World Regional Geography Book Series

What does Finland mean to a Finn, Sichuan to a Sichuanian, and California to a Californian? How are physical and human geographical factors reflected in their present-day inhabitants? And how are these factors interrelated? How does history, culture, socio-economy, language and demography impact and characterize and identify an average person in such regions today? How does that determine her or his well-being, behaviour, ambitions and perspectives for the future? These are the type of questions that are central to The World Regional Geography Book Series, where physically and socially coherent regions are being characterized by their roots and future perspectives described through a wide variety of scientific disciplines. The Book Series presents a dynamic overall and in-depth picture of specific regions and their people. In times of globalization renewed interest emerges for the region as an entity, its people, its landscapes and their roots. Books in this Series will also provide insight in how people from different regions in the world will anticipate on and adapt to global challenges as climate change and to supra-regional mitigation measures. This, in turn, will contribute to the ambitions of the International Year of Global Understanding to link the local with the global, to be proclaimed by the United Nations as a UN-Year for 2016, as initiated by the International Geographical Union. Submissions to the Book Series are also invited on the theme 'The Geography of…', with a relevant subtitle of the authors/editors choice. Proposals for the series will be considered by the Series Editor and International Editorial Board. An author/editor questionnaire and instructions for authors can be obtained from the Publisher.

This book series is published in cooperation with the International Geographical Union (IGU). The IGU is an international, non-governmental, professional organization devoted to the development of the discipline of Geography. The purposes of the IGU are primarily to promote Geography through initiating and coordinating geographical research and teaching in all countries of the world.

The World Regional Geography Book Series is now indexed in SCOPUS.

Norm Catto · Guy R. Vincent

Canada: Land and People

A Physical and Human Geography

Springer

Norm Catto
Department of Geography (ret.)
Memorial University
St. John's, NL, Canada

Guy R. Vincent
Faculté des Arts et des Sciences Sociales
Département d'Histoire et de Géographie
Université de Moncton
Moncton, NB, Canada

ISSN 2363-9083 ISSN 2363-9091 (electronic)
World Regional Geography Book Series
ISBN 978-3-032-04390-0 ISBN 978-3-032-04391-7 (eBook)
https://doi.org/10.1007/978-3-032-04391-7

Picture Credit: Anton Balazh - Fotolia.com

This Springer imprint is published by the registered company Springer Nature Switzerland AG
The registered company address is: Gewerbestrasse 11, 6330 Cham, Switzerland

To my late parents Juliette and André

— Guy R. Vincent

Preface

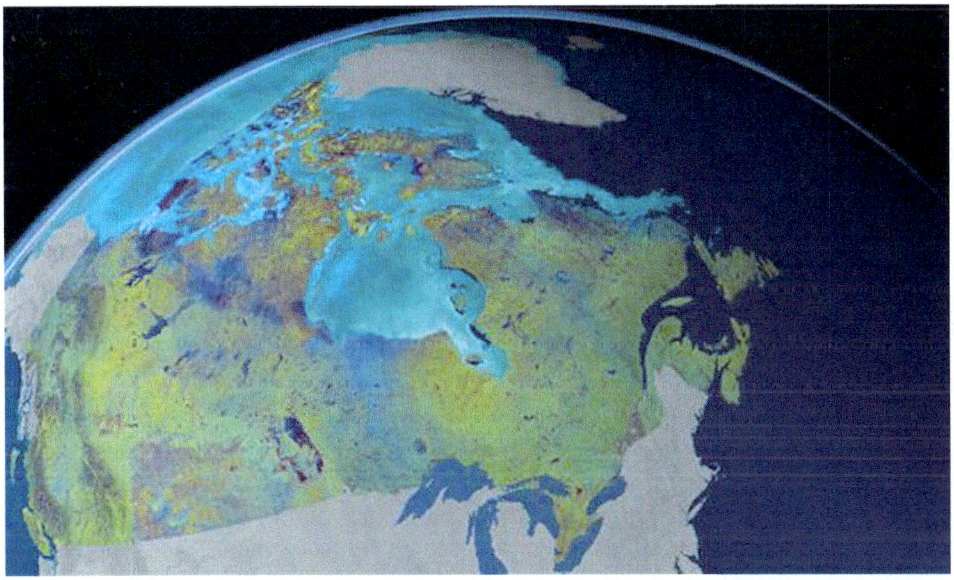

(Credit: RADARSAT-2 Data and products © MDA Geospatial Services Inc. 2014—All rights reserved. RADARSAT is an official mark of the Canadian Space Agency.)

Canada: Land and People forms part of the Springer-Nature series of country studies of Physical and Human Geography. With more than 40 million people (in 2025) spread across almost 10 million km^2, there are many aspects of Canadian geography of potential interest to international readers, including university geographers. Here, we've attempted to apply the geographical focuses on environment-human interactions and sense of place to an investigation of Canada. This volume emphasizes connections between land and people from a variety of perspectives, considering the country's increasingly urban and diverse demographies.

There are many aspects of Canadian Geography in addition to those discussed in these chapters. Canada's geographers continue to vigorously research innumerable topics in Physical and Human Geography. The Canadian Association of Geographers, Royal Canadian Geographical Society, and Canadian Geographical Education, among other academic and professional organizations, activity promote geographical research and education to Canadians. A single volume cannot possibly address all geographical issues of interest, even in an introductory manner. Researchers interested in Canadian Geography, therefore, can take this as a starting point for further detailed and rewarding study.

Canada, "village" or "settlement" in Haudenosaunee (Huron Iroquois), sits in northern North America and is bathed by three oceans. It has the USA as its southern neighbour and Russia on the other side of the North Pole. Ice coverage through several glacial ages contributed to shape a varied natural landscape which would eventually welcome human life: first by nomadic people migrating from Asia through Beringia following and hunting their prey to survive in hostile environments; and secondly by waves of European explorers and settlers who would transform the way humans interact with the land. Blessed by its geography, Canada has evolved into a developed country and a modern society.

Three themes will be recurrent throughout the book: continentality, North-South trends and East-West outcomes. Firstly, Canada's history and development is closely related and, to a certain degree, influenced by that of the United States because of its proximity. Both share a large part of a continent (if we include Mexico) that is diverse and rich in resources. Events south of the border will have consequences "up North". Secondly, because of its sheer size, North America is inherently subject to North-South directional tendencies, both in physical and human terms. Dominant physiographic features, for instance, are generally aligned in such an axis. Human contacts are more prone to be developed and maintained between people from either side of the borders than across the continent. Some North-South alignments, economic, cultural, or social will then appear almost naturally. Finally East-West currents have been also a common theme. The American "Manifest destiny", or the "Go West young man" mantra, resulted in the hasty establishment of the international boundary along the forty-ninth parallel from the prairies to the Pacific coast that served to artificially cut the more natural North-South flows.

Traditionally, overviews of Canadian geography have taken a regional approach. Usually, the regions chosen are the political provincial and territorial divisions: Atlantic Canada, Québec, Ontario, Prairie Canada, British Columbia, and the Northern Territories. The same regions are commonly invoked to analyse politics, provide advice to prospective immigrants, discuss music, visual, and performing arts, consider mitigation and adaptation measures for climate change, and many others.

However, the boundaries of the provinces and territories have changed repeatedly over the years, with none remaining unaltered. Several proposals include separating the largest Census Metropolitan Areas (CMA) from the surrounding provinces, establishing a National Capital district around Ottawa-Gatineau, merging New Brunswick, Nova Scotia, and Prince Edward Island into a single Maritime Union entity, detaching Northern Ontario and Nunatsiavut as separate provinces, among others.

Many differences within each of these regions are greater than those among them. Toronto, Montreal, and Vancouver are not merely larger versions of Timmins, Métis-sur-Mer, and Valemount, respectively, either in their natural environments or their human geographies. Renaming Newfoundland to Newfoundland and Labrador in 2001 recognized some of the differences within that province, although the geographies of Atlantic Boreal Newfoundland, the Boreal and Taiga Canadian Shield ecozones of southern and central Labrador, and the Arctic Cordillera of northern Nunatsiavut are very distinct.

Ecozones, based on climate, vegetation, fauna, soils, terrain, and geology, represent a physically based system of regions, cutting across provincial and territorial boundaries. The ecozonal organization of the first chapters of this volume recognizes the connections between environments and landscapes, and human demographics, activities, and cultures. The differences in demographics, economics, and transportation choices among the largest CMA, and between the CMA, and the smaller cities are reflected in the organization of the urban chapters of the volume.

Norm: The research and writing for this volume reflects 35 years of university life as a professional geographer in Canada. Over that time, I have taught more than 40 different courses and more than 175 course sections at Memorial University of Newfoundland and Labrador and elsewhere. I have conducted more than 4000 university classes, seminars, laboratory sessions, and field excursions, for the more than 14,000 students who I've had the pleasure of working for and with. I've learned from thousands of colleagues who I've had the privilege of interacting with and learning from in research papers, conferences, academic and professional websites, laboratories, lectures, field excursions, classrooms, and conferences, especially my colleagues in the Department of Geography at Memorial.

Guy: Professional geographer for 30 years at Université de Moncton, a francophone university, at Moncton campus. I have prepared and taught over twenty different courses on very varied topics to several thousand students. Curious about the world we live from a very young age (Montreal's Expo67 opened my eyes to the world!), I have developed a deep interest for the different aspects of geography and a desire to always convey my passion better. I always have been keen to learn more through interacting with colleagues, reading and researching, regularly attending and participating at conferences, travelling and discovering. Geography is and has been so rewarding!

St. John's, Canada — Norm Catto
Moncton, Canada — Guy R. Vincent

Contents

1 Introduction

Abstract

Canada: Land and people discuss the physical and human geography of Canada. The introduction includes an overview of background information, including classification of climate, physiography, geology, ecozones, and soils, the elements of the natural environments and landscapes. This chapter considers ecozones and regional environments, including climate, geology, botany, zoology, soils, water bodies and wetlands, and landscapes, based on the classification of Environment Canada and Agriculture and Agrifoods Canada with modifications. Chapter 2 discusses the Atlantic Ecozone, encompassing the Appalachian Mountains and coastal waters. Chapter 3 looks at the Mixedwood Plains Ecozone in the St. Lawrence-Great Lakes Platform. Chapter 4 discusses the Boreal Ecozones from southern Labrador to northeastern British Columbia. Chapter 5 concerns the Prairie Grasslands Ecozone. Chapter 6 deals with the Montane, Boreal, and Taiga Cordilleran Ecozones. Chapter 7 focuses on the Pacific Maritime Ecozone and coastal Pacific Ocean waters. Chapter 8 discusses the Subarctic and Taiga Ecozones. Chapter 9 considers the Arctic Ecozones. Chapter 10 discusses parks and conserved areas, focusing on the evolving trend towards park establishment and operation based on more holistic appreciations linking people and land, and the implications for conservation. The second part of this volume focuses on the people of Canada, including population, demography, urban centres, political organization and governance, historical geography and territorial evolution, and settlement patterns. Chapter 11 considers the historical context and present conditions involving migration and immigration to Canada, and the current ethnic composition of the country. Chapters 12 and 13 focus on the urban areas, concentrating on the Census Metropolitan Areas, with emphasis on population, growth, urban transit and commuting, environmental stresses, urban vegetation, and inequality in urban systems. Chapter 14 discusses transportation, including marine, rail, road, and air. Chapter 15 discusses forestry, focusing on the national aspects of harvesting, disturbances, production, consumption, and GHG emissions. Chapter 16 looks at Agriculture and Food across Canada. Chapter 17 discusses Energy and Electricity. Aspects of coal, petroleum, and natural gas production and exploitation are considered, along with nuclear, wind, solar, and tidal energies. Chapters 18, 19, and 20 discuss the economic development, industrialization, and urbanization of Canada; winter life in Canadian cities; and boundary definitions and cadastral layout. The final chapter is a short conclusion and commentary.

Keyword

Canada · Land · Environment · People · Demography · Settlement patterns

1.1 Introduction

Part 1 of this volume focuses on the land and environments of Canada (Fig 1.1). In the first section of this chapter, we'll give an overview of background information, including classification of climate, physiography, geology, ecozones, and soils, the elements of the natural environments and landscapes. Eight subsequent chapters will look at ecozones and regional environments, including climate, geology, botany, zoology, soils, water bodies and wetlands, and landscapes. We'll consider these factors as parts of a whole environmental package, allowing comparisons to those of adjacent areas.

The volume of material involved in a study of the environments and landscapes of Canada seems daunting at first glance. As Canada is the second largest country in the world by area, and one of the most diverse environmentally, a wide

N. Catto and G. R. Vincent, *Canada: Land and People*, World Regional Geography Book Series,
https://doi.org/10.1007/978-3-032-04391-7_1

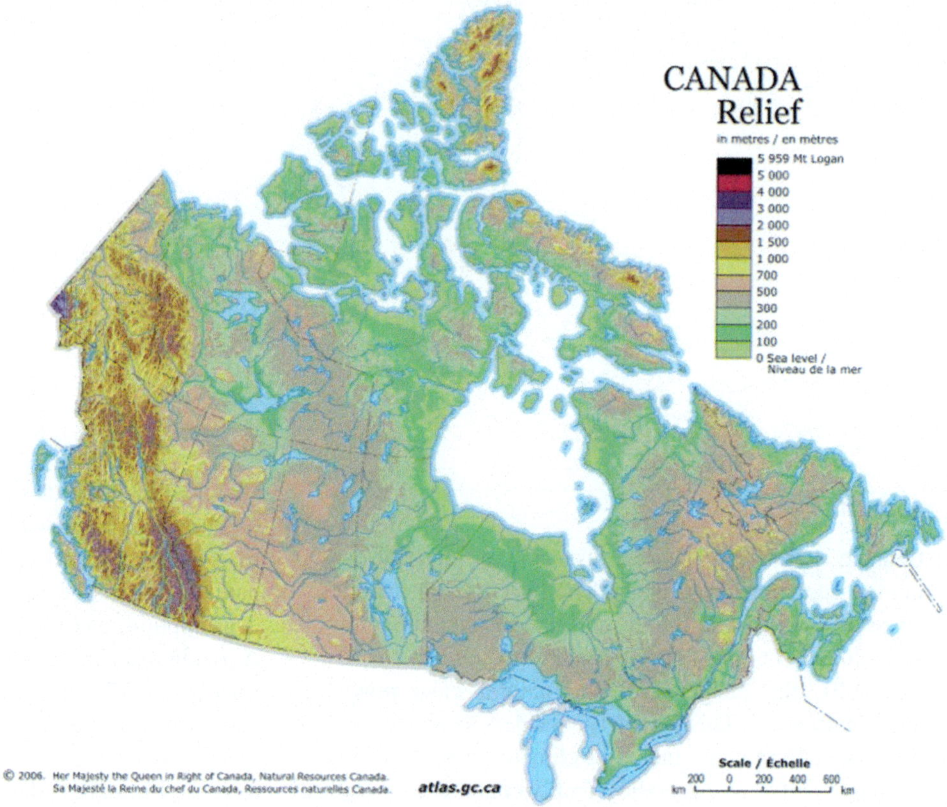

Fig. 1.1 Relief map of Canada (Government of Canada)

variety is inevitable. Our goal is to give you some impressions of the environments and landscapes, as you go through the chapters. In evaluating environments, consideration of the role of *Homo sapiens* is necessary, especially as humans have modified regions to considerable degrees over the past 400 years.

Ecozones (Fig 1.2) used here are based on the classification of Environment Canada and Agriculture and Agrifoods Canada (1995), with modifications (Table 1.1). Chapter 2 discusses the Atlantic Ecozone, encompassing the physiographic Appalachian Mountains and coastal waters. Ecological and landscape subdivisions include the Atlantic Highlands (Gaspé, Beauce, and northern New Brunswick), Atlantic Maritime Plain (southern New Brunswick, Nova Scotia, Prince Edward Island, and Iles-de-la-Madeleine), and the coastal waters of the North Atlantic Ocean (Scotian Shelf, Gulf of St. Lawrence, and Newfoundland and Labrador Shelf). The Newfoundland Boreal landscape is also discussed in this chapter.

Chapter 3 looks at the Mixedwood Plains Ecozone in the physiographic St. Lawrence Platform, including the Great Lakes. The St. Lawrence Platform includes three parts: eastern (Mingan Archipelago, Anticosti, Strait of Belle Isle), central (southern Québec-eastern Ontario), and western (southern Ontario). The Great Lakes–St. Lawrence River drainage system is also discussed.

Chapter 4 discusses the Boreal Ecozones extending across central Canada from southern Labrador to northeastern British Columbia. The Boreal Shield Ecozone includes the Grenville Geological/Structural Province (Laurentides; Québec and central Ontario), the Superior and Keweenawan Geological/Structural Provinces (northern and northwestern Ontario), and the southwestern part of the Churchill Geological/Structural Province (Manitoba, Saskatchewan, and Alberta) of the southern Canadian Shield. The Boreal Plains Ecozone includes the areas with Köppen-Geiger Dfb and transitional Dfc climates with boreal forest vegetation, developed on terrain of the Interior Platform, extending from Manitoba to northeastern British Columbia.

Chapter 5 concerns the Prairie Grasslands Ecozone of Manitoba, Saskatchewan, and Alberta. Physiographically, this southern part of the Interior Platform is subdivided by the Manitoba Escarpment and Missouri Coteau into the three successive westward steppes of the Manitoba, Saskatchewan, and Alberta Plains.

Chapter 6 deals with three Cordilleran Ecozones. The Montane Cordillera includes the southern Rocky Mountains (Foreland Structural Belt, Eastern Cordillera System physiographic subregion), and the Selkirk and Omineca Mountains,

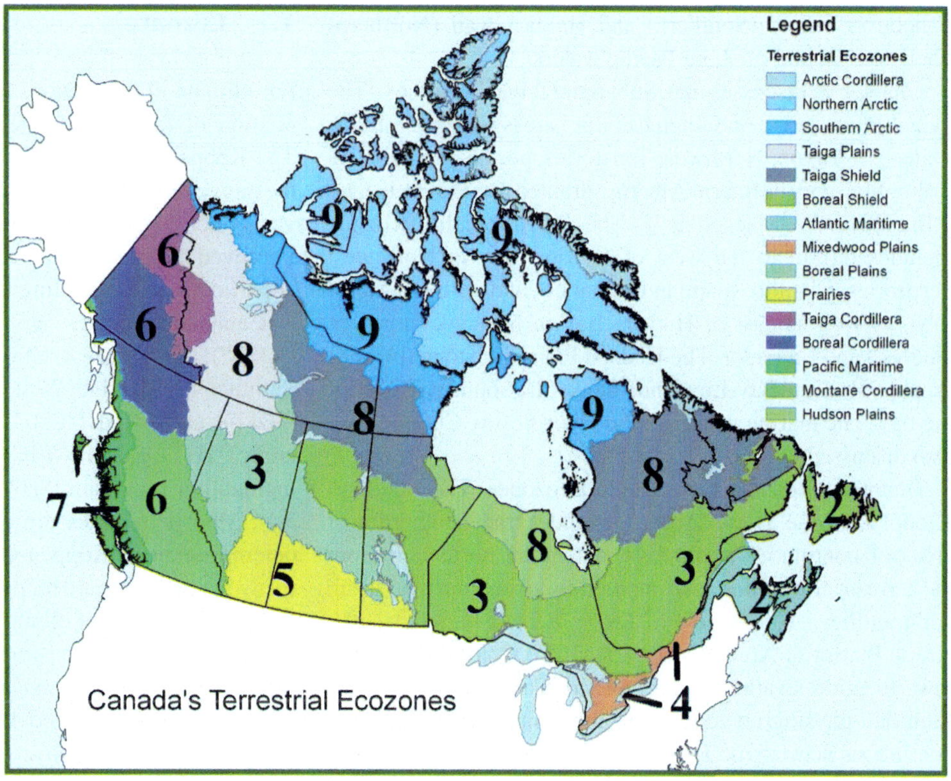

Fig. 1.2 Ecozones of Canada. Areas covered in Chaps. 2–9 are labelled. (Environment and Climate Change Canada and Agriculture Canada, 1995). {https://www.nrcan.gc.ca/sites/nrcan/files/forest/Canada_terrestrial_ecozones_lg_e.jpg}

Table 1.1 Layout of Chaps. 2–9

Ch	Terrestrial ecozones	Marine ecozones	Köppen-Geiger climates	Physiographic regions	Geological provinces
2	Atlantic maritime; boreal newfoundland	Gulf of St. Lawrence; Scotian Shelf; Nfld. Shelf	Dfb, Dfc	Appalachian mountains; maritime plain	Appalachian orogen
3	Mixedwood plains	Great lakes	Cfa, Dfa, Dfb	St. Lawrence lowlands	St. Lawrence platform
4	Boreal shield; boreal plains	–	Dfb	Canadian shield; interior plains	Canadian shield; interior platform
5	Prairie grasslands	–	BSk, Dfb	Interior plains	Interior platform
6	Montane cordillera; boreal cordillera; taiga cordillera	–	BSk, Dfb, Dfc, H	Cordilleran mountains	Cordilleran orogen
7	Pacific maritime	Pacific offshore; Pacific shelf; Salish sea	Cfb, Cfc, Csb, H	Cordilleran mountains	Cordilleran orogen
8	Taiga plains; taiga shield; Hudson plains	Hudson bay complex; labrador shelf	Dfc, ET	Interior plains; Canadian shield; Hudson plains	Interior platform Canadian shield; Hudson platform
9	Southern Arctic; Northern Arctic; Arctic cordillera	East. Arctic; West. Arctic; Arctic Basin; Arctic Archipelago	ET, EF, H	Arctic plains; Canadian shield; Innuitian mountains; Arctic coastal plain	Arctic platform; Canadian shield; Innuitian orogen

Interior Plateau, and the Okanagan Valley (Omineca and Intermontane Structural Belts, Interior Cordillera System physiographic subregion). The Boreal Cordilleran Ecozone includes the northern Rocky Mountains, the Interior Cordilleran landscapes of central Yukon (Omineca and Intermontane Structural Belts), and the Kluane and St. Elias Mountains in southwestern Yukon (Insular Structural Belt, Western Cordillera System physiographic subregion). The most northerly Taiga Cordillera Ecozone includes the Mackenzie, Richardson, and Ogilvie Mountains, and northern Yukon (Foreland Structural Belt, Eastern Cordillera System).

Chapter 7 focuses on the Pacific Maritime Ecozone and coastal Pacific Ocean waters. The Pacific Maritime Ecozone includes the Coast, Cascade, Vancouver Island, and Haida Gwaii ranges, as well as intermontane valleys, fjords, and coastal plains. Coastal Pacific waters include the Salish Sea,

Vancouver Island (Southern) and Haida Gwaii (Northern) Shelf waters, and offshore Pacific waters.

Chapter 8 discusses the Subarctic/Taiga Ecozones. The Taiga Plains Ecozone includes the areas of the Interior Plateau of Yukon, Northwest Territories, northeastern British Columbia, and northern Alberta, drained by the Deh Cho ("Big River", "Mackenzie"). The Taiga Shield Ecozone includes terrain to the west of Hudson Bay in Northwest Territories, Kivalliq (mainland Nunavut), and northern Manitoba, as well as east of Hudson Bay in Nunavik (northern Québec) and Labrador. The Hudson Plains Ecozone (physiographic Hudson Bay Lowland) flanks the southern coast of the bay. The marine waters of the Hudson Bay Complex are also discussed.

Chapter 9 considers the Arctic Ecozones. The Southern Arctic Ecozone (Low Arctic) includes the southernmost areas of continuous permafrost, extending northward from the terrestrial tree line, and marked by dwarf upright shrub communities. The Northern Arctic Ecozone includes the Arctic Platform, Arctic Ocean (Beaufort) Coastal Plain, and low- to mid-elevation regions of the Canadian Shield and Innuitian physiographic regions, and is marked by deep and continuous permafrost. Higher elevation terrain on Umigmak Nuna (Ellesmere Island), Qikiqtaaluk(Baffin Island), and in Torngat Mountains National Park (NL) lies within the Arctic Cordillera Ecozone. Marine Ecozones include the Eastern Arctic, Western Arctic, Arctic Archipelago waters, and the Arctic Basin waters of the Beaufort Sea.

Chapter 10 discusses parks and conserved areas, focusing on the evolving trend towards park establishment and operation based on more holistic appreciations linking people and land, and the implications for conservation. Canada has seen accelerated growth in the number of federal, provincial, and co-managed protected areas, both terrestrial and marine, over the past twenty years. Positive factors, such as the increased involvement of many communities in park management, are unfortunately counterbalanced by negative factors, including declines in ecological integrity within parks and overall environmental challenges affecting broader regions of Canada.

Recent years have seen the emergence of a stronger Indigenous presence in the use of place names for terrain features in Canada. This long overdue acknowledgement of the reality of Canadian Indigenous culture can complicate place name recognition for English, French, and international readers. We have included a partial list of equivalences between English and Indigenous names for terrain features (Table 1.2). In the chapters of Part 1, we will use both English and Indigenous names for landscape features, particularly for those in Northwest Territories, Nunavut, and Yukon, although we will emphasize the Indigenous terms.

1.2 Climate

For climate classification, this volume uses the 2006 modification of the Köppen-Geiger system (Beck et al. 2006). The Köppen-Geiger system evaluates climates according to the ranges of temperature and precipitation throughout the year. Each subsequent edition since 1922 has incorporated improved measurements, more climate stations, and changes at particular sites resulting from climate change.

Canadian climates are classified as Steppe or Prairie ("BSk"); Temperate ("Cfa"); Marine West Coast ("Cfb"); Summer dry Marine West Coast ("Csb"); southern Boreal ("Dfa"); mid-boreal ("Dfb"); northern boreal or taiga ("Dfc"); arctic ("ET"); glacial ("EF") and Maritime Polar ("EM"), and highland or mountain ("H", for terrain in excess of 1500 m asl). Map boundaries are approximate (Fig. 1.3). Locally, determinations of Köppen-Geiger zones vary, and some (relatively small) zones are recognized in some variants but not in others. Discussion of climates also includes Climate Normal data for winds, clouds, fog, humidity, and sunshine (Environment Canada 2020; *see also* Phillips 1990), key elements which are not accounted for in the Köppen-Geiger zones. Climate-related hazards are also discussed.

1.3 Physiographic Regions and Geological Provinces

The physiographic subdivision of Canada used here is based on the system summarized by Bostock (2014). Considering the broad characteristics of the landscapes, Canada can be divided into nine major physiographic regions (Fig. 1.4).

Geologically, the country has been divided into 13 terrestrial and 4 marine geological (also called structural) provinces (Natural Resources Canada 2009; also see Wheeler et al. 1996). As (Fig. 1.5) the terrestrial physiographic regions and geological provinces overlap, it is convenient to discuss them together.

The geological history of Canada is marked by numerous episodes of mountain formation, each of which is termed an orogeny. Igneous intrusions (such as granites) associated with younger orogenies frequently penetrate and metamorphose older rocks, producing a complex pattern of igneous and metamorphic units of differing ages. After much research, Canada has been divided into geological provinces, each of which is defined on the basis of the ages of rock assemblages and on the nature of folding and faulting. The prevalence of igneous intrusions means that each geological province contains rocks of several different ages, particularly in the Canadian Shield.

Table 1.2 Indigenous terminology for selected place names in Canada

Indigenous	English Canadian	*Indigenous*	English Canadian
Aqvituq	Hopedale	*Nunavik*	Nouveau Quebec
Arviat	Eskimo Point	*Paulatuk*	Paulatuk
Arviliit	Ottawa Islands	*Pingualuit*	Nou. Quebec Crater
Ausuiktuq	Grise Fiord	*Qaisualuk*	Mount Thor
Ausuituu Tariunga	Jones Sound	*Qamani'tuaq*	Baker Lake
Behchok`ǫ	Rae-Edzo	*Qausuittuq*	Resolute
Deh Cho; Deh Gah Got'ine	Mackenzie River	*Qausuittuq*	Bathurst Island
Igluigaarjuk	Chesterfield Inlet	*Qikiqtaaluk*	Baffin Island
Ikahuak	Sachs Harbour	*Qikiqtaq*	King William Island
Ikpiarjuk	Arctic Bay	*Qikiqtarjuaq*	Broughton Island
Ilagiit nunagivaktangit	Foxe Basin	*Qikiqtaryuk*	Herschel Island
Ilulliq	Melville Island	*Qipuqqaq*	Postville
Inukjuak	Inoucdjouac	*Qulitasaniakvik*	Schwarzenbach Falls
Iqaluit	Frobisher Bay	*Saknirutiak Imanga*	Baffin Bay
Iqaluktuuttiaq	Cambridge Bay	*Salliq*	Coral Harbour
KakKasuak	Mealy Mountains	*Salluit*	Sugluk
Kangiqluruluk	Oliver Sound	*Sanikiluaq*	Belcher Islands
Kangiqsualujjuaq	Fort George	*Sanirajak*	Hall Beach
Kangiqsujuaq	Maricourt (Wakeham)	*Sanirajak*	Melville Peninsula
Kangiqsujuaq	Wakeham Bay	*Sarvarjuaq*	North Water
Kangiqtiniq	Rankin Inlet	*Shartoo*	Wales Island
Kangiqtualuk	Cumberland Sound	*Shugliaq*	Southampton Island
Kangiqtugaapik	Clyde River	*Sivanitirutinguak*	Mount Asgard
K'asho Got'ine	Fort Good Hope	*Sǫ`ǫmbak'è*	Yellowknife
Katannilik	Meta Incognita Penin	*St'át'imcets*	Lillooet
Kikiak	Rigolet	*T'saatine nda ghe'in'deh*	Lesser Slave Lake
Killiniq	Cape Chidley	*Tallurutiup Timanga*	Lancaster Sound
Kimmirut	Lake Harbour	*Taloyoak*	Spence Bay
Kingailik	Prince of Wales Island	*Tasiuja*	Eclipse Sound
Kingait	Cape Dorset	*Tatlurutit*	Devon Island
Kingngailap Nunanga	Boothia Peninsula	*Teetł'it Zheh*	Fort McPherson
Kingoak	Bathurst Inlet	*Thebacha*	Fort Smith
Kitlineq	Victoria Island	*Tikiraqjuaq*	Whale Cove
Kivalliq	Keewatin	*Tłeg´ǫhłı̨*	Norman Wells
Kugaaruk	Pelly Bay	*Tsiigèhnjik*	Arctic Red River (river)
Kuganajuup Qikiqtang	Somerset Island	*Tsiigehtchic*	Arctic Red River (town)
Kugluktuk	Coppermine	*Tu Nedhe*	Great Slave lake
Kuujjuaq	Fort Chimo	*Tunnunik*	Prince Albert Crater
Kuujjuaraapik	Poste-de-la-Baleine	*Ukkusiksalik*	Wager Bay
Kuujjuarapik	Great Whale River	*Ulukhaktok*	Holman
Łíídlı̨ Kų́ę́	Fort Simpson	*Umingmak Nuna*	Ellesmere Island
Marruuvik	Makkovik	*Umingmat Nunaat*	Axel Heiberg Island
Mittimatalik	Pond Inlet	*Ungava kangiqluk*	Ungava Bay
Nalluata imanga	Navy Board Inlet	*Usqsuqtuuq*	Gjoa Haven
Naujaat	Repulse Bay	*Wînipekw*	Hudson Bay

(continued)

Table 1.2 (continued)

Indigenous	English Canadian	*Indigenous*	English Canadian
Nirjutiqavvik	Cobourg Island	*Wînipekw*	James Bay
Nitassinan	Labrador	*Xátł'odehchee*	Hay River
Nunajnguk	Nain	*Zhahti Kue*	Fort Providence

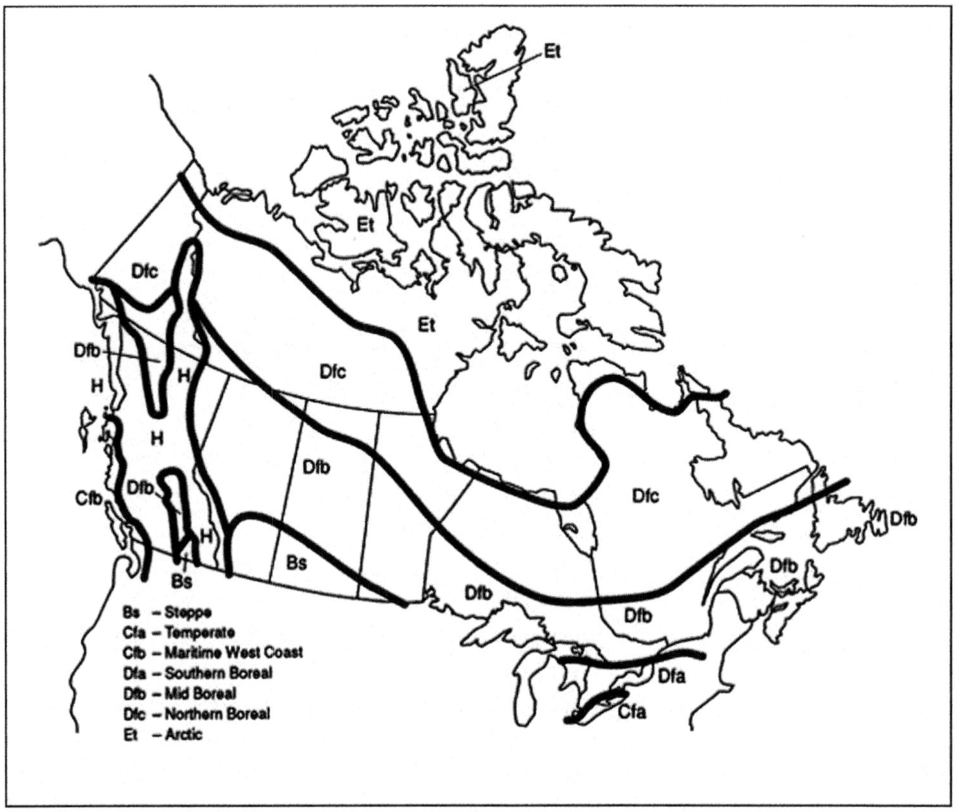

Fig. 1.3 Köppen-Geiger climate zones. Some small areas are not illustrated

The Appalachian Mountain System is a diverse physiographic region, encompassing the Atlantic Provinces (except Labrador) and eastern Québec. Atlantic Canada includes the provinces of New Brunswick, Newfoundland and Labrador, Nova Scotia, and Prince Edward Island. Maritime Canada (or the Maritimes) includes New Brunswick, Nova Scotia, and Prince Edward Island, but not Newfoundland and Labrador.

In the Appalachian Mountain System, topography ranges from rugged ridges to flat plains. Geologically, the region is included with the Appalachian Orogen province. Ages of rock units range from latest Precambrian to Mesozoic.

The St. Lawrence Platform is marked by low relief. This physiographic region is the most densely populated in the country and has undergone the most human modification. It is split into three parts by southward extensions of the Grenville (Laurentides) region of the Canadian Shield. The geological St. Lawrence Platform is predominantly underlain by Palaeozoic sedimentary rocks.

The Hudson Platform (Hudson Bay Lowland) consists of gently sloping plains surrounding Hudson Bay, developed on Palaeozoic sedimentary bedrock, and covered by marine sediment formed when the waters of Hudson Bay stood at higher elevations than the present sea level, forming the Tyrrell Sea following the last deglaciation.

The Interior Platform (Interior Plains) includes the southern Prairies, Alberta Plateau, and Mackenzie (Deh Cho) Valley. Relief is varied, but the topography is generally rolling, rather than flat. Sedimentary rocks dominate, with ages from Palaeozoic through Cenozoic.

The Cordilleran Mountain System includes the mountain ranges of western Canada, including the Rocky, Mackenzie, Coast, Cascade, Selkirk, and St. Elias Mountains, and the intervening plateaux. Bedrock units vary widely in lithology and range in age from P:recambrian to Cenozoic.

Geologically, the Cordilleran Orogen Structural Province is divided into the Eastern, Interior, and Western Systems, oriented southeast-northwest. The Eastern System is marked by the Foreland Structural Belt, including the Rocky Mountains of Alberta and British Columbia and the Mackenzie Mountains of NWT. The Interior (also called Central) System

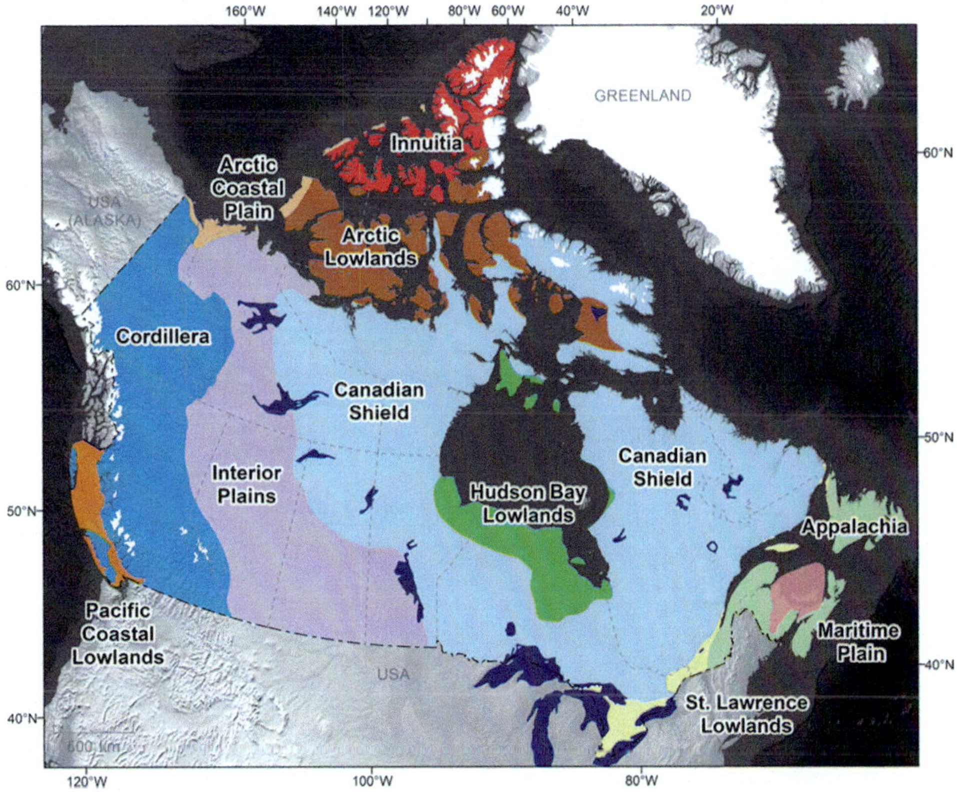

Fig. 1.4 Physiographic regions of Canada, after Bostock (2014). (Natural Resources Canada; doi:10.4095/293408) {Adapted from "Landscapes and Landforms of Western Canada"}

consists of the Omineca (Selkirk, Purcell, and Omineca Mountains) and Intermontane (Okanagan Valley, Cariboo Plateau) Structural Belts. The Western System includes the Coastal (Coast, Cascade, St. Elias Mountains) and the Insular (Vancouver Island, Haida Gwaii) Structural Belts.

The Arctic Coastal Plain and Arctic Platform are marked by gentle topography, with low relief. The rock units are sedimentary, later Mesozoic and Cenozoic in age. Geologically, the Arctic Platform Structural Province, including the Arctic Coastal Plain, is flanked to the south and east by the Bear, Tu Nedhe, and Churchill provinces of the Canadian Shield, to the north by the Innuitian Orogen, and to the west by the Arctic Coastal Plain.

The Innuitian Mountain System is diverse topographically, including mountains, plateaux, rolling hills, and flat plains. Geologically, this region is assigned to the Innuitian Orogen Structural Province. Successive folding, faulting, and sedimentary deposition during the mid-Palaeozoic through Mesozoic have resulted in complex landscapes. On eastern Bathurst Island, for example, the terrain consists of a series of sharply defined ridges and valleys aligned north–south, reflecting the folded terrain of the mid Devonian Cornwallis fold belt. In the western, central, and northern parts of the island, the ridges and valleys are more sharply defined than in the east and are aligned west-southwest–east-northeast, with a series of deep, linear embayments. These ridges and valleys were created during the Late Devonian Parry Islands event and reflect both folding and faulting. The terrain in the area where the fold belts intersect, in the eastern third of Bathurst Island, is marked by irregular rolling hills and hummocks.

1.3.1 Canadian Shield

The Canadian Shield is marked by rolling topography, and developed as the Precambrian bedrock was gradually eroded. Virtually all of the erosion occurred prior to Quaternary glaciation.

Geologically, the Canadian Shield is complex, with Precambrian rocks ranging in age from Proterozoic (541–2500 million years old) to Archean (more than 2500 million years old). Numerous subdivisions have been proposed. Developments and improvements in chronological techniques have resulted in several terminological changes and boundary adjustments for geological provinces of the Canadian Shield since 1970. In this volume, we'll designate eight Structural Provinces (Fig. 1.6): Bear, Churchill, Grenville, Keweenawan (also called Southern), Makkovik, Nain, Superior, and Tu Nedhe (formerly termed Slave).

The Bear Province represents the northwestern extent of the Canadian Shield. Exposed west of Coronation Gulf, it includes the terrain from Kugluktuk southwest to Great Bear Lake. These rocks are Proterozoic metamorphosed sedimentary and metamorphosed basic volcanic units, and range in age

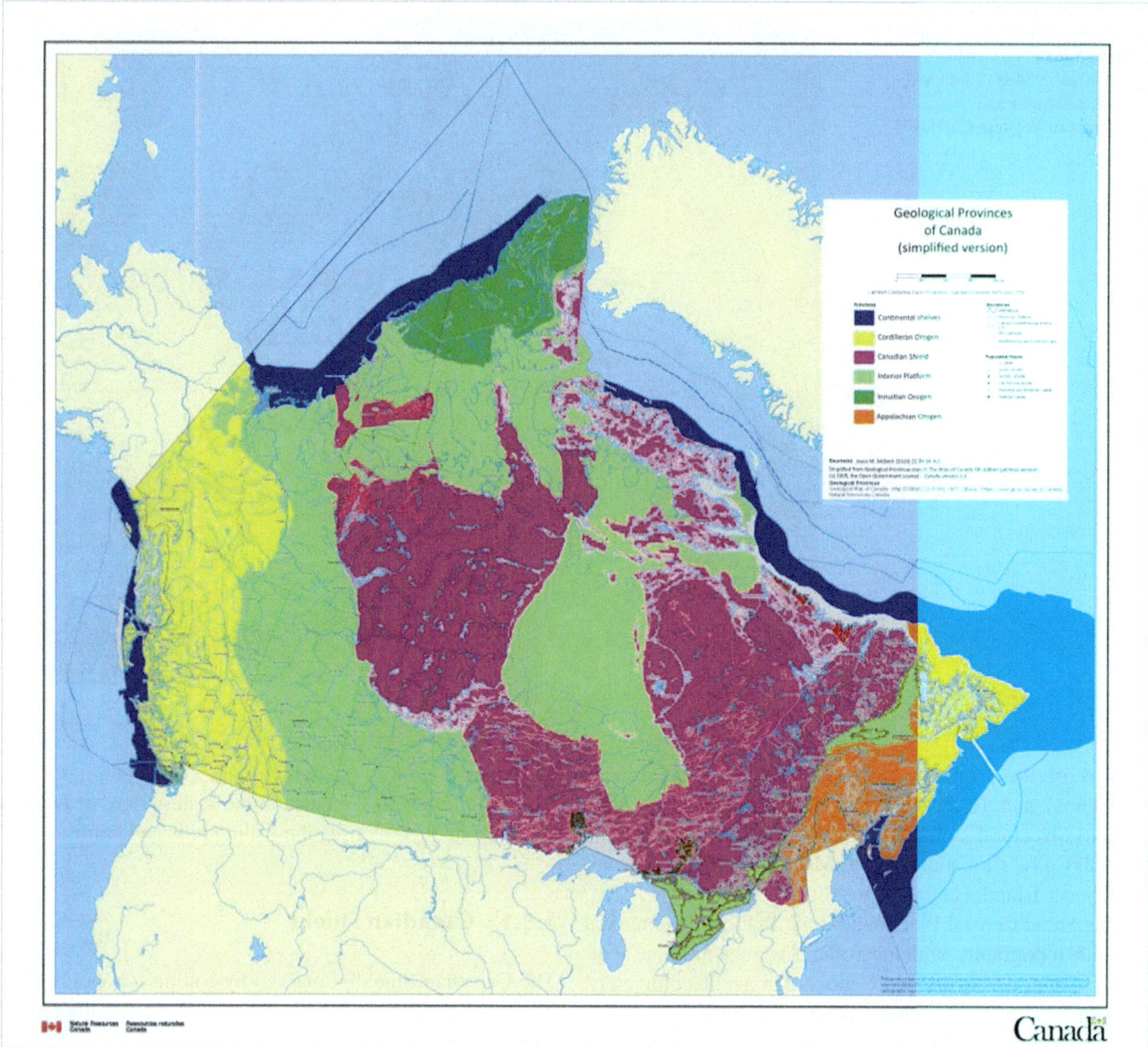

Fig. 1.5 Geological provinces of Canada. (Natural Resources Canada) {https://open.canada.ca/}

from 2200 to 1400 million years. Volcanic greenstone belts have been exploited for copper by the Kilusiktormiut people at Kugluktuk. Quartz sandstones and quartzites form resistant units. Younger diabase dykes and sills (~1000 million years) have intruded the older rocks, forming resistant ridges and ledges. The numerous small islands in Coronation Gulf are exposed pieces of the diabase sills.

The Churchill Province, defined and named for the Churchill River of northern Manitoba, extends across the central part of the Canadian Shield. The province as originally defined contained a large number of geological units, ranging in age from Archaean to Neo-Proterozoic. This is the most heterogeneous of the geological provinces, and efforts at reclassifying, subdividing, and renaming the Churchill Province have been ongoing for many years.

The Grenville Province comprises the southeastern part of the Canadian Shield, extending from Goose Bay to Georgian Bay, and including the core of the Long Range Mountains. Most of the province is dominated by igneous acidic rocks (granites and related types) and acidic, highly metamorphosed rocks (gneiss and schist) derived from the metamorphism of pre-existing sedimentary and igneous rocks during several Precambrian orogenic events. The Grenville Province is considered to be among the oldest parts of the Canadian Shield by geologists, but it has been so intensively modified by metamorphism during the orogenic events that determining the original age of the rocks is essentially impossible. Orogenic activity began in the Grenville Province approximately 1600 million years ago, and culminated in the Ottawan

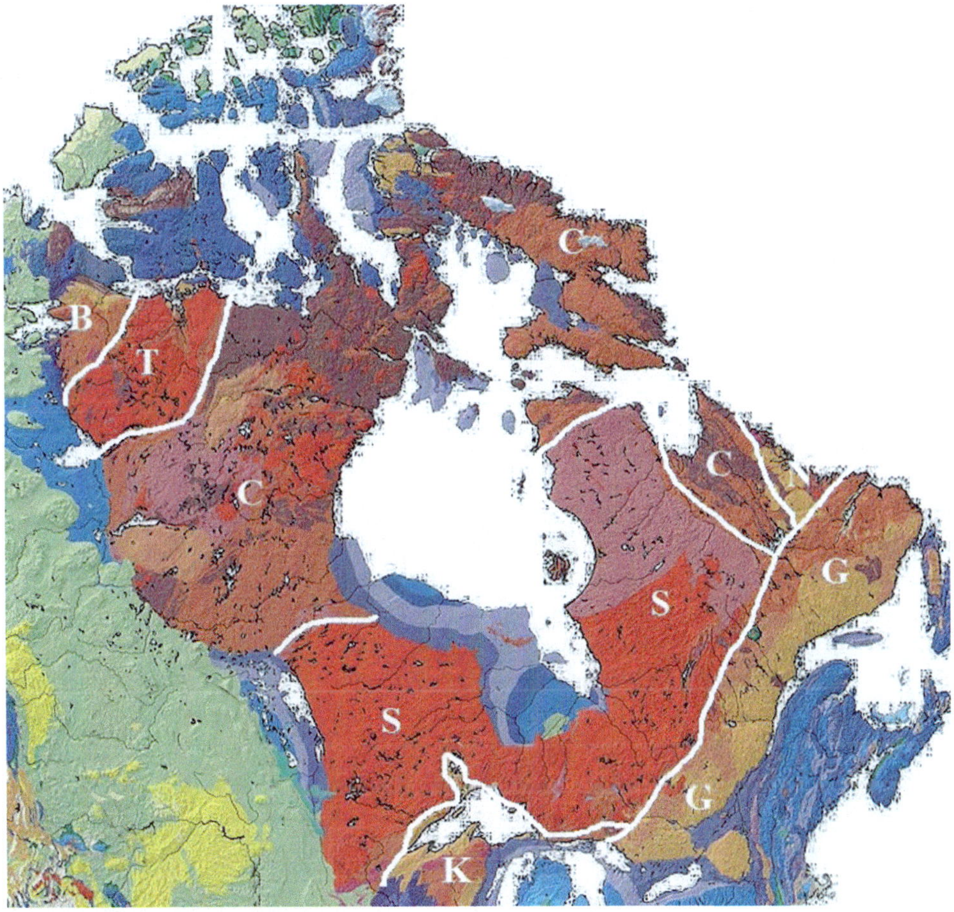

Fig. 1.6 Geological provinces of the Canadian Shield. B = Bear; C = Churchill; G = Grenville; K = Keweenawan; S = Superior; T = Tu Nedhe. (Natural Resources Canada {from Wikimedia Commons})

Orogeny between 1100 and 1000–985 million years ago, during the Meso-Proterozoic era of the Precambrian.

The Keweenawan Province consists of sedimentary and metamorphosed sedimentary rocks of Palaeo-Proterozoic age exposed along Lake Superior, which were welded to the Superior Province and deformed during the Meso-Proterozoic, 1370 million years ago. In Canada, Keweenawan Province rocks are relatively weak basalts, gneisses, schists, slates, argillites, and phyllites. Consequently, they have been eroded evenly along the Lake Superior coast, which has resulted in the gently curving Superior shoreline, devoid of any harbours. These rocks underlie much of the eastern half of Lake Superior, and their erodability accounts for the location of the lake and its great depth.

The Makkovik Province occupies a wedge in coastal Labrador, flanked by the Grenville province to the south and by Arctic Cordilleran landscapes of the Nain province to the west and north. Rock units range from Archaean to Palaeo-Proterozoic and consist of a variety of granitic plutons, metamorphosed sediments, and mylonite shear zones. The coastline is a complex of indentations and bedrock islands and skerries.

The Nain Province extends along the Labrador coastline from Nachvak Bay southward to the Kanairktok River. Inland, its limits are defined by younger deformed rocks and intrusions of the Churchill Structural Province. Rocks range from Archean to Meso-Proterozoic (3800 to 1305 million years old). Rock types include gneisses, basalts, layered intrusions of basic norite, gabbro, and anorthosite, sandstones, and chert (microcrystalline silica) beds. The anorthosite units, formed almost exclusively of labradorite plagioclase feldspar, can be used as facing stone and ornamental flooring in buildings and also are valuable for jewellery production. Ramah chert has been used for the manufacture of artefacts throughout the human occupation of Labrador.

The Superior Province forms a broad arc from the Nelson River basin in northern Manitoba across northern Ontario, through northern Québec and Nunavik along the eastern shore of Hudson Bay. The Superior Province is flanked by the Grenville Province to the southeast, and by the Churchill Province to the west and north. The rocks of the Superior Province were formed during the Archaean Eon, and were deformed during the Kenoran Orogeny, 2480 million years ago. Alternating belts of resistant gneiss/granite and weaker volcanic "greenstone" bedrock, have produced a characteristic corrugated topography marked by ridges and valleys accentuated by faulting throughout the Superior Province.

The Tu Nedhe Province (formerly "Slave") in the northwestern Canadian Shield is named for the large lake in the southwestern NWT. Rock types include granite, gneiss, migmatite, metamorphosed volcanic rocks, and metamorphosed sedimentary rocks and schists, formed during the Archean Eon and deformed ca. 2400 million years ago. Younger Proterozoic basalt sills, metamorphosed volcanic units, and dolomites are also present. Kimberlite pipes have intruded the Precambrian rocks. Kimberlite, a basic rock containing olivine and phlogopite mica, also can contain diamonds (Fig. 1.6).

1.4 Quaternary Glaciation

The Quaternary is the most recent period of geological time, extending from 2.6 million years ago (2.6 Ma) to the present. In Canada, the Quaternary is characterized by repeated glacial advances, separated by warmer interglacial periods. The Quaternary period of geological time is defined in the Northern Hemisphere to encompass the time from the initiation of continental glaciation in North America, up to the present instant.

The onset of Quaternary glaciation in the Northern Hemisphere saw the formation of glaciers ranging in size from individual cirques confined to a single mountain, to continent-wide complexes of glaciation (inlandsis). Alpine glaciers developed in all the mountain range throughout Canada. Discussion of the Quaternary history of each ecozone is presented in the following chapters, with a brief summary of glacial history here.

Several phases of continental glaciation affected Canada during the Quaternary (e.g. Dyke et al. 2003; Dyke 2004; Stokes et al. 2015) (Fig. 1.7 and Fig. 1.8). Glacial deposits pre-dating the last interglacial (130,000 years ago) have been identified only in isolated locations, including central Yukon, western NWT, southern Alberta and Saskatchewan, southern Ontario, southern Québec, Qikiqtaaluk (Baffin Island), Maritime Canada, and Newfoundland. With the exception of central Yukon and western NWT, the deposits are exposed only in sections, without associated surface landforms.

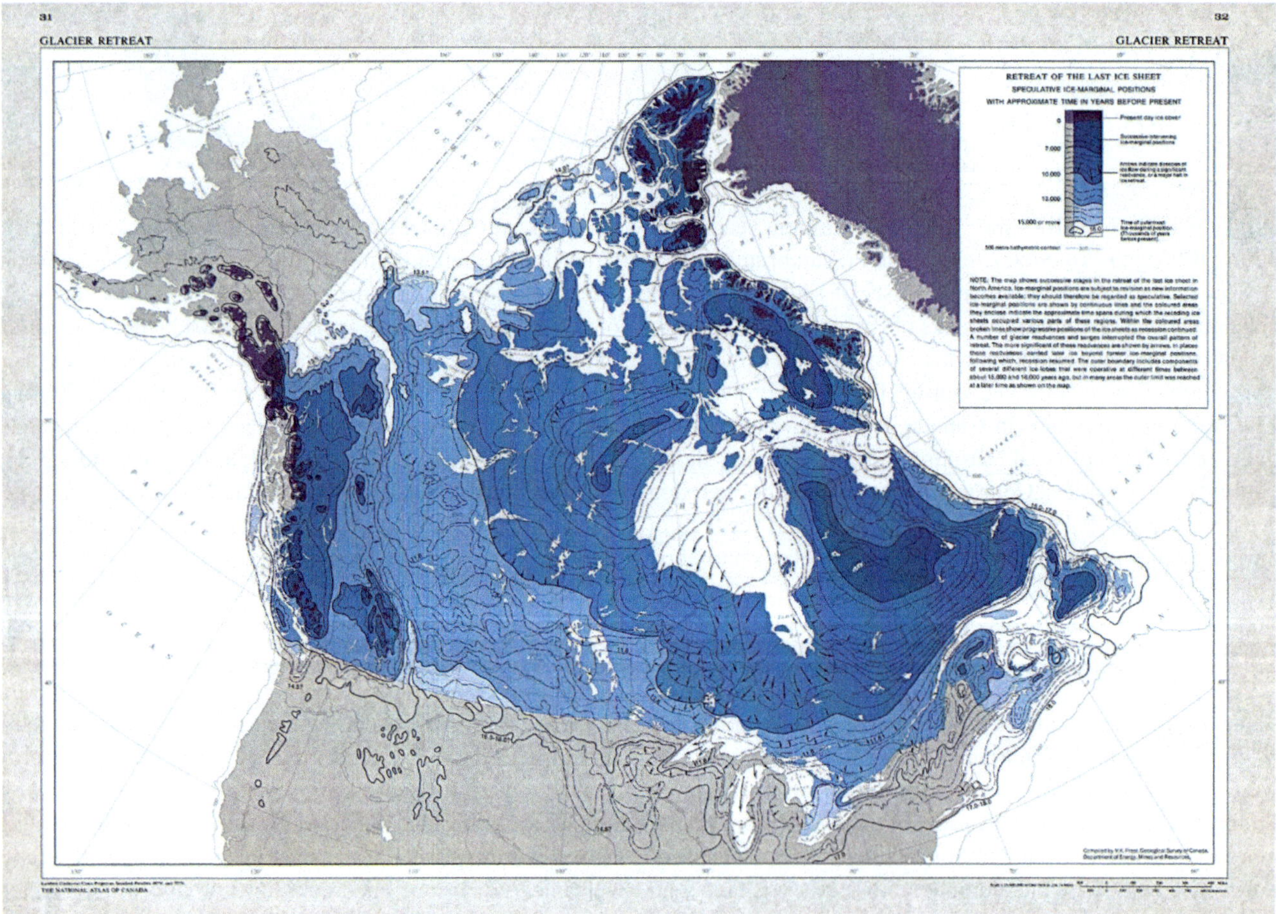

Fig. 1.7 Glaciation of Canada: retreat of the last ice sheet. (Natural Resources Canada, Atlas of Canada; https://open.canada.ca/)

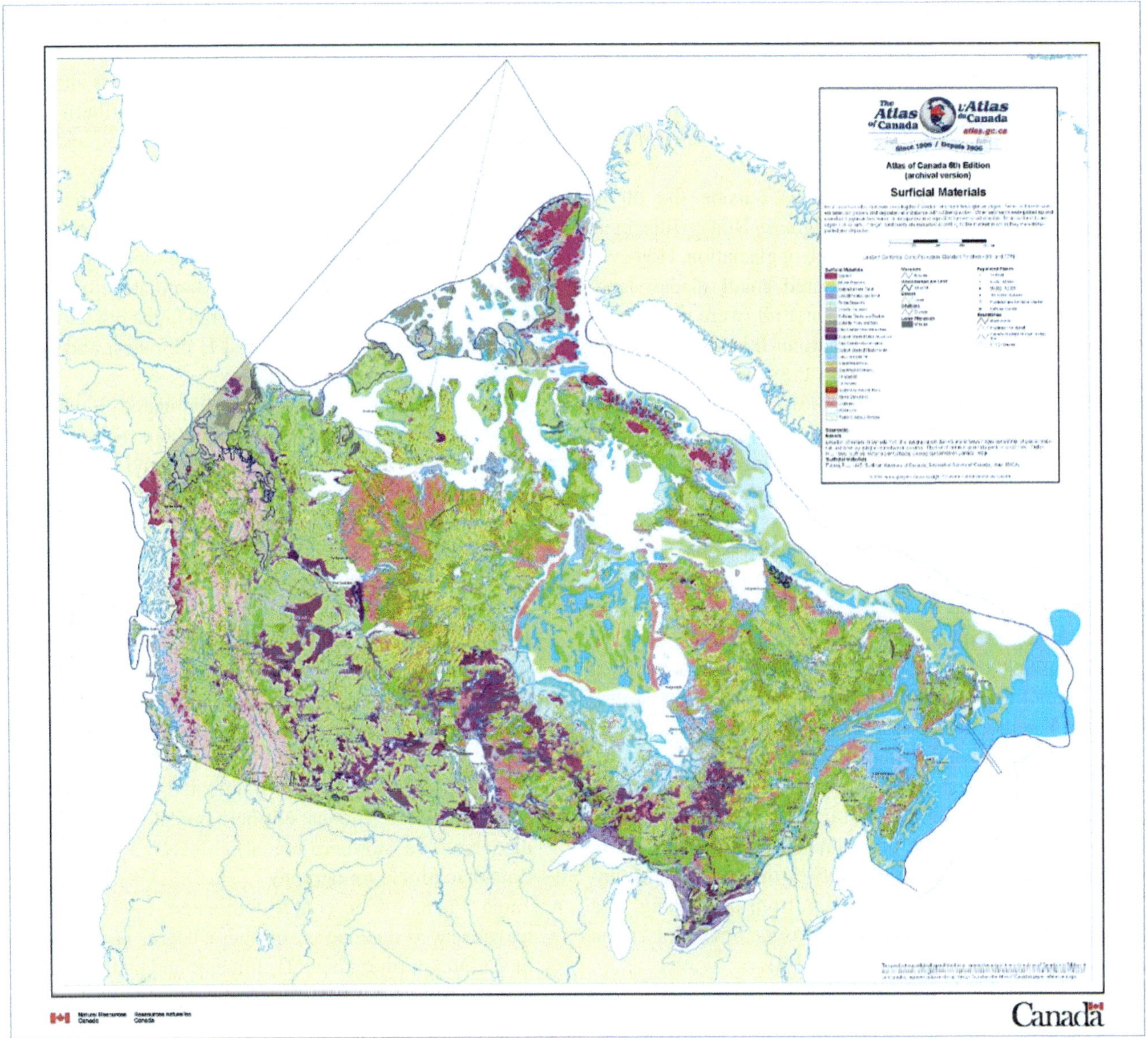

Fig. 1.8 Glacial deposits and landforms. Geological Survey of Canada Map 1880A (Fulton 1995). (Natural Resources Canada, https://doi.org/10.4095/205040)

Most of the evidence preserved in landscapes relates to the most recent glacial event, between 29,000 and 11,700 years ago (e.g. Fulton 1995; Catto 2015; Slaymaker 2017; Slaymaker and Catto 2020). The term "Laurentide" is applied to all continental glaciation which originates from any area of the Canadian Shield and dates to the most recent glaciation, with final deglaciation during the initial phases of the Holocene. The Laurentide Inlandsis included several centres, domes, sectors, and ice sheets, which coalesced to produce the inlandsis. Ice from a glacial accumulation area centred west and north of Hudson Bay in mainland Nunavut (Kivalliq) expanded to the south and west to cover mainland Northwest Territories and Nunavut, and most of the western prairies. Another accumulation centre formed over Foxe Basin and southern Qikiqtaaluk. Ice from a third accumulation area centred east of Hudson Bay in western Labrador and adjacent Nunavik (northern Québec) extended eastward into the Labrador Sea, northward to block Hudson Strait and coalesce with ice over Qikiqtaaluk and Foxe Basin, westward across northern Québec and Hudson Bay to coalesce with ice in eastern Kivalliq and northern Manitoba, and southward to Ontario and Maritime Canada.

The Cordilleran Inlandsis formed over central British Columbia, originating in the Prince George area. Ice radiated southwards and westwards to cover British Columbia and extend into Washington and Idaho, northwards to central Yukon, and eastwards to the Rocky Mountains. A tongue of Cordilleran ice moved through Yellowhead Pass in Jasper

National Park, coalesced with westward-moving Laurentide ice between Edson and Hinton, Alberta, and then flowed south along the Rocky Mountain Foothills. Coalescence of the two ice masses is indicated by the "Erratics Train", including the Okotoks erratic, marking the suture line between ice from east and west across southwestern Alberta.

In northern Nunavut, many areas outside the alpine glaciated high mountains of Umigmak Nuna (Ellesmere Island) show relatively little evidence of glaciation. However, the presence of erratic and scattered small glacial landforms indicates that most areas, if not all, were covered by glacial ice at some time. The Innuitian Inlandsis expanded across northern Nunavut, reaching the continental shelf of the Beaufort Sea (England et al. 2006).

Local ice caps grew and coalesced throughout Atlantic Canada, blocking intrusion of the Laurentide ice from the north (Shaw et al 2006; Slaymaker and Catto 2020). Laurentide ice coalesced with local ice caps to form ice streams through the Gulf of St. Lawrence, which was glaciated due to marine drawdown to 120 m below present sea level. Offshore, the Grand Banks, Scotian Shelf, and Georges Bank were exposed by lower sea level. Locally ice streams extended along the major embayments to the continental shelf. Glacial flow patterns and deglaciation were complex, but after Younger Dryas readvances throughout the region, rising sea level resulted in rapid deglaciation.

Northern and western Yukon escaped glaciation during the Quaternary, and is the largest tract of unglaciated terrain within Canada. The cold, dry climate initiated in the early Quaternary as a result of the uplift of the St. Elias Mountains to the south prevented glaciers from developing, and also prevented the Cordilleran Inlandsis ice of central Yukon from expanding northwestward (Fig. 1.7).

1.5 Ecozones

Terrestrial Ecozones, including several smaller ecoregions, are based on combinations of vegetation, fauna, climate, and terrain (Environment Canada and Agriculture and Agrifoods Canada 1995). Fifteen terrestrial ecozones are currently recognized (Fig. 1.2):

- Atlantic Maritime, with mixed broadleaved deciduous–narrowleaved deciduous–coniferous forests; southern- to mid-boreal fauna; Dfb (mid-boreal) climates; and rolling mountainous to planar terrain.
- Mixedwood Plains, with broadleaved deciduous forests including narrowleaved deciduous and conifers; southern- to mid-boreal fauna; Cfa (temperate), Dfa (southern boreal), and Dfb (mid-boreal) climates; and plains.
- Boreal Shield, with confer-dominated forest including narrowleaved deciduous; mid-boreal to northern boreal fauna; moist-to-moderate Dfb (mid-boreal) climates; and undulating Canadian Shield topography.
- Boreal Plains, with confer-dominated forest including narrowleaved deciduous; mid-boreal to northern boreal fauna; dry-to-moist Dfb (mid-boreal) climates; and plains to rolling hills.
- Prairies, with tallgrass, midgrass. Shortgrass, and prairie parkland assemblages; prairie fauna; BSk (steppe) to dry Dfb (mid-Boreal) climates; and prairie topography.
- Montane Cordillera, with confer-dominated forests and alpine tundra vegetation; Cordilleran fauna reflecting altitudinal zonation; BSk, Dfb, and H climates; and mountainous topography with plateaux.
- Boreal Cordillera, with confer-dominated forests and alpine tundra vegetation; Cordilleran fauna reflecting altitudinal zonation; Dfc and H climates; and mountainous topography with plateaux.
- Taiga Cordillera, with confer-dominated forests and alpine tundra vegetation; Cordilleran and southern Arctic fauna; Dfc, Dfd, and H climates; and mountainous topography with plateaux.
- Pacific Maritime, with warmth-loving coniferous forests; southern Cordilleran fauna; Cfb (maritime West Coast), Cfc (cool maritime west coast), and Csb (summer dry maritime West Coast) climates; and mountainous topography.
- Taiga Shield, with confer-dominated forest including narrowleaved deciduous; northern boreal fauna; moist-to-moderate Dfc (northern boreal) climates; and undulating Canadian Shield topography.
- Taiga Plains, with confer-dominated forest including narrowleaved deciduous; northern boreal fauna; dry-to-moist Dfc (northern boreal) climates; and plains to rolling hills.
- Hudson Plains, with tundra vegetation; northern boreal to Arctic fauna; Dfc to ET (tundra) climates; and plains with abundant wetlands.
- Arctic Cordillera, with alpine tundra vegetation; limited fauna; ET, EF, and H climates; and mountainous topography.
- Southern Arctic, with tundra vegetation; northern boreal to Arctic fauna; ET (tundra) climates; and flat to undulating topography.
- Northern Arctic, with tundra vegetation; Arctic fauna; ET (tundra) climates; and flat to undulating topography.

Throughout the volume, we have used both scientific (Linnaean) and local common English names. Using both naming systems is necessary, because people living in different parts of Canada have tendencies to use different names for the same species, or the same name for different species. This is especially true for plants. Tables 1.3 and 1.4 list Linnaean and common names for selected plant and faunal species, respectively (Fig. 1.6).

Table 1.3 Scientific and common names for plant species in Canada

Scientific name	Common names	Typical habitats
Conifers—Cedars		
Thuja occidentalis	Cedar, Eastern White Cedar, Arbor vitae	Eastern Boreal forest
Thuja plicata	Western Red Cedar	Pacific Coast;
Callitropsis nootkatensis	Yellow Cypress; Nootka Cypress	Pacific Coast
Junipers		
Juniperus communis	Common juniper	Boreal forest
Juniperus horizontalis	Trailing Juniper	Boreal forest
Juniperus scopulorum	Rocky Mountain (Western) Juniper	Montane BC
Juniperus virginiana	Eastern Red Cedar	Southern Great Lakes
Pines		
Pinus albicaulis	Whitebark Pine	Subalpine Cordillera
Pinus banksiana	Jack Pine; Princess Pine	Boreal forest
Pinus contorta	Lodgepole Pine; Shore Pine	Cordillera
Pinus flexilis	Limber Pine	Southern Cordillera
Pinus monticola	Western White Pine	Pacific; Columbian
Pinus ponderosa	Ponderosa Pine	Southern Interior BC
Pinus resinosa	Red Pine	SE Boreal
Pinus rigida	Pitch Pine	Thousand Islands NP
Pinus strobus	Eastern White Pine	SE Boreal
Firs		
Abies amabilis	Amabilis Fir; Cascades Fir; Pacific Silver Fir	Pacific Coast
Abies balsamea	Balsam Fir; Var	Boreal Forest
Abies grandis	Grand Fir	Pacific; Columbian
Abies lasiocarpa	Alpine Fir; Rocky Mountain, Fir	Cordillera
Hemlocks		
Tsuga canadensis	Eastern Hemlock	Acadian; Temperate
Tsuga heterophylla	Western Hemlock	Pacific; Columbian
Tsuga mertensiana	Mountain Hemlock	Cordillera
Douglas Firs		
Pseudotsuga menziesii menziesii	(coastal) Douglas Fir	Pacific Coast
Pseudotsuga menziesii glauca	Interior Douglas Fir; Blue Douglas Fir	SW Cordillera
Spruces		
Picea engelmannii	Engelmann Spruce	Rocky Mountains
Picea glauca	White Spruce; Cat Spruce; Northern Spruce	Boreal Forest
Picea mariana	Black Spruce	Boreal Forest
Picea rubens	Red Spruce	Acadian
Picea sitchensis	Sitka Spruce	Pacific Coast
Larches		
Larix laricina	Tamarack; Larch	Boreal forest
Larix lyallii	Alpine Larch	Southern Rocky Mts
Larix occidentalis	Western Larch	S. Columbian
Yews		
Taxus brevifolia	Western Yew	Pacific Maritime
Taxus canadensis	Canada Yew; Palm	Boreal Forest
Deciduous—Ashes		
Fraxinus americana	White Ash	Temperate forest

(continued)

Table 1.3 (continued)

Scientific name	Common names	Typical habitats
Fraxinus nigra	Black Ash	Acadian; Temperate
F. pennsylvanica austini	Northern Red Ash	Southern Boreal Forest
F. pennsylvanica asunintegerrina	Prairie Green Ash	Assiniboine Valley
Fraxinus quadrangulata	Blue Ash	Carolinan Forest
Basswood		
Tilia americana	Basswood; lime	Temperate
Beeches		
Fagus grandifolia	Beech	Temperate; Acadian
Castanea dentata	Chestnut	Temperate; Carolinan
Oaks		
Quercus alba	White Oak	Temperate
Quercus bicolor	Swamp White Oak	Carolinan
Quercus garryana	Garry Oak	Southern Pacific Coast
Quercus macrocarpa	Bur Oak	Eastern Prairies
Quercus muehlenbergii	Chinquapin Oak	Carolinan
Quercus palustris	Pin Oak	Carolinan
Quercus prinus	Chestnut Oak	Carolinan
Quercus rubra	Northern Oak; Red Oak	Temperate; Acadian
Quercus shumardi	Shumard's Oak	Temperate
Quercus velutina	Black Oak	Temperate; Carolinan
Maples		
Acer circinatum	Vine Maple	Southern Pacific Coast
Acer glabrum	Douglas Maple	Southern Montane
Acer macrophyllum	Bigleaf Maple	Pacific Coast
Acer negundo	Manitoba Maple; Box-Elder	Prairies; temperate
Acer nigrum	Black maple	Temperate
Acer pensylvanicum	Striped Maple; Moosewood	Temperate; Acadian
Acer rubrum	Red Maple	Temperate; Acadian
Acer saccharinum	Silver Maple	Temperate; Carolinan
Acer saccharum	Sugar Maple	Temperate; Acadian
Acer spicatum	Mountain Maple	Eastern Boreal forest
Apples		
Malus coronaria	Wild Crab Apple	Carolinan
Cherry family		
Prunus americana	Wild Plum	Carolinan
Prunus emarginata	Bitter Cherry	Southern Pacific Coast
Prunus nigra	Canada Plum	Carolinan
Prunus pennsylvanica	Pin Cherry	Temperate; S Boreal
Prunus serotina	Black Cherry	Temperate
Prunus virginiana	Choke Cherry	Temperate; S Boreal
Birch Family		
Alnus incana rugosa	Grey or Hoary Alder	Boreal
Alnus rubra	Red alder	Pacific Coastal
Alnus viridis (alnobetula) crispa	Green Alder; Mountain Alder	Boreal
Alnus viridis sinuata	Sitka Alder	Pacific Coastal
Betula alleghaniensis	Yellow Birch	Temperate; Acadian

(continued)

Table 1.3 (continued)

Scientific name	Common names	Typical habitats
Betula caerulea	Blueleaf Birch	Acadian
Betula cordifolia	Mountain White Birch	Boreal; Acadian
Betula glandulosa	Dwarf Birch, resin birch, shrub birch	Northern Taiga
Betula lenta	Cherry (sweet) Birch	Port Dalhousie ON
Betula nana nana	Arctic Birch	E Baffin Island
Betula neoalaskana	Alaska Birch	Taiga
Betula occidentalis	Water Birch	Prairies; W Boreal
Betula papyrifera	White Birch; Paper Birch	Boreal; Temperate
Betula populifolia	Grey Birch	Acadian
Corylus cornuta	(beaked) Hazel	Temperate; Acadian
Carpinus caroliniana	Hornbeam; Bluebeech	Temperate; Acadian
Ostrya virginiana	Ironwood; Hop-hornbeam	Temperate; Acadian
Dogwoods		
Cornus alternifolia	Alternate-leaved dogwood	Temperate; Acadian
Cornus canadensis	Bunchberry	Boreal Forest
Cornus florida	Flowering Dogwood	Carolinan
Cornus nuttalii	Pacific Dogwood	Pacific Coast
Elm family		
Ulmus americana	White Elm; American Elm	Temperate
Ulmus rubra	Red Elm	Temperate
Ulmus thomasii	Rock Elm	Temperate
Walnut family		
Carya cordiformis	Bitternut Hickory	Temperate
Carya glabra	Pignut Hickory	Carolinan
Carya laciniosa	Big Shellbark Hickory	Carolinan
Carya ovata	Shagbark Hickory	Temperate
Juglans cinerea	Butternut	Temperate
Juglans nigra	Black walnut	Temperate
Mountain Ash/Dogberry		
Sorbus americana	Dogberry, rowan, Mountain ash	Boreal forest
Sorbus decora	Showy Mt. Ash, Dogberry	Boreal forest
Willows and Poplars	*(more than 75 species nationally)*	
Populus angustifolia	Narrowleaf cottonwood	SW Prairies
Populus balsamifera	Balsam Poplar	Boreal Forest
Populus deltoides deltoides	Eastern Cottonwood	Temperate
Populus deltoides monilifera	Plains Cottonwood	Prairie Parkland
Populus grandidentata	Large-toothed Aspen	Temperate
Populus tremuloides	Trembling Aspen	Boreal Forest
Populus trichocarpa	Black Poplar; Western Balsam Poplar	Pacific; montane
Salix amygdaloides	Peachleaf Willow	Temperate
Salix bebbiana	Beaked Willow	Boreal Forest
Salix discolor	Pussy Willow	Boreal Forest
Salix hookeriana	Hooker's Willow	Pacific Coast
Salix lasiandra	Pacific Willow	Pacific Coast
Salix lucida	Shining Willow	Boreal Forest
Salix nigra	Black Willow	Temperate; Acadian

(continued)

Table 1.3 (continued)

Scientific name	Common names	Typical habitats
Salix scouleriana	Scouler's Willow	Prairie; interior BC
Salix candida	Sageleaf willow	
Other Deciduous Trees and Shrubs		
Arbutus menziesii	Arbutus; Madrone	S Pacific Coast
Asimina triloba	Pawpaw	Carolinan
Celtis occidentalis	Hackberry	Carolinan
Cercis canadensis	Redbud	Carolinan
Crategus sp.	Hawthorn	Temperate
Gleditsia triacanthos	Honey-locust	Carolinan
Gymnocladus dioicus	Kentucky Coffee-tree	Carolinan
Hamamelis virginiana	Witch Hazel	Carolinan
Liriodendron tulipfera	Tulip-tree	Carolinan
Magnolia acuminata	Cucumber-tree	Carolinan
Magnolia macrophylla	Bigleaf Magnolia	Carolinan
Morus rubra	Red Mulberry	Carolinan
Nyssa sylvatica	Black Gum; Tupelo	Carolinan
Platanus occidentalis	Sycamore	Carolinan
Ptelea trifoliata	Hop-Tree; skunk bush	Carolinan
Rhus typhina	Staghorn Sumac	Temperate
Toxicodendron vernix	Poison Sumac	Temperate
Sambucus nigra	Blue Elder	Interior BC
Sambucus pubens	Eastern Red Elderberry	Temperate
Sassafras albidum	Sassafras	Carolinan
Toxicodendron diversilobium	Pacific Poison Oak	Pacific Coast
Viburnum lentago	Blackhaw; Nannyberry; Viburnum	Temperate
Zanthoxylum americanum	Prickly Ash	Temperate
Herbs, Shrubs, Grasses, Ferns, Lichens, Mosses	**Taxa are listed in alphabetical order by scientific name, does not imply taxonomic affinity**	
Aconitum delphinifolium	Mountain Monkshood	Pacific Cordilleran
Adiantum pedatum	Northern maidenhair Fern	Boreal
Agropyron smithii	Western Wheat Grass	Prairie
Alopecurus alpinus	Foxtail	Arctic
Amelanchier	Chuckley-pear	Widespread
Amelanchier alnifolia	Saskatoon	Prairie
Ammophila brevigulata	Marram Grass	Coastal
Andromeda polifolia	Andromeda; Bog-rosemary	Boreal; Taiga
Andropogon	Bluestem grasses	Prairie
Andropogon scoparius	Little Bluestem Grass	Prairie
Androsace septentrionalis	Andosace	Tundra
Anemone parvifolia	Anemone	Tundra; Taiga
Anemone patens wolfgangiana	Prairie Crocus, pasqueflower	Prairie (Manitoba)
Anemonella thalicitores	Rue-anemone	Thousand Islands NP
Antennaria lanata	Everlasting	Alpine
Aquilegia jonesii	Jones' Blue Columbine	Alpine
Arctagrastis latifolia	Arctagrastis	Tundra
Arctostaphylos alpina	Alpine bearberry	Montane; Boreal

(continued)

Table 1.3 (continued)

Scientific name	Common names	Typical habitats
Arctostaphylos rubra	Red Bearberry	Boreal
Arctostaphylos uva-ursi	Bearberry, kinnikinnick	Widespread
Arenaria	Sandwort	Boreal
Armeria maritima	Sea Thrift; sea pink	Coastal; saline
Artemisia tridentata	Sagebrush	Widespread
Asplenium virida	Green spleenwort	Boreal
Aster conspicuus	Showy aster	Montane; Boreal
Aster yukonensis	Yukon Aster	Boreal
Astralagus alpinus	Alpine Milk-vetch	Arctic; Cordillera
Astralagus williamsii	Milk-vetch	Cordillera
Azalea loiseleuria	Northern Azalea	Taiga; Tundra
Baptisia tinctoria	Wild Indigo	Temperate
Beckmannia syzigachne	Prairie Slough Grass	Temperate
Bouteloua spp.	Grama Grass	Prairie
Braya purpurescens	Braya	Arctic
Calamagrostis purpurescens	Reed Bentgrass	Subalpine
Calamagrostis rubescens	Pinegrass	Montane
Campanula rotundiflora	Bluebell	Prairie
Carex spp.	Sedges	Widespread
Cassiope mertensiana	Alpine Heather	Alpine
Cassiope tetragona	Arctic White Heather	Tundra
Castilleja	Paintbrush; Prairie Fire	Prairie
Cerastium alpinum	Mouse-eared Chickweed	Alpine
Chamerion angustifolium	Narrow-leaved fireweed	Boreal; Taiga
Chamerion latifolium	Broad-leaved fireweed	Boreal; Taiga
Cicuta douglasii	Western Water Hemlock	Cordillera
Cicuta muculata	Water Hemlock	Temperate
Cirsium pitcheri	Pitcher's Thistle	Pukaskwa NP
Cladonia rangifera	Caribou Moss	Tundra; Taiga
Claytonia sarmentosa	Alaska Springbeauty	Pacific Cordilleran
Clintonia borealis	Clintonia	Boreal
Clintonia unifolia	Queen's-cup lily	Cordillera
Cochlearia	Scurvy-grass	Coastal
Conopholis americana	Cancer-root	Acadian
Coptis trifolia	Goldthread	Acadian
Corallorhiza maculata	Spotted coral-root	Acadian
Corallorhiza trifida	Early coral-root	Acadian
Corydalis pauciflora	Blue Corydalis	Pacific Cordilleran
Cypripedium acaule	Pink Lady's-slipper	Acadian
Cypripedium calceolus	Yellow Lady's-slipper	Boreal carbonate
Cypripedium franklinii	Franklin's Lady's-slipper	Pukaskwa
Cypripedium montanum	Yellow Mountain Orchid	Subalpine
Cypripedium passerinum	Sparrow-egg Lady's-slipper	Boreal carbonate
Cypripedium reginae	Showy Lady's-slipper	Acadian; Temperate
Cystopteris fragilis	Fragile Fern	Taiga

(continued)

Table 1.3 (continued)

Scientific name	Common names	Typical habitats
Cystopteris montana	Mountain Bladder Fern	Alpine; Taiga
Cytisus scoparius	Scotch Broom	Pacific non-native
Danthonia	Oatgrass	Prairie
Delphinium	Delphinium; Larkspur	Widespread
Deschampsia brevifolia	Hairgrass	Arctic
Deschampsia cespitosa	Tuft-Grass	Alpine
Diapensia lapponica	Diapensia	Tundra; Taiga
Dodecatheon conjugens	Shooting-Star	Saline
Draba	Draba	Taiga; Tundra
Drosera	Sundew	Boreal
Dryas drummondii	Yellow Dryas	Boreal
Dryas integrifolia	Mountain Avens	Boreal; Taiga
Dryas octopetala	Western Mountain Avens	Alpine
Dryopteris disjunta	Oak Fern	Acadian
Dryopteris fragrans	Fragrant Shield Fern	Boreal; Taiga
Dupontia fisherii	Dupontia	Arctic
Elaeganus argentea	Silverberry	Prairie
Elymus	Beach Rye	Pacific Coastal
Elymus innovatus	Wild Hairy Rye	Montane
Empetrum eamesii	Pink Crowberry	Taiga; Tundra
Empetrum nigrum	Crowberry, Curlewberry	Taiga; Tundra; Boreal
Epigaea repens	Mayflower; Trailing Arbutus	Temperate; Acadian
Equisetum arvense	Horsetail	Widespread
Erigeron	Fleabane	Widespread
Eriogonum	Buckwheat	Prairie
Eriophorum	Cottongrass	Taiga; tundra
Erysimum inconspicuum	Small-flowered Rocket	Boreal carbonate
Escobaria vivipara	Purple Ball Cactus	Prairie
Festuca	Fescue	Prairie
Fragaria virginiana	Strawberry	Temperate
Gaultheria shallon	Salal	Pacific Coastal
Gentianella	Gentian	Boreal
Gentianella detonsa	Island Gentian	Boreal
Geranium robertanium	Robert's Geranium	Temperate; S Boreal
Geum triflorum	Geum; Prairie Avens	Prairie; S Boreal
Goodyera tesselata	Rattlesnake Plantain	Temperate
Gymnocarpium robertanium	Limestone Polypody	Boreal CO_3
Habernaria dilatata	Bog Orchid	Boreal CO_3; Arctic
Habernaria hookeri	Hooker's Orchid	Boreal
Helianthus	Sunflower	Prairie
Heracleum lanatum	Cow-Parsnip	Montane
Hesperostipa comata	Spear Grass	Prairie
Hesperostipa spartea	Porcupine Grass	Prairie
Hierochloe	Holygrass	Arctic
Hydrocotyle umbellata	Water Pennywort	Temperate
Hylocomium splendens	Feathermoss	Montane

(continued)

Table 1.3 (continued)

Scientific name	Common names	Typical habitats
Ilex	Holly	Pacific Coastal
Iris versicolor	Purple Iris	Boreal
Isotria medeoloides	Small Whorled Pegonia	Carolinan
Isotria verticillata	Large Whorled Pegonia	Carolinan
Japonicus glaber	Beach Pea	Coastal
Juncus	Rushes	Widespread
Juncus arcticus	Northern Bog Rush	Boreal; Taiga
Kalmia angustifolia	Sheep Laurel	Boreal
Kalmia polifolia	Bog Laurel	Boreal
Koeleria cristata	June Grass	Prairie
Leucanthemum vulgare	White Daisy; Ox-eyed Daisy	Non-native
Lilium philadelphicum	Prairie Lily	Prairie
Linnaea borealis	Twinflower	Boreal; Taiga
Linum lewisii	Flax	Prairie
Listera australis	Southern Twayblade	Acadian
Listera borealis	Twayblade	Boreal
Listera convallariodes	Broadlipped twayblade	Boreal
Lobelia kalmii	Kalm's Lobelia	Boreal
Lonicera involucrata	Black Twinberry	Pacific Coastal
Lupinus	Lupin	Widespread
Luzula nivalis	Luzula	Arctic
Lychnis alpina	Alpine catchfly	Boreal
Lycopodium	Club Moss	Widespread
Lysichiton americanum	Skunk Cabbage	Cordillera
Lysimachia maritima	Sea Milkwort	Saline
Mahonia (Berberis) repens	Oregon Grape	Pacific Cordilleran
Malus fusca	Pacific Crabapple	Pacific Coastal
Menziesii ferruginea	False Azalea	Subalpine
Menziesii glabella	False Azalea	Subalpine
Mertensia maritima	Oyster leaf	Coastal
Myrica gale	Sweet Gale	Boreal
Oplopanax horridus	Devil's Club	Cordillera
Opuntia humifusa	Eastern Prickly Pear	Carolinan
Opuntia polyacantha	Prickly Pear	Prairie
Oxyria	Sorrel	widespread
Oxytropis	Locoweed	Prairie
Panicum virgatum	Panic Grass	Prairie
Papaver cornwalliensis	Poppy	Arctic
Papaver nudicale	Arctic Poppy	Arctic
Papaver pygmaeum	Alpine Poppy	Alpine
Papaver radicum	Poppy	Arctic
Parrya arctica	Parrya	Arctic
Phacelia lyalli	Lyall's Scorpion Weed	Subalpine
Phippsia algida	Phippsia	Arctic
Phyllodoce empetriformis	Pink Mountain Heather	Subalpine

(continued)

Table 1.3 (continued)

Scientific name	Common names	Typical habitats
Phyryma leptostachya	Lopseed	Temperate carbonate
Plantago rugelli	Rugell's Plantain	Temperate
Poa secunda	Sandberg's Blue Grass	Prairie
Poaceae	Grasses	Widespread
Polemonium viscosum	Sky Pilot	Alpine
Polygana incarnata	Pink Milkwort	Carolinan
Polygonum	Knotweed	Widespread
Polystichum lonchitis	Northern Holly Fern	Widespread
Potentilla	Cinquefoil	Widespread
Potmamogeton	Pondweed	Widespread
Primula egaliksensis	Greenland Primrose	Taiga; Tundra
Prunella vulgaris	Heal-all	Pacitic Coastal
Pteridum aquilinum	Bracken	Boreal
Puccinella	Goose-Grass	Taiga; Tundra
Pyrola	Wintergreen	Widespread
Ranunculus	Buttercup	Widespread
Rhacomitrium	Rhacomitrium	Boreal
Rhizocarpon geographicum	Rhizocarpon	Widespread
Rhododendron albiflorum	White-flowered Rhododendron	BC Cordillera
Rhododendron canadense	Rhodora	Boreal
Rhododendron decumbens	Northern Labrador Tea	Taiga
Rhododendron groenlandicum	Labrador Tea	Boreal; Taiga
Rhododendron lapponicum	Lapland Rosebay	Boreal; Taiga
Rhus copallina	Winged Sumac	Thousand Islands NP
Ribes	Gooseberry	Widespread
Rosa acicularis	Prairie Rose	Prairie
Rosa arkansana	Low Prairie Rose	Prairie
Rosa nutkana	Nootka Rose	Pacific Coastal
Rubus canadensis	Blackberry	Widespread
Rubus chamaemorus	Bakeapple; cloudberry	Boreal; taiga
Rubus ideaus	Raspberry	Boreal; Temperate
Rubus parviflorus	Thimbleberry	Montane
Rubus spectablis	Salmonberry	BC Cordillera
Rudbeckia serotina	Black-eyed Susan	Widespread
Salicornia rubra	Samphire	Saline
Salvia	Sage	Widespread
Sanguisorba canadensis	Burnet; bottlebrush	Boreal; taiga
Sarracenia purpurea	Pitcher Plant	Boreal; Taiga
Saxifraga purpurescens	Purple Saxifrage	Arctic
Scirpus cespitosus	deergrass	Boreal
Sedum acre	Stonecrop	Boreal
Seliginella wallaceii	Spikemoss	Pacific Coastal
Senecio psuedoarnica	Seabeach Senecio	Saline
Shepherdia canadensis	Soapberry, Buffalo berry	Widespread
Silene	Campion	Boreal; Taiga; Arctic
Solidago hispida	Goldenrod	Widespread

(continued)

Table 1.3 (continued)

Scientific name	Common names	Typical habitats
Solidago ptaemicoides	White Aster	Temperate carbonate
Spartina pectinata	Cord Grass	Prairie
Sphagnum	Sphagnum	Widespread
Streptopus roseus	Rose-twisted stalk	Temperate
Suaeda	Sea Blite	Saline
Symphoricarpos	Snowberry	Montane
Thalictrum alpinum	Alpine Meadow-Rue	Boreal carbonate
Toxicodendron radicans	Poison Ivy	Temperate; S Boreal
Toxicoscordion venenosum	Death Camas	Prairie
Triantha glutinosa	False Asphodel	Arctic
Trientalis borealis	Starflower	Boreal
Trifolium pratense	Red Clover	Non-native
Triglochum maritima	Seaside Arrowroot	Taiga; Arctic
Trillium cernuum	Nodding Trillium	Acadian
Trillium erectum	Red Trillium; Wake-Robin	Temperate
Trillium grandiflorum	White Trillium	Temperate
Trillium undulatum	Painted Trillium	Temperate
Trisetum spicatum	Spike False-Oak	Alpine
Trollius laxus	Globeflower	Subalpine
Typha latifolia	Cattail	Wetlands
Ulex	Gorse	Pacific Coastal
Utriculata vulgaris	Bladderwort	Widespread
Vaccinium angustifolium	Low Blueberry	Boreal
Vaccinium boreale	Dwarf Blueberry	Boreal; Taiga
Vaccinium caespitosum	Dwarf Bilberry	Montane
Vaccinium macrocarpon	Cranberry	Boreal; Taiga
Vaccinium myrtiloides	Blueberry	Boreal; Temperate
Vaccinium oxycoccos	Bog Cranberry	Boreal
Vaccinium scoparium	Grouseberry	Subalpine
Vaccinium stamineum	Deerberry; huckleberry	Thousand Islands NP
Vaccinium uliginosum	Bilberry	Boreal; Taiga; Arctic
Vaccinium vitis-idaea	Partridgeberry; lingonberry	Boreal; Taiga
Verbena	Verbena	Pacific Coastal
Viburnum dentatum	Southern Arrowwood	Thousand Islands NP
Vicia gigantia	Giant Vetch	Pacific Coastal
Viola culcullata	Marsh Blue Violet	Temperate; Boreal
Viola nephrophylla	Northern Bog Violet	Boreal
Viscum album	Mistletoe	Pacific Coastal
Vitis aestivalis	Summer Grape	Thousand Islands NP
Vitis labrusca	Fox Grape	Temperate
Woodsia	Woodsia	Arctic
Xanthoria elegans	Xanthoria lichen	Widespread
Xerophyllum tenax	Bear Grass	Waterton Lakes NP
Yucca	Yucca	Dry BC interior
Zizania aquatica	Wild Rice	Boreal

Table 1.4 Scientific and common names for faunal species in Canada

Scientific name	Common name(s)	Typical habitats
Mammals		
Alces alces	Moose	Boreal/Taiga
Antilocarpa americana	Pronghorn	Grasslands
Aplodontia rufa	Mountain beaver	Okanagan
Bison bison athabascensis	Wood bison	Boreal
Bison bison bison	Plains bison	Grasslands
Canis latrans	Coyote	Widespread
Canis lupus	Wolf	Boreal/Taiga
Castor canadensis	Beaver	Widespread
Cervus canadensis	Wapiti	Boreal/Cordillera
Cervus dama	Fallow deer	James I, Gulf Islands
Clethrionomys gapperi	Gapper's red-backed vole	Taiga/mid-Boreal
Clethrionomys rutilus	Northern red-backed vole	Taiga
Cynomys ludovicianus	Black-tailed prairie dog	Grasslands
Dicrostonyx hudsonicus	Ungava/Nunavik lemming	Northern Lanrador
Dicrostonyx torquatus	Collared lemming	Arctic
Didelphis virginiana	Opossum	Southern Boreal
Dipodomys ordii	Ord's kangaroo rat	Grasslands
Enhydra lutris	Sea otter	Pacific Coast
Erethizon dorsatum	Porcupine	Boreal
Felis concolor	Cougar, mountain lion	Boreal/Cordillera
Glaucomys sabrinus	Northern flying squirrel	Boreal
Glaucomys volans	Southern flying squirrel	Southern Boreal
Gulo gulo	Wolverine	Boreal/Taiga
Lasionycteris noctivagens	Silver-haired bat	Boreal
Lasiurus borealis	Red bat	Boreal
Lasiurus cinereus	Hoary bat	Boreal
Lemmus sibiricus	Brown lemming	Taiga/Arctic
Lepus americanus	Snowshoe hare	Boreal/Taiga
Lepus arcticus	Arctic hare	Arctic
Lepus townsendii	Jackrabbit	Grasslands
Lontra canadensis	Otter	Widespread
Lynx canadensis	Lynx	Boreal/Taiga
Lynx rufus	Bobcat	Southern Boreal
Marmota caliguta	Hoary marmot	Cordillera
Marmota flaviventris	Yellow-bellied marmot	Cordillera
Marmota monax	Groundhog, woodchuck	Boreal
Marmota vancouverensis	Vancouver island marmot	Vancouver Island
Martes americana	Pine marten	Boreal/Taiga
Martes pennanti	Fisher	Boreal
Mephitis mephitis	Skunk	Widespread
Mustela erminea	Ermine, weasel	Widespread
Mustela nigripes	Black-footed ferret	Grasslands
Mustela vison	Mink	Widespread
Myotis evotis	Long-eared bat	Boreal
Myotis lucifugus	Little brown bat	Widespread

(continued)

Table 1.4 (continued)

Scientific name	Common name(s)	Typical habitats
Myotis septentrionalis	Northern long-eared bat	Southern Boreal
Myotis volans	Long-legged bat	Boreal
Ochontona princeps	Pika	Cordillera
Odocoileus hemonius	Mule/black-tail deer	Grasslands
Odocoileus virginiana	White-tail/red deer	Boreal
Ondrata zibethica	Muskrat	Widespread
Oreamnos americanus	Mountain goat	Cordillera
Ovibos moschatus	Musk ox	Arctic
Ovis canadensis	Bighorn sheep	Cordillera
Ovis dalli	Dall's sheep	Cordillera
Procyon lotor	Raccoon	Widespread
Rangifer tarandus arcticus	Barren-ground caribou	Arctic
Rangifer tarnadus caribou	Woodland caribou	Taiga
Rangifer tarandus pearyii	Peary caribou	Arctic
Scalopus aquaticus	Eastern mole	Pelee
Sciurus carolinensis	Grey/black squirrel	Boreal
Sciurus niger	Fox SQUIRREL	Pelee
Sorex cinereus	Common/masked shrew	Boreal
Sorex gaspensis	Gaspe shrew	Gaspe
Sorex palustris	Water shrew	Boreal
Sphermophilis sp.	Ground squirrels	Widespread
Spilogale putorius	Spotted skunk	Southern Boreal/Grasslands
Sylvilagus floridanus	Cottontail	Southern Boreal
Sylvilagus nuttalli	Nuttall's cottontail	Grasslands
Tamias striatus	Chipmunk	Boreal
Tamiasciurus hudsonicus	Red squirrel	Boreal
Taxidea taxus	Badger	Boreal/Grasslands
Thomomys talpoides	Northern pocket gopher	Grasslands
Urocyon cineroargenteus	Grey fox	Southern Boreal
Ursus americanus	Black bear	Boreal/Cordillera
Ursus arctos horribilis	Grizzly bear	Cordillera
Ursus maritimus	Polar bear	Arctic
Vulpes lagopus	Arctic fox	Arctic
Vulpes velox	Swift fox	Extirpated
Vulpes vulpes	Red fox	Widespread
Amphibians		
Acris crepitans blanchardi	Blanchard's cricket frog	Southern Boreal
Ambystoma gracile	Northwestern salamander	Pacific Coastal
Aneides ferreus	Clouded salamander	Pacific Coastal
Bufo americana	American toad	Widespread
Hyla regilla	Pacific tree frog	Pacific Coastal
Rana aurora	Red-legged frog	Pacific Coastal
Rana catesbeiana	Bullfrog	Southern Boreal
Rana pipens	Leopard frog	Southern Boreal
Rhyacotritum olympicus	Olympic salamander	Pacific Coastal
Tarichia granulosa	Rough-skinned newt	Pacific Coastal

(continued)

Table 1.4 (continued)

Scientific name	Common name(s)	Typical habitats
Reptiles		
Chalydra serpentina	Snapping turtle	Boreal
Charina bottae	Rubber boa	Okanagan
Chrysemys picta	Eastern painted turtle	Southern Boreal
Columber constrictor	Racer	Southern Boreal
Crotalus horridus	Timber rattlesnake	Extirpated
Crotalus viridus	Prairie rattlesnake	Grasslands
Dermochelys coriacca	Leatherback turtle	Coastal
Elaphe obsoleta	Black rat snake	Southern Boreal
Elaphe vulpina	Fox snake	Southern Boreal
Eucemes faciatus	Five-lined skink	Pelee
Eucemes septentrionalis	Northern prairie skink	Spruce Woods
Graptemys geographica	Map turtle	Southern Boreal
Heterodon nascius	Western hognosed snake	Grasslands
Heterodon platyrhinos	Eastern hognosed snake	Southern Boreal
Hypsiglena torquata	Desert night snake	Okanagan
Lampropektis triangulatum	Milk snake	Southern Boreal
Nerodia siphedon	Northern water snake	Southern Boreal
Opheodrys vernalis	Smooth Green Snake	Southern Boreal
Phyrnosoma douglasii	Horned Lizard	Okanagan
Sistrurus catenatus	Massassauga Rattlesnake	Southern Boreal
Thamnophis radix	Plains garter snake	Grasslands
Thamnophis sirtalis	Eastern garter snake	Boreal
Sternotheris odoratus	Musk turtle, stinkpot	Southern Boreal
Fish		
Acipenser oxyrhynchus	Atlantic sturgeon	Atlantic drainage
Alosa pseudoharengus	Alewife	Great Lakes
Ambloplites rupestris	Rock bass	Great Lakes
Angulla rostrata	Eel	Atlantic drainage
Aplodinous grunniens	Drum	Boreal
Coregonus culpeaformis	Whitefish	Boreal/Taiga
Esox lucius	Northern pike	Boreal/Taiga/Grasslands
Esox masquinongy	Muskellunge	Boreal
Fundulus heteroclitus	Mummichog	Gulf of St. Lawrence
Ictalrus punctatus	Channel catfish	Great Lakes
Lota lota	Ling, burbot	Atlantic drainage
Microgadus tomcod	Tomcod	Atlanic Coastal
Micropterus dolomieu	Largemouth bass	Boreal
Micropterus salmoides	Smallmouth bass	Boreal
Morone americana	White perch	Lake Ontario
Morone chrysops	White bass	Red River
Nocomis micropogon	River chub	Bruce Peninsula
Notropis heterodon	Blackchin shiner	Bruce Peninsula
Oncorhynchus gorbuscha	Pink salmon	Pacific Coastal
Oncorhynchus keta	Chum salmon	Pacific Coastal
Oncorhynchus kisutch	Coho salmon	Pacific Coastal

(continued)

Table 1.4 (continued)

Scientific name	Common name(s)	Typical habitats
Oncorhynchus nerka	Sockeye salmon	Pacific Coastal
Oncorrhynchus tshawytscha	Tyee salmon	Pacific Coastal
Osmerus mordax	Smelt	Boreal
Perca flavescens	Yellow perch	Boreal
Petromyzon marinus	Lamprey	Great Lakes
Salmo clarki	Cuttthroat trout	BC Interior
Salmo gairdneri	Steelhead trout	BC Interior
Salmo salar	Atlantic salmon	Atlantic drainage
Salvelinus alpinus	Arctic char	Taiga/Arctic
Salvelinus fontinalis	Brook trout	Boreal
Salvelinus namaycush	Lake trout	Boreal/Taiga
Stenodus leichthyus mackenzii	Inconnu	Yukon
Stizostedion vitreum	Pickerel, walleye	Boreal
Thymallus signifer	Arctic grayling	Arctic

1.6 Soils

Canadian geographers generally use the Canadian System of Soil Classification (CSSC), developed by Agriculture Canada (Soil Classification Working Group 1998, 2001). The CSSC involves the recognition of specific horizon types, and analysis of the chemistry of the soils. Research and refinement of the CSSC has been concerned primarily with soils in agricultural regions. Soils in Canada are divided into 10 Orders, which in turn are subdivided into Great Groups" and "Subgroups" (Fig. 1.9). Equivalent terms in the World Database of the FAO are presented in Table 1.5.

Brunisol soils characteristically have weak horizon development, with a moderately to poorly defined Bm horizon at least 5 cm thick. Brunisols may form under taiga forests (Köppen-Geiger Dfc climates), in moderately drained sites, or where vegetation clearance (natural or artificial) or other factors restrict pedogenic processes.

Chernozem soils are defined in association with Köppen-Geiger BSk steppe climates and can only develop in prairie regions underlain by calcareous sediment. *Cherny* is the Ukrainian word for "black". Chernozems are characterized by a thick (>10 cm) humic-rich Ah surface layer, underlain by a Cca horizon marked by accumulation of calcium carbonate through groundwater movement and precipitates, and a Ck parent material (lake sediment, glacial, or limestone bedrock) inherently rich in calcium. Chernozems are alkaline soils with high pH ($OH^- > H^+$). Grasses thrive on the natural lime in the soil, and extend their roots down to the Cca horizon to obtain it. As the grass vegetation develops, calcium carbonate is cycled through the system repetitively, with each cyclic event improving the soil for grass growth. Consequently, grassland soils are fertile and stable, as long as the grasses are not disturbed. Removal of the original grass vegetation can result in a loss of soil fertility, which must be compensated for by fertilization.

Cryosol soils are subject to extensive frost disturbance. Cryosols are developed in the northern parts of the Discontinuous Permafrost one, and in the Continuous Permafrost zone. The thickness of the cryosol is determined by seasonal freezing and thawing of the "active layer", marking the depth to which vascular plant roots can penetrate.

Gleysol soils form in poorly drained areas, including some wetlands. They are mineral soils (not including peats, or other soils with high organic content), marked by gley: mottles developed from alternating oxidation and reduction of minerals, particularly those containing iron. Iron exposed to oxygen turns red (Fe^{3+} ions). Iron which is inundated in water is reduced, turning greenish-grey (Fe^{2+} ions). In a gleysol, fluctuations of the water table alternately expose and reduce iron-bearing minerals. As conversion is not instantaneous, a varying water table produces a mottled pattern, with some iron being reduced and other ions oxidized, resulting in the mottling.

Luvisol soils form by eluviation of material from an overlying Ae to an underlying horizon, but the material involved is clay, rather than iron. A Luvisolic Bt horizon has an enhanced concentration of clay. Luvisols may have moderate to high pH. They develop on clay-rich glacial or lake deposits, on shales and fine sandstones, or on limestones. Luvisols are characteristic of drier, interior boreal forested areas, along the Aspen Parkland ecotone between boreal forest and grassland in Prairie Canada, and under mixedwood vegetation on carbonate-bearing bedrock.

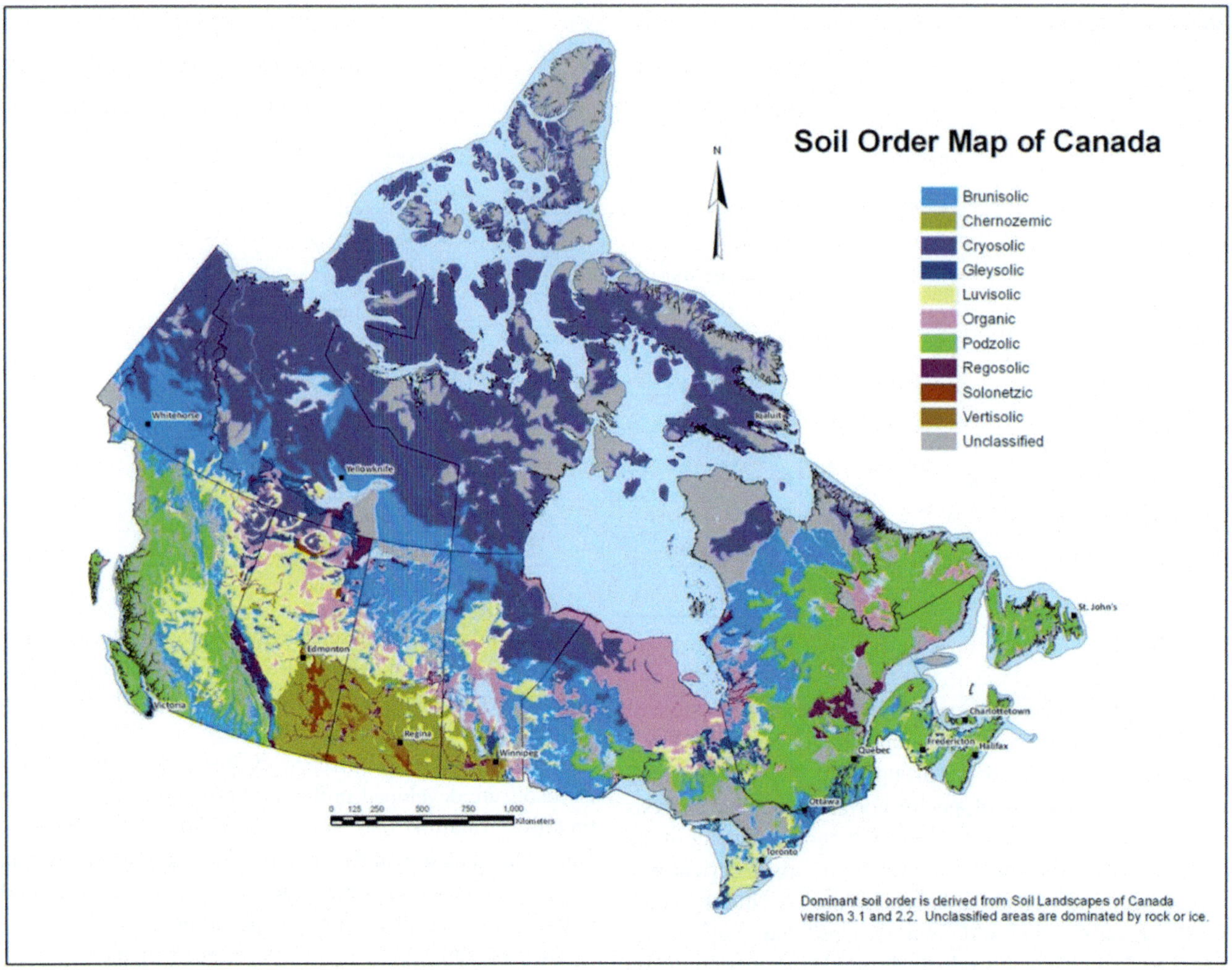

Fig. 1.9 Soil orders of Canada. (Agriculture and Agri-Food Canada 2004). {https://sis.agr.gc.ca/cansis/publications/maps/soc/all/soils/index.html}

Organic soils are dominated by organic matter, with no or few mineral grains. The Von Post scale measures organic decomposition or humification on a scale of 1 to 10, with 1 being least decomposed and 10 being the most decomposed. Organic soils are divided by the degree of decomposition of organic matter, ranging from fibrisols (least decomposed, Von Post 1–5) through mesisols (6–7) to humisols (8–10). Weakly decomposed fibric peats characterize spruce- and sedge-dominated fen areas, where decomposition only began following the last deglaciation and is interrupted by freezing each autumn.

Podzol soils typically develop under boreal forests, over quartz-feldspar rich bedrock or Quaternary sediments, in mid-boreal (Dfb) climates. Podzols are coarse-grained and acidic (pH 3.8–5.8). Decreased air and water infiltration results from compaction, and evidence of frost disturbance is common. The characteristic soil horizons are an eluviated Ae, marked by downward leaching of nutrients by water, underlain by the diagnostic Podzolic B. A Podzolic B horizon must be black or red, at least 10 cm thick, coarser than clay, with a low pH, and with either high Fe^{3+} or high combined Fe^{3+} and Al^{3+} concentrations.

Regosols are thin soils, less than 10 cm thick, over bedrock. Regosols lack distinctive soil horizons.

Solonetz soils are sodium-rich, developed in saline areas under Koppen-Geiger BSk climates. Solonentz soils contain a characteristic Bn horizon, with strongly developed vertical columns, capped by white sodium salts ($NaNO_3$ and Na_2SO_4). The column caps are commonly indurated and extremely difficult to break or plough, and the high sodium levels frequently render the soil infertile for many cereal crops. Solonetzic soils develop naturally in areas where saline bedrock occurs near the surface, or where groundwater flow brings sodium salts to the soil.

Vertisol soils develop where montmorillonite and other swelling clays are subjected to seasonally wetting and drying under Köppen-Geiger BSk steppe climates. In Canada, Vertisols are uncommon, confined to small areas of southern Saskatchewan and Alberta (Fig. 1.9).

Table 1.5 Comparison of soil classification. CSSC: Canadian System of Soil Classification; WRB: World Reference Base for Soils, Food and Agriculture Organization, United Nations

CSSC	WRB (FAO)
Brunisol, dystric	Cambisol, dystric
Brunisol, eutric	Cambisol, eutric
Brunisol, melanic	Cambisol, melanic
Brunisol. sombric	Cambisol, umbric
Chernozem, black	Chernozem
Chernozem, brown	Kastanozem, aridic
Chernozem, dark brown	Kastanozem, haplic
Chernozem, dark grey	Greyzemic phaeozem
Cryosol	Cryosol
Gleysol	Gleysol
Luvisol	Luvisol
Organic	Histosol
Podzol, ferro-humic	Podzol, orthic
Podzol, humic	Podzol, humic
Podzol, humo-ferric	Podzol, orthic
Regosol	Regosol
Solod	Solodic planosol
Solodized solonetz	Haplic solonetz
Solonetz	Solonetz
Vertisol	Vertisol

1.7 People

Part 2 of this volume focuses on the people of Canada. Here, we'll give an overview of background information, including population, demography, urban centres, political organization and governance, historical geography and territorial evolution, and settlement patterns. Chapter 11 considers the historical context and present conditions involving migration and immigration to Canada, and the current ethnic composition of the country.

Chapters 12 and 13 focus on the urban areas. Chapter 12 concentrates on the Census Metropolitan Areas (CMA), the largest urban centres, with emphasis on population, growth, urban transit and commuting, environmental stresses, urban vegetation, and inequality in urban systems. Chapter 13 looks at the complexities of Canadian cities as shaped by their geographical, social, economic, and cultural diversity.

Chapter 14 discusses transportation, including marine, rail, road, and air. The St. Lawrence Seaway is discussed in Chap. 4 (Great Lakes–St. Lawrence Mixedwood Plains). Rail and highway routes through the Cordillera are considered in Chap. 6, and navigation challenges in Subarctic and Arctic waters are considered in Chaps. 8 and 9.

Chapter 15 discusses forestry, focusing on the national aspects of harvesting, disturbances, production, consumption, and GHG emissions. Wildland fires are also discussed in Chap. 15. Specific details of forest assemblages and tree species are covered in Chaps. 2 (Atlantic), Chapter 3 (Boreal), Chap. 4 (Mixedwood), Chapter 6 (Cordillera), and Chapter 7 (Pacific Maritime).

Chapter 16 looks at agriculture and food across Canada. The relationships between soils, climate, and agricultural production for each Ecozone are discussed in Chaps. 2–8.

Chapter 17 discusses energy and electricity. Aspects of coal, petroleum, and natural gas production and exploitation are considered, along with nuclear, wind, solar, and tidal energy.

Chapter 18 discusses the economic development and industrialization and urbanization of Canada. Chapter 19 discusses winter life in Canadian cities., as well as outdoor winter recreation as pursued mainly by urban residents. Chapter 20 considers patterns and lines on and around the land (boundary definitions and cadastral layout). Chapter 21 presents a short conclusion and commentary.

We decided that the linkages between aquatic organisms and the marine and lacustrine environments of Atlantic, Great Lakes, Pacific, and Arctic waters warranted considering the fisheries of each Ecozone in the appropriate chapters. You will find details of Atlantic fisheries in Chap. 2; freshwater species and fisheries in Chaps. 3, 4, 5, 6, and 8; Pacific Maritime fisheries in Chap. 7; and marine subarctic and arctic fisheries in Chaps. 8 and 9.

Similarly, rather than present a single chapter concerning mining, we have incorporated the discussions of metallic minerals and ores in the appropriate chapters, based on mine location. Removing ores from their geological settings would leave them out of context. Hydrocarbon developments are discussed in Chap. 17 (Energy), as well as in Chap. 5 (Prairies).

Unless otherwise stated, statistical data is taken from the 2021 Census of Canada (Statistics Canada 2021–2024).

1.8 Population Distribution

Among the world's countries, Canada ranks second in area (9,984,670 km^2), but approximately 37th in population, with 0.5% of earth's human residents (Table 1.6) (Statistics Canada 2021–2024, World Bank Open Data 2024, World Factbook 2023). The 2021 census population of 36,991,981 represented an increase of approximately 2 million (5.2%) since 2016. Almost all of that increase is due to immigration: the low rate of natural population increase (0.1% p.a.) contrasts with the total population growth rate of 0.77 per year (Table 1.7). Statistics Canada estimated the total population at 39,566,248 on 1 January 2023, following an increase of 1,050,110 people throughout 2022. International migration in 2023 accounted for 95.9% of this almost unprecedented growth rate of 2.7% per year, the highest among the G-7 countries and comparable to rates in several less economically developed countries. At that rate, the total population of Canada would double in approximately 26 years, compared to 75 years for 1991–2001.

Growth rates and population densities vary substantially across Canada. The vast terrain includes areas almost unoccupied by humans, along with intensely developed

Table 1.6 Area and population data for Canadian provinces and territories. (Statistics Canada, Census of Canada 2021)

	Area (km^2)	Pop. (2021)*	% Change 2016–21*	Population/km^{2*}	CMA Pop. (%)**	% Indigenous [i]	Date entered confederation
Canada	9,984,670	36,991,981	5.2	3.70	27,281,119 (73.7%)	4.9	
Newfoundland and Labrador (NL)	405,212	510,550	−1.8	1.26	212,579 (41.6%)	9	1949
Prince Edward Island (PEI; PE)	5,660	154,331	8.0	27.27	0 [c]	2	1873
Nova Scotia (NS)	55,284	969,383	5.0	17.53	465,703 (48.0%)	6	1867
New Brunswick (NB)	72,908	775,610	3.8	10.64	396,940 (51.2%)	4	1867
Québec (QC)	1,542,056	8,501,833	4.1	5.51	6,083,107 (71.6%)	2	1867
Ontario (ON)	1,076,395	14,223,942	5.8	13.21	11,795,482 (82.9%)	3	1867
Manitoba (MB)	647,797	1,342,153	5.0	2.07	834,678 (62.2%)	18	1870
Saskatchewan (SK)	651,036	1,132,505	3.1	1.74	566,697 (50.0%)	16	1905
Alberta (AB)	661,848	4,262,635	4.8	6.44	3,124,615 (73.3%)	7	1905
British Columbia (BC)	944,735	5,000,879	7.6	5.29	3,801,318 (76.0%)	6	1873
Northwest Territories (NWT; NT)	1,346,106	41,070	−1.7	0.03	0 [y]	51	1870***
Yukon (YT)	482,443	40,232	12.1	0.08	0 [w]	23	1898
Nunavut (NU)	2,093,190	36,858	2.5	0.02	0 [iq]	86	1999

* 2021 Census of Canada, Statistics Canada
** Within 41 largest Census Metropolitan Areas (CMA) in Canada, 2021, [i] % self-identified as Indigenous, 2021 [c] 51.1% in Charlottetown (78,858) [y] 49.5% in Yellowknife (20,340) [w] 79.3% in Whitehorse (31,913) [iq] 20.2% in Iqaluit (7,429)
*** Subsequently divided into Alberts, Saskatchewan, Yukon, Nunavut, and parts of Manitoba, Ontario, and Québec

Table 1.7 Demographic data summary. See text for discussion

Parameter	2021 census of Canada
Crude birth rate	10.2/1000
Crude death rate	8.1/1000
Total population growth rate	0.77% per annum
Rate of natural increase	0.1% per annum
Total fertility rate	1.4
Infant mortality rate	4.4/1000
Mean childbearing age	31.4 y
Mean life expectancy	82.2 y
Female: male ratio at birth	0.95
Female: Male ratio overall	1.02
Median age	41.0 y
Population 0–14 y	15.6%
Population $\geq$ 65 y	18.8%
Immigrant population	23.0%
Net migration rate	5.5/1000

large and economically dominant cities. Northwest Territories, Nunavut, Yukon, have fewer than 1 person /km^2, as do the northern parts of Québec (Nunavik), Manitoba, Saskatchewan, British Columbia, and Newfoundland and Labrador (Nunatsiavut). The populations of the Boreal Shield regions of Ontario and Québec, and the northern Boreal Plains of Alberta, are concentrated in mining- and transportation-related cities along the southern margin, with abundant open space between towns and to the north. In contrast, the Maritime Provinces (NB, NS, PEI), the Appalachian region of Québec, and the southern and central Prairies (AB, SK, MB) have population concentrations in urban and agricultural areas, including the large Prairie cities of Calgary, Edmonton, Regina, Saskatoon, and Winnipeg. The most densely populated regions are the Great Lakes–St. Lawrence Mixedwood Plains of Ontario (including Toronto and Ottawa) and Québec (including Montreal and Québec), and British Columbia's Lower Mainland (including Vancouver) and southern Vancouver Island (Victoria), with the Pacific Maritime ecozone.

All seven of the largest political provinces traverse multiple ecozones, with very different economic contexts, resources, rates of urbanization, and population densities. Ontario includes large areas of the Boreal Shield and Hudson Platform, together with the densely populated Mixedwood Plains along the St. Lawrence River and lower Great Lakes. Québec includes areas of Mixedwood Plains, Appalachian Highlands, Boreal Shield, Taiga Shield, and Arctic environments. British Columbia's population is largely concentrated in the most southerly 10% of the Pacific Maritime ecozone, while the province includes large areas of the Cordillera and Boreal Plains. Less than half of the land area of the Prairie provinces of Alberta, Saskatchewan, and Manitoba is Prairie Grassland. Consequently, discussions of many demographic, socio-economic, and resource issues are better framed considering the ecozones outlined above, rather than regions defined by political provincial boundaries. Political viewpoints and public policies enunciated by the provinces tend to be dominated by the perspectives of the most populous urban areas, which can leave other regions being or feeling under-represented in discussions.

1.9 Demographics

Canada is characterized by a low crude birth rate (approximately 10.2/1000 population in 2021; Table 1.7). From the high levels following World War II (27.4/1000 in 1950), CBR declined sharply to 15.6/1000 in 1973. After a small increase between 2004 and 2008, since 2009 (11.1/1000), the CBR has slowly declined. Higher CBR are characteristic of the largest urban centres (e.g. Toronto, 20.6/1000 in 2022), and regions with greater percentages of Indigenous people (e.g. Nunavut, 21.5/1000 in 2022). Lower CBR (7.5–8.5/1000) occur in areas with statistically older populations, such as the island of Newfoundland.

The overall crude death rate was 8.1/1000 population in 2021. After a period of decline from 1950 (8.8/1000) to 1984 (7.0/1000), the CDR has increased steadily over the past 40 years, a trend reflecting the gradually increasing median age of the population. However, the Standardized Mortality Rate, the ratio between the observed number of deaths and the number of deaths would be expected, based on the age- and sex-specific rates and the population size using the same age/sex groups, steadily declined from 2007 (9.5/1000) to 2019 (7.7/1000). The SMR rose during the impact of COVID-19, reaching 8.6/1000 (2022), largely reflecting deaths in the elderly population. Higher CDR for males results in an overall shift in the female: male ratio from 0.95 at birth to 1.02 in the general population of all ages. Life expectancy in 2021 was 82.2 years, with females outliving males by approximately 6 years. The slow decline in recent years (81.8 in 2023) is attributed in part to the effects of COVID-19 (Fig. 1.10).

Statistically, deaths during the winter months exceed those in summer months by 10–15%. The main causes of death during the past decade were various forms of cancer and heart disease, together accounting for almost half of total mortality (Fig. 1.11). COVID-19 (including complications arising) accounted for 3–5% of all deaths during peak activity, primarily among those aged 80 and over (Fig. 1.12).

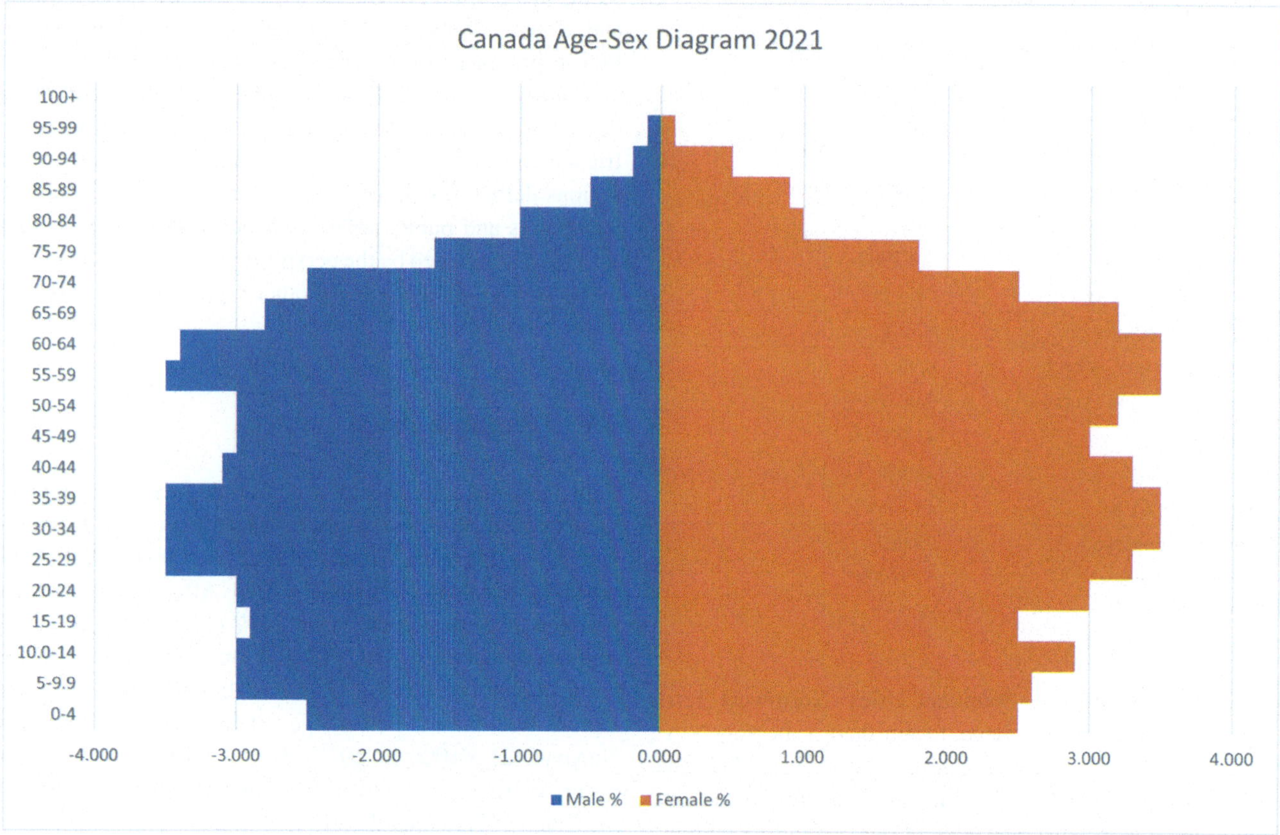

Fig. 1.10 Age-sex distribution in Canada, 2021. (Data from Statistics Canada, Census of Canada 2021)

Fig. 1.11 Causes of death in Canada, 2015–2019. (Data from Statistics Canada)

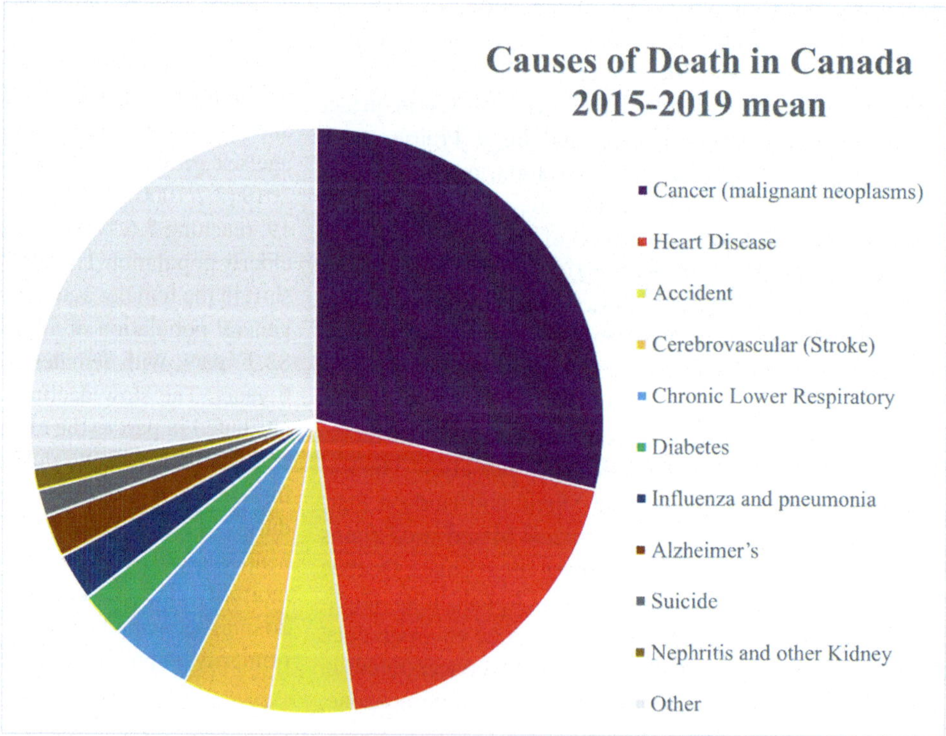

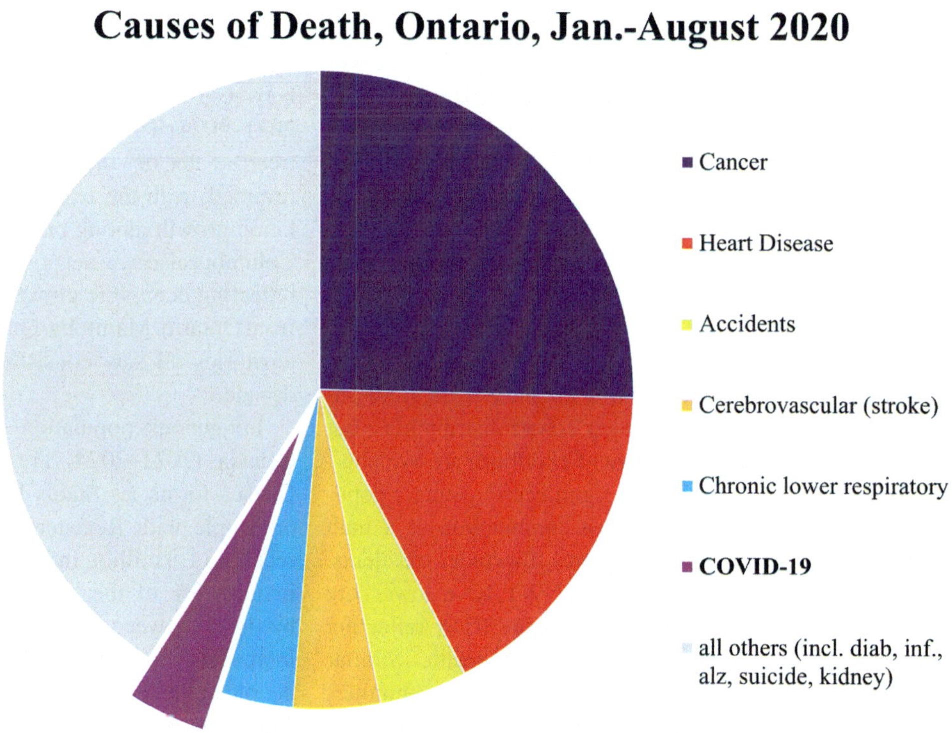

Fig. 1.12 Causes of death in Ontario, January–August 2020. (Data from Government of Ontario)

In 2021, the infant mortality rate was 4.4 per 1000 live births (Statistics Canada 2021–2024). For international comparison purposes, adjusted infant mortality rates for Canada exclude live births with a birth weight of less than 500 g. The adjusted infant mortality rate in Canada is lower than the averages for European Union and OECD countries. Approximately, 75% of infant deaths occur with 1 month of birth. IMR in Canada fell from 41.3/1000 in 1950 to 9.0/1000 in 1983 and 5.3/1000 in 2000.

IMR in Canada varies substantially by parental income and education levels, by province and territory, and by ethnic origin (Public Health Services 2019; Statistics Canada 2021–2024). Across the country, IMR are 1.9 times higher than the national average for Métis, 2.3 times higher for all First Nations Indigenous groups, and 3.9 times higher for Inuit. Some of the disparities can be attributed to socio-economic status, availability of prenatal and premature birth intensive care, and traumas induced by past colonial practices including forced relocation and residential schools. IMR are generally higher for males in the individual provinces, with the exception attributable to low birth numbers in lower populations. Conspicuous differences exist in the three territories, however, with markedly higher male IMR.

In Nunavut, where the IMR has been approximately 3 times the national average since 2007, the rate for male deaths has exceeded female deaths by a factor of almost 4. A potential cause identified by public health workers in Nunavut is bed sharing, where the parents or mother, in particular, shares the same bed with the child, which may die as a result of accidental suffocation. Infections and smoking by parents also have been recognized as problems. Genetic hypoglycemia tends to differentially affect males, and genetic birth defects have also been recognized as a problem. Since 2016, the government of Nunavut has distributed baby boxes to all mothers of newborns, with the hope that this infant mortality rate, especially among males, can be reduced.

The total fertility rate in 2021 was 1.4, well below the threshold for total population stability (Statistics Canada 2021–2024; also see World Factbook 2023, World Bank Open Data 2024). The TFR has declined markedly from 1.7 since 2008. For the cohort between ages 20 and 29, the fertility rate was halved between 1990 and 2021. Nunavut is the only jurisdiction with a TFR greater than 2.1 (2.23 in 2022). The average child-bearing age increased from 27.8 in 1990 to 31.4 in 2021.

The median age in 2021 was 41.0 years. With 15.6% of the population aged 0–14, and 18.8% aged 65 or over, Canada has had an increasing dependency ratio since 1981. In the 1970s, the relatively high dependency ratio reflected the number of children. In 2021, the increasing dependency ratio of approximately 65% results from having more senior citizens.

In a Canadian context, the dependency ratio is somewhat arbitrary. The assumption that at age 15, a person is no longer a family dependent and is expected to be a contributor to working society is not accurate for most. At the other end of the scale, many who are technically seniors are still working. Which age qualifies as youth or senior really depends on socio-economic status and societal conditions. Both boundaries have shifted in the older direction in Canada. For employment purposes, Statistics Canada considers a youth to be a person between ages 15 to 24, suggesting that 24 might be more appropriate as the lower boundary, rather than 15. The average age for first childbirth for Canadian women is 29.6, and the average age for first formal marriage is 31.6.

The age-sex diagram (Fig. 1.10) indicates relatively equal distribution of age segments younger than 75 y for both sexes (Statistics Canada 2021–2024). The latter part of the post-World War II "Baby Boomer" population is reflected in the bulge of the 55–64 y cohorts. The recent increase in immigration has begun to alter the younger cohort distributions, but the lower TFR and increasing age of childbirth are evident.

Nunavut is characterized by a significantly younger population than the national average, a combination of a high birth rate and lower life expectancy for older residents. Inuit communities in Nunavut and NWT have lower life expectancies (e.g. 70.74 y in Nunavut in 2021), reflecting socio-economic determinants influencing health. Similar patterns categorizing indigenous First Nations communities in other Canadian jurisdictions, including northern Manitoba, Saskatchewan, and Alberta, and Yukon, influence overall demographics. Lower life expectancies in indigenous communities are unfortunately typical across Canada: the Innu community of Natuashish had a median age of death of 37 in 2021, more than 35 years below the provincial average for Newfoundland and Labrador. The totals in Alberta also reflect interprovincial migration of younger people.

The four Atlantic Provinces have higher median ages and proportions of senior citizens, and lower percentages of children and youth. The ageing demographics have become a concern for people and political figures alike. Within the provinces, regional differences are pronounced, with rural areas having more seniors, fewer childbirths, and enhanced out-migration of youths and young adults. In Newfoundland and Labrador, the median age of 42.4 for the St. John's Census Metropolitan Area (CMA) contrasts with 52.6 for the province as a whole, and 21.2 for Natuashish. Approximately, 15% of the indigenous population of NL is aged 15–24, compared to 9.8% of the non-indigenous population.

Rural–urban differences are common in all jurisdictions. Rural communities generally tend to have lower percentages of children than do urban communities. Smaller urban centres (5000 or less) tend to have lower percentages of young people than do larger urban centres. However, for the largest metropolitan centres, the percentage of children tends to be less than provincial averages in the urban core and comparable to the average in suburban and exurban areas. Comparison of the 2021 figures for the City of Toronto with the broader Toronto CMA indicates lesser percentages of children under the age of 15 (13.7% vs. 15.7%) and higher percentages of people aged 20–35 (25.9% vs. 24.4%).

Internal migration also results in demographic change. In 2022, more than 393,000 people were recorded as migrants between Canadian provinces and territories (Statistics Canada 2021–2024) Through the decade 2012–22, the long-standing trend of net out-migration from the Atlantic Provinces has reversed, with the result that PEI had the most rapid population growth among provinces in 2021. Alberta and British Columbia also saw net positive internal migration in 2021–23, reflecting economic growth, in contrast to net out-migration from Ontario. Manitoba, Québec, Saskatchewan, and the three territories all saw consistent out-migration throughout the decade.

Indigenous population numbers recorded by Statistics Canada (2021–2024) are based on self-identification on Census forms, as Status First Nations, Inuk (Inuit), Métis, or people with Registered Indian status. The 2021 census recorded 1.8 million Indigenous people, representing approximately 5% of the total population. The population grew by 9.7% between 2016 and 2021, with Status First Nations increasing at the slowest rate (4.1%) and Registered Status people at the fastest (27.2%). Particularly noteworthy are the increasing percentages of Indigenous people living in CMA and CA. Indigenous people form majorities in Nunavut and Northwest Territories, and substantial percentages in Yukon, Manitoba, and Saskatchewan.

Throughout this volume, we will emphasize the terms used by Indigenous peoples for self-identification. Table 1.8 lists the Indigenous peoples mentioned in this volume, along with formerly used English-Canadian terms.

1.10 Urban Centres

Despite its reputation and mythology as a country dominated by wilderness, Canada is highly urbanized. Using the internationally accepted definition of urban (population centre of at least 1000 people), the percentage increased from 69.67% in 1961 to 81.65% in 2021. Under this definition, the majority of the population in all provinces and territories would be classified as urban.

A more useful approach is to consider urban centres with a population of 100,000 or more. Statistics Canada defines a census metropolitan area (CMA) as having a total population of at least 100,000, with at least 50,000 living in the metropolitan core. In 2021, Statistics Canada (2021–2024) designated 41 CMA. Centres with populations between 10,000 and 100,000 are designated as Census Agglomerations (CA). Municipalities adjacent to the core are included within a CMA or CA based on the degree of integration with the core, as estimated from daily home-workplace commuter traffic. This designation assumes that working from home is not significant. The 41 CMA accounted for the majority of the 5.2% increase in population from 2016 to 2021.

Table 1.8 Indigenous peoples discussed in the text

Indigenous	English Canadian
A'aninin	Gros Ventre
Algonquian	Algonquin
Dene	Athabaskan, Athapaskan
Denésoliné; Denesulin	Chipewyan
Eeyou	Cree (East)
Gayogo̱hó:nǫ'	Cayuga
Haida; X̱aayda	Haida
Haudenosaunee	Iroquois
Ililiw	Cree (Moose)
Innu	Montagnais
Innu	Naskapi
Inuit	Inuit
Kainai	Blood
Kanien'kehá:ka	Mohawk
Lakota	Sioux
Maliseet; Wolastoqiyik	Maliseet
Mi'kmaq; Mi'kmaw	Mi'kmaq, Micmac
Nakota	Assiniboine
Nehinaw	Cree (Plains)
Nuu-Chah-nulth	Nootka
Odawa	Ottawa
Ojibwe	Chippewa, Ojibwa, Salteaux
Onödowá'ga	Seneca
Onontaerrhonon	Onondaga
Onyota'a: ka	Oneida
Piikani	Piegan
Siksika	Blackfoot
Sivullirmiut	Dorset
Taissumanialungmiut	Thule
T'atsaot'ine	Yellowknives
Tionontati	Petun
Tlicho	Dogrib
Tsúùtínà	Sarcee
Wendat; Wyandot	Huron
Wolastoqiyik	Maliseet
X̱aayda	Haida

Within Canada, 73.7% of residents lived in a CMA in 2021. The most heavily urban provinces are Ontario (Toronto, Ottawa, Hamilton, Kitchener, London), British Columbia (Vancouver, Victoria), Alberta (Calgary, Edmonton), and Québec (Montreal, Québec). The single CMA of Winnipeg (MB), Halifax (NS), and St. John's (NL) dominate their respective provinces. In the provinces without a CMA, smaller CA form significant population concentrations in Charlottetown (PEI, 51.1%, Yellowknife (NWT, 49.5%), and Whitehorse (YT, 79.3%).

City growth and relative population rank since Confederation reflects the westward expansion of white settlement, coupled with economic development. Saint John (NB), Halifax, and Kingston (ON) were gradually supplanted by Vancouver, Calgary, Edmonton, and Winnipeg. Toronto rose from the 3rd largest population centre in 1871 to 2nd in 1891 and 1st in 1991, surpassing Montreal and Québec. The Ottawa-Gatineau CMA (including Ottawa in Ontario and Gatineau in Québec) maintained its position throughout the transition from a commercial lumber town to a governmental and IT community. Historically, commercial growth and immigration allowed Montreal to overtake Québec in population in the early 1800s.

1.11 Political Organization and Governance

Canada is a confederation of 10 provinces and 3 territories. Provinces are allocated power and responsibility for certain aspects of governance, as listed in the Constitution Act (formerly the British North America (BNA) Act; Dodek 2021). Other aspects of governance are assigned to the Federal Parliament, based in Ottawa (Library of Parliament 2022, Parliament 2024). Provincial legislatures are considered to be co-sovereign with the Federal government (Brooks 2007; Hueglin 2021). A change in responsibility for an aspect of governance or public policy requires a constitutional amendment, necessitating consent of the Federal government and at least 7 of the 10 existing Provincial legislatures from provinces with at least 50% of the total Canadian population.

Territories, in contrast, have no sovereign powers, except those delegated to each by the Federal Parliament. In recent years, increasing devolution of powers from the Federal government to the territories has occurred, coupled with the establishment of self-governance in Nunavut and Northwest Territories. Devolution is accomplished by Acts of Parliament, and does not require constitutional amendments.

Historically, 8 of the 10 provinces were admitted to Confederation directly as provinces. Alberta and Saskatchewan were detached from the Northwest Territories and granted provincial status in 1905 (see below), based on their increasing populations and economic significance, under a simpler process. The three current territories all have small populations and low population densities. The ongoing transference of powers from the Federal government has limited any desire to apply for provincial status.

Under the Constitution Act (Stanley 1969), the Federal Government has power over the following aspects:

- Public debt and property.
- Regulation of trade and commerce.
- Unemployment insurance.

- Direct and indirect federal taxation.
- Postal service.
- Census and statistics.
- Defence.
- Navigation and shipping.
- Quarantine for disease control.
- Marine and inland commercial fisheries.
- Interprovincial and international ferries and railways.
- Currency, banking, and bankruptcy.
- Weights and measures.
- Patents and copyrights.
- Crown-Indigenous relations.
- Citizenship, marriage, and divorce.
- Criminal law and federal penitentiaries.
- Infrastructure connecting provinces; or within a single province but to the advantage of more than one province.

Provinces have jurisdiction over the following:

- Direct taxation within province.
- Management and sale of provincial public lands.
- Prisons.
- Hospitals.
- Municipalities.
- Formalization of marriage; property and civil rights.
- Administration of civil/criminal justice.
- Education.
- Incorporation of companies.
- Natural resources.
- Matters of a merely local or private nature.

Aspects with powers shared between the federal and provincial governments are:

- Old age pensions.
- Immigration.
- Agriculture.

Contemplation of these lists suggests a problem which has confounded governance in Canada since 1867: overlapping, uncertain, and unspecified areas of federal and provincial jurisdiction (e.g. Stanley 1969; Hueglin 2021). In some instances, compromises have developed. Provinces issue marriage licenses and birth certificates. Prison sentences of two years or more are served in federal penitentiaries, whereas shorter sentences (e.g. "two years less a day") are served in provincial jails. Federal and provincial income taxes are collected simultaneously, utilizing the services of the federal Revenue Canada. Provincial and federal departments of agriculture generally work effectively together. For other aspects, matters have not always been considered harmoniously. Three aspects of particular concern to geographers are health, natural resource management, and environment.

"Health" was not specifically mentioned in the original BNA or 1982 Constitution Acts. Became "hospitals" were listed as a provincial responsibility, the provinces have responsibility for providing healthcare services (including prescription drugs); medical and dental care; training, licensing, and regulation of physicians, dentists, and pharmacists; health insurance; and occupational health and safety. However, the federal jurisdiction over criminal matters (e.g. those pertaining to reproductive rights) and "quarantine" (evident during COVID-19) overlaps with provincial responsibilities. Significantly, because health care forms a steadily increasing budget cost for all provinces, which rely on the transfer of funds raised through federal taxation and borrowing, the Federal government has significant leverage: it can limit or refuse financial contributions to provinces which do not follow the federal guidelines established in the Canada Health Act (1984). A key provision of the Canada Health Act precludes physicians from charging patients additional fees patients fees beyond the sum collected from the province for an insured service, or establishing many forms of private clinics. The Canada Health Act does not apply to dentists or pharmacists.

In 1867, "natural resources" primarily involved logging and extraction of solid minerals. Fisheries were allocated to the Federal government, and agriculture is a shared responsibility. The development of large-scale energy industries has brought provincial jurisdictions into conflict with the Federal government (and other provinces) over the construction of pipelines and power lines, financing of large projects, regulation of trade and commerce, federal taxation (e.g. for climate change mitigation), and initiation of new or expanded operations.

"Environment" was not mentioned in the BNA Act. All provinces and territories have separate ministries of the environment, along with Environment and Climate Change Canada. Climate change policy and response is not explicitly indicated as a particular government's responsibility, leading to federal-provincial tensions. Offshore petroleum resources are beyond the provincial territorial limits, but offshore exploration and drilling are licensed by the provinces, in the same spaces managed by the federal ministries of Environment and Climate Change, Fisheries and Oceans, and Natural Resources. Another consequence of the jurisdictional overlap is that while Canada's mitigation policies for climate change are led by Environment and Climate Change Canada, adaptation efforts and policies are led by Natural Resources Canada.

Canada is a bicameral system under the general form of British parliamentary government (Library of parliament 2022, Parliament 2024). The official Head of State is the Monarch of the Commonwealth, represented at the federal level by the Governor General and by Lieutenant Governors in each province. These positions are strictly ceremonial and limited term, with holders appointed by the monarch upon recommendation from the Prime Minister (federal) or provincial heads of government.

The primary responsibility for government at the federal level rests with the House of Commons. Members of Parliament are elected by universal suffrage (aged 18 and over), with the person receiving the greatest number of eligible votes winning the seat. Most elections are won by plurality. Members typically represent designated political parties, although independent candidates can run. The party receiving the greatest number of seats in the House of Commons usually forms the government, with the party leader becoming Prime Minister and Head of Government. Minority governments or coalitions result when no single party captures a majority of seats. Elections are held at a maximum of every five years.

Allocation of seats is based on population, using an electoral quotient of 121,891 eligible voters per seat (as of 2022). Modifications account for provinces and territories with lower populations, and also use the principle that no province or territory can have fewer seats in the House of Commons than it had prior to each reassessment and reallocation (most recently in 2022). Of the total of 343 seats allocated in 2022 for the subsequent 2025 election, individual provinces have between 4 (PEI) and 122 (Ontario). Each of the 3 territories is allocated 1 seat. Together, the four most populous provinces (Ontario, Québec, British Columbia, and Alberta) account for 280 of the 343 seats.

Review of electoral boundaries in each province is the responsibility of independent 3-person electoral boundary commissions, chaired by a provincial judge. Sitting federal or provincial politicians are barred from serving. Typically, the other commission members are long-serving provincial servants or experts in demographics, sometimes including professors of Geography. In addition to attempting to have similar-sized populations in each electoral district, commission members also must consider social and geographical factors, including communities of interest or identity.

The Senate consists of members appointed by the Governor General on recommendation by the sitting Prime Minister: senators are not elected. Seat allocations are based on population, modified to allow one senator from each of the three territories and also using the principle that no province can have fewer seats in the Senate than it had prior to reallocation. Senators represent a province or territory, not individual constituencies. At present, there are 105 Senators. Senators can be appointed to represent political parties, or may be independent. Senators have no fixed terms, but must retire at age 75. The Senate reviews and debates bills proposed by the House of Commons. Less commonly, Senators can initiate some legislation, although not that involving spending or revenue generation. Numerous proposals to reform or abolish the Senate have been made in the past 50 years.

Provincial and territorial legislatures consist of a single elected chamber, with no Senate, elected by universal suffrage. Provincial and territorial elections are held at a maximum of every five years, but never close to a federal election date. In the provinces and Yukon, the party receiving the greatest number of seats usually forms the government, with the party leader becoming Head of Government.

The Northwest Territories and Nunavut have consensus governments, without parties. Elected members meet and select one of their numbers to become Premier, who then assigns ministerial portfolios and responsibilities. Each vote requires a majority to pass. Consensus government incorporates indigenous community values of cooperation, effective use of human resources, and common accountability. Consensus government is also used in the Inuit self-governing community of Nunatsiavut (Labrador).

Many of the comprehensive land claims agreements signed by the Government of Canada since 1975, in addition to others signed by provinces, have included provisions for Indigenous self-government (Crown-Indigenous Relations and Northern Development 2023a, b). There are currently 25 self-government agreements within Canada. Self-government involves local control over resource management issues, the delivery of government programmes and services, education, economic development, and determining which laws and policies are exclusively or jointly administered by federal, provincial, or Indigenous jurisdictions. Examples within provinces include Nunatsiavut (NL), Nunavik (Québec), Nisga'a (BC), and Sioux Valley Dakota Nation (MB). In 2024, self-government agreements covered all of Nunavut, northern Northwest Territories, and most of Yukon.

Municipal governments operate in cities, towns, and villages. They operate at the "pleasure" of the provincial or territorial government. The powers allocated to them are determined by the province or territory under the provisions of the BNA and Constitution Acts. Municipalities can be created, merged (amalgamated), or disbanded by the province at any time. Canada's first municipal government was established at Saint John (NB) in 1785. Typically, municipalities are responsible for fire protection; local streets; local public transportation; planning and development; provision of water, sanitation, and other public services; parks and recreational facilities; and snow clearing. Municipalities typically are responsible for management of primary and secondary education through school boards, although curricula and teacher training, licensing, and staffing usually are the responsibility of the provincial government. Larger cities have their own police forces, but most communities are policed by the provincial forces (Ontario, Québec) or by the RCMP under contract to the province. The large majority of funding for municipal programmes and services comes directly from property taxes and fees paid by residents.

Unincorporated areas are outside any municipal boundary. Limited services in these communities are provided either by a local board partially supported by the province, or directly by the province. Unincorporated communities typically do not have planning, recreational, local boards of education, water supply, or police services, and may lack fire protection as well. Residents in unincorporated areas generally do not pay municipal taxes.

"Crown land" is land owned directly by the Federal or provincial governments (Alberta 1997, Crown-Indigenous Relations and Northern Development 2020). When the colonial administrations were established in future provinces, land which was unowned (by settlers) at the time or had not been granted by the monarch was designated as the property of the Crown. Federal crown land, representing 41% of the terrain of Canada, is largely located in the territories and is administered through self-government agreements by indigenous peoples. Federal crown land within the provinces is largely confined to national parks, military bases, and indigenous reserves. Provincial crown land represents 48% of the terrain of Canada and largely remains under the direct control of the provinces. Percentages vary from a minimum of 12% (PEI) to more than 90% (BC), with additional areas within provinces covered by self-government agreements (e.g. Nunavik, Nunatsiavut).

Crown land remains the property of the Crown (in practice, the provincial and federal governments), and cannot be unilaterally sold. Leasing of Crown lands for logging and mineral extraction rights (including petroleum production) and royalties derived from production provide resource-based provinces with substantial revenue. Individuals may also lease pieces of crown land for various terms for their own use, particularly the construction of seasonal residences.

The Royal Canadian Mounted Police (RCMP) originally was constituted as the North-West Mounted Police in 1873, charged with policing in the (then-) Northwest Territories. The force merged with the federal Dominion Police, operating in eastern Canada, in 1921. In 2025, the RCMP is responsible for policing rural and most smaller urban centres under contracts to provincial governments, with the exceptions of Ontario (Ontario Provincial Police), Québec (Sûreté du Québec), and parts of Newfoundland and Labrador (Royal Newfoundland Constabulary). Today, RCMP officers do not wear red serge or Stetson hats, or ride horses, except on ceremonial occasions. As with many police forces, questions have arisen concerning accountability, sensitivity, relationships with racialized, indigenous, and 2SLGBTQI+ communities, and treatment of female and racialized officers, as acknowledged in public opinion reports commissioned for the RCMP (Ekos Research Associates 2023) and others.

The Supreme Court of Canada consists of 9 judges, appointed by the Governor General on recommendation from the Prime Minister (Supreme Court of Canada 2024). In Canada, all judges are appointed based on their legal expertise and previous service in lower courts, including provincial supreme courts, and are assumed to be politically impartial. The Supreme Court Act requires the appointment of 3 judges from Québec. Typically, 2 or 3 judges come from Ontario, two from western Canada, with the others from Atlantic or northern Canada. Judges serve until mandatory retirement at age 75, unless they resign or are deceased. All are required to be functionally bilingual in English and French.

The criminal justice system throughout Canada is based on the "Common Law" system inherited from the original British colonial administrations (Atkinson and Atkinson 2020; Canada 2024). Criminal law is a federal responsibility, as designated in the BNA Act. Indigenous laws and practices are gradually being incorporated in Canadian courts. Unfortunately, indigenous people form a disproportionately large segment of those incarcerated.

Québec has a hybrid system, with civil matters addressed using the *Code Napoléon* modification of French civil law, as guaranteed in both the original conditions associated with the Treaty of Paris (1763) and the BNA Act. The Supreme Court of Canada may agree to hear matters arising from either Common or Civil Law proceedings.

1.12 Historical Geography and Territorial Evolution

Numerous volumes focusing on national, provincial, regional, and local Canadian histories, written from a variety of perspectives, are available (e.g. McNaught 1988; Brown 2012; Conrad et al. 2012; Crowley et al 1993; Hayes 2006). From the geographers' perspective, we are primarily interested in how historical events shaped the establishment and development of communities, political geography, and territorial evolution.

Selected events prior to 1766 ce are summarized in Table 1.9. After initial exploration and unsuccessful attempts at colonization by Europeans, terrestrial bases were established along the Atlantic coast at Red Bay (1530), Port-Royal (1605), Québec (1608), and Cupid's (1610). Initial European arrivals encountered Indigenous populations with differing degrees of socio-political organization, notably the Haudenosaunee Confederacy ("Iroquois", see Table 1.8), established ca. 1142 ce, together with Mi'kmaq, Odawa, Omàmiwinini, Wabanaki, Wyandot, and numerous other cultures (e.g. Canadian Geographic 2018). Inter-cultural relations varied from mutual avoidance (Beothuk of insular Newfoundland), to cooperation, to confrontation and conflict. Along the St. Lawrence valley and Lower Great Lakes, Omàmiwinini-Wyandot-Innu trade agreements with Québec and Haudenosaunee agreements with the Dutch based at Fort Nassau (now Albany NY) accentuated existing rivalries and conflicts, resulting in the destruction of the Wyandot culture (1649 ce).

Table 1.9 Historical timeline of selected events, 1015 ce–1766 ce

Year (ce)	
1015	Leif Eirikson arrives at L'Anse-aux-Meadows
ca. 1142	Haudenosaunee Confederacy established
1494	Doctrine of Discovery proclaimed by Pope Alexander VI
1497–1583	Explorations by Cabot (1497), Corte Real (1500), Cartier (1534), and Gilbert (1583)
1530–1620	Basque whaling station at Red Bay, Labrador (NL)
1604–1608	Champlain founds settlements at Ile-Ste-Croix (NB, 1604, subsequently abandoned), Port-Royal (NS, 1605), and Québec (1608)
1610	Guy founds settlement at Cupid's (NL, subsequently abandoned); initial occupation of Renews (NL) ca. 1615
1610–1692	"Beaver Wars" French—Omàmiwinini (Algonquin)—Wyandot -Innu vs. Haudenosaunee conflicts
1615–1664	Haudenosaunee—Dutch trading arrangements
1621	Calvert founds Ferryland (NL)
1629–1632	Kirke captures Québec (1629), returned to France (1632). Kirke receives Ferryland in compensation
1648–1649	Haudenosaunee—Wyandot conflict
1664	Dutch admiral de Ruyter captures and burns St. John's
1670	Rupert's Land allotted to Hudson's Bay Company by Charles II (Great Britain)
1678	Dutch renounce all claims in eastern North America, including Acadia
1689–1697	King William's War; Battle of Québec (1690); d'Iberville attacks Br. settlements in NL (1696–7); ends w. Treaty of Ryswick (1697)
1701	Great Peace of Montreal between 39 Indigenous leaders and French colony of Québec ends "Beaver Wars"
1702–1713	Queen Anne's War (Spanish Succession) ends w. Treaty of Utrecht (1713). Acadia to Britain; Ile Royale (Cape Breton I), Ile-St-Jean (PEI) to France. NB remains in dispute
1722–1725	Dummer's (Father Rale's) War; Mi'kmaq-British conflict at Canso and Annapolis Royal
1749–1753	Mi'kmaq-French alliance defeated by British in Acadia
1754–1763	Seven years' war. Expulsion of Acadians (1755). British capture Louisbourg (1758), Québec (1759), Montreal (1760). French attack St. John's (1762). Wabanaki-Mi'kmaq–British "Peace and Friendship" Treaties agreed at Halifax (1760). War ends w Treaty of Paris (1763): Ile Royale, Ile St-Jean (Saint John's I.), Québec to Britain; St-Pierre-et-Miquelon to France; Anticosti I. & Iles-de-la-Madeleine to Newfoundland (Britain). British establish colonies of Québec, Nova Scotia, Cape Breton, and Newfoundland
1763–1766	Odawa (led by Obwaandi'eyaag (Pontiac))—British conflict

Throughout the 17th and 18th centuries (ce), Britain, France, Holland, Portugal, and Spain used North American colonies and fishing bases as bargaining chips in negotiating European conflicts, a process which began with the abandonment of the Red Bay station (1620) and continued through the nineteenth century ce. Kirke, de Ruyter, d'Iberville, La Parouse, and others attacked settlements and people with little influence on ultimate political results. The series of European treaties and territorial exchanges ended with the 1763 Treaty of Paris, which confirmed British control of Québec (including modern southern Ontario), Nova Scotia (including modern NB and PEI), Cape Breton (as a separate colony), and Newfoundland. St-Pierre-et-Miquelon remained in possession of France, which also retained fishing rights along the west coast of Newfoundland (until 1904). Residents of the regions, both indigenous and non-indigenous, were rarely consulted in the decision-making processes. In many cases, notably Lawrence's Expulsion of the Acadians (1755), political positions and hostile intent were assumed without warrant.

Events whose significance lay somewhat latent until the twenty-first century involved Indigenous-colonial relations (Crown-Indigenous Relations and Northern Development 2023a, b). In 1701, the "Great Peace of Montreal" agreement between 39 Indigenous leaders and nations, including representatives of the Haudenosaunee, Mississauga, Nipissing, Omàmiwinini, Odawa, Ojibwe, Wabanaki, Wyandot, and others, and Louis-Hector de Callière for the French colony of Québec, ended the "Beaver Wars". The agreement allowed free travel and trade throughout the region from the St. Lawrence Valley westward to the Great Lakes drainage basin, along with a cessation of Indigenous-French hostilities. Indigenous leaders hoped that the agreement would constrain advances across the Appalachians from the British colonies along the Atlantic seaboard. Failure of this objective led to subsequent Odawa-British (1763–66) and Indigenous-USA (e,g. Tippecanoe, Ohio, 1811) conflicts. Importantly, the Great Peace did not entail relinquishment of land rights or termination of indigenous sovereignty: the terrain throughout the region remained unceded.

Similarly, British colonial authorities and Indigenous peoples agreed to a series of "peace and Friendship" Treaties in the Maritimes, notably the Wabanaki-Mi'kmaq-British Treaty at Halifax (1760). As with the Great Peace of Montreal, these treaties ended hostilities and allowed free trade and travel, but did not entail surrender of sovereignty or resources (Crown-Indigenous relations and Northern Development Canada 2023a, b). The British commitment to establish a "truck house" (trading post) resulted in a guarantee of Indigenous rights to fish and hunt throughout the region to maintain a moderate livelihood, rights upheld in court decisions in the 1980s and since. Thus, the terrain from Gaspé through the Maritimes to the south coast of Newfoundland remained unceded as well.

In 1670, Charles II of Great Britain granted exclusive trading rights to a British syndicate over all the terrain draining into Hudson Bay. With the king's cousin Prince Rupert as patron, the terrain was royally designated as "Rupert's Land". Subsequently, British governments held that the Hudson's Bay Company rights extended to all mineral resources, and that the Company was entitled to exercise sovereignty over the terrain. This situation prevailed in Rupert's Land, and in the adjacent North-Western Territory (transferred to the Hudson's Bay Company in 1821) until after Confederation.

Selected events between 1763 and 1867 ce are summarized in Table 1.10. After assuming political control after 1763, the British government extended the boundary of Québec over most of the terrain covered by the Great Peace of Montreal. However, the subsequent USA-British War (1776–1783) resulted in the establishment of the USA east of the Mississippi River, the migration of pro-British residents (United Empire Loyalists, UEL) and pro-British indigenous groups to the regions of Nova Scotia, Québec, and Ontario, the establishment of New Brunswick as a distinct colony (with UEL settlement), the Mississauga purchase by Britain to provide land along the Grand River for pro-British Haudenosaunee (Haldimand Tract) and land along Lake Ontario for British settlers, and the attack on Prince of Wales Fort by La Perouse in support of the USA. Between 1784 and 1812, fourteen additional British-Indigenous treaties covered lands in southern Ontario west of the existing community of Katarokwi (renamed Kingston, 1788), all involving cessation of land rights for compensation (Crown-Indigenous relations and Northern Development Canada 2023a, b).

The 1791 division of Québec into Lower and Upper Canada marked the first time that these two regions were politically separate, and also formally introduced "Canada" as a political designation. The division reflected socio-economic and linguistic differences between the longer-established French settlements of Lower Canada, and the very recently acquired lands of the Mississauga purchase of Upper Canada, including the communities of Newark (later Niagara-on-the-Lake, est. 1792), York (later Toronto, est. 1793), and Malden (later Amherstburg, est. 1796).

From a sporting perspective, the British–Indigenous–USA War (1812–1815) ended with an overtime tie (with the final battle fought after the Treaty of Ghent was signed). USA attacks on Lower Canada were blunted by local fighters and the leadership of Salaberry de Valleyfield. In Upper Canada, the initial USA attack across the Detroit River was defeated by the brilliant psychological warfare of the team led by Brock (Br) and Tecumseh (Shawnee). Unfortunately, the team did not hold together: Brock was killed when he prematurely attacked at Queenston Heights without waiting for his main body of troops (1812). Tecumseh was unable to persuade the successive British commanders to work together effectively with the Shawnee, leading to defeat and his death in 1813. The local Upper Canada colonists initially were less than active participants in the hostilities, but that changed following the burning of farms and homes by USA troops. The war convinced the British authorities that defence against the USA was necessary, and that this could be best achieved by uniting the colonies of British North America, an objective pursued from 1815 to 1873.

Along the Pacific coast, the voyages of the British captain Vancouver (following up on Cook) and the Spanish captain Bodega y Quadra led to the Nootka Incident dispute between them. The five-cornered dance also involved Indigenous peoples of the Salish Sea, and potential interventions from USA traders and Russian voyagers from Sitka (AK). Eventually, Vancouver and Bodega y Quadra reached an agreement, also with cooperation of Indigenous groups, which left the renamed Vancouver's-Bodega y Quadra's Island open to both British and Spanish traders, without conveying sovereignty to either, or removing it from the Salish Sea communities. This agreement and subsequent negotiations paralleled the earlier "Peace and Friendship" treaties. Eventually, the HBC administered British interests, while the Spanish (Mexican) interests were effectively transferred to the USA after the latter annexed California, and the Russian claims finally formally lapsed with the purchase of Alaska by the USA (1867).

Gradual expansion of British colonial activity on the island and adjacent Lower Mainland led to a series of Indigenous-British conflicts, notably with Hwistesmetxe'qen in the Okanagan region. The Oregon Boundary Treaty (1846) established the British-USA boundary at 49°N, with the British retaining all of Vancouver's Island. The colony of New Caledonia was established, renamed British Columbia (for the river) in 1858. Vancouver's Island was a separate colony from 1849–1866. The only settlement of land claims and indigenous rights occurred in a series of treaties between 1850 and 1854, covering only small areas of Vancouver Island (Crown-Indigenous relations and Northern Development Canada 2023a, b). Interior BC remained as unceded territory, increasing occupied by British colonial settlements.

Interior western Canada remained under the control of the Hudson's Bay Company. In 1811, in part as a response

Table 1.10 Historical timeline of selected events, 1769 ce–1866 ce

Year (ce)	
1769	Saint John's I detached from NS as separate British colony; named Prince Edward Island 1798
1774	Quebec Act. Québec boundary extended to Ohio R, including Iles-de-la-Madeleine. Anticosti assigned to Québec 1774; to Newfoundland 1809; to Québec 1825
1782	Last armed intervention by France within Canada; La Perouse attacks and destroys Prince of Wales Fort, Hudson Bay coast, MB
1784	NB detached from NS as separate British colony; Mississauga (Grand River) purchase by Britain (ON)
1789–90	Nootka Incident. Vancouver–Bodega y Quadra agreement leaves BC coast open to British and Spanish traders. Spain's claims subsequently transferred to USA
1791	Québec divided into Lower and Upper Canada
1811	Red River colony land grant to Selkirk by Britain
1812–14	British—Indigenous—USA War of 1812 in Lower and Upper Canada. Ended with Treaty of Ghent (1815)
1816	Pemmican conflicts (MB); Hudson's Bay Company (HBC), North West Company (NWC), Bois-Brules (Metis), Salteaux-Ojibwe (Anihšināpē)
1818	49°N established as British-USA boundary from Lake-of-the-Woods to "Oregon Country". HBC administers British-claimed territory
1820	Cape Breton colony attached to NS
1821	NWC merges with HBC; Britain transfers administration of North-Western Territory to HBC
1825	Britian and Russia establish 141°W as North-Western Territory-Alaska boundary
1829	Death of Shanawdithit, last Beothuk
1830–1850	Hwistesmetxe'qen (Nicola)—British conflict, Okanagan region
1837–1840	Papineau and Mackenzie lead rebellions in Lower and Upper Canada (1837–38). Fenian raids from USA (until 1870). Durham report leads to political reform, merger into single British colony of Canada, divided into Canada East and Canada West (1840)
1838–1842	British-USA Aroostock conflict (NB). 1842 Ashburton-Webster Treaty establishes NB-Maine boundary, assigns NW Angle (Lake of the Woods) to USA
1846	Britain-USA Oregon Boundary Treaty; New Caledonia colony established; renamed British Columbia (1858). Vancouver's I. was separate colony 1849–66
1850	Indigenous-British (Robinson) treaties in Upper Great Lakes region, Ontario
1850–1854	Indigenous-British (Douglas) Treaties, southern Vancouver Island
1855	Bytown designated as capital of Canada colony; renamed Ottawa
1859	HBC Administration of North-Western Territory separated from Rupert's Land
1864–1866	Conferences in Charlottetown, Québec, and London (England) discuss political organization and governance of British North America

to suffering induced by the colonial policy of Clearances of Scots Highlanders, Lord Selkirk was granted land in the Red River valley of southern Manitoba for a Scots colony. The colonists arrived as the HBC establishment was confronted by commercial competition from the more dynamic North West Company (NWC). Employing the services of Haudenosaunee, Métis, Omàmiwininì, Odawa, Ojibwe, and Québecois voyageurs, operating more effectively along the trade route through the Ottawa River, Upper Great Lakes, and Lake-of-the-Woods, and disregarding the official legal monopoly held by the HBC, the NWC rapidly provided both effective trade competition and political and inter-community tensions. The Pemmican series of conflicts in 1816 involved HBC, NWC, Bois-Brules (Métis), Salteaux-Ojibwe (Anihšināpē), and Scots. To restore order, the HBC and NWC were merged in 1821 to form a revised HBC, although former NWC personnel dominated the new entity and NWC trade practices were adopted. The British government transferred administration of the North-Western Territory, the drainage basins of the Deh Cho (Mackenzie) and upper Yukon Rivers, where NWC traders such as Alexander Mackenzie had been operating, to the HBC. The boundaries of the HBC-administered terrain were established as 49°N and 141°W by treaties with the USA (1818) and Russia (1825) respectively. The northern boundary was the Arctic Ocean and the drainage divide north and east of Hudson Bay.

Meanwhile, residents of Upper and Lower Canada were becoming increasingly dissatisfied with the colonial administrators. While the 1837 Upper Canada "rebellion" led by WL Mackenzie was poorly conceived and rapidly failed, conflict in Lower Canada between the colonial authorities and the Patriotes led by L-J Papineau was more extensive, lengthy, and significant. In response to the justified grievances of the residents, the British government through the Durham report led to political reform, with the merger of Lower and Upper Canada into a single British colony of Canada, divided into Canada East and Canada West (1840). Although the merger addressed the British concerns of defence and maintenance of order, heightened by sporadic Fenian raids and some agitation for merger with the USA, it did not satisfy many residents. During this time, a series of Indigenous-British (Robinson) Treaties were negotiated, involving transfer of sovereignty and resources over the terrain adjacent to Lakes Huron and Superior in western and northern Ontario. From the perspective of Canada, the treaties were intended to regular colonial settlement along Lake Huron, and secure access to copper and silver along Lake Superior.

Several factors, including perceived threats of USA invasion, ongoing boundary and fishery assess disputes, logistics and costs of railway construction, and a desire to increase efficiency and reduce costs of government administration, led to discussion of the political organization and governance of the British North American colonies. Initial proposals to re-unify the formerly united Maritime colonies of NS, NB, and PEI were modified when representatives of Canada, led by JA Macdonald, arrived in Charlottetown to propose an all-encompassing union. Newfoundland sent observers to the meetings, but did not fully participate. Indigenous peoples were not consulted. Subsequent meetings in Québec and London revealed disagreement between establishing a strong central government (as favoured by Macdonald) or having the provinces retain most responsibilities and powers (as favoured by PEI, along with some of the representatives from NB, NS, and "Canada East"). The division of powers in the British North America Act (discussed above) represented a compromise among these positions, although PEI declined to accept the terms. 1 July 1867 saw Confederation of the provinces of New Brunswick, Nova Scotia, Ontario, and Québec (split from the former 1840 province) as Canada. The St. Lawrence-Hudson Bay drainage divide formed the northern boundaries of Ontario and Québec. The British Government retained some aspects of political sovereignty, not fully relinquished until passage of the Statute of Westminster (1931).

Selected events following 1867 ce are summarized in Table 1.11 (see Natural Resources Canada 2015, Crown-Indigenous relations and Northern Development Canada 2023a, b). Following Confederation, the British government continued to encourage political consolidation of territory, along with Canadian politicians (particularly in Ontario, with a developing shortage of agricultural land). The HBC relinquished political control of Rupert's Land and North-Western Territory, designed Northwest Territories (NWT). A delay in the transfer of authority from HBC to Britain to Canada from Britain resulted in a period of political interregnum. In the Red River valley, Métis residents under Louis Riel assumed control in the political vacuum, with the intention of negotiating with Canada. However, failure of the negotiations led to conflict, with Riel's group being suppressed with minimal violence but substantial social upheaval for the Métis communities. A square around the Red River Valley became the province of Manitoba in 1870, with the boundaries subsequently extended (1881, 1889, 1912).

The recently reunited province of British Columbia joined Confederation in 1871, with more enthusiasm among the mainland population than on Vancouver Island. Settlement of the San Juan Islands (1872) and Alaska boundary disputes (1903) established the present boundaries of British Columbia. In contrast to the Northwest Territories, no extensive land claim or treaty negotiations were conducted with Indigenous peoples in most of BC until the late twentieth century.

Prince Edward Island had remained as a separate colony since 1867. Under financial pressure from railway construction and the necessity to purchase property rights from absentee landlords, coupled with the desire for increased commercial ties, PEI opted to join Confederation in 1873 as the seventh province.

Beginning in 1871, Canada undertook negotiations with Indigenous peoples in the Northwest Territories (Crown-Indigenous relations and Northern Development Canada 2023a, b). A series of 11 numbered treaties were negotiated covering land claims across the region, beginning with areas now part of Manitoba (#1, #2), northwestern Ontario (then part of NWT, #3), Saskatchewan, Alberta, northern Manitoba, and northeastern BC (#4, #5, #6, #7, #8), northern Ontario (then part of NWT, #9), and northeastern Saskatchewan (#10). With the exception of #10, the other treaties involved negotiations between the federal government and the Indigenous peoples, before the establishment of provincial authority in the land. Negotiations for Treaty #10 began before the establishment of the province of Saskatchewan. Treaty #11 (1921) covered the terrain of Mackenzie District of NWT and southeast Yukon. The numbered treaties all involved establishment of Canadian governmental sovereignty in exchange for compensation and establishment of Reserves. Political subdivisions of the former NWT resulted in the establishment of districts (1876–1895).

Construction of a transcontinental railway was among the terms negotiated by BC for entry into Confederation. The subsequent establishment of the North-West Mounted Police, negotiation of Treaties #1 through #7, plummeting of the *Bison* population, and subdivision of the NWT into districts all predated large-scale European settlement of the prairies. Ongoing tensions between Métis (led by Riel and Dumont) and Néhinaw (led by Pitikwahanapiwiyin and Mistahimaskwa) and Canadian authorities in (present-day)

Table 1.11 Historical timeline of selected events from 1867 ce

Year (ce)	
1867	Confederation of provinces of New Brunswick, Nova Scotia, Ontario, and Québec as Canada. St. Lawrence-Hudson Bay drainage divide forms northern boundaries of Ontario and Québec. Terms of Confederation are established in the British North America Act of parliament of Great Britain
1868–1870	HBC relinquishes political control of Rupert's Land and North-Western Territory, designed Northwest Territories (NWT). Delay in transfer to Canada from Britain results in political interregnum. Failed negotiation with Manitoba residents leads to Riel conflict. Riel's group suppressed and MB becomes 5th province (1870); boundaries extended 1881
1871	BC joins Confederation as 6th province. Canada-Indigenous Treaty 1 (southeast MB) and Treaty 2 (west MB)
1872	San Juan Islands dispute resolved by Wilhelm I (Germany) in favour of USA
1873	PEI joins Confederation as 7th province. Canada-Indigenous Treaty 3 (now northwest ON). North West Mounted Police est
1874	Canada-Indigenous Treaty 4 (now south SK). Western boundary of Ontario extended
1875	Canada-Indigenous Treaty 5 (central and north MB)
1876	Canada-Indigenous Treaty 6 (central SK and AB). Keewatin District established within NWT
1877	Canada-Indigenous Treaty 7 (southwest AB)
1880	Arctic Islands transferred from Britain to Canada
1882	Assiniboia, Saskatchewan, Alberta, Athabaska Districts established within NWT; *Bison* population plummets
1885	Métis and Néhinaw—Canada conflict (Northwest Rebellion). CPR completed. Establishment of Residential Schools
1889	Ontario boundary extended westward to Lake of the Woods
1895	Ungava, Mackenzie, Yukon, Franklin Districts established within NWT
1896–1898	Klondike Gold Rush. Yukon established as separate Territory. Québec boundary extended (1898)
1899	Canada-Indigenous Treaty 8 (northwest SK, north AB, northeast BC)
1903	Alaska boundary dispute settled
1905	Canada-Indigenous Treaty 9 (north Ontario). Alberta and Saskatchewan become 8th and 9th provinces
1906	Canada-Indigenous Treaty 10 (northeast SK)
1912	Manitoba boundary extended to 60°N; Ontario and Québec boundaries extended to Hudson Bay
1918	Female Suffrage proclaimed
1921	Canada-Indigenous Treaty 11 (NWT (Mackenzie District), southeast Yukon)
1923	Canada—Chippewa–Mississauga Treaty (Williams) resolves issues along Lake Ontario from 1783–84
1925	Canada extends territorial claim to North Pole
1927	Canada-Newfoundland Labrador boundary dispute adjudicated by Privy Council of Great Britain
1931	Statute of Westminster transfers all political powers to the Governments of Canada and Newfoundland
1934	Newfoundland declares bankruptcy, ceases to be independent; administered by Britain through Commission of Government
1949	Newfoundland joins Confederation as 10th province; Official name changed to Newfoundland and Labrador in 2001
1960	Indigenous suffrage proclaimed
1982	Constitution Act repatriates British North America Act as an Act of Canadian parliament
1999	Nunavut devolves from NWT to become separate Territory
2008	Truth and Reconciliation Commission established

central Saskatchewan led to conflict with human bloodshed (Northwest Rebellion), suppressed by armed force in 1885. The Canadian Pacific Railway was completed in the same year.

During this time, the Government of Canada supported expansion of church-operated systems of Residential Schools, involving the support of all political parties and many members of Canadian academic communities. Persisting until 1996, Residential Schools attempted to facilitate the assimilation of Indigenous children into Canadian society, often employing cruel and unethical methods. At least 150,000 children were forced to attend, of whom at least 6,000 died at the Residential Schools. Discussed extensively by the Truth and Reconciliation Commission (2012), the Residential School systems are widely considered to have been genocidal. Indigenous people did not receive the right to vote in Canada until 1960.

As a result of the Klondike Gold Rush, Yukon was detached from NWT to form a separate Territory in 1898 (Natural Resources Canada 2015). Alberta and Saskatchewan were formed as provinces in 1905. Areas of the NWT as it existed in 1870 were detached to Manitoba, Ontario, and Québec, with the provinces reaching their present boundaries in 1912. The Arctic Islands north of the Hudson Bay drainage divide were transferred to Canada from Britain in 1880. With the extension of its territorial claims to the North Pole in 1925, Canada had reached its present boundaries, with the exception of the area of Newfoundland and Labrador.

Newfoundland had rejected Confederation in 1866, and Canadian Prime Minister Thompson's efforts to entice it to join ended with his death in 1892. On the North American mainland, Canada (on behalf of Québec) claimed all the terrain drained by rivers entering the Gulf of St. Lawrence west of 57°06′30″ W, including the headwaters of the Mecatina, Petit-Mecatina, Magpie, and Romaine rivers to the north of 52° N. Newfoundland exercised jurisdiction over the coastal communities of the Labrador Sea and Strait of Belle Isle, and in the Lake Melville area. Western Labrador was considered by Canada to belong to the Northwest Territories, while Newfoundland considered it to be part of Newfoundland. As the two parties could not agree, the dispute was adjudicated by the Privy Council of Great Britain (1927), ruling that the "Coast of Labrador" had been defined to extend to the drainage divide between Hudson Bay and the Labrador Sea. Canada did receive a small tongue of terrain between the Romaine and Magpie Rivers extending to the north of 52°N. This decision was accepted by the federal government of Canada with relatively little interest, although politicians in Québec were less pleased. Although Canada accepted the arbitration, Québec never formally has.

In 1934, a combination of the effects of the Great Depression and political mismanagement resulted in the bankruptcy of Newfoundland, forcing the island to revert to Commission of Government under more direct British rule. At that time, Newfoundland offered to sell Labrador to Canada for $110,000,000, a substantial discount from the original $150,000,000 offer made in late 1927. Canada, however, was also feeling the effects of the economic depression in 1934, and declined to purchase Labrador on both occasions, for reasons that have never been thoroughly explained. One hypothesis suggests that Canada anticipated Newfoundland's eventual entry into Confederation; another suggestion is that Canada was waiting for a further discount. Following World War II, and faced with the refusal of Great Britain to continue with the Commission of Government, Newfoundland narrowly voted to join Canada in 1949. The province officially changed its name to Newfoundland and Labrador in 2001.

In 1999, Nunavut devolved from NWT to form a separate Territory (Natural Resources Canada 2015). The establishment of self-government in Nunavut and subsequent agreements with Canada have resulted in the transfer of territorial rights and resource management, rectifying their exclusion from the initial transfer of the Arctic Islands to Canada in 1880.

1.13 Settlement Patterns and Resources, Historical, and Present

Generally, people will settle in areas offering suitable living conditions: not too hot or cold, availability of water and food resources, favourable topography among other factors. Settlement in Canada forms a mosaic along the southern border, where there are arable lands!

Being the first to establish permanent colonies, France proceeded by hesitation in successive attempts (not many, if at all, between Cartier and Champlain). The years of the early seventeenth century were turbulent times in France. The focus was rather on internal affairs than exploration and colonization. Nevertheless, colonization slowly picked up, but as the settlers, labourers, soldiers, known as "engagés" (engaged) were mostly males, natural increase remained low. Although France's population was the highest in Europe, fearing exodus, immigration was controlled and remained relatively low. Instead, King Louis XIV then relied on approximately one thousand "filles du roy" (King's daughters) who were sent to encourage family formation from 1663 to 1673. Consequently, natural increase grew significantly (Lacroix 2019) but the French colony population remained modest compared to the British colonies. Domestic markets for emerging businesses remained very limited.

Great Britain, it seems, had much more conviction and (possibly) logic? To start with, even though France was first to establish permanent colonies in North America, British immigration was considerably more important and the population increase was faster. From 250,000 in 1700, Colonial America's population grew tenfold to 2.5 million in less than eight decades (Lemon 2001). By comparison, around 70,000 people of French descent lived in the St-Lawrence valley at the end of the French regime (Hayes 2006). When comparing their respective spatial strategies, British colonies were typically more clustered settlements along the Atlantic coast, and their population increase initiated a westward advance (see Chap. 11). French possessions, on the other hand, were much more scattered. The fur trade routes, and the control of main accesses to the sea, appear to have been the centre piece of their extensive possessions, a territory difficult to defend for such a small population. However, there were several concentrations along the St-Lawrence River, notably around Québec City, Trois-Rivières, and Montreal.

The current pattern of Canadian settlement has 99% of the population living on 20% of its territory (DeKoninck 2004); 66% lives within 100 km from the American boundary on 4% of the total land (Statistics Canada 2021–2024). Canada is an urban nation as 80% of the population lives in urban areas (with a population of at least 1,000 and no fewer than 400 persons km^{-2}). Additionally, this large proportion of the population lives on less than 1% (0.25%) of Canada's land mass. Statistics Canada categorizes urban areas into the larger 41 census metropolitan areas (CMA), which accounts for 71.9% of the total population, and 119 census areas (CA) which represent 12%, the remaining percentage being the rural population (Statistics Canada 2021–2024). Furthermore, almost half of the Canadian population lives in the six largest urban regions, including 35.7% in the three largest. However, these urban regions and the remaining rural regions are sprinkled unevenly.

1.13.1 Another Way to Look at Canada

Harris (2006) has compared Canada to an archipelago, a land with groupings of people expressing a host of regionalisms, hence a difficult way to forge a common identity. DeKoninck (2004), inspired by Hamelin (1969), added to the idea by dividing the settlement pattern into three types: three continuous clusters, three strings of discontinuous settlement strips, and the dispersed margins:

- Three main clusters ("Main Street", ALSAMA (Alberta, Saskatchewan, Manitoba), Lower Fraser Valley).
- Connecting Strips (Rockies, "North central", Atlantic).
- The margins (by boat, by road or trail, by air (FIFO, Fly-in-Fly out)).

This typology of Canadian human geography shows how population is distributed and their (approximate) physical settings. The interactive Historical Atlas of Canada (2024) is a helpful visualization of the spatial evolution of population and the political boundaries of Canada between 1851 and 1961. Canada's Minerals and Mining Map (Natural Resources Canada 2024) is also a good source for more detailed information (Fig. 1.13).

1.13.1.1 Three Main Nodes or Clusters: "Where Food Grows"

The three main nodes or population clusters show a strong relation with the types of ecoregions where they concentrate. "Main Street" (the Québec-Windsor axis), the prairie provinces "triangle" and the Lower Fraser Valley in British Columbia are fertile regions suitable for agriculture, and the relative flatness of the facilitates population presence. These three main clusters account for around thirty million people, or close to eighty per cent of the Canadian population in 2021: "Main Street" (20 million), ALSAMA (7 million) and the Lower Fraser Valley (3 million). The Québec-Windsor Axis has a much higher population density, as the land area is smaller than the Prairies' estimated 490,000 km^2 (Natural Resources Canada). Although the measuring methods can vary, Yeates (1975) estimated the axis land area to be close to 175,000 km^2, with 298 km^{-2} population density on "Main Street" vs 14 km^{-2} in the Prairies triangle.

"Main Street" or the "Québec-Windsor axis" is the axis along the St-Lawrence River to the Great Lakes (Yates 1975). It is also the main initial colonization basin of Canada, subsequently gradually penetrating the axis on farmlands from East to West. A long string of poles developed along the axis progressively from East to West. Of Québec City (1608), Trois-Rivières (1634), and Montreal (1642), the latter became the early metropolis of Canada where population, businesses and industry congregated for several decades before being doubled by Toronto by the second half of the twentieth century (see Chap. 12).

Agriculture, the main reason why people settled along the St-Lawrence axis in the first place, is still present although it has suffered from urban growth and sprawl, especially in the Montreal and Toronto regions. In some places, traditional agriculture has converted to specialty exploitation including goat, sheep, ostrich, buffalo farming, niche-market grains, and viniculture. Agriculture has made its way to the city as some community gardening initiatives have sprouted with some ventures using greenhouses on building roofs to grow food. Lufa Farms Inc., in Montreal, is an example.

The population along this axis is not as continuous as its name might indicate. Nevertheless, around half of the Canadian population (20 million) lives in this cluster and it constitutes Canada's economic core where manufacturing, administrative and financial activities concentrate. Although Ontario and Québec have respectively attracted heavier (e.g. auto, machinery, steel) and lighter industry (e.g. textile, clothing, agri-food), their economy has recently been moving into high-technologies and R&D. Of these poles, a trio has emerged as metropolises in their distinctive arena: Toronto as the financial metropolis, Montreal has seen its role shift from Canadian metropolis to Québec and francophone world metropolis, and Ottawa as the national capital.

The "ALSAMA triangle" is unmistakable when looking at Canada's land cover. The region is large and has a population of around seven million people that appears evenly dispersed.

Settlement of the prairies has occurred in a succession of migration waves, of which the most important was from mainly European immigrants at the turn of the twentieth century (see Chap. 11). Farming became the economic base of the prairies but with some geographic distinctive characteristics.

Of the three provinces, Alberta has, with more than half, the largest population within this region. Several urban service centres, and the provincial capitals, are sprinkled in an almost regular fashion among which a handful of metropolitan centres emerged as important poles in their distinctive areas. Three of these, Winnipeg, Edmonton, and Calgary, roughly delimit this triangle. Winnipeg had gained importance as being a regional metropolis and "The gateway to the West", a rail transportation hub and especially the location of the Grain Exchange trading market between 1897 and 2018, making

Fig. 1.13 Types of settlement patterns

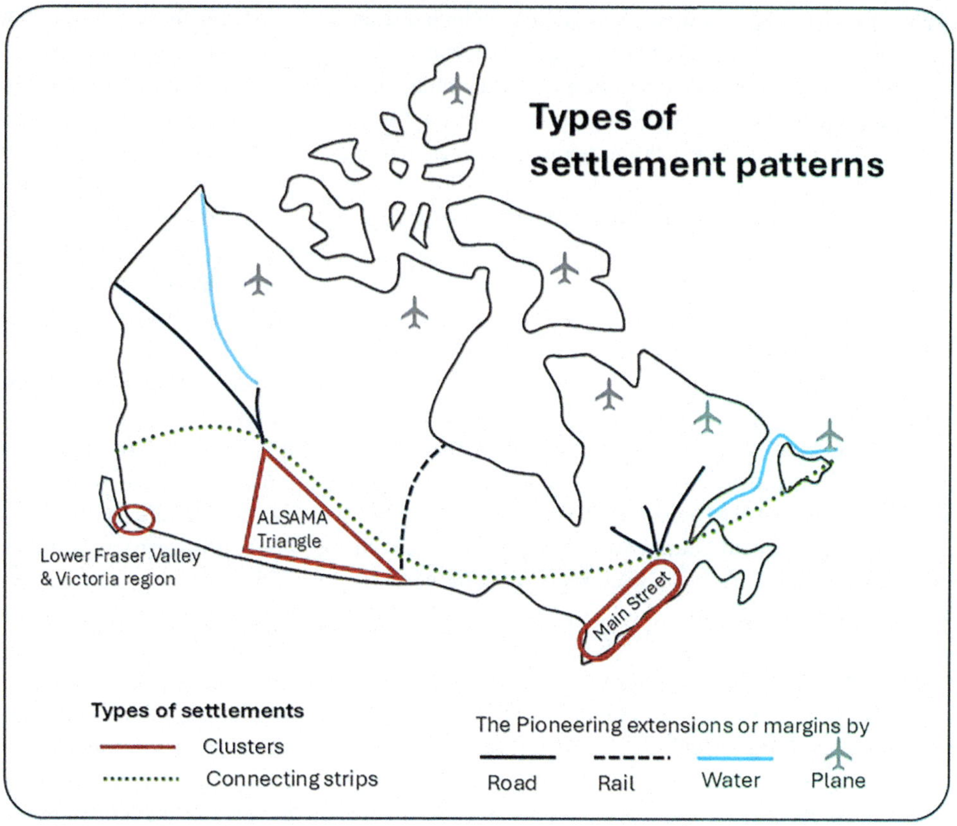

Adapted from De Koninck, 2004

2021 approximate population totals for settlement patterns types (compiled from Statistics Canada)

Clusters	Main Street ALSAMA Triangle Lower Fraser Valley	20 millions 7 millions 3 millions	30 millions
Connecting strips	East Coast Boreal forests South cordillera	2.5 millions 1.5 millions 1 million	5 millions
Pioneering extensions or Margins	East West	26,000 10,300	36,300

it the wheat capital of Canada. Winnipeg has lost some of its lustre over time while Calgary, initially because of the ranching industry, and Edmonton, gained momentum as the demand for oil reached new heights. However, reliance on commodities such as oil, cattle and grain that are conditioned by external forces is subject, as they have been before, to considerable value fluctuations. Consequently, this vulnerability has led these cities to diversify their economic base.

These five metropolitan centres account for at least half of their province's population. None of the combinations of two CMAs in Saskatchewan nor Alberta dominate their province population as Winnipeg does. Winnipeg represents 62% of Manitoba's population while Edmonton (33.3%) and Calgary (34.7%) combined just top Winnipeg by 6%, and Saskatoon (28%) and Regina (22%) total 50% of Saskatchewan's population.

The remainder of the Prairies' population is scattered throughout in regional oriented service towns, villages, and rural areas, displaying a remarkable hierarchical symmetry from diversified large metropolitan cities to locally geared service centres along the road corridors, thus being reminiscent of the central place theory. However, some changes have been noticeable during the last few decades.

Migration data show that population in most rural areas is decreasing while largest metropolitan areas and job producing regions show positive variations. But, again, these trends will vary according to world market. For instance, Alberta has seen its migration data swing one way or the other when oil value peaked then fell dramatically during the 2007–2009 recession, the Fort McMurray 2016 wildfire, and the COVID-19 crisis.

Illustrative of an economic reorganization and farms consolidation is the gradual replacement of iconic small wooden grain elevators that grace the skyline of almost every town by much larger and modern ones made of concrete that respond to a different spatial logic. Because they require more space, these integrated "prairie giants" are built outside towns and are more centrally located to serve several towns.

The "Lower Fraser Valley" (or Georgia Basin) is isolated from the rest of Canada by mountain ranges and graced by a mild climate and rich and fertile valley soils. Thus, vegetation grows and appears lush almost year-round in a spectacular natural setting. The Georgia Basin sits on a comparably small piece of land on the edge of the continent and is surrounded by snowy peaks, water, protected agricultural lands and the United States borders. Nonetheless, essentially Vancouver focused, the region, that also includes the southern part of Vancouver Island, is the undisputable and energetic West coast financial and commercial centre and home to around 3 million people. Like Winnipeg, its share of the provincial population is unbalanced at 60%, the Okanagan Valley is a distant second with a little over a tenth of the Lower Fraser Valley's share.

British Columbia was connected to Canada when the Pacific Canadian Railway chose Vancouver as its western terminal in 1885. Similarly as with the United States and California, this reach for the West coast precipitated its annexation as a province to the Canadian federation with central and eastern Canada.

The location of the region makes it the perfect and picturesque interface between "sea to sky", as the name of the highway from Vancouver to Whisler indicates. It is also the link between the rest of Canada and the Pacific world. It has become the Canadian gateway to Asia. Culturally, on a British background, and initially Chinese, more diverse Asian later, immigration has added a cosmopolitan flavour to its population makeup, so much so that it gave the region the combination of an "Asian lotus and Californian laid back feel" to it. Commercially, the Vancouver port has grown into an exportation and importation hub at an international level. It gave the opportunity to export British Columbia's rich variety of natural resources (i.e. forest products, coal, copper, fruits, fish, and sea food), but also agricultural crops and natural resources from the Prairies and finished goods from the Canadian industrial core.

With limited space for expansion, the Vancouver region also shows its singularity by its consciousness on planning. Since its founding, it has experienced a cycle of political and planning regimes balancing back and forth from a permissive British model that reacts to market forces to a restrictive and regulatory USA model aimed at controlling those forces (Langford 2012). Conscious Contemporary urban planning in Vancouver appears more connected today to cities such as Seattle, Portland OR, and San Francisco interacting with comparable natural settings and often regarded as example of avant-garde urbanism.

Being so far away from the rest of the country "back East", residential architecture was influenced early on by currents from American West coast cities. Thus, early twentieth-century American Craftsman and Californian bungalows are still very present in Vancouver single-family neighbourhoods. Even the "Vancouver Special" is a "local creation" from the mid-twentieth century.

The Lower Fraser Valley region has sprawled considerably from its initial core. The Greater Vancouver Regional District (GVRD), now Metro Vancouver Regional District, a provincial entity, was created in 1967. It encompasses most of the Lower Fraser Region. It has worked on the shared principle of livability among its federate members (metrovancouver.org). Evocative of the GVRD mission was the title of its first document: the Livable Region Strategic plan (Greater Vancouver Regional District 1996). A second entity, The Fraser Valley Regional district was created in 1995 and contains the remaining area of the Lower Fraser Valley Region.

1.13.1.2 Strips of Discontinuous Settlement: "Where Trees Grow"

These strips of discontinuous settlement are adjacent to the tree line on the southern part of the country and connect the abovementioned clusters. Population totaling approximately 5 million is dispersed in small pockets on land, regularly along circulation axis, and where agriculture is marginal at best. However, the diversity of resources found near those strips explain human presence. The Eastern coastal region is the most populous of these with close to 2.5 million, while the population of the "mixed forest margin" totaling 1.5 million is strewn from Sept-Iles to the Manitoba border line, and 1 million people are spread out in the Southern Cordilleras.

Not only is the "Eastern coastal" population is distributed in discontinuous settlements, it is also fragmented into four provinces (five regions including Gaspésie), two of which are islands, one is a peninsula and the other is contiguous to both the south shore of Gaspésie and the state of Maine. It could be said that "Geography kills the Maritimes… and Atlantic Canada!". The dislocated aspect of the region makes circulation challenging. Even the term "region", defined as territory with specific characteristics that give it unity, may seem inadequate. In that sense, Eastern coastal Canada is rather a patchwork of dissimilar regions. Easternmost settlements have relied on fishing, as some still do, others on logging in

New Brunswick, some farmed potatoes on Prince Edward Island. Even Acadie is not as uniform as it may seem as Arseneault's map shows (see Chap. 11).

Access to distant Newfoundland requires 6 to 14 h on a ferry from Cape Breton, or a trip to Halifax to get a direct flight, as (or more) expensive as flying from Halifax to Vancouver. The road to PEI has only been "connected" to the continent since the opening of the Confederation Bridge in 1997. Travel to the Iles-de-la-Madeleine (Magdalen Islands) necessitates a long ferry ride from Souris PEI. Gaspésie is far from everywhere in Atlantic Canada. Relatively, so is Central Canada from Atlantic Canada!

Population is dispersed and mainly rural, close to half in Nova Scotia and PEI, 40% in New Brunswick and Newfoundland, and 72% in Gaspésie. There are several urban centres, most of them being peripheral. Moncton is centrally located within the Martimes, but Halifax is the easiest gateway to Newfoundland by air.

History has not been kind to Atlantic Canada. Wooden shipbuilding enjoyed growth for a while but declined with the advent of larger iron hulled vessels (ca. 1870–1880). Early coal mining benefited from tariff barriers under the Federal Government National Policy at the end of the nineteenth century but slowly faded afterwards with nearer Pennsylvania coal supplying Ontario steelworks. Railway transportation and maintenance also had a significant impact, notably in Moncton where the CN shops were located. However, more reliable technology coupled with competition from road transportation led to their closure in 1986. Although some manufacturing industry developed early, the weakness and the fragmented nature of the regional market, but also the distance from larger markets, has often resulted in the closing or the purchase by larger firms and moving to Central Canada. When New Brunswick, Nova Scotia, and later, Prince Edward Island became part of the Canadian federation, economic decline was already evident.

The contemporary region's economy is mostly tied with natural resources such as agriculture, primarily potatoes in PEI and New Brunswick, fishing, logging, and offshore oil extraction in Newfoundland. Secondary activities such as paper and wood mills, oil refining, fish packing plans derive from the extracting process. The service sector has allowed a few urban regions, notably Halifax and Moncton, to attract businesses and population. However, most of the population growth has been fuelled by rural out-migration. Conversely though, the last few Canadian censuses (Statistics Canada 2021–2024) revealed that the three Maritime provinces showed a notable population increase while Newfoundland has struggled to stabilize its population.

The Mixed Forest Margin and Southern Cordillera are relatively similar. Both regions are natural resources rich, those resources being mainly minerals, forest, and hydropower: *Blue gold, black gold, green gold, white gold. Ho! Gold too!* Therefore, extracting and primary processing activities dominate. These activities benefit from the use of efficient technologies and do not require large labour forces. Nonetheless, their presence has fuelled permanent settlements, although sometimes remote, and a communication network within the region and/or with their main node (i.e. "Main Street", ALSAMA, Lower Fraser Valley). Some are service centres; others are industrial towns where resources undergo initial processing. Regional main centres are more diversified. Examples include Sept-Iles and Rouyn-Noranda in Québec, Sudbury in Ontario, Fort McMurray in Alberta, and Prince Rupert in British Columbia.

The "mixed forest margin" discontinuous settlement strip runs along the mixed forest margins on the Canadian Shield contour and more precisely from Sept-Iles to Ontario-Manitoba boundary. The regional population totals approximately a little under 1.5 million.

In the early stages of industrialization on "Main Street", minerals were found nearby in sufficient quantities to fuel nascent industry. Large-scale mining in boreal Canada is intertwined with the construction of the Canadian Pacific Railway as several deposits of minerals were found when cutting rock on the Canadian Shield to lay the tracks on. These discoveries led to a fury of prospection throughout Canada (Cranstone 2002). Mining developed in Northen Ontario and spread later in neighbouring Abitibi. Gold (ON, QC), silver (QC), nickel, copper, platinum group metals, gold, silver, cobalt, selenium, and tellurium (around Sudbury) mining were eventually established (see Chap. 3).

Forest (green gold) commercial operations have long been related to shipbuilding and lumber trade with Great Britain along the St-Lawrence axis. Pulp and paper industry then developed, manufacturing a large variety of products. The industry has, however, experienced setbacks due to international competition since the 1970s and, more recently, with the advent of electronic media which reduced demand on newspaper and book printing (see Chap. 15). A recent segment of the industry has been the development of bioenergy to reduce dependence on fossil fuels. Moreover, such practices as precision forestry, forest monitoring, management, and sustainability, cutting planning, and the use of efficient harvesting technology have made their way into the Canadian forest industry. Nonetheless, wildfires and clearcutting are ongoing concerning issues, as discussed further in Chap. 15.

Water (blue gold) is a resource with many possibilities that have been used for a long time, for instance to move machinery mechanically through a large wheel activated by water current. However, hydroelectric power (white gold) began to be used by manufacturing companies needing a large quantity of energy in the process. Such enterprises were willing to invest in building hydropower dams to cater to their needs. The excess production of power would be sold to domestic use. Aluminium making is a typical example. Arvida

QC (now part of Saguenay) in the mixed forest margin region, and Kitimat BC, in the Southern Cordillera, among others, were ALCAN creations (now owned by Rio Tinto).

The "Southern Cordillera" region is basically the area south of the Yellowhead highway that runs from the Rocky Mountains to Prince Rupert. The population count is around 1 million (Statistics Canada 2021–2024). The mountainous topography of the region partly explains the lower population number. Several towns are dispersed throughout. The Okanagan Valley, confined between mountain ranges, includes a string of cities which total a population of a little over 350,000 people living in a dry and mild climate.

What differs the Southern Cordillera from the mixed forest margin is its spectacular setting where mountain ranges, plateaux and valleys combine to offer a variety of economic activities. However, the environment is more suitable for agriculture, especially for fruits orchards, vineyards and wine making, particularly in the Okanagan Valley. "Beautiful British Columbia", the slogan on vehicle license plates, or "Supernatural" on tourism publicity, relies heavily on tourism, as well. Resources based in the Pacific Maritime temperate conifer forest, which offers higher value products, and minerals such as coal, copper, zinc, and silver contribute to British Columbia's economy.

1.13.1.3 Pioneering Extensions: The Resource Trail or "Where the Resources Are"

Differing from the previous types of region, "pioneering extensions" are almost exclusively resource motivated. Their locations are very sporadic and often remote. They are almost exclusively in the resource rich "Far North". Canada possesses "Base" and "Precious" metals, including Uranium (SK), diamonds (NWT, QC), Iron (QC), as well as rare earth elements reserves (Natural Resources Canada 2024). Some mines might be short-lived until the resource is exhausted. The extensions were developed by water routes, by roads, railways or by air for the express purpose of connecting them to the main nodes. Among the communities that were developed using water transport, some have been connected by road subsequently (see Chap. 14).

We can recognize two axes of "trappers and fishers extensions" on opposite sides of the Canadian North. One follows Newfoundland's Great Northern Peninsula and Québec's Lower North Shore (Basse-Côte-Nord) in the East, and another along the Mackenzie River (Deh Cho) in the West. The first one totals a population of approximately 26,000 stretched along the two shores. There are 70 communities with population ranging from 20 to 1000 plus St. Anthony, at the northern end of the Viking Trail road, with a comparably soaring 2,258 inhabitants. Communities' names containing "River", "Harbour", "Port", "Cove", "Point", "Brook" or "Bay" are evocative of their location and their economic base, with fisheries, some logging, and tourism, as at L'Anse- aux-Meadows historic site. Québec's Basse-Côte-Nord population is a little over 9,000. A long stretch of communities between Kegaska and Vieux-Fort is as of 2024 only accessible by water.

Approximately, 10,000 people live along the Mackenzie (6,700) and Slave (3,600) rivers. Six out of 15 communities' English names contain the word "Fort", evocative of its exploration and fur trade past.

"Roads to the adjacent margins… and to connect the dots!" represent a type of pioneering extension. This could also be referred to as the close periphery, or a step beyond the margins, that has been connected by road. In that sense, the "trappers and fishers extensions" could fit that description in cases where roads connect those settlements that were initially pioneered by water transport and still exist. Chibougamau, Mattagami, Fort McMurray, Flin Flon, and Peace River would perfectly fit this definition. The huge hydropower projects undertaken in the 1960s, (Manicouagan) and 1970s (Baie James) by Les bâtisseurs d'eau ("Water builders") are also good examples (see Chap. 17).

Several ore exploitation operations will necessitate the establishment of a "temporary" or "boom" town until the closing of the mine, for instance. This type is similar to the previous one with the difference that the railway connection is essentially aimed at carrying material from deposits to place of export or processing. Examples include the lines from Winnipeg to Churchill MB, and from Fermont and Labrador City to Port-Cartier and Sept-Iles (see Chap. 14).

An example of a short-lived mining operation is the case of Gagnon QC, a mining town established by Québec Cartier Mining Company to extract iron ore. The mine closed in 1984. Population went from 3,360 in 1981 to 5 five years later (Statistics Canada 1986 Census of Canada).

"Fly in fly out" (FIFO) has emerged as an alternative work arrangement to building towns where exploitation does not necessarily require a large working force in remote areas, or supplement existing accommodations. Mary River iron mine in Nunavut is one example of these conditions. Camps of varying sizes are built or carried to the site to accommodate workers who will work a number of days and return home to rest, typically one week at work and one week at home.

References

Alberta (1997) About public lands. Queen's Printer for Alberta

Atkinson P, Atkinson D (2020) The Canadian justice system: an overview. LexisNexis Canada

Beck C, Grieser J, Kottek M, Rubel F, Rudolf B (2006) Characterizing global climate change by means of Koppen climate classification. http://koeppen-geiger.vu-wien.ac.at/pdfs/beck_et_al_ksb_2006.pdf

Bostock HS (2014) Physiographic regions of Canada. In: Geological survey of Canada map 1254A (2nd edn), scale 1:5,000,000. https://www.nrcan.gc.ca/maps-tools-and-publications/maps/geology-and-geosciences/16876. Accessed 12 July 2022

Brooks SF (2007) Canadian democracy: an introduction. Oxford University Press, Don Mills
Brown C (2012) Illustrated history of Canada. McGill-Queen's Press
Canada (2024) Justice: Canada's system of justice. https://www.justice.gc.ca/eng/csj-sjc/
Canadian Geographic (2018) Indigenous peoples' Atlas of Canada. Royal Canadian Geographical Society, Ottawa
Catto N (2015) Geomorphology: landscapes we live in. Oxford University Press, Don Mills, Ontario
Conrad M, Finkel A, Fyson D (2012) Canada: a history. Toronto: Pearson
Cranstone DA (2002) A history of mining and mineral exploration in Canada and outlook for the future. Natural Resources Canada
Crowley TA, Crowley T, Murphy R (1993) The essentials of Canadian history: pre-colonization to 1867—the beginning of a nation. Res Educ Assoc
Crown-Indigenous Relations and Northern Development (2020) Federal crown land. Queen's Printer for Canada
Crown-Indigenous Relations and Northern Development (2023a) History of indigenous peoples, aboriginal affairs and northern development Canada and the treaty relationship https://www.rcaanc-cirnac.gc.ca/eng/1338907166262/1607904846325
Crown-Indigenous relations and Northern Development Canada 2023b. Treaties and agreements. https://www.rcaanc-cirnac.gc.ca/eng/1100100028574/1529354437231
DeKoninck R (2004) Le monde à la carte. Cahier d'exercices. Québec, Presses Inter Universitaires, 6e éd
Dodek A (2021) The Canadian constitution. Parliament. https://www.parl.ca/
Dyke AS (2004) An outline of North American deglaciation with emphasis on central and Northern Canada. In: Ehlers J, Gibbard PL (eds) Quaternary glaciations—extent and chronology, Part II: North America. Elsevier, Amsterdam, pp 371–424
Dyke AS, Moore A, Robertson L (2003) Deglaciation of Canada. Geol Surv Can Open File 1574. Thirty-two digital maps at 1:7,000,000 scale with accompanying digital chronological database and one poster (two sheets) with full map series; 1 CD-ROM.
England J, Atkinson N, Bednarski J, Dyke AS, Hodgson DA, O'Cofaigh C (2006) The Innuitian ice sheet: configuration, dynamics and chronology. Quatern Sci Rev 25:689–703
Environment Canada (2020) Canadian climate normals 1981–2010. https://climate.weather.gc.ca/climate_normals/index_e.html
Environment Canada and Agriculture and Agrifoods Canada (1995). Terrestrial Ecozones and Ecoregions of Canada. https://sis.agr.gc.ca/cansis/publications/maps/eco/all/regions/eco_all_regions_7m_canada_en.jpg. Accessed 12 July 2022
Fulton RJ (1995) Surficial materials of Canada. In: Geological survey of Canada map 1880A, Scale 1:5 000 000. https://www.nrcan.gc.ca/maps-tools-and-publications/maps/geology-and-geosciences/16876. Accessed 12 July 2022
Greater Vancouver Regional District (1996) Livable region strategic plan. https://metrovancouver.org/about-us/Documents/livable-region-strategic-plan.pdf
Hamelin L-E (1969) Le Canada. Presses Univ de France, Paris
Harris RC (2006) Canada and nature: some reflections at the scale of a country. Annales De Géographie 649:259–269
Historical Atlas of Canada (2024) Interactive maps. http://www.historicalatlas.ca/website/hacolp/
Hayes D (2006) Historical atlas of Canada. Douglas & McIntyre
Hueglin TO (2021) Federalism in Canada: contested spaces and uneasy balances. University of Toronto Press, pp 286–319
Lacroix P (2019) Promises to keep: French Canadians as revolutionaries and refugees, 1775–1800. J Early Am His 9–1:59–82
Langford W (2012) Is Sutton Brown god? Planning expertise and the local state in Vancouver, 1952–1973. BC Stud 173. https://doi.org/10.14288/bcs.v0i173.2371
Lemon JT (2001) Colonial America in the 18th Century. In: McIlwraith TF, Muller EK (eds) North America: the historical geography of a changing continent. Lanham: Rowman & Littlefield, pp 143–164
Library of Parliament (2022) Overview of the Canadian parliamentary system|Our Country, Our Parliament. lop.parl.ca
McNaught K (1988) The penguin history of Canada. Penguin books
Natural Resources Canada (2009) Geological provinces of Canada. https://www.nrcan.gc.ca/maps-tools-and-publications/maps/geology-and-geosciences/16876. Accessed 12 July 2022
Natural Resources Canada (2015) Territorial evolution. The Atlas of Canada. https://atlas.gc.ca/ette/en/
Parliament (2024) Parliament. https://www.parl.ca/
Natural Resources Canada (2024) Canada's minerals and mining map. https://atlas.gc.ca/mins/en/index.html
Phillips D (1990) The climates of Canada. Environ Canada 176
Public Health Services (2019) Infographic: inequalities in infant mortality in Canada https://www.canada.ca/en/public-health/services/publications/science-research-data/inequalities-infant-mortality-infographic.html
Shaw J, Piper DJW, Fader GBJ, King EL, Todd BJ, Bell T, Batterson MJ, Liverman DJE (2006) A conceptual model of the deglaciation of Atlantic Canada. Quaternary Sci Rev 25:2059–2081
Slaymaker O (2017) Landscapes and landforms of Western Canada. Springer International Publishers, Cham, Switzerland
Slaymaker O, Catto NR (2020) Landscapes and landforms of Eastern Canada. Springer International Publishers, Cham, Switzerland
Soil Classification Working Group (1998) The canadian system of soil classification, 3rd edn. Publication, Agriculture and Agri-Food Canada, p 1646
Soil Classification Working group (2001) Soils of Canada. Agriculture and Agri-foods Canada, Research Branch. Scale 1:6,500,000 map. https://sis.agr.gc.ca/cansis/publications/maps/soc/all/soils/soc_all_soils_2004en.png. Accessed 12 July 2022.
Stanley GFG (1969) A short history of the Canadian constitution. Ryerson, Toronto
Statistics Canada (2021–2024) Census of Canada 2021
Stokes CR, Tarasov L, Blomdin R, Cronin TM, Fisher TG, Gyllencreutz R, Hättestrand C, Heyman J, Hindmarsh RCA, Hughes ALC, Jakobsson M, Kirchner N, Livingstone SJ, Margold M, Murton JB, Noormets R, Peltier WR, Peteet DM, Piper DJW, Preusser F, Renssen R, Roberts DH, Roche DM, Saint-Ange F, Stroeven AP, Teller JT (2015) On the reconstruction of palaeo-ice sheets: recent advances and future challenges. Quatern Sci Rev 125:15–49
Supreme Court of Canada (2024) Supreme court: welcome. https://www.scc-csc.ca/home-accueil/index-eng.aspx
Truth and Reconciliation Commission of Canada (2012) Truth and reconciliation commission of Canada: calls to action
Wheeler JO, Hoffman PF, Card KD, Davidson A, Sanford BV, Okulitch AV, Roest WR (1996) Geological map of Canada. Geological Survey of Canada, Map D1860A https://ftp.maps.canada.ca/pub/nrcan_rncan/raster/atlas_4_ed/eng/environment/land/025_26.jpg. Accessed 12 July 2022
World Bank Open Data (2024) Canada
World Factbook (2023) The World FactBook—Canada
Yeates M (1975) Main street: Windsor to Quebec City. Macmillan of Canada, Toronto

2 Atlantic and Boreal Newfoundland Ecozones

Abstract

The Atlantic Ecozone encompasses the physiographic Appalachian Mountains and coastal waters. Ecological and landscape subdivisions include the Atlantic Highlands, Atlantic Maritime, Newfoundland Boreal, and the coastal waters of the North Atlantic Ocean. Most areas of Atlantic Canada have mid-boreal climates, marked by relatively cool conditions and seasonally consistent precipitation, with humid and perhumid moisture regimes. Geologically, Appalachian Canada includes rocks of the Canadian Shield Grenville Province, the St. Lawrence Platform, the Appalachian Mountain System, and the Maritime Plain. Quaternary glaciation involves Laurentide ice from the accumulation centre in western Labrador, numerous local ice caps, and alpine glaciers in the Long Range Mountains and Monts-Notre Dame. All parts of the region were glaciated at some time during the Quaternary, and all but the highest areas were covered by ice during the most recent glaciation. Offshore areas, such as the Grand Banks and Sable Island Bank, saw consistently rising relative sea levels from the Last Glacial Maximum low of 110 m below present sea level 18,000 years ago. Regions adjacent to the Laurentide Inlandsis to the north, such as the tip of the Northern Peninsula and the North Shore of the Gulf of St. Lawrence and Strait of Belle Isle, saw consistent relative sea level fall from deglacial maxima. Intermediate areas have more complex patterns of relative sea level change. The presence of multiple ice centres, with different timing of deglaciation and hence subsequent rebound, further complicates the situation. Some regions have experienced periods with relative sea level alternately higher and lower than at present. Vegetation assemblages in Atlantic Canada include the Acadian Deciduous Forest, New Brunswick-Gaspé Boreal Forests, Newfoundland Boreal Forest, and Barrens communities. Forest soil orders of Atlantic Canada include podzols, brunisols, and gleysols. Thin, poorly developed regosols form over bedrock. Organic order soils are associated with wetlands. Faunal assemblages represent a transition between the Atlantic Maritime (mid-boreal and transitional southern boreal), Atlantic Highlands (mid-boreal), and Newfoundland boreal (mid-boreal to northern boreal) communities. Atlantic Canada is washed by coastal currents derived from the St. Lawrence and other river systems, and by the Gulf Stream and Labrador Current. The main flow of the Gulf Stream occurs beyond the shelf-slope break, although interaction influences Maritime coastlines and southern Newfoundland west of Cape Race. The Grand Banks cod (Gadus morhua) fishery has been central to Newfoundland's economy, culture, and psychology. Cold waters from the Labrador Current and the island of Newfoundland interact with the warm Gulf Stream to create upwelling, nutrient blooms, and suitable habitats for plankton, invertebrates, pelagic, and benthic species.

Keywords

Atlantic Ecozone · Atlantic Highlands · Atlantic Maritime · Newfoundland Boreal · North Atlantic coastal waters

2.1 Introduction

This chapter discusses the Atlantic Ecozone, encompassing the physiographic Appalachian Mountains and coastal waters (Fig. 2.1). Ecological and landscape subdivisions include the Atlantic Highlands (Gaspé, Beauce, and northern New Brunswick), Atlantic Maritime (southern New Brunswick, Nova Scotia, Prince Edward Island, and Iles-de-la-Madeleine), Newfoundland Boreal, and the coastal waters of the North Atlantic Ocean (Scotian Shelf, Gulf of St. Lawrence, and Newfoundland and Labrador Shelf).

N. Catto and G. R. Vincent, *Canada: Land and People*, World Regional Geography Book Series,
https://doi.org/10.1007/978-3-032-04391-7_2

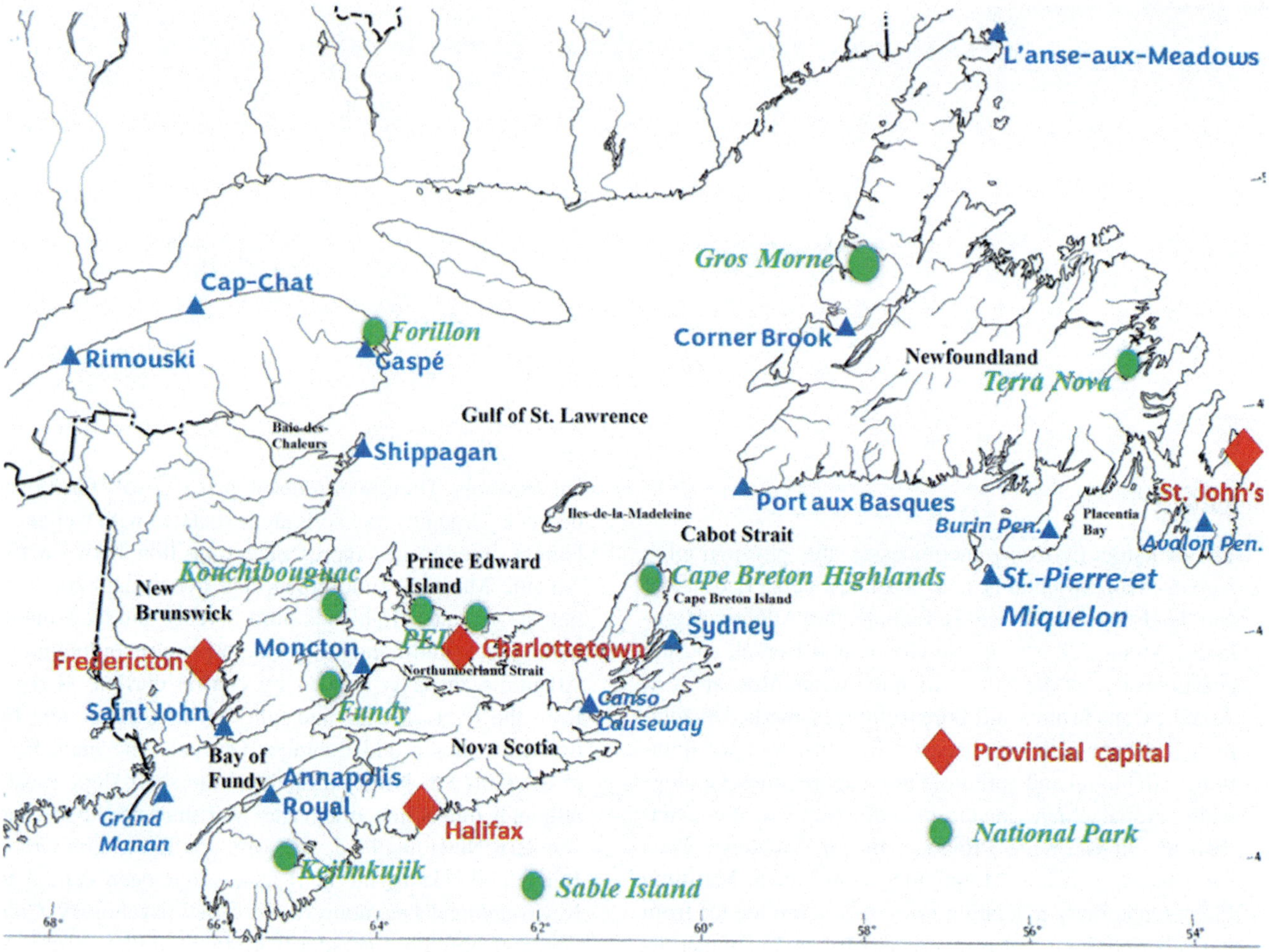

Fig. 2.1 Atlantic Canada, indicating locations discussed in the text

By convention in Canada, "Atlantic Canada" includes the provinces of New Brunswick, Nova Scotia, Prince Edward Island, and Newfoundland and Labrador. "Maritime Canada" (or "Maritime Provinces") includes only New Brunswick, Nova Scotia, and Prince Edward Island, reflecting the pre-1949 political configuration. Gaspé, Beauce, and Iles-de-la-Madeleine, within Québec, are excluded from Atlantic Canada politically. However, the similarity in natural attributes has led us to include these regions in this chapter. In a similar vein, the northern boreal, taiga, and tundra regions of Labrador (see Chaps. 4, 8, and 9) are not considered here.

2.2 Climate

Most areas of Atlantic Canada have mid-boreal climates (Köppen-Geiger Dfb), marked by relatively cool conditions and seasonally consistent precipitation, with humid and perhumid moisture regimes (Banfield 1981a, b; Phillips 1990, Beck et al. 2006). Climates are controlled by the dominant westerly winds and the proximity of the relatively cold waters of the North Atlantic Ocean and freshwater influx from the St. Lawrence Estuary. Local factors, such as topography and the prevalence of onshore and offshore breezes, create distinct mesoclimatic and microclimatic regimes in many locations (Environment Canada 2020). Halifax International Airport receives almost 1.5 times the snowfall of downtown Halifax, 25 km away. Some examples of regional climates in Atlantic Canada are presented here.

2.2.1 Avalon Peninsula

The Avalon Peninsula, including St. John's, is the area of Newfoundland showing the most marked maritime influence (Banfield 1981a, b; Fig. 2.2). Summers are short, cool, and wet (normally, the driest and hottest month is August). Winters are moderately mild and wet. Long springs (March through June) and relatively short autumns (September through mid-October) are normal. Influenced by southwesterly winds

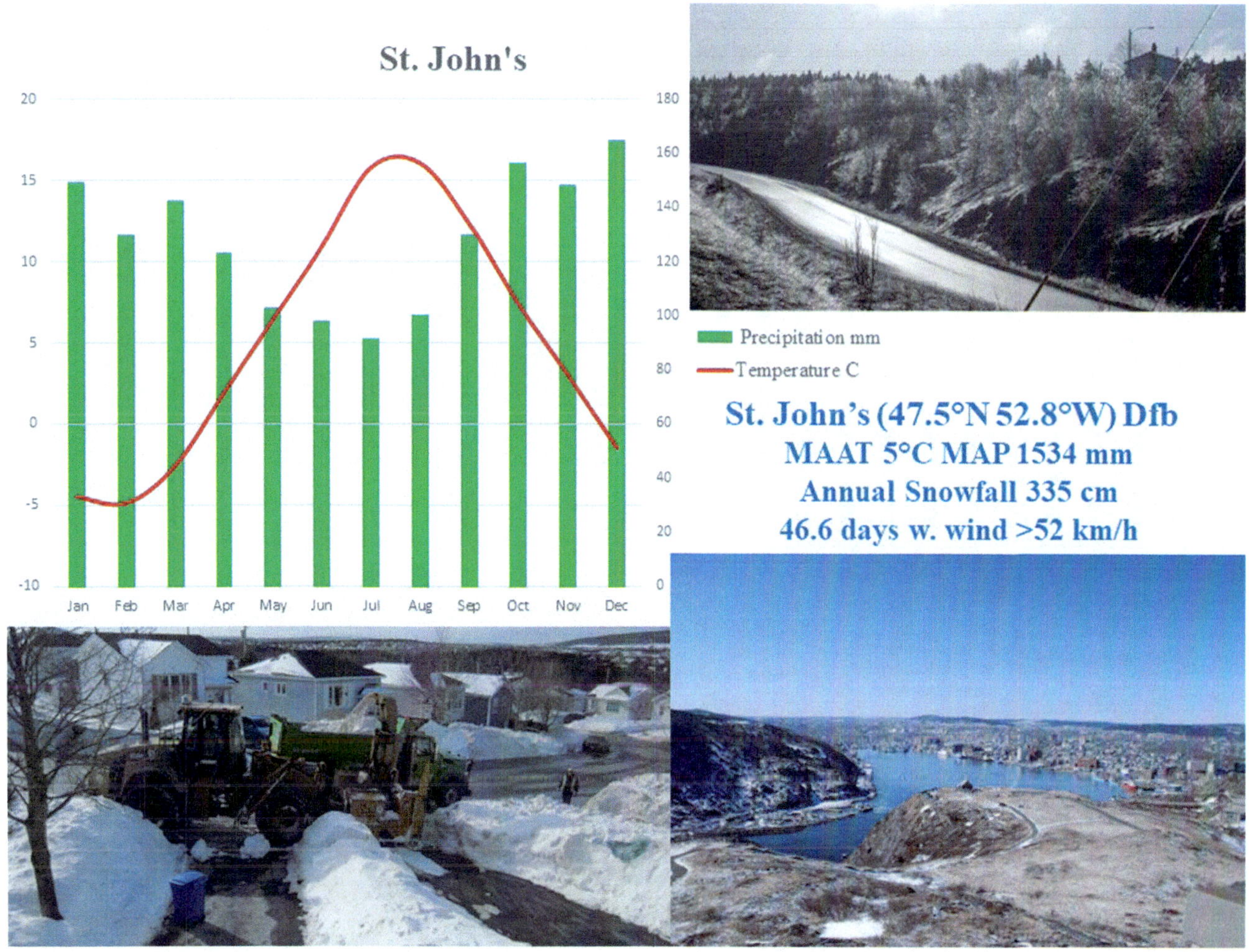

Fig. 2.2 St John's NL has a mid-boreal (Dfc) climate, with abundant snowfall and freezing rain events. Climate Data from Environment and Climate Change Canada (2020)

blowing landward, interior areas are 1–2 °C colder than adjacent coastal sites. Along Conception Bay, January and February mean temperatures also are slightly higher in coastal locations sheltered from northeast winds, in comparison to those exposed to the northeast. Sites exposed to direct south-westerly winds are somewhat more variable in temperature than are sheltered areas (Catto and St. Croix 1998, Catto et al. 2003).

Freeze–thaw cycles are generally numerous from mid-December to early April, and frost events may occur at any time from early September to June. In exposed coastal head-lands, several freeze–thaw cycles may occur daily during the early and late winter. Shoreline areas receive less snowfall and more freezing rain and drizzle than do interior locations, although rainfall events in coastal areas may be associated with freezing precipitation events inland. In coastal sites, typically 15–25% of the precipitation falls as snow, although in exposed regions subject to onshore winds the proportion of snowfall may be less than 10%.

Fog is common along the coastline. In the western Avalon Peninsula, Argentia, the foggiest weather station in Canada, averages 206 days y^{-1} with at least 1 h of fog. Fog is less ubiquitous on the Conception Bay coast than along the open Atlantic shoreline adjacent to St. John's.

St. John's receives 335 cm of snow annually, with some years seeing more than 600 cm (Environment Canada 2020). During the mid-latitude cyclonic storm of 17–18 January 2020, some areas of the metropolitan region received more than 90 cm, with drifts exceeding 15 m in height. Wind gusts can exceed 170 km h^{-1}, with waves exceeding 10 m. The mid-latitude cyclone of 14–15 February 1982 resulted in the loss of all 84 people aboard the offshore oil rig *Ocean Ranger*.

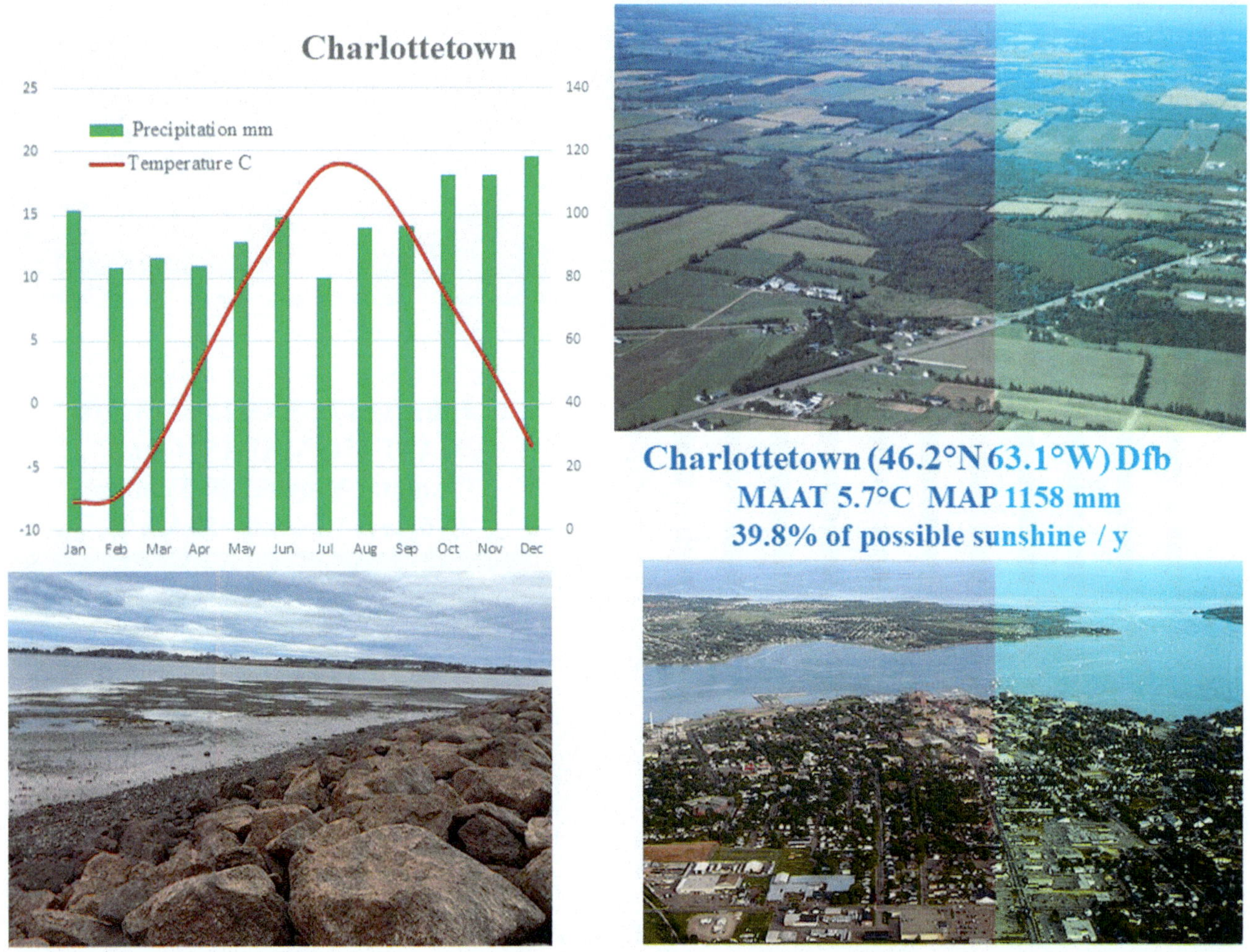

Fig. 2.3 Climate, Charlottetown PEI. Climate Data from Environment and Climate Change Canada (2020)

2.2.2 Prince Edward Island

Although Prince Edward Island is generally considered to have a mild summer climate, the seasonal temperature values differ little from those of other mid-boreal regions (Fig. 2.3). At Charlottetown, January and July mean temperatures are −7.7 °C and 18.7 °C, respectively (Environment Canada 2020). The area receives approximately 1100–1200 mm of precipitation in a typical year, of which 15–20% falls as snow. The combination of temperature and precipitation produces a humid moisture regime, with adequate moisture in the soil for plant growth throughout most of the growing season. The frost-free period is ~200 days along most of the north coast, allowing agriculture to flourish.

Fog is relatively uncommon along the north coast. Prince Edward Island National Park receives 1900 h of sunshine per year, with the sun shining for at least 1 h on 280 days in a typical year. This abundant sunshine, combined with warm summer temperatures (110 days of high summer/year), few thunderstorms (10/year), and warm coastal waters for swimming contributes to the attractiveness of the area for tourists. August sea surface temperatures average 17 °C). Even during the coldest month of February, water temperatures seldom fall below 0 °C, frequently keeping the north shore ice-free throughout the winter (although drift ice from the Gulf of St. Lawrence can be propelled onshore). In contrast, surface water temperatures in Northumberland Strait fall to −2 °C in February.

The prevailing winds are westerlies across PEI. However, local onshore northwesterly breezes are common along the north coast, and are significant in forming, shaping, and maintaining the coastal dunes (Catto et al. 2002, Catto and Catto 2009). Storm winds are associated with hurricanes from the southwest during the summer and autumn. During the winter, northerly and southwesterly storm winds affect the north coast of PEI. These storms are particularly effective at modifying the coastline during years where offshore ice cover is limited or develops late in the winter. Storms occurring during winters marked by more persistent snow cover inland and persistent, regionally extensive sea ice cover, are ineffective at modifying the coastline.

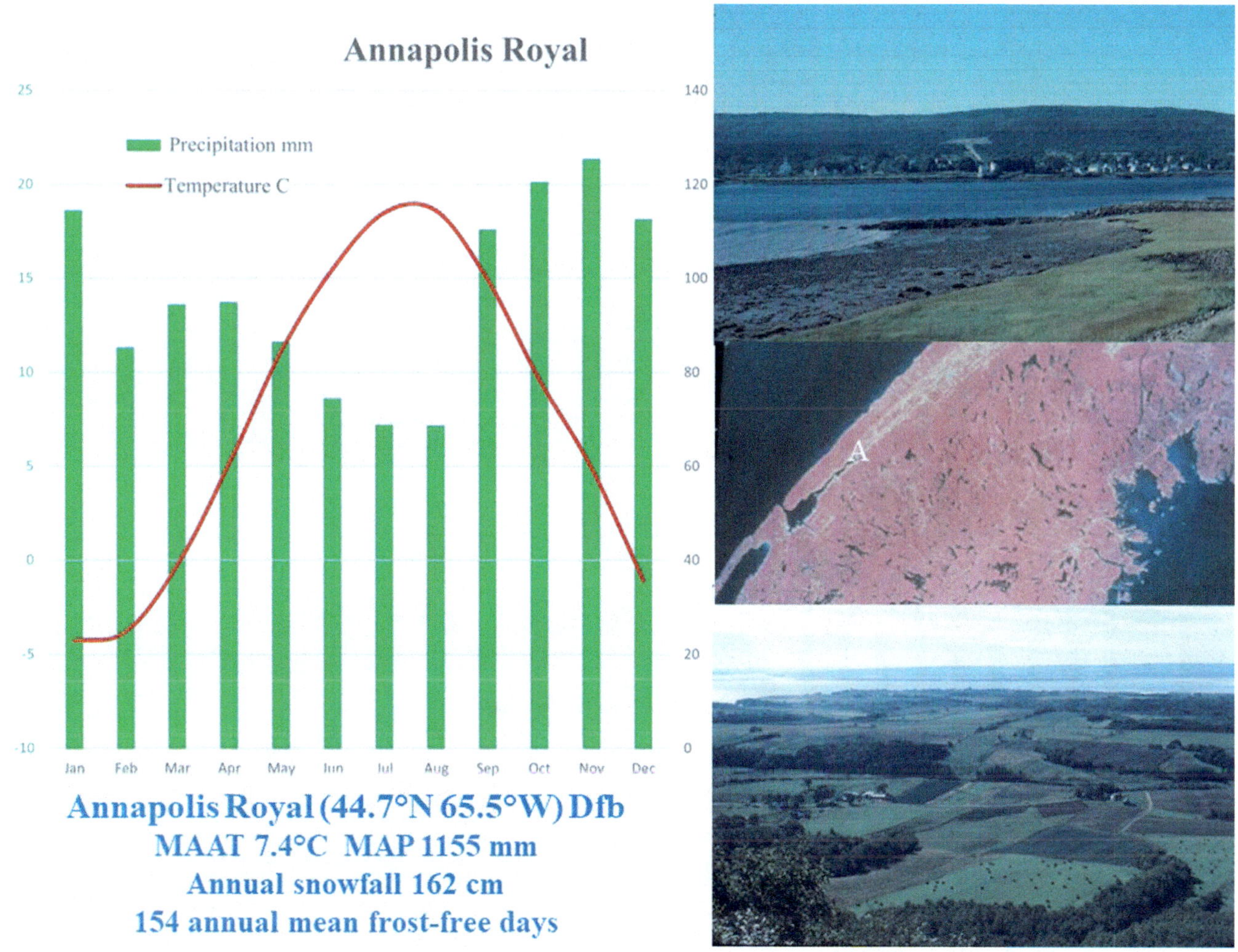

Fig. 2.4 Annapolis Royal NS (A on Landsat image, centre right) is located in the agriculturally productive Annapolis Valley. Hypertidal conditions are present in the Bay of Fundy. Climate Data from Environment and Climate Change Canada (2020)

2.2.3 Annapolis Valley

Physiography has a major effect on the climate of the Annapolis Valley (Fig. 2.4). North Mountain effectively isolates the Annapolis Valley from the effects of the Bay of Fundy, thus reducing the maritime influences in the interior. This effect is most marked during January, when the prevailing winds are northerly. Winds travelling across the Bay of Fundy bring cold, wet air to the few communities on the north slope of North Mountain. The topographic obstacle forces the winds to rise, and thus, they effectively pass over the Annapolis Valley in the lee of the mountain. Warmer air is trapped within the valley, limiting the number of freeze–thaw cycles and the amount of freezing rain, and thus allowing the introduced apple trees to survive. The northerly winds affect the precipitation and temperature patterns of South Mountain, bringing freezing rain and cold temperatures. The Annapolis Valley thus represents a winter thermal oasis (January mean −4.3 °C) between the two upland areas (Environment Canada 2020).

During the summer months, the prevailing winds are southwesterly, travelling along the length of the valley from Digby to Wolfville. At this time of the year, the differences in climate between the uplands and valley are less pronounced.

2.2.4 Gaspé

Gaspé is influenced by northern boreal (modified Köppen Dfc) climates (Beck et al. 2006, Environment Canada 2020). Winters are cold and snowy, with January daily mean temperatures of −10 °C and 480–500 cm of snowfall each year at interior sites in Gaspésie (Fig. 2.5). Forillon National Park, more exposed, receives 320 cm of snowfall in a typical year. Measurable amounts of snow fall on 49 days annually, and in a typical year blowing snow hinders travel and human activity

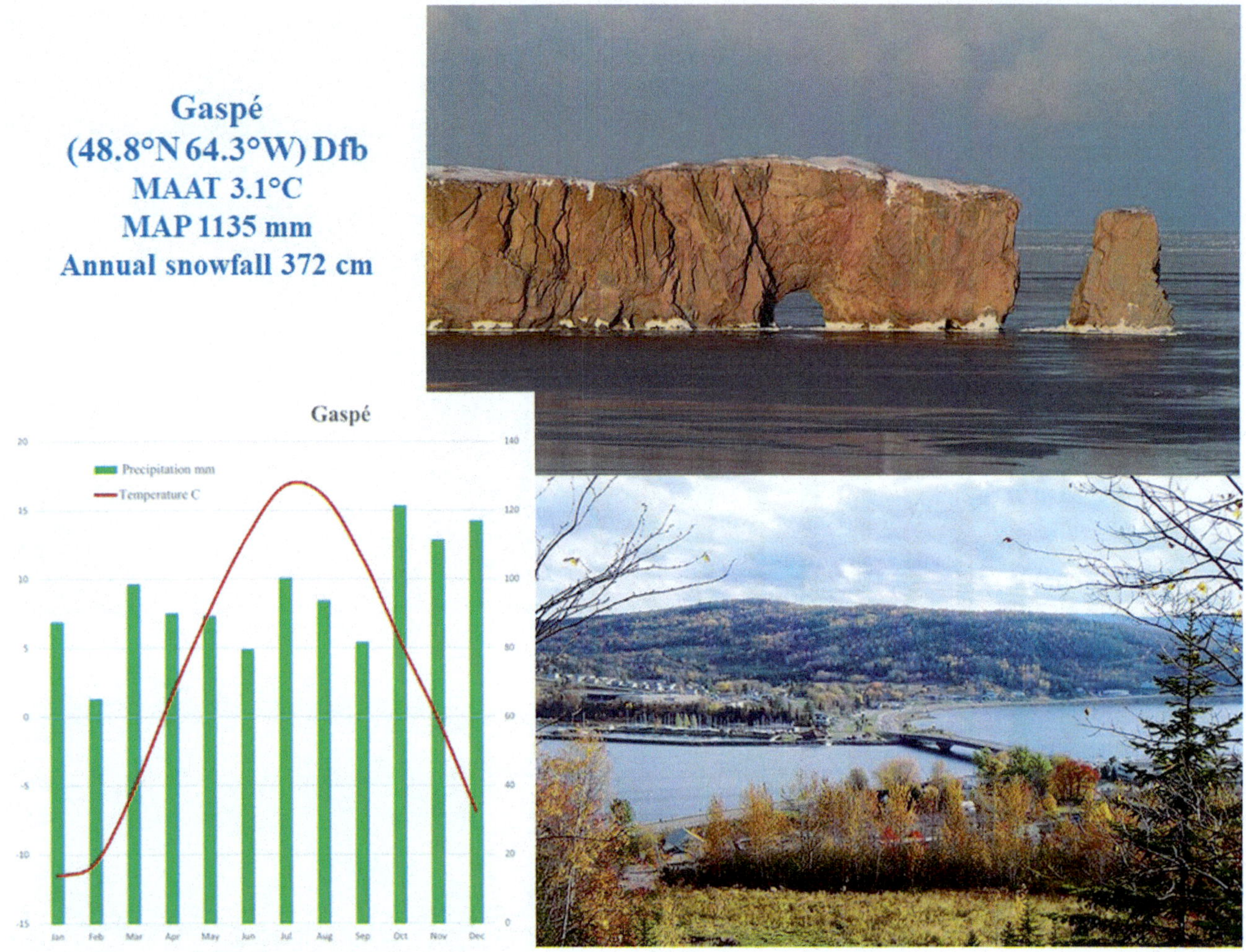

Fig. 2.5 Climate, Gaspé, Québec. Climate Data from Environment and Climate Change Canada (2020)

on 25 days. Snow cover persists for approximately 165 days in forested inland areas and for 100–120 days at Forillon. January winds are light on many days, due to the development of a strong maritime Polar pressure system in the Gulf of St. Lawrence; when strong winds do occur, they are generally onshore breezes. Freezing rain commonly occurs during late March and early April.

Summer temperatures are cool (July daily mean 16 °C on the coast, 18 °C at Forillon park headquarters). The growing season varies from 180 days in sheltered areas adjacent to the Restigouche to 160 days on the exposed northern shore. July winds are southeasterly, swinging up the Gulf of St. Lawrence and striking the peninsula on its southeastern margin. The topographic barriers aligned at 90° to the prevailing wind direction create windward and leeward zones across the peninsula, leading to lower amounts of precipitation (±1200 mm annually) on the northern shore (e.g. Cap-Chat) than on the southern shore (±1350 mm annually). On the southern shore, precipitation falls on 200 days y^{-1}, creating a humid moisture regime, and fog is common (50 days y^{-1}). Thunderstorms are less frequent than in the southern Maritimes, with a mean of 7 occurring annually.

The general climate picture is complicated by local, microclimate effects. Along the valley inland from Anse-au-Griffon, on the north side of Forillon, microclimate effects lead to higher daytime and lower night time summer temperatures than on the adjacent coastline. The predominant south easterly winds, blowing from land to water, limit the influence of diurnal onshore-offshore breezes, giving the valley a more continental climate than its surroundings. During the winter, the high-pressure system minimizes wind activity, and there is no temperature difference between Anse-au-Griffon valley and its surroundings.

2.3 Bedrock

Geologically, Appalachian Canada (including Gaspé) includes rocks of the Canadian Shield Grenville Province, the St. Lawrence Platform, the Appalachian Mountain System, and the Maritime Plain (Natural Resources Canada 2009; Bostock 2014). The Long Range Mountains of Newfoundland, including the core of Gros Morne National Park, are comprised of gneiss and granite interspersed with granite plutons. The Long Ranges represent part of the Grenville Province of the Canadian Shield, exposed from southern Labrador to Georgian Bay in Ontario (see Chap. 3). These igneous rocks are the oldest exposed in Appalachian Canada (Fig. 2.6).

The narrow Newfoundland Coastal Plain, underlain by thrusted sedimentary bedrock and covered by Holocene marine sediments, discontinuously borders the western margin of the Long Range from the Port-au-Port Peninsula to the Strait of Belle Isle (Colman-Sadd et al. 1990; Bostock 2014). This coastal fringe, below approximately 75–90 m above sea level (rising towards the north) is part of the eastern St. Lawrence Platform of Anticosti Island and the Mingan shore of Québec (see Chap. 4). The boundary between the coastal plain and the Long Range highlands coincides roughly with the limit of postglacial marine inundation following deglaciation.

2.3.1 Appalachian Mountain System

The Appalachian Mountain System contains rocks ranging from later Precambrian (Neo-Proterozoic) to Carboniferous. These units have been deformed by three major orogenies—Avalonian, Taconic, and Acadian—into folded, faulted, and metamorphosed ridges, uplands, valleys, and plateau (Fig. 2.7).

2.3.1.1 Avalonian Orogeny

The Avalonian Orogeny occurred 650–500 million years ago, and affected the regions which today form the Avalon, Burin, Connaigre, and Bonavista Peninsulas of Newfoundland, as well as areas in eastern mainland Nova Scotia (O'Brien 1998, Murphy et al. 2001; Shellnutt et al. 2019; Landing et al. 2022). On the Avalon Peninsula, Neo-Proterozoic and Ediacaran units were deformed into a series of northeast-southwest trending folds, accompanied by faulting (King 1988). The major faults are aligned NNW-SSE to NNE-SSW. The heights of land follow the axes of the geological structure, with

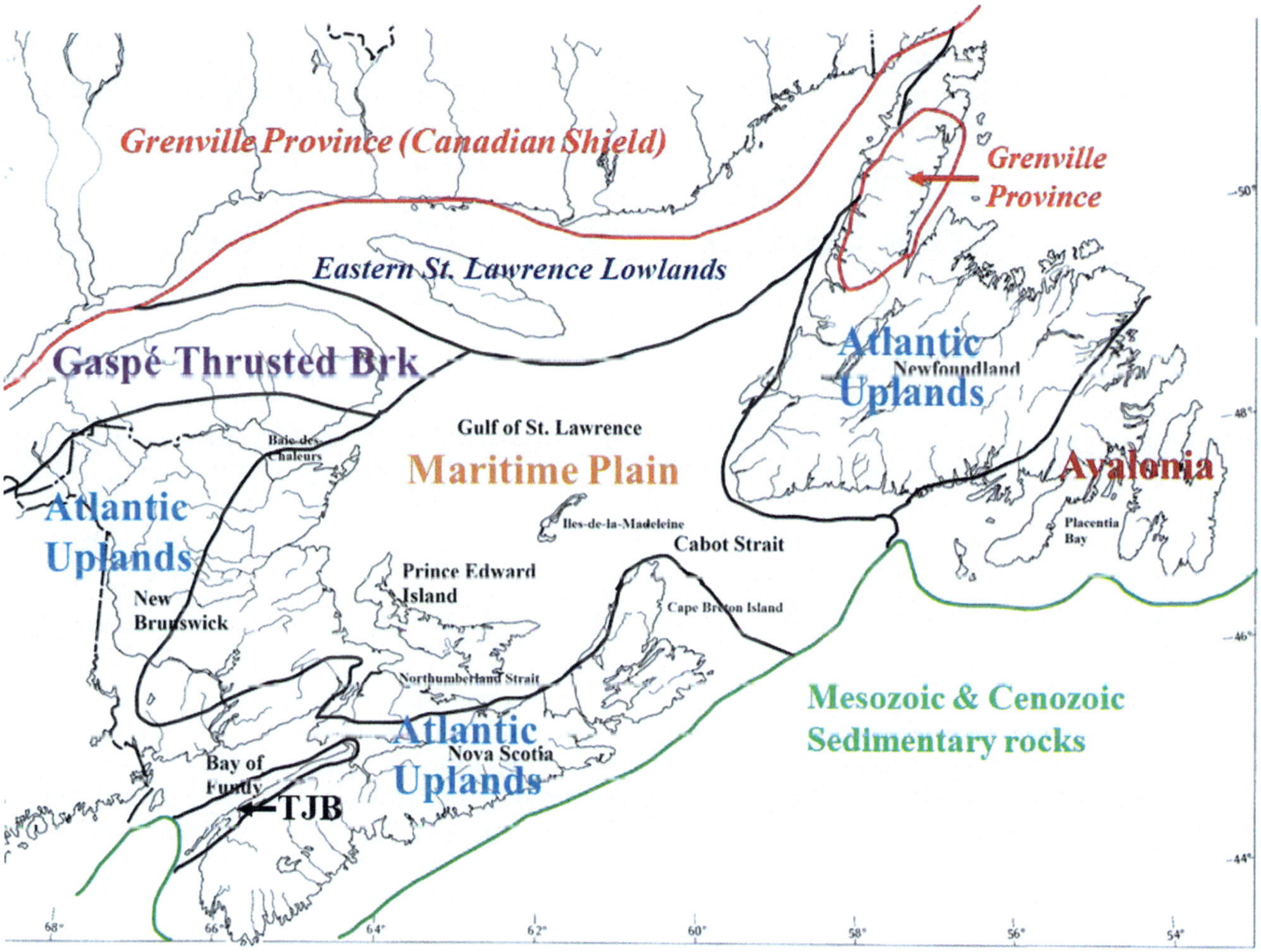

Fig. 2.6 Major bedrock zones of Appalachian Atlantic Canada

Geological Period (ages)	Event	Regions of Exposure	Sites
Quaternary (0-2.6 Ma)	glaciation	throughout Atlantic Canada	numerous
Cenozoic Era (2.6-66 Ma)	Offshore marine sediment deposition	Scotian Shelf Grand Banks	numerous
Cretaceous (66-144 Ma)			
Jurassic-Triassic (144-252 Ma)	Bay of Fundy rifting	North Mountain NS; Grand Manan NB	
Permian-Carboniferous (252-359 Ma)	Pangaea; terrestrial redbeds	Maritime Plain (Iles-de-la-Madeleine; PEI; NB; NS)	PEI NP Joggins NS Hopewell R. NB Kouchibouguac NP
Devonian (359-419 Ma)	Acadian Orogeny (350-415 Ma)	Atlantic Uplands (NB-NL-NS) Gaspé Peninsula	Forillon NP Miguasha QC Cape Breton H. NP Kejimkujik NP
Silurian (419-444 Ma)			
Ordovician (444-485 Ma)	Taconic Orogeny (440-480 Ma)	Atlantic Uplands (NB-NL-NS) Gaspé Peninsula E. St. Lawrence L.	Tablelands NL Rawdon Hills NS Lunenburg NS Fundy NP
Cambrian (485-541 Ma)			
Ediacaran (541-635 Ma) Neo-Proterozoic Era (to 1000 Ma)	Avalonian Orogeny (500-650 Ma)	Avalon, Bonavista, Burin Pen. NL	Mistaken Point NL
Meso-Proterozoic Era (1000-1600 Ma)	Grenville Prov., Canadian Shield	Long Range Mountains NL	Gros Morne NP

Fig. 2.7 Geological events in Atlantic Canada

maximum elevations in excess of 330 m above sea level (asl). Prominent landmarks, such as Signal Hill and Cape Spear, are underlain by resistant units.

Structural geology dominates over glacial overdeepening in St. Mary's Bay (>210 m depth), Conception Bay (>280 m), Placentia Bay (>300 m), and Trinity Bay (>580 m). In smaller embayments and coves, the coastal geomorphology of the Avalon Peninsula is largely controlled by differential resistance to erosion. Most prominent cliffs are supported by resistant conglomerates and coarse arkose (feldspar-rich) sandstones. In contrast, harbours are generally associated with weaker interbedded greywackes (silty sandstones) and argillites and shales, locally accentuated by faulting. North of Brigus, northeast-southwest trending faults superimposed on SSW-plunging anticlines and synclines result in the development of deep northeast-southwest trending embayments and the excellent harbours of Bay de Grave, Bay Roberts, Spaniards Bay, and Harbour Grace (Catto 2020b).

Mistaken Point Ecological Reserve, a UNESCO World Heritage site, conserves some of the oldest known and most diverse assemblages of large, biologically complex mid-Ediacaran (580–560 Ma) organisms (Clapham et al. 2003; Narbonne 2005; Liu et al. 2011; Fig. 2.8). These soft-bodied marine animals were buried and preserved in exceptional detail by influxes of volcanic tephra. Ongoing erosion due to the interaction between structural geology and wave impact has exhumed more than 100 fossil sea floor surfaces. Foot traffic due to increased visitation also represents a threat to the fossil assemblages.

The Avalonian rocks of eastern Newfoundland formed in the Iapetus Ocean, ancestor to the North Atlantic, adjacent to what is now the coastline of Portugal. Gradual closure of Iapetus through plate tectonics was accompanied by volcanic eruptions, intrusion of granite and gabbro plutons, and metamorphism of the original marine sediments to argillites and slates. When the Iapetus Ocean basin closed completely during the subsequent Taconic Orogeny, eastern Newfound-

Fig. 2.8 Mistaken Point Ecological Reserve, Avalon Peninsula. Gently dipping Ediacaran rocks of the main "D" and "E" surfaces (left) display a complex assemblage of multi-cellular soft-bodied fossils (lower right). The steeply dipping strata at Pigeon Cove, 10 km west (upper right), feature a different fossil assemblage (middle right)

land was attached to North America for the first time along the Dover-Hermitage Fault.

To the east of what is now Gros Morne National Park, Avalonian tectonic activity resulted in the formation of a mid-ocean rift valley system. Volcanoes erupted along the widening rift system, forming basalts.

Sediments from the Grenville province of the Canadian Shield to the west began to gradually fill the widening ocean basin, and were subsequently lithified to form limestones, shales, and conglomerates. These sedimentary rocks, of Cambrian and Ordovician ages, contain brachiopods, graptolites, and trilobites. During the process of sediment accumulation, slumping and flows occurred along the edges of sediment banks, forming disturbed masses of broken material. These sediments, when lithified into rocks, formed breccia. At Cow Head, extensive exposures of the Cow Head breccia represent the collapsed margin of a coral reef that grew when Gros Morne lay in the Southern Tropics, ca. 500 Ma (James and Stevens 1986; Fig. 2.9).

2.3.1.2 Taconic Orogeny

When the Iapetus Ocean basin closed completely during the Ordovician, the impact caused the development of the Taconic Mountains, and Appalachia began to form (Prave et al. 2000, Science Direct 2022). Slabs of Iapetus Ocean crust were detached and pushed up during the collision, forming the Tablelands and other ophiolite complexes in western Newfoundland (Berger et al. 1992; Brookes 1993; Staal et al. 2007). These rocks consist of peridotite and dunite rich in nickel, chromium, and vanadium (Fig. 2.10). Olivine is the most common mineral in unmodified peridotite and dunite rock, but most has been altered by atmospheric exposure to soft orange-yellow serpentine and chlorite. Pillow basalts, formed by submarine eruptions of volcanoes, are exposed in the Green Gardens area.

In Nova Scotia, Taconic (Cambrian-Ordovician) metamorphosed sedimentary rocks extend from the Rawdon Hills (east of Windsor) to the southern shore at Halifax and Lunenburg (Keppie 2000). These rocks are easily

Fig. 2.9 Ordovician breccia is exposed along the Gulf of St. Lawrence shore at Cow Head, western Newfoundland. Upper left: on the Cow Head tombolo, the shore faces northwest into the prevailing storm winds and waves (Govt. of NL airphoto). Right: wave and storm activity, combined with frost action, constantly expose and erode the breccia. Bottom left: the breccia is composed of rubble from the seaward flank of Ordovician coral reefs

eroded, and form gently sloping plains and low, rolling hills. Quartzites contain small quantities of gold, which formerly was mined in the Rawdon Hills area and from sea cliffs at the Ovens and Cunard Cove, south of Lunenburg (Fig. 2.11). Samuel Cunard, of shipping fame, started by extracting gold from family land at Cunard Cove.

2.3.1.3 Acadian Orogeny

Deformation of the central and western parts of the Appalachian Mountain System continued during the subsequent Devonian Acadian Orogeny (Bradley 1983; Scotese and McKerrow 1990; Murphy and Keppie 2005). Granite intrusions formed conspicuous landforms in Terra Nova National Park (NL), and Cape Breton Highlands National Park, Kejumkujik National Park, South Mountain, and Peggy's Cove (NS) (Colman-Sadd et al. 1990; Macdonald 1994; Keppie 2000; Barr et al. 2021; Fig. 2.12). The granite plutons have survived environmental modification and glaciation to form irregularly undulating terrain, relatively resistant to erosion. Many of these granite plutons are readily recognizable by their coarse pink potassium feldspar and clear quartz crystals, some of which are more than 4 cm long. Linear black dykes of gabbro and andesite also intruded the pre-existing rocks. Many of these dykes are aligned parallel to faults and joints, particularly in coastal areas.

Subsequently, during the latter Devonian and early Carboniferous, sandstones and shales were deposited over the metamorphosed sediments and granites. In most areas, these sediments have been removed under the influence of climate and environment over millions of years, but they remain as a fringe covering the older rocks at elevations less than 200 m around the western margin of Cape Breton (Barr et al. 2021). In the Avon River Valley(NS), these deposits also contain

Fig. 2.10 Tablelands, Gros Morne National Park. Top left: On this false-colour Landsat image, the almost-unvegetated Tablelands is a conspicuous green area, flanked by boreal forest (red) and high-altitude barrens vegetation (brown). Lower left: Ophiolite rock limits vegetation cover to species which can tolerate high nickel concentrations. Upper right: the Tablelands contrast strongly with the boreal forest flanking Bonne Bay. Lower right: Frost and gravity combine to produce talus cones and streams

economic quantities of salt, anhydrite, and gypsum. Gypsum (calcium sulphate) and anhydrite (dehydrated gypsum), used to manufacture wallboard, are mined at several large quarries near Brooklyn. The limestone units contain fossils of gastropods, corals, cephalopods, brachiopods, bivalves, bryozoa, and rare trilobites. The sedimentary rocks are easily erodable by river and wave activity.

As in other areas of Appalachian Canada, the physiography of Gaspé is controlled in large measure by the type of bedrock exposed (Bourque et al. 2000; Brisebois and Nadeau 2003). The ridges represent the results of extensive deformation of sedimentary bedrock, comparable to that associated in later geological times with the construction of the Rocky Mountains (Fig. 2.13). The basic principles of thrust fault development, central to our understanding of the deformation which accompanies collision of tectonic plates, were initially proposed from study of the rocks of Gaspé and vicinity by Sir William Logan, the first director of the Geological Survey of Canada, in the 1850s.

The rocks in Gaspé are limestones, shales, and sandstones of Ordovician, Silurian, and Devonian age. These units were initially deposited as flat-lying sediments in tropical seas, and contain trilobites, brachiopods, and corals. Graptolites are present in the shales.

Subsequently, the rocks were subjected to thrusting from the south-southeast during the Palaeozoic Taconic Orogeny (480–440 million years ago) and then during the Acadian Orogeny (415–350 million years ago). The strata were pushed towards the north-northwest, where the Canadian Shield north of the St. Lawrence River formed an undeformable barrier (Malo and Bourque 1993; Malo and Kirkwood 1995; Jutras et al. 2003). As the strata were thrust to the north, they were stacked atop each other as the leading edges of each unit buckled against the Appalachian Structural Front. Weaker

Fig. 2.11 The Ovens, Lunenburg NS, form in Taconic (Cambrian-Ordovician) metamorphosed sedimentary rocks (upper left). Upper right: view seaward from the cave interior, indicating the inclined strata. The Taconic units, modified by glacial erosion, are exposed along the south coast of Nova Scotia from Bridgewater and Lunenburg (lower left) to Halifax and Dartmouth (lower right)

units, such as shales, failed first and were subject to additional folding. More resistant units, especially limestones, moved as blocks over the deformable shales. In many areas, repeated thrusting pushed older, more resistant limestones over younger, weaker shales, thus creating a series of ridges oriented at right angles to the direction of thrusting, with gentle slopes indicating the source of the deformation. The presence of older rocks overlying younger rocks is characteristic of thrust faulted terrains. The differential resistance of the limestones and shales is evident along the coastline, where coves represent areas of shale outcrops and high cliffs mark limestone exposures. The picturesque sites of Percé Rock, Ile Bonaventure, and the cliffs of Forillon are all associated with limestone exposures.

Miguasha Natural UNESCO World Heritage Site, located within the parc national Miguasha, is a paleontological site on the southern coast of the Baie des Chaleurs (Miguasha National Park 2007, Cloutier 2013). Dating from 370 million years ago, its rocks contain five of the six fossil fish groups associated with the Upper Devonian "Age of fishes". Minguasa has the highest number and best-preserved fossil specimens of the lobe-finned fishes that gave rise to the first four-legged, air-breathing terrestrial vertebrates.

2.3.2 Maritime Plain

In eastern New Brunswick, northern Nova Scotia, Prince Edward Island, and Iles-de-la-Madeleine, the Appalachian rocks are overlain by soft, gently sloping "red-bed" sedimentary rocks of the Maritime Plain (van der Poll 1989, Keppie 2000, Lavoie 2008, New Brunswick 2008, Bostock 2014, Hétu et al. 2020). These units, dominantly sandstones, formed in a series of braided streams and alluvial fans which developed at the feet of mountains during the later Carboniferous and Permian periods, 360–270 million years ago (Fig. 2.14).

Fig. 2.12 Granite units in Atlantic Canada formed during the Acadian Orogeny. Upper: Peggy's Cove. All of the unvegetated granite is regularly overwashed by storm waves. Lower: Burgeo NL (left), South Mountain NS (centre), and Black Brook, Cape Breton Highlands National Park NS (right) show differential crystallization and joint structure, influencing erosion and landform development

Prince Edward Island lay in the heart of the supercontinent Pangaea, and was subjected to a dry continental climate. All the sediments were exposed to the air for prolonged periods following deposition, and consequently most were oxidized extensively. The age of the rocks decreases successively from southwest to northeast, with the oldest Carboniferous rocks in coastal New Brunswick and the youngest Permian rock exposed on Iles-de-la-Madeleine.

The Permian rocks on Iles-de-la-Madeleine were deformed by salt-diapirs approximately ~300 Ma. The sandstones and shales are subject to erosion along the Gulf of St. Lawrence and Northumberland Strait coasts, and form relatively weak, crumbling cliffs.

Joggins Fossil Cliffs UNESCO World Heritage Site contains an extensive Carboniferous terrestrial fossil record of 96 genera, including reptile and amphibian tracks and upright fossilized trees (Falcon-Lang et al. 2006; Kosters and Grey 2021). Plant fossils include *Calamites* (distantly related to the modern horsetail), and *Lepidodendron* (distantly related to the modern club-moss). Unlike their modern descendants, both *Calamites* and *Lepidodendron* were large plants, exceeding 10 m in height. Fossils of the Carboniferous bolosaurian parareptile *Erpetonyx arsenaultorum* and the Permian reptile *Eryops*, a crocodile-like animal 2.5 m long, have been found in Prince Edward Island (Modesto et al. 2015).

The coastal plain southeast of Fundy National Park is underlain by younger red-bed clastic sedimentary rocks, predominantly sandstones and conglomerates, of Carboniferous and Triassic age (New Brunswick 2008). The sedimentary rocks are subject to erosion and form crumbling cliffs along the Fundy shore, notably at Hopewell Rocks where Carboniferous conglomerates crop out. Similarly aged sedimentary rocks underlie the Annapolis Valley. The Wolfville Formation contains dinosaur bones and tracks, primarily early sauropods such as *Plateosaurus* and early the codonts such as *Stagonoleopsis*.

Fig. 2.13 Gaspé Peninsula. Resistant ridges of limestone thrust towards the north-northwest during the Taconic and Acadian Orogenies form curved ridges along the northern side of the peninsula. Strata were stacked as the leading edges of each unit buckled against the Appalachian Structural Front. (Landsat Image, NOAA)

2.3.3 Triassic and Jurassic Basalts

The youngest rocks are the North Mountain Triassic and Jurassic basalts along the Nova Scotia Fundy coast, formed in mid-ocean rifts and transform faults associated with the Fundy Aulacogen (Kontak 2008; Kontak and Webster 2010; Fig. 2.15). During the latter part of the Triassic, the supercontinent of Pangaea began to break up, and a spreading mid-ocean ridge formed where the Bay of Fundy is today. Basalts poured out from submarine volcanoes, covering the ocean floor. Spreading continued until the early part of the Jurassic, and the resulting volcanic rocks subsequently hardened into the North Mountain Basalt and uplifted to their present elevation. The basalts are characterized by strong vertical joints, often forming hexagonal columns. These rocks form the southern shoreline of the Bay of Fundy from Cape Split to Digby Neck and Brier Island. Similar basalts are exposed on Grand Manan Island, forming 100–120 m-high cliffs along the northwestern shore (McHone 2011; Fyffe 2014). Strong vertical joints, commonly forming hexagonal columns, mark most of the basalts. Frost wedging along these joints in coastal regions and along escarpments has resulted in numerous rockfalls.

2.4 Quaternary

2.4.1 Glaciation

Quaternary glaciation in Appalachian Canada involves both Laurentide ice from the accumulation centre in western Labrador, numerous local ice caps and small sheets, and alpine glaciers in the Long Range Mountains and Monts-Notre Dame. All parts of the region were glaciated at some time during the Quaternary, and all but the highest areas were covered by ice during the most recent glaciation, beginning 30,000 years ago (Rampton et al. 1984; Grant 1989; Stea et al. 2011; Catto et al. 2020a).

The extent of Laurentide glaciation is determined by glacially transported erratics from the Canadian Shield, including granites, gneisses, and anorthosites. Earlier Quaternary Laurentide glaciations covered all of Gaspé, New Brunswick, Prince Edward Island, Iles-de-la-Madeleine, mainland Nova Scotia, and the west coast of Newfoundland, not reaching eastern Cape Breton and Newfoundland east of the Long Range Mountains. Aside from the erratics, these left few traces upon the landscapes.

Fig. 2.14 The Maritime Plain is underlain by Pennsylvanian and Permian sedimentary bedrock. Upper right: Arch at Cap Egmont PEI shows differential erosion of friable shale and more resistant sandstone. Glacial deposits crown the arch. Upper right: *Calamites* fossil, Joggins NS. Lower left: Falls and topples in sandstone triggered by frost action and wave undercutting, Skmaqn–Port-la-Joye–Fort Amherst National Historic Site PEI. Lower right: coastal erosion at Cape Kildare PEI

2.4.1.1 Gaspé

During the most recent glaciation, 30,000–10,000 years ago, penetration of Laurentide ice was obstructed by the buildup of alpine glaciers and ice caps in Gaspé, and by the Escuminac Ice Centre in the Gulf of St. Lawrence (Olejczyk and Gray 2007; Stea et al. 2011). The high elevation areas of the Chic-Chocs and Notre Dame Mountains served as ideal regions for the initiation and growth of ice caps during the Quaternary (Fig. 2.16). Local glaciers from the mountains spread radially over Gaspé, extending south to the Baie des Chaleurs and north to the St. Lawrence River, intersting southward-moving Laurentide ice (Morrisette et al. 2020).

In Forillon, two cirques represent the source areas for alpine glaciers, and many others are evident throughout Gaspé. These local glaciers covered the peninsula, finally receding from the coastal areas about 13,000 years ago. The last glaciers ablated from the Chic-Chocs prior to 10,000 years ago, after a final readvance during the Younger Dryas episode. Along the margins of the glaciers, belts of permafrost terrain developed, and traces of the ice-wedge casts and frost-heave structures are evident today.

2.4.1.2 New Brunswick and Western PEI

Eastern New Brunswick, the western part of Prince Edward Island, and the Iles-de-la-Madeleine were influenced by glacial ice which formed in the area currently submerged by the Gulf of St. Lawrence (Grant 1989; Dredge et al. 1992; Josenhans and Lehman 1999; Stea et al. 2011; Catto et al. 2020a; Hétu et al. 2020). Formation of the various Quaternary glaciers required the removal of vast volumes of water from the oceans, resulting in a general lowering of sea level in the northern North Atlantic. Glacial ice of the Escuminac Ice Centre (named for Escuminac NB on Miramichi Bay) expanded southeast from the Gulf of St. Lawrence, crossing

Fig. 2.15 Triassic and Jurassic basalts. Upper right: Cape Split NS. Upper left: Columnar forms at Westport (Digby Neck) NS. Lower: vertical cliffs at Dark Harbour (left) and Southwest Head (right), Grand Manan NB

the PEI coast at Cabot Beach Provincial Park and covering all of Prince County and western Queen's County (Fig. 2.17). This glacier brought erratics and sediments from the Gulf of St. Lawrence and deposited them inland across Prince County (Prest 1973; Catto 1998a). The ice sheets fluctuated in extent during the last glaciation, finally departing from the north coast of PEI approximately 13,000 years ago.

Central and western New Brunswick were influenced by glaciation from the north and northwest, with ice moving across the terrain from the Appalachian Mountains (Rampton et al. 1984; Grant 1989; Stea et al. 2011; Todd et al. 2016). Glaciation extended along the Bay of Fundy to the Gulf of Maine, influenced by topography and interacting with ice caps from Nova Scotia. North of Fundy National Park, the Caledonian Hills formed a barrier to glacial flow, and thus the main ice flows were diverted around the park. At various times, ice flowed along the Petitcodiac River valley to Shepody Bay, following the modern coast to the southwest and filling Chignecto Bay. A second major glacial flow occurred west of the park, flowing south down the Saint John River valley and over St. Martins to the Bay of Fundy and Grand Manan Island. Later flows approached the Fundy Park region from the west and north, approximately 25,000 years ago. The area was covered by thin, non-erosive cold-based ice which was frozen to the ground during the last glacial event. The cold-based ice was incapable of erosion or deposition, and thus no glacial landforms were developed at this time. Grand Manan Island was deglaciated about 15,000 years ago, coastal Fundy about 14,000 years ago, and Kouchibouguac NP about 13,000 years ago, with complete deglaciation of New Brunswick following shortly thereafter (Shaw et al. 2006).

2.4.1.3 Nova Scotia and Eastern PEI

Southern Nova Scotia was influenced by several distinct ice caps, each of which left a record of its passing by forming drumlins and other landforms, by depositing distinctive

Fig. 2.16 The high elevation areas of the Chic-Chocs and Notre Dame Mountains served as ideal regions for the initiation and growth of ice caps during the Quaternary. (Wikimedia/Jean-Philippe Caron)

erratic boulders, and by eroding striations in the bedrock (Stea et al. 1992, 2011; Grant 1989; Shaw et al. 2006, Todd et al. 2016). Initially, the region was covered by ice flowing from the west-northwest, across the Bay of Fundy from New Brunswick. This ice transported basaltic pebbles and boulders from North Mountain and deposited them in the Annapolis and Avon Valleys. The next ice advance came from the north, extending from the Escuminac Ice Centre. Pebbles from North Mountain were carried south across the Annapolis Valley and deposited over the granitic rocks of South Mountain. The presence of these basaltic pebbles, which had to be carried uphill to reach their current resting places on South Mountain, indicates that glacial activity was responsible.

During the third phase of glacial activity, an ice cap built up over Kejimkujik National Park and South Mountain and expanded to the north, northeast, and southeast. Granitic pebbles from South Mountain were carried north and deposited in the valleys and on North Mountain. A final phase of glacial activity involved westward flow from glaciers centred in the area south of Antigonish. The picture of glacial activity thus involves multiple competing flows, alternatively covering the area from each direction. The most distinctive glacial features on the landscape are drumlins, reaching 50 m in length and 15 m in height (Fig. 2.18). Drumlin fields extend southwest to the coastline at Bridgewater, Lunenburg, and Halifax, indicating ice flow towards the southeast.

The Annapolis and Avon Valleys and the southern shore of Nova Scotia were deglaciated approximately 13,000 years ago. The subsequent Younger Dryas cold climate event (approximately 11,000 years ago) brought cold conditions to the area, limiting biological activity in the lakes, but the remnant glaciers over South Mountain did not readvance.

Western King's County and Queen's County, PEI, including Stanhope and Rustico, was covered by glacial ice which advanced northwards from Nova Scotia, crossing Northumberland Strait (Prest 1973; Grant 1989; Shaw et al. 2006; Stea et al. 2011; Catto et al. 2020a). This ice advance can be recognized through landforms and striations, and also through the presence of boulders of older sedimentary and igneous rock derived from the Antigonish-New Glasgow area of Nova Scotia. Eastern King's County (Souris, St. Peter's Bay, East Point) may have remained ice-free at the beginning of the glaciation, but were subsequently covered by ice.

2.4.1.4 Cape Breton

In contrast to other upland areas of Atlantic Canada, glaciation had very little effect on the northern Cape Breton landscape (Grant 1989). Striations and glacially transported erratics indicate that the Highlands were affected by several glacial events involving ice originating from the west and southwest early in the Quaternary, and all areas were covered by ice at some time. During the last interglacial, 125,000 years ago, relatively warm conditions prevailed

Fig. 2.17 Glaciation in the Malpeque Bay region of PEI forms gently undulating topography underlain by clay-rich till. Inset: Glacial deposits overlying Pennsylvanian bedrock, Cap Egmont PEI

throughout Cape Breton. Fossil plant remains are preserved at Mabou Harbour and Bay St. Lawrence.

During the most recent glaciation, ice from the Bras d'Or Lakes area was confined to the region south of Mabou Harbour and Wreck Cove. The highlands were glaciated by cold-based ice, which remained frozen to the underlying sediment and rock and did not produce landforms.

2.4.1.5 Newfoundland: Northern Peninsula

In Newfoundland, Gros Morne was glaciated by ice originating from the Long Ranges, resulting in the formation of the fjord topography (Grant 1989; Berger et al. 1992; Shaw et al. 2006; Fig. 2.19). Western Brook Pond is the most-photographed and best-developed fjord. The total relief in the area is in excess of 750 m, and the fjord exceeds 150 m in depth throughout most of its length. It is isolated from marine waters by a submarine fan complex, which has subsequently been inundated, exhumed, and modified by aeolian (wind) activity. The fjord is floored with marine sediment containing mollusc shells and flanked by interlobate moraines and kame terraces.

Trim lines, erosional features, and differential zones of modification indicate that three distinct glacial events have occurred in the region. The oldest tapers seaward from the inland flanks of Western Brook Pond (700 m asl) to 400 m at the termination of the cliffs at the fjord mouth. The age of this glaciation is unknown, but it is considered to pre-date the last interglacial, prior to 125,000 years ago. Two younger glacial events have carved trimlines which also taper seaward, at elevations 200–270 m and 300–350 m below the glacial maximum. The younger of these two events is considered to represent the most recent glaciation, forming at some time between 30,000 and 10,000 years ago. The middle event is of uncertain age.

Southward-flowing Laurentide ice merged with the Northern Peninsula glaciers near Ten Mile Lake, on the

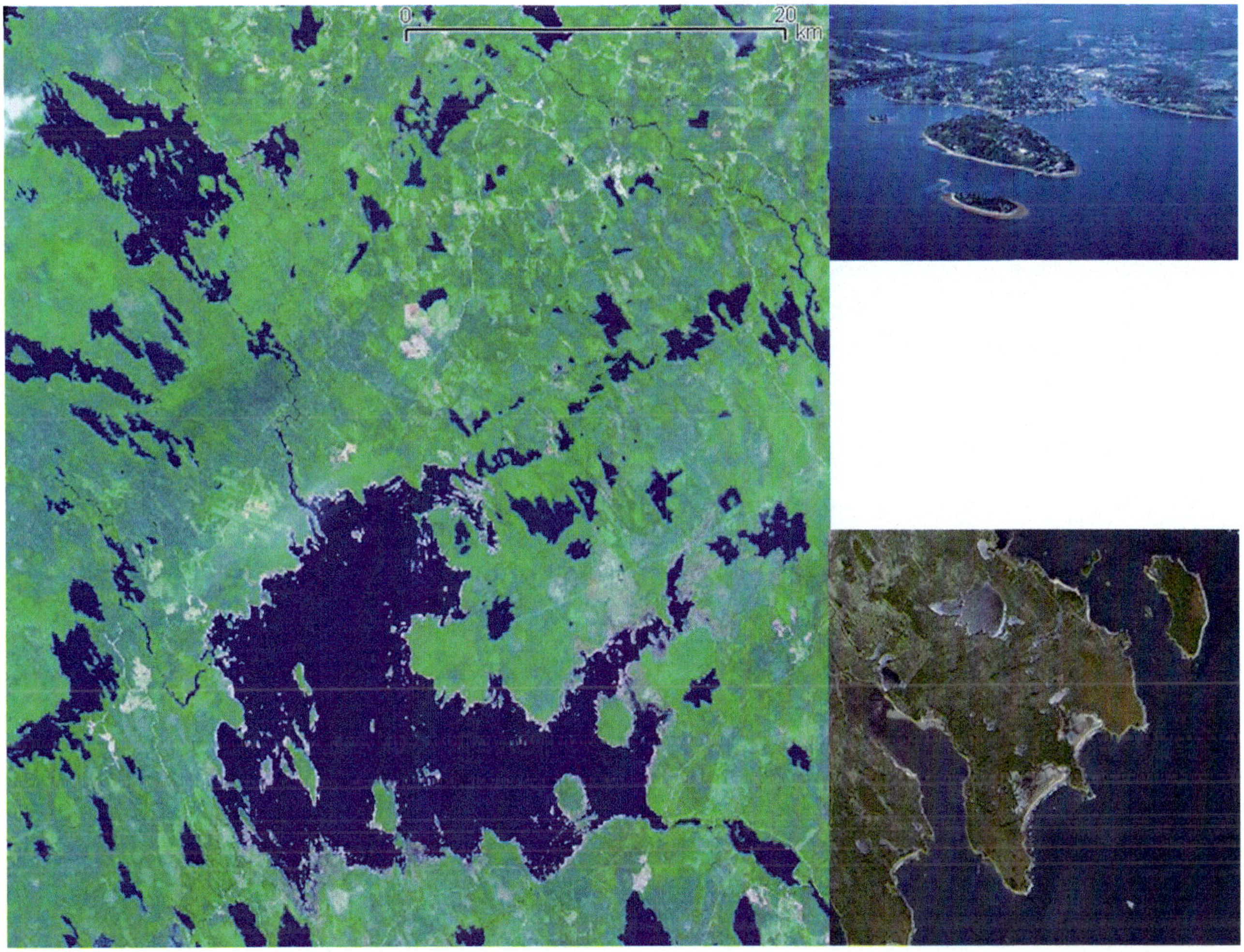

Fig. 2.18 Kejimkujik National Park NS displays a landscape marked by deposition of glacial drumlins over South Mountain granite (left). Right: Drumlins indicating ice flow towards the southeast are located along the coast from Mahone Bay southwest to Bridgewater

Northern Peninsula east of St. Barbe. The combined glacier extended south into the Gulf of St. Lawrence, being augmented by ice from Gros Morne. Granitic boulders could come from either the Grenville rocks in the core of the Long Range Mountains, or from Labrador to the northeast. An emerged delta present at the mouth of Shoal Brook suggests that ice persisted in the heads of the fjords for ±3000 years after the outer coast was deglaciated between 13,000 and 12,000 years ago.

2.4.1.6 Central Newfoundland

East of the Long Range Mountains, independent ice centres developed in central Newfoundland. Ice covered the entire area from the Long Ranges to the Isthmus of Avalon, extending offshore into the Gulf of St. Lawrence and Cabot Strait in the southwest, along the entire south coast and Burin Peninsula in the south, and through White Bay, Hamilton Sound, Bonavista Bay, and the Bonavista Peninsula in the north. Large ice streams developed in Fortune, Placentia, and Trinity Bays, with ice extending offshore across the exposed margins of the northern North Atlantic (Shaw et al. 2006; Brushett et al. 2007; McHenry and Dunlop 2016).

The Terra Nova region was glaciated by eastward-moving ice from the Middle Ridge area southwest of the park (Fig. 2.20). The park was ice-covered throughout the last glaciation. At its maximum extent 20,000 years ago, glacial ice-covered all of Bonavista Bay and extended more than 100 km offshore. Erosional roche moutonées and striations, and deposits of glacial till containing granite boulders from Terra Nova village, indicate that the ice flowed northeastward and eastward towards the coast.

As the Laurentide glaciers in Labrador began to melt, sea level rose around Newfoundland. The glaciers along the coastline were destabilized as a result and began to melt rapidly. By 13,000 years ago, the coastlines of Bonavista, Trinity, and Placentia Bays were largely free of glacial ice, although ice

Fig. 2.19 Western Newfoundland was glaciated both by local ice from the Long Ranges, and Laurentide ice moving southwestward along the Gulf of St. Lawrence coast. Upper left: Winter Gulch, Gros Morne National Park, channelled Long Range ice to the Gulf of St. Lawrence. Upper right: glacial erratic at Robinsons indicates southwest flow. Lower left: cirque formed by local glaciation, Tablelands, Gros Morne National Park. Lower right: Western Brook Pond is a fjordal lake formed by Long Range glaciation

remained inland covering Terra Nova National Park. Eskers and other glacifluvial sand and gravel features formed along the Terra Nova River and other streams as subglacial rivers carried sediments towards the sea, and delta deposits formed at the river mouths.

By about 11,500 years ago, Terra Nova National Park had been deglaciated, and the ice had retreated to Terra Nova village. At that time, the marine waters had fallen from their highest level, and much of the coastal areas were bare of vegetation and exposed to frost action. Isolated areas of permafrost developed on the exposed ground, allowing formation of frost-heave structures and ice-wedge casts (Liverman et al. 2000).

2.4.1.7 Avalon Peninsula

The independent nature of Avalon Peninsula glaciation is indicated by geomorphic ice flow indicators, striations, and the absence of the distinctive Palaeozoic granite clasts derived from the Terra Nova region to the northwest (Catto 1998b, 2020b). Multiple small ice caps developed on the Avalon Peninsula, with sequential shifts in formation and ice flow direction (Fig. 2.21).

In the westernmost Avalon, ice from the main part of Newfoundland flowed southeastward into northern Placentia Bay, carrying Palaeozoic granitic clasts from the Terra Nova and Swift Current areas. Two distinct ice streams developed, with one flowing southward into Placentia Bay and across Merasheen, Long, and Red Islands. The other ice stream flowed across the western Isthmus of Avalon into Trinity Bay, carrying Palaeozoic granitic clasts and silty marine sediment from Placentia Bay.

Three major phases of glaciation occurred during the period 30,000–10,000 years ago, each involving several distinct ice centres. The initial phase involved glacial accumu-

Fig. 2.20 Glaciated landscape, Terra Nova National Park

lation at several centres along the heights of land. Ice flowed seaward, following pre-existing structural features, and all parts of the Avalon Peninsula were glaciated.

The Last Glacial Maximum was the dominant geomorphic glacial event on the Avalon. Ice from the St. Mary's Bay centre flowed across the southern Avalon Peninsula to Placentia Bay, St. Mary's Bay, and the open Atlantic Southern Shore. Additional ice centres persisted over the Heart's Content and Witless Bay Barrens, as well as on the Isthmus.

The concluding phase of glaciation began about 15,000 years ago, when sea level reached −65 m, the controlling depth of St. Mary's Bay (Catto 1998b). The resulting destabilization of the St. Mary's Bay ice centre initiated deglaciation throughout the Avalon. The major ice centres collapsed, punctuated by transitory short-lived centres on heights of land. The effect on the landscape was limited in most instances. Deglaciation of the Avalon Peninsula was completed before 10,000 years ago.

2.4.2 Relative Sea Level History

When considering geomorphology, erosion, and coastal hazards, it is the relative sea level with respect to the terrain (or a harbour) that is important, rather than the absolute volume of marine water in the ocean. Changes in sea level are measured relative to a point, such as a wharf or cliff. A rise in relative sea level results in flooding and inconvenience, regardless of the total amount of water in the basin. From a relative sea level perspective, it does not matter if the change involves an increase in the amount of water, or subsidence of the land, or a combination of both: the net result is a rise in relative sea level.

As deglaciation began, the combined effect of water influx to the ocean with removal of the weight of the Laurentide glacier, and to a lesser extent melting of the local glaciers, produced a complex pattern of changes of relative sea level (Grant 1989; Shaw et al. 2002; Catto 2011; Slaymaker et al. 2020a). Offshore areas, such as the Grand Banks and Sable Island Bank, saw consistently rising relative sea levels from the Last Glacial Maximum low of 110 m below present sea level 18,000 years ago. Sea level rise offshore caused destabilization of ice tongues and streams at tidewater, leading to calving and rapid retreat. In contrast, regions adjacent to the Laurentide Inlandsis to the north, such as the tip of the Northern Peninsula and the North Shore of the Gulf of St. Lawrence and Strait of Belle Isle, saw consistent relative sea level fall from deglacial maxima of 120–170 m asl (above modern sea level) to present.

Intermediate areas, including most of Atlantic Canada, have more complex patterns of relative sea level change. Immediately following deglaciation, the land was glaci-isostatically depressed due to the weight of the ice, and high relative sea level results. Over time, the land rebounds due to the loss of ice cover, but the volume of ocean water is simultaneously increasing. Thus, the potential drop in relative sea level due to rebound is countered by the potential rise from increased water volume. In Atlantic Canada, the presence of multiple ice centres, with different timing of deglaciation and hence subsequent rebound, further complicates the situation. Some regions have experienced periods with relative sea level alternately higher and lower than at present.

Fig. 2.21 The Avalon Peninsula was glaciated from local ice caps throughout the late Quaternary. Upper left: Glacial erosion of Ediacaran bedrock, Mad Rocks, Bay Roberts. Lower left: Straited and grooved roche moutonnée indicating northeasterly flow, Roaches Line. Upper right: Middle Cove bluff displays glacial till at the base, capped by reworked debris flow deposits. The bluff top is receding at ~10 cm y^{-1}. Lower right: glacial deposits at Colinet, with $1 coin for scale

2.4.2.1 Early Deglacial Relative Sea Level

About 12,000 years ago, as deglaciation proceeded, lower relative sea level due to glaciation had exposed the Grand Banks, Sable Island Bank, and Georges Bank (Shaw et al. 2022). Glaci-isostatic depression flooded land surrounding Malpeque and Badeque Bays (PEI) to 20 m asl (above modern sea level). Prince County consisted of two islands at this time: one extending from North Cape to O'Leary, and a second surrounding Mount Pleasant and Wellington, in addition to small areas of land surrounding Kensington and Kinkora (Prest 1973; Shaw et al. 2002).

Simultaneously, higher relative sea levels occurred in northern Cape Breton Island (5 m asl), eastern New Brunswick (45 m asl in Kouchibouguac); Forillon National Park (25 m asl), Cap-Chat, Gaspé (100 m asl); Conception Bay (10–20 m asl), Deer Lake (50 m asl), Trout River NL (70 m asl; Fig. 2.22), Cow Head (100 m asl), and the tip of the Northern Peninsula (150 m asl). Generally, higher relative sea levels occurred in more northern sites, closer to the Laurentide Inlandsis. Glacial ice remained in central Newfoundland, the Chic-Choc and Appalachian Mountains, and in northern Québec and Labrador.

In southern and eastern areas, the period of raised sea levels was very short. Relative sea having dropped to its present position by 10,000 years ago in eastern and southwestern Newfoundland, on Grand Manan Island, and in southern Gaspé. Relative sea level fall below its modern position exposed much of the southern Gulf of St. Lawrence and Northumberland Strait. On the Iles-de-la-Madeleine, it reached −17 m asl (Hétu et al. 2020). Prince Edward Island was attached by land to New Brunswick and Nova Scotia (Shaw et al. 2002).

Fig. 2.22 Postglacial higher relative sea level, west coast of Newfoundland. Upper: Trout River terrace (right) and Old Man Stack (left) indicate relative sea level ~50 m above present. Lower left: marine sediment exposed on land at the head of Bonne Bay. Lower right: glacial sediment underlies the coastal plain flanking the Long Range Mountains at Gros Morne

2.4.2.2 Early to Mid-Holocene

Throughout the early Holocene (10,000–8000 years ago), sea level fell steadily in most areas. In eastern Prince Edward Island, drowned stumps of deciduous trees mark former forested areas developed at that time (Grant 1989). Submerged deltas and wave-cut terraces in Placentia Bay indicate lower sea levels offshore of Swift Current (−8 m asl) and Argentia (−19.6 m asl). Along the Burin Peninsula, the postglacial lowstand varies from −12.4 m asl at Long Harbour (Fortune Bay), to −19.5 m asl at Marystown Harbour. Similar lowstands have been recorded at Port-au-Choix (~3 m below present 3000 years ago, Stephenville, and Burgeo (Shaw et al. 2002; Catto et al. 2003; Catto 2011; Renouf 2011; Slaymaker et al. 2020a).

The trend reversed during the mid-Holocene. By that time, the Grand Banks and Georges Bank were completely inundated, only Sable Island remained exposed over Sable Island Bank, Prince Edward Island was an island, and the coastlines of New Brunswick, Cape Breton Island, Gaspé, and Newfoundland had essentially reached their present form. Between 3000 and 8000 years ago, depending on location, the land then began to subside from the over-compensated positions, resulting in renewed relative sea level rise. Evidence of transgression due to relative sea level rise is reflected by enhanced erosion along many beaches, and inundation of terrestrial peat deposits and trees.

2.4.2.3 Latest Holocene and Present Trends

At present, rates of relative sea level rise vary from about 1 mm y^{-1} in northern localities such as L'Anse-aux-Meadows and Cap-Chat to 3.0–3.5 mm y^{-1} in southern locations such as Halifax, Charlottetown, and St. John's (Shaw et al. 1999, Debaie et al. 2006, Vasseur and Catto 2008, Catto et al. 2003, Catto 2011, 2012, 2020a, 2020b, Slaymaker et al. 2020a, Warren and Lulham 2021; Fig. 2.23). Although these rates appear slow in human terms (fingernails grow 10–30 times more rapidly), they are considered very rapid geologically. They are also sufficiently rapid to result in the inundation of archaeological sites throughout Atlantic Canada, including Ferryland (NL), Fort Beausejoir (NB), and Louisbourg (NS).

When the fortress at Louisbourg, south of Sydney, was constructed by France beginning in 1721, sea level was approximately 2.5 m below present. Mooring rings used by French vessels now are exposed only at low tide. Relative sea level rise of approximately 3.5 mm y^{-1} has resulted in substantial erosion of the historic remains at Louisbourg. The appearance of the harbour today differs substantially from that of the mid-1700s, as broad tidal flats have become inundated by the rising marine waters.

Future changes in sea level will be determined by combinations of ongoing glaci-isostatic adjustment and increases in volume of the oceans due to glacial melting. For New Brunswick, Nova Scotia, Prince Edward Island, and southern Newfoundland, application of recent global mean sea level projections has high end values exceeding 2 m above the present position by 2100, due to accelerated melting of polar ice sheets. The suggested rates of change of 10–20 mm y^{-1} are substantially in excess of current rates.

2.5 Vegetation

Vegetation assemblages in Atlantic Canada include the Acadian Deciduous Forest, New Brunswick—Gaspé Boreal Forests, Newfoundland Boreal Forest, and Barrens communities (CNVC 2016).

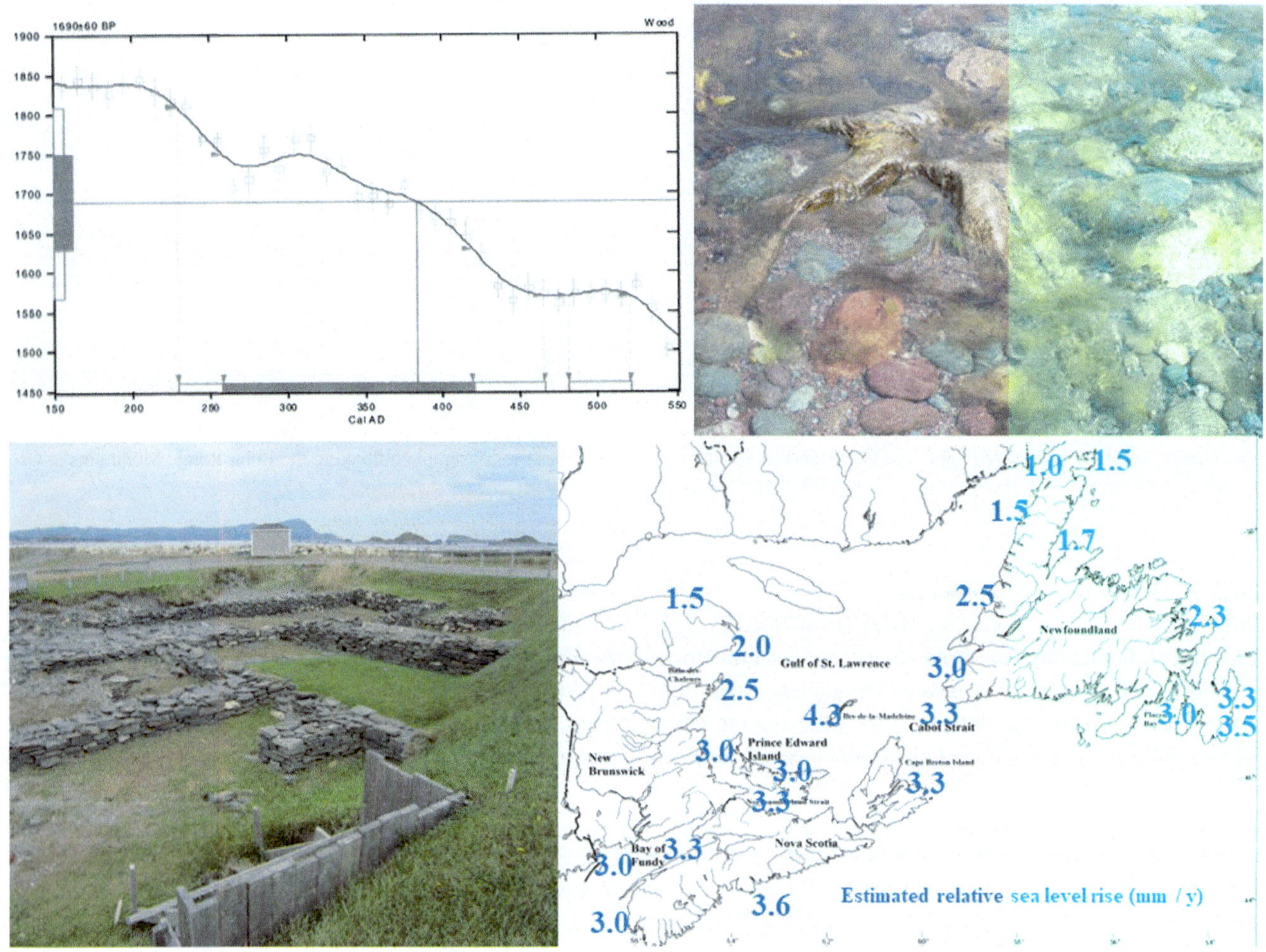

Fig. 2.23 Holocene sea level. Upper: Inundated terrestrial vegetation indicates rising relative sea level. ^{14}C dating (left) of submerged black spruce (*Picea mariana*) stump, Broad Cove, Avalon Peninsula (right) indicates that relative sea level was at least 2 m below present ca. 380 ce. Lower left: Archaeological investigations at Ferryland indicate that relative sea level has risen since ca. 1610 ce. Lower right: Estimates of current rates of relative sea level rise, Atlantic Canada

2.5.1 Acadian Deciduous Forest

Prior to extensive logging and agricultural clearance ca. 1800 ce, Prince Edward Island; the Annapolis Valley and sheltered inland areas of Nova Scotia; the Maritime Plain, lower Saint John River valley, and lower Restigouche valley of New Brunswick; and the most sheltered sites of Gaspé, were covered by the Acadian Deciduous Forest (Lescarbot 1609, Denys 1672; Stewart 1806; Gaudet 1979; Johnson 1986; Loo and Ives 2003). Broad-leaved deciduous trees, such as *Fagus grandifolia* (beech), *Acer saccharum* (sugar maple), *Quercus rubra* (northern red oak, the official tree of PEI), *Fraxinus americana* (white ash), and *Fraxinus nigra* (black ash) dominated the assemblage (Figs. 2.24 and 2.25). *Ulmus americana* (white elm) was also present. *Fagus grandifolia* appears to have covered more than 50% of the inland areas of PEI National Park and the higher ground surrounding Malpeque Bay in 1800. Other southern boreal species such as *Betula alleghaniensis* (yellow birch, the official tree of Québec), *Picea rubens* (red spruce, the official tree of NS), *Pinus strobus* (white pine), and *Tsuga canadensis* (hemlock) were also common. More northerly species including *Picea glauca* (white spruce), *Abies balsamea* (balsam fir), and *Acer rubrum* (red maple) were present in smaller numbers. *Betula papyrifera* (white birch) and *Pinus banksiana* (jack pine) were both uncommon.

Today, little of these natural assemblages remain. The woodlots and the wooded areas of the national and provincial parks contain combinations of native species in altered proportions (generally with more spruce, birches, maples, and poplars (*Populus balsamifera, P. tremuloides*) than originally, and fewer white pine, beech, oak, ash, and elm), mixed with introduced species from Ontario, Europe, and Asia. The most common co-dominants are yellow birch and sugar maple (Rowe 1972, Zinck 1998, Munro et al. 2014, CNVC 2016, Baldwin et al. 2019). In many groves, introduced species outnumber the native trees. The only area of Prince Edward Island where a remnant of the Acadian deciduous forest is present is located on the easternmost tip, 2–10 km west of East Point. Here, poor agricultural conditions have limited human settlement. The poor soil, however, also limited the growth of native deciduous trees, and this forest remnant contains few broad-leaved specimens.

2.5.1.1 Deciduous Species

Northern red oak may survive more than 300 years in favourable circumstances, and trees reach 25 m in height where not shaded by other trees (particularly beech). Oaks prefer well-drained, sandy, nutrient-rich soils, and are hindered in riverbank settings, although they are relatively salt-tolerant (Gaudet 1973; Martin 1977; Hosie 1979). Changes in flood frequency and magnitude along rivers due to logging, with decreased mean discharge and increased flood runoff and freshet, along with increased beaver populations, tend to favour *Acer pennsylvanicum* (moosewood) and *Acer rubrum* (red maple) over northern red oak (Johnson 1986). Northern red oak is more freezing rain tolerant than beech, but less than birches or maples.

Fig. 2.24 Modified remnants of the Acadian deciduous forest vegetation ca. 1800 ce are present in PEI. Broad-leaved *Fagus grandifolia* (beech), *Acer saccharum* (sugar maple), and *Quercus rubra* (northern red oak) dominated the assemblage. *Betula papyrifera* (white birch) was uncommon ca. 1800 ce, but forms a larger component of today's forest

Fig. 2.25 Acadian deciduous forest species. Upper left: *Betula alleghaniensis* (yellow birch). Upper right: *Acer saccharum* (sugar maple). Lower left: *Quercus rubra* (red oak) bark and leaves. Lower right: *Fagus grandifolia* (beech), Kouchibouguac National Park

White ash (*Fraxinus americana*) grows to 30 m height in favourable situations. It is called "white" because the wood is lighter than that of the "black" ash (*Fraxinus nigra*). It prefers slightly acidic to neutral soils, does not grow on peat, and is not wind or saltspray tolerant. It is a slower-growing species than the other North American ashes (Hosie 1979). Although ash leaves can be steeped to make a tonic to dissolve kidney stones, this would not have been common in Acadia. White ash is the preferred wood for making hockey sticks; black ash wood is too brittle.

Yellow birch, named for its conspicuous bark, grows to 20 m height and may live for 200 years (Gaudet 1973; Hosie 1979; Hinds 2000; Drobyshev et al. 2014, Munro et al. 2014). It does not grow in shaded or dry sites and requires at least 20 years before seed production commences, delaying its response to changes in climate. As a favourite browse species, increased yellow birch numbers are often associated with declining deer populations (Benson and Dodds 1977). Late frosts can be very detrimental (Vasseur and Catto 2008).

Yellow birch is used to manufacture doors, furniture, flooring, cabinets, and plywood. It is the preferred wood species in North America for toothpicks. In Mi'kmaq ethnobotany, the sap is used as a diuretic. Although more difficult to split and dry than white birch, yellow birch is useful for firewood, as it burns slowly, produces few shards and firebrands, and is an effective heat producer (5.5 million kilocalories per cord {3.62 m^3}).

2.5.1.2 Coniferous Species

Today, conifers form a lesser proportion of the Acadian Deciduous Forest (generally less than 30%), although disturbances due to previous logging continue to the assemblages (Fig. 2.26). White pines are confined to isolated shoreline areas where they could not be logged. Individual pines are as much as 75 years old, but many of the older specimens have suffered over the past 15 years due to acidic precipitation.

Tsuga canadensis are the oldest and largest trees in lesser-disturbed areas such as Kejimkujik National Park. Eastern

Fig. 2.26 Acadian forest coniferous species. Left: *Tsuga canadensis* (eastern hemlock). Right: *Thuja occidentalis* (eastern white cedar)

hemlock is shade-tolerant and prefers areas with lower fire frequency (Hosie 1979; Zinck 1998, Munro et al. 2014). As fire suppression generally has been practised over the last 100 years, hemlock has expanded more rapidly than jack pine. It thrives in the deciduous forests of sugar maple and yellow birch, and thus tends to supplant white pines in these environments. Under natural conditions marked by few fires, as was the situation from ca. 4000 to approximately 50 years ago, hemlock was able to gradually expand its niches in Kejimkujik, as well as in Cape Breton Highlands National Park (Nova Scotia 2021), as its northern extent in Québec simultaneously contracted. Several groves of hemlock with trees in excess of 300 years in age are present in Kejimkujik (Parks Canada 2021).

Eastern hemlock is shunned by loggers, due to its hardness and its tendency to break in shards when cut (Hosie 1979). It is a poor choice for firewood, as it tends to throw off shards explosively when burned. Consequently, eastern hemlock can expand its range or share of the forest assemblages of eastern Canada following commercial logging, as long as the soil remains moist and shaded, and the deer population is low. Ice storm damage slows the rate of hemlock expansion by reducing shade.

Red spruce (*Picea rubens*) is a slow-growing tree, reaching maximum heights of 20–25 m over 300 years (Hosie 1979; Hinds 2000). Red spruce is shade-tolerant and prefers moist, well-drained soils with moderate pH and some calcium content (Fig. 2.27). Its intolerance of acidic precipitation led to significant dieback in areas adjacent to coal-burning operations between 1900 and 1960. With a relatively shallow rooting system, it is subject to wind throw, leading to declines in logged areas and sites adjacent to roads (Cooper and Clay 1994). Red spruce hybridizes with black spruce (*Picea mariana),* and the latter species dominates in cooler, moister, and lower pH sites. Red spruce is primarily used for pulpwood, and also finds employment in guitars and fiddles.

Thuja occidentalis (eastern white cedar; arborvitae) is present in southern New Brunswick, western PEI, and the Isthmus of Chignecto and the Annapolis Valley of Nova Scotia (Rowe 1972, Earle 2018). In Atlantic Canada, it grows to 15 m height, and is very slow-growing, locally surviving for 500 years. Prior to agricultural clearance and logging, cedar was largely confined to nutrient-rich swamps with higher pH and high calcium content, and in areas with exposed limestone (c.f. Johnston 1990). Subsequently, while wetland drainage reduced habitat, cedar was able to spread and become naturalized as the use of agricultural lime increased and eutrophic conditions developed in riparian settings. In agriculturally less-productive upland sites, cedar colonizes abandoned fields and pastures. Common use as an ornamental has also facilitated spread of cedar in exurban areas.

Cedar is logged in New Brunswick to produce planking, siding, shakes, and shingles. Terpenes can be refined to make cleansers, disinfectants, and insecticides. Foliage is rich in Vitamin C, and has been used as a cure for scurvy. Cedar serves as nursery habitat and food source for snowshoe hares and squirrels.

Fig. 2.27 *Picea rubens* (red spruce) is shade-tolerant and prefers moist, well-drained soils with moderate pH and some calcium content

Ongoing climate change poses several issues for cedar, particularly in Nova Scotia where it is listed as a threatened species in the wild (Lemieux 2010). Cedar requires moist, but not saturated soils, rendering it vulnerable to both summer drought and excessive snowmelt, freshet, and autumn thunderstorms.

2.5.1.3 Understory Vegetation

Understory vegetation of note in sugar maple-yellow birch forests typically includes *Epigaea repens* (trailing arbutus or mayflower, provincial flower of Nova Scotia), marsh blue violet (*Viola cucullatta,* provincial flower of New Brunswick), starflower (*Trientalis borealis*), rose twisted-stalk (*Streptopus roseus*), cancer-root (*Conopholis americana*), and oak fern (*Dryopteris disjuncta*). The spotted coralroot (*Corallorhiza maculata*) is present only in hemlock groves (Zinck 1998, Hinds 2000, Munro et al. 2014, Atlantic Canada Conservation Data Centre 2022).

The provincial flower of Prince Edward Island is *Cypripedium acaule*, the pink lady-slipper orchid, which prefers moist, acidic peaty or conifer needle-duff soil (Martin 1977). Originally, the more spectacular *Cypripedium reginae* (showy lady's-slipper) was designated as the provincial flower, but during the preparations for the 1967 Canadian centennial, it was noted that *C. reginae* was extremely uncommon, whereas *C. acaule* was more readily visible along island roadsides and trails. Consequently, in 1965 the provincial legislature designated *C. acaule* as the new provincial flower. Unfortunately, ongoing agricultural development since 1965 has severely limited the numbers of all the island's orchids. Transplantation of *Cypripedium* is extremely difficult, hampering efforts to re-populate otherwise suitable coniferous glades (Fig. 2.28).

Cicuta maculata (water hemlock), not related to the eastern hemlock *Tsuga canadensis*, is found in marshy areas. It stands 30–90 cm high, has lance-shaped dark green leaves 5–20 cm long, arranged in groups of three, and bears a compound cluster of 10–20 small white flowers. Water hemlock is poisonous to all mammals (Schep et al. 2009).

Toxicodendron radicans (poison ivy) is widespread in shady deciduous woodlands (Zinck 1998; Hinds 2000, Munro et al. 2014). This erect shrub, commonly 10–30 cm high, is easily recognizable by its bright green leaves, with conspicuous veins, arranged in groups of three (hence the saying, "Leaves of three, let it be"). Contact with the leaves produces skin irritation in most people, but the effects are seldom more severe and rarely last longer than a week. Cattle and deer are able to eat poison ivy with impunity. Cats and dogs are also immune, as the oil which causes irritation in humans remains on the outer fur, unless a human decides to pet the animal. Poison ivy responds well to elevated CO_2 levels (Mohan et al. 2006), indicating that it will become more prevalent under warming climates in Maritime Canada.

2.5.2 New Brunswick—Gaspé Boreal Forests

Transitions from Mixed Acadian Deciduous Forest to Boreal Forest assemblages occur with elevation in Cape Breton and Gaspé, with latitude in central New Brunswick, and with distance from the coast in Fundy National Park (Rowe 1972;

Fig. 2.28 *Cypripedium* species

Deschênes et al. 2010; CNVC 2016, Baldwin et al. 2019). The Fundy Coastal Forest is dominantly coniferous, with *Picea mariana* (black spruce), *Picea glauca* (white spruce), *Abies balsamea* (balsam fir), *Larix laricina* (tamarack), *Betula papyrifera* (white birch), and raised *Sphagnum* moss bogs (Hinds 2000). This assemblage reflects the exposure of the area to coastal winds and freezing drizzle, both of which hinder the establishment of broad-leaved deciduous trees. The presence of *Acer rubrum* (red maple) along low lying river valleys, especially at the mouths of the streams, reflects the tolerance of this species to periodic flooding (Hosie 1979).

Gaspé provides a good example of how the combination of elevation and orientation to the prevailing winds has produced a series of distinctive vegetation zones (Grondin et al. 1998, Forêt Québec 2002, Desponts et al. 2004, CNVC 2016; Fig. 2.29). The exposed southern coast supports a *Abies balsamea–Picea glauca* dominated assemblage (balsam fir-white spruce). The spruce trees of the coastal fringe area produce the best spruce beer in Gaspé. Species diversity varies with bedrock type. Areas with limestone bedrock support more beech and cedar, whereas white and red pines prefer the more acidic soils associated with sandstones and shales. Red maples preferentially inhabit the floodplains, as this species is tolerant of periodic inundation, whereas pines prefer drier ground.

The inland areas on the southern flank of the Monts-Notre-Dame are milder than the coastal regions, and lack freezing rain. Trees include beech, cedar, yellow birch, sugar maple, and balsam fir. Sugar maples growing on limestone bedrock have distinctive sap and syrup, but formerly were a good source of roots for root beer production. The numerous maple sugar producers ("érabliere" or "sugar bushes") are established on bedrock of differing chemistry, producing maple syrups which can be distinguished by connoisseurs. With increasing elevation (approximately 100 m) and distance from the coast, the forest assemblage changes, with *Abies balsamea* and *Pinus banksiana* (jack pine) replacing the cedar, beech, ash, and elm. The spread of Dutch Elm Disease has also reduced the elm population, with most of the vacant spots occupied by ash and sugar maple.

At elevations above approximately 500 m asl (above sea level), *Betula papyrifera* dominates the deciduous assemblage, and the broad-leaved trees and *B. alleghaniensis* are absent (Grondin et al. 1998). Balsam fir (var) is the dominant tree of this assemblage, and pure stands of fir are present. The decrease in the numbers of fires (some of which were set by the Mi'kmaq prior to European settlement to encourage deer habitation, and for agricultural clearance) has favoured the expansion of balsam fir at the expense of the pines.

At elevations higher than approximately 750 m asl, the Gaspé Boreal assemblage is replaced with a more northerly *Abies balsamea–Picea mariana* assemblage, lacking pines and most deciduous trees. Scrubby *Betula papyrifera, Populus tremuloides* (trembling aspen), *Populus balsamifera* (balsam poplar), and *Alnus sp.* (alders) are present in small numbers. Above approximately 1000 m asl, forest vegetation cannot survive, and the summits of Mont Jacques-Cartier and other peaks are occupied only by alpine tundra vegetation.

Fig. 2.29 Inland areas in northern New Brunswick and on the southern flank of Gaspé are milder than the coastal regions, and lack freezing rain. Trees include beech, cedar, yellow birch, sugar maple, and balsam fir

Permafrost features are actively forming on these summits (Gray et al. 2016).

2.5.3 Newfoundland Boreal Forest

The combination of cool summers, freezing rain, strong winds, humid to perhumid moisture regimes, and the Gulf of St. Lawrence, restrict the vegetation of Newfoundland to boreal forest assemblages (Rowe 1972, Damman 1983, Ryan 1995, CNVC 2016, Baldwin et al. 2019). Heathland assemblages are present in coastal and barrens and at altitude. Unique vegetation assemblages are associated with the ophiolite Tablelands of the west coast.

Typical boreal forest assemblages (Fig. 2.30) are dominated by black spruce (*Picea mariana*) and balsam fir (*Abies balsamea*). White spruce (*Picea glauca*) is most prevalent in well-drained and coastal sites. White birch (*Betula papyrifera*), balsam poplar (*Populus balsamifera*), trembling aspen (*Populus tremuloides*), and green alder (*Alnus viridis crispa*) are common deciduous species. Broadleaf species such as sugar maple (*Acer saccharum*), northern red oak (*Quercus rubra*), and beech (*Fagus grandifolia*) are not native to Newfoundland, due in part to their inability to tolerate multiple freeze–thaw cycles and freezing rain. *Betula alleghaniensis* (yellow birch) is only present in the most sheltered areas, in small numbers.

2.5.3.1 Coniferous Species

On drier, coarser soils, larch (*Larix laricina*), *Juniperus communis* (common juniper), and *Juniperus horizontalis* (trailing juniper) are important constituents, with larches forming nearly homogeneous stands and junipers preferring open, sometimes unstable terrain (Ryan 1995). White pine (*Pinus strobus*) was formerly an important component in central Newfoundland, but was preferentially logged during the 19th and early twentieth centuries. Red pine (*Pinus resinosa*) was widespread throughout the central part of Newfoundland during the mid-Holocene, but climatic cooling and moisture increases beginning 4000 years ago, have caused its range to be severely contracted. Red pine is currently confined to 12 sites in sheltered locations in the Glovertown and Gander areas, in West Brook Ecological Reserve near Springdale, and surrounding the village of Terra Nova (Catto 2006a, b). Jack pine (*Pinus banksiana*), a dominant in many eastern Canadian boreal forests, is absent from the island of Newfoundland.

Balsam fir (*Abies balsamea*) grows to 25 m tall at maturity (more commonly 15 m). It lives up to 70 years at sheltered sites and is a very fast-growing species (Ryan 1995). Needles are flat, in contrast to spruce, and are retained by trees for longer periods during drought than are spruce needles (so balsam firs are more suitable as Christmas trees). Bark can be distinguished from spruce bark by the presence of resin blisters.

Fig. 2.30 Newfoundland boreal forest assemblages are dominated by *Picea mariana* (black spruce), *Abies balsamea* (balsam fir), and *Picea glauca* (white spruce), with wetland vegetation flanking the numerous ponds

Balsam fir is widely distributed throughout Newfoundland and Labrador south of 54°N. Balsam fir prefers forested areas and commonly forms dense, almost pure stands. Its shallow root system means that it is vulnerable to being toppled by strong winds, and so an isolated balsam fir cannot survive in an exposed, windswept area, although dense stands can by relying on each other for mutual support. It prefers acidic mineral soil but can survive on peaty soil.

Balsam fir is a key species for the pulpwood and paper industry. In the past, it was used for a wide variety of medical remedies, including first aid for cuts, as a disinfectant, and as a cure for scurvy (although spruce is more effective). Refined balsam resin ("Canada balsam") was formerly used in a number of industrial applications but has been replaced by artificial petroleum-based compounds. Balsam fir stands are under pressure from balsam fir sawfly and balsam wooly adelgid, and particularly from moose browsing (McLaren et al. 2004; Gosse et al. 2011).

2.5.3.2 Deciduous and Understory Species

White birch (*Betula papyrifera*) is a common boreal species, growing to 20 m in height in favourable situations, but more commonly 10 m (Ryan 1995). In a sheltered situation, a white birch can live for more than 100 years. White birch seldom forms pure stands in Newfoundland, but is commonly present on the fringes of spruce and fir forests. Birch often colonizes newly burned or cleared areas, is often found along the edges of old homesteads or farms, and has increased in numbers as a result of land clearance. Birch was a useful tree for early settlers, both for birch bark and as a source of medicine and flavouring (for birch candy and birch beer). It was also useful for furniture manufacture.

Trembling Aspen (*Populus tremuloides*) and balsam poplar (*Populus balsamifera*) prefer well-drained, moist, sandy, or gravelly soils with small concentrations of silt and clay ("loamy soils"). They are fast growing but short-lived generally less than 50 years. In Newfoundland, they tend to form almost pure stands on the edges of forests. Over time, a grove of trembling aspens may be gradually replaced by white birch.

Green alder (*Alnus viridis crispa*) colonizes recently cut areas, and is especially common along the fringes of the highway where other trees have been cleared. Hoary or grey alder (*Alnus rugosa*) is less common in Newfoundland, preferring sheltered areas.

Showy dogberry (*Sorbus decora*) and common dogberry (*Sorbus americana*) are called Mountain Ash or Rowan elsewhere in Canada. These species are common in many boreal forests throughout Canada, reaching maximum heights of 12 m. Showy dogberry is more resistant to winter temperatures than is the common dogberry, and hence extends further up north-facing slopes (Ryan 1995). Despite its name, showy dogberry generally produces fewer berries than does common dogberry. Dogberry wine was a staple of settlers from early times, although it is virtually unknown throughout the remainder of boreal Canada.

Chuckley-pear (*Amelanchier laevis*, serviceberry) grows to 10 m in height. Chuckley-pear prefers moderate to well-drained, coarse to sandy soils. The wood can be used for making shovel handles and tools, as it is hard and durable, but the trees are too small for use as lumber. In western Canada,

Fig. 2.31 Understory vegetation, Newfoundland boreal forest. Boreal forest understory assemblages include *Cornus canadensis* (bunchberry), *Andromeda polifolia* (andromeda), sphagnum moss, and *Abies balsamea* seedlings. Inset: *Linnaea borealis* (twinflower)

berries from the related species *Amelanchier alnifolia* (Saskatoon) are used to make jam, but this has never been common in Newfoundland.

Boreal forest understory assemblages include *Cornus canadensis* (bunchberry, crackerberry), *Linnaea borealis* (twinflower), *Andromeda polifolia* (andromeda), *Clintonia borealis* (clintonia), *Trientalis borealis* (starflower), and *Viola nephrophylla* (northern bog violet)(Fig. 2.31). Ombrotrophic, oligotrophic bogs support assemblages dominated by bakeapple (*Rubus chamemorus*) and Pitcher Plant (*Sarracenia purpurea).*

Vegetation in exposed coastal areas is dominated by "tuckamore" (krummholz white spruce, *Picea glauca;* Fig. 2.32). Tuckamore trees characteristically form a low, dense tangle of gnarled trunks and interlocking branches, aligned in response to the prevailing coastal wind. In many cases, the upper branches are shaped into wave forms. Exposed coastal bluffs and fringes of salt marshes are occupied by sea thrift (sea pink, *Armeria maritima*), Lapland rosebay (*Rhododendron lapponicum*), and *Rhacomitrium* moss.

2.5.4 Barrens Assemblages

In Newfoundland, "barrens" refers to open landscapes dominated by heathland assemblages, with few or no shrubs or trees (Fig. 2.33). Although marked by thin regosols or organic veneers over coarse glacial deposits or bedrock, barrens typically have 90% or more vegetation cover. Barrens serve as important ecological areas, as well as providing important cultural and food resources for many Newfoundland residents (see Weber 2022). Typical barrens assemblages (Damman 1983, Ryan 1995) include crowberry (*Empetrum nigrum*), Labrador Tea (*Rhododendron groenlandicum*), partridgeberry (*Vaccinium vitis-idaea*), dwarf blueberry (*Vaccinium boreale*), bilberry (*Vaccinium uliginosum*), low blueberry (*Vaccinium angustifolium*), bearberry (*Arctostaphylos uva-ursi*), bakeapple (*Rubus chamemorus*), Saxifrage (*Saxifraga*), and feather moss (*Raecomitrium lanuginosum).*

Empetrum nigrum (crowberry in southern and central Newfoundland; blackberry in the Northern Peninsula and Labrador) is the dominant species in most undisturbed headlands. It is commonly used as a food resource in northern communities, but much less so elsewhere in Newfoundland.

The name "partridgeberry" for *Vaccinium vitis-idaea* (lingonberry) was in use in Newfoundland and Labrador and Cape Breton Island before 1770 (Story et al. 1997). "Partridges" (rock ptarmigan, *Lagopus muta*) consume partridgeberries and crowberries during late summer and early autumn. Because immature berries host the larva of the partridgeberry fruitworm (*Grapholite libertine*), they should not be harvested until touched by first frost, usually in late September–mid–October.

The name "bakeapple" for *Rubus chamemorus* (cloudberry) comes from the experiences of early French residents of Newfoundland. The bakeapple is not native to France, so

Fig. 2.32 Tuckamore, Great Barachoix NL. This tree is approximately 65 years old and has reached its maximum height. Common tuckamore species are *Picea glauca* and *Abies balsamea* (foliage inset)

when settlers first encountered it, they had no idea what the plant was. As an unknown berry, it was referred to as "Beie Qu'Appelle" (What is this berry called?). Over time, "beie qu'appelle" became "bakeapple".

Many barrens adjacent to human coastal communities ('outports') reflect domestic woodcutting between 1650 and 1850 (Catto 2020b; Catto and Catto 2012). Strong winds subsequently precluded re establishment of trees, and grazing by sheep reduced vegetation cover and facilitated frost action. Active and relict periglacial features are present in areas that show evident disturbance of heathland vegetation, or where introduced European species are not present.

Ground hugging barrens assemblages can be severely damaged through tourist foot traffic. At several sites, including Cape Spear, Mistaken Point Ecological Reserve, Mad Rocks Bay Roberts, and Cape Bonavista, the amounts of human foot traffic and rates of erosion along paths are directly correlated. At Cape St. Marys, restriction of foot traffic led to decreased erosion and partial recovery of the coastal vegetation. In the absence of similar action elsewhere, as tourism continues to impact coastal sites, vegetation damage and erosion will increase (Catto et al 2020b).

Disturbed heathland areas contain mixtures of native and introduced vegetation (Fig. 2.34). Species such as red clover (*Trifolium pratense*), white daisy (ox-eye daisy, *Leucanthemum vulgare*) burdock (*Arctium minus*), butter and eggs (*Linaria vulgaris*), yarrow (*Achillea millefolium*), and orange hawkweed (*Hieracium aurantiacum*) are not native, but have spread invasively. Native *Kalmia angustifolia* (sheep laurel), *Kalmia polifolia* (bog laurel), *Rhododendron lapponicum* (Lapland Rosebay), and *Rhododendron canadense* (rhodora) also take advantage of cleared barrens sites. *Rosa nitida* (shining rose), *R. rugosa* (rugose), *Juniperus communis* (common juniper), and *J. horizontalis* (trailing juniper) often indicate former sheep pasturing. The presence of raspberry (*Rubus ideaus*), which prefers well-drained sandy to gravelly mineral soils, in heathland assemblages also indicates spread from adjacent outport communities.

Heath and moss barrens also occur at higher interior elevations, with scattered krummholz *Picea mariana* or *Picea glauca* forming treeline. *Juniperus*. Labrador Tea (*Rhododendron groenlandicum*), rhodora (*Rhododendron canadense*), and dwarf bilberry (*Vaccinium boreale*) are common. Some of the coastal heathland species are replaced by highland equivalents, such as alpine azalea (*Kalmia procumbens*), alpine bearberry (*Arctostaphylos alpina*), and pink crowberry (*Empetrum eusmesii*). *Diapensia lapponica* (cushion plant) forms disjunct populations as far south as the Avalon Wilderness Area and the Hawke Hills, west of St. John's (Baehre 2022).

As in the coastal settings, the barren conditions at many summits are not truly reflective of the local climate. At the popular tourist site of Gros Morne mountain, as much as 75% of the vegetation formerly present on the summit has been inadvertently destroyed by human feet.

Fig. 2.33 Barrens species. Clockwise from top left: *Rubus chamaemorus* (bakeapple), *Vaccinium angustifolium* (low blueberry), *Vaccinium boreale* (dwarf bilberry), *Rhododendron groenlandicum* (Labrador Tea), and *Empetrum nigrum* (crowberry). *Sarracenia purpurea* (pitcher plant) occurs in wet sites

2.5.4.1 Tablelands

The unique ecology of the Tablelands area of Gros Morne is due to the combination of periglacial conditions marked by cryoturbation and strong wind activity, with ophiolite (peridotite) bedrock containing concentrations of magnesium, chromium and nickel that are toxic to most plants. These harsh conditions restrict the flora, and many locations are completely devoid of vascular plants (Bouchard et al. 1986; Berger et al.1992).

Areas which are somewhat less exposed are dominated by restricted heath communities including *Lychnis alpina* (alpine catchfly), *Arenaria* (sandwort), *Solidago hispida* (goldenrod), *Rhacomitrium* moss, and prostrate *Juniperus horizontalis*. Areas where snowpacks persist have additional species due to the moister conditions, including *Sanguisorba canadensis* (burnet), *Scirpus cespitosus* (deergrass), and *Adiantum pedatum* (northern maidenhair fern). Carnivorous *Sarracenia purpurea* (pitcher plant) and *Drosera rotundifolia* (sundew) also are present (Fig. 2.35).

2.6 Soils

Forest soil orders of Atlantic Canada include podzols, brunisols, and gleysols. Thin, poorly developed regosols form over bedrock. Organic order soils are associated with wetlands (Soil Classification Working Group 1998, 2001).

2.6.1 Podzols

Podzols are acidic (pH 3.8–5.8), developed on granite, quartz-rich sandstone, sedimentary rocks with feldspar, gneiss, or glacial sediment derived from these bedrock types (Sanborn

Fig. 2.34 Disturbed barrens. Upper right: Many small barrens adjacent to human coastal communities reflect forest clearance between 1650 and 1850, coupled with strong winds which precluded re establishment of trees. Mad Rocks, Bay Roberts. Clockwise from bottom right: typical species indicating human disturbance include *Rosa sp.* (rose), *Hieracium aurantiacum* (Orange Hawkweed), *Rubus idealis* (raspberry), and *Aster novae-angliae* (aster)

et al. 2011). Podzols in Atlantic Canada form where boreal forest vegetation has persisted for hundreds to several thousand years, prior to any clearance for agriculture (Fig. 2.36).

Better conditions for conifer growth, with warmer temperatures, longer frost-free seaspons, and more precipitation, favour the production of humic matter and hence influence podzol development. Podzols in Atlantic Canada tend to be better developed in Köppen-Geiger Dfb (boreal) than in Dfc (taiga) climates (other factors being equal). Atlantic Canada's most agriculturally productive soils are podzols.

Soils are the critical element in the agricultural development of Prince Edward Island (PEI Soil Survey 1988). The climate of the agricultural areas does not differ greatly from many other regions in Atlantic Canada, but agricultural productivity is much higher. The soils, developed on glacial materials derived from iron-rich sandstones and shales, were richer in both humic matter and calcium carbonate than most other podzolic soils in Atlantic Canada. These differences in soil chemistry can be attributed to the original deciduous forest cover, which added substantial quantities of leaf mould and mulch to the soil yearly, hence providing humic matter and nutrients. In the areas surrounding Malpeque Bay, higher sea levels following deglaciation (Prest 1973; Shaw et al. 2002) allowed carbonate-bearing silts and sands (and locally, mollusc shells) to be deposited, adding calcium carbonate to the soil and raising its pH. The sandy texture of the soils is highly conducive to potato farming. With the removal of the beech-oak-sugar maple forest, the supply of humic matter to the soil has gradually been declining, and farmers have compensated by adding humus and nutrients artificially.

Locally, resistant horizons have developed in the humo-ferric podzols. Thin, black cemented glass-like placic horizons develop in areas where iron and manganese oxides accumulate and bond with organic molecules to form metal–organic or sesquioxide complexes. These horizons are imper-

Fig. 2.35 Sundew (*Drosera*) inhabits poorer soils on barrens and Tablelands areas throughout Newfoundland. Hearts Content Barrens, courtesy Carissa Brown

meable to groundwater, and if regionally extensive can impede tree growth or agriculture by preventing drainage.

The soils of the agricultural Annapolis Valley, middle Saint John River Valley, and Humber Valley are also dominantly podzolic (NS Soil Survey 1972, Kirby 1988, Holmstrom and Thompson 1989, Fahmy et al. 2010). In the low-lying regions adjacent to the rivers, periodic sediment deposition by flooding and the presence of high amounts of organic material washed down by the streams resulted in the formation of different soils. Cumulic regosols, marked by alternating layers of river sediment and thin organic-rich horizons, occur in the most frequently flooded locations. Areas which were less regularly flooded developed brunisols. Although these soils will eventually develop additional horizons and become more chemically mature, they are suitable for supporting plant growth. Human regulation of the river systems has effectively curtailed the continuing formation of flood-influenced soils (Fig. 2.37).

Highland areas such as Fundy National Park and Gaspé also contain Fe^{3+}-rich podzols, generally low in organic content and with pH values ranging from 4 to 5.5. The productivity of these soils is relatively low, making the areas unsuitable for most forms of agriculture, with farming limited to root crops and cabbage. The poor soil is one factor responsible for delaying and hindering the spread of European settlement. In Gaspé, the rigourous boreal climate limits podzol development, except in the areas which support the Restigouche deciduous forests (Bureau d'aménagement de l'est du Québec 1966).

Fig. 2.36 Podzolic soils dominate in agricultural regions of PEI, such as Miscouche. Right: podzolic soil profile, indicating surface LFH, Ah (dark grey), eluviated Ae (white), iron-rich Bf (red-brown), and parent C glacial horizon (brown)

Fig. 2.37 Agricultural landscape, Canning, Annapolis Valley NS

2.6.2 Brunisols

Brunisols in the CSSC are subdivided by pH and the thickness of the humic Ah horizon (Smith et al. 2011). In Atlantic Canada, dystric brunisols, with low pH (acidic) and thin Ah horizons, are typical under boreal conifers on acidic bedrock. Dystric brunisols develop where more rigourous climates (Dfc), insufficient time since deglaciation, shorter frost-free seasons, thin sediment cover, or other factors preclude podzol formation. Some sites with exposed limestone support calcareous eutric brunisols.

Brunisols are typical throughout Newfoundland (Fig. 2.38), including regions with boreal (Dfb) climates such as the Avalon Peninsula and Exploits River Valley, and those with more rigourous taiga (Dfc) climates, such as the Northern Peninsula and Annieopsquotch Mountains (e.g. Heringa 1981, Woodrow and Heringa 1987). Acidic, thin, and coarse, these soils generally would not be considered suitable for agriculture, but that has not always been recognized.

2.6.3 Gleysols

Poorly drained areas, including some wetlands, are characterized by gleysols. Gleysols develop in discharge zones or depressions where repeated fluctuations of the water table (typically several per year) alternately expose and inundate iron-bearing minerals. As oxidation and reduction are not instantaneous, a varying water table produces a mottled pattern in the soil. Variations in water table position reflect snowmelt, summer drought and thunderstorm activity, river flooding, freezing and thawing, formation of elevated seasonal frost tables (tjäle), fragipan horizons, and/or withdrawal of groundwater by humans. Gleysols are generally agriculturally poor.

2.6.4 Organic Soils

Organic Order soils are dominated by organic matter with no or few mineral grains. In Atlantic Canada, fibrisol peats characterize conifer- and sedge-dominated fens, and sedge- and sphagnum-dominated bogs. These soils form in a variety of topographic settings, ranging from riparian margin fens to summit-covering blanket bogs.

Ombrotrophic blanket bogs drape the underlying topography, and may develop on plateaux and slopes (Warner and Rubec 1997). Blanket bog formation requires perhumid environments, with annual precipitation in excess of 1000 mm, at least 160 days with precipitation per year, and the warmest month having a mean temperature less than 15 °C. In Atlantic Canada, blanket bog development occurs in coastal areas of eastern Newfoundland and southern Cape Breton Island.

Catena sequences are common in upland areas. In Cape Breton, the plateau areas contain fibrisols developed in blanket bogs, interspersed with brunisols developed on inorganic (mineral-rich) sediments (Cann et al. 1963, NS Soil Survey 1972). The flanks of the plateau commonly display catena sequences, with ferro-humic podzols on the upper slopes, gleyed ferro-humic podzols in midslope positions, and fera gleysols marked by periodic saturation at the bases of the slopes. Iron in the soils on the upper slopes is oxidized to the

Fig. 2.38 Dystric Brunisol, Bauline, Avalon Peninsula

ferric state (Fe^{3+}), turning the soil red. On the lower slopes, the iron is present in the reduced state (Fe^{2+}), resulting in grey-green tones. All of these soils are acidic and iron-rich and are relatively poor in organic matter. None are suitable for agricultural development. Small areas of soils near the communities are used for root crops.

2.7 Fauna

Faunal assemblages of Atlantic Canada also represent a transition between the Atlantic Maritime (mid-boreal and transitional southern boreal), Atlantic Highlands (mid-boreal), and Newfoundland boreal (mid-boreal to northern boreal) communities (Banfield 1981a, b; Naughton 2012). Ongoing climate change is resulting in northward and altitudinal movement of species in Nova Scotia, New Brunswick, and Gaspé. The Newfoundland fauna are largely isolated from those of the Atlantic Maritime and Highlands, although migration across the Strait of Belle Isle and human introductions has occurred.

2.7.1 Atlantic Maritime

Today, a typical Atlantic Maritime faunal community includes *Alces alces* (moose), *Odocoileus virginianus* (white-tailed deer), *Castor canadensis* (beaver), *Mephitis mephitis* (skunk), *Martes americana* (pine marten), *Procyon lotor* (raccoon; Fig. 2.39), *Erethizon dorsatum* (porcupine), *Lepus americanus* (snowshoe hare), *Vulpes vulpes* (red fox), and *Ursus americanus* (black bear). Predators have been eliminated (*Canis lupus* {wolf}) or greatly reduced in numbers (*Lynx canadensis* {lynx} and *Martes americana*). Woodland caribou (*Rangifer tarandus caribou*) were extirpated prior to 1900. The existence of *Puma concolor* (cougar, locally called "eastern panther"), rumoured in New Brunswick for many years, has been demonstrated by genetic evidence (Lang et al. 2013). Although isolated cougars still live in the region, the species no longer is able to play a significant role in the ecosystem.

Some species have adapted well to the combination of reduced predator pressure and increased human activity. Both *Procyon lotor* (raccoon) and *Erethizon dorsatum* (porcupine) have adjusted to the new conditions, along with some smaller animals such as the water shrew (*Sorex palustris*). Locally, increased porcupine activity has resulted in increased tree losses, as porcupines can damage or kill trees by eating the bark at ground level. Porcupine numbers and damage, however, are kept in check by silvicultural practices, as porcupines prefer the taste of pine bark to that of spruce or fir. Porcupines are frequent victims of automobiles, as they have a fatal weakness for the salt which accumulates on highways.

Raccoons are gradually increasing in numbers, as they have learned to forage from the garbage discarded by humans. Many designs of waste disposal containers have been tried in efforts to stop the raccoons from raiding, but the intelligence of the raccoons has usually proven sufficient to defeat the containers over time. Contrary to popular belief, raccoons do not need to rinse their food in water prior to eating. Most raccoons in the wild do not rinse their food, and those in captivity tend only to rinse certain foods, such as fish. The behaviour seems to be related to that of domestic cats, who will often play with dead food before eating it, probably to simulate hunting. Raccoons fishing in the wild seldom waste effort, and rarely place a wounded fish back in the water to give it a chance to escape.

Extensive logging has occurred in most areas of New Brunswick and Nova Scotia, along with agricultural clearance. With the shift from timber harvesting to pulp-and-paper operations, and with the forestry industry preference for monoculture stands of identical characteristics, reforestation efforts since 1960 have concentrated on developing uniform stands of fast-growing conifers (particularly *Picea* and *Abies*) for market. This practice has led to a loss of habitat for many of the animal species. Reductions in the population of *Martes*

Fig. 2.39 *Procyon lotor* (raccoon) has adjusted well to human modification of landscapes

americana (Pine Marten) and *Glaucomys sabrinus* (Northern flying Squirrel) are evident. The relatively small protected areas are insufficient to provide many of the forest species with the amount of territory which they need to maintain successful populations (Samson et al. 2010).

Smaller animals have also been affected by the changes in forest composition. The population of Eastern Painted Turtles (*Chrysemys picta*) has decreased in New Brunswick (COSEWIC 2018), due to loss of habitat and infilling of ponds with debris and river systems with silt generated by logging. In addition, Eastern Painted Turtles were formerly collected in substantial numbers for sale as pets, despite the species' very short life span in captivity. Fundy National Park lies at the eastern extremity of the range of the Eastern Painted Turtle, as their ability to breed is controlled in part by water temperature (Cook et al. 1984).

Agricultural development and human predation have virtually eliminated the native fauna of Prince Edward Island. Black bear, wild red fox, lynx, and wolf were eliminated before World War I; the larger herbivores such as moose, deer, and porcupine disappeared shortly thereafter. Although some foxes have escaped from commercial fur farms and exist in the wild, no native wild species larger than a house cat currently inhabit the island with the exception of raccoons (*Procyon lotor*). Other native species still present include snowshoe hare (*Lepus americanus*), mink (*Neogale vison*, some of which are descended from escapees from fur farms), and muskrat (*Ondrata zibethica*). Beaver was extirpated prior to 1700 ce, and those present today are descendants from New Brunswick re-introductions (Curley et al. 2019).

The ubiquitous coyote (*Canis latrans*) arrived in eastern PEI, near Souris, in 1983. Since that time, the species has spread across the island, and now occupies all suitable habitats. Although coyote hunting is permitted, the population grew rapidly in the 1980s and 1990s, almost doubling in some years. Subsequently, the rate of population increase has slowed, as the available habitat was occupied. Today, the population appears to have stabilized at about 2000 (PEI 2003).

Combinations of climate change and human activity have facilitated the spread of some species. In the Annapolis Valley, *Odecoileus virginianus* (white-tailed deer) populations have recently begun to increase, as the deer have taken advantage of the introduced apple trees. The Avon River lies approximately 50 km to the east of the pre-1600 ce range limit of *Odecoileus virginianus* (Benson and Dodds 1977, Patton 1991). On Cape

Fig. 2.40 *Rangifer tarandus caribou* (woodland caribou) herd, Portugal Cove South, NL

Breton Island, the coastal fauna dominated by southern boreal species such as snowshoe hare, porcupine, and red fox were joined by the non-native raccoon (ca. 1956), bobcat (*Lynx rufus*, ca. 1960), and coyote (ca. 1980) which have spread along the coast from the entry point of the Canso Causeway (Parks Canada 2018). Caribou were extirpated, and replaced by introduced white-tail deer ca. 1900–1910.

During nineteenth- through twentieth-century extirpation, *Lynx canadensis* survived in the snowier, more isolated highlands. The southern boreal species *Lynx rufus* has steadily expanded its range throughout the past 30 years. Simultaneously, the range and numbers of *Lynx canadensis* have declined systematically. These changes indicate that the climate of the region has been gradually warming, improving conditions for *Lynx rufus* while posing increasing problems for *Lynx canadensis*. Similarly, climatic changes during the past 100 years have allowed *Lynx rufus* to colonize Kouchibouguac National Park (Parks Canada 2020a).

Kejimkujik National Park is the only haunt in Atlantic Canada of the Southern Flying Squirrel (*Glaucomys volans*). Unlike its northern relative, the southern flying squirrel prefers oak trees to conifers, usually selecting an abandoned woodpecker hole (COSEWIC 2006a, b). Consequently, it is a difficult life for southern flying squirrels in boreal forests, although they are also found in the deciduous forests of Southern Ontario. This nocturnal animal is highly sociable, and commonly lives in dense communal groups with as many as 3 flying squirrel adults per hectare. Flying squirrels are active throughout the winter.

Kejimkujik National Park is also the only place in Atlantic Canada where Blanding's Turtle (*Emydoidea blandingii*) and the Northern Ribbon Snake (*Thamnophis sauritus septentrionalis*) survive (Cook et al. 1984, Ernst and Barbour 1989). Small populations of both species also occur in Southern Ontario, but both are considered to be threatened in Canada. The total number of Blanding's Turtles in Nova Scotia is less than 150 (Environment and Climate Change Canada and Parks Canada Agency 2019). Nesting beaches at Heber's Meadow are routinely disturbed by egg-eating raccoons, and unfortunately by increased foot traffic from hikers. The park also supports the only population of snapping turtles (*Chalydra serpentina*) in Maritime Canada. Eastern Painted Turtles (*Chrysemys picta*) are also present.

In addition to the Northern Ribbon snake, Kejimkujik supports four other snake species: Maritime Garter (*Thamnophis sirtalis pallidula*), Northern Red-Bellied (*Storeria occipitomaculata occipitomaculata*), Northern Ring-Necked (*Diadophis punctatus edwardsi*), and Eastern Smooth Green snakes (*Opheodrys vernalis*). The Northern Ring-Necked snake is present in smaller numbers than the other three species, and all snakes within the park have suffered in the past from indiscriminate killing by panicky *Homo sapiens*.

Nova Scotia has 13 species of amphibians: five salamanders, seven frogs, and one toad (Nova Scotia Herpetofaunal Atlas n.d.). All of these amphibians, with the possible exception of the Blue-spotted salamander (*Ambystoma laterale*), live in Kejimkujik National Park. The Maritime Boreal ecoclimatic conditions allow amphibian and reptile species that could not survive elsewhere in Nova Scotia to overwinter here. Amphibian populations have been declining in Kejimkujik, in common with most other areas of boreal and temperate North America.

Sea mink (*Mustela macrodon*) was driven to extinction by human hunting ca 1880. It was approximately twice the size of the terrestrial mink. Little is known about the ecology of this marine mammal. It was harvested for food by the Mi'kmaq people of the Fundy region, and may have ranged northeastward to the south coast of Newfoundland (Black et al. 1998). However, as almost everything about this extinct species, this is uncertain.

2.7.2 Atlantic Highlands

Atlantic Highland faunal assemblages have most of the species of the Atlantic Maritime regions, although with larger populations of colder- and snow-tolerant species such as moose, lynx, black bear, and woodland caribou (Fig. 2.40), and fewer representatives of raccoon and skunk. In Gaspé, fauna is more boreal in character than any other area of Appalachian Canada with the exception of Gros Morne (which does not have the faunal diversity of Gaspé). Recent climatic change has had relatively little impact on the faunal assemblage to date, but human pressures are increasing on all species. In addition, the recent spread of coyotes (*Canis latrans*) to the peninsula has resulted in a decline in caribou numbers, due to winter kill of calves.

In contrast to the general situation in Atlantic Canada, the population of white-tailed deer (*Odocoileus virginianus*) is declining in Forillon National Park. Suggested causes include coyote predation, the effect of confinement of the animals by Route 197, and the gradual vegetation changes in some areas of the park. The Anse-au-Griffon valley, formerly an agricultural area prior to park establishment, was a favourite haunt of white-tailed deer. Since agricultural activity has ceased, however, the valley is being reclaimed by coniferous forests, more favourable for moose than for deer. Thus, the moose population in Forillon is rising at the same time as deer numbers fall. An additional related factor is the steady increase in the beaver (*Castor canadensis*) population, which has resulted in an increase in wetlands caused by damming, providing terrain more suitable for moose than for deer. The black bear population has also increased in recent years.

Some species of animals survive in Gaspé as a consequence of the extensive limestone beds which mark the major ridges of the Notre Dame Mountains. The peninsula formerly supported a diverse population of bats, which take advantage of the numerous caves developed in the limestone. *Myotis lucifugus* (little brown bat) was the most common of the bat species. Mature individuals commonly have a wingspan of approximately 15–20 cm and a body length of 6–10 cm, and may live for thirty years (Banfield 1981a, b). Bats form a valuable component of the ecosystem, as they effectively control the flying insect population during the summer (a useful service in interior Gaspé, with its shallow ponds and marshy riverbanks suitable for mosquito production). Unfortunately, white-nose syndrome has affected Gaspé bats, and the populations have become endangered (COSEWIC 2013).

Another species which has adapted to life in Gaspé limestones is the Gaspé shrew, *Sorex gaspensis* (Fig. 2.41). Although Gaspé is its main population base, *Sorex gaspensis* also occurs in the Mount Carleton and Woodstock areas of New Brunswick and in limestone areas in central Cape Breton Island, south of the national park (COSEWIC 2006a, b). With a body length of 5 cm at maturity (not including the 6–8 cm long tail), the shrew is small in stature but large in its importance as an insectivore and arachnidovore (unfortunately, there is no elegant word to describe an organism that eats spiders), and as a source of food for small carnivores. *Sorex gaspensis* has evolved for life in limestone terrains with narrow crevices and caves by developing a flattened skull with a long snout, a slender body with very flexible leg joints, and short legs with relatively long toes.

2.7.3 Newfoundland

During the Quaternary, the island of Newfoundland was entirely covered by glacial ice. When the ice began to melt, approximately 14,000 years ago, the relative sea level was higher than at present. As a result, the only species that could reach Newfoundland were those that could swim the (at the time) 30 km wide Strait of Belle Isle, those that could fly, or those that could endure the cold conditions of the Québec North Shore, cross the frozen Strait of Belle Isle during the winter months, and make their way along the Northern Peninsula. Species that did reach Newfoundland tended to evolve in isolation, and thus, several distinctive Newfoundland races or subspecies have developed)(Snow 1996).

Prior to 1600 ce, Newfoundland had only 14 terrestrial mammal species (Snow 1996, Newfoundland and Labrador n.d.). Of these, 9 were predators, insectivores, or scavengers. Some, such as the Newfoundland black bear, were both predator and scavenger. The pine marten (*Martes americana*) was confined to the densest interior forests, and the Newfoundland black bear (*Ursus americanus hamiltonii*) does not appear to have lived in exposed coastal areas. Carnivores and scavengers included lynx (*Lynx canadensis*),

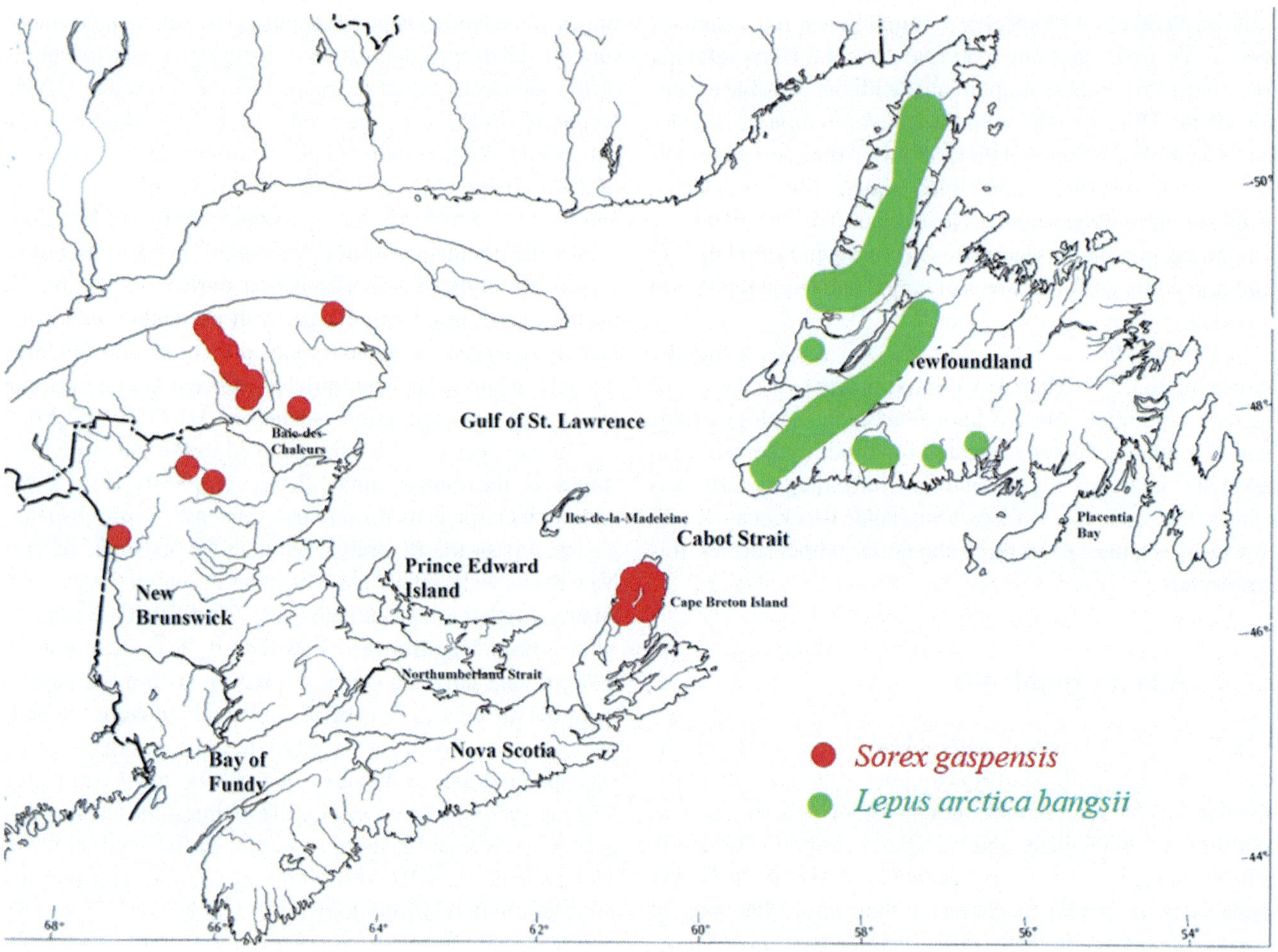

Fig. 2.41 Current occurrences of *Sorex gaspensis* (Gaspé shrew) and *Lepus arcticus bangsii* (Arctic hare) in Atlantic Canada. See text for discussion

red fox (*Vulpes vulpes*), river otter (*Lontra canadensis*), ermine (*Mustela richardsonii*), and Newfoundland wolf (*Canis lupus beothucus*). Small native herbivores and insectivores include the Little Brown Bat *(Myotis lucifugus*), the Northern Long-eared Bat (*Myotis septentrionalis*), the beaver (*Castor canadensis*), the Meadow Vole (*Microtus pennsylvanicus*), and the Newfoundland muskrat (*Ondrata zibethicus*). Maritime species such as white-tail deer, moose, porcupine, skunk, raccoon, squirrels, snakes, and amphibians were not present.

The native Newfoundland wolf was eliminated by human hunting ca. 1926 (Maunder 1982). Although five Labrador wolves and 13 Labrador wolf/coyote hybrids, presumably arriving on sea ice, have been genetically confirmed in Terra Nova National Park and the Baie Verte Peninsula since 2010, there is no evidence that a breeding population has been established. Attempts to use artificial wolf urine (produced in Sweden) to deter moose from approaching the Trans-Canada Highway were not effective.

In most areas of Canada, the lynx's primary food is the snowshoe hare (*Lepus americanus*). The pattern in Newfoundland was somewhat different in 1600, however, because snowshoe hares were not present prior to importation in the mid-1800s. Prior to that time, the lynx had to survive by eating birds, and the few native small herbivores. When snowshoe hares were introduced to Newfoundland, the population of lynx jumped, and it rapidly replaced the wolf as the most common predator until the middle of the twentieth century. Lynx will prey on sheep, and hence many sheepherders have attempted to control the population. This, along with the spread of urban and suburban areas, has resulted in a drop in the lynx population.

The Newfoundland black bear has been expanding its range across the island in recent years. The crossing of the Isthmus of Avalon by the black bears suggests that relatively milder climatic conditions have prevailed since 1990. In the mid-1990s, the black bear reached the easternmost parts of the Avalon Peninsula. Shy and secretive animals, they do not frequent urban areas. The Newfoundland populations are marked by slightly larger body masses (typical mass 135 kg; maximum 270 kg) and skull sizes than mainland populations (Banfield 1981a, b, Newfoundland and Labrador n.d.).

The higher altitudes of the Long Ranges support a small population of *Lepus arcticus bangsii* (arctic hare, Newfoundland subspecies; Hearn and Species Status Advisory Committee 2012). This disjunct population has become separated from the remainder of the arctic hares on the Labrador mainland. The hares probably reached Gros Morne shortly after deglaciation. As the surrounding areas were influenced by progressively milder climates, the higher parts of the Long Ranges became the only region suitable for the arctic hare.

2.7.3.1 Introduced and Invasive Species

The Newfoundland faunal community has been augmented by the arrivals of the black rat (*Rattus rattus*, mid-1600 s), Norwegian rat (*Rattus norvegicus*, ca. 1775), snowshoe hare (*Lepus americanus*, introduced 1864), moose (*Alces alces*, introduced 1878 and 1904), brown trout (*Salmo trutta*, introduced 1883), mink (*Neogale vision*, escapees from Markland, 1937), masked shrew (*Sorex cinereus*, 1958), chipmunk (*Tamias striatus*, introduced 1962–64–68), red squirrel (*Tamiasciurus hudsonicus*, introduced 1964), coyote (ca. 1985), Gapper's red-backed vole (*Myodes gapperi*, 1998), and garter snake (*Thamnophis sirtalis*, ca. 2014). Snowshoe hare and brown trout were deliberately introduced as government policy to provide food sources and entertainment (Fig. 2.42). By providing alternate food sources, the invasions of masked shrew and Gapper's red-backed vole have benefitted predatory species such as pine marten (Strong and Leroux 2014), just as lynx benefitted from the introduction of snowshoe hare.

The lack of natural predators for moose (*Alces alces*) has led to locally severe over-population problems, and excessive browsing by the moose (Strong and Leroux 2014). Terra Nova National Park's relatively small area is insufficient to insulate most large species from the effects of surrounding

Species	Source	Date Arrived
Black Rat	Europe	Mid-1600s (inv)
Norwegian Rat	Europe	<1775 (inv)
Norway Maple	New England	<1800 (int)
Snowshoe Hare	Ontario	1864 (int)
Moose	Maritimes	1878, 1904 (int)
Brown trout	Scotland	1883 (int)
Japanese Knotweed	Ontario	~1890 (int)
Forget-me-not	Ontario	~1900 (int)
Mink	Farmed animals at Markland	1937 (escaped)
Masked Shrew	Maritimes	1958 (inv)
Chipmunk	Ontario	1962-64-68 (int)
Red Squirrel	Maritimes	1964 (int)
Coyote	Labrador	~1985 (inv)
Gapper's Red-Backed vole	Maritimes	1998 (inv)
Garter Snake	Maritimes	2014-16 (inv)

Fig. 2.42 Invasive (inv) and introduced (int) species on the island of Newfoundland. Photographs clockwise from top right: *Alces alces* (moose); Japanese knotweed (*Fallonica japonica*); Forget-me-not (*Myosotis sylvatica*); garter snake (*Thamnophis sirtalis*), Gapper's red-backed vole (*Myodes gapperi*); masked shrew (*Sorex cinereus*)

logging and human settlement. Both Terra Nova and Gros Morne National Parks have initiated public consultations concerning moose culling and management on several occasions.

On the island of Newfoundland, more than 520 moose-vehicle collisions typically occur annually, costing the lives of 2 humans and ~520 moose (NL 2022). A male moose can run at 60 km h^{-1}, covering 16.7 m in one second. Considering a typical human reaction time of slightly more than one second, stopping a typical mid-size car travelling at 75 km h^{-1} on dry asphalt requires approximately 60 m. At 100 km h^{-1}, the stopping distance is approximately 100 m, and the time required about 6.5 s.

In the higher regions of the Long Range Mountains, the grazer *Rangifer tarandus caribou* (woodland caribou; Fig. 2.40) has been able to maintain its population against the competition of the introduced moose). Moose did not arrive in Gros Morne until 1925 and required some time to displace caribou from their mutual range. Moose were not common in the region until after 1945, and were outnumbered by caribou until the mid-1970s. They tend to frequent elevations below 500 m asl, with the exception of tuckamore areas and open coastal barrens. Approximately, 2000 moose live within the Gros Morne region. In the area between 100 and 400 m asl, the modal density of the moose population is about 2 moose /km^2. As in Terra Nova, the boreal forest is suffering greatly from over-browsing by the introduced moose, which has led to the legalized hunting of the species within the park boundary, beginning in 2011 (Parks Canada 2020b).

2.7.4 Freshwater Fish

Salmon (*Salmo salar*) has been a major food source for predators and humans throughout the postglacial history of Atlantic Canada. Salmon spawn in October–November, preferring fine gravel in riffle areas or shallow pools (Scott and Crossman 1998). Deep pools are preferred feeding sites, and hence are the areas where salmon are generally caught, but spawning fish require faster water and coarser sediment. The gestation period is approximately 110 days, and hatching usually coincides with water temperatures of 4 °C or greater. The young fish remain in the gravel after hatching for three to six weeks, emerging in late May–early June. Maturation from hatchling to seabound parr requires 3–4 years in Atlantic Canada. The salmon return, as grilse, after 2–3 years. Although Atlantic salmon can live up to 10 years, and make as many as 4 return journeys to their spawning grounds, most do not spawn more than once. Less than 20% of Atlantic salmon around the northern North Atlantic succeed in spawning more than once.

Over the past 60 years, salmon numbers in many streams have declined due to local overfishing and loss of habitat resulting from river regulation by dams and disturbances of flow and river bottom sediment related to logging activity (Fig. 2.43). Logging disturbances and local over-harvesting are the primary causes of the decline in salmon populations. Despite some enhancement programmes, the productive capacity of potential Atlantic salmon habitats has declined by almost 20% across Atlantic Canada since 1870. However, salmon populations in some rivers have increased, frequently due to joint community-NGO-governmental stocking initiatives.

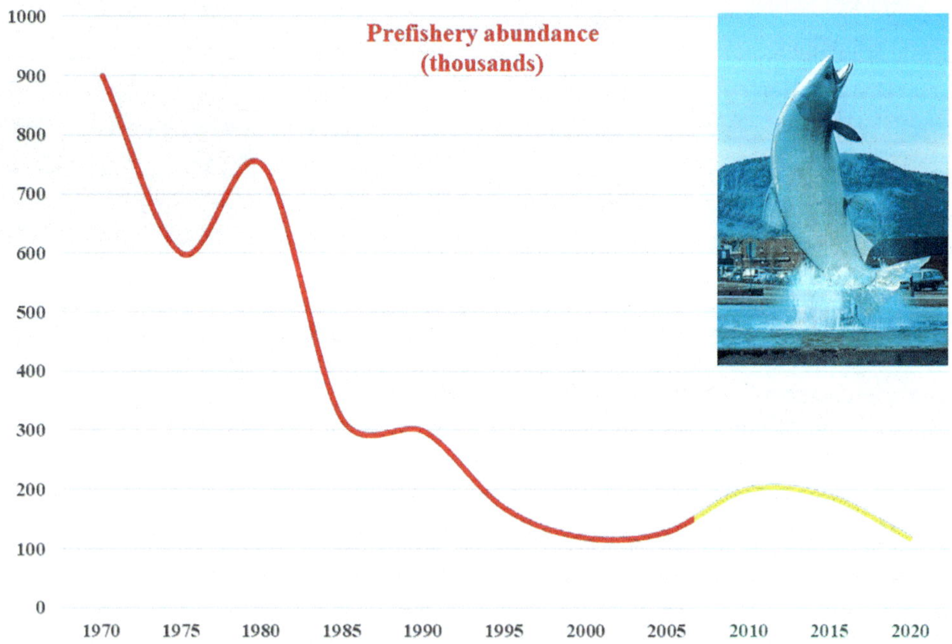

Fig. 2.43 Estimates of prefishery Atlantic salmon (*Salmo salar*) population, Atlantic Canada. Data from Atlantic Salmon Federation reports. Inset: Salmon angling is an important recreational and tourist activity in Atlantic Canada. This statue is located in Campbellton NB

In addition to salmon, native freshwater fish species include Arctic char (*Salvelinus alpinus*; confined to Newfoundland and rare), brook trout (*S. fontinalis*; common), rainbow smelt (*Osmerus mordax*), and eel (*Anguilla rostrata*) (Scott and Crossman 1998). Brown trout (*Salmo trutta*) and steelhead trout (*Salmo gairdneri*) have been introduced.

Brackish waters surrounding Prince Edward Island are inhabited by the Mummichog *(Fundulus heteroclitus)*, a small fish 3 cm in length (Scott and Crossman 1998; Finley et al. 2013). The mummichog cannot survive in water temperatures below 1 °C, and hence, its presence indicates the extent of the warm coastal waters derived from the Gulf Stream. In Atlantic Canada, it is also present along the west coast of Cape Breton Island, along the Iles-de-la-Madeleine, in the Bay of Fundy, in Port-au-Port Bay, and on Sable Island, but populations are largest along the north coast of Prince Edward Island.

2.8 Marine Environments

Atlantic Canada is washed by coastal currents derived from the St. Lawrence and other river systems, and by the Gulf Stream and Labrador Current. The main flow of the Gulf Stream occurs beyond the shelf-slope break, although interaction influences Maritime coastlines and southern Newfoundland west of Cape Race.

2.8.1 Currents

The Gulf Stream is the strongest current in the North Atlantic, with a mean discharge of 55 million m^3 s^{-1} (Stommel 1966, Colling 2001; Knauss and Garfield 2016; Garrison and Ellis 2017). The surface pressure gradient slopes northeast from the Caribbean, so that mean absolute sea level (as measured from space is ±0.5 m higher in the Caribbean than in the Barents Sea. This provides impetus for northeastward water flow. The Gulf Stream meanders, with a width of 50–75 km and a peak velocity of 15 km h^{-1}. August sea surface temperatures average 17–18 °C off Maritime Canada, with February SST of 9–10 °C. Mean surface salinity approximates 35 psu.

As the Gulf Stream meanders, it forms rings, analogous to oxbows in a terrestrial river. Typically, 18 to 22 rings form each year (Auer 1987). Rings detached on the northwest (landward) margin have warm cores, allowing more southerly species such as tuna to survive. Cold core rings develop on the southeastern (oceanic) margin. The rings drift southwestwards at 8–12 km d^{-1}, and may persist for as long as 5 years after formation (Silver et al. 2021).

In the North Atlantic Gyre, approximately 15% of the discharge forms the Norwegian Current, gradually diverted to the west along the Kalaallit Nunaat (Greenland) coast. Water circulates along the eastern shores of Davis Strait and Baffin Bay, to approximately 75°N (North Water, see Chap. 9). Augmented by cold, fresh water from glaciers, flow returns southward as the Labrador Current.

In comparison to the Gulf Stream, the Labrador Current carries much less water (~6 million m^3 s^{-1}), is 10–20 km wide, and is much slower (1 km h^{-1} at Strait of Belle Isle, 2–3 km h^{-1} at Cape Race). One of slowest currents in the world ocean, the Labrador Current's southward flow is hindered by the pressure gradient. On occasion, the current can reverse to flow northward, or not flow with measurable velocity (see Chap. 8). August SST off St. Anthony and Cape Race are typically 8-9 °C and 12–13 °C, respectively, with mean salinity of 32 psu.

2.8.2 Offshore Regions

The open North Atlantic offshore of Nova Scotia (Scotian Shelf) and southern Newfoundland consists of a series of shallow banks (35–120 depth) separated by deep submarine canyons (Breeze and Horsman 2005; Fig. 2.44). Georges Bank, the furthest southwest, is divided politically between the USA and Canada. Browns, German, Baccaro, LaHave, Emerald, Western-Sable Island, and Banquereau Banks flank the Nova Scotia coastline. The deep Laurentian Channel (to 550 m) extends southeast from the Gulf of St. Lawrence, separating the Scotian Shelf from St. Pierre Bank and the Grand Banks to the east (Todd 2016c). The banks and canyons provide habitat for a wide variety of marine species. Several canyons, included within marine protected areas, have assemblages of cold-water corals and sponges.

The geomorphology of the continental shelf and slope results from the bedrock geology, in combination with Quaternary glacial and sea level history (Piper et al. 1990, Todd et al. 1999; Harris et al. 2014; Normandeau et al. 2020). The shelf-slope break has remained in the same position since the later Jurassic (Pe-Piper et al. 2007). Glaciation reached the edge of the Continental Slope along the Scotian Shelf (Todd 2016a). Glacial landforms including DeGeer moraines are preserved on the surface of German Bank, to depths of 100 m below modern sea level (Todd 2016b). Submarine canyons, developed where slopes exceed 3°, are marked by slope failures along the margins. During deglaciation and consequent sea level rise, sediments were reactivated in downslope turbidity currents and along-slope contour currents (Campbell and Mosher 2016; Mosher et al. 2017). Although transport of sediments in the canyons continues, there is no evidence of active turbidity currents at present.

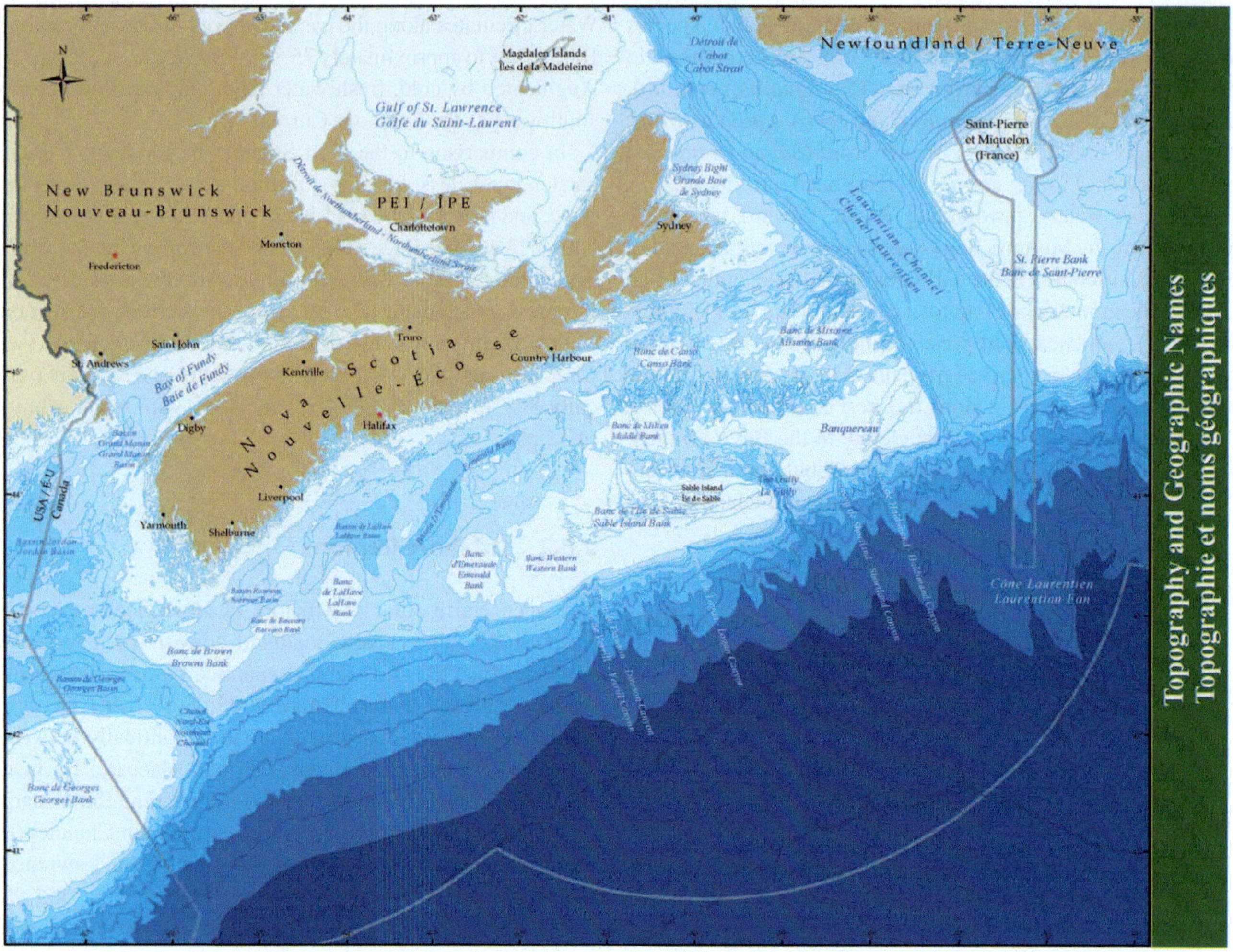

Fig. 2.44 Bathymetry of the Scotia Shelf region. (Fisheries and Oceans Canada)

2.8.2.1 Sable Island

Sable Island National Park Reserve (established 2013) is located 300 km southeast of Halifax. The island (42 km long, 1.5 m maximum width) is the only spot on the offshore banks that remains above relative sea level, although the island is currently undergoing relative sea level rise and erosion (Fig. 2.45). Sable Island developed on the surface of a terminal moraine, formed under lower relative sea level during the Last Glacial Maximum.

Visitation is tightly controlled, requiring a permit from Parks Canada (Parks Canada 2019). Sable Island is noted for its history of shipwrecks (at least 350 since 1583) and its population of feral horses, descended from those confiscated from Acadian settlers during the deportations, beginning in 1755 (Christie 1995, de Villiers and Hirtle 2004). The island is occupied by a transverse coastal dune complex, subject to erosion due to storm action and overgrazing. Production of natural gas from offshore wells ceased in 2018.

2.8.2.2 Georges Bank

Georges Bank was the subject of a prolonged boundary dispute between Canada and the USA, as both countries wished to control the lucrative scallop *(Placopecten magellanicus)* fishery, as well as reserving submarine terrain for hydrocarbon exploration (Van der Zwaag 1983, Chircop et al. 1984, 1995, IJC 1984, Russell 1992, Byers and Østhagen 2016; Fig. 2.46). Scallops undergo optimal growth and spawning in water temperatures of 10–18 °C, and live at depths of less than 150 m (Stewart and Arnold 1994; Packer et al. 1999). The shallow banks, bathymetrically isolated from the warmer Gulf Stream waters to the southeast and having low turbidity, provide ideal scallop habitat.

Traditionally, the Georges Bank scallop beds were fished by vessels based in Georges Bank scallop beds fished by vessels based in Digby and Yarmouth (NS), and in Gloucester and New Bedford (Massachusetts). Digby and Gloucester were dominated by family-owned enterprises,

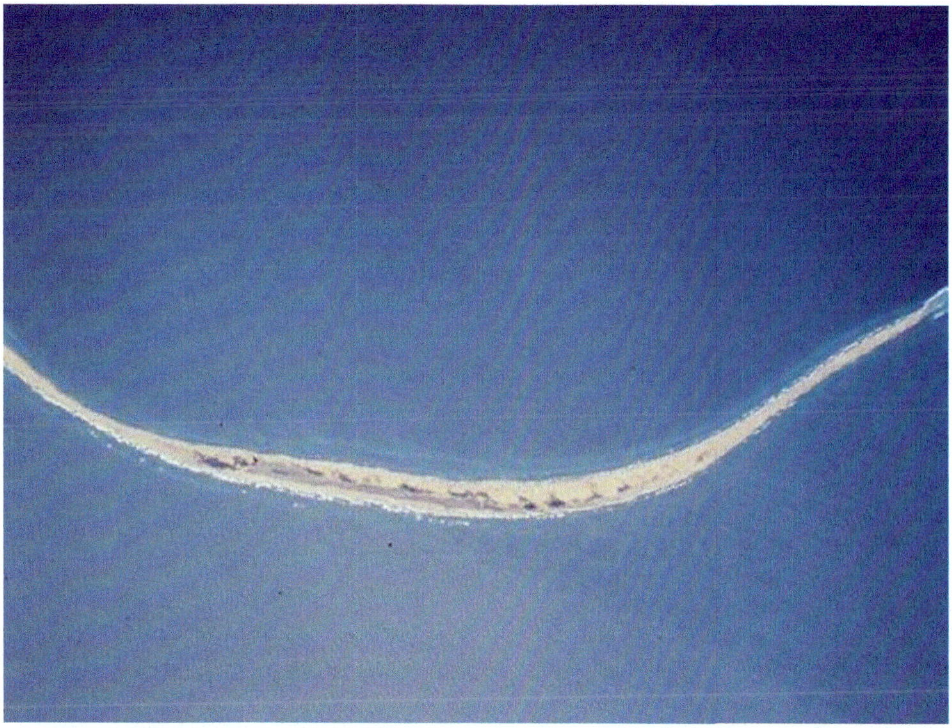

Fig. 2.45 Sable Island National Park Reserve. (NASA)

while Yarmouth and New Bedford supported larger commercial vessels. After the two countries filed overlapping Exclusive Economic Zone (EEZ) claims, the dispute was referred to the International Court of Justice (IJC) in 1979.

The 1984 verdict (IJC 1984) indicated that geographic adjacency better expressed the link between a State and its submarine entitlements than did "natural prolongation" (i.e., the configuration of the continental shelf). The court rejected the historical ties and vital economic interests arguments raised by both countries, as well as rejecting Canada's contention that Nantucket and Cape Cod were "geographical anomalies" that could be discounted, and the USA contention that Georges Bank should be considered as a non-divisible ecological region. The demarcation line was based on the principle of equidistance modified by the length of coastline of each country, setting a precedent for delineation of marine boundaries elsewhere. Canada, which had claimed 33% of the bank, received approximately 16%. The USA had claimed the entire Georges Bank.

Canada and USA had agreed to a common starting point for the offshore boundary southwest of the small Machias Seal Island, which was thus excluded from the IJC decision. The 700 km^2 of lobster habitat surrounding Machias Seal Island ("Grey Zone") remains in dispute.

The Gulf of St. Lawrence includes the shallow St. Lawrence Platform (50–100 m depth) on the southwest, flanked by the deeper Laurentian Channel extending from the mouth of the St Lawrence Estuary to the Continental Shelf, including Cabot Strait (104 km minimum width, 480 m maximum depth) between Nova Scotia and Newfoundland (Breeze and Horsman 2005; Normandeau et al. 2015). The deep-water 4000 m Laurentian Fan, including a Holocene slope failure of more than 14 000 km^2, marks the end of the channel at the Continental Slope (Todd 2016c). Northumberland Strait, separating New Brunswick and Prince Edward Island, has a minimum width of 13 km and a maximum depth of 65 m. The Abegweit (formerly Confederation) Bridge was completed across the strait in 1997. To the northeast of the Laurentian Channel, the tributary Anticosti and Esquiman Channels separate shallower banks along the western Newfoundland and eastern Anticosti Island coasts.

2.8.2.3 Northern Right Whale

The main shipping lane across the Gulf of St. Lawrence coincides with the spring and summer feeding areas of the critically endangered Northern Right Whale (*Eubalaena glacialis*). The Northern Right Whale was slow, social, contained large quantities of oil, preferred shallow coastal waters, and floated when harpooned (Kenney 2009; Baumgartner et al. 2017). All of these attributes made it the "right" whale to kill, and it invariably was the first species eliminated as whalers moved into new territories. Never numerous, with an estimated population off New England of 1000–2000 ca. 1600 ce (Monsarrat et al. 2015), it was no longer economic to pursue Right Whales in the region by 1750. With an estimated population of 100, the right whale was the first whale species to be protected from whaling, in 1937, but numbers have not shown any significant recovery.

Fig. 2.46 Georges Bank was the subject of adjudication between Canada and the United States of America by the International Court of Justice in 1984. The map shows the final boundary settlement, along with the location of the major family (outline star) and large-scale commercial (closed star) scallop ports of Digby NS, Yarmouth NS, Gloucester MA, and New Bedford (MA). Inset top left: Machias Seal Island (United States Coast Guard). Right: *Placopecten magellanicus* (Atlantic scallop), showing life position (upper), shell (middle), and fate as human food (lower)

Vessels are currently restricted to 10 knots (18.5 km h^{-1}) maximum speed.

2.8.3 Tidal Features

Tidal fronts move counterclockwise in the Gulf of St. Lawrence (Dohler 2017). Mixed, mainly semi-diurnal mesotidal conditions dominate the southern Gulf of St. Lawrence coastlines of New Brunswick and Prince Edward Island, and the northern part of the Northern Peninsula. Semi-diurnal microtidal regimes are prevalent along most of the west coast of Newfoundland, and in eastern Northumberland Strait. Diurnal regions are located in western Northumberland Strait, and to the southwest of the Iles-de-la-Madeleine in the Gulf of St. Lawrence. The Gulf of St. Lawrence semi-diurnal nodal point is located to the west of the Iles-de-la-Madeleine.

2.8.3.1 Southern Gulf of St. Lawrence Coastline

The New Brunswick southern Gulf of St. Lawrence coastline is marked by long, shallow mesotidal embayments. "Kouchibouguac" can be translated from the Mi'kmaq as "river of long tides". Salt marshes are developed in most tidally influenced lagoons and estuaries (Fig. 2.47).

Typically, the seaward margin of a saltmarsh is dominated by vascular salt-tolerant plants such as *Salicornia* (samphire), *Spartina* (saltwater cord-grass), *Festuca* (fescue), and *Sueda* (sea blight). The higher areas support *Carex* (sedge), *Myrica gale* (sweet gale), *Juncus* (rush), and *Iris*. Saltmarsh vegetation can reduce incoming wave heights by up to 70%, and tidal energy by more than 90%, resulting in deposition of sediment,

Fig. 2.47 Salt marsh, Alberton, Cascumpec Bay PEI

gradual buildup of the marsh, and insulation of the shoreline from anomalous high tides and storm surges (Ashall et al. 2016; Silinski et al. 2016, Catto 2022c).

Saltmarsh development is influenced by sediment influx, the rate of sea level change, ice scouring, and human activity (Chmura and Hung 2004, Catto 2022c). Organic detritus and sediments accumulate in saltmarshes at rates of 1–10 mm per year. If sea level rises at a comparable rate and sediment influx is maintained, the saltmarsh will continue to develop, as long as there is available and suitable terrain inland. However, if the sediment influx is reduced, the saltmarsh will begin to erode along the seaward edge. Removal of the vegetation will accentuate the process of erosion. If the saltmarsh is isolated from tidal influx (e.g. by constructing an embankment), the lack of sediment influx and tidal flushing will result in the demise of the marsh.

The other diagnostic coastlines along the southern Gulf of St. Lawrence are the barrier island and linear spit complexes and protected lagoons (Dubois and Grenier 1993, Ollerhead and Davidson-Arnott 1995; Catto 2015). Barrier islands develop as sediment is transported parallel to the shore by littoral drift (Fig. 2.48). Sand is brought to the beach by incoming swash waves, which are then refracted seaward as backwash. As swash and backwash occur repeatedly, the sand gradually moves parallel to the shoreline. Barrier island development requires gentle offshore bathymetric slopes, a large supply of sand, and gradually rising sea level. Examples are present at Kouchibouguac NP and Les dunes-de-Bouctouche (NB); Cascumpec Bay, Pītaweikek (Sandhills), Malpeque Bay, and Prince Edward Island National Park (PEI), and Iles-de-la-Madeleine (Québec; Fig. 2.49).

The mesotidal barrier islands along the Gulf of St. Lawrence are flanked by deep tidal channels. The strong tidal currents produce a cellular pattern of littoral drift, rather than a continuous conveyor. The sequence is split into individual barrier islands with relatively high dunes, which shelter the lagoon from storm activity. When storm surges occur, they tend to be funnelled into the tidal channels, further deepening these passes. The net result is that the tidal channels do not migrate, allowing them to serve as consistent passageways from lagoon to open ocean for sediments, tidal currents, fish, and human vessel traffic.

Transverse dunes are formed at approximately 90° azimuth (normal) to the direction of wind motion. As the winds moving the sand and building the dunes are primarily blowing onshore, there is a limited influx of sand currently available for dune replenishment.

Development of the dune-backed coasts is controlled by a combination of littoral current and sediment movement, aeolian activity, sea level rise, and human actions. A system of coastal barrier beaches depends upon having an ample supply of sand to replace that transported laterally by littoral drift. Sand supply has been curtailed in some areas by the erection of docks and breakwaters, interrupting the course of littoral drift. Rising sea level is also resulting in coastal erosion. Some of the sand necessary to maintain the beaches

Fig. 2.48 Barrier island coastlines, PEI. Upper left: Cascumpec spit and lighthouse; Upper right: tidal inlet, Conway sand flats; Lower right: overwash fans, New London Bay, PEI National Park; Lower right: attached spit, Malpeque

is stored temporarily in the foredune systems. Thus, any loss of sand from the foredunes, or transport of sand inland, has a negative impact on the health of the beaches. Ongoing relative sea level rise has gradually limited the scope for inland expansion, simultaneously eroding the dunes on the seaward side, while restricting migration on the landward side.

2.8.3.2 Pītaweikek

Flanking the northern side of Malpeque Bay, the Pītaweikek (Sandhills) barrier island and dune complex is one of the least disturbed coastal dune complexes in Atlantic Canada (Catto and Catto 2009; Fig. 2.50). Pītaweikek ("the place where tea is found") has great traditional importance for Mi'kmaq cultural heritage. The terrain is valued as a source of berries and medicinal plants, as a base for fishing, for camping, and for ceremonial uses. The cultural importance of the Pītaweikek landscape has encouraged interest in its conservation, particularly as many other areas of the north coast of PEI have been rapidly modified by development. Pītaweikek has been proposed as a national park.

Comparison of the relatively undisturbed Pītaweikek complex with other more disturbed dune-backed coastal areas of Prince Edward Island and elsewhere in Atlantic Canada (Catto 2002, 2006b; Catto et al. 2020b) indicates that the terrain has retained most of its natural landscape. Dunes at the adjacent Cabot Beach Provincial Park have receded more than 3 m since 1990, predominantly due to erosion generated by human activity (Catto et al. 2002). Increased tourism without appropriate forethought will result in significant damage from erosion of the dune complex. The example provided by the Greenwich unit of Prince Edward Island National Park indicates that the potential does exist to simultaneously promote tourism and allow beach access without causing significant damage to the dunes, conserving the ecological integrity of the terrain.

Fig. 2.49 Iles-de-la-Madeleine, Québec, showing barrier dunes connecting Permian sandstone islands. (NASA Earth Observatory)

2.8.3.3 Bay of Fundy

In Atlantic Canada, the semi-diurnal tide progresses southwestward, requiring approximately 4 h to travel from Cape St. Francis to Yarmouth (Dohler 2007). The Bay of Fundy is remarkable for its tides, among the highest in the world. The name "Fundy" comes from the French word "fendu", meaning "split", derived from Cape Split which separates the Minas Basin from the main, western extent of the bay.

The funnel-like configuration of the shores and bathymetry of the bay focus the incoming tide, causing the tidal range to rise progressively through the bay, from approximately 6 m at the bay mouth to 12–14 m in the Minas Basin (Greenberg et al. 2012; Archer 2013; Catto 2015). The record tidal range in the Bay of Fundy is 31.6 m, observed at Burntcoat Head, Nova Scotia. The hypertidal activity in the bay has a major influence on the region's geomorphology and human utilization of the coast (Catto 2020c; Fig. 2.51).

The tidal bore of the Petitcodiac River estuary (NB) can be seen along the length of the estuary, but is most pronounced in the narrowest reach, adjacent to Moncton. The height of the bore varies with the strength of the tide, but generally is between 50 cm and 1 m. Construction of a causeway across the Petitcodiac River at Moncton has interfered with the water circulation in the upper part of the estuary, causing an overall reduction in bore height (Locke et al. 2003; Gerwing et al. 2017).

Tidal currents result in erosion and transportation of unconsolidated sediments when flow exceeds 10 cm s^{-1} (0.36 km h^{-1}). As tidal velocities in confined channels along Fundy regularly exceed this value, channel bottoms are scoured. Channels frequently are flanked by broad tidal flats, coated with sand and silt, which extend seaward for tens of metres to kilometres. On Fundy's tidal flats, sediment is provided primarily from offshore and coastal sources, rather than fluvial influx. Deposition consists of highly fluid mixtures of sand, mud, and water. The sandy substrates are very suitable habitats for burrowing molluscs (soft-shell clams, quahogs), and the flats provide important resting and feeding areas for seabirds during low tide.

Erosion by the tidal currents at Cape Hopewell, where the incoming currents are forced to turn to the northwest up the estuary, has resulted in the formation of Hopewell Rocks (Trenhaile et al. 1998; Catto 2020c). These pillars, up to 25 m high, are composed of relatively easily eroded, red Carboniferous conglomerate (Fig. 2.52). Erosion by frost wedging along a system of interconnected fractures aligned

Fig. 2.50 Pītaweikek PEI is a conserved area managed by the Mi'kmaq Confederacy. Upper left: relatively undisturbed coastal sand dune landscape north of Georges Island. Upper right: Saucer blowout in parabolic dune, resulting from natural aeolian activity. Lower left: small parabolic dune landscape is valued as a source of berries and medicinal plants. Lower right: deflation hollow resulting from excessive foot traffic indicates the sensitivity of the landscape to human usage. Access to Pītaweikek courtesy of Mi'kmaq Confederacy, Lennox Island

both parallel to and at 90° to the coast resulted in the gradual detachment of the pillars from the cliff, giving the pillars their characteristic blocky, rectangular cross-sections. Tidal action then eroded the bases of the pillars, producing the characteristic "flowerpot" shape with blocky columns supported on slender, hourglass-shaped bases 8–10 m high. Tidal erosion in the area has been ongoing almost since deglaciation.

Although the bases appear fragile, the rate of erosion of the bases of the pillars is relatively slow. Incoming tidal currents are largely focused on the opposite (southeastern) bank of the estuary, and outgoing currents tend to flow most strongly in the centre of the channel. "Decapitated" pillars are very uncommon. The rock falls present along the bases of the cliffs are largely the products of frost wedging on the slopes above, rather than being due to destruction of flowerpot pillars by erosion at high tide level.

Exploitation of potential tidal power resources of Atlantic Canada has been discussed for many years. Although proposals have been floated for developments at East Point (PEI), the Bras d'Or Lakes (NS), and the Strait of Belle Isle, Rigolet, and Cartwright (NL), most of the interest and controversy has centred on the Bay of Fundy (Karsten et al. 2008; Catto 2020c). The experimental 50 GWh/y barrage facility at Annapolis Royal (NS), operational since 1984, is the only functioning power plant. Numerous proposals have been made to exploit tidal energy in the Western Passage of Passamaquoddy Bay (NB-Maine) and in the Minas Channel (NS).

Although initial proposals involved barrage construction, the ecological and financial consequences are considered excessive, and current interest focuses on anchored in-stream turbines. At Annapolis Royal, an existing dam was retrofitted to allow increased tidal influx and generate power. Modi-

Fig. 2.51 Tidal features, Bay of Fundy. Upper: Tidal flat at low tide, Blue Beach NS (left); detail of sandy sediment suitable for mollusc breathing holes (right). Right: tidal channels at Kennetcook NS show steep banks and meanders. Larger submarine dunes and sand waves (lower right) form during storm events. Lower left: tidal channel, flat, and ebb-tide delta, Port Williams NS

fied tidal activity resulted in the harbour and the ingrown Annapolis River, which flows through weak Triassic clastic bedrock mantled by Quaternary silt.

2.8.4 Strait of Belle Isle

The Strait of Belle Isle (length 125 km, minimum width 17.4 km) separates Newfoundland from Labrador. Maximum depths exceed 60 m. Sea ice is generally present for 9 months in a typical year, and ferry navigation between St. Barbe (NL) and Blanc-Sablon (QC) is generally restricted to 8 months from May through late December (Fig. 2.53). A submarine cable linking the Muskrat Falls Hydroelectric project to Newfoundland has been constructed. Proposals for a road or rail tunnel ("Fixed Link") across the strait have been suggested at various times since 1949.

The Strait of Belle Isle shoreline is dominated by erosion under the combined actions of sea ice and tidal action (Catto 2011, 2020a). Tidal range along the central Northern Peninsula coast varies from 2.5 to 3.5 m. Offshore, a platform underlain by Ordovician rock and covered by gravel slopes towards the northwest at a gentle angle. The development of the rock platform offshore was primarily the result of frost weathering during intervals of lower sea level when the rock was exposed, rather than being the product of direct abrasion by waves, tides, or sea ice.

Seasonal ice shove is common. Warmer winters in the southern Gulf of St. Lawrence, resulting in part from strong southwesterly winds, lead to restricted ice cover. However, southwesterly winds tend to push sea ice northeastwards along the Northern Peninsula, resulting in persistent ice cover north of St. Barbe, even during years when ice cover has been restricted or non-existent south of Bellburns.

Fig. 2.52 Cape Hopewell NB displays tidally scoured Carboniferous sandstone pillars. The mean high tide position is indicated by the top of the smoothed supporting column. The rock above high tide is subject to frost wedging. The intertidal zone (lower left) is dominated by coarse sand and gravel

Sea ice can act as an erosional agent. In tidally influenced areas, sea ice can be driven onshore on each cycle, resulting in the production of distinctive geomorphic features. Repeated thrusting of slabs of sea ice results in remobilization of pebbles, cobbles, and boulders, acting to remove sediment and expose wide rock platforms. In some locations, the influence of seasonal ice shove results in the construction of a line of boulders (a boulder rampart), or the emplacement of individual boulders perched on the rock substrate.

L'Anse aux Meadows at the southeastern edge of the Strait of Belle Isle is the only definitively authenticated Viking occupation site in North America, although many other candidates have been suggested. The short Norse period of occupation is currently estimated at ca. 1015 ce. L'Anse aux Meadows as a UNESCO World Heritage Site (Bell et al. 2003, Slaymaker et al. 2020b).

2.8.5 Offshore Newfoundland

Burgeo Bank and St. Pierre Bank, rising to within 30 m of the sea surface, flank the Laurentian Channel and Hermitage Channel (343 m maximum depth) to the east (Fig. 2.54). The major fishing grounds of St-Pierre Bank are separated from St-Pierre-et-Miquelon by the 140 m deep St-Pierre Channel (Simms et al. 2007).

St-Pierre-et-Miquelon (Fig. 2.55) has been under French control, with the exception of the period 1713–1763, since its initial settlement in the mid-1600 s. The territory was traded to Britain by France under the 1713 Treaty of Utrecht, in return for the west coast of Newfoundland (the "French Shore"). In 1763, France regained St-Pierre-et-Miquelon in the Treaty of Paris. With its population of approximately 6000, and a total area of 240 km^2, St-Pierre-et-Miquelon is a political division of France, a "collectivité territorial" (since

Fig. 2.53 Sea Ice, Gulf of St. Lawrence, 11 February 2013. (NASA Earth Observatory)

1985). The designation was created specifically for St-Pierre-et-Miquelon, to solve political problems for France while attempting to answer the needs of the St-Pierrois.

St-Pierre's location adjacent to Newfoundland resulted in a series of marine boundary disputes and discussions about fishing rights which were somewhat resolved by mediated settlement in 1992–1994 (Day 1992). Disputes still exist, however, in particular with respect to the seaward extent of the economic zone, and the exploitation of pelagic fish stocks. Britain had never considered St-Pierre to be part of either Newfoundland or mainland British North America, and hence Canada has no possible claim to St-Pierre-et-Miquelon. The attitude of successive Canadian federal governments since 1949 has basically been one of minimal interference, few political initiatives, and "damage-control" style reaction to any problems raised by St-Pierre or France.

The Grand Banks have water depths of 15 to 91 m (Simms et al. 2007). Exposed during the Last Glacial Maximum (Shaw et al. 2002), the surface is marked by oxidized gravel and relict periglacial features. However, the offshore position and harsh glacial climate prevented the Grand Banks from serving as a biological refugium. Never glaciated and isolated from terrestrial sediment influx, the sediment overlying the Grand Banks is limited to deposition from suspension in the Labrador Current. The margins are reworked by contour currents (Rashid et al. 2017). The shallow waters also

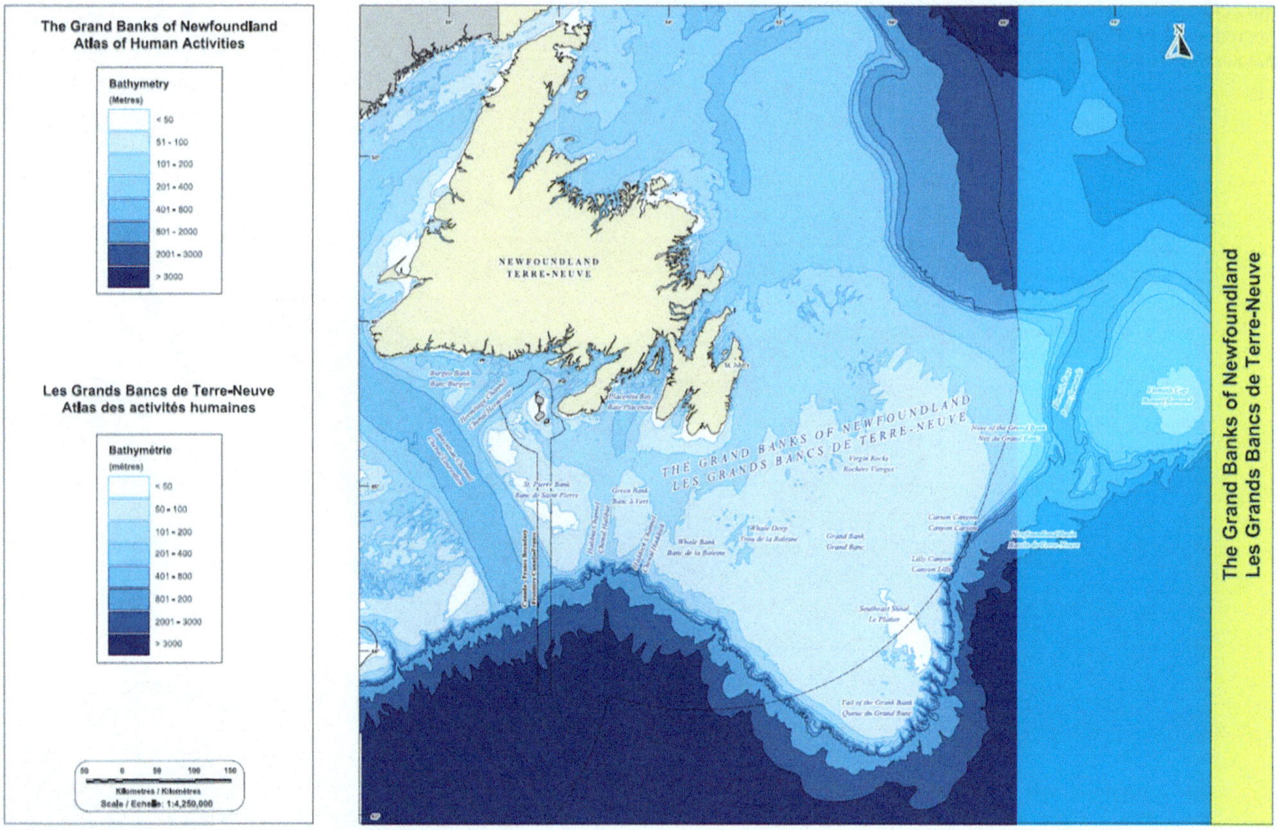

Fig. 2.54 Bathymetry of the Grand Banks. (Fisheries and Oceans Canada)

limit iceberg movement, with the bergs passing offshore of the Banks, over the resting place of RMS Titanic (Cochonat et al. 1989).

The majority of the Grand Banks lies within Canada's proclaimed EEZ (370 km or 200 nautical miles; Simms et al. 2007; Fig. 2.53). Two areas, the "Nose" (eastern extremity) and "Tail" (southern extremity) lie beyond the EEZ limit. Both are currently fished by several states (including Canada) under the auspices of the Northwest Atlantic Fisheries Organization (NAFO). Canada has laid claim to the Nose and Tail, arguing that they are parts of the contiguous Continental Shelf. Flemish Cap (122 m minimum depth) is located offshore of the Nose, separated from the Grand Banks by Flemish Pass (Piper and Campbell 2005). Orphan Knoll (1600 m minimum depth) lies outside the EEZ to the north, separated from the Grand Banks by the Orphan Basin. Further north, the Bonavista Shelf, cut by glacial outlet valleys extending from Trinity, Bonavista, and White Bays, flanks northeastern Newfoundland offshore.

Newfoundland and Labrador's offshore petroleum industry involves exploitation of Jurassic and Cretaceous units in the Flemish Pass, Jeanne d'Arc Basin, and Orphan Basin, all within the EEZ. Production began in 1998 and is administered by the joint Canada-Newfoundland and Labrador Offshore Petroleum Board (C-NLOPB).

2.8.6 Marine Fisheries

The Grand Banks cod (*Gadus morhua)* fishery has been central to Newfoundland's economy, culture, and psychology since the initial exploitation by Basque, Portuguese, and Spanish fish harvesters (*Bacalao*). In Newfoundland, the word "fish" refers to cod: other species (e.g. capelin, halibut, salmon) are identified separately. Cold waters from the Labrador Current and the island of Newfoundland interact with the warm Gulf Stream to create upwelling, nutrient blooms, and suitable habitats for plankton, invertebrates, pelagic, and benthic species.

Initially, fishing involved small vessels, with local processing (Ryan 1990, Hutchings and Myers 1995, Wright 1998, Rose 2003, Higgins 2007, Rowshvra et al. 2020). Most commonly, fish were split from nose to tail into slabs, and sun-dried on wooden platforms ("flakes") in waterside sites ("rooms"). Salt was in short supply, and generally was used sparingly, if at all. This "dried" processing method distinguished Newfoundland cod from those packed whole in salt ("wet fishery") for transport. It also resulted in a large store of dried fish awaiting transport each autumn, committing the local fish harvesters to much work throughout the summer with an uncertain return. As Newfoundland-based merchants took over the business of supplying necessary goods in Spring

Fig. 2.55 St-Pierre-et-Miquelon is a collectivité territorial of France. (International Space Station Image, NASA, USA)

and buying dried fish from harvesters in Fall, both groups became enmeshed in the "Truck" system of revolving credit.

The inshore and nearshore cod fishery harvested only mature fish. These fish, born in coastal waters, migrated north after birth, and did not return until reaching maturity, a process requiring 8–12 years. Although the odds against maturation are high (an estimated 1 of every 2 million eggs develops into a mature cod), fishing of only mature individuals allowed the spawning stock to be maintained.

Development of larger-scale operations after World War II and Confederation (1949), accompanied by Canadian factory trawlers, onshore cold storage and freezing plants, and increased fishing by foreign vessels, resulted in fundamental changes to the fishery, rippling through economy, culture, and demographics. Overfishing was the primary cause of cod stock depletion, culminating in the imposition of the Northern Cod Moratorium in July 1992. The Moratorium represented a major turning point in the geography of Newfoundland and Labrador, both cultural and physical.

Fisheries mismanagement has been ubiquitous, attributable to overfishing, overcapacity of fish plants, competition between jurisdictions, political pressure, lack of scientific knowledge, and especially lack of communication between the participants involved in all aspects of managing the cod fishery. Atlantic cod and haddock catches declined 69% and 80% respectively between their historic highs in the late 1960s and 1992. The discard rates are estimated at 40% for deep-water fisheries. Canada had an overcapacity in fisheries vessels and gear of approximately 35% in 1992.

The difficulties and lack of a fully integrated management strategy are compounded by the vagrancies of changing temperatures and oceanological fluctuations. By 1991, the

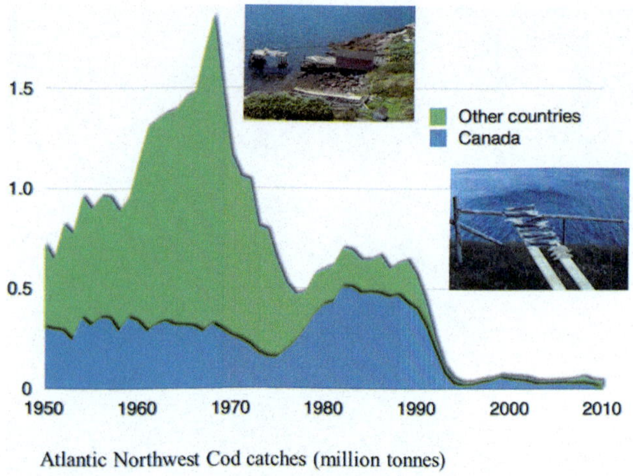

Fig. 2.56 Atlantic Northwest Cod (*Gadus morua*) catches, 1950–2010. The Northern Cod Moratorium was imposed in July 1992. (Fish Statistics Database, FAO, United Nations.). Inset right: cod drying on flake, Bacon Cove. Avalon Peninsula, July 1991. Inset upper: Abandoned and unmaintained flake and stage, Placentia Bay, July 2006

stock had crashed, and signs of recovery have remained limited since (Fig. 2.56). Management of the cod stocks is complicated by the numerous small stocks present in isolated embayments and estuaries. These "bay cod" are non-migratory, and their numbers reflect conditions and fishing effort within each individual bay. In some bays, stocks did not decline, leading to frustration by local fish harvesters at restrictions.

Other groundfish species include *Melonogrammus aeglefinus* (haddock) and *Pollachius virens* (pollock). Ocean climate variations have led to expansion and contraction of the ranges of these species (Chmura et al. 2008). Flatfish species include *Glyptocephalus cynoglossus* (witch flounder) and *Hippoglossoides platessoides* (plaice). *Hippoglossus hippoglossus* (halibut) inhabits waters over the shallow banks (70–90 fathoms, or 140–180 m depth). *Limanda ferruginea* (yellowtail flounder) and *Pseudopleuronectes americanus* (winter flounder) are also present (Scott and Scott 1988).

Reinhardtius hippoglossoides (turbot) has gradually replaced halibut and plaice as a significant commercial target, as numbers of the former species have declined. Turbot is now a highly prized species in Europe (Bowering and Brodie 1995). The preference for small fish has led to concerns that the harvesting of immature turbot will damage the breeding stock.

Clupea harengus (Atlantic herring) are particularly susceptible to changes in water temperature and salinity (Scott and Scott 1988). Salinity levels below 35.2 ppt result in almost total mortality of pelagic herring larvae.

Scomber scombrus (mackerel) have not traditionally been regarded as a major food source or a desirable fish by Newfoundlanders and Labradorians. However, the decline in other fish stocks has recently stimulated interest. Mackerel prefer water depths of less than 200 m, but are not found in estuarine or brackish water areas (Scott and Scott 1988). Salinity fluctuations affect mackerel and herring equally severely. In mackerel fisheries elsewhere in the North Atlantic, loss of a single year class due to adverse salinity conditions has triggered recurring problems in subsequent years.

Other species found throughout the coastal waters include *Sebastes marinus* (redfish), *Myoxocephalus scorpiodes* (Arctic sculpin, or kanayuk), and *Cyclopterus lumpus* (lumpfish). Pressure on these "underutilized" species has grown as cod and flatfish numbers have dropped.

One of the keystone fish species is capelin (*Mallosus villosus*). Requiring water spawning temperatures of 5–8 °C, capelin is the characteristic pelagic finfish of northern North Atlantic waters and is a primary food source for cod, haddock, seabirds, and seals. As a major prey species, capelin stocks thus have a significant impact on cod numbers (Carscadden et al. 2000; Anderson et al. 2002; Rose 2005). Newfoundland and Labrador capelin represent one of four distinctive stocks in the northern North Atlantic and do not intermingle with the Barents, West Greenland, and Icelandic stocks. Newfoundland capelin spawn onshore, preferring granular gravel beds. Freshwater input creates suitable summer conditions for the fish, with low salinity, and highly nutrient-rich water with abundant copepod populations, the main food of capelin.

Numerous invertebrate species are also present. The most important of these from a fisheries viewpoint, snow crab (*Chionoecetes opilio*) and northern shrimp (*Pandalus borealis, P. montagui*, are primarily harvested in offshore waters of the North Atlantic and Labrador Sea.

2.8.7 Burin Tsunami 1929

Submarine slope failures triggered by earthquake activity may generate a pair of tsunami, with one progressing seaward, acting as a far-field event for distant shorelines. The more deadly backflow tsunami will move landward, causing a near-field event.

Shortly after 5:00 pm on 18 November 1929, an earthquake with estimated magnitude M_S 7.2 occurred where the Laurentian Channel reaches the edge of the continental shelf, 340 km south of the Burin Peninsula (Ruffman 1991, Hyndman et al. 2008, Earthquakes Canada n.d.; Fig. 2.57). The earthquake at 400 m water depth triggered a submarine slope failure, sending 200 km^3 of slumping sediments, turbidity currents, and a tsunami seaward. Eventually, a small tsunami was reported at Lisbon and Casablanca.

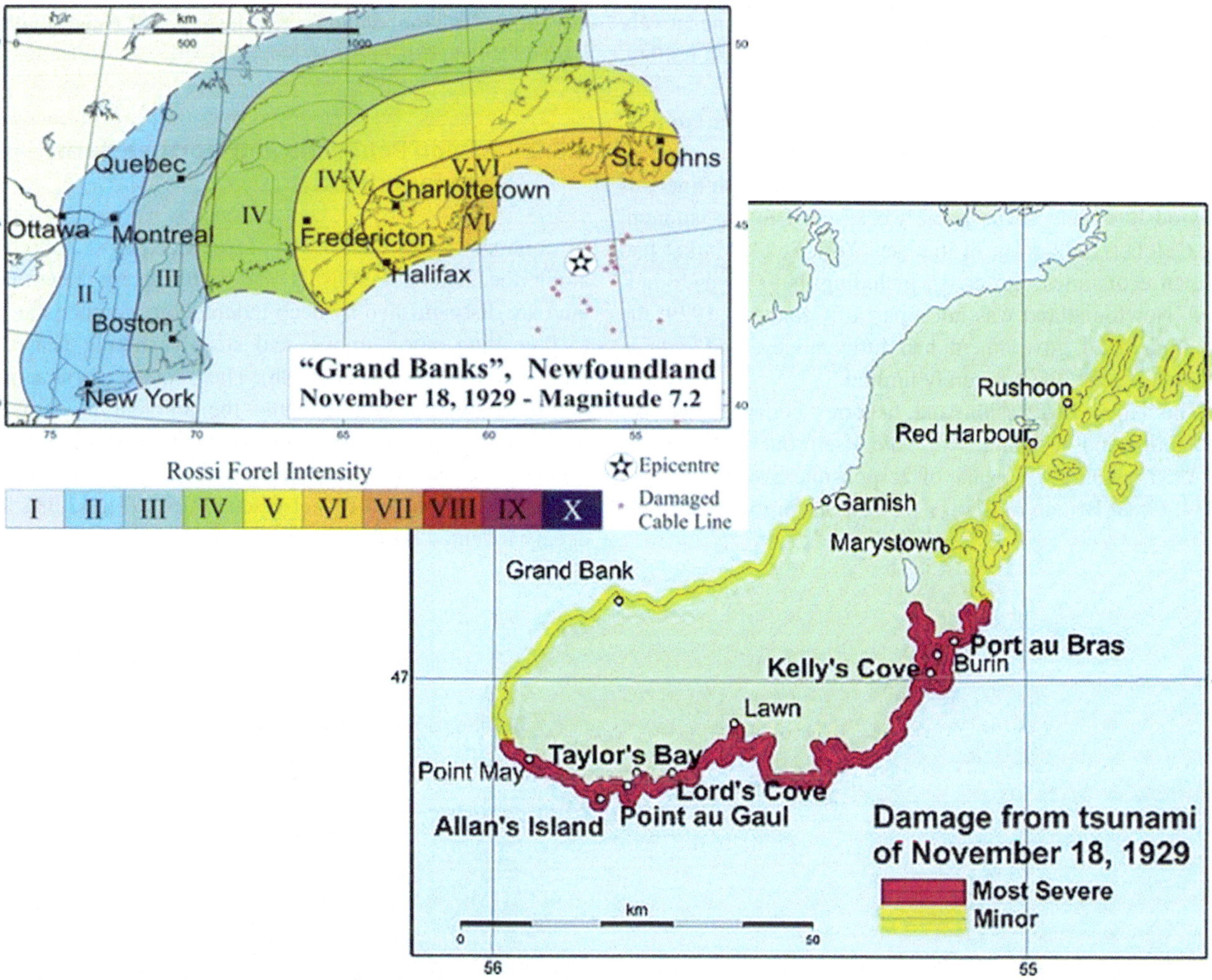

Fig. 2.57 Burin Peninsula tsunami, 18 November 1929. Upper left: earthquake epicentre (star), estimated Modified Mercalli damage intensities (Roman numerals), and locations of breakages in submarine cables (red dots). Lower right: Areas suffering damage from the tsunami. (Earthquakes Canada, Natural Resources Canada)

Propelled by the released earthquake energy, the turbidity currents moved downwards along the continental slope at 60–80 km h^{-1}, eroding the underlying sediment. En route, the turbidity currents snapped the Trans-Atlantic telephone cables, laid on the ocean floor, in succession. The speeds of the turbidity currents could be determined from the time that telephone service was interrupted. Today, fibre optical cables connect North America and Europe across the Atlantic Ocean along the same routes. Modern fibre-op cables are armoured and have an internal nested construction, designed to prevent service interruptions in the event of future turbidity current activity.

Simultaneously, a backflow tsunami began to move northward and westward, with a velocity of 225 km h^{-1}. Initial impact on the fishing communities of the southern Burin Peninsula occurred about 6:30 pm, after nightfall. Twenty-nine people were killed, the largest loss of life documented for an earthquake within Canadian territory. The tsunami travelled through relatively deep (~100 m) coastal waters until nearing landfall, allowing it to retain significant energy. Locally, tsunami height reached 6 m, with run-up heights reaching 27 m and inundation extending more than 1200 m inland.

Losses included homes, buildings, ships, fishing vessels, and more than 120,000 kg of prepared salted cod (Ruffman 1991; Hyndman et al. 2008). The estimated monetary cost was equivalent to approximately $25 million in 2022. This numerical value does not convey the extent of damage to the livelihood of small communities: the price of the lumber, nails, and paint was far less than the effort required to rebuild the fishery. The loss of the salt cod, ready for shipping, meant that many residents had lost the results of the year's fishing effort for no return. Disturbance to the seafloor had negatively affected the habitats for groundfish such as cod and halibut, and affected the squid fishery.

Recognition of the disaster and relief were slow in coming. In 1929, there were no road, rail, air, or telephone connections to the Burin Peninsula, and the sole telegraph line was previously under repair. Three weeks prior to the tsunami, the October 1929 crash of the New York Stock Market had thrown economies into crisis, including that of Newfoundland. Newfoundland was not a part of Canada in 1929, the Newfoundland government had little money, and outside sources of relief were severely limited.

The Great Depression and fisheries collapses led the Newfoundland legislature to vote itself out of existence in 1933, ending 79 years of responsible government. In 1934, Great Britain took over the responsibility for directly governing Newfoundland, which lasted until Newfoundland joined Canada in 1949.

2.8.8 Avalon Peninsula and Northeastern Newfoundland

In contrast to the Gulf of St. Lawrence, most shorelines along the Avalon Peninsula and northeastern Newfoundland are distinguished by deep indentations, aligned parallel to prevailing storm tracks, and steeply sloping nearshore bathymetry (Catto 2011, 2020b). The frequent storm activity in all seasons, long fetch, and the ambient moderate to high energy, in combination with the coastal configurations, produces shorelines dominated by swash activity, shore-normal transport, and reflective conditions (Fig. 2.58). The deeply indented Avalon coast is marked by coves with wave

Fig. 2.58 Avalon Peninsula coastline. Upper left: Marine planation terrace, New Melbourne. Upper right: Self-organized cusps on reflective shoreline, Topsail Beach. Lower right: wave-dominated shore-normal reflective shoreline, Middle Cove. Lower left: Baymouth bar developed on glacial terminal moraine, Great Barachoix, Placentia Bay

action laterally confined by bedrock, focusing energy at approximately 90° to the beach and generating shore-normal transport. Swash-backwash motion dominates, producing features which extend seaward (cuspate spits, tombolos) or indicate that waves move sediment normal to the shore (self-organized cusps and overwash fans).

Directly offshore of most Avalon beaches, steeply sloping bathymetry is common. Most waves reach the vicinity of the shore before breaking, producing relatively high energy conditions along the beach, strong seaward return flows (including rip currents), and reflective energy regimes (Catto 2020b).

Ice foot development is commonly a major factor in the geomorphic development of shorelines north of Spaniards Bay. Formation of an ice foot largely precludes winter erosion of beach sediments (Forbes and Taylor 1994; Catto 2015). The southerly extent of persistent ice foot development coincides with the position of the −0.5 °C February SST isotherm. Adhering ice is common on all beaches that are not anthropogenically disturbed.

Offshore pack ice persists from February to April, and icebergs are present along the coastline throughout the spring and early summer. Formation of the ice foot begins in late December during most winters, and several beaches commonly retain an ice foot until late March.

References

Anderson JT, Dalley EL, O'Driscoll RL (2002) Juvenile capelin (*Mallotus villosus*) off Newfoundland and Labrador in the 1990s. ICES J Mar Sci 59:917–928

Archer AW (2013) World's highest tides: hypertidal coastal systems in North America, South America, and Europe. Sed Geol 284–285:1–25

Ashall LM, Mulligan RP, van Proosdij D, Poirier E (2016) Application and validation of a three-dimensional hydrodynamic model of a macrotidal salt marsh. Coast Eng 114:35–46

Atlantic Canada Conservation Data Centre (2022) Illustrated flora of PEI. http://accdc.com/peiflora/s1.htm

Auer SJ (1987) Five-year climatological survey of the Gulf Stream system and its associated rings. J Geophys Res 92:11709–11726

Baehre H (2022) Hierarchies of habitat: *Diapensia lapponica* on the Avalon Peninsula. MSC thesis, Department of Geography, Memorial University

Baldwin K, Allen L, Basquill S, Chapman K, Downing D, Flynn N, MacKenzie W, Major M, Meades W, Meidinger D, Morneau C, Saucier J-P, Thorpe J, Uhlig P (2019) Vegetation zones of Canada: a biogeoclimatic perspective. [Map] Scale 1:5,000,000. Natural Resources Canada, Canadian Forestry Service, Sault Ste. Marie, ON

Banfield C (1981b) The climatic environment of Newfoundland. In: Macpherson AG, Macpherson JB (eds) The Natural Environment of Newfoundland Past and Present. Memorial University, St. John's, pp 83–153

Banfield AWF (1981a) The mammals of Canada. National Museum of Natural Sciences, National Museums of Canada, p 438

Barr SM, White CE, Fisher BE, McKinnon JS, Barras AL, Hapgood DS (2021) DP ME 433, digital geological data generated as part of the bedrock geological mapping compilation project for Cape Breton Island, Nova Scotia. Nova Scotia Geological Survey

Baumgartner MF, Wenzel FW, Lysiak NSJ, Patrician MR (2017) North Atlantic right whale foraging ecology and its role in human-caused mortality. Mar Ecol Prog Ser 581:165–181. https://doi.org/10.3354/meps12315

Beck, C. Grieser J, Kottek M, Rubel F, Rudolf B (2006) Characterizing global climate change by means of köppen climate classification. http://koeppen-geiger.vu-wien.ac.at/pdfs/beck_et_al_ksb_2006.pdf

Bell T, Macpherson JB, Renouf MAP (2003) "Wish you were here..." a thumbnail portrait of the Great Northern Peninsula Environment 1000 AD. In: Lewis-Simpson SM (ed.), Vinland revisited: the norse world at the turn of the first millennium. selected papers from the viking millennium international symposium 15–24 September 2000, Newfoundland and Labrador. Historic Sites Association of Newfoundland and Labrador, St. John's

Benson DA Dodds DG (1977) The deer of Nova Scotia. Department of Lands and Forests, Province of Nova Scotia, Halifax, pp 92

Berger AR, Bouchard A, Brookes IA, Grant DR, Hay S, Stevens RK (1992) Gros Morne National Park, Newfoundland—geology, topography, vegetation. Geological Survey of Canada Miscellaneous Report 54

Black DW, Reading JE, Savage HG (1998) Archaeological records of the extinct sea mink, *Mustela macrodon* (Carnivora: Mustelidae), from Canada. Canadian Field-Naturalist 112:45–49

Bostock HS (2014) Physiographic regions of Canada. Geological Survey of Canada Map 1254A (2nd edition), scale 1:5,000,000. https://www.nrcan.gc.ca/maps-tools-and-publications/maps/geology-and-geosciences/16876. Accessed 8 April 2024

Bouchard A, Hay S, Gauvin C, Bergeron Y (1986) Rare vascular plants of Gros Morne National Park, Newfoundland, Canada. Rhodora 88:481–502

Bourque P-A, Malo M, Kirkwood D (2000) Paleogeography and tectono-sedimentary history at the margin of Laurentia during Silurian-earliest Devonian time: the Gaspé Belt. Geological Society of America Bulletin 112, Québec, pp 4–20

Bowering WR, Brodie WB (1995) Greenland halibut (*Reinhardtius hippoglossoides*). A review of the dynamics of its distribution and fisheries off eastern Canada and Greenland. In: Hopper AG (ed) Deep-water fisheries of the North Atlantic Oceanic Slope. Kluwer, Dordrecht, pp 113–160

Bradley DC (1983) Tectonics of the acadian orogeny in New England and adjacent Canada. J Geol 91:381–400

Breeze H Horsman T (2005) The Scotian shelf: an atlas of human activities. Fisheries and Oceans Canada. https://www.dfo-mpo.gc.ca/oceans/publications/scotian-atlas-ecossais

Brisebois D, Nadeau J (2003) Géologie de la Gaspésie et du Bas-Saint-Laurent. Ministère des Ressources naturelles et de la Faune du Québec, DV 2003–08

Brookes IA (1993) Table mountain, Gros Morne National Park, Newfoundland: Canadian landform examples 26. Can Geogr 37:60–73

Brushett D, Bell T, Batterson MJ, Shaw J (2007) Ice-flow History of Placentia Bay, Newfoundland: multibeam seabed mapping. Newfoundland and Labrador Department of Natural Resources, Geological Survey Report 07-1, pp 215–228

Bureau d'aménagement de l'est du Québec 1966. Sols du bas Saint-Laurent, de la Gaspésie et des Îles-de-la-Madeleine. Atlas rêgional du Bas-St.-Laurent, de la Gaspésie, et des îles-de-la-Madeleine

Byers M, Østhagen A (2016) Why does Canada have so many unresolved maritime boundary disputes? Canad Yearbook Int Law 54 https://doi.org/10.1017/cyl.2017.14

Campbell DC, Mosher DC (2016) Geophysical evidence for widespread Cenozoic bottom current activity from the continental margin of Nova Scotia, Canada. Mar Geol 378:237–260

Cann DB, MacDougall JI, Hilchey JD (1963) Soil survey of Cape Breton Island NS. Nova Scotia Soil Survey Report 12

Carscadden JE, Frank KT, Leggett WC (2000) Ecosystem changes and the effects on capelin (*Mallotus villosus*), a major forage species. Can J Fish Aquat Sci 58:73–85

Catto NR (1998a) Comparative analysis of striations and diamicton fabrics as ice flow direction indicators, Malpeque Bay region, Prince Edward Island. Boreas 27:259–274

Catto NR (1998b) The pattern of glaciation on the Avalon Peninsula of Newfoundland. Géog Phys Quatern 52:23–45

Catto NR (2002) Anthropogenic pressures on Coastal Dunes, southwest Newfoundland. The Canadian Geographer 46:17–32

Catto NR (2006a) More than 16 Years, more than 16 stressors: evolution of a high energy reflective beach, 1989–2005. Géog Phys Quatern 60:49–62

Catto N (2015) Geomorphology: landscapes we live in. Oxford University Press, Don Mills, Ontario

Catto NR (2020a) Labrador and the Northern Peninsula of Newfoundland: foreshore and flanking geomorphology. In: Slaymaker O, Catto NR (eds) Landscapes and landforms of Eastern Canada. Springer International Publishers, Cham, Switzerland, pp 179–210

Catto NR (2020b) The avalon of Newfoundland: geomorphology, people, and landscape. In: Slaymaker O, Catto NR (eds) Landscapes and landforms of Eastern Canada. Springer International Publishers, Cham, Switzerland, pp 333–368

Catto NR (2020c) Atlantic Canada's tidal coastlines: geomorphology and multiple resources. In: Slaymaker O, Catto NR (eds) Landscapes and landforms of Eastern Canada. Springer International Publishers, Cham, Switzerland, pp 401–430

Catto NR, Catto G (2012) Landscape response to human impact in coastal Newfoundland, Canada: 29,000 km of "untouched" coastline. Landscape Archaeology Conference, LAC 2012. Topoi, Special 3:225–227

Catto NR, MacQuarrie K, Hermann M (2002) Geomorphic response to late holocene climate variation and anthropogenic pressure, northeastern Prince Edward Island, Canada. Quatern Int 87:101–118

Catto NR, Slaymaker O, Kovanen DJ (2020a) Quaternary glacial, glacimarine, and glacilacustrine history. In: Slaymaker O, Catto NR (eds) Landscapes and landforms of Eastern Canada. Springer International Publishers, Cham, Switzerland, pp 49–78

Catto NR, Slaymaker O, Kovanen DJ (2020b) Human-modified landscapes. In: Slaymaker O, Catto NR (eds) Landscapes and landforms of Eastern Canada. Springer International Publishers, Cham, Switzerland, pp 475–508

Catto NR, Catto G (2009) Geomorphology, sedimentology, and management issues, Hog Island (Pemamgiag) sandhills, PEI. Report to the Mi'kmaq Confederacy of Prince Edward Island, p 99

Catto NR, St. Croix L (1998) Urban geology of St. John's, Newfoundland. In: Karrow PF, White OL (eds.), Urban geology of Canadian Cities. Geological Association of Canada Special Paper 42, pp 445–462

Catto, NR, Scruton, DA, Ollerhead, LMN (2003) The coastline of Eastern Newfoundland. Canadian Technical Report Fisheries and Aquatic Sciences 2495

Catto NR (2006b) Impacts of climate change and variation on Provincial Parks, NL. Ministry of Environment and Conservation, Newfoundland and Labrador, p 160

Catto NR (2011). Coastal erosion in Newfoundland. Ministry of Environment and Conservation, Newfoundland and Labrador, p 144

Catto NR (2012) Natural hazard and vulnerability assessment in Atlantic Canada: review, progress, and challenges. In: Haque CE, Etkin D (eds.), Disaster risk and vulnerability. McGill-Queen's University Press, pp 235–274

Chircop A, Van der Zwaag D, Mushkat P (1995) The gulf of maine agreement and action plan. Mar Policy 19:317–333

Chmura G, Pohle G, Van Guelpen L, Page F, Costello M, (2008) Climate change and thermal sensitivity of commercial marine species. Climate Change Impacts and Adaptations Directorate report, http://www.geog.mcgill.ca/climatechange/index.htm

Chmura G, Hung G (2004) Controls on salt marsh accretion: a test in salt marshes of Eastern Canada. Estuaries 27:70–81

Christie BJ (1995) The horses of Sable Island. Pottersfield Press, Lawrencetown Beach, N.S.

Clapham ME, Narbonne GM, Gehling JG (2003) Paleoecology of the oldest known animal communities: Ediacaran assemblages at Mistaken Point, Newfoundland. Paleobiology 29:527–544

Cloutier R (2013) Great Canadian Lagerstätten 4. The Devonian Miguasha Biota (Québec): UNESCO World Heritage Site and a time capsule in the early history of Vertebrates Geoscience Canada 40, pp 149–163. https://doi.org/10.12789/geocanj.2013.40.008

CNVC (2016) Canadian National Vegetation Classification. http://cnvc-cnvc.ca. Accessed 2 April 2024

Cochonat P, Ollier G, Michel JL (1989) Evidence for slope instability and current-induced sediment transport, the RMS Titanic wreck search area, Newfounfland Rise. Geo-Marine Lett 9:145–152

Colling A (2001) Ocean circulation. Butterworth-Heinemann, Open University, Boston

Colman-Sadd SP, Hayes JP, Knight I (1990) Geology of the island of Newfoundland. Newfoundland and Labrador, Department of Mines and Energy, Geological Survey Branch, Map 90–01

Cook FR Johnston JA, Douglas C, Cook JC (1984) Introduction to Canadian amphibians and reptiles. National Museums of Canada, National Museum of Natural Sciences, Ottawa

Cooper L, Clay D (1994) An historical review of logging and river driving in Fundy National Park. Research Notes of Fundy National Park. No. FUN94–05. Parks Canada, p 73

COSEWIC (2006a) COSEWIC assessment and update status report on the Gaspé shrew *Sorex gaspensis* in Canada. Committee on the Status of Endangered Wildlife in Canada, Ottawa

COSEWIC (2006b) COSEWIC assessment and update status report on the southern flying squirrel *Glaucomys volans* (Atlantic (Nova Scotia) population and Great Lakes Plains population) in Canada. Committee on the Status of Endangered Wildlife in Canada, Ottawa

COSEWIC 2013. COSEWIC assessment and status report on the Little Brown Myotis *Myotis lucifugus*, Northern Myotis *Myotis septentrionalis* and Tri-colored Bat *Perimyotis subflavus* in Canada. Committee on the Status of Endangered Wildlife in Canada. Ottawa.

COSEWIC 2018. Assessment and Status Report on the Midland Painted Turtle (*Chrysemys picta marginata*) and the Eastern Painted Turtle (*Chrysemys picta picta*) in Canada. www.registrelep-sararegistry.gc.ca/default_e.cfm.

Curley R, Keenlyside DL, Kristmanson HE, Dibblee RL (2019) A review of the historical and current status of American Beaver (*Castor canadensis*) on Prince Edward Island, Canada. Can Field-Nat 133:332–342

Damman AW (1983) An ecological subdivision of the island of Newfoundland. In: South GR (ed) Biogeography and ecology of the island of Newfoundland. DW Junk, The Hague, pp 163–206

Day D (1992) The Saint Pierre and Miquelon Dispute: towards a further redefinition of French fishing rights in the Northwest Atlantic. Ocean Coast Manag 18:371–403

de Villiers M, Hirtle S (2004) A Dune Adrift: The Strange Origins and Curious History of Sable Island. McClelland & Stewart, Toronto

DeBaie L, Murphy M, Austen A, Fitzgerald N, Daigle R (2006) Bouctouche case Study: evaluating integrated effects upon tourism in Bouctouche. In: Daigle R, Forbes DL, Parkes G, Ritchie H, Webster T, Bérubé D, Hanson A, DeBaie L, Nichols S, Vasseur L (eds.) Impacts of sea level rise and climate change on the coastal zone of Southeastern New Brunswick. Environ Canada, Ottawa

Denys N (1672) Description géographique et historique des costs de l'Amérique septentionale, avec l'histoire naturelle du pais. https://www.canadiana.ca/view/oocihm.34784/3. Accessed 12 Aug 2022

Deschênes J, Gosselin C, Hardy L (2010) Portrait territorial du Bas-Saint-Laurent. Ressources naturelles et Faune, Gouvernement du Québec, Québec, p 117

Desponts, Mireille, Brunet G, Bélanger L Bouchard M (2004) The eastern boreal old-growth balsam fir forest: a distinct ecosystem. Can J Bot 82:830–849

Dohler G 2007. Tides in Canadian waters. Canadian Hydrographic Service. Fisheries and Oceans Canada

Dredge LA, Mott RJ, Grant DR (1992) Quaternary stratigraphy, paleoecology, and glacial geology, Iles de la Madeleine, Québec. Can J Earth Sci 29:1981–1996

Drobyshev I, Guitard M-A, Asselin H, Genries A, Bergeron Y (2014) Environmental controls of the northern distribution limit of yellow birch in eastern Canada. Can J for Res 44:720–731

Dubois J-MM, Grenier A (1993) The Magdalen Islands, Gulf of Saint Lawrence. Coastlines of Canada. In: Proceedings, 8th symposium on coastal and ocean management. American Society of Civil Engineers, July 19–23, pp 170–182

Earle CJ (2018) *Thuja occidentalis*. The Gymnosperm Database. https://www.conifers.org/cu/Thuja_occidentalis.php. Accessed 13 Aug 2023

Earthquakes Canada n.d. Earthquakes Canada. Natural Resources Canada

Environment and Climate Change Canada and Parks Canada Agency (2019) Action plan for the Blanding's Turtle (*Emydea blandingii*), Nova Scotia Population, in Canada. Species at Risk Act Action Plan Series. Environment and Climate Change Canada and Parks Canada Agency, Ottawa

Environment and Climate Change Canada (2020) Canadian Climate Normals 1981–2010. https://climate.weather.gc.ca/climate_normals/index_e.html. Accessed 8 Aug 2023

Ernst CH, Barbour RW (1989) Snakes of eastern North America. George Mason University Press, Lanham, MD

Fahmy SH, Hann SWR, Jiao Y (2010) Soils of New Brunswick, the second approximation. Research Branch Agriculture and Agri-Food Canada. Fredericton, New Brunswick

Falcon-Lang HJ, Benton MJ, Braddy SJ, Davies SJ (2006) The Pennsylvanian tropical biome reconstructed from the Joggins Formation of Nova Scotia, Canada. J Geol Soc 163:561–576

Finley MA, Courtenay SC, Teather KL, Hewitt LM, Holdway DA, Hogan NS, van den Heuzel MR (2013) Evaluating cumulative effects of anthropogenic inputs in Prince Edward Island estuaries using the mummichog (*Fundulus heteroclitus*). Integr Environ Assess Manag 9:496–507

Forbes DL, Taylor RB (1994) Ice in the shore zone and the geomorphology of cold coasts. Prog Phys Geogr 18:59–89

Forêt Québec (2002) Ministère des Ressources naturelles du Québec,. 2002+. Les guides de reconnaissance des types écologiques. Gouv. du Québec, Québec, QC. http://www.mffp.gouv.qc.ca/forets/inventaire/guide-types-ecologiques-carte.jsp

Fyffe LR (2014) The Grand Manan Terrane of New Brunswick: tectonostratigraphy and relationship to the Gondwanan margin of the Iapetus Ocean. Geosci Can 41:483–501

Garrison T, Ellis R (2017) Essentials of oceanography. Cengage Learn

Gaudet JF (1973) Native trees of Prince Edward Island and the more common woodland shrubs. Dept. of Agriculture and Forestry, Charlottetown, Prince Edward Island, p 98

Gaudet JF (1979) Forestry past and present on Prince Edward Island. Department of Agriculture and Forestry, Prince Edward Island, p 69

Gerwing TG, Hamilton DJ, Barbeau MA, Haralampides K, Yamazuki G (2017) Short-term response of a Downstream Marine System to the partial opening of a Tidal-River Causeway. Estuaries Coasts 40:717–725

Gosse J, Hermanutz L, McLaren B, Deering P, Knight T (2011) Degradation of boreal forests by nonnative herbivores in Newfoundland's national parks: recommendation for ecosystem restoration. Nat Arch J 31:331–339

Grant DR (1989) Quaternary geology of the Atlantic Appalachian region of Canada. In: Fulton RJ ed., Quaternary geology of Canada and Greenland. Geological Survey of Canada, Geology of Canada, pp 393–440

Gray J, Davesne G, Fortier D, Godin E (2016) The thermal regime of Mountain Permafrost at the summit of Mont Jacques-Cartier in the Gaspé Peninsula, Québec, Canada: a 37 year record of fluctuations showing an overall warming trend. Permafrost Periglac Process 28:266–274

Greenberg DA, Blanchard W, Smith B, Barrow E (2012) Climate change, mean sea level, and high tides in the Bay of Fundy. Atmos Ocean 50(3):261–276

Grondin P, Blouin J, Racine P, D'Avignon H, Tremblay S (1998) Rapport de classification écologique du sous-domaine bioclimatque de la sapinière à bouleau blanc d l'est. Foret Québec, Direction des inventaires forestiers, Ministère des Resources naturelles du Québec

Harris PT, Macmillan-Lawler M, Rupp J, Baker EK (2014) Geomorphology of the oceans. Mar Geol 352:4–24

Hearn B, Species Status Advisory Committee (2012) The Status of Arctic Hare (*Lepus arcticus bangsii*) in Insular Newfoundland. Species Status Advisory Committee report 26

Heringa PK (1981) Soils of the Avalon Peninsula, Newfoundland. Newfoundland Soil Survey report 3

Hétu B, Rémillard AM, Bernatchez P, St-Onge G (2020) Landforms and landscapes of the Magdalen Islands: between geology and climate. In: Slaymaker O, Catto NR (eds) Landscapes and land forms of Eastern Canada. Springer International Publishers, Cham, Switzerland, pp 431–452

Higgins J (2007) Heritage NL. 19th century cod fisheries. https://www.heritage.nf.ca/articles/economy/19th-century-cod.php. Accessed 13 Aug 2022

Hinds HR (2000) Flora of New Brunswick. University of New Brunswick, Department of Biology, Fredericton, NB

Hodych JP, King AF (1989) Geology of Newfoundland and Labrador. Newfoundland J Geol Educ 10

Holmstrom DA, Thompson BL (1989) Soils of the Annapolis Valley area of Nova Scotia. Nova Scotia Soil Survey, report 22. Agriculture Canada

Hosie RC (1979). Native trees of Canada. Canada Department of Forestry, Don Mills, Ontario

Hutchings J, Myers R (1995) The biological collapse of atlantic cod off Newfoundland and Labrador: an exploration of historical changes in exploitation, harvesting technology, and management. In: Arnason R, Felt L, eds., The North Atlantic fisheries: successes, failures, and challenges, vol. 3, p. 37–93. Memorial University of Newfoundland

Hyndman D, Hyndman D, Catto NR (2008) Natural Hazards and Disasters. Nelson

ICJ 1984. Delimitation of the Maritime Boundary in the Gulf of Maine Area (Canada/United States of America). ICJ Report 246 [Gulf of Maine]. 23 ILM 1197

James NP, Stevens RK (1986) Stratigraphy and correlation of the Cambro-Ordovician Cow Head Group, western Newfoundland. Geological Survey of Canada Bulletin 366

Johnson RS (1986) Forests of Nova Scotia: a history. Nova Scotia Department of Lands and Forests. Four East Publications, Halifax, Nova Scotia, p 407

Johnston WF (1990) *Thuja occidentalis* L. Northern white-cedar. In. Burns RM, Honkala BH, Silvics of North America, Vol. 1, Conifers. United States Department of Agriculture Forest Service, Agricultural Handbook 654, pp 580–588

Josenhans HW, Lehman S (1999) Late glacial stratigraphy and history of the Gulf of St. Lawrence, Canada. Can J Earth Sci 36:1327–1345

Jutras P, Prichonnet G, McCutcheon S (2003) Alleghanian deformation in the eastern Gaspé Peninsula of Québec, Canada. Geol Soc Am Bull 115:1538–1551

Karsten RH, McMillan JM, Lickley MJ, Haynes RD (2008) Assessment of tidal current energy in the Minas Passage, Bay of Fundy. In: Proceedings, IMechE 222A: journal of power and energy, https://doi.org/10.1243/09576509JPE555

Kenney RD (2009) Right whales *Eubalaena glacialis, E. japonica,* and *E. australis.* In: Perrin WF (ed.), Encyclopedia of marine mammals. elsevier science & technology, pp 962–971

Keppie JD (2000) Geological map of the province of Nova Scotia Map ME 2000–001. Nova Scotia Geological Survey

King AF (1988) Geology of the Avalon Peninsula, Newfoundland (parts of 1K, 1L, 1M, 1N and 2C). Map 88–001. Newfoundland and Labrador Geological Survey

Kirby DA (1988) Soils of the Pasadena—Deer Lake area, Newfoundland Report No. 17 Newfoundland Soil Survey

Knauss JA, Garfield N (2016) Introduction to physical oceanography. Waveland Press

Kontak DJ (2008) On the edge of CAMP: geology and volcanology of the Jurassic North Mountain Basalt, Nova Scotia. Lithos 101:74–101

Kontak DJ Webster TL (2010) OFM ME 2010-007: overview map for the bedrock geology maps of basaltic rocks of the North Mountain Formation from Brier Island to Cape Split. Nova Scotia Geological Survey

Kosters EC, Grey M (2021) From scratch: building the governance structure of the Joggins. Foss Inst Geoheritage 13(1). https://doi.org/10.1007/s12371-021-00531-5

Landing E, Keppie JD, Keppie DF, Geyer G, Westrop SR (2022) Greater Avalonia—latest Ediacaran-Ordovician "peribaltic" terrane bounded by continental margin prisms ("Ganderia," Harlech Dome, Meguma): review, tectonic implications, and paleogeography. Earth Sci Rev 224:103863. https://doi.org/10.1016/j.earscirev.2021.103863

Lang LD, Tessier N, Gauthier N, Wissink R, Jolicoeur H, Lapointe F-J (2013) Genetic confirmation of Cougars (*Puma concolor*) in Eastern Canada. Northeast Nat 20:383–396

Lavoie D (2008) Appalachian Foreland basin of Canada. Sedimentary Basins of the World 5C, Elsevier, pp 65–103

Lemieux MJ (2010) A management plan for native occurrences of Eastern White Cedar (*Thuja occidentalis* L.) in Nova Scotia. Nova Scotia Department of Natural Resources

Lescarbot M (1609), translated by Erondelle P (1928). In: Ross ED, Power E, ed., Nova Francia, a Description of Acadia 1606. Harper and Brothers, London, p 346

Liu AG, McIlroy D, Antcliffe JB, Brasier MD (2011) Effaced preservation in the *Ediacara Biota* and its implications for the early Macrofossil record. Palaeontology 54:607–630

Liverman DGE, Batterson MJ, Catto NR, Mackenzie C, Munro-Stasiuk M, Scott S, Sommerville A (2000) Evidence of late-glacial permafrost in Newfoundland. Quatern Int 68(71):163–174

Locke A, Hanson JM, Klassen GJ, Richardson SM, Aubé CI (2003) The damming of the petitcodiac river: species, populations, and habitats lost. Northeast Nat 10:39–54

Loo J, Ives N (2003) The Acadian forest: historical condition and human impacts. For Chron 79:462–474

Macdonald MA (1994) Map ME 1994–001: geological map of the South Mountain Batholith, Western Nova Scotia [Parts of 11D and 21A] [1:250 000]. Nova Scotia Geological Survey

Malo M, Bourque P-A (1993) Timing of the deformation events from Late Ordovician to mid-Devonian in the Gaspé Peninsula. In: Roy DC, Skehan SJ (eds.) The Acadian Orogeny: recent studies in New England, Maritime Canada and the Autochthonous foreland: Geological Society of America, Special Paper 275, p 101–122

Malo M, Kirkwood D (1995) Faulting and progressive strain history of the Gaspé Peninsula in Post-Taconian time: a review. In: Hibbard JP, van Staal CR, Cawood PA, eds., Current Perspectives in the Appalachians-Caledonian Orogen: Geological Association of Canada, Special Paper 41, p 267–282

Martin K (1977) Island Woodland Plants. Prince Edward Island, Ministry of Environment, Energy, and Forestry

Maunder J (1982) The Newfoundland wolf. Newfoundland Museum, Historic Resources Division, Information sheets from the Newfoundland Museum, No. 8

McHenry M, Dunlop P (2016) The subglacial imprint of the last Newfoundland Ice Sheet, Canada. J Maps 12:462–483

McHone JG (2011) Triassic basin stratigraphy at Grand Manan, New Brunswick, Canada. Atlantic Geol 47:125–137

McLaren BE, Roberts BA, Djan-Chekar N, Lewis KP (2004) Effects of overabundant moose on the Newfoundland landscape. Alces 40:45–59

Miguasha National Park 2007. Vertebrates. https://miguasha.ca/mig-en/vertebrates.php

Modesto SP, Scott DM, MacDougall MJ, Sues H-D, Evans DC, Reisz RR (2015) The oldest parareptile and the early diversification of reptiles. Proc Royal Soc B 282:20141912. https://doi.org/10.1098/rspb.2014.1912

Mohan JE, Ziska LH, Schlesinger WH, Thomas RB, Sicher RC, George K, Clark JS (2006) Biomass and toxicity responses of poison ivy (*Toxicodendron radicans*) to elevated atmospheric CO_2.In: Proceedings of the National Academy of Sciences 103, pp 9086–9089

Monsarrat S, Pennino MG, Smith TD, Reeves RR, Meynard CM, Kaplan DM, Rodrigues AS (2015) A spatially explicit estimate of the prewhaling abundance of the endangered North Atlantic right whale. Conserv Biol 30:783–791

Morissette A, Hétu B, Lajeunesse P (2020) Quaternary heritage, glacial landforms, and depositional assemblages in the Témiscouata-Madawaska Valley (Eastern Québec). In: Slaymaker O, Catto NR (eds) Landscapes and landforms of Eastern Canada. Springer International Publishers, Cham, Switzerland, pp 381–399

Mosher DC, Campbell DC, Gardner JV, Piper DJW, Chaytor JD, Rebesco M (2017) The role of deep-water sedimentary processes in shaping a continental margin: the Northwest Atlantic. Mar Geol 393:245–259

Munro MC Newell RE Hill NM (2014) Nova Scotia plants. Nova Scotia Museum, Halifax

Murphy JB, Keppie JD (2005) The Acadian Orogeny in the Northern Appalachians. Int Geol Rev 47:663–687

Murphy JB, Pisarevsky SA, Nance RD, Keppie JD, Jessell MJ (2001) Animated history of Avalonia in Neoproterozoic—early Proterozoic. General contributions. J Virtual Explor 3:45–58

Narbonne GM (2005) The Ediacara Biota: neoproterozoic origin of animals and their ecosystems. Annu Rev Earth Planet Sci 33:421–442

Natural Resources Canada (2009) Geological provinces of Canada. https://www.nrcan.gc.ca/maps-tools-and-publications/maps/geology-and-geosciences/16876 Accessed 12 July 2023

Naughton D (2012) The natural history of Canadian mammals. Canadian Museum of Nature, Ottawa

New Brunswick Dept of Natural Resources (2008) Bedrock geology of new brunswick. Miner Policy Plan Div 1(500):000

Newfoundland and Labrador n.d. Mammals. https://www.gov.nl.ca/ffa/wildlife/snp/programs/education/animal-facts/mammals/

Newfoundland and Labrador n.d. Black bears our traditional predator. https://www.gov.nl.ca/ffa/wildlife/all-species/bear/

Newfoundland and Labrador 2022. Newfoundland and Labrador moose management plan 2022–2026. https://www.gov.nl.ca/ffa/files/22282-Moose-Management-Report-July-12.pdf

Normandeau A, Lajeunesse P, St-Onge G (2015) Submarine canyons and channels in the Lower St. Lawrence Estuary (Eastern Canada): morphology, classification and recent sediment dynamics. Geomorphol 241:1–18

Normandeau A, Piper DJW, Shaw J, Todd BJ, Campbell DC, Mosher DC (2020) The seafloor of Southeastern Canada. In: Slaymaker O, Catto

NR (eds) Landscapes and landforms of Eastern Canada. Springer International Publishers, Cham, Switzerland, pp 453–474

Nova Scotia Herpetofaunal Atlas n.d. NS Herp Atlas Project. Department of Biology, Acadia University, Wolfville NS. https://discovery.acadiau.ca/herpatlas/guide.htm

Nova Scotia Soil Survey (1972) Soils of Nova Scotia. Agriculture Canada

O'Brien SJ (1998) Geology of the Connaigre Peninsula and adjacent areas, southern Newfoundland (parts of NTS 1M/5, 6, 11, 12 & 14 and 11P/9 & 9). Geological Survey Branch, Dept. of Mines and Resources, Government of Newfoundland and Labrador, Map 98-02

Olejczyk P, Gray JT (2007) The relative influence of Laurentide and local ice sheets during the last glacial maximum in the eastern Chic-Chocs Range, northern Gaspé Peninsula, Quebec. Can J Earth Sci 44:1603–1625

Ollerhead J, Davidson-Arnott RGD (1995) The evolution of Buctouche Spit, New Brunswick, Canada. Mar Geol 124:215–236

Packer DB, Cargnelli LM, Griesbach SJ, Shumway SE (1999) Essential fish habitat source document: sea scallop, *Placopecten magellanicus*, life history and habitat characteristics. NOAA Technical Memorandum NMFS-NE No.134, p 21

Parks Canada (2018) Cape Breton Highlands National Park: Mammals. https://www.pc.gc.ca/en/pn-np/ns/cbreton/decouvrir-discover/faune-animals/mammiferes-mammals

Parks Canada (2019) Sable Island National Park Reserve of Canada management plan. https://www.pc.gc.ca/en/pn-np/ns/sable/info/plan/gestion-management-2019

Parks Canada (2020a) Kouchibouguac National Park mammals. https://www.pc.gc.ca/en/pn-np/nb/kouchibouguac/nature/faune-wildlife/mammiferes-mammals

Parks Canada (2020b) Forests at risk in Newfoundland and Labrador. https://www.pc.gc.ca/en/nature/science/conservation/forets-forests/grosmorne

Parks Canada (2021) Kejimkujik National Park. https://www.pc.gc.ca/en/pn-np/ns/kejimkujik

Patton A (1991) Deer history. Nova Scotia. https://novascotia.ca/natr/wildlife/conserva/deer-history.asp

Pe-Piper G, Piper DJW, Jansa LF, De Jonge A (2007) Early Cretaceous opening of the North Atlantic Ocean: implications of the petrology and tectonic setting of the Fogo Seamounts off the SW Grand Banks, Newfoundland. Geol Soc Am Bull 119:712–724

Phillips D (1990) The climates of Canada. Environment Canada, p 176

Piper DJW, Campbell DC (2005) Quaternary geology of Flemish Pass and its application to geohazard evaluation for hydrocarbon development. Geol Assoc Can Spec Pap 43:29–43

Piper DJW, Mudie PJ, Fader GB, Josenhans HW, Maclean B, Vilks G (1990) Quaternary geology. In: Keen MJ, Williams GL, eds., Geology of the continental margin of Eastern Canada. Geol Surv Can 2:475–607

Prave AR, Kessler LG, Malo M, Bloechl WV, Riva J (2000) Ordovician arc collision and foredeep evolution of the Gaspe Peninsula, Quebec: the Taconic Orogeny in Canada and its bearing on the Grampian Orogeny in Scotland. J Geol Soc 157:393–400

Prest VK (1973) Surficial deposits, Prince Edward island. Geological Survey of Canada Map 1366 A

Prince Edward Island Soil Survey (1988) Soils of Prince Edward Island. Agriculture Canada

Prince Edward Island, Ministry of Environment Energy and Forestry (2003) Coyote policy. https://www.princeedwardisland.ca/sites/default/files/publications/coyote_policy.2003.pdf

Rampton VN, Gauthier RC, Thibault J, Seaman AA (1984) Quaternary geology of New Brunswick. Geological Survey of Canada, Memoir 416

Rashid H, Mackillop K, Sherwin J, Piper DJW, Marche B, Vermooten M (2017) Slope instability on a shallow contourite-dominated continental margin, southeastern Grand Banks, eastern Canada. Mar Geol 6:1–13

Renouf MAP (2011) The cultural landscapes of Port au Choix: Precontact hunter-gatherers of northwestern Newfoundland. Springer, New York

Rose GA (2005) Capelin (*Mallotus villosus*) distribution and climate: a sea "canary" for marine ecosystem change. ICES J Mar Sci 62:1524–1530

Rose GA (2003) Fisheries resources and Science in Newfoundland and Labrador: an independent assessment. Royal Commission on Renewing and Strengthening Our Place in Canada. https://www.gov.nl.ca/publicat/royalcomm/research/Rose.pdf

Rowe JS (1972) Forest regions of Canada. Canadian Forestry Service Publication 1300. Ottawa

Ruffman A (1991) The 1929 'Grand Banks' Earthquake and the historical record of Earthquakes and Tsunamis in Eastern Canada. Proceedings, geological survey of Canada Workshop on Eastern Seismicity Source Zones for the 1995 Seismic Hazard Maps, March 18–19, Ottawa, Geological Survey of Canada Open File 2437, 193

Russell D (1992) International ocean boundary issues and management arrangements'. In: Van der Zwaag D (ed) Canadian Ocean Law and Policy. Butterworths, Toronto, pp 463–507

Ryan S (1990) History of the Newfoundland cod fishery. Inshore Fisheries Workshop, Gander, Newfoundland. https://www.cdli.ca/cod/history5.htm

Ryan AG (1995) Native trees and shrubs of Newfoundland and Labrador. Park Interpretation Publication 14, Parks and Natural Areas Division, Newfoundland and Labrador

Samson C, Forget P, Pelletier A-M, Villeneuve F (2010) Population dynamics of the American marten (*Martes americana*) in an industrial forest of northwestern New Brunswick. https://www.fundymodelforest.net/images/pdfs/WS_2010_Jan28_Samson.pdf

Sanborn P, Lamontagne L, Hendershot W (2011) Podzolic soils of Canada: genesis, distribution, and classification. Can J Soil Sci 91:843–880

Schep LJ, Slaughter RJ, Becket G, Beasley DM (2009) Poisoning due to water hemlock. Clin Toxicol 47:270–278

Science Direct (2022) Taconic orogeny. https://www.sciencedirect.com/topics/earth-and-planetary-sciences/taconic-orogeny. Accessed 12 Aug 2022

Scotese CR, McKerrow WS (1990) Palaeozoic, palaeogeography and biogeography. Geol Soc Mem 12:1–21

Nova Scotia (2021) Forest vegetation types SH2—Hemlock—White pine/Sarsaparilla

Scott WB, Crossman EJ (1998) Freshwater fishes of Canada. Galt House, Oakville. Canada. Fisheries Research Board of Canada. Bulletin 184

Scott WB, Scott MG (1988) Atlantic fishes of Canada. Canadian Bulletin of Fisheries and Aquatic Sciences 219

Shaw J, Gareau P, Courtney RC (2002) Palaeogeography of Atlantic Canada 13–20 kyr. Quatern Sci Rev 21:1861–1878

Shaw J, Piper DJW, Fader GB, King EL, Todd BJ, Bell T, Batterson MJ, Liverman DGE (2006) A conceptual model of the deglaciation of Atlantic Canada. Quatern Sci Rev 25:2059–2081

Shaw J, Taylor RB, Forbes DL, Solomon S, Ruz, M-H (1999) Sensitivity of the coasts of Canada to Sea-level Rise. Geological Survey of Canada, Bulletin 505

Shellnutt JG, Owen JV, Yeh M-W, Dostal J, Nguyen DT (2019) Long-lived association between Avalonia and the Meguma terrane deduced from zircon geochronology of metasedimentary granulites. Sci Rep 9:4065. https://doi.org/10.1038/s41598-019-40673-9

Silinski A, Heuner M, Troch P, Puijalon S, Bouma TJ, Schoelynck L, Schröder U, Fuchs E, Meire P, Temmerman S (2016) Effects of contrasting wave conditions on scour and drag on pioneer tidal marsh plants. Geomorphology 255:49–62

Silver A, Gangopadhyay A, Gawarkiewicz G, Nishchitha E, Silva S, Clark J, (2021) Interannual and seasonal asymmetries in Gulf Stream Ring Formations from 1980 to 2019. Scientific Reports 11, Article number: 2207. https://doi.org/10.1038/s41598-021-81827-y

Simms J, Coates C, Coughlan G Mercer D (2007) The grand banks of Newfoundland: atlas of human activities. Oceans and Habitat Management Branch, Fisheries and Oceans Canada (Newfoundland and Labrador Region). https://www.dfo-mpo.gc.ca/oceans/publications/nfld-atlas-tnl. Accessed 13 Aug 2023

Slaymaker O, Catto NR, Kovanen DJ (2020a) Holocene geomorphic processes in Eastern Canada. In: Slaymaker O, Catto NR (eds) Landscapes and landforms of Eastern Canada. Springer International Publishers, Cham, Switzerland, pp 79–119

Slaymaker O, Catto NR, Kovanen DJ (2020b) Protecting geodiversity in eastern Canada. In: Slaymaker O, Catto NR (eds) Landscapes and landforms of Eastern Canada. Springer International Publishers, Cham, Switzerland, pp 557–582

Smith CAS, Webb KT, Kenney E, Anderson A, Kroetsch D (2011) Brunisolic soils of Canada: genesis, distribution, and classification. Can J Soil Sci 91:695–717

Snow D (1996) Land mammals of Newfoundland and Labrador. St. John's, Newfoundland Dept. Tourism, Culture and Recreation

Soil Classification Working Group (1998) The Canadian system of soil classification, 3rd ed. Agriculture and Agri-Food Canada, Publication 1646

Soil Classification Working group (2001) Soils of Canada. Agriculture and Agri-foods Canada, Research Branch. Scale 1:6,500,000 map. https://sis.agr.gc.ca/cansis/publications/maps/soc/all/soils/soc_all_soils_2004en.png. Accessed 25 July 2022

Staal CR van, Whalen JB, McNicoll VJ, Pehrsson S, Lissenberg CJ, Zagorevski A, Breemen O. van, Jenner GA (2007) The Notre Dame arc and the Taconic orogeny in Newfoundland. Geol Soc Am Mem 200:511–552

Stea RR, Conley H, Brown Y, (1992) Surficial geology of Nova Scotia Map ME 1992–003. Nova Scotia Geological Survey

Stea R, Seaman AA, Pronk T, Parkhill MA, Allard S, Utting D (2011) The Appalachian glacier complex in Maritime Canada. In: Ehlers J, Gibbard PL, Hughes PD (eds.), Quaternary glaciations—extent and chronology: a closer look. Dev Quat Sci 15:631–659

Stewart PL, Arnold SH (1994) Environmental requirements of the sea scallop (*Placopecten magellanicus*) in eastern Canada and its response to human impacts. Can Tech Rep Fish Aquat Sci 2005:1–36

Stewart J (1806) An account of Prince Edward Island in the Gulph of St. Lawrence, North America. Wakefield. S.R. Publishers, England; Johnson Reprint Corporation, New York 1967

Stommel H (1966) The gulf stream. University of California Press

Story GM Kirwin WJ Widdowson JDA (1997) Dictionary of Newfoundland English. https://www.heritage.nf.ca/dictionary/#3277

Strong JS, Leroux SJ (2014) Impact of non-native terrestrial mammals on the structure of the terrestrial mammal food web of Newfoundland. Canada Plos ONE 9(8):e106264. https://doi.org/10.1371/journal.pone.0106264

Todd BJ (2016a) Recessional moraines on the southern Scotian Shelf of Atlantic Canada. Geol Soc Lond Mem 46:257–258

Todd BJ (2016b) De Geer moraines on German Bank, southern Scotian Shelf of Atlantic Canada. Geol Soc Lond Mem 46:259–260

Todd BJ, Fader GBJ, Courtney RC, Pickrill RA (1999) Quaternary geology and surficial sediment processes, Browns Bank, Scotian Shelf, based on multibeam bathymetry. Mar Geol 162:165–214

Todd BJ, Shaw J, Valentine PC (2016) Submarine glacial landforms on the Bay of Fundy–northern Gulf of Maine continental shelf. Geol Soc Lond Mem 46:429–436

Todd BJ (2016c) The Laurentian channel: a major cross-shelf trough in Atlantic Canada. In: Dowdeswell J, Canals M, Jakobsson M, Todd BJ, Dowdeswell EK, Hogan KA (Eds) Atlas of submarine glacial landforms: modern, quaternary and ancient. Geol Soc Lond Mem 46:161–162

Townsend CA, Gault I (1984) The making of a maritime boundary: the Gulf of Maine case'. Oil Gas Law Tax Rev 17:173–181

Trenhaile AS, Pepper DA, Trenhaile RW, Dalimonte M (1998) Stacks and notches at hopewell rocks, New Brunswick, Canada. Earth Surf Proc Land 23:975–988

Tsuga canadensis—Pinus strobus/Aralia nudicaulis. https://novascotia.ca/natr/forestry/veg-types/sh/sh2.asp

Van de Poll HW (1989) Lithostratigraphy of the Prince Edward Island redbeds. Atlantic Geol 25:23–35

Van der Zwaag D (1983) The fish feud: The US and Canadian boundary dispute. Lexington Books, Lexington, MA, pp 89–94

Vasseur L, Catto N (2008) Atlantic Canada. In: Lemmen DS, Warren FJ, Lacroix J, Bush E (eds) From impacts to adaptation: canada in a changing climate 2007. Government of Canada, Ottawa, pp 119–170

Warner BG Rubec CD. 1997. The Canadian wetland classification system. Wetlands Research Centre, University of Waterloo, Waterloo, p 68

Warren F, Lulham N (2021) Canada in a changing climate: national issues report. Government of Canada, Ottawa, ON

Weber JT (2022) Traditional uses and beneficial effects of various species of berry-producing plants in eastern Canada. Botany 100:175–182

Woodrow Ef Heringa Pk (1987) Pedoclimatic Zones of the Island of Newfoundland Report 32 Newfoundland Soil Survey

Wright M (1998) The shadow of the past: an historical perspective on the newfoundland cod fishery, 1950–1992. World Wildlife Fund's Endangered Seas Campaign—Series Title, pp 68–82. https://scholar.uwindsor.ca/historypub/100

Zinck M (1998) Roland's flora of Nova Scotia. Nimbus Publishing, Halifax

3 Boreal Shield and Boreal Plains Ecozones

Abstract

The Boreal Shield extends from the coastline of southern Labrador, across the Laurentides of Québec, north of Georgian Bay and Lake Superior, westward to north-central Manitoba and northern Saskatchewan. The border between the Boreal Shield and Boreal Plains ecozones is defined by the easternmost outcrops of Palaeozoic and Mesozoic sedimentary rock. Climates in the Boreal Shield ecozone generally fall with the Köppen-Geiger Dfb (mid-boreal) regime. In the Boreal Plains, climates vary from Köppen-Geiger Dfb (mid-boreal) to southernmost Dfc (northern boreal — taiga). Generally, precipitation decreases from east to west across the ecozones. Geologically and physiographically, the rolling topography of the Precambrian Canadian Shield is flanked by plains and platforms of progressively younger sedimentary rock of Palaeozoic and Mesozoic ages. In Western Canada, the Boreal Plains lie within the Interior Platform physiographic region, extending west to the Foothills of the Cordillera. The landscapes reflect Quaternary glaciation between 100,000 and 10,000 years ago, combined with sediments deposited in ice-dammed lakes during glacial retreat, meltwater channels, and modern geomorphic processes. The natural vegetation assemblages of the Boreal Shield ecozone varied from mixed, deciduous-dominated in the southernmost areas, to mixed but conifer-dominated in central regions, to conifer-dominated with lesser deciduous trees in the northern and western regions. Logging and the cessation of anthropogenic burning have resulted in substantial changes to the vegetation assemblages in the southernmost and central areas. In the Boreal Plains, the Boreal Forest communities are dominated by Populus tremuloides (trembling aspen), Picea glauca (white spruce), Betula papyrifera (white birch), and Pinus banksiana (jack pine). Moister areas support Abies balsamea (balsam fir), Picea mariana (black spruce), and Larix laricina (tamarack). Wetlands are important elements in the landscape. Organic wetlands dominated by peat accumulation represent more than 90 percent of the total wetland area. Soils in low-lying areas are poorly drained, leading to the development of seasonally saturated, iron-rich Fera Gleysols. In well-drained situations, soils range from weakly developed regosols and brunisols to moderately developed humo-ferric podzols. Soils of the Boreal Plains ecozone reflect the underlying bedrock and Quaternary sediment. Southern to mid-boreal faunal assemblages typically include porcupine, moose, and white-tailed deer. Smaller herbivores include muskrat, beaver, snowshoe hare, chipmunk, woodchuck, red squirrel, and northern flying squirrel. Omnivores include raccoon, red fox, black bear, and skunk. Wolverine is present, albeit in small, threatened populations. Carnivores include lynx, mink, North American ermine, otter, and pine marten.

Keywords

Boreal shield · Boreal plains · Boreal forest · Wetlands · Canadian shield · Interior plateau

3.1 Introduction

This chapter discusses the natural environments and landscapes of the Boreal Shield and Boreal Plains ecozones (Fig. 3.1). The Boreal Shield extends from the coastline of southern Labrador, across the Laurentides of Québec, north of Georgian Bay and Lake Superior in Ontario, westward to north-central Manitoba and northern Saskatchewan. Although it is boreal climatically and ecologically, the island of Newfoundland is discussed in the Atlantic Chap. 2. Protected areas include La Mauricie, St. Lawrence Islands, Georgian Bay Islands, and Pukaskwa National Parks, provincially administered Parc-des-Laurentides, Mont-Tremblant, La Verendrye, Killarney, Algonquin, Quetico, and Lac La Ronge parks, and Pimachiowin Aki World UNESCO Heritage Site.

N. Catto and G. R. Vincent, *Canada: Land and People*, World Regional Geography Book Series,
https://doi.org/10.1007/978-3-032-04391-7_3

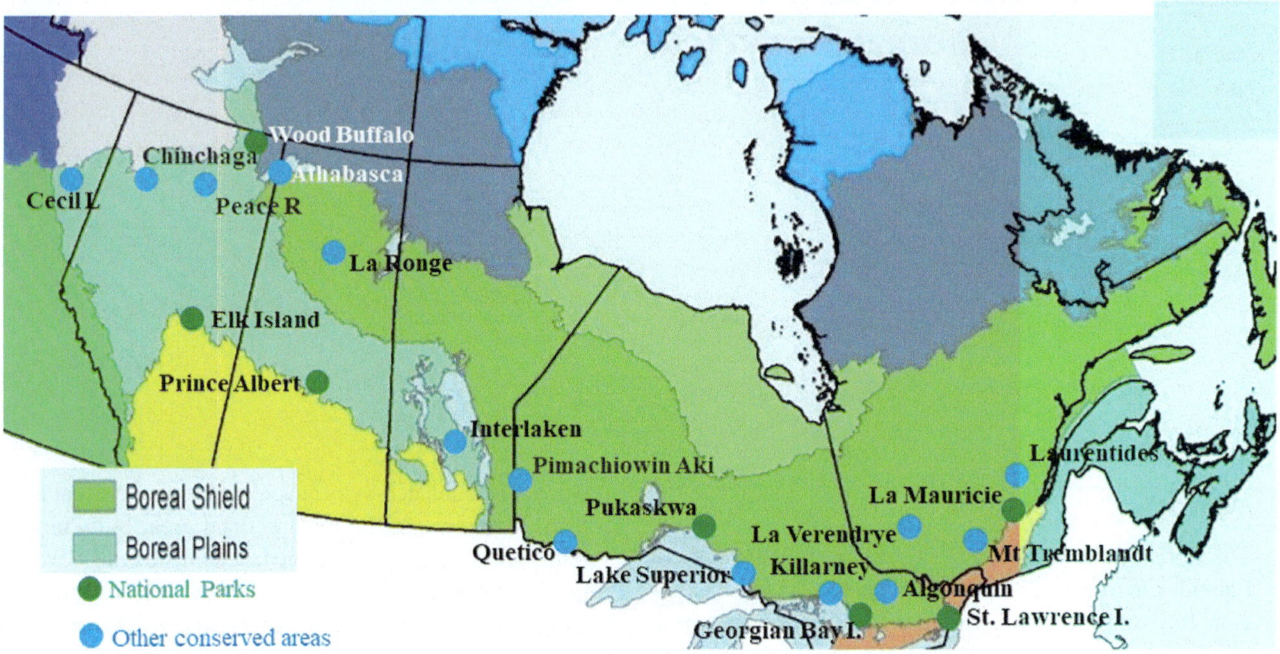

Fig. 3.1 Boreal plains and Boreal shield ecozones, with locations discussed in the text

The border between the Boreal Shield and Boreal Plains ecozones is defined by the easternmost outcrops of Palaeozoic and Mesozoic sedimentary rock. The Boreal Plains ecozone is bordered to the north by Taiga Plains (discussed in Chap. 8) and to the west by the Cordillera (Chap. 6). To the south, the Boreal forest grades ecologically into the Aspen Parkland and Mixed Grassland zones. In this volume, we've considered the Aspen Parkland and the ecotonal regions straddling the theoretical border, including Riding Mountain and Prince Albert National Parks, within the Prairie grasslands Chap. 5. Protected areas within the Boreal Plains include Wood Buffalo National Park (AB-NWT), Cecil Lake Ecological Reserve (BC), and Athabasca Sand Dunes (SK), Peace River Wildlands (AB), and Chinchaga Wildlands (AB) Provincial Parks. Parks Canada has proposed the establishment of a national park in the Interlake area of Manitoba.

3.2 Climate

Climates in the Boreal Shield ecozone generally fall with the Köppen-Geiger Dfb (mid-boreal) regime (Beck et al. 2006; Figs. 3.2, 3.3, 3.4, 3.5, 3.6, 3.7, 3.8 and 3.9). In the Boreal Plains, climates vary from Köppen–Geiger Dfb (mid-boreal) to southernmost Dfc (northern boreal—taiga). Generally, precipitation decreases from east to west across the ecozones.

3.2.1 Boreal Shield

January mean temperatures for the mid-boreal climates of the Boreal Shield range between −8 °C and −16 °C (Environment and Climate Change Canada 2020; Figs. 3.2 and 3.3). For July, the means range from 15 °C (along the shore of Lake Superior) to 23 °C (inland). Inland areas display large ranges between summer and winter temperatures, producing a moderate continental climatic effect. Extreme temperatures throughout the year differ by as much as 65 °C.

Along the shore of Georgian Bay, temperatures in August and early September are as much as 5 °C cooler than those 10 km inland, and onshore breezes are common. During the fall, the gradual decline in surface water temperatures, in contrast to the rapid cooling of the land, reverses the temperature gradient. Lakefront temperatures are warmer than those inland throughout the autumn and remain warmer until the margins of the islands freeze (January). The central part of Georgian Bay does not freeze, however, and onshore breezes bring cold air and blizzards across the entire region.

The mean surface temperature of the nearshore areas of Lake Superior during July, its warmest month, is 9 °C. The lake does not act to moderate the winter climate. The north shore of Superior freezes in late December-early January and remains frozen into mid-April, with ice cover influenced by the prevailing winds. Variations in timing have shown no consistent trend since the first records were kept in the early

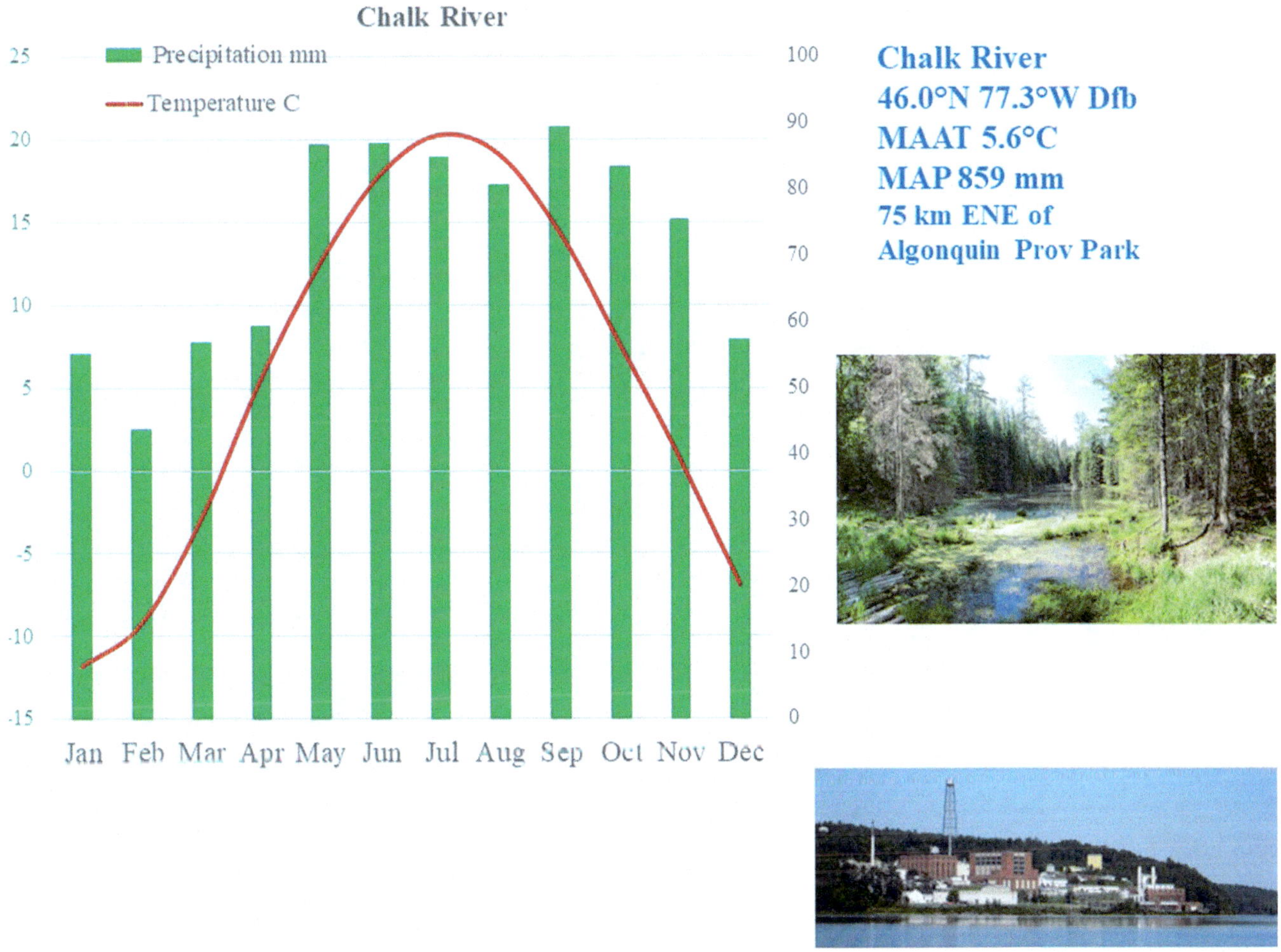

Fig. 3.2 Chalk river laboratories, located 180 km northwest of Ottawa along the Ottawa river, and 75 km ENE of Algonquin provincial park, has a Köppen-Geiger Dfb climate. (Climate data from Environment and Climate Change Canada 2020)

1850s. The lake last froze over completely in Winter 1996, and in some years, coverage was less than 10% (NOAA 2022), but there is no consistent trend.

Mean annual precipitation in the Dfb Boreal Shield varies between 900 and 1300 mm, with higher totals characteristic of areas subjected to onshore breezes from Georgian Bay and Lake Superior (Fig. 3.6). Thunderstorms occur regularly (10–20 per year), but seldom attain the violence characteristic of those of southern Ontario (Phillips 1990). Typically, 25–30% of precipitation falls as snow. Snow cover persists for 140–170 days, and the mid-winter snowpack commonly reaches 120 cm depth in forested areas. Northwest winter winds come from the Arctic High-pressure system centred over Kivilliq, west of Hudson Bay. Snowfall produces suitable conditions for both downhill and cross-country skiing. Many of Québec's ski resorts are located in the southern Laurentides (see Fig. 3.5).

3.2.1.1 Algonquin Provincial Park

Despite the precipitation totals, summers and autumns are generally sunny over most of the Boreal Shield, with a mean of 2100 h of sunshine/year in Algonquin Provincial Park and vicinity (Environment and Climate Change Canada 2020). Without anthropogenic influences, forest fires might be expected to burn each hectare of Algonquin every 70–100 years. However, the fire cycle between 1636 and 1779 was artificially reduced due to lesser indigenous use of the forest, and accelerated after 1780 due to logging and camping, reaching 1 per 17 years by 1940 (Guyette and Dey 1995). After 1945, the Ontario Ministry of Natural Resources (and its predecessors) aggressively practised a total fire suppression policy, only recently partly modified. In Algonquin, the changes in fire frequency say much about human usage and policy—reduced due to disease among the indigenous population after 1636, followed by enhanced pressure,

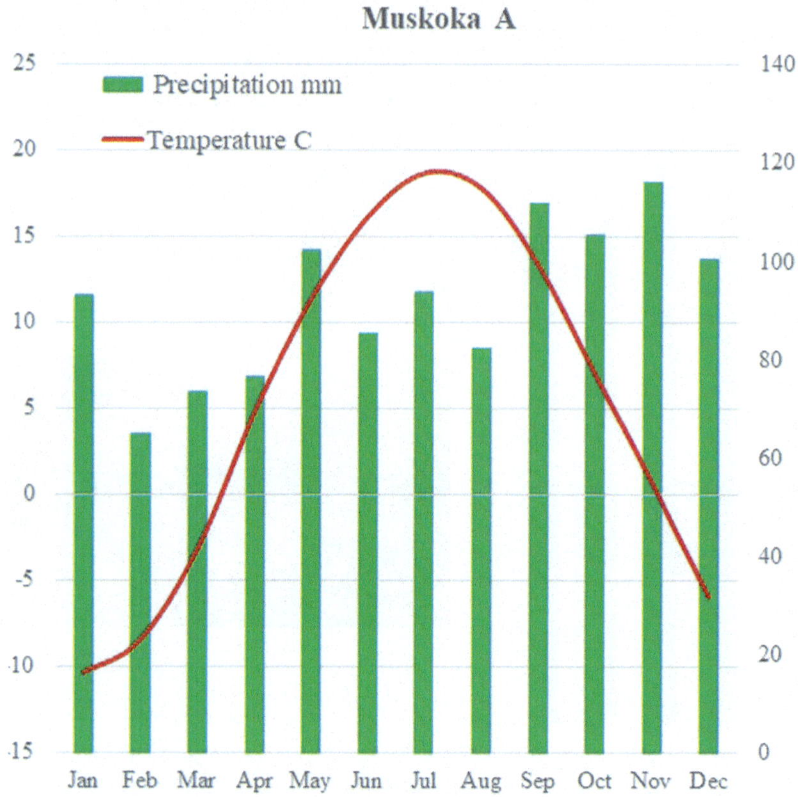

Muskoka Airport
45.0°N 79.3°W Dfb
MAAT 5.2°C
MAP 1105 mm
80 km SW of Algonquin Prov Pk; 40 km ENE of Georgian Bay Islands National Park

Fig. 3.3 Muskoka airport ("A"), located midway between Gravenhurst and Bracebridge 140 km north of Toronto, has a Köppen-Geiger Dfb climate. Muskoka is a major recreational area for southern Ontario. (Climate data from Environment and Climate Change Canada 2020)

followed by recent total suppression—and relatively little about changing climate or forest composition. Fire danger generally decreased in the Boreal Shield of Ontario and Québec between 1901 and 2002 (Girardin et al. 2013), making it difficult to predict future effects of climate change based on historical data. In most areas, there is no purely natural baseline for comparison throughout the past 500 years.

The major environmental problem in Algonquin, and to a lesser extent elsewhere in the Laurentide Grenville Province, in the past 50 years has been acidic precipitation, which poses significant problems for organisms (Neary et al. 1990, Ouimet et al. 2006; Fig. 3.4). Under non-industrialized conditions, precipitation in eastern Canada would have a pH of approximately 5.0 (slightly acidic), a value close to the pH of most natural lake waters in the region. Organisms have adjusted over time to these conditions. The lake waters lack any substantial buffering capacity, due to the granitic and gneissic bedrock in which they sit, and thus the acidity of the lakes quickly reflects the acidity of the precipitation.

Industrialization in adjacent regions lowered the mean pH of rainfall in Algonquin Park to 4.2–4.0 by the early 1990s (Environment Canada 2013), thus increasing the concentration of hydrogen and sulphide ions (and acidity) in the rainfall by approximately 800–1000%. Rainfall of this acidity is sufficient to cause serious harm to aquatic organisms. Individual rainfall events had much lower pH values, and the impact of acid rain was thus severe.

Wind direction is important in determining how much acid rain reaches the park throughout the year, and the magnitude of its effect on particular organisms. During the winter months, winds are relatively light and originate from the northwest. The areas to the northwest of Algonquin are not heavily settled or industrialized, and precipitation during this period generally does not contain an excess of hydrogen and sulphide ions or acidity.

The winds gradually shift during the early spring, moving successively from northwest to west to southwest by early summer. During this period, the shifting winds pass over

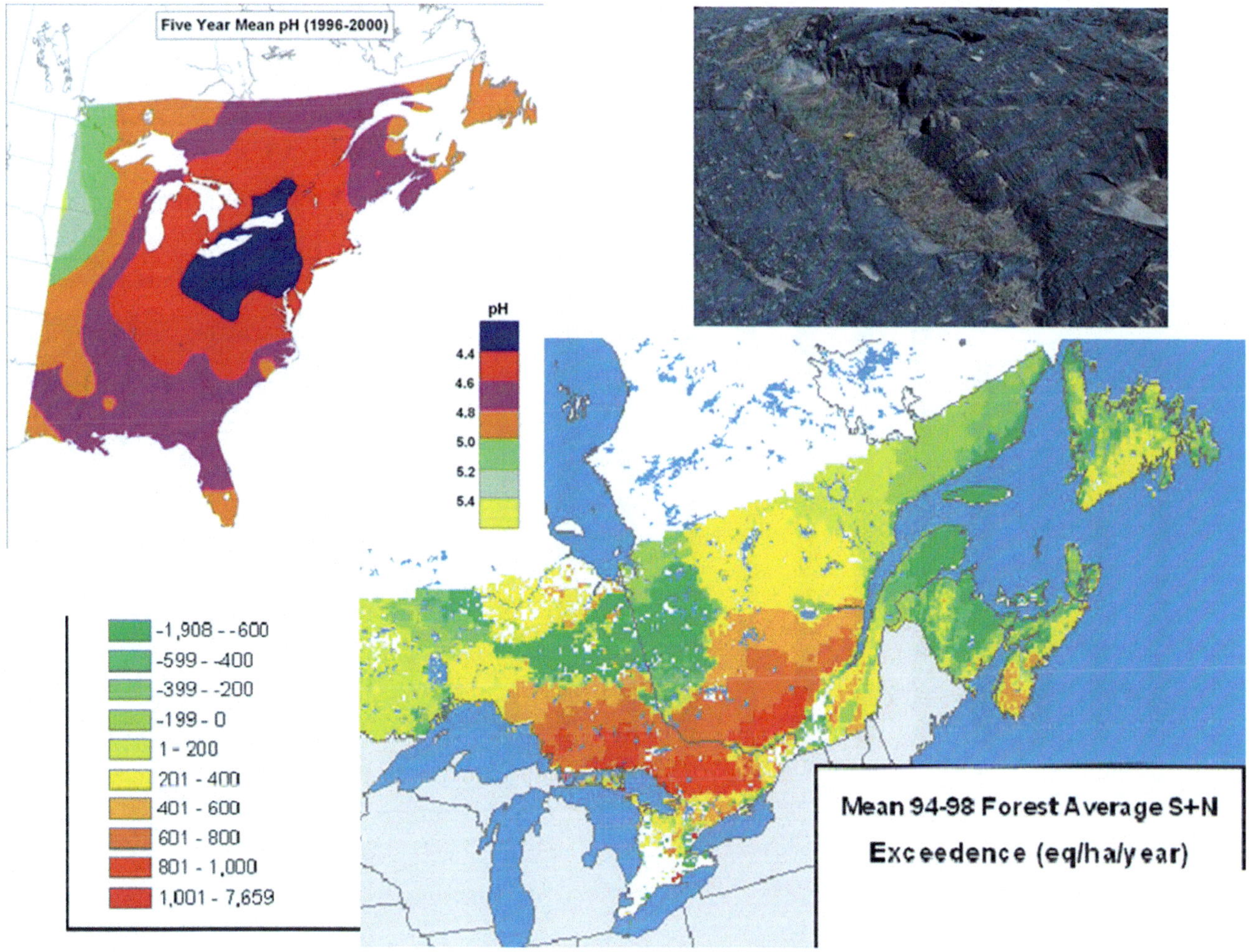

Fig. 3.4 Acid precipitation is a significant environmental issue in the Boreal Shield of eastern Canada. Left: mean pH values for 1996–2000. Low pH rainfall has a greater adverse effect on the acidic bedrock of the Grenville geological province than on the carbonate bedrock north of Lake Ontario. Lower right: Mean exceedance values for sulphur (sulphuric acid) and nitrogen (nitric acid) in precipitation, 1994–1998. High values characterize the Boreal Shield region from Sudbury to Algonquin Park and La Mauricie, resulting in stress on boreal forests (Environment and Climate Change Canada 2022). Inset: bedrock contaminated by acidic precipitation resulting from early twentieth century open roasting of nickel ore, Sudbury

a series of industrialized and acid pollution-emitting areas. These winds can bring acidic precipitation to the park, and hence any organism attempting to breed or grow during the mid-spring to early summer period is subject to interference. During mid-summer, the winds begin to undergo a reciprocal shift, and by mid-autumn, the northwesterly pattern is re-established. Organisms are thus potentially subject to severe doses of acid precipitation throughout the primary growing season.

The 1991 United States-Canada Air Quality Agreement was initiated to address the problem of acid precipitation, by limiting emissions of sulphur dioxide and nitrous oxides. Substantial progress has been made in the past thirty years: total Canadian SO_2 emissions dropped by 78% between 1990 and 2020 (Environment and Climate Change Canada 2022). Annual precipitation-related sulphate deposition declined in Algonquin Park from approximately 20 kg ha^{-1} y^{-1}in 1990 to approximately 10 kg ha^{-1} y^{-1} in 2010, with a similar rate of decline in precipitation-related nitrate deposition over the same period. Chemical analyses show that Algonquin's lakes are showing reduced concentrations of sulphates and nitrates and increasing pH, along with increases in dissolved organic carbon content, indicating that some recovery is under way. Unfortunately, ecological damage from previous episodes of acidic precipitation remains (c.f. Hadley et al. 2016) and will require substantial time to be ameliorated or reversed.

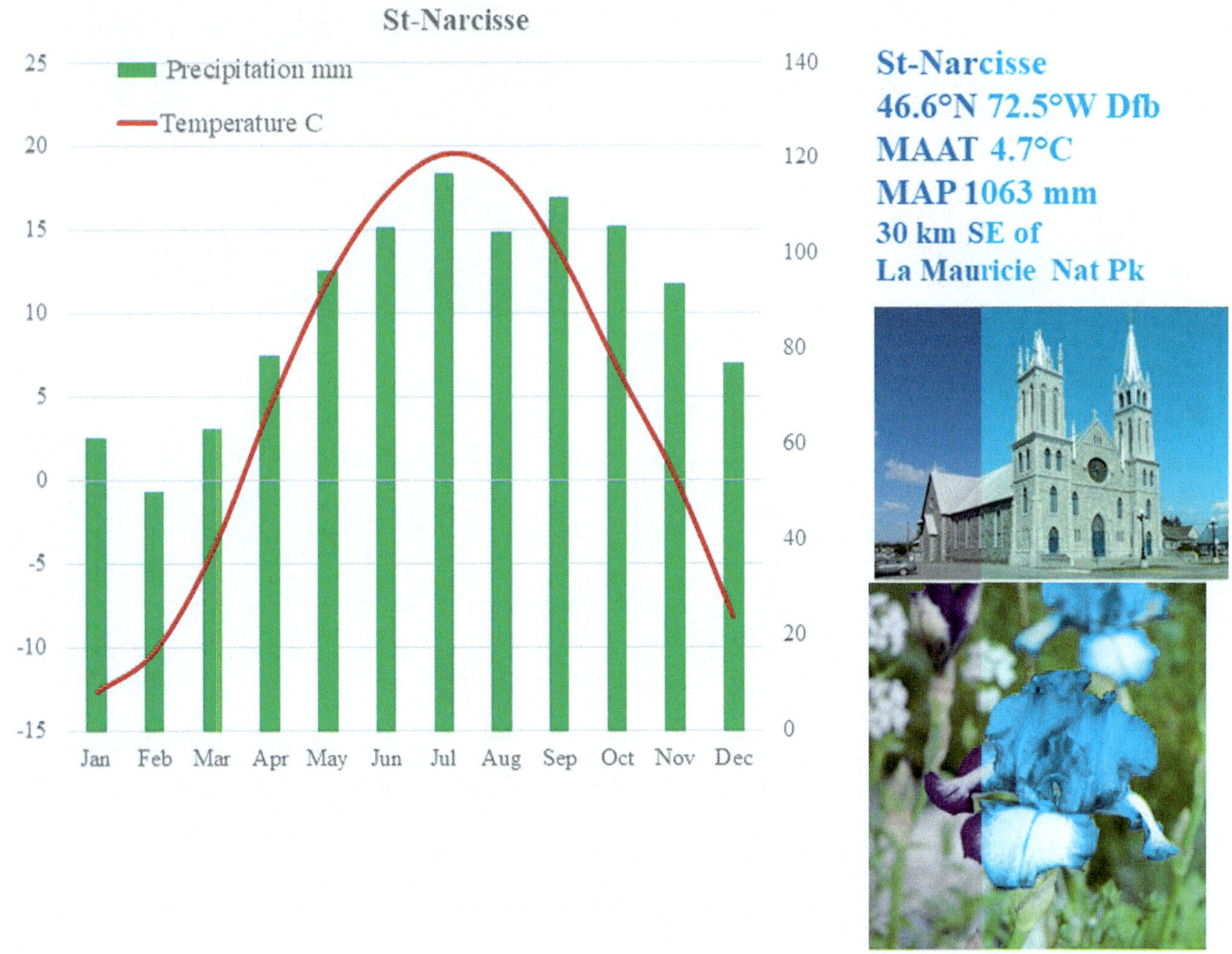

Fig. 3.5 St-Narcisse, Québec, located 30 km southeast of La Mauricie, has a Dfb climate. Inset: northern blue flag iris (*Iris versicolor*), official flower of Québec since 1999. (Climate data from Environment and Climate Change Canada 2020)

3.2.2 Boreal Plains

The Boreal Plains ecozone is relatively dry, receiving 350–600 mm of precipitation per year (Fig. 3.10). Violent weather events, such as thunderstorms and hailstorms, are much less common than across the southern prairies, and tornadoes are extremely rare. Approximately 30% of the precipitation falls as snow, and snow cover persists for 150–190 days, dependent upon aspect and elevation. Prevailing winds are westerly (Phillips 1990; Environment and Climate Change Canada 2020).

Wood Buffalo lies in the transition zone between the mid-boreal (Köppen-Geiger Dfb) and northern boreal (Köppen-Geiger Dfc) climatic regimes (Fig. 3.11). Daily mean temperatures are −24 °C in January and 17 °C in July. The large temperature range, indicative of a continental climatic setting, puts great stress on plant life. The area characteristically receives 25–30 high summer days per year, and about an equal number of days with minimum temperatures less than −35 °C in the winter. Despite the small amount of precipitation, the cold winter temperatures limit evaporation, and the area has a positive net moisture budget and an overall humid moisture regime.

Along the western margin of the Boreal Plains, the Peace River Lowland is marked by cold winters (January mean −18 °C) and warm summers (July mean 17 °C). Annual precipitation is 600 mm, 200 cm of which falls as snow. The lowland has a humid moisture regime (MI > 0), a positive net moisture budget, and prevailing westerly winds. Topographic effects along the deeply incised Peace River valley have created a ' "thermal oasis", where temperatures regularly exceed those of the surrounding uplands and prairies by 5 °C (Environment and Climate Change Canada 2020). Unfortunately, warmer summer temperatures have led to local water shortages for agriculture.

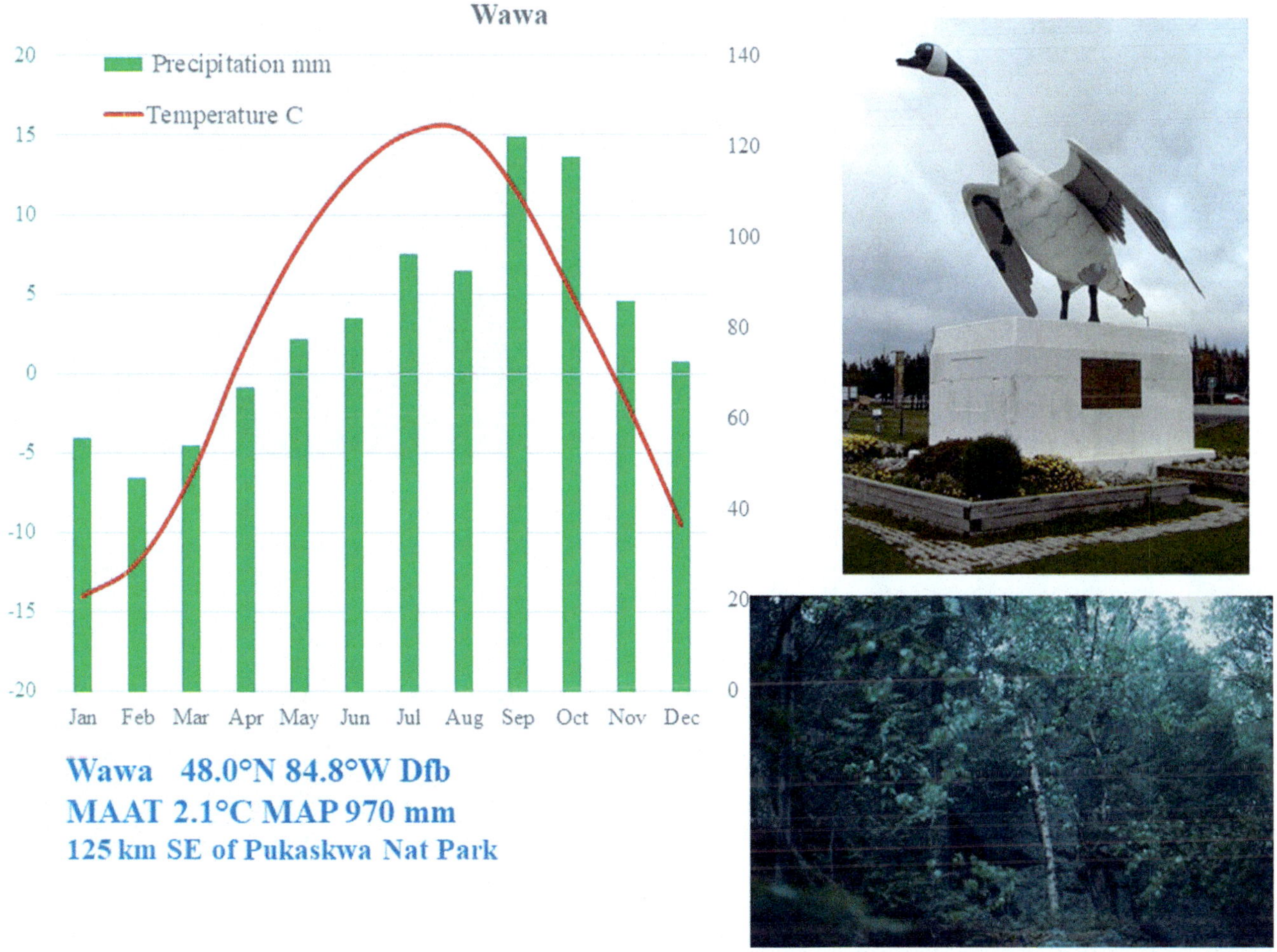

Fig. 3.6 Wawa, inland from Lake Superior and 125 km southeast of Pukaskwa National Park, has a Köppen-Geiger Dfb climate. The general westward trend for reduced precipitation is apparent in comparing Muskoka, Wawa, and Kenora. (Climate data from Environment and Climate Change Canada 2020)

3.3 Bedrock

Geologically and physiographically, the rolling topography of the Precambrian Canadian Shield is flanked by plains and platforms of progressively younger sedimentary rock of Palaeozoic and Mesozoic ages (Bostock 2014). In eastern Canada, the younger rocks underlie the St. Lawrence Platform (discussed in Chap. 4). In Western Canada, the Boreal Plains lie within the Interior Platform physiographic region, extending west to the Foothills of the Cordillera.

3.3.1 Boreal Plains

Palaeozoic limestones and dolomites form lake cliffs, flat platforms, and karst topography in the Interlake region of the Boreal Plains, between Lakes Winnipeg and Winnipegosis (Natural Resources Canada 2009; Fig. 3.12). Hecla/Grindstone Provincial Park displays examples of these landscapes (Manitoba 2022). Subterranean dissolution by groundwater has produced caves and tunnels, linked to the surface by near-vertical oval and dumbbell-shaped dolines. Erosion of limestone has produced karst topography with collapse dolines, such as Little Limestone Lake (Ford 2017).

3.3.1.1 Wood Buffalo

Karst topography exposed in the eastern part of Wood Buffalo National Park is developed on mid Devonian interbedded limestones, shales, and evaporite gypsum, anhydrite, halite, and sylvite (Mossop and Shetsen 1994, Ford 1997). Alexandra Falls (Hatto deh Naili, 32 m high) is located where the Hay River tumbles over resistant limestone, 30 km southwest of Tu Nedhe (Great Slave) Lake.

The evaporite bedrock units are subject to chemical dissolution by groundwater and by rainwater (Fig. 3.13). Surface exposures of halite (NaCl) and sylvite (KCl) dissolve rapidly,

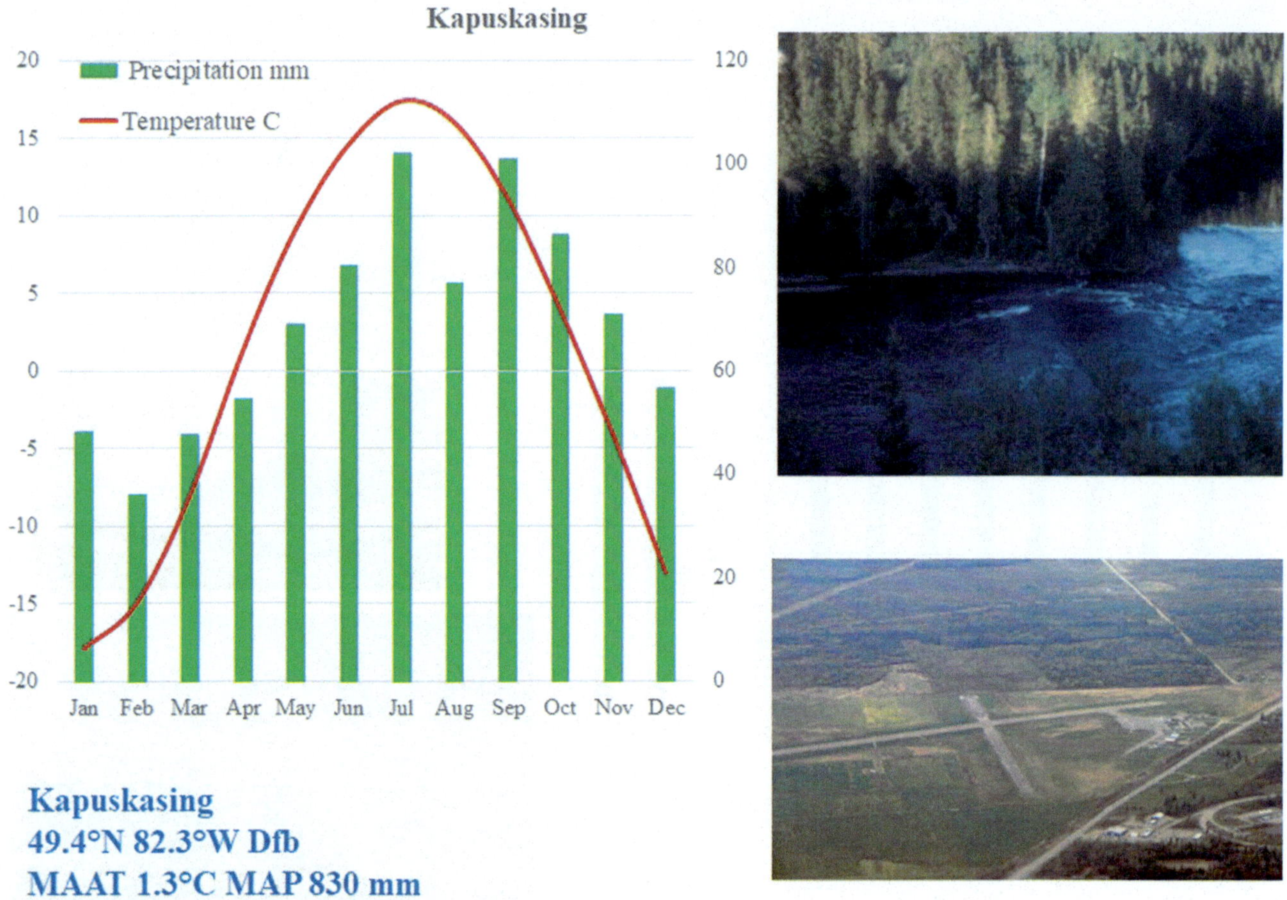

Fig. 3.7 Kapuskasing lies near the northern limit of the mid-boreal Dfb climate zone. Vegetation is dominated by spruce-fir coniferous forests. (Climate data from Environment and Climate Change Canada 2020)

producing saline flats and shallow saucer-shaped basins, such as the potholes of Sass River.

Cold saline springs are present at the bases of most limestone escarpments. The discharge of sodium nitrate, calcium sulphate, sodium chloride, potassium nitrate, and other salts at the bases of the escarpments onto the flat terrain surrounding the Salt and Little Buffalo Rivers has produced large salt flats.

The surface sediments are calcareous, and the lake bottoms are floored with soft calcium carbonate marl and mud. The lakes are rimmed with *Picea mariana* (black spruce), *Larix laricina* (tamarack), and *Salix candida* (sageleaf willow; Cody 1995). The shallowest potholes are the most alkaline (pH 7.6–8.3) and support *Scirpus validus* (bulrush). Potholes deeper than 1.5 m are less alkaline (pH 7.2–7.3) and support *Typha latifolia* (cat-tail) and *Carex aquatilis* (sedge) instead of bulrush.

Local areas of saline terrain are present, where discharging groundwater brings salts from the Devonian bedrock to the surface. In these terrains, solonetz soils support a halophyte assemblage with *Salicornia rubra* (samphire), *Glaux maritima* (sea milkwort), and *Dodecatheon sp.* (shooting star). Arboreal vegetation cannot survive in these areas. The saline marshlands and shallow ponds provide suitable habitats for several small crustaceans, which are a major food source for the region's bird life.

The evaporate deposits are overlain sequentially by Upper Devonian limestones and shales, and conglomerates and sandstones of the Cretaceous McMurray Formation. The petroleum-bearing strata of the Athabasca Oil Sands are associated with the middle sandstone unit of the McMurray Formation (Mossop and Shetsen 1994). This formation is exposed along the entire extent of the Athabasca River from Fort McMurray north to Wood Buffalo.

3.3.1.2 Western Boreal Plains

The western parts of the Boreal Plains are underlain by shales and sandstones of late and mid-Cretaceous age (Fig. 3.14). Sandstones form some resistant hills, such as the Birch Hills and Caribou Hills, and shales crop out in low-lying regions, such as the Peace River valley. Along the southern margin

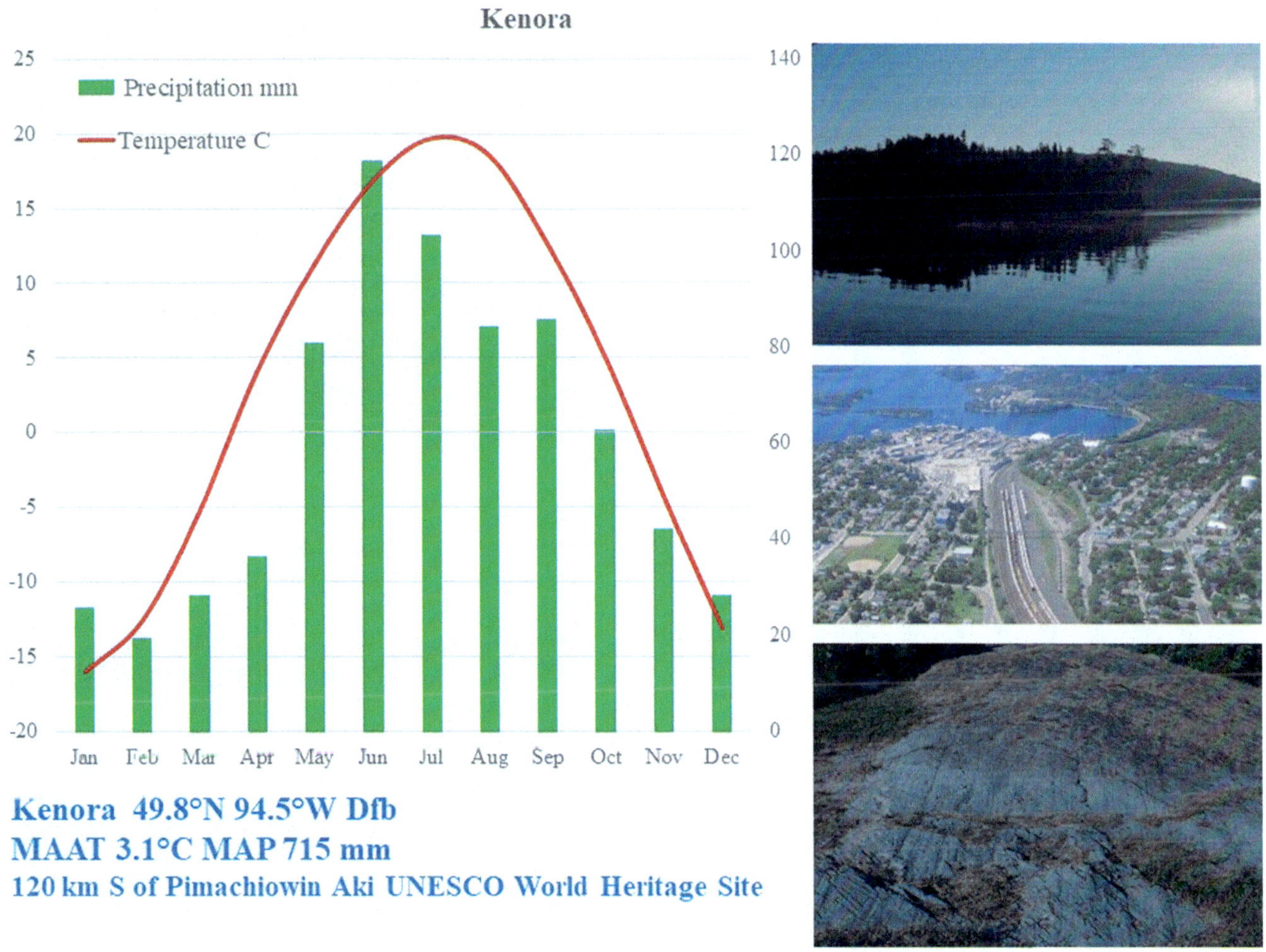

Fig. 3.8 Kenora, in westernmost Ontario, has a Köppen-Geiger Dfb climate. The region is marked by glaciated landscapes of the Superior geological province. (Climate data from Environment and Climate Change Canada 2020)

of the Canadian Shield, shale beds are particularly susceptible to glacial ice thrusting. The weaker substrate shale was ripped out and thrust along failure planes beneath the glacial sole. Detached slabs adhere to the glacier base (Aber and Ber 2007; Catto 2015). Many shale areas subject to ice thrusting are marked by deeper lakes, such as Waskesiu in Prince Albert National Park. Throughout much of the western Boreal Plains, the bedrock does not form a dominant influence on the topography, as it is generally covered by thick glacial deposits.

To the southwest, the bedrock has been affected by northeast-directed thrusting activity from the Rocky Mountains, and eastward-dipping strata are common. Even here, the bedrock structure has little influence on the topography or geomorphic development of the region. The high concentrations of montmorillonite and bentonite in the strata, however, together with their poorly consolidated nature, greatly increase the susceptibility of these rocks to slope failure following fluvial or anthropogenic undercutting or overloading.

3.3.2 Boreal Shield

From west to east, the Boreal Shield contains rocks associated with the Churchill, Superior, Keweenawan, and Grenville geological provinces (Natural Resources Canada 2009).

3.3.2.1 Churchill Province

The Churchill Geological Province is defined and named for the Churchill River (Nehinaw: Missinipi), which extends from northeastern Alberta across Saskatchewan, reaching Hudson Bay at Churchill, Manitoba (Fig. 3.15). The complexes of igneous and metamorphic geological units, ranging in age from Archaean to Neo-Proterozoic, produce landscapes marked by rectangular drainage patterns marked by 90° bends and confluences. Related contorted patterns develop in areas where contorted metamorphic or folded sedimentary bedrock is exposed. Both drainage patterns indicate strong structural control from the underlying bedrock, with little or no sedi-

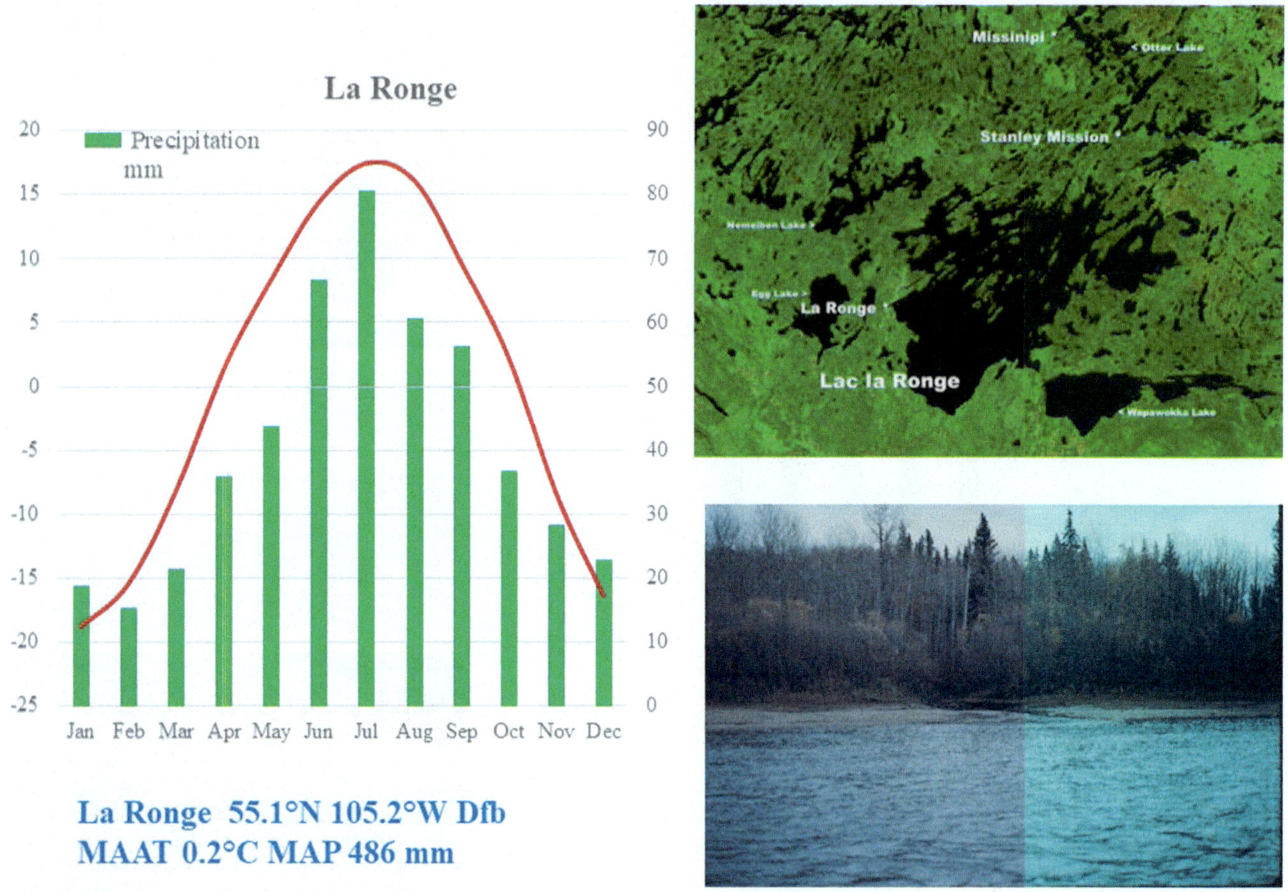

Fig. 3.9 La Ronge, located along the southern margin of the Churchill geological province, has a classical interior Köppen-Geiger Dfb climate marked by dry winters and wet summers. (Climate data from Environment and Climate Change Canada 2020) The Orbis (NASA) satellite image illustrates the contrast between the exposed Churchill metamorphosed terrain in the north and the sedimentary rock of the Boreal Plains to the south

ment cover, on gentle to moderate slopes (Catto 2015). Both the smallest glint lakes, and the largest such as Wollaston and Reindeer, are marked by deeply indented shorelines and small islands and skerries. The low relief causes Wollaston Lake to display bifurcated drainage, with 90% of the water draining into the Churchill drainage basin and 10% draining west into the Fond-du-Lac River and Athabasca drainage basin: the "continental divide" between Arctic Ocean and Hudson Bay drainage passes through Wollaston Lake. The waters of Reindeer Lake, artificially controlled by the Whitesand Dam, drain into the Reindeer and Churchill rivers. Deep Bay, near the southern end of Reindeer Lake, is a 13 km wide and 220 m deep circular feature, formed by a meteorite impact approximately 99 million years ago (Ma) (Grieve 2001).

3.3.2.2 Superior Province

Rocks of the Superior geological province are exposed across central Manitoba, northern Ontario, and northwestern Québec, including at Pimachiowin Aki World UNESCO Heritage Site, Quetico Provincial Park, and Pukaskwa National Park (Fig. 3.16). The rocks of the Superior Province, the oldest area of the Canadian Shield, were formed during the Archaean Eon, and were deformed during the Kenoran Orogeny, 2480 Ma (Card 1990; Hocq 1994a; Hocq and Verpaelst 1994, Percival 2007). The Superior province landscape shows more varied relief than does the Churchill geological province, although it is generally less rugged and more rolling than the Grenville (Laurentian) region of the Canadian Shield (Bostock 2014).

Pukaskwa straddles the boundary between two geological provinces of the Canadian Shield. The majority of the park lies within the Superior Geological Province, consisting of a series of belts of alternating resistant acidic gneiss and granite, locally cut by diabase dykes, and weaker volcanic (primarily basalt) and sedimentary rocks derived from the debris produced by explosive volcanic eruptions ("volcaniclastics"). The weaker volcanic rocks erode to produce northeast-southwest oriented valleys, separated by gneiss and

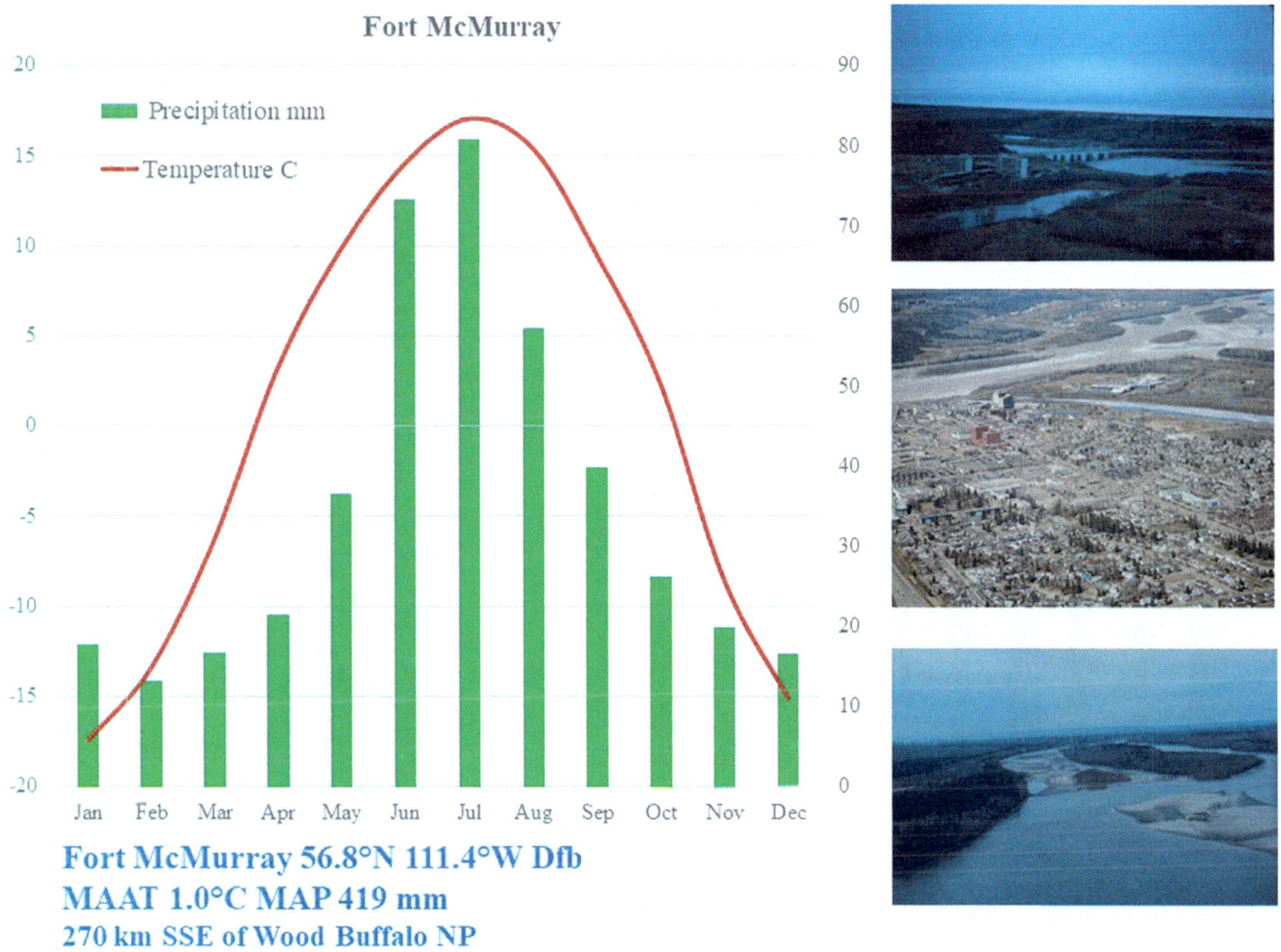

Fig. 3.10 Fort McMurray, along the Athabasca River 270 km SSE of Wood Buffalo National Park, is the largest city in the Boreal Plains. It lies near the northern limit of the mid-boreal Dfb climate zone. (Climate data from Environment and Climate Change Canada 2020)

granite ridges, giving a corrugated appearance to the landscape. Faults have accentuated the pre-existing ridge-and-valley topography.

This style of alternating volcanic and gneiss/granite bedrock units is common throughout the Superior Province (Percival 2007). Chlorite and amphiboles in the volcanic rocks commonly weather to a dark green on the surface, and the units are referred to as "greenstone belts". The basalts and volcaniclastic sediments of the greenstone belts formed during eruptions of island arc volcanoes, associated with Archaean plate tectonic trenches. The mineralogy of the Archaean greenstones, differing from that associated with more recent volcanic activity in the Palaeozoic, reflects the immature nature of Earth's crust 2500 Ma ago. Accompanying greenschists, formed during low temperature regional metamorphism, also contains the green minerals chlorite, epidote, and serpentine, along with quartz, potassium feldspars, and actinolite and hornblende amphiboles.

In many areas of the Superior Province, the greenstone belts contain economic minerals. Hydrothermal activity during and after volcanism resulted in the migration and concentration of gold in a greenstone belt sequence at Hemlo, north of Pukaskwa along the Trans-Canada Highway, allowing several large mines to operate since 1985. The original wealth of Rouyn-Noranda, Timmins, Kirkland Lake, Red Lake, Val d'Or, and other communities in the Superior province is based on the exploitation of volcanogenic ores of gold (Sandlos and Keeling 2021).

In contrast to surface placer mining used to exploit fluvial deposits of gold, such as in the Klondike of Yukon, extraction of gold from Superior Province rocks involves underground, technological , energy-, and capital-intensive operations. "Hard-rock" mining involves extraction of gold from the typically sulphur-bearing greenstones, processing 100–200 kt daily to produce a few grammes of gold (e.g. Lakshmanan and Gorain 2019). Gold concentrations are very low, and as a result, substantial waste rock is produced as tailings. Underground mining, as generally practised today in Canada, generates about 10% of the waste rock per kg of gold produced compared to surface open-pit operations.

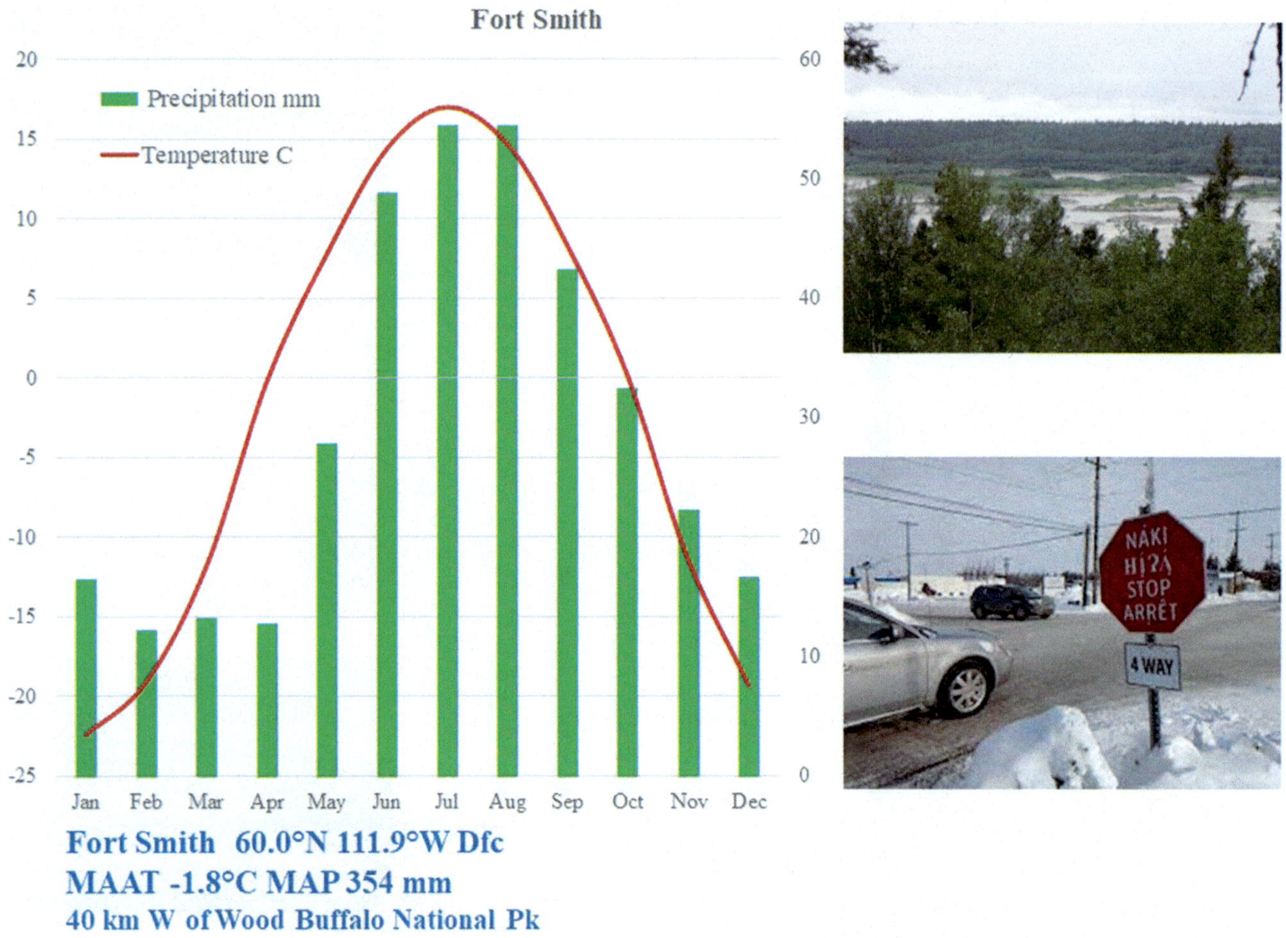

Fig. 3.11 Fort Smith (Thebacha), with a mean annual air temperature of −1.8 °C, has a Köppen-Geiger Dfc climate. Drivers are required to stop in Nehinaw (Cree), Chipewyan (Denesulin), English, or French. (Climate data from Environment and Climate Change Canada 2020)

The standard procedure, used in approximately 90% of Canadian mines, involves cyanide leaching to extract the gold (see Eisler and Wiemeyer 2004). This produces wastewater which must be neutralized or stored in holding ponds. Production of 1 g of gold requires approximately 2500 l of water, and generates 2500 kg of tailings and 20 kg of CO_2 emissions. Large-scale gold mining is also responsible for approximately 10% of the annual mercury emissions worldwide.

Copper, zinc, nickel, and cobalt have been mined at sites throughout the Superior province (Sandlos and Keeling 2021). The Sudbury Basin (Irruptive), resulting from a meteorite impact along the Superior-Keeweenawan province boundary 1850 Ma (Rousell and Card 2009; Fig. 3.17), has been mined commercially since 1886. Ores are extracted from norites and gabbros from layered igneous intrusions within the Sudbury Igneous Complex. The primary minerals, chalcopyrite, pyrrhotite, pentlandite, and millerite, are all sulphides. Extraction of the metals requires extensive processing, sulphide leaching, and smelting, generating waste and hydrogen sulphide gases. Previous mining activity has resulted in acidification of soils and water bodies, leading to ecosystem damage and altering more than 1000 km^2 of the local landscape (Dirszowsky 2020), in addition to producing downwind damage from acidic precipitation. Refinement of mineral processing has reduced, although not eliminated environmental effects.

Initially, mining at Sudbury was above ground, as is typical for copper and silver mining of igneous layered intrusions and porphyries. This technique produces large open pits and tailings piles, as are evident at Rouyn-Noranda, Temagami, and Cobalt–Haileybury–New Liskeard. The gradual transition to nickel mining after 1910 at Sudbury, and subsequently to platinum group elements, led to the development of underground mining, later used to exploit the nickel ores at Thompson. Copper mining at the other Superior province sites gradually became uneconomic after World War II.

Other elements mined include cesium and tantalum (Tanco MB), palladium (Lac des Iles ON), lithium (Val d'Or region QC), and cobalt (Cobalt ON, Thompson MB). Cesium is used in petroleum drilling muds, atomic clocks, and thermionic generators. Tantalum is used in electronic capacitors and

Fig. 3.12 Bedrock and glacial deposits, Boreal Plains, Manitoba. Lonely Lake (centre) and Ebb-and-Flow Lake (east, right side; Ojibwe *Gaa-gwekwekejiwang*) are developed in karst associated with mid-Devonian carbonate rock. Variations in water level and colour are associated with both karst groundwater and surface water activity. Glacial Lake Agassiz deposits, including NNW-SSE oriented iceberg scour tracks, blanket the bedrock. (International Space Station Image 15E06985, NASA, USA

resistors, and in carbide alloys for high-temperature applications. Palladium is used in ceramic capacitors and automobile catalytic convertors. Lithium mining of igneous pegmatites in Canada must compete with large-volume sources in Australia, China, and Chile, although the increasing use of lithium-ion batteries is a spur to domestic production.

Cobalt and chromium are not currently produced in Canada. Interest in both elements has rapidly increased.

Fig. 3.13 Evaporite deposits and karst, Wood Buffalo area. Upper left: Collapse doline, McLelland Lake, Alberta. Upper right: Salt flats, Sass River, Wood Buffalo National Park. Lower left: Sodium nitrate, calcium sulphate, sodium chloride, and potassium nitrate precipitate to form the flats. Lower right: residual deposits of non-evaporite material remain after dissolution

Cobalt is used in gas turbines, jet engines, hip replacements, radial tyres, and in lithium-ion batteries used in automobiles and cellphones. Global production has increased from ~50 kt in 2000 to ~150 kt in 2020. Of this tonnage, 70% is produced in the Democratic Republic of Congo, and shipped to China for processing and refinement. Social and environmental issues associated with open-pit mining of cobalt in the DRC, combined with political instability and supply-chain concerns, have heightened interest in production from Canadian sources. Production of chromium, used in stainless steels, is dominated by South Africa, Kazakhstan, and India.

Recent interest in development of mines for chromium and cobalt production has focused on the McFaulds Lake (Ring-of-Fire) region of the northernmost extent of the Superior province in northern Ontario. Environmental and socio-economic concerns of the Ojibwe residents of the Attawapiskat area would have to be addressed and satisfactorily resolved before any development could proceed.

Lake Superior Provincial Park, 200 km south of Pukaskwa along Highway 17, conserves a Boreal Shield ecological assemblage over Superior Province granites and gneisses. Agawa Rock, a Superior Province granite cliff, displays pictographs depicting canoes and several animals (Conway 2010; Fig. 3.18). Mishipeshu, known as the Great Lynx, Underwater Wildcat, and the Ojibwe spirit of the waves of Lake Superior, is a key element. The pictographs most probably date from 1650 to 1760. Some researchers attribute the pictographs to the Anishinaabe Ojibwe leader Shingwaukonce who led opposition to copper mining at Mamainse Point to the south (Sandlos and Keeling 2021). Mishipeshu is also the traditional Ojibwe protector of copper.

3.3.2.3 Keweenawan Province

The southern coastal area of Pukaskwa lies in a younger geological subdivision of the Canadian Shield. This Keweenawan Province (also termed the "Southern Province") consists of sedimentary and metamorphosed sedimentary rocks of Palaeo-Proterozoic age, which were welded to the Superior Province and deformed during the Meso-Proterozoic, 1370 Ma (Natural Resources Canada 2009; Slaymaker et al. 2020a).

The Keweenawan Province rocks are relatively weak basalts, gneisses, schists, slates, argillites, and phyllites and thus do not form the ridge-and-valley topography charac-

Fig. 3.14 Western Boreal Plains bedrock is dominated by Cretaceous sedimentary units. Upper: Cretaceous sandstone and shale, capped by glacial till and Glacial Lake Peace sediments, flank the Peace River valley of British Columbia. Lower: the Kiskatinaw River downcuts through weak Cretaceous shale, siltstone, and bentonite

teristic of the Superior Province (Fig. 3.19). Consequently, they have been eroded evenly along the coast, which has resulted in the gently curving Superior shoreline, devoid of any harbours. These rocks underlie much of the eastern half of Lake Superior, and their erodability accounts for the location of the lake and its great depth. Similar rocks underlie the low-relief Nipigon Plain. More resistant diabase and gabbro sills locally form ridges, notably the Thunder Bay Hills (northwest shore of Lake Superior; Zaniewski et al. 2020) and Penokean Hills (north shore of North Channel, Lake Huron), with relief locally exceeding 280 m.

A short-lived silver mine operated at Silver Islet near Thunder Bay (Sandlos and Keeling 2021). The Penokean Hills region surrounding Elliot Lake and Blind River was the scene of intense uranium mining in the 1950s and early 1960s (Sandlos and Keeling 2021). Uranium was extracted from uraninite contained in quartz pebble conglomerate beds. Rare earth elements were produced in small quantities as byproducts. Mining continued until 1997, although the health effects of silicosis and inhalation of radon gas on miners will persist for longer. Elliot Lake currently promotes itself as an inexpensive retirement community.

3.3.2.4 Grenville Province

The Grenville province comprises the southeastern part of the Canadian Shield, extending from Happy Valley-Goose

Fig. 3.15 Churchill geological province. Left: Western part of the Churchill Province in northeastern Alberta (west, left edge), northern Saskatchewan (east), northwestern Manitoba (east, far right edge), and Northwest Territories (north), showing Lake Athabasca, Wollaston, Reindeer, and Kasba Lake, July 2001. Lake and river geomorphology reflect structural geology and tectonic deformation. Smoke plumes from forest fires are moving northeastward. (NASA). Upper right: Glacially eroded Athabasca sandstone bedrock, Livingstone Lake, northern Saskatchewan, indicating ice flow towards the southwest. (NAPL). Lower right: O-Pipon-Na-Piwin (Southern Indian) Lake, northern Manitoba, is part of the Churchill River. The Missi Falls control dam at the south end has raised the lake level by ~3 m. The southern part of the lake is bordered by resistant granite and pegmatite. Less resistant metamorphic rocks flank the northern part, marked by denser vegetation (red tones) on this false-colour satellite image (NASA)

Bay to Georgian Bay, and including the core of the Long Range Mountains of Newfoundland (Hocq 1994b, Wheeler et al. 1996, Natural Resources Canada 2009, Catto et al. 2020, Slaymaker et al. 2020a). Most of the province is dominated by igneous acidic rocks (granites and related types) and acidic, highly metamorphosed rocks (gneiss and schist) derived from the metamorphism of pre-existing sedimentary and igneous rocks during several Precambrian orogenic events (Fig. 3.20). The Grenville Province is considered to be among the oldest parts of the Canadian Shield by geologists, but it has been so intensively modified by metamorphism during the orogenic events that determining the original age of the rocks is essentially impossible.

Orogenic activity began in the Grenville Province approximately 1600 Ma, and culminated in the Ottawan Orogeny between 1100 and 1000–985 Ma, during the Meso-Proterozoic era of the Precambrian. At this time, a now-destroyed continent to the southeast (where the Appalachians of Beauce are now) was thrust against the Grenville Province rocks, causing metamorphism and mountain-building. The style of deformation was in some respects similar to that occurring in the modern Himalayan Mountains of India and Nepal. The resultant mountains were high, similar to today's Himalaya.

Subsequently, 1000 million years of erosion has ground away the peaks, leaving only the roots of the mountain system

Fig. 3.16 Superior province basaltic rocks are exposed at Pukaskwa National Park. Inset: quartz-feldspar pegmatite dyke cutting fine-crystalline granite, Jackfish, Ontario

exposed as the Grenville Province. Locally, thicknesses up to 10 kms of rock have been removed by erosion, leaving the gneiss and granites which formed the subsurface core of the ranges exposed. Although erosion took place throughout the past 1000 million years, most material was probably removed during the initial stages, when the mountains stood at their greatest elevations above the Precambrian seas. Quaternary glaciation during the past 2.588 million years removed very little material (Catto 2015). In many areas, the topography is very similar to what it was prior to Quaternary glaciation. Algonquin, Georgian Bay Islands, Thousand Islands, and La Mauricie provide examples of Grenville province landscapes.

Algonquin Provincial Park is a low relief area (approximately 200 m), marked by many steep cliffs, numerous lakes with ragged shorelines, and isolated hills. This irregular topography gives travellers using surface transport the illusion of passing through much more rugged terrain. The configuration and alignment of the lakes and rivers is almost exclusively a product of bedrock geology. Algonquin is influenced by two major fault systems (Catto 2015). The primary system trends WNW-ESE, and the major rivers (such as the Petawawa and Bonnechere) are developed along this orientation, as is the Ottawa River further north. A secondary system of faults and joints has developed at right angles to the primary system, producing the dog-leg pattern characteristic of the Petawawa and Ottawa Rivers, and the angular shorelines of lakes such as Opeongo. Although some lakes, such as Lake of Two Rivers, are impounded in part by glacial sediment, analysis of remotely sensed images indicates the dominant role of bedrock on physiography.

The Ottawa-Bonnechere Graben is bounded by the normal Petawawa (south) and Mattawa (north) Faults. The graben, dropped between the faults, forms the Ottawa River valley. The major rivers have not carved these valleys by fluvial action: the valleys formed as the hanging-wall blocks dropped, creating flat valleys filled with preserved Cambrian and Ordovician sedimentary rocks, and subsequently with Quaternary sediments. Earthquake activity continues along

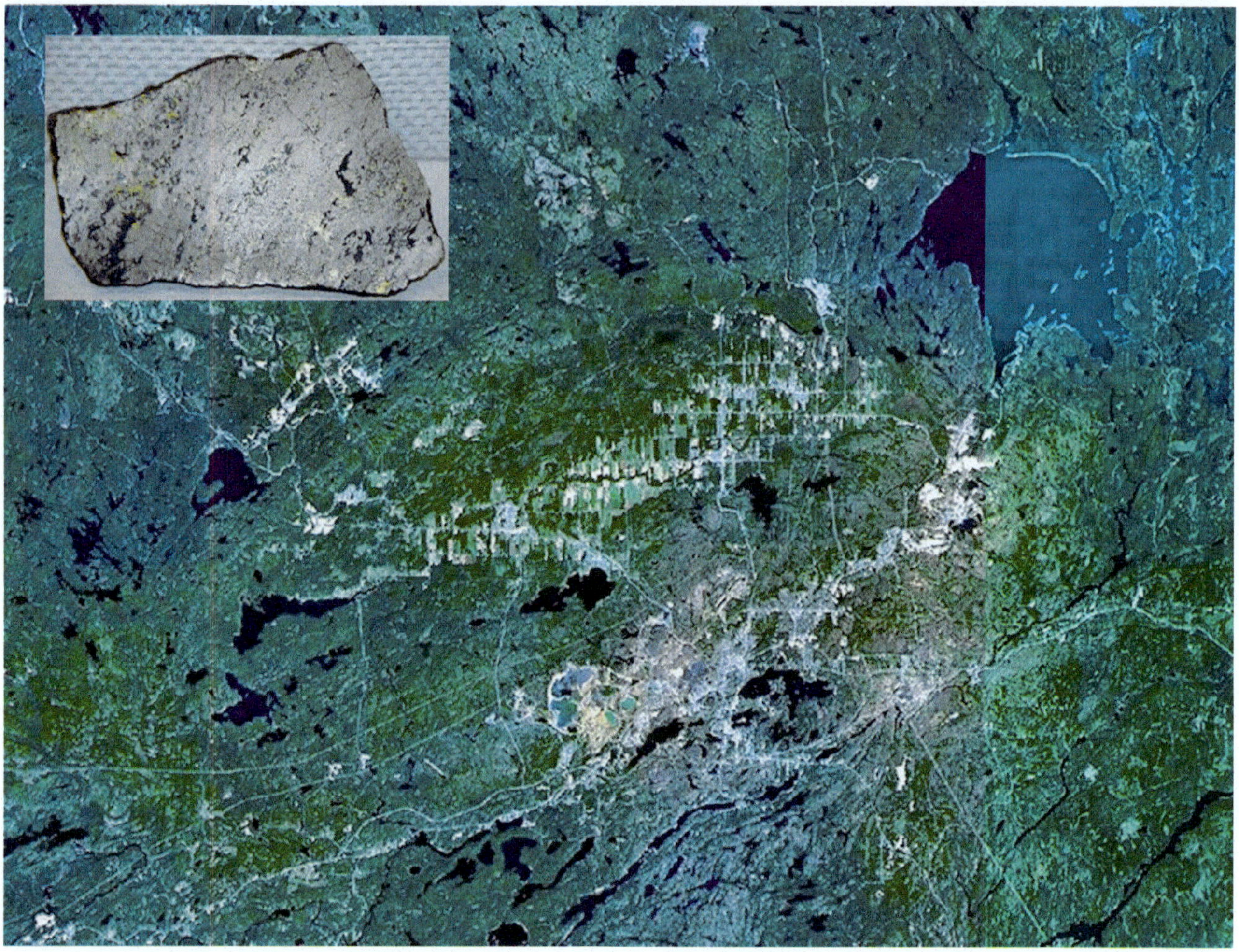

Fig. 3.17 Sudbury Impact structure (1850 Ma; 60 × 30 km). The oval Nickel Irruptive is marked by numerous mines with associated surface disturbances, notably along the southeastern margin. The central Sudbury Basin is infilled with younger (1800–1400 Ma) Precambrian sedimentary and volcanic rocks. Keeweenawan province rocks flank the Impact Structure to the south and west, with Superior province granites and gneisses to the east and northeast. Wanapitei Lake (upper right) is a younger (~37 MA) impact feature. (NASA). Inset: pentlandite-pyrrhotite nickel-copper norite ore, Sudbury

the series of faults associated with all of these grabens (Earthquakes Canada 2022).

Most of the 59 islands within Georgian Bay Islands National Park, and the shoreline of the bay north of the mouth of the Severn River, are included within the eroded terrain of the Grenville province. The islands are too small to contain lakes or surface streams. Relief on the islands is muted, as the features eroded into the bedrock are generally small in size and locally are blanketed with lacustrine clay deposited during postglacial high phases of Lake Huron (see Chap. 4). Inland, relief approximates 150 m.

The Georgian Bay Mnidoo Gamii UNESCO Biosphere Reserve (GBBR 2022) encompasses the eastern coast of Georgian Bay, from the Severn River north to the French River, including the numerous islands composed of Grenville Canadian Shield rock (Fig. 3.21). Casually, this largest freshwater archipelago on Earth is known as the 30,000 islands. Depending on the water level in the bay, and what you wish to consider an "island" (does an individual exposed boulder count?), 30,000 may be a considerable underestimate.

The Thousand Islands are developed in a narrow belt of Canadian Shield granites and gneisses that extends from the main area of the Grenville Province southeast across the St. Lawrence River to connect with the Adirondack Hills of New York. This Frontenac Axis resembles the remainder of the Grenville Province and Laurentian Highlands physiographically, marked by rolling to rugged topography, isolated hills and depressions, and steep cliffs. Relief is approximately 150 m (Fig. 3.22). Regardless of how large a rock exposed above the St. Lawrence has to be in order to be considered an

Fig. 3.18 Agawa Rock pictographs depict canoes (inset, upper left) and Mishipeshu, along with other animals (right). The pictographs most probably date from 1650 to 1760

"island", the total number of exposed pieces of landfalls far short of 1000. There are less than 100 named points of land in Ontario and New York combined.

La Mauricie lies in the Laurentian Highlands of the Laurentian region of the Canadian Shield (Bostock 2014), generally referred to as "Les Laurentides". The park is dominated by a system of southeast-northwest trending valleys developed along faults, occupied by the major streams and elongated lakes. Relief in the area is approximately 300 m. The terrain between the fault valleys generally consists of rolling to irregularly rugged isolated hills, with interspersed gently rolling plateaux.

After 985 Ma, the fringes of the Grenville Province were covered by Cambrian (541–485 Ma) and Ordovician (444–485 Ma) sedimentary rocks. Although most have been eroded, a few small areas have been preserved (Wheeler et al. 1997).

Brent Crater, in the northwestern part of Algonquin, is a meteor impact crater (Grieve 2001). At 3.8 km in diameter, it is among the smallest of Canada's 24 documented meteor impact structures. The meteor impact approximately 450 Ma excavated Cambrian and Ordovician sedimentary rocks, remnants of which are preserved in the crater.

Manicouagan, 180 km northwest of Sept-Iles, is one of the largest and best-preserved impact structures on Earth (Spray et al. 2010). The 100 km-diameter impact structure, approximately 214 million years old, is currently occupied by a 65-km-diameter hydroelectric reservoir (Fig. 3.23).

Most areas of the Grenville province can be classified as diverse and interesting from a mineralogical viewpoint, but relatively unproductive and uneconomic from a mining viewpoint. The southern fringes adjacent to the population centres of the Great Lakes–St. Lawrence Lowlands rapidly attracted prospectors, in many cases revisiting sites previously noted by indigenous inhabitants, but these mining ventures tended to be short-lived (Sandlos and Keeling 2021).

Industrial iron mining in Canada began at the Forges du Saint-Maurice, north of Trois Rivieres, in 1730 (Samson 1998). Other iron deposits formed by igneous intrusions, and metamorphosed marbles and skarns, along the southern margins of the Grenville province have been exploited since 1820, notably at Marmora (magnetite) and Lac Tio (ilmenite). Titanium is also produced at La Tio. Uranium mines operated in the Bancroft area from the 1950s through 1982.

Fig. 3.19 Keweenawan province rocks underlie Michipicoten Island (upper). The relatively weak basalts, gneisses, schists, slates, argillites, and phyllites erode evenly, precluding the development of deep harbours (NASA). Lower right: Keeweenawan basalt and metamorphic rocks, Coastal Trail, Lake Superior Provincial Park. (Wikipedia/Adam Kahtava). Inset lower left: copper-bearing conglomerate ore

Niobium, used in structural steel and superalloys, is produced at St-Honoré, along with rare earth elements.

3.4 Quaternary

Although earlier Quaternary glaciations covered all of the Boreal Shield and Boreal Plains, few traces of these events remain. The landscapes reflect ice activity between 100,000 and 10,000 years ago, combined with sediments deposited in ice-dammed lakes during glacial retreat, meltwater channels, and modern geomorphic processes (Fulton 1995; Dyke et al. 2003; Stokes et al. 2015; Catto et al. 2020, Slaymaker and Catto 2020)

3.4.1 Boreal Shield

In Québec and Ontario, the Boreal Shield was glaciated by the Labrador centre of Laurentide Inlandsis, which advanced from western Labrador and adjacent Québec (Schefferville-Labrador City area; Klassen et al. 1992; Veillette et al. 2017). Ice flowed south across La Mauricie and Gatineau, south-west across Algonquin and Pukaskwa, and west-southwest over Pimachiowin Aki and Quetico. In easternmost Manitoba, Labrador centre ice coalesced with southward-flowing ice from Kivalliq, in mainland Nunavut. Kivalliq ice flowed westward and southwestward across the Churchill geological province (Stokes 2020) and to the southwest and west over the Boreal Plains (Catto et al. 1996, Hartmann et al. 2018).

Fig. 3.20 Grenville geological province. Upper: Igneous activity, metamorphism, deformation, and Quaternary glaciation have combined to produce the intricate shorelines of Lake Muskoka (left) and Lake Opeongo, Algonquin Provincial Park (right). Lower left: polished Grenville granite displays quartz, orthoclase feldspar, and mafic iron-bearing minerals. Lower right: Grenville gneiss exposed at Mary's Harbour, Labrador

The Boreal Shield regions remained ice-covered throughout the period from approximately 100,000 to 12,000 years ago (Dyke 2004; Occhietti et al. 2011).

The resistant rocks of the Churchill, Superior, and Grenville provinces did not yield easily to glacial erosion. In most places, the ice flowed parallel to the geological structure, accentuating topography such as the ridge-and-valley terrain of the Superior province which existed previously.

The basic topography of the Boreal Shield, including Lake Superior, was present prior to Quaternary glaciation. The effects of glacial erosion are confined to slightly modifying the topography, rather than creating it.

Deposition of sediment requires previous erosion. Unconfined glaciers flowing over hard bedrock are generally ineffective at erosion, so although erosion of the hard granitic and high-grade metamorphic gneiss of the Canadian Shield is extremely slow, with total erosion during the Quaternary generally amounting to less than 2 m, very little material is produced for potential deposition. As a result, hard-rock glaciated landscapes are generally dominated by erosional features, such as roches moutonnées and glacial grooves. These features are relatively small, and the maximum thickness of rock locally removed by glacial erosion throughout the Quaternary in La Mauricie, Algonquin, Pukaskwa, Quetico, and Pimachiowin Aki is less than 20 m. As areas without thick sediment cover also lack the necessary source material to generate glacial depositional landforms, depositional features are largely confined to areally restricted zones.

The style of erosion, and the resulting assemblage of landforms, is also related to the resistance, structure, and fracture patterns of the underlying bedrock (Figs. 3.24 and 3.25). Rock units with limited resistance to erosion, but without conspicuous joint systems, are more likely to be evenly eroded

Fig. 3.21 The Georgian Bay Biosphere Reserve extends north from Georgian Bay Islands National Park to Killarney Provincial Park. Inset: George Lake, Killarney. (Wikimedia Commons/Kristopher Kinsinger)

into smooth features polished by glacial abrasion. Conversely, harder units with closely spaced joints are more susceptible to erosion by plucking.

3.4.1.1 Glacial Erosional Features

Glacial flow modifies bedrock obstructions, smoothing the up-ice sloping stoss side into a curved surface which minimizes friction resistance. The down-ice lee side is subject plucking due to regelation. Suitable rock type, bedrock fracture pattern, and orientation with respect to glacial flow can produce a gently smoothed stoss side and a more irregular or jagged lee, indicating the direction of glacial flow. Stoss-and-lee landforms with heights of metres to 10 s of metres and lengths of metres to 10 s of metres are commonly referred to as roches moutonnées (Catto 2015).

If the bedrock obstruction is aligned at an oblique angle to the direction of glacial flow, both the stoss and lee sides may undergo abrasion and subglacial meltwater erosion, producing whalebacks (metres to 10 s of metres), and rock drumlins (10 s to hundreds of metres). Although these features are double-faceted, the stoss up-ice side slopes at a gentler angle than does the lee down-ice side.

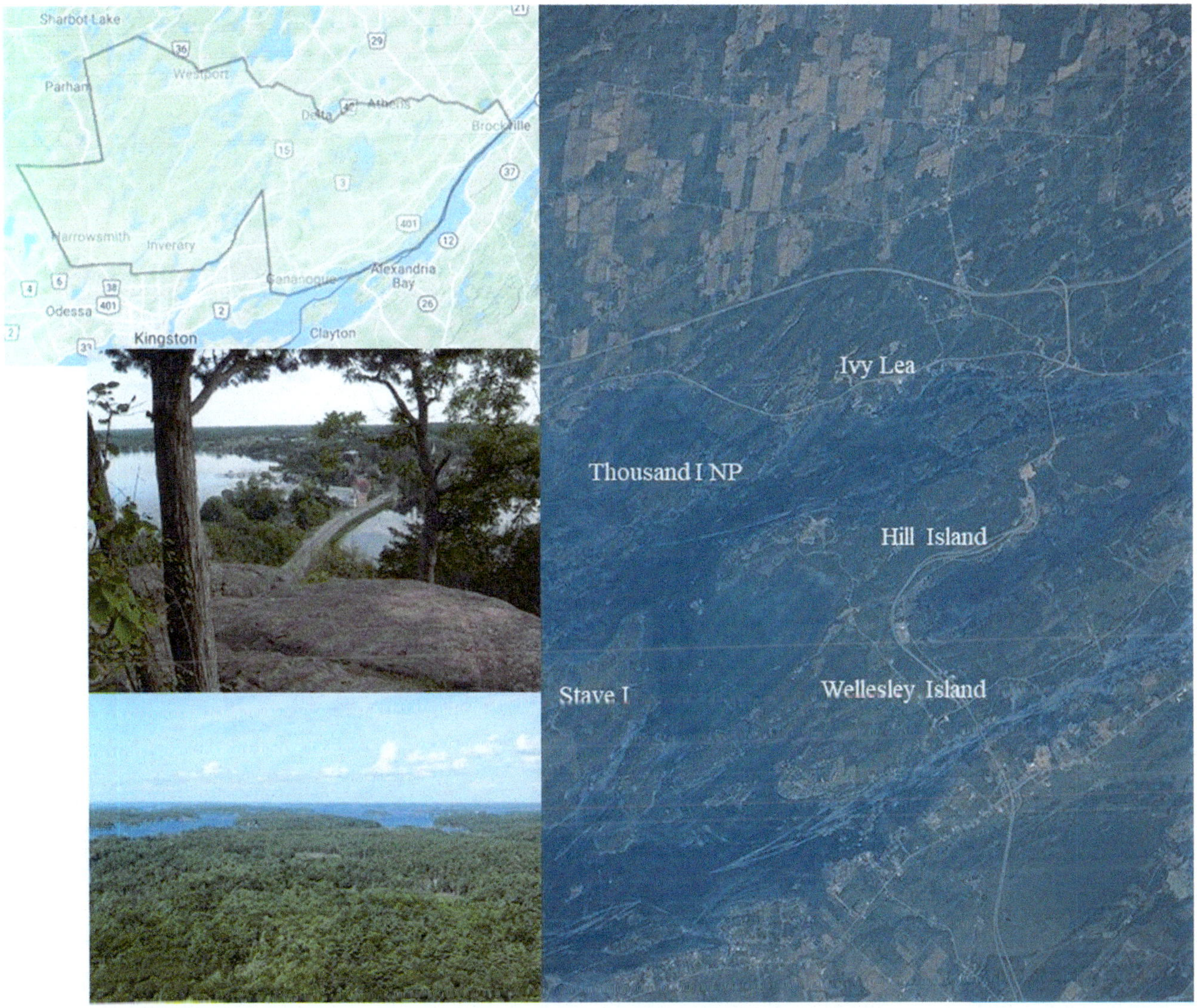

Fig. 3.22 The Frontenac Arch UNESCO Biosphere Reserve includes Thousand Islands National Park. Upper left: area of Frontenac Arch UNESCO Biosphere Reserve. Centre left: Foley Mountain overlook of Westport, along the Rideau Canal waterway at the northern edge of the Reserve (Wikimedia Commons/Petersent). Lower right: View of the Thousand Islands (Wikimedia Commons/Ad Meskens). Right: satellite image of the Ivy Lea-Thousand Islands National Park area. The Canada-USA border follows International Rift between Hill and Wellesley Islands (NASA)

After deglaciation, bedrock which is sufficiently resistant to further erosion and environmental modification may allow preservation of micro-erosional features, such as striations. Striations on sediment-covered bedrock can be preserved for long periods following deglaciation. However, if the bedrock is exposed, modification and erosion rapidly obscure or destroy striations.

Striations are linear, 10 s of centimetres to metres in length, 1–10 mm deep, and 0.2–3 mm wide, formed by clasts dragged along the bedrock substrate beneath the sole of a warm-based glacier. Production of striations requires a very thin film of subglacial water, a supply of clasts for abrasion, and suitable underlying bedrock. A glacier flowing at a speed of 12 m/y would move 1 m in 1 month, suggesting that individual striations represent a few days to months of glacial flow. Commonly, a striated surface is the only record of a glaciation that lasted thousands to tens of thousands of years.

Elongated erosional glacial grooves also can develop in suitable bedrock. Grooves vary from straight, linear features with symmetrical cross-sections, to irregularly curved features with variations in width and depth, to features which show cubic parabolic cross-sections. Cavettos are saucer-shaped erosional features in bedrock, produced by subglacial meltwater under high pressure.

Fig. 3.23 Manicouagan impact structure, formed approximately 214 Ma, is currently occupied by a 65-km-diameter hydroelectric reservoir (NASA)

3.4.1.2 Younger Dryas Glacial Readvance

Retreating ice exposed the southern fringes of the Boreal Shield between 13,000 and 11,000 years ago. Overall ice retreat was punctuated by several local readvances, involving relatively thin ice largely constrained by local topography. Although some events were entirely local, a significant regional readvance between 11,700 and 11,000 years ago has been linked to essentially simultaneous glacial readvances throughout much of the world, the "Younger Dryas" event.

Along the northern flank of the St. Lawrence and Ottawa Valleys, a short-lived retreat 12,000 years ago was followed by a sharp re-advance of glacial ice, between 11,700 and 11,000 years ago (Lasalle and Shilts 1993; Occhietti 2007; Fig. 3.26). This glacial advance built the large St.-Narcisse moraine, north of the St. Lawrence River, extending from La Malbaie to Thurso (east of Ottawa). The relatively short duration of this event (in comparison to other glacial fluctuations) and its widespread areal extent mark it as an unusual episode in the deglacial history of Canada. Correlative readavances also occurred in Algonquin Provincial Park, along the Ottawa River north of Pembroke, and near The Pas, Manitoba (Catto et al. 2020). Elsewhere in eastern Canada, Younger Dryas glacial advances are recorded in the area north of Parrsboro and other localities in Nova Scotia, Prince Edward Island, the Northern Peninsula of Newfoundland (Ten Mile Lake moraine), and in eastern Labrador (Sebaskachu moraine).

The Younger Dryas event initially was recognized by studying fossil pollen assemblages in Scandinavia, where colder episodes were marked by an increase in the proportion of the pollen of the alpine species *Dryas octopetala* (mountain

Fig. 3.24 Evidence for glaciation in the Boreal Shield is commonly preserved as micro-erosional features. Clockwise from top left: Striations (mm depth) and grooves (cm to 10 cm depth), Sudbury; two sets of striations showing multiple glaciation on a roche moutonnée, Kenora; suchelwannen indicating abrasion by high-pressure subglacial water flow, Pukaskwa; cavettos indicating and plucking by subglacial water, Pukaskwa

avens). It corresponds to a similar areally extensive but short duration glacial readvance in Norway, Sweden, and Finland. Analysis of cores taken from the glacial ice of Kalaallit Nunaat (Greenland) has also revealed evidence for a short, sharp glacial re-advance *ca.* 11,700 years ago. Other effects of the Younger Dryas have been recorded in numerous locations worldwide, including some unglaciated areas marked by a period of colder conditions.

The chronological similarity between all these glacial readvances suggests that they were triggered by a common circumstance which affected the entire North Atlantic Ocean. The cause of the Younger Dryas event is at present uncertain, but it is believed to be related to changes in oceanic circulation and water temperatures, triggered by the sudden arrival of large quantities of fresh, cold water into the Arctic and Atlantic Oceans. The initiation of the Younger Dryas has been related to melting of the Laurentide Glacier, or discharges of meltwater from Glacial Lake Agassiz in prairie Canada, or both in combination (e.g. Leydet et al. 2018). Large volumes of meltwater entering the North Atlantic Ocean could have resulted in the development of a surface cap of cold, fresh water. This would act to cool the margins of the North Atlantic, triggering glacial readvances. When the glaciers readvanced, the supply of meltwater was curtailed, and the oceanic circulation returned to its pre-Younger Dryas status.

It is this possibility which has led to speculation that meltwater from Kalaallit Nunaat could trigger a reduction in the temperature or velocity of the Gulf Stream, impacting the climate of northwestern Europe. Evidence of past episodes of freshwater discharge, as indicated by distinctive sediments carried from terrestrial sources, is preserved in marine deposits and has been widely correlated to climate changes in numerous areas. However, the effect of future meltwater discharge from the Kalaallit Nunaat glaciers cannot be

Fig. 3.25 Roche moutonnées indicate the direction of ice flow. Upper left: subglacial abrasion smooths the up-ice (stoss) side into a streamlined form. Ice flow from lower left to upper right. Lower left: Down-ice (lee) side abrasion can also occur, producing double-faceted features (whalebacks). Ice flow direction from right to left. Upper right: Surfaces of larger roche moutonnées developed in less resistant rock units may show grooves and subglacial meltwater features. Lower right: Roche moutonnée are ubiquitous in Boreal Shield landscapes. Ice flow direction was left to right

directly predicted by comparison to the events of the Younger Dryas.

At the start of the Younger Dryas, large areas (including all of the Boreal and Taiga Shields) were still covered by glacial ice, which influenced local climates and provided a basis for glacial readvance. Ocean temperatures were cooler than those prevalent today. The volume of meltwater discharge from the ablating North American glaciers was much larger than that potentially produced from coastal Kalaallit Nunaat. It was directed into the Arctic Ocean from several sources, notably the Mackenzie River, and thus would be expected to have a different impact than would input from Kalaallit Nunaat. The available evidence suggests that the impact of meltwater from Kalaallit Nunaat could result in cooling of the surface waters of the Labrador Sea, but will not result in cooling of the Gulf Stream to the same degree of magnitude as characterized the Younger Dryas.

3.4.1.3 Renewed Glacial Recession

Following the Younger Dryas, progressive glacial retreat resumed. Around the southeastern margin of the Boreal Shield, deglaciation of the outlet of the St. Lawrence River at Québec City allowed marine waters to flood the St. Lawrence Valley, and Georgian Bay Islands NP was inundated by the glacilacustrine waters of Lake Algonquin (see Chap. 4). Stranded blocks of slowly ablating ice persisted in the centre of Algonquin Park for up to 1,000 years after the main deglaciation took place, immediately following the conclusion of the Younger Dryas. These isolated ice masses produced hummocky moraines, short eskers, and

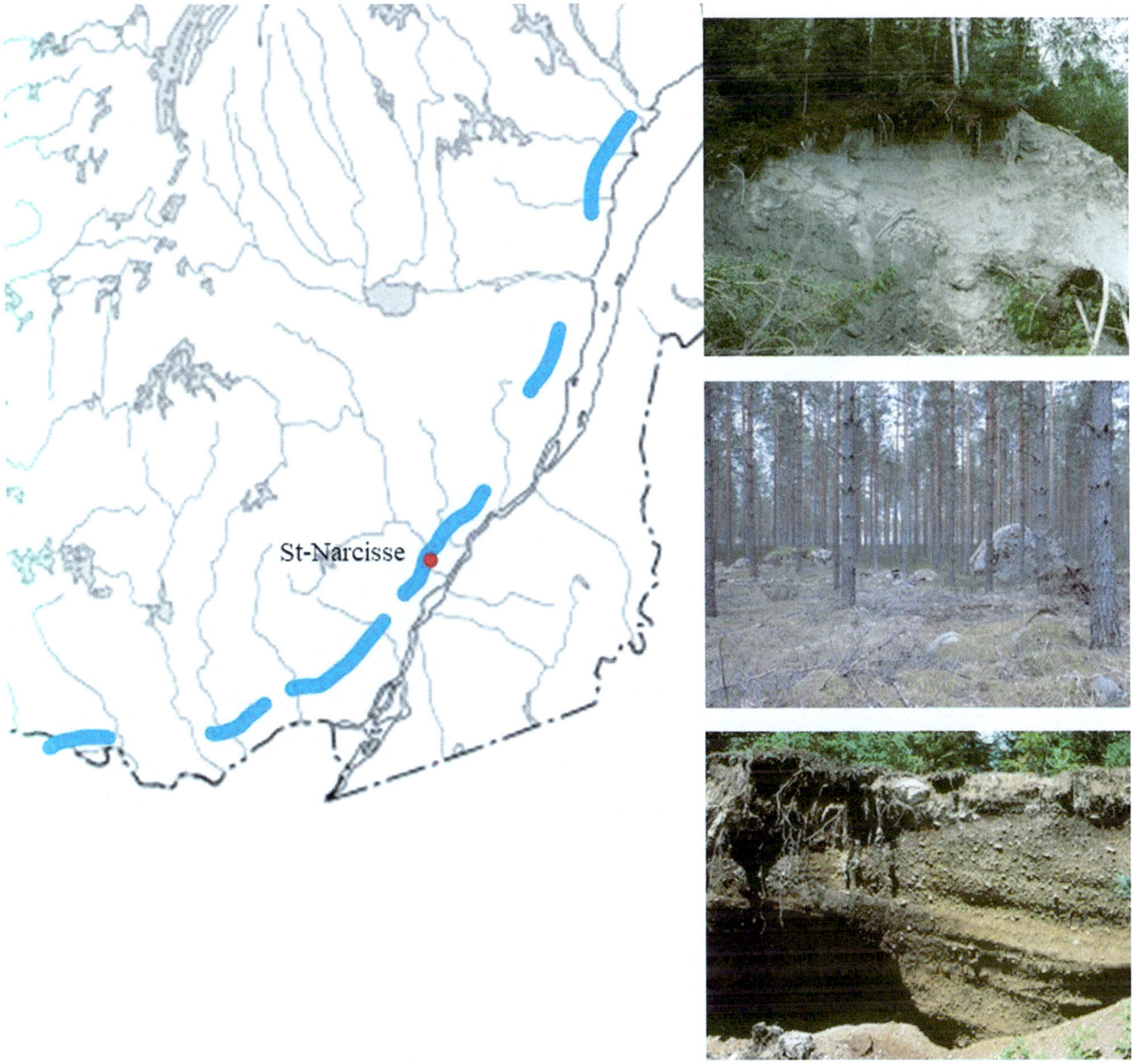

Fig. 3.26 The St. Narcisse Moraine, and its eastward extension into Labrador, represents the Younger Dryas ice frontal position in eastern Canada. Left: Location of moraine segments in Québec. Right: sediments associated with the moraine include glacial till and flanking marine deposits (upper), boulder-covered surfaces (centre), and glacimarine and glacifluvial gravel (lower)

gravel kames throughout the central park region. Deglaciation of the park, and the glaci-isostatic depression which followed, allowed the waters of the Lake Huron basin to the west to drain directly across Algonquin into the Ottawa Valley, *ca.* 11,500 years ago.

Further west, deglaciation of Pukaskwa National Park did not begin until shortly before 11,000 years ago, when the ice left the northern shore of Lake Superior. The lake basin was occupied by a later phase of Lake Minong, which received water from Lake Agassiz to the west for a short time and drained into the Lake Huron basin to the east. The shoreline of Pukaskwa remained ice-free, and the Superior basin continued to drain east into the Huron basin and through the North Bay Outlet into the Ottawa River. Pukaskwa, Quetico, and Pimachiowin Aki were not completely free of glacial ice until about 10,000 years ago.

3.4.1.4 Glacial Lake Ojibway

One of the last areas of the Boreal Shield to be deglaciated was the Cochrane–Kapuskasing region of northeastern Ontario. By 10,000 years ago, the ice had retreated northwards through the upper Ottawa Valley, past Temiscaming. Ice

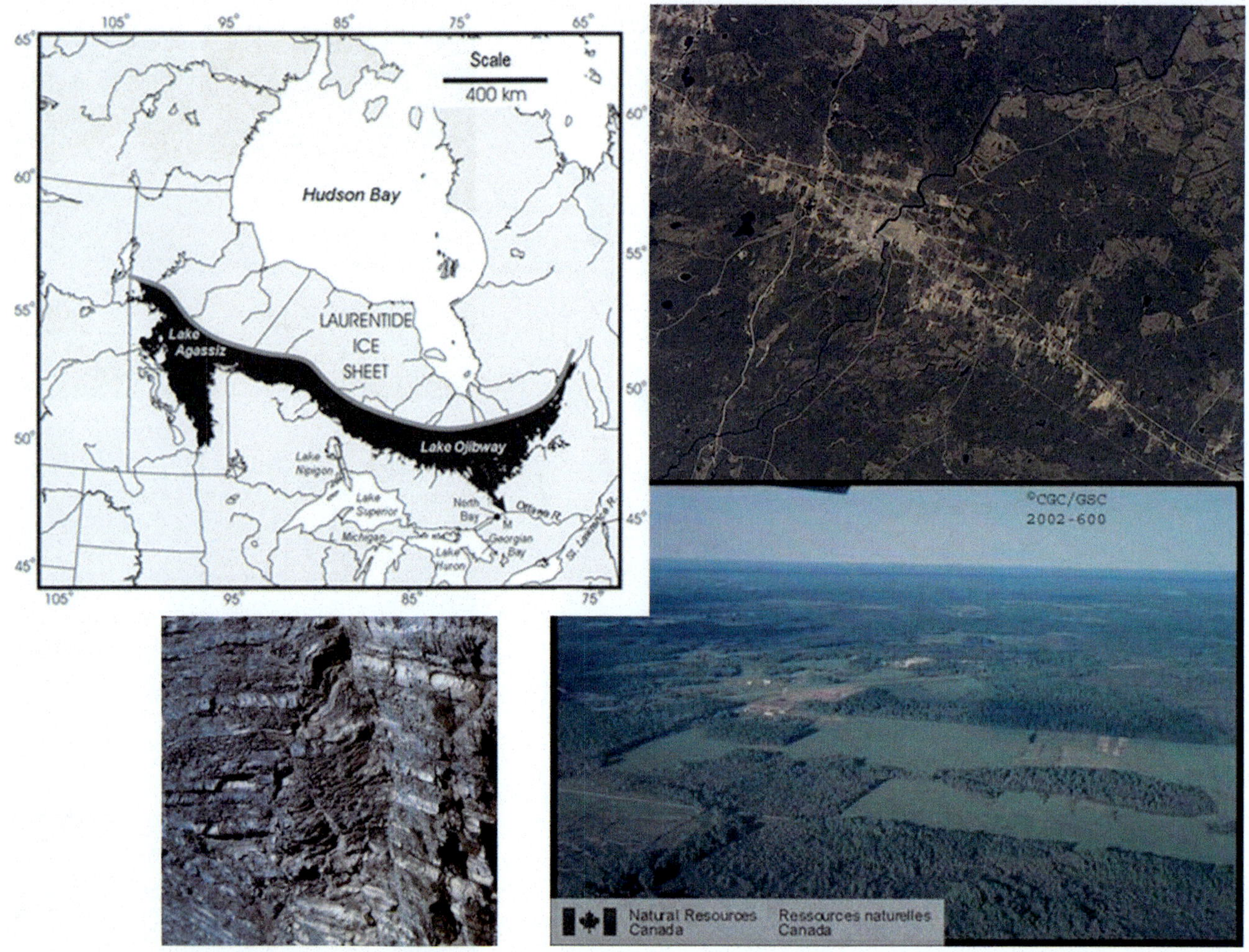

Fig. 3.27 Glacial Lake Ojibway developed as northward drainage to Hudson Bay was blocked by retreating Laurentide ice. Upper left: The maximum extent of Glacial Lake Ojibway was contiguous with Glacial Lake Agassiz to the west. Drainage was southward through the Ottawa River. Lower right: Varves produced by seasonal sedimentation. Left: satellite and aerial photographs of the Glacial Lake Ojibway clay belt landscape near Kapuskasing (upper) and Timmins (lower)

stood across northern Ontario and Québec, blocking northward drainage and impounding lakes along the margins of the retreating glacier (Fig. 3.27). As the ice continued to retreat, the lakes amalgamated into one body, now termed Lake Ojibway, extending westward to the headwaters of the Albany River. The Clay Belt is a product of northern Ontario's glacilacustrine history.

Initially, the amalgamated Lake Ojibway continued to drain south into the Ottawa River system through Lake Temiscaming, as ice blocked the natural drainage northwards to James Bay (Catto et al. 2020; Brouard et al. 2021). Discharge from Glacial Lake Agassiz to the west (see Chap. 5) also contributed. Seasonal discharge of meltwater plumes from the ice front allowed deposition of varves, yearly rhythmites of alternating layers of silt (spring runoff) and clay (summer settling). Some of the varves of Glacial Lake Ojibway were disrupted by earthquake disturbances, which can be traced across the lake basin (Brooks 2020).

About 8,000 years ago, marine waters flooded Hudson Strait, causing the rapid withdrawal of the ice to the north, and the demise of Glacial Lake Ojibway. Less than 500 years may have been required to completely destabilize the ice mass in Hudson Bay.

3.4.2 Boreal Plains

Throughout the Boreal Plains, the presence of Canadian Shield granite and gneiss erratics indicate that the area was glaciated by ice which moved uphill, against the current regional drainage (Fenton et al. 2013, Slaymaker 2017). In some locations, as the ice moved against the topographic

Fig. 3.28 Boreal Plains glaciation. Upper: hummocky terrain west of Grande Prairie, Alberta. Lower left: Laurentide glacial sediment (buff) overlain by vegetated Glacial Lake Peace clay, Halfway River area, BC. Lower right: Sand lenses showing ice-thrust features are common in the silt- and clay-rich tills of northern Alberta

slope, ice-thrust features were developed. The soft Cretaceous bedrock was particularly susceptible to glacial thrusting, and many lake basins represent the areas of blocks of bedrock gouged out intact by the advancing glaciers. In some cases, the blocks of rock were deposited adjacent to the lake basins, and it is possible to recognize that the rock exposed in the hills to the southwest of the lake was once on the lake bottom.

In the Boreal Plains of Saskatchewan, ablation of the Kivalliq ice mass began approximately 13,000 years ago, leading to the formation of hummocky terrain (Fig. 3.28). These low-relief features dominate many areas of the modern Boreal Plains landscape. Sinuous eskers, ridges of gravel deposited subglacially by streams flowing in tunnels beneath the ablating ice, were also formed. Deglaciation was complete approximately 11,500 years ago.

Prior to glaciation, the streams of northern Alberta drained largely to the north-northeast. Gravels of preglacial Quaternary age are readily recognized throughout Alberta because

Fig. 3.29 Glacial and postglacial sedimentation, Peace River valley, BC. Left: Silt and clay deposits of Glacial Lake Peace are subject to repeated slope failures, induced by undercutting of the Cretaceous siltstone and Quaternary glacial deposits by the Peace River. Right: Glacifluvial deposits can contain fossils useful for 14C dating. Upper: extracting a *Bison* femur from glacifluvial gravels, Fort St. John area, with the aid of a front-end loader. Lower: the extracted femur was 14C dated, indicating an age of approximately 10,000 years

they lack glacially derived granitic and gneissic clasts from the Canadian Shield (Catto et al. 1996). Part of northern Alberta drained into the proto-Mackenzie basin drainage system, whereas the northeastern part of the province was drained by an eastward-flowing river paralleling the modern Fond-du-Lac River in Saskatchewan.

The Boreal Plains of northern Alberta (including Wood Buffalo NP) and the Peace River lowland were also glaciated by southwestward-flowing ice from Kivalliq, advancing up the topographic slope, and thus impounded waters during both the advance and retreat phases. Although the ice flow direction during the most recent phase of glaciation was generally southwestward, local southeastward-moving lobes developed as well, in areas where topographic obstructions were aligned at 90° to the overall flow direction (southeast-northwest).

3.4.2.1 Glacial Lake Peace

The Peace River lowland, extending west to the foothills of the northern Rocky Mountains, was affected by glacial ice from two source areas during the Quaternary (Figs. 3.29 and 3.30). Laurentide (Kivalliq) ice advanced from the east to cover the eastern part of the Peace River lowland at some time after 22,000 years ago (Catto et al. 1996, Hartmann et al. 2018). Cordilleran ice from west of the Rocky Mountain Trench advanced eastward to form the Portage Mountain Moraine, adjacent to the WAC Bennett Dam, after 16,000 years ago. Although all parts of the area were covered at one time or another during the last glacial maximum, the two ice sheets did not advance simultaneously and never actually coalesced. The area between the shifting ice fronts was occupied by a succession of proglacial lakes, and the Portage Mountain Moraine was built as a glacifluvial feature into the waters

Fig. 3.30 Glacial Lake Peace. Left: typical landscapes of areas inundated by Glacial lake Peace, northwest of Grand Prairie (upper) and west of Fort St. John (lower). Upper right: laminated and folded sediments, near Bear Flat, BC. Centre right: dropstone in varved sediment, near Watino, Alberta. Lower right: debris flow failure involving Glacial Lake Peace sediments, Halfway River, BC

of Glacial Lake Peace, impounded by the Laurentide glacier to the east as it blocked the Peace River drainage (Hickin et al. 2015). Various phases of Glacial Lake Peace followed the retreating ice margin from the Dawson Creek area of British Columbia to Lake Athabasca between 15,000 and 12,000 years ago. Conditions did not become suitable for human habitation until the Laurentide glacier retreated and Glacial Lake Peace lowered, about 11,000 years ago. Around this time, people occupied the Charlie Lake cave archaeological site, adjacent to the shoreline of Glacial Lake Peace (Driver 1999).

Glacial Lake Peace clay is particularly susceptible to rotational sliding (Catto 2015). Undercutting of the base and removal of sediment from the toe causes repetitive rotational slides. Each successive event forms a minor scarp below the main scarp. Transverse (tensional) and radial (extensional) cracks are commonly developed in the sediment.

Rotational slides are transitional to other styles of mass movement. As the sediment moves, it tends to gradually lose internal cohesion, transitioning to a flow. Destabilization of a slide toe in contact with a river, lake, or ocean at the bottom of the undercut slope encourages loss of cohesion and flow. Constant erosion at the base locally oversteepens the slope, triggering renewed movement. Undercutting of the toe is a significant factor in most rotational slides. As a result, most rotational slide features combine block movement and cohesion-reduced flow. Lubrication of the sediment due to rainfall, snowmelt, or groundwater is also commonly involved.

3.4.2.2 Athabasca River and Glacial Lake McConnell

The largest glacial lake in western Canada was Glacial Lake Agassiz, formed as a result of the blockage of drainage northeastward from the Red River to Hudson Bay. Glacial

Lake Agassiz existed from 15,000 to 7,500 years ago, in a variety of configurations, and with drainage routes varying with the progressive retreat (and occasional re-advance) of the Laurentide glacier from Kivalliq (see Chap. 5).

The Clearwater River spillway, currently occupied by the Clearwater River and the Athabasca River north of Fort McMurray, was formed by an immense discharge of water northwestward from Glacial Lake Agassiz (Fisher et al. 2009). This discharge to the northwest coincided with a termination of flow to the east, via Lake Superior, and resulted in rapid drawdown of Lake Agassiz from its maximum level of 484 m asl (above sea level) in western Saskatchewan. The spillway model accounts for the difference in cross-sectional profile between the more typical form of the Athabasca River west of Fort McMurray and the broad, steep-walled valley below Fort McMurray. The outflow followed the Clearwater River and only reached the Athabasca at the modern confluence. Following the cessation of drainage from Lake Agassiz, the Athabasca River gradually evolved into its present configuration.

Today, the Athabasca and Peace Rivers meet in the Peace-Athabasca Delta at the western end of Lake Athabasca (Wolfe et al. 2005). During initial deglaciation, the area lay under the easternmost extension of Glacial Lake Peace, about 12,000 years ago. The lower Athabasca River valley formed a subordinate arm of the lake, receiving drainage from the outwash plain to the northeast, the deglaciated terrain of central Alberta to the south, and the drainage basin of the Clearwater River to the southeast.

With continued deglaciation to the north, the upper Mackenzie Valley was flooded under the waters of Glacial Lake McConnell, extending southwards to the Athabasca River (Smith 1994; Wolfe et al. 2017; Fig. 3.31). The subsequent withdrawal of Lake McConnell exposed the sands deposited on its beaches to wind. Large sand dunes developed along the Tu Nedhe (Slave) River and along the southern shore of Lake Athabasca (especially in Saskatchewan; Lamb and Guedo 2012; Fig. 3.32). These sand dune regions have undergone several periods of development since their initial formation, and re-activation occurs whenever the pine forests atop them are destroyed by fire.

Dynamic ice-jam flooding has occurred repeatedly in Fort McMurray, with more than 25 floods since 1875 (Fig. 3.33). The most recent major flood occurred 26 April–2 May 2020, displacing more than 13,000 people and causing property damage in excess of $230 million. Flooding is triggered by dynamic ice-jamming along the Athabasca River, accentuated by rain-on-snow events and spring freshet. The Athabasca–Clearwater confluence forms a choke point in the system. Dynamic ice movement is less obstructed along the broad Clearwater-Athabasca north valley, but there are numerous points where ice can ground. Thus, more ice can move through the broader channel more rapidly, leading to large dams. Ice-jamming is much less common along the southwestern reach of the Athabasca upstream of Fort McMurray.

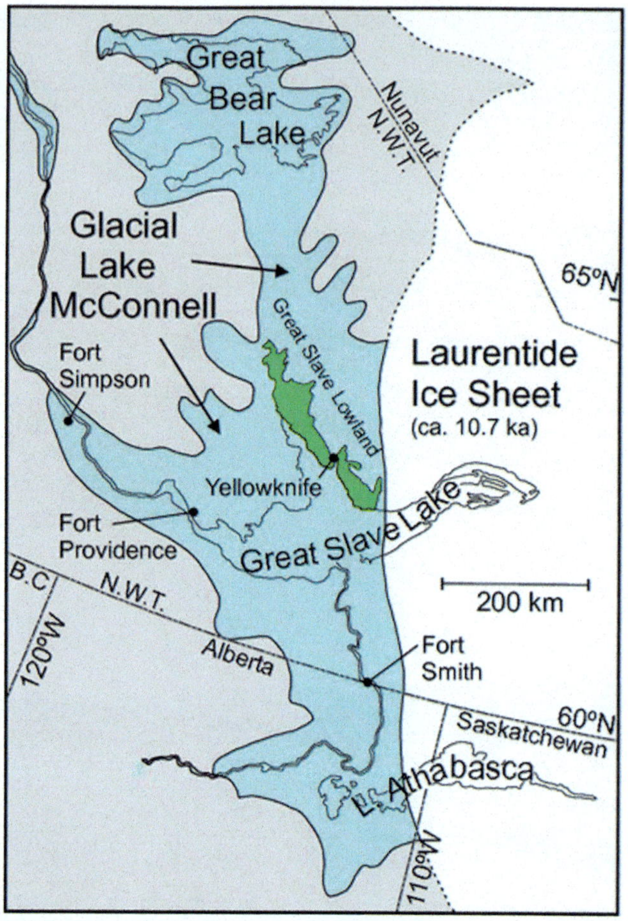

Fig. 3.31 Glacial Lake McConnell extended from Lake Athabasca north to Great Bear Lake, fed by meltwater from the Laurentide Ice Sheet and by Glacial Lake Peace to the southwest. (Wolfe et al. 2017)

3.5 Vegetation Assemblages

3.5.1 Boreal Shield

Prior to 1800 ce, the natural vegetation assemblages of the Boreal Shield ecozone varied from mixed, deciduous-dominated in the southernmost areas (e.g. Thousand Islands), to mixed but conifer-dominated in central regions (e.g. La Mauricie, Algonquin), to confer-dominated with lesser deciduous trees in the northern and western areas (e.g. Cochrane, Pukaskwa, Pimachiowin Aki; Brandt et al. 2013). Logging and the cessation of anthropogenic burning has resulted in substantial changes to the vegetation assemblages in the southernmost and central areas (Quinn 2004, Puric-Mladenovic and Pinto 2005), with lesser effects in the north and west.

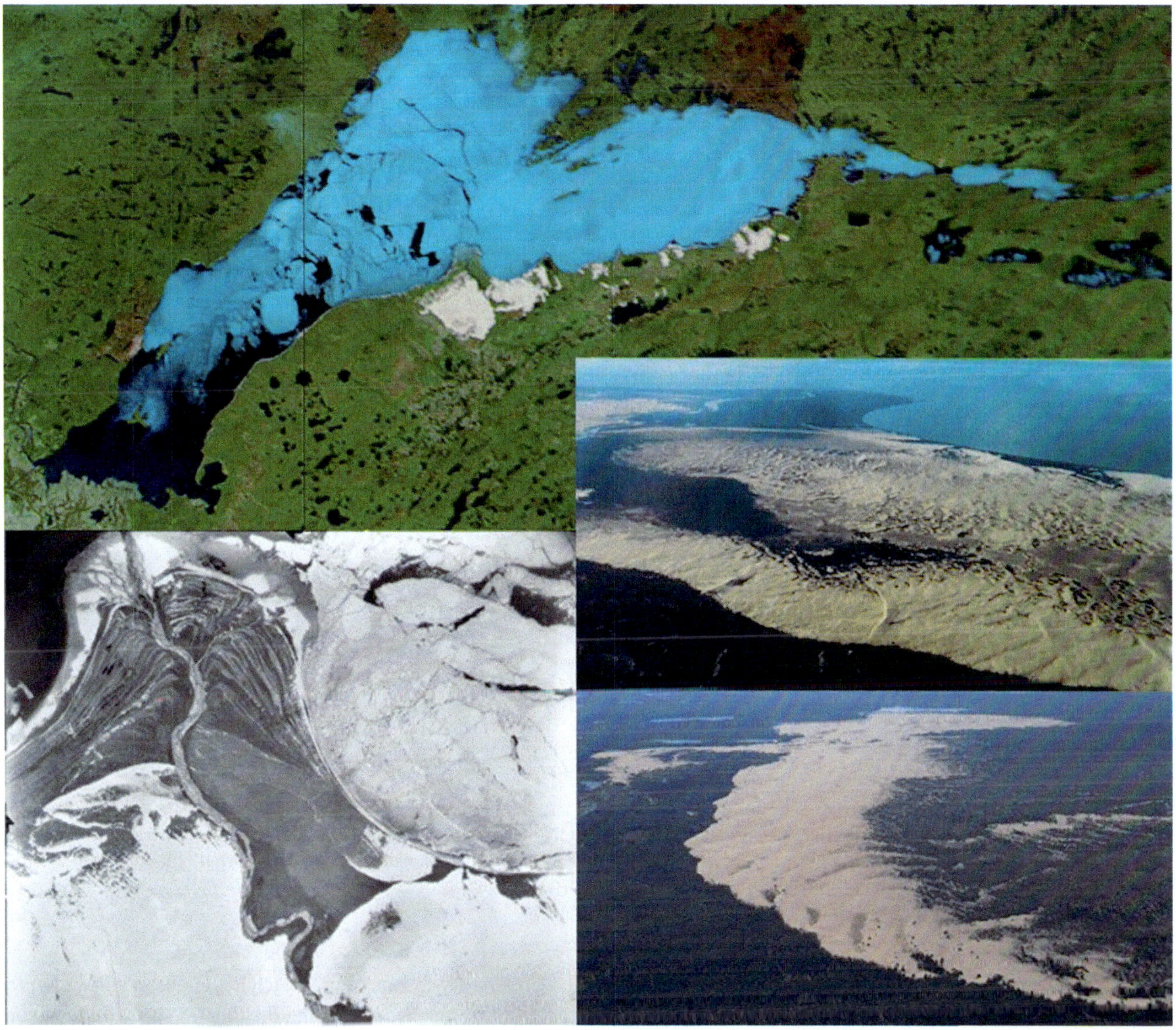

Fig. 3.32 Athabasca sand dunes. Upper: Ice-covered Lake Athabasca. Sand dune areas are light brown patches along the southern margin of the lake. Ridges to the south indicate former shorelines of Lake Athabasca and Glacial Lake McConnell. (NASA). Lower left: William River delta. Sand dune area is the light terrain to the south. Right: Parabolic and linear dunes in the William River area. (Wikipedia/Tim Beckett)

3.5.1.1 Mixed, Deciduous-Dominated Thousand Islands

The natural forest cover of the Thousand Islands is a mixed deciduous-dominated assemblage (Rowe 1972, Baldwin et al. 2019, CNVC 2016), with *Acer saccharum* (sugar maple) as the most conspicuous tree. Other common deciduous species are *Fagus grandifolia* (beech), *Tilia americana* (basswood), *Fraxinus pennsylvanica* (red ash), *Fraxinus americana* (white ash), *Betula alleghaniensis* (yellow birch), *Quercus rubra* (red oak), *Acer pennsylvanicum* (striped maple), *Ostrya virginiana* (ironwood), and *Prunus serotina* (black cherry). Ash trees in this region sometimes develop yellow leaves ("ash yellows") as a result of weather-induced stress (usually drought in summer months). Coniferous species are less numerous, and are dominated by *Pinus strobus* (eastern white pine; where not removed by logging), *Pinus resinosa* (red pine), and *Tsuga canadensis* (eastern hemlock). As elsewhere in eastern Canada, the combination of logging, agricultural disturbance, and cessation of burning by the indigenous Haudenosaunee (Iroquois) population (see Blarquez et al. 2018) has produced major changes in the forest assemblage.

One unique conifer species present is *Pinus rigida* (pitch pine; Fig. 3.34). This southern temperate species does not grow naturally anywhere else in Canada, except in isolated enclaves along the St. Lawrence Valley (Cody 1975; Hosie

Fig. 3.33 Fort McMurray is subject to flooding resulting from ice-damming of the Athabasca and Clearwater Rivers. Upper left: The Clearwater River flowing in the meltwater channel from Glacial Lake Agassiz joins the Athabasca River at Fort McMurray. Upper right: view looking north along the Athabasca River (Wikipedia/Tallgirl). Lower left: ice fills the Athabasca (Wikipedia/MiguelB). Lower right: the wandering Athabasca River downstream of Fort McMurray is subject to dynamic ice-jamming

1979). In Thousand Islands National Park, it is confined to southwest-facing slopes. Pitch pine is a slow-growing species, with individuals requiring 200 years or more to reach 20 m in height. It prefers dry sites, where other trees (particularly other pines) cannot establish themselves, and is favoured by frequent forest fires. Pitch pine can produce new shoots from dormant buds contained within its bark, and thus can quickly regenerate after a fire, even if the trunk is charred and the needles removed. During the Haudenosaunee-induced controlled burning of the forests, pitch pine was thus a major beneficiary. The Haudenosaunee also used the pitch for external treatment of cuts and skin ailments.

In the absence of forest fires, pitch pines find it difficult to reproduce, and many specimens do not grow in areas where fires have not occurred in the past 60 years. With the effective fire suppression resulting from European settlement, and the ongoing suppression practised in the intensely used recreational areas of the Thousand Islands, the future for pitch pine does not appear positive. Substantial populations of pitch pine exist in the United States, however, and thus it does not face global extinction.

Understory species unique to the Thousand Islands area include *Rhus copallina* (winged sumac), *Anemonella thalicitores* (rue-anemone), *Vitis aestivalis* (summer grape), and *Viburnum dentatum* (southern arrowwood), all of which survive only in sheltered locations and are at their northern limits (Cody 1975). *Vaccinium stamineum* (deerberry or huckleberry), extremely rare in Canada, grows at the western end of Grenadier Island. Substantial populations of huckleberries exist in the eastern United States.

During a typical year, the St. Lawrence River freezes around most of the islands by early January, and all channels are ice-covered sufficiently to support animal traffic by late January. In warmer years, initial freezing did not occur until mid- to late-January. In contrast, colder years were marked by freezing in early December. The duration of frozen conditions has a critical influence on animal migration, and years where freeze-up occurs earlier are also marked by increased desiccation of sensitive plants. In a typical year, the growing season approximates 165 days in the most exposed areas and up to 20 days longer in locally sheltered zones (Environment and Climate Change Canada 2020).

Fig. 3.34 *Pinus rigida* (Pitch pine)

3.5.1.2 Mixed, Conifer-Dominated

The mixed conifer-dominated boreal forest assemblage is dominated by the conifers *Picea glauca* (white spruce), *Abies balsamea* (balsam fir), *Pinus banksiana* (jack pine), *Pinus strobus* (eastern white pine), *Pinus resinosa* (red pine), and *Thuja occidentalis* (white cedar; Rowe 1972, Ministère des Ressources naturelles du Québec 2002, CNVC 2016; Fig. 3.35). Temperate conifers such as *Tsuga canadensis* (eastern hemlock) are generally rare, and do not occur north of the north shore of Lake Huron. Eastern hemlock and white cedar prefer soil with higher levels of alkalinity, developed from limestone bedrock, and thus do not thrive in the acidic podzols derived from granitic and gneissic rocks (Hosie 1979). Deciduous species are a lesser component of the assemblage, and are dominated by *Acer saccharum* (sugar maple), *Acer rubrum* (red maple), *Populus balsamifera* (balsam poplar), *Populus tremuloides* (trembling aspen), *Betula papyrifera* (white birch), *Betula alleghaniensis* (yellow birch), and *Fraxinus americana* (white ash).

Removal of the formerly dominant *Pinus strobus* by logging (see Chap. 4) allowed other trees to re-colonize the vacated areas. *Pinus banksiana* (jack pine), formerly an insignificant component of the assemblage growing only in the areas of poorest soil, was able to replace the logged *Pinus strobus*. *Pinus banksiana* is a slow-growing, relatively short-lived (50 years) species, which requires clear areas to expand. Although it is favoured by repetitive fires, fires occurring less than 15 years apart will result in the destruction of a jack pine forest. As a "hard" pine, and one with a characteristically twisted and knotted trunk, jack pine was considered inferior as a source of timber. Its dominance in the modern environment of Algonquin Park and other southern Boreal Shield regions was only possible as a result of removal of *Pinus strobus*. The spread of jack pine benefited Kirtland's Warbler (*Dendroica kirtlandii*), a species that nests only in jack pine stands of 30 hectares or more, where all trees are less than 6 m high (Sykes and Clench 1998). Other species, including *Acer saccharum* (sugar maple), *Acer rubrum* (red maple), *Populus*

Fig. 3.35 Mixed, conifer-dominated assemblages dominate the southern parts of the Boreal Shield. Dominant conifers are *Abies balsamea* (balsam fir), *Picea glauca* (white spruce), *Pinus banksiana* (jack pine), and *Pinus strobus* (white pine, where not removed by logging). Narrow-leaved deciduous species include *Betula papyrifera* (white birch) and *Populus balsamifera* (poplar). Broad-leaved deciduous trees such as *Quercus rubra* (red oak, lower right) are subordinate components. Many areas have been altered by logging and recreational development

tremuloides (trembling aspen), and *Betula papyrifera* (white birch), were also able to occupy areas forcibly vacated by *Pinus strobus*.

The understory species present include *Arctostaphylos uva-ursi* (bearberry), *Rhododendron groenlandicum* (Labrador Tea), *Vaccinium vitis-idaea* (bog cranberry), *Vaccinium uliginosum* (bilberry), and *Rubus acaulis* (raspberry). The floral emblem of Québec is the purple iris (*Iris versicolor*).

3.5.1.3 Conifer-Dominated

Conifer-dominated assemblages mark the northern and western areas of the Boreal Shield, including Pimachiowin Aki and Pukaskwa (Rowe 1972, Saucier 2008, Baldwin et al. 2019, CNVC 2016; Fig. 3.36). *Pinus banksiana* (jack pine) dominates on poor soils, gravelly beach ridges, and in exposed locations. More sheltered areas are occupied by *Abies balsamea* (balsam fir), *Picea glauca* (white spruce), *Larix laricina* (tamarack), and *Picea mariana* (black spruce). Pukaskwa lies near the northern limit of *Pinus strobus* (eastern white pine). Deciduous trees form a much smaller component of the assemblage, and are dominated by *Betula papyrifera* (white birch), *Populus tremuloides* (trembling aspen), and *Populus balsamifera* (balsam poplar), with shrubby *Salix sp.* (willow), *Alnus incana rugosa* (hoary alder), *Sorbus decora* (showy mountain ash), and *Corylus* (hazel). The understory vegetation contains boreal species such as *Rhododendron groenlandicum* (Labrador Tea; Dampc and Luczkiewicz 2015), *Vaccinium vitis-idaea* (bog cranberry), and *Vaccinium uliginosum* (bilberry). Unusual understory species in Pukaskwa include Pitcher's Thistle (*Cirsium pitcheri*) and Sparrow's egg Lady Slipper (*Cypripedium passerinum*), the latter currently under threat (Irvine and Patterson 2021).

Fig. 3.36 Conifer-dominated vegetation assemblage, Kaministiquia River, west of Thunder Bay, typical of the northern parts of the Boreal Shield

3.5.2 Boreal Plains

In the Boreal Plains, the Boreal Forest communities are dominated by *Populus tremuloides* (trembling aspen), *Picea glauca* (white spruce), *Betula papyrifera* (white birch), and *Pinus banksiana* (jack pine; Fig. 3.37). Broad-leaved deciduous species are absent. *Picea glauca* is typically the tallest species in the area, locally growing to 25 m in height. On well-drained sites, *Pinus banksiana* (jack pine) or *Pinus banksiana x Pinus contorta* (naturally hybridized jack pine–lodgepole pine cross) are significant (Lusebrink et al. 2013). The hybridization of the pines proceeded following deglaciation, as the jack pine spread from the east and the lodgepole pine spread from the southwest.

Moister areas support *Abies balsamea* (balsam fir), *Picea mariana* (black spruce), and *Larix laricina* (tamarack). Bogs are dominated by *Rhododendron groenlandicum* (Labrador tea), *Betula glandulosa* (dwarf birch), *Rubus chamaemorus* (cloudberry), *Sphagnum* moss, *Eriophorum* (cottongrass), and *Carex* (sedges; Cody 1995, CNVC 2016, Crispo et al. 2021). Most bogs and rivers are fringed by *Salix* (willow) and *Alnus viridis crispa* (green alder). Drier slopes support *Vaccinium uliginosum* (blueberry) and *Arctostaphylos uva-ursi* (bearberry). The understory vegetation consists of typical boreal species such as *Arctostaphylos uva-ursi* (bearberry), *Rhododendron groenlandicum* (Labrador Tea), *Vaccinium vitis-idaea* (lingonberry), *Vaccinium uliginosum* (bog cranberry), *Rubus chamaemorus* (cloudberry), *Salix* (willow), *Alnus viridis crispa* (green alder), and *Betula glandulosa* (dwarf birch).

3.5.3 Wetlands

Wetlands are important elements in the landscape in many areas of Canada, representing more than 14 per cent of the exposed terrain of the country, including the Boreal Shield and Boreal Plains. Canada currently has about 14 per cent of the worldwide total wetland area. Organic wetlands dominated by peat accumulation represent more than 90 per cent of the total wetland area (National Wetlands Working Group 1988).

Wetland terminology has become confused and confusing over the years. Some terms, such as *muskeg*, have been employed in so many different ways by different students of Canadian Geography that they have lost all usefulness. We will define a wetland as an exposed surface area that has standing water or saturated sediment at the surface, but is not an open-water body. Although wetlands include coastal marshes (saltwater and freshwater) and riparian marshes and swamps, here we'll consider organic wetlands developed in interior boreal ecoclimatic regimes.

Oligotrophic wetlands, with low nutrient concentrations, develop because the availability of nutrients is limited or restricted. Eutrophic conditions result when the supply of nutrients is in excess of the requirements of the vegetation. Eutrophic conditions can result from local oversupply of

Fig. 3.37 Boreal Plains vegetation assemblages are dominated by *Picea mariana* (black spruce), with *Populus tremuloides* (aspen) and *Betula papyrifera* (white birch). Clockwise from top left: Firebag River, Alberta; Hudson Hope, BC; Kiskatinaw River, BC; Halfway River, BC

nutrients (e.g. human discharge of nutrient-rich sewage or runoff from agricultural manure) or from lack of water circulation causing nutrients to be retained in the wetland system. A eutrophic wetland is characterized by large concentrations of algae or other forms of rapidly growing vegetation. Wetlands with intermediate amounts of nutrients are termed mesotropic.

Sources and characteristics of the nutrient load are also considered. In an ombrotrophic wetland, nutrients are locally derived from organic decay, with no outside source of clastic mineral material. The wetland is not connected to a drainage network. All water provided to an ombrotrophic wetland comes directly from precipitation, and the moisture regime is directly linked to the precipitation at that specific location.

Minerotrophic wetlands receive inputs of clastic minerals to supplement the organically derived components through a variety of mechanisms, primarily running water from streams. Water levels reflect conditions in the entire upstream drainage basin, rather than strictly local conditions. During a dry interval, a minerotrophic wetland may retain standing water, due to influx from wetter areas upslope.

3.5.3.1 Bogs and Fens

Combinations of nutrient concentrations, sources, and characteristics, together with vegetation types and moisture content in sediments, categorize the various wetland types (Catto 2015; Slaymaker et al. 2020b). A bog is an ombrotrophic wetland, without flowing water, deriving its water supply exclusively from precipitation. Most bogs have very limited nutrient concentrations and therefore are also oligotrophic (Fig. 3.38).

Bogs can have water impounded in a small isolated basin beneath the surface layer of organic matter. These floating or quaking bogs develop where the accumulating precipitation is precluded from draining readily, due to the presence of impermeable material under the basin and the lack of a surface outlet. Most bogs do not have standing water: the surface of the peat may be seasonally saturated, and water can be

Fig. 3.38 Boreal wetlands include bogs (top left) and fens (main photograph). *Picea mariana* (black spruce) is the dominant conifer, along with deciduous shrubs and *Populus tremuloides* (aspen)

squeezed out of it, but no pools of water are present on the surface.

Fens receive water from both overland and groundwater flow and are minerotrophic. The surfaces are depressed below the surrounding terrain and serve as areas of surface water influx and groundwater discharge. Fens are generally mesotrophic or slightly eutrophic, with abundant quantities of nutrients, permitting tree growth, along with sedges in the wetter areas.

Typically, fen peat has undergone limited modification, contains woody fibres, and is locally mixed with clastic silt and fine sand. In deeper-water fens, where flowing water and mineral influx dominate over organic detritus, mixed organic-mineral gyttja accumulates. Gyttja, which also accumulates in many shallow boreal and taiga lakes, contains higher concentrations of nutrients but lower carbon concentrations than does peat. Fens can be subdivided by form, extent of standing water, and vegetation type, into a variety of categories.

A precipitation event has very different results on fen and bog surfaces. In a fen, the water level reflects not only precipitation on site but also precipitation occurring upstream in the drainage basin. The fen, including its vegetation cover, acts to slow the rate of discharge of upstream water, reducing flooding downstream by spreading the discharge volume over a longer period. Water is absorbed into the unsaturated surface layer, which varies substantially in depth throughout the year. In some fens, evapotranspiration can exceed local precipitation, with the water levels retained due to the influx of water from upstream (Waddington et al. 2009). However, rain falling on a bog surface can only be retained until it is completely saturated. No infiltration to depth is possible, as this layer is already saturated. An ombrotrophic bog has little excess capacity for holding water, and plays little role in reducing flooding. The storm precipitation tends to run directly across the surface of the bog, locally carving shallow rills.

3.6 Soils

3.6.1 Boreal Shield

In the Boreal Shield ecozone, a variety of soil types are present, controlled primarily by the chemical and textural composition of the parent material and by the state of drainage (Soil Classification Working Group 1998, 2001 Wilpert 2022). Soils in low-lying areas are poorly drained, leading to the development of seasonally saturated, iron-rich Fera Gleysols. These soils are marked by a distinct mottling pattern, involving patches of grey-green gleyed zones with ferrous iron ions where the soils have been saturated below the water table.

In well-drained situations, soils range from weakly developed regosols and brunisols to moderately developed humo-ferric podzols (Fig. 3.39). In areas with thicker accumulations of glacial sediment, brunisols and podzols are present. These soils are highly acidic (typical pH values of 4.5–5.5), organic-poor, and are unsuitable for agriculture.

During the period from 1840 to 1910, after agricultural settlements had occupied the suitable land of the St. Lawrence and Ottawa Valleys, many fruitless efforts were made to expand farming north to the Grenville province. The Government of Ontario promoted a system of "Colonization Roads", and engaged citizens to construct these and clear the land (often in conjunction or collusion with logging operations in *Pinus strobus* forests), with the idea that the people would become farmers, working the thin, acidic soils. The only legacies of these efforts were increases in the proportions of *Populus* and *Acer* (as these trees re-occupied the labouriously cleared land upon agricultural failure), and the creation of groups of frustrated would-be farmers, many of whom saw their land for the first time as they arrived to begin farming it. After 70 years of failure, the Hastings Colonization Road (Ontario Highway 62) was described by surveyor Charles Aylsworth as "The Trail of Broken Hearts" (Kirby 2014).

Evidence of failure did not prevent the Government of Ontario from trying again. The Glacial Lake Ojibway Clay Belt from Kapuskasing to Cochrane was the scene of one of

Fig. 3.39 Acidic dystric brunisols (left) and podzols (right) are typical soils developed under Dfb boreal climates and coniferous vegetation over acidic bedrock. These soils are agriculturally limited

the last efforts to entice homesteaders to farm in Canada (Kent 1966; Vanderhill 1988). Following the discovery of silver at Cobalt, Haileybury and New Liskeard to the south, and findings of gold at Kirkland Lake and Val D'Or to the southeast and Timmins to the southwest, the governments of Canada and Ontario decided that the new mining communities would benefit from the establishment of agricultural areas on the soils of the Clay Belt. Further impetus to agricultural development was provided by the construction of the Temiskaming and Northern Ontario railway north from North Bay to service the mining districts, eventually extended northward to Moosonee on James Bay. Additional railroad construction involved the east–west National Transcontinental Railway. This line, intended by Wilfred Laurier as an alternative to the CPR, linked Québec City to Winnipeg via La Tuque (QC) and Cochrane and Nakina (ON), passing north of Lake Superior. Agricultural settlement along the Québec part of the line, in Rouyn-Noranda and Abitibi, was strongly promoted by the Catholic Church in Québec (Gourd 1973).

Unfortunately, although additional mineral finds continued (e.g. at Rouyn-Noranda), agricultural activity was not successful. The Clay Belt offered suitable soil that experienced farmers could have successfully exploited, but the climate was not conducive to development of farms, and the immigrants from urban Ontario and Québec generally were not experienced farmers. Most of the would-be homesteaders rapidly abandoned farming, turning instead to mining, forestry, or emigration to meet their economic needs. Today, climate change is renewing interest in agricultural development of the Clay Belt, albeit driven by personal rather than government initiative.

3.6.2 Boreal Plains

Soils of the Boreal Plains ecozone reflect the underlying bedrock and Quaternary sediment (Soil Classification Working Group 1998, 2001) In the Peace River Lowland and the Grande Prairie region, the soils are dominantly Dark Grey Luvisols (under boreal forests and flanking Aspen Park lands, commonly developed on glacial till; Fig. 3.40) and Dark Grey to Black Chernozems (under grasslands and areas cultivated since ca. 1900 ce, commonly developed on glacilacustrine sediment). Although initial surveys in the 1870s identified these areas as agriculturally suitable, decisions to route railway construction to the south delayed agricultural development. Adjacent upland areas and forested terrain to the north dominantly support dark grey luvisols and brunisols. Solonetz soils develop where sodium-rich bedrock is exposed, as in Wood Buffalo National Park.

Fig. 3.40 Dark Gray Luvisol developed under mixed *Populus tremuloides* (aspen)-conifer assemblage. A thick Ah horizon overlies the clay-rich Bt showing columnar structure

3.7 Fauna

Southern to mid-boreal faunal assemblages typically include the herbivores *Erethizon dorsatum* (porcupine), *Alces alces* (moose), and *Odocoileus virginianus* (white-tailed deer). Smaller herbivores include muskrat (*Ondatra zibethicus*), beaver (*Castor canadensis*), snowshoe hare (*Lepus americanus*), chipmunk (*Tamias striatus*), woodchuck (*Marmota monax*), red squirrel (*Tamiascurus hudsonicus*), and northern flying squirrel (*Glaucomys sabrinus*), along with several bats, moles, and voles (Banfield 1981; Naughton 2012). Omni vores include *Procyon lotor* (raccoon), *Vulpes vulpes* (red fox), black bear (*Ursus americanus*), masked shrew (*Sorex cinereus*), and *Mephitis mephitis* (skunk). The scavenger *Gulo gulo* (wolverine) is present, albeit in small, threatened populations.

Fig. 3.41 *Lynx canadensis*. (United States Fish and Wildlife Service)

Carnivores include *Lynx canadensis* (lynx; Fig. 3.41), mink (*Neogale vison*), North American ermine (*Mustela richardsonii*), otter (*Lontra canadensis*), and pine marten (*Martes americana*). *Puma concolor* (cougar), formerly native to the southern part of the Boreal Shield, has been largely eliminated by human activity, although isolated populations remain (e.g. in Pukaskwa). Cougars inhabit the northern part of the Boreal Shield ecozone.

Canis lupus (wolf) was formerly present in substantial numbers throughout the region, but was largely eliminated by human action as well (Rempel et al. 2021). Eastern wolves (*Canis lupus lycaon*; total number uncertain) have inhabited La Mauricie since 2010, but are not sufficiently numerous to play a substantial ecological role (Parks Canada 2021b). Algonquin also harbours a population of wolves. The Algonquin wolf is a hybrid among the eastern wolf, grey (timber) wolf, and the coyote (COSSARO 2016). An adult male stands approximately 60–70 cm high at the shoulder, and has a mass of 25–30 kg. Eastern wolves form small packs (typically 4–6 animals), and range over 150-500 km^2. Estimated population in the park is less than 500, and the species is classified as threatened.

The effective removal of the largest predator species from the ecosystem has allowed other predators to assume the vacant ecological niches, and to increase their numbers. In particular, *Canis latrans* (coyote) has expanded its range and numbers considerably as wolves have been eliminated. In 1919, coyotes were not present throughout Ontario and Québec, but they have successively extended their range eastward. Throughout the southern parts of the Boreal Shield, the coyote has assumed a position as the dominant predator of many species (Parker 1995; Patterson and Messier 2000), including white-tailed deer (Messier et al. 2011).

3.7.1 *Odocoileus virginianus* (White-Tailed Deer)

Odocoileus virginianus (white-tailed deer) populations generally increased over the past 70 years across most of the southern Grenville province, in response to lessening of predation pressure from both the now-departed *Puma concolor* and from *Homo sapiens* (Deer Friendly n.d.). From a low point ca. 1900 ce, *Odocoileus virginianus* has increased its population to reach approximately the number of deer in 1500 ce (Fig. 3.42). Population "explosions" of whitetails periodically occur (VerCauteren et al. 2011), leading to winter kill and forcing human intervention in the form of rescue operations and feeding programmes. Cold winters, such as 2008–09 and 2014–15, have resulted in local population declines of 30% in boreal Ontario and Québec. In the southern Boreal Shield, pressure from hunters in the surrounding areas and local culls necessitated by over-population in predator-deprived areas have acted to check the growth of the white-tailed deer population within protected areas.

Deer prefer forest areas with mature trees and limited underbrush (Waller and Alverson 1997). Indigenous Algonkians and Ilnu (Nehilaw) periodically burned areas of forest, which served to remove underbrush, inhibit sapling growth, and simultaneously enriched the soil and encouraged development of green shoots (Blarquez et al. 2018). These practices favoured the deer, which served as a major food

Fig. 3.42 *Odocoileus virginianus* (white-tailed deer) is present throughout the Boreal Shield and Boreal Plains ecoregions. Left: male; right: female. (US Department of Agriculture)

source for the people, in addition to favouring the development of a forest dominated by mature white pines and fringing areas with alders and jack pines. When burning stopped following European settlement, logged tracts of forest were replanted with white spruce, forming dense thickets of low trees that the deer found unattractive. As well, the absence of periodic fertilization resulting from burning led to declines in soil quality and productivity. Many cleared areas were occupied by sugar maples, and sugar maple "bushes" were established to utilize the sap. These maple groves were also unsuitable as deer habitats, but were preferred to the white spruce plantations by the deer. Paradoxically, the population of white-tailed deer in woodlot areas adjacent to agricultural terrain thus locally increased, while simultaneously populations in the reforested areas steadily dropped. With the cessation of logging and plantation of white spruce monocultures in La Mauricie, the situation may improve for the white-tailed deer. Long-term improvement, however, will be contingent on re-establishment of the natural fire frequency within the park, as well as regulation of hunting in the environs.

Cervid-vehicle collisions are significant problems across Canada (e.g. Laliberté and St-Laurent 2020; Fig. 3.43). A typical mature male moose has a body mass of 600 kg, compared to 100 kg for a large white-tail deer and 1400 kg for a typical mid-sized car.

Areas with the most deer or moose have the greatest likelihood of collisions, although the effect tends to decrease as human population density increases (e.g. Jiang et al. 2012). Accidents are most likely to occur in the evening, either because deer and moose are more active at these times, or because humans and animals are less able to see each other at those times. Deer in isolated, protected, or non-hunting areas show more daytime activity than those in human-influenced areas, suggesting adaptation to human pressure. Typically, collision rates increase at the beginning of the plant-growing season, suggesting a connection with animal movement in search of food (Steiner et al. 2014). Collision rates tend to be higher on roads with high traffic volumes, either throughout the year or seasonally. Roads through forested areas or with curves, requiring lower speed driving, have lower collision rates (Eldegard et al. 2012; Neumann et al. 2012).

Deer and moose perceive objects primarily through motion, and have only limited ability to focus on a particular object, seeing a 300° arc of vision essentially equally. Moving vehicles, which don't look, sound, or smell like familiar predators, may not be perceived as threats. Deer are more likely to perceive and react to shorter wavelengths (ultraviolet, blue) than longer wavelengths (yellow, red) (Cohen et al. 2014). Deer vision has adapted to detect predator movement at a distance, particularly at dawn or dusk, somewhat similarly to the Purkinje effect (dark adaptation) noted in changed human perception of colours from daylight to dusk. A potential consequence of this is that deer and moose perceive vehicle headlights (particularly tungsten halogen lights, with greater red and longer-wave emissions than incandescent lights) differently than humans do. High-intensity discharge headlights produce a larger proportion of visible light, particularly shorter-wave blue light.

A male white-tail deer can run at 45 km h^{-1}, covering 12.5 m in one second. A male moose can run at 60 km h^{-1} covering 16.7 m in one second. Considering a typical human reaction time of slightly more than one second, for a typical mid-size car travelling at 100 km h^{-1}, the stopping distance is approximately 100 m, and the time required about 6.5 s.

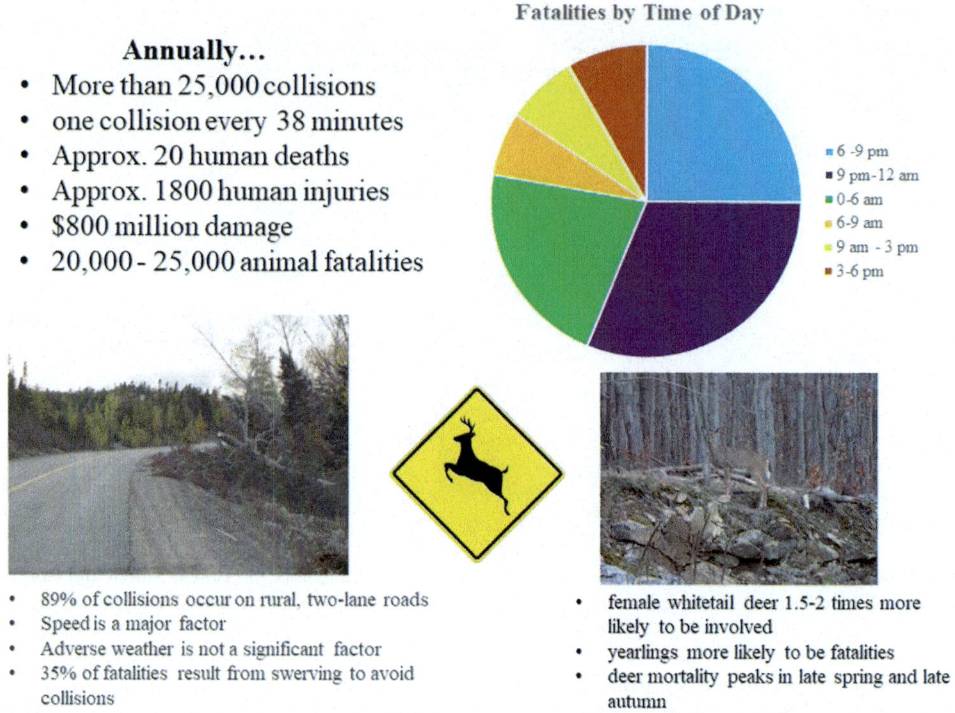

Fig. 3.43 Deer-vehicle interactions have significant economic, human, and faunal costs throughout boreal Canada. (Data sources: Traffic Injury Research Foundation, Transport Canada; Desjardins Insurance)

Discussions of potential mitigation strategies such as roadside fencing may focus on financial cost–benefit-perceived risk analysis. Conover et al. (1995) balanced an estimated $19.7 billion in benefits (money spent by wilderness users and hunters, value of venison) against $1 billion in vehicle collision damage, browse damage to timber and agricultural crops, and losses caused by deer transmittal of Lyme disease. Despite the human injuries and deaths resulting from deer-vehicle collisions, the net conclusion was that deer have an overall high positive value.

In contrast to white-tailed deer, populations of *Rangifer tarandus caribou* (woodland caribou) are generally under pressure in the Boreal Shield, primarily due to habitat loss caused by logging and road construction (Rempel et al. 2021). Pimachiowin Aki is the largest protected area which provides suitable habitat for woodland caribou.

3.7.2 Boreal Plains

The faunal assemblage of the Boreal Plains ecozone represents a transition between the transitional Grassland–Aspen Parkland fauna to the south, including wapiti (*Cervus canadensis*), mule deer (*Odocoileus hemionus*), and *Bison* (see Chap. 5), and the Boreal Shield and Taiga Shield fauna to the north and east. In areas subjected to agricultural disturbance, most of these species have been extirpated or reduced greatly in numbers. The only species which have been truly successful in resisting human penetration are coyote and mule deer. The latter have taken advantage of the removal of wolves and cougars from the region to expand their numbers (Burkholder et al. 2018), and locally pose a serious problem for farmers.

Prairie species have responded to general climatic warming by moving northwards into the Boreal Plains. In areas without agricultural activity, typical midgrass prairie residents such as badger (*Taxidea taxus*), jackrabbit (*Lepus townsendii*), mule deer (*Odocoileus hemionus*), white-tail deer (*O. virginianus*), and wapiti (*Cervus Canadensis*) are expanding their ranges northwards, along with ground squirrel (*Citrullus sp.*) and pocket gopher (*Thomomys sp.*) species. Southward migration of northern boreal species occurs during the winter months, including woodland caribou (*Rangifer tarandus caribou*). Free-roaming bison live in Wood Buffalo National Park (see Chap. 5).

3.7.3 Freshwater Fish

Boreal freshwater fish are significant ecologically, culturally, and recreationally (Scott and Crossman 1998). Fish health is a key indicator of ecological conditions, and one with extremely high public and political visibility. "Algonquin" is an Ojibwe word which can be translated as "the place where we spear fish". The fish referred to here are probably pickerel, pike, and perch. Concerns about acidic precipitation focused on its effects on the fish populations, and specifically on the species of interest to anglers. In particular, the emotional commitment

Fig. 3.44 Black bass. Upper: *Micropterus salmoides* (largemouth). Lower: *Micropterus dolomieu* (smallmouth)

that many Ontario residents feel to black bass, combined with the long history of angling and the traditionalism involved, meant that their decline was keenly felt.

The two most desirable species for most anglers are the black bass *Micropterus dolomieu* (smallmouth) and *M. salmoides* (largemouth; Fig. 3.44). Bass angling has been mentioned in glowing terms in Ontario accounts since 1816. Largemouth bass tend to be heavier (up to 6 kg) and longer (up to 40 cm) than smallmouth, but in the Boreal Canadian Shield the latter are better "fighters" and are more highly prized. Both species spawn in the spring, with the males sweeping out a 1 m radius of sediment on the lake bottom for a nest, into which the female lays in excess of 5000 eggs (Brown et al. 2009a, 2009b). The largemouth prefers shallower, warmer lakes (generally <6 m deep, water temperatures 25–28 °C) than the smallmouth (optimum water temperature 16 °C). Lakes tend to support only one black bass species, depending on their bathymetry; throughout the southern Laurentian Highlands, smallmouth bass are more common than largemouth.

The spring spawning habits of these species, combined with their intolerance of pollution and acidity (especially true for the smallmouth), rendered them susceptible to attack from acidic precipitation. Many lakes suffered drastic declines in smallmouth bass numbers, and several are completely "dead" (although reintroductions have been achieved since 1995). Locally, introduced species such as *Ambloplites rupestris* (rock bass) have added to the pressure by consuming eggs and bass fry, due to their earlier hatching times.

Salvelinus fontinalis (brook trout) is a common inhabitant of fast-flowing, gravel-bottomed rivers throughout Boreal Canada (Scott and Crossman 1998; Hasnain et al. 2010). It requires well-oxygenated, cold (<20 °C), clear waters, and does not flourish in streams choked by logging debris. Brook trout are commonly 25–30 cm in length and may have a mass of 5 kg at maturity (although 2 kg is more common). They are carnivorous, eating up to 50% of their body weight each week. As they spawn in the fall, they are less affected by acidic precipitation in Algonquin during this critical season.

Salvelinus namaycush (lake trout) prefers deeper, colder (less than 10 °C) well-oxygenated lake waters (Behnke 2002). Lake trout are larger than brook trout, commonly 38–50 cm in length and 5 kg mass at maturity (the record is 46 kg), and have great longevity (commonly more than 20 years). They are carnivorous, preying upon smaller fish such as cisco (*Leucichthys sp.*) and sculpins, and are vulnerable to declines in numbers of their preferred prey. Lake trout are also fall spawners and are migratory.

Coregonus clupeaformis (whitefish) is present as two distinct populations in Algonquin Provincial Park (Crossman and Mandrak 1992). A "normal-sized" population (25–40 cm length, 2–10 kg in mass) and a dwarf population (5–10 cm in length, less than 1 kg mass) exist in adjacent lakes, but do not co-habit. These distinct populations appear to have developed separately as a result of modification of the drainage patterns of the park during deglaciation and glaci-isostatic rebound, and have remained separate as the lakes developed. Whitefish cannot move readily through rapids or shallow waters and require cold conditions. They spawn on gravelly or stony substrates in late November and early December and require water temperatures of less than 8 °C for successful reproduction (the mortality of the eggs reaches 99% at a water temperature of 10 °C). These three salmonid species, therefore, are among the least affected by acidic precipitation during the spawning season.

In contrast, the species of the pike family, including *Esox lucius* (northern pike), *Esox masquinongy* (muskellunge), *Sander vitreum* (walleye; also called doré and pickerel), and *Perca flavescens* (yellow perch), are all spring spawners (Fig. 3.45). Environmental preferences and times of spawning may favour some species over others (Scott and Crossman 1998). *Esox lucius* (northern pike) generally spawn at spring breakup, in clear, warm, shallow water, and young pike grow rapidly in the early spring. *Esox masquinongy* (muskellunge) generally spawn 4–6 weeks after northern pike, and grow relatively slowly. Thus, the muskellunge eggs hatch in time to provide the young, aggressive, hungry pike with a suitable meal. In lakes which contain substantial numbers of both species, the muskellunge tends to be outcompeted by the pike. Muskellunge is also less tolerant of environmental variations in water chemistry and will hybridize with pike in some circumstances. Consequently, the range and numbers of muskellunge are steadily contracting, as habitats are gradually taken over by the northern pike. Whereas northern pike are found south of the treeline in all areas of Canada east of the Rocky Mountains and north of the Appalachians, muskellunge are confined to two disjunct areas of the southern Canadian Shield, west of La Mauricie and east of the Mani-

Fig. 3.45 Boreal freshwater fish. Upper left: *Esox lucius* (northern pike; Wikipedia/Josyan Pierson). Centre left: *Sander vitreum* (walleye; NOAA). Lower Left: *Esox masquinongy* (muskellunge; Eric Enbretson, United States Fish and Wildlife Service). Right: statues of muskellunge in Gananoque (upper) and Kenora (lower) attest to its popularity for sport angling

toba border, and are not found in streams draining into Lake Superior (Mandrake and Crossman 1992).

Perca flavescens (yellow perch) and *Sander vitreus* (walleye) spawn in late April and early May, respectively, in Algonquin Provincial Park. Perch is less demanding in its spawning and living requirements, tolerating shallow, reedy or brush-filled water and thus can thrive in areas which have been logged. Walleye require clear, fast-running water for spawning, and are extremely vulnerable to (illegal) human predation during the mass migrations which characterize the spawning season. All of the pike family spawn during the peak acidic precipitation period, but the "environmental tolerance" of northern pike and perch have enabled them to withstand this environmental threat to a greater degree than have muskellunge and walleye. All species, however, are under some pressure as a result of previous acidic precipitation.

A few lakes and large rivers support lake sturgeon (*Acipenser fulvescens*). The presence of lake sturgeon in Pimachiowin Aki and lakes in the Clay Belt indicate former connections to the Great Lakes–St. Lawrence and Mississippi drainage basins resulting from postglacial lake formation and glaci-isostatic deformation. Lake sturgeon prefers turbid waters with suspended sediment and hence often occupies lakes underlain with glacilacustrine clays. They are carnivores, feeding on crustaceans, molluscs, gastropods, insects, and eggs of other fish species. An adult male may reach 2.5 m in length and exceed 150 kg. Unusual feeding behaviours include swimming upside down, jumping, and backwards tail walking, with the fishes' body above the water level (Pimachiowin Aki World Heritage Project 2012).

Combinations of overfishing (1850–1950), loss of habitat due to dam construction, and warming water due to climate change have placed severe pressure on lake sturgeon populations. In most areas, including Pimachiowin Aki, only indigenous First Nations may harvest lake sturgeon.

3.7.4 Birds

3.7.4.1 Boreal Shield

About half of North America's birds spend time in the boreal forest, and more than 3 billion breed there annually. Typical birds of the Boreal Shield include flycatchers, grosbeaks, kinglets, siskins, thrushes, vireos, and warblers, along with waterfowl, raptors, woodpeckers, and ptarmigan (Boreal Songbird Initiative 2015).

Pukaskwa National Park has adopted the common loon (*Gavia immer*) as its symbol. Loons arrive immediately following spring breakup, and proceed to nest (Godfrey 1986, Bird 2017). Only one pair nests on all but the largest lakes. Loons have a tendency to remain with the same partner throughout life. They are graceful in the water and under it, somewhat ungainly in flight, and embarrassingly clumsy on land. Loons cannot take off from land or from a tree; they require a running start over water, sometimes as much as 500 m on calm days. They leave the park shortly before freeze-up, and winter in marine waters. Pukaskwa's loon population, with other Ontario loons, heads for the South coast of Newfoundland, Nova Scotia, and New England. Loons are negatively influenced by acidic precipitation, as well as by avian botulism.

Fig. 3.46 *Meleagris gallopavo* (turkey) has been re-introduced to parts of the meadow area of the Thousand Islands. (Wikipedia/Riki7 photo)

The "wild" turkey (*Meleagris gallopavo*) was extirpated from southern Ontario by 1902, but introduced birds have repopulated parts of the meadow area of the Thousand Islands (Bowman 2007; Fig. 3.46). Although some of the wild turkeys may be descendants of escapees from farms prior to 1945, most are descendants of introduced birds. Today's domestic turkey, bred for human uses, is incapable of flight. Turkeys suffer from cold winters, and declines in their populations were noted during the cold years of 1810–1818 (particularly after the disastrous "year without a summer" of 1816), and again in 1840–1845. During severe winters, such as that of 1842, more than 75% of Canadian turkeys died.

3.7.4.2 Boreal Plains

Wood Buffalo is noted for its whooping crane (*Grus Americana*) population (Godfrey 1986, Parks Canada 2021a; Fig. 3.47). The whooping crane, a large wading bird with a wingspan of 2.5 m and standing 1.2 m high, was never an

Fig. 3.47 *Grus americana* (whooping crane) (US Department of Agriculture)

abundant species in North America. It is dependent on specific summer and winter nesting sites for its propagation, requiring abundant supplies of bulrushes for nest construction and alkaline waters to provide nutrition. The major summer nesting ground remaining for whooping cranes is the Sass River area of Wood Buffalo, and only the shallowest pothole lakes can be used by the birds. Small populations exist in other pothole areas, but these birds congregate in the Sass River area prior to their late October migration to their remaining winter nesting ground, the Aransas National Wildlife Refuge on the Gulf of Mexico coast of Texas.

The population of birds declined throughout the early 1900s, although they were not preferentially hunted. Some of the decline was due to disturbance of the winter nesting grounds, but most seems to be due to the natural changes in the summer breeding grounds, induced by climatic fluctuations. The dry 1930s were particularly difficult for the whooping crane, as potholes dried and bulrushes died. By 1941, the population had declined to only 15 birds. Preservation efforts begun in the mid-1930s were continued strenuously after World War II, and the population has slowly risen. Current estimates suggest that the total population is between 200 and 220 birds.

The Whooping Crane is one of a very few species whose demise cannot be blamed entirely on *Homo sapiens*. It thus stands in sharp contrast to other bird species hunted to extinction in North America, notably the Passenger Pigeon and the Great Auk. Changes in wetland hydrology resulting from dam construction on the Peace River, however, could have a negative impact on the Whooping Crane.

Prince Albert and Wood Buffalo support colonies of White Pelicans (*Pelecanus erythrorhynchos*). White pelicans are fish-eaters but prefer to catch their fish while wading or swimming, rather than diving from above (Green and Elmberg 2014). With a wingspan of 3 m, a flock of these highly gregarious birds is an impressive sight. Other significant bird species include ruffed grouse (*Bonasa umbellus*), canvasback duck (*Aythya valisineria*), mallard duck (*Anas platyrhynchos*), and night heron (*Nycticorax nycticorax*). A variety of raptors are present, including Great Grey Owl (*Strix nebulosa*, left), Great Horned Owl (*Bubo virginianus*), barred owl (*Strix varia*), Long-eared owl (*Asio otus*), short-eared owl (*Asio flammeus*), goshawk (*Accipiter gentilis*), sharp-shinned hawk (*Accipiter striatus*), red-tailed hawk (*Buteo jamaicensis*), Swainson's hawk (*Buteo swainsoni*), and ferruginous hawk (*Buteo regalis*).

3.7.5 Reptiles and Amphibians

Georgian Bay Islands National Park in the southern Boreal Shield contains the greatest diversity of reptiles and amphibians of any Canadian national park. Species include eastern snapping turtle (*Chalydra serpentina;* Fig. 3.48), map turtle (*Graptemys geographica*), musk turtle or stinkpot (*Sternotherus odoratus*), spotted turtle (*Clemmys guttata*), massasauga rattlesnake (*Sistrurus catenatus*), eastern fox snake (*Elaphe vulpina*), milk snake (*Lampropeltis triangulum*), hognose snake (*Heterodon platirhinos*), bullfrog (*Rana catesbeiana*), leopard frog (*Rana pipiens*), and American toad (*Bufo americana*). Environmental changes have caused substantial declines in frog and toad numbers, and snakes are periodically harassed by visitors (particularly the Massasauga), but turtle numbers are stable.

Fig. 3.48 *Chelydra serpentina* (eastern snapping turtle). (Bruce Hallman, United States fish and Wildlife Service)

Chelydra serpentina (eastern snapping turtle) is a species which prefers quiet, mud-bottomed lakes (Cook et al. 1984). Mature snapping turtles are 40–45 cm in length, not including the long, serrated, prehensile tail. Unlike most turtles, the tail cannot be fitted under the carapace. Typical mass for a mature male is 39 kg, but a record specimen had a mass of 143 kg. Despite their size and somewhat fearsome reputation, the snapping turtle is often rather passive, reacting only when prodded or taunted. It eats a mixed diet, usually consisting of approximately 30% plant matter and 70% fish and invertebrates, often scavenging its meat. It is particularly fond of eating fish caught by anglers that are left hanging over the sides of boats or docks. Leaving a string of fish in the water will result in a string of fish heads. Snapping turtles require warm waters to move and feed, and will not eat when water temperatures drop below 16 °C.

After a somewhat unusual mating ritual, highlighted by the male repeatedly biting the female's neck, snapping turtles nest in nearshore gravel banks above the water level. The young hatch in early October, and frequently immediately begin their winter hibernation.

Attempts made to augment the natural population with laboratory-bred individuals were initially unsuccessful, as repeated efforts to hatch eggs under controlled conditions produced only female turtles. The sex of the unborn turtles appears to be partially controlled by the incubation temperature, with temperatures of ± 30 °C required for females and 25 °C–28 °C for males.

The massasauga rattlesnake (maximum length 90 cm) survives only in the most isolated, inaccessible areas of southern Ontario (Fig. 3.49). Its venom induces local destruction of the capillaries and surrounding tissues, interrupting circulation and oxygen supplies and rapidly causing death in the mice, voles, and shrews that it favours as prey (Ernst and Barbour 1989). The venom's effect on humans is much slower and less sure, usually requiring three hours or more to cause serious injury or death. Since 1956, the total number of human fatalities in Ontario is 2, one of whom had been playing with the snake before being bitten. Rattlesnakes use their rattles to advertise their presence to potential threats such as deer, in order to avoid being stepped upon.

The Thousand Islands also support a diverse assemblage of reptile and amphibian species. Notable among these are *Pantherophis spiloides* (grey rat snake), Canada's largest snake, with a maximum length of 2.5 m, and *Nerodia sipedon* (northern water snake), which preys on small fish and amphibians. Both snakes are harmless to humans.

The shallow leads and ponds support *Lithobates catesbeiana* (bullfrog). Bullfrog numbers, in common with other amphibian species, are currently in decline throughout southern Ontario (Berrill et al. 1992). This decrease in population matches those of amphibians in virtually all areas surrounding the Great Lakes and is linked to environmental degradation. Mild summer evenings in the islands formerly featured the croaking of bullfrogs, but an estimated less than half of the 1985 population of frogs was present ten years later,

Fig. 3.49 *Sistrurus catenatus* (massasauga rattlesnake) survives in reduced numbers in restricted locations along Georgian Bay. Left: ca. 1800 ce, the massasauga was present throughout southwestern Ontario. Currently, with the exception of enclaves at Ojibwa Prairie (near Windsor) and Wainfleet Bog along Lake Erie, it is confined to the northern Bruce Peninsula, Manitoulin Island, and the Georgian Bay biosphere reserve. (Data source: Committee on the status of species at risk in Ontario). Right: *Sistrurus catenatus*. (United States Geological Survey)

and numbers continue to decline. The scarcity of frogs has begun to affect the population of Great Blue Herons (*Ardea herodias*).

In the eastern and southern Boreal Plains, small snakes such as the plains garter snake (*Thamnophis radix*) are common in less-modified areas, but have suffered from human activities elsewhere. The Narcisse snake den in the Interlake, Manitoba, area is noted for the large concentration of overwintering garter snakes (Nature North n.d.; Fig. 3.50). Huge Spring "mating balls" involve more than 100 male snakes for every female. Previously, more than 70,000 plains garter snakes were collected annually, primarily for sale as pets, prior to a major crash in the population in the early 1990s. Today, the population is gradually recovering.

3.8 Boreal Shield and Plains Human Communities

In 2024, approximately 1.3 million people lived in the 1.7 million km^2 Boreal Shield Ecozone discussed in this chapter. The majority live in the southern part of the Shield regions of Ontario, concentrated in the mining- and transportation-related cities of Kenora, Thunder Bay, Sault Ste. Marie, Sudbury, and North Bay; other mining-related communities (e.g. Kirkland Lake, Timmins); and communities with recent growth based on recreation and tourism (e.g. Bracebridge, Huntsville, Parry Sound). In Québec, mining (e.g. Rouyn-Noranda, Val d'Or), forestry (e.g. La Tuque), recreation and tourism (e.g. Mont-Laurier), and proximity to the major urban centres of Gatineau, Montreal, and Québec account for most of the population. In Manitoba, centres are associated with mining (Thompson), forestry (The Pas), and hydroelectric development (Gillam). Abundant open space is present between towns, increasing northward.

Boreal Shield communities developed as transportation pushed westward and northward between 1850 and 1950. Hard-rock mining generally followed transportation, rather than preceding it. Town layouts thus were influenced by the planning concepts of the time. Rather than developing in linear forms along a single transportation axis, or agglomerated patterns in response to personal land ownership, expediency, or practicality, the initial boreal Shield towns were laid out in grid patterns, followed by curvilinear patterns in newer communities (e.g. Thompson). Even in the oldest mining communities (e.g. Cobalt), where a single company did not dominate operations in the initial phases, subsequent economic consolidation, sometimes facilitated by fires, resulted in grid street patterns. Towns were imposed upon the pre-exiting landscape, rather than developed in harmony with it.

Communities initially developed to service forestry, such as Bracebridge and Parry Sound, underwent changes from initial agglomerated layouts as the economic focus shifted after 1860. Parry Sound sequentially became a trans-shipment

Fig. 3.50 *Thamnophis radix* (garter snakes), Narcisse Snake Dens, Manitoba. (Travel Manitoba)

port for prairie grain when the Canada Atlantic Railway was completed across Algonquin Park and port facilities were constructed at Depot Harbour (1898), followed by a recreation and tourism based community when the port was abandoned (1953). Summer recreational activity is largely responsible for the growth of Bracebridge from population 6903 (1971) to 17,305 (2021, Census of Canada).

Boreal Plains communities in northern Alberta and northeastern British Columbia have about 450,000 people, within the total of 737,287 km^2. The largest centres of Fort McMurray, Grande Prairie, Cold Lake, Fort St. John, Dawson Creek, and Peace River account for approximately two-thirds of the total population. Petroleum, agriculture, and forestry provided the economic impetus for the establishment and growth of these communities. The initial towns were laid out in grid patterns, with curvilinear patterns in newer communities.

References

Aber JS, Ber A (2007) Glacitectonism. Developments in quaternary science series 6. Elsevier, London, 246 p

Baldwin K, Allen L, Basquill S, Chapman K, Downing D, Flynn N, MacKenzie W, Major M, Meades W, Meidinger D, Morneau C, Saucier JP, Thorpe J, Uhlig P (2019) Vegetation zones of Canada: a biogeoclimatic perspective. scale 1:5,000,000. Natural Resources Canada, Canadian Forestry Service, Sault Ste. Marie, ON

Banfield AWF (1981) The mammals of Canada. National Museum of Natural Sciences, National Museums of Canada, 438 p

Beck C, Grieser J, Kottek M, Rubel F, Rudolf B (2006) Characterizing global climate change by means of Köppen climate classification. http://koeppen-geiger.vu-wien.ac.at/pdfs/beck_et_al_ksb_2006.pdf

Behnke RJ (2002) Trout and salmon of North America. Free Press, New York, p 359

Berrill M, Bertram S, Tosswill P, Campbell V (2012) Is there a bull frog decline in Ontario? In Bishop C, Pettit K (eds) Declines in amphibian populations: designing a national monitoring strategy. Canadian Wildlife Service, Occasional Paper #76

Bird DM (2017) Birds of Canada. DK publishing, London

Blarquez O, Talbot J, Paillard J, Lapointe-Elmrabi L, Pelletier N S, Pierre CG (2018) Late Holocene influence of societies on the fire regime in southern Québec temperate forests. Quaternary Science Reviews 180, 63–74

Boreal Songbird Initiative (2015) Comprehensive guide to Boreal birds. https://www.borealbirds.org/comprehensive-boreal-bird-guide. Accessed 2 April 2024.

Bostock HS (2014) Physiographic regions of Canada. Geological Survey of Canada Map 1254A (2nd edn), scale 1:5,000,000. https://www.nrcan.gc.ca/maps-tools-and-publications/maps/geology-and-geosciences/16876. Accessed 25 July 2022.

Bowman J (2007) Wild Turkey. In: Cadman MD, Sutherland DA, Beck GG, Lepage D, Couturier AR (eds) Atlas of the breeding birds of Ontario, 2001–2005. Bird Studies Canada, Environment Canada, Ontario Field Ornithologists, Ontario Ministry of Natural Resources, and Ontario Nature, Toronto, Ontario, pp 132–133

Brandt JP, Flannigan MD, Maynard DG, Thompson ID, Volney WJA (2013) An introduction to Canada's boreal zone: ecosystem processes, health, sustainability, and environmental issues. Environ Rev 21:207–226. https://doi.org/10.1139/er 2013 0040

Brooks GR (2020) Evidence of a strong paleoearthquake in ~9.1 ka cal BP interpreted from mass transport deposits, western Quebec—northeastern Ontario, Canada. Quat Sci Rev 234:106250. https://doi.org/10.1016/j.quascirev.2020.106250

Brouard E, Roy M, Godbout PM, Veillette JJ (2021) A framework for the timing of the final meltwater outbursts from glacial Lake Agassiz—Ojibway. Quat Sci Rev 274:107269. https://doi.org/10.1016/j.quascirev.2021.107269

Brown TG, Runciman B, Pollard S, Grant ADA (2009a) Biological synopsis of largemouth bass (*Micropterus salmoides*). Canadian technical report, fisheries and aquatic sciences 2884, 27 p

Brown TG, Runciman B, Pollard S, Grant ADA, Bradford MJ (2009b) Biological synopsis of smallmouth bass (*Micropterus dolomieu*). Canadian technical report, fisheries and aquatic sciences 2887, 50 p

Burkholder EN et al (2018) To jump or not to jump: mule deer and white-tailed deer fence crossing decisions. Wildl Soc Bull 42:420–429

Canadian National Vegetation Classification (CNVC) (2016) Eastern North American Temperate Hardwood—Conifer Forest; Eastern North American Boreal Forest. http://cnvc-cnvc.ca. Accessed 2 Apr 2024

Card KD (1990) A review of the superior province of the Canadian shield, a product of Archean accretion. Precambr Res 48:99–156

Catto N (2015) Geomorphology: landscapes we live in. Oxford University Press, Don Mills, Ontario

Catto NR, Liverman DGE, Bobrowsky PT, Rutter NW (1996) Laurentide, Cordilleran and Montane glaciers in the western Peace River—Grande Prairie region, Alberta and British Columbia, Canada. Quat Int 32:21–32

Catto N, Slaymaker O, Kovanen DJ (2020) Quaternary glacial, glacimarine, and glacilacustrine history. In: Slaymaker O, Catto NR (eds) Landscapes and landforms of eastern Canada. Springer International Publishers, Cham, Switzerland, pp 49–78

Cody WJ (1975) Plants of the thousand islands. Parks Canada. http://www.oliverkilian.com/ecology/thousand-islands/island-insights/plants/plants.html. Accessed 25 July 2022

Cody WJ (1995) Vascular plants of Wood Buffalo National Park. Centre for Land and Biological Resources Research, Agriculture and Agri-food Canada, Ottawa, ON

Cohen BS, Osborn DA, Gallagher GR, Warren RJ, Miller KV (2014) Behavioral measure of the light-adapted visual sensitivity of white-tailed deer. Wildl Soc Bull 38:480–485

Conover M, Pitt W, Kessler K, DuBow T, Sanborn W (1995) Review of human injuries, illnesses, and economic losses caused by wildlife in the United States. Wildl Soc Bull 23:407–414

Conway T (2010) Spirits on stone: lake superior Ojibway History, Legends, and the Agawa Pictographs. Privately published, ISBN 978-1-878094-01-8

Cook FR Johnston JA Douglas C Cook JC (1984) Introduction to Canadian amphibians and reptiles. National Museum of Natural Sciences, National Museums of Canada

COSSARO (2016) Ontario species at risk evaluation report for Algonquin Wolf (*Canis sp.*), an evolutionarily significant and distinct hybrid with *Canis lycaon, C. latrans*, and *C. lupus* ancestry Committee on the Status of Species at Risk in Ontario (COSSARO). http://cossaroagency.ca/wp-content/uploads/2017/06/Accessible_COSSARO-evaluation-Algonquin-Wolf.pdf Accessed 25 July 2022

Crispo M Jean M Fenton NJ Leduc A Bergeron Y (2021) Factors explaining the composition and diversity of vascular plant understories along a transcontinental gradient in the Canadian boreal forest. J Veg Sci 32:e13058. https://doi.org/10.1111/jvs.13058

Crossman EJ, Mandrak NE (1992) Fish distribution and community analysis, Algonquin Park: annual report for 1991 and completion report, 1989–1991. Department of Ichthyology and Herpetology, Royal Ontario Museum. Unpublished Report

Dampc A, Luczkiewicz M (2015) Labrador tea—the aromatic beverage and spice: a review of origin, processing, and safety. J Sci Food Agric 95:1577–1583

Deer Friendly (n.d.) Decline of deer populations. https://www.deerfriendly.com/decline-of-deer-populations. Accessed 25 July 2022.

Dirszowsky RW (2020) Natural and human-impacted landscapes of the sudbury mining region. In: Slaymaker O, Catto NR (eds) Landscapes and landforms of eastern Canada. Springer International Publishers, Cham, Switzerland, pp 525–541

Driver JC (1999) Raven skeletons from paleoindian contexts, Charlie Lake Cave, British Columbia. Am Antiq 64:289–298

Dyke AS (2004) An outline of North American deglaciation with emphasis on central and northern Canada. In: Ehlers J, Gibbard PL (eds) Quaternary glaciations—extent and chronology, part II: North America. Elsevier, Amsterdam, pp 371–424

Dyke AS, Moore A, Robertson L (2003) Deglaciation of Canada. Geological Survey of Canada, Open File 1574

Earthquakes Canada (2022) https://www.earthquakescanada.nrcan.gc.ca/index-en.php

Eisler R, Wiemeyer SN (2004) Cyanide hazards to plants and animals from gold mining and related water issues. Rev Environ Contam Toxicol 183:21–54

Eldegard K, Lyngved JT, Hjeljord O (2012) Coping in a human-dominated landscape: trade-off between foraging and keeping away from roads by moose (*Alces alces*). Eur J Wildl Res 58. https://doi.org/10.1007/s10344-012-0640-4

Environment Canada (2013) Acid rain. https://www.ec.gc.ca/air/default Accessed 25 July 2022

Environment and Climate Change Canada (2020) Canadian climate normals 1981–2010. https://climate.weather.gc.ca/climate_normals/index_e.html. Accessed 25 July 2023

Environment and Climate Change Canada (2022) Environmental indicators. https://www.canada.ca/en/environment-climate-change/services/environmental-indicators/air-pollutant-emissions.html. Accessed 25 July 2022

Ernst CH, Barbour RW (1989) Snakes of eastern North America. George Mason University Press, Lanham, MD

Fenton MM, Waters EJ, Pawley SM, Atkinson N, Utting DJ, Mckay K (2013) Surficial geology of Alberta. Alberta Geological Survey, AER/AGS Map 601

Fisher TG, Waterson N, Lowell TV, Hajdas I (2009) Deglaciation ages and meltwater routing in the Fort McMurray region, northeastern Alberta and northwestern Saskatchewan, Canada. Quat Sci Rev 28:1608–1624

Ford D (2017) Little limestone lake: a beautiful marl lake in the interlake region, Manitoba. In Slaymaker O (ed.) Landscapes and landforms of western Canada, Springer, Cham, Switzerland, pp 123–130

Fulton RJ (1995) Surficial materials of Canada. Geological Survey of Canada Map 1880A, Scale 1:5000000. https://www.nrcan.gc.ca/maps-tools-and-publications/maps/geology-and-geosciences/16876. Accessed 25 July 2022

GBBR (n.d.) Georgian bay biosphere reserve. https://www.gbbr.ca/. Accessed 2 Aug 2022

Girardin MP, Ali AA, Carcaillet C, Gauthier S, Hély C, Le Goff H, Terrier A, Bergeron Y (2013) Fire in managed forests of eastern Canada: Risks and options. Forest Ecology and Management 294, 238–249

Godfrey WE (1986) The Birds of Canada. National Museum of Natural Sciences, National Museums of Canada, p. 595

Gourd B-B (1973) La colonisation des Clay Belts du Nord-Ouest québécois et du Nord-Est ontarien. Revue D'histoire De L'amérique Française 27:235–256

Green AJ, Elmberg J (2014) Ecosystem services provided by waterbirds. Biol Rev 89:105–122

Grieve RAF (2001) Impact cratering on Earth. In: Brooks GR (ed) A synthesis of geological hazards in Canada. Geol Surv Canada Bull 548:207–222

Hadley KR, Paterson AM, Reid RA, Rusak JM, Somers KM, Ingram R, Smol JP (2015) Altered pH and reduced calcium levels drive near extirpation of native crayfish, Cambarus bartonii,in Algonquin Park, Ontario, Canada. Freshwater Science 34, 918–932

Hartman GMD, Clague JJ, Barendregt RW, Reyes AV (2018) Late Wisconsinan Cordilleran and Laurentide Glaciation of the Peace

River valley east of the rocky mountains, British Columbia. Can J Earth Sci 55. https://doi.org/10.1139/cjes-2018-0015
Hasnain SS, Minns CK, Shuter BJ (2010) Key ecological temperature metrics for Canadian freshwater fishes. Climate change Research Report 17, Applied Research and Development Branch, Ontario, Ministry of Natural Resources. Sault Ste. Marie ON, p 54
Hickin AS, Lian OB, Levson VM, Cui Y (2015) Pattern and chronology of glacial Lake Peace shorelines and implications for isostacy and ice-sheet configuration in northeastern British Columbia, Canada. Boreas 44:288–304. https://doi.org/10.1111/bor.12110.ISSN0300-9483
Hocq M (1994a) La Province du Supérieur. Geologie du Quebec. Ministère des Ressources naturelles. Québec, MM 94-01:7–20
Hocq M (1994b) La Province de Grenville. Geologie du Quebec. Ministère des Ressources naturelles. Québec, MM 94-01:75–94
Hocq M, Verpaelst P (1994) Les sous-provinces de l'Abitibi et du Pontiac. Geologie du Quebec. Ministère des Ressources naturelles. Québec, MM 94-01:21–38
Hosie RC (1979) Native trees of Canada. Canada Department of Forestry, Don Mills, Ontario
Irvine CC, Patterson LD (2021) Status and declining trend of Sparrow's-egg Lady's-slipper (*Cypripedium passerinum*) orchids in Pukaskwa National Park, Ontario, Canada. Can Field Nat 135. https://doi.org/10.22621/cfn.v135i4.2723
Jiang X, Qiu Y, Lyles RL (2012) Hazard assessment of deer-vehicle collisions in Michigan. Hum Ecol Risk Assess 19(4) https://doi.org/10.1080/10807039.2012.691817
Kent J (1966) Agriculture in the clay belt of Northern Ontario. Can Geogr 10:117–126
Kirby P (2014) The trail of broken hearts; surveying, building and settling the hastings colonization road. Kirby Books
Klassen RA, Paradis S, Bolduc AM, Thomas RD (1992) Glacial landforms and deposits, Labrador (Newfoundland) and eastern Québec. Geological Survey of Canada Map 1814A
Lamb EG, Guedo DD (2012) The distribution, abundance, and environmental affinities of the endemic vascular plant taxa of the Athabasca Sand Dunes of northern Saskatchewan. Ecoscience 19:161–169
Laliberté J, St-Laurent MH (2020) In the wrong place at the wrong time: Moose and deer movement patterns influence wildlife-vehicle collision risk. Accid Anal Prev 135:105365. https://doi.org/10.1016/j.aap.2019.105365
LaSalle P, Shilts WW (1993) Younger Dryas-age readvance of Laurentide ice into the Champlain Sea. Boreas 22:25–37
Leydet DJ, Carlson AE, Teller JT, Breckenridge A, Barth AM, Ullman DJ, Sinclair G, Milne GA, Cuzzone JK, Caffee MW (2018) Opening of Glacial Lake Agassiz's eastern outlets by the start of the Younger Dryas cold period. Geology 46:155–158
Lusebrink I, Erbilgin N, Evenden ML (2013) The lodgepole × jack pine hybrid zone in Alberta, Canada: a stepping stone for the mountain pine beetle on its journey East across the boreal forest? J Chem Ecol 39:1209–1220
Mandrak NE, Crossman EJ (1992) A checklist of Ontario freshwater fishes annotated with distribution maps. Royal Ontario Museum Life Sciences Miscellaneous Publication, Toronto
Manitoba (2022) Hecla-grindstone provincial park. https://www.gov.mb.ca/sd/pubs/parks-protected-spaces/park_info/hecla_grindstone_pp.pdf
Messier F, Barrette C, Pettorelli J (2011) Coyote predation on a white-tailed deer population in southern Quebec. Can J Zool 64:1134–1136
Ministère des Ressources naturelles du Québec, Forêt Québec (2002) Les guides de reconnaissance des types écologiques. Gouv. du Québec, Québec, QC. http://www.mffp.gouv.qc.ca/forets/inventaire/guide-types-ecologiques-carte.jsp. Accessed 13 Aug 2022
Mossop GD, Shetsen I (1994) Geological atlas of the western Canada sedimentary basin. Canadian Society of Petroleum Geologists and Alberta Research Council, Calgary, AB, Canada
National Wetlands Working Group (1988). Wetlands of Canada. Ecological Land Classification Series #24. Environment Canada, Ottawa and Polyscience Publications, Montreal
Natural Resources Canada (2009) Geological Provinces of Canada. https://www.nrcan.gc.ca/maps-tools-and-publications/maps/geology-and-geosciences/16876. Accessed 12 July 2022
Nature North (n.d.) Narcissse Snake Dens. http://www.naturenorth.com/spring/creature/garter/Narcisse_Snake_Dens_Videos.html Accessed 2 Feb 2024
Naughton D (2012) The natural history of Canadian mammals. Canadian Museum of Nature
Neumann W, Ericsson G, Dettki H, Bunnefeld N, Keuler NS, Helmers DP, Radeloff VC (2012) Difference in spatiotemporal patterns of wildlife road-crossings and wildlife-vehicle collisions. Biol Cons 145:70–78
NOAA (2022) Annual maximum ice cover, Lake Superior. National Ocean and Atmosphere Administration. https://www.glerl.noaa.gov/data/ice/glicd/AMIC/Superior.png. Accessed 25 July 2022.
Occhietti S (2007) The Saint-Narcisse morainic complex and early Younger Dryas events on the southeastern margin of the Laurentide ice sheet. Géog Phys Quat 61:89–117
Occhietti S, Parent M, Lajeunesse P et al (2011) Late Pleistocene-early Holocene decay of the Laurentide ice sheet in Québec-Labrador. In: Ehlers J, Gibbard PL, Hughes PD (eds) Quaternary glaciations—extent and chronology, part II: a closer look. Developments in quaternary science, vol 15, pp 601–630
Ouimet R, Arp PA, Watmouth SA, Aherne J, Demerchant I (2006) Determination and Mapping critical loads of acidity and exceedances for upland forest soils in Eastern Canada. Water, Air, and Soil Pollution 172: 57–66
Parker GR (1995) Eastern coyote: the story of its success. Nimbus Publishing, Halifax, NS
Parks Canada (2021a) Whooping cranes. https://www.pc.gc.ca/en/pn-np/nt/woodbuffalo/decouvrir-discover/science_nature/Cranes_Grue/point-update. Accessed 25 July 2022
Parks Canada (2021b) Conserving the eastern wolf population in La Mauricie National Park. https://www.pc.gc.ca/en/nature/science/especes-species/reussite-success/itm11e. Accessed 25 July 2022
Patterson BR, Messier F (2000) Factors influencing kill rates of white-tailed deer by Coyotes in eastern Canada. J Wildl Manag 64:721–732
Percival JA (2007) Geology and Metallogeny of the Superior Province, Canada. in Goodfellow, W.D., ed., Mineral Deposits of Canada: A Synthesis of Major Deposit-Types, District Metallogeny, the Evolution of Geological Provinces, and Exploration Methods: Geological Association of Canada, Mineral Deposits Division, Special Publication
Phillips D (1990) The climates of Canada. Environment Canada, p 176
Puric-Mladenovic D, Pinto F (2005) Pre-settlement vegetation modeling in Algonquin Park and Temagami. https://uwaterloo.ca/library/geospatial/collections/digital-projects/historicalsoils-veg/index-artus-franscecut-heidenreich-puric-mladovich-and/danijela-puric-mladenovic. Accessed 25 July 2022
Quinn NWS (2004) The presettlement hardwood forests and wildlife of Algonquin Provincial Park: a synthesis of historic evidence and recent research. For Chron 80(6):7–17
Rempel RS, Carlson M, Rodgers AR, Shuter JL, Farrell CE, Cairns D, Stelfox B, Hunt LM, Mackereth RW, Jackson JM (2021) Modeling cumulative effects of climate and development on moose, wolf, and caribou populations. J Wildl Manag 85:1355–1376. https://doi.org/10.1002/jwmg.22094
Rowe JS (1972) Forest regions of Canada. Canadian Forest Service, Department of the Environment Publication 1300, Ottawa
Samson R (1998) The Forges du Saint-Maurice: beginnings of the iron and steel industry in Canada (1730–1883). Presses de l'Université Laval

Sandlos J, Keeling A (2021) Mining Country: a history of Canada's mines and miners. James Lorimer & Company, Toronto

Saucier JP (2008) Defining the Boreal in the ecological land classification for Québec. In: Proceedings of the fourth annual workshop, conservation of arctic flora and fauna, technical report 15, pp 53–57

Scott WB, Crossman EJ (1998) Freshwater fishes of Canada: fisheries research board of Canada. Bulletin 184

Slaymaker O (2017) Landscapes and landforms of western Canada. Springer International Publishers, Cham, Switzerland

Slaymaker O, Catto NR (2020) Landscapes and landforms of eastern Canada. Springer International Publishers, Cham, Switzerland

Slaymaker O Catto N Kovanen DJ (2020a) Eastern Canadian landscapes as a function of structure, relief and process: the big picture. In: Slaymaker O, Catto NR (eds) Landscapes and landforms of eastern Canada. Springer International Publishers, Cham, Switzerland, pp 3–48

Slaymaker O Catto N Kovanen DJ (2020b) Holocene geomorphic processes in eastern Canada. In: Slaymaker O, Catto NR (eds) Landscapes and landforms of eastern Canada. Springer International Publishers, Cham, Switzerland, pp 79–117

Smith DG (1994) Glacial lake McConnell: paleogeography, age, duration and associated river deltas, Mackenzie River Basin, western Canada. Quat Sci Rev 13:829–843

Soil Classification Working Group (1998) The Canadian system of soil classification, 3rd edn. Publication, Agriculture and Agri-Food Canada, p 1646

Soil Classification Working Group (2001) Soils of Canada. Agriculture and Agri-foods Canada, Research Branch. Scale 1:6,500,000 map. https://sis.agr.gc.ca/cansis/publications/maps/soc/all/soils/soc_all_soils_2004en.png. Accessed 25 July 2022

Spray JG, Thompson LM, Biren MB, O'Connell-Cooper C (2010) The Manicouagan impact structure as a terrestrial analogue site for lunar and Martian planetary science. Planet Space Sci 58:538–551

Steiner W, Leisch F, Hackländer K (2014) A review on the temporal pattern of deer-vehicle accidents: impact of seasonal, diurnal and lunar effects in cervids. Accid Anal Prev 66:168–181

Stokes CR (2020) Glacial landform assemblages of mainland Nunavut, west of Hudson Bay, and their palaeo-glaciological significance. In: Slaymaker O, Catto NR (eds) Landscapes and landforms of eastern Canada. Springer International Publishers, Cham, Switzerland, pp 121–144

Stokes CR, Tarasov L, Blomdin R, Cronin TM, Fisher TG, Gyllencreutz R, Hättestrand C, Heyman J, Hindmarsh RCA, Hughes ALC, Jakobsson M, Kirchner N, Livingstone SJ, Margold M, Murton JB, Noormets R, Peltier WR, Peteet DM, Piper DJW, Preusser F, Renssen R, Roberts DH, Roche DM, Saint-Ange F, Stroeven AP, Teller JT (2015) On the reconstruction of palaeo-ice sheets: recent advances and future challenges. Quat Sci Rev 125:15–49

Sykes PW, Clench MH (1998) Winter habitat of Kirtland's Warbler: an endangered nearctic/neotropical migrant. Wilson Bull 110:244–261

Vanderhill BG (1988) Agriculture's struggle for survival in the great clay belt of Ontario and Quebec. Am Rev Can Stud 18:455–464

Veillette JJ, Roy M, Paulen RC, Ménard M, St-Jacques G (2017) Uncovering the hidden part of a large ice stream of the Laurentide Ice Sheet, northern Ontario, Canada. Quat Sci Rev 155:136–158

VerCauteren, KC, Anderson CW, Van Deelen TR, Drake D, Walter WD, Vantassel SM, Hygnstrom SE (2011) Regulated commercial harvest to manage overabundant white-tailed deer: an idea to consider? Wildl Soc Bull 35:185–194

Waddington JM, Quinton WL, Price JS, Lafleur PM (2009) Advances in Canadian Peatland Hydrology, 2003-2007. Canadian Water Resources Journal 34, 139–148

Waller WM, Alverson WS (1997) The white-tailed deer: a keystone herbivore. Wildl Soc Bull 25:217–226

Wheeler JO, Hoffman PF, Card KD, Davidson A, Sanford BV, Okulitch AV, Roest WR (1996) Geological map of Canada. Geological Survey of Canada Map D1860A. https://ftp.maps.canada.ca/pub/nrcan_rncan/raster/atlas_4_ed/eng/environment/land/025_26.jpg. Accessed 12 July 2022

Wilpert K (2022) Forest soils—what's their peculiarity? Soil Syst 6(5). https://doi.org/10.3390/soilsystems6010005

Wolfe BB, Kart-Riddoch TL, Vardy SR, Falcone MD, Hall RI, Edwards TWD (2005) Impacts of climate and river flooding on the hydroecology of a floodplain basin, Peace-Athabasca Delta, Canada since A.D. 1700. Quatern Res 64:147–162

Wolfe SA Morse PD Kokelj SV Gaanderse AJ (2017) Great slave lowland: the legacy of Glacial Lake McConnell. In Slaymaker O (ed) Landscapes and landforms of Western Canada. Springer, Cham, Switzerland, pp 87–96

Zaniewski K, Phillips B, Dean F (2020) Northwestern Ontario—the Thunder Bay region. In: Slaymaker O, Catto NR (eds) Landscapes and landforms of Eastern Canada. Springer International Publishers, Cham, Switzerland, pp 159–177

4 Mixedwood Plains; Great Lakes–St. Lawrence Lowlands

Abstract

The Mixedwood Plains ecozone surrounds the St. Lawrence River and Great Lakes. The ecozone includes the eastern (Mingan Archipelago, Anticosti, Strait of Belle Isle), central (southern Québec-eastern Ontario), and western (southern Ontario) zones. The 540-km-long St. Lawrence connects Lake Ontario at Kingston to the Atlantic Ocean northeast of Québec. Including the Great Lakes, the St. Lawrence drainage basin represents about 25% of Earth's surface freshwater resource. The Great Lakes include Superior, Huron, Erie, and Ontario, divided between the USA and Canada, and Michigan, entirely within the USA. Most locations within the Mixedwood Plains ecozone have mid-boreal (Dfb) climates. Limited zones with southern boreal (Dfa) and moist temperate (Cfa) climates are present in southwestern Ontario. The St. Lawrence Platform is underlain by Palaeozoic sedimentary bedrock. The lowlands were glaciated several times during the Quaternary. During the last major glaciation, Laurentide ice advanced from western Labrador and adjacent Québec to cover all of the St. Lawrence-Great Lakes lowland. The Laurentide ice retreated in stages, punctuated by short-lived readvances. Final deglaciation of the northernmost Mixedwoods ecozone occurred shortly after 11,000 years ago. As glacial retreat progressed, ice continued to block the Great Lakes' drainage through the Niagara and St. Lawrence Rivers. In combination with glaci-isostatic depression of newly exposed terrain, this resulted in a complex pattern of shifting lake configurations and changing outlets over the period from 16,000 to 4,000 years ago. The Mixedwood Forest includes both deciduous and coniferous trees and represents a transition between the temperate and the Boreal Shield forests. Although much of the Mixedwood Forest has been removed through agricultural clearance and urbanization, new species have been introduced, and climate change has altered species ranges, some indications of the natural assemblages still are evident. The more southerly boreal areas support an assemblage dominated by deciduous trees. Point Pelee's climate supports the temperate Eastern Deciduous "Carolinian" Forest. Soil types are controlled primarily by the type of Quaternary sediment which they are developed upon. Grey-brown luvisols, marked by clayey horizons rich in calcium carbonate, developed on glacilacustrine and glacimarine silts and clays under deciduous vegetation. The fauna is transitional between boreal and temperate assemblages. Faunal population densities throughout the Ottawa and St. Lawrence Valleys appear to have been relatively low prior to eighteenth and nineteenth centuries' logging and settlement, and the effects of agricultural and urban development have reduced numbers still further. Predators in particular have suffered large declines in numbers.

Keywords

Mixedwood plains · St. Lawrence river · Great lakes · Quaternary glaciation · Boreal-temperate vegetation

4.1 Introduction: St. Lawrence River and Great Lakes

The Mixedwood Plains ecozone surrounds the St. Lawrence River and Great Lakes. Environment Canada and Agriculture and Agrifoods Canada (1995). The ecozone includes three parts: eastern (Mingan Archipelago, Anticosti, Strait of Belle Isle), central (southern Québec-eastern Ontario), and western (southern Ontario) (Fig. 4.1).

Initially named "Grande rivière de Canada" by Jacques Cartier in 1535, the name "Fleuve St-Laurent" was first used by Samuel de Champlain in 1604. Indigenous names for the river include Ktsitekw (Wαpánahki), Kaniatarowanénhne (Kanien'kehá:ka), Kahnawáʔkye (Haudenosaunee), Laooendaooena (Wendat), Magtogoek

N. Catto and G. R. Vincent, *Canada: Land and People*, World Regional Geography Book Series,
https://doi.org/10.1007/978-3-032-04391-7_4

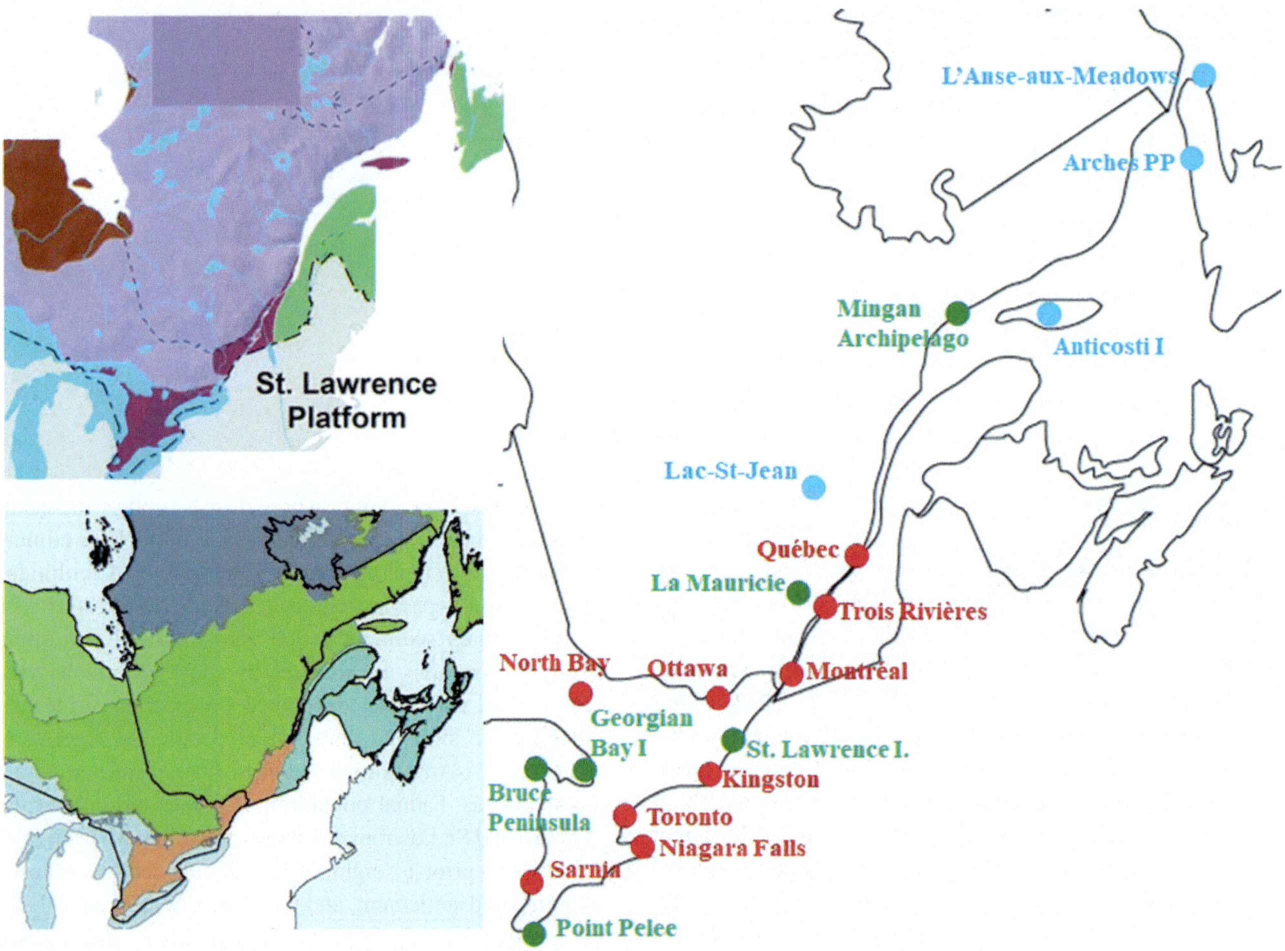

Fig. 4.1 Mixedwood Great lakes–St. Lawrence Lowland, showing localities discussed in text. Green: national parks; Red: human communities; Blue: other significant locations. Inset upper left: St. Lawrence Platform areas in purple. Inset lower right: Mixedwood ecoregion (orange)

(Algonquian-Omàmiwininiwag), Micta sipi (Atikamekw), and Wepìstùkwiyaht sīpu (Innu), all of which can be translated as "great river" or "great pathway". Humans have been using the river as an artery of communication and travel for 9000 years.

4.1.1 Upper St. Lawrence

The 540 km-long St. Lawrence connects Lake Ontario at Kingston to the Atlantic Ocean at Ile d'Orleans, northeast of Québec (Figs. 4.2 and 4.3). Including the Great Lakes, the St. Lawrence drainage basin encompasses 1,344,200 km^2 (839,200 km^2 in Canada), representing about 25% of Earth's surface freshwater resource (Fisheries and Environment Canada 1978; Great Lakes Basin 2003). The St. Lawrence River at Québec has a mean discharge of more than 9850 m^3/s, and the discharge during freshet conditions can exceed 14,000 m^3/s. During floods, peak discharge rates can be thousands of times larger than the average discharge values. Human modifications along the route have had substantial impacts on discharge and water levels (Assani 2016). Major tributaries include the Ottawa (1271 km length), Richelieu (124 km, draining Lake Champlain), St-Maurice (563 km), St-François (218 km), and Yamaska (160 km).

After exiting the Thousand Islands, the mean gradient of the 200 km reach of the St. Lawrence River between Brockville and Montreal was 0.00034, prior to impoundments for the St. Lawrence Seaway. The river was gravel-dominated and wandering, with low sinuosity and a single dominant channel. Seaway construction involved locks and impoundments, including Lakes St. Lawrence, Saint-François, and Saint-Louis, to accommodate the 68 m drop between Brockville and Montreal (Macfarlane 2014; Clamen and Macfarlane 2018; Great Lakes–St. Lawrence Seaway 2022). One result was enhancement of fringing riparian wetlands, notably the Lac Saint-François National Wildlife Area, which continues to evolve (Hudon et al. 2018).

Fig. 4.2 The St. Lawrence River extends from Lake Ontario (SW) to Québec (NE). Major tributaries include the Ottawa, confluence at Montreal; and the Richelieu, flowing north from Lake Champlain to confluence at Sorel. Agricultural and urban development occupy the central part of the Great Lakes–St. Lawrence Lowlands, flanked by the Boreal Shield ecoregion (Grenville geological province) to the north and west, by the Appalachian Mountains of eastern Québec and Gaspé, and the Adirondack Mountains of New York to the south. Also visible are the western part of the St. Lawrence Platform north of Lake Ontario, the Lac-St-Jean–Saguenay region in the northeast, and suspended sediment in the St. Lawrence Estuary downstream of Québec. (Envisat, European Space Agency)

Replacement of the original Long Sault Rapids and St. Lawrence River by the 250 km^2 Lake St. Lawrence flooded 9 villages and farmland, displacing 6500 people (International Lake Ontario-St. Lawrence River Board 2022). Although farmers and residents of the "Lost Villages" were compensated, disputes occurred, with some residents being forcibly removed. Most residents were relocated to newly constructed communities at South Stormont, Ingleside, and Iroquois. Several heritage homes and buildings were transported from the area prior to inundation, and collected to form the historical tourist attraction of Upper Canada Village.

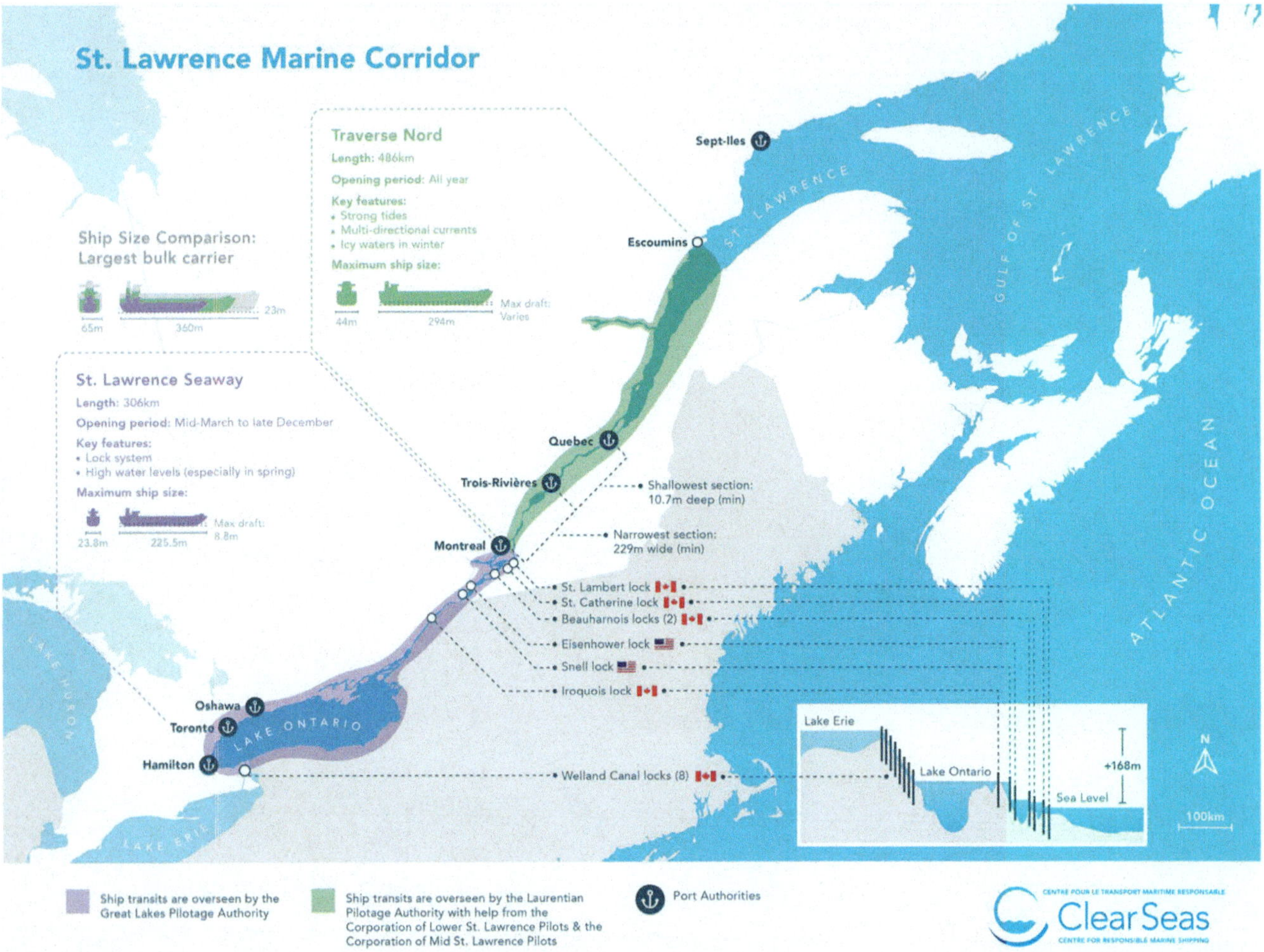

Fig. 4.3 St. Lawrence Marine Corridor (courtesy Clear Seas). The Great Lakes–St. Lawrence waterway extends from Lake Superior to Québec. Canals include Sault Ste. Marie (Superior to Huron), Welland (Erie to Ontario), and the St. Lawrence Seaway, with 7 locks between Iroquois and Montreal. Port authorities are discussed in Chap. 14

The last obstacles to navigation along the St. Lawrence are the 4.8 km-long Lachine Rapids, in the western part of Montreal. Montreal's development reflects its position at the head of navigation. "Lachine" is a derogatory reference to the original French explorers' belief that the St. Lawrence River would eventually lead to China. The rapids are characterized by large standing waves, formed over resistant bedrock ridges. The geomorphology and the urban development of Montreal meant that canal construction was required, rather than impoundment. After initial efforts beginning in 1689 failed, the first Lachine Canal north of the river was completed in 1825 (Parks Canada 2019). The modern St. Lawrence Seaway uses the South Shore Canal to bypass the rapids, the only ones still visible along the St. Lawrence. The lock size, as well as those of the Welland Canal, limits larger marine vessels to the river below Montreal (Great Lakes–St. Lawrence Seaway 2022).

4.1.2 Ottawa River

The Ottawa River (Algonquian *Kitchissippi)* joins the St. Lawrence through Lac-de-Deux Montagnes and three channels at Montreal. Tributaries include the Mattawa (76 km), Petawawa (187 km), Bonnechere (145 km), Madawaska (230 km), Mississippi (170 km), Rideau (150 km), and South Nation (175 km) in Ontario, and the Coulonge (240 km), Gatineau (390 km), and Lièvre (330 km) in Québec (Fig. 4.4).

The Ottawa and its major tributaries are susceptible to ice-jamming and spring freshet flooding (Watt 1989; Buttle et al. 2016; Catto et al. 2020; Public Safety Canada 2022). The river from Pembroke to Lac-de-Deux Montagnes is gravel-dominated and wandering, with low sinuosity and one or two dominant channels. Cascades hinder navigation at Rapides-des-Joachims, Chats Falls, Remic Rapids, Chaudière Falls,

Fig. 4.4 Ottawa and Gatineau flank the Ottawa River. Major tributaries include the Gatineau River from the north, and the Rideau River from the south. Grenville gneiss and granite of the Canadian Shield is exposed above the 213 m marine limit in Gatineau Park, northwest of Gatineau. Ottawa is underlain by Cambrian sandstone and Ordovician limestone. Most areas below 213 m asl are covered by thick marine clay deposits of the Champlain Sea, resulting in planar topography and permitting development of the grid pattern of streets. Dams at Hog's Back Falls and Chaudière Falls impound the Rideau and Ottawa Rivers respectively. The Rideau Canal was completed in 1832, bypassing Rideau Falls. Dow's Lake, an artificial impoundment along the canal, is the remnant of Dow's Great Swamp (geomorphically, marsh and fen), which extended to the Ottawa River upstream of Chaudière Falls. (SPOT image, used in Slaymaker-Catto 2020 volume)

Fig. 4.5 Archipelago of Lac-Saint-Pierre, formed by the St. Lawrence at the southwest end, contains 14 channels, and 103 islands (NASA, Landsat)

and Long Sault, among other locations. The Long Sault Rapids were bypassed by the Carillon, Chute-à-Blondeau, and Grenville Canals in 1833 (Parks Canada 2021a), but a deep-water canal was never constructed along the Ottawa River.

The St. Lawrence Lowlands area adjacent to the Ottawa River from Pembroke to Lac-de-Deux Montagnes is commonly referred to as the Ottawa Valley, although it was not carved by the Ottawa River. The lowlands extend between 5 and 20 km north of the Ottawa River. The region is a gently northward-sloping plain, with very little relief in many areas (<100 m). Most of the relief is related to the Ottawa River, and to bedrock escarpments which cross the region from southeast to northwest. In Ottawa, dams at Hog's Back Falls and Chaudière Falls impound the Rideau and Ottawa Rivers respectively. The Rideau Canal was completed in 1832, bypassing Rideau Falls (Watson 2007; Catto et al. 2020; Parks Canada 2021b). Dow's Lake, an artificial impoundment along the canal, is the remnant of Dow's Great Swamp (geomorphically, marsh and fen), which extended to the Ottawa River upstream of Chaudière Falls.

Tributary streams which enter the Ottawa River east of Ottawa from the south (e.g. South Nation River) commonly have their headwaters within a few kilometres of the St. Lawrence. The position of the drainage divide, closer to the St. Lawrence than to the Ottawa, and the flatness of the lowland terrain are both consequences of the glacial history of the region, and the development of the Champlain Sea following deglaciation (Catto 2015).

The southern margin of the Ottawa Valley is marked by the Frontenac Axis at Brockville, and by the Canadian Shield's Grenville geological province at communities such as Smith's Falls and Calabogie. The division between the Ottawa and St. Lawrence Valleys is cultural rather than physical. Residents of towns along the St. Lawrence generally consider themselves to live in the St. Lawrence Valley if the town is engaged in activities connected to the St. Lawrence Seaway (such as Cornwall). Residents of towns without significant Seaway industries (such as Prescott and

Morrisburg) are more likely to consider themselves as living in the Ottawa Valley. Those residents living north of the barely perceptible height-of-land (in communities such as Winchester and Spencerville) are firmly within the Ottawa Valley cultural landscape.

4.1.3 Lower St. Lawrence

Between Montreal and Québec, the St. Lawrence River drops only 6 m over its 250 km length. Although it is not generally tidal, marine vessels can navigate throughout without requiring locks or canals, aided by ice-breaking in winter. Ice-jamming occurs along the section between Montreal and Lac-Saint-Pierre, notably at the confluence with the Richelieu River at Sorel-Tracy (SODES 2019). This reach is marked by one or two low-sinuosity channels and large longitudinal bar-islands, including Ile-Bouchard and Ile-St-Ours (bird sanctuary).

The Richelieu River flows north from Lake Champlain, crossing the limestone ridge at Saint-Jean-sur-Richelieu. The channel, confined by bedrock and flanked by a narrow floodplain, has no room for the expansion necessary to accommodate increased discharge without flooding the margins (Saad et al. 2015). Spring freshet flooding in 2011 displaced 25,000 people and caused $400 million damage (2022 equivalent; Public Safety Canada 2022).

A delta initially formed during the recession of the post-glacial Champlain Sea (see below) has been progressively modified to form the 46 km^2, 14 channels, and 103 islands of the Archipelago of Lac-Saint-Pierre, at the southwest margin of the lake (Fig. 4.5). Most of the islands are flooded during spring freshet, and the lake margins occasionally are inundated by storm surge from the St. Lawrence Estuary. Erosion of the silt-dominated Ile-de-Grace flanking the shipping channels results from vessel traffic, augmented by dredging.

The archipelago, lake, and fringing wetlands together form a Ramsar Biosphere Reserve, including the Grande-Ile Wildlife Refuge and Nicolet Bird Sanctuary (IBA 2021). Grande-Ile is an important site for the Great Blue Heron (*Ardea herodias*). Annually, more than 500,000 snow geese (*Anser caerulescens*) and 100,000 Canada geese (*Branta canadensis*) pass through Nicolet, along with about 290 other bird species. The sites also support 90 fish species and 27 rare plant species.

Lac-Saint-Pierre's shallowness and ecological importance have sparked water quality concerns (Québec 2013). In the past 40 years, substantial improvement has resulted from municipal sewage treatment, regulation of domestic septic tank systems, and enhanced waste management. Unfortunately, use of the area for military activities between 1950 and 2000 has resulted in unexploded ordnance on the lake bottom (Canada National Defence 2017). The process of recovery and disposal is ongoing.

Downstream, from the mouth Lac-Saint-Pierre past Trois Rivières to Québec and Ile-de-Orléans, the St. Lawrence forms a low-sinuosity channel carrying sand-and-gravel bedload, with large longitudinal bars. Although high storm surge penetrates the reach on occasion, tidal geomorphic features are not present. Rivière Saint-Maurice joins the St. Lawrence at Trois Rivières (Karrow 1998). Between 1650 and 1750, the seigneurial practice of dividing the land into elongated "long lots" with short river frontage, subdivided at approximately 90° to the river upon inheritance, resulted in influx of Champlain Sea silt and clay to the St. Lawrence through the numerous property-line drainage ditches. To provide each landowner with access to the river through progressively narrowing frontages, clearance of vegetation from the bank increased erosion and sediment influx (Fig. 4.6).

4.1.4 St. Lawrence Estuary

Ile-de-Orléans marks the inland limit of the St. Lawrence Estuary, where the river widens from 1 km at Québec to 19 km (Fisheries and Environment Canada 1978). An arbitrary seaward boundary can be set at 50 km width, east of Godbout and Matane, as the estuary transitions to the open Gulf of St. Lawrence. The estuary extends for more than 600 km east of Québec. Subaqueous tidally modified longitudinal bars are present in the upper estuary, west of Rivière-du-Loup and Tadoussac, at the mouth of the Saguenay River. Mesotidal and macrotidal semi-diurnal tides mark the estuary, with mean tidal ranges of 4.9 m at Ile-de-Orléans and 3.8 m at Kamouraska and Rivière-du-Loup (Dohler 2017; Matte et al. 2019).

The St. Lawrence Estuary is a modified, partially mixed at depth river-dominated estuary (Fig. 4.7). The saltwater wedge saline front is generally present offshore of Ile-de-Orléans, except when modified by spring freshet or storm activity. Fluvial sediment discharge and tidal action combine to produce a zone of high turbidity and suspended sediment concentration, typically extending 70-90 km seaward from Ile-de-Orléans (Rossignol 1998; Beauchesne et al. 2020; Fabris et al. 2021). Turbulence, upwelling, disruption of the halocline, and caballing flow combine to produce favourable conditions for many planktonic, molluscan, fish, and seabird species (Nozères et al. 2010).

Seabirds use the northern margins of the estuary and Gulf of St. Lawrence, particularly at Mingan Archipelago National Park Reserve, as breeding, resting, and feeding grounds (Morris et al. 2012; Parks Canada 2020). Bird

Fig. 4.6 Grondines, Québec. The traditional agricultural pattern involved long lots oriented at 90° to the St. Lawrence River, allowing each farmer frontage to the major transportation artery. Successive division of inherited property between siblings resulted in reduced river frontage for each person. (International Space Station image, NASA)

Fig. 4.7 St. Lawrence Estuary, from Québec downstream past Ile d'Orléans to Isle-aux-Coudres, southeast of Baie-Saint-Paul. Plumes with high turbidity and suspended sediment concentration, generated by fluvial input and tidal action, are apparent. Note the difference in water quality between the south (main shipping channel) and north sides of Ile-d-Orléans, and the enhanced sediment input from Montmagny and Saint-Jean-Port-Joli south of Isle-aux-Grues (NASA)

species include Atlantic Puffin (*Fratercula arctica*), Eider-duck (*Somateria mollissima*), Black-legged Kittiwake (*Rissa tridactyla*; a species in decline in Mingan), Razorbill (*Alca torda*, present in low numbers but slowly increasing), and Semipalmated Sandpiper (*Calidris pusilla*). More than 15,000 nesting pairs of Herring Gulls are residents of the park, and increased numbers have recently led to destruction of the native flora, including unusual species such as the Sparrow's Egg Lady's-Slipper. Parks Canada is currently wrestling with the dilemma of whether to interfere with this natural increase of herring gulls.

4.1.5 Saguenay River

The Saguenay River flows 170 km from Lac-Saint-Jean to meet the St. Lawrence Estuary at Tadoussac. The lower 100-km reach, Saguenay Fjord, extends from St-Fulgence to Tadoussac and is conserved in Parc National Saguenay Fjord (Fig. 4.8). At the confluence is the Saguenay–St. Lawrence Marine Park, a National Marine Conservation Area co-managed by Parks Canada and Société des établissements de plein air du Québec (Sépaq). Recently, pollution, vessel traffic, and fluctuating water temperatures in the Estuary and Gulf of St. Lawrence have reduced the populations of marine mammal species such as Beluga (*Delphinapterus leucas*; Fisheries and Oceans Canada 2012; Ménard et al. 2022). Although belugas have suffered extensively from pollution, particularly in the important feeding areas around the mouth of the Saguenay River, other marine mammals such as seals and porpoises have been increasing in number. A casual visitor to the archipelago is Narwhal (*Monodon monoceros*), arriving during anomalously cold winters marked by severe ice conditions in the Gulf of St. Lawrence.

Fig. 4.8 Parc National Saguenay Fjord. Upper: Saguenay Fjord. (Wikimedia/Thomson MG). Lower left: Cap Trinité, 50 km inland from Tadoussac, is a monzonite (quartz-poor granitic rock) modified by glacial erosion. (Wikimedia/M Thomson M). Lower right: Satellite image of Parc National Saguenay Fjord, showing confluence of the Saguenay at St. Lawrence at Tadoussac (International Space Station image, NASA)

Flooding of north-flowing Saguenay tributaries in July 1996 caused $750 million damage (2022 equivalent; Brooks and Lawrence 1998; Buttle et al. 2016; Public Safety Canada 2022). Excessive rainfall over previously logged terrain, obstruction of flow by bridge abutments, riverbank slope failures, and failure of the concrete wing wall of the Rivière aux Sables dam all contributed.

4.1.6 Great Lakes

The Great Lakes include Superior, Huron, Erie, and Ontario, divided between the USA and Canada, and Michigan, entirely within the USA (Figs. 4.9 and 4.10). Basic data are presented in Table 4.1 (Fisheries and Environment Canada 1978; Great Lakes Basin 2003; USEPA 2006).

4.1.6.1 Lake Ontario

Lake Ontario (Wendat: *great lake*; Haudenosaunee; *katarokwi)* has a surface area of 19,000 km^2 (10,388 km^2 in Canada), making it the seventh largest lake in Canada (Fig. 4.11). The elevation of the lake surface varies seasonally and from year-to-year, but has averaged 75 m above sea level during the past decade. Although Ontario is the smallest of the Great Lakes, it is also the second deepest, with a maximum depth of 244 m (175 m in Canadian waters). The northern part of the lake basin is divided bathymetrically into several subdivisions, including the Kingston Basin to the north of the subaqueous sill joining Duck and Galloo Islands, and the shallow Bay of Quinte. Water depths in the bay increase from west to east, from 1 to 3 m at Trenton, 5 m at Belleville, and 14 m at the mouth of the bay at Glenora Ferry. The pattern of increasing depths is not uniform, however, as the bay is divided by a shallow sill at Telegraph Narrows, near

Table 4.1 Great Lakes data. Lake levels fluctuate seasonally, yearly, and over longer periods, causing variations in these parameters

Lake	Superior	Michigan	Huron	Erie	Ontario
Water Area (km^2)	84283	58000	63000	25812	19000
Maximum Depth (m)	405	245	229	64	244
Elevation asl (m)	183	176	176	174	75
Volume (m^3)	12100	4900	3500	484	1640
Mean residence Time (y)	196	62	22	3	6
Outlet	St. Mary R. (Sault Ste. Marie)	Strait of Mackinaw	St. Clair River (Sarnia)	Niagara River (Fort Erie)	St. Lawrence (Kingston)

Fig. 4.9 Great Lakes from space. The Niagara Escarpment, southern margin of the Grenville province of the Canadian Shield, and plumes in shallow water formed by sediment and waste discharges from communities are apparent (NASA)

Deseronto. Water depths in North Channel, which connects the bay to the main expanse of Lake Ontario, exceed 60 m. The majority of Lake Ontario lies in a west-east structural basin, with a narrow shelf along the northern shoreline and consistent water depths offshore of 60–140 m.

Construction of the Oswego branch of the Erie Canal from the Mohawk River drainage *via* Rome NY to enter Lake Ontario at Oswego (1828) allowed some marine species to enter the basin, including sea lamprey (*Petromyzon marinus) ca. 1835* (GLFC 2022), *and Morone americana* (white perch) after 1950 (Mills et al. 1993). Sea lamprey is a primitive cartilage fish, resembling an eel in shape (Fig. 4.12). It preys upon black bass, lake trout, and whitefish, by latching onto the host fish and systematically sucking the flesh and blood of its victim. In marine environments, *Petromyzon marinus* is a minor predator, and many marine fish have adapted to its presence. In the lacustrine environment, however, fish have not yet developed a natural defence, and the introduction of the predator into the Great Lakes has had a devastating effect. Intensive effort since 1970 has resulted in the development of mechanisms that effectively have controlled and locally eliminated the lamprey (GLFC 2022). Lampreys reproduce during a very short period of time, and eradication of the eggs has proven effective.

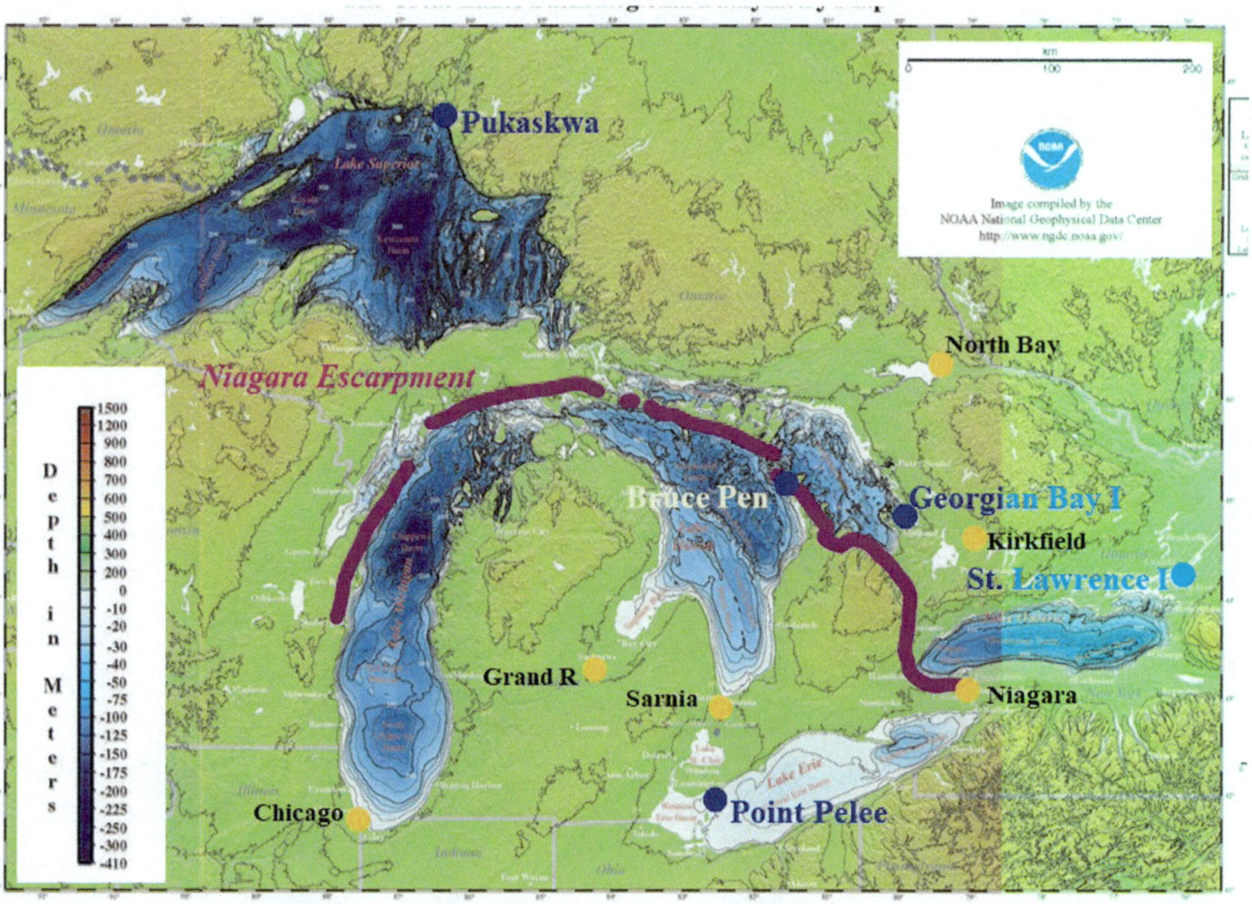

Fig. 4.10 Bathymetry of the Great Lakes. Blue dots: National Parks. Yellow dots: Glacial lake outlets. Purple Line: Niagara Escarpment (NOAA base map)

Deep water and strong outflow through the St. Lawrence River at Kingston result in short water residence times of approximately 6 years. Despite the limited urbanization along the Canadian shoreline east of Bowmanville, cumulative environmental stress on Lake Ontario is high throughout, reflecting the urban concentration of the Niagara-Toronto-Bowmanville "Golden Horseshoe", agricultural runoff inputs, and contributions from New York State (GLEAM 2015). The cumulative effects of agricultural, industrial, and urban activity had converted much of Lake Ontario from its originally oligotrophic nature to a mesotrophic condition. The shallowest parts of the bay became eutrophic and unsuitable for most fish life.

4.1.6.2 Lake Erie

Lake Erie (Haudenosaunee: *erielhonan,* "long tail") is the smallest (484 km^3 of water), shallowest (maximum depth 64 m, offshore of Long Point), and second-smallest in terms of area (25,812 km^2) of the Great Lakes (Fig. 4.13). The areas surrounding Point Pelee are much shallower than the remainder of the lake, due to a submerged glacial moraine. The small basin west of Pelee and south of the Detroit River is generally <20 m deep. Lake Erie drains north through the Niagara River across Niagara Falls.

Water circulates rapidly through the Erie basin because of its shallowness. Typically, individual water molecules can traverse the entire basin in <3 years. This short residence time has greatly facilitated efforts to clean the lake of pollutants, because once new pollutants are stopped from entering the lake, those in residence are quickly flushed out.

Although Lake Erie was one of the most severely polluted bodies of water anywhere in the world in 1969 (the last time the Cuyahoga River caught fire; Stradling and Stradling 2008), efforts at clean-up have proven at least partially successful (GLEAM 2015). Two major factors were the ban on phosphate detergents, and changing economic conditions which discouraged further expansion of heavy industry in northern Ohio. Although it was fashionable in the 1970s to blame the United States for the pollution problems, owing to the lack of heavy industry and large cities on the Cana-

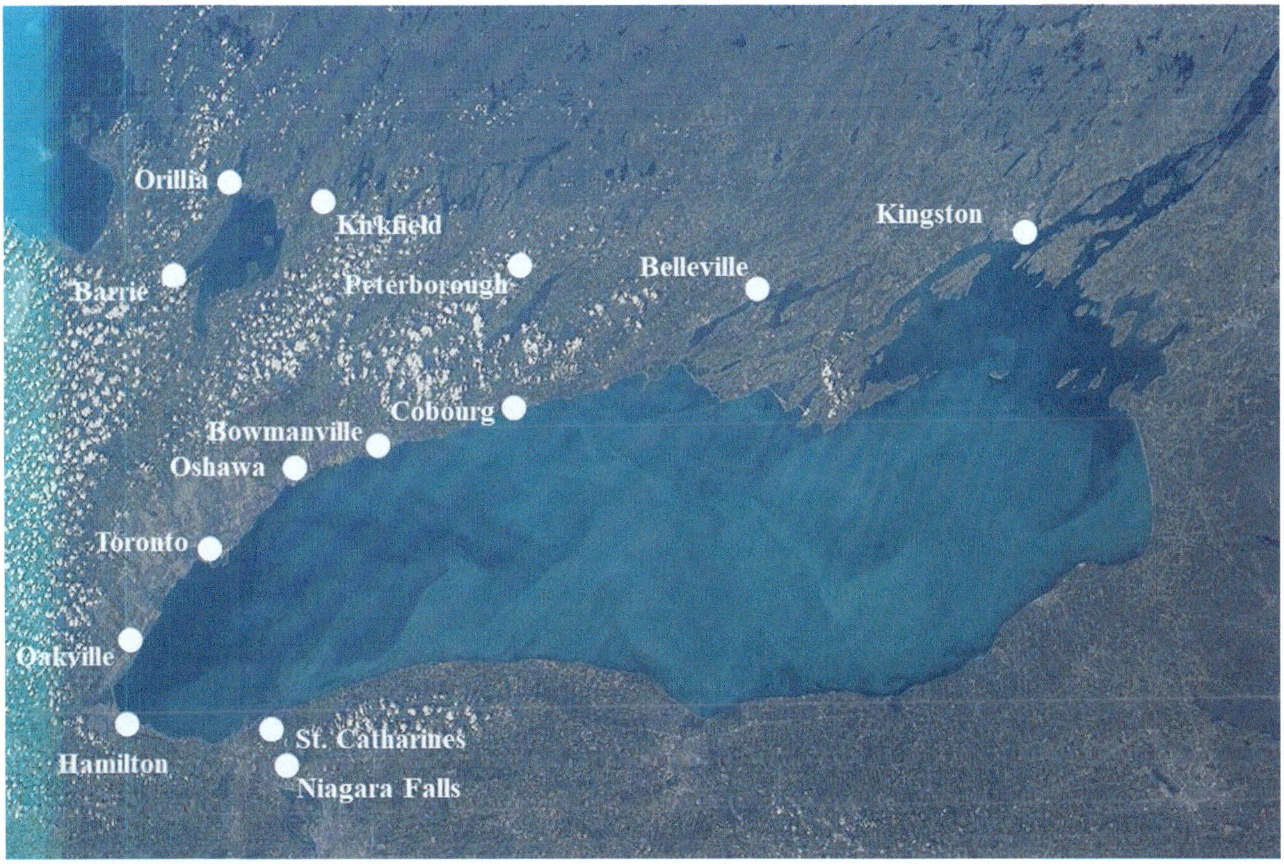

Fig. 4.11 Lake Ontario. Whitish plumes are produced by carbonate precipitation (whiting) in warmer waters, combined with fine sediments (NASA)

dian shore of Lake Erie, little real difference existed in the effectiveness of the two countries in combatting pollution. Local pollution problems in Lake Erie were produced in Canada (GLEAM 2015), particularly by agricultural fertilizers (Leamington), chemical plants (Dunnville), and toxic spills in river systems (Grand River south of Brantford). A proposed industrial complex at Nanticoke would have added further potential for pollution. Fortunately, however, efforts to eliminate pollution sources proved successful in the 1980s, and the rapid flow-through and short residence time characteristic of Lake Erie allowed remediation efforts to show rapid results.

The initial Welland Canal bypassing Niagara Falls was constructed in 1829 (Jackson 1991). Later versions in 1833, 1848, 1887, 1932, and 1972 shortened and deepened the channel (Fig. 4.14). The 1932 iteration was designed to last until 2030, although no new construction was planned as of 2022. The Welland Canal currently can accommodate vessels up to 225.5 m length, 24 m beam, 8.08 m draft, and 35.5 m height above water line. The canal is 34.5 km long, has 8 locks, and typically takes 11 hours to transit. It is closed to traffic between late December and late March (Great Lakes–St. Lawrence Seaway 2019).

Lake Erie is fed by southward drainage from Lake Huron through the St. Clair River, Lake St. Clair, and the Detroit River. The St. Clair River Delta is among the largest freshwater deltas in North America (Thomas et al. 2006; Catto 2015). Lake St. Clair has a modal depth of only 3 m, with limited wave activity, forming a passive basin for fluvial discharge. Annual dredging of a 10 m deep channel through the delta and across the lake, to facilitate ship traffic, has resulted in the extension of the dredged distributaries into the lake, greatly accentuating the digitate configuration of the delta. The Detroit River International Wildlife Refuge, established in 2001, conserves several islands and fragments of riparian marshlands (USFWS n.d.).

4.1.6.3 Lake Huron

Lake Huron (Wendat: *karegnondi)* is the second largest of the Great Lakes in area (63,000 km^2) and has a maximum depth of 229 m. Although Lakes Huron and Michigan form a single contiguous water body at ~176 m asl, by convention they are

Fig. 4.12 *Petromyzon marinus* (sea lamprey) first entered Lake Ontario in 1835, and has now spread to the Upper Great Lakes. The lamprey attaches to its prey, such as Salvelinus namaycush (lake trout) with its suction-like mouth (Upper: T. Lawrence, Great Lakes Fishery Commission; Lower: United States Fish and Wildlife Service)

designated as separate lakes along the Strait of Mackinaw. The bathymetry of Lake Huron reflects the structural geology, with SE-NW ridges (notably extending offshore of Point Clark) separating basins (Figs. 4.15 and 4.16).

Georgian Bay (area 15,111 km^2, volume 660 km^3, maximum depth 165 m) is divided bathymetrically from the main area of Lake Huron by the sublacustrine extension of the Niagara Escarpment, between Tobermory at the northern end of the Bruce-Saugeen Peninsula and Little Baymouth on Manitoulin Island (Sly and Munawar 1988). The first British surveyor, Captain Owen, designated the water body as a separate "Lake Manitoulin" in 1825, but ten years later the present nomenclature was adopted (Barry 1968). This division primarily influences deep-water circulation, and has no effect on fish migration. The escarpment forms the southern flank of Manitoulin and Cockburn Islands, extending west into Michigan. Manitoulin Island (2786 km^2) is the largest island within a lake on Earth. North Channel (3950 km^2, 90 km^3, maximum depth 85 m) separates Manitoulin Island from the Canadian Shield terrain to the north. The bathymetry of Georgian Bay and North Channel reflect the underlying structural geology of the Grenville and Keeweenawan provinces of the Canadian Shield.

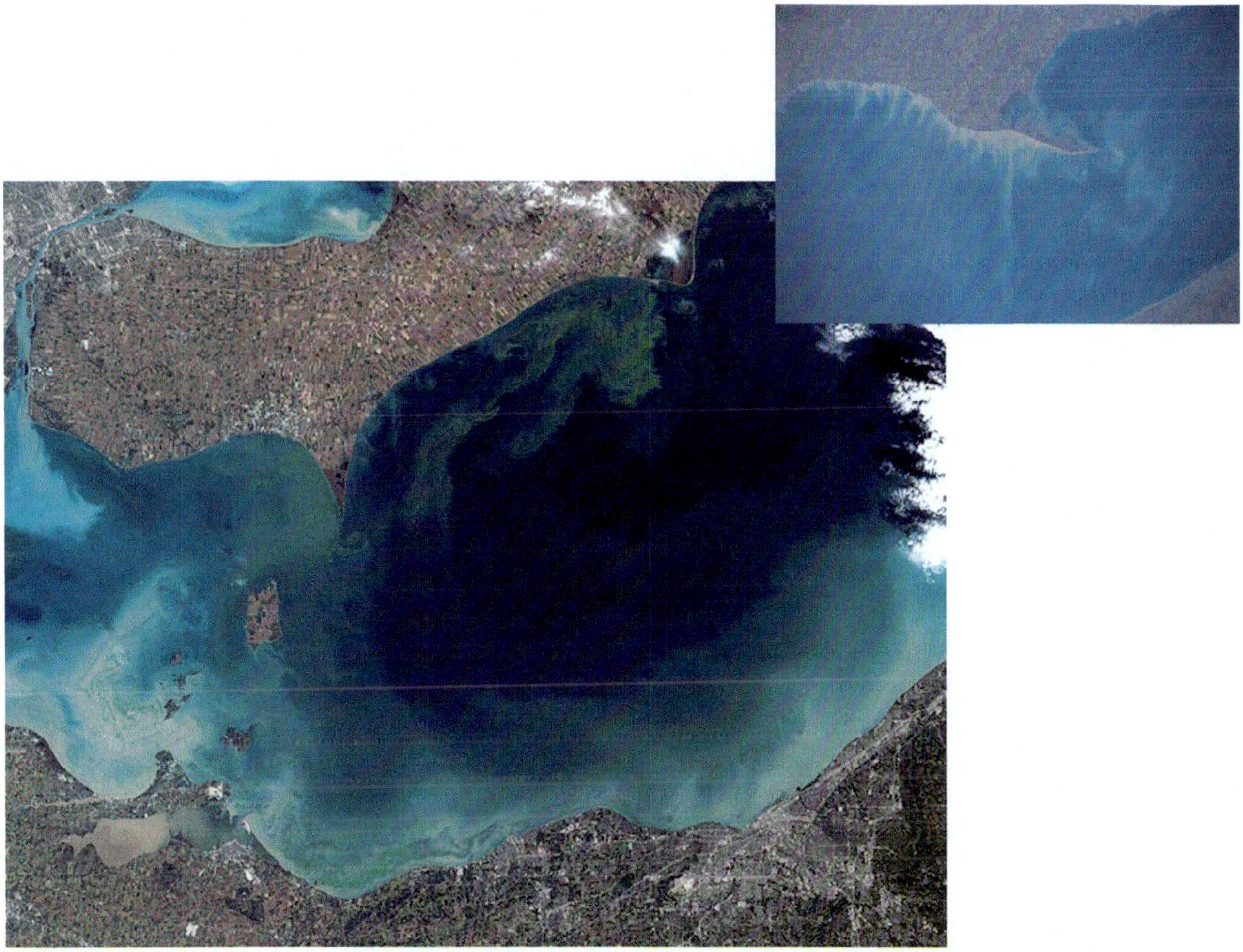

Fig. 4.13 Lake Erie. Left: view of western Lake Erie. The Pelee Moraine extends southward from Point Pelee National Park to Pelee Island (ON) and Sandusky (USA). Intense agricultural development marks the region from Windsor (upper left, west) to Rondeau Provincial Park (right). White sediment plume extends from the mouth of the Detroit River at Amherstburg. Algal concentrations are visible along the north shore from Point Pelee to Rondeau. (NASA) Right: sediment plumes extending from Long Point, 150 km ENE of Rondeau. (International Space Station, NASA)

Lake Huron, Georgian Bay, and North Channel have been subjected to lesser ecological stress than the lower Great Lakes (including Michigan), and are still largely oligotrophic (GLEAM 2015). However, local mesotropic zones are present adjacent to communities including Grand Bend, Goderich, and Port Elgin (Huron), Owen Sound, Collingwood, and Wasaga Beach (Georgian Bay), and Algoma Mills and Thessalon (North Channel).

Introduced and invasive species reach Lake Huron from both Sarnia and Chicago. In addition to the sea lamprey, more recent marine invaders (ca. 1985) are the zebra mussel (*Dreissena polymorpha*) and Quagga mussel (*D. bugensis*). Accidentally imported from the Black Sea on the hulls and in the bilge water of ocean-going freighters (Fisheries and Oceans Canada 2021), these mussels have outcompeted the native molluscan fauna, and are multiplying throughout the drainage basin. Eradication programmes are under way, but no effective method of control has yet been devised. Various species of Asian carp (*Cyprinus sp.*), present in the Chicago Ship Canal, an obsolete transportation route connecting Lake Michigan to the Mississippi River, have the potential to be highly invasive and damaging.

In the terrestrial areas of the Niagara Escarpment, the relative lack of game fish in ponds (no bass or trout) led to attempts to introduce new species for angling. Populations of Atlantic Salmon (*Salmo salar*) and hybridized "splake" (Speckled trout-lake trout crossbreeds) were introduced to the Bruce-Saugeen Peninsula on several occasions. None of these introductions was very successful, particularly the numerous "splake" attempts which all resulted in non-viable

Fig. 4.14 Welland Canal. Extending from Port Colbourne on Lake Erie to St. Catharines on Lake Ontario, the canal bypasses Niagara Falls on the west. An older segment of the canal through Thorold connects to the Welland River. (NASA). Inset: Allanburg through-truss vertical lift bridge. The 60.9-m span was built in 1930 and refurbished in 2001 following a vessel collision

Fig. 4.15 Lake Huron Coast. Left: Coastline from Kincardine (S) to Bruce Nuclear Generating Station (N). Agricultural terrain is underlain by silt-dominated glacial till and Devonian limestone. Blue dots mark the raised beach of Glacial Lake Algonquin; areas with thick lake sediments along the shoreline support deciduous vegetation and are typically used recreationally. (ISS/NASA). Upper right: Ordovician limestone flanks the south coast of Midland Bay, eroded by southwestward-flowing glaciation to form the harbours of Penetanguishene (W), Midland (C), and Port McNicoll (E). The north shore, including Beausoleil Island in Georgian Bay Islands National Park, is Grenville gneiss and granite, resistant rocks without good harbours. (NTS maps 31D/12 and 31 D/13). Lower right: Several sandy recreational beaches occur along Lake Huron and southern Georgian Bay

populations of genetically altered and unstable fish. The introduced fish, however, did survive long enough to prey on rare native chub species such as River Chub (*Nocomis micropogon*) and Blackchin Shiner (*Notropis heterodon*). Further pressure on the native chubs, shiners, and daces has resulted from escapee bait minnows, and from goldfish, carp, and other introduced species deliberately or accidentally added to streams and ponds (Parks Canada 2016a). In the semi-integrated drainage of karst areas, a single introduced fish can decimate an entire native species within its lifetime.

4.1.6.4 Lake Superior

Lake Superior (*Ojibwe Gichigami*) is connected to North Channel through the St. Mary River at Sault Ste.-Marie. Superior is the largest freshwater body in the world, with a surface area of 84,243 km^2 and a mean water volume of 12,100 km^3, exceeding that of all the other Great Lakes combined. Water entering the lake has an average residence time of 196 years. With a maximum depth of 405 m, the bathymetric relief rivals that above the water surface. The floor of Lake Superior is marked by a dendritic system of channels, which are not directly related to the present

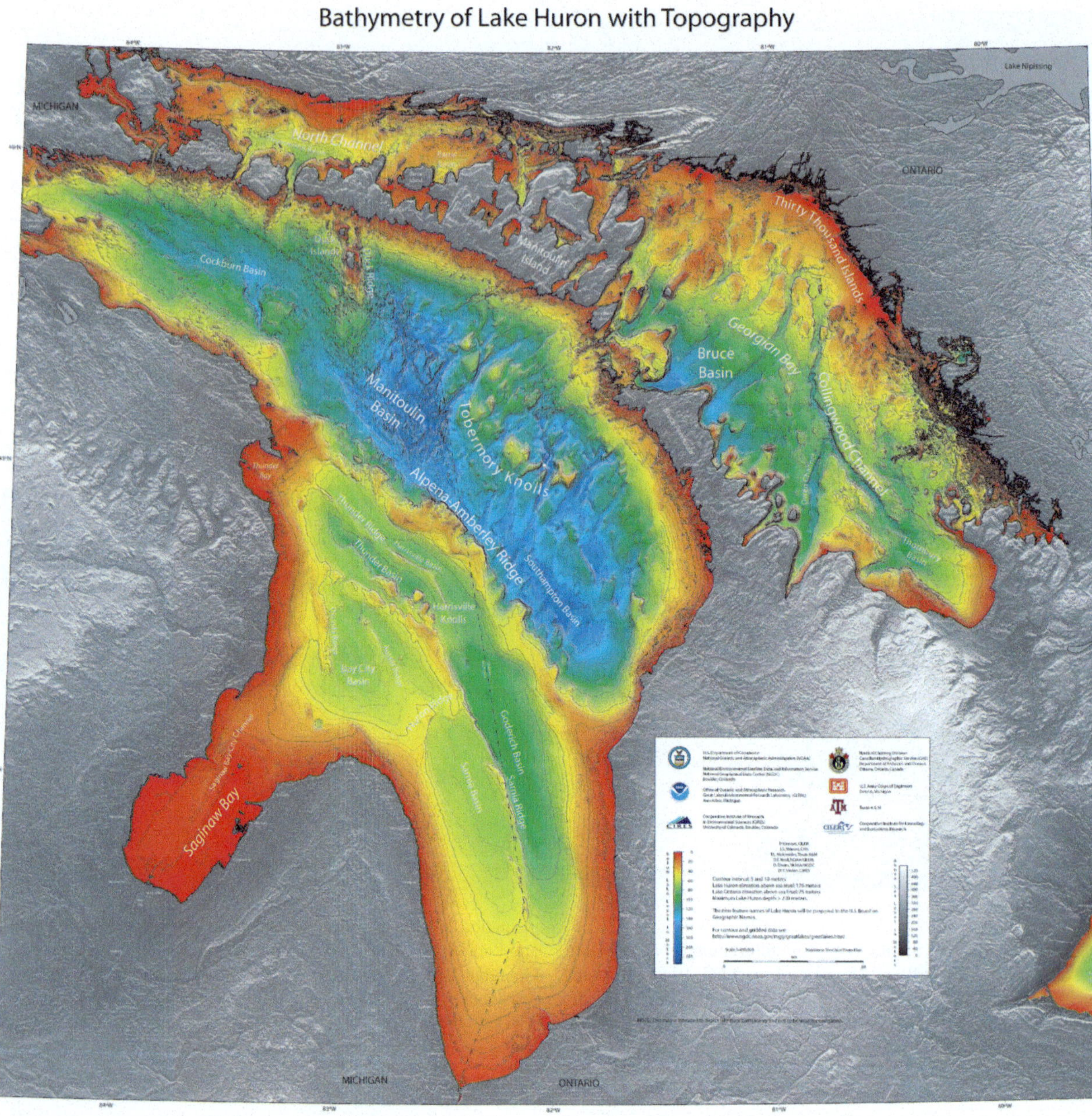

Fig. 4.16 Lake Huron bathymetry. The deep Manitoulin and Southampton Basins were occupied by Glacial Lake Stanley during the low North Bay Outlet phase. The underwater Niagara Escarpment separates Lake Huron from Georgian Bay. Glacial Lake Hough occupied the Bruce Basin during the low North Bay Outlet phase. The deep channel offshore of Owen Sound was excavated by glacial and fluvial activity during the Quaternary (NOAA)

patterns of lake surface currents. The mean water level of Lake Superior is 183 m above sea level. Superior is largely oligotrophic, with areas of environmental stress in Canada confined to Thunder Bay and Sault Ste. Marie (GLEAM 2015). Superior is discussed further in Chap. 3.

4.2 Climate

Most locations within the Mixedwood Plains ecozone have mid-boreal (modified Köppen-Geiger Dfb) climates. Limited zones with southern boreal (Dfa) and moist temperate (Cfa) climates are present in southwestern Ontario (Phillips 1990; Beck et al. 2006; Environment and Climate Change Canada 2020).

4.2.1 St. Lawrence-Ottawa Valley (Figs. 4.17 and 4.18)

Ottawa's climate is classified as modified Köppen-Geiger Dfb (Fig. 4.17). Winters are typically rigorous, with January daily mean temperatures of −10.3 °C (Environment and Climate Change Canada 2020). Mean annual snowfall is 224 cm, and snow cover persists for 130 days. Winter winds come dominantly from Arctic sources to the west-northwest. During January, cold, dry westerly winds commonly alternate with warmer, moisture-laden southwesterlies, bringing daily shifts in precipitation patterns. Freezing rain occurs on 15 days in a typical year. The National Capital Commission (NCC n.d.) maintains the Rideau Canal, which traverses the city, as an outdoor skating rink during the winter. The tradition of skating on the canal dates back to shortly after its construction in 1832, and skating is regarded as a very significant part of Ottawa life (see Chap. 13).

Spring brings a shift in wind pattern, and the southwesterly winds originating from the Gulf of Mexico dominate throughout the spring, summer, and early autumn. Springs are pleasant (although moist), but summers are typically hot and very humid. The July daily mean temperature is 21 °C, and thunderstorms are common (23 y^{-1} on average). Ottawa typically receives 2000 hours of sunshine and 125 high summer days per year. Comfortable sightseeing conditions for the average tourist occur on 80 % of days in May (when Ottawa holds its annual Festival of Spring, featuring tulips) and 85 % of October days (when the deciduous tree leaves have turned colour), but only during 68 % of days in July.

In human hazard terms, an extreme heat event is a period of unusually high atmospheric-related heat stress that may have adverse health consequences for the affected population. Although a heat event represents an extended period of temperatures above the normal value, there is no gener-

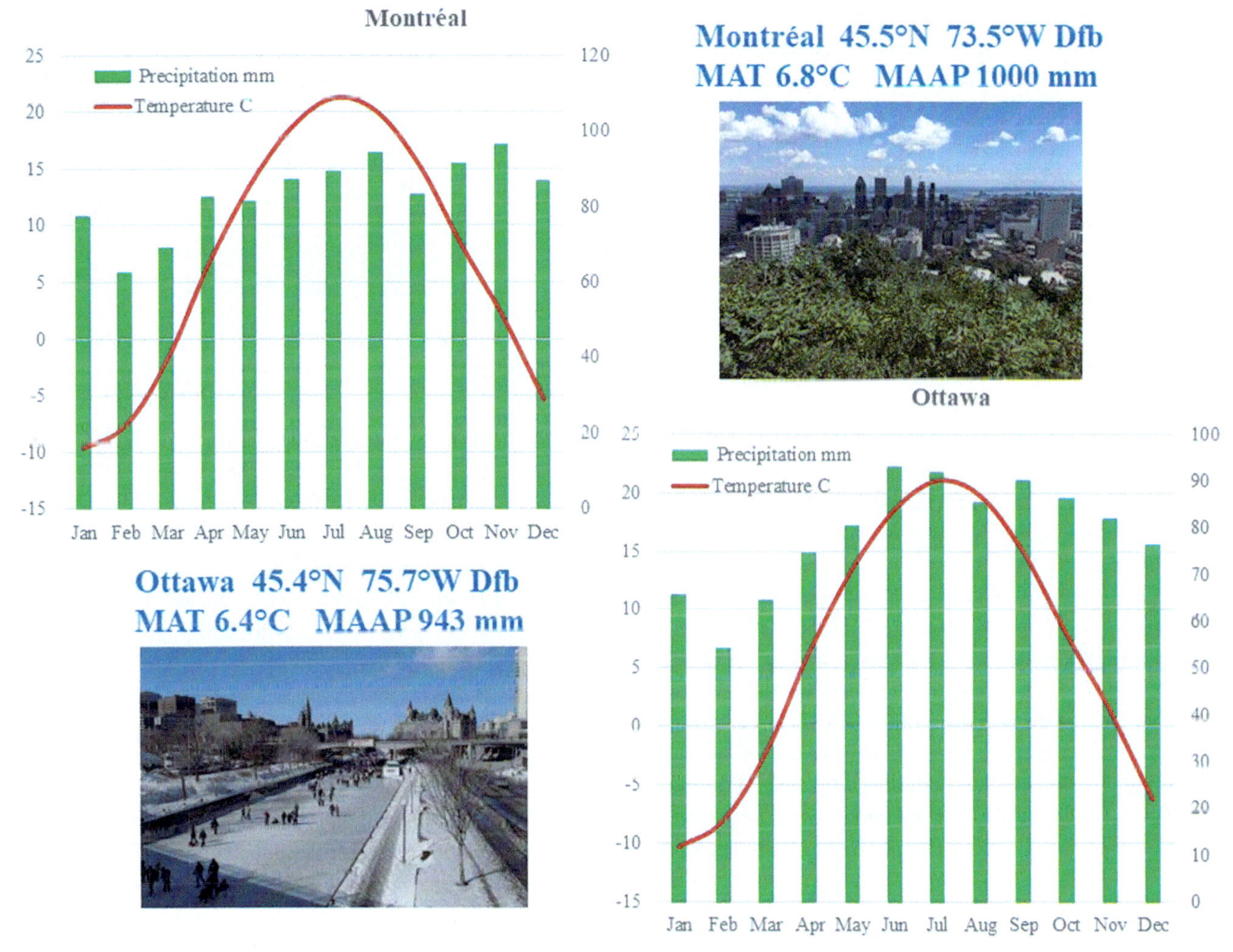

Fig. 4.17 Montreal and Ottawa have Köppen-Geiger Dfb climates, with hot, humid summers and cold winters suitable for outdoor skating. (*Data* Environment and Climate Change Canada 2020)

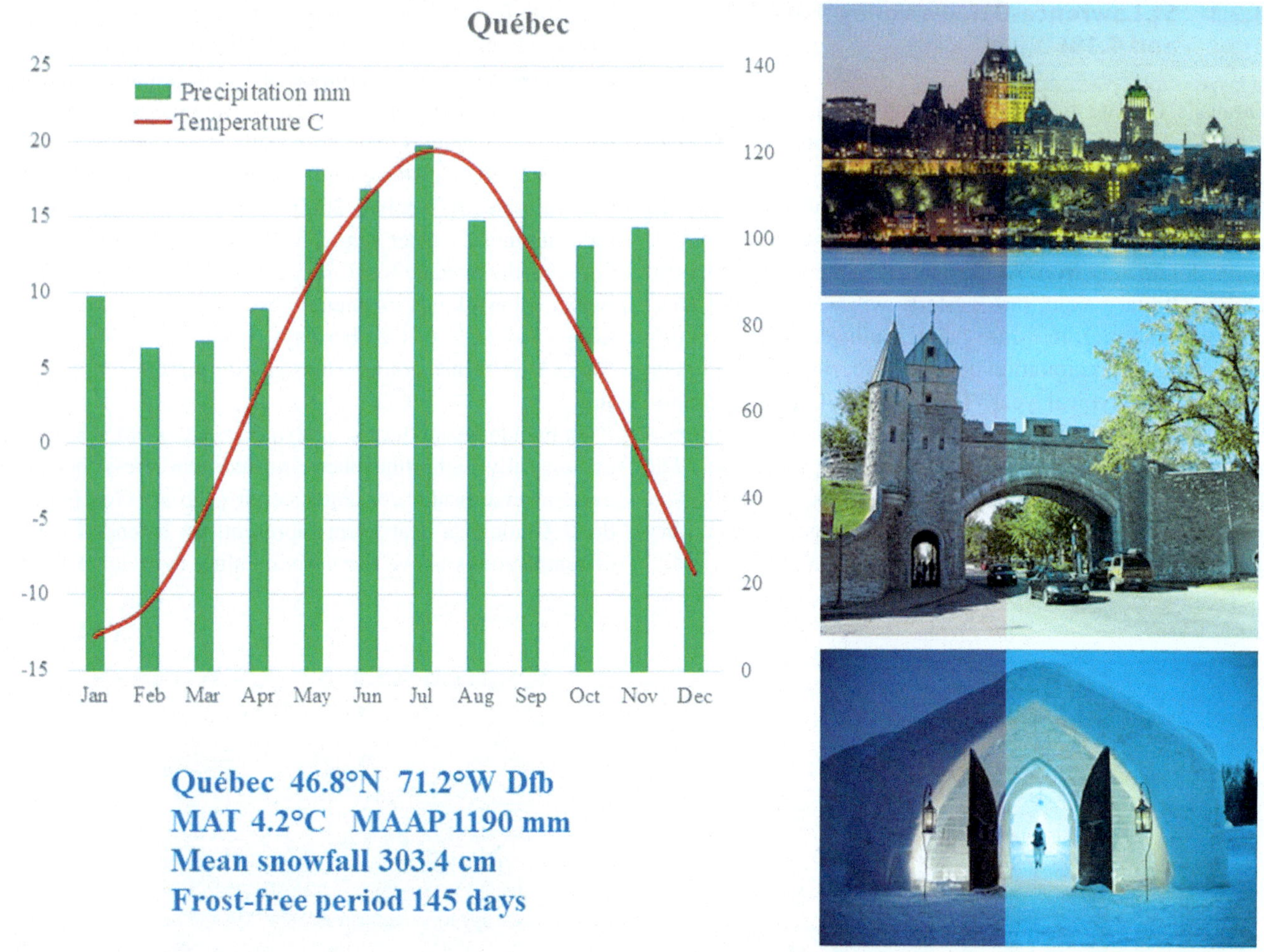

Fig. 4.18 Québec's Dfb mid-boreal climate is marked by colder winters and more snowfall than is Montreal (*Data* Environment and Climate Change Canada 2020)

ally accepted definition of for how long and how high above normal should qualify. Environment Canada currently defines an extreme heat event as involving three consecutive days with maximum temperature of at least 32 °C (Bishop-Williams et al. 2015).

In Canada, increased numbers of extreme heat events have been recorded since 1953, as measured by both daytime and nighttime humidex values (Warren and Lulham 2021). The Humidex is a common measure of the combined effects of temperature and humidity (Environment Canada 2019). Designed initially for Canadian weather forecasting, the Humidex compares temperatures with relative humidity values to estimate equivalent temperatures. Higher relative humidity does not necessarily imply higher humidex values. Although the relative humidity in Ottawa typically is 50–60% in summer, the high temperatures make Ottawa summers hot and sticky. High humidex values are typical for Montreal, with more than 40 summer days exceeding 30 (Environment Canada 2019). Nighttime increases are more common, as also noted in many locations. Projections based on climate modelling suggest that the number and intensity of extreme heat events will continue to increase in future.

4.2.1.1 Freezing Rain and Verglas

Freezing rain falls in liquid form and freezes on contact with a surface. Other forms of winter precipitation, including ice pellets, graupel (soft hail), and small hail, fall and arrive in solid form (Phillips 1990; Stewart 1992; Cortinas et al. 2004; Ahrens et al. 2015). Sleet, a term for a variety of assemblages of mixed rain and snow, is not used by Environment Canada. Freezing drizzle involves droplets <0.5 mm diameter. As water descends from clouds at temperatures close to 0 °C, partial melting and refreezing are common. The largest solid particles (ice pellets and snow) tend to melt more slowly. The smallest droplets are thus most likely to arrive at the ground in liquid form. Colder air directly above the ground surface may act to decrease the amount of freezing rain, favouring the formation of solid particles, which generally pose a lesser hazard. Most freezing rain events involve only liquid droplets, rather than mixtures of liquid and solid precipitation.

As water droplets fall, the kinetic energy generated by movement partially counteracts the effect of air temperature. Supercooled droplets are liquid, but have temperatures <0 °C. Freezing drizzle may remain liquid at −14 °C, whereas larger freezing rain droplets typically require temperatures of −5 °C or warmer to avoid freezing prior to contact. Once the droplet encounters a surface, movement ceases and the droplet freezes on contact. The resulting frozen glaze can be beautiful, but also may constitute a significant hazard. Precipitation amounts can exceed 120 mm, and accumulations of glaze can exceed 90 mm.

Freezing rain events require a thermal inversion, with warmer air aloft trapping air at or below 0 °C within 800–2000 m of the ground surface. Liquid droplets encounter a relatively thin zone of air <0 °C: passing through a thicker cold air mass would result in partial or total refreezing before arrival on surface, producing graupel or ice pellets. Particular freezing rain events result from the specific conditions that allow the thermal inversion to develop, coupled with the presence of abundant atmospheric moisture. Predictions of both the total amount of precipitation and the proportions of freezing rain and solid particles can be difficult. These complexities not only mean that not all freezing rain events are identical, but that the process of accumulation of glaze differs, with consequences for infrastructure.

Typically, freezing rain events are initiated shortly before sunrise, when air and dew point temperatures are at their daily minimum. Most events are very short-lived, typically <1 h: only about 5% of events last for 6 h or more. Cold surface air temperatures tend to persist for several hours after the event, allowing the adhered ice to remain.

The 4–10 January 1998 verglas currently ranks with the Western Canadian drought of 2000–2001, the southern Alberta flooding of 2013, and the Fort McMurray wildfire of 2016 as one of the most costly disasters recorded in Canada (Lecomte et al. 1998, Statistics Canada 2011; Public Safety Canada 2022). Extending from eastern Ontario through New Brunswick, the six-day verglas resulted in local accumulations of freezing rain in excess of 100 mm (Environment Canada 1998; Henson et al. 2011; Fig. 4.19).

The combination of glaze loading and winds caused the collapse of numerous high-tension electrical towers, resulting in loss of power to most of southern Québec, including Montreal, and much of eastern Ontario (Statistics Canada 2011; Rezaei et al. 2016). In the worst-affected "Triangle of Darkness" between Longueil and St-Jean-sur-Richelieu, 23,000 towers and poles and 80 km of power lines were destroyed. More than 4.7 million people lost power, and 2.6 million were impeded from employment, representing 19% of the Canadian workforce (Environment Canada 1998; Lecomte et al. 1998). The prompt distribution of financial compensation in communities such as Montérégie may have led to reduced social stresses and decreased rates of property crime (Maltais et al. 2001). The interruption of power caused significant losses to dairy farmers, as unrefrigerated milk had to be dumped (Statistics Canada 2011). Total losses in Canada exceeded $3 billion. Interruption to power supply has become the greatest socio-economic impacts of verglas events (Call 2010).

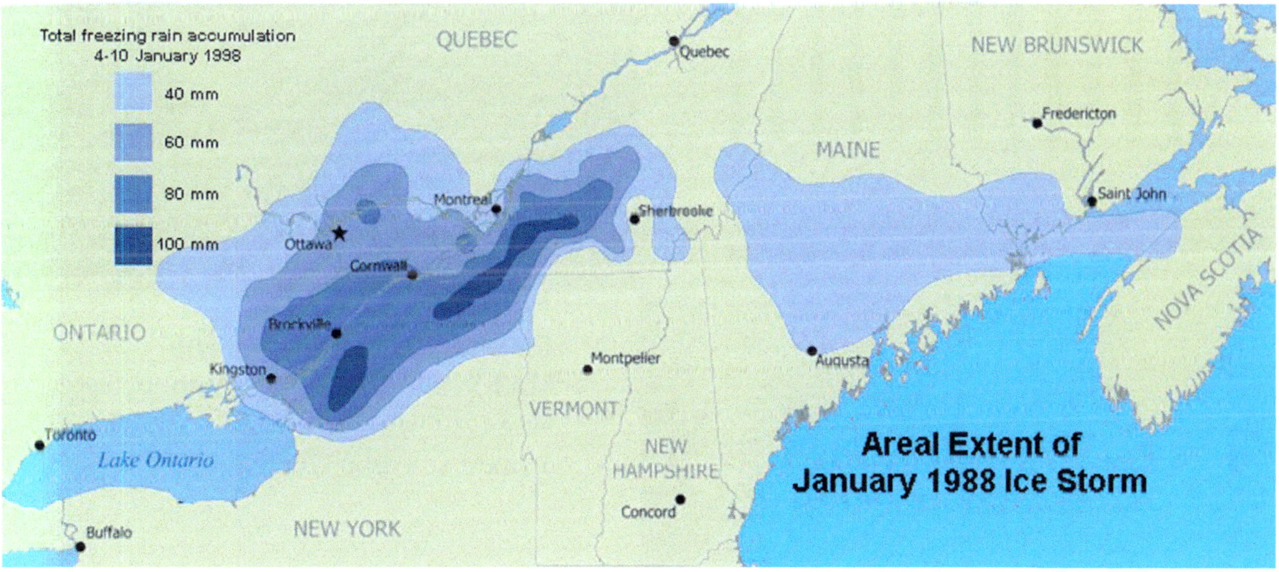

Fig. 4.19 Area impacted by the January 1988 verglas (ice storm) (Environment Canada and Public Safety Canada)

Trees played a significant role in damage, both through falling branches and disruptions to the maple syrup industry. Failure of tree branches is due to both glaze loading and wind. Freezing rain events are a major cause of mortality of mature hardwood deciduous trees in eastern Canada (Proulx and Greene 2001; Deschênes et al. 2019). The forest structure also plays a role. The broad canopies in beech-dominated forests can reduce damage to smaller trees, in contrast to the larger amounts of damage suffered in more open oak- or sugar maple-dominated groves. Isolated trees, particularly in urban areas, are susceptible to damage. Although most damaged trees do recover, they may initially respond by producing more foliage at the expense of stem growth (see Pisaric et al. 2008). In a sugar maple grove, this results in more leaves, but lesser trunk and branch growth, with lesser maple sap.

Maple syrup production is a significant industry in this region (Statistics Canada 2011). Québec maple syrup production generates more than 12,000 full time jobs, and $610 million GDP. During the verglas, many maple syrup operations in Québec and Ontario suffered significant damage. In eastern Ontario, more than 150,000 trees were affected, representing 25% of Ontario maple syrup production (Kidon et al. 2001). For many maple syrup operators and dairy farmers, the immediate losses were of lesser concern than the long-term financial impacts of interrupted operation.

4.2.2 Toronto

Toronto (Fig. 4.20) is a highly developed city economically, with high automobile usage, and relatively light winds, with a population of approximately 4 million. In downtown Toronto, the present urban heat island effect produces a 2–4 °C increase in temperature compared to surrounding rural areas (Natural Resources Canada 2016). Seasonally, the greatest differences occur in late summer and in winter, with the minimum differences in May (Moshin and Gough 2012). Differences of 1–2 °C are evident between downtown

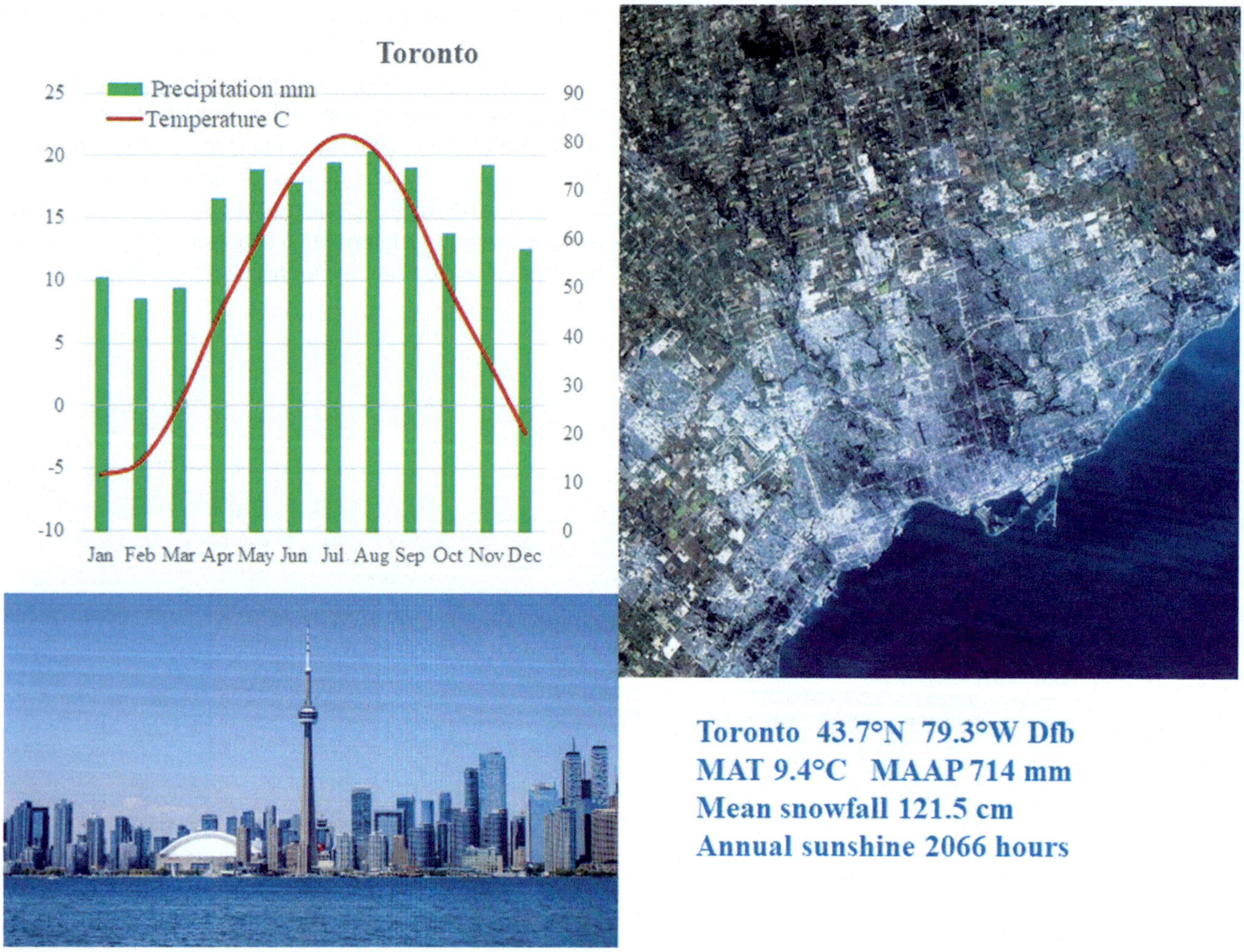

Fig. 4.20 Toronto is located near the southern margin of the Köppen-Geiger Dfb climate zone in Canada, and receives 40% of the snowfall of Québec. The effects of urban development are evident on the satellite image. (*Data* Environment and Climate Change Canada 2020; image: NASA)

Toronto and the temperatures recorded at Pearson International Airport, the principal weather station of Environment Canada for metropolitan Toronto (Moshin and Gough 2012). Substantial variations exist throughout Toronto, resulting from local land use and infrastructure.

Toronto Public Health estimates that extreme heat contributes to at least 120 deaths annually (c.f. Smoyer et al. 2000; Pengelly et al. 2007; Health Canada 2011). Mortality due to extreme heat or heat stroke is predominantly associated with cardiovascular, respiratory, and cerebrovascular causes (Health Canada 2012; Li et al. 2015; Song et al. 2017). Combinations of factors are generally responsible for extreme heat event mortality (Åström et al. 2011; Benmarhnia et al. 2015; Laverdière et al. 2015; Xu and Tong 2017; see Chap. 12 for further discussion).

Tornadoes are uncommon in southern and eastern Ontario, but a destructive event can be expected to occur approximately every 5-8 years (Sills 2012; Public Safety Canada 2022). More rarely, hurricanes modified by interaction with eastward-moving storms to form extratropical transitions track across the Appalachians and arrive in southern Ontario with devastating effects (Robinson and Cruikshank 2006). During the night of 15–16 October 1954, Extratropical Transition Hazel II caused intense flash flooding along the sand-and-gravel–dominated meandering Don, Rouge, and Humber Rivers in Toronto, killing 81 people and causing more than $900 million (2024 $) in damage. Subsequently, the river valleys and their tributary ravines were designated as conservation areas. Rouge National Urban Park, established in 2015, encompasses 40 km^2 of river and adjacent ravine terrain in eastern metropolitan Toronto.

4.2.3 Niagara Peninsula

Niagara Falls and the lower plain along the Lake Ontario shores at St. Catharines, Niagara-on-the-Lake, and Grimsby enjoy a very mild boreal (modified Köppen-Geiger Dfb) climate (Environment and Climate Change Canada 2020; Fig. 4.21). January daily mean temperatures for the area as a

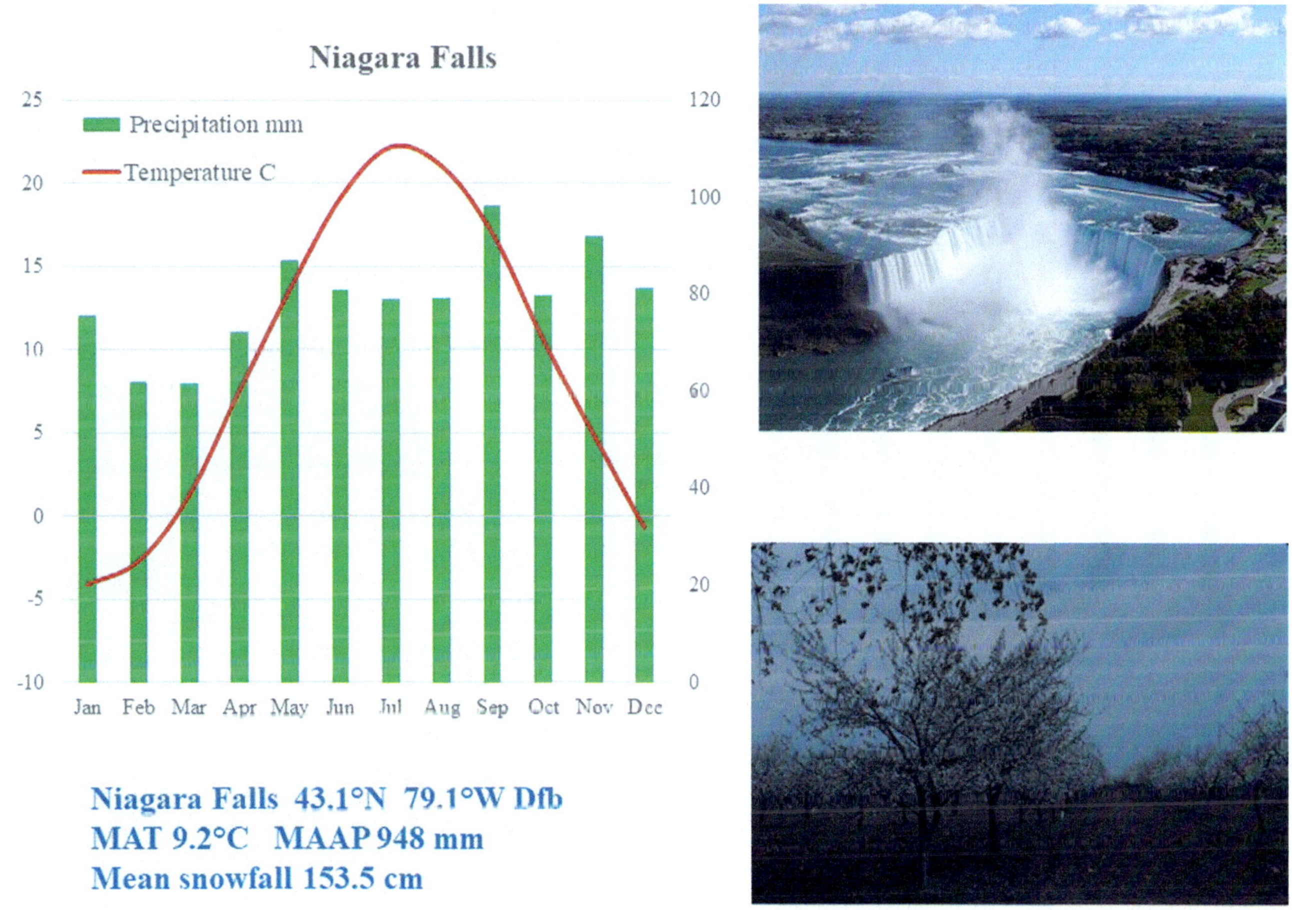

Fig. 4.21 Niagara Falls also is located near the southern margin of the Köppen-Geiger Dfb climate zone. The area south of the Niagara Escarpment receives more snowfall than the Lake Ontario shoreline communities of St. Catharines and Grimsby, in part due to moist southwesterly winds from Lake Erie. (*Data* Environment and Climate Change Canada 2020)

whole approximate −4 °C, and there are small microclimatic areas where the January means are >0 °C. July mean temperatures approximate 24 °C. Temperatures on the southern, higher plain are slightly lower in January, and are slightly higher or similar in July.

Precipitation throughout the area averages 800–850 mm annually. The northern plain typically receives 30 cm less snowfall per year than does the southern, due to the prevailing southwesterly winds which blow across Lake Erie and bring more snow (110 cm annually) to the south. The differences are accentuated along the Lake Erie shores, with Buffalo NY frequently receiving almost twice as much snow through the winter as does St. Catharines. High humidity throughout the summer months contributes to the large numbers of thunderstorms (30 y^{-1}) and lightning flashes (more than 200 y^{-1}) characteristic of the area.

4.2.4 Warm Moist Temperate Dfa/Cfa Climates: Point Pelee

Isolated areas along the northwestern shore of Lake Erie, including Point Pelee National Park and vicinity are classified as possessing a warm moist temperate (Köppen-Geiger Dfa; Fig. 4.22) climate. This area is the mildest in southern Ontario, and only the Gulf of Georgia Islands in British Columbia has a milder climate within Canada (Phillips 1990; Environment and Climate Change Canada 2020). The mean annual temperature is 10 °C.

The area receives 800 mm of precipitation in a typical year, of which about 10% falls as snow. Commonly, measurable amounts of snow fall on as few as 15 days y^{-1}. Snow in the region tends to be dry and light, resulting in blizzards and white-out conditions on the rare occasions when it occurs,

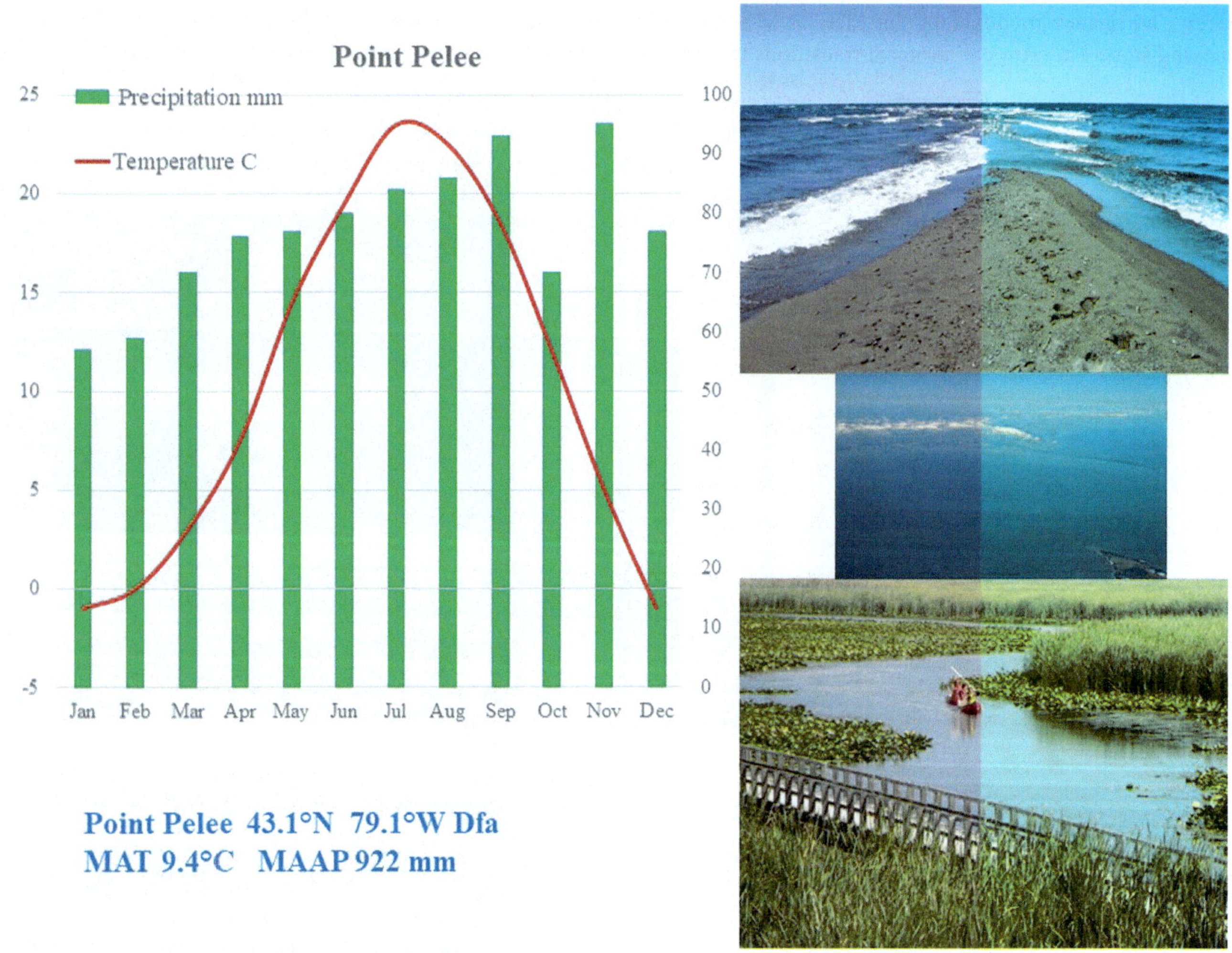

Fig. 4.22 Point Pelee National Park and adjacent Leamington have Dfa (southern boreal) Köppen-Geiger climate. Cfa climate is present on Pelee Island to the south. (*Data* Environment and Climate Change Canada 2020; aerial image: Wikimedia/Aaron Headly)

but temperatures remain relatively high and frost penetration relatively low. Short, sharp snowfall events are characteristic. Snow cover commonly persists for >2 months in all but the most densely wooded areas. In sheltered regions, snow cover may last for <20 days y^{-1}. Freezing rain is rare, typically occurring on 6–9 days y^{-1}. The frost-free period lasts between 180 and 230 days, depending upon aspect.

During the summer months, warm moist air is brought to Pelee by the prevailing southwesterly winds, blowing across Lake Erie. Violent thermals can occur in the late afternoons, and the area receives more thunderstorms than any other part of Canada, 35 annually. Tornadoes occur throughout southwestern Ontario, with about one newsworthy event every two years, but none has struck at Pelee over the past 40 years. The climate, therefore, is not typical of that of the remainder of Canada.

4.3 Bedrock

The St. Lawrence Platform is divided into three regions (Natural Resources Canada 2009; Bostock 2014; Fig. 4.23). The Eastern region includes the coastal plain along the northeast shore of Newfoundland, Anticosti Island, and Mingan Archipelago, flanked by the Canadian Shield to the north and the Appalachians to the south. The Central region is the St. Lawrence and Ottawa Valleys, bordered by the Frontenac Axis of the Canadian Shield to the southwest (Thousand Islands area), and by the eastern Québec Uplands of Appalachia to the east. The western region includes the Great Lakes lowlands, forming a triangle flanked by lakes Ontario, Erie, and Huron. All three regions are underlain by Palaeozoic sedimentary bedrock.

4.3.1 Eastern Region

In the Eastern region, the igneous and metamorphic Grenville Province rocks are flanked by the Ordovician limestones of the Mingan Archipelago, the northernmost and oldest units of the succession evident in the Gaspé Peninsula (Wheeler et al. 1996; see Chap. 2). These Ordovician limestones have been folded into a series of very gentle anticlines and synclines. Fossils present include trilobites, gastropods, and the cephalopod molluscs *Orthoceras*, a distant ancestor of squid, octopus, and the modern Pacific

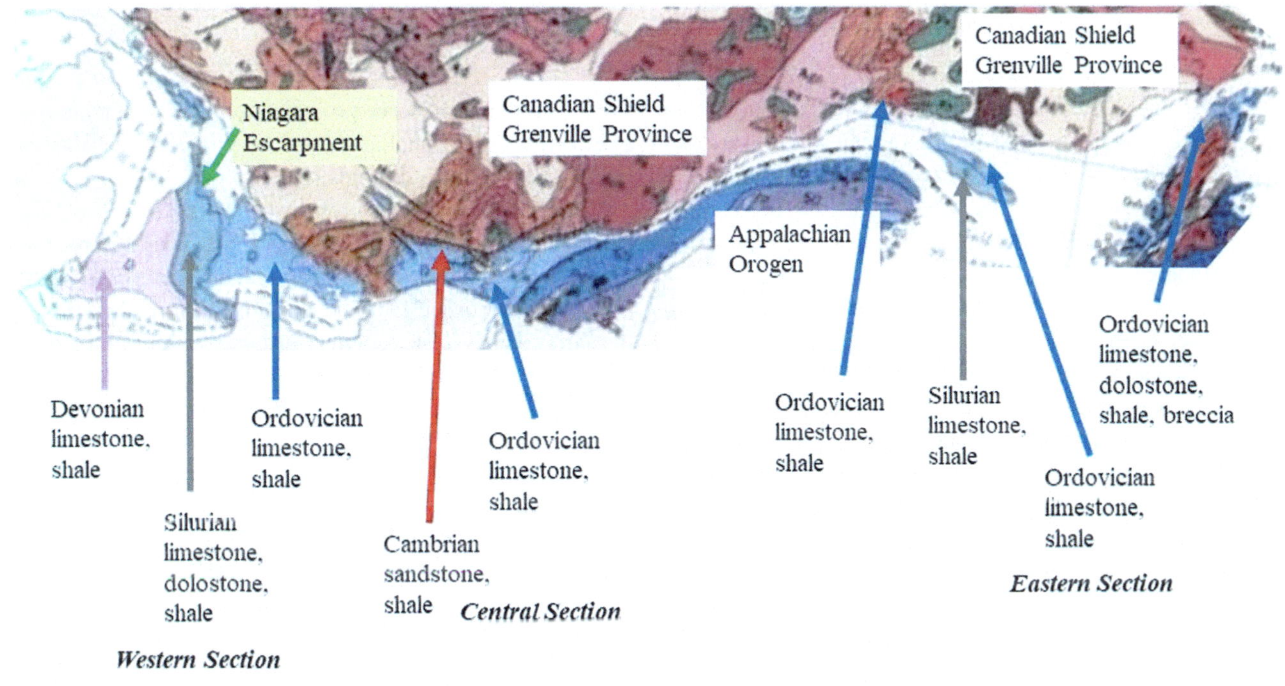

Fig. 4.23 St. Lawrence Platform Bedrock overview. Western section: successively younger sedimentary rocks from northeast to southwest, with minimal structural deformation. Central section: Cambrian (Nepean) sandstone and shale along the northern margin adjacent to the Canadian Shield, with Ordovician limestone and shale in the Ottawa and St. Lawrence Valleys, with minimal structural deformation. Eastern section: Ordovician sedimentary rocks at Mingan Archipelago (minimal deformation), northern Anticosti (deformation), and western Newfoundland (significant thrust faulting), with Silurian rock in southern Anticosti. (Base map: Geological map of Canada, Natural Resources Canada)

Fig. 4.24 Anticosti Island. The northern shore is composed of Ordovician reef limestones and shales, subject to erosion along faults and sedimentary boundaries between reefs. The central and southern parts are composed of Silurian sedimentary rocks, more evenly eroded to produce a coastline without indentations. Parallel drainage patterns with straight streams mark the Silurian landscapes (Landsat, NASA)

Nautilus. Younger Silurian limestones, dolostones, and shales form a monoclinal SSW-dipping cuesta on Anticosti Island (Innu-aimun *Notisquan)*, located between Jacques-Cartier and Honguedo straits (Chen et al. 2016; Beaupré 2020). The official Boundary Stratotype for the transition between the youngest Ordovician (Gamachian) and Silurian strata occurs at Cap de la Vache-qui-Pisse, on Anticosti Island (Fig. 4.24).

The Gamachian strata provide important clues to the progress of evolution between the Ordovician and Silurian geological periods (Barnes 1988). During this geological event, the region was located in the Southern Hemisphere, and was subject to changes in sea level induced by continental glaciation in what is now the western part of the Sahara Desert. The fluctuations in water level caused species to become extinct, and others to evolve, and the evidence for these changes is preserved in the rock and fossil records. Subsequently, Anticosti and Mingan were attached to the North American continent as the Iapetus Ocean closed.

Along the Gulf of St. Lawrence in northern Newfoundland, a narrow coastal plain, underlain by thrusted early Palaeozoic sedimentary bedrock, covered by Holocene marine sediments, discontinuously borders the western margin of the Long Range (Colman-Sadd et al. 1990; Catto 2020). Table Point Ecological Reserve protects the mid-Ordovician trilobite, orthocerid cephalopod, and brachiopod fossil assemblage. Bryozoans, crinoids, gastropods, ostracods, and porifera fossils are also present to the north. Burnt Cape Ecological Reserve and Watts Point Ecological Reserve were established to preserve the unique plant assemblages developed on the exposed Ordovician limestone and dolostone of the Strait of Belle Isle region (Hermanutz et al. 2002).

4.3.2 Central Region

The St. Lawrence and Ottawa Valleys are underlain by Cambrian and Ordovician sedimentary rocks, which dip gently towards the south (Globensky 1987; Ontario Geological Survey 1991). One of the oldest of these, the Cambrian Nepean sandstone, forms the steep cliffs of Parliament Hill in Ottawa (Bélanger 1998). Younger Ordovician limestones, dolomites, and shales, containing trilobites, brachiopods, and corals, crop out to the south of the Ottawa River and across the St. Lawrence Valley (Macdonald and Harrison 1979; Fig. 4.25). SE-NW trending normal faults mark the Ottawa Valley from Montreal northwest to Pembroke, locally forming 10-25 m high steep escarpments.

Fig. 4.25 Ottawa bedrock. Left: Ottawa is largely underlain by Ordovician limestones and shales. Waterfalls develop over resistant limestones, including flat-lying units (Rideau Falls, upper) and folded strata (Hog's Back Falls, lower). Right: Cambrian Nepean sandstone, primarily exposed in quarries, was extensively used as facing stone in Ottawa's public buildings, although it is subject to environmental modification and must be periodically refurbished. Upper: Centre Block, Parliament Buildings (photo taken prior to ongoing restoration). Lower: Pretoria Bridge, following restoration

The ten isolated summits of the Monteregian Hills, including Mont-St-Bruno (218 m asl), Mont-Mégantic (1105 m asl), and Mont Royal in Montreal (233 m asl), form conspicuous landscapes in the southern St. Lawrence valley (Boyer et al. 1985). Aligned in a SE (oldest) to NW (youngest) chain, the Monteregian Hills formed as the North American tectonic plate slid over the New England Hotspot between 140 and 118 million years ago (Fig. 4.26). The peaks are formed of Cretaceous intrusions of gabbro with associated syenite and flanks of metamorphosed hornfels (Roulleau and Stevenson 2013). The combination of basic mineralogy, elevation, and unsuitability for farming has produced unique landscapes, partially conserved as natural areas (e.g. O'Neill Sanger et al. 2021).

The flat-lying sedimentary rocks of the St Lawrence Valley grade eastwards through the somewhat deformed early Palaeozoic units of the Eastern Québec Uplands to the strongly deformed Notre Dame (northeast), Mégantic (centre), and Sutton-Green (southwest) Appalachian mountains. The Notre Dame Mountains extend southwestwards from Gaspé (see Chap. 2). Parc régional du massif du sud, 65 km ESE of Québec, represents an orographic landscape of the Notre Dame Mountains, including Mont-du-Midi, Mont-St-Magloire, and Mont-Chocolat.

The Mégantic Hills formed when folded Silurian and Devonian limestones and sandstones were intruded by Devonian granites. Parc-Mont-Mégantic and the Samuel-Brisson ecological reserve include these units, in addition to the Monteregian Mont-Mégantic. With representative upper Appalachian ecosystems and critical bird conservation sites (Québec Environment n.d.) Mont-Mégantic is also the world's first International Dark Sky Reserve.

The Sutton Mountains, forming the northern extension of the Green Mountains of Vermont, consist of anticlinally folded Cambrian and Ordovician sedimentary rocks with a core of deformed metamorphic schists and volcanic units (Fig. 4.27). Relief exceeds 1000 m. The Hereford Community Forest, surrounding Mont-Hereford (864 m), conserves a forest landscape representative of the ecological changes that the Sutton Hills have undergone in the past 300 years (Foret Herford n.d).

Fig. 4.26 The SE-NW chain of Monteregian Hills consists of ten Cretaceous gabbro intrusions formed between 140 and 118 Ma. Upper left: Mont Yamaska (right, ca. 140 Ma); Mont Rougemont (centre, ca. 137 Ma); Mont-St-Hilaire (left, 135 Ma). Upper right: Mont Shefford (ca. 130 Ma), near Granby (Wikimedia/uncivil fire). Lower: Silhouette of Mont St-Hilaire (Wikimedia/Krestavilis)

4.3.3 Western Region

The Great Lakes lowland consists of two very low-relief plains: a lower plain adjacent to Lake Ontario (Baker et al. 1998); and a broad, gently undulating upper plain, sloping almost imperceptibly to the southwest, extending south to the shore of Lake Erie. The two plains are separated by the Niagara Escarpment, which averages 100 m in relief and varies from a vertical cliff to a series of steep slopes within a 5 km-wide belt (Karrow and Morin 1998; Cowell 2020; Desloges et al. 2020).

Bedrock along the Lake Ontario shoreline is a succession of Ordovician limestones and shales, very gently dipping to the north or flat-lying (Ontario Geological Survey 1991). The limestone units contain trilobites and brachiopods, and are parent materials for calcareous melanic brunisols and grey-brown luvisols. The most resistant limestone and mixed limestone-dolomite units underlie ridges. Valley areas are underlain by weaker limestones and intercalated limestones and shales.

4.3.3.1 Niagara Escarpment

Niagara Falls represents the most spectacular part of the Niagara Escarpment, which extends to the west from Niagara to Hamilton and Dundas (where it is called "The Mountain"), and from there northwest to the Saugeen-Bruce Peninsula (Wiarton and Bruce Peninsula National Park) that separates Lake Huron from Georgian Bay (Fig. 4.28). Numerous small waterfalls cascade over the escarpment along its length.

The Niagara Escarpment is the product of differential physical modification of bedrock units over time. Faulting is not involved in its formation. Prior to the development of the automobile, the only practical route over the escarpment in southwestern Ontario was through the Dundas Gap, west of Hamilton. This transportation problem led to the construction of a succession of Welland Canals to bypass the falls.

Fig. 4.27 Lac Spruce, Parc d'environnement naturel de Sutton, 35 km SW of Magog, Sutton Mountains. (Wikimedia/Nicholas Raymond)

The lower plain, along the shores of Lake Ontario, contains the escarpment's oldest rocks. These units, the Lorraine shale, Grimsby (or Cataract) sandstone, and Rochester (or Clinton) shale, are Ordovician and Silurian age (exposed in Niagara Gorge). All are readily eroded, especially the Rochester shale beds (Johnson et al. 1992; Brett et al. 1995).

These units are overlain by the extremely resistant Lockport dolomite (top of Niagara Gorge succession). It formed during the Silurian, as limestone was gradually hardened and converted into dolomite by the substitution of magnesium (Mg^{2+}) ions for approximately half of the calcium (Ca^{2+}) ions originally present in the rock. This substitution, which occurred underground as magnesium-rich groundwaters slowly percolated through the rocks, hardened the dolomite and created cavities (vugs) in the rocks where some changes in crystal structure resulted. The Lockport dolomite forms the escarpment. The upper plain is developed on younger Silurian rocks, the Guelph dolomite and limestone and the Salina limestone and shale, which formed still later in the Silurian and overlie the Lockport dolomite.

As the rocks gradually were exposed on the surface and weathered over the past 100 million years, the Salina and Guelph units were removed from the area of the escarpment, and the Rochester shale was eroded below. As rivers ran over the unyielding Lockport dolomite, water infiltrated the joints in the rock. The waters froze in the Rochester shale, prying it apart, and the Lockport dolomite was undermined. As the shale was removed, the dolomite formed a flat-topped rampart with a near-vertical cliff standing above the area formerly occupied by the now-eroded shale, and the Niagara Escarpment was developed. The rates of physical weathering involved in this erosional activity indicate that the escarpment required millions of years for its formation, and it thus developed in much its present form prior to the Quaternary glaciation of the area.

4.3.3.2 Southwestern Ontario

The rocks underlying the terrain to the west and south of the Niagara Escarpment are mid-Devonian shales, limestones, and dolomites (Johnson et al. 1992). Younger than the sequence exposed at Niagara Falls, these rocks are generally deeply buried by Quaternary sediments, and do not crop out or have a great influence on the physiography of southwestern Ontario. Sedimentary rocks near Petrolia and Oil City east of Sarnia contained petroleum. The first well was drilled in the area in 1858, and the region produced

Fig. 4.28 Niagara Escarpment. Upper left: Shaded Relief Model, based on processed satellite data. The Dundas Gap, west of Hamilton on Lake Ontario, is a critical transportation route across the escarpment. Also visible is the escarpment developed in older Ordovician rocks along Midland Bay to the north (see Fig. 4.15). (NASA). Upper right and lower left: Views of the Niagara Escarpment south of Collingwood. Small waterfall is supported by the Silurian Lockport dolostone. Lower right: The Escarpment near Grimsby, Ontario. (Wikimedia/Devastator)

petroleum throughout the nineteenth and early twentieth century. It was these petroleum deposits, now long exhausted economically, that led to the establishment of refineries and petrochemical manufacturing industries at Sarnia (see Chap. 17).

4.3.3.3 Karst (Carbonate) Landscapes

Exposed carbonate bedrock in several parts of the St. Lawrence Lowlands has allowed karst landscapes to develop. On the Saugeen-Bruce Peninsula, karst topographic features include doline lakes, formed from the dissolution of limestone by both surface waters (recently accentuated by acid precipitation) and ground waters (Cowell 2020; Fig. 4.29). Dolines dissolved by surface waters tend to be broad and shallow, with saucer-like cross-sections. Small wetlands develop along the fringes of these dolines, and in areas where the water can slowly drain from the centre of the depression. Dolines formed from subsurface dissolution are oval with near-vertical sides. Small erosional karst karren form ridges, depressions, and irregular pinnacles <2 cm high across exposed flat-lying limestones. Lake cliffs of limestone, up to 90 m high, are eroded at the base by the cold water of Lake Huron and Georgian Bay, forming dissolved notches and caves at water level. The narrow bases of Flowerpot Island and its neighbours are also the result of dissolution of limestone by lake water.

Mingan Archipelago initially developed as the Ordovician rocks were subject to dissection, forming a plain cut by valleys parallel to the bedding of the rocks and following faults and joints (Brisebois and Nadeau 2003; Fig. 4.30). When sea level rose following deglaciation, the dissected plain was flooded, creating the archipelago. Relief is gener-

Fig. 4.29 Karst at Bruce Peninsula National Park. Upper: Erosional Flowerpot developed in Silurian dolostone (Wikimedia/Borbrav). Lower: The Grotto and adjacent beach are a popular tourist attraction (Wikimedia/WikiPedant)

Fig. 4.30 Mingan Archipelago, Québec features a chain of about 40 islands that formed as the continent rose after the last glaciation. Over the course of the last few thousand years limestone rocks eroded to form the largest group of monoliths in Canada. *Photograph* M. Villeneuve/Wikimedia Commons, CC BY 2.0. (Slaymaker and Catto ch 1)

ally low, approximately 100 m, and the remnants of the plain slope to the southeast. The northern coast of Anticosti Island represents another facet of the pre-existing planar topography (Beaupré 2020).

The erosion of the limestone has created a complex network of pillars, steep bluffs, flowerpot structures (with narrow bases, widening upwards), sea caves, and sea arches. Many of these features have attractive descriptive local names, such as the Tete-du-chat (Cat's Head) cliff; the "Full Dome", developed where a gentle anticlinal fold of limestone is washed by the sea; "Fallen Dome", marked by erosion of the anticline's crown; Cap-St-Charles and False Pass Pillars; and "The Chalice", "The Hen", and "The Pottery" flowerpot monoliths.

The development of these features, sea caves, and of arches such as Ile de Quarry arch, is a result of frost weathering in the cool oceanic climate (Catto 2015). Breaking waves send seaspray into fractures and crevices in the rocks. In the cold climate, freezing of the spray wedges the fractures and crevices apart, inducing failure. Additional erosion results from the chemical dissolution of the limestone, a process which proceeds more rapidly under cool conditions than in warmer environments. The development of the features is thus a reflection of the climate, rather than reflecting the energy

Fig. 4.31 The Arches Provincial Scenic Attraction, NL, is composed of Ordovician dolomitic limestone. Fracture patterns are related to thrust faulting, augmented by secondary jointing and frost wedging. Mean high tide reaches the base of the supporting pinnacle. Beach gravels include local Ordovician carbonate and clastic pebbles, and granitic and gneissic clasts derived from glacial transport from the Grenville Province of the Canadian Shield

level of the coast. Direct erosion by waves is unimportant in the development of these arches, pillars, and stacks.

Along the Newfoundland Coastal Plain, The Arches Provincial Scenic Attraction preserves two sea arches eroded from Ordovician dolomitic limestone (Catto 2020; Fig. 4.31). The arches are controlled by the pattern of fractures and joints, with the main fracture patterns paralleling the coast. The joint orientation also controls the shape of each arch. Frost wedging is the primary process involved, with wave action responsible for removal of disaggregated material.

4.3.4 Earthquake Activity

Consistent intraplate earthquake activity occurs in the St. Lawrence and Ottawa Valleys, along the former plate boundary which continues to undergo tectonic stresses (Fig. 4.32). Moderate seismic hazards characterize Ottawa, Montreal, and Québec (Earthquakes Canada n.d.). In 1944, an Mw 6.6 earthquake occurred southeast of Ottawa, with its epicentre in the Cornwall, Ontario, area. The 2010 Mw 5.0 Val-des-Bois earthquake produced noticeable shaking with some panic in Ottawa. Slope failures triggered by seismic activity occur in glacimarine sediments (Quinn et al. 2011; Lévy et al. 2012).

The Charlevoix seismic zone is the most active seismic region in eastern North America (Lamontagne et al. 2003; Earthquakes Canada n.d.). Since the founding of Québec in 1608, notable earthquakes have occurred in 1663, 1791, 1860, 1870, 1925, and 1988. The largest, the 5 February 1663 La Malbaie earthquake, estimated Richter MW 7.2 ± 0.2, triggered numerous slope failures along the St. Lawrence and Saguenay valleys. The estimated MMI of VI resulted in significant damage to chimneys, masonry walls, and a brewery. Damage was recorded in Boston.

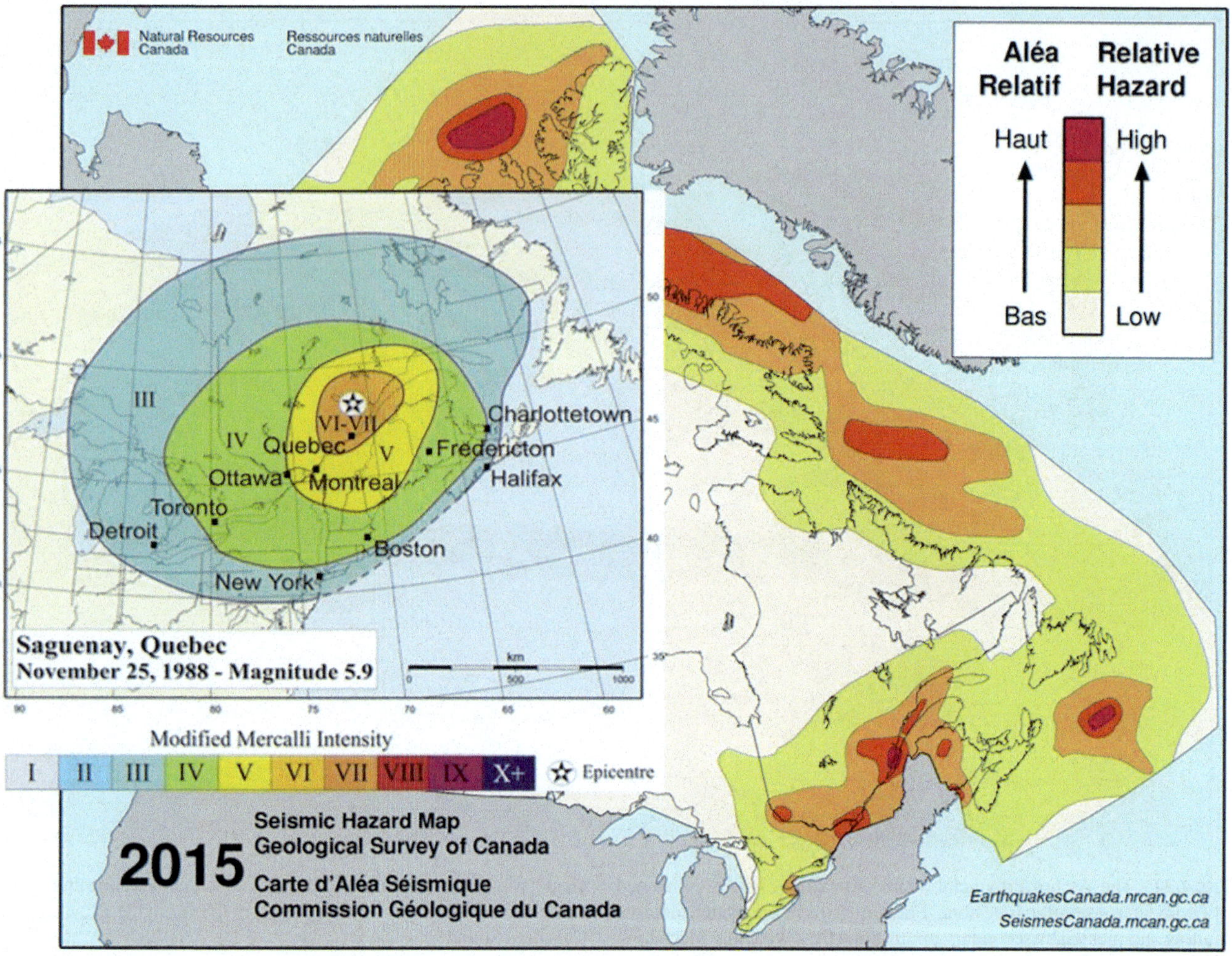

Fig. 4.32 Seismic hazard in the St. Lawrence Valley is greatest in the Charlevoix Seismic Zone (northeast of Québec) and in the Montreal-Ottawa region. Inset: Intensity map for the 1988 earthquake centred in the Charlevoix Seismic Zone. (Earthquakes Canada, Natural Resources Canada

4.4 Quaternary

4.4.1 Glacial History

The St. Lawrence-Great Lakes lowlands were glaciated several times during the Quaternary (Fig. 4.33). Successions of Quaternary deposits dating to more than 200,000 years old are preserved in Rouge River Park (Scarborough Bluffs, Toronto; Fig. 4.34), and along the north shore of Lake Erie (Karrow 1990). To the north of Niagara Falls, older gorges cut by stream systems developed prior to the last glaciation, such as St. David's Gorge, have been buried by younger glacial sediments (Hobson and Terasmae 1969).

During the last major glaciation, Laurentide ice advanced from western Labrador and adjacent Québec to cover all of the St. Lawrence-Great Lakes lowlands (Dyke et al. 2003; Dyke 2004; Ontario Geological Survey 2010; Mahaney et al. 2014; Catto et al. 2020; Desloges et al. 2020). Ice flowed south-southwest along the Gulf of St. Lawrence coastline of Newfoundland, south across the Québec North Shore, southeast across the central and lower St. Lawrence Valley, southwest across Ottawa, and southeast, following the upper Ottawa River. Once the ice reached the upper St. Lawrence River, it swung to the southwest, following the general trend of the St. Lawrence River towards the Thousand Islands. North of Toronto, thick deposits of the Oak Ridges Moraine formed where ice moving through Lake Ontario met ice coming south from Algonquin Park. Other Laurentide ice streams flowed south through the Georgian Bay, Simcoe, and Lake Huron basins (Sookhan et al. 2018). The Lake Erie lobe of the Laurentide ice sheet advanced to the southwest approximately 27,000 years ago, covering the entire Lake Erie basin and extending southwest into Ohio.

North of Lake Ontario, and in the region inland from Lakes Simcoe, Huron, Erie, and Ontario, glacial deposits with mixtures of clasts derived from local Palaeozoic bedrock and Canadian Shield granitic and metamorphic rocks are exposed, forming sandy and gravelly substrates,

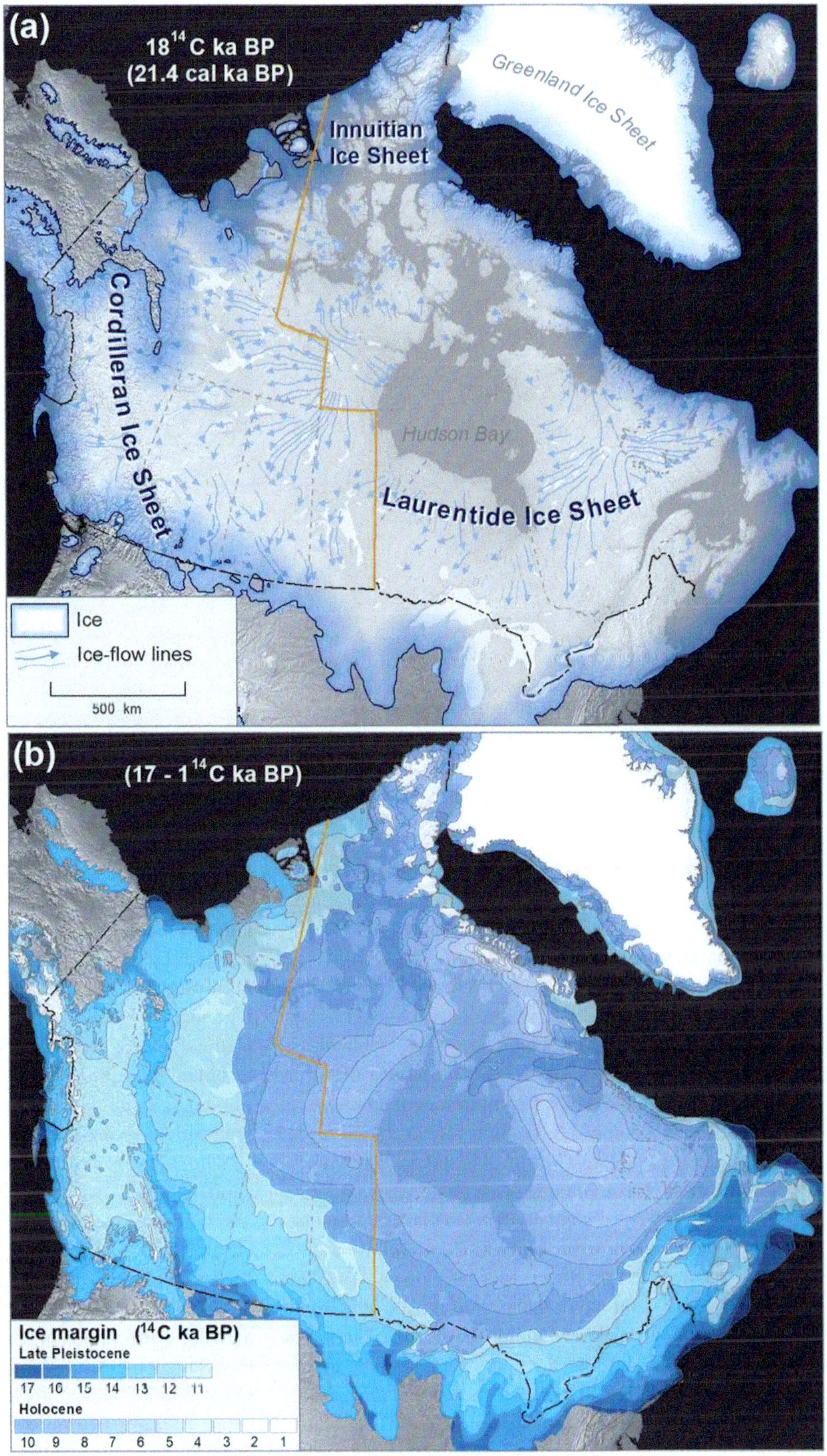

Fig. 4.33 Ice masses in Canada. Upper: Distribution of Cordilleran, montane, plateau and continental ice coverage and generalized ice flow direction (modified from Shaw et al. 2020) of most recent (MIS 2) glaciation, beginning ca. 30,000 a. Lower: Extent of ice coverage at various stages of deglaciation. Ice margins from Dyke et al. (2003) (Slaymaker and Catto ch 2)

Fig. 4.34 At Bluffers Park in eastern Toronto, adjacent to Rouge River National Urban Park, the Scarborough Bluffs displays a thick succession of Quaternary sediments spanning 150,000 years

along with glacifluvial eskers and kames. In most other areas of the Great Lakes–St. Lawrence lowlands, the glacial deposits are blanketed by thick accumulations of lake clay (southern Ontario) or glacimarine clay (Ottawa Valley, St. Lawrence Valley, and Gulf of St. Lawrence shorelines).

More than 3000 drumlins are present in the Peterborough drumlin field north of Lake Ontario (Maclachland and Eyles 2013; Marich 2016; Fig. 4.35). The largest drumlins in the Peterborough field exceed 3 km in length and 40 m in relief. The field is a composite feature, formed by advances from the Simcoe Lobe (west), from Algonquin Park (northeast), and from the Lake Ontario Lobe (southeast), as determined from the origins of glacially faceted clasts. The slopes of many of the Peterborough drumlins do not show a clear difference between the up-flow (stoss) and down-flow (lee) sides. The highest point along the axes of most of the drumlins is approximately halfway along the profile. The slope angles of the stoss and lee sides are very similar, and in many cases, the lee side shows a steeper slope. Differences between adjacent drumlins are common, both in relative slope angles and overall shape. As recognized in fields elsewhere, drumlin morphology is not a reliable indicator of the direction of flow (Spagnolo et al. 2010).

A group of 16 tunnel valleys, formed during glaciation, are currently occupied by the Kawartha Lakes in Ontario (Russell et al. 2003; Fig. 4.36). Tunnel valleys, marked by cubic parabolic cross-sections and carved by high-pressure subglacial meltwater flows, are associated with eskers. Depths may exceed 50 m, and the flat-bottomed channel bases and steep sides suggest flow under highly turbulent conditions, with rapid bursts of water fed by impounded subglacial lakes along the margin of the Canadian Shield to the north.

The Laurentide ice retreated in stages, punctuated by short-lived readvances (Karrow and Calkin 1985; Karrow 1989; Dyke et al. 2003; Catto et al. 2020; Desloges et al. 2020). In western Lake Erie, the Pelee moraine formed approximately 16,000 years ago. Progressive retreat deglaciated the Québec North Shore (including Mingan) and the north shore of Lake Erie about 14,000 years ago; the Saugeen-Bruce Peninsula, the north shore of Lake Ontario, and Québec about 13,000 years ago; and the north shore of Lake Huron, Ottawa, and the northern flank of the St. Lawrence Valley about 12,000 years ago. Along the northern flank of the St. Lawrence and Ottawa Valleys, a glacial readvance between 11,700 and 11,000 years

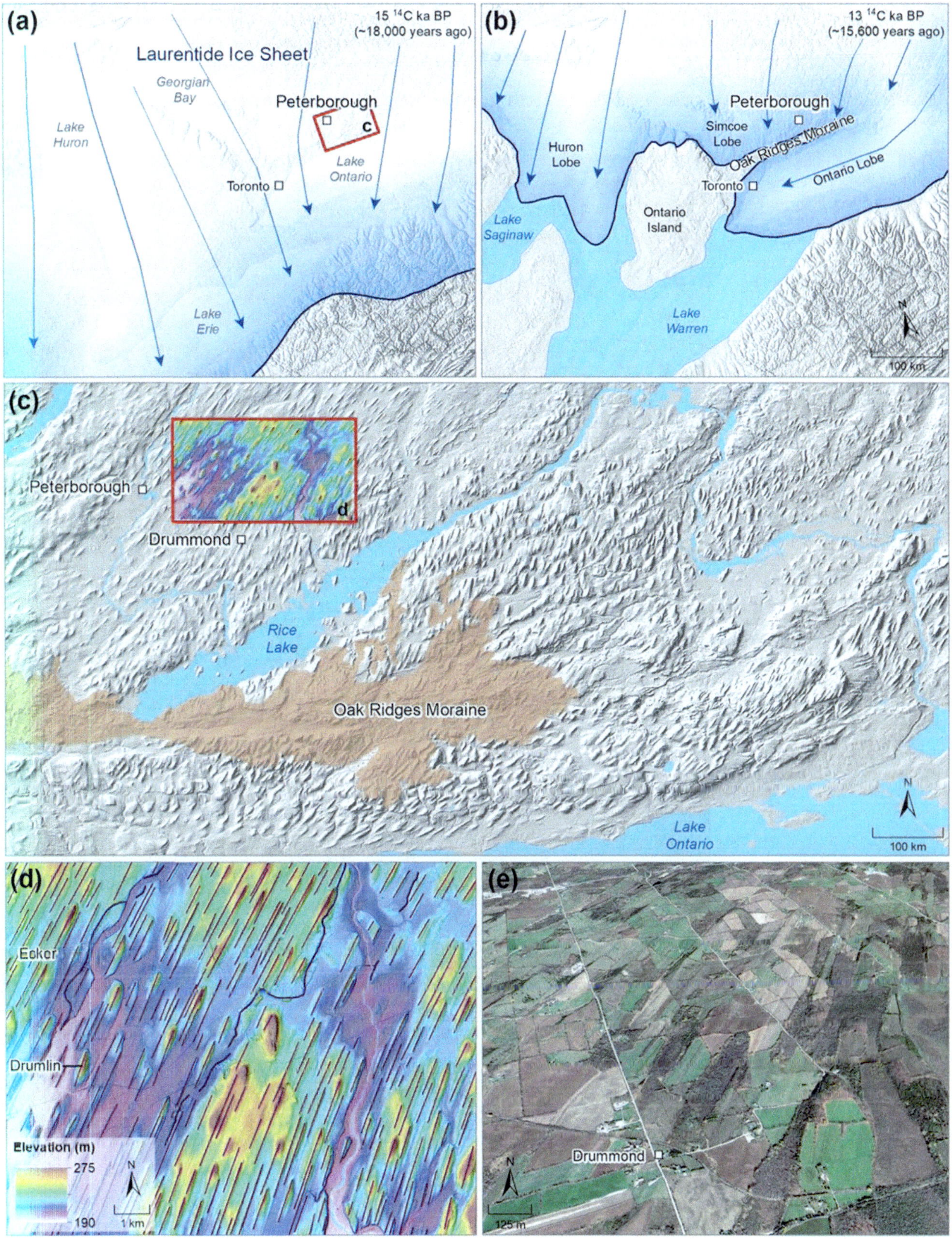
(a)
15 ¹⁴C ka BP
(~18,000 years ago)
Laurentide Ice Sheet
Georgian Bay
Lake Huron
Peterborough
c
Lake Ontario
Toronto
Lake Erie
(b)
13 ¹⁴C ka BP
(~15,600 years ago)
Peterborough
Simcoe Lobe
Huron Lobe
Oak Ridges Moraine
Toronto
Ontario Lobe
Lake Saginaw
Ontario Island
Lake Warren
N
100 km
(c)
Peterborough
d
Drummond
Rice Lake
Oak Ridges Moraine
N
Lake Ontario
100 km
(d)
Esker
Drumlin
Elevation (m)
275
190
N
1 km
(e)
Drummond
N
125 m

◂**Fig. 4.35** The Peterborough Drumlin Field is an example of a Laurentide glaciated landscape in southern Ontario. The field's approximately 3000 drumlins covers the region from Lake Ontario to north of Peterborough. **a** and **b** Landforms created by glacial advance, retreat, stagnation, and final melting of the Simcoe Ontari, and Huron Lobes of the Laurentide Ice Sheet (adapted from Maclachlan and Eyles 2013). **c** Drumlin orientation in the northern and eastern parts of the drumlin field record southward flow of the Simcoe lobe. The southern part of the field records northwestward movement of the Ontario lobe from the Lake Ontario basin. *Source* shaded relief image from MRD128-REV; © Queen's Printer for Ontario 2010. **d** and **e** Drumlins can range from approximately 45 m up to >5 km in length, a few metres 500 m wide, and from a few metres to >40 m relief. *Source* **d**—Lidar from Ontario Geological Survey (2010); **e**—Google Earth: Landsat/Copernicus, DigitalGlobe 2019. Centred on about 44°16′12″ N, 78°12′36″ W. (Slaymaker and Catto ch 2)

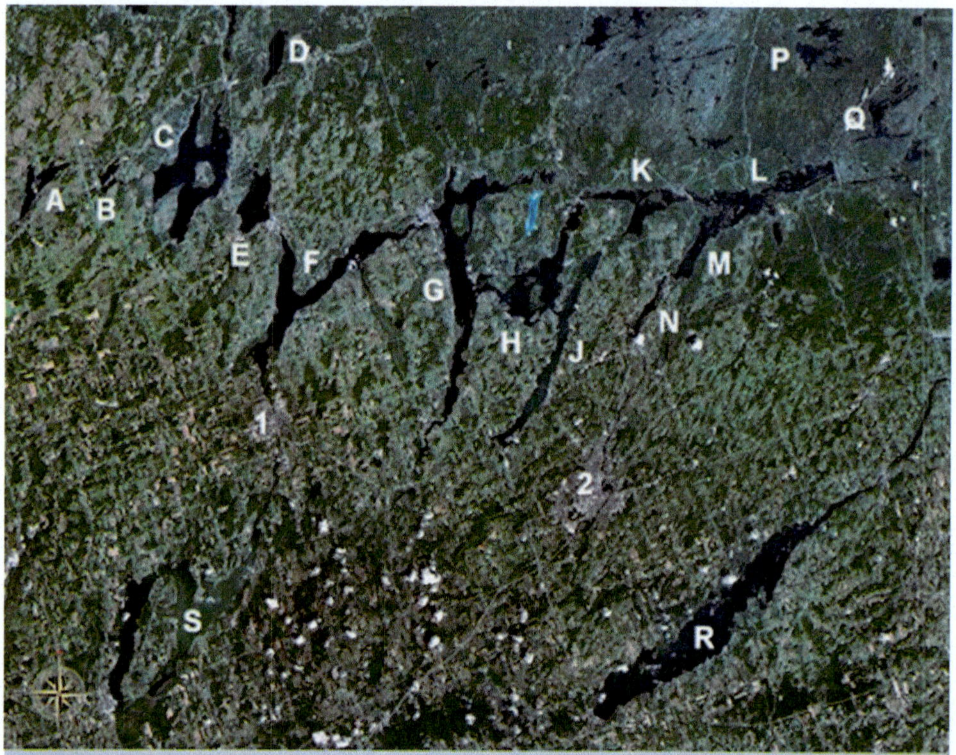

Fig. 4.36 Tunnel valleys are formed by subglacial discharges of meltwater under high pressure. Segments with glacifluvial sand-and-gravel deposits may be present. Upper: The Kawartha Lakes represent a series of tunnel valleys (labelled A through S) formed by southwestward discharge from the Laurentide Inlandsis. Erosion was concentrated in weaker limestone and shale bedrock, following structural weaknesses. (NASA). Lower: Tunnel valley at Ranney Gorge, occupied by the Trent River at Campbellford, 25 km east of Rice Lake (Wikimedia/Robert Taylor). Tunnel valleys in upper image: A, Canal L; B Mitchell L, C Balsam L, D Four Mile L, E Cameron L, F Sturgeon L, G Pigeon L, H Buckhorn L, J Chemong L, K Lower Buckhorn L, M Clear L, N Katchewanooka L, P Jack L, Q Kassabog L, R Rice L, S Lake Scugog. 1: Lindsay; 2: Peterborough

ago formed the St.-Narcisse moraine (see Chap. 3). Final deglaciation of the northernmost Mixedwoods ecozone occurred shortly after 11,000 years ago.

4.4.2 Great Lakes Deglacial History

As glacial retreat progressed, ice continued to block the Great Lakes' drainage, rendering them unable to drain through the Niagara and St. Lawrence rivers (Karrow and Calkin 1985; Karrow 1989; Catto et al. 2020; Desloges et al. 2020). In combination with glaci-isostatic depression of newly exposed terrain along the ice margin, this resulted in a complex pattern of shifting lake configurations and changing outlets over the period from 16,000 to 4,000 years ago (Fig. 4.37). Initially, lakes drained to the south. As glacial retreat opened new, lower outlets along the ice margin, lake configurations shifted northwards. Continued glaci-isostatic recovery eventually led to the abandonment of northerly outlets, with the reconfigured Great Lakes draining through Sarnia, Niagara Falls, and the St. Lawrence River, as they do today.

4.4.2.1 Lake Erie Basin

When the ice stood at the Pelee moraine 16,000 years ago, Lake Erie could not drain eastward. The lake ponded against the ice front, flooding the Toledo (Ohio) area southwest of the lake, and drained southwestward through several channels into the Ohio River, and eventually into the Mississippi. This highest phase of Lake Erie is referred to as Lake Maumee. As the ice gradually retreated northward, progressively lower lakes (Whittlesey, Warren, Wayne, Arkona) were formed, each one being dammed at its eastern end by the retreating ice. The later stages of this sequence extended north into the southern part of the Lake Huron basin, which had also been deglaciated by then. These lower lakes drained to the west *via* the Grand River of Michigan into Lake Michigan, and thence south through Chicago into the Mississippi system.

By 14,000 years ago, the glacial front stood midway between Fort Erie and Niagara-on-the-Lake. A small lake was ponded against the retreating ice, but was unable to drain north because the falls and Lake Ontario basin were still ice-covered. When the ice did retreat to the escarpment edge *ca.* 13,700 years ago, a deep lake quickly developed between the escarpment and the ice, and the resulting production of icebergs by calving caused the glacier to retreat rapidly through the entire Lake Ontario basin.

4.4.2.2 Lake Ontario Basin

Approximately 13,000 years ago, the ice front remained in the Thousand Islands area, blocking the eastern outlet of Lake Ontario. Waters in the basin were ponded against the ice front to the east, forming a high-level lake stage in the basin (±140 m above modern Lake Ontario), referred to as Lake Iroquois (Karrow and Calkin 1985; Karrow 1989; Catto et al. 2020; Lewis and Anderson 2020). As Lake Iroquois was prevented from draining to the northeast along the St. Lawrence Valley by the retreating glacial ice, it drained to the east through an outlet at Rome, NY, USA. From Rome, the waters followed the Mohawk River valley to Albany (along the route currently occupied by the Erie Canal), and then flowed south along the Hudson River to New York City. Lake Iroquois was fed by waters flowing from Niagara Falls to the southwest, and more importantly from the Lake Huron basin to the northwest *via* the Kirkfield outlet and the Trent River system.

4.4.2.3 Glacial Lake Algonquin

As the glacier retreated northward, the ice-impounded Glacial Lake Algonquin developed in the Lake Huron, Georgian Bay, and Lake Michigan basins (Karrow and Calkin 1985; Larsen 1987; Karrow 1989; Teller 1995; Catto et al. 2020). Initially, Glacial Lake Algonquin drained to the Mississippi River, through the outlet at Chicago. Continued glacial retreat opened outlets for the upper Great Lakes which followed the ice front. These northern outlets, such as those at Kirkfield and (later) North Bay, were lower than the outlet at the south end of Lake Huron (Sarnia), and the south end of Lake Michigan (Chicago), because they were still isostatically depressed, due to the weight of glacial ice which had recently covered them. Isostatic rebound had simultaneously raised the levels of the Sarnia and Chicago outlets, terminating their use. The main period of drainage through the Kirkfield outlet into the Lake Ontario basin occurred between 12,500 and 11,700 years ago.

As a result, Glacial Lake Algonquin drained eastward, rather than southward into the Lake Erie basin. Niagara Falls was thus deprived of most of its water (Tinkler et al. 1994). The falls shrank to a relative trickle, and the rate of erosion at the lip, controlled by the volume of water passing over the Lockport dolomite, was extremely small. During the period between 13,000 and 8,000 years ago, when almost all of the drainage from Huron, Superior, and Michigan bypassed Niagara Falls, the falls and gorge remained at essentially the same position.

As the ice front retreated northeastwards, lower lake levels formed in the Lake Ontario basin as new outlets opened to the northeast. The water surface dropped in stages from the Lake Iroquois level to the Lake Frontenac positions (±90–110 m above modern lake level), the Lake Belleville position (±30 m above modern lake level), and finally, about 12,500 years ago, to levels ±45 m below that of the modern Lake Ontario, approximately 30 m above sea level.

After the Younger Dryas readvance 11,700 years ago (LaSalle and Shilts 1993; Occhietti 2007), renewed northward glacial retreat in the Lake Huron and Georgian Bay basins opened lower northern outlets for Glacial Lake Algonquin, initially through Algonquin Park (at Fossmill and Barron

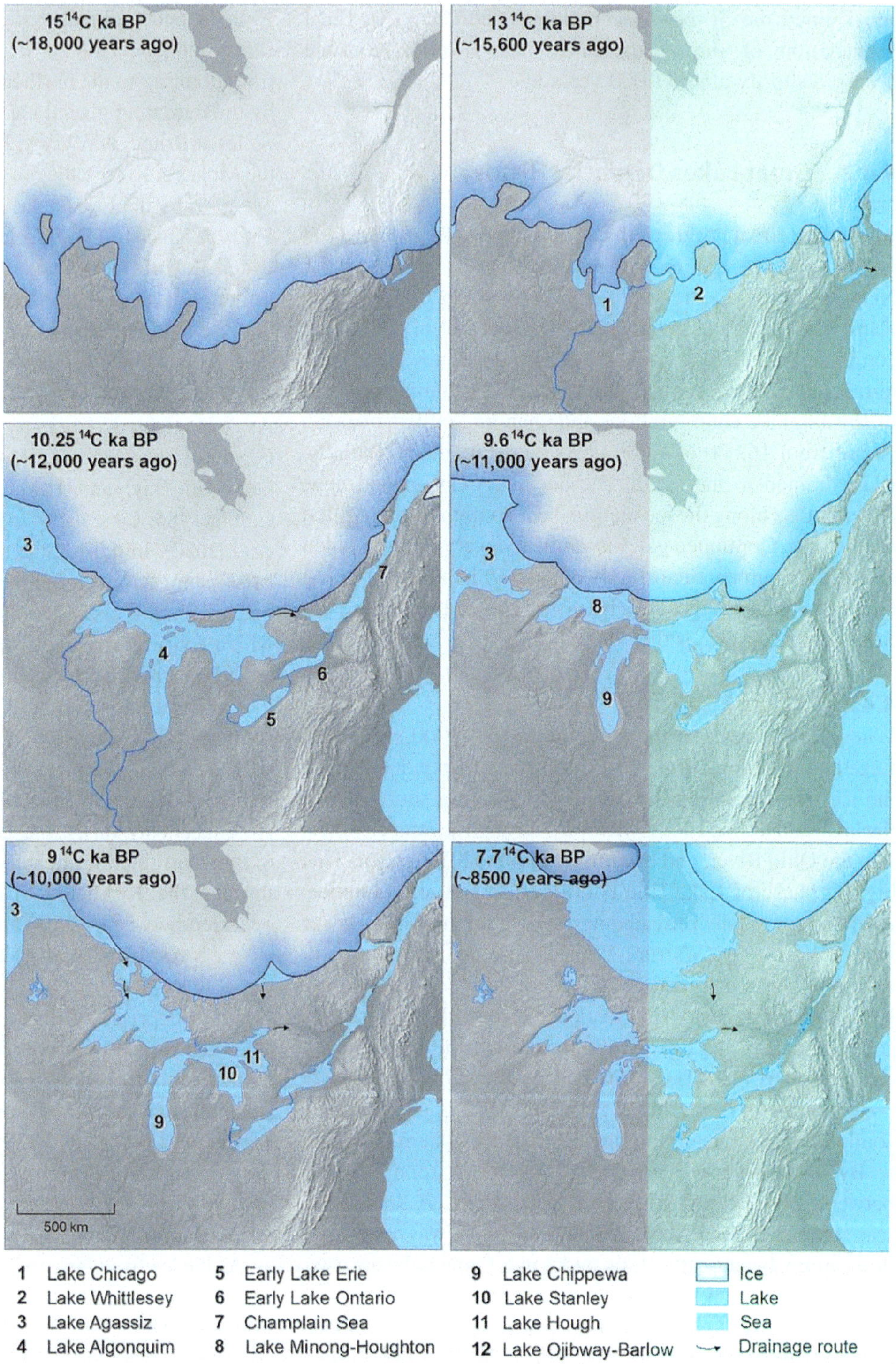

Fig. 4.37 Stages of development of the Great Lakes. Ice margins from Dyke et al. (2003). See text for discussion. (Slaymaker and Catto ch 2)

Canyon) and subsequently through North Bay, into the Ottawa Valley. Throughout these events, Glacial Lake Algonquin formed high-level shorelines along Lake Huron and southern Georgian Bay.

Diversion of the waters of the upper Great Lakes bypassed the Lake Ontario basin. The dry areas of the basin were subjected to frost action, and ice-wedge polygons developed around the margins of the reduced lake (Sly and Prior 2011). Although Lake Ontario had retreated to a small area in the deepest part of the basin, it continued to drain northeast into the St. Lawrence Valley. Recent research suggests that the Champlain Sea, which had formed to the northeast, may have extended as a tongue of water below the modern level of Lake Ontario, as far west as offshore of Hamilton. Water level began to rise in the Lake Ontario basin after 12,000 years ago. Low water levels persisted in Lake Erie.

4.4.2.4 North Bay Outlet

By 11,000 years ago, the North Bay Outlet and Mattawa River (Algonquian "meeting of waterways") carried the entire discharge of the Lake Huron, Georgian Bay, and Michigan basins, and the deglaciated part of Lake Superior (Karrow and Calkin 1985; Catto et al. 2020). The French River and Mattawa River Gorges formed part of the North Bay Outlet, connecting the Huron Basin with Lake Nipissing and the Ottawa River (Fig. 4.38). Today, the volume of water discharged at Sarnia from Lake Huron is 90–95% of the volume of the St. Lawrence at Cornwall (Fisheries and Environment Canada 1978). Consequently, when the North Bay outlet opened, the St. Lawrence was reduced to about 5–10% of its current volume, and the Upper Great Lakes waters were diverted to flow past Ottawa.

By 10,000 years ago, drainage through North Bay caused water levels to drop in the Huron and Michigan basins, and the Georgian Bay basin became isolated from the main body of Lake Huron. These low-level lakes, termed "Lake Hough" (Georgian Bay), "Lake Stanley" (Huron), and "Lake Chippewa" (Michigan), remained in the basins as the outlet at North Bay gradually rose. By approximately 8400 years ago, the water level in the Lake Huron Basin stood at 176 m above sea level (at the present elevation of the lake), and Georgian Bay and Lake Huron were joined together again. Additional water entered the North Bay Outlet from Lake Minong, in the deglaciated part of the Lake Superior basin, and for a short time from Lake Agassiz in northwesternmost Ontario (see Chap. 5).

Fig. 4.38 North Bay Outlet, a main drainage route from Glacial Lake Algonquin. Upper: Mattawa River Gorge; Lower: French River Gorge. These two channels formed part of the North Bay Outlet

Fig. 4.39 Niagara Falls (NASA)

Water levels continued to rise as the North Bay Outlet rebounded. By 8000 years ago, the water level in the Huron, Michigan, and Superior Basins stood at the same elevation, approximately 186 m above sea level. A single large lake, termed the "Nipissing Great Lakes", developed in the basins (Karrow and Calkin 1985; Larsen 1987; Catto et al. 2020). Shorelines developed during the Nipissing Great Lakes phase are present at approximately 10 m above the present level of the lake in Georgian Bay Islands and Bruce Peninsula National Parks. As the North Bay Outlet continued to rise isostatically, new outlets for the Nipissing Great Lakes developed at Chicago and Sarnia, and all three outlets were functioning simultaneously about 6000 years ago.

Lakes Erie and Ontario continued to rise throughout the Holocene, a process which was accelerated after the Niagara River again received the majority of the drainage of the Upper Great Lakes after 5500 years ago (Tinkler et al. 1994). By 6000 years ago, Niagara Falls was receiving approximately 40 % of its modern water discharge, with the remainder draining eastward from Lake Huron *via* North

Bay and southward from Lake Michigan *via* Chicago into the Mississippi River.

Eventually, the combination of isostatic recovery at North Bay and enhanced downcutting at Sarnia caused the gradual abandonment of the outlets at North Bay (*ca.* 5500 years ago) and Chicago (*ca.* 4500 years ago). The Lake Huron–Lake Michigan basin drained southward through Sarnia, and became isolated from the Lake Superior basin at Sault Ste. Marie about 3500 years ago. The closure of the competing outlets at 4500 years ago (North Bay) and 4500 years ago (Chicago) eventually forced all of Lake Huron's discharge to travel south past Sarnia, into Lake Erie and to Niagara Falls. By approximately 4500 years ago, the modern drainage pattern of the Great Lakes was completely established.

4.4.2.5 Niagara Falls

The modern Niagara Falls is <4500 years old (Tinkler et al. 1994; Fig. 4.39). Water has continued to course over the falls almost uninterrupted and undiverted, except by the requirements of hydroelectric power plants, for the past 4,500 years. On March 29, 1848, spring breakup in the Niagara River and strong winds combined to completely dam the river with broken ice floes. As a result, the supply of water to the falls downstream was cut off, and the falls were dry for approximately 30 h (Phillips 1993; Pope 2016). The present rapid rates of erosion of the lips of the falls, so severe as to necessitate the temporary diversion of water away from the American Falls in order to cap the escarpment with concrete, reflect only the activity of the past 4500 years.

Erosion by the Niagara River has resulted in a steady southward recession of the falls, as the river has cut a deep 8 km-long gorge extending from the escarpment margin at Queenston Heights to the current position of the falls. The total difference in elevation between the upper and lower plain here is approximately 100 m, of which the falls themselves account for 55–57 m, the remaining 45 m being represented by the descent through the Whirlpool Rapids of the Gorge. The Niagara River diverges upstream of the Falls around Goat Island, with the main body of water travelling over the Horseshoe or Canadian Falls to the west of the island, and a smaller volume of water travelling to the east of Goat Island and descending over the American Falls. The Canada-US border divides Horseshoe Falls.

The total discharge of the Niagara River entering Lake Ontario is approximately 8000 m^3 s^{-1} (Fisheries and Environment Canada 1978). The volume of water passing over the falls is considerably less than this figure, as most is used for hydroelectric power generation at the Adam Beck generating station of Ontario Hydro, and by US power stations. The amount of water diverted to each country's turbines is regulated by treaty, so that the "scenery" is not impaired during daylight hours. Often, the volumes of water diverted during the night greatly exceed the daytime totals. Although the idea of preservation in a national park came to North America too late to totally preserve the falls, the co-existence of power generation and conservation of scenery here represented a positive step. More than 96 % of the volume passing into Lake Ontario initially came from the upper Great Lakes *via* Sarnia and Windsor, so that diversion of this water would cause Niagara Falls to be reduced to a relative trickle.

4.4.3 Champlain Sea

About 13,000 years ago, deglaciation of the outlet of the St. Lawrence River at Québec City allowed marine waters to flood the St. Lawrence Valley (Occhietti 1989; Fulton 1995; Occhietti et al. 2001; Catto et al. 2020). This marine event, termed the Champlain Sea, resulted in the deposition of great thicknesses of silt and clay (locally exceeding 70 m) and numerous fossils of seals, whales, herring, molluscs, and other marine organisms (Fig. 4.40). The resulting Champlain Sea embayment of the Atlantic Ocean reached an elevation of 213 m above modern sea level in Gatineau Park, and blanketed most of the terrain below this elevation with marine silt and clay. The northwestern arm of the sea extended along the Ottawa Valley to Chalk River, Ontario (Catto et al. 1981). The southwestern arm extended along the St. Lawrence southwest into the Lake Ontario Basin to offshore of Hamilton (Lewis and Todd 2019).

As the land gradually rebounded from the weight of the glacial ice, the Champlain Sea was forced to contract from northwest to northeast. Simultaneously, areas in the southwest (which were deglaciated first) rebounded, forcing the sea to retreat towards Québec and causing the future tributary streams of the Ottawa River to flow north, down the newly developing slope and away from the St. Lawrence River (Catto 2015).

By approximately 10,000 years ago, the combined effects of rebound at Ottawa and the recovery of the outlet at Québec City had reduced the former Champlain Sea to a small body of water (elevation 75–80 m a.s.l.) which was isolated from the Gulf of St. Lawrence. This Lampsilis Lake (named after a fossil mollusc found in its sediments) persisted in the Ottawa and St. Lawrence Valleys for another 1000 years, until continued rebound forced the lake to dissipate completely. By 9000 years ago, the ancestral Ottawa and St. Lawrence Rivers flowed through the valleys.

As the Champlain Sea withdrew and shrank in the Ottawa Valley, the Ottawa River extended into the basin from the west. By 11,000 years ago, the Ottawa Valley was receiving waters from the upper Great Lakes *via* North Bay. The Ottawa River expanded greatly as a result, carving the additional channels south of the current stream that would eventually be occupied by the Mer Bleue and other peat bogs.

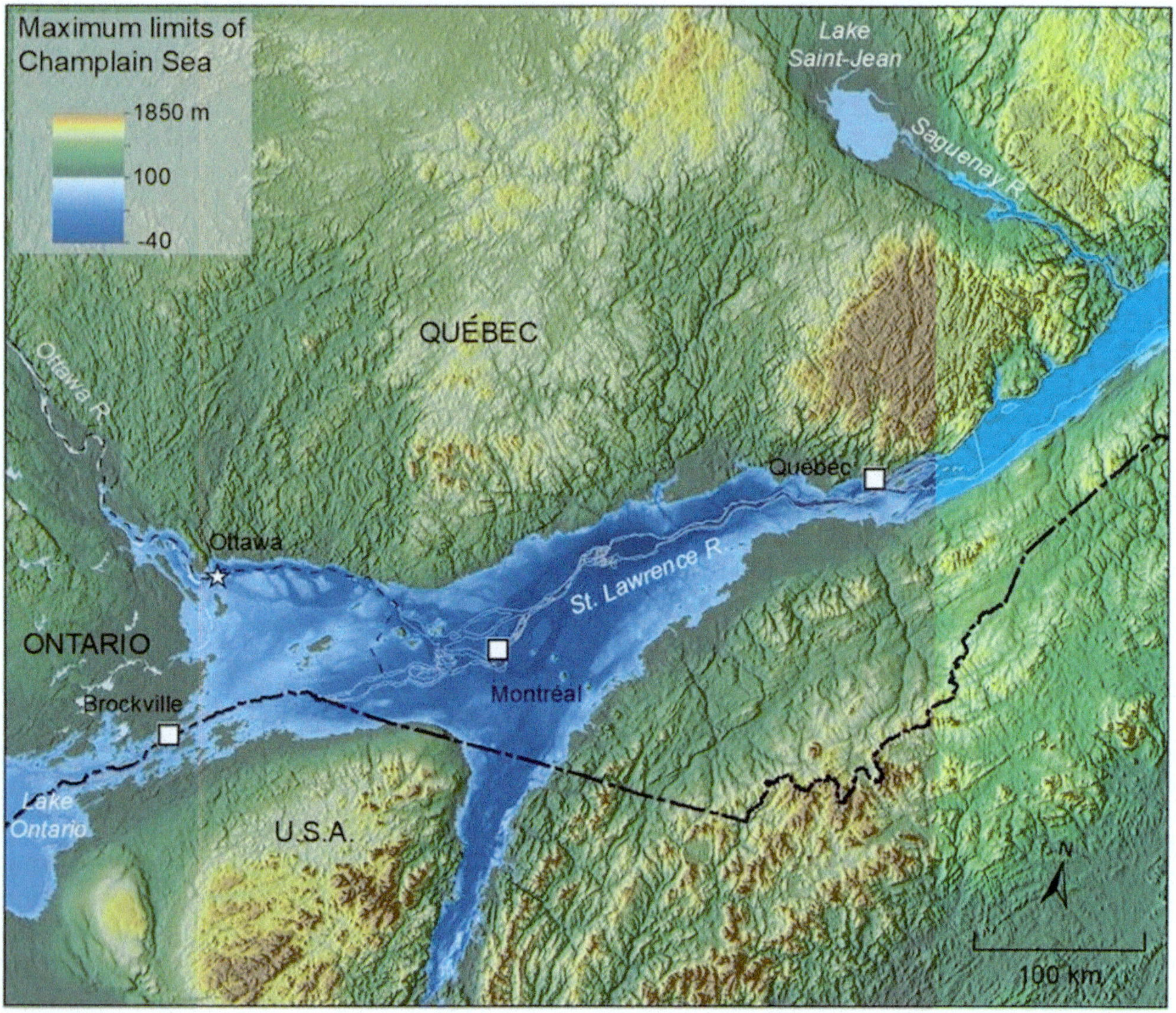

Fig. 4.40 The Champlain Sea occupied the glaci-isostatically depressed land of the St. Lawrence Lowland between Québec City and Brockville, and extended into Lake Ontario, the Ottawa River valley and the Saguenay Fjord from 13,000 to 10,000 years ago. The modern Ottawa River evolved as the ancestral river and its tributaries adjusted to the retreat of the Champlain Sea. (Slaymaker and Catto 2020)

Drainage through the North Bay Outlet continued throughout the early and mid-Holocene. As North Bay gradually rose isostatically, the volume of water passing through the outlet to the Ottawa River was steadily reduced. Final closure of the North Bay Outlet to Great Lakes drainage isolated Lake Nipissing and the French River from the Ottawa Valley and caused the Ottawa River to recede to its present position and volume of discharge.

Together with numerous fossils of seals, whales, herring, molluscs, and other marine organisms, the Champlain Sea deposited thicknesses of marine and brackish water silt and clay (commonly termed "Leda clay"), locally exceeding 70 m. *Leda glacialis* is a fossil mollusc. Initially, this fossil was identified in the clay, but subsequent work demonstrated that the shells represented a different species, *Portlandia arctica*. Thus, the sediment is named after a fossil that is not present in it.

4.4.3.1 Champlain Sea Clay Slope Failures

Champlain Sea clay is a highly unstable material, maintaining its strength only when undisturbed or dry (Lefebvre 2017). When disturbed by a sudden shock (earthquake or blasting) or when subjected to heavy rainfall, the clay fails and liquefies, causing rotational slides and rapid mud- and debris flows. Scars showing the locations of past slope failures are present along many streams, and any riverbank exposure of the sediment is prone to failure (Fig. 4.41).

Fig. 4.41 Slope failures in glacimarine sediments. Upper: St-Jean-Vianney. Lower: Site of Lemieux. See text for discussion (Geological Survey of Canada, Natural Resources Canada)

The first recorded fatal mass movement in Canada occurred on 29 Sept 1771, when Champlain Sea clay failed at St-Pierre-de-Montmagny, Québec. At Notre-Dame-de-la-Salette, a rotational slide along a cutbank blocked the Lièvre River, flooding the town and drowning 34 people in 1908 (Locat et al. 2017a). Another rotational slide in Champlain Sea clay killed four people in St-Jude, Québec, on 11 May 2010 (Locat et al. 2017b).

At St-Jean-Vianney, Québec, a rotational slide destroyed a community built on a cutbank overlooking the Saguenay River along a cutbank, killing 31 people on 4 May 1971 (Evans 2001; Nutz et al. 2020). The failure resulted from a combination of heavy rains, river undercutting, and slope overloading due to construction. Following the disaster, the remaining homes were abandoned, and the former community designated as a memorial site. After 1971, communities throughout Québec and eastern Ontario prevented new construction along cutbanks. Repetitive "fingerprint" rotational slides were identified in many areas of eastern Ontario.

The setting of Lemieux, Ontario, resembled that of St-Jean-Vianney, and a previous rotational slide had blocked the adjacent South Nation River. Consequently, the Geological Survey of Canada recommended the abandonment of Lemieux. All buildings were purchased by the Government of Ontario, residents relocated and compensated, and Lemieux was abandoned in 1991. On 20 June 1993, the abandoned Lemieux was completely destroyed when a rotational slide and flow displaced 2.8 million m^3 of Champlain Sea clay into the South Nation River. The relocation of Lemieux represents a proactive mitigation response to a mass movement hazard (Brooks et al.1994; Lawrence et al. 1996).

Champlain Sea Clay can also fail if heavy structures are placed upon it without proper re-enforcement. One of the most embarrassing architectural incidents involved the Victoria Memorial Museum, currently occupied by the National Museum of Natural History and constructed as the headquarters of the Geological Survey of Canada, that opened in 1911. This building, constructed on infilled land overlying Champlain Sea Clay, was designed with imposing outer walls of thick Nepean sandstone and a ten-storey high frontal tower. Even before the official opening, the outer walls began to sink into the clay, and the tower commenced leaning backwards over the body of the structure. The "leaning tower of Ottawa" was pulled down in 1915, and a series of repairs begun on the structure of the building, a process which has continued periodically ever since.

4.4.4 Gulf of St. Lawrence Postglacial Marine Events

Along the Gulf of St. Lawrence, glaci-isostatic depression allowed marine waters to cover the coastal plains following deglaciation. Sea level during this Goldthwait Sea phase rose to ±215 m a.s.l., depositing marine clays inland along the embayments and river valleys of the mainland shore, and forming marine deltas (Brouard et al. 2016; Dietrich et al. 2020). As isostatic rebound progressed, the sea gradually receded from the land, and the archipelago began to develop. The erosional features evident in the limestone cliffs were developed under conditions of modern sea levels, and relict features reflecting former marine shorelines are present at higher elevations.

A glacimarine arm of the Gulf of St. Lawrence extended along the Saguenay River to 50 km northwest of the modern north shore of Lac-St.-Jean, forming Golfe Laflamme (Nutz et al. 2020; Fig. 4.42). After its initial formation

Fig. 4.42 Lac-St-Jean and upper Saguenay River. Golfe Laflamme covered all of the terrain currently used for agriculture. (NASA). Inset: agricultural terrain in Saguenay. (Wikimedia/Adqproductions)

about 13,000 years ago, rapid glaci-isostatic uplift caused Golfe Laflamme to recede. Lac-St.-Jean formed when the Kenogami bedrock sill emerged at its southeastern margin, about 9000 years ago.

Deglaciation of the Newfoundland Gulf of St. Lawrence coastline proceeded between *ca.* 13,500 and *ca.* 12,000 years ago. Glaci-isostatic depression resulted in marine incursion, with elevations increasing from south to north, from 135 m above modern sea level at St. Barbe, 140 m at Watts Point, and approximately 150 m above modern sea level at Burnt Cape Ecological Reserve (Catto 2020). Isostatic recovery resulted in consistent sea level fall until ca. 2000 years ago, followed by slow rise.

4.5 Vegetation

4.5.1 Mixedwood Forest

The Mixedwood Forest not only includes both deciduous and coniferous trees but represents a transition between the temperate forests to the south in the USA and the Boreal Shield forests (see Chap. 3) to the north and east (Rowe 1972; Elliott 1998; Gagnon 2004; CNVC 2016; Ministère des Forêts 2018). Although much of the Mixedwood Forest has been removed through agricultural clearance and urbanization, new species have been introduced, and climate change has altered species ranges, some indications of the natural assemblages still are evident (e.g. Butt et al. 2016; Foret Hereford 2015; Terrail et al. 2019; Danneyrolles et al. 2021).

Typical mid-boreal deciduous species include *Acer rubrum* (red maple), *Betula papyrifera* (white birch), *Populus tremuloides* (trembling aspen), *Populus balsamea* Balsam poplar), *Alnus incana rugose* (green alder), and *Crataegus* (hawthorn). Conifers include *Abies balsamea* (balsam fir), *Picea mariana* (black spruce), *Picea glauca* (white spruce), *Pinus banksiana* (jack pine), *Pinus resinosa* (red pine), *Pinus strobus* (eastern white pine), and *Thuja occidentalis* (eastern white cedar).

Southern boreal deciduous species include *Fagus grandifolia* (beech), *Betula alleghaniensis* (yellow birch), *Tilia americana* (basswood), *Quercus rubra* (northern red oak), *Acer saccharum* (sugar maple), *Prunus virginiana* (choke cherry), and formerly *Ulmus americanus* (elm). Beech requires undisturbed conditions, and forms relatively pure stands with dense canopies of leaves, shading other species (Hosie 1979; Kershaw 2001). In Southern Ontario, individual beech trees may live in excess of 200 years, but pure beech stands are relatively rare. The Haudenosaunee (Iroquois), Wyandot (Huron), and Ojibwe people utilized beech (*gawe'mîc*), including nuts, leaves, and bark, and may have aided in the preferential spread of beech (Kuhnlein and Turner 1991). Beech is not resistant to fire. Expansion of sugar maple has had negative impacts on beech (St-Jean et al. 2021)

4.5.1.1 *Pinus strobus* (Eastern White Pine)

Prior to 1750, the dominant tree throughout much of eastern North America was *Pinus strobus* (eastern white pine; provincial tree of Ontario), which locally represented more than 60 % of some stands (Fig. 4.43). It prefers well-drained, well-developed sandy podzol soil (Hosie 1979; Kershaw 2001). *Pinus strobus* was spiritually significant to the indigenous Algonquian, Haudenosaunee, and Mi'kmaq peoples. Periodic burning removed the understorey to favour deer populations, facilitating growth of white pine seedlings.

Following European settlement, eastern white pine rapidly became the most valuable commercial species (e.g. Gillis 1975; Claveau and Laframboise 1983; Gaudreau 1999; Dick et al. 2014). It was preferentially targeted for logging, due to its straight trunk (more than 30 m high for a 200-year old tree), ideal for construction of ship's masts. White pine floats well and is water resistant. As a "soft" pine, white pine is a very easy wood to work with, easy to shape and polish, producing no splinters or shards, flexible and resilient, not susceptible to warping, and can be used for most woodworking purposes.

Commercial exploitation of eastern white pine in eastern North America began in the 1720s. In eastern Canada, exploitation accelerated after 1795 when the supply of Baltic timber to Britain was interrupted by the Napoleonic Wars, combined with the decline in USA and Maritime Canadian supplies. Consequently, progressive northwestward—moving logging from USA to the boreal Canadian Shield virtually eliminated throughout much of eastern Canada in the 19th century. *Pinus strobus* is also very intolerant of acidic precipitation.

The effects of clearance of the eastern white pine forests are well-documented in the Ottawa Valley. Ottawa was founded as a logging town, and logging of white pines continued until the last major stands in the headwaters of the Ottawa River and its tributaries were cut in the 1890s (Dick et al. 2014). Other species which were logged preferentially were *Fagus grandifolia* (beech) and *Quercus rubra* (northern red oak). Removal of northern red oak was a significant factor in the extinction of the Passenger Pigeon (*Ectopistes migratorius*) (Novak et al. 2018).

Removal of *Pinus strobus* allowed other species to occupy the cleared ground, such as *Acer rubrum* (red maple), *Acer saccharum* (sugar maple), *Betula papyrifera* (white birch), *Pinus resinosa* (red pine), and *Populus tremuloides* (trembling aspen). These species, together with a mix of southern and northern trees such as *Thuja occidentalis* (white cedar), *Betula alleghaniensis* (yellow birch), *Tilia americana* (bass-

Fig. 4.43 *Pinus strobus* (eastern white pine). Left: foliage; Right: trunk. See text for discussion

wood), *Abies balsamea* (balsam fir), and *Picea glauca* (white spruce), form the present tree cover of the Ottawa Valley (CNVC 2016), Drobyshev et al. (2014)

One species which initially benefited from the clearance of white pine, beech, and red oak was the American White Elm, *Ulmus americana*. Free from the interference of the large shade trees and the dense white pine stands, elms expanded rapidly, forming almost pure stands in some areas (Kershaw 2001). White elm wood was the preferred material for the manufacture of hockey sticks until the early 1960s. Elms were also widely planted as ornamental trees, especially during the period prior to and shortly after World War I as Ottawa struggled to make the transition from rollicking lumber town to national capital. Elms were preferred over beech and oak, as they grew faster than either, provided more shade than oak, and were thought to be more symmetrical (and more elegant) than beech.

Unfortunately, North American elms were totally unexposed naturally to the chief natural elm pathogen, the fungus *Ceratocystis ulmi*. Accidently introduced from Europe after 1930, the "Dutch Elm Disease" quickly travelled from elm to elm by riding on the elm bark beetle. The fungi and beetle are specific to elms, and in a natural forest assemblage, the presence of other tree species would hinder the spread of the beetle and fungus. The planted elm stands and those which developed on cleared land, however, were dominated by elms, and formed perfect environments for the spread of the fungus. The vast majority of the elms planted in Ottawa and throughout Ontario have succumbed to the disease. Efforts continue to assist elm recovery (Elm Recovery Project n.d.).

4.5.1.2 Sugar Maple (*Acer saccharum*)

Another beneficiary of the removal of eastern white pine was the Sugar Maple (*Acer saccharum*; Fig. 4.44). Prior to 1800, sugar maples were subordinate components of the Mixedwood forest, hindered from expansion by pine forests, forest fires, and high deer populations. Removal of the pines and white-tail deer, reduced fire frequency, and agricultural development all favoured sugar maple (e.g. Puric-Mladenovic 2004; 2011; Bannon et al. 2015; Terrail et al.

Fig. 4.44 *Acer saccharum* (sugar maple). See text for discussion

2019; Copenheaver et al. 2020; St-Jean et al. 2021). The open terrain allowed enhanced distribution of maple pollen by wind (Roussy and Kavan 2014), as maples tend to have infrequent, unpredictable rapid mass flowering early in spring. The bee and insect populations critical for pollen distribution also increased with agriculture. Maple is faster-growing than beech, oak, or ash and is also more tolerant of the lesser-nutrient podzols and brunisols that previously supported pines. In consequence, "pure" maple stands began to form in the 1800s, eventually leading to the establishment of "sugar bushes" suitable for commercial maple syrup production. Currently, sugar maple is the dominant component of the Mixedwood forests of Southern Ontario and Québec (Kershaw 2001, CNVC 2016).

Sugar maple's enhanced visibility in the physical landscape, especially in autumn, led to its increased visibility culturally. First used as a symbol in the early 1800s, sugar maple was adopted by Lower Canada (Quebec) and Upper Canada (Ontario) patriots in 1837. By 1848, it was coupled with the beaver as a Canadian symbol, was officially recognized 1859 by the Prince of Wales, appeared on the "Red Ensign" flag ca. 1890, and on the national Coat of Arms in 1921 (Swan 1977; Canadian Heritage n.d.).

The present northern limit of *Acer saccharum* coincides with the northern boundary of Canada in 1867 (in addition to PEI). Although its range is expanding northwards with climate change, the political extent of Canada has expanded much more rapidly. Today, the range of *Acer saccharum* represents about 8% of the area of Canada. Locally, sugar maple is now under some pressure from the invasive Norway Maple (*Acer platanoides*), introduced from New England prior to 1800 (Ontario Invasive Plant Council 2021). Climate change represents an additional stressor (Moreau et al. 2020). *Acer saccharum* has thus both been assisted and harmed by human activity.

Sugar maple forests provide regulating, provisioning, and cultural ecosystem services (DePratto and Kraus 2017). A 2013 study (Group AGÉCO 2013) estimated the value of the regulating services, the ability of ecosystems to maintain conditions conducive to human life, provided by sugar maples, including water filtration, reduction of erosion, and sequestration of GHG emissions to exceed $1.9 billion annually. GHG sequestration is estimated to represent ~25% of vehicle emissions in Québec. Provisioning services, direct services and materials provided by maple forests, including food production, medicinal products, and ornamentals, represents $520 million annually. Intangible cultural benefits, including leisure, tourism, and aesthetic and cultural values related to education and heritage total $326 million annually.

4.5.1.3 Trillium

The provincial flower of Ontario is the white trillium (*Trillium grandiflorum*; Fig. 4.45). This brilliant three-petalled flower requires shaded woodland conditions and humus-rich soil, preferably fed by both pine duff and deciduous leaf mould (Case and Case 1997). These requirements are met in the complex forests developed in the area, and hence this was one of the most productive trillium habitats in the province, both for *Trillium grandiflorum* and for its red-tinged relative, *Trillium erectum* (wake-robin). Clearance of the original forests for agriculture, and the re-colonization of abandoned areas by pioneer species such as *Populus tremuloides* (aspen poplar), has led to a loss of trillium habitat. *Trillium grandiflorum* is a difficult species to transplant successfully.

4.5.2 Southern Boreal Deciduous Forest

The more southerly boreal areas along Lake Ontario and southwestern Ontario south of Lake Simcoe and central Lake Huron supported an assemblage dominated by deciduous trees (Rowe 1972; Elliot 1998; CNVC 2016; Fig. 4.46). Species include *Fraxinus americana*, *Fraxinus pennsylvanica* (red ash), *Fraxinus pennsylvanica austini* (northern red ash), *Carya glabra* (red hickory), *Hamamelis virginiana* (witch-hazel), *Carpinus caroliniana* (hornbeam), *Ostrya caroliniana* (ironwood), *Rhus typhina* (staghorn sumac), *Populus grandidentata* (largetooth aspen), *Sambucus pubens* (eastern red elderberry), *Acer nigrum* (black maple), *A. saccharinum* (silver maple), *Quercus alba* (white oak), *Ulmus thomasii*

Fig. 4.45 *Trillium grandifolium*

Fig. 4.46 Southern boreal deciduous trees include *Fraxinus americana*, *Fraxinus americana* (white ash), *Acer saccharinum* (silver maple), and *Quercus alba* (white oak). Lower left: *Juglans nigra* (black walnut) foliage. Lower right: *Ulmus americana* (white elm) foliage

(rock elm), *Quercus bicolor* (swamp white oak), and *Toxicodendron vernix* (poison sumac). Most of these species are at their northern limits, and many can only survive in the most sheltered areas (Hosie 1979; Kershaw 2001). Losses in forest cover have occurred due to introduced pest species such as Asian Long-horned beetle, which attacks maples (*Acer sp.*), and Emerald Ash Borer (*Agrilus planipennis)*, which attacks white ash (*Fraxinus americana*). The Emerald Ash Borer, introduced near Detroit (2002), has killed more than 50 million White Ash in eastern North America (Knight et al. 2013).

Prunus serotina (black cherry), *Juglans nigra* (black walnut), *Juglans cinerea* (butternut), *Carya cordiformis* (bitternut hickory), *Carya ovata* (shagbark hickory), and *Ulmus rubra* (slippery elm) can be used as food sources and for medicines (Kuhnlein and Turner 1991). *Carya ovata* stands near Wiarton and Owen Sound may represent descendants of those intentionally established by the Tionontati (Petun) First Nation.

Black walnut and butternut prefer well-drained, fertile, calcareous melanic brunisols, in areas with minimal shade cover from overhanging vegetation (Hosie 1979; Kershaw 2001). They characteristically colonize abandoned beaches formed by previous high stands of the glacial lakes. Both species release juglone, a toxin, from fallen leaves and roots, which acts to inhibit the growth of other trees. Utilization of walnuts and butternuts by the Haudenosaunee allowed these trees to spread, as did the practice of periodic burning to accommodate agriculture and to provide deer runs. *Platanus occidentalis* (sycamore), *Quercus nigra* (black oak), and *Ulmus rubra* (slippery elm) prefer growing on poorly drained clayey soils, and are shade-tolerant. These species occupy the former lake-bottom areas between the ridges. In large quantities, *Quercus nigra* acorns are poisonous to livestock.

Some trees reach their northern limits in the Niagara area. *Betula lenta* (cherry birch) is among the rarest trees in Canada, being represented by one grove at Port Dalhousie, near St.

Catharines (Hosie 1979; Zoladeski and Hayes 2013). The presence of this species only in this small area of the lower plain attests to a temperate microclimatic environment at Port Dalhousie. However, as more than half of the specimens in this single grove died between 1967 and 2005, the tree remains under extreme pressure from shoreline erosion along Lake Ontario. Cherry birch has not been successful at colonizing additional areas, possibly due to the lack of genetic diversity in this very disjunct population. The nearest wild population is located 70 km to the east (against the prevailing wind direction), in New York state.

4.5.2.1 *Castanea dentata* (chestnut)

A species which was formerly abundant in the region is *Castanea dentata* (chestnut; not to be confused with the ornamental horse-chestnut or Ohio buckeye (*Aesculus hippocastanum*), frequently planted in cities and native to Walpole Island in Lake St. Clair). This broad-leaved, graceful deciduous tree was a co-dominant in the region, locally outnumbering all other species, especially on well-drained sites (Fig. 4.47). Mature chestnut trees were up to 35 m high and 100 cm in trunk diameter. It was generally not harvested for wood, as its nuts were considered a valuable food source and its wood inferior to oak for construction or burning (Hosie 1979; Kershaw 2001).

Unfortunately, as with its co-inhabitant *Ulmus americana*, the chestnut was unexposed to its chief European pathogen until a fungus was introduced in the 1920s from New York State, having spread from the initial accidental introduction into North America in 1904 from China (Anagnostakis and Hillman 1992; Boland et al. 2012). The "chestnut blight", marked by the development of bulbous orange cankers in the cambrium of the tree, is caused by the fungus *Endothia parasitica.* Unlike the elm blight, *Endothia parasitica* is not specific to chestnuts, and can be spread by a variety of insects, on bird's feet, and even by splashing raindrops. Consequently,

Fig. 4.47 *Castanea dentata* (chestnut). Formerly a dominant species, chestnut saplings (foreground left) are killed by "chestnut blight" before reaching maturity. Top right: foliage. Bottom right: bulbous orange cankers caused by the fungus *Endothia parasitica.* (*J. O'Brien, United States Forestry Service*)

it did not depend on encountering pure stands of chestnuts for its spreading and propagation. Between 1925 and 1950, virtually every chestnut tree in southern Ontario died, and no new saplings survived. Initial attempts to develop fungicides failed completely (there still is no suitable fungicide), and little optimism remained in the late 1950s that the species could return.

At present, chestnut trees attacked by the blight grow as saplings to approximately 3 m height and 10 cm in diameter. When they reach this size, the blight causes death in virtually all instances. As the chestnut reproduces vegetatively by sprouting from a stump, in addition to producing pollen, trees need not survive to maturity to generate a new crop of saplings. However, mature trees are required to enable the species to spread. Thus, although chestnut saplings are found over much of its former range, mature trees are extremely rare. Unable to develop an effective fungicide, some chestnut researchers have focused on the prospect of creating a competing fungus, which would replace *Endothia parasitica* and out-compete it, while simultaneously being less virulent to the chestnuts. Progress in this endeavour has been slow.

Attempts to develop blight-resistant chestnut trees involve "backcrossing", introducing genetic material from related chestnut species native to China, which have evolved along with the blight and developed some evolutionary immunity. Progress requires multiple backcrossing efforts, harvesting the chestnuts produced by each tree, and then introducing more genetic material.

4.5.2.2 Coniferous Species

Coniferous species, including *Juniperus virginiana* (eastern red cedar), *Thuja occidentalis, Pinus strobus*, and *Abies balsamea*, are subordinate in most areas. *Thuja occidentalis* flourishes in poorly drained, calcareous areas. In Bruce Peninsula National Park, old-growth *Thuja occidentalis* are found in cliff-top settings over poorly developed highly calcareous soils (and locally, growing directly on limestone). The oldest, stunted cedars are more than 700 years in age, and survived the nineteenth century logging blitz due to their cliff-top positions and their gnarled trunks (Kelly and Larson 1997, 2007). Today, the cliff-top cedars are under threat from tourists who come to admire them, and inadvertently destroy the thin soils and trample and compress the near-surface roots.

4.5.3 Carolinan Forest

Point Pelee's climate allows a temperate forest to survive which resembles that of North Carolina, commonly referred to as the "Carolinian Forest" (Rowe 1972; Allen et al. 1990, Waldron 2003). More technically, it is referred to as the Eastern Deciduous Forest by botanists (CNVC 2016). The small area which the forest is able to occupy, however, is a severe handicap, and many of its species are endangered or threatened in Canada (although large populations exist south of Lake Erie).

The Eastern Deciduous Forest lacks conifers except for *Pinus strobus* (eastern white pine), which survives only in areas of open, sandy soil which is stable and not subject to wave reworking or the formation of dunes. Originally, white pines were present in coastal areas along Lake Erie, but they have not recolonized the terrain subsequent to logging. Today, white pines grow only where the deciduous species cannot become established. Species common in the boreal forest, such as *Pinus banksiana* (jack pine), *Picea glauca* (white spruce), *Picea mariana* (black spruce), *Acer rubrum* (red maple), *Sorbus decora* (mountain ash), and *Betula papyrifera* (white birch) are absent.

In addition to the broad-leaved deciduous trees found elsewhere in southwestern Ontario, such as *Platanus occidentalis* (sycamore), *Carya cordiformis* (bitternut hickory), *Fagus grandifolia* (beech), *Ulmus rubra* (slippery elm), *Acer saccharinum* (silver maple), and *Quercus nigra* (black oak) among others, the Eastern Deciduous Forest contains a number of unusual species (OMNR 2011; Figs. 4.48 and 4.49). Although these trees all occur together in a fully developed Carolinian forest, many are represented by only a few stands of individuals in southwesternmost Ontario, due to habitat destruction, removal by cutting, and the various microclimates operating at the northern fringes of their respective ranges.

The most common tree (or large shrub) is hackberry, *Celtis occidentalis*, which reaches a maximum height of 15 m but is generally smaller at Point Pelee. The crown tends to grow with irregular masses of fine twigs, called "witches" brooms. It is a shade-tolerant species, which grows on a variety of soils (Kershaw 2001) Under natural circumstances, hackberry would be an understory shrub in a dense deciduous forest, and would not be able to compete with white pines on open sandy sites. Removal of many of the competing trees, however, has opened sites to colonization by hackberry.

Several of the other trees, such as *Asimina triloba* (pawpaw), *Sassafras albidum* (sassafras), *Malus coronaria* (native crabapple), and *Gleditsia triacanthos* (honey locust), were important as food sources for animals in the region, and also proved to be of medicinal value to indigenous residents and newly arriving settlers (Kuhnlein and Turner 1991). *Ptelea trifoliata* (hop-tree), found growing on exposed beach sites (OMNR 2011; Parks Canada 2012), provided berries which could be used as a substitute for hops in beer manufacture. Species favoured by wildlife uninterested in alcohol, such as pawpaw, sassafras, and *Nyssa sylvatica* (black gum, also called tupelo), tended to live in moist bottomland areas or on the fringes of swampy ground, and hence were not

Fig. 4.48 Eastern deciduous (Carolinian) Forest trees. Left: *Acer-Carya-Quercus-Platanus* assemblage. Upper right: *Quercus nigra* (black oak) and *Acer saccharinum* (silver maple) are common. Lower right: *Liriodendron tulipifera* (tulip-tree)

greatly disturbed by the initial 19th century wave of European settlers. Some species, including *Carya laciniosa* (shellbark hickory), *Carya cordiformis* (bitternut hickory), *Ulmus rubra* (slippery elm), *Prunus nigra* (Canada plum), *Prunus americana* (American Plum), and *Prunus serotina (*black cherry), appear to have been deliberately cultivated and encouraged by the indigenous Caldwell First Nation (Anishinaabek).

The Eastern Deciduous Forest species also include other trees unique to southernmost Ontario (Kershaw 2001; OMNR 2011; Fig. 4.49). Although not all species grow within Point Pelee National Park, due to its small size, the deciduous trees formerly covered all of the plains bordering Lake Erie. Species include *Morus rubra* (red mulberry), *Liriodendron tulipifera* (tulip-tree), *Cercis canadensis* (redbud; only on Pelee Island), and *Cornus florida* (flowering dogwood tree). Also present are *Enonymus atropurpureus* (Burning bush), *Populus deltoides deltoides* (broad-leaved eastern cottonwood), *Viburnum lentago* (blackhaw), *Fraxinus quadrangulata* (blue ash), *F. profunda* (pumpkin ash), *Quercus ellipsoidalis* (northern pin oak), *Q. palustris* (pin oak), *Q. bicolor* (swamp white oak), *Q. muelenbergii* (chiquapin oak), *Q. shumardi* (shumard oak), and *Zanthoxylum americanium* (prickly ash, also called toothache tree).

Several variants of *Magnolia*, including the very rare cucumber-tree (*Magnolia acuminata*) added pleasant contrasts and aromas to the landscape. These trees formed a dense broad-leaved forest cover, with canopy levels ranging from 15 to 25 m above the forest floor, and prevented understory vegetation from being established in many areas (Environment Canada 2015).

Gymnocladus dioicius (Kentucky coffee-tree) also grows in the area. A shade-intolerant species, reaching a maximum height of 35 m, it prefers woodland fringe areas and often grew in proximity to the hop-tree. Although some early (perhaps desperate) settlers roasted its berries to make a substitute for coffee, the berries are toxic to humans in large quantities. Animals seldom consume any part of the Kentucky coffee-tree. Trees have limited seed production, with male and female

Fig. 4.49 Eastern deciduous (Carolinian) Forest plants. Upper left: *Morus rubra* (red mulberry). Upper right: *Nyssa sylvatica* (black gum). Lower Left: *Celtis occidentalis* (hackberry). Lower right: *Magnolia*

flowers on separate trees, with seeds dispersed by floatation in floodwaters. Many populations in Ontario are unisexual and rely on vegetative reproduction.

The Kentucky coffee-tree is currently threatened in southern Ontario through a combination of habitat loss, clearance of land for agriculture, and flood-control measures. In addition, the arrival of the Double crested Cormorant (*Nannopterum auritum*) has impacted the trees, and the bird droppings are toxic (Hebert et al. 2014; OMNR 2017).

In isolated wetland areas, "swamp forests" have developed where a combination of elevated water tables (often resulting from human modification of drainage patterns) and Dutch Elm Disease have killed the groves of *Ulmus americana, Ulmus rubra,* and *Ulmus thomasii* that originally were present (Fig. 4.50). These areas also commonly support sycamore, silver maple, pin oak, swamp white oak, and chinquapin oak, which are immune from Dutch Elm Disease and are more tolerant of elevated water tables once they reach maturity (Parks Canada 2016b; Canadian Wildlife Service 2001). The dead trees of the "swamp forest" have formed ideal habitats for tree swallows, woodpeckers, great horned owls, and a variety of warblers, including the Prothonotary Warbler, *Protonotaria citrea.*

Removal of the Eastern Deciduous Forest assemblage during agricultural clearance was essentially complete across most of southwestern Ontario *ca.* 1900, with only isolated groves remaining. Similar habitat destruction threatens several understory plants of the Carolinian forest with extinction in Canada (Lamb and Rhynard 1994), including *Opuntia humifusa* (eastern prickly pear cactus, found only at Point Pelee and Fish Point provincial nature reserve); *Isotria verticillata* (large whorled pegonia, found at only 2 localities, and not observed since 1996); *Polygala incarnata* (pink milkwort, found at only two localities); and *Isotria medeoloides* (small whorled pegonia orchid, found only at Catfish Creek Swamp, and not observed since 1998).

Further problems resulted from the arrival of invasive, non-native species, especially purple loosestrife (*Lythrium sali-*

Fig. 4.50 Swamp forests developed over elevated water tables are dominated by silver maple and oaks

caria) in wetland areas. Of the 70 species of vascular plants currently growing within the boundaries of Pelee National Park, 40% are non-native invaders. The tree assemblage survived in a more intact condition on Pelee and Middle Islands, due to isolation. In Point Pelee National Park, the sandy sediments of the spit precluded the establishment of these trees. Consequently, the spit was dominated by *Pinus strobus,* as are several other sandy spits extending into Lake Erie (such as Point Rondeau and Long Point).

4.5.4 Exposed Limestone Assemblages

The limestones of Mingan Archipelago and the Gulf of St. Lawrence coast of Newfoundland support southern boreal, mid-boreal, taiga, and arctic species (Fig. 4.51). *Cypripedium calceolus* (flat-petalled yellow lady's-slipper) and *Erysimum inconspicuum* (small-flowered rocket) are both found in moorland coastal areas (Couillard et al. 1987). The bouldery beach inhabitant *Taraxacum laurentianum* (Laurentian dandelion) is also present in Mangan, and unlike more common dandelions elsewhere in Canada, is a rare and preserved species. *Erysimum inconspicuum coarctatum* (small-flowered rocket) is also unique to Mingan. Exposed coastal areas also support halophytes such as Oyster leaf (*Mertensia maritima*) and Seabeach Senecio (*Senecio pseudo-arnica*).

Several arctic and taiga species reach their southern limits in the Mingan Archipelago, including the cliff-dwelling *Cystopteris montana* (mountain bladder-fern), and limestone plateaux species such as *Primula egaliksensis* (Greenland primrose) and *Cerastium alpinum* (mouse-eared chickweed). These species, along with boreal lime-loving plants such as *Asplenium virida* (green spleenwort), *Gymnocarpium robertianum* (limestone polypody), *Cypripedium passerinum* (sparrow's-egg Lady's-slipper), *Botychium minganense* (Mingan moonwort), *Draba aurea* (gilded draba), *Dryas drummondii* (yellow dryas), *Erigeron lonchophyllus* (spear-leaved fleabane), and *Gentianella detonsa* (island gentian), inhabit exposed regions and grow directly on the lime-

Fig. 4.51 Limestone barrens support unique and restricted vegetation. Burnt Cape NL

stone bedrock. The swamp thistle (*Cirsium scariosum*) is present along the shorelines of the four islands nearest to Harve-St-Pierre.

Thalictrum alpinum (alpine meadow rue) is present as two disjunct populations in North America. Displaced to the fringes of the continent during glaciation, separate populations survive along the fringes of the subarctic Pacific coast, and in subarctic and northern boreal Atlantic shorelines; Mingan represents one of its most southerly growing sites.

Several southern boreal species reach their northern limits in the Mingan Archipelago, prevented from expanding north by unfavourable climates and unsuitable bedrock. Among these species are *Geranium robertianum* (Robert's geranium), *Habernaria hookeri* (Hooker's orchid), *Listeria convallarioides* (broad-lipped twayblade orchid), and *Lobelia kalmii* (Kalm's loebelia). All of these species prefer more sheltered conditions, preferably along the fringes of forested terrain, and require humic-rich, organic or conifer duff soils.

Enedimic calciphile plants present at Burnt Cape and Watts Point Ecological Reserves and other limestone barrens of coastal Newfoundland include Arctic bladderpod (*Lesquerella arctica*), alpine arnica (*Arnica alpina*), dwarf hawksbeard (*Crepis nana*), and Burnt Cape cinquefoil (*Potentilla usticapensis*) (Catto 2020). Fernald's braya (*Braya fernaldii*) is listed as "threatened", and Long's braya (*Braya longii*) and the barrens willow (*Salix jejuna*), both confined to Watts Point Ecological Reserve, are listed as "endangered" (Hermanutz et al. 2002).

In Bruce Peninsula National Park, calcareous bedrock and exposed coastal environment allow a vegetation assemblage unique to the tip of the peninsula to develop, with 43 orchid species and other calciphile plants (Parks Canada 2016a). Among the more unusual species are dwarf lake iris (*Iris lacustris*), Rugel's plantain (*Plantago rugelli*), prairie slough grass (*Beckmannia syzigachne*), and lopseed (*Phryma leptostachya*). Other species include stonecrop (*Sedum acre*), Prairie avens (*Geum triflorum*), white aster (*Solidago ptarmicoides*), and Indian paint brush (*Castilleja coccinea*). Diversity is more limited on the non-saline coast of Georgian Bay than along the Atlantic seaboard. Tourist pressure has also been felt here, with many of the rarer orchid species suffering from unauthorized bouquet collection and transplantation efforts. Virtually any attempt to grow these orchids in environments different from their calcareous coastal settings will fail.

4.6 Soils

Several types of soils are present in the Mixedwood Plains ecozone, controlled primarily by the type of Quaternary sediment which they are developed upon. The most common type is the grey-brown luvisol, marked by clayey horizons rich in calcium carbonate, developed on glacilacustrine and glacimarine silts and clays under deciduous vegetation (Soil Classification Working Group 1998, 2001; Lavkulich and Arocena 2011; Fig. 4.52). Grey-brown luvisols formerly supported oak, elm, and beech forests. These soils are generally fertile and humic-rich, enriched by humic matter and leaf mould formerly contributed by the deciduous trees. In glacimarine areas, clayey soils frequently show poor drainage, limiting the range of economic crops.

Melanic brunisols (FAO eutric cambisols) develop on carbonate and sandstone bedrock and sandier glacilacustrine and glacial sediment. A typical melanic brunisol is neutral to slightly basic, with high concentrations of calcium carbonate, poor horizon differentiation, and abundant humus, and is well-drained, easy to work, and has great agricultural potential

Fig. 4.52 Mixedwood soils. Left: Melanic brunisol, with Ah horizon over modified Bm horizon. Right: Luvisol, with strongly developed Bt horizon. This luvisol has been altered by agricultural ploughing (Ap) on the surface (Canadian System of Soil Classification, Agriculture Canada)

(Soil Classification Working Group 1998, 2001; Smith et al. 2011). The melanic brunisols in the Niagara Peninsula and those surrounding Leamington in southwestern Ontario are among the most fertile soils in Canada, supporting large-scale corn and tomato agriculture and viticulture.

Clearance of the land for agricultural development has had a major impact on the soil, as well as the biological communities. Removal of the deciduous vegetation means that extensive and persistent fertilization (either chemical or organic) commonly is necessary to maintain high crop yields. As the water tables drop in the region due to extraction of the groundwater for human requirements, nutrients migrate downwards in the soils, producing eluviated melanic brunisols. Ongoing eluviation transforms melanic brunisols into grey-brown luvisols, with accumulation of transported clay at depth. Maintenance of luvisolic soils would require the presence of broadleaved deciduous vegetation. In the absence of forest cover, heavy fertilization is the only method of maintaining the chemistry of these soils.

Poorly drained areas are characterized by gleysols (Soil Classification Working Group 1998, 2001). More exposed or northerly regions dominated by conifers support brunisols, podzols, and regosols. Humo-ferric and ferric podzols are developed in areas where granitic, gneissic, or sandstone bedrock is present. Dystric brunisols and regosols develop on the olivine-rich bedrock of the Monteregian Hills. Aeolian dunes with pine groves are overlain by thin, acidic regosols,

Organic soils derived from peat accumulations are present in bogs and fens throughout the region. The Mer Bleue peat bog, located southeast of Ottawa, occupies 50 km^2. Developed in three abandoned former channels of the Ottawa River, the peat has accumulated in the Mer Bleue since the early Holocene (Fig. 4.53). The area supports a diverse bird fauna, and a large population of leopard frogs (*Rana pipiens*). During late July and early August, the frogs migrate throughout

Fig. 4.53 Mer Bleue peat bog in summer (upper) and winter (lower). Mer Bleue formed in abandoned channels of the Ottawa River southeast of Ottawa

the bog, frequently interrupting traffic and making driving hazardous on Anderson Road, which crosses through the bog.

Holland Marsh (varying from fens in the south to a freshwater marsh in the north) was an 18 km long, 8500 ha wetland flanking the Holland River north of Toronto (Catto et al. 2020). In the mid-1950s, a systematic programme of polder construction and drainage was undertaken. The soils are primarily organic Mesisols (Mesic Histosols) and Fibrisols (Fibric Histosols), supporting bok choy, endive, and onions in large commercial farms. Due to the relatively dry summers, the friable fibrous organic material in drained fields is subject to aeolian erosion.

4.7 Fauna

The fauna of the Mixedwood Plains also is transitional between boreal and temperate assemblages. Faunal population densities throughout the Ottawa and St. Lawrence valleys appear to have been relatively low prior to 18th and 19th century logging and settlement, and the effects of agricultural and urban development have reduced numbers still further. Predators in particular have suffered large declines in numbers.

Currently, small mammals such as *Sciurus carolinensis* (grey squirrel), *Procyon lotor* (raccoon), *Mustela richardsonii* (North American ermine), *Neogale vison* (mink), *Vulpes vulpes* (red fox), and *Erethizon dorsatum* (porcupine) dominate the assemblage (Banfield 1981; Naughton 2012). All of the larger predators, including *Ursus americanus* (black bear), *Puma concolor* (cougar), *Canis lupus* (timber wolf), and *Lynx canadensis* (lynx) were eliminated prior to 1900. An eastern subspecies of wapiti, *Cervus canadensis canadensis*, was also exterminated prior to 1850. Among the few species to increase in population recently is *Odocoileus virginianus* (white-tailed deer), primarily due to the elimination of its major predators (Deer Friendly n.d.). Deer thrive in predator-free environments: on Anticosti Island, the introduced population of 220 in 1897 currently numbers approximately 150,000, or 18.9/km^2 (Béaupre 2020).

Marmota monax (woodchuck; groundhog) is found throughout southern Ontario and Québec (Fig. 4.54). This rodent generally reaches a weight of 1-2 kg at maturity, and is notorious for its ability to dig up meadows and lawns and its ability to rapidly reproduce in the absence of predators (CWF 1991).

One particular representative of this species, "Wiarton Willie", achieved national fame for his annual performances every February 2nd. The superstition that if the groundhog sees his shadow the area will receive six more weeks of winter, is not substantiated by climatic statistics. In a regional context, six weeks of winter beyond February 2 would represent an anomalously hard year, as winter conditions are generally over in Wiarton by the beginning of March. *Marmota monax*, as most burrowing rodents, perceives colours and tones vastly differently than humans do. Wiarton Willie responds most readily to infrared light, and prefers nocturnal activity. Under these circumstances, one can wonder if he ever "sees his shadow" at all.

Some new species of mammals arrived after human settlement. *Canis latrans* (coyote) was not native to eastern Ontario, but has been able to survive in the area due to the elimination of its potential competitors *Canis lupus, Lynx canadensis*, and *Puma concolor* (Parker 1995). Fox squirrels (*Sciurus niger*) were introduced to Pelee Island *ca.* 1890, and have done well in the light of limited competition. *Sylvilagus floridanus* (cottontail) has expanded its range steadily northwards since the mid-1800s, helped by both warmer climates and agricultural clearance of the deciduous forests (Godin 1977). It arrived in southwestern Ontario about 1885. During the past 90 years, its range has expanded and contracted in accordance with yearly and decadal shifts in climate. Currently, the northern limit of its range is located approximately 40 km north of Lake Ontario.

Fig. 4.54 *Marmota monax* (groundhog)

4.7.1 *Didelphis virginiana* (Opossum)

A southern species particularly sensitive to climatic shifts is *Didelphis virginiana* (opossum), the only marsupial native to North America (Banfield 1981; Naughton 2012; Walsh and Tucker 2017; Fig. 4.55). Opossums advance northward into southern Ontario during warm, dry winters and retreat into the US during cold, wet winters, on a decadal cycle. During the relatively cold 1920s, the St. Lawrence River and northern Lake Ontario froze during the winter months, and opossums retreated southward. A subsequent warm period in the 1930s permitted the marsupials to return prior to 1934, and winter freeze-up was minimal. This climatic shift had consequences for human affairs as well. During the 1920s, vehicles were able to drive from Canada to the United States across the river and lake ice, which greatly facilitated smuggling of liquor into the USA in the winter season. Repeal of Prohibition in the USA in 1933 corresponded with warmer climates, which would have made winter smuggling much more difficult to accomplish, had the market not already "dried up".

Similar advances in the opossum range occurred during the early 1960s (especially during the dry period from 1961 to 1966) and 1980s, but each period was followed by colder weather and a retreat of the marsupials. During the period from 1987 to 1991, the opossum advanced northward again, but another retreat was forced by the winters of 1992 and 1993. After 1996–1997, the winters have been sufficiently mild to allow the opossum to expand its range yet again (Walsh and Tucker 2017). During "El Niño" years (e.g. 1981–1982, 1997–1998), conditions are more favourable for opossums already living in forested regions, but the absence of snow cover leads to frostbite if the animals venture onto open ground. In addition, the delayed freezing of the river hinders migration of new individuals across the Thousand Islands. Mature opossums captured in Ontario almost invariably have frost-bitten ears and tails.

4.7.2 *Urocyon cinereoargenteus* (Grey Fox)

Another species formerly characteristic of Southern Ontario was *Urocyon cinereoargenteus* (grey fox) (Banfield 1981; Naughton 2012; Fig. 4.56). The grey fox should not be confused with the commercially farmed "silver fox". Silver foxes used for furs are zoologically the same as red foxes (*Vulpes vulpes*) but have been bred for their distinctive fur colour. Grey foxes were adapted to life in the deciduous forest woodlands of Quinte and southern Ontario, with an ability to climb trees. Their favourite prey were cottontail rabbits (*Sylvilagus floridanus*), rodents including grey squirrels (*Sciurus carolinensis*), red squirrels (*Tamiasciurus hudsonicus*), and southern flying squirrels (*Glaucomys volans*). In the dense deciduous forests, *Urocyon* was able to compete successfully (and frequently out-compete) *Vulpes vulpes*, the common red fox.

Fig. 4.55 *Didelphisvirginiana* (opossum) (US Fish and Wildlife Service)

Fig. 4.56 *Urocyon cineroargentus* (grey fox) (Wikimedia/Alan Schmierer)

The advent of European settlement, and in particular the arrival of United Empire Loyalists displaced from New England after the American Revolution, caused difficulties for the grey fox. Agricultural clearance favoured the more adaptable red fox, which could survive in meadows and grasslands, over the forest-requiring grey fox (Statham et al. 2012).

Grey foxes were either completely or virtually extirpated from Canada sometime between the late-1800s to 1905. The present Canadian populations represent foxes which have expanded northward from the United States, taking advantage of shifts in agricultural style, changing public attitudes towards predators, and ongoing climate warming (COSEWIC 2015). Three separate grey fox populations now exist in Canada, each of which represents a different subspecies. *Urocyon cinereoargenteus borealis* occupied the Bay of Quinte region during the late 1930s-early 1940s, and has maintained small numbers since. The other two subspecies, *Urocyon cinereoargenteus cinereoargenteus* and *Urocyon cinereoargenteus ocythous,* inhabit the Lake Erie fringe of southern Ontario and the Red River Valley of southern Manitoba, respectively. The Southern Flying Squirrel, which prefers beech forests and is also found in areas with sugar maple, silver maple, oak, and hickory, was also greatly reduced in numbers as a result of agricultural clearance (COSEWIC 2006). Small populations currently live throughout southern Ontario, but they are not genetically distinct.

4.7.3 Birds at Point Pelee

Point Pelee supports a tremendous variety of bird species, with more than 350 recorded. More than 120 species routinely call at the point throughout the year, and many more "casuals" arrive in the wake of severe storms (iNaturalist n.d.). Southern, inland species such as cardinals (*Cardinalis cardinalis*), scarlet tanagers (*Piranga olivacea*), orchard orioles (*Icterus spurius*), and northern or Baltimore orioles (*Icterus galbula*) mingle with northern species such as Canada Jays (*Perisoreus canadensis*), blue jays (*Cyanocitta cristata*), and Canada geese (*Branta canadensis*), and marine species such as Arctic terns (*Sterna paradisaea*).

4.7.4 Reptiles and Amphibians

The southern boreal and temperate regions supported a diverse population of snakes. The Niagara Peninsula formerly had two species of rattlesnakes—the timber rattlesnake (*Crotalus horridus*) and the massasauga (*Sistrurus catenatus*). The timber rattlesnake is almost certainly extinct in the region, as none have been observed since 1941. The massasauga has also been extirpated from Niagara (Ernst and Barbour 1989).

Numerous other snakes, all non-venomous, are present in the area (Ernst and Barbour 1989; Fig. 4.57). All have suffered

Fig. 4.57 Reptiles of southwestern Ontario. Upper: *Heterodon platirhinos* (Eastern Hognose snake) (Wikimedia/ Barry Giles). Lower: *Lampropeltis triangulum* (Milk Snake) (Wikimedia Will Brown)

human persecution in the past, but attitudes have changed among the farming community in particular as the beneficial effects of the neighbourhood snake have become more known. Among the more common are *Heterodon platirhinos* (hognose snake), a harmless snake with a taste for toads (*Bufo americana*). This 115-cm maximum length snake is naturally defenceless against predators. Its only escape mechanism when cornered or surprised is to imitate the massasauga rattlesnake by moving its tail in a threatening fashion—an effective defence against an alert, experienced predator, but an unfortunate choice when facing a inexperienced, panicky *Homo sapiens*.

Lampropeltis triangulum (milk snake) is another potentially helpful reptile to humans, as it is particularly fond of rats and mice. It reaches a maximum length of 135 cm. The milk snake's colour pattern of light brown mottles and rings on a beige background of scales is responsible for its name. Early farmers in southern Ontario believed that the colour pattern developed because the milk snake drank cow's milk, stolen from the udders of the farmers' cows by night-raiding, milk-sucking milk snakes. Considering the udder size of a typical Holstein cow, and the stomach size of a typical milk snake, draining one cow's udder of a noticeable amount of milk would require the combined efforts of several thousand milk snakes. Modern professional farmers have long since moved beyond such folklore, and few would persecute milk snakes today.

With its warm winter climate, Pelee is a suitable habitat for reptiles (27 species) and amphibians (20 species) (Parks Canada 2016b). One extremely rare subspecies, *Acris crepitans blanchardi* (Blanchard's Cricket Frog), may survive only on Pelee Island, although there are no confirmed records since the 1970s. The five-lined skink (*Plestiodon faciatus*) is also confined within Canada to the Pelee area. Unfortunately, as elsewhere in Canada, amphibian and reptile populations are declining. Blanding's Turtle (*Emydoidea blandingi*), formerly common, is now considered threatened, and several other reptiles and amphibians are suffering population losses.

4.8 Human Impact

The 170,000 km^2 area of the central and western parts of the Mixedwood Plains ecozone was home to approximately 21,500,000 people in 2021, about 56% of the population of Canada. Three of the four largest census metropolitan areas: Toronto (6.2 million), Montreal (4.3 million), and Ottawa-Gatineau (1.5 million), and 10 of the largest 15 are located here, along with much of the country's industrial and commercial development (see Chap. 12). The concentration of people and infrastructure has greatly affected the natural environment, both spatially and in species composition and populations (Fig. 4.58). Introduced and invasive species are significant components of the urbanized areas.

Agriculture, with the associated land clearance, and forestry dominated both economically and in ecological impact between 1600 and 1867. The 1871 Census indicated that more than 80% of the total population lived in rural areas. Settlement was largely agricultural, with soil quality and accessibility to water transportation as the dominant factors. Along the St. Lawrence and Great Lakes shorelines, towns grew in association with the local agricultural resources, rather than in association with mining or manufacturing. Forests were relatively quickly exhausted of the most valuable species, particularly eastern white pine, with lumbering communities such as Ottawa surviving through imported wood from the Boreal Shield to the north. The initial pattern of town development and transportation linkages persisted after industry and commerce evolved in the latter 1800s.

Land currently in agricultural production occupies approximately 85,000 km^2 in the Mixedwood Plains, of which about 60% is used as cropland. Based on potential soil productivity and climate, the most suitable terrain (Canada Land Inventory Classes 1–3) represents about 110,000 km^2. Much of the land with agricultural potential is currently used for other purposes, particularly urban development, or is not in production. Approximately 50% of Canada's Class 1 farmland is located within a 50 km radius of downtown Toronto. Removal of agricultural land

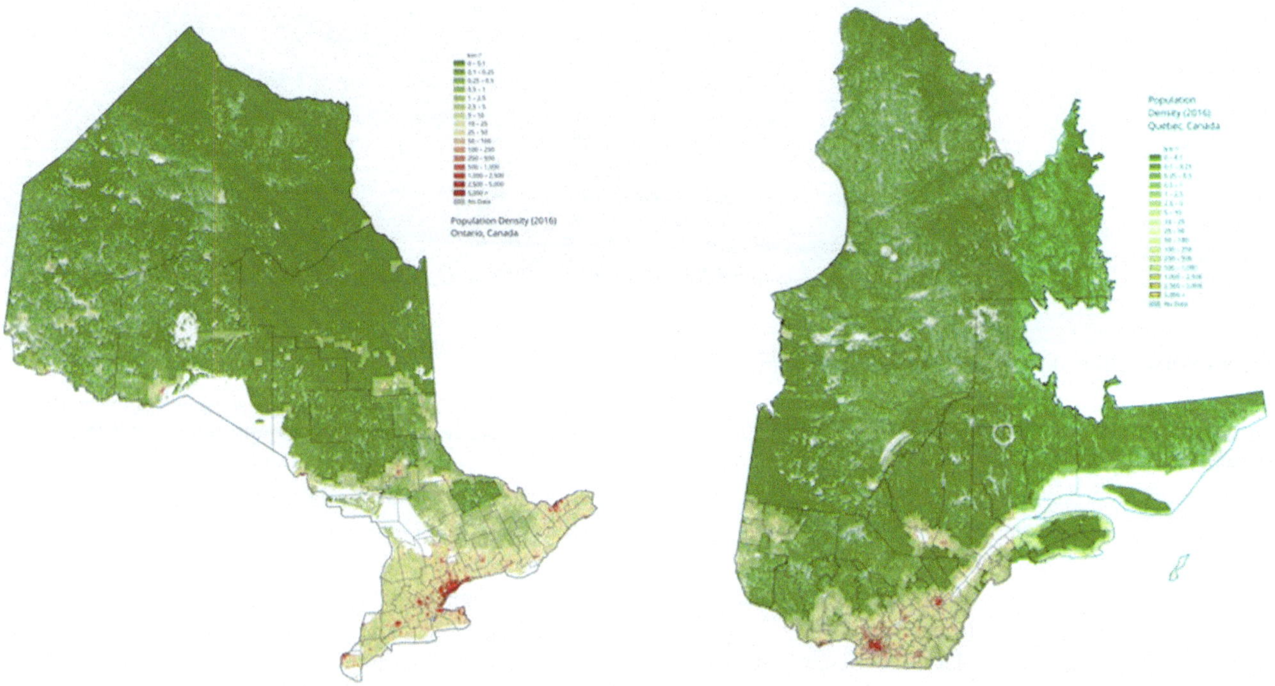

Fig. 4.58 Population densities for Ontario (left) and Québec (right) indicate the human impact on the Mixedwood Ecoregion (Wikimedia/awmcphee)

for urban and suburban development is a significant issue in the Niagara Peninsula and southwestern Ontario, as well as other locations in the Mixedwood Plains.

The density of urban and exurban development limits opportunities for establishment of more natural regimes and conservation areas. Introduced and invasive species have fundamentally altered vegetation assemblages and faunal populations throughout the ecozone, including within national and provincial parks. The climate has undergone anthropogenically induced changes. Overall, the environmental regime of the Mixedwood Plains represents a hybrid of effects, reflecting human activity extending to its initial occupation by people.

References

Ahrens CD, Jackson P, Jackson C (2015) Meteorology today: an introduction to weather, climate, and the environment. Nelson Cengage

Allen GM, Eagles PFJ, Price SD (1990) Conserving Carolinian Canada. University of Waterloo Press, Ontario

Anagnostakis SL, Hillman B (1992) Evolution of the chestnut tree and its blight. Arnoldia 52:2–10

Assani AA (2016) Analysis of the impacts of man-made features on the stationarity and dependence of monthly mean maximum and minimum water levels in the Great Lakes and St. Lawrence River of North America. Water 8:485. https://doi.org/10.3390/w8110485

Åström DO, Forsberg B, Rocklöv J. (2011) Heat wave impact on morbidity and mortality in the elderly population: a review of recent studies. Maturitas. 2011 Jun; 69(2):99–105

Baker CL, Lahti LR, Roumbanis DC (1998) Urban geology of Toronto and surrounding area. In Karrow PF, White OL (eds) Urban geology of Canadian cities. Geological Association of Canada Special Paper, vol 42, pp 323–352

Banfield AWF (1981) The mammals of Canada. National Museum of Natural Sciences, National Museums of Canada, 438 p

Bannon K, Delagrange S, Bélanger N, Messier C (2015) American beech and sugar maple sapling relative abundance and growth are not modified by light availability following partial and total canopy disturbances. Can J for Res 45:632–638

Barry JP (1968) Georgian Bay: the sixth great lake. Boston Mills Press

Barnes CR (1988) Stratigraphy and palaeontology of the Ordovician-Silurian boundary interval, Anticosti Island, Quebec, Canada. Bull Br Mus nat Hist (Geol) 43, 195–219

Beauchesne D, Daigle RM, Vissault S, Gravel D, Bastien A, Bélanger S, Bernatchez P, Blais M, Bourdages H, Chion C, Galbraith PS, Halpern BS, Lavoie C, McKindsey CW, Mucci A, Pineault S, Starr M, Ste-Marie A-S, Archambault P (2020) Characterizing exposure to and sharing knowledge of drivers of environmental change in the St. Lawrence system in Canada. Front Marine Sci 7:2020.00383. https://doi.org/10.3389/fmars.2020.00383

Beaupré M (2020) Anticosti Island: geomorphological context of a unique Karst area. In: Slaymaker O, Catto NR (eds) Landscapes and landforms of Eastern Canada. Springer International Publishers, Cham, Switzerland, pp 229–244

Beck C, Grieser J, Kottek M, Rubel F, Rudolf B (2006) Characterizing global climate change by means of Köppen climate classification. http://koeppen-geiger.vu-wien.ac.at/pdfs/beck_et_al_ksb_2006.pdf

Benmarhnia T, Deguen S, Kaufman JS, Smargiassi A (2015) Review Article: Vulnerability to Heat-related Mortality: A Systematic Review, Meta-analysis, and Meta-regression Analysis. Epidemiology. 2015 Nov; 26(6): 781–93

Bélanger JR (1998) Urban geology of Canada's National Capital area. In Karrow PF, White OL (eds) Urban geology of Canadian cities. Geological Association of Canada Special Paper 42, pp 365–384

Bishop-Williams KE, Berke O, Pearl DL, Kelton DF (2015) A spatial analysis of heat stress related emergency room visits in rural Southern Ontario during heat waves. BMC Emerg Med. 2015 Aug 6;15:17

Boland GJ, Ambrose J, Husband B, Elliott KA, Melzer MS (2012) Recovery strategy for the American chestnut (Castanea dentata) in Ontario. Ontario recovery strategy series. Prepared for the Ontario Ministry of Natural Resources, Peterborough, Ontario

Bostock HS (2014) Physiographic regions of Canada. Geological survey of Canada Map 1254A (2nd edition), scale 1:5,000,000. https://www.nrcan.gc.ca/maps-tools-and-publications/maps/geology-and-geosciences/16876. Accessed 12 July 2022

Boyer L, Bensoussan A, Durand M, Grice RH, Berard J (1985) Geology of Montreal, Quebec, Canada. Bull Assoc Eng Geol 22:333–394

Brett CE, Tepper WM, Goodman WM, LoDuca ST, Eckert B-Y (1995) Revised stratigraphy and correlations of the Niagaran provincial series (Medina, Clinton, and Lockport groups) in the type area of western New York. United States Geological Survey Bulletin 2086

Brisebois D, Nadeau J (2003) Géologie de la Gaspésie et du Bas St-Laurent. DV 2003–08, 1:250 000 map. Ministère des Ressources Naturelles de la Faune et des Parcs, Québec

Brooks GR, Aylsworth GM, Evans SG, Lawrence DM (1994) The Lemieux landslide of June 20, 1993, South Nation Valley, southeastern Ontario—a photographic record. Geological Survey of Canada, Miscellaneous Report 56

Brooks GR, Lawrence DE (1998) Geomorphic effects and impacts of severe flooding. photographic examples from the Saguenay area, Québec. Geological Survey of Canada, Miscellaneous Report 62

Brouard E, Lajeuenesse P, Cousineau PA, Govare E, Locat J (2016) Late Wisconsinan deglaciation and proglacial lakes development in the Charlevoix region, southern Québec, Canada. Boreas 45:754–772

Butt S, Ramprasad P, Fenech A (2016) Changes in the landscape of Southern Ontario, Canada, since 1750: impacts of European colonization. In: Fenech A MacIver D Aulld H Hansell R (eds) Integrated mapping assessment. Environment Canada. Paper 6. https://www.yumpu.com/en/document/view/31426684/paper 6 changes-in-the-landscape-of-southern-upei-projects

Buttle JM, Allen DM, Caissie D, Davison B, Hayashi M, Peters DL, Pomeroy JW, Simonovic S, St-Hilaire A, Whitfield PH (2016) Flood Processes in Canada: regional and spatial aspects. Canadian Water Resources Journal 41:7–30

Call DM 2010 Changes in Ice Storm Impacts over Time: 1886–2000. Weather, Climate, and Society 2, 23–35. Lecomte EL Pang AW Russell JW 1998. Ice Storm '98. Institute for Catastrophic Loss Reduction

Canadian Heritage (n.d.) Timeline: the Maple Leaf. https://www.canada.ca/content/dam/pch/documents/services/unofficial-symbols-canada/Timeline_MapleLeaf_EN.pdf. Accessed 2 Aug 2022

Canada National Defence (2017) Practicing UXO safety at Lac Saint-Pierre, Quebec.https://www.canada.ca/en/department-national-defence/services/uxo/uxo-locations/practicing-uxo-safety-lac-saint-pierre.html

Canadian Wildlife Service (2001) Point Pelee National Park, Ontario. Information Sheet on Ramsar Wetlands Environment Canada, Ottawa. https://rsis.ramsar.org/RISapp/files/RISrep/CA368RIS.pdf. Accessed 2 Aug 2022

Case FW, Case RB (1997) Trilliums. Timber Press, Portland, Oregon

Catto N (2015) Geomorphology: landscapes we live in. Oxford University Press, Don Mills, Ontario

Catto NR (2020) Labrador and the Northern Peninsula of Newfoundland: foreshore and flanking geomorphology. In: Slaymaker O, Catto NR (eds) Landscapes and Landforms of Eastern Canada. Springer International Publishers, Cham, Switzerland, pp 179–210

Catto NR, Patterson RJ, Gorman WA (1981) Late Quaternary marine sediments at Chalk River, Ontario. Can J Earth Sci 18:1261–1267

Catto NR, Slaymaker O, Kovanen DJ (2020) Quaternary Glacial, Glacimarine, and Glacilacustrine History. In Slaymaker O., Catto NR., ed., Landscapes and Landforms of Eastern Canada. Springer International Publishers, Cham, Switzerland, 49–78

Chen DL, Jiang C, Duchesne MJ, Malo M (2016) Geological characteristics and petroleum resource assessment of the Macasty formation, Anticosti Island, Quebec, Canada. Geological Survey of Canada Open File 8018

Clamen M, Macfarlane D (2018) Plan 2014: the historical evolution of Lake Ontario–St. Lawrence River Regulation, Canadian Water Resour J 43:416–431

Claveau P, Laframboise Y (1983) L'exploitation forestière destinée au commerce du bois au 19e siècle, Ethnotech inc., 443 p

Colman-Sadd SP Hayes JP, Knight I (1990). Geology of the island of Newfoundland. Newfoundland and Labrador, Department of Mines and Energy, Geological Survey Branch, Map 90–01.

Copenheaver CA, Shumaker KL, Butcher BM, Haha G, Perkins JL, Dukes CJ, Thompson EG, Pisaric MFJ (2020) Dendroclimatology of sugar maple (Acer saccharum): Climate-growth response in a late-successional species Dendrochronologia 63:125747. https://doi.org/10.1016/j.dendro.2020.125747

COSEWIC (2006) COSEWIC assessment and update status report on the southern flying squirrel Glaucomys volans (Atlantic (Nova Scotia) population and Great Lakes Plains population) in Canada. Committee on the Status of Endangered Wildlife in Canada. Ottawa

COSEWIC (2015) COSEWIC assessment and status report on the Gray Fox *Urocyon cinereoargenteus* in Canada. Committee on the Status of Endangered Wildlife in Canada, Ottawa

Cortinas JV, Bernstein BC, Robbins CC, Strapp JW (2004) An Analysis of Freezing Rain, Freezing Drizzle, and Ice Pellets across the United States and Canada: 1976–90. Weather and Forecasting 19, 377–390.

Couillard L, Grondin P, Pelletier D (1987) Les plantes rares de l'Archipel de Mingan. Environment Canada

Cowell DR (2020) Karst geomorphology of the Saugeen-Bruce Peninsula, Ontario. In: Slaymaker O, Catto NR (eds) Landscapes and Landforms of Eastern Canada. Springer International Publishers, Cham, Switzerland, pp 277–296

CNVC (2016) Canadian national vegetation classification.http://cnvc-cnvc.ca. Accessed 2 Aug 2022

CWF (1991) Hinterland: Who's Who: Woodchuck. Canadian Wildlife Federation. https://www.hww.ca/en/wildlife/mammals/woodchuck.html. Accessed 2 Aug 2022

Danneyrolles V, Cyr D, Gitardin MP, Gauthier S, Asselin H, Bergeron Y (2021) Influences of climate fluctuations on northeastern North America's burned areas largely outweigh those of European settlement since AD 1850. Environ Res Lett 16:114007. https://doi.org/10.1088/1748-9326/ac2ce7

Deer Friendly (n.d.) Decline of deer populations. https://www.deerfriendly.com/decline-of-deer-populations. Accessed 2 Aug 2022

Deschênes E, Brice M-H, Brisson J (2019) Long-term impact of a major ice storm on tree mortality in an old-growth forest. For Ecol Manage 448:386–394

Desloges JR, Phillips RTJ, Byrne M-L, Cockburm JMH (2020) Geomorphology of the Great Lakes Lowlands of Eastern Canada. In: Slaymaker O, Catto NR (eds) Landscapes and landforms of Eastern Canada. Springer International Publishers, Cham, Switzerland, pp 259–276

Dick M, Porter TJ, Pisaric MFJ, Wertheimer È, deMontigny P, Perreault JT, Robillard K-L (2014) A multi-century eastern white pine tree-ring chronology developed from salvaged river logs and its utility for dating heritage structures in Canada's National Capital Region. Dendrochronologia 32:120–126

DePratto B, Kraus D (2017) Putting a value on the ecosystem services provided by forests in Canada: case studies on natural capital and conservation. TD Bank Group (TD) and the Nature Conservancy of Canada (NCC). https://www.natureconservancy.ca/assets/documents/nat/Natural-Capital_2017_draft.pdf. Accessed 2 Aug 2022

Dietrich P, Normandeau A, Lajeunesse P, Ghienne J-F, Schuster M, Nutz A (2020) Deltaic complexes of the Québec North Shore. In: Slaymaker O, Catto NR (eds) Landscapes and landforms of Eastern Canada. Springer International Publishers, Cham, Switzerland, pp 245–258

Dohler G 2017. Tides in Canadian waters. Canadian Hydrographic Service.

Drobyshev I, Guitard M-A, Asselin H, Genries A, Bergeron Y (2014) Environmental controls of the northern distribution limit of yellow birch in eastern Canada. Can J For Res 44:720–731

Dyke AS (2004) An outline of North American deglaciation with emphasis on central and northern Canada. In: Ehlers J, Gibbard PL (eds) Quaternary glaciations—extent and chronology, Part II: North America. Elsevier, Amsterdam, pp 371–424

Dyke AS, Moore A, Robertson L (2003) Deglaciation of Canada. Geological Survey of Canada Open File 1574

Earthquakes Canada (n.d.). Natural resources Canada. http://earthquakescanada.nrcan.gc.ca/. Accessed 2 Aug 2022

Elliott KA (1998) The forests of southern Ontario. For Chron 74:850–854

Environment Canada and Agriculture and Agrifoods Canada (1995) Terrestrial ecozones and ecoregions of Canada. https://sis.agr.gc.ca/cansis/publications/maps/eco/all/regions/eco_all_regions_7m_canada_en.jpg. Accessed 12 July 2022

Environment Canada (1998) Ice Storm 98—Worst ice storm in Canadian history. Environment Canada Report, http://www1.tor.ec.gc.ca/events/icestorm98/icestorm98_the_worst_e.html#top. Accessed 2 Aug 2022

Environment Canada (2015) Action plan for the Cucumber Tree (Magnolia acuminata) in Canada. Species at risk act action plan series. Environment Canada, Ottawa

Environment Canada (2019) Humidex. https://www.canada.ca/en/environment-climate-change/services/seasonal-weather-hazards/warm-season-weather-hazards.html#toc7. Accessed 2 Aug 2022

Environment and Climate Change Canada (2020) Canadian Climate Normals 1981–2010. https://climate.weather.gc.ca/climate_normals/index_e.html. Accessed 2 Aug 2022

Ernst CH, Barbour RW (1989) Snakes of eastern North America. George Mason University Press, Lanham, MD

Evans SG (2001) Landslides. In: Brooks GR (ed) A synthesis of geological hazards in Canada, vol 548. Geological Survey of Canada Bulletin, pp 43–80

Fabris A-S, Larouche P, Montero-Serrano, J-C (2021) Characterization of suspended matter size and composition in the St. Lawrence Estuary (eastern Canada). Reg Stud Marine Sc 45:101838. https://doi.org/10.1016/j.rsma.2021.101838

Fisheries and environment Canada (1978) Hydrological atlas of Canada. National Research Council of Canada, Ottawa

Fisheries and Oceans Canada (2012) Recovery strategy for the Beluga Whale (Delphinapterus leucas) St. Lawrence Estuary Population in Canada. Species at risk act recovery strategy series. Fisheries and Oceans Canada, Ottawa

Fisheries and Oceans Canada (2021) Zebra mussel. https://www.dfo-mpo.gc.ca/species-especes/profiles-profils/zebramussel-moulezebree-eng.html. Accessed 2 Aug 2022

Foret Hereford (2015) Fiche d'information sur l'évolution historique du paysage forestier de la Forêt communautaire Hereford. www.forethereford.org/fr/foret-conservation/documents/Fiche_info_foret_naturelle_FHI_2015.pdf. Accessed 2 Aug 2022

Fulton RJ (1995) Surficial materials of Canada. Geological Survey of Canada Map 1880A

Gagnon D (2004) La Forêt Naturelle du Québec, un survol. Groupe de recherche en écologie forestière interuniversitaire Université du Québec à Montréal. https://www.bibliotheque.assnat.qc.ca/depotnumerique_v2/affichagenotice.aspx?idn=9634. Accessed 2 August 2022

Gaudreau G (1999) La récolte des forêts publiques au Québec et en Ontario, 1840–1900. McGill-Queen's University Press, Montréal and Kingston

Gillis SJ (1975) The timber trade in the Ottawa Valley, 1806–54. Parks Canada, Manuscript Report 153

Globensky Y (1987) Géologie des Basses-Terres du Saint-Laurent. Ministère de l'Énergie et des Ressources, Québec, MM 85–02

Godin AJ (1977) Wild mammals of New England. Johns Hopkins University Press, Baltimore

Great Lakes Basin (2003) Natural Resources Canada. www.atlas.gc.ca. Accessed 2 Aug 2022

Great Lakes Environmental Assessment and Mapping (GLEAM) Project (2015). https://www.greatlakesnow.org/2015/10/u-of-m-researchers-launch-great-lakes-map/

Great Lakes Fishery Council (n.d.). Sea lamprey. http://www.glfc.org/sea-lamprey.php. Accessed 2 Aug 2022

Great Lakes–St. Lawrence Seaway System (n.d.). https://greatlakes-seaway.com/en/. Accessed 2 Aug 2022

Great Lakes–St. Lawrence Seaway System (2019). Welland Canal. https://greatlakes-seaway.com/wp-content/uploads/2019/10/welland.pdf. Accessed 2 Aug 2022

Groupe AGÉCO (2013) Le capital écologique du Grand Montréal: une évaluation économique de la biodiversité et des écosystèmes de la Ceinture verte, report prepared by Groupe AGÉCO for the David Suzuki Foundation and Nature-Action Québec, February 2013

Health Canada (2011) Communicating the health risks of extreme heat events: toolkit for public health and emergency management officials. Health Canada, Ottawa, 60 p

Health Canada (2012) Heat alert and response systems to protect health: best practices guidebook. Health Canada, Ottawa, 92 p

Hebert CE, Pasher J, Weseloh DVC, Dobbie T, Dobbyn S, Moore D, Minelga V, Duffe J (2014) Nesting cormorants and temporal changes in Island Habitat. J Wildl Manag 78:307–313

Hermanutz L, Mann H, Anions MF, Ballam D, Bell T, Brazil J, Djan-Chékar N, Gibbons G, Maunder J, Meades SJ, Nicholls W, Smith N, Yetman G (2002) National recovery plan for Long's Braya (Braya longii Fernald) and Fernald's Braya (Braya fernaldii Abbe). National recovery plan no. 23. Recovery of Nationally Endangered Wildlife (RENEW), Ottawa

Hobson GD, Terasmae J (1969) Pleistocene geology of the buried St. Davids gorge, Niagara Falls, Ontario. Geophysical and palynological studies. Geological Survey of Canada, Paper 68-67

Hosie RC (1979) *Native Trees of Canada. Canada Department of Forestry, Don Mills, Ontario.* Hudon C, M. Jean, G. Létourneau 2018. Temporal (1970–2016) changes in human pressures and wetland response in the St. Lawrence River (Québec, Canada) C. Sci Total Environ 643:1137–1151

IBA (Important Bird Areas) (2021) La Grande Ile, Berthierville, Québec. https://www.ibacanada.org/site.jsp?siteID=QC118. Accessed 2 Aug 2022

iNaturalist (n.d.) Birds of Point Pelee National Park of Canada. https://www.inaturalist.org/guides/4992. Accessed 2 Aug 2022

International Lake Ontario-St. Lawrence River Board (2022). https://ijc.org/en/loslrb/lake-st-lawrence. Accessed 2 Aug 2022

Jackson JN (1991) The construction and operation of the first, second, and third Welland canals. Can J Civ Eng 18:472–483

Johnson MD, Armstrong DK, Sanford BV, Telford PG, Rutka MA (1992) Palaeozoic and Mesozoic geology of Ontario. In: Thurston PC, Williams HR, Sutcliffe RH, Stott GM (eds) Geology of Ontario. Ontario geological survey, Special vol 4, issue 2, pp 907–1008

Karrow PF (1989) Quaternary geology of the Great Lakes sub-region. In: Fulton RJ (ed) Quaternary Geology of Canada and Greenland. Geological survey of Canada series, no. 1. Geological Survey of Canada, Ottawa, pp 326–350

Karrow PF (1990) Interglacial beds at Toronto, Ontario. Géog Phys Quatern 44:289–297

Karrow PF (1998) Urban geology note: Trois-Rivières, Quebec. In Karrow PF, White OL (ed) Urban geology of Canadian cities. Geological Association of Canada Special Paper 42, pp 397–400

Karrow PF, Calkin PE (1985) Quaternary evolution of the Great Lakes. Geological Association of Canada, Special Paper 30

Karrow PF, Morin FJ (1998) Urban geology note: Hamilton, Ontario. In Karrow PF, White OL (eds) Urban geology of Canadian cities. Geological Association of Canada Special Paper 42, pp 279–286

Kelly PE, Larson DW (1997) Effects of rock climbing on populations of presettlement eastern white cedar (*Thuja occidentalis*) on cliffs of the Niagara Escarpment, Canada. Conserv Biol 11:1125–1132

Kelly PE, Larson DW (2007) The last stand: a journey through the ancient cliff-face forest of the Niagara escarpment. Natural Heritage Books, 176 pp

Kershaw L (2001) Trees in Ontario: including tall shrubs. Lone Pine Publishing. Edmonton, AB, p 240

Kidon J, Fox G, McKenney D, Rollins K (2001) Economic impact of the 1998 ice storm on the eastern Ontario maple syrup industry. Forestry Chronicle 77 (4)

Knight KS, Brown JP, Long RP (2013) Factors affecting the survival of ash (*Fraxinus* spp.) trees infested by emerald ash borer (*Agrilus planipennis*). Biol Invasions 15:371–383

Kuhnlein HV, Turner NJ (1991) Traditional plant foods of Canadian indigenous peoples: nutrition, botany, and use. Gordon and Breach, Amsterdam

Lamb L, Rhynard G (1994) Plants of Carolinian Canada. Federation of Ontario Naturalists

Lamontagne M, Keating P, Perrault S (2003) Seismotectonic characteristics of the Lower St. Lawrence Seismic Zone, Quebec: insights from geology, magnetics, gravity, and seismics. Can J Earth Sci 40:317–336

Larsen CE (1987) Geological history of glacial Lake Algonquin and the Upper Great Lakes. U.S. Geological Survey Bulletin 1801

LaSalle P, Shilts WW (1993) Younger Dryas-age readvance of Laurentide ice into the Champlain Sea. Boreas 22:25–37

Laverdière É, Généreux M, Gaudreau P, Morais JA, Shatenstein B, Payette H (2015) Prevalence of risk and protective factors associated with heat-related outcomes in Southern Quebec: A secondary analysis of the NuAge study. Can J Public Health. 106: 315—21

Lavkulich LM, Arocena JM (2011) Luvisols of Canada: Genesis, distribution, and classification. Can J Soil Sci 91:781–806

Lawrence DE, Aylsworth JM, Morey CR (1996) Sensitive Clay Flows Along the South Nation River, Ontario, Canada and their Impact on Land Use. In Proceedings of the 7th International Symposium on Landslides, Trondheim, Norway, June 1996, edited by K Senneset . A.A. Balkema Ltd., Rotterdam, 479–484

Lefebvre G (2017) Sensitive clays of Eastern Canada: from geology to slope stability. In: Thakur V, L'Heureux J-S, Locat A (eds) Landslides in sensitive clays. Advances in natural and technological hazards research, vol 46. Springer International Publishing

Lévy S, Jaboyedoff M, Locat J, Demers D (2012) Erosion and channel change as factors of landslides and valley formation in Champlain Sea Clays: the Chacoura River, Quebec, Canada. Geomorphology 145–146:12–18

Lewis CFM, Anderson TW (2020) A younger glacial lake Iroquois in the lake Ontario basin, Ontario and New York: Re-examination of Pollen stratigraphy and radiocarbon dating. Can J Earth Sci 57:453–463

Lewis CFM, Todd BJ (2019) The early Lake Ontario barrier beach: evidence for sea level about 12.8–12.5 cal. ka BP beneath western Lake Ontario in eastern North America Boreas, vol 48, pp 195–214

Li T, Horton RM, Bader DA, Huang G, Sun Q, Kinney PL (2015) Heat-related mortality projections for cardiovascular and respiratory disease under the changing climate in Beijing, China. Sci. Rep., 5, 11441

Locat A, Locat P, Demers D, Leroueil S, Robitaille D, Lefebvre G (2017a) The Saint-Jude landslide of 10 May 2010, Quebec, Canada: investigation and characterization of the landslide and its failure mechanism. Can Geotech J 54:1357–1374

Locat J, Turmel D, Locat P, Therrien J, Létourneau M (2017b) The 1908 Disaster of Notre-Dame-de-la-Salette, Québec, Canada: Analysis of the Landslide and Tsunami. In: Thakur V, L'Heureux J-S, Locat A (eds) Landslides in sensitive clays: from research to implementation. Advances in Nature and Technological Hazards Research No. 46. Springer, Switzerland, pp 361–371

MacDonald G, Harrison JE (1979) Geological survey of Canada, "A" Series Map 1508A

Macfarlane D (2014) Negotiating a River: Canada, the US, and the creation of the St. University of British Columbia Press, Vancouver, Lawrence Seaway

Maclachland JC, Eyles CH (2013) Quantitative geomorphological analysis of drumlins in the Peterborough Drumlin Field, Ontario, Canada. Geografiska Annalar A 95:125–144

Mahaney WC, Hancock RGV, Milan A, Pulleyblank C, Costa PJM, Milner MW (2014) Reconstruction of Wisconsinan-age ice dynamics and compositions of southern Ontario glacial diamictons, glaciofluvial/lacustrine, and deltaic sediment. Geomorphology 206:421–439

Maltais D Lalonde C Bellerose C Robichaud S Simard A Fortin M Lebeau A Mayer R (2001) Les consequences de la tempête de verglas sur la santé des indiïidus, des interwenants et des communautés. Université du Québec à Chicoutimi.

Marich AS (2016) Quaternary geology of the Lindsay and Peterborough areas, southern Ontario. Ontario Geological Survey Open File Report 6321

Matte P, Secretan Y, Morin J (2019) Drivers of residual and tidal flow variability in the St. Lawrence fluvial estuary: Influence on tidal wave propagation. Cont Shelf Res 174:158–173

Ménard N, Turgeon S, Conversano M, Martins CA (2022) Sharing the waters: application of a marine spatial planning approach to conserve and restore the acoustic habitat of endangered beluga whales (*Delphinapterus leucas*) in and around the Saguenay–St Lawrence Marine Park. Marine Pollut Bull 175:113325. https://doi.org/10.1016/j.marpolbul.2022.113325

Mills EL, Leach JH, Carlton JT, Secor CL (1993) Exotic species in the Great Lakes: a history of biotic crisis and anthropogenic introductions. J Great Lakes Res 19:1–54

Ministère des Forêts, de la Faune et des Parcs Québec 2018. Zones de végétation et domaines bioclimatiques du Québec.https://mffp.gouv.qc.ca/les-forets/amenagement-durable-forets/inventaire-ecoforestier/. Accessed 2 Aug 2022

Moreau G, Achim A, Pothier D (2020) An accumulation of climatic stress events has led to years of reduced growth for sugar maple in southern Quebec, Canada. Ecosphere 11(7):e03183. https://doi.org/10.1002/ecs2.3183

Morris RD, Pekarik C, Moore DJ (2012) Current status and abundance trends of common terns breeding at known coastal and inland nesting regions in Canada. Waterbirds 35:194–207

Mohsin T, Gough WA (2012) Characterization and estimation of urban heat island at Toronto: impact of the choice of rural sites. Theoret Appl Climatol 108:105–117

National Capital Commission (NCC) (n.d.) Rideau Canal Skateway. https://ncc-ccn.gc.ca/places/rideau-canal-skateway. Accessed 2 Aug 2022

Natural Resources Canada (2009) Geological Provinces of Canada. https://www.nrcan.gc.ca/maps-tools-and-publications/maps/geology-and-geosciences/16876. Accessed 12 July 2022

Naughton D (2012) The natural history of Canadian mammals. Canadian Museum of Nature

Novak BJ, Estes JA, Shaw HE, Novak EV Shapiro B (2018) Experimental investigation of the dietary ecology of the extinct passenger pigeon, Ectopistes migratorius. Front Ecol Evol 6:20. https://doi.org/10.3389/fevo.2018.00020

Nozères C, Archambault D, Chouinard P-M, Gauthier J, Miller R, Parent E, Schwab P, Savard L, Dutil J-D (2010) Identification guide for marine fishes of the estuary and northern Gulf of St. Lawrence and sampling protocols used during trawl surveys between 2004 and 2008 Regional Science Branch Fisheries and Oceans Canada Maurice Lamontagne Institute Mont-Joli (Québec)

Nutz A, Lajeunesse P, Ghienne J-F, Schuster M, Brouard E, Dioetrich P, Bouchette F, Roquin C, Cousineau PA (2020) Deglacial to paraglacial history of the Lake Saint-Jean Lowlands: a geomorphological perspective. In: Slaymaker O, Catto NR (eds) Landscapes and landforms of Eastern Canada. Springer International Publishers, Cham, Switzerland, pp 145–158

Occhietti S (1989) Quaternary geology of St. Lawrence valley and adjacent Appalachian subregion. In: Fulton RJ (ed) Quaternary geology of Canada and Greenland. Geology of Canada series, no. 1. Geolgical Survey of Canada, Ottawa, pp 350–379

Occhietti S (2007) The Saint-Narcisse morainic complex and early Younger Dryas events on the southeastern margin of the Laurentide ice sheet. Géog Phys Quatern 61:89–117

Occhietti S, Parent M, Shilts WW, Dionne J-C, Govare E, Harmand D (2001) Late Wisconsinan glacial dynamics, deglaciation, and marine invasion in southern Quebec. Geological Society of America Special Paper 351

O'Neill Sanger CE, St-Jacques J-M, Peros MC, Schwartz KA (2021) Reconstructed high-resolution forest dynamics and human impacts of the past 2300 years of the Parc national de Mont-Orford, southeastern Québec, Canada. Holocene 31:1019–1032

Ontario Geological Survey (1991) Bedrock geology of Ontario, southern sheet. Ontario Geological Survey, Map 2544

Ontario Geological Survey (2010) Surficial geology of southern Ontario. Miscellaneous Release—Data 128-REV

OMNR (2011) Ontario Ministry of Natural Resources: Ontario Tree Atlas. http://www.mnr.gov.on.ca/en/Business/ClimateChange/2ColumnSubPage/267027.html. Accessed 2 Aug 2022

Ontario Invasive Plants (2021) Norway Maple. https://www.ontarioinvasiveplants.ca/wp-content/uploads/2021/05/BMP_NorwayMaple_April2021_WEB.pdf. Accessed 2 Aug 2022

Ontario Ministry of Natural Resources and Forestry (2017) Recovery strategy for the Kentucky Coffee-tree (Gymnocladus dioicus) in Ontario. Ontario Recovery Strategy Series. Prepared by the Ontario Ministry of Natural Resources and Forestry, Peterborough, Ontario

Parks Canada (2016a) Multi-species action plan for Bruce Peninsula National Park and Fathom Five National Marine Park of Canada. Species at risk act action plan series. Parks Canada Agency, Ottawa

Parks Canada (2016b) Multi-species action plan for point Pelee National Park of Canada and Niagara National Historic Sites of Canada. Species at risk act action plan series. Parks Canada Agency, Ottawa. iv + 39 pp

Parks Canada (2019b) Canal Lachine. https://www.pc.gc.ca/en/lhn-nhs/qc/canallachine/culture/histoire-history/histoire-history. Accessed 2 Aug 2022

Parks Canada (2020) Mingan Arch National Park Reserve: Marine Birds. https://www.pc.gc.ca/en/pn-np/qc/mingan/nature/faune-wildlife/oiseaux-birds/marins-marines. Accessed 2 Aug 2022

Parks Canada (2021a) Canal Carillon. https://www.pc.gc.ca/en/lhn-nhs/qc/canalcarillon. Accessed 2 Aug 2022

Parks Canada (2021b) Rideau Canal.https://www.pc.gc.ca/en/lhn-nhs/on/rideau. Accessed 2 Aug 2022

Parks Canada Agency (2012) Recovery strategy for the common Hoptree (*Ptelea trifoliata)* in Canada. Species at risk act recovery strategy series. Parks Canada Agency. Ottawa

Parker GR (1995) Eastern Coyote: the story of its success. Nimbus Publishing, Halifax, NS

Pengelly LD, Campbell ME, Cheng CS, Fu C, Gingrich SE, Macfarlane R (2007) Anatomy of heat waves and mortality in Toronto: lessons for public health protection. Can J Public Health 98:364–368

Phillips D (1990) The climates of Canada. Environment Canada, 176 p

Phillips D (1993) The Day Niagara Falls Ran Dry: Canadian weather facts and Trivia. KeyPorter Toronto

Pisaric MFJ, King DJ, MacIntosh AGM, Bemrose R (2008) Impact of the 1998 ice storm on the health and growth of sugar maple (Acer saccharum Marsh.) dominated forests in Gatineau Park, Quebec. Journal of the Torrey Botanical Society 135, 530–539

Pope A (2016) Throwback Thursday: The Day Niagara Falls Ran Dry. https://canadiangeographic.ca/articles/throwback-thursday-the-day-niagara-falls-ran-dry. Accessed 2 Aug 2022

Proulx OJ, Greene DF (2001) The relationship between ice thickness and northern hardwood tree damage during ice storms. Can J For Res 31, 1758–1769

Puric-Mladenovic D (2004) Historic forest cover vision for the future. The Regional Municipality of York, ON. http://forests-settled-urban-landscapes.org/Presettlement/files/York-2004. Accessed 2 Aug 2022

Puric-Mladenovic D (2011) Pre-settlement vegetation mapping for the Greater Toronto Area, including the Regions of Hamilton, Halton, Peel and York and the Credit Valley Watershed. Faculty of Forestry, University of Toronto. https://academic.daniels.utoronto.ca/forestry/wpcontent/uploads/sites/4/2016/04/Final_Report.pdf. Accessed 2 Aug 2022

Public Safety Canada (n.d.). Canadian Disaster Database. https://www.publicsafety.gc.ca/cnt/rsrcs/cndn-dsstr-dtbs/index-en.aspx. Accessed 2 Aug 2022

Québec Environnement (2013) Le Lac Saint-Pierre, Un joyau à restaurer. https://www.environnement.gouv.qc.ca/eau/lac-st-pierre/doc-synthese.pdf. Accessed 2 Aug 2022

Québec Environnement (n.d.) Réserve écologique Samuel-Brisson. https://www.environnement.gouv.qc.ca/biodiversite/reserves/samuel_brisson/res_15.htm. Accessed 2 Aug 2022

Québec, Ressources naturelles. Vegetation Zones and Bioclimatic Domains in Québec. https://mern.gouv.qc.ca/english/publications/forest/publications/zone-a.pdf. Accessed 2 Aug 2022

Quinn PE, Hutchinson DJ, Diederichs MS, Rowe RK (2011) Characteristics of large landslides in sensitive clay in relation to susceptibility, hazard, and risk. Can Geotech J 48:1212–1232

Rezaei SN, Chouinard L, Langlois S, Légeron F (2016) Analysis of the effect of climate change on the reliability of overhead transmission lines. Sustainable Cities and Society 27, 137–144

Robinson D, Cruikshank K (2006) Hurricane Hazel: Disaster Relief, Politics, and Society in Canada, 1954-55. Journal of Canadian Studies 40, 37–70.

Rossignol A (1998) L'estuaire maritime et le golfe du Saint Laurent: Carnet d'océanographie. Rimouski: INRS-Oceanology & Pêches et Océans Canada, p 12

Roulleau E Stevenson R (2013) Geochemical and isotopic (Nd–Sr–Hf–Pb) evidence for a lithospheric mantle source in the formation of the alkaline Monteregian Province (Quebec). Can J Earth Sci 50: 650–666.

Roussy AR Kevan PG (2014) Pollination and reproduction in Sugar Maple Trees. Ontario Arborist, ISAOOctNov_Pollination_SugarMaple.pdf. Accessed 2 Aug 2022

Rowe JS (1972) Forest regions of Canada. Can For Serv, Dept Environ Pub 1300. Ottawa, ON. 172 p. + map.

Russell HAJ, Arnott WC, Sharpe DR (2003) Evidence for rapid sedimentation in a tunnel channel, Oak Ridges Moraine, southern Ontario, Canada. Sed Geol 160:33–55

Saad C, El Adlouni S, St-Hilaire A, Gachon P (2015) A nested multivariate copula approach to hydrometeorological simulations of spring floods: the case of the Richelieu River (Québec, Canada) record flood. Stoch Env Risk Assess 29:275–294

Shaw J, Barrie JV, Conway KW, Lintern DG, Kung R (2020) Glaciation of the northern British Columbia continental shelf: the geomorphic evidence derived from multibeam bathymetric data. Boreas, 49, 17–37

Sills D (2012) Map of all verified tornadoes in Canada between 1980 and 2009. Environment Canada

Slaymaker O, Catto NR (2020) Landscapes and landforms of Eastern Canada. Springer International Publishers, Cham, Switzerland

Sly PG, Munawar M (1988) Great Lake Manitoulin: Georgian Bay and the North Channel In: Munawar M (ed) Limnology and fisheries of Georgian Bay/North Channel Ecosystems. Hydrobiologia, vol 163. Kluwer, pp 1–19

Sly P, Prior J (2011) Late glacial and postglacial geology in the Lake Ontario basin. Can J Earth Sci 21:802–821

Smith CAS, Webb KT, Kenney E, Anderson A, Kroetsch D (2011) Brunisolic soils of Canada: genesis, distribution, and classification. Can J Soil Sci 91:695–717

Smoyer K, Rainham D, Hewko J (2000) Heat-stress-related mortality in five cities in Southern Ontario: 1980–1996. Int J Biometeorol 44: 190–197

SODES (St Lawrence Economic Development Council) (2019). https://www.st-laurent.org/en/ice-jam-in-lake-saint-pierre-the-shortage-of-icebreakers-threatens-the-economy-and-public-safety/. Accessed 2 Aug 2022

Soil Classification Working Group (1998) The Canadian system of soil classification, 3rd edn. Publication, Agriculture and Agri-Food Canada, p 1646

Soil Classification Working group (2001) Soils of Canada. Agriculture and agri-foods Canada, Research Branch. Scale 1:6,500,000 map. https://sis.agr.gc.ca/cansis/publications/maps/soc/all/soils/soc_all_soils_2004en.png. Accessed 12 July 2022

Solymar B (n.d.) Stewardship guide to Grasslands in Southern Ontario: an introduction for farmers and rural landowners. Ontario Barn Owl Recovery Project. http://www.bsc-eoc.org/regional/barnowl.html. Accessed 2 Aug 2022

Song X, Wang S, Hu Y, Yue M, Zhang, T, et al. (2017) Impact of ambient temperature on morbidity and mortality: An overview of reviews. Science of the Total Environment 586:241–254.

Sookhan S, Eyles N, Arbelaez-Moreno L (2018) Converging ice streams: a new paradigm for reconstructions of the Laurentide Ice Sheet in southern Ontario and deposition of the Oak Ridges Moraine. Can J Earth Sci 55:373–396

Spagnolo M, Clark CD, Hughes ALC, Dunlop P (2010) The topography of drumlins; assessing their long profile shape. Earth Surf Proc Land 36:790–804

Statham MJ, Sacks BN, Aubry KB, Perrine JD, Wisely SM (2012) The origin of recently established red fox populations in the United States: translocations or natural range expansions? J Mammal 93:52–65

Statistics Canada (2011) Census of Canada. https://www12.statcan.gc.ca/census-recensement/2011/as-sa/98-310-x/2011003/fig/fig3_2-1-eng.cfm

Stewart RE (1992) Precipitation types in the transition region of winter storms. Bull, Am Meteorol Soc 73:287–296

St-Jean A, Meunier S, Nolet P, Messier C, Achim A (2021) Increased levels of harvest may favour sugar maple regeneration over American beech in northern hardwoods. For Ecol Manage 499:119607. https://doi.org/10.1016/j.foreco.2021.119607

Stradling D, Stradling R (2008) Perceptions of the Burning River: deindustrialization and Cleveland's Cuyahoga River. Environ Hist 13:515–535

Swan C (1977) Canada: symbols of sovereignty. University of Toronto Press

Teller JT (1995) History and drainage of large ice-dammed lakes along the Laurentide Ice Sheet. Quaternary Internattional 28:83–92

Terrail R, Dupuis S, Danneyrolles V, Fortin M-J, Boucher Y, Arsenault D (2019) Reorganization of tree assemblages over the last century in the northern hardwoods of eastern Canada. Appl Veg Sci 22:474–483

Thomas RL, Christensen MD, Szalinska E, Scarlat M (2006) Formation of the St. Clair River Delta in the Laurentian Great Lakes system. J Great Lakes Res 32:738–7548

Tinkler KJ, Pengelly JW, Asselin G, Parkins WG (1994) Postglacial Recession of Niagara Falls in Relation to the Great Lakes. Quatern Res 42:20–29

University of Guelph Arboretum (n.d.) Elm Recovery Project. https://arboretum.uoguelph.ca/researchandstewardship/elmrecovery. Accessed 2 Aug 2022

US EPA (2006) Great Lakes, physical facts. United States Environmental Protection Agency. http://www.epa.gov/glnpo/physfacts.html. Accessed 2 Aug 2022

US Fish and Wildlife Service and Detroit River International Wildlife Refuge (n.d.). https://www.fws.gov/refuge/detroit-river. Accessed 2 Aug 2022

Waldron G (2003) Trees of the Carolinian Forest. Boston Mills Press, Erin, Ontario

Wheeler JO, Hoffman PF, Card KD, Davidson A, Sanford BV, Okulitch AV, Roest WR (1996) Geological Map of Canada. Geological Survey of Canada, Map D1860A

Watson KW (2007) The Rideau route: exploring the pre-canal waterway. Privately Published, Elgin, Ontario

Watt WE (1989) Hydrology of floods in Canada: a guide to planning and design. National Research Council of Canada, Associate Committee on Hydrology, Ottawa

Walsh LL, Tucker PK (2017) Contemporary range expansion of the Virginia opossum (*Didelphis virginiana*) impacted by humans and snow cover. Can J Zool 96:107–115

Warren F, Lulham N (2021) Canada in a changing climate: National Issues Report; Government of Canada, Ottawa, ON

Xu Z Tong S (2017) Decompose the association between heatwave and mortality: Which type of heatwave is more detrimental? Environmental Research 156, 770–774

Zoladeski, C, Hayes K (2013) Recovery strategy for the Cherry Birch (Betula lenta) in Ontario. Ontario Recovery Strategy Series. Prepared for the Ontario Ministry of Natural Resources, Peterborough, Ontario, vi + 12 pp

Prairie Grasslands Ecozone

5

Abstract

The Prairie Grassland regions form the southern part of the Interior Platform of Canada. They represent unique landscapes developed under continental climates. The southern Interior Platform is divided into three major physiographic plains, separated by escarpments. The prairies include a broad region with cold steppe climate (Koppen-Geiger BSk). The true prairie is flanked by a transitional belt of mid-boreal climate (Dfb) in the Aspen Parkland to the north and west. The Cypress Hills form a boreal enclave, the result of altitude. In general, the climate becomes systematically drier from east to west, influenced by the Rocky Mountains of the Cordillera. Vegetation, fauna, and soils show corresponding changes. The easternmost part of the Manitoba Plain supports tallgrass prairie, with humic-rich black soils. Western Manitoba and southeastern Saskatchewan are marked by midgrass prairie vegetation, and greyer, less humic soils. The driest areas of southwestern Saskatchewan and southern Alberta support a shortgrass prairie flora, and brown, humic-poor soils. A similar transition from midgrass to shortgrass prairie is evident from north to south, between Edmonton and Calgary, and between Saskatoon and Regina. Distinctive soils characterize the Tallgrass, Midgrass, and Shortgrass prairies, and the parkland areas. Soils continue to evolve under changing land uses, moisture regimes, and climates. BSk soils are dominantly Chernozems, with some Solonetz and minor areas of Vertisols. The Interior Platform is underlain by Palaeozoic, Mesozoic, and Palaeogene sedimentary rocks, with ages generally decreasing from east to west. Most strata are flat lying to very gently dipping towards the southwest. Although the prairies were glaciated multiple times during the Quaternary, almost all of the landforms and landscapes are associated with the most recent glacial event, beginning 28,000 years ago. Most of the region was covered by glacial ice originating from Kivalliq (mainland Nunavut). Simultaneously, Cordilleran glacial ice expanded from the Yellowhead Pass area in Jasper National Park, and flowed to the south along the Rocky Mountain Front. As the regional topography sloped towards Hudson Bay, the pre-existing drainage patterns were blocked by the glacial fronts. Prior to 1500 CE, the area was dominated by large grazing mammals, notably Bison bison bison. Ungulate populations included Cervus canadensis and Odocoileus virginianus in the parkland and northern midgrass regions, Odocoileus hemionus in the central areas, and Antilocapra americana in the shortgrass prairie. The primary predatory species were Canis lupus, Puma concolor, Lynx canadensis, Lynx rufus, Taxidea taxus, and Canis latrans. Ursus americanus and Ursus arctos horribilis were also present. The impact of newer faunal arrivals was considerable and changed the assemblages significantly.

Keywords

Prairie Grasslands · Cold steppe climate · Tallgrass · Midgrass · Shortgrass · Bison

5.1 Introduction

The Prairie regions form the southern part of the Interior Platform of Canada. They represent unique landscapes developed under continental climates. Although the region has the reputation in some (non prairie resident) quarters as being flat, monotonous, and uniform, a single trip across the prairies should convince the observer that there is much to see and many differences from place to place. Many long-time prairie residents develop deep emotional ties to the landscape.

In general, the climate of the prairies becomes systematically drier from east to west, influenced by the Rocky Mountains of the Cordillera. Vegetation, fauna, and soils show corresponding changes. The easternmost part of the Manitoba Plain supports tallgrass prairie, with humic-rich black

N. Catto and G. R. Vincent, *Canada: Land and People*, World Regional Geography Book Series,
https://doi.org/10.1007/978-3-032-04391-7_5

soils. Western Manitoba and southeastern Saskatchewan are marked by midgrass prairie vegetation, and greyer, less humic soils. The driest areas of southwestern Saskatchewan and southern Alberta support a shortgrass prairie flora, and brown, humic-poor soils. A similar transition from midgrass to shortgrass prairie is evident from north to south, between Edmonton and Calgary, and between Saskatoon and Regina.

The relatively rapid agriculturally based settlement from 1870 through 1910, and the perceived lack of "spectacular" tourist attractions, delayed establishment of national parks in the prairies. Currently, there are four national parks in the southern Interior Plains, of which only Grasslands occupies a truly open prairie area (Fig. 5.1).

5.2 Physiographic Subdivisions

The southern Interior Platform is divided into three major physiographic plains, separated by escarpments (Mossop and Shetson 1994, Bostock 2014; Rennie 2017; Fig. 5.2). Eastern and central Manitoba occupy the lowest of these steppes, the First Prairie Level or Manitoba Plain. The First Prairie Level is bordered to the west and southwest by the Manitoba Escarpment. The central third of the prairies is occupied by the Second Prairie Level or Saskatchewan Plain, which extends from the Manitoba Escarpment in southwestern Manitoba to the Missouri Coteau escarpment in central Saskatchewan. The third and highest steppe is the Third Prairie Level or Alberta Plain, extending from the Missouri Coteau to the Rocky Mountains.

5.2.1 First Prairie Level (Manitoba Plain)

The Red River Valley is the first area of the Prairies that westbound travellers encounter. It represents a substantial contrast to the rolling, heavily forested terrain of the Canadian Shield evident along the shores of Lake-of-the-Woods and in Whiteshell Provincial Park, and most visitors are aware that a sharp physiographic, climate, and environmental boundary has been crossed (Fig. 5.3). As the first part of the prairies to be settled by agriculturally inclined Europeans, the Red River Valley has experienced the greatest human impact on the environment (Welsted et al. 1996).

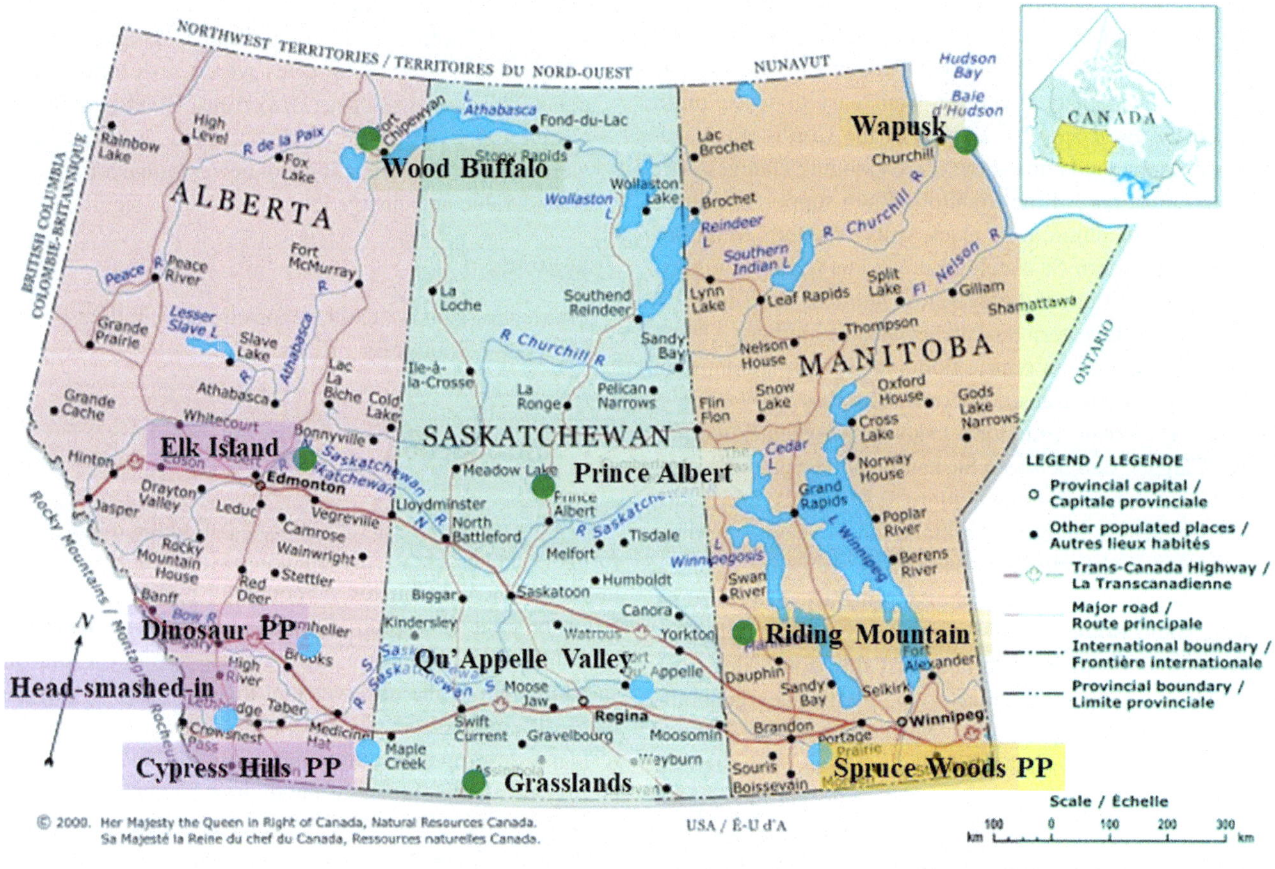

Fig. 5.1 Prairie provinces, showing locations discussed in text. Green dots: national parks; blue dots: other protected areas

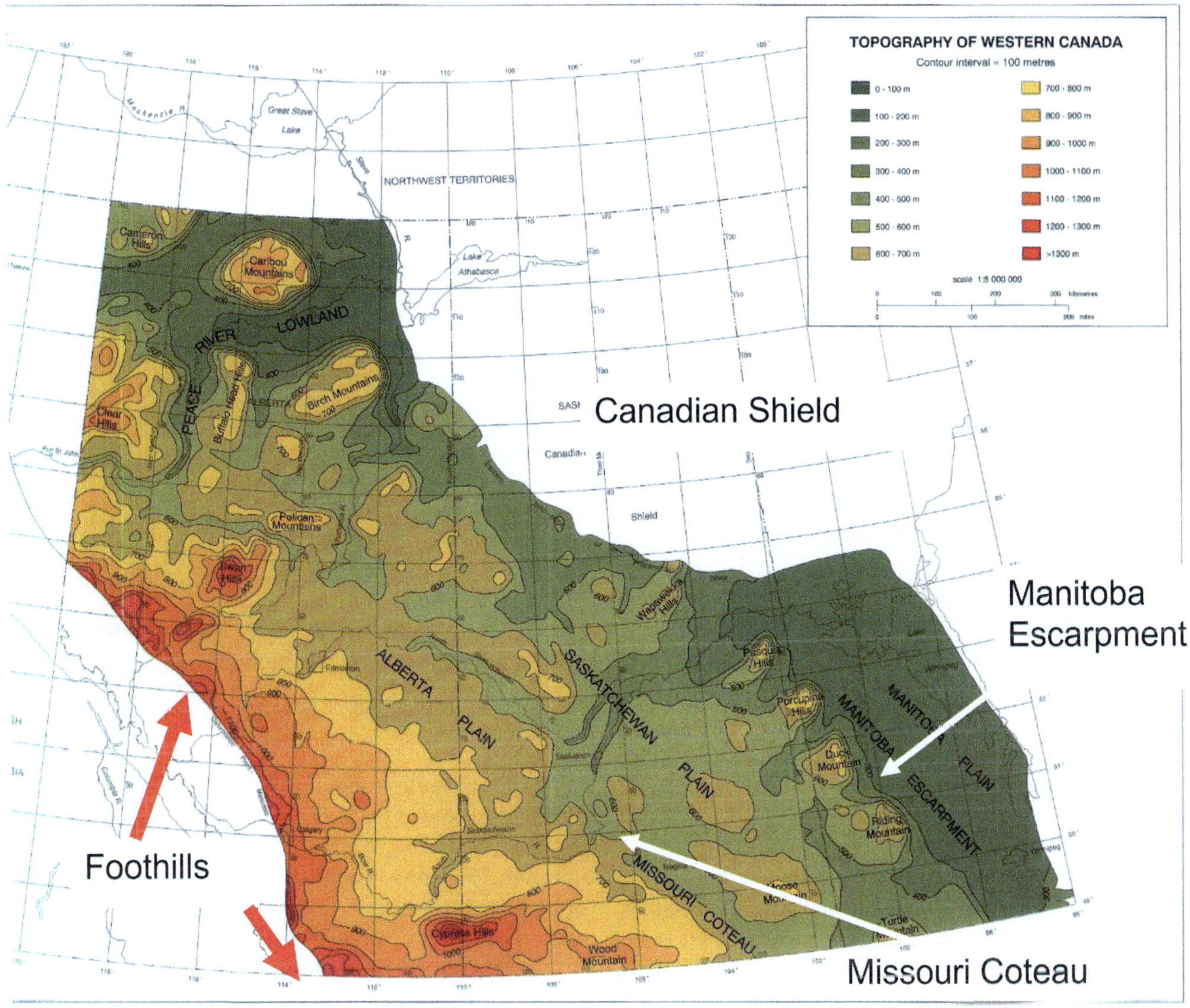

Fig. 5.2 Physiography of Prairie Canada, showing the major physiographic divisions. Geologic Atlas of Western Canada, Alberta Geological Survey

The Red River flows north across the First Prairie Level, the lowest, most easterly, and flattest steppe of the Prairies (Fig. 5.4). The highest elevations in Winnipeg are less than 15 m above the "normal" flood stage of the Red River. Winnipeg is situated at the confluence of the Red River and its major tributary, the Assiniboine River. Like most major prairie communities, it developed on the most productive agricultural land in the vicinity (Kaye 1996; Rannie 1996). Prior to European settlement, this area supported the richest and most diverse floral and faunal community of the Canadian Prairies (Henderson and Koper 2014; Hamel 2015).

Relief in this area is extremely low, generally less than 50 m, making this region atypically flat in comparison to the remainder of the prairies (Brooks 2017). The Red River has not incised a deep canyon through the plain, and the valley is a broad, gently sloping arcuate depression up to 100 kms wide. The Red River is located in the centre of this depression, and meanders northwards to flow into Lake Winnipeg at Netley.

Lake Winnipeg appears to be huge on topographic maps, with a surface area of 24,387 km^2 (larger than Lake Ontario), but its maximum depth is a mere 18 m (Fisheries and Environment Canada 1978). Lake Manitoba, north of Portage-la-Prairie, is even shallower, with a maximum depth of 4 m. The low relief of the area renders it highly susceptible to flooding. "Winnipeg" can be translated from Nehinaw as "big muddy water".

Flooding, triggered by snowmelt and freshet, moves progressively northwards (Warkentin 1997; Rannie 2015; Buttle et al. 2016; Brooks 2017). During severe winters, such as 1996–97, snowfall in southern Manitoba exceeds 2 m (Environment and Climate Change Canada 2020). Spring

Fig. 5.3 Chernozem soils developed on Glacial Lake Agassiz sediment mark the flattest eastern region of the Manitoba Plain

comes rapidly, with mean temperatures rising from −7.0 °C in March to 11.6 °C in May, producing rapid snowmelt and large freshets.

Initially, the CPR line was to cross the Red River at Selkirk, north of Winnipeg, where 10 m high bluffs restrict flooding. The residents of Winnipeg, however, exerted sufficient political pressure to force the line to be diverted to the south, into the area most susceptible to flooding. The Red River has flooded several times since Fort Garry was constructed, notably in 1881 (when the newly completed CPR bridge at Winnipeg was destroyed) and in 1950, when much of riverfront Winnipeg was submerged (Bumsted 1993; Rashid 2011).

The spring 1997 flood saw water levels exceeding those of 1950, with discharges more than 50 times the average value (Warkentin 1997; Haque 2000; Rannie 2015; Brooks 2017). On April 19, more than 90% of Grand Forks (ND, USA), was under water, requiring evacuation of the entire city and severely damaging the water and sanitation systems. Flooding progressed northwards into Manitoba in early May, flooding more than 1800 km^2 and 20 small communities, and displacing 27,400 people. The estimated cost of the flood exceeded $1.5 billion (2024 equivalent).

To attempt to avoid a repeat of the economic and human disruption of the 1950 flood, the "Red River Floodway" was constructed to bypass Winnipeg to the east, connecting the Red south and north of the city. Flooding in 1997 upstream of Winnipeg exceeded the 1950 level, with the peak discharge reaching 3400 $m^3 s^{-1}$ (compared to the normal 75 $m^3\ s^{-1}$). However, diversion of the floodwaters around Winnipeg saved an estimated $10 billion in flood damage costs in 1997, a substantial payoff of the $300 million construction cost (City of Winnipeg 1997, Haque 2000; Catto 2015).

Although communities in southern Manitoba were devastated by the flood of May 1997, the city of Winnipeg suffered relatively minor damage. Even though the floodwaters rose higher in 1997 than they had in 1950, the Red River Floodway had successfully diverted most of the water around the city (Rannie 2015; Brooks 2017). Had Winnipeg suffered inundation comparable to that of 1950, direct costs could easily have exceeded $10 billion. The Red River Floodway has thus proven to be a good investment.

Ten subsequent floods occurred along the Red River in southern Manitoba between 1998 and 2019, approximately one every two years. After the 1997 flood, an additional $750 million (2022 equivalent) was spent on infrastructure upgrades in Winnipeg. South of the city, dykes were constructed around 8 communities, and funding provided to elevate other buildings and homes on pads. Selkirk, north of the Red River Floodway, suffered flood damage in 2009 (Wazney and Clark 2015).

Fig. 5.4 The Red River flows north through the Manitoba Plain. Upper left: Winnipeg is built at the confluence of the Red and Assiniboine Rivers. Downtown is located north of the Assiniboine River. (International Space Station/NASA image). Lower left: the Red River Floodway was built in 1967 to divert flood waters east of Winnipeg. (Wikipedia/J. Park). Right: Red River near Morris MB shows tight meanders and oxbow lakes. Glacial Lake Agassiz sediment overlies glacial features oriented NNW-SSE. (National Air Photo Library)

5.2.2 Second Prairie Level (Saskatchewan Plain)

The Qu'Appelle Valley, formed by glacial meltwater discharge and hosting a major tributary of the Assiniboine, lies in the heart of the Second Prairie Level above the Manitoba Escarpment (Klassen 1975, 1989; Rannie 2017; Saskatchewan Geological Survey 1999; Fig. 5.5). Relief is generally low on the upland areas, approximately 70–80 m from valley bottom. The upland prairies are marked by rolling hills and irregular mounds and ponds, created by glacial hummocky terrain. Many of the ponds are saline or brackish, lacking outlets and often drying up during the summer. Ephemeral saline ponds, locally called "sloughs" or "potholes", are readily recognizable from the air by the presence of whitish rings of precipitated salts and the distinctive saline-loving vegetation which surrounds them (Last and Ginn 2005).

The Qu'Appelle Valley is bordered to the north by the Touchwood Hills (Mossop and Shetson 1994; Bostock 2014). This area of rolling hills is characterized by locally exposed bedrock, some of which has been thrust and pushed by advancing glacial ice, and a thin cover of glacial sediments. This area has a relief of 200–250 m, and is not as suitable for agriculture as are the surrounding regions. The Touchwood Hills serve as a refuge for fauna, and as a groundwater recharge area.

To the north, the Second Prairie Level extends across north-central Saskatchewan and northeastern Alberta. Topography is marked by rolling hills, belts of hummocky terrain, and arcuate ridges of ice-thrust, disturbed bedrock, aligned northwest–southeast (Fulton 1995; Fenton et al. 2013). Relief is generally less than 100 m, but no areas of flat topography comparable to the Red River valley or the Regina Basin exist. The elevation of the plain rises gradually from east to west.

Fig. 5.5 The Qu'Appelle valley is a deeply incised trough through the Saskatchewan Plain, with steep (locally vertical) sides 100–150 m high and a flat base 2–5 km wide. The valley was created by rapid discharges of meltwater from glacial lakes impounded against the retreating ice sheet. (NASA, Space shuttle image, Chris Hatfield). Inset: The modern Qu'Appelle carries less than 5% of the peak volume of these great meltwater floods, and it forms a tiny misfit stream along the centre of the deep canyon

The undulating escarpment of the Missouri Coteau (Fig. 5.6), trending southeasterly from Last Mountain Lake and Moose Jaw to Radville and the 49th parallel west of Estevan, forms the boundary between the Saskatchewan and Alberta Plains (Fenton et al. 1994; Bostock 2014), becoming less distinct north of the North Saskatchewan River valley. The Coteau varies from 15 to 40 km in width, with relief of 100–150 m from the Second Prairie level to the summit. Hummocky terrain marked by sloughs is common.

5.2.3 Third Prairie Level (Alberta Plain)

The Third Prairie level includes southwestern Saskatchewan and southern and central Alberta (Fig. 5.7). In southwestern Saskatchewan, Wood Mountain rises to 1048 m, flanked by ice-thrust topography to the east and northeast (Fenton et al. 1994; Christiansen and Sauer 1997; Aber 2007). Rivers in this area are permanent, but may shrink to 1% of their spring discharges during the late summer. The major streams, such as the Frenchman River, are ingrown (cut through sediment) or incised (cut through bedrock) meandering systems, with tight meanders and depths characteristically less than 2 m. Although these rivers are easy to ford, their deeply entrenched valleys are up to 3 km wide. Many are former meltwater discharge channels, and steep slopes subject to failure are ubiquitous (Klassen 1992). The river valleys thus form obstacles to animal migration and human movement across the terrain.

The Third Prairie Level in the Fort Macleod-Claresholm region is cut by the Oldman River, a braided stream which has developed a broad valley with steeply sloping walls 30 to 60 m high. Slumps and flows involving bedrock and glacial sediment are common along the valley walls. The surrounding terrain is a gently rolling plain, with hummocky topography and ice-thrust ridges of bedrock. Relief in this area beyond the valley is approximately 100 m.

South-central Alberta is a first glance a region of some uniformity. The rolling Third Prairie Level stretches to the horizon, the Rocky Mountains are slightly beyond view, and the topography is generally subdued. The traveller driving northeast from Calgary passes kilometres of hummocky terrain, broad wheat fields, fallow land, and more broad wheat fields. Approximately 10 kms from downtown Drumheller, however, a feature becomes noticeable on the horizon. The Red Deer River has incised a valley with almost vertical walls, 150 m deep and 10 km wide, marked by isolated pinnacles, blind side canyons, and scalloped topography. Legends of people riding off the prairie on black nights directly over the edge of the valley seem almost believable. The badlands of the Red Deer River at Drumheller are the largest example of badland topography in western Canada, encompassing 800 km^2 (Bryan et al. 1987; Campbell 1987). The upland surfaces are typically hummocky, marked by gently rolling hills. Isolated blocks of ice-thrust bedrock, notably the Hand Hills, provide some relief (Klassen 1989; Evans 2000). The overall relief of both prairie and Red Deer River valley approximates 250–300 m.

The Cypress Hills form a major topographic feature amidst the southern Third Prairie Level (Klassen 1992; Bostock 2014; Fig. 5.8)). The hills consist of three plateaux, the East, Centre, and West Blocks. Both Alberta and Saskatchewan have established provincial parks in the Cypress Hills, with Alberta's encompassing the northern part of the West Block and Saskatchewan's preserving parts of the Centre and East Blocks.

The northern escarpment of the three hill blocks rises above the prairie as a steep, locally near-vertical wall, with slump and sediment slide features developed from its crest. The relief increases from 150 m in the east to 400 m in the Elkwater, Alberta, area at the western end of the Cypress Hills. The plateaux are very gently southwesterly sloping surfaces, with less than 5 m relief over most of the flattest plateau, surmounting the West Block. The southern slopes of the Cypress Hills are irregularly terraced, marked by mass movements, and do not form pronounced escarpments in most areas (Klassen 1989).

The Cypress Hills form a striking contrast to the general appearance of the surrounding Shortgrass Prairie. The elevation difference has allowed the formation of a boreal enclave within the steppe, colonized by species that have survived since deglaciation, and those which have been able to spread from the Rocky Mountains to the west.

Fig. 5.6 The Missouri Coteau region in southwest Saskatchewan is marked by shortgrass vegetation assemblages and hummocky terrain

Fig. 5.6 (continued)

Fig. 5.7 High Alberta Plain south of Vulcan

Fig. 5.8 Cypress Hills escarpment, looking north towards Elkwater Lake

5.3 Climate

The prairies include a broad region with cold steppe climate (Koppen-Geiger BSk; Figs. 5.9 and 5.10). The true prairie is flanked by a transitional belt of mid-boreal climate (Dfb) in the Aspen Parkland to the north and west (Beck et al. 2006, Environment and Climate Change Canada 2020). The Cypress Hills form a boreal enclave, the result of altitude.

5.3.1 Manitoba

At the eastern margin, the Red River Valley (including Winnipeg) lies physiographically within the prairies, although its combination of mean annual precipitation and temperature pattern places it in the steppe-boreal climatic transition. Daily mean January and July temperatures at Winnipeg are −16.4 °C and 19.4 °C, respectively

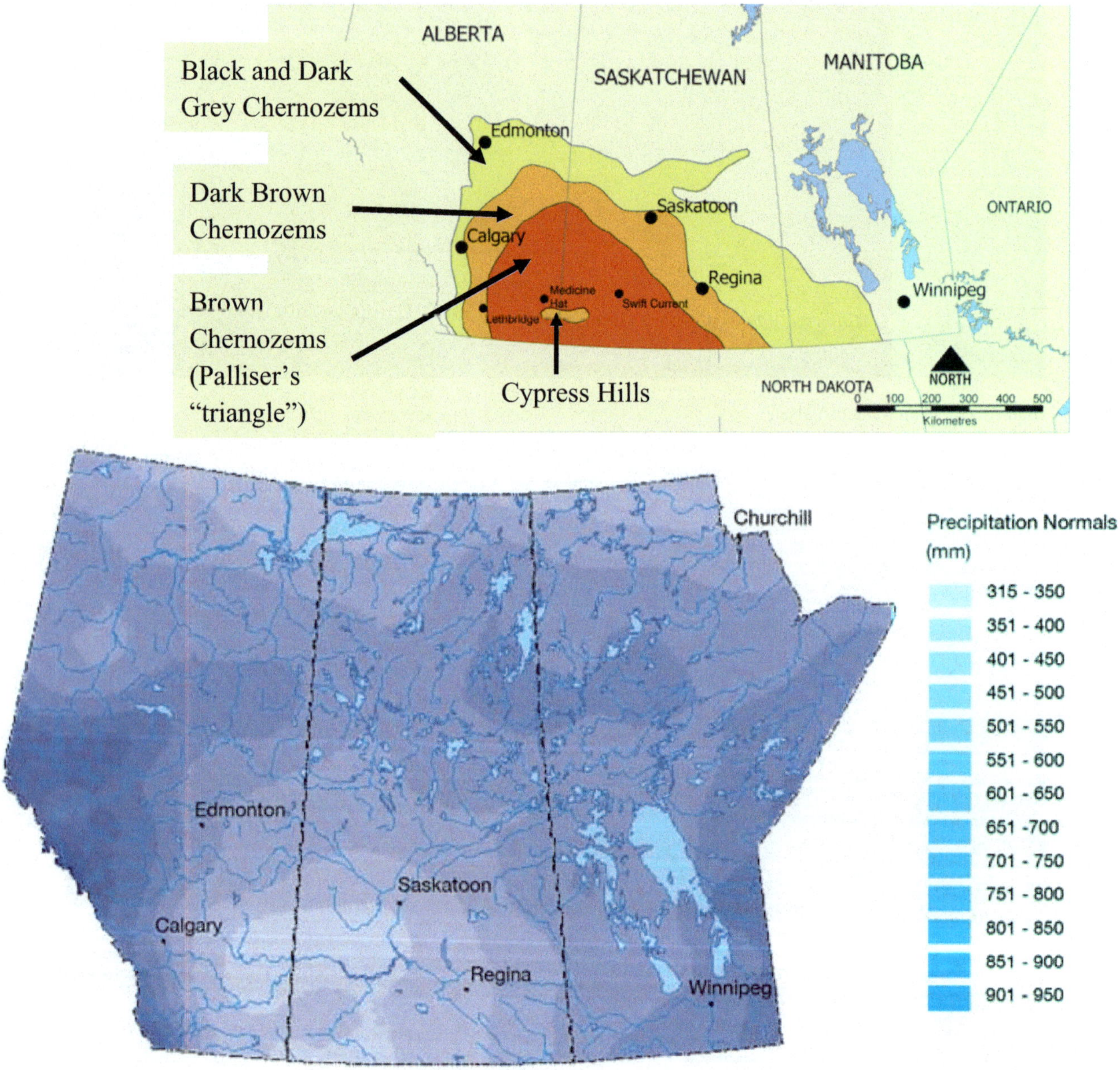

Fig. 5.9 Broad Climate Pattern across the grasslands ecoregion. Upper: soil types and general drier climates (red tones) show east–west and south–north patterns. Lower: Precipitation normals show more complexity, but do not indicate runoff or evapotranspiration. (Environment and Climate Change Canada)

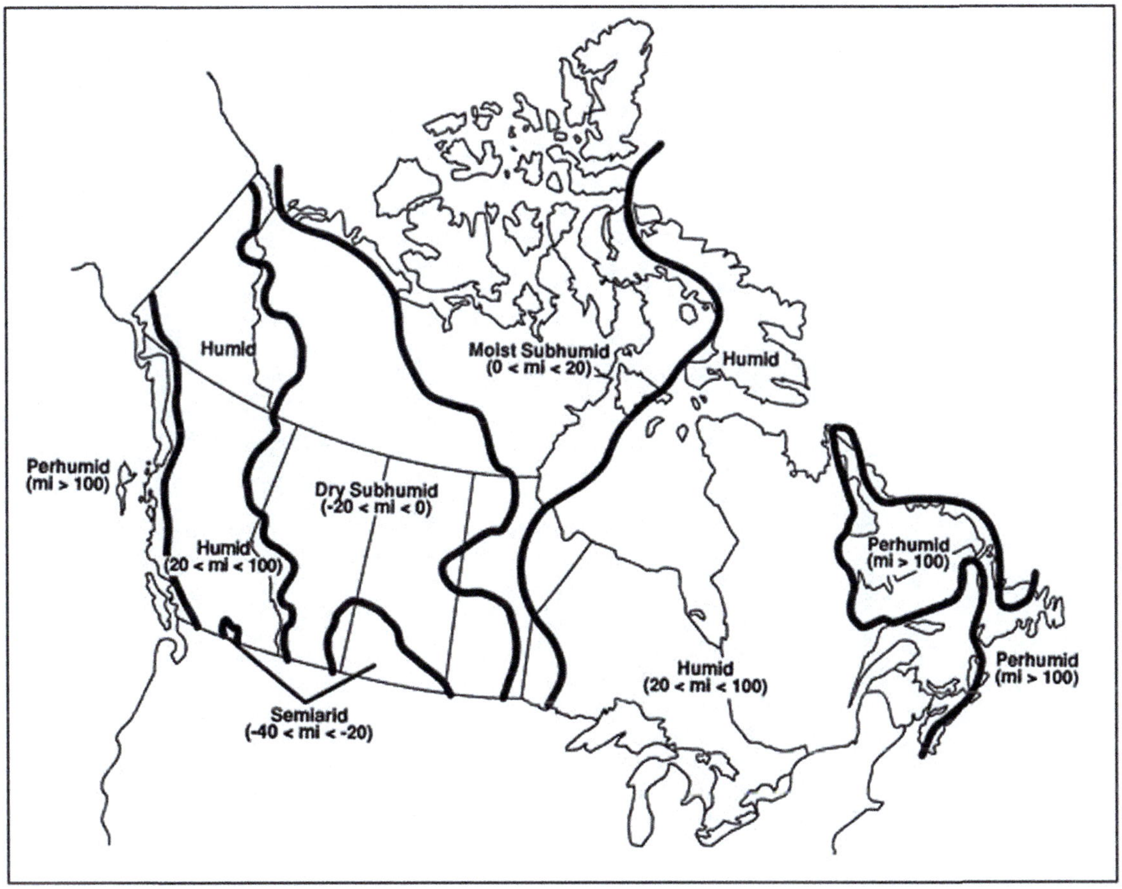

Fig. 5.10 Moisture Regimes in Canada. The Moisture Index (mi) is calculated as: mi = 100 × (S − D)/PE, where S is the water surplus, D is the water deficit, and PE is the potential evapotranspiration (mm). Much of the Grasslands ecoregion is classified as Semi-arid, with moisture indices less than −20 indicating overall water deficit

(Environment and Climate Change Canada 2020; Fig. 5.11). The large temperature ranges are characteristic of inland, "continental" climate regions (Phillips 1990). The typical summer is marked by 120 high summer days, abundant sunshine (2300 h y^{-1}), and ~22 thunderstorms. Winters can be bitterly cold, with temperatures below −20 °C for at least 1 h during 65 days of a typical year, and wind-chill exceeding −40 °C during 10% of January hours. The growing season is short, with a mean of 110 frost-free days per year: there is a 50% chance of frost after May 25, and a similar probability of frost before Sept 20.

Precipitation is high (500–550 mm y^{-1}) by prairie standards. The overall moisture regime is dry subhumid (MI = −5), so more water is lost through evapotranspiration and runoff than is replenished through precipitation (Cohen et al. 2015; Sauchyn et al. 2020). In the long run, therefore, Winnipeg will eventually run out of water, even if groundwater is utilized, despite periodic floods. At present rates of consumption and anticipated population growth, however, a severe water shortage (as opposed to the occasional summer drought) will not occur for two hundred years. Approximately 20% of the precipitation falls as snow, and freezing rain is very rare. Snow cover persists from 110–130 days, but is frequently patchy due to removal by strong winds. Winds are persistently westerly, and contribute to the human perception of Winnipeg winters. Secondary southerly winds originating from the Gulf of Mexico occur during the summer months, rarely bringing tornadoes to the area.

Climate differences between Manitoba Escarpment (e.g. Riding Mountain National Park) and midgrass Assiniboine Valley areas (e.g. Brandon), and between slopes which face north and those facing south, are common throughout the area. Generally, slopes which face north are covered with snow for longer periods each year, are less influenced by wind, are moister, and are more prone to mudflows and slumps than are south-facing slopes.

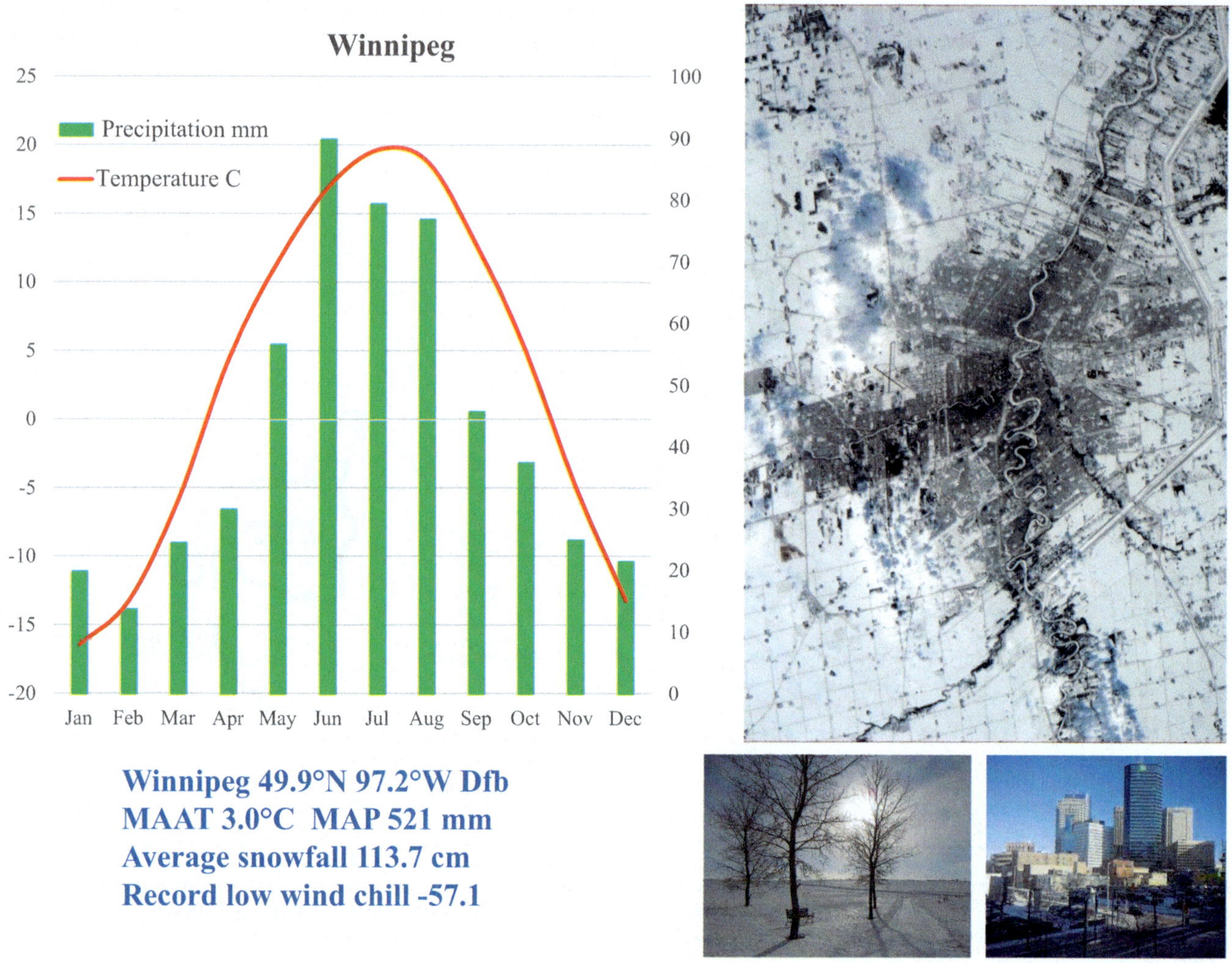

Fig. 5.11 Winnipeg has a Köppen-Geiger Dfb climate, marked by warm summers and cold winters. (Satellite image: NASA) (Data: Canadian Climate Normals, Environment and Climate Change Canada)

Daily mean temperatures in the Spruce Woods Provincial Park region along the Assiniboine valley are −15 °C (January) and 18 °C (July). In Riding Mountain, temperatures average 2 °C cooler. Mean annual precipitation varies between 550 mm (south-facing slopes) and 750 mm (north-facing slopes in Riding Mountain). Snowfall varies between 100 and 120 cm y^{-1}, and snow cover persists up to 165 days on sheltered, north-facing slopes. Riding Mountain and most north-facing slopes have a neutral moisture budget, but the south-facing slopes in the southern part of the area have dry subhumid moisture regimes (MI = −15), making them locally susceptible to drought. Summers are hot, sunny (2050–2100 sunshine hours), and marked by thunderstorms (23–27 per year) and occasionally destructive hailstorms (typically 1–2 per year). Winds are consistently westerly (Environment and Climate Change Canada 2020).

5.3.2 Saskatchewan

To the west, the midgrass Qu'Appelle valley of Saskatchewan has a climate on the border between BSk and Dfb (Fig. 5.12). Daily mean temperatures in January and July are −14 °C and 18 °C, respectively (Environment and Climate Change Canada 2020). Winter days with temperatures <−30 °C and summer temperatures >30 °C are common. Precipitation is low, averaging 500 mm y^{-1}, of which 20–25% falls as snow. Northern (south-facing) slopes tend to receive less overall precipitation and especially less snowfall than do north-facing slopes. Snow cover persists for 130–140 days on south-facing slopes, and up to 30 days longer on north-facing slopes and in the Touchwood Hills. Summer evaporation generally slightly exceeds precipitation, and the area has an annual dry subhumid moisture regime (mi = −5). Winds are predominantly westerly throughout the year.

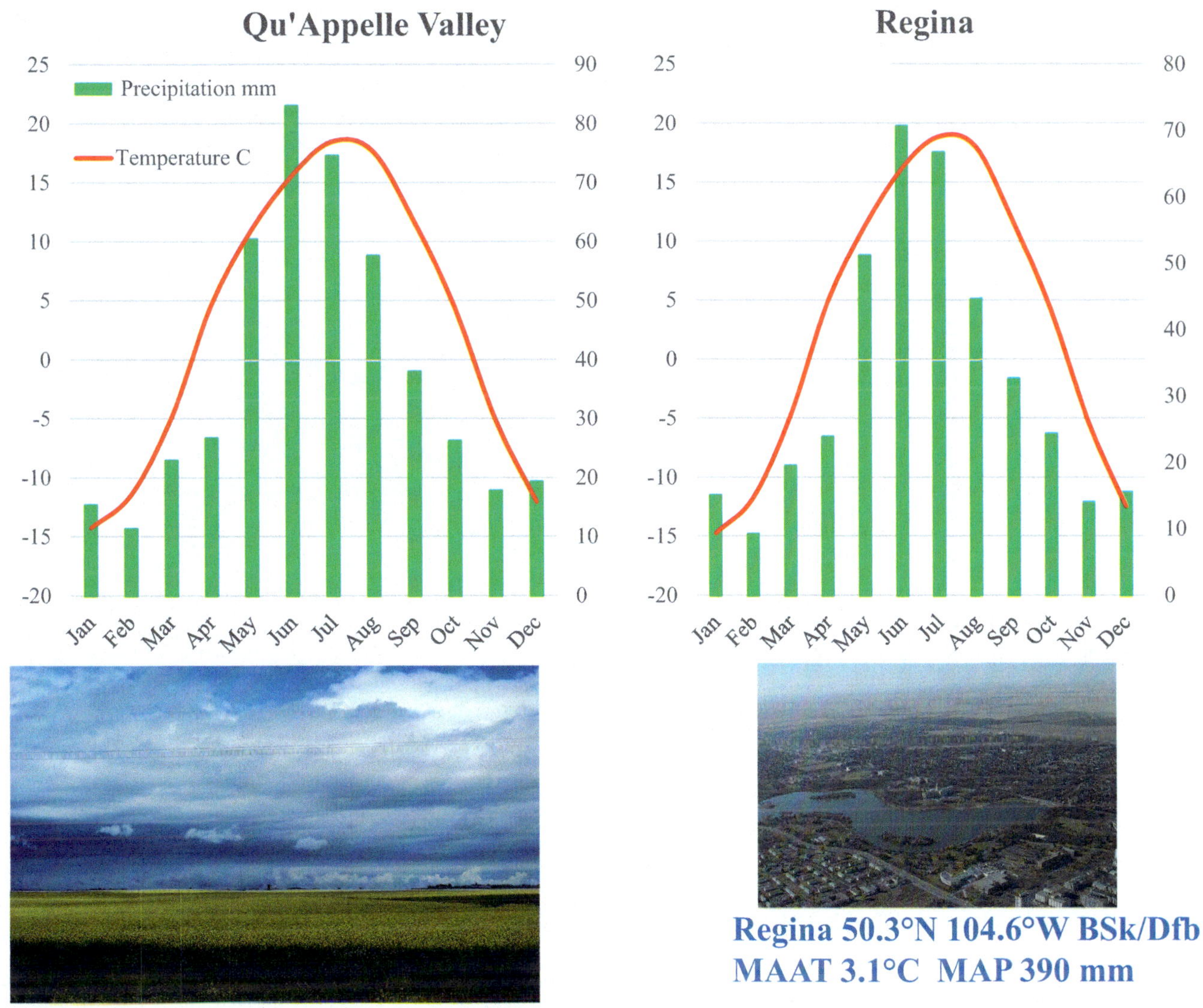

Fig. 5.12 Climagraphs for Qu'Appelle Valley and Regina, regions in the transition zone between BSk /and Dfb Köppen-Geiger climates. (Data: Canadian Climate Normals, Environment and Climate Change Canada)

The Shortgrass Prairie area is statistically the warmest in Canada during the summer (Phillips 1990; Environment and Climate Change Canada 2020; Sauchyn et al. 2020). July daily mean temperatures vary from 23 °C to 25 °C across the Grasslands region (Fig. 5.13). Temperatures of 45 °C were recorded in Midale and Yellow Grass, Saskatchewan, near the base of the Missouri Coteau, on July 5, 1937. During the previous summer of 1936, a severe heat wave and drought was responsible for at least 500 deaths across the southern prairies (Phillips 1990; Bonsal et al. 2011a, b). In prairie Canada, 5–10 deaths have been attributed to excessive heat in each year from 1989, with increased tolls since 2015 (Sauchyn et al. 2020). During a typical summer in Weyburn, the daytime high exceeds 30 °C on 20 days. The area is among the sunniest in Canada, receiving 2400–2500 h of sunshine/year. Winter temperatures are brisk, with typical January daily means of −12 to −15 °C, and temperatures below −40 °C rare but not unheard of.

The heart of the Shortgrass Prairie, including Grasslands National Park, is one of the driest areas in southern Canada. Precipitation totals vary between 220 and 320 mm per year, of which approximately one third falls as snow. Moisture budgets are uniformly negative and have been since ca. 8000 years ago. Recently, notable droughts occurred in 1933–1939, 1961, 1976–1977, 1988, and 2000–2002. Soil cracks develop by combination of desiccation and frost action under the BSk climate.

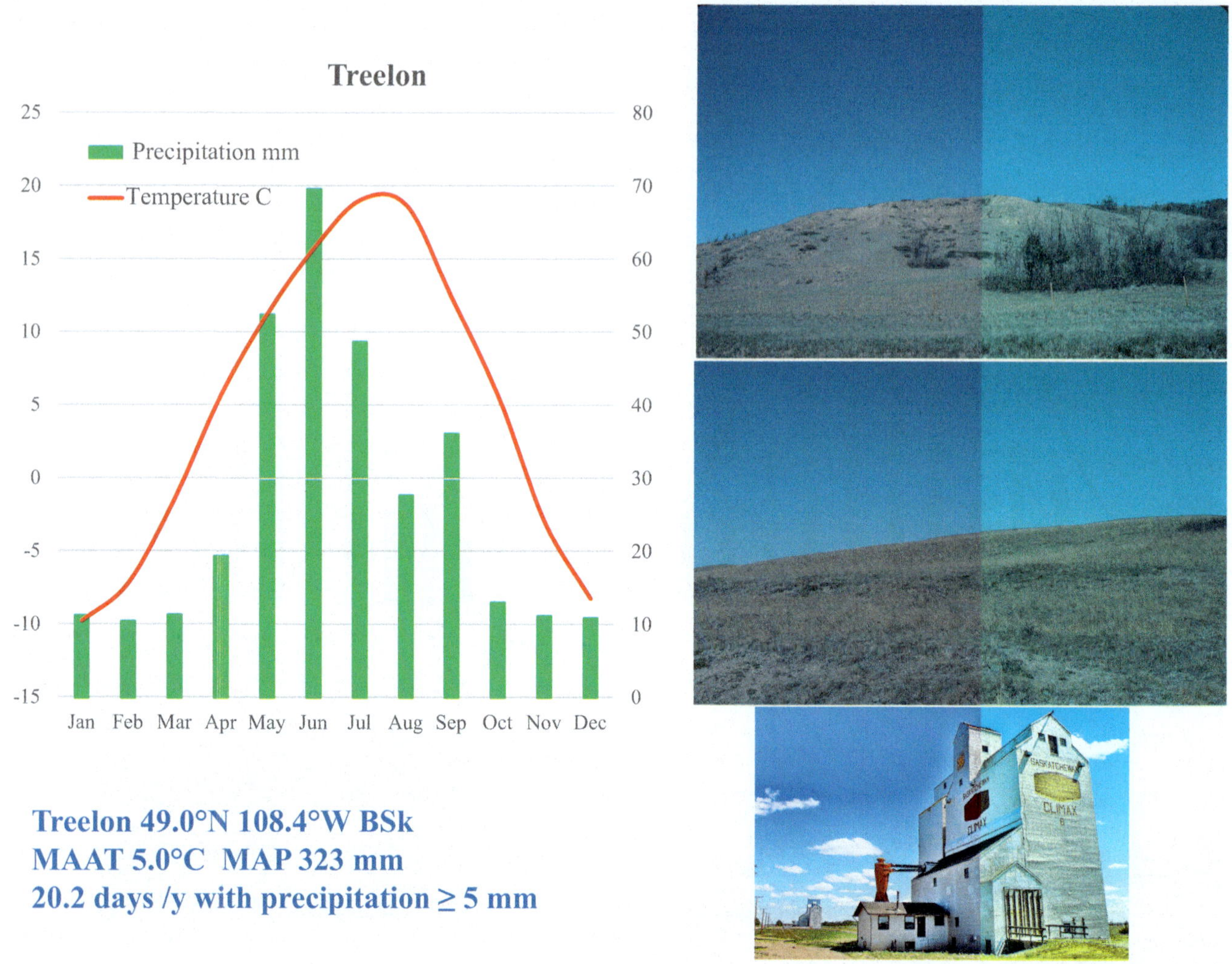

Fig. 5.13 Treelon SK, on the 49th parallel 60 km ESE of Grasslands National Park, has a dry cold steppe BSk climate. Shortgrass prairie is characteristic. Inset: Wooden grain elevators, such as that formerly present at Climax 22 km to the north, have largely disappeared from the prairie landscape. (Data: Canadian Climate Normals, Environment and Climate Change Canada)

Summer thunderstorms accompanied by hail and excessive runoff may nominally end or pause a meteorological drought, but may not provide water useful for crop growth. Accumulated snow gradually melts to produce soil moisture and recharge groundwater aquifers, whereas winter rain may run across frozen surfaces and not provide replenishment.

5.3.3 Alberta

Along the western margin of the prairies, each altitudinal zone, from the lowest steppe (BSk) areas at Fort Macleod and Claresholm, to the boreal (Dfb) summits of the Porcupine Hills at 1500 m asl, differs climatically and ecologically (Figs. 5.14 and 5.15). Daily mean temperatures at the Head-Smashed-In Bison Jump Historic Site are −10 °C (January) and 15 °C (July); mean temperatures in Fort Macleod (230 m lower) may be as much as 5 °C higher (Environment and Climate Change Canada 2020). Mean annual precipitation at Head-Smashed-In is approximately 300 mm, of which 40% falls as snow. Snow cover typically persists for 130–140 days in sheltered areas, but is rapidly removed by winds in most locations. The moisture regime at Head-Smashed-In is classified as dry subhumid (mi = ± −5). Higher elevation areas are generally colder and snowier than Head-Smashed-In, and lower regions tend to be warmer and drier.

During the winter months, dry winds descending the eastern flanks of the Rocky Mountains absorb large quantities of moisture and snow. Sublimation occurs, no water is produced, and the terrain remains dry. In the adjacent foothills and southwestern prairies from between Calgary and Fort Macleod, these "Chinook" winds result in the removal of snow throughout the winter, drying the uppermost soil layers in the low-lying areas (Nkemdirim 1997, Macdonald et al.

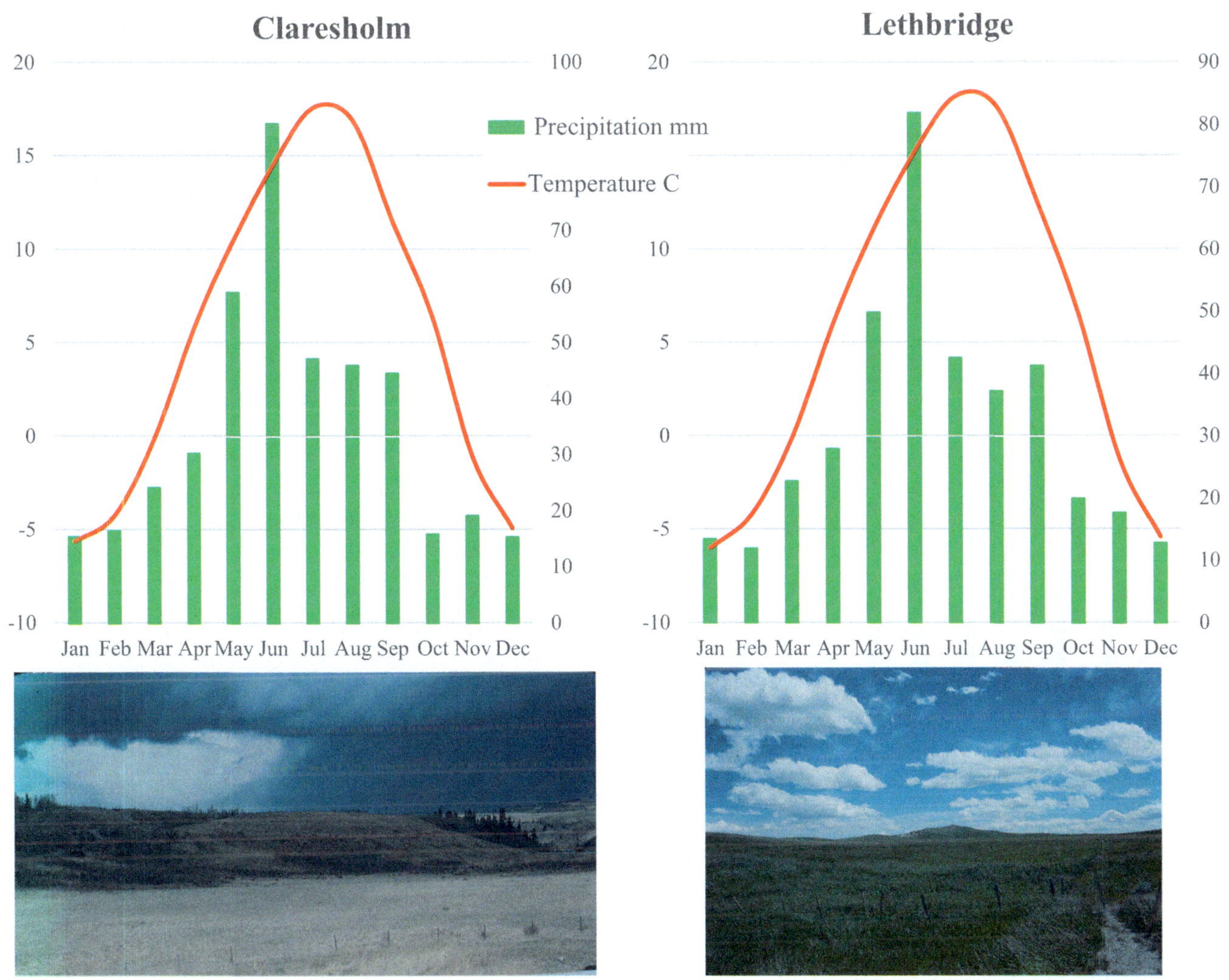

Fig. 5.14 Climagraphs for Claresholm and Lethbridge, southern Alberta. (Data: Canadian Climate Normals, Environment and Climate Change Canada)

2018). The Chinooks allow some snow-intolerant species to survive in the region, but also increase the depth of frost penetration, deterring others. Species which cannot tolerate open, cold conditions, such as *Lynx rufus* (Bobcat), cannot survive under the influence of Chinooks, but snow-tolerant species such as *Lynx canadensis* (Lynx), also find the area unfavourable. Thus, the area west of Claresholm is one of the few regions of Canada south of the treeline which does not support a natural *Lynx* population (Koen et al. 2014).

To the northwest, Drumheller also lies within the cold steppe realm (BSk). Daily mean temperatures in January and July are 12 °C and 20 °C, respectively, and extremes are common in both summer and winter (Environment and Climate Change Canada 2020). Precipitation is low, averaging 450 mm y^{-1}. Rain occurs both as sporadic cloudbursts associated with late afternoon summer thunderstorms, and as minor episodes of drizzle throughout the spring and fall. During the summer, a common phenomenon is virga, where rain falling from above evaporates before striking the ground.

The area receives more sunshine than any other in Canada, with up to 2600 h y^{-1}. On average, the sun shines for at least 1 h on 340 days y^{-1}. Humidity is very low, with completely cloudless skies occurring during 35% of the afternoon hours between late June and late September. Winter snowfall is low, averaging 105 cm, and much snow is removed by the prevailing westerly winds. The area has a dry subhumid moisture regime, locally necessitating irrigation. Sightseers have a 97% chance of suitable weather for at least 4 h on any July day.

Along the northern grassland-boreal prairie parkland ecotone, Edmonton and Elk Island National Park lie in the transitional zone between the mid-boreal Dfb and cold steppe Bsk climatic regions (Environment and Climate Change Canada 2020). Daily mean temperatures in July and January

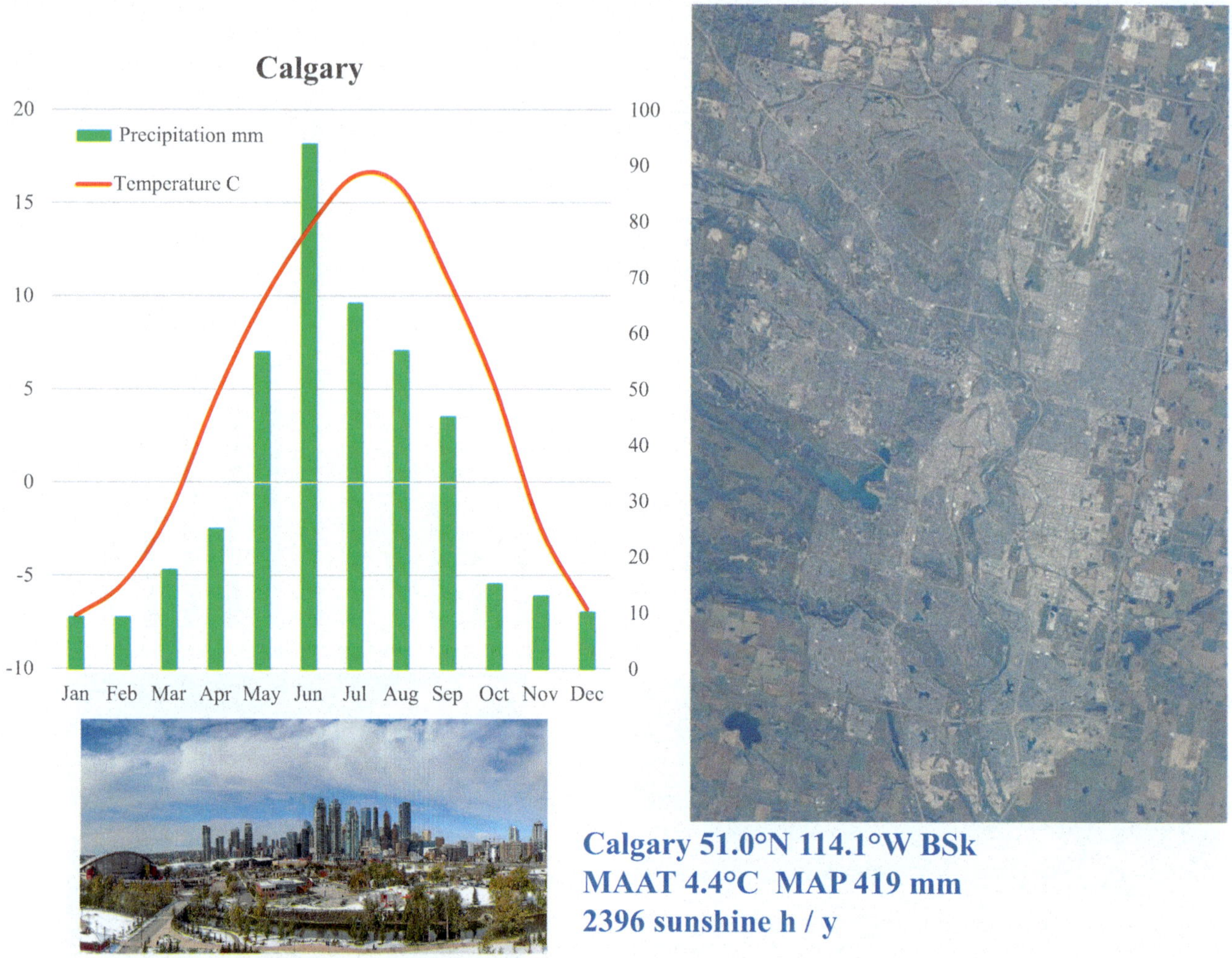

Fig. 5.15 Calgary has a Köppen-Geiger BSk (cold steppe) climate, marked by dry winters and abundant sunshine. (Satellite image: NASA). (Data: Canadian Climate Normals, Environment and Climate Change Canada)

are 18 °C and −14 °C, respectively. Minimum temperatures drop below −20 °C on 50 days during a typical year, and intervals where the daytime temperature does not rise above −30 °C for several days are not uncommon. Mean annual precipitation is 500 mm, of which 20–25% falls as snow (110 cm mean). Snow cover reaches a mean of 45 cm depth and persists for 180 days in wooded areas, but prevailing westerly winds rapidly remove snow from exposed regions. The area has a dry subhumid moisture regime, and receives 100 days of high summer in a typical year.

5.4 Bedrock Geology

The Interior Platform is underlain by Palaeozoic, Mesozoic, and Palaeogene sedimentary rocks, with ages generally decreasing from east to west (Mossop and Shetsen 1994; Wheeler et al. 1996; Fig. 5.16). Most strata are flat-lying to very gently dipping towards the southwest.

The Red River Valley is underlain predominantly by Ordovician and Silurian limestones, dolomites, and shales (Matile et al. 1998). These sediments are flat-lying to gently dipping, and are generally deeply buried beneath Quaternary deposits. One of the few rock quarries in the area, at Tyndall to the northeast of Winnipeg, produces distinctly mottled brown limestones containing numerous coral, brachiopod, bryozoan, algal, and arthropod fossils. This Tyndall limestone is extensively used for building facades and interiors throughout Canada, including the main foyer and other parts of the Parliament Buildings in Ottawa.

A small region south of Winnipeg is underlain by younger Jurassic rocks, predominantly sandstones and siltstones, deposited in shallow seas. These rocks contain fossil

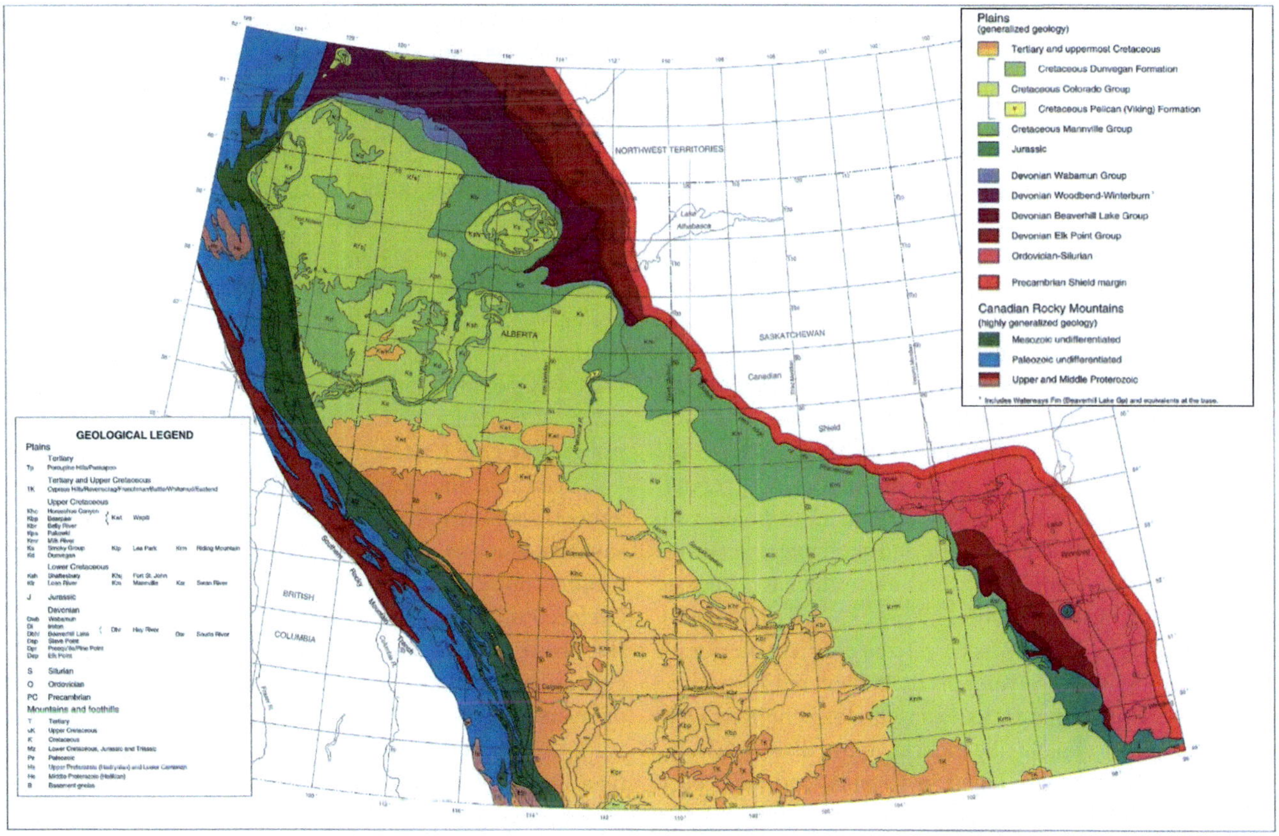

Fig. 5.16 Bedrock geology of Prairie Canada, showing the major physiographic divisions. Geologic Atlas of Western Canada, Alberta Geological Survey

plesiosaurs and other marine reptiles. The Jurassic sediments are soft, and easily eroded.

The rocks to the northeast of the Manitoba Escarpment in Riding Mountain National Park are mid-Devonian dolomites and limestones (Bannatyne et al. 1979). Unlike the Niagara Escarpment, the position of the Manitoba Escarpment is not controlled by any single rock type. To the south of Riding Mountain, Mid-Cretaceous and Late Cretaceous sandstones and shales crop out locally, but are commonly buried by thick accumulations of Quaternary sediment. All of the rock units are relatively easily erodible, and all contain calcium carbonate.

The Second and Third Prairie Levels are underlain by Late Cretaceous sandstones and shales, which dip gently to the west-southwest (Mossop and Shetsen 1994; Saskatchewan Geological Survey 1999, Prior et al. 2013). River systems commonly develop by erosion of soft strata at ± 90° to the orientations of the bedding planes. In most areas, however, the bedrock is covered by Quaternary sediment, so it does not influence the local details of relief.

In central Alberta, the bedrock has little effect on the physiography of the region, but is significant economically. The Cretaceous rocks contain coal seams and natural gas deposits, and they overlie Devonian reef rocks which contain petroleum. These Devonian reef limestones and dolomites represent the first rocks which provided significant quantities of petroleum and natural gas in Alberta, and thus fuelled the Alberta economy.

Saline soils and groundwaters result from the erosion of older Devonian Prairie Evaporite Complex salt strata, which underlie the Cretaceous rocks throughout the area (Mossop and Shetsen 1994). Several types of halides are present in the Devonian beds, including halite (commercial rock salt, $NaCl$) and sylvite (commercial potash, KCl). Saskatchewan's potash and potassium-fertilizer industry is based on the mining of these deposits. Although pure sylvite is colourless, Saskatchewan deposits generally contain some iron oxide which colours the potash bright ochre or orange, and produces a distinctive, highly pervasive coloured dust. Potash mining areas can usually be spotted at some distance from the mine, simply by observing the colour of gravel roads and highway shoulders.

The southern prairies are underlain by Late Cretaceous sandstones and shales, both marine and non-marine (Mossop and Shetsen 1994; Saskatchewan Geological Survey 1999, Prior et al. 2013). These clastic rocks are assigned to several geological units, including the Bearpaw Formation (marine shales) and the Edmonton Group (deltaic sandstones, shales, and coal seams). These rocks are very poorly consolidated, and can be dug with a hand shovel in most areas. Many burrowing rodents find the Cretaceous sandstones easier to tunnel through than the overlying glacial deposits.

In Saskatchewan, although a variety of Late Cretaceous terrestrial dinosaur fossils such as *Triceratops* are found in the area, the most exciting finds are those of *Tyrannosaurus rex*. The area extending from Grasslands to the *T. rex* Discovery Centre in Eastend has produced several *T. rex* fossils, including rarely preserved skin casts, bones with preserved internal structures, teeth, and skull material (Royal Saskatchewan Museum 2022). Further to the east, Jurassic marine shales with plesiosaur fossils are exposed (Sato 2003). The community of Ponteix has erected a concrete plesiosaur at its village limit (les Auvergnois de Ponteix n.d.).

The Late Cretaceous rocks exposed along the Red Deer River in Alberta contain the richest, most diverse assemblage of dinosaur fossils found in Canada, and have been "mined" for both fossils and coal for the last hundred years (Fig. 5.17). Dinosaur collecting in this region is now strictly regulated. Many of the best finds are on display at the Royal Tyrrell Palaeontological Museum, west of Drumheller in the Red Deer valley, and new excavations continue at Dinosaur Provincial Park, near Duchess.

Dinosaur Provincial Park has been declared a UNESCO World Heritage Site on the basis of its fossil faunal assemblage (Currie and Koppelhus 2005, Slaymaker et al. 2017). The dinosaur assemblage in the Red Deer Badlands is dominated by grazing herd genera, including *Triceratops* and hadrosaurs. Predatory species are relatively rare, and with the exception of *Albertosaurus* (a close relative of *Tyrannosaurus rex*) most are small, possibly scavengers. The stegosaurs and sauropods (such as *Diplodocus*) had become extinct millions of years prior to the Late Cretaceous.

Triceratops and the other ceratopsian dinosaurs were social animals, and travelled in large herds across the Cretaceous savanna, much as bison travelled over the prairies in the nineteenth century. Although grasses had not yet developed in the Cretaceous, *Triceratops* was primarily a grazer, foraging in river-bank areas. Many of the best specimens in the Tyrrell collection are *Triceratops* that were caught by sudden flooding of rivers as they attempted to cross. The largest ceratopsians were 15 m in length and 4,000 kg in mass, making them larger than bison. The horns appear to be primarily designed for rooting out bulbs and tubers, and for mating displays, rather than for defence against predators.

Fig. 5.17 Badlands overview from south of Drumheller. Dark grey: coal strata; grey: montmorillonite-bentonite siltstone and shale; light grey-white: siltstone and sandstone

5.4.1 Badlands

Badlands are areas of dissected topography with narrow ravines, sharp crests and ridges, isolated columns (hoodoos), and pinnacles, developed on shales, mudstones, and sandstones under arid to sub-humid conditions (Figs. 5.17 and 5.18). The badlands of the Red Deer River are the largest example of badland topography in western Canada (Bryan et al. 1987; Campbell 1987; DeLugt and Campbell 1992; Evans 2000; Catto 2015).

The Late Cretaceous siltstones contain bentonite, in a variety of hues from dark brown to purple to pale green to bright yellow, produced from alteration of volcanic tephra over 65–70 million years. Bentonite has the ability to absorb water up to 16 times its volume, producing a characteristic swelling on the surface that resembles popcorn. As the bentonite swells, it becomes slicker and more prone to sliding and mass movement.

Prolonged dry conditions produce deep desiccation cracks, further widened by frost, microorganisms, and burrowing animals. Grassland cover generally reduces the extent of shrinkage and desiccation crack development. Rainstorms during the summer months recharge the groundwater, and cause surface swelling of the bentonite, sealing the surface and trapping the groundwater. In winter, the trapped groundwater freezes, initiating frost creep and heave.

In spring, the weakened and saturated sediments creep, flow, and slide towards the valley bottom. Groundwater movement through cylindrical pipes is critical in triggering badland development (Bryan et al., 1987; Campbell, 1987; DeLugt and Campbell 1992). Piping requires loose, weak sediment, trans-

Fig. 5.18 Badlands morphology. Cretaceous sandstones, siltstones, and coals exposed at Grasslands SK (upper left), Manyberries AB (upper right), Drumheller, and other localities, are soft and easily eroded. The siltstones contain substantial quantities of the clay mineral bentonite, in a variety of hues from dark brown to purple to pale green to bright yellow, produced from the environmental modification of volcanic tephra since initial deposition. Bentonite is particularly efficient at retaining water and has the ability to absorb water up to 16 times its volume, producing a characteristic swelling on the surface that resembles popcorn (lower left and right). As the bentonite swells, it becomes slicker and more prone to sliding and mass movement

ported grain by grain at very low velocity. The surrounding sediment must be sufficiently cohesive to allow the vertical pipes to develop without failure. Saturated materials fail by subsidence, creep, viscous grain flow, slow glide, and high-viscosity debris flow.

Hoodoo development requires a resistant sandstone caprock over the underlying supporting pillar.

Continued disturbance causes the valley walls to recede laterally from the river, widening the valley without deepening it. The extensive development of the badlands along the Red Deer River valley indicates that a large proportion of the badland development may have occurred prior to the most recent glaciation (Evans 2000). Henderson and Tanke (2010) suggested that approximately 60% of the original volume of rock in Dinosaur Provincial Park has been lost to erosion, along with more than 10,000 skeletons of large dinosaurs.

5.4.2 Porcupine and Cypress Hills

In southwestern Alberta, the Porcupine Hills flanking the prairie are underlain by yellow–brown Palaeocene sandstones (Prior et al. 2013). This Porcupine Hills sandstone is resistant to weathering and was used to construct the Alberta Legislature Building in Edmonton. The sandstone forms the major bluffs of the hills, such as Head-Smashed-In (Slaymaker et al. 2017; Fig. 5.19). Fossils of deciduous tree leaves, redwood needles, and freshwater mollusca are common.

The Palaeocene sandstone represents deposition in braided river systems which developed across Alberta as the last marine seas receded, 65–60 million years ago. The rivers drained northeastwards to the Beaufort Sea. This drainage pattern remained essentially intact until central Alberta was glaciated. The glaciers blocked the drainage, infilled many pre-existing valleys with sediment, and created the easterly drainage channels currently occupied by the Oldman, Bow, Red Deer, and North Saskatchewan Rivers (Klassen 1989; Fenton et al. 1994, 2013).

The base of the Cypress Hills is composed of Cretaceous sedimentary rock (Prior et al. 2013). Bentonitic shales, containing a mixture of marine muds, deltaic silts, and volcanic ashes, formed at the margins of a shallow sea which extended northwards from the Gulf of Mexico to cover eastern Alberta ca. 70–66 million years ago. The sea advanced and retreated several times, allowing both marine animals (such as ammonites and ichthyosaurs) and terrestrial animals (such as hadrosaurs and *Triceratops*) to flourish in turn. Thin coal seams indicate periods when terrestrial vegetation, including the ancestor to today's true Bald Cypress trees of the southeastern United States, was able to colonize land.

The Cypress Hills are capped by younger Oligocene gravels. These gravels, approximately 30 million years old, contain fossil rhinoceros-like teeth, relics of the last rhinoceros to inhabit North America. They were deposited

Fig. 5.19 East flank of the Porcupine Hills west of Fort MacLeod

in river channels which originated in the Belt Mountains of Montana, and flowed northeasterly. In the intervening time, most of the original river sediment has been eroded away, leaving the Cypress Hills as isolated relict segments of the former river channels.

5.5 Quaternary

5.5.1 Glaciation

Although the prairies were glaciated multiple times during the Quaternary, almost all of the landforms and landscapes are associated with the most recent glacial event, beginning 28,000 years ago (Klassen 1989; Fulton 1995; Fullerton et al. 2000; Matile and Keller 2004a, b; Fenton et al. 2013; Slaymaker and Kovanen 2017a, b; Fig. 5.20). Deposits of older glacial and interglacial events are present along the Qu'Appelle, Oldman, Bow, and South Saskatchewan river valleys. Many of the large mammals that inhabited North America prior to the last glacial retreat, including mammoth, sabre-toothed cat, American cheetah, American lion, camel, and horse are represented by fossilized bones in these deposits.

The only nunataks remaining above the glacial inundation were the West and Centre Blocks of the Cypress Hills (Klassen 1989, 1992). The elevation of the north-facing escarpment, combined with the great distance from the Cypress Hills to the source area of the glaciers in Kivalliq, prevented glacial ice from overtopping the summits. The ice flowed around the plateaux, leaving them exposed. Wind-blown sand and silt accumulated on the summits, forming a blanket of loess and subduing the previous relief (Catto 1983). During periods of intensive permafrost activity when the hills were surrounded by glacial ice, cracking of the frozen ground allowed water to penetrate and freeze, forming wedges of ice 20 cm across and 1 m deep. When these wedges of ice melted, the cracks were filled by wind-blown loess, allowing ice-wedge casts to be preserved along the escarpment east of Elkwater.

Glacial ice advanced across the prairies from two sources (Klassen 1989; Dyke et al. 2003; Dyke 2004, Evans et al. 2014; Slaymaker and Kovanen 2017a, b). Most of the region was covered by glacial ice originating from Kivalliq (mainland Nunavut). The Kivalliq ice initially advanced across the Canadian Shield, eroding granitic and gneissic clasts, before extending southward into Manitoba and southwestward across Saskatchewan and Alberta (Fig. 5.21). As the Kivalliq ice advanced against the topographic gradient from the Cordillera across the drainage into Hudson Bay, it blocked the ancestral Saskatchewan, Red, and Nelson streams. The presence of the Canadian Shield clasts thus indicates transport upslope, against gravity, implying glaciation rather than fluvial activity.

In the Red River Valley, the ice flowed south through what is now Lake Winnipeg to North Dakota, eroding the carbonate bedrock and depositing calcium carbonate-rich till (Matile et al. 1998; Fullerton et al. 2000; Matile and Keller 2004a). An esker formed beneath the ice mass is preserved today as Bird's Hill Provincial Park, northeast of Winnipeg. Further west, the ice advanced almost directly upslope along the eastern flank of the Manitoba escarpment (Matile and Keller 2004b). Ice flow was largely diverted to the southeast, forming a strongly flowing lobe which extended into South Dakota and Des Moines, Iowa. Locally, glacial flutings, such as at Katepwa in the Qu'Appelle valley, indicate ice flow direction (Klassen 1975).

The upslope movement of glacial ice against the Missouri Coteau resulted in the formation of ice-thrust features, composed of deformation structures including folds and faults involving sediment and/or bedrock, created by glacial thrusting (Christiansen and Sauer 1997; Aber 2007; Aber and Ber 2007). In the Dirt Hills, arcuate ridges of weak sedimentary bedrock were produced by glacial thrusting, adjacent to a temporary terminus. Locally, meltwater channels erode and accentuate the depressions. Ice-thrust features, drag folds, and other glacial deformations involving both bedrock and sediment are common in glacial deposits throughout Alberta (c.g. Catto 1984; Fig. 5.22). Ice from Kivalliq extended south to Great Falls, Montana, and west to the margins of the Foothills of the Rocky Mountains, covering almost all of the prairies of Canada.

Along the western margin of the prairies from Edson south to Fort Macleod, continental glaciation flowing southwestwards from Kivalliq covered the eastern area with glacial deposits containing Canadian Shield (Laurentide) erratics. Kivalliq ice was diverted towards the southeast as it encountered the topography of the Foothills, including the Porcupine Hills (Jackson et al. 2008). Simultaneously, Cordilleran glacial ice expanded from the Yellowhead Pass area in Jasper National Park, flowed to the south along the Rocky Mountain Front, and covered the Porcupine Hills. In southwestern Alberta, the division between the areas influenced by the two glaciers lies to the west of Claresholm and Fort Macleod, and to the east of Head-Smashed-In.

Coalescence of the Cordilleran and Kivalliq ice masses is indicated by the Erratics Train, a string of Cambrian quartz sandstone boulders derived from the Mount Edith Cavell area of Jasper Park (Jackson et al. 2008; Jackson 2017; Fig. 5.23). The Erratics Train extends southwards from Calgary to the St. Mary River. indicating the suture line between ice from east and west. The largest of these erratics is located at Okotoks, and has a mass of 22,000 kg. "Okotoks" is a Siksika word, which translates as "Big Rock". The Okotoks erratic has long been regarded as sacred to the Siksika.

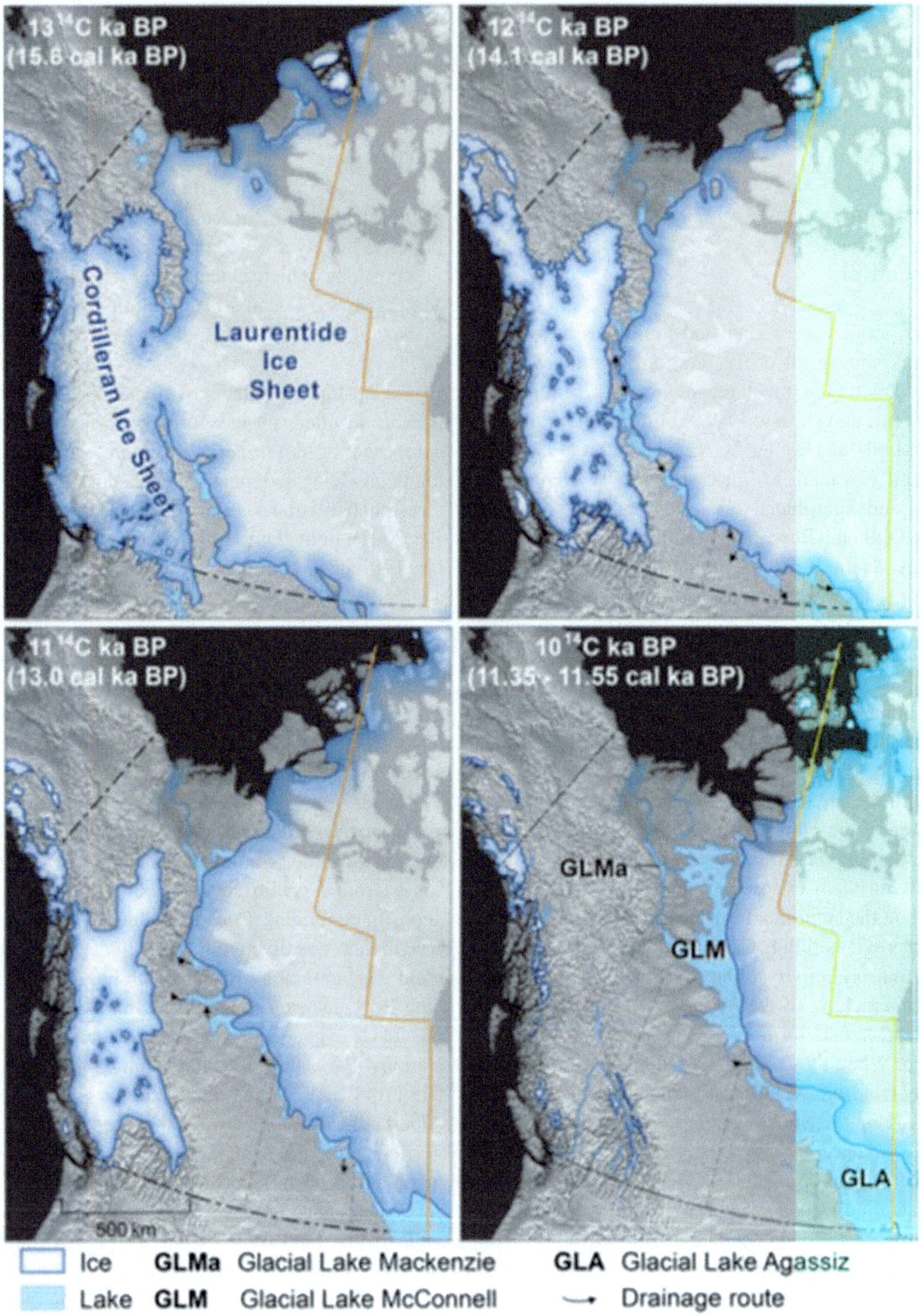

Fig. 5.20 Pattern of glaciation and glacial lakes in western Canada (Slaymaker 2017)

Fig. 5.21 Glacial features. Upper: Erratics derived from the Churchill province of the Canadian Shield indicate transport by Laurentide glacial ice moving against the topographic slope in western Canada. These examples are found in Grasslands. Lower: ice thrust ridge developed in glacial sediment, Medicine Hat. Flow was from the right

Fig. 5.22 Ice thrusting of Quaternary sediment. Left: drag-fold deformation of underlying gravel by local glacial advance towards the right (east), Edmonton Convention Centre. Upper right: deformation of ground squirrel burrow ca. 30,000 years old by ice thrusting towards the left, Wetaskiwin area, AB. Lower right: ice-thrust deformations in intertill sand lenses and glacial till, Clover Bar, Edmonton

The presence of the Erratics Train indicates that the ice sheets coalesced during the last glaciation, approximately 18,000 to 14,000 years ago. These results, in conjunction with other data suggesting that humans had arrived in South America *ca.* 14,000 years ago, indicate that the glaciers had coalesced during the initial stages of human migration to the Americas (see Catto et al. 1996; Hartman et al. 2018). Therefore, the suggestion that an "Ice-Free Corridor" existed between the Cordilleran and Kivalliq glaciers, potentially allowing human migration along the Rocky Mountain Foothills, is now regarded as incorrect. The early arrival of humans in South America indicates that people used a different route to bypass the glaciated terrain. Most archaeologists now believe that the initial migration followed the Pacific Coast of North America, across Haida Gwaii, during a period when sea level was much lower than at present (see Chap. 7).

5.5.2 Ice Retreat and Glacial Lakes

5.5.2.1 Hummocky Terrain

About 18,000 years ago, the Kivalliq glacial ice began to recede northeastwards and northwards (Dyke et al. 2003). Across most of the prairies, the low slope of the receding ice resulted in regional stagnation, rather than stepwise retreat with well-defined recessional moraine positions. Between 15,000 and 12,000 years ago, the southwestern sides of the Manitoba Escarpment and Missouri Coteau were flanked by broad regions of stagnating ice. The resulting undulating topography evident in the Saskatchewan and Alberta Plains west of the escarpments is marked by well-developed belts of hummocky terrain (Fulton 1995; Fig. 5.24).

Hummocky terrain is marked by apparently random assemblages of low hummocks, mounds, knobs, and ridges of glacial diamicton (till) and disturbed deposits of various types,

Fig. 5.23 Glacially transported Cambrian quartz sandstone boulders form the Erratics Train in southwestern Alberta, indicating the interaction between Cordilleran and Laurentide ice during the Last Glacial Maximum. Left: Okotoks (Sikskia: "Big Rock"), largest of the erratics. Right: Erratic west of Fort MacLeod, which was utilized as a bison rubbing stone

interspersed with lenses and beds of sorted fluvial sediment, and kettles, depressions, and basins frequently containing lacustrine sediment (Benn and Evans 2010; Catto 2015). Typically, relief in an area of hummocky terrain does not exceed 20 m. Features are highly variable in dimension, but most are small, less than 100 m in diameter. In relatively dry areas of the Canadian Prairies, the grass-covered hummocks flank shallow dry depressions or small, brackish ponds (sloughs or potholes). Many hummocky terrain belts are extensively farmed, with the sloughs representing the only parts left uncultivated. In dry prairie areas, hummocks are conspicuous features. In more humid prairie areas, hummocky terrain modified by farming has a mottled appearance. In boreal regions flanking the Aspen Parkland, hummocky terrain with conifer vegetation is marked by interspersed wetlands.

Interior structures and sediment composition vary widely. Most hummocky terrain is dominated by diamicton, with at least some glacial signature evident, but some hummocky areas are composed of sorted glacifluvial sand and gravel. Cross-sections through hummocks reveal diamictons of all types, including ice-thrusts, squeezed diapirs, subglacial traction deposits, passive meltout units, debris flow beds, and collapse features.

The variety in internal composition and external forms suggests that hummocky terrains can form as a result of several processes. Previously, most hummocky terrain was interpreted as the result of passive deposition, primarily on a gradually ablating glacier surface, during widespread stagnation of continental ice, over periods of hundreds of years. Ice sheets which had advanced upslope against the regional topography, as throughout much of prairie Alberta and Saskatchewan, are considered particularly susceptible: once the ice ceased advancing, widespread stagnation would ensue across the central and western prairies, leading to the deposition of the extensive belts of hummocky terrain (some with areas of hundreds of km^2).

Study of rates of glacial melting in areas where hummocky terrain is currently forming suggest that mean annual air temperature is the most important climate variable involved, but also emphasize the role played by topography. This research suggests that the southern prairies of Canada during deglaciation, as summer temperatures increased, would be particularly susceptible to widespread stagnation and the consequent formation of hummocky terrain. However, the widespread distribution of hummocky terrain in climate zones ranging from dry, warm, summer prairie to taiga to cool, maritime environments also could indicate that different processes may be involved.

5.5.2.2 Drainage Diversions

As the regional topography sloped towards Hudson Bay, the pre-existing drainage patterns were blocked by the glacial fronts (Klassen 1989; Dyke et al. 2003). In southernmost Alberta, glacial meltwater and streamflow from the Cordillera were diverted into the Milk River drainage, connected to the Missouri and Mississippi Rivers. Deep dry meltwater chan-

Fig. 5.24 Hummocky terrain is characteristic of glaciation in Prairie Canada. Upper left: typical hummocky landscape showing hummocks as mottles in cultivated fields, slough ponds, and shallow meltwater channels, eastern Saskatchewan (National Air Photo Library). Lower left: hummocky terrain with meltwater channel, southwestern Alberta. Right: Structures in hummocky terrain near Cooking Lake Alberta indicate deposition in supraglacial ponds (upper), ice thrusting (centre), and transport of friable blocks of coal (lower)

nels, locally called "coulees", represent former routes taken by glacial meltwater during the progressive retreat. Largely abandoned coulees along the north and west flanks of the Cypress Hills, and another connecting Lake Pakowki to the Milk River, developed during deglaciation (Klassen 1992). Dissipation of the meltwater, glacial retreat, and the advert of drier climates after 8,000 years ago isolated Lake Pakowki, forming an Internal drainage pattern with endorheic inflowing steams (Catto 2015).

Across Alberta and western Saskatchewan, blockage of the previous drainage routes along the South and North Saskatchewan valleys resulted in a series of short-lived glacially impounded lakes, with drainage channels flanking the ice fronts and oriented towards the northwest or southeast (Klassen 1989; Saskatchewan Geological Survey 1999; Kehew and Teller 1994; Slaymaker and Kovanen 2017a, b). Deglaciation began approximately 13,000 years ago. The presence of the hummocky topography indicates that deglaciation proceeded slowly (Dyke et al. 2003; Fisher et al. 2009), and that many blocks of ice persisted, buried by sediment, until about 9000 years ago. Decaying ice blocks remained for approximately 3,000 years after the first humans occupied the region.

During the progress of deglaciation in the Edmonton area, the North Saskatchewan River was temporarily blocked by ice, resulting in the deposition of thick layers of lacustrine clay over the low-lying terrain (Fenton et al. 2013). Subsequently, the Glacial Lake Edmonton clay supported the development of rich black chernozemic soils, lush midgrass—Aspen Parkland vegetation, and ideal habitat for *Bison*. When glacial retreat

Fig. 5.25 Meltwater channels drained ice-impounded glacial lakes. Upper: Gwynne Outlet from the ground (left) and air (right), the major drainage channel for Glacial Lake Edmonton. Lower left: the Qu'Appelle Valley drained Glacial Lake Regina. Lower right: Glacial Lake Regina landscape

progressed, the rapid drainage of Glacial Lake Edmonton formed the Gwynne Outlet meltwater channel to the south (Fig. 5.25). Subsequent glacial retreat opened drainage routes northwards to Glacial Lake McConnell, formed in the Boreal Plains region of northern Alberta, and eventually to the Beaufort Sea (see Chap. 3).

Similarly, glacial lake water was impounded against the retreating Kivalliq ice in the Qu'Appelle valley 12,000 years ago (Klassen 1975). The Qu'Appelle valley is a deeply incised trough through the Saskatchewan Plain, with steep (locally vertical) sides 100–150 m high and a flat base 2–5 km wide. The valley was created by rapid eastward discharges of meltwater from glacial lakes impounded against the retreating ice sheet (Kehew et al. 2009). The modern Qu'Appelle carries less than 5% of the peak volume of these great meltwater floods, and it forms a tiny misfit stream along the centre of the deep canyon. Today, the reservoir dammed in the base of the valley gives some indication of the appearance of the original drainage channel.

5.5.2.3 Glacial Lake Agassiz

In Manitoba, the retreating ice blocked northward drainage of the Red and Assiniboine rivers into Hudson Bay. The meltwater ponded against the ice formed Glacial Lake Agassiz (e.g. Teller and Clayton 1983; Teller and Leverington 2004; Dilworth and Fisher 2018; Leydet et al. 2018; Teller et al. 2018). This lake initially drained southwards through North Dakota and Minnesota into the Mississippi River (Fig. 5.26). As glacial retreat progressed, interrupted by short re-advances, Lake Agassiz gradually grew northward and progressively covered all of the Red River Valley, depositing great thicknesses of clay and silt (locally > 100 m). The topography, sediments, and soils of the Red River Valley are primarily the result of Lake Agassiz, and Winnipeg sits near the deepest part of the lake. The periodic floods of the Red River represent an echo of its former lacustrine expanse over southern Manitoba.

Lake Agassiz developed new outlets as the glacial ice retreated. At various times between 15,000 and 8,000 years

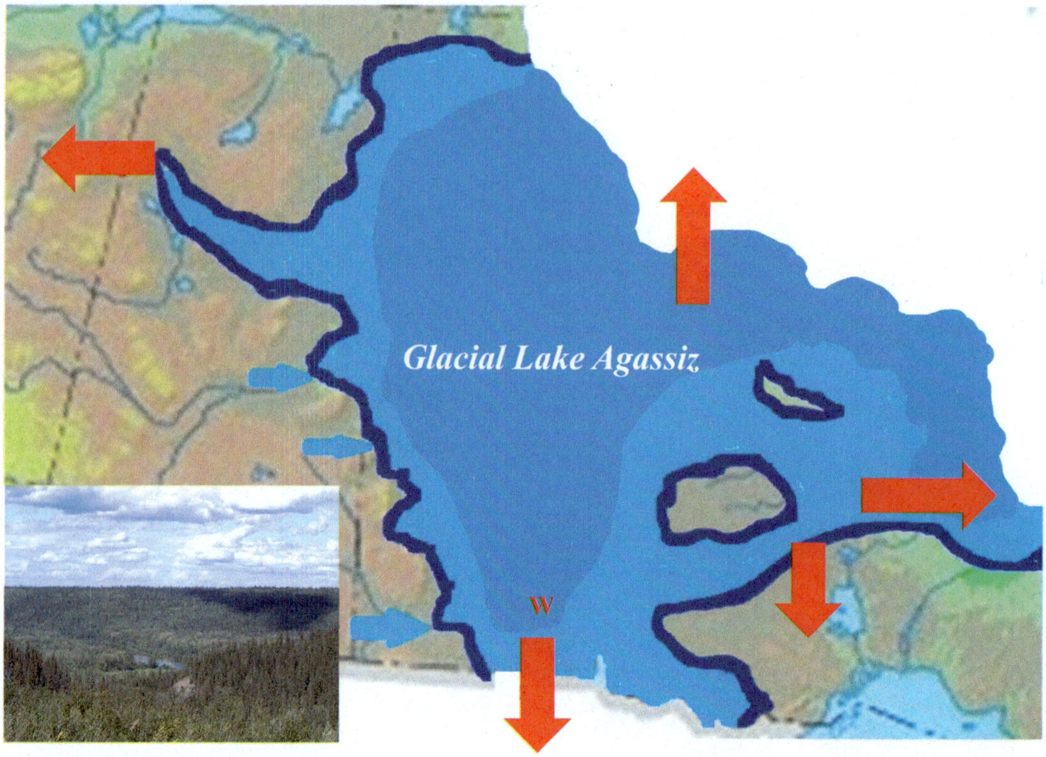

Fig. 5.26 Total extent of terrain covered by Glacial Lake Agassiz. The lake existed as several phases, expanding to the northeast as the Laurentide Inlandsis (white) retreated. Red arrows represent outlets open during various phases: Lake Traverse south to Mississippi Drainage; Clearwater River west to Fort McMurray (inset); northern Ontario outlets east to Great Lakes Basin and Glacial Lake Ojibway (see Chap. 3), and finally northward drainage to the opening Tyrrell Sea in Hudson Bay (see Chap. 8). Blue arrows are major inflows, including the Assiniboine, Qu'Appelle, and Saskatchewan Rivers. W: modern location of Winnipeg. (will redraw)

ago, the lake drained southward; southeastward into the Lake Superior basin at Duluth, Minnesota; eastward into Lake-of-the-Woods through Whiteshell Provincial Park; eastward into the Lake Nipigon basin, and thence via the Ottawa River into the Gulf of St. Lawrence; and northwestward via the Clearwater River of Saskatchewan into the Athabasca at Fort McMurray, Alberta, and thence into Glacial Lake McConnell and eventually to the Beaufort Sea. Each change in level allowed new gravel and sand beaches to form around the margins of the new lakes. These gravel beaches provide the major sources of aggregate for construction in the Winnipeg area today.

During the lifetime of the lake, icebergs calved from the glacial front to the north drifted across the water, grounding as they reached the shallow southern end. Ridges of pushed sediment and imprints made by the keels of these grounding icebergs can be seen on aerial photographs of the Red River Valley. These iceberg keel structures were studied intensely by petroleum development engineers, prior to the establishment of oil drilling platforms in the ice-laden waters of the Beaufort Sea and offshore NL.

Eventually, continued northward retreat allowed Lake Agassiz to drain directly into Hudson Bay, causing the southern part of the lake to rapidly dry up. The site of Winnipeg emerged from the lake waters about 9,000 years ago. The current Lake Winnipeg and Lake Manitoba thus represent remnant puddles of Lake Agassiz.

5.5.2.4 Assiniboine Delta and Spruce Woods Provincial Park

Where the Assiniboine River entered Lake Agassiz, a complex of delta sediments and alluvial underflow fan deposits developed (Rannie et al. 1989; Rannie 2017). The waters of Lake Agassiz had receded from the upper Assiniboine Valley by 9,500 years ago. The "Assiniboine Delta" sediment was subsequently reworked by westerly winds as Lake Agassiz shrank, producing the Spirit Sands of Spruce Woods Provincial Park (Rannie 2017; Fig. 5.27).

.The dunes of Spruce Woods are parabolic features, crescents of windblown sand with arms pointing into the prevailing wind, towards the west. Parabolic dune development involves a combination of limited sand supply, efficient sediment retention and surface adhesion, and consistently oriented winds. At Spruce Woods, sediment retention is critical in their establishment and maintenance. Sediment supply was limited by the presence of ephemeral standing water in ponds, and by snow. The Assiniboine River limits eastward dune migration, contributing to sediment retention and forcing dunes to increase in height and allowing stacking of successive generations of landforms. Deposition is influenced by sporadic wind gusts. The highest and most laterally extensive parabolic dunes, locally coalesced into transverse forms, are located along the margin of the Assiniboine. Depositional episodes were transitory in the dune field, and were separated by periods of stability not associated with marked climate change.

Parabolic dunes in prairie areas, including Spruce Woods, the Great Sand Hills (Saskatchewan), and Grande Prairie (Alberta), respond to climate warming and drying by becoming more active (Munyikwa et al. 2011; Wolfe 2013, 2017; Wolfe et al. 2013). Pollen assemblages indicate that warmer conditions than those of the present existed between 8,000 and 5,000 years ago across western Canada. At that time, the prairie grasslands expanded, forests contracted, lakes and sloughs dried up, and parabolic dunes underwent enhanced migration.

Over the past 3,000 years, the general tendency favoured forest expansion, and contraction of the grassland areas, although local exceptions existed (Wolfe et al. 2007). Many dunes were occupied by vegetation and stabilized. Since ca. 1860, increased warming of the climate has resulted in a reversal of this trend in some areas of the southern prairies. More recent dry episodes, accompanied by renewed sand mobilization and dune migration, occurred in the early 1880s, 1910s, 1930s, and 1950s. In the Great Sand Hills and elsewhere in the southern Canadian Prairies, aridification was a significant factor triggering dune migration, but human disturbances and practices (local desertification) also were impor-

Fig. 5.27 The Assiniboine Delta formed as the Assiniboine River entered Glacial Lake Agassiz southeast of Brandon. Subsequent drainage allowed parabolic and dome dunes to develop. Today, Spruce Woods Provincial Park (Spirit Sands) supports a riparian Aspen-Bur Oak Parkland assemblage in wetter areas, with scattered mid-grass vegetation and conifers on the dune surfaces

tant in triggering movement. In undisturbed areas with vegetation cover, the dunes remain stabilized (see Hugenholtz et al. 2010; Wolfe 2013, 2017).

Spruce Woods Provincial Park provides an interesting case study of the impact of human disturbance of parabolic dune vegetation. The sand exposed by foot traffic in the park is susceptible to wind transport. As the dunes are currently promoted by the Government of Manitoba as a major tourist attraction, complete stability (with the dunes immobile and vegetation-covered) is as undesirable as completely mobile conditions. When aeolian erosion became severe in some parts of Spruce Woods in the late 1980s, foot traffic was banned and people were urged to remain on the designated paths. Two relatively wet summers in the early 1990s resulted in increased stabilization and vegetation growth, significantly reducing the "desert-like" feel of the park, and hence its attractiveness to tourists. This represents an extremely interesting problem in park management: it appears that, at least under some weather conditions, human foot traffic contributes to the creation of a landscape which humans desire to see. In the long term, however, the dunes will become de-stabilized if current overall warming and drying trends continue, regardless of the number of human feet and the regulatory measures of the park wardens.

5.6 Vegetation Assemblages

In conjunction with climates, vegetation assemblages vary from east to west, and also north–south (Environment Canada and Agrifoods Canada 1995; Alberta Parks 2015; CNVC 2016). Aspen (also called Prairie) Parkland assemblages mark the ecotones between BSk and mid-boreal (Dfb) climates. No intervening "C" climate zone is present. In broad terms, the vegetation zones include (Fig. 5.28).

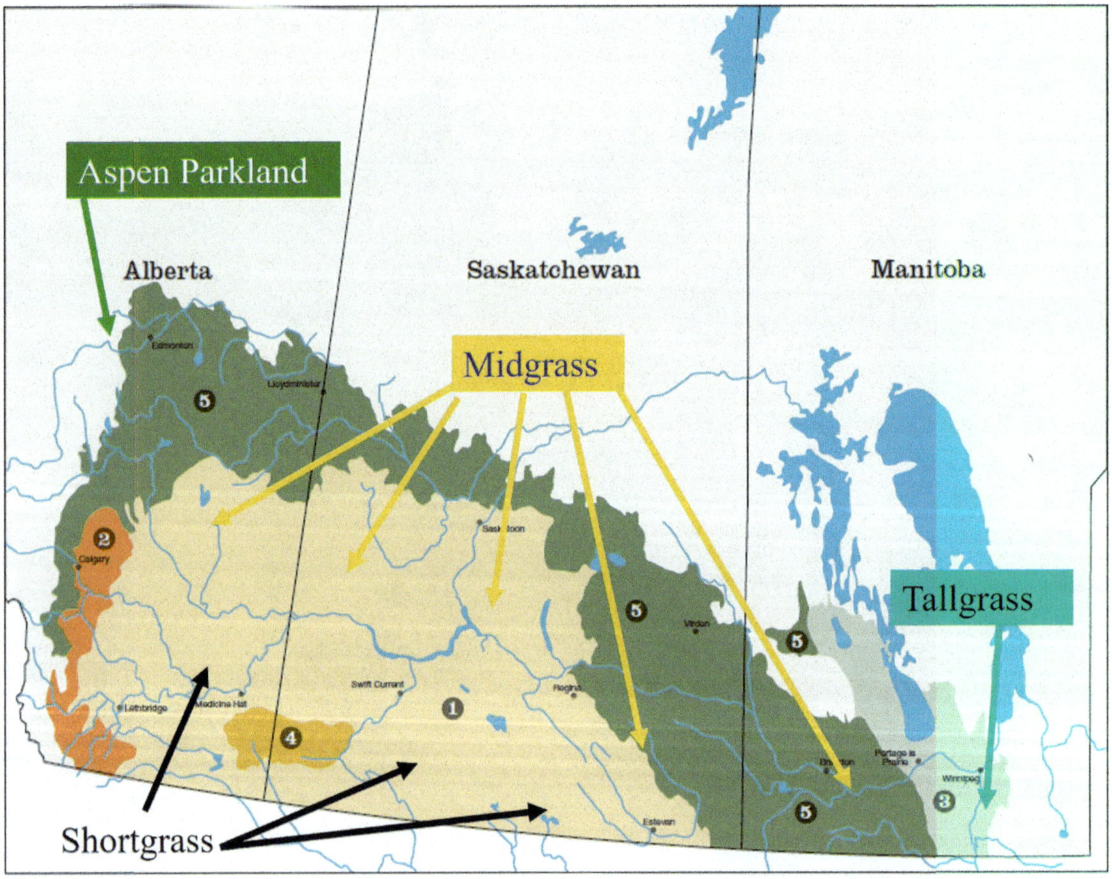

Fig. 5.28 Vegetation assemblages of Prairie Canada. 1 Mixed Grass Prairie: includes Shortgrass dominated assemblages in south and west, with higher proportions of C4 grasses; Midgrass assemblages in north and east, with higher proportions of C3 plants. 2: Fescue Prairie: Shortgrass (south) and Midgrass (north) assemblages dominated by fescue. 3: Tallgrass Prairie (south) with Prairie Parkland groves. 4: Fescue Prairie with Prairie Parkland. 5: Midgrass Prairie/Prairie Parkland transition zone. See text for discussion

(a) Tallgrass Prairie, with interspersed Aspen Parkland groves and riparian assemblages (Red and lower Assiniboine river valleys, and small regions of southern Manitoba and southeastern Saskatchewan);
(b) Midgrass Mixed Grass Prairie (western Manitoba, eastern and central Saskatchewan, north-central and western Alberta);
(c) Midgrass Prairie/Aspen Parkland ecotone (northern flank of Midgrass Mixed Grass Prairie in Manitoba, Saskatchewan, and north-central Alberta);
(d) Fescue Prairie ecotone (western flank of Midgrass Mixed Grass Prairie in western Alberta); and
(e) Shortgrass Prairie, including areas with halophyte and xerophyte vegetation (southwestern Saskatchewan, southeastern Alberta).

5.6.1 Tallgrass Prairie

The Tallgrass Prairie of Canada represents the northern extension of the biome, developed on the wettest Black Chernozems of southern Manitoba and southeastern Saskatchewan (Fig. 5.29). Prior to 1770, the assemblage was dominated by grasses to 2-m height, including a mixture of C3 and C4 types. Although most of the Tallgrass Prairie has been ploughed for agriculture, small patches are preserved in provincial and municipal parks (Henderson and Koper 2014; Pearn and Hamel 2014; Hamel 2015).

The dominant plants of the Tallgrass Prairie were the grasses *Andropogon* (bluestem), *Hesperostipa spartea* (porcupine or needle-grass), *Spartina pectinata* (cord grass), and *Panicum virgatum* (panic grass, also called millet); and the forbs *Helianthus* (sunflower), *Rudbeckia hirta* (black-eyed Susan), *Pulsatilla nuttalliana* (prairie crocus or pasqueflower,

Fig. 5.29 Tallgrass Prairie occupied southern Manitoba and southeastern Saskatchewan ca. 1870 ce, but was removed during agricultural development. Inset: Tallgrass Prairie Memorial in eastern Winnipeg marked the northern limit ca. 1870. Upper right: *Andropogon* (big bluestem grass). Lower left and centre: Tallgrass Prairie assemblage at Tallgrass Prairie Preserve, near Tolstoi MB (Wikipedia/Joe Passe). Lower right: *Panicum virgatum* (panic grass)

the provincial flower of Manitoba), *Hypoxis hirsuta (*eastern yellow stargrass), and *Symphyotrichum* spp.(asters)(Looman and Best 1994). The grasses covered the terrain to a height of 1 m or more prior to agricultural disturbance, and their densely interwoven root systems frequently extended more than 2 m into the soil. Typically, more than 50% of the biomass was below the ground surface. "Sod-busting" in the Tallgrass Prairie involved considerably more than removing a thin, green carpet.

This assemblage has been substantially modified by agricultural clearance, fire suppression, removal of fauna, and the spread of imported plants throughout the prairie. Species such as red clover (*Trifolium pratense*) and white daisy (ox-eye daisy, *Chrysanthemum leucanthemum*) are not native to the Tallgrass Prairie of North America, but have spread invasively and now dominate many areas. Escaped agricultural plants such as wheat hybrids, oats, buckwheat, and canola (rapeseed), along with Kentucky bluegrass (*Poa pratensis)* and bentgrass (*Agrostis stolonifera*) are also now common throughout the area (Hammermeister et al. 2001).

Along the river systems, sufficient moisture existed to allow trees to become established. The fringes of trees along the Red and Assiniboine Rivers, in some low-lying moist areas, and along the margins of the Canadian Shield to the east, represent the Aspen (Prairie) Parkland assemblage (CNVC 2016). In eastern Manitoba, the dominant trees of the Parkland are *Populus tremuloides* (trembling aspen), *Populus deltoides* (cottonwood), and *Quercus macrocarpa* (bur oak; Fig. 5.30). Other deciduous species such as *Acer negundo* (Manitoba Maple; also called box elder) and *Betula papyrifera* (white birch) are present in small numbers (Hosie 1979). Conifers are naturally absent from the Aspen-Bur Oak Parkland, but have been introduced during the past 100 years.

The balance between the Aspen-Bur Oak Prairie Parkland and the Tallgrass Prairie is controlled by moisture availability and human interference. During the past 200 years, the

Fig. 5.30 *Quercus macrocarpa* (bur oak), Spruce Woods Provincial Park. Lower right: Detail of foliage. Insets show *Rudbeckia hirta* (black-eyed Susan, left) and *Pulsatilla nuttalliana* (prairie crocus or pasque-flower, the provincial flower of Manitoba, centre), characteristic species of the Bur Oak-Aspen Parkland and adjacent Tallgrass Prairie

southern prairies of Manitoba have been progressively drying out, as the climate grew warmer (Thorpe 2011; Sauchyn et al. 2020). This drying trend would favour expansion of the Tallgrass Prairie at the expense of the Parkland. Local human activity, however, has disrupted the pattern. By preferentially clearing the Tallgrass Prairie areas for agriculture, altering the natural assemblages by introductions (such as clover and daisy), suppressing fire, and changing the drainage patterns of the region, human activity has favoured the expansion of the Aspen-Bur Oak Parkland into marginal agricultural regions (Kaye 1996; Welsted et al. 1996).

Microclimatic variations control the development of different vegetation assemblages on slopes of different aspect. On north-facing slopes, and along the lager rivers, the Aspen-Bur Oak Parkland assemblage dominates. Dominant trees here are *Quercus macrocarpa* (bur oak), *Populus tremuloides* (aspen), *Populus deltoides monilifera* (cottonwood), and *Acer negundo* (Manitoba Maple). The Assiniboine Valley marks the western limit for some broad-leaved species, such as *Tilia americana* (basswood). Others, such as *Fraxinus pennsylvanica subintegerrina* (prairie green ash) and *Ulmus americana* (white elm) are confined to the wettest, lowest valley areas (Hosie 1979). Dutch Elm disease reached western Manitoba approximately ten years ago, and virtually all of the elms are affected. The disease has thus succeeded in spreading over the entire range of *Ulmus Americana* (see Chap. 4).

5.6.2 Midgrass Mixed Grass Prairie

The Midgrass mixed grass Prairie was the dominant vegetation assemblage throughout the southern and central prairies prior to 1870. Over the past 150 years, the original assemblage has been modified by agricultural activity, fire suppression, introduced vegetation species, removal of bison and other fauna, grazing animals (including horses), and climate change (Figs. 5.31 and 5.32). Midgrass assemblages developed on Black, Dark Brown, Dark Grey, and Grey Chernozems, dominated by C3 grasses to 0.5-m height. The suitability of these soils for cereal crops led to the rapid alteration of most of the midgrass prairie between 1880 and 1910, and there are no national or large provincial parks preserving true midgrass assemblages.

Dominant plants in a typical midgrass assemblage include *Hesperostipa comata* (Needle-and-thread grass, provincial grass of Saskatchewan), *Hesperostia curtiseta* (porcupine

Fig. 5.31 Much of the midgrass mixed grass prairie region has been converted to agriculture, as in this canola field near Yorkton SK. A line of orange *Lilium philadelphicum* (western prairie lily, provincial flower of Saskatchewan) is visible in the foreground. Trees are Acer negundo (box elder), artificially planted as a windbreak

Fig. 5.32 Dauphin Lake region, western Manitoba, showing contrast between agriculturally modified midgrass mixed grass prairie developed on glacial and lacustrine sediments with the less agriculturally productive terrain flanking Lake Winnipegosis to the east. Boreal forested terrain is present to the west (Mossey Hills) and south-west (northern boundary of Riding Mountain National Park). North–south ridges on the western side of the image, marked by changes in agricultural patterns, are former beaches of Glacial Lake Agassiz. (NASA)

grass, notably in moister areas), *Elymus lanceolatus* (thick-spike wild rye), and *Pascopyrum smithii* (western wheat grass; Fig. 5.33). Other components include *Andropogon scoparius* (little bluestem grass), *Bouteloua* (grama grass), *Nassella viridula* (green needlegrass), *Calamagrostis montanensis* (plains reedgrass), *Linum lewisii* (flax), *Cornus canadensis* (bunchberry), and *Erigeron sp.* (fleabane) (Moss 2000; Government of Alberta 2010; Looman and Best 1994; Alberta Parks 2015; Alberta Environment and Parks 2016; CNVC 2016).

Needle-and-thread is a C3 species of Poaceae. One of the bunchgrasses, it grows in small clumps (bunches), with straight stems and slender leaves (Looman and Best 1994). Seeds are typically 2 cm long, with a sharp tip and long awn (tail), resembling a needle and thread. Excellent forage for grazing animals, the seeds of needle-and-thread grass can represent a hazard if ingested by some species, particularly dogs.

Little Bluestem Grass is a C4 species of Poaceae. Growing in a bunch habit, with open spaces between plants, it prefers drier soil and can tolerate alkaline conditions and some sodium content (Looman and Best 1994; Moss 2000). When touched by frost, its leaves turn dark orange, giving it the alternative name "blaze".

Fig. 5.33 Midgrass species. Upper left: *Hesperostipa comata* (needle-and-thread grass; National Park Service USA). Upper right: *Hesperostia curtiseta* (porcupine grass; Krista Lundgren, USFWS). Lower left: *Nassella viridula* (green needle grass; National Park Service USA). Lower right: *Bouteloua* (grama grass; National Park Service USA)

Also present are several species of sedge, *Rudbeckia* (black-eyed Susan), *Oenothera pilosella* (sundrop), *Artemisia frigida* (sagebrush), *Phlox hoodii* (moss phlox), *Rosa acicularis* (prairie rose), and *Lilium philadelphicum* (western prairie lily, provincial flower of Saskatchewan). *Lilium philadelphicum* prefers moister settings, where its orange-red flower make conspicuous showings in late June and early July. It commonly spreads in ditches along roads and adjacent to watered agricultural fields, and has become more prominent in the Saskatchewan landscape as a result of human disturbance.

Invasive species of concern include Kentucky bluegrass, crested wheat grass, smooth brome, quackgrass, and purple loosestrife (Hammermeister et al. 2001). All five species spread rapidly and result in significant declines in biodiversity in affected areas, although livestock and native grazers can utilize crested wheat grass. Quackgrass (*Agropyron repens*), introduced in imported grain, is ubiquitous in agricultural areas. The Saskatchewan government has active campaigns attempting to eradicate both quackgrass and purple loosestrife.

Much of the midgrass prairie is characterized by belts of glacial hummocky terrain, with many semi-permanent lakes and sloughs, generally less than 1 km in diameter and less than 2 m deep even after heavy rains. These "prairie potholes" vary in water quality, ranging from fresh to saline, and from clear to extremely clay-laden.

Fig. 5.34 Aspen (*Populus tremuloides*) parkland is found in the transition from midgrass to boreal assemblages (upper) and in riparian sites (Souris River, Manitoba, lower right). Inset: *Acer negundo* (box elder, Manitoba maple) is a subordinate component of Parkland assemblages in the eastern prairies

Potholes which retain water throughout the summer area dominated by *Schoenoplectus* sp. (bulrushes), *Typha* sp. (cat-tails), and *Carex sp.* (sedges)(Galatowitsch and van der Valk 1996; Put 2013). Seasonally wet areas may support Scirpus (club-rush), *Glyceria striata* (mannagrass), *Calamagrostis canadensis (*bluejoint*),* and *Juncus balticus (*Baltic rush). Saline potholes are habitats for halophytes such as *Salicornia rubra* (samphire), *Glaux maritima* (sea milkwort), and *Dodecatheon sp.* (shooting star).

Regardless of water quality, all the water and vegetation was usable by the native fauna (particularly bison), and provided important breeding, resting, and feeding areas for waterfowl and migratory birds. The health of the prairie potholes is an important measure of the overall states of the prairie, both in terms of biological diversity and of the effects of climate change (Adamus 1996). Restoration efforts of prairie pothole habitats, particularly those drained for agriculture between 1965 and 1980, are a major focus of organizations such as Ducks Unlimited.

Climate change in the region has had a significant influence on the distribution of grassland and adjacent parkland areas. Over the past 3,000 years, the general tendency favoured forest expansion, and contraction of the grasslands. Aeolian dunes (e.g. Maidstone SK) were stabilized by *Calamovilfa longifolia (*prairie sandreed), *Sporobolus cryptandrus* (sand dropseed), *Artemisia cana* (silver sagebrush), and *Juniperus horizontalis* (trailing juniper). Recently, however, increased warming of the climate has resulted in a reversal of this trend. Anticipated future trends include decreases in midgrasses with corresponding increases in shortgrasses, and increased prevalence of C4 species (Thorpe 2011; Sauchyn et al. 2020).

Agricultural development has replaced more than 80% of the original midgrass prairie across western Canada, with nearly total replacement in the most densely utilized areas of black chernozemic soil (such as those adjacent to Edmonton and Regina). In addition to cereal cultivation, railway and road construction and residential development have resulted in extensive habitat fragmentation.

5.6.3 Midgrass Prairie/Aspen Parkland Ecotone

Along the northern margin of the mixed midgrass prairie from Riding Mountain northwest to Prince Albert and Elk Island national parks, the ecotone transition to the boreal forest is dominated by Aspen (*Populus tremuloides*; Figs. 5.34 and 5.35). Parkland assemblages also occupy wet riparian situations where human modification has not been extensive (Johnson et al. 1995; Moss 2000, ABMI 2015, Alberta Environment and Parks 2016).

The natural vegetation is dominated by *Populus tremuloides* (trembling aspen), with lesser amounts of *P. balsamifera* (balsam poplar). *Acer negundo* (Manitoba Maple or Box Elder) and *Quercus macrocarpa* (bur oak) are not native to the region west of the Assiniboine River valley (Hosie 1979). However, *Acer negundo* has encroached due to human planting to the southwest, for both ornamental purposes, and to establish windbreaks. Shrubs associated with the parkland include *Rosa acicularis* (wild prairie rose), *Potentilla sp.* (cinquefoil), *Elaeagnus argentea* (silverberry), *Shepherdia canadensis* (soapberry), *Rubus sp.* (raspberry), *Ribes sp.* (gooseberry), *Amelanchier sp.* (saskatoon), *Prunus virginiana* (choke cherry), and *Prunus pensylvanica* (pin cherry). Scattered conifers including *Picea glauca* (white spruce), *P. mariana* (black spruce) and *Larix laricina* (tamarack), and areas of midgrass prairie with fescue, forbs, and sedges are also present. In areas that were initially cleared for agriculture, and then allowed to overgrow, *Populus balsamifera* (balsam poplar), *Salix* (willow), and *Alnus sp.* (alder) are common.

Amelanchier alnifolia (saskatoon berry), low (30- 70 cm high) bushy plant, a relative of the chuckley-pear, represented an important food source to the fauna of the region, and was used by early travellers as a staple ingredient in pemmican (dried and jerked bison meat; Ngapo et al. 2021). It was also used by early settlers as a substitute for blueberries in pies and jams and for wine manufacture, although it is rather dry, and it remains popular today.

Much of the Aspen Parkland ecotone has been substantially disturbed by human intervention, and the vegetation assemblages do not resemble a natural community, containing more pure stands of *Populus tremuloides* than would be expected under normal conditions. The development of pure stands of *Populus tremuloides* has allowed the insects which feed on this species to greatly increase their numbers. The most ubiquitous of these is the tent cater-

Fig. 5.35 Dark grey luvisol developed under aspen parkland vegetation. Clay is translocated from the Ahe to the Bt horizon

pillar (*Malacosoma disstria*). Although as an adult *Malacosoma disstria* is an innocuous moth, as a caterpillar the species is voracious. Colonies of tent caterpillars hatch from pale white, gossamer-like tents woven in the branches of the aspens, and proceed to strip the trees of leaves. During bloom years, entire groves of aspens may be completely stripped of leaves, leaving a ghostly assemblage of naked trees by early July. Although tent caterpillars occur in most areas of southern Canada, concentrations of aspen make this region particularly vulnerable to their infestation (Hogg and Shwarz 1999; Brandt et al. 2003).

5.6.4 Fescue Prairie Ecotones

Along the western margin of the shortgrass prairie, flanking the conifer-dominated eastern slopes of the Porcupine Hills, a narrow ecotone belt dominated by Plains Rough fescue (*Festuca*) is present (CNVC 2016). Plains Rough Fescue zones also border the north flank of the coniferous Cypress Hills, and the southern areas of Prince Albert National Park, transitional to Aspen Parkland (Fig. 5.36). Plains Rough Fescue grasslands were once common across the northern and western parts of the prairies. The fescue grassland areas are currently shrinking, due to agricultural development in Alberta, and Parkland forest encroachment in Saskatchewan (Pavlick and Looman 1984; Holcroft Weerstra 2003; Parks Canada 2011; Guedo and Lamb 2013; Alberta Parks 2015).

In southwestern Alberta, the transition zone from Shortgrass to Rough Fescue-midgrass to forest transition zone is narrow, less than 30 km in width in many areas, and is most sharply developed in the area west of Fort Macleod and Claresholm, where the Alberta Plain grades into the foothills of the Rocky Mountains, the easternmost part of which is the Porcupine Hills. Vegetation assemblages differ markedly with elevation. At the lowermost elevations (e.g. Fort Macleod) a shortgrass prairie assemblage is present, dominated by *Bouteloua* (grama grass), *Artemisia* (sage), *Stipa comata* (spear grass), Cyperaceae (sedges), *Festuca* (fescue), and *Castilleja* (Prairie Fire). Succulents such as *Opuntia polyacantha* (prickly pear) and *Escobaria vivipara* (purple ball cactus) are present on south-facing slopes (Looman and Best 1994; Moss 2000).

At slightly higher elevations (e.g. Claresholm), Rough Fescue-midgrass prairie is present. Along with *Festuca*, genera such as *Andropogon* (bluestem grass), *Agropyron* (western wheat grass), Erigeron (fleabane), *Bouteloua* (grama grass), and *Amelanchier alnifolia* (saskatoon berry) dominate. The Rough Fescue prairie belt is relatively narrow, and is completely absent from the area west of Fort Macleod, in part due to agriculture.

The lower slopes of the Porcupine Hills boreal zone are occupied by coniferous forest (Alberta Parks 2015; CNVC 2016). The dominant species are *Pinus contorta* (lodgepole pine) and *Pseudotsuga menziesii glauca* (interior Douglas fir). The Douglas Firs are relatively small (10–15 m), and are the interior variety developed in stressful conditions, with twisted, small needles and 5 cm long, somewhat gnarled cones. Minor amounts of *Picea glauca* (white spruce) are also present. Deciduous trees form a lesser component of the assemblage, dominated by *Populus tremuloides* (trembling aspen) and *Betula papyrifera* (white birch).

The uppermost elevations in the Porcupine Hills lack *Populus* and *Betula*. The conifer assemblage is dominated by alpine species, including *Picea glauca* (white spruce), *Picea engelmannii* (Engelmann spruce), *Pinus flexlis* (limber pine), and *Abies lasiocarpa* (alpine fir), in addition to *Pinus contorta*. *Pseudotsuga* is not present in the highest areas. Deciduous shrubs (e.g. Salix) and understory vegetation are also very similar to the assemblages of the boreal forests.

The northern flank of the Cypress Hills supports Rough Fescue Prairie, grading upwards to a *Pinus contorta* (lodgepole pine) dominated forest (Fig. 5.37), with minor amounts of *Populus tremuloides* (trembling aspen). This forest can only survive in the coldest and snowiest north-facing slopes on the northern flanks of the plateaux. The pines grow with straight trunks 10–20 m high, with few lower branches, and are very suitable for constructing lodgepoles for portable Siksika lodges. Lodgepole pine is native to the Rocky Mountains, and its pollen was transported to the Cypress Hills by the prevailing westerly winds. *Pinus contorta* closely resembles the northeastern boreal pine species *P. banksiana* (jack pine), and is capable of hybridizing with it (Lusebrink et al. 2013). The French voyageurs who discovered the Cypress Hills named them for the *cyprés,* the Québecois term for jack pine. Grey luvisols are developed under the arboreal vegetation.

The relatively small area in Prince Albert National Park represents 35% of Canada's total fescue grassland (Parks Canada 2011; Guedo and Lamb 2013). The expansion of forest cover is largely due to the former policy of suppressing forest fires, which originally acted to restrict the spread of trees in the Aspen Parkland areas and thus permitted the fescue grasslands to develop. In the park, total fire suppression was practiced from *ca.* 1930 to the mid-1960s, but this policy has been replaced with a more "natural" approach. Fires are now permitted to burn unchecked, unless they threaten human infrastructure or sensitive faunal zones (such as the pelican nesting grounds). Fire frequency is low in the area, however, with any particular hectare of the forest likely to burn once every 200–250 years. Thus, a 30-year non-suppression period is statistically unlikely to curtail forest expansion, and the forests have continued to steadily encroach.

Fig. 5.36 Fescue prairie, shortgrass assemblages, southern Alberta

5.6.5 Shortgrass Prairie

True shortgrass assemblages are restricted to the driest areas of southwestern Saskatchewan and southeastern Alberta (Hargrave 2022; Government of Alberta 2010; ABMI 2015; Alberta Parks 2015; CNVC 2016). In more northerly, wetter, or riparian settings, shortgrass and midgrass species combine to form mixedgrass assemblages.

Shortgrass Prairie is dominated by grasses which characteristically grow less than 30 cm high, and in many areas do not exceed 10 cm in height (Fig. 5.38). The dominant species are *Bouteloua sp.* (grama grass), *Hesperostipa comata* (spear grass), *Koeleria cristata* (June grass), *Poa secunda* (Sandberg's blue grass), several species of Cyperaceae (sedges), *Selaginella densa* (spikemoss), and *Artemisia sp.* (sagebrush) (Looman and Best 1994; Moss 2000).

Fig. 5.37 *Pinus contorta* (lodgepole pine) is the dominant tree species of the Cypress Hills. Lower: podzolic soil profile developed under *Pinus contorta* on loess and Palaeocene gravel, Cypress Hills

In moderately acidic areas (pH 6.2–6.8), two species of roses grow: the wild prairie rose, *Rosa acicularis*, the provincial emblem of Alberta; and the low pink prairie rose, *Rosa arkansana*. The low pink prairie rose is more adapted to dry conditions, but is less snow-tolerant than the wild prairie rose. Humic-poor brown chernozems characterize these dry regions. The terrain is too dry to support mid-grass prairie species, such as the Prairie Lily.

In dry areas, cacti and other xerophytic species are present (Fig. 5.39). The two dominant cactus species are *Opuntia polyacantha* (prickly pear cactus), with bright yellow flowers during late June and early July; and *Escobaria vivipara* (purple ball cactus), with small magenta flowers in June. *Castilleja* (Prairie Fire or paintbrush), with its distinctive spikes of upright red flowers from June to September, also is common on dry sites.

Fig. 5.38 Shortgrass prairie is dominated by grasses which characteristically grow less than 30 cm high. The dominant species are *Bouteloua sp.* (grama grass), *Hesperostipa comata* (spear grass), *Koeleria cristata* (June grass), *Poa secunda* (Sandberg's blue grass), several species of Cyperaceae (sedges), *Selaginella densa* (spikemoss), and *Artemisia sp.* (sagebrush). Lower right: Soil cracks developed by combination of desiccation and frost action in shortgrass, Eastend area, SK

The saline slough and pothole areas are marked by belts of Cyperaceae (sedges) and halophytic plants such as *Salicornia rubra* (red samphire) and *Dodecatheon pulchellum* (saline shooting star; right). Trees are not present around even the most stable prairie potholes. Along the Frenchman River in Grasslands National Park, isolated stands of *Populus angustifolia* (narrowleaf cottonwood) and naturalized *Acer negundo* (box elder) represent the only arboreal vegetation.

Poisonous species of vegetation are relatively rare, and cause little difficulty for the native fauna. Introduced cattle and sheep, however, are not sufficiently adapted to the region to recognize the potentially lethal species. The two most frequently encountered poisonous plants are *Oxytropis sp.* (locoweed) and *Toxicoscordion venenosum* (death camas; Campbell et al. 1956). The leaves of both species are poisonous to cattle and sheep, and the tubers of death camas are especially lethal to the close-cropping sheep. Both species tend to grow on dry hillslopes, generally south-facing. Although cattle losses due to locoweeds are not commonly significant, death camas can cause serious problems to sheep-herding operations, especially in late spring. Bison are relatively unaffected by locoweed and do not eat death camas; cattle and sheep do not discriminate.

5.7 Soils

Soils in Prairie Canada have developed under the influences of the cold steppe climate, bedrock geology, Quaternary history, grassland and prairie parkland vegetation assemblages, faunal activity, and human influences over the past 10,000 years. Distinctive soils characterize the Tallgrass, Midgrass, and Shortgrass prairies, and the parkland areas. Soils continue to evolve under changing land uses, moisture regimes, and climates. BSk soils are dominantly Chernozems, with some Solonetz and minor areas of Vertisols

Fig. 5.39 Shortgrass Prairie species. Upper: *Opuntia polyacantha* (prickly pear cactus) and Lower left: *Toxicoscordion venenosum* (Death Camas; Jennifer McNew, Bureau of Land Management USA) grow in south-facing xerophytic sites. Lower right: *Oxytropis sericea* (white locoweed; NRCS Montana, US Department of Agriculture), along with other locoweed species, grows in a variety of settings in the shortgrass region

5.7.1 Chernozems

Initial establishment of the Tallgrass Prairie and subsequent agricultural development both required suitable soil. The combination of the continental prairie climate, carbonate-rich bedrock and sediments, and the gradual development of the grassland assemblages allowed the formation of humic-rich clayey Black Chernozems, the ideal soil for growing cereal crops (Fig. 5.40).

True Black Chernozems can only develop in steppe climate regions underlain by calcareous sediments. They are characterized by a thick (>10 cm) humic-rich Ah surface layer, containing 1–17% organic carbon and with carbon: nitrogen ratios greater than 1:17 (Soil Classification Working Group 1998, 2001; Pennock et al. 2011). Chernozems are alkaline soils with high pH ($OH^- > H^+$). The underlying Bm horizon is commonly dark brown, and nutrients from this horizon are moved vertically upwards (by the grass roots) and downwards (by water infiltration). The underlying Cca horizon is marked by an accumulation of calcium carbonate, resulting from groundwater movement and accumulated precipitates, giving it a grey-white appearance. The lowermost Ck horizon is the parent material (lake sediment, glacial, or limestone bedrock), inherently rich in calcium.

Grasses thrive on the natural lime in the soil, and extend their roots down to the Cca horizon to obtain it. As the grass vegetation develops, calcium carbonate is cycled through the system repetitively, with each cyclic event improving the soil for grass growth. Consequently, grassland soils are fertile and stable, as long as the grasses are not disturbed. Removal of the original grass vegetation can result in a loss of soil fertility, which must be compensated for by fertilization.

Fig. 5.40 Chernozems. Left: Brown chernozem, characteristic of semi-arid shortgrass regions. Centre: Dark Brown chernozem, characteristic of drier midgrass prairie. Left: Black Chernozem, characteristic of moister midgrass and tallgrass prairie. (Canadian System of Soil Classification, Agriculture Canada)

In the riparian zones, the Aspen-Bur Oak Prairie Parklands are developed on Luvisolic soils, marked by lower humic levels, downward translocation of clays, and (generally) lower calcium carbonate concentrations (Soil Classification Working Group 1998; 2001; Lavkulich and Arocena 2011). Encroachment of Aspen Parkland vegetation onto a chernozemic area will result in the gradual transformation of the soil from chernozem to luvisol, but the process requires several hundred years. Thus, recent encroachments of Aspen-Bur Oak Parkland can be recognized by observing areas where trees grow on chernozemic soils.

Midgrass regions are characterized by dark brown and dark grey chernozems, which have lower humic contents (2–5% C) than the black chernozems of the Tallgrass Prairie, but are similarly calcareous. Black Chernozems are found in favourable sites under the best-developed midgrass prairie vegetation. The black chernozems which underlie the Edmonton area are among the richest soils in the country, and were a major factor leading to Edmonton's growth.

Soil types in most midgrass regions are controlled by two principal factors: vegetation type and position on the slopes. All of the sediments and bedrock in the region contain substantial amounts of calcium carbonate, and hence all soils (except peat successions developed in bogs) are alkaline. Soil assemblages which differ due to slope position are referred to collectively as catenary sequences. The Prairie Parkland areas are characterized by luvisol soils, generally dark grey luvisols (Fig. 5.35).

The soils were initially developed under a more arboreal vegetation assemblage in many midgrass locations, and are still in transition to true grassland soils. The area of forested land was expanding prior to 1880, and conditions which would have favoured forest expansion existed between 1990–1925 and in the 1960s. As soil development requires as much as 600–800 years after a stable vegetation assemblage develops, many areas covered by grassland vegetation still are characterized by luvisol soils.

Under arboreal vegetation at the summits of hills, dark grey luvisols are developed (Lavkulich and Arocena 2011). Dark grey luvisols are marked by the downward translocation of clay particles, forming clay-rich Bt horizons at depth with strongly columnar structures. The soil develops where groundwater is able to percolate downwards, and thus forms in relatively permeable areas on crests.

At the bases of the slopes, soils are water-saturated throughout much of the year. In these groundwater discharge areas, with groundwater moving vertically upwards or laterally out of the slopes, iron within the soils is periodically oxidized and reduced, as water tables rise and fall in response to precipitation. This fluctuation produces a mottled or "gleyed" pattern in the soils, and these sites thus support Luvic Gleysols. Changes in water table position from year to year can affect soil development, with luvic gleysols developing from luvisols as water tables are raised and reverting to luvisols as water tables lower. Soil transformation requires several tens of years, however, so that individual years of drought or excessive precipitation are insufficient to cause the soils to change significantly.

Shortgrass regions commonly support Brown Chernozem soils (Soil Classification Working Group 1998, 2001; Pennock et al. 2011). Although Brown Chernozems are calcium-carbonate rich and form similarly to Dark Brown or Black Chernozems, their Ah horizons contain much lower concentrations of organic carbon (1–2%), a consequence of the drier climate and reduced vegetation productivity. Brown Chernozems have relatively low agricultural productivity, particularly for cereal crops. In many areas, the best usages are for rangeland or leaving the soils unexploited agriculturally. The more northerly shortgrass areas support Grey Chernozems, more organic-rich and agriculturally productive than the Brown Chernozems, but with less humic matter than the Dark Grey and Black Chernozemic soils found to the north and east.

5.7.2 Solonetz

In the Shortgrass and the drier areas of the Midgrass Prairie, saline areas are marked by the development of sodium-rich solonetz soils (Soil Classification Working Group 1998, 2001; Miller and Brierley 2011). Solonetz soils contain a characteristic Bn horizon, with strongly developed vertical columns, capped by white sodium salts ($NaNO_3$ and Na_2SO_4). The column caps are commonly indurated and extremely difficult to break or plough, and the high sodium levels frequently render the soil infertile for many cereal crops. The underlying Csk hozion reflects the high concentration of sodium in the parent material.

Solonetz soils develop naturally in areas where saline bedrock occurs near the surface, or where groundwater flow brings sodium salts to the soil (Miller and Brierley 2011; Nachshon et al. 2013; Oliveira Filho et al. 2020. Two hundred years ago, solonetz soils were relatively rare in the eastern midgrass prairie, as the area is underlain by thick Quaternary deposits. Irrigation in surrounding regions, combined with climate change and local changes in water tables, has resulted in the recent development of salinized areas and solonetz soils. The consequent losses to agriculture are considerable. The groundwater system of the eastern prairies is complex, and transport of groundwater several tens or even hundreds of kilometres from catchment to discharge areas is not uncommon. Thus, the saline groundwaters discharged today in the region may originate from substantial distances.

Solonetz soils undergo three stages of development (Miller and Brierley 2011; Fig. 5.41). In the first, Solonetz, stage, Na^+ migrates upward in groundwater discharge zones from Na^+-rich parent material. As Na^+ gradually accumulates in Bn horizons, it inhibits root penetration, and begins to form an impermeable cap, hampering drainage. When the $NaNO_3$ cap hardens completely, the soil reaches the Solodized Solonetz stage. Only halophyte vegetation can persist, and the land is agriculturally unproductive. The Ah horizon breaks down due to the lack of nutrient input, forming a sandy Ae horizon which is subject to wind erosion. The resulting exposure of the water-soluble NaNO3 cap causes it to gradually break down, resulting in the third stage of development, a Solod. Partial dissolution of the $NaNO_3$ cap and the downward migration of Na^+ allows non-halophytes to occupy the site. Humus begins to accumulate, forming a thicker Ah horizon.

5.7.3 Vertisol

Vertisols develop where seasonal wetting and drying of deformable (swelling) montmorillonite clays occurs (Soil Classification Working Group 1998, 2001). Surface cracking from seasonal desiccation and frost action facilitates vertisol development in areas with limited vegetation cover.

5.8 Fauna

Analysis of the changes in the faunal assemblages in the Canadian Prairies reveals the impact which removal or introduction of a very few species can have upon the ecosystem as a whole. Prior to 1500 ce, the area was dominated by large grazing mammals, notably *Bison bison bison* (plains bison). Ungulate populations included *Cervus canadensis* (wapiti, or elk) and *Odocoileus virginianus* (white-tailed deer) in the

Fig. 5.41 Stages of solonetz development. (Upper right photo: Canadian System of Soil Classification, Agriculture Canada)

parkland and northern midgrass regions, *Odocoileus hemonius* (mule deer) in the central areas, and *Antilocarpa americana* (pronghorn antelope) in the shortgrass prairie. Communities of rodents developed in harmony with these intermediate feeders. The primary predatory species were *Canis lupus* (wolf), *Puma concolor* (cougar), *Lynx canadensis* (lynx), *Lynx rufus* (bobcat), *Taxidea taxus* (badger), and *Canis latrans* (coyote). Black bears (*Ursus americanus*) and Plains Grizzly Bears (*Ursus arctos horribilis*) were also present.

In time, however, other species made their home on the range, including *Homo sapiens* and a cast of associated animals. The impact of these new arrivals was considerable, and has changed the faunal assemblage significantly.

5.8.1 Bison

The fauna prior to *ca.* 1500 ce was dominated by *Bison*. Two variants are currently recognized: *Bison bison bison* (plains bison), and *Bison bison athabascae* (wood bison; Fig. 5.42).

Plains bison were the largest herbivores on the prairies (McDonald 1981; Lott 2002; Naughton 2012; Krumbiegel and Sehm 1989; Gates and Jung. 2013). Mature males stand 1.4–1.6 at the shoulder (higher than most tallgrass prairie vegetation), and typically exceed 700 kg, with some individuals reaching 1000 kg. Mature females typically are 400–450 kg. Both females and males have curved, black horns (typically 58–64 cm for males), retained throughout the year. The characteristic muscular hump between the shoulder blades supports bison heads during feeding and fighting. Typical lifespan in the wild is 12–15 years. Females begin breeding at age 3, have a single calf in each breeding season, and typically produce 4–5 offspring during a lifetime.

B. b. athabascae is distinguished from the plains variant, *B. b. bison*, by being somewhat larger (mature males stand 1.8 m at the shoulder), heavier (1400 kg), and having a larger head with a broader horn span (typically 68 cm for mature males). These adaptations reflect the tendency for animals in colder environments to become heavier and more compact than their southern counterparts (more mass per body surface area). Although there is controversy as to whether *B. b. athabascae* was truly a subspecies, or whether it was simply a population which has developed different physical characteristics in response to a particular environment (an "ecotype"), it was a distinctive organism which evolved in response to a particular ecological regime (Geist 1991; van Zyll de Jong et al. 1995; Chiodini 2021).

Bison are noted for their poor eyesight, extremely strong sense of smell (detecting aromas from more than 3 km distance), speed (sprints to 65 km/h, capable of sustained travel at 30 km/h), and vocal repertoire. Males in particular can bellow loudly during mating season, although there is some evidence suggesting that the loudest males are not necessarily as successful as are the strong, quieter types (Wyman et al. 2008).

Diet varies seasonally and by location. Grasses and sedges usually dominate, but herbs, woody shrubs, aspen, willow, and poplar leaves, asters, sunflowers, and daisies are also consumed. Bison tend to consume high-quality protein plants in spring, gradually reducing the protein content of their diets throughout autumn and winter. They seek out recently burned areas to take advantage of newly growing green foliage.

Eighteenth- and early nineteenth century reports cite fast-travelling herds numbering several thousand, which required several hours to pass a given point (e.g. Isenberg 200; Lott 2002). Mass drownings occurred where bison attempted to cross rivers in flood, as still occurs today in the Peace-Athabasca Delta of Wood Buffalo National Park. The grazing

Fig. 5.42 Bison ecotypes. Left: Plains bison, *Bison bison bison*. Right: Wood bison, *B. b. athabascae* is distinguished from the plains variant, *B. b. bison*, by being somewhat larger (mature males stand 1.8 m at the shoulder), heavier (1400 kg), and having a larger head with a broader horn span (typically 68 cm for mature males). (Photo: Wikipedia/Rufus46)

bison aerated the soil, and re-fertilized the grazed areas, allowing plant growth to proceed in approximate equilibrium. Bison do not generally graze to soil level, leaving a fringe of vegetation to recover after the herd has vacated, and the animals migrated constantly. Thus, the vegetation was not over-grazed at any particular locality, and soil erosion was minimized.

Bison also grazed throughout the winter months, exposing the vegetation below the snow by rolling their massive heads and sweeping the snow from side to side. In the drier areas, this periodic removal of snow facilitated frost penetration, which broke down the structure of the soil and allowed more effective infiltration of spring precipitation. Aeration of the soil by the bison's activities favoured burrowing rodents.

Bison seek out shallow muddy or damp depressions to wallow, covering their fur with sediment (Fig. 5.43). Wallowing fulfils a variety of functions, including removal of excess fur (especially around the head), removing lice and ticks, reducing itching from insect bites, body cooling, relieving muscle aches and bruises, social demonstrations, and play. Wallowing acts to deepen the depressions, aerating soil and increasing moisture infiltration, providing shallow water sources for plants, insects, and small animal species. Wallowing also adds biomass from discarded fur and skin, and spreading seeds and burrs.

Bison also used "rubbing stones" to remove excess winter fur and visiting insects (Carne 2009. Glacial erratics from the Canadian Shield were preferred, due to their rougher, crystalline surfaces (Saskatchewan 1986). The vicinity of

Fig. 5.43 Rubbing stones and wallows. Left: Head-rolling serves to remove excess fur and parasites, as well as to clear snow during the winter. Upper right: Quartz sandstones of the Erratics Train, resistant with rough edges, were suitable rubbing stones. Centre right: bison wallows along rivers can suffer from over-use where herds are non-migratory. Lower right: with migratory herds, wallows are organically enriched and trap snow during winter, providing distinctive micro-habitats for vegetation

the rubbing stone was also an area of enhanced biological productivity, due to the additional fur, skin cells, and other waste.

In the absence of a suitable rubbing stone, bison removed insects and excess fur by rubbing against upright shrubs and trees. In the Aspen Parkland, this practice hindered the growth and spread of *Populus tremuloides*, favouring grass and fescue. The relative extent of open grassland and parkland groves thus reflected bison populations, with increasing bison use corresponding with decreasing tree cover. In many areas, shifts in proportions of tree cover thus reflects changes in bison population, rather than changes in climate.

5.8.1.1 Bison Jumps

Bison played a vital role in the cultures of indigenous people across the prairies. In several locations, bedrock escarpments served as locations where bison could be herded and killed (Brink 2008). Bison herds were stampeded towards the cliffs, often by young men dressed in bison robes, along "drive lanes" marked by stones. The herd, once stampeded, would continue to run blindly over the escarpment. The majority of the people would camp below the cliff face, killing any wounded animals and butchering the casualties. Several of these "bison jumps" were routinely used between 7,000 and 1,000 years ago in southern and central Alberta. Head-Smashed-In, the largest of these jumps in Canada, is located on the eastern margin of the Porcupine Hills 10 km northwest of Fort Macleod, and has been designated as an Alberta Provincial Historic Site, Archaeological reserve, and UNESCO World Heritage Site (Head-Smashed-In n.d.; Brink 2008; Slaymaker et al. 2017; Fig. 5.44). The site is managed and operated by the Piikani (Peigan) First Nation.

Repetitive use of bison jumps was underway by 5,000 years ago. Bison formed the basis of the Plains economy in southern Alberta, though other species contributed significantly to the diet. Bison provided not only food (meat, gristle, bone grease, marrow, blood) but also raw materials for tools (bone, sinew), shelter (hides, bone pegs), paints (fats), ceremonial trappings (horn headdresses, hides, bone), dog food (bones, offal), and other uses.

The Piikani used the Head-Smashed-In jump for the last 1000 years. Prior to this, both ancestral Piikani and Kutenai used the site. Excavations in the major kill deposits located at the south end of the complex indicated this jump's use history extends at least 5700 years before the present. Please see Brink (2008) for a complete discussion; a summary is provided here.

The complex consists of a series of buffalo jumps along the edge of the hills where bison were run off the edge of Porcupine Hills Formation sandstone cliffs. Below the cliffs lie bone deposits, usually situated on slump blocks. Below these blocks are a series of bison processing campsites usually located adjacent to springs. The complex association of cliffs, kills and camps stretches from 10 km along the eastern base of the hills.

To the west of the hills and butchering camps are the highlands and basin of Olsen Creek. From here the bison were gathered and driven towards the various jumps. The drive directions were controlled by a complex system of 12 interlocking drive lane complexes which were marked by regularly spaced small stone piles ("cairns").

Operation of this complex was a highly sophisticated affair involving the close cooperation and control of a large number of people (perhaps 1000) in the driving, killing and processing of the bison. The fall drives, along with the spring drives in locales where the bison calved (south of the Oldman River) were two focal points of the Piikani seasonal round. Large numbers of bison were hunted and processed and the meat dried, pounded and mixed with marrow and grease from the broken and boiled bones along with Saskatoon berries to make pemmican, a highly nutritious food (Ngapo et al. 2021) for use in winter or summer when the hunts failed or when other necessary activities—the summer Sundance, cutting lodgepoles, quarrying stone, digging roots, and picking berries—placed demands on the peoples' time and schedules. Often these resources were not where the bison were at these seasons of the year.

Bison driving was a totally integrated part of the social, political and spiritual of the Piikani. The spiritual significance of the site is indicated by the presence of "buffalo stones" or "In'imskins" (the Upper Cretaceous ammonite *Baculites sp.*), and the practice of placing the skulls in special places near the jumps. The dead bison's spirits would call to their children asking them to come to the jump and be killed for the peoples' benefit.

5.8.1.2 Post- 1500 ce Changes

Beginning shortly after 1500 ce, a series of unfortunate events drastically impacted the bison, and all of the residents of the grasslands and parklands.

One of the first new arrivals was the domestic horse, *Equus*. Horses became extinct in North America approximately 6,000 years ago (Murchie et al. 2021), along with mammoths and other megafauna. Recently, evidence of human utilization of horses during the Last Glacial Maximum has been found in northern Yukon (Bourgeon and Burke 2021).

Horses spread rapidly over the prairie, possibly reaching the southern Canadian Prairies by no later than 1700 ce, and soon began to compete with the pronghorn and mule deer for forage (Davies and Boyd 2019). Unlike these native grazing species, horses cropped the grasses and forbs closely, hindering reproduction of the vegetation and beginning the processes of soil erosion. They were also more difficult for the large predators, wolf and cougar, to kill, and hence horse populations tended to expand as mule deer and pronghorn

Fig. 5.44 Head-smashed-in Buffalo Jump is a UNESCO World Heritage site, administered by the Piikani First Nation. Upper: sandstone cliff marking the edge of the jump. Lower left: Bison view of the fescue shortgrass prairie directly below the jump

populations slowly dropped. Horses were rapidly employed by Piikani, Siksika, and other indigenous people for transportation and bison hunting (Piikani Nation n.d.), leading to breeding and increased numbers of escapees. Horse populations did have some influence on other prairie fauna in the 1700s and early to mid-1800s.

Although there currently is no fossil evidence indicating that *Equus* was present in North America between 6,000 years ago and 1519 ce, there are genetic relationships between the former and present North American populations (Murchie et al. 2021). It can be argued that *Equus* is native to North America, but simply disappeared for a time before reoccupying the terrain. In that sense, horses would not be considered invasive or alien, which may have implications for wildlife management (Zomorodi and Walker 2019). Alberta Environment does consider *Equus* to be feral and potentially invasive. With the exception of the herd on Sable Island (NS), there is no official protection for wild horses in Canada.

Today, wild horses are essentially free of predators. Overgrazing on sandy soils can cause extreme amounts of erosion, necessitating re-location or culling of herds (Davies and Boyd 2019). Although uncontrolled expansion of the population is a serious problem in some areas of the USA, the total wild horse population in the Alberta Foothills was estimated at 1314 in 2021 (AEP 2022).

The next, much larger, impact of *Homo sapiens* resulted from the impetus to colonization and settlement from the east. In 1616, Samuel de Champlain received intelligence from Nishnaabeg (Ojibwe, Nipissing) people about *Bison*, including hides, obtained from trade with people in Minnesota or Manitoba (Biggar 1929). Champlain assigned the French term "buffles" to the animal, which was subsequently anglicized to "buffalo". At that time, *Bison bison bison* ranged across western Canada from the western edge of the Canadian Shield north to northern Alberta and west to the Rocky Mountains. In the USA, Bison were present south of the Great Lakes as far east as Buffalo NY (Hornaday 1889). The total North American population is estimated as at least 60 million animals in 1780 ce (Bureau of Sport Fisheries and Wildlife 1965).

Although much of the substantial population crash is often attributed to railway construction, most of the contraction of *Bison* range and population reduction occurred prior to completion of the first US transcontinental railway in 1869 (Hornaday 1889; Bureau of Sport Fisheries and Wildlife 1965; Shaw 1995; Isenberg 2000; Lott 2002; Fig. 5.45). Clearance and fencing of agricultural land, and emphasis on use and export of hides, were greater factors in extirpation. In most areas, bison disappeared before railways arrived. In 1870, the North American population was between 5 and 6 million animals. The eastern border of the *Bison* range had retreated to central North and South Dakota, and *Bison* was no longer present in Manitoba or central Saskatchewan north of the North Saskatchewan River. These declines reflected overall losses in the migrating herds, rather than local over-exploitation.

In the USA, cases where trains were delayed by several hours due to herds on the tracks were noted in the 1860s and early 1870s. It appears that railway executives, railway engineers, and railway passengers alike were not impressed with the time "wasted" watching bison cross the tracks. Some railways claimed that tracks needed repair after the passage of large bison herds, which is not totally improbable. Hunting for meat or "sport" killing were much less significant than the hide trade (Taylor 2011), with 1.1 million bison hides exported from the USA in 1875. Pronghorns and mule deer were also targeted to a lesser degree.

By 1881, when the CPR began to build across the Canadian Midgrass Prairie, the bison population had already been seriously weakened. As the species migrated extensively across the prairies (without respecting the 49th parallel), losses in the U.S. adversely affected the "Canadian" population. The "Canadian" population crashed in 1883. Total North American bison population dropped from 395,000 in 1880 to less than 1,000 in 1890, with free-ranging Bison confined to the Boreal forest of northern Alberta. *Bison bison bison* would have become at least locally extinct had not individual animals been previously conserved in farms and private rangelands. The removal of *Bison* from the ecosystem was coupled with the drop in pronghorn and mule deer numbers.

These changes had immediate impacts on many other species. The aeration and fertilization functions performed by grazing, wallowing, and rubbing, the lack of winter snow clearance by massed bison heads, and the replacement of migrating grazers with stationary close-cropping cattle and sheep, resulted in a decline in botanical productivity. The effects of dry years became much more pronounced, as the weakened and stressed vegetative community struggled with the cyclically unfavourable climate. The conditions for burrowing rodents became unfavourable, and populations of these species dropped. Pressure on the predatory species was increased by human persecution of both predators and prey. Rattlesnakes and swift foxes, as well as wolves, cougars, and grizzly bears, were targeted for direct elimination.

5.8.1.3 Modern Bison Populations

Today, the bison within Wood Buffalo National Park in the Boreal Plains (see Chap. 3) are Canada's largest free-ranging herd, but they have not existed without human interference (Mckennirey and Richardson 1985). When the park was designated in 1922, the *Bison bison athabascae* variant was relatively pure. Free-ranging plains bison had been eliminated from the southern prairies, precluding the possibility of interbreeding. A preserve for plains bison was established at Wainwright, Alberta, and designated "Buffalo National Park". The bison for this park were imported from the herd in Yellowstone National Park, U.S.A. Subsequently, the area of Buffalo National Park (now Canadian Forces Base Wainwright) was requisitioned for tank manoeuvres, and the plains bison were evacuated to several locations, including Wood Buffalo. The plains and wood variants interbred freely, and consequently no "pure" examples of *B. b. athabascae* remain (Gates and Jung 2013; Environment and Climate Change Canada 2018). The population is estimated at 4,000–5,000 animals. Subsequently, some of these "hybrid" bison have been sent to other areas, including Elk Island National Park.

Reintroduction of plains bison in Canada is largely dependent on the herd currently resident in Elk Island National Park, descendants from those transferred from the herd at Wood Buffalo National Park. Unlike their northern counterparts, they are believed to be tuberculin-free at present. The

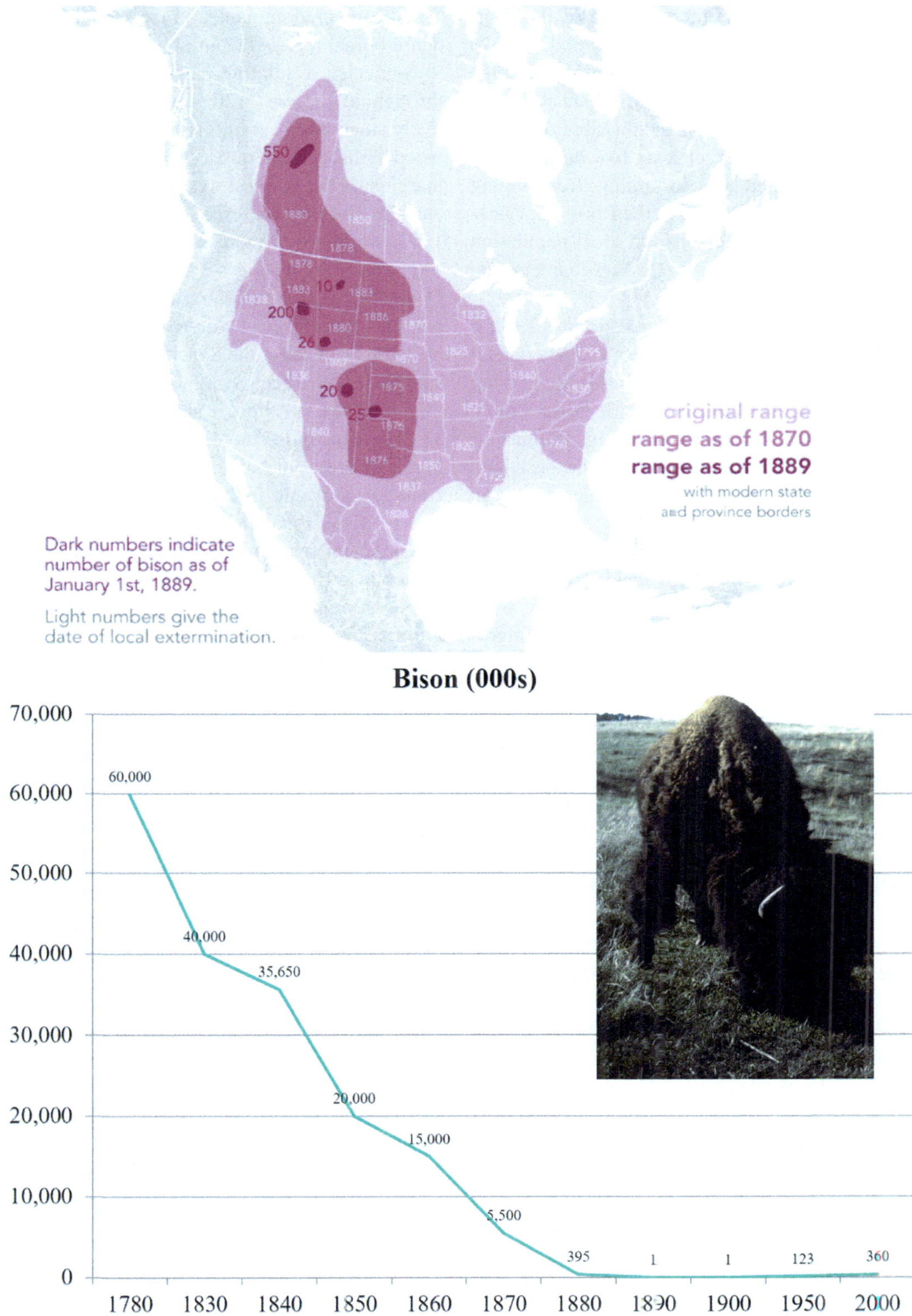

Fig. 5.45 Bison decimation in the nineteenth century was documented by WT Hornaday in 1889 (upper). Numbers continued to fall until the early 1900s. Current population is estimated at 360,000 to 400,000. (Data: Gates and Jung 2013, and other sources listed in text)

animals represent the results of interbreeding of the original *Bison bison athabascae* population of Wood Buffalo with the *B. b. bison* population imported from Yellowstone (Wyoming) and Wainwright, Alberta, and hence are not a pure subspecies. The long-term population and distribution objective is to ensure the existence of at least five disease-free, genetically diverse, connected, self-sustaining, free-ranging local populations distributed throughout their original Canadian range, with a minimum size for each local population of 1,000 animals (Gates and Jung 2013; Environment and Climate Change Canada 2018).

The utilization of the bison as a domestic species has recently become popular throughout western Canada (Canadian Bison Association 2022). Bison spend less time grazing than do cattle, and are able to digest the prairie grasses more effectively, although cattle are somewhat less selective in their feeding habits. Grazing bison appears to be the best land management strategy (not considering economic factors) in large land areas marked by homogeneous use, where limited human labour is available, and where highly sensitive or riparian woodlands are present. Cattle ranching is favoured under the opposite suite of conditions, particularly in areas where competing land uses are practised. Although the migratory habits of bison must be curbed, the present generations of bison are descended from several captive generations and tend to be more sedentary. Keeping large herbivores confined exerts a significant impact on the terrain and vegetation: one cow nursing a calf will eat as much vegetation as 300 prairie dogs (Miller et al. 1994).

5.8.2 Wapiti- Whitetail-Mule-Pronghorn

With the removal of the bison, four large herbivores occupy the vegetation zones across the boreal forest-parkland- grasslands spectrum (Fig. 5.46). The range of *Odocoileus virginianus* (white-tail deer) extends westward from the Boreal Shield and Mixedwood ecozones along the boreal forest-parkland transition (Banfield 1981, Naughton 1992). As browsers, the southwestern limit of whitetail deer generally coincides with the edge of the Aspen Parkland, but the elimination of the bison has allowed it to expand into some rough fescue and midgrass areas, locally facilitated by the encroachment of aspens no longer subjected to pressure from bison rubbing.

The other large herbivore in open boreal forested and parkland areas is *Cervus canadensis* (wapiti, also called elk). The Nehinaw (Cree) word "wapiti" refers to the distinctive white rump. Mature males typically are 150–500 kg, standing 3–4 m high at the shoulder, allowing them to navigate forested, parkland, and tallgrass landscapes (Banfield 1981). Only males have antlers, shed each winter as the testosterone levels drop. Wapiti are predominantly grazers, and do not occupy the dense boreal forests favoured by moose (*Alces alces*).

Subjected to hunting pressure and habitat fragmentation throughout the past 120 years, wapiti are present in relatively small numbers in forested enclaves such as the Touchwood Hills of eastern Saskatchewan, as well as along the margins of the boreal forest in Prince Albert National Park and the Cordilleran forest of the Porcupine Hills. Prior to 1900, wapiti were more widely disturbed across the rough fescue and mixed midgrass prairie, extending south to the Red Deer River valley in central Alberta. Wapiti were also resident in southwestern Ontario until the early 1800s, and have been reintroduced to the Bancroft area (Rosatte 2017). Several subspecies have been recognized, but in some instances these may represent ecotypes, rather than true subspecies.

One of the mild curiosities about Elk Island National Park is that it does not have any wild representatives of *Cervus canadensis.* Wapiti in this region were eliminated by hunting and by the habitat destruction resulting from agricultural development surrounding the park, and the park's 194 km^2 area is insufficient to sustain a free-ranging population. The wapiti currently living in the park are descendants of reintroduced animals.

The dominant large herbivore in midgrass and mixed grassland areas is *Odocoileus hemionus* (mule or black-tailed deer). Mule deer are a gregarious species noted for its slightly smaller size than the white-tail (mature males ~100 kg, 1 m shoulder height) and distinctive black stripe and black-tipped tail. Mule deer antlers grow by bifurcation, producing a fork-like pattern, unlike the white-tail's prongs emanating from a single beam. Only male deer have antlers, which are regrown in spring and summer (Geist 1998; van Zwoll 2004).

Mule deer have a unique gait, moving in a series of short bounds with all four feet landing together. This "stotting" or "pronking" produces a characteristic bobbing motion which distinguishes mule deer from the more smoothly running white-tail. Stotting is an evolutionary adaptation to the open fescue and midgrass ecology.

Mule deer both browse and graze, the diet reflecting available resources. In prairie Canada, favourite foods include *Elymus lanceolatus* (thick-spike wild rye), *Hesperostipa comata (*Needle-and-thread grass, *Pascopyrum smithii* (western wheat grass), Festuca sp., Artemisia sp. (sagebrush), *Erigeron sp.* (fleabane), and Populus tremuloides (aspen). They have also adapted to consume introduced vegetation, including Kentucky bluegrass (*Poa pratensis),* red clover (*Trifolium pratense*, Quackgrass (*Agropyron repens*), and dandelion (*Taraxacum officinale*). This eclectic diet, combined with the ability to thrive in diverse open landscapes including agricultural and exurban areas, has resulted in stable or expanding populations of mule deer in Canada, rebounding from low numbers in the mid-1900s.

Fig. 5.46 Deer species. Upper left: *Odocoileus virginianus* (whitetail). Upper right: Current northern range limit of *Odocoileus hemionus* (mule deer) (Data: Banfield 1981; Governments of Alberta, British Columbia, Saskatchewan, Yukon). Lower left: *Odocoileus hemionus* (mule deer; US Fish and Wildlife Service). Lower right: *Cervus canadensis* (wapiti)(US National Park Service)

The dominant large herbivore in shortgrass landscapes is *Antilocapra americana* (pronghorn, also called antelope; Fig. 5.47). With half the mass of a mule deer (mature males 40–50 kg), but standing 1 m at the shoulder, pronghorns have evolved to travel across open shortgrass (O'Gara and Yoakum 2004). Pronghorns can reach 95 km/h, and can maintain 80 km/h for more than 1 h. However, pronghorns cannot jump (see Robb et al. 2022). Binocular vision is another useful adaptation to the open shortgrass landscape. Horns are shed annually.

Prior to the introduction of barbed-wire fencing and habitat fragmentation, pronghorns migrated as far as 800 km annually, in response both to vegetation growth and seasonal snow cover (Jakes et al. 2018). Unlike bison and wapiti, pronghorns are not capable of clearing snow effectively to access vegetation. Although pronghorns are gregarious, the relative scarcity of vegetation resources in shortgrass landscapes means that each mature male requires access to about 250 hectares to survive. Diets include a mixture of forbs, grasses, and shrubs, including *Bouteloua sp.* (grama grass), *Hesperostipa comata* (spear grass), Cyperaceae (sedges), *Selaginella densa* (spikemoss), *Castilleja* (Prairie Fire), *Artemisia sp.* (sagebrush), and *Opuntia polyacantha* (prickly pear cactus).

In addition to *Homo sapiens*, the primary predator of pronghorn is *Canis latrans* (coyote). Coyotes are threats only to new-borne pronghorns (O'Gara and Yoakum 2004). Within two weeks, the young are able to detect coyotes with their binocular vision and outrun them, within the protection of the herd. In an evolutionary adaptation, all young in a herd are commonly born within a 2-week period. The solitary hunting coyotes are thus presented with an abundance of prey, but only for a very short time: once the coyote has killed, eaten, and digested its first meal, the remaining pronghorn young have matured sufficiently to evade its next effort.

Fig. 5.47 *Antilocapra americana* (Pronghorn). Upper left and lower left: males (Gary Kramer, US Department of Agriculture; James Leupold, US Fish and Wildlife Service). Upper right: female. (James Leupold, US Fish and Wildlife Service). Lower right: Current northern range limit. (Data: IUCN, Government of Alberta, Government of Saskatchewan)

Sharp declines in pronghorn population in the early twentieth century prompted the 1922 establishment of Wawaskesy National Park, northeast of Medicine Hat, as a preserve (see Chap. 10). Population increases throughout 1920–1945, coupled with the 1941 requisition of the park as a military training ground and chemical weapons testing area ("Suffield Block"), led to the degazetting of Wawaskesy in 1947. The landscape was designated as the Suffield National Wildlife Area in 2003, although military activities continue, and the land is open to petroleum exploitation.

5.8.3 Shortgrass Fauna

In addition to bison and pronghorn, the Shortgrass and drier Midgrass prairies supported a unique Canadian fauna prior to human intervention. Some representatives of this assemblage remain.

Cynomys ludovicianus (black-tailed prairie dog) reaches its northern limit in the southernmost Canadian Prairies (Fig. 5.48). At present, its only Canadian locality is Grasslands National Park and vicinity (Parks Canada 2021). Male black-tailed prairie dogs are up to 1.5 kg in mass, and usually live for three years on a diet of forbs, grasses, and carrion. Black-tails are extremely social animals, living together in large colonies or "dog towns", and sharing responsibilities such as tunnel construction, foraging, and keeping sentry duty for their numerous predators—rattlesnakes, hawks, owls, badgers, coyotes, swift foxes (*Vulpes velox*), and the black-footed ferret, *Mustela nigripes*. The burrowing activities of blacktails would rapidly destroy the soil and vegetation in areas where they were allowed to spread unchecked, but the combined effects of predation and bison re-fertilization allowed all members of the community to be adequately supported by the land.

Fig. 5.48 *Cynomys ludovicianus* (Black-tailed Prairie Dog). Upper: "dog towns" are marked by abundant burrowing. Without predation and bison re-fertilization, net degradation can result from increased populations. Lower: *Cynomys ludovicianus* reaches its northern range in Grasslands National Park. (Elise Smith, US Fish and Wildlife Service)

The most specialized of the region's predators was the black-footed ferret, *Mustela nigripes.* This relative of the weasel lived almost exclusively on black-tailed prairie dogs. Its body is elongated, generally about 30 cm in length, and its legs are extremely short. The black-footed ferret's attack strategy was generally direct confrontation, usually by burrowing into the prairie dog's den. Its typical body mass of ±1 kg was often less than the mass of its prey.

The black-footed ferret was once believed to be extinct in the wild throughout North America, and the only individuals known to be alive were those kept for breeding at a U.S. facility near Laramie, Wyoming, and their descendants. In 1988, these numbered exactly 23. The Wyoming facility, however, successfully bred and re-introduced the black-footed ferret to nineteen areas, including Grasslands (Tuckwell and Everest 2009).

Vulpes velox (swift fox, also sometimes erroneously called kit fox) preys on jackrabbits, burrowing rodents, carrion, and grasshoppers (Michalsky et al. 2020). The periodic plagues of grasshoppers common to all grassland regions were warmly received by most of the predatory species. *Vulpes velox* is adapted to shortgrass prairie life, with its low silhouette and its powerful legs, capable of propelling the fox at 55 km/h. *Vulpes velox* currently is threatened throughout the shortgrass prairie. After being extirpated in Canada in the 1930s, *Vulpes velox* has been re-introduced (Michalsky et al. 2020).

Rodent species inhabiting this region include *Lepus townsendii* (jackrabbit), Nuttall's cottontail rabbit (*Sylvilagus nuttallii*), and *Spermophilus sp.* (ground squirrels). Although western residents tend to call any small or medium-sized rodent a "gopher", including ground squirrels and black-tails, technically only *Thomomys* belongs in this category. *Thomomys talpoides* (northern pocket gopher), along with other pocket gophers, evolved rapidly in the grasslands during the last 4 million years, living on the forbs fuelled by bison activity. A small rodent, less than 8 cm long, it prefers moist soils. *Dipodomys ordii* (Ord's Kangaroo Rat), a bipedal hopper capable of bounds up to 2 m, similar to the Asian gerbil, survives on a diet of seeds and grasshoppers and took advantage of the larger rodent and bison activity in aerating sandy soil.

Crotalus viridus (western or prairie rattlesnake; Fig. 5.49) was also a common predator, thriving on the smaller rodents (pocket gophers and kangaroo rats). The rattles evolved as a warning device for the benefit of grazing bison and pronghorn antelope (*Antilocarpa americana*). Unlike the swift fox, *Crotalus viridus* is a slow-moving (5 km/hour maximum) predator, and thus prefers to attack from ambush. It grows a new rattle segment each time it sheds its skin, which can occur several times in one season, and older rattle segments frequently break off. A large prairie rattler is about 1 m in length, and may live up to 20 years (Cook et al. 1984). Although its venom effectively paralyses small rodents, it is much less effective on large animals. Approximately 3% of rattlesnake bites to humans result in human fatalities, usually in cases where treatment has been delayed for more than two hours. A much higher proportion of rattlesnake bites to humans result in rattlesnake fatalities.

The Prairie rattlesnake has expanded its range northward up to 100 km since construction of the CPR in 1882–3. The old superstition—"snakes won't cross tracks"—has been superseded as climate change continued. As many farmers do not encourage snake range expansion, climate change is the most reasonable explanation.

Fig. 5.49 *Crotalus viridis* (prairie rattlesnake). Upper: Mature male (US National Park Service). Lower: Current northern range limit of *Crotalus viridis*. (*Data source* COSEWIC)

Other unique shortgrass species include greater short-horned lizard (*Phrynosoma hernandesi;* COSEWIC 2018), burrowing owl (*Speotyto cunicularia*), sage grouse (*Centrocercus urophasianus*), ferruginous hawk (*Buteo regalis*), Bullock's oriole (*Icterus bullockii*), prairie chicken (*Tympanuchus cupido*), Magpie (*Pica pica*), and Sprague's Pipit (*Anthus spragueii*), the latter known for its airborne courtship rituals (Godfrey 1986; Bird 2017).

5.8.4 Tallgrass Fauna

The presence of the relatively moist Red River valley allowed species to migrate north from North Dakota and Minnesota that could not survive in the harsher environments of the Canadian Shield to the east and the Midgrass Prairie to the west. Consequently, the Red River valley organisms were often genetically distinct from those of surrounding areas. Although similar species lived in the milder climates of Southern Ontario, the individuals in the Red River valley developed distinctive and separate genetic characteristics as they became more isolated. Although these differences were not recorded in the larger mammal populations prior to their extirpation, study of the smaller mammals has revealed distinct Red River subspecies and variants. Among these are *Sylvilagus floridanus similis* (Red River cottontail), *Lepus americanus phaeonotus* (Manitoba snowshoe hare), *Sciurus carolinensis hypophaeus* (Red River grey squirrel), and *Tamiasciurus hudsonicus minnesota* (Minnesota red squirrel).

Similarly, fish species such as white bass (*Morone chrysops*) and chestnut lamprey (*Ichthyomyzon castaneus*) present in the Red River are absent from surrounding areas (Scott and Crossman 1998). Both fish migrated up the Red River valley during deglaciation, following the retreating shoreline of Lake Agassiz northward.

5.8.5 Midgrass-Parkland-Boreal Fauna

As elsewhere on the prairies, the fauna of the midgrass, parkland, and boreal transitional zones has suffered from *Homo sapiens'* attentions since the 1870s. In addition to bison, wapiti, and white-tailed (parkland) or mule deer (midgrass), typical fauna would have included *Vulpes vulpes* (red fox), *Lynx canadensis* (lynx), *Felis concolor* (cougar), *Ursus arctos horribilis* (plains grizzly bear), *Ursus americanus* (black bear), *Canis lupus griseoalbus* (wolf), *Canis latrans* (coyote), *Mephitis mephitis* (striped skunk), *Erethizon dorsatum* (porcupine), *Lepus americanus* (snowshoe hare), *Castor canadensis* (beaver), and *Taxidea taxus* (badger). Adjacent boreal forests supported populations of *Alces alces* (moose), *Rangifer caribou* (woodland caribou), pine marten (*Martes americana*), and fisher (*Martes pennanti*).

Of this fauna, the bears, cougars, lynx, wolves, pine marten, and caribou have been extirpated by hunting and habitat fragmentation. Red fox and coyote, the smaller and more adaptable predators, have managed to thrive in the vacated ecological niches. Confined to landscapes west of the Canadian Shield margin in 1890, coyote has been steadily

Fig. 5.50 Left: *Spilogale putorius* (spotted skunk) is confined to southernmost Manitoba in Canada, although it has an extensive range in the United States. (US National Parks Service). Right: *Mephitis mephitis* (striped skunk) is common throughout the Grassland, Mixedwood, and southern Boreal ecoregions. (John Collins, US Fish and Wildlife Service)

expanding its range and increasing in population throughout Canada south of the taiga since 1920, including urban environments. Skunk and porcupine have also adapted to human activity and are increasing their populations.

Taxidea taxus (badger) has also managed to survive in the region, despite unwelcome human attention (COSEWIC 2012). This large (75 cm long) striped member of the rodent family is adept at burrowing, and preys upon the shrews, voles, ground squirrels, snakes, and other small animals of the grassland region. It is particularly fond of eggs. Local increases in the populations of small rodents resulting from agricultural development and the elimination of large predators has favoured the badger, as badgers fill ecological niches forcibly vacated by other predatory species.

Disjunct populations exist in the coniferous forest and dunes of Spruce Woods Provincial Park.

Spilogale putorius (eastern spotted skunk) is an inhabitant of the coniferous forest area (Fig. 5.50). This skunk is distinguished from the more common striped skunk (*Mephitis*) by its smaller size and its pattern of irregular white mottles, especially on the head. Typically, *Spilogale putorius* had a maximum body length of 50 cm in Spruce Woods, with a 30–40 cm long tail. For unknown reasons, it has become extremely rare across its entire range (Gompper and Jachowski 2016).

The coniferous forests also provide a suitable habitat for *Glaucomys sabrinus* (northern flying squirrel). This common inhabitant of the boreal forest is capable of gliding for 80 m supported by flaps of skin between its forelimbs and lower torso, but generally prefers to take much shorter trips between densely packed conifers (Wells-Gosling and Heaney 1984). It has a maximum body length of 30 cm (although most individuals are somewhat smaller), and is a nocturnal feeder on conifer cones and seeds.

The dune areas allow reptiles to survive, as here they have suitable sediment to burrow in, can take advantage of the sand's ability to reflect heat during the summer, and are relatively free of human predation. An unusual reptile present within Canada only in Spruce Woods Provincial Park and in the adjacent Canadian Forces Base Shilo to the west is *Plestiodon septentrionalis septentrionalis* (northern prairie skink). The Spruce Woods population is disjunct, with the main range of the species located in the Dakotas. The northern prairie skink is a small lizard (20 cm long maximum), marked by alternating dark green and light olive-brown stripes along the length of the body (Cook et al. 1984). The subspecies' confinement to the dune areas leaves it potentially vulnerable to climatic changes or human interference.

5.9 Environmental Impact on Human Residents: Droughts

Droughts lasting between 8 months and more than 5 years are repetitive occurrences in Prairie Canada, with more than 40 significant events since 1800 (Bonsal et al. 2011a, b; Fleming and Sauchyn 2013; Lapp et al. 2013; Sauchyn and Ilich 2017; Sauchyn et al. 2020). Droughts affecting different areas occurred in 1858–1872, 1878–1884, 1912–1914, 1930–1937, 1958–1963, 1999–2005, 2010–2011, and 2015–2017. The severe drought conditions in 2001–2002 extended from British Columbia to Nova Scotia, resulting in lost agricultural productivity in excess of \$3.6 billion, and Canada's gross domestic product decreased by ~\$5.8 billion (Barrie et al. 2011, Hanesiak et al. 2011).

Over the past 100 years, a general trend towards drying is apparent in the southern Canadian Prairies (Asong et al. 2018; Sauchyn et al. 2020). Less than 2% of Canadian agricultural production comes from irrigated land, with most water obtained from rivers, rather than groundwater. In Prairie Canada, meteorology, agriculture, and surface water hydrology are closely linked, although effects become evident over time.

Understanding the climate and meteorological parameters responsible for one drought can aid prediction of the next. If several droughts exhibit the same pattern of parameters, a model for drought occurrence can be developed. Droughts in the Canadian Prairies from 1858 (documented by the British North American Expedition, led by Captain John Palliser; see Wolfe et al. 2013), through 1930–1937 (Great Depression), to 1963 appear to have originated from a common set of circumstances. Unfortunately, the 1999–2005 drought did not conform to the previously documented pattern (Barrie et al. 2011, Hanesiak et al. 2011).

In western Canada, moisture originates from air masses flowing across the North Pacific Ocean. The Pacific—North American teleconnection (PNA) relates atmospheric pressure over the North Pacific Ocean with airflow and precipitation across North America. During the PNA positive phase, barometric pressures are higher than normal over Hawai'i and Prairie Canada. The jet stream is strengthened and moves eastward, carrying moisture around the high pressure areas towards the southeast. This normally would result in warmer, drier weather across Prairie Canada from autumn through spring, but has relatively little influence on summer temperatures. During the PNA negative phase, barometric pressures are lower than normal over Hawai'i and the prairies, the jet stream weakens and moves westward, and winter precipitation generally increases.

The Pacific Decadal Oscillation (PDO) also influences western Canada's weather (Fig. 5.51). PDO is defined by changes in sea surface temperature in the North Pacific Ocean. During a PDO positive phase, Pacific coastal waters of the Alaska (northbound) and California (southbound) Currents become relatively warmer, while the central and western North Pacific become cooler. Temperature and precipitation both increase over coastal British Columbia, but both tend to decrease over Prairie Canada. A PDO negative phase is characterized by wetter winter in Prairie Canada.

Both the PNA and PDO are influenced by complex and variable teleconnections to ENSO. Positive PNA and PDO tend to be associated with El Niño phases of ENSO, especially stronger events, but can develop under neutral or weak La Niña conditions. As both PNA and PDO operate over longer periods (several years to decade) than does ENSO (months to years), the three oscillations are not synchronous.

The combination of a positive PNA, positive PDO, and some El Niño phases would suggest decreased winter and spring precipitation, higher summer temperatures (increased evapotranspiration), and meteorological drought over the Canadian Prairies, progressing to surface hydrological drought. This "amplified warm" dry, warm summer pattern, accompanied by high pressure blocking atmospheric ridges, strong winds, soil erosion, and blowing dust, occurred during the droughts of 1858–1872 and 1930–1937. The duration of each drought spanned several El Niño—La Niña oscillations, but ENSO phases were highly variable throughout both episodes. Similar, although less extreme, conditions marked the droughts of 2010–2011 and 2015–2017 (Szeto et al. 2016). The strong 2015–16 El Niño may have contributed to the latter drought.

5.9.1 Drought of 1999–2005

The drought of 1999–2005 differed substantially from this pattern (Barrie et al. 2011, Bonsal et al. 2011a, b; Hanesiak et al. 2011). Initial development in southern Alberta was marked by La Niña conditions, along with negative PNA and PDO. In 2001–2002, extreme drought conditions developed under a relatively strong La Niña, a moderately negative PNA, and a strongly negative PDO. High pressure blocking ridges ("omega" highs) did not occur over the prairies. Drought in the central and northern prairies was not accompanied by heat. At Edmonton and Saskatoon, northwesterly air flow produced drier conditions accompanied by below-average temperatures: a cold drought, with no extremely hot days and little precipitation. Precipitation deficits marked all seasons, with reductions in excess of 30% in some areas. Most precipitation occurred as minor events (<10 mm), and convective thunderstorms were uncommon.

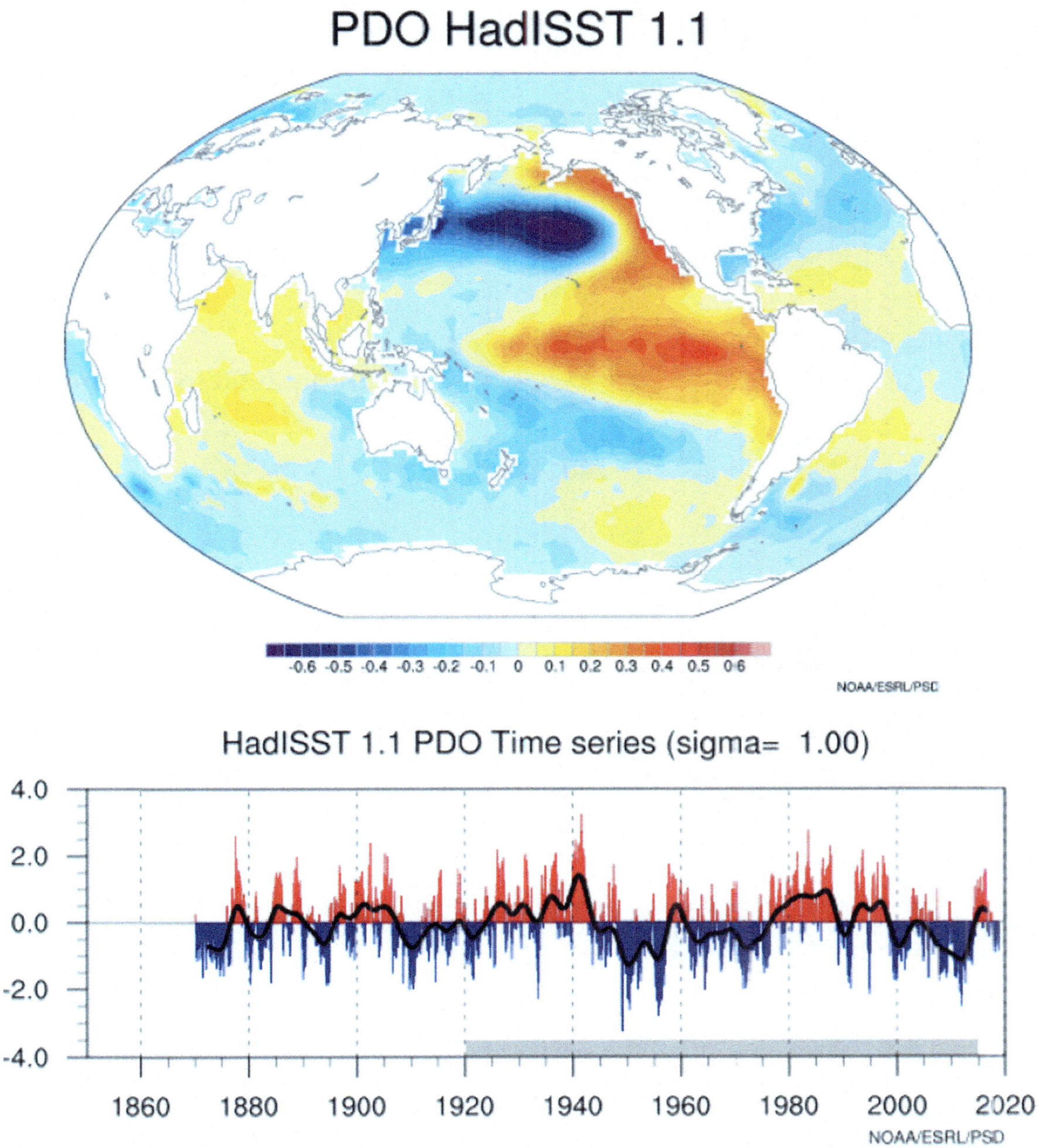

Fig. 5.51 Pacific Decadal Oscillation. Upper: Positive phase of PDO based on Hadley climate centre data. Cooler temperatures in the northwest Pacific Ocean During a PDO positive phase, temperature and precipitation both increase over coastal British Columbia, but tend to decrease over Prairie Canada. A PDO negative phase is characterized by wetter winter in Prairie Canada. Lower: Variation of PDO, 1870–2019. (NOAA)

El Niño conditions dominated between 2003 and 2005, although negative PDO and PNA phases persisted. The affected areas shifted to the north and east, extending from northern Alberta to central Manitoba.

Shifting atmospheric conditions and affected areas demonstrated that the 1999–2005 drought was highly complex (Hanesiak et al. 2011). The concept of a persistent, single "amplified warm" drought with dominant high pressure ridges and positive PNA, with a single origin and uniform effects required modification (Figs. 5.52 and 5.53).

The drought is now understood to involve episodes of three types of upper atmospheric flow conditions: "amplified warm"; "amplified cold", with significant cold-air advection and reduced temperatures; and "zonal", west-to-east air movement involving downsloping flow from cool, dry air masses over the Cordillera. The influences of PNA, PDO,

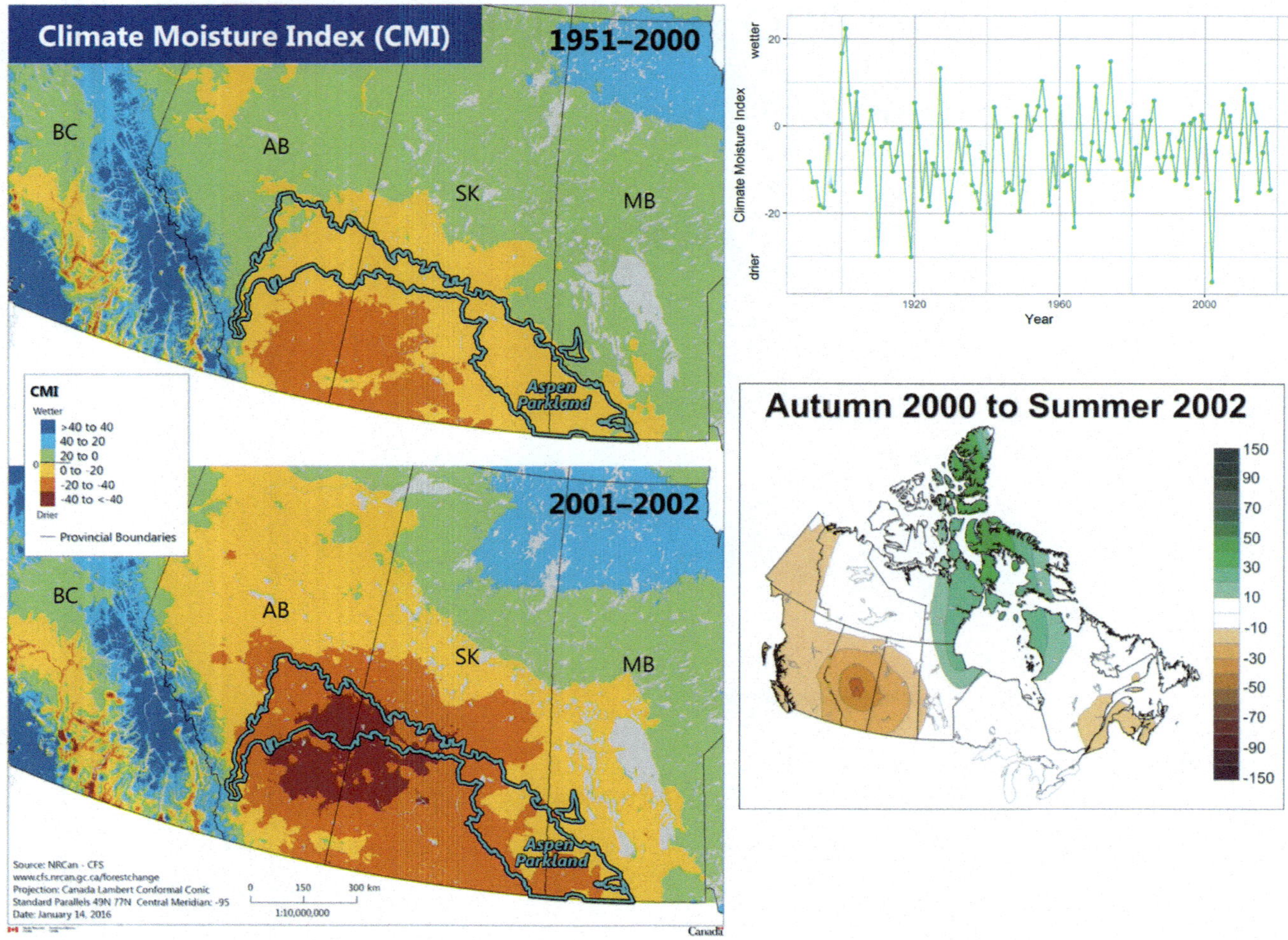

Fig. 5.52 The 1999–2005 drought had substantial ecological, financial, and societal impacts in Prairie Canada. Left: comparison of the mean Climate Moisture Index (CMI) over 1950–2000 compared to 2001–2002. Upper right: variation in CMI over time. Lower right: Differences in precipitation compared to climate normal values, Autumn 2000 to Summer 2002. (Environment and Climate Change Canada)

and ENSO phases on amplified cold and zonal events are limited. Atmospheric circulation patterns changed during the six years of the drought, and none of the regimes was dominant throughout.

Across the prairies, the limited precipitation in autumn 1998 and snow cover in the winter of 1998–1999 may have contributed to the drought by reducing recharge of soil moisture. Winter snow cover remained below average in most affected regions throughout the drought. Typical hydrological lags were 1–3 years for lakes and reservoirs, and one to five years for groundwater aquifers.

In addition to impacting agriculture, the drought accentuated wildfire hazards in the northern prairies, and affected hydroelectricity generation by reducing streamflow. Reduced generating capacity is a significant concern for power companies across western Canada. Lower river discharge is often accompanied by reduced input from snowmelt in the Rocky Mountains, as winter precipitation decreases. Snowpack thicknesses have consistently decreased in the Northern Cordillera since 1979, with much of the decrease attributed to rising winter temperatures.

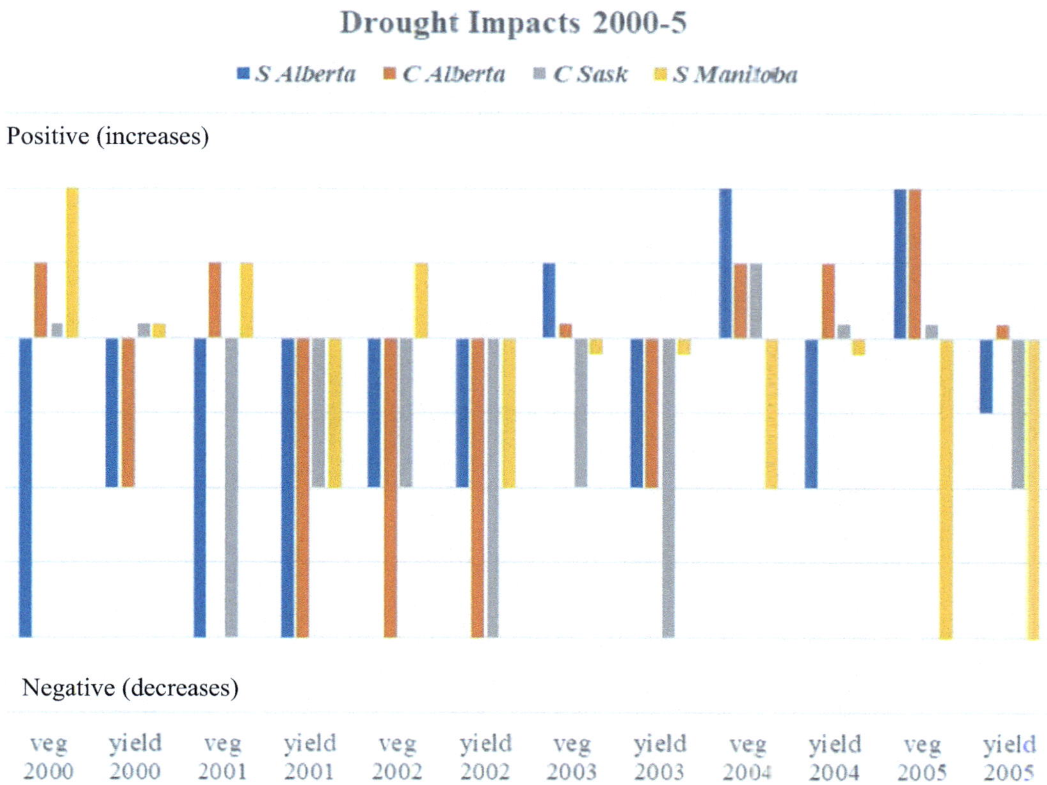

Fig. 5.53 Impacts of 2000–2005 drought, Prairie Canada. Forest vegetation health and productivity (veg) and crop yields of spring wheat, barley, canola, and field peas (yield) varies spatially and temporally. (Data source: Hanesiak et al. 2011, and work cited therein)

5.9.2 Implications

The prospects and potential impacts of future drought and water scarcity are of concern across the southern prairies (Sauchyn et al. 2020; Fig. 5.54 and 5.55). Surface water allocation has generally proceeded on the seniority principle, with existing patterns and users having priority (Wheater and Gober 2013; Gober and Wheater 2014). In many areas of the Saskatchewan River Basin, all of the available water has already been allocated, and there is no excess capacity (Corkal et al. 2011; Kamar Ali and Klein 2014). Alternative policies prioritizing supply to people, particularly urban residents, or proportional reductions for all users, have also been suggested. In southern Alberta, the seniority principle favours existing irrigation over supply to urban residents of Calgary. Proportional reductions would affect all users, but would allow irrigation to continue.

Water conservation measures are in place in Calgary, although they are less common or comprehensive in other prairie communities (e.g. Kevinsen et al. 2014). Green lawns and shade trees require abundant irrigation to live in a prairie climate. In prairie urban areas, where most residents water lawns, maintenance of outdoor vegetation typically accounts for at least 30% of residential summer water use (Statistics Canada 2015).

The environmental sustainability of prairie agriculture would be affected by future droughts. Increased soil erosion, salinization, contamination, reduction in water quality and quantity, and desertification are all likely (Sauchyn et al. 2020).

Salinization impacts vegetation, aquatic ecosystems, wildlife, and agricultural production (Nachshon et al. 2013). Replacement of the natural shortgrass and midgrass vegetation assemblages with annual crops of wheat and canola, and drainage of wetlands and sloughs, impact evapotranspiration and water movement, concentration sodium and potassium salts and contributing to salinization over longer terms. conversion of prairie grasslands to annual dryland crops and drainage of wetlands. More than 40% of prairie wetlands have been drained for agriculture and urban development since 1900 (IISD 1994).

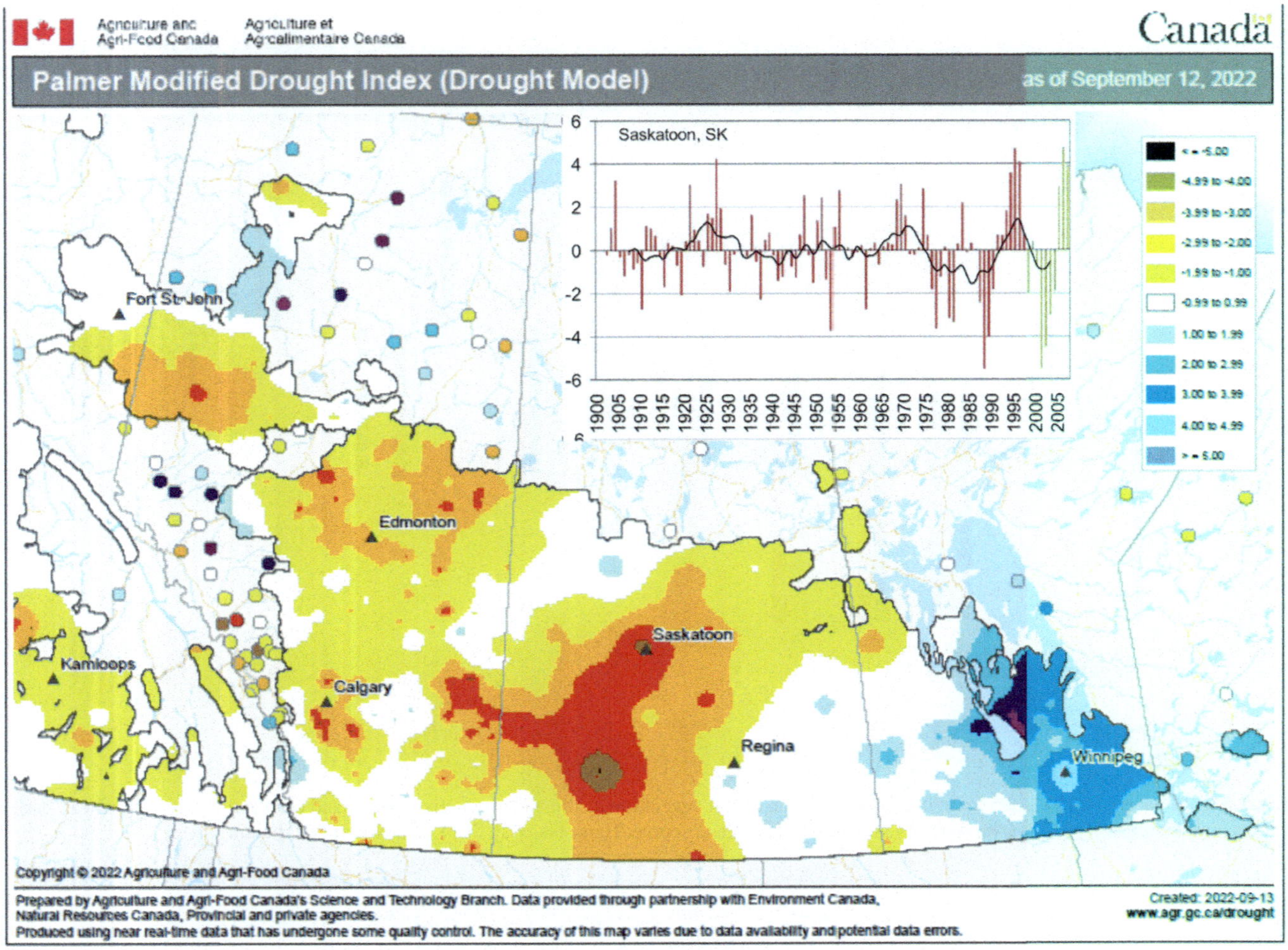

Fig. 5.54 Palmer Modified Drought Index for western Canada, 12 September 2022. (Agriculture and Agri-food Canada). Inset: PDSItime series for Saskatoon. (Environment and Climate Change Canada)

Historically, repetitive efforts to establish agricultural settlement in the southern prairies were influenced by the variability in the moisture regime, both in location and timing (Lapp et al. 2013; Sauchyn et al. 2015a, b, 2020). Palliser's expedition of 1857–1860 crossed the driest area of the southern prairies ("Palliser's Triangle") during a drought (Wolfe et al. 2013). Strongly influenced by his difficulty in travelling through active sand dunes of the Middle Sand Hills of southeastern Alberta, he declared that the region would "forever be comparatively useless" for agriculture. Simultaneously, expeditions led by Henry Youle Hind noted conditions more favourable for agriculture on the northern and eastern prairies (Morton 1980).

When the initial efforts made to establish homesteaders along the Canadian Pacific Railway in western Saskatchewan failed in 1883–4, and the federal government was on the point of abandoning the effort, heavy rain and snowfall replenished the area in the late 1880s. John Macoun, the CPR-employed botanist who had encouraged settlement of the region without actually visiting it, thus was able to describe the region as suitable for settlement. This boom-and-bust pattern continued until 1930–1937, when the severity of the drought overbore all other arguments. Migration of rural residents to more fertile terrain in the northern prairies and elsewhere resulted, influenced by debt, social networks, and government relief. As noted in previous and subsequent droughts, farm abandonment was greatest in areas with sandy soil, regardless of the amount of precipitation. Farms on clayey soil tended to be less vulnerable.

In meteorological terms, the 1999–2005 drought was not as severe as the droughts of the 1850s and 1930s (Barrie et al. 2011, Bonsal et al. 2011a, b; Hanesiak et al. 2011). Precipitation was reduced for a shorter period, primarily during the summer, and the effects were less persistent. Summer temperatures were not extremely hot and were locally cooler than average, reducing evapotranspiration and the Palmer Drought Severity Index. Socio-economic effects, however, were substantial: although the degree of exposure was less,

Fig. 5.55 An evening of virga at Tompkins SK. In drier climates marked by high-altitude clouds, much of the moisture in clouds does not reach the ground, re-evaporating en route as virga

sensitivity and vulnerability were high. The difference in meteorological conditions between the 1999–2005 and earlier events indicates that the Canadian Prairies are exposed to drought from a variety of causes. Societal vulnerability to drought has been accentuated due to urban and economic growth, combined with changes in agribusiness.

In a drought-sensitive region, assessment of both past and potential future extremes is valuable. Trees respond to stresses by growing more slowly. If a tree is not killed, the record of droughts is preserved in its annual rings. Dendrochronology, the use of tree rings to date and assess changes in climate, is a valuable tool in reconstructing the history of droughts. In western Canada, dendrochronological records allow climate and drought reconstructions over more than 1400 years (Sauchyn et al. 2015a, b; Sauchyn and Ilich 2017).

In the North and South Saskatchewan river basins, two sustained periods prior to 1880 indicate droughts which would overwhelm current water use and management policies. In contrast, 1890–1915 saw abundant, although fluctuating, water availability, the most anomalous period of the past 1000 years, coinciding with expansion of European settlement and the initial allocations of water to users. Recent discussions of possible changes to water allocation in the Canadian Prairies are set within this legacy. Protecting water at the watershed scale has been a focus, but few Indigenous communities are involved in the process. The Athabasca River watershed dendrochronological record indicates the influence of PDO and PNA, and reveals a larger range of flows and severity of hydrologic deficits than do the shorter instrumental records (Sauchyn et al. 2015a, b).

References

Aber JS (2007) Ice-thrust terrains of the Missouri Coteau. In: Thraves BD, Lewry ML, Dale JE, Schlichtmann H (eds) Saskatchewan: geographical perspectives. Department of Geography, University of Regina, Regina, pp 28–31

Aber JS, Ber A (2007) Glacitectonism. Developments in quaternary science series 6. London: Elsevier, 246 pp

Adamus PR (1996) Bio-indicators for assessing ecological integrity of prairie wetlands. EPA/600/R-96/082. USEPA Environmental Research Laboratory, Corvallis, OR. http://www.epa.gov/owow/wetlands/wqual/ppaindex html

Alberta Biodiversity Monitoring Institute (ABMI) (2015) The status of biodiversity in the grassland and parkland regions of Alberta. Preliminary Assessment. Accessed 4 August 2022

Alberta Environment and Parks (AEP) (2016) Principles for minimizing surface disturbances in native grassland—principles, guidelines, and tools for all industrial activity in native grassland in the prairie and parkland landscapes of Alberta. Edmonton, Alberta

Alberta Environment and Parks (AEP) (2022) Feral horse numbers. https://www.alberta.ca/feral-horse-management.aspx#jumplinks-3. Accessed 4 August 2022

Alberta Parks (2015) natural regions and subregions of Alberta. A framework for Alberta's Parks. Alberta tourism, parks and recreation. Edmonton, Alberta, 72pp

Asong ZE, Wheater HS, Bonsal B, Razavi S, Kurkute S (2018) Historical drought patterns over Canada and their teleconnections with large-scale climate signals. Hydrol Earth Syst Sci 22:3105–3124

Banfield AWF (1981) The mammals of Canada. National Museum of Natural Sciences, National Museums of Canada

Bannatyne BB, McCabe HR, McRitchie WD, Norris W, Scoates RFJ, Weber W (1979) Geology of Manitoba. Manitoba Geological Survey

Beck C, Grieser J, Kottek M, Rubel F, Rudolf B (2006) Characterizing global climate change by means of Köppen Climate Classification. http://koeppen-geiger.vu-wien.ac.at/pdfs/beck_et_al_ksb_2006.pdf

Benn DI, Evans DJA (2010) Glaciers and glaciations. Arnold, London, p 816

Biggar HP (1929) The works of Samuel de Champlain, vol 3. Champlain Society, Toronto

Bird DM (2017) Birds of Canada. DK publishing, London

Bonsal BR, Wheaton EE, Chipanshi AC, Lin C, Sauchyn DJ, Wen L (2011a) Drought research in Canada: a review. Atmos Ocean 49:303–319

Bonsal BR, Wheaton EE, Meinert A, Siemens E (2011b) Characterizing the surface features of the 1999–2005 Canadian prairie drought in relation to previous severe twentieth century events. Atmos Ocean 49(4):320–338

Bostock HS (2014) Physiographic regions of Canada. Geological survey of canada map 1254A (2nd edition), scale 1:5,000,000. https://www.nrcan.gc.ca/maps-tools-and-publications/maps/geology-and-geosciences/16876. Accessed 25 July 2022

Bourgeon L, Burke A (2021) Horse exploitation by Beringian hunters during the Last Glacial Maximum. Quatern Sci Rev 269:107140. https://doi.org/10.1016/j.quascirev.2021.107140

Brandt JP, Cerezke HF, Mallett KI, Volney JA, Weber JD (2003) Factors affecting trembling aspen (*Populus tremuloides* Michx.) health in the boreal forest of Alberta. Saskatchewan Manitoba Forest Ecol Manag 178:287–300

Brink JW (2008) Imagining head-smashed-in: aboriginal buffalo hunting on the northern plains. AU Press

Brooks GR (2017) Red River valley, Manitoba: the geomorphology of a low-relief, flood-prone prairie landscape. Slaymaker O (2017) Landscapes and Landforms of Western Canada. Springer International Publishers, Cham, Switzerland, pp 143–156

Bryan RB, Campbell IA, Yair A (1987) Postglacial geomorphic development of the Dinosaur Provincial Park badlands, Alberta. Can J Earth Sci 24:135–146

Bumsted JM (1993) The MB flood of 1950: an illustrated history. Watson and Dwyer Publishing

Bureau of Sport Fisheries and Wildlife (1965) The American Buffalo. Conservation Note, 12

Buttle JM, Allen DM, Caissie D, Davison B, Hayashi M, Peters DL, Pomeroy JW, Simonovic S, St-Hilaire A, Whitfield PH (2016) Flood Processes in Canada: regional and spatial aspects. Canadian Water Resour J 41:7–30

Campbell IA (1987) Canadian landform examples—3: Badlands of Dinosaur Provincial Park, Alberta. Canadian Geogr 31:82–87

Campbell JB, Lodge RW, Budd AC (1956) Poisonous Plants of the Canadian prairies. Research Branch, Canada Department of Agriculture Publication 900

Canadian Bison Association n.d. https://www.canadianbison.ca. Accessed 4 August 2022

Canadian National Vegetation Classification (CNVC) (2016) Accessed 4 August 2022

Carne S (2009) Buffalo rubbing stone and Wisdom. Ezine Articles. Accessed 4 August 2022

Catto NR (1983) Loess in the Cypress Hills, Alberta, Canada. Can J Earth Sci 20:1159–1167

Catto NR (1984) Glacigenic deposits at the Edmonton convention centre, Edmonton, Canada. Can J Earth Sci 21:1473–1482

Catto NR (2015) Geomorphology: landscapes we live in. Oxford University Press, Don Mills, Ontario

Catto NR, Liverman DGE, Bobrowsky PT, Rutter NW (1996) Laurentide, Cordilleran and montane glaciers in the western peace river-Grande prairie region, Alberta and British Columbia, Canada. Quatern Int 32 21–32

Chiodini R (2021) The elusive and mysterious wood bison. http://www.ozarkbisons.com/articles/wood-bison.pdf. Accessed 4 August 2022

Christiansen EA, Sauer EK (1997) The Dirt Hills structure: an ice-thrust feature in southern Saskatchewan, Canada. Can J Earth Sci 34:76–85. https://doi.org/10.1139/e17-007

Cohen S, Koshida G, Mortsch L (2015) Climate and water availability indicators in Canada: Challenges and a way forward. Part III—Future scenarios. Canadian Water Resour J 40:160–172

Cook FR Johnston JA Douglas C Cook JC (1984) Introduction to Canadian amphibians and reptiles. National Museum of Natural Sciences, National Museums of Canada

Corkal DR, Diaz H, Sauchyn D (2011) Changing roles in Canadian water management: a case study of agriculture and water in Canada's South Saskatchewan River Basin. Int J Water Resour Dev 27:647–664

COSEWIC (2012) COSEWIC assessment and status report on the American Badger *Taxidea taxus* in Canada. Committee on the Status of Endangered Wildlife in Canada. Ottawa. xviii + 63 pp

COSEWIC (2018) COSEWIC assessment and status report on the Greater Short-horned Lizard *Phrynosoma hernandesi* in Canada. Committee on the Status of Endangered Wildlife in Canada. Ottawa. xi + 64 pp

Currie PJ, Koppelhus EB (2005) Dinosaur Provincial Park: a spectacular ancient ecosystem revealed. Indiana University Press, Bloomington, IN, USA

Davies KW, Boyd CS (2019) Ecological effects of free-roaming horses in north American Rangelands. Bioscience 69:558–565

De Lugt J, Campbell IA (1992) Mass movements in the badlands of Dinosaur Provincial Park, Alberta, Canada. Catena Supplement 23:75–100

Dilworth J, Fisher TG (2018) Determining the Lake Agassiz Moorhead Phase lowstand elevation from compaction ridges and newly identified strandlines in the Red River Valley. USA Geomorphology 319:216–225

Dyke AS, Moore A, Robertson L (2003) Deglaciation of Canada. Geological Survey of Canada, Open File 1574

Environment Canada and Agriculture and Agrifoods Canada (1995) Terrestrial Ecozones and ecoregions of Canada. https://sis.agr.gc.ca/cansis/publications/maps/eco/all/regions/eco_all_regions_7m_canada_en.jpg. Accessed 12 July 2022

Environment Canada (2018) Recovery strategy for the wood Bison (*Bison bison athabascae*) in Canada. Species at risk act recovery strategy series. Environment and climate change Canada. Ottawa. viii + 59 pp

Environment and Climate Change Canada (2020) Canadian Climate Normals 1981–2010. https://climate.weather.gc.ca/climate_normals/index_e.html. Accessed 25 July 2022

Evans DJA (2000) Quaternary geology and geomorphology of the Dinosaur Provincial Park area and surrounding plains, Alberta, Canada: the identification of former glacial lobes, drainage diversions and meltwater flood tracks. Quatern Sci Rev 19:931–958

Evans DJA, Nathaniel JP, Young, Cofaigh CÓ (2014) Glacial geomorphology of terrestrial-terminating fast flow lobes/ice stream margins in the southwest Laurentide Ice Sheet Geomorphology 204:86–113

Fenton MM, Schreiner BT, Nielsen E, Pawlowicz IG (1994) Quaternary geology of the western plains. In: Mossop GD, Shetsen I (eds) Geological atlas of the Western Canada sedimentary basin. Canadian Society of Petroleum Geologists and Alberta Research Council. http://www.ags.gov.ab.ca/publications/wcsb_atlas/atlas.html.

Fenton MM, Waters EJ, Pawley SM, Atkinson N, Utting DJ, Mckay K (2013) Surficial geology of Alberta. Alberta Geological Survey, Map 601

Fisher TG, Waterson N, Lowell TV, Hajdas I (2009) Deglaciation ages and meltwater routing in the Fort McMurray region, northeastern Alberta and northwestern Saskatchewan, Canada. Quatern Sci Rev 28:1608–1624

Fisheries and Environment Canada (1978) Hydrological atlas of Canada. NRC of Canada, Ottawa. https://sis.agr.gc.ca/cansis/publications/maps/eco/all/zones/eco_all_zones_30m_canada.png

Fleming SW, Sauchyn DJ (2013) Availability, volatility, stability, and teleconnectivity changes in prairie water supply from Canadian Rocky Mountain sources over the last millennium. Water Resour Res 49:64–74

Fullerton DS, Ringrose SM, Clayton L, Schreiner BT, Goebel JE (2000) Quaternary geologic map of the Winnipeg 4° × 6° Quadrangle, United States and Canada. Quaternary geologic atlas of the United States

Fulton RJ (1995) Surficial materials of Canada. Geological Survey of Canada Map 1880A

Galatowitsch SM, van der Valk AG (1996) The vegetation of restored and natural prairie wetlands. Ecol Appl 6:102–112

Gates CC, Jung TS (2013) COSEWIC assessment and status report on the plains bison (*Bison bison bison*) and the wood bison (*Bison bison athabascae*) in Canada. Committee on the Status of Endangered Wildlife in Canada, Ottawa

Geist V (1991) Phantom subspecies: the wood bison *Bison bison "athabascae"* Rhoads 1897 Is Not a Valid Taxon, but an Ecotype. Arctic 44:283–300

Geist V (1998) Deer of the world: their evolution, behaviour, and ecology. Stackpole Books, Mechanicsburg, PA

Godfrey WE (1986) The Birds of Canada. National Museum of Natural Sciences, National Museums of Canada, 595 p

Gober P, Wheater HS (2014) Socio-hydrology and the science–policy interface: a case study of the Saskatchewan River basin. Hydrol Earth Syst Sci 18:1413–1422

Gompper M, Jachowski D (2016) *Spilogale putorius*. IUCN red list of threatened species. https://doi.org/10.2305/IUCN.UK. Accessed 4 August 2022

Government of Alberta (2010) What is the Grassland Vegetation Inventory (GVI)? Grassland Vegetation Inventory (GVI) Specifications, Alberta Sustainable Resource Development, Government of Alberta 5th Edition. Accessed 4 August 2022

Guedo DD, Lamb EG (2013) Prescribed burning has limited long-term effectiveness in controlling Trembling Aspen (*Populus tremuloides*) encroachment into fescue grassland in Prince Albert National Park. Canadian Field-Naturalist 127:50–56

Hamel C (2015) Tall grass prairie natural area plan summary. The Nature Conservancy of Canada, Winnipeg, MB. 19 pp. https://www.natureconservancy.ca/assets/documents/mb/Public-NACP-TGPNA-2015.pdf. Accessed 4 August 2022

Hammermeister A, Gauthier MD, McGovern K (2001) Saskatchewan native prairie: statistic of a vanishing ecosystem and dwindling resource. Native Plant Society of Saskatchewan, Saskatoon. Accessed 4 August 2022

Hargrave AN (2022) Field guide: identification of common range plants of Southern Saskatchewan. Agriculture and Agri-Food Canada's Greencover Canada Program and administered through the Saskatchewan Forage Council (SFC). https://www naturesask.ca/rsu_docs/common-range-plants-southern-sask---sask-forage-council.pdf. Accessed 4 August 2022.

Hartman GMD, Clague JJ, Barendregt RW, Reyes AV (2018) Late Wisconsinan Cordilleran and Laurentide glaciation of the peace river valley east of the rocky mountains, British Columbia. Canadian J Earth Sci 55. https://doi.org/10.1139/cjes-2018-0015

Hanesiak JM, Stewart RE, Bonsal BR, Harder P, Lawford R, Aider R Amiro BD, Atallah E, Barr AG, Black TA et al (2011) Characterization and summary of the 1999–2005 Canadian prairie drought Atmos Ocean 49:421–452

Haque CE (2000) Risk assessment, emergency preparedness and response to hazards: the case of the 1997 red river valley flood Canada. Nat Hazards 21:225–245

Head-Smashed-In (n.d.) Head-smashed-in UNESCO world heritage site Accessed 4 August 2022.

Henderson DC, Koper N (2014) Historic Distribution and ecology of tall-grass prairie in Western Canada. North American Prairie Conf Winnipeg Manitoba 23:40–49

Henderson DM, Tanke DH (2010) Estimating past and future dinosaur skeletal abundances in Dinosaur Provincial Park, Alberta, Canada Can J Earth Sci 47:1291–1304

Hogg EH, Shwarz AG (1999) Tree-ring analysis of declining aspen stands in west-central Saskatchewan. Aspen Bibliography. Paper 1056. https://digitalcommons.usu.edu/aspen_bib/1056 Accessed 4 August 2022

Holcroft Weerstra (2003) Plains rough fescue (*Festuca hallii*) grassland mapping—central parkland natural Subregion of Alberta resource data branch, Alberta sustainable resource development, Edmonton Alberta Biota Consultants. Accessed 4 August 2022

Hornaday WT (1889) The Extermination of the American Bison to 1889 Government Printing Office, Washington

Hosie RC (1979) Native trees of Canada. Canada Department of Forestry. Don Mills, Ontario

Hugenholtz CH, Bender D, Wolfe SA (2010) Declining sand dune activity in the Southern Canadian Prairies: Historical context, controls and ecosystem implications. Aeol Res 2:71–82

IISD (1994) Sustainability of Canada's Agri-Food system-a prairie perspective. International institute for sustainable development (IISD) for the faculty of agricultural and food sciences, University of Manitoba

Isenberg AC (2000) The destruction of the bison: an environmental history, 1750–1920. Cambridge University Press

Jackson LE Jr (2017) The Foothills Erratics Train Region. Slaymaker O. Landscapes and Landforms of Western Canada. Springer International Publishers, Cham, Switzerland, pp 157–165

Jackson LE Jr, Leboe ER, Little EC, Holme PJ, Hicock SR, Shimamura K, Nelson FEN (2008) Quaternary stratigraphy and geology of the Foothills, southwestern Alberta. Geolo Surv Canada Bull 583:19–46

Jakes AF, Gates CC, Descesare NJ, Jones PF, Goldberg JF, Kunkel KE, Hebblewhite M (2018) Classifying the migration behaviors of pronghorn on their Northern range. J Wildl Manag 82:1229–1242

Johnson D, Kershaw L, MacKinnon A, Pojar J (1995) Plants of the Western boreal forest and aspen Parkland. Lone Pine Publishing and Canadian Forest Service, Edmonton (AB), p 392

Kaye M (1996) Historical development of cultural landscapes of Manitoba to 1870. In: Welsted J, Everitt J, and Stadel C (eds), The Geography of Manitoba, The University of Manitoba Press, Winnipeg, pp 79–89

Kamar Ali M, Klein KK (2014) Implications of current and alternative water allocation policies in the Bow River Sub Basin of Southern Alberta. Agric Water Manag 133:1–11

Kehew AE, Lord ML, Kozlowski A, Fisher TG (2009) Proglacial megaflooding along the margins of the Laurentide Ice Sheet. In: Burr D, Carling PA, Baker VR (eds) Megaflooding on earth and mars. Cambridge University Press, Cambridge, pp 104–127

Kehew AE, Teller JT (1994) History of late glacial runoff along the southwestern margin of the Laurentide Ice Sheet. Quatern Sci Rev 13:859–877

Kevinsen J, Patrick RJ, Bharadwaj LA (2014) A framework for assessing effective urban water management: lessons from the Canadian Prairie. Water Int 39:113–127

Klassen RW (1975) Quaternary geology and geomorphology of Assiniboine and Qu'Appelle valleys of Manitoba and Saskatchewan. Geol Surv Canada Bull 228

Klassen RW (1989) Quaternary geology of the southern Canadian Interior Plains. In: Fulton RJ (ed) Quaternary geology of Canada and Greenland. Geological Survey of Canada, Geology of Canada, no. 1. pp 138–174

Klassen RW (1992) Surficial geology, Wood Mountain, Saskatchewan geological survey of Canada, "A" Series Map 1802A

Koen EL, Bowman J, Lalor JL, Wilson PJ (2014) Continental-scale assessment of the hybrid zone between bobcat and Canada lynx. Biol Cons 178:107–115

Krumbiegel I, Sehm GG (1989) The geographic variability of the plains bison. A reconstruction using the earliest European illustrations of both subspecies. Arch Nat History 16:169–190

Lapp SL, St. Jacues J_M, Sauchyn DJ, Vantose JR (2013) Forcing of hydroclimatic variability in the northwestern Great Plains since AD 1406. Quat Int 310:47–61

Last WM, Ginn FM (2005) Saline systems of the great plains of Western Canada: an overview of the limnogeology and paleolimnology. Saline Syst 1:1–10

Lavkulich LM, Arocena JM (2011) Luvisols of Canada: genesis, distribution, and classification. Can J Soil Sci 91:781–806

Les Auvergnois de Ponteix, n.d. https://lesauvergnoisdeponteix.ca/. Accessed 4 August2022

Leydet DJ, Carlson AE, Teller JT, Breckenridge A, Barth AM, Ullman DJ, Sinclair G, Milne GA, Cuzzone JK, Caffee MW (2018) Opening of Glacial Lake Agassiz's eastern outlets by the start of the Younger Dryas cold period. Geology 46:155–158

Looman J Best KF (1994) Budd's flora of the Canadian Prairie Provinces. Research Branch Agriculture Canada Publication 1662

Lott DF (2002) American Bison: a natural history. University of California Press

Lusebrink I, Erbilgin N, Evenden ML (2013) The lodgepole × jack pine hybrid zone in Alberta, Canada: a stepping stone for the mountain pine beetle on its journey East across the boreal forest? J Chem Ecol 39:1209–20

Matile GLD Keller GR (2004a) Surficial geology of Neepawa map sheet NTS 62 J MB. Manitoba Geol Survey SG-62 J

Matile GLD, Keller GR (2004b) Surficial geology of southern Manitoba (south of 53°); Manitoba industry, economic development and mines, Manitoba geological survey, surficial geology compilation map series, SG-SMB, scale 1:500 000

Matile GLD, Thorleifson LH, Grant N, Burt A, Mann J (1998) Geology of the Winnipeg region NATMAP project (NTS 62H/W, 62I and 52L/W). Manitoba Energy and Mines, Geological Services, Report of Activities 1998:161–171

McDonald JN (1981) The North American Bison: their classification and evolution. University of California Press

McKennirey M, Richardson B (1985) The Great Buffalo Saga. National Film Board/Parks Canada video, https://www.nfb.ca/film/great_buffalo_saga/

Miller B, Ceballos G, Reading R (1994) The Prairie Dog and biotic diversity. Conserv Biol 8:677–681

Miller JJ, Brierley JA (2011) Solonetzic soils of Canada: genesis, distribution, and classification. Can J Soil Sci 91:889–902

Morton WL (1980) Henry Youle Hind. University of Toronto Press

Moss EH (2000) Flora of Alberta. (2nd Ed) Revised by Packer JG. Toronto (ON): University of Toronto Press

Mossop GD Shetsen I (1994) Geological Atlas of the Western Canada Sedimentary Basin; Canadian Society of Petroleum Geologists and Alberta Research Council: Calgary, AB, Canada

Munyikwa K, Feathers JK, Rittenour TM, Shrimpton HK (2011) Constraining the Late Wisconsinan retreat of the Laurentide ice sheet from western Canada using luminescence ages from postglacial aeolian dunes. Quat Geochronol 6:407–422

Murchie TJ, Monteath AJ, Mahony ME et al (2021) Collapse of the mammoth-steppe in central Yukon as revealed by ancient environmental DNA. Nat Commun 12:7120. https://doi.org/10.1038/s41467-021-27439-6

Nachshon U, Ireson A, van der Kamp G, Wheater H (2013) Sulfate salt dynamics in the glaciated plains of North America. J Hydrol 499:188–199

Naughton D (2012) The natural history of Canadian mammals. Canadian Museum Nat

Ngapo TM, Champagne C, Chilian C, Dugan MER, Gariepy Vahmani SP, Bilodeau LP (2021) Pemmican, an endurance food: past and present. Meat Sci 178:108526

Nkemdirim LC (1997) On the frequency and sequencing of Chinooks. Phys Geogr 18:101–113

O'Gara BW, Yoakum JD (2004) Pronghorn: ecology and management. University Press of Colorado, Boulder CO

Oliveira Filho JS, Pinheiro CR, Pereira MG, Valldares GS, Camara R (2020) Sodification and solodization processes: Pedogenesis or natural soil degradation? J S Am Earth Sci 104:102909

Parks Canada (2011) Restoration and maintenance of fescue grasslands in Prince Albert National Park. Native plant society of Saskatchewan meeting February 12th, 2011

Parks Canada (2021) Recovery strategy and action plan for the black-tailed prairie dog (*Cynomys ludovicianus*) in Canada [Proposed]. *Species at Risk Act* Recovery Strategy Series. Parks Canada Agency. Ottawa. vi + 49 pp

Pavlick LE, Looman J (1984) Taxonomy and nomenclature of rough fescues, Festuca altaica, F. campestris (F. scabrella var. major) and F. hallii, in Canada and the adjacent part of United States. Can J Bot 62:1739–1749

Pearn M, Hamel C (2014) Proceedings of the Manitoba Tall grass prairie preserve research symposium. Nature conservancy of Canada, Winnipeg, Manitoba. 49 p

Pennock D, Bedard-Haughn A, Viaud V (2011) Chernozemic soils of Canada: genesis, distribution, and classification. Can J Soil Sci 91:719–747

Phillips D (1990) The climates of Canada. Environment Canada, 176 p

Prior GJ, Hathway B, Glombick PM, Pana DI, Banks CJ, Hay DC, Schneider CL, Grobe M, Elger R, Weiss JA (2013) Bedrock geology of Alberta. Alberta Geological Survey Map 600

Piikani Nation n.d. https://piikanination.wixsite.com/piikanination/about-us. Accessed 4 August 2022

Put MO (2013) Comparison of the Vegetation and Seed Banks of natural and restored wetlands in the Prairie Pothole region, Saskatchewan. Unpublished MSc thesis, Department of Biology, University of Regina

Rannie WF (1996) Evolution of the lower Assiniboine River. In: Welsted J, Everitt J, Stadel C (eds) The Geography of Manitoba. The University of Manitoba Press, Winnipeg, pp 28–39

Rannie WF (2015) The 1997 flood event in the Red River basin: Causes, assessment and damages. Canadian Water Resour J 41:45–55

Rannie WF (2017) Landscapes of the Assiniboine River Watershed. In: Slaymaker O (ed) Landscapes and landforms of Western Canada. Springer International, Cham, Switzerland, 131–142

Rannie WF, Thorleifsen LH, Teller JT (1989) Holocene evolution of the Assiniboine River paleochannels and Portage la Prairie alluvial fan. Can J Earth Sci 26:1834–1841

Rashid H (2011) Interpreting flood disasters and flood hazard perceptions from newspaper discourse: tale of two floods in the Red River valley, Manitoba, Canada. Appl Geogr 31:35–45

Robb BS, Merkle JA, Sawyer H, Beck JL, Kauffman MJ (2022) Nowhere to run: semi-permeable barriers affect pronghorn space use. J Wildl Manag 2022:e22212. https://doi.org/10.1002/jwmg.22212

Rosatte R (2017) Home Ranges and Movements of Elk (Cervus canadensis) restored to Southern Ontario, Canada. Canadian Field-Naturalist 134. https://doi.org/10.22621/cfn.v130i4.1926

Royal Saskatchewan Museum (2022) *T. rex*. https://royalsaskmuseum.ca/trex. Accessed 4 August 2022

Saskatchewan (1986) Buffalo rubbing stone provincial historic site. Canada's Historic Places. Accessed 4 August 2022

Saskatchewan Geological Survey (1999) Geological Map of Saskatchewan. Scale 1:1 000 000

Sato T (2003) *Terminonatator ponteixensis*, a new elasmosaur (Reptilia:Sauropterygia) from the Upper Cretaceous of Saskatchewan. J Vertebr Paleontol 23:89–103

Sauchyn D, Davidson D, Johnston M (2020) Prairie Provinces. In: Warren FJ, Lulham N, Lemmen DS (eds) Canada in a changing climate: regional perspectives report. Government of Canada, Ottawa, Ontario

Sauchyn D, Ilich N (2017) Nine hundred years of weekly streamflows: stochastic downscaling of ensemble tree-ring reconstructions. Water Resour Res 53:9266–9283

Sauchyn DJ, St-Jacques J-M, Luckman BH (2015) Long-term reliability of the Athabasca River (Alberta, Canada) as the water source for oil sands mining. Proceedings, National Academy of Sciences. https://doi.org/10.1073/pnas.1509726112

Sauchyn DJ, Vanstone J, St-Jacques J-M, Sauchyn R (2015b) Dendrohydrology in Canada's western interior and applications to water resource management. J Hydrol 529:548–558

Scott WB, Crossman EJ (1998) Freshwater fishes of Canada. Galt House, Oakville. Canada. Fisheries Research Board of Canada. Bulletin 184

Shaw JH (1995) How many bison originally populated western rangelands? Rangelands 17:148–150

Slaymaker O (2017) Landscapes and landforms of Western Canada. Springer International Publishers, Cham, Switzerland

Slaymaker O Kovanen DJ (2017a) Pleistocene landscapes of Western Canada. In: Slaymaker O (ed) Landscapes and landforms of Western Canada. Springer International, Cham, Switzerland, 27–48

Slaymaker O Kovanen DJ (2017b) Holocene and anthropocene landscapes of Western Canada. In: Slaymaker O (ed) Landscapes and landforms of Western Canada. Springer International, Cham, Switzerland, pp 49–75

Soil Classification Working Group (1998) The Canadian system of soil classification, 3rd edn. Publication, Agriculture and Agri-Food Canada, p 1646

Soil Classification Working group (2001) Soils of Canada. Agriculture and agri-foods Canada, Research Branch. Scale 1:6,500,000 map. https://sis.agr.gc.ca/cansis/publications/maps/soc/all/soils/soc_all_soils_2004en.png. Accessed 25 July 2022

Szeto K, Zhang X, White RE, Brimelow J (2016) The 2015 extreme drought in western Canada. Bull Ameri Meteorol Soc. https://doi.org/10.1175/BAMS-D-16-0147.1

Slaymaker O, Catto NR, Kovanen D (2020) Geomorphological landscapes and Conservation. In: Slaymaker O (ed) Landscapes and landforms of Western Canada. Springer International, Cham, Switzerland, pp 413–435

Statistics Canada (2015) Outdoor water use. https://www150.statcan.gc.ca/n1/pub/16-508-x/16-508-x2017002-eng.htm. Accessed 4 August 2022

Taylor MS (2011) Buffalo Hunt: international trade and the virtual extinction of the North American Bison. Amer Econ Rev 101:3162–3195. https://doi.org/10.1257/aer.101.7.3162

Teller JT Clayton L (1983) Glacial Lake Agassiz. Geol Assoc Canada Special Paper 26

Teller JT, Leverington DW (2004) Glacial Lake Agassiz: a 5000 yr history of change and its relationship to the d18O record of Greenland. Geol Soc Am Bull 116:729–742

Teller JT, McGinn RA, Rajaparac HM, Shuklac AD, Singhvi AK (2018) Optically stimulated luminescence ages from the Lake Agassiz basin in Manitoba. Quatern Res 89:478–493

Thorpe J (2011) Vulnerability of prairie grasslands to climate change prepared for prairies regional adaptation collaborative (PRAC) Saskatchewan Research Council Publication, 12855–2E11

Tuckwell J, Everest T (2009) Management plan for the black-tailed prairie dog (*Cynomys ludovicianus*). Species at Risk Act Management Plan Series. Parks Canada Agency, Ottawa, ON

van Zwoll W (2004) Mule deer. Mule Deer Foundation, Reno, NV

van Zyll de Jong CG, Gates C, Reynolds H, Olson W (1995) Phenotypic variation in remnant populations of North American bison. J Mammalogy 76:391–405

Warkentin AA (1997) The Red River flood of 1997: An overview of the causes, predictions, characteristics and effects of the Red River Flood of the Century. In Proceedings of the Red River Valley '97 Flood symposium. Winnipeg, MB: Canadian Water Resources Association

Wazney L, Clark SP (2015) The 2009 flood event in the Red River basin: causes, assessment, and damages. Canad Water Resour J. https://doi.org/10.1080/07011784.2015.1009949

Wells-Gosling N, Heaney LR (1984) Glaucomys sabrinus. Mammal Species 22:1–8

Welsted J, Everitt J, Stadel C (1996) The geography of Manitoba. The University of Manitoba Press, Winnipeg

Wheater H, Gober P (2013) Water security in the Canadian Prairies: science and management challenges. Philos Trans Royal Soc A 371:20120409. https://doi.org/10.1098/rsta.2012.0409

Wheeler JO, Hoffman PF, Card KD, Davidson A, Sanford BV, Okulitch AV, Roest WR (1996) Geological map of Canada. Geol Surv Canada Map D1860A

Wolfe SA (2013) Cold-climate aeolian environments. In: Shroder JF (Editor-in-chief), Lancaster N, Sherman DJ, Baas ACW (eds) Treatise on geomorphology, vol 11, Aeolian Geomorphology. Academic Press, San Diego, pp 375–394

Wolfe SA (2017) The Prairie Dune Archipelago. In: Slaymaker O (ed) Landscapes and landforms of Western Canada. Springer International Publishers, Cham, Switzerland, pp 167–176

Wolfe SA, Hugenholtz CH, Evans C, Huntley DJ, Ollerhead J (2007) Potential aboriginal occupation-induced dune activity, Elbow Sand Hills, northern Great Plains, Canada. Great Plains Res 17 173–192

Wolfe SA, Hugenholtz CH, Lian OB (2013) Palliser's triangle: reconstructing the 'central desert' of the southwestern Canadian prairies during the late 1850s. The Holocene 23:699–707

Wyman MT, Mooring MS McCowan B, Penedo MCT, Hart LA (2008) Amplitude of bison bellows reflects male quality, physical condition and motivation. Anim Behav 76:1625–1639

6 Cordilleran Ecozones

Abstract

The Cordillera encompasses the mountains and intervening plateaus of western Alberta, western Northwest Territories, Yukon, and British Columbia. Ecological divisions within the Cordillera reflect the physiography and geological structure, which produces a combination of northwest-southeast transit routes for flora and fauna, superimposed on latitudinal and latitudinal transitions. Each of the physiographic divisions contains distinctive botanical and zoological communities and has a distinctive geological and Quaternary history. Geologically, the Cordilleran Structural Province is divided into 5 NW-SE trending "Belts", reflecting differences in geological history. Within each belt are several distinctive geological units or terranes. Altitude plays a strong role in Cordilleran climates. Cold steppe (Köppen-Geiger BSk), Boreal (Köppen-Geiger Dfb), and taiga (Dfc) climates are succeeded by tundra (ET) climates in lower relief or more northerly settings, and by alpine (Köppen-Geiger H) climates above 1500 m asl. Foreland Structural Belt rocks span geological time from the Precambrian to the Palaeocene. Although rock units generally increase in age from east to west, complex thrust faulting has locally resulted in younger rock being overlain by displaced older units, in addition to creating the characteristic northwest-southeast oriented pattern of ranges and valleys. The Omineca Structural Belt includes two distinctive groups of mountain ranges: the Columbia Mountains in south-central British Columbia, and a second extending across northern BC and Yukon. The boundary with the Foreland Structural Belt is marked by the Rocky Mountain Trench in British Columbia, and in part by the Tintina Trench in central Yukon. Three distinctive styles of Quaternary glaciation affected the Canadian Cordillera. The eastern flank of the Foreland Structural Belt, including the Rocky Mountain Foothills and Richardson Mountains, was affected by the maximum extent of Laurentide glaciation originating from Kivalliq in the Canadian Shield. Numerous alpine glaciers developed at higher altitudes in all the major ranges. The Cordilleran Inlandsis expanded from the Chilcotin and Nechako Plateaus of the Intermontane Structural Belt to cover British Columbia (with the exception of the highest mountains and the Peace River Lowland) and southern and central Yukon. The region north of the Ogilvie Mountains and west of the Richardson Mountains in northern Yukon remained unglaciated throughout the Quaternary. In montane regions, altitudinal variations in climate cause differences in vegetation assemblages and soil development. However, vegetation is not distributed along straight contour lines. Faunal distribution in the Cordillera is also controlled by altitude. Human pressure has severely affected the fauna in the lower (montane and boreal) areas, especially in Banff, but the assemblages in the subalpine and alpine ecoclimatic regions have been less affected.

Keywords

Cordillera · Structural belts · Quaternary glaciation · Vegetation assemblages

6.1 Introduction

The Cordillera encompasses the mountains and intervening plateaus of western Alberta, western Northwest Territories, Yukon, and British Columbia. The Cordillera is one of the most scenic regions of Canada, and probably the most highly publicized from the viewpoint of international tourism (Fig. 6.1).

Ecological divisions within the Cordillera reflect the physiography and geological structure, which produces a combination of northwest–southeast transit routes for flora and fauna,

N. Catto and G. R. Vincent, *Canada: Land and People*, World Regional Geography Book Series,
https://doi.org/10.1007/978-3-032-04391-7_6

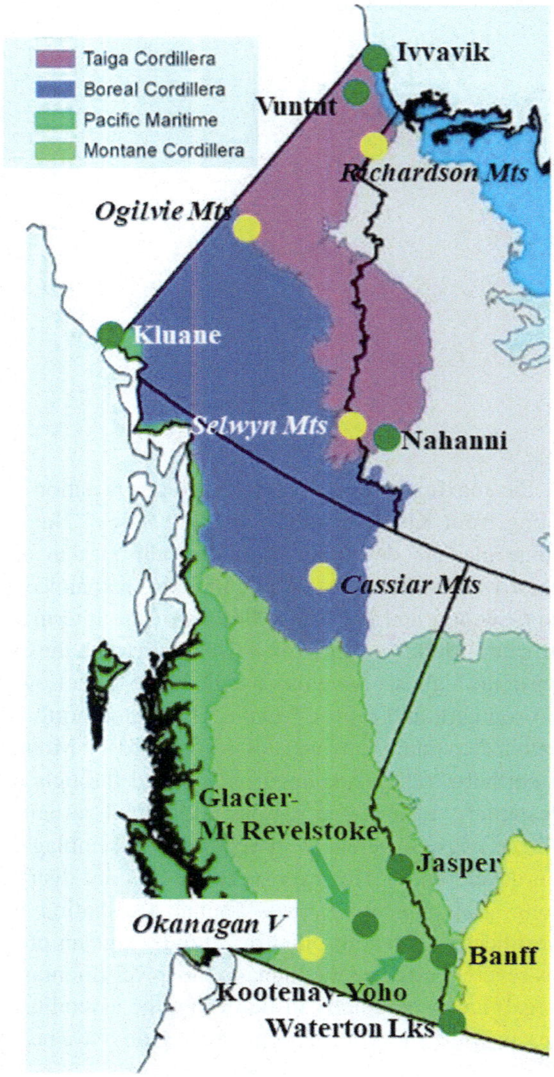

Fig. 6.1 Locations in the Boreal, Montane, and Taiga Cordillera discussed in text. Green circles are national parks. Base map: Ecoregions of Canada (Environment and Climate Change Canada)

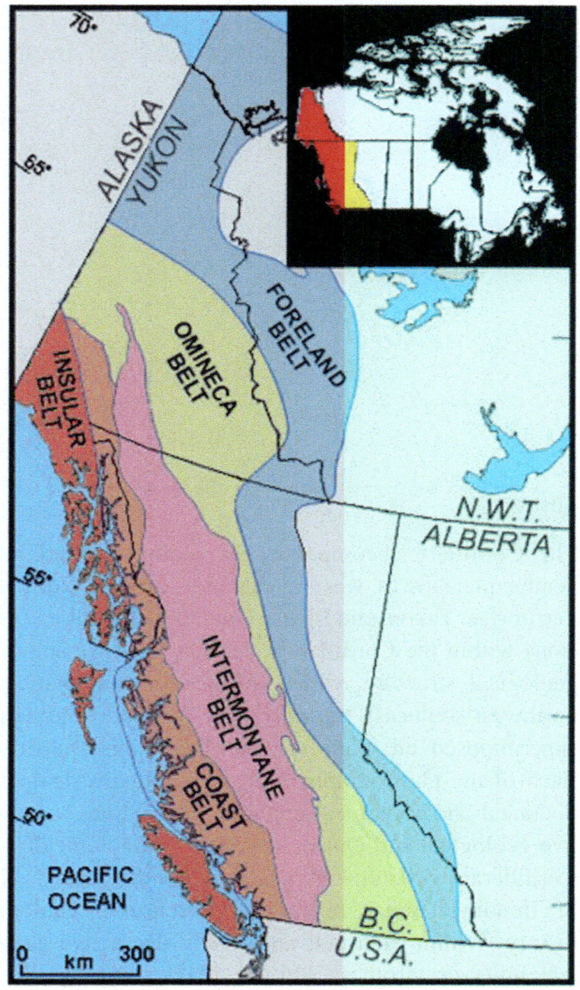

Fig. 6.2 Structural Belts of the Canadian Cordillera (Wikipedia)

superimposed on latitudinal and latitudinal transitions. Each of the physiographic divisions contains distinctive botanical and zoological communities and has a distinctive geological and Quaternary history.

Geologically, the Cordilleran Structural Province is divided into 5 NW–SE trending "Belts", reflecting differences in geological history (Gabrielse and Yorath 1991; Wheeler and McFeely 1991; Wheeler et al. 1996, Price 1994; Haggart et al. 2006; Nelson and Colpron 2007; Fig. 6.2). Within each belt are several distinctive geological units or terranes. A Terrane (spelled that way) is a connected series or group of rocks, with a similar tectonic or geologic history. We can recognize connections between the geological structural belts and broad physiographic systems. Three structural belts will be discussed in this chapter.

The Foreland Structural Belt corresponds to the Eastern Physiographic System (Bostock 2014; Slaymaker et al. 2017). This includes the Rocky Mountains of Alberta and easternmost British Columbia, the Mackenzie and Selwyn Mountains of the Northwest Territories, and the Richardson Mountains of Yukon. Banff, Jasper, Waterton Lakes, Kootenay, Yoho, and Nahanni National Parks lie within the Eastern System. The "Rocky Mountains" are just one of several mountain chains in the Eastern System of the Cordillera. The Eastern system is bordered on the east by the Foothills, and on the west by the Rocky Mountain Trench. Ecozones within the Foreland Structural Belt show a transition from Montane Cordillera (southern and central Rocky Mountains), Boreal Cordillera (northern Rocky Mountains), Taiga Cordillera (Mackenzie and Werencke Mountains), to Tundra Cordillera (Richardson Mountains). Northwest Yukon (Porcupine Plateau) is also included within the Taiga Cordillera ecozone.

The Omineca and Intermontane Structural Belts (Price 1994) are parts of the Interior (or Central) Physiographic System (Bostock 2014; Slaymaker and Kovanen 2017). The Omineca structural belt includes the Columbia Mountains of southern British Columbia, subdivided into the Selkirk, Purcell, Monashee and Cariboo Mountains (Church and Ryder 2010), the Omineca and Cassiar Mountains of northern BC, and the Pelly and Ogilvie Mountains of Yukon (Smith et al. 2004). The Interior System is marked by less rugged topography than the Eastern System. Ecozones within the Omineca structural belt include Montane Cordillera (southern and central British Columbia) and Boreal Cordillera (Omineca and Cassiar Mountains, and Yukon). National Parks include Glacier and Mount Revelstoke. For convenience, we will also include discussion of the Taiga Cordillera in the St. Elias Range of southwestern Yukon, including Kluane National Park.

The Intermontane Structural Belt includes plateaux and valleys between the Omineca and the Coast and Insular Structural belts to the west (Church and Ryder 2010). These include the Okanagan Valley and Thompson Plateau (southern British Columbia); Chilcotin and Nechako Plateaux and Lowlands (central BC); and Liard Lowlands, Yukon Plateau, and Yukon River valley (northern BC and southern Yukon). The Semi-arid Plateaux ecozone (Environment Canada 1995) includes the Okanagan Valley, the most ecologically distinctive area, and the Thompson Plateau. Montane Cordillera ecologies characterize central British Columbia, transitional to Boreal Cordillera to the north.

The Coast and Insular Structural Belts are included in the Western Physiographic System. This System includes the Coast and Cascade Ranges of British Columbia, Vancouver Island, Haida Gwaii, and the St. Elias Mountains of Yukon. This System lies within the Pacific Maritime Ecozone, discussed in Chap. 7.

6.2 Climate

Altitude plays a strong role in Cordilleran climates. Cold steppe (Köppen-Geiger BSk), Boreal (Köppen-Geiger Dfb) and taiga (Dfc) climates are succeeded by tundra (ET) climates in lower relief or more northerly settings, and by alpine (Köppen-Geiger H) climates above 1500 m asl.

6.2.1 Rocky Mountains

At Banff National Park, the mean annual precipitation increases westward and with altitude (Holland and Coen 1983), from 500 mm at the east gate, to 1200 mm at Banff townsite (1350 m asl) and 1300 mm at Lake Louise (Environment and Climate Change Canada 2020; Fig. 6.3). The proportion falling as snow varies from 35% at the townsite to 80% at elevations above 2800 m. Snow cover persists for 220 days at Banff, and the snow pack depth averages 190 cm. Typically, Banff townsite has daily mean temperatures of −15 °C in January and 14 °C in July. The overall moisture regime is humid, due to the low summer temperatures. Summers in Banff townsite are marked by relatively warm, sunny weather. Typically, the community receives 1700 h of sunshine per year. Thunderstorms occur at lower elevations (18 per year at Banff) but are rarer at higher elevations.

To the west of the Continental Divide, the climates strongly influenced by the prevailing westerly winds (Phillips 1990). Climatic parameters for any particular site are controlled primarily by its altitude and its aspect with respect to the prevailing winds, with west-facing sites receiving substantially more moisture than leeside sites facing east (Fig. 6.4). In Kootenay National Park, mean annual precipitation ranges from 270 mm to more than 1000 mm, with montane sites being drier than subalpine and alpine locations (Achuff et al. 1984; Rodengen et al. 2022). The proportion of precipitation falling as snow varies from a low of 30% in montane, east-facing sites to highs in excess of 80% on alpine, west-facing sites. Snow cover in most areas persists for a minimum of 180 days. Snow pack depths range from less than 30 cm at Radium Hot Springs to more than 200 cm in alpine areas.

Daily mean temperatures also vary with altitude and aspect. Typical January and July daily means at Golden, B.C., are −14 °C and 14 °C, respectively (Environment Canada 2020). The difference between the extreme high and low temperatures in Yoho and Kootenay is not as great as in Banff, but the parks are somewhat cooler overall throughout the year.

6.2.2 Selkirk Mountains

Westward, the Selkirk Mountains are moister than the Rockies, due to their position athwart the prevailing westerly winds from the Pacific Ocean. Mean annual precipitation is 1300–2000 mm, of which 30–70% falls as snow (dependent upon elevation; Environment and Climate Change Canada 2020; Fig. 6.5). The driest areas, such as Revelstoke town (1300 mm per year) are located at the base of east-facing slopes. Wetter areas, such as the summit of Mount Revelstoke (2000 mm) tend to be higher in altitude and face west, into the prevailing winds.

Snow arrives at the summit of Mount Revelstoke in late September, and the avalanche season begins in November. In contrast, Revelstoke town does not generally receive substantial snowfall until December. Heavy snowfalls usually mark

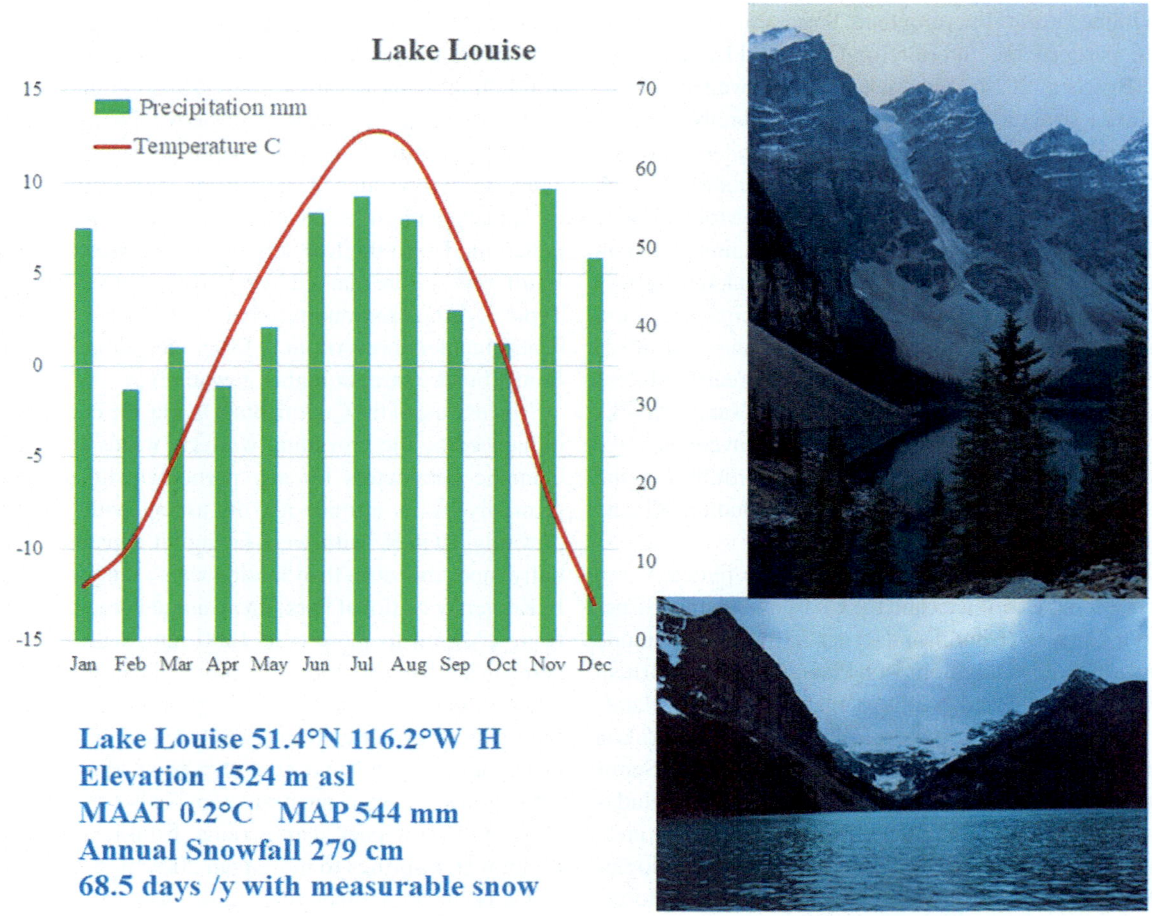

Fig. 6.3 Lake Louise, 1524 m asl, has a Köppen-Geiger H climate (Data: Canadian Climate Normals, Environment and Climate Change Canada 2020)

January and February in the town, with sleet in March and the last snow in April. Within Mount Revelstoke National Park, snow cover varies from 80 days (lowest elevations) to more than 200 days (summit of Mount Revelstoke). Similar variations categorize Glacier National Park, although snow cover durations are generally 10–15 days longer throughout, due to the higher elevation. In Rogers Pass, the snowpack commonly exceeds 300 cm in depth by mid-March. Glaciers are present in Glacier National Park, although they are currently in recession.

During El Niño years, snowfall and snow cover duration are substantially reduced throughout the region, regardless of elevation and aspect (McClung 2013). The weather during an El Niño January resembles normal March weather. Under these circumstances, snow tends to be heavy and mixed with freezing rain (Hägeli and McClung 2003).

Avalanches may be triggered following periods of individual heavy snowfall events, or longer periods of above-average snowfall (Campbell et al. 2007, Haeberli et al. 2014). Alternating periods of freezing and thawing destabilize the snowpack, producing layering. Hoar layers, consisting of recrystallized, approximately spherical particles of modified snow, are particularly weak. Hoar is produced as fresh precipitation interacts with deposited snow, allowing reorientation and recrystallization of snowflakes (Horton and Jamieson 2016). Snow overlying the hoar layer rests on an unsteady foundation, subject to motion if disturbed. Disturbance is most likely when the hoar layer is slightly below 0 °C.

Wind is the most common natural disturbance. Gravitational failure of an overhanging cornice or failure due to loading by additional snow, ice, or rock from above can also occur. Avalanches can also be artificially triggered by planted explosives or noise. In Rogers Pass, cannons are used to initiate pre-emptive small failures, using either solid projectiles or bursts of compressed air.

Avalanches vary in origin, flow style, speed, and impact on humans (Haeberli et al. 2014). They commonly change dynamically as they progress from the source zone (origin), along the track, to termination in the run-out zone (Fig. 6.6). Powder avalanches begin as dry snow (with little or no liquid

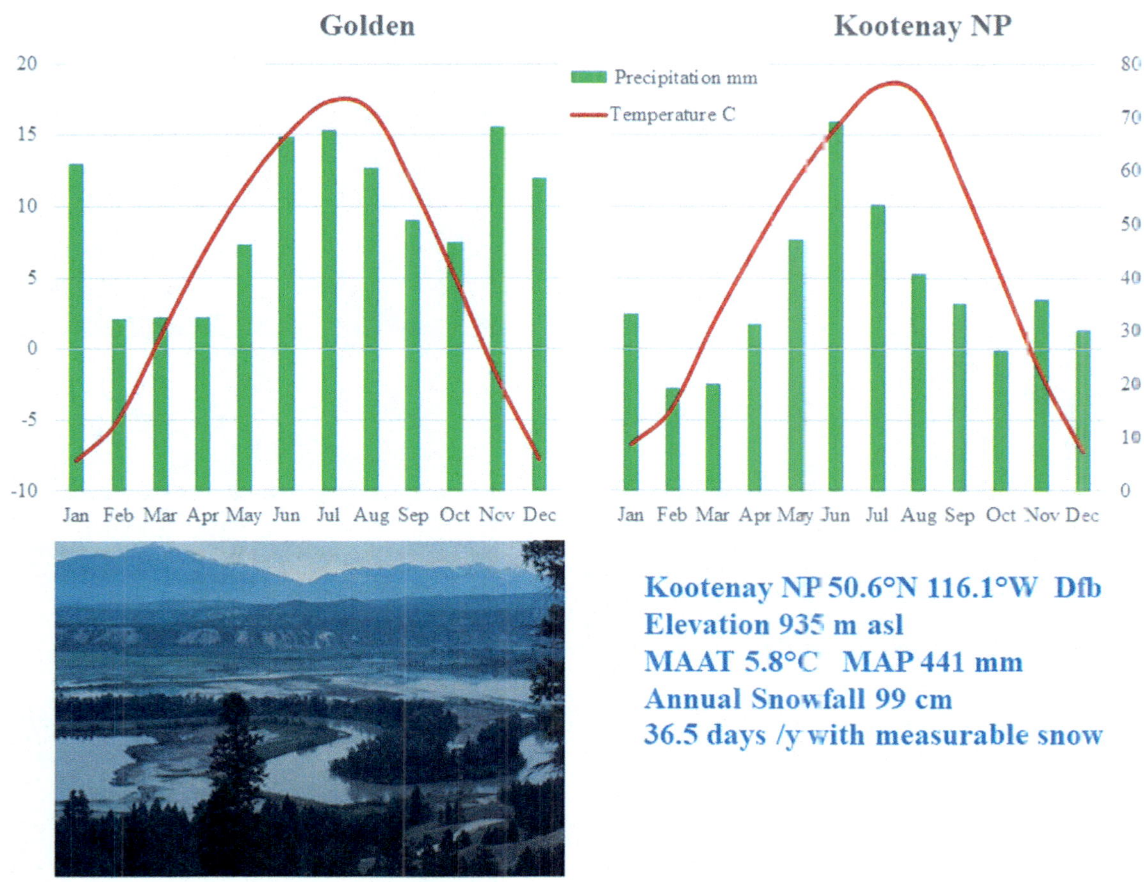

Fig. 6.4 Golden (west of Yoho National Park) and Kootenay National Park both have Dfb climates Precipitation differences are related to aspect. Inset: Columbia River near Radium Hot Springs (Data: Canadian Climate Normals, Environment and Climate Change Canada 2020)

water) fails near or above tree line, incorporating air during movement downslope. Powder avalanches can flow at more than 200 km/h, but usually stop in heavily treed areas. Once the powder ceases moving, snow particles settle and form a tight pack, with limited passages for oxygen to circulate from outside. Powder avalanches form the greatest hazard to recreational users of high backcountry areas, including skiers and snowboarders. The attraction of experiencing fresh powder is directly associated with the hazard.

Hard-slab avalanches involve the movement of a partially hardened tabular mass above a hoar layer. Hard slabs are formed by sintering, as snowflakes and ice pellets are compressed and recrystallized at temperatures slightly below freezing. The detached slab initially slides, but internal stresses rapidly break it apart into individual flowing fragments. Hard-slab avalanches can exceed 80 km/h, and frequently cut tracks into forests. Tracks formed by previous avalanches are commonly reutilized. Inactive tracks may be distinguished by differences in vegetation and lack of recent erosion.

Wet-slab avalanches also initially move by sliding. They include some liquid water, and are launched at temperatures at or above 0 °C. Wet-slab avalanches can begin in forested areas, or on relatively short slopes in hilly terrain. Although they are the slowest-moving, typically at 20–35 km/h, they are faster than most humans, especially on snow-covered slopes. Slab avalanches are characteristic of lower elevation areas, and fatalities may include snowmobilers and community residents.

Hard-slab and wet-slab avalanche snow has undergone consolidation and recrystallization prior to movement. When the avalanche comes to rest, the pre-existing structure may result in less settling and reorganization, allowing passages between particles to remain. Unfortunately, dense, compacted snow contains few passages allowing circulation of oxygen.

A person encased in snow will suffer from oxygen deprivation, or asphyxia. Most avalanche fatalities result from asphyxia, with many of the remainder attributable to trauma (e.g. McIntosh et al. 2007; Haegeli et al 2011).

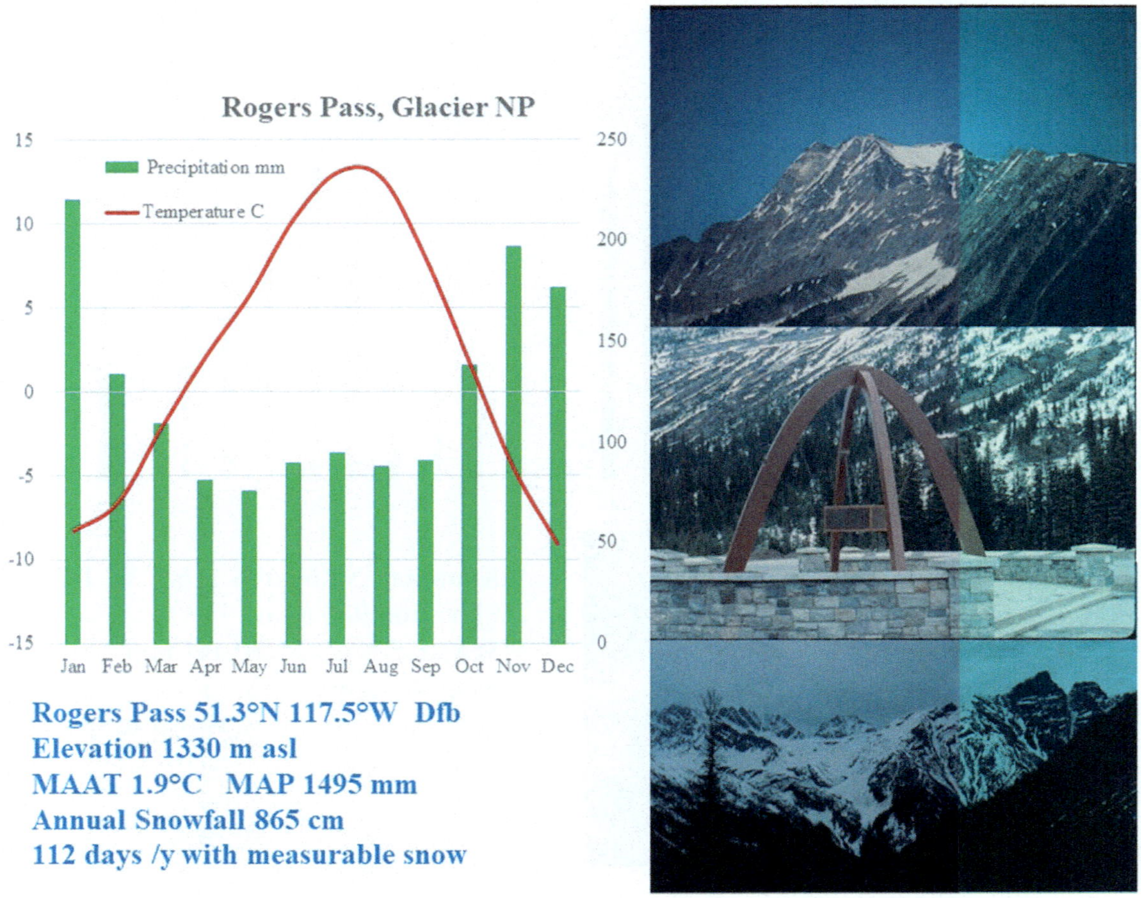

Fig. 6.5 Rogers Pass, Glacier National Park (Data: Canadian Climate Normals, Environment and Climate Change Canada 2020)

Snow involved in avalanches varies in temperature, density, and packing. A person immersed in dense or tightly packed snow is particularly susceptible to rapid asphyxia. Survival rates for Canadian and Swiss victims completely buried in snow are similar (~46–47%), although the average rescue time in Canada (18 min) is half that in Switzerland (35 min). Avalanche victims in British Columbia and Atlantic Canada are more likely to be buried in denser, compacted snow than those in the drier and colder Swiss Alps (Haegeli et al. 2011).

Annually, avalanches result in 10 to 15 fatalities in Canada (Jamieson et al. 2010; Haegeli 2019). The demographic profiles of avalanche victims have changed significantly in the past 150 years. Initially, most were killed at work, typically in mining or railroad operations, or in communities at the base of slopes. In Canada, the largest avalanche death toll occurred when 58 railroad workers were killed in Rogers' Pass in 1910 (Greenaway 2009), ultimately leading to the replacement of the above-ground line with the Connaught and Mt. Macdonald tunnels.

Between 1900 and 1960, the annual toll of avalanche victims generally declined in North America, associated with changes in occupational patterns and lifestyles, and improved weather forecasting, warning systems, and rescue efforts (Page et al. 1999). Since 1960, increased recreational use of mountainous areas has resulted in increased fatalities and injuries (Boyd et al. 2009).

The Canadian Avalanche Association and the Alpine Club of Canada have conducted demographic analysis of avalanche victims in Canada (Boyd et al. 2009; Jamieson et al. 2010). The typical avalanche fatality is a male, between 20–29 years old, killed between noon and 2 pm, under clear, calm conditions. Most fatalities involved dry powder or hard-slab slides, above or near tree line, in Alberta and British Columbia. Most victims were engaged in recreational activities, including skiing and snowboarding or snowmobiling. Unfortunately, perhaps 90% or more of these victims were killed by avalanches accidentally started by themselves or their companions. Human error is the dominant presumptive cause of a large majority of Canadian alpine accidents.

Fig. 6.6 Avalanches leave visible traces on landscapes. Upper left: Avalanche tracks in vegetation, Banff National Park. Upper right: Debris deposited in avalanche run-out zone, Yoho National Park (Mel Reasoner). Lower left: Depositional fans in the run-out zone form from avalanche activity and summer water flow. Peyto Lake, Banff National Park. Lower right: snowsheds and ramparts constrain avalanche flow to protect highways and railways. Field area, Yoho National Park

In a typical year, the daily mean January temperature in Glacier town is -16 °C (Environment and Climate Change Canada 2020). Corresponding values for Revelstoke town and the summit of Mount Revelstoke are −12 °C and −20 °C, respectively. Winter temperatures during El Niño years are up to 10 °C higher on average, and days above freezing are common in valley bottoms. The net effects of El Niño conditions are to reduce moisture recharge, enhance slope erosion through water runoff and avalanching, and increase the likelihoods of summer drought and forest fires.

Summers are relatively warm, especially in the lower, more western locales in Mount Revelstoke. The daily mean July temperature is approximately 16 °C in Revelstoke town, 2–3 °C cooler in Glacier town, and is slightly above 0 °C on the highest summits and active glaciers. El Niño has little or no influence on summer weather. The lowland areas of the parks receive 100–110 high summer days per year. Thermal inversions, with cool air trapped below warm, are common through the summer, along with thunderstorms (15 per year) at lower elevations. The area is considered to have a Highland climate in the Köppen-Geiger classification system.

6.2.3 Okanagan Valley

The driest area of the southern Cordillera is the floor of the southern part of the semi-arid Okanagan Valley (Phillips 1990; Environment Canada 1995; Fig. 6.7). The prevailing westerly winds dominate the climate of the Okanagan Valley, producing moist conditions on the west-facing eastern slope of the valley (e.g. along the front of the higher parts of the Monashee Mountains) and drier conditions on the east-facing slopes at the foot of the Cascades (Rayne and Forest 2016; Hewer and Gough 2021). Mean annual precipitation varies from 1000 mm on the wettest slopes to less than 300 mm on the driest east-facing slopes at Vaseux Lake Provincial Park and Osoyoos (Environment and Climate Change Canada 2020). Snowfall also varies considerably with aspect, from 80–140 cm. Snow cover persists for approximately 90 days in a typical year. West-facing slopes are subhumid, whereas east-facing slopes have a semiarid moisture regime. Both sides of the larger Okanagan Valley show individual examples of orographic precipitation differences. This climatic division based on aspect exerts a dominant control on the vegetation and soils throughout the Okanagan.

Daily mean temperatures are among the mildest in Canada, with the January daily mean at approximately 0 °C and the July value at 24 °C. The mild winters allow the cultivation of soft fruit trees (Quamme et al. 2010) and viticulture (Senese et al. 2012) and permit the survival of cold- and snow-intolerant reptile species. The area typically receives between 200 and 220 growing days per year, a requirement for successful soft fruit production. Ongoing climate change may have impacts on viticulture, with more heat stress but reduced risk of frost damage (Hewer and Gough 2021).

The Okanagan Valley is classified as a steppe (modified Köppen-Geiger BSk) region, based primarily on the relatively low rainfall (Beck et al. 2006). The adjacent montane areas of the Cascade and Columbia Mountains are considered to have boreal and highland climates.

6.2.4 Nahanni National Park Reserve

In the Mackenzie Mountains, Nahanni National Park Reserve lies along the transition between the northern boreal (Köppen-Geiger Dfc) climate zone, and the highland (Köppen-Geiger H) zone, defined to occur at 1500 m asl. The northern side of the Nahanni valley is generally wetter and warmer than the southern (Fig. 6.8). Daily mean temperatures at Tungsten, NWT, are −24 °C in January and 11 °C in July (Environment and Climate Change Canada 2020). In contrast, valley bottom temperatures at Virginia Falls are 5 °C warmer in winter, and 2–3 °C warmer in summer. Temperature differences between southwest-facing and northeast-facing slopes regularly exceed 5 °C.

Similar differences exist in precipitation patterns. At Tungsten, the mean annual precipitation is 645 mm, in contrast to the 600 mm total at Virginia Falls and the up to 800 mm totals on some higher, west-facing slopes of the Mackenzie Mountains. Snowfall comprises 45% of total precipitation at Tungsten, slightly less at Virginia Falls, and up to 85% in the highest areas of the Mackenzie Mountains. South-facing slopes receive 10–20% more precipitation than north-facing slopes, and marginally less snow. The extreme winter temperatures, however, largely preclude the development of easily recognizable vegetation differences between the slopes.

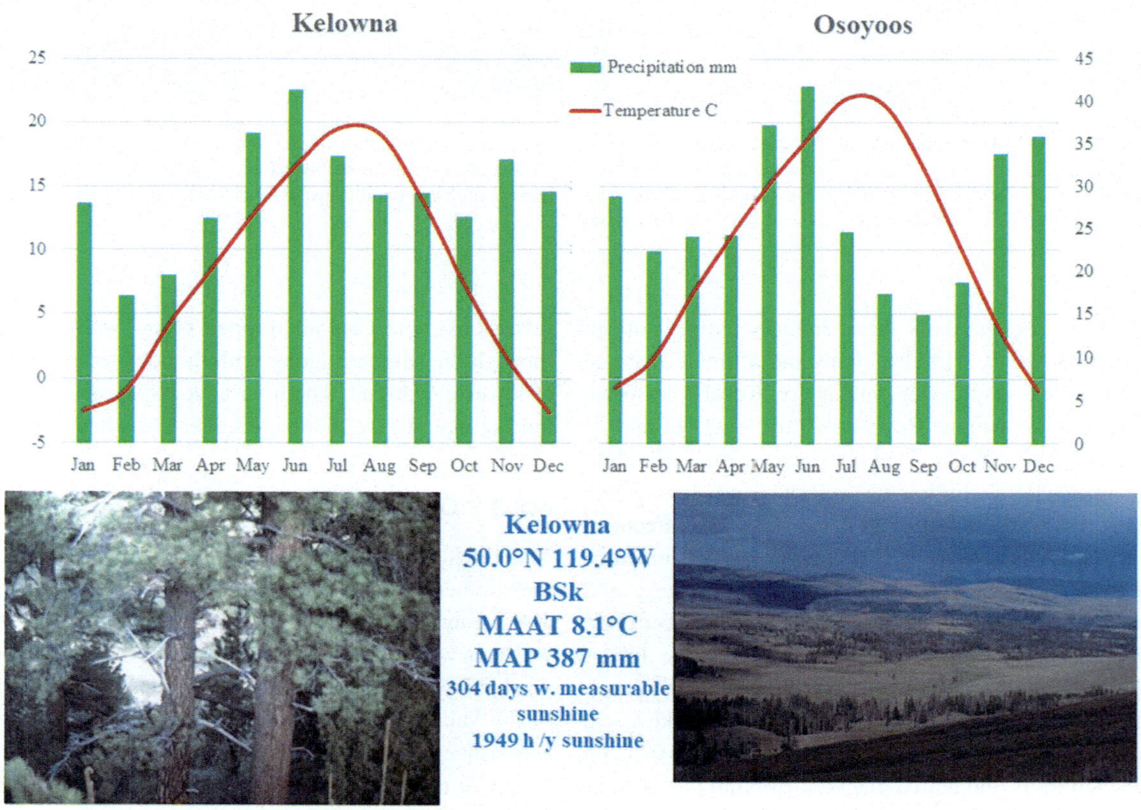

Fig. 6.7 Okanagan Valley climates at Kelowna and Osoyoos are BSk (steppe). (Data: Canadian Climate Normals, Environment and Climate Change Canada 2020)

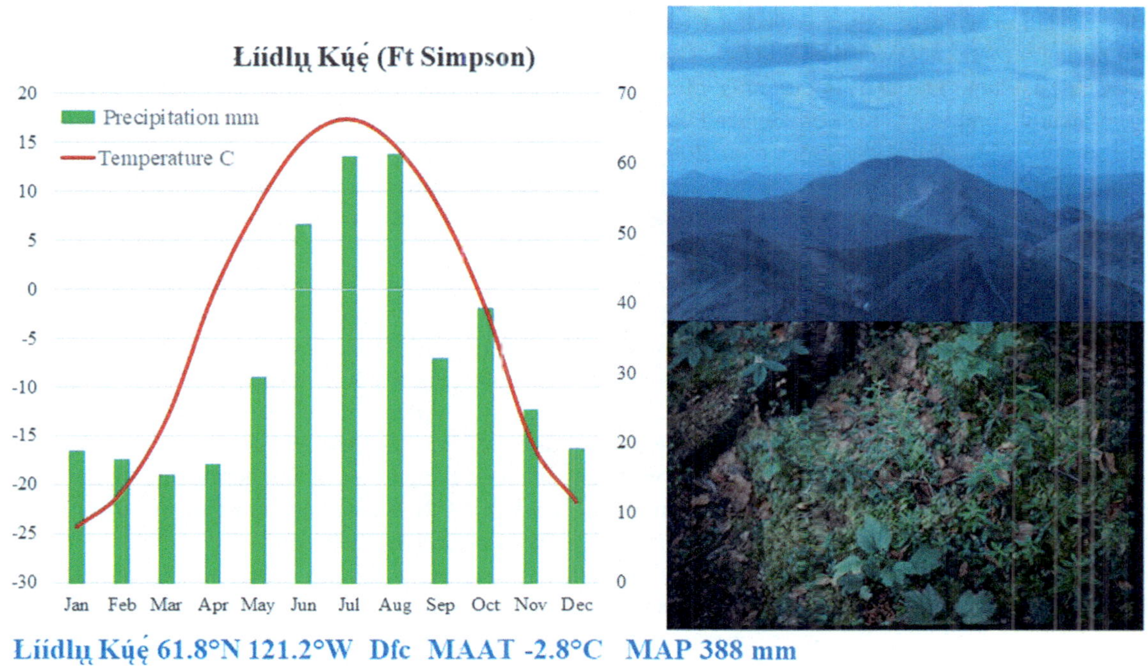

Fig. 6.8 Łíídlı̨ı̨ Kų́ę́ (Fort Simpson), adjacent to Nahanni National Park, has a taiga (Dfc) Köppen-Geiger climate. Right: Topography and vegetation, Mackenzie Mountains (Data: Canadian Climate Normals, Environment and Climate Change Canada 2020)

The higher elevation of the Mackenzie Mountains compared to the relatively low Selwyn Ranges results in colder and snowier winters in the Mackenzies. Snow cover persists for up to 260 days in the higher parts of the Mackenzie Mountains. The moisture regime is classified as humid. The prevailing winds are southwesterly. Discontinuous permafrost marks elevations below treeline, with continuous permafrost present at elevations above 1220 m a.s.l.

6.2.5 Kluane National Park

The lowest parts of Kluane National Park have a northern boreal (modified Köppen-Geiger Dfc) climate, and lie within the Alpine North Pacific Cordilleran ecoclimatic region (Wahl et al. 1987; Environment Canada 1995; Environment and Climate Change Canada 2020; Fig. 6.9). The climate is considered to be among the most rigourous in Canada. The January daily mean temperature at Burwash Landing on Kluane Lake is −27 °C, and temperatures below −40 °C are common in mountainous areas. More than 120 days per year have minimum temperatures less than −20 °C, and the frost-free period at lake level is 35 days. The now-abandoned community of Snag, northwest of Kluane near the Yukon-Alaska border, holds the Canadian record for the coldest temperature ever reliably recorded, at −63 °C (Phillips 2016).

The July daily mean temperature at Kluane Lake is 5 °C. Burwash Landing has 25 high summer days and 1300 h of sunshine in a typical year, along with 6 thunderstorms. Mean annual precipitation ranges from a low of 1500 mm at Burwash Landing to 2900–3200 mm in the core of the Icefield Ranges. Burwash Landing receives 400 cm of snow each year (±25% of the total precipitation), but virtually all of the precipitation above 3000 m asl falls as snow. The snowpack at Burwash Landing exceeds 180 cm by mid-March, and snow cover persists at this altitude for 220 days. The high precipitation and cold temperatures produce a perhumid moisture regime. Discontinuous permafrost is present at Burwash Landing, and alpine permafrost influences all areas above ±1100 m asl. The prevailing winds are westerlies, originating from the relatively warm Alaska Current southwest of the St. Elias Mountains.

6.2.6 Taiga Cordillera: Northern Yukon Interior

Northern Yukon developed a continental climate during the Quaternary, as the St. Elias Range rose to the southwest and intercepted the Pacific air masses (Winkler 2000; Finzel et al.

2011). The prevailing wind is westerly, driven by thermal convection during the summer over the Old Crow Basin and the Arctic Coastal Plain. The area is currently marked by extreme continental conditions (Köppen-Geiger Dfc at Old Crow and Dawson; Wahl et al. 1987; Beck et al. 2006). The Ogilvie Mountains have a subalpine to alpine climate (Dfc to H), but generally show less variation in temperature and have more precipitation than the lower-lying Yukon Plateau. Westerly winds are prevalent throughout the year in the Ogilvie Ranges. Even more extreme conditions (Dfd) prevail in the core of the Richardson Mountains. As most of the summits are below 1500 m asl, the terrain is too low to be classified within the Highland (H) climate zone.

At Dawson, January and July daily mean temperatures are −26 °C and 16 °C, respectively (Environment and Climate Change Canada 2020; Fig. 6.10). During the summer, at least one day each year has a high temperature in excess of 30 °C, and 30 high summer days occur in an average year. Winter nights below −50 °C are not uncommon. Temperatures in the Yukon River valley are more extreme than those on the plateau uplands, due to the common thermal inversions and the overpassing of warm Pacific air which has surmounted the St. Elias Range and thus bypasses Dawson at high altitudes.

Dry conditions prevail throughout most of the year (Wahl et al. 1987). The mean annual precipitation in the vicinity of Dawson City is 300–400 mm, with slightly more precipitation in the plateau regions. Snowfall accounts for about 40% of the total precipitation, generally falling in autumn and spring. Little precipitation falls between November and February. Snow cover persists for 170–180 days in plateau regions. In August, thunderstorms occur (5 per year), but tend to be of short duration and are much less intense than those affecting the prairies or southern Ontario. The long summer days at 64°N account for most of the 1700 h of sunshine Dawson receives each year.

Prevailing winds are westerlies throughout the year. The bypassing effect results in very light winds at ground level

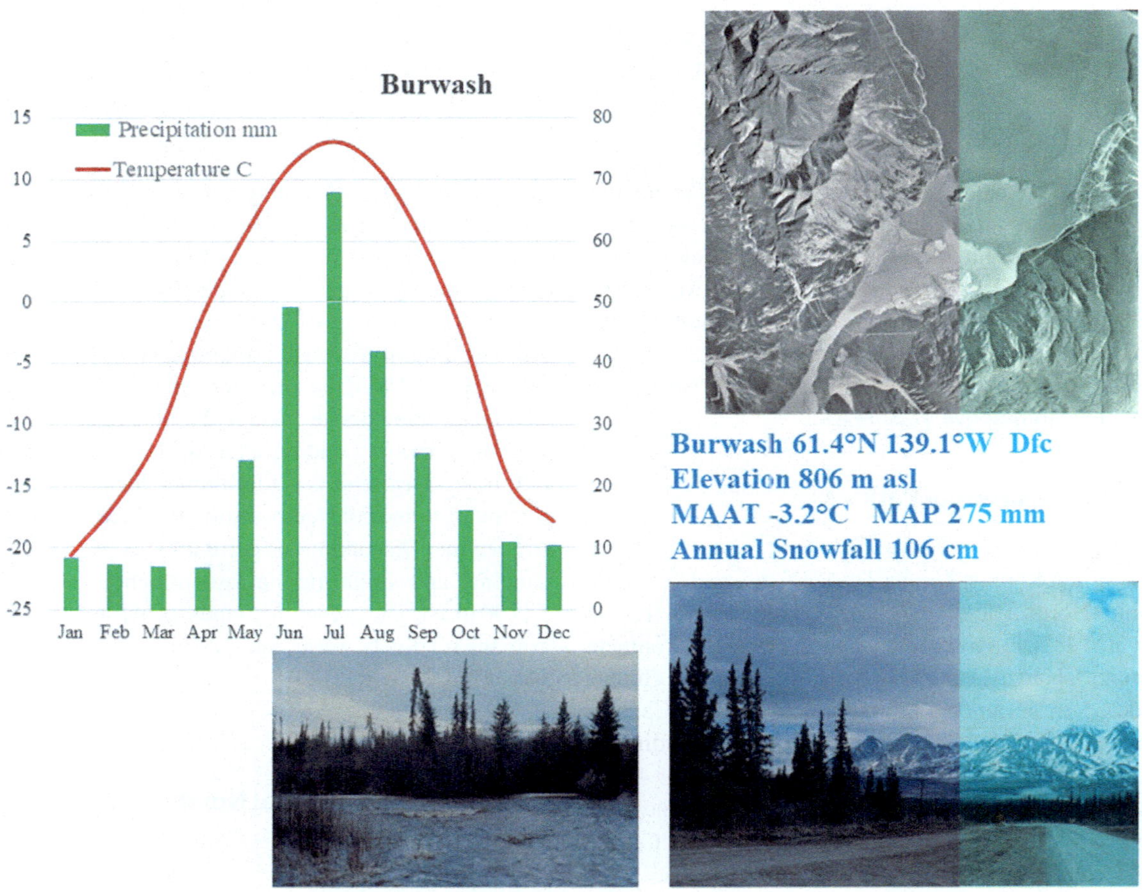

Fig. 6.9 Burwash, adjacent to Kluane National Park, has a Dfc climate with summer precipitation maximum. Air photo shows southern end of Kluane Lake (NAPL) (Data: Canadian Climate Normals, Environment and Climate Change Canada 2020)

throughout the winter months. Calm conditions (without measurable winds) occur during 75% of daylight hours in January, and during 85–90% of the night hours. In the summer, warming of the land surface generates strong thermally driven winds, often funnelled along the Yukon River valley and Tintina Trench.

6.3 Bedrock

6.3.1 Foreland Structural Belt

6.3.1.1 Rocky Mountains

The Rocky Mountains and adjacent foothills form the southern part of the Foreland Structural Belt (Gabrielse and Yorath 1991; Wheeler and McFeely 1991; Price 1994; Haggart et al. 2006; Nelson and Colpron 2007). Rocks exposed span geological time from the Precambrian to the Palaeocene (Fig. 6.11). Although rock units generally increase in age from east to west, complex thrust faulting has locally resulted in younger rock being overlain by displaced older units, in addition to creating the characteristic northwest–southeast oriented pattern of ranges and valleys.

The oldest Precambrian units form the northwest - southeast trending Main Ranges (Leckie 2017), including Mount Robson west of Jasper (Fig. 6.12). In Yoho and Kootenay National Parks, there are three main chains: an eastern chain along the Continental Divide, including the Waputik, Bow, and Ball Ranges; a chain through western Yoho and central Kootenay, including the Van Horne, Ottertail, and Vermilion Ranges; and a western chain in southwestern Kootenay, consisting of the Brisco Ranges. The eastern and central chains are separated by glacially modified valleys occupied by the Vermilion and Ottertail Rivers, and the Brisco Ranges are separated from the Vermilion Range by the valley of the Kootenay River.

The bedrock in Yoho and Kootenay Parks represents the base of the succession found in Banff, Waterton Lakes, and Jasper to the east (Massey et al. 2005; AGS 2015; Figs. 6.13 and 6.14). The oldest rocks are Neo-Proterozoic sediments

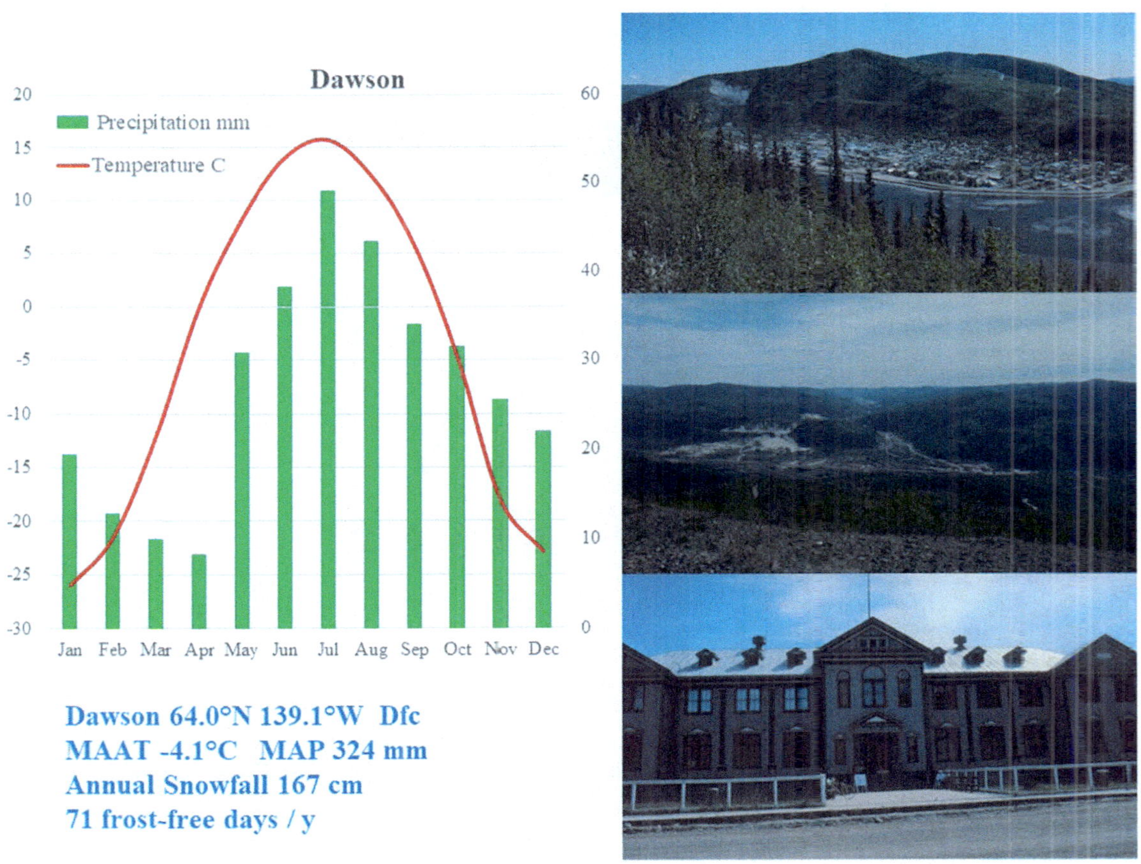

Fig. 6.10 Dawson, in the unglaciae western Yukon, has a Dfc climate (Data: Canadian Climate Normals, Environment and Climate Change Canada 2020)

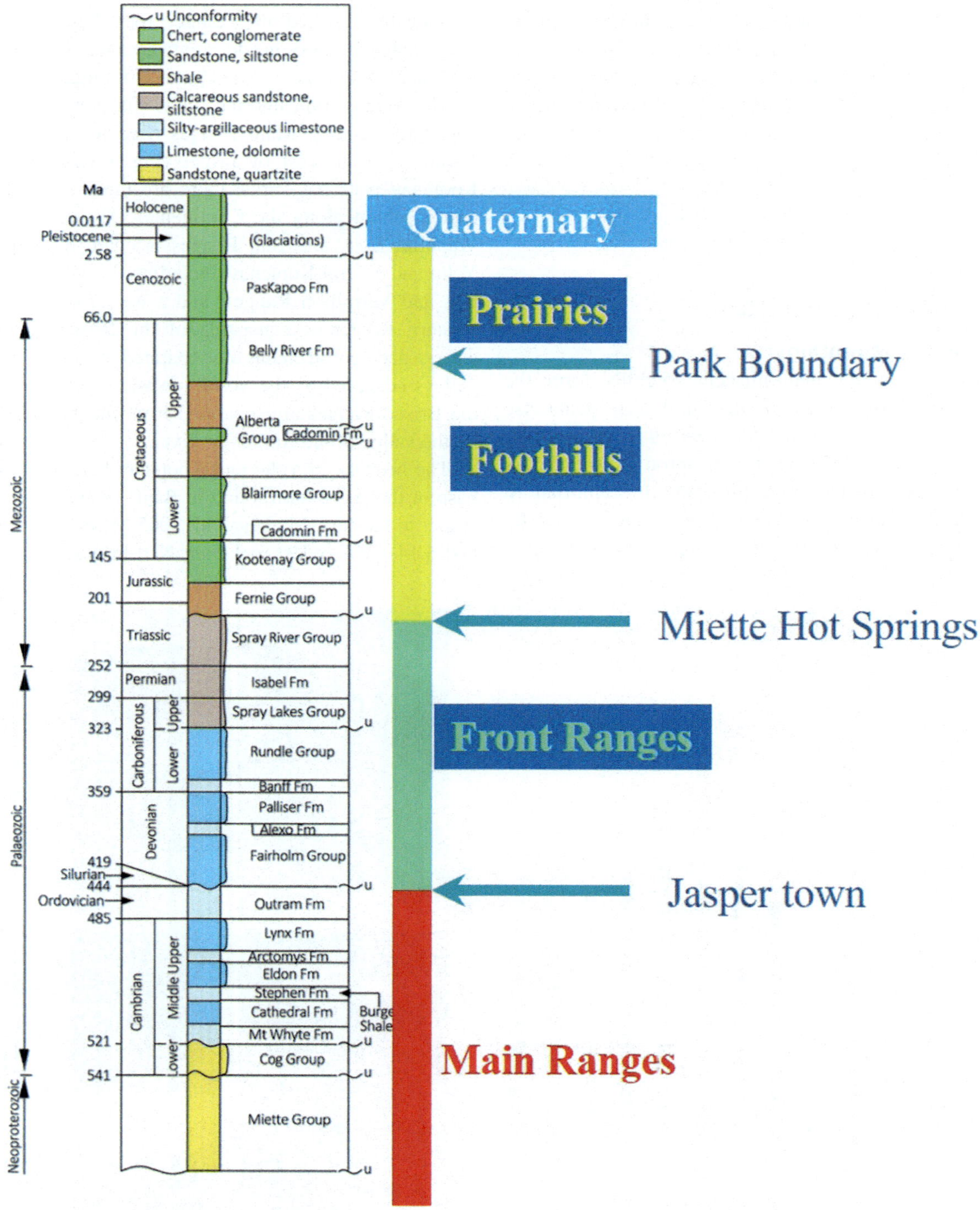

Fig. 6.11 Bedrock stratigraphy of the southern Rocky Mountains. Left column: Stratigraphic formations (Geologic Atlas of Western Canada (Mossop and Shetsen 1994), Alberta Geological Survey (Prior et al. 2013)). Right column: regions of exposed rock in the Foreland Structural Zone of Banff, Jasper, and Waterton Lakes National Parks. Arrows indicate boundaries between zones in Jasper National Park

Fig. 6.12 Mount Robson BC is the highest peak in the Canadian Rockies (Vic Levson)

of the Windermere Supergroup (Miette Group), 1325–900 million years old, consisting of shales, siltstones, metamorphosed slates, and Precambrian glacial deposits long since lithified after being formed by glacial ice 900 million years ago (Smith et al. 2011). Other units of the Windermere Supergroup contain stromatolites, fossilized mounds of blue-green algae formed in shallow marine waters during the Ediacaran.

Younger strata in the Main Ranges include Cambrian Gog Group sandstones and Cambrian and Ordovician limestones and shales (Desjardins et al. 2010; Fig. 6.13). Silurian sediments, absent from the Banff succession, have been preserved in Yoho in thrust-faulted assemblages of shales and limestones.

Karst features produced from the chemical dissolution of carbonate limestone and dolostone are present in the Main Ranges, including Castleguard Cave in the Columbia Icefields (Ford and Smart 2017). Dissolution of exposed limestone by surface streams, in combination with fluvial erosion by transported clasts, can produce meandering channels marked by overhanging walls. The initial focus of surface dissolution may be provided by a fault or structural weakness, or may be due to confinement of a steeply dipping bed by other rock types. Eroding streams produce assemblages of open areas and covered tunnels, forming narrow canyons ("bogaz") and limestone "natural bridges". An eroding stream can "disappear" from the surface in a ponor, excavated in limestone by the stream as it flows from the surface to the subsurface. Maligne Canyon in Jasper National Park and Mistaya Canyon in Banff National Park are among the examples in the Main Ranges (Leckie 2017).

The most famous of the geological units is the mid-Cambrian Burgess Shale, a thin stratum located in central Yoho (Morris and Whittington 1985; Coppold and Powell 2014; Fig. 6.15). The Burgess shale is remarkable for its fossil assemblage of soft-bodied forms (without shells). Mud and marine organisms were deposited in moderately deep water at the front of a mid-Cambrian algal reef, and accumulated under anaerobic conditions. The absence of oxidation and bacteria meant that the muscle tissues and internal organs of the organisms were preserved. Thus, when the mud lithified into the Burgess shale, the entire organisms were preserved, rather than only the shells as is normal. Over 120 species of Cambrian organisms have been identified, many of which are found in no other rock units throughout the world. The uniqueness of the Burgess shale has long been recognized by geologists, and has led to the designation of Yoho as a UNESCO World Heritage Site (Slaymaker et al. 2017). Fossil collecting from the shale is strictly prohibited, and park wardens frequently inspect the packs of returning hikers.

The establishment of Kootenay National Park in 1920 was centred around Radium Hot Springs. The hot pool at Radium Hot Springs has a temperature of 40 °C, and the water rising from 2 km depth in the Earth's crust contains calcium sulphate, epsom salt (magnesium sulphate), sodium sulphate,

Fig. 6.13 The Main Ranges in Yoho National Park are capped by Gog Group quartz sandstone (Mel Reasoner)

calcium bicarbonate, and hydrogen sulphide, in addition to radium.

Radioactivity was relatively new to the human public, and it was widely believed that radium salts could possibly cure almost anything. During the 1920s, Radium Hot Springs exaggerated the amount of radium in the waters, in order to draw more customers! At this time, it was common for people to take tonics with radioactive salts, and a substantial industry developed to prepare these products. Workers purified the salts by blowing vapours away using straws over open vats, while inhaling large quantities of gaseous radon and other radioactive substances. The demand for radium products began to die in the early 1930s, which also was the period when the first workers and customers began to die of radiation poisoning. Fortunately, however, the actual concentration of radium and other radioactive elements in the waters of Radium Hot Springs is relatively low, and it is unlikely that anyone would suffer as a result of being exposed to the low levels of radiation in the spring waters, much less than those received directly from Sun.

Kootenay Paint Pots are produced where three springs bring iron-rich sulphate acidic water to the surface (Fig. 6.16). Ochre (goethite) was collected by the indigenous Ktunaxa people. The waters originate from a pyrite deposit at depth. The distinctive microbial community has been suggested as a possible analogue for microbial communities on Mars (Grasby et al. 2013).

The western part of Banff is occupied by the Main (Bow and Waputik) Ranges, which consist of sharp-crested peaks which have been somewhat more eroded than the Front Ranges further east. Glacial troughs, cirques (such as Lake Louise), and hanging valleys are common, and most valleys show signs of glacial modification. Northeastern slopes are generally steep, whereas the southwestern slopes are gentle. Cross-cutting links are rare. The summits of the Main Ranges correspond to the Continental Divide between

Fig. 6.14 Mount Edith Cavell, Jasper National Park, shows thrusting of sedimentary bedrock characteristic of the Main Ranges

Atlantic (Hudson Bay) drainage and Pacific drainage, and also mark the provincial boundary between Alberta and British Columbia.

The Main Ranges are composed of resistant Cambrian sandstones (Gog Group) which overlie Precambrian shales and siltstones of the Windermere Supergroup (Pana and Elgr 2013; AGS 2015). The major peaks of the Kicking Horse Pass and Lake Louise areas in Banff, Mount Robson (*Yexyexéscen*) in BC (Fig. 6.12), and Mount Edith Cavell in Jasper (Fig. 6.13), are made predominantly of Gog sandstone, as are the steep cliffs of Athabasca Falls (Leckie 2017). Tracks of various coastal organisms are preserved in the Gog sandstones (Gibb et al. 2017). Gog Group sandstone from the vicinity of Mount Edith Cavell was transported to southern Alberta by Cordilleran glacial ice and deposited as the Erratics Train (see Chap. 5) are composed of Gog Group sandstones.

Castle Mountain (Siksika: *Miistukskoowa*; Fig. 6.17) in Banff has been the subject of considerable controversy since 1946, when Mackenzie King decided that this peak, commonly considered to resemble a castle battlement, was better named for Dwight Eisenhower. This began a running argument involving Park officials, who routinely erected signs announcing the elevation of "Mt. Eisenhower", and local zealots who just as routinely removed them and substituted others identifying the peak as "Castle Mountain". Finally, in 1981, Parks Canada conceded, and agreed that the entire range could be designated "Castle Mountain", although the very uppermost peak was still designated "Eisenhower Peak".

Miistukskoowa and the surrounding area of Siksika land was added to the park in 1911. In 2017, the Government of Canada recognized that the transfer was illegal (see Chap. 10). The Siksika were awarded financial compensation and acquired the rights to manage the hostel and other commercial enterprises in the *Miistukskoowa* area. Interpretive signage detailing the history was erected at the site, and the process of reinstalling the name *Miistukskoowa* for the physical feature begun.

The central and eastern parts of Banff and Jasper display the Front Ranges (Sawback, Sundance, Goat, Vermilion, Palliser, and Fairholme Ranges), sharp-crested, serrated peaks with steep northeastern slopes and gently sloping southwestern slopes. In Banff, rocks of all geological periods from Cambrian through Cretaceous, with the exception of the Silurian, are exposed in the Front Ranges (Pana and Elgr 2013; Leckie 2017). The youngest strata (Cretaceous) are located at the eastern borders of the parks (Caplan 1998), and the ages of the rocks progressively increase from east to west.

The Front Ranges are composed of resistant limestones and dolomites, primarily of Devonian age (Fairholme and Palliser groups) and of younger Early Carboniferous age (Rundle Group), interspersed with easily erodible shales

Fig. 6.15 Yoho National Park is marked by northwest–southeast valleys and the Main Ridges of the Rocky Mountains, formed by thrust faulting from the southwest. Inset: *Ottoia* fossil, Burgess shale (Smithsonian Institution, National Museum of Natural History)

(AGS 2015). In Jasper, the Devonian Fairholme and Palliser limestone units, containing coral, brachiopod, stromatoporoid, bryozoan, and trilobite fossils, form strongly castellate and picturesque rock faces at Roche Miette, Roche Ronde, and Roche-a-Perdrix (Leckie 2017; Fig. 6.18). The Early Carboniferous Rundle Group limestones, dolomites, and shales (containing coral, brachiopod, mollusc, crinoid, bryozoan, and trilobite fossils) are on display at Mount Rundle (Banff).

Valleys cutting the Front Ranges are deep glacial troughs, with minor steep-sided tributaries, and are developed along weak rock units and thrust faults. Most major valleys run parallel to the trend of the mountain ranges, with narrow cross-cutting links aligned at 90° to the ridges.

The topography of the Front Ranges is largely the result of thrust faulting (Wheeler and McFeely 1991; Price 1994). During the latest Cretaceous and early Palaeogene, the terrain to the west of the Rocky Mountain Trench of British Columbia was thrust against the landmass of North America to the northeast. As the land masses were sutured together, thrusting from the southwest pushed the strata into northwest–southeast trending ridges, aligned at right angles to the direction of the applied thrusting pressure.

In several areas, much older rocks were overthrust across younger units. In Waterton Lakes National Park, the entire succession of Proterozoic rocks has been displaced 60 km to the east-northeast, coming to rest overlying the younger Cretaceous rocks. The succession was displaced as a single

Fig. 6.16 Kootenay paint pots, Kootenay National Park (Wikipedia/Membeth)

slab, above a thrust fault known as the Lewis Overthrust (Velden and Cook 1994). The absence of any folding within the Proterozoic succession indicates that the block was moved as a solid body, without internal deformation.

Subsequent erosion has isolated a small remnant of the Overthrust as *Nínaiistáko* (Chief Mountain), an island of Proterozoic rock within Glacier National Park (USA) directly south of Waterton Lakes, flanked and underlain by Cretaceous sandstone (Fig. 6.19). *Nínaiistáko* was initially thrust eastward along the Lewis Thrust, and then detached from the adjacent mountain front by erosion of the thinnest part of the thrust sheet, creating a hole or "fenster".

6.3.1.2 Rocky Mountain Foothills

Further east, the Front Ranges are flanked by the Rocky Mountain Foothills, composed of Jurassic, Cretaceous, and Palaeocene sandstones, shales, and coal seams (Pana and Elgr 2013; AGS 2015). These units, soft and easily eroded, contain plant fossils, and some contain molluscan, insect, and dinosaur remains. These strata indicate that marine waters advanced and receded over western Alberta several times during the Cretaceous. The foothills are rounded to rolling, with gentle southwestern slopes and irregularly eroded northeastern slopes. Valleys between the foothill ridges are broad, with gentle slopes where eroded through shale and steeper slopes in the sandstone areas, but have not been glacially overdeepened. Most valleys trend northwest–southeast, at approximately 90° to the direction of glacial flow. Generally, the Foothills region shows little sign of glacial modification, except for the presence of perched glacial erratics and some areas of glacially eroded fluted sandstone.

6.3.1.3 Northern Foreland Structural Belt

The Mackenzie Mountains form the northern extension of the Rocky Mountains, but are separated from the main ranges of the Rockies by the Liard Lowland (Intermontane Plateaux), a broad basin excavated by the tributaries of the Liard River (Smith et al. 2004; Leslie-Panek and McMechan 2021; Fig. 6.20). The Selwyn Mountains (Fig. 6.21) also are part of the Foreland Structural Belt, but they differ substantially from the Mackenzie Mountains in physiography and geology, and are usually not considered as part of the Rocky Mountains (Gabrielse and Yorath 1991; Price 1994; Bostock 2014; Ford and Hamilton 2017).

The Mackenzie Mountains, well-displayed in Nahanni National Park Reserve (Ford 2017), form a series of northwest–southeast trending folded and thrust-faulted ridges very similar to the Front Ranges and Main Ranges of Banff and

Fig. 6.17 Miistukskoowa (Castle Mountain), Banff National Park

Jasper. Relief reaches 1600 m, and the topography is rugged. Resistant late Precambrian to Carboniferous carbonate rocks form steep northeastern castellate faces. Younger Permian to Triassic sedimentary rocks cap the sequences (Fallas and MacNaughton 1994; Wright et al. 2007).

In Nahanni National Park Reserve, the South Nahanni River divides the Mackenzie and Selwyn Mountains. The river valley is deep, and trends northwest–southeast. The drainage pattern of the tributaries of the South Nahanni River indicates that the stream has pirated the headwaters of former rivers (ancestors of the Redstone and Keele Rivers) which flowed directly to the northeast, across the Mackenzie Mountains in large trunk valleys (Ford 2017). Tectonic activity during the Palaeogene caused the Mackenzie Mountains to rise, aiding in the southward diversion. The effects are still evident, as the Nahanni region is among the most seismically active in Canada. The new drainage pattern trends towards the southeast, diverting the water through the South Nahanni valley and around the southern flank of the Mackenzie Mountains. The South Nahanni is a tributary of the Liard River, which in turn flows into Deh Cho (Mackenzie River) at Łíídlı̨ı̨ Kų́ę́ (Fort Simpson).

Karst topography is prominent in Nahanni (Brook and Ford 1975; Ford 2017; Ford and Hamilton 2017; Fig. 6.22). The dominant processes, climate, previous geological history, and bedrock type all play roles in the formation of karst terrain and its hazards.

Dissolution features, 10 s of centimetres to 10 s of metres in depth, develop along prominent joints or fractures. The alignment of these cutters or grikes (pronounced "greeks") reflects the jointing pattern. In Nahanni, local dissolution has produced an interlocking labyrinth of features. Labyrinth grikes develop where climate conditions restrict vegetation growth and periodic high-energy events or rapid snowmelt clear the surfaces for further denudation.

Prolonged surface dissolution of limestone can produce large, flat-bottomed depressions, 10–50 km long, 6–8 km wide, to 200 m depth, flanked by steep walls. These features are termed polje (pronounced "pol-yay"). Development of a polje usually reflects an underlying near-surface structural geological influence. In Nahanni National Park, polje develop where water is temporarily impounded in depressions overlying limestone, either due to blockage of the ponor outlets by seasonal ice jams, or during extreme summer precipitation events.

The Nahanni region is currently tectonically active (Earthquakes Canada n.d.). Earthquakes occur regularly throughout the park region, and the Nahanni valley is currently the most tectonically active area in Canada, with earthquakes to magnitude of 6.5 on the Richter scale. These seismic events represent the ongoing adjustment of the welding of North America to Wrangellia and the other exotic terranes west

Fig. 6.18 Front range mountains, Jasper National Park, are composed of Palaeozoic sedimentary rocks. Upper: Roche Miette Lower: Roche Ronde

Fig. 6.19 Ninaistako (Chief Mountain), Glacier National Park USA, is a conspicuous feature of the Lewis Overthrust visible from Waterton Lakes National Park directly to the north (National Parks Service, USA)

of the Tintina Trench and Rocky Mountain Trench, ongoing from the Jurassic to the Neogene (Gabrielse and Yorath 1991; Goes 2013; Yukon Geological Survey 2021b).

The ongoing tectonic activity also is responsible for the genesis and circulation of hydrothermal fluids. Surface deposition of calcium carbonate can occur around mineral springs, usually where hot mineral-rich hydrothermal water under pressure reaches the surface. Precipitation of calcium carbonate tufa results in the formation of small terraces and flat sheets of sediment surrounding the hot springs. White mounds and terraces of actively forming tufa are accompanied by areas with concentrations of ferric iron (brown-stained) or manganese (black) ions, surrounded by grey areas of inactive tufa deposits. Tufa is actively forming at Gahnhthah (Rabbitkettle Hot Springs) (Gulley 1993; Fig. 6.23).

The Selwyn Mountains are composed of sandstones, shales, and conglomerates of late Precambrian to Devonian age (Yukon Geological Survey 2021a). These rock units are susceptible to erosion, marked by rounded summits and inter-ridge plateaux rather than by deeply carved valleys. Relief generally is less than 1000 m, and the mountains generally lack steep slopes and castellate faces. Glacial erosional features are less prominent throughout the Selwyn Ranges. The height of land in the Selwyn Mountains forms the border between Yukon and NWT, in addition to representing the drainage divide between the Liard and Yukon River watersheds.

The Richardson Mountains, including the British Mountains to the northwest, are considered as the northernmost part of the Foreland Structural Belt (Price 1994; Yukon Geological Survey 2021a). The core of the Richardson Mountains consists of an anticline of Cambrian sedimentary rocks, flanked by younger sedimentary rocks of Ordovician through Cretaceous age (Norris 1986; Catto 1996; Pinet 2021; Fig. 6.24). A series of north–south trending ridges, generally smoothed and flat-topped, are separated by north–south trending relatively broad valleys. North–south trending thrust faults and normal faults parallel the axis of the anticline. Resistant quartzites and sandstones form the ridges, and weathered siltstones and shales form the valleys. Relief throughout the range is generally less than 500 m, and the topography resembles that of foothills rather than a major mountain chain. The mountain belt is 35 km wide in the headwater area of the Caribou River and narrows to

Fig. 6.20 Mackenzie Mountains. Upper: glacial cirques are evident in the headwaters of Tsiigèhnjik (Arctic Red River). Lower: Nailicho (Virginia Falls), Nahanni (Wikipedia/Paul Gierszewski)

Fig. 6.21 Selwyn Mountains. Left: Resistant granite-cored Itsi Range (peaks at 2500 m), looking to the southeast across the valley of South Macmillan River (Geological Survey of Canada, Natural Resources Canada). Right: Pelly River, western flank of the Selwyn Mountains, is flanked by glacifluvial kame terraces. Round pits mark locations of former ice blacks trapped in glacifluvial gravel (National Air Photo Library)

15 km in the vicinity of the headwaters of the Vittrekwa River. The belt widens to the north of the Vittrekwa River, and is 80 km wide along 68°N.

6.3.1.4 Porcupine Plateau

The Porcupine Plateau, including the Bell and Bonnet Plume Basins, and Old Crow Flats, is also assigned to the Foreland Structural Belt (Fig. 6.25). At the southern end of the Richardson Mountains, the Peel River has exploited structural weaknesses to cut a pair of narrow west–east canyons, Aberdeen Canyon to the west and Peel River Canyon to the east. The area between the canyons is occupied by the almost featureless Bonnet Plume Basin (North Yukon Planning Commission 2006; Bostock 2014).

To the west, the Bell Basin, surrounding the Bell River, forms part of the Porcupine Plateau. The Bell River unites with rivers flowing westward from the Richardson Mountains, and with the Eagle River flowing northward from the Wernecke Mountains, to form the Porcupine River, the major tributary of the Yukon.

To the south of the British Mountains is the unglaciated terrain of Old Crow Flats (French 2017), known to the Gwich'in people as Vuntut. In 1993, the northern part of this geologically and topographically interesting area was preserved as Vuntut National Park.

Although the village of Old Crow is more than 2000 km from the Bering Sea by river, it stands at only 300 m above sea level. The terrain boasts more than 2000 lakes, most less than 2 m in depth and less than 3 km in diameter, formed by ablation of the underlying permafrost (Roy-Léveillée and Burn 2017). These thermokarst lakes are generally oval or circular in outline.

The Porcupine Plateau is underlain by flat-lying to gently dipping sandstones, shales, and lignite coals, ranging in age from Early Cretaceous to Neogene (North Yukon Planning Commission 2006; Yukon Geological Survey 2021a). These sedimentary rocks are soft and are readily eroded.

6.3.2 Omineca Structural Belt

The Omineca Structural Belt includes two distinctive groups of mountain ranges: the Columbia Mountains in south-central British Columbia, and a second extending across northern BC and Yukon (Church and Ryder 2010; Webster et al. 2020). These ranges are separated by the Nechako Plateau and Fraser River drainage. The boundary with the Foreland Structural Belt is marked by the Rocky Mountain Trench in British Columbia, and in part by the Tintina Trench in central Yukon.

Fig. 6.22 Karst features are present in Nahanni National Park Reserve, and in the Deh Cho (Mackenzie) drainage south to Wood Buffalo National Park. Left: Karst Labyrinth topography, Nahanni (Paul Sanborn, University of Northern British Columbia). Upper right: Collapse doline, north of Fort Vermilion, Alberta. Centre right: Solution doline, caused by dissolution of limestone on the surface. Lower right: Surface dissolution karren features

The southern part of the Rocky Mountain Trench represents a major physiographic and geological boundary separating the Rocky Mountains of Yoho and Kootenay from the Selkirk Mountains of Glacier and Mount Revelstoke (van der Velden and Cook 1996), formed by normal faulting. The areas to the west lie within the zone of terranes adhered to North America during the tectonic thrusting and orogenies responsible for creating the North American Cordillera. These rocks were not originally part of the North American continent, and have been sutured to the Rocky Mountain Trench and the Rocky Mountains during a series of events throughout the Mesozoic and early Cenozoic Eras (Gabrielse and Yorath 1991; Goes 2013).

From south to north, the southern part of the Rocky Mountain Trench is occupied by the south-flowing upper Kootenay River, the north-flowing Columbia River (Fig. 6.26), the south-flowing Canoe River, and the northwest-flowing upper reaches of the Fraser River from Tete-Jaune-Cache. These rivers are marked by anabranching and braided reaches (Makaske et al. 2009; Eaton et al. 2010; Catto 2015), and were not navigable by vessels larger than canoes. Near Columbia Lake, the head of the Columbia River, the south-flowing Kootenay is separated from the north-flowing Columbia by 1.9 km terrace of sandy gravel. An ill-advised, expensive, and ultimately abortive attempt to construct a canal to join two unnavigable rivers is commemorated at Canal Flats (Welwood 2003). North of Donald, the Columbia River and the tribu-

Fig. 6.23 Gahnhthah (Rabbitkettle) Hot Springs, Nahanni National Park, is forming as tufa actively accumulates (Nat Rutter)

tary Canoe Rover are submerged by the Kinbasket Reservoir impounded by the Mica Dam.

In British Columbia, the terrain to the west of the Rocky Mountain Trench is assigned to the Omineca Structural Belt of the Cordillera (Price 1994). In southern and central BC, several mountain ranges including the Purcell (west of the upper Kootenay River, east of the lower Kootenay River), Selkirk (bordered by the Columbia and lower Kootenay Rivers), Monashee (west of the Columbia River at Revelstoke and the Arrow Lakes, and east of the Okanagan Valley), and Cariboo Mountains (headwaters of the Fraser River west of the Rocky Mountain Trench) encompass the Columbia Mountains (Church and Ryder 2010). The Kootenay River flows south along the Rocky Mountain Trench, leaves British Columbia at Roosville (south of Cranbrook), turns north in the USA, re-enters British Columbia at Rykerts (south of Creston), flows north through Kootenay Lake reservoir, and finally turns west to join the Columbia River at Castlegar. The Columbia River flows north from Columbia Lake along the Rocky Mountain Trench past Radium Hot Springs, Golden, and Donald; in the Big Bend around the Selkirk Mountains through the Kinbasket Reservoir, and south along the western flank of the Selkirk Mountains through the Lake Revelstoke and Arrow Lakes reservoirs to the USA border south of Trail.

The Selkirk Mountains in Glacier National Park have rounded to serrated summits, but lack the prominent castellate peaks characteristic of the Rocky Mountains (Fig. 6.27). Ridges are not preferentially aligned in a consistent orientation, and the drainage pattern is dominantly rectilinear on a broad scale. Glacial features are present but are not as prominent in most areas as in the Rocky Mountains, with the exception of some well-developed horns such as Mount Sir Donald and Cheops in the vicinity of Rogers Pass (Mathews 1989).

The rocks of the Selkirk Ranges are predominantly igneous and high-grade metamorphic (Massey et al. 2005). Granites, gabbros, granodiorites, volcanic rocks, gneisses, and schists predominate. Isolated areas of limestones and marbles are present, and form cave systems such as the 5.9 km long Nakimu Caves between the upper and lower Cougar Valley areas of Glacier National Park (Morris 2002). Nakimu (*Secwépemc*: grumbling spirit) refers to the noise made by Cougar Brook as it enters the ponor. Ages of the rock units vary, but almost all are Precambrian. These rocks were formed during deformations and mountain-building orogenies in Precambrian times, and were transported by plate tectonic action and attached to North America hundreds of millions of years later. During the attachment processes, additional stresses and heating caused more metamorphism and alter-

Fig. 6.24 Richardson Mountains. Upper left: unglaciated felsenmeer terrain along drainage divide, MWT-Yukon border. Upper left: alluvial fan dominated by periglacial mass movement, MacDougall Pass area. Lower left: Northern Richardson Mountains, north of Rat River NWT. Lower right: Exposed vertically aligned strata, southern Richardson Mountains, Snake River region

ation of the already badly deformed units. These complexly deformed igneous and metamorphic rocks thus bear no resemblance to the thrust sedimentary rocks of Yoho and Kootenay National Parks, east of the Rocky Mountain Trench. Similar rocks comprise the Purcell and Monashee Mountains (Massey et al. 2005).

The Cariboo Mountains, along the southern and eastern flanks of the Fraser River, form the northern part of the Columbia Mountains (Church and Ryder 2010). Rocks include Precambrian igneous and metamorphic units, along with isolated areas of carbonate rocks with karst topography and caves (Hickson and Vigouroux 2014), similarly to the other Columbia ranges. However, the Cariboo Mountains also contain several areas of Quaternary volcanic activity, notably in Wells Gray Provincial Park west of the North Thompson River (Hickson et al. 1995; Hickson and Hollinger 2014; Neave 2015; Fig. 6.28).

Flat-topped tuya volcanoes result from eruptions under glacial ice sheets. In British Columbia, several tuyas that developed under the Cordilleran Inlandsis between 20,000 and 740,000 years ago are now exposed, including Hyalo Ridge in Wells Gray Provincial Park. Tuya formation begins with a basaltic eruption, which melts part of the overlying ice and produces pillow basalt as the lava congeals in a subglacial lake. Water leaks downwards from the lake into the volcano, destabilizing the magma and increasing the content of volatile dissolved gas. These hydrovolcanic eruptions tend to erupt explosively. Eventually, the volcano breaks through the glacier surface, causing the pyroclastic eruptions to become less violent. Basaltic lava spreads across the newly exposed surface, producing the flat-topped tuya.

The northern part of the Omineca Structural Belt includes the Omineca, Cassiar, Pelly and Ogilvie Mountains. The Omineca and Cassiar Mountains are dominated by granitic

Fig. 6.25 Unglaciated landscapes of the Porcupine Plateau. Upper left: periglacial topography resulting from pronounced frost modification under dry, cold climate. Upper right: thermokarst lakes, Wind River Basin. Lower left: Meandering Old Crow River, near Old Crow. Lower centre: Frost-shattered tor. Lower right: slope modified by gelifluction

and volcanic rocks of ages from Precambrian to Mesozoic (Massey et al. 2005). The terranes were attached to North America along the Rocky Mountain Trench between 180 and 115 million years ago. Volcanic activity continued in the Cassiar Mountains until about 5 million years ago. Younger eruptions occurred along the southwestward flank of the Cassiar and Pelly Mountains (Yukon Geological Survey 2021a). In the Fort Selkirk area in central Yukon, Ne Ch'e Ddhawa pyroclastic cone last erupted about 15,000 years ago, and Volcano Mountain erupted only a few hundred years ago (Jackson et al. 2012). Other volcanic features are present throughout southern and central Yukon. About 8.4 million years ago, slowly congealing basaltic lava formed hexagonal columns, making conspicuous ramparts at Miles Canyon, along the Yukon River north of Whitehorse (Hart and Villeneuve 1999).

The Pelly Mountains consist of NW–SE oriented slabs of Palaeozoic limestones, shales, and volcaniclastic rocks, and Carboniferous metamorphosed volcanic sediments, separated by faults indicating thrusting to the southwest (Yukon Geological Survey 2021a). Several lead–zinc deposits have been exploited in the past. The highest peak, Fox Mountain, is 2404 m asl. Glacial erosional features are prominent throughout the mountains. The range is bordered on the northwest by the Tintina Trench (Bostock 2014).

In contrast to the other ranges of the Omineca Structural Belt, the Ogilvie Mountains are located on the north side of the Tintina Trench, separating the Intermontane Yukon Plateau region around Dawson City (Fig. 6.29) from the Foreland Old Crow Basin and Porcupine Plateau (Yukon Geological Survey 2021a). Rolling to rugged topography characterizes the Ogilvie Mountains, which rise to 2213 m a.s.l. and are often warmer and moister than the Dawson City lowlands, as they are able to intercept the westerly Pacific winds (Wahl et al. 1987).

The Ogilvie Mountains are younger than the ranges to the southeast, and are composed of Cretaceous granitic rocks of the Tombstone Suite, exposed at Tombstone Territorial Park (Yukon Geological Survey 2021a, b). Local glaciers developed in the Ogilvie Mountains during the last glaciation, but were confined to the higher elevations (Beierle 2002). In combination with the granitic lithologies, the result is a series of rounded domed peaks, such as Tombstone Mountain, in contrast to the more rugged Pelly Mountains. The Tr'ondëk Hwëch'in First Nation and the Yukon Government jointly administer the park (Tombstone Territorial Park 2009).

Fig. 6.26 Anabranching reach of Columbia River in the Rocky Mountain Trench west of Kootenay National Park. Left: Anabranching reach near Castledale, 50 km northwest of Radium Hot Springs. The abundant influx of coarse sediment fed to the Columbia from the surrounding mountains overwhelms the ability of the river to transport it, so a mismatch exists between sediment supply and stream power. The channels are confined by the vegetated islands and small levees, so the excess sediment cannot be dispersed across the floodplain. The sediment accumulates in the channel bottoms, gradually aggrading the channels and triggering avulsion. (National Air Photo Library). Right: Anabranching reach of the Columbia River near Radium Hot Springs

6.3.3 Intermontane Structural Belt

The Okanagan Valley (Fig. 6.30) lies within the Intermontane Structural Belt, bordered to the east by the Monashee Range of the Columbia Mountains (Omineca Structural Belt), a rolling to rugged complex of igneous and metamorphic rocks with a net relief of 1400 m (Church and Ryder 2010). To the west, the Okanagan is bordered by the sharp, conical summits of the volcanic peaks of the Cascade Mountains (Coast Structural Belt), encompassing Cathedral and Manning Provincial Parks, with relief of 1600 m (Church and Ryder 2010). The Cascades intercept the prevailing westerly winds, bringing semi-arid conditions to the western side of the Okanagan Valley. Mount Boucherie, west of Kelowna, is a Palaeogene volcano in the Cascade Mountains (Greenough and Roed 2004). The east face is composed of columnar dacite. The north and south flanks are composed of andesite and rhyolite.

The Okanagan valley itself is underlain by Carboniferous and Permian sedimentary rocks (Okulitch 2013). In the Osoyoos area, Permian sandstones and shales crop out. Younger Eocene (Cenozoic) sediments, containing broad-leaved and conifer needle fossils, are present on the flanks of the Cascade Ranges (particularly in chert in the Princeton area; Dillhoff et al. 2013). The flat-lying bedrock geology is not evident in the physiography of the Okanagan Valley.

The central parts of the Thompson, Chilcotin, and Nechako Plateaux further north, are underlain by volcanic rocks, including plateau ("flood") basalts, flat sheets blanketing the underlying topography (Church and Ryder 2010). The basalts represent a sequence of eruptions between 31 and 1.1 million years ago (Massey et al. 2005; Bordet et al. 2014). The Grand Canyon of the Nechako formed as the river cut down through the basalt sheets. Cheslatta Falls is located at the mouth of the Cheslatta Valley, which hangs above the main canyon.

The Liard Plateau forms most of the 277,017 km^2 drainage basin of the Liard (Nêt'ıł Tué') River, a major tributary of Deh Cho (Bednarski 1998; Smith et al. 2004; Culpterry et al. 2005). The plateau interrupts the mountain ranges of both the Omineca and Foreland Structural Belts, and hence formed a convenient lowland transit route, marked by boreal vegetation. Rather than being Scottish, "Liard" comes from the voyageur word for cottonwood (*Populus deltoides*,

Fig. 6.27 Landscapes of the Selkirk Mountains in Glacier National Park are dominated by granitic and metamorphic summits with rounded profiles influenced by alpine glaciation. East-facing slopes receive less precipitation but have cooler temperatures. Glaciers throughout the Selkirks ae currently in retreat. Lower right: the Beaver River carries sediment discharge from Rogers Pass east to the Columbia

which actually does not grow here). The plateau is underlain by Devonian and Cretaceous sedimentary rock (Yukon Geological Survey 2021a).

6.3.3.1 Yukon Plateau, Dawson City

Farther north, Dawson City is situated in the Yukon Plateau (Smith et al. 2004; Fig. 6.31). Relief throughout the plateau is low, approximating 100 m. Discontinuous permafrost is present throughout the plateau. The plateau is rolling to undulating, without significant features or deep river valleys, with one exception.

The Yukon River slices through the Yukon Plateau along a deep valley system carved 150–350 m below the plateau level. West of Dawson City to the Alaskan border, the river follows the Tintina Trench, which bisects the plateau along a straight northwest–southeast line (Gabrielse and Yorath 1991; Nelson and Colpron 2007; Yukon Geological Survey 2021b).

Late Proterozoic metamorphosed sedimentary rocks dominate the area surrounding Dawson City and the southern Yukon Plateau but are not present north of the Tintina Fault system. These Late Proterozoic rocks are the original sources of the gold in the Klondike. Gold was concentrated in the Late Proterozoic rocks by heated groundwater in hydrothermal vein deposits, and then was subsequently eroded by streams and re-concentrated in gravel and sand placer deposits in the White Channel gravels and other fluvial units.

Jurassic and Cretaceous sandstones and shales, which do not contain any gold, are present north of the Tintina Trench. The Tintina Fault is lined with Cenozoic sands and gravels, formed from the erosion of the older rocks on either side of the fault.

Dawson City lies in the unglaciated region of western Yukon, the largest tract of unglaciated terrain within Canada

Fig. 6.28 Kostal Lake Pyroclastic Cone, Wells Gray Provincial Park (National Air Photo Library) Inset: Volcanic Block, Wells Gray Provincial Park (Photo courtesy Paul Metcalfe)

(Fulton 1995; Dyke et al. 2003; Slaymaker et al. 2017). Cold, dry climatic conditions, initiated in the early Quaternary as a result of the uplift of the St. Elias Mountains to the south (Wahl et al. 1987; Winkler 2000; Finzel et al. 2011), prevented continental glaciers from developing in the Dawson vicinity, and also prevented the Cordilleran glaciers of central Yukon from expanding northwestward to the region. As a result, Dawson City lacks the glacial features which form a part of most other Canadian landscapes (French 2017).

It seems paradoxical that Dawson should escape glaciation. However, the terrain adjacent to the town is not suitable for glacier accumulation, and the climatic conditions did not allow enough snow to accumulate on the glaciers to the southeast. As the St. Elias Mountains in southwest Yukon continued to rise throughout the Quaternary, as they are still doing today, conditions became progressively less favourable for glacial development.

The lack of glaciation has played a vital role in the Dawson economy. The absence of glacial erosion and the long period throughout the Cenozoic available for hydrothermal groundwater modification of the underlying bedrock has resulted in the development of gold deposits. The gold was produced hydrothermally within the rocks, much as calcareous scale accumulates on hot water pipes, and then was concentrated in stream beds with gravel as placer gold deposits. These placers have been preserved in the hills surrounding Dawson, as the streams (such as the "White Channel") which made them have shifted position throughout the Quaternary. Their preservation, however, is due to the lack of Quaternary glaciation in the region.

6.3.4 St. Elias Mountains

The St. Elias Mountains tectonically are part of the Western System of the Cordillera (Gabrielse and Yorath 1991; Price 1994; Haggart et al. 2006; Nelson and Colpron 2007). However, their physical separation from coastal British Columbia, and their taiga and tundra ecology in strong contrast to the Pacific Maritime Ecoregion, make it convenient to discuss this region here.

The strata incorporated in the St. Elias Mountains, including Kluane National Park, span from Cambrian to Jurassic time (Muller et al. 1966; Israel and Cobbett 2008; Fig. 6.32). Most of the rocks are sedimentary, with sandstones, shales, and limestones dominating. Metamorphism has altered some of the strata to schists and slates, and volcanic rocks are also present. During the later orogenic events, Cretaceous granite plutons were intruded into the sequence.

Fig. 6.29 High taiga landscape west of Dawson. *Picea glauca* (white spruce) colonizes dry terrain, usually growing on gravels, sands, or disturbed bedrock. The forests are open, with trees widely spaced

Kluane's rocks were assembled in their present geographic position during the Mesozoic. The park is part of the "suspect" or "exotic terrane" referred to as Wrangellia (after Wrangell, Alaska). Slabs of continental rock moved across the Pacific Ocean basin, driven by plate tectonic processes, and collided with North America. During the accretion events, each slab was attached or "docked" into North America, resulting in the gradual build-up of the land mass of southwestern Yukon and western British Columbia. All of Kluane's sedimentary rocks originated on the Asian/Australian side of the Pacific Ocean, and the fossils contained in them are similar to fossils found in the western part of the Pacific Rim. The granite plutons developed as a result of the pressure induced as the rocks of Wrangellia docked against North America.

The southeast-northwest trending faults responsible for the valleys in northeastern Kluane are strike-slip (transcurrent) faults. The largest of these systems in Kluane is the Denali Fault, which extends westward to Denali Mountain in Alaska. The Denali Fault, and the other parallel systems to the south, reflect the movement of slices of rock throughout Wrangellia. These systems also parallel the Tintina Trench system present to the east and north. A Mw 7.9 earthquake occurred along the fault in November 2002 (Eberhart-Phillips et al. 2003).

The development of the St. Elias Mountains is a very recent event. Occurring only 5 million years ago, it post-dates all of the accretionary events and the formation of the granites. After the Wrangellia rocks were firmly attached to North America, continued plate movement caused the Pacific Plate to begin to slide under the North American Plate, creating an oceanic trench along the Aleutian Islands of Alaska and the Alaskan coast. This subduction was responsible for the uplift of the St. Elias Mountains along the North American margin, as the rocks were buckled (Winkler 2000; Finzel et al. 2011). Plate movement in the area continues today, and is responsible for the many large earthquakes felt along the Alaskan shoreline and in Yukon.

6.4 Quaternary Glaciation

Three distinctive styles of Quaternary glaciation affected different regions of the Canadian Cordillera (Slaymaker et al. 2017). The eastern flank of the Foreland Structural Belt, including the Rocky Mountain Foothills and Richardson Mountains, was affected by the maximum extent of Laurentide glaciation originating from Kivalliq in the Canadian Shield (Dyke 2004; England et al. 2009). Numerous alpine glaciers developed at higher altitudes in all the major ranges. In addition, the Cordilleran Inlandsis expanded from the Chilcotin and Nechako Plateaux of the Intermontane Structural Belt to cover British Columbia (with the exception of the highest mountains and the Peace River Lowland)

Fig. 6.30 Okanagan Valley landscapes. Upper left: dry east-facing slope along Okanagan Lake near Penticton (Laura Byrne). Upper right: Penticton, situated between Okanagan Lake (north) and Skaha Lake (NASA). Lower left: Valley landscape near Kelowna. Lower right: Vaseux Lake, showing glacially eroded trough margin

and southern and central Yukon. Lobes of the Cordilleran Inlandsis extended through Yellowhead Pass, along the Peace River Valley, and across the Liard Lowland to interact with the Laurentide glacier to the east. The region north of the Ogilvie Mountains and west of the Richardson Mountains in northern Yukon remained unglaciated throughout the Quaternary (Catto 1996; North Yukon Planning Commission 2006; French 2017).

6.4.1 Laurentide Glaciation

Interaction between Laurentide and Cordilleran ice in western Alberta produced the Erratics Train, made of quartzite boulders from Jasper National Park (see Chap. 5), and also reached the easternmost part of Waterton Lakes National Park. Laurentide ice also blocked the drainage of the Peace, Athabasca, and Liard Rivers, producing Glacial Lakes Peace and McConnell (see Chap. 3). The eastern areas of Nahanni

Fig. 6.31 Upper: Dawson is built along he braided Yukon River. Lower: The Tintina Trench extending across central Yukon, north and west of Dawson City, is a transcurrent fault system. Ogilvie Mountains are visible to the north

Fig. 6.32 Mount Logan, Saint Elias Range, Kluane National Park, is the highest peak in Canada (NOAA)

National Park Reserve were covered by Kivalliq ice on three occasions, as indicated by glacial deposits, most recently between 30,000 and 12,000 years ago (Ford 2017).

In the Richardson Mountains, the Quaternary record is marked by three distinct glacial events (Catto 1996). All of the glacial deposits were formed by glaciers originating from the Canadian Shield to the southeast, as the Richardson Mountains were too low in elevation to support glaciers during most of the Quaternary. An early glacial phase is represented by sediments containing clasts derived from the Canadian Shield, transported by ice advancing towards the west and northwest. The Peel River formed as a result of glacial obstruction of the pre-existing regional drainage pattern at this time. A second glacial event deposited till along the Snake River and in the southern and central parts of the Peel Plateau, impounded a lake in the Bonnet Plume Basin, flooding Peel River Canyon, and extended to the glacial maximum limit in McDougall Pass (Fig. 6.33) and along the Beaufort Sea coast, reaching Ivvavik National Park. The timing of this glaciation is not firmly established, but it pre-dated 40,000 years ago.

The third and most recent glaciation reached the Bonnet Plume Basin about 25,000 years ago. In the central and northern Peel Plateau, including the Rat River valley and McDougall Pass, this most recent glacial event terminated 25–40 km west of the Peel River, and thus did not reach the pre-existing glacial maximum.

6.4.2 Alpine Glaciation

Alpine glacial landforms and sequences are well-studied in the southern mountain parks and the St. Elias Range. Multiple events occurred during the Quaternary, with each glaciation typically being less extensive than its predecessor. Most of the evidence is preserved as erosional landforms, including cubic parabolic troughs, cirques, aretes, and horns typical of alpine glaciated landscapes worldwide (Benn and Evans 2010; Catto 2015). Hanging valleys are common, and spectacular waterfalls such as the 360 m high Helmet Falls (several steps and chutes) in the Vermilion Range and the 254 m high Takakkaw Falls in the Yoho River valley of the Waputik Range (Leckie 2017; Fig. 6.34) are wonderful evidence of glacial modification. Takakkaw Falls is the most spectacular in the Cordillera.

Most of the landforms and sediments were created during the most recent glaciations. The erosional nature of the landscapes means that timing of the glacial events is still somewhat uncertain. Small terminal moraines mark maximum ice front positions. Lateral and terminal moraines also can receive sediment from gravitational mass movements from the surrounding mountains.

Sturzstroms (rock avalanches) can deposit large quantities of debris on alpine glacier surfaces. The resulting accumulations of sediment tend to be dominated by highly angular

Fig. 6.33 McDougall Pass represents the maximum western extent of glaciation in the Richardson Mountains. Ice flow was towards the left (west)

debris originating from a single bedrock unit. However, in many cases they can be extremely difficult to distinguish from "normal" terminal moraines formed by glacial flow under suitable climate conditions. Moraine Lake in Banff could be interpreted as a "normal" terminal moraine or as a deposit formed by a mass movement from the flanking bedrock wall onto glacial ice (Catto 2015; Fig. 6.35).

Glaciated landscapes are important tourist attractions: more than a third of Canada's national parks include either active or former areas of alpine glaciations. Alpine landscapes involve more than glaciers. Bedrock type and rock structure, such as bedding planes, folds, and faults, exert major influences on locations and rates of glacial erosion. Accumulation of snow, a necessary initial step in the formation of an alpine glacier, requires a suitable hollow in the bedrock. Snow accumulation and ice preservation are also influenced by the direction of prevailing winds, and the distance from major sources of moisture.

Over long periods, uplift of mountain ranges resulting from plate tectonic activity can progressively increase elevation, leading to enhanced alpine glaciation. Rates of uplift influence snow accumulation and glacial development. The alpine glacial history of a location marked by rapid tectonic uplift (such as Kluane) will differ from those in areas with slower uplift (such as Jasper and Banff) or areas where no uplift is occurring (such as Torngat Mountains and Qikiqtaaluk (see Chap. 9)).

6.4.2.1 Banff's Alpine Glaciers

The sequence at Banff National Park is among the most studied (Rutter 1972; Bobrowsky and Rutter 1992). During the most recent major glaciation, 30,000 to 15,000 years ago, ice derived from the coalescence of local montane glaciers and expanded icefields filled the Bow River Valley, extending east to Canmore. Glacial till hardened by calcium carbonate deposition forms conspicuous hoodoos east of Banff townsite (Fig. 6.36). The glaciers began to retreat at approximately 15,000 years ago, and by 13,000 years ago Banff townsite was deglaciated. The ice continued to rapidly retreat up the Bow Valley beyond Banff, and had receded to the vicinity of the Continental Divide at the commencement of the Holocene, 11,700 years ago. In both Banff and Jasper, glaciers were less extensive 10,000 years ago than they are today (Levson and Rutter 1989; Luckman 2017; Luckman et al. 2020).

While the glacial ice blocked the course of the Bow River east of Banff during the retreat, river systems developed along the edges of the ice mass (Rutter 1972). East of Banff, the river flowing south of the trunk glacier was unable to negotiate the narrow, ice-choked pass, and so was forced to cut a new route south of the original valley. When the glacier retreated from Banff, the new channel remained as the route of the Bow River, and the old glaciated valley was left as a dry channel to the north of the Bow River. This dry valley is now occupied by the Trans-Canada Highway and the Canadian Pacific Railway, and the isolated hill between the two

Fig. 6.34 Takakkaw Falls, Yoho National Park, is located at the mouth of a hanging valley

channels is referred to as Tunnel Mountain. Initially, the CPR had intended to drill a tunnel directly through Tunnel Mountain, as the original surveys had failed to reveal the dry valley to the north.

After substantial retreat during the Early Holocene, glaciers in the southern Cordillera readvanced at least twice. The most extensive, longest, and most regionally widespread advance occurred during the "Little Ice Age", which began ca. 1400 ce and culminated approximately 1845 ce (Demuth and Horne 2017; Luckman 2017). During this "Little Ice Age" event, most glaciers in Banff and Jasper advanced between 1 and 2 km (Fig. 6.37).

6.4.2.2 Athabasca Glacier, Jasper National Park

The recent Quaternary history of Jasper's Athabasca Glacier is well-studied (Luckman et al. 2020; Figs. 6.38 and 6.39). In the Wilcox Pass area, adjacent to the Athabasca Glacier, pollen evidence indicates that tundra vegetation grew in this area ca. 10,500 years ago, and thus the glaciers had withdrawn from the region by this time (Beaudoin and King 1990). The current vegetation of the area is subalpine, and hence the climate was slightly colder ca. 10,500 years ago than it is today. As the climate warmed through the early Holocene, the glaciers retreated, and new vegetation arrived at Wilcox Pass The tundra plants were succeeded by *Pinus contorta* (lodgepole pine, complete with mistletoe parasites), and then by *Pinus albicaulis* (whitebark pine), *Abies lasiocarpa* (alpine fir), and *Picea engelmannii* (Engelmann spruce), indicating a climate warmer than that at present. The Mazama ash was deposited about 7,000 years ago, near the conclusion of this mid-Holocene warm period. Gradual climatic cooling throughout the late Holocene is indicated in the pollen record by a reversion to the subalpine forest, and finally ca. 3000 years ago to the treeline conditions that currently categorize the Wilcox Pass area.

Fig. 6.35 Moraine Lake, Banff National Park. The accumulation of disorganized angular sediment impounding the lake has been variously interpreted as a supraglacial moraine, a slope failure deposit, or a combination of both. See text for discussion

Fig. 6.36 Hoodoos east of Banff town, composed of glacial till cemented by calcium carbonate

Fig. 6.37 Peyto Lake, Banff National Park, showing the characteristic turquoise produced by suspended rock flour

The glacier re-advanced as the climate cooled. Terminal moraines at the Athabasca Glacier, dated by tree-ring analysis, indicated that the glacier was actively moving forward at 1843–1844 ce, reaching its most extensive position in more than 7,000 years (Demuth and Horne 2017; Luckman 2017; Luckman et al. 2020).

Shortly thereafter, however, the ice started to retreat, and in the ensuing years receded to its current position. This rapid retreat indicates an extreme fluctuation in the climate, representing a complete reversal of the climatic trend of the previous 7000 years. The abrupt retreat suggests that climate change, induced at least in part by human activities, is responsible. The Athabasca Glacier and others in Jasper provide one of the strongest indications of the reality of climate warming. Many cirque glaciers in Banff, Waterton Lakes, Yoho, and Glacier National Park have completely ablated since 1860 ce (Warren and Lulham 2021).

When Banff was established as a national park in 1885, most of the glaciers were still close to their most recent advance limits (ca. 1845 ce). Rapid retreat since 1900 has resulted in significant recession. The Chateau Lake Louise, built to view the Albert Glacier, is now out of visual range of the ice. The glaciers today have retreated to the approximate positions at which they were situated during the early Holocene, about 11,000 years ago.

6.4.2.3 Selkirk and Cariboo Alpine Glaciers

Glaciers in the Selkirk and Cariboo Ranges advanced during the "Little Ice Age", at some time between 1200 and 1850 ce. Following general climatic warming, these glaciers also started to retreat, and most have receded more than 3 km since 1850 (e.g. Beedke et al. 2015). The Illecillewaet Glacier's rapid retreat represents one example of this general recession. The retreat continues to the present, and several Columbia Mountains Glaciers have disappeared within the past 100 years (Warren and Lulham 2021).

6.4.2.4 St. Elias Alpine Glaciers

Alpine glaciation also developed in the St. Elias Mountains, as they rose during the Quaternary and the climate in the Northern Hemisphere cooled (Turner et al. 2016). Several glacial advances from the Icefield Ranges north to the Kluane Ranges and Kluane Lake are indicated by terminal and lateral moraines in the main valleys (Rampton 1980a, b). During the Macauley Glaciation ice extended north of the park boundary and coalesced with the Cordilleran Inlandsis McConnell glacier to the east, near Whitehorse. The Macauley glacial ice retreated during the latter part of the last glaciation.

Recent advances during the past 2000 years have also occurred in Kluane National Park, and some glaciers continue to advance today (Fig. 6.40). Many Kluane glaciers, such

Fig. 6.38 Athabasca Glacier, Jasper National Park. Upper left: glacially transported debris can be deposited after glacial ablation. Upper right: supraglacial stream entering ice mass through moulin. Lower left: Clear glacial ice without sediment. Lower right: discharge of subglacial stream, carrying reworked glacifluvial sediment

as the Steele, Kaskawulsh, Lowell, and Donjek, are subject to extremely rapid glacial advances, called surges. During surges, a glacier may move forward several metres in one day, an extremely high velocity (although most surges only last a matter of days in total). The narrow, deep valleys flanked by glaciers may become blocked by ice if the glaciers cross the valleys at right angles. Surged glaciers are frequently marked by contorted medial moraines.

Surging can cause drainage routes to be blocked by ice, and lakes impounded against the glaciers (Flowers et al. 2014; Shugar et al. 2017). Advances of the Lowell glacier across the Alsek River blocked the southward drainage of the Alsek to Alaska, flooding the valley to the north and creating Neoglacial Lake Alsek. Analysis of tree-ring data and radiocarbon dates indicate that Lake Alsek formed and drained several times during the last 2000 years, including one event at some time between 1736 and 1832, a major event at some time between 1848 and 1891, and a minor event after 1891 but before 1917. During the last glacial maximum, an even larger glacial lake impounded by the ice flooded much of the low-lying terrain east of the park, producing glacial Lake Champagne, named for a now-abandoned townsite. Unfortunately, Lake Champagne appears to have contained only water. At present, several smaller ice-dammed lakes exist in the Kluane Ranges, impounded by surged glaciers.

The ice-dammed lakes are inherently unstable (Clarke 2014). Tunnels are eroded by the lake waters in the ice dam, and when the tunnel is completed the entire impounded lake basin drains rapidly, often in a few hours. When the basin has drained, the ice dam re-forms, and the basin once again becomes impounded as the waters fill up the enclosure. Drainage occurs when the water pressures and depths in the lake are sufficient to erode tunnels and float the ice dam. Hazard Lake, impounded by the Steele Glacier, drained and filled repeatedly between 1968 and the 1990s.

The ouflow of water from an ice-dammed lake is called a jökulhlaup, an Icelandic term. Jökulhlaups move extremely rapidly (in excess of 50 km/h), and cause widespread alter-

Fig. 6.39 Alpine glaciation, Jasper National Park. Upper left: Athabasca Glacier. Upper right: lateral moraine of Athabasca Glacier, formed ca. 1843 ce. Centre right: Marker indicating frontal position of Athabasca Glacier in summer 1965. Lower left: Small alpine glaciers throughout Jasper National Park are in retreat. Lower right: Glacial sediment from Cavell Glacier, solidified by carbonate cement in hoodoos, was deposited ca. 12,000 years ago

ation and erosion in the valley downstream of the impoundment. Erosional features carved by Neoglacial jökulhlaups are visible in the Alsek and other valleys. Potentially, large ice-dammed lakes and jökulhlaups could develop if glacial surging continues, as several major streams in Kluane are flanked by actively moving ice.

6.4.3 Cordilleran Inlandsis

The Cordilleran Inlandsis spread from its accumulation areas in northern British Columbia to reach the Coast Ranges, central Yukon, western Alberta, and southernmost British Columbia and Puget Sound (Clague and Ward 2011; Ferbey et al. 2013; Brennand and Perkins 2017; Huntley et al. 2017; Fig. 6.41).

Southern and Central Yukon were glaciated on at least four occasions during the Quaternary. Cordilleran ice flowed northwestward along the Yukon and Pelly River valleys and the Tintina Trench, reaching its maximum extent west of Fort Selkirk (Stroeven et al. 2014; Fig. 6.42). Cordilleran ice also coalesced with ice from the Selwyn Mountains to the east, flowing along the Stewart River's headwaters. The most recent, McConnell Glaciation, reached its maximum sometime before 20,000 years ago. The Yukon River valley below Fort Selkirk, including Dawson City, remained unglaciated throughout the Quaternary.

Cordilleran ice originating from the Nechako Plateau of central British Columbia also flowed eastward through Yellowhead Pass (Plouffe 2000; Stumpf et al. 2004; Blais-Stevens and Clague 2007). Radiocarbon dates indicate that the last glacial advance to override Jasper townsite occurred

Fig. 6.40 Alpine glaciation, Kluane National Park. Upper left: Mount St. Elias, Chugach Mountains (National Parks Service). Upper right: Lateral and Medial moraines on Logan Glacier (National Parks Service). Lower left: Surface of surged Steele Glacier shows crevassing and accumulation of debris at snout (Nat Rutter) Lower right: Debris-covered surface of Steele Glacier (Dave Liverman)

at some time after 30,000 years ago, and the ice had retreated from Jasper ca. 13,000 years ago (Levson and Rutter 1989).

Yellowhead Pass is the lowest route across the southern Rockies, at 1132 m asl. The pass is lower than the Obed Summit drainage divide between the McLeod and the Athabasca drainage systems in the foothills east of Jasper National Park. The very low elevation (500 m lower than Kicking Horse Pass west of Banff) not only facilitates biological migration, but it allowed the pass to serve as an outlet for the Cordilleran Inlandsis.

Cordilleran ice also flowed to the south along both arms of the Columbia Valley, covering all but the highest points of the Selkirk, Monashee, and Gold Ranges. Revelstoke and the lower parts of Glacier Park were probably deglaciated about 11,000 years ago.

The Okanagan Valley was glaciated at least twice during the Quaternary (Greenough and Roed 2004; Stumpf et al. 2013; Nichol et al. 2015). After an advance some time prior to 45,000 years ago, the Cordilleran glacier retreated, allowing the development of forests and the deposition of river and lake sediments (termed the "Bessette" and "Olympia" deposits), along with ash from the Cascade volcanoes to the southwest. The glacial ice readvanced over the area about 30,000 years ago. Ice retreated from the area at the conclusion of the last glaciation, standing at Penticton 11,700 years ago. As the ice continued to retreat northwards from Penticton to Kamloops, a series of glacial lakes were impounded in the narrow valleys. Lake Okanagan represents merely the most recent of this system of long, fjordal lakes.

6.4.4 Unglaciated Northern Yukon

The terrain north of the Ogilvie Mountains, including Old Crow and Vuntut National Park, remained unglaciated throughout the Quaternary (Fig. 6.42). The rising St. Elias Mountains to the south blocked the moist Pacific winds, producing a dry periglacial climate (modified Köppen-Geiger Dwd) which did not allow glacial expansion. Together with adjacent Alaska, this terrain is referred to as Beringia (Zazula

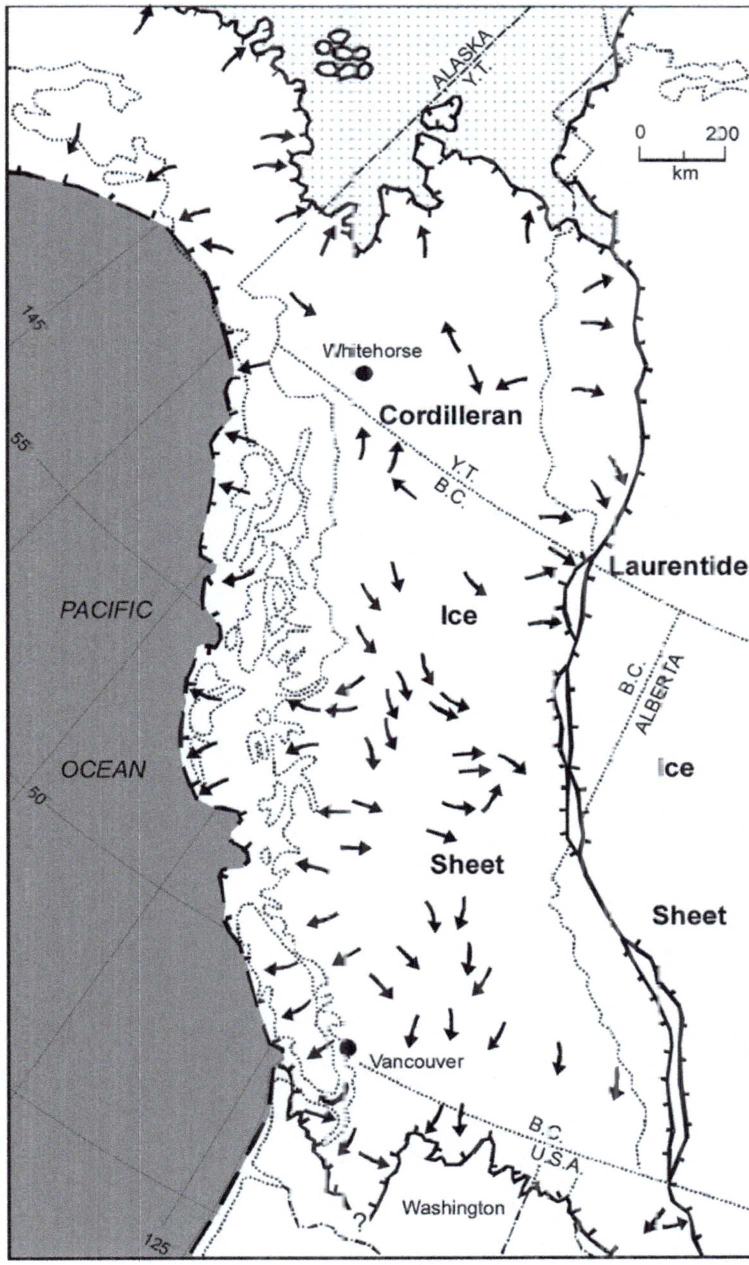

Fig. 6.41 Cordilleran Inlandsis glaciation (Courtesy Brent Ward, Simon Fraser University)

et al. 2003; Froese et al. 2009; Edwards et al. 2018; Hoffecker et al. 2020).

Beringia supported a unique vegetation assemblage, unlike that existing anywhere in North America today. Compression of the biomes by glaciation resulted in an assemblage with alpine, subalpine, taiga, boreal forest, prairie parkland, and southern prairie steppe species growing together. This "steppe-tundra" assemblage provided an ideal habitat for large grazing and browsing animals and their predators. Many faunal remains have been preserved in the fine river-bank Yedoma sediments ("mucks") and windblown silts (loess) present throughout western Yukon and adjacent Alaska (Schirrmeister et al. 2013).

The unglaciated terrain provided a suitable home for the large mammals of northern North America during the glacial episodes (Harington 2011; MacDonald et al. 2012; Williams 2022). A continuous series of climatic changes affected the evolution and distribution of the fauna, especially the "megafauna" (animals larger than 44 kg).

Mammuthus primigenus (wooly mammoth) is the best known of the grazers. Mammoths travelled in herds, as do modern caribou, and seem to have grazed similarly. A full-

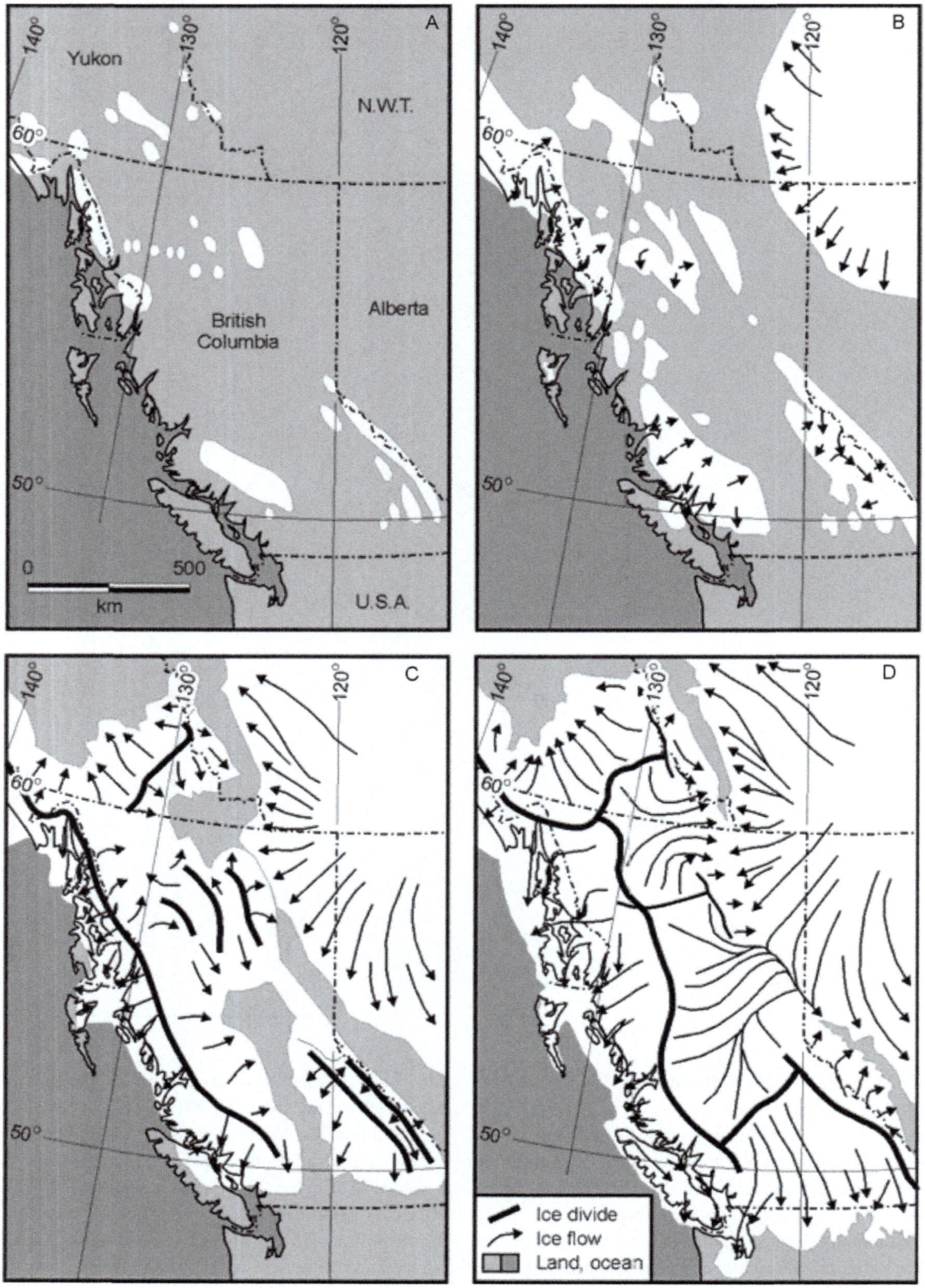

Fig. 6.41 (continued)

Fig. 6.41 (continued)

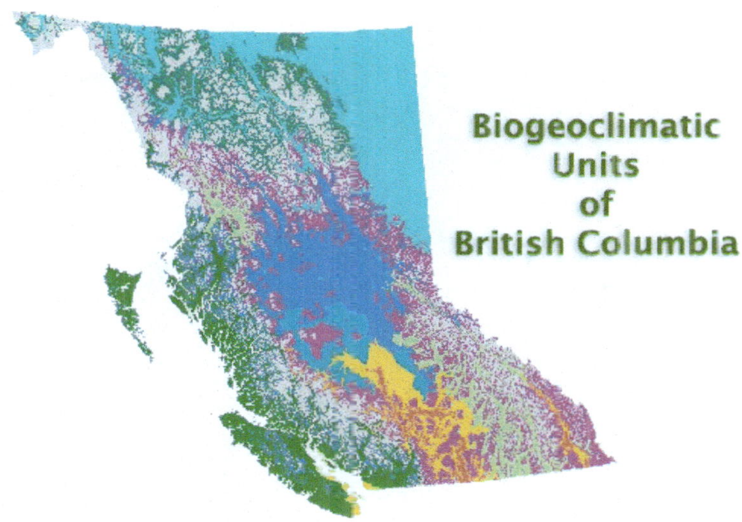

grown male mammoth stood 2.8 m high at the shoulder, and had a mass of 5 metric tonnes. Mammoths preserved in permafrozen muck deposits have been studied intensively, and their internal organs and stomach contents analysed. The results indicate that the mammoth enjoyed a varied diet, taking full advantage of all the different vegetation available in the steppe-tundra region. Other grazers present included *Equus* (horse), *Camelops* (camel), *Rangifer* (caribou), and *Bison*.

Woodland areas also supported large herbivores. Chief among these browsers was *Mammut* (mastodon), which appeared to be solitary throughout most of the year, much as the modern moose does. The faunal remains of *Mammut* recovered indicate that its preferred vegetation was alder and birch browse, but many other plants were consumed as well. Other woodland browsers included *Symbos* (woodland musk ox), *Platygonus* (peccary), *Casteroides* (giant beaver, approximately the size of a modern black bear), and Tapirs.

Predatory species were also common in the region. The chief predators were *Canis dirans* (dire wolf, larger, heavier, and less agile than modern *Canis lupus*), *Panthera leo atrox* (cave lion), *Arctodus simus* (short-faced bear, 50% larger than the grizzly), *Acinonyx* (cheetah), and *Smilodon* (sabre-toothed cat). All of these predators were built to ambush prey, rather than attempt to chase it down as modern wolves and lions attempt to do. Many were specific to one particular prey

Fig. 6.42 Northern Yukon landscapes. Upper: Terminal moraine along Keele River marks the limit of glaciation. Ice advanced from the left. Lower: Aberdeen Canyon, Peel River, lies within the unglaciated area of northeastern Yukon

species: *Smilodon*, for example, appears to have survived primarily by dining on young mammoths.

This community thrived throughout the last glacial period in Yukon. Less than 2,000 years after the commencement of the Holocene, however, virtually all of these species had disappeared from North America. In some instances, the evolution of newer, more agile species (especially predators such as *Ursus arctos horribilis* and *Canis lupus*) was responsible for the extinction. Other predatory species disappeared due to the elimination of their prey, as happened to *Smilodon* when mammoths became extinct. For the large grazers and browsers, two possible reasons for the extinctions have been suggested: climatic change following deglaciation, and human over-hunting.

The climatic shift following deglaciation resulted in the development of latitudinal belts of vegetation, with prairie species migrating southward and arctic species moving northward. Organisms which required a balanced diet consisting of all these species could not survive the breakup of the steppe-tundra vegetation communities. This vegetation change would have particularly affected the large grazers, such as *Mammuthus*.

Simultaneously, the arrival of *Homo sapiens* in the area brought the grazers in contact with a sophisticated group of hunters. The naive fauna of the area would have been easy prey initially. Although it is unlikely that human hunting by itself could have eliminated these species, it may have severely affected organisms which were already suffering the effects of the shifts in vegetation types and abundance.

6.5 Vegetation

In montane regions, altitudinal variations in climate cause differences in vegetation assemblages and soil development. However, vegetation is not distributed along straight contour lines, as other factors are involved (Demarchi 2011).

Bedrock lithology is a significant control: some plants prefer carbonate-rich substrate, whereas others thrive on acidic soil produced from quartz sandstone. In thrust-faulted ranges of the Foreland Belt, alternating limestone and sandstone strata result in vegetation distributions not strictly controlled by altitude. Differences also exist, particularly in understory vegetation, over the granitic rocks of the Omineca Structural Belt.

Bedrock structure is also significant. In thrust-faulted ranges, the gently sloping upthrust sides support more regular transitions in altitudinal vegetation assemblages than do the steeply sloping downthrust faces. Castellate mountains may show "missing" altitudinal zones, where steep faces preclude vegetation establishment in otherwise suitable climate regimes.

Aspect with respect to insolation, and orientation with respect to prevailing winds, are also important controls. In this Northern Hemisphere environment, Sun's rays strike south-facing slopes more frequently. Topography also can provide local shading. As a result, it is not uncommon for altitudinal transitions in vegetation to vary by several tens to a few hundred metres on opposite sides of a mountain valley. The prevailing wind direction, generally westerly in the Cordillera but with local topographic funnelling, also influences vegetation. Avalanches allow deciduous shrubs to occupy tracks cleared of conifers.

Fire, both natural and anthropogenic, also influences Cordilleran vegetation assemblages (McGee et al. 2015; Coogan et al. 2021; Prichard et al. 2021). Natural fire frequencies are highest at lower altitudes and in boreal forest assemblages. Assemblages dominated by pine, spruce, or interior Douglas Fir are more susceptible. The eastward and upslope spread of the Mountain Pine Beetle (*Dendroroctorus ponderosae*) has produced swaths of dead and dying conifers, which burn more readily.

In the southern Rocky Mountains and Selkirk Mountains, anthropogenic fire frequency was abruptly increased by railroad construction and operation of steam locomotives (1885–1950). Mining, logging, and development of wildland-urban interface communities, such as Kelowna in the Okanagan Valley, also play significant roles.

In the Rocky Mountains, the fire frequency has been greatly affected by European settlement and the fire suppression policy formerly favoured by Parks Canada (Rhemtulla et al. 2002; Pyne 2007; Coogan et al. 2021). Whereas prior to 1850, fires were set by the Nehinaw (Cree) First Nation in lowland areas in Banff, total fire suppression became the policy after World War 1. Currently, fires are at 10% of the estimated "historic" level defined as that where records exist, beginning immediately after the arrival of Hudson's Bay Company personnel who kept notes.

In order to approach a more natural fire cycle, Parks Canada has initiated the practice of "prescribed burns", under which parts of the forest (preferably those remote from high-visibility sites) are burned. The goals are to reduce the potential damage from hazardous large wildland fires, and to mimic the effects of natural fire on forest ecology. In some respects, prescribed burning is the modern equivalent to the repeated use of fire by previous generations of humans. Although many rural residents burn brush or grass each spring to rejuvenate the land, most prescribed burning is conducted on government-owned or—administered lands.

Among the faunal winners under the prescribed burn regimes are hares, lynx (preying on the hares), and woodpeckers, which may increase in population up to 50 times following a fire. Parks Canada estimates suggest the cost of a prescribed burn is $ 100/ha in Jasper (although it can be as high as $1200/ha), which contrasts with the up to $ 70,000 per hectare cost involved in fighting an uncontrolled fire.

Prescribed burning remains controversial (McFarlane et al. 2008). The overall ecological benefits may conflict with both immediate biodiversity concerns and perceived threats to human health. In addition, any burning is regarded as unacceptable under a total suppression policy. Fear of liability or backlash in the event of problems can be a strong deterrent. Public attitudes to prescribed burning tend to reflect personal views on human-environmental relations in general.

Animal activity can also influence vegetation assemblages in the Cordillera. Examples include terrain modification by beaver (*Castor canadensis*), rubbing of deciduous trees at lower elevations by wapiti (*Cervus canadensis*), white-tailed deer (*Odocoileus virginanus*), and formerly bison (*Bison bison bison*). At higher altitudes, bighorn sheep (*Ovis canadensis*), mountain goat (*Oreamnos americanus*), black bear (*Ursus americanus*) and grizzly bear (*Ursus arctos horribilis*) also influence vegetation distribution. Marmots (*Marmota* sp.) and pika (*Onchonta princeps*) distribute seeds in subalpine and high alpine environments, respectively.

As a result of all these factors, the altitudes of the tree line (altitude limit of upright woody-stemmed vegetation) and the timber line (altitudinal limit of "practically useable" timber) will vary significantly within a particular area. Neither "tree" nor "timber" lines are true contour lines, both showing significant ecotonal variations over short distances. Prostrate woody-stemmed vegetation may extend to altitudes substantially above tree line. Similar variations occur in all altitudinal vegetation successions in the Cordillera.

6.5.1 Banff: Altitudinal Vegetation Succession

In the southern Rocky Mountains, the vegetation succession on the east-facing slopes at Banff National Park, from the Bow River Valley (1200 m asl) to Mount Rundle (2984 m asl), provides an example (Holland and Coen 1982; Parks Canada 2017a; see Fig. 6.43). At the lowest elevations in the Bow River Valley, the vegetation is an Aspen Parkland assemblage, dominated by *Populus tremuloides* (trembling aspen). Common understory species include *Campanula rotundi-*

Fig. 6.43 Altitudinal vegetation succession, Kananaskis, southern Rocky Mountains. Lakeside: Upper Foothills vegetation assemblage, dominated by *Pinus contorta* (lodgepole pine) and *Picea engelmannii* (Engelmann spruce). Midslope: Eastern Slope Subalpine Forest with *Picea engelmannii, Abies lasiocarpa* (alpine fir), and *Larix lyallii* (alpine larch). Upper slope: Upper Subalpine forest with *Abies lasiocarpa* and *Pinus albicaulis* (whitebark pine). Summits: alpine tundra vegetation

flora (bluebell), *Rosa acicularis* (wild rose), *Berberis repens* (Oregon grape, with waxy-blue berries), and *Symphoricarpos albus* (snowberry). Luvisols develop in the forested areas wherever erosion does not obliterate the horizons. Directly adjacent to the Bow Valley and up some of the side valleys, soils of the Brunisolic Order can be identified. These soils appear as transition soils between the Chernozemic soils and the Luvisolic soils to the east and the Podzolic soils to the west (Soil Classification Working Group 2001).

The Upper Foothills vegetation assemblage occupies terrain to approximately 1350–1450 m asl, including Banff townsite (Fig. 6.44). The assemblage is dominated by the boreal species *Pinus contorta* (lodgepole pine) and *Picea engelmannii* (Engelmann spruce). Other conifers include *Pseudotsuga menziesii glauca* (interior Douglas Fir), with lesser *Pinus flexilis* (limber pine) and *Abies lasiocarpa* (alpine fir). These three species are more prevalent further south, at Waterton Lakes National Park. To the north in Jasper National Park, *Pinus flexlis* does not occur, and *Pseudotsuga* is very restricted (Parks Canada Agency 2017a). Dominant conifers here include the boreal species *Picea glauca* (white spruce) and *Picea mariana* (black spruce) at lower altitudes, replaced by Engelmann spruce (or Engelmann-white spruce hybrids) at elevation.

Deciduous trees form a lesser component of the assemblage, dominated by *Populus tremuloides* (trembling aspen), *P. trichocarpa* (black cottonwood), and *Betula papyrifera* (white birch). Other species include *Acer glabrum* (Douglas maple, a small shrub), *Shepherdia canadensis* (buffaloberry), *Rubus idaeus* (raspberry), *Rubus parviflorus* (thimbleberry), *Clematis columbiana* (blue clematis), *Heracleum lanatum* (Cow-parsnip, despised by cattle), and *Clintonia uniflora* (Queen-cup lily). The soils developed under the Upper Foothills Assemblage are dominantly humo-ferric podzols and dystric brunisols.

Eastern Slope Subalpine Forest assemblages occupy terrain to 1800–2000 m asl (Fig. 6.45). The dominant tree is *Picea engelmannii*, with associated *Pinus flexilis*, *Abies lasiocarpa* (alpine fir), and *Larix lyallii* (alpine larch). Douglas Fir and tree birch are absent from the subalpine zone. Shrub birch, *Salix* (willow), *Juniperus scopulorum* (Rocky Mountain juniper), and *Alnus* (alder) are present in small numbers. Other plants in the subalpine zone include *Ranunculus* sp. (buttercup), *Gentian* sp. (gentian), *Cypripedium montanum* (Mountain orchid), and *Trollius laxus* (globe flower).

Eastern Slope Subalpine Forest assemblages in Waterton Lakes National Park (Kujit 1982; Parks Canada 2017b) also include *Xerophyllum tenax* (bear grass), which in Canada is

Fig. 6.44 Upper Foothills vegetation assemblage Left: *Pinus contorta* (lodgepole pine). Right: *Populus tremuloides* (trembling aspen)

Fig. 6.45 *Picea engelmannii* (Englemann spruce). Upper: Tangle Falls, Jasper National Park, is near the upper altitudinal limit. Lower: *Picea englemannii* near its lower altitudinal limit, along a tributary of the Athabasca River

confined to southwesternmost Alberta and adjacent British Columbia, and does not grow in any other national park. Bear grass is actually not a grass, but is a lily. It is a favourite food of wapiti and migrating mountain goats, but bears seldom eat it or pay any attention to it. Two other species confined to Waterton are *Aquilegia jonesii* (Jones' blue columbine) and *Phacelia lyalli* (Lyall's scorpion weed). Both survived in Waterton during glaciation on isolated, uncovered flanks of mountains, while all the surrounding terrain was buried beneath the advancing ice. Soils are dominantly dystric brunisols and regosols (Coen and Holland 1976).

In Banff, terrain between ±1450 and ±1600 m asl is considered to lie within the Montane Southern Cordilleran ecoclimatic region; that above ±1600 m is classified as Subalpine Southern Cordilleran. The timber line, the upper limit of "practically useable" timber, occurs at 1600–1750 m.

At ±1800 m asl, the assemblage becomes more subalpine in character. *Larix lyallii, Pinus flexilis*, and the deciduous shrubs are not present, and *Picea engelmannii* is a lesser component. *Abies lasiocarpa* is a prominent component of the assemblage between 1850–2000 m asl. The dominant tree is *Pinus albicaulis* (whitebark pine; Fig. 6.46), a gnarled, twisted tree which usually grows in a krummholz (tuckamore) habit and survives for 50–100 years (Rowe 1972; Bosie 1979; Resler et al. 2014). At elevations above 2000 m, *Pinus albicaulis* is the only tree species present, and it forms the tree line, the uppermost position of upright, woody-stemmed vegetation, at ±2200–2400 m in Banff, ±2300–2450 m asl in Waterton Lakes, and ±1900–2000 m asl in Jasper. Soils are poorly developed, thin regosols.

Whitebark pine has a symbiotic relationship with the alpine bird species *Nucifraga Columbiana* (Clark's nutcracker). The nutcracker is the primary seed disperser for the whitebark pine: areas without Clark's nutcracker lack whitebark pine, and vice versa. Any threat to either species will thus greatly imperil the other.

Fig. 6.46 *Pinus albicaulis* (whitebark pine) is the treeline conifer species in the southern Canadian Rocky Mountains. Lower right: Whitebark pine has a symbiotic relationship with the alpine bird species *Nucifraga Columbiana* (Clark's nutcracker) (Jen Hook, National Park Service)

At present, the whitebark pine is threatened by the eastward expansion of the mountain pine beetle, *Dendroctorus ponderosae* Introduced into British Columbia in the early 2000s, the mountain pine beetle initially attacked the Ponderosa pines of southern BC, but gradually spread across the province as it fed on other conifers (Chen 2014). By 2006, most conifer stands of central and eastern British Columbia (although not the Pacific Coastal forests) were infected. The beetle has continued to spread eastward and northeastward into Alberta. Although several conifer species are at risk, the attacks are most severe for slow-growing species, particularly whitebark pine.

Spread of the mountain pine beetle is facilitated by warmer winter temperatures. Although the optimum summer temperature for growth of all four stages of the beetle's life (egg, larvae, pupae, and adult) is between 20 and 25 °C (hence, warmer than the typical temperatures prevailing in the subalpine zone), growth continues as long as the summer temperature remains above 5 °C. The critical winter (November–February) temperature is approximately −27 °C: at this temperature, significant mortality of the larvae begins. Temperatures of −35 °C are sufficient to result in complete mortality of the larvae, which would effectively eliminate the species. However, warmer winter temperatures throughout south-central British Columbia and adjacent Alberta since 2009 have allowed the larvae to over-winter, facilitating the spread of the outbreak.

Alpine tundra vegetation occupies the terrain above tree line. Typical alpine tundra plants include *Arctostaphylous rubra* (alpine bearberry), *Cassiope* (arctic heather), *Chamerion latifolium* (broad-leaved fireweed), *Potentilla* (cinquefoil), *Pyrola* (wintergreen), *Saxifraga* (saxifrage), *Silene* (moss campion), *Polemonium viscosum* (sky pilot), and *Vaccinium uliginosum* (bilberry). The bright yellow to apricot alpine poppy (*Papaver pygmaeum)* is confined to Waterton Lakes National Park, and also represents a survivor of glaciation. Static cryosols and regosols are the only soil types present at these altitudes.

6.5.2 Selkirk Mountains

The differences in vegetation between the slopes of the Rocky Mountains (Coen and Kuchar 1982; Parks Canada 2017b; Parks Canada Agency 2017c, d) and Glacier and Mount Revelstoke National Parks in the Selkirk Mountains reflect the moister climate of the western parks (Haber and Soper 1980; Douglas et al. 1998–2002; BC Ministry of Forests 2016; Parks Canada Agency 2017d; Klinkenberg 2020). The lowermost areas of Glacier and Mount Revelstoke are influenced by a Montane Southern Cordilleran ecoclimatic regime, and support the Interior Montane Columbian Forest (Rowe 1972; Fig. 6.47). The dominant trees are *Tsuga heterophylla* (western hemlock) and *Thuja plicata* (western red cedar). Other trees commonly present include *Pseudotsuga menziesii glauca* (interior Douglas Fir), *Taxus brevifolia* (western yew), and *Picea glauca* (white spruce). Deciduous trees are uncommon, and *Populus tremuloides* (trembling aspen) is the only deciduous species to form dense stands.

Rare wetland areas are present along the lowest reaches of the Illecillewaet River, with cattail (*Typha latifolia*), sedges (Cyperaceae), water hemlock (*Cicuta douglasi*), and skunk cabbage (*Lysichiton americanum*). Elsewhere in the southern Cordillera, these areas have been cleared in the process of railroad or highway construction, or have been inundated by hydroelectric reservoirs.

The understory vegetation includes dense thickets of *Oplopanax horridum* (Devil's Club; Fig. 6.48), an extremely irritating species which is difficult to walk through, produces a characteristic musky odour when contacted, and readily adheres to clothing and skin (Klinkenberg 2020). Devil's Club severely restricts mobility of all medium and large mammal species in the region. Drier east-facing slopes support *Pinus monticola* (western white pine) and minor amounts of *Pinus ponderosa* (ponderosa pine), the latter species at its northern limit. The soils are predominantly humo-ferric podzols and dystric brunisols.

Above 1450 m (Glacier) and 1500 m (Mount Revelstoke), the Columbian forest is replaced by a lower subalpine forest (Rowe 1972). The dominant tree species here are *Picea engelmannii* (Engelmann Spruce), *Abies lasiocarpa* (alpine fir), *Tsuga heterophylla*, and *Tsuga mertensiana* (mountain hemlock). Junipers are present in open areas. In areas influenced by recent forest fires (within ±300 years) and on coarse soil, *Pinus contorta* (lodgepole pine) is a co-dominant.

Above 1900–2000 m a.s.l., the lower subalpine forest is replaced by an upper subalpine assemblage, dominated by *Picea engelmannii* and *Abies lasiocarpa*, without the hemlocks and pines. Understory vegetation includes Western Spring Beauty (*Claytonia lanceolata*), alpine lily (*Lloydia serotina*), and Tiger Lily (*Lilium columbianum*). Fireweed (*Chamerion latifolium* and *Chamerion angustifolium*) and paintbrush (*Castellija* spp.) occupy clearings and meadow areas. Soils in both subalpine zones are dominantly regosols.

The tree line is located at ±2200 m in Mount Revelstoke and ±2300 m in Glacier, although local variations due to aspect, rock type and slope, and other climatic factors are common. The alpine vegetation of both parks is dominated by *Cassiope mertensiana* (alpine heather), *Antennaria lanata* (everlasting), *Potentilla* (cinquefoil), *Phyllodoce glandiflora* (heather), *Vaccinium scoparium* (grouseberry), and *Dryas octopetala* (western mountain avens). The only shrubs present are *Salix* (willow) and *Betula glandulosa* (dwarf birch), both commonly confined to the lowermost 100–250 m of the alpine zone. Cryoturbation is rare in the parks, and the soils are static cryosols and regosols.

Fig. 6.47 Interior Montane Columbian forest, Selkirk Mountains, dominated by *Tsuga heterophylla* (western hemlock) and *Thuja plicata* (western red cedar). *Populus tremuloides* (trembling aspen) is the only deciduous species to form dense stands (lower right)

Fig. 6.48 *Oplopanax horridus* (devils club) (National Parks Service)

6.5.3 Okanagan Slopes Vegetation Assemblages

Distinctive vegetation assemblages resulting from precipitation differences associated with westerly prevailing winds and orographic uplift are well-displayed across the Okanagan. Three distinctive vegetation assemblages are present in the region (Rowe 1972; Demarchi 2011; British Columbia Ministry of Forests 1990, 2016). The west-facing slopes support a moist, Southern Columbian Forest. The drier east-facing slopes support a Southern Montane Forest, and grassland and xerophytic vegetation is present in the driest sections of the area, at Vaseux Lake and Osoyoos (Parish et al. 1996; Stewart and Hebda 2000; Klinkenberg 2020; *also see* Gayton 2003; Bai et al. 2004; Lee et al. 2014; Hamilton et al. 2022).

The Southern Columbian Forest (Fig. 6.49) is dominated by *Tsuga heterophylla* (western hemlock) and *Thuja plicata* (western red cedar). These species form an essentially continuous canopy during the summer, and grow to 30 m in height, effectively shading the forest floor and limiting understory vegetation (Rowe 1972; Hosie 1979). One of the most ubiquitous understory species, Devil's Club (*Oplopanax horridum*), is also among the most unpopular with humans. Fire frequency here is low, due to the subhumid moisture regime, and any particular hectare of the forest would burn approximately every 500 years under natural circumstances. Associated species include *Pseudotsuga menziesii menziesii* (coastal Douglas Fir) and *Pinus monticola* (western white pine).

The Southern Montane Forest (Fig. 6.50) is dominated by *Pinus ponderosa* (ponderosa pine). This forest is more open than the Columbian Forest, and lacks *Oplopanax*

Fig. 6.49 Southern Columbian forest conifer assemblages are dominated by *Tsuga heterophylla* (western hemlock) and *Thuja plicata* (western red cedar), with associated *Pseudotsuga menziesii menziesii* (coastal Douglas Fir)

Fig. 6.50 Southern montane forest landscape, Okanagan Valley. Inset: *Pinus ponderosa* (ponderosa pine)

horridum. *Pinus ponderosa* grows to 25–30 m in height in this region, and is readily distinguished by its straight trunk with brightly mottled brown and orange bark (Hosie 1979; Klinkenberg 2020). Ponderosa pine forests make spectacular backdrops for films, due to this handsome colour scheme and the general openness of the forest. Many western movies have been filmed in Ponderosa terrains, regardless of the intended geographical setting of the production. Associated tree species in the Southern Montane Forest include *Pseudotsuga menziesii menziesii* (coastal Douglas Fir), *Pinus contorta* (lodgepole pine), and *Larix occidentalis* (western larch). Deciduous trees are minor constituents of the Montane Forest, and the most common species is *Populus tremuloides* (trembling aspen). Broad-leaved species are absent.

In contrast to the Columbian forest, the Ponderosa pine-dominated Montane forest is highly susceptible to wildland fire (Filmon 2004; McGee et al. 2015). Lightning is the most common cause, accounting for about two-thirds of wildland fires. In contrast to human-caused fires, lightning can strike anywhere, including very remote areas.

Much of the human losses from fires in southern British Columbia is the result of people's choices to construct houses in forests. These "interface communities" commonly consist of single-family detached homes. Lower rates of taxation and the feel of "living in the forest" attract many who wish to escape urban congestion and enjoy nature. They like to be surrounded by trees and shrubs: fire fuel.

6.5.4 Xerophytic Shortgrass Assemblage

The driest Semi-Arid Interior areas at Osoyoos support a shortgrass assemblage, with many xerophytic plants (Redpath 1990; Durance 1992; Scudder 2000; Klinkenberg 2020; Fig. 6.51). Typical of the shortgrass areas are *Bouteloua* (grama grass), *Andropogon* (bluestem grass), *Erigeron* (fleabane), *Artemisia* (sage), *Stipa comata* (spear grass), *Castilleja* (Prairie Fire), *Opuntia polycantha* (prickly pear), and *Escobaria vivipara* (purple ball cactus). Halophytes are not present. The semi-arid shortgrass landscape, including the Nk'mip conservation area, is not a true desert in the climate sense (Köppen-Geiger BSk), but it is the driest area in Canada.

The Okanagan Valley of south-central British Columbia has become one of the most popular vacation and retirement spots for western Canadians. Its mild climate, convenient access to major centres, and interesting scenery has drawn increasing numbers of people, creating a real estate boom

Fig. 6.51 Xerophytic shortgrass assemblages, southern Okanagan valley, are dominated by *Bouteloua* (grama grass), *Andropogon* (bluestem grass), *Erigeron* (fleabane), *Stipa comata* (spear grass), and *Artemisia* sp. (sage). Lower left: valley bottom grey chernozem landscape. Upper right: K'li'lx (Spotted Lake). Lower right: *Artemisia* (sage)

throughout the area. This residential development, combined with the agricultural development (chiefly orchards and vineyards) which has already occurred, has substantially altered the natural environments and landscapes of the Okanagan.

The rapid pace of development has hindered Parks Canada's efforts to establish a national park representative of this Semi-Arid Intermontane Plateau terrain. One of the few undeveloped areas is located at Churn Creek, on the east side of Okanagan Lake south of Kelowna. Negotiations to establish a reserve have been ongoing, slowly, with the British Columbia government. The Osoyoos Desert Centre conserves part of the former Nk'mip (Okanagan) drylands (Scudder 2000; Bresner 2014).

K'li'lx (Spotted Lake) is located along Highway 3 northwest of Osoyoos. The spots are produced over summer, as lake waters evaporate to leave hyperconcentrated deposits of magnesium, calcium, and sodium sulphates. Changing compositions, reflecting different amounts of evaporation and rainfall, cause changes in spot size, number, and colour. As evaporation progresses, the rims harden to further isolate the spots (Renaut and Long 1989).

The soils also show the influences of climate and vegetation type. The moist west-facing slopes are characterized by humic podzols at lower elevations, developed on the thickest and most organic-rich sediments. The upper parts of the Columbian Forest zone are developed on dystric brunisols. Both the podzols and brunisols are thick, productive soils, suitable for both coniferous forest growth and agriculture, although they are acidic.

The drier, east-facing slopes have melanic brunisols as the most common soil type at higher elevations, grading downslope to grey chernozems. Areas of regosols and dystric brunisols are also present. These soils have lower organic contents and are less productive than the moister west-facing slope units.

The valley bottom soils are grey chernozems, with minor areas of luvisols (Soil Classification Working Group 1998, 2001). Most of these soils are moderately productive. The driest parts of the region support brown chernozems, the least productive soils of the region.

6.5.5 Northern Foreland Structural Belt Vegetation Assemblages

Northern areas of the Foreland Structural Belt, including the Mackenzie and Richardson Mountains, show similar altitudinal vegetation transitions, albeit with lesser effects of rainshadowing and human intervention (Rowe 1972; CNVC 2016; Baldwin et al. 2019; Montesano et al. 2020; Fig. 6.52). In Nahanni National Park, below ±1220 m a.s.l., taiga forest occupies the montane climatic region (Gimbarzevsky et al. 1979). The dominant tree species are *Picea glauca* (white spruce) and *Picea mariana* (black spruce). *Larix laricina* (tamarack) is present only in the most sheltered areas, and *Pinus contorta* (lodgepole pine) reaches its northwestern limit at the eastern edges of the park (Porsild and Cody 1980).

Deciduous species are subordinate throughout the region. The most common is *Betula neoalaskana* (northern white birch), with lesser amounts of *Populus tremuloides* (trembling aspen). Riverbank areas are colonized by shrubby birch, *Salix* (willow), *Alnus incana rugosa* (hoary alder), and *Alnus viridis crispa* (green alder). Understory and shrubs include *Betula glandulosa* (dwarf birch), *Menziesia glabella* (false azalea), *Vaccinium scoparium* (grouseberry), *Hylocomium splendens* (feathermoss) and *Phyllodoce empetri-*

Fig. 6.52 Northern Foreland vegetation is dominated by *Picea glauca* (white spruce) and *Picea mariana* (black spruce) (Upper left: Rat River, NWT). Upper right: Understory assemblage with *Rhododendron groenlandicum* (southern Labrador Tea), *Rhododendron decumbens* (northern Labrador Tea), *Rubus chamaemorus* (cloudberry), *Vaccinium uliginosum* (bilberry), and *Sphagnum*. Lower left: Areas of *Chamerion* sp. (fireweed) indicate areas recently burned by tundra fires in peat. Lower right: segregated ice lens underlying wetland vegetation with scrubby *Picea mariana* (black spruce), *Betula glandulosa* (dwarf birch), and *Salix* sp. (willow)

formis (mountain heather). Gleysolic and regosolic soils dominate, with organic soils present in bog areas.

Above the tree line, an alpine assemblage is present. The only shrubs present are scattered *Betula glandulosa* (dwarf birch) and *Salix* (willow). The herbs and smaller plants of the alpine region include *Vaccinium vitis-idaea* (bog cranberry), *Rubus chamaemorus* (cloudberry), *Pyrola* (wintergreen), *Artemisia* (sage), *Chamerion angustifolium* (narrow-leaved fireweed), *Chamerion latifolium* (broad-leaved fireweed), *Silene* (moss campion), *Saxifraga* (saxifrage), *Vaccinium uligniosum* (bilberry), *Eriophorum* (cottongrass), *Sphagnum* moss, and various species of Cyperaceae (sedge) and Poaceae (grasses). *Cladonia rangifera* (reindeer or caribou "moss", actually a lichen) is also present. Static and turbic cryosols and regosols are present.

The Richardson Mountains and the southern part of Vuntut National Park support a taiga forest (Rowe 1972; Porsild and Cody 1980). The dominant trees are *Picea mariana* (black spruce) in moist areas and *Picea glauca* (white spruce) in dry areas. *Larix laricina* (tamarack) is present on moist sites in the southern parts of the Peel Plateau. The most northerly trees are *Picea glauca*, which reproduce vegetatively (without forming pollen) approximately two of every three years. The belt of transition corresponds approximately to the division between "discontinuous" and "continuous" permafrost.

Chamerion latifolium (broad-leaved fireweed) occupies recently burned areas. Deciduous shrubs include *Alnus viridis crispa* (green alder), *Betula glandulosa* (dwarf birch), and *Populus tremuloides* (trembling aspen). Deciduous trees, such as *Betula papyrifera* (paper birch) and *Populus tremuloides* taller than 4 m, are not present in the northern parts of the taiga.

Herbaceous plants include *Empetrum nigrum* (crowberry), *Eriophorum* (cottongrass), *Rhododendron groenlandicum* (southern Labrador Tea), *Rhododendron decumbens* (northern Labrador Tea), *Rubus chamaemorus* (locally called cloudberry), *Shepherdia canadensis* (buffaloberry), *Vaccinium uliginosum* (bilberry), and *Vaccinium vitis-idaea* (bog cranberry). Turbic and static cyrosols and regosols dominate the soil assemblage.

6.5.6 Kluane National Park

Altitude is also the major factor controlling the vegetation assemblages of Kluane National Park in the St. Elias Mountains. In the lowermost areas, between 770 m and approximately 1100 m asl, a montane parkland assemblage is present (Cody 1996; Cody et al. 2005). The dominant trees are *Populus tremuloides* (trembling aspen), *Picea glauca* (white spruce), *Populus balsamifera* (balsam poplar), *Betula kenaica* (Kenai birch), and *Betula neoalaskana* (Alaska birch, a close relative of white birch {*Betula papyrifera*}). *Picea mariana* (black spruce) is extremely rare, and *Larix laricina* (tamarack) does not occur. Associated understory species include *Puccinellia deschampsiodis* (goose grass), *Aster yukonensis* (Yukon aster), *Chamerion angustifolium* (narrow-leaved fireweed, the territorial floral emblem of Yukon), *Chamerion latifolium* (broad-leaved fireweed), *Juniperus communis* (juniper), *Dryas drummondii* (yellow dryas), and *Shepherdia canadensis* (buffaloberry).

Moist areas support *Salix* (willow), *Alnus incana tenufolia* (northern alder), and *Juncus arcticus* (bog-rush). Dystric brunisols are the most common and best-developed soils.

Between 1100 and 1500 m asl (depending upon aspect), the lower subalpine community is present (Fig. 6.53). The largest plants are shrubby and krummholz *Picea glauca, Betula glandulosa* (dwarf birch), and *Salix*. Understory vegetation includes Poaceae (grasses), *Artemisia* (sage), *Juniperus horizontalis* (trailing juniper), *Arctostaphylous uva-ursi* (bearberry), *Calamagrostis purpurasens* (reed-bentgrass), *Astragalus williamsii* (milk-vetch), *Potentilla* (cinquefoil), *Pyrola* (wintergreen), and *Linnaea borealis* (twinflower). The dominant soils are regosols and dystric brunisols.

Above 1500 m asl, the upper subalpine community is present on some sheltered slopes, marked by an open sclerophyll shrub assemblage, such as *Salix richardsonii*. These areas lack *Picea glauca*.

Alpine tundra occurs at elevations above 1500 m to 2000 m asl, depending upon aspect. *Dryas integrifolia* (dryas) is the most characteristic species of the St. Elias tundra. Other common constituents include *Carex* (sedge), *Artemisia* (sage), *Oxytropis*, *Potentilla, Salix reticulata, Arctostapylous alpina* (alpine bearberry), *Cassiope tetragona* (arctic white heather), *Silene acaulis* (campion), and *Trisetum spicatum* (tuft-grass). Vegetation is present up to ±3000 m asl. Soils are poorly developed static cryosols and regosols.

6.6 Fauna

Faunal distribution in the Cordillera is also controlled by altitude. Human pressure has severely affected the fauna in the lower (montane and boreal) areas, especially in Banff, but the assemblages in the subalpine and alpine ecoclimatic regions have been less affected. Most species undertake seasonal altitudinal migrations, moving downwards in the winter and upwards during the summer months.

6.6.1 Rocky Mountains

The 4,130,081 paying visitors who swarmed through Banff in 2022–2023 (not including those who drive through along the Trans-Canada Highway) have a large impact on the natural environment. In addition, the park's position straddling the

Fig. 6.53 Subalpine vegetation, Kluane. Foreground: open sclerophyll shrub assemblage, including *Salix richardsonii*. Background: Alpine tundra occurs at elevations above 1500 m to 2000 m asl, depending upon aspect

major transportation artery from Calgary to Vancouver results in large amounts of through traffic. Traffic volumes, with over 11,000,000 people per year, have grown to the point where the Trans-Canada Highway through the park was converted to a four-lane expressway.

Construction of overpasses for fauna, involving concrete bridges covered with soil, coupled with fencing of the entire highway, was designed to solve this problem (Clevenger et al. 2009; Fig. 6.54). The various species showed different degrees of comfort with the crossings: wapiti used the routes immediately (even as construction proceeded), whereas grizzlies, wolves, wolverine, and lynx required several years before venturing onto the overpasses. Whereas most species prefer high, wide, and short overpasses, black bear and cougar prefer routes that are low-slung, long and narrow.

The lowermost elevations are dominated by species also present in the foothills and the higher, colder prairie regions (Banfield 1981; Holland and Coen 1983; Naughton 2012; Parks Canada 2017a). The largest grazer in Banff is *Cervus canadensis* (wapiti). *Odocoileus hemonius* (mule deer), and *O. virginianus* (white-tail deer), formerly confined to adjacent prairie areas, have expanded their ranges westward with decreases in winter snow cover. *Alces alces* (moose) are present in the valleys and lower subalpine elevations in Jasper, along with woodland caribou (*Rangifer tarandus*).

Bison bison bison was present prior to 1880 (reintroduced as a free-ranging herd in Banff in 2017). In Waterton Lakes, a small herd was re-introduced in the paddock area in 1952, and remained until the 2017 Kenow wildland fire. A new herd was introduced in February 2021. These animals, however, are not

Fig. 6.54 Animal overpass, Trans-Canada Highway, Banff National Park

free-ranging, and must be fed in winter. A small free-ranging herd of *Bison bison athabascae* (?) was introduced in Jasper in 1978.

Formerly, the predators *Canis lupus* (wolf) and *Puma concolor* (cougar) were also present in valleys and lower subalpine elevations, but they have been eliminated. The resulting increase in the wapiti population has led to local overgrazing problems. Smaller predators such as *Canis latrans* (coyote), *Lynx canadensis* (lynx), and *Vulpes vulpes* (red fox) are still present in these areas, as are *Lepus americanus* (snowshoe hare) and *Castor canadensis* (beaver).

The population of beaver has increased throughout the lower elevations of Banff, with the cessation of trapping pressure and the reduction in predatory species. Warm summers throughout the past decade have led to the warming of numerous beaver ponds, encouraging the growth of intestinal parasites which produce an intestinal disturbance in humans commonly referred to as "beaver fever". Although beaver fever is more of a nuisance than a severe health hazard for most people, its effects can be debilitating and are always unpleasant. The parasite does not depend on beavers to breed or spread, but it finds a suitable environment in shallow ponds such as those created by beaver damming activity.

Banff is famous for its bear populations (Raine and Kanasa 1990; Parks Canada 1997; Herrero et al. 2021). Both the black bear (*Ursus americanus*) and the grizzly bear (*Ursus arctos horribilis*) are present (Fig. 6.55). The grizzly bear is larger (350 kg) than the black bear, and has a distinctive "hump" over its shoulder blades. The grizzly tends to inhabit higher, densely wooded areas, and shuns human contact whenever possible. Both bears are present in Jasper, and generally are more numerous (100–120 grizzlies) than in Banff.

The black bear is more gregarious, and is found at lower elevations. Typically, the black bear's diet consists largely of vegetable matter (about three-quarters), supplemented by carrion and insects. Live mammals and birds make up less than 1% of a typical bear's menu.

Although black bears are also generally not fond of human contact, some individuals are willing to raid human waste disposal sites and food supplies throughout the Banff region. Unfortunately, such bears frequently become nuisances, and must be tranquilized and transported by helicopter to remote sections of the park; return visitors are generally destroyed. The number of bear attacks on humans is grossly exaggerated by many commentators, but unfortunately in any population of 4,100,000 humans there will always be some individuals willing to put themselves and others in danger by performing thoughtless actions, such as taunting bears, attempting to photograph them at close range, or disturbing mothers with young cubs in tow.

Fig. 6.55 Grizzly (*Ursus arctos*) and black (*Ursus americanus*) bears inhabit the Cordilleran parks. Left: current ranges of bear species in western Canada. Upper right: *Ursus arctos* (Wikipedia/Bobisbob). Lower right: *Ursus americanus*, Mount Revelstoke National Park

The subalpine environment supports a different animal community. The most conspicuous species is the bighorn sheep (*Ovis canadensis*), a favourite of tourists travelling the Banff-Jasper Icefields Parkway (Fig. 6.56). *Ovis canadensis* usually travels in small herds, and is particularly fond of the salt accumulations on park highways (a habit which creates monumental traffic jams on occasion). Male bighorns are up to 160 kg (Banfield 1981; Naughton 2012). Although the general removal of predatory species (particularly wolves and cougars) has contributed to an increase in numbers, locally triggering overgrazing, human poaching has recently become a severe problem.

Bighorn sheep also carry a parasite detrimental to *Homo sapiens*. The wood tick prefers to live on sheep, but will attach itself to any convenient animal or human host. The most efficient method of acquiring a tick (should you wish to) is to pet a sheep, a rather foolhardy manoeuvre; the next most efficient way is to walk through a juniper thicket. Ticks burrow into human skin head first, but can usually be convinced to vacate by applying a lit match to their bodies. The most severe disease carried by the tick is Rocky Mountain Spotted Fever, a virulent (and formerly fatal) condition which is now readily diagnosed and treated (Public Safety Canada 2010). Banff National Park represents the northern limit of the influence of Rocky Mountain Spotted Fever.

Distinctive rodent species also inhabit the subalpine environment. *Marmota caligata* (hoary marmot), a 20–30 cm long, 1 kg herbivore is present throughout the Cordillera. Its close relative *Marmota flaviventris* (yellow-bellied marmot) is less restrictive in its diet and more likely to establish contact with Homo sapiens in a campground situation, but is confined to the southern Rockies and rarely is found north of Bow Pass in Banff. Males of both marmot species are polygamous, which is perhaps a useful evolutionary adaptation considering the male: female ratio of 1:4 in a typical marmot population.

Fig. 6.56 *Ovis canadensis* (bighorn sheep) are common at lower elevations in Jasper National Park. Increased populations can lead to overgrazing

Canis lupus (wolf) is a permanent resident of the higher subalpine zone in Banff, and throughout the subalpine of Jasper. Other species include *Puma concolor* (cougar), *Gulo gulo* (wolverine), *Martes americana* (pine marten), *Martes pennanti* (fisher), and *Mustela erminaea* (ermine).

Oreamnos americanus (mountain goat) inhabits the alpine terrain (Fig. 6.57). Mature males have masses to 120 kg. At these altitudes, predators are rare visitors, and human predation is more difficult. The population of mountain goats is apparently not subject to the same fluctuations in number which have recently affected bighorn sheep and wapiti. The distinctive rodent at these altitudes is *Ochotona princeps* (American pika), a 30–60 g inhabitant of rocky talus slopes.

Although the faunal assemblages of Yoho and Kootenay are similar to those of Banff and Jasper (Parks Canada 2017a; Parks Canada Agency 2017a, b, c), the high elevations of Vermilion and Kicking Horse Passes preclude direct migration of valley and lower subalpine species such as wapiti, white-tailed deer, and moose. These animals reached Kootenay and Yoho Parks by travelling northwards along the Columbia and Kootenay River valleys, rather than directly from Banff to the east. Consequently, genetic differences between the populations in British Columbia and Alberta exist.

6.6.2 Selkirk Mountains

In the Selkirk Mountains of Glacier and Mount Revelstoke, the dense snow cover and isolation of the park interiors limits the faunal assemblage (Parks Canada 2010; Parks Canada Agency 2017d; Clevenger et al. 2014). The only two ungulates to survive throughout the year are *Oreamnos americanus* (mountain goat) and *Rangifer tarandus caribou* (caribou; considered by some to be a separate ecotype "mountain caribou"). Summer visitors include *Cervus canadensis nelsoni* (wapiti), *Odocoileus hemonius* (mule deer), *O. virginianus* (white-tail deer), and *Alces alces* (moose), but these species migrate out of the parks in late September. As migrants, all of these ungulates are subject to pressures resulting from habitat loss due to logging and highway construction. Bison formerly were present in the Columbia Valley. Other valley-dwelling species include *Castor canadensis* (beaver), *Ondrata zibethica* (muskrat), *Erethizon dorsatum* (porcupine), pine marten (*Martes americana*), and snowshoe hare (*Lepus americanus*).

The scarcity of large ungulates limits the predatory species. *Canis lupus* (wolf), *Canis latrans* (coyote), and *Puma concolor* (cougar) are transitory visitors. *Neogale vison* (mink) is extremely rare, and there is only one record of

Fig. 6.57 *Oreamnos americanus* (mountain goat) (Marreen Farrelly, Public Domain)

Fig. 6.58 Okanagan species. Upper: *Charina bottae* (rubber boa) (United States Department of Agriculture). Lower: Aplodontia rufa (mountain beaver) (US National Parks Service)

Lontra canadensis (otter) since the parks' establishment. The most common predators are *Gulo gulo* (wolverine) and *Mustela erminea* (weasel). *Marmota caligula* (hoary marmot) is the most common moderate-sized rodent species. *Ursus americanus* (black bear) and *Ursus arctos horribilis* (grizzly bear) are relatively abundant in the lower subalpine and upper subalpine zones, respectively. Other subalpine residents include pine marten, snowshoe hare, Columbian Ground Squirrel (*Spermophilus columbianus*), red squirrel (*Tamiascirus hudsonicus*), and porcupine. Above treeline, in addition to grizzly, caribou and mountain goat, hoary marmot (*Marmota caligula*), golden-mantled ground squirrel (*Spermophilus lateralis*), and pika (*Ochotona princeps*) are present.

These parks are home to numerous bats, particularly in the caves (Parks Canada 2022). Species include long-eared bat (*Myotis evotis*), long-legged bat (*Myotis volans*), silver-haired bat (*Lasionycteris noctivagans*), Townsend's big-eared bat (*Corynorhinus townsendii*, ileft), big brown bat (*Eptescius fuscus*), and hoary bat (*Lasiurus cinereus*). As elsewhere in North America, white nose syndrome has affected the bats.

The lowland areas of Mount Revelstoke and Glacier National Parks provide suitable nesting grounds for numerous bird species. In addition to local species such as the white-tailed ptarmigan (*Lagopus leucurus*) and Clark's nutcracker (*Nucifraga columbiana*), seasonal inhabitants include some migrants from the north (Steller's jay, *Cyanocitta stelleri*) and others from neotropical and tropical areas, including pine siskin (*Spinus pinus*), Cassin's finch (*Carpodacus cassinii*), and house finch (*Carpodacus mexicanus*; Godfrey 1986; Bird 2017). Populations of summer-resident tropical birds have declined in Mount Revelstoke, due primarily to habitat loss in their winter rain forest residences.

6.6.3 Okanagan Valley

The faunal assemblage of the Okanagan valley has been severely altered as a result of the intensive agricultural development of the region. Prior to agricultural and subsequent urban expansion, the area supported an interior Cordilleran ungulate-dominated community, with *Ovis canadensis* and *Cervus canadensis*, and *Oreamnos americanus* in higher altitude regions. Predators included *Canis lupus, Puma concolor*, and *Ursus arctos horribilis*. Today, little of this faunal assemblage remains.

In the driest area surrounding Osoyoos and Vaseux Lake Provincial Park, a unique dry-land fauna is present (Fig. 6.58). *Charina bottae* (rubber boa), the only native Canadian boa constrictor, reaches a length of 70 cm in the Okanagan, and is a nocturnal predator on small mammals (Cameron and St. Clair 2003). Other snake species present include *Crotalus viridis oreganus* (Oregon rattlesnake) and *Coluber constrictor mormon* (western yellow-bellied racer). *Hypsiglena torquata* (desert night snake), a small, secretive snake, which preys on frogs, salamanders, and toads, is also present in the Osoyoos region and the Lower Similkameen Valley (BC Ministry of Enviornment 2015).

Lizards, salamanders, frogs, and toads are common in the grassland regions. *Phrynosoma douglasii douglasii* (pygmy short-horned lizard, often called a "horned toad"), is probably extirpated (Environment and Climate Change Canada 2017). Bird species include calliope hummingbird (*Stellula calliope*), western bluebird (*Siala mexicana*), western tanager (*Piranga ludoviciana*), and pygmy owl (*Glaucidium gnoma*) (Godfrey 1986; Bird 2017).

The mountain beaver (*Aplodontia rufa*) is a mammal species unique in Canada to the Cascade Mountain fringes of the Okanagan Valley (BC Ministry of Environment 2013). Although a rodent, it is not related to modern beavers, muskrats, weasels, or other common rodents of Canada. It appears to represent an evolutionary branch of Rodentia which developed in the Late Jurassic, and thus can be considered as a "living fossil". The mountain beaver resembles a tailless muskrat, 30–40 cm long, with small ears and a body designed for burrowing. It prefers riparian habitats and can inhabit any terrain below timberline (up to 2900 m in the Okanagan region). Loosely colonial, the mountain beaver does not build dams, and prefers a diet of forbs, grasses, and young conifers.

Although the mountain beaver is an unusual and little-known species, it is not endangered. It is able to survive well in logged terrain, and has adjusted to human penetration of its environment. In some areas, it has actually attained "pest" status, due to its habit of ringing conifer saplings of their bark (and hence killing them). Mountain beavers also enjoy strawberries, raspberries (including canes), cabbages, parsnips, turnips, and other crops, and are not fondly regarded by fruit-growers and market gardeners in the Okanagan.

6.6.4 Northern Fauna

The northern Rocky Mountains, Mackenzie Mountains, and St. Elias Mountains support similar altitudinal faunal assemblages to those further south, with differences resulting from colder climates, isolation resulting from glaciation, and lesser (albeit some) human pressure (Banfield 1981; Danby 2003; Naughton 2012). Snow-intolerant species such wapiti and deer cannot survive in many areas. Summer migrations of most subalpine species occur, involving movement of *Rangifer tarandus caribou, Ursus americanus,* and *Ovis dalli* to alpine regions.

Ovis dalli (Dall's sheep) is characteristic of the Mackenzie and St. Elias Mountains (Fig. 6.59). This northern relation of the bighorn sheep (*Ovis canadensis*) of the southern Rockies is heavier, and has a larger head and whiter fur than its southern counterpart. Dall's sheep inhabits almost all of Yukon below treeline. The populations of Dall's sheep have suffered some poaching lately.

Deh Cho has been an effective transit route for fish moving both north and south (Scott and Crossman 1998; Stephenson et al. 2005). Among the species represented are Dolly Varden (*Salvelinus malma*), Lake trout (*Salvelinus namaycush*), Char (*Salvelinus alpinus*), Whitefish (*Coregonus clupeaformis*), Grayling (*Thymallus arcticus*), Goldeye (*Hiodon alosoides*), Yellow Perch (*Perca flavescens*), Walleye (*Stizostedion vitreum*), and Chum Salmon (*Oncorhynchus keta*). Most of these species have navigated southward to reach Nahanni. The southern species goldeye, lake trout, Dolly Varden, and yellow perch appear to have come via Glacial Lake McConnell.

When the first Europeans descended Deh Cho northward under the leadership of Alexander Mackenzie in 1789, they relied primarily on fish as their food source. Although most of the species encountered were familiar, such as whitefish (*Coregonus clupeaformis*), arctic grayling (*Thymallus arcticus*), and arctic char (*Salvelinus alpinus*), one unfamiliar large species (typically 2–5 kg) was encountered. The voyageurs named the fish "*inconnu*", which literally translates as "unknown", and local residents still refer to the species as

Fig. 6.59 *Ovis dalli* (Dall's sheep) (US National Parks Service)

inconnu (or "conny"). Later investigation revealed that the inconnu is a subspecies (*Stenodus leucichythys mackenzii*) of a freshwater fish found in the Caspian Sea drainage of western Asia, extending to Siberia and Kamchatka. Its presence in northern Yukon and Alaska, but nowhere else in North America, indicates that a freshwater connection over land existed between Siberia and Alaska during the Quaternary. Inconnu, therefore, are thus tangible evidence for the existence of a land route across the terrain now flooded by the Bering Sea.

6.7 Human Transportation

Substantial facets of human impact in the Canadian Cordillera involve tourism and establishment of protected areas (discussed in Chap. 10) and transportation infrastructure.

6.7.1 Southern Cordillera

Topographic constraints largely channelled transportation into north–south routes until the 1850s. The looping courses of the Columbia and Kootenay Rivers made east–west foot traffic across the Selkirk and Purcell Mountains unnecessary. The relatively low Eagle Pass (550 m) provided a connection across the Monashee Mountains between the Columbia (Revelstoke) and the South Thompson-Fraser (Kamloops) drainage systems.

Travel across the Rocky Mountains was limited to passes that were most suitable for foot travel, rather than being strictly constrained by elevation. Kicking Horse Pass (1627 m), with abundant large woody debris and a fast-flowing river, was a more difficult route on foot than the more circuitous Ottertail (2140 m) or Vermilion (1680 m) alternatives. For westbound fur trade-related transport and eastbound bison-hunting related transport by the Ktunaxa people, Howse Pass (1539 m) lay on the direct route between the Columbia and North Saskatchewan drainage systems. However, the North West Company abandoned Howse Pass for the fur trade after 1811, and changes in bison abundance and Ktunaxa population gradually rendered the pass impassable. By the early 1880s, accumulated large woody debris along the Blaeberry and North Saskatchewan rivers precluded regular foot traffic.

Similarly, Athabasca Pass (1753 m) was the primary route for westbound fur trade traffic from central Alberta, in preference to the lower but more difficult foot routes through Yellowhead (1132 m) and McGregor (1555 m) Passes. Athabasca Pass formed a more direct route between the Athabasca and Columbia River drainages, in contrast to Yellowhead and McGregor, which opened into the Fraser River drainage system. Prior to 1848, westbound traffic was directed towards the Columbia River and the main outlet for furs at Fort Vancouver (Washington). Use of Yellowhead for foot traffic accelerated after the Oregon dispute settlement established the USA border at 49 N, leaving Fort Vancouver in the USA, as well as the Cariboo Gold Rush (1860–1868), resulting in redirection of traffic to the North Thompson and Fraser River valleys. In northern British Columbia, traffic followed the Peace River through the Rocky Mountains, only occasionally using the overland route through Pine Pass (875 m).

With abundant archaeological material, the oldest dating from approximately 11,000 years ago, Crowsnest Pass (1358 m) contains ample evidence of human utilization. Crowsnest Pass provides better access to the bison-hunting region of southern Alberta, and to the Kootenay River drainage, than the more circuitous route through Brown's Pass, Montana (1907 m). Continuous human occupation kept Crowsnest Pass in use, unlike Howse Pass, allowing it to function as a transportation artery.

The Columbia, Kootenay, and Fraser Rivers, although navigable by canoe, were not suitable for steam vessels. The Cariboo Road was constructed inland of the Fraser Valley from Lytton to Alexandria. The advent of increased east–west trade, developing railway technology and socio-economic implications and impacts, and the political connection of British Columbia to eastern Canada (1871) spurred rail construction. After prolonged debate, railways were constructed through Kicking Horse, Rogers, and Eagle Passes (Canadian Pacific Railway 1884–1885; Longworth 2009), Crowsnest Pass (CPR 1898) and Yellowhead Pass (Grand Trunk Pacific 1911, Canadian Northern Railway 1913), the choices dictated by combinations of low gradient and directness of east–west routing. In particular, the CPR lines did not follow pre-existing trade or settlement patterns. Subsequent road construction (Crowsnest Pass 1917; Kicking Horse Pass 1928; Yellowhead Pass 1944; Pine Pass 1951; Eagle Pass 1952; Rogers Pass 1962), along with construction along the Fraser and Thompson Rivers, followed the rail routes. Additional routes were constructed across the Purcell and Monashee Mountains (Kootenay Pass, 1774 m; Blueberry-Paulson Summit, 1535 m) and through the Coquihalla Pass (1244 m), utilizing gradients impossible for railways.

Engineering and maintenance concerns include avalanches, slope failures (debris flows, rock falls, sturzstroms in Crowsnest Pass), river flooding, wildland fires, and snowfalls and blizzards. The routes which did not follow traditional trails are in general more susceptible to interruptions and require enhanced maintenance. The scarcity of transit routes increases vulnerability in the event of closure for any notable period.

6.7.2 Northern Cordillera

With the exception of the White Pass and Yukon railway (1900) and subsequent highway (1979), modern highway routes across the drainage divides in the northern Cordillera do not follow traditional trails. Rapid wartime construction of the Alaska Highway across Summit (1305 m) and Muncho (1095 m) in 1942, and of the Canol Road across Macmillan Pass (1465 m) in the Mackenzie Mountains

Fig. 6.60 Summer night on the Dempster Highway, Richardson Mountains, Taiga Cordillera

in 1943 used the most expedient methods preceded by limited surveying. Slope failures and permafrost posed engineering difficulties which were not previously anticipated. Repair, reconstruction, and some re-routing of the Alaska Highway alleviated these issues. The Canol Road, built to service the abortive attempt at oil extraction (1944–1945) at Norman Wells NWT, rapidly fell into disrepair due to permafrost-related failures and bridge collapses, and is currently impassable to vehicle traffic through the Northwest Territories.

The most recent road constructed across the Cordillera is the Dempster Highway (1979) across the Richardson Mountains (927 m height of land), connecting Dawson City to Inuvik (Fig. 6.60). The road passes through Widespread Discontinuous Permafrost, dominated by cryoturbation and cryogenic-gravitational features. The major concern with river icing currently focuses on the interruption to the ferry services across Deh Cho at Tsiigehtchic (Arctic Red River) and across the Peel River south of Teetł'it Zheh (Fort McPherson). Truck traffic on the Dempster Highway is interrupted until the rivers freeze solidly.

References

Achuff PL, Holland WD, Coen GM, van Tighem K (1984) Ecological land classification of Kootenay National Park, BC. Alberta Institute of Pedology Publication M 84-10

Alberta Geological Survey (2015) Alberta table of formations; Alberta energy regulator. http://ags.aer.ca/document/Table-of-Formations.pdf. Accessed 8 Aug 2022

Bai Y, Broersma K, Thompson D, Ross TJ (2004) Landscape-level dynamics of grassland-forest transitions in British Columbia. J Range Manag 57:66–75

Baldwin K, Allen L, Basquill S, Chapman K, Downing D, Flynn N, MacKenzie W, Major M, Meades W, Meidinger D, Morneau C, Saucier J-P, Thorpe J, Uhlig P (2019) Vegetation zones of Canada: a biogeoclimatic perspective. Scale 1:5,000,000. Natural Resources Canada, Canadian Forestry Service, Sault Ste. Marie, ON

Banfield AWF (1981) The mammals of Canada. National Museum of Natural Sciences, National Museums of Canada

BC Ministry of Environment (2013) Management plan for the Mountain Beaver (*Aplodontia rufa*) in British Columbia. B.C. Ministry of Environment, Victoria, BC, 28 pp

BC Ministry of Environment (2015) Habitat atlas for wildlife at risk night snake *Hypsiglena torquata deserticola*. https://www.env.gov.bc.ca/okanagan/esd/atlas/species/night_snake.html. Accessed 8 Aug 2022

BC Ministry of Forests, Lands, and Natural Resource Operations (2016) Biogeoclimatic zones of British Columbia. Map, 1:2,000,000

Beaudoin AB, King RH (1990) Late quaternary vegetation history of Wilcox Pass, Jasper National Park, Alberta. Palaeogeogr Palaeoclimatol Palaeoecol 80:129–144

Beck C, Grieser J, Kottek M, Rubel F, Rudolf B (2006) Characterizing global climate change by means of Köppen climate classification. http://koeppen-geiger.vu-wien.ac.at/pdfs/beck_et_al_ksb_2006.pdf

Bednarski JM (1998) Quaternary geology of Fort Liard map area, Northwest Territories. Geol Surv Bull 596

Beedke MJ, Menounos B, Whete R (2015) Glacier change in the Cariboo Mountains, British Columbia, Canada (1952–2005). Cryosphere 9:65–80

Beierle BD (2002) Late quaternary glaciation in the Northern Ogilvie Mountains: revised correlations and implications for the stratigraphic record. Can J Earth Sci 39:1709–1717

Benn DI, Evans DJA (2010) Glaciers and Glaciation. Hodder Arnold

Bird DM (2017) Birds of Canada. DK Publishing, London

Blais-Stevens A, Clague JJ (2007) Surficial geology, southeast portion of the Prince George Map Area, British Columbia. Open File 5274, Geological Survey of Canada. https://doi.org/10.4095/224162

Bobrowsky PT, Rutter NW (1992) The quaternary geologic history of the Canadian rocky mountains. Géographie physique et Quaternaire 46:5–50

Bordet E, Mihalynuk MG, Hart CJR, Mortensen JK, Friedman RM, Gabites J (2014) Chronostratigraphy of Eocene volcanism, central British Columbia. Can J Earth Sci 51:56–103

Bostock HS (2014) Physiographic regions of Canada. Geological Survey of Canada Map 1254A, 2nd edn, scale 1:5,000,000. https://www.nrcan.gc.ca/maps-tools-and-publications/maps/geology-and-geosciences/16876. Accessed 6 Aug 2022

Boyd J, Haegeli P, Abu-Laban RB, Shuster M, Butt JC (2009) Patterns of death among avalanche fatalities: a 21-year review. Canadaian Medical Assoc Journal 180(5):507–12. https://doi.org/10.1503/cmaj.081327

Brennand TA, Perkins AJ (2017) The deglacial landscape of the Southern Fraser Plateau, British Columbia. In: Slaymaker O (ed) Landscapes and landforms of Western Canada. Springer International Publishers, Cham, Switzerland, pp 277–290

Bresner K (2014) Sharing identity through indigenous tourism: Osoyoos Indian Band's Nk'Mip Desert Cultural Centre. Anthropologica 56:135–150

British Columbia Ministry of Forests (1990) A guide to site identification and interpretation for the Kamloops Forest Region Part 1. B.C. Ministry of Forests, Victoria, B.C.

Brook GA, Ford DC (1975) The karstlands of the South Nahanni region, N.W.T. Contract 72-32C, Parks Canada, 501 pp

Cameron M, St. Clair R (2003) COSEWIC status report on the rubber boa *Charina bottae* in Canada. Committee on the Status of Endangered Wildlife in Canada, Ottawa, 14pp

Campbell C, Bakermans L, Jamieson B, Stethem C (2007) Current and future snow avalanche threats and mitigation measures in Canada. Canadian Avalanche Centre. Prepared for Public Safety Canada. http://www.ucalgary.ca/asarc/files/asarc/AvalancheThreatsMitigationCanada.pdf. Accessed 8 Aug 2022

Caplan ML (1998) Sedimentology and sequence stratigraphy of Devonian-Carboniferous strata, southern Alberta. Bull Can Pet Geol 46:487–514

Catto N (1996) Richardson Mountains, Yukon-Northwest territories: the Northern portal of the postulated ice-free corridor. Quatern Int 32:3–20

Catto N (2015) Geomorphology: landscapes we live in. Oxford University Press, Don Mills, Ontario

Chen H (2014) A spatiotemporal pattern analysis of historical mountain pine beetle outbreaks in British Columbia, Canada. Ecography 37:344–356

Church M, Ryder JM (2010) Physiography of British Columbia. In: Pike RG, Redding TE, Moore RD, Winkler RD, Bladon KD (eds) Compendium of forest hydrology and geomorphology in British Columbia, vol 1. Land Management

Clague JJ, Ward B (2011) Pleistocene glaciation of British Columbia. Dev Quatern Sci 15:563–573

Clarke GKC (2014) A short and somewhat personal history of Yukon glacier studies in the twentieth century. Arctic 67:1–21

Clevenger AP, Ford AT, Sawaya MA (2009) Banff wildlife crossings project: integrating science and education in restoring population connectivity across transportation corridors. Final report to Parks Canada Agency, Radium Hot Springs, British Columbia, Canada, 165pp. (Accessed 8 Aug 2022)

Clevenger AP, Sawaya MA, Landguth EL, Dorsey BP (2014) Mitigating multi-species mortality and fragmentation on the Trans-Canada highway through Mount Revelstoke and Glacier National Parks, British Columbia. Final report to Parks Canada Agency, Revelstoke, British Columbia. (Accessed 8 Aug 2022)

CNVC (Canadian National Vegetation Classification) (2016) Canadian national vegetation classification. http://cnvc-cnvc.ca. Accessed 8 Aug 2022

Cody WJ (1996) Flora of the Yukon territory. National Research Council (NRC) Press, Ottawa, Ontario, Canada

Cody WJ, Bennett BA, Caswell P (2005) New records of vascular plants in the Yukon Territory VII. Can Field-Naturalist 119:417–436

Coen GM, Holland WD (1976) Soils of Waterton Lakes National Park, Alberta. Alberta Institute of Pedology, S 73–33

Coen GM, Kuchar P (1982) Biophysical (ecological) land inventory of Yoho NP, BC. Alberta Institute of Pedology M-82-2

Coogan SCP, Daniels LD, Boychuk D, Burton PJ, Flannigan MD, Gauthoer S, Kafka V, Park JS, Wotton BM (2021) Fifty years of wildland fire science in Canada. Can J For Res 51:283–302. https://doi.org/10.1139/cjfr-2020-0314

Coppold M, Powell W (2014) A geoscience guide to the burgess shale. Burgess Shale Geoscience Foundation

Culpterry JM, Prowse D, Luiker EA (2005) Mackenzie river basin. Rivers of North America, pp 804–850

Danby RK (2003) Birds and mammals of the St. Elias mountain parks: checklist evidence for a biogeographic convergence zone. Can Field-Nat 117:1–18

Demarchi DA (2011) An introduction to the ecoregions of British Columbia. Ecosystem Information Section. Ministry of Environment Victoria, British Columbia

Demuth MN, Horne G (2017) Decadal-centenary glacier mass changes and their variability, Jasper National Park of Canada, Alberta, including the Columbia Icefield region. Geological Survey of Canada, Open File 8229, 32 p. https://doi.org/10.4095/304236

Desjardins PR, Buatois LA, Pratt BE, Mágano MG (2010) Stratigraphy and sedimentary environments of the Lower Cambrian Gog Group in the southern rocky mountains of western Canada. Bull Can Pet Geol 58:403–439

Dillhoff RM, Dillhoff TA, Greenwood DR, DeVore ML, Pigg KB (2013) The Eocene Thomas Ranch flora, Allenby Formation, Princeton, British Columbia, Canada. Botany 91:514–529. https://doi.org/10.1139/cjb-2012-0313

Douglas GW, Straley GB, Meidinger D, Pojar J (1998–2002) Illustrated flora of British Columbia. B.C. Ministry of Sustainable Resource Management and B.C. Ministry of Forests, Victoria, B.C., 8 volumes

Durance E (1992) Vanishing desert of the Okanagan. Borealis 3(3):16–21

Dyke AS (2004) An outline of North American deglaciation with emphasis on central and northern Canada. In: Ehlers J, Gibbard PL (eds) Quaternary glaciations—extent and chronology, Part II: North America. Elsevier, Amsterdam, pp 371–424

Dyke AS, Moore A, Robertson L (2003) Deglaciation of Canada. Geological Survey of Canada Open File 1574

Earthquakes Canada (n.d.) Natural resources Canada. http://earthquakescanada.nrcan.gc.ca/. Accessed 2 Aug 2022

Eaton BC, Millar RG, Davidson S (2010) Channel patterns: braided, anabranching, and single-thread. Geomorphology 120:353–364

Eberhart-Phillips D, Haeussler PJ, Freymueller JT, Frankel AD, Rubin CM, Craw P, Ratchkovski NA, Anderson G, Carver GA (2003) The 2002 Denali fault earthquake, Alaska: a large magnitude, slip-partitioned event. Science 300:1113–1118

Edwards ME, Lloyd A, Armbruster WS (2018) Assembly of Alaska-Yukon boreal steppe communities: testing biogeographic hypotheses via modern ecological distributions. J Syst Evol 56:466–475

England JH, Furze MFA, Doupe JP (2009) Revision of the NW laurentide ice sheet: implications for paleoclimate, the northwest extremity of Beringia, and Arctic Ocean sedimentation. Quatern Sci Rev 28:1573–1596

Environment and Climate Change Canada (2017) Recovery strategy for the Pygmy short-horned Lizard (*Phrynosoma douglasii*) in Canada. Species at risk act recovery strategy series. Environment and Climate Change Canada, Ottawa

Environment and Climate Change Canada (2020) Canadian Climate Normals 1981–2010. https://climate.weather.gc.ca/climate_normals/index_e.html. Accessed 8 Aug 2022

Environment Canada and Agriculture and Agrifoods Canada (1995) Terrestrial ecozones and ecoregions of Canada. https://sis.agr.gc.ca/cansis/publications/maps/eco/all/regions/eco_all_regions_7m_canada_en.jpg. Accessed 12 July 2022

Fallas KM, MacNaughton RB (2014) Geology, Carcajou Canyon (northwest), northwest territories. Geological Survey of Canada, Canadian Geoscience Map 94, Scale 1:100,000

Ferbey T, Arnold H, Hickin AS (2013) Ice-flow indicator compilation, British Columbia. Open-File 2013-06, Ministry of Energy, Mines and Natural Gas, British Columbia Geological Survey, Victoria, BC, Scale 1:1 650,000

Filmon G (2004) Firestorm 2003 provincial review. http://www.2003firestorm.gov.bc.ca/firestormreport/FirestormReport.pdf. Accessed 15 July 2022

Finzel ES, Trop JM, Ridgway KD, Enkelmann E (2011) Upper plate proxies for flat-slab subduction processes in southern Alaska. Earth Planet Sci Lett 303:348–360. https://doi.org/10.1016/j.epsl.2011.01.014

Flowers GE, Copland L, Schoof CG (2014) Contemporary glacier processes and global change: recent observations from Kaskawulsh Glacier and the Donjek Range, St. Elias Mountains. Arctic 67:22–34. https://doi.org/10.14430/arctic4356

Ford DC (2017) The many landscapes of South Nahanni National Park. In: Slaymaker O (ed) Landscapes and landforms of Western Canada. Springer International Publishers, Cham, Switzerland, pp 205–226

Ford DC, Hamilton JP (2017) Cold climate, Canyons and Karst along the Canol Road Corridor, Mackenzie Mountains, NWT. In: Slaymaker O (ed) Landscapes and landforms of Western Canada. Springer International Publishers, Cham, Switzerland, pp 187–204

Ford DC, Smart CC (2017) Castleguard Cave and Karst, Columbia Icefield, Alberta and British Columbia. In: Slaymaker O (ed) Landscapes and landforms of Western Canada. Springer International Publishers, Cham, Switzerland, pp 227–240

French HM (2017) The Northern Interior Yukon: an example of periglaciation. In: Slaymaker O (ed) Landscapes and landforms of Western Canada. Springer International Publishers, Cham, Switzerland, pp 257–266

Froese DG, Zazula GD, Westgate JA, Preece SJ, Sanborn PT, Reyes AV, Pearce NJG (2009) The Klondike goldfields and Pleistocene environments of Beringia. GSA Today 19(8) 4–10

Fulton RJ (1995) Surficial materials of Canada. Geological Survey of Canada Map 1880A

Gabrielse H, Yorath CJ (1991) Geology of the Cordilleran orogen in Canada. Geological Survey of Canada, Geology of Canada Series 4. https://doi.org/10.4095/134069

Gayton DV (2003) Forest research extension partnership British Columbia grasslands: monitoring vegetation change. http://www.forrex.org/publications/FORREXSeries/FS7_Part1.pdf. Accessed 6 Aug 2022

Gibb S, Pemberton SG, Chatterton BDE (2017) Arthropod trace fossils of the upper lower Cambrian Gog Group, southern Rocky Mountains of Canada. Ichnos 24:91–123

Gimbarzevsky P, Peaker JP, Addison P, Talbot S (1979) Nahanni National Park, north west territories; integrated survey of biophysical resources. Parks Canada

Godfrey WE (1986) The birds of Canada. National Museum of Natural Sciences, National Museums of Canada, 595 p

Goes S (2013) Western North America's jigsaw. Nature 496(7443):25–27

Grasby SE, Richards BC, Sharp CE, Brady AL, Jones GM, Dunfield PF (2013) The paint pots, Kootenay National Park, Canada—a natural acid spring analogue for Mars. Can J Earth Sci 50:94–108

Greenaway JE (2009) March 4, 1910 Rogers Pass Avalanche. http://jccabulletin-geppo.ca/march-4-1910-rogers-pass-avalanche/. Accessed 8 Aug 2022

Greenough JD, Roed MA (2004) Okanagan geology. Kelowna Geology Committee. Sandhill Book Marketing

Gulley AL (1993) Rabbitkettle hotsprings, Nahanni National Park Reserve: a Hydrological Study. MSc Earth Sciences Carleton. https://curve.carleton.ca/2b2c47c3-f9fa-4395-9531-8eed5d6a8a79. Accessed 4 Aug 2022

Haber E, Soper J (1980) Vascular plants of Glacier NP BC. National Museum of Natural Sciences, National Museums of Canada. Syllogeus #24

Haeberli W, Whiteman CA, Shroder JF (2014) Snow and ice-related hazards, risks, and disasters. Academic Press, Waltham, MA

Haegeli P, Falk M, Brugger H, Etter H-J, Boyd J (2011) Comparison of avalanche survival patterns in Canada and Switzerland. Can Med Assoc J 2011:789–795. https://doi.org/10.1503/cmaj.101435

Haegeli P (2019) Snow avalanches. https://blog.alpineclubofcanada.ca/state-of-the-mountains/2019/1/14/snow-avalanches. Accessed 8 Aug 2022

Hägeli P, McClung DM (2003) Avalanche characteristics of a transitional snow climate—Columbia Mountains, British Columbia, Canada. Cold Reg Sci Technol 37:255–276

Haggart J, Enkin RJ, Monger JWH (2006) Paleogeography of the North American cordillera. Geological Association of Canada, Special Publication 46

Hamilton NP, Sanborn PT, Burton PJ (2022) Effects of grazing on plant communities and successional processes vary along an aridity gradient at a northern temperate grassland. Plant Ecol 223:151–170

Harington CR (2011) Pleistocene vertebrates of the Yukon Territory. Quatern Sci Rev 30:2341–2354

Hart CJR, Villeneuve M (1999) Geochronology of Neogene alkaline volcanic rocks (Miles Canyon basalt), southern Yukon Territory, Canada: the relative effectiveness of laser 40Ar/39Ar and K-Ar geochronology. Can J Earth Sci 36:1495–1507

Herrero S, Roulet J, Gibeau M (2021) Banff National Park: science and policy in Grizzly Bear Management. Ursus 12:161–168

Hewer MJ Gough WA (2021) Climate change impact assessment on grape growth and wine production in the Okanagan Valley (Canada). Clim Risk Manag 33:100343. https://doi.org/10.1016/j.crm.2021.100343

Hickson C, Hollinger J (2014) Wells Gray rocks. Thompson Rivers University, Kamloops, BC

Hickson CJ, Vigouroux N (2014) Volcanism and glacial interaction in the Wells Gray-Clearwater volcanic field, east-central British Columbia. GSA Field Guides 38:169–191. https://doi.org/10.1130/2014.0038(08)

Hickson CJ, Moore JG, Calk L, Metcalfe P (1995) Intraglacial volcanism in the Wells Gray-clearwater volcanic field, east-central British Columbia, Canada. Can J Earth Sci 32:838–851

Hickson CJ, Pollack J, Struik L, Hollis L, Yonge C (2020) Discovery of a significant cave entrance in stripe karst, Horsethief Creek Group, Wells Gray Provincial Park, British Columbia, Canada. Can J Earth Sci 57:662–670

Hoffecker JF, Elias SA, Potapova O (2020) Arctic Beringia and Native American origins PaleoAmerica 6:158–169

Holland WD, Coen GM (1982) Ecological (biophysical) land classification of Banff and Jasper National Parks Vol. II: soil and vegetation resources. Alberta Institute of Pedology Publication No. SS-82-44

Holland WD, Coen GM (1983) Ecological (biophysical) land classification of Banff and Jasper National Parks volume I: summary. Alberta Institute of Pedology Publication No M-83

Horton S, Jamieson JB (2016) Modelling hazardous surface hoar layers across western Canada with a coupled weather and snow cover model. Cold Reg Sci Technol 128:22–31

Hosie RC (1979) Native trees of Canada. Canada Department of Forestry, Don Mills, Ontario

Huntley DH, Hickin AS, Lian OB (2017) The pattern and style of deglaciation at the Late Wisconsinan Laurentide and Cordilleran ice sheet limits in northeastern British Columbia. Can J Earth Sci 54:52–75

Israel S, Cobbett R (2008) Kluane Ranges bedrock geology, White River area (Parts of NTS 115F/9, 15 and 16; 115G/12 and 115K/1, 2). In: Emond DS, Blackburn LR, Hill RP, Weston LH (eds) Yukon exploration and geology 2007, Yukon Geological Survey, pp 153–167

Jackson LE, Nelson FE, Huscroft CA, Villeneuve M, Barendregt RW, Storer JE, Ward BC (2012) Pliocene and Pleistocene volcanic interaction with Cordilleran icesheets, damming of the Yukon River and vertebrate paleontology, Fort Selkirk volcanic group, west-central Yukon. Quatern Int 260:3–30

Jamieson B, Haegeli P Gauthier D (2010) Avalanche accidents in Canada volume 5 1996–2007. CAA. www.avalancheassociation.ca/resource/resmgr/files/avalancheaccidentsv5.pdf. Accessed 7 Aug 2020

Klinkenberg B (2020) E-Flora BC: electronic atlas of the plants of British Columbia. Department of Geography, University of British Columbia. Accessed 8 Aug 2022

Kuijt J (1982) A flora of Waterton Lakes NP. Hignell Printing, Winnipeg

Leckie D (2017) Rocks, ridges, and rivers: geological wonders of Banff, Yoho, and Jasper National Parks. Steeling Home Publications, Calgary

Lee RN, Bradfield GE, Krzic M, Newman RF, Cumming WFP (2014) Plant community–soil relationships in a topographically diverse grassland in southern interior British Columbia, Canada. Botany 92:837–845

Leslie-Panek J, McMechan M (2021) A seismic structural overview of Liard Basin, Northeast British Columbia, Canada. Interpret J Subsurf Char 9:T523–T532. https://doi.org/10.1190/INT-2020-0078

Levson VM, Rutter NW (1989) Late Quaternary stratigraphy, sedimentology, and history of the Jasper townsite area, Alberta, Canada. Can J Earth Sci 26:1325–1342

Longworth HA (2009) Tracks, tunnels and trestles: an environmental history of the construction of the Canadian Pacific Railway. MA thesis, Department of History, University of Victoria

Luckman BH, Sperling BJR, Osborn GD (2020) The Holocene history of the Columbia icefield. Can Quatern Sci Rev 242:106436. https://doi.org/10.1016/j.quascirev.2020.106436

Luckman BH (2017) Glacier landscapes in the Canadian Rockies. In: Slaymaker O (ed) Landscapes and landforms of Western Canada. Springer International Publishers, Cham, Switzerland, pp 241–246

MacDonald GM, Beilman DW, Kuzmin YV, Orlova LA, Kremenetski KV, Shapiro B, Wayne RK, Van Valkenburgh B (2012) Pattern of extinction of the woolly mammoth in Beringia. Nat Commun 3. https://doi.org/10.1038/ncomms1881

Makaske B, Smith DG, Berendsen HJA, deBoer AG, van Nielen-Kiezebrink MF, Locking T (2009) Hydraulic and sedimentary processes causing anastomosing morphology of the upper Columbia River, British Columbia, Canada. Geomorphology 111:194–205

Massey NWD, MacIntyre DG, Desjardins PJ, Cooney RT (2005) Geology of British Columbia. BC Geological Survey, Map 2005-5

Mathews WH (1989) Development of Cordilleran landscapes during the Quaternary. In: Fulton RJ (ed) Quaternary geology of Canada and Greenland. Geological Survey of Canada, Ottawa, ON, pp 32–34

McClung DM (2013) The effects of El Nino and La Nina on snow and avalanche patterns in British Columbia, Canada, and central Chile. J Glaciol 59:783–792

McFarlane BL, Watson DOT, McGee T (2008) Vegetation management in Banff NP: a survey of local residents. Report for Banff NP. http://research.eas.uaberta.ca/hdhresearch. Accessed 7 Aug 2022

McGee T, McFarlane B, Tymstra C (2015) Wildfire: a Canadian perspective, in wildfire hazards, risks, and disasters. https://doi.org/10.1016/B978-0-12-410434-1.00003-8

McIntosh SE, Grissom CK, Olivares CR, Kim HS, Tremper B (2007) Cause of death in avalanche fatalities. Wildern Environ Med 18:293–297

Montesano PM, Neigh CSR, Macander M, Feng M, Noojipady P (2020) The bioclimatic extent and pattern of the cold edge of the boreal forest: the circumpolar taiga-tundra ecotone. Environ Res Lett 15:105019. https://doi.org/10.1088/1748-9326/abb2c7

Morris M (2002) Nakimu caves. National Park feature articles. Columbia Mountains Institute of Applied Ecology. https://cmiae.org/national-park-feature-article/nakimu-caves/. Accessed 8 Aug 2022

Morris SC, Whittington HB (1985) Fossils of the Burgess Shale, a national treasure in Yoho National Park, British Columbia Miscellaneous Report 43

Mossop GD, Shetsen I (1994) Geological atlas of the Western Canada Sedimentary Basin; Canadian Society of Petroleum Geologists and Alberta Research Council. https://ags.aer.ca/reports/atlas-western-canada-sedimentary-basin. Accessed 8 Aug 2022

Muller JE, Christie RL (1966) Geology Kluane Lake, Yukon Territory. Geological Survey of Canada, "A" Series Map 1177A

Naughton D (2012) The natural history of Canadian mammals. Canadian Museum of Nature

Neave R (2015) Exploring wells Gray Park, 6th edn. Wells Gray Tours, Kamloops, BC

Nelson J, Colpron M (2007) Tectonics and metallogeny of the British Columbia, Yukon and Alaskan Cordillera, 1.8 Ga to the present. In: Goodfellow WD (ed) Mineral deposits of Canada: a synthesis of major deposit-types, district metallogeny, the evolution of geological provinces, and exploration methods: geological association of Canada, Mineral Deposits Division, Special Publication No. 5, pp 755–791

Nichol C, Monahan P, Fulton R, Ping J, Wei X, Thomson S (2015) Quaternary stratigraphy and evidence for multiple glacial episodes in the north Okanagan Valley, British Columbia. Can J Earth Sci 52:338–356

Norris DK (1986) Geology of the northern Yukon and northwestern District of Mackenzie. Geological Survey of Canada, Map, 1581 A

North Yukon Planning Commission (2006) Simplified geology and ecodistricts, North Yukon planning region. www.nypc.planyukon.ca. Regional ecosystems and ecoregions. Accessed 6 Aug 2022

Okulitch AV (2013) Geology, Okanagan Watershed, British Columbia. Geological Survey of Canada, Open File 6839. https://doi org/10.4095/292220

Page CE, Atkins D, Shockley LW, Yaron M (1999) Avalanche deaths in the United States: a 45-year analysis. Wildern Environ Med 10:146–151

Pana D, Elgr R (2013) Geology of the Alberta rocky mountains and foothills. Alberta Geological Survey Map 560

Parish, R. Coupé R, Lloyd D (1996) Plants of the southern interior of British Columbia. BC Ministry of Forests and Lone Pine Publishing, Edmonton, Alta

Parks Canada Agency (2017a) Multi-species action plan for Jasper National Park of Canada. Species at risk act action plan series. Parks Canada Agency, Ottawa, iv + 21 pp

Parks Canada Agency (2017b) Multi-species action plan for Yoho National Park of Canada. Species at risk act action plan series. Parks Canada Agency, Ottawa, iv + 18 pp

Parks Canada Agency (2017c) Multi-species action plan for Kootenay National Park of Canada. Species at risk act action plan series. Parks Canada Agency, Ottawa, iv + 21 pp

Parks Canada Agency (2017d) Multi-species action plan, Mount Revelstoke National Park of Canada and Glacier National Park of Canada. Species at risk act action plan series. Parks Canada Agency, Ottawa. iv + 19 pp

Parks Canada (1997) Banff National Park management plan. Ministry of Canadian Heritage, Ottawa, Ontario, Canada

Parks Canada (2010) Management plan 2010—Mount Revelstoke and Glacier National Parks and Rogers Pass national historic site of Canada. Parks Canada, Ottawa, Ontario

Parks Canada (2017a) Multi-species action plan for Banff National Park of Canada. Species at risk act action plan series. Parks Canada Agency, Ottawa. iv + 27 pp

Parks Canada (2017b) Multi-species action plan for Waterton Lakes National Park of Canada and Bar U Ranch National Historic site of Canada. Species at risk act action plan series. Parks Canada Agency, Ottawa, iv + 31 pp

Parks Canada (2022) Bats in the mountain national parks. https://www.pc.gc.ca/en/pn-np/mtn/chauvessouris-bat. Accessed 8 Aug 2022

Phillips D (1990) The climates of Canada. Environment Canada, 176 p

Phillips D (2016) Canada's coldest day. https://www.canadashistory.ca/explore/environment/canada-s-coldest-day. Accessed 5 Aug 2022

Pinet N (2021) Structural geology of the eastern Richardson Mountains, Yukon and Northwest territories: some field observations and a note of caution for palinspastic reconstructions. In: MacFarlane KE (ed) Yukon exploration and geology 2020. Yukon Geological Survey, pp 1–18

Plouffe A (2000) Quaternary geology of the Fort Fraser and Manson River map areas, central British Columbia. Geol Surv Canada Bull 544. Ottawa

Porsild AE, Cody WJ (1980) Vascular plant flora of continental NWT, Canada. National Museum of Natural Sciences, National Museums of Canada

Price RA (1994) cordilleran tectonics and the evolution of the Western Canada sedimentary basin in geological atlas of the Western Canada sedimentary basin. In: Mossop GD, Shetsen I (comp) Canadian Society of Petroleum Geologists and Alberta Research Council. https://ags.aer.ca/atlas-the-western-canada-sedimentary-basin/chapter-2-cordilleran-tectonics. Accessed 8 Aug 2022

Prichard SJ et al (2021) Adapting western North American forests to climate change and wildfires: 10 common questions. Ecol Appl 31(8):e02433. https://doi.org/10.1002/eap.2433

Prior GJ, Hathway B, Glombick PM, Pana DI, Banks CJ, Hay DC, Schneider CL, Grobe M, Elgr R, Weiss JA (2013) Bedrock geology of Alberta. Alberta Geological Survey Map 600

Public Safety Canada (2010) Rocky mountain spotted fever. https://www.canada.ca/en/public-health/services/laboratory-biosafety-biosecurity/pathogen-safety-data-sheets-risk-assessment/rickettsia-rickettsii.html. Accessed 8 Aug 2022

Pyne SJ (2007) Awful splendour: a fire history of Canada. UBC Press

Quamme HA, Cannon AJ, Neilsen D, Caprio JM, Taylor WG (2010) The potential impact of climate change on the occurrence of winter freeze events in six fruit crops grown in the Okanagan Valley. Can J Plant Sci 90:85–93

Raine RM, Kanasa JL (1990) Black bear seasonal food habits and distribution by elevation in Banff National Park, Alberta: Bears: their biology and management. In: Eighth international conference on bear research and management, Victoria, BC, pp 297–304

Rampton VN (1980a) Surficial materials, Kluane National Park, Yukon Territory. Geological Survey of Canada, Preliminary Map 13-1979

Rampton VN (1980b) Quaternary landforms, Kluane National Park, Yukon Territory. Geological Survey of Canada, Preliminary Map 14-1979

Rayne S, Forest K (2016) Rapidly changing climatic conditions for wine grape growing in the Okanagan Valley region of British Columbia, Canada. Sci Total Environ 556:169–178

Redpath K (1990) Identification of relatively undisturbed areas in the South Okanagan and Similkameen valleys, British Columbia. Canadian Wildlife Service, Pacific and Yukon Region, Delta, BC. Technical Report 108

Renaut RW, Long PR (1989) Sedimentology of the saline lakes of the Cariboo Plateau, Interior British Columbia, Canada. In: Renaut RW (ed) Sedimentology and diagenesis of evaporites. Sediment Geol 64:239–264

Resler LM, Shao Y, Tomback DF, Malanson GP (2014) Predicting functional role and occurrence of Whitebark Pine (*Pinus albicaulis*) at alpine treelines: model accuracy and variable importance. Ann Assoc Am Geogr 104:703–722

Rhemtulla JM, Hall RJ, Higgs ES, Macdonald SE (2002) Eighty years of change: vegetation in the montane ecoregion of Jasper National Park, Alberta, Canada. Can J For Res 32:2010–2021

Rodengen TJ, Pellatt MG, Kohfeld KE (2022) Paleoecological investigation of vegetation, climate and fire history in, and adjacent to, Kootenay National Park, Southeastern British Columbia, Canada. Front Ecol Evol 9:768785. https://doi.org/10.3389/fevo.2021.768785

Rowe JS (1972) Forest regions of Canada. Canadian Forest Service, Department of the Environment Publication 1300. Ottawa

Roy-Léveillée P, Burn CR (2017) Old crow flats: thermokarst lakes in the forest–Tundra transition. In: Slaymaker O (ed) Landscapes and landforms of Western Canada. Springer International Publishers, Cham, Switzerland, pp 267–276

Rutter NW (1972) Geomorphology and multiple glaciation in the area of Banff, Alberta. Geological Survey of Canada Bulletin 206

Schirrmeister L, Froese D, Tumskoy V, Grosse G, Wetterich S (2013) Yedoma: Late Pleistocene ice-rich syngenetic permafrost of Beringia. In: Encyclopedia of quaternary science, pp 542–552

Scott WB, Crossman EJ (1998) Freshwater fishes of Canada. Fisheries Research Board of Canada. Bulletin 184. Galt House, Oakville

Scudder GGE (2000) The Osoyoos Desert society: experimental studies on ecological restoration of the shrub-steppe habitat. In: Scudder GGE, Darling LM (ed) Proceedings of a conference on the biology and management of species and habitats at risk, Kamloops, B.C., 15–19 Feb., 1999, vol 2. B.C. Ministry of Environment, Lands and Parks, Victoria, B.C. and University College of the Cariboo, Kamloops, B.C., pp 797–800. Accessed 8 Aug 2022

Senese DM, Wilson W, Momer B (2012) The Okanagan wine region of British Columbia, Canada. In: Dougherty PH (ed) The geography of wine: regions, terroir and techniques. Springer, pp 81–94

Shugar DH, Clague JJ, Best JL, Schoof C, Willis MJ, Copland L, Roe GH (2017) River piracy and drainage basin reorganization led by climate-driven glacier retreat. Nat Geosci 10:370–375. https://doi.org/10.1038/ngeo2932

Slaymaker O (2017) Landscapes and landforms of Western Canada. Springer International Publishers, Cham, Switzerland

Slaymaker O, Catto NR, Kovanen D (2017) Geomorphological landscapes and Geoparks. In: Slaymaker O (ed) Landscapes and landforms of Western Canada, chap 29. Springer

Smith CAS, Meikle JC, Roots CF (2004) Ecoregions of the Yukon territory: biophysical properties of Yukon landscapes. Agriculture and Agri-Food Canada, PARC Technical Bulletin No. 04-01, Summerland, British Columbia, 313 p

Smith MD, Arnaud E, Arnott RWC, Ross GM (2011) The record of neoproterozoic glaciations in the Windermere Supergroup, southern Canadian Cordillera. Geol Soc Lond Memoirs 36:413–424. https://doi.org/10.1144/M36.37

Soil Classification Working Group (1998) The Canadian system of soil classification, 3rd edn. Agriculture and Agri-Food Canada, Publication 1646

Soil Classification Working Group (2001) Soils of Canada. Agriculture and agri-foods Canada, research branch. Scale 1:6,500,000 map. https://sis.agr.gc.ca/cansis/publications/maps/soc/all/soils/soc_all_soils_2004en.png. Accessed 12 July 2022

Stephenson SA, Burrows JA, Babaluk JA (2005) Long-distance migrations by Inconnu (*Stenodus leucichthys*) in the Mackenzie River System. Arctic 58:21–25

Stewart H, Hebda RJ (2000) Grasses of the Columbia basin of British Columbia. B.C. Ministry of Forests, Victoria, B.C.

Stroeven AP, Fabel D, Margold M, Clague JJ, Xu S (2014) Investigating absolute chronologies of glacial advances in the NW sector of the Cordilleran Ice Sheet with terrestrial in situ cosmogenic nuclides. Quatern Sci Rev 92:429–443

Stumpf AJ, Ferbey T, Plouffe A, Clague JJ, Ward BC, Paulen RC, Bush ABG (2013) Discussion: "Streamlined erosional residuals and drumlins in central British Columbia, Canada" by J. Donald McClenagan, (2013). Geomorphology 189:41–54

Stumpf AJ, Broster BE, Levson VM (2004) Glacial stratigraphy of the Bulkley River region: a depositional framework for the late Pleistocene in central British Columbia. Géographie Physique et Quaternaire 58:217–228

Tombstone Territorial Park Management Plan (2009) https://yukon.ca/sites/yukon.ca/files/env/env-tombstone-territorial-park-management-plan.pdf. Accessed 7 Aug 2022

Turner DG, Ward BC, Froese DG, Lamothe M, Bond JD, Bigelow NH (2016) Stratigraphy of Pleistocene glaciations in the St Elias Mountains, southwest Yukon, Canada. Boreas 45:521–536

Van der Velden AJ, Cook FA (1994) Displacement of the Lewis thrust sheet in southwestern Canada: new evidence from seismic reflection data. Geology 22:819–822

Van der Velden AJ, Cook FA (1996) Structure and tectonic development of the southern Rocky Mountain Trench. Tectonics 15:517–544

Wahl HE, Fraser DB, Harvey RC, Maxwell JB (1987) Climate of Yukon environment Canada. Atmospheric Environment Service. Climatol Stud 40. En57-7-40-eng.pdf

Warren F, Lulham N (2021) Canada in a changing climate: national issues report. Government of Canada, Ottawa, ON

Webster RA, Archibald DA, Pattison DRM, Pickett JA, Jansen JC (2020) Tectonic domains and exhumation history of the Omineca Belt in southeastern British Columbia from $^{40}Ar/^{39}Ar$ thermochronology. Can J Earth Sci 57. https://doi.org/10.1139/cjes-2019-0131

Welwood RJ (2003) Baillie-Grohman's diversion. British Columbia Historical News, p 2, Fall 2003. https://open.library.ubc.ca/viewer/bch/1.0190653. Accessed 7 Aug 2022

Wheeler JO, McFeely P (1991) Tectonic assemblage map of the Canadian Cordillera and adjacent parts of the United States of America. Geological Survey of Canada, Map 1712A

Wheeler JO, Hoffman PF, Card KD, Davidson A, Sanford BV, Okulitch AV, Roest WR (1996) Geological map of Canada. Geological Survey of Canada Map D1860A

Williams JW (2022) Bottom-up versus top-down megafauna–vegetation interactions in ancient Beringia. Proc Natl Acad Sci 119(5):2121734119. https://doi.org/10.1073/pnas.2121734119

Winkler GR (2000) A geologic guide to Wrangell-Saint Elias National Park and Preserve, Alaska. United States Geological Survey Professional Paper 1616

Wright DF, Lemkow D, Harris JR (2007) Mineral and energy resource assessment of the Greater Nahanni Ecosystem under consideration for the expansion of the Nahanni National Park Reserve, Northwest Territories. Geological Survey of Canada, Open File 5344, pp 327–365. https://doi.org/10.4095/224562

Yukon Geological Survey (2021a) Yukon digital bedrock geology. Yukon Geological Survey. http://data.geology.gov.yk.ca/Compilation/3#InfoTab. Accessed 6 Aug 2022

Yukon Geological Survey (2021b) A digital atlas of terranes for the northern Cordillera. Yukon Geological Survey. https://data.geology.gov.yk.ca/Compilation/2. Accessed 6 Aug 2022

Zazula GD, Froese DG, Schweger CE, Mathewes RW, Beaudoin AB, Telka AM, Harington CR, Westgate JA (2003) Ice-age steppe vegetation in east Beringia. Nature 423:603

Pacific Maritime Ecozone

7

Abstract

The Pacific Maritime Ecozone encompasses the west coast of Canada. The "Salish Sea" encompasses all the waters lapping eastern and southern Vancouver Island and the Lower Mainland of British Columbia. Pacific Maritime Ecozone climates are dominated by proximity to the Pacific Ocean, including the relatively warm northbound currents, and the prevailing westerly winds' interaction with coastal topography. Westerly wind influences continue eastward to the summits of the Coast Range, and through the valleys of the Skeena and Stikine Rivers. Geologically, the region lies within the Insular Structural Belt (Vancouver Island and Haida Gwaii, along with narrow coastal plains) and the Coast Structural Belt (Cascade Mountains; Coast Ranges, including the Pacific, Kitimat, and Boundary Ranges; and Stikine Plateau, with numerous subdivisions). Earthquake activity characterizes the offshore Pacific continental shelf and adjacent islands. Recently and currently active volcanoes occur in the Cascade Mountains and Stikine Plateau. Quaternary glaciation involved ice from two sources. Cordilleran ice expanded from interior British Columbia to reach the Pacific Coast along the major valleys, deepening fjords, and flowed westward onto Vancouver Island and Haida Gwaii. Although glacial ice reached the continental shelf in some locations (Shaw et al. 2020), recent evidence suggests that the northwestern tip of Vancouver Island may not have been ice-covered during the last glaciation, providing a biological refugium 20,000 years ago. Retreat began about 18,000 years ago, with rapid deglaciation once climate warming began. The defining trees of the Coastal Southern Pacific Cordilleran ecoclimatic region are large, warmth-loving conifers. Along the west coast of Vancouver Island, the perhumid moisture regime allows the development of a dense coniferous forest. The Northern Pacific Coast Forest is dominated by Tsuga heterophylla, Abies amabilis, Thuja plicata, Picea sitchensis, and Callitropsis nootkatensis. Further inland in northern British Columbia, the eastern slopes, and areas at higher elevations, support a vegetation assemblage adapted to colder and drier conditions. Faunal assemblages in interior sites include species from both the Pacific Maritime and the Interior Cordilleran Ecoclimatic regions. Terrestrial faunal assemblages of the islands are very limited. Species of Pacific salmon include Oncorhynchus gorbuscha, O. keta, O. kisutch, O. nerka, and O. tshawytscha. The dominant ocean surface current is the Alaska Current, which flows northward from Oregon to Alaska, bringing relatively warm water along open Pacific shorelines. Approximately 80% of British Columbia's human population lives in the coastal zone of the Pacific Maritime Ecozone.

Keywords

Pacific Maritime · Salish Sea · Insular Structural Belt · Coast Structural Belt · Sea level variations · Conifer-dominated forests

7.1 Introduction

The Pacific Maritime Ecozone encompasses the west coast of Canada, including coastal British Columbia, the Cascade Ranges, Coast Ranges, Skeena, and Stikine regions, offshore islands including Vancouver Island and Haida Gwaii, and the adjacent waters overlying the Pacific Continental Shelf and Slope (Fig. 7.1). The St. Elias Mountains of SW Yukon, which are ecologically and climatically taiga and high alpine, are discussed in the Cordillera Chap. 6.

The "Salish Sea" encompasses all the waters lapping eastern and southern Vancouver Island and the Lower Mainland of British Columbia (Freelan 2009; Tucker and Rose-Redwood 2015; Fig. 7.2). Salish Sea is replacing terms such as "Strait of Georgia" and "Gulf of Georgia" (named for George

N. Catto and G. R. Vincent, *Canada: Land and People*, World Regional Geography Book Series,
https://doi.org/10.1007/978-3-032-04391-7_7

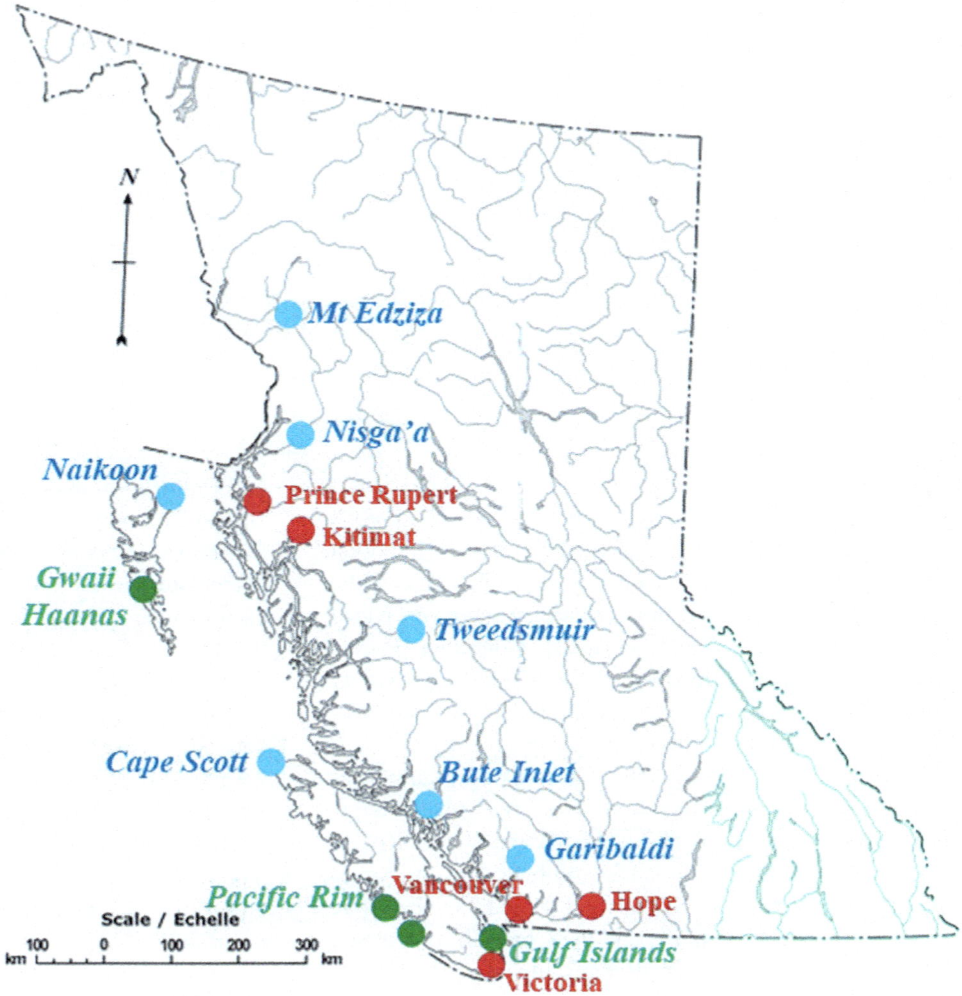

Fig. 7.1 Location of sites discussed in text. Green dots: National Parks. Blue dots: other conserved areas and sites discussed in text. Red dots: communities

III of Britain) and "Juan de Fuca Strait" (named for a Greek navigator who never visited the area). The Gulf [of Georgia] Islands lie between the British Columbia Lower Mainland, including Vancouver, and Vancouver Island. Considering the waters as a single unified Salish Sea makes more sense, both culturally and oceanographically (Jones et al. 2021; Trimbach et al. 2021).

7.2 Climate

Pacific Maritime Ecozone climates are dominated by proximity to the Pacific Ocean, including the relatively warm northbound currents, and the prevailing westerly winds' interaction with coastal topography (Fig. 7.2). Westerly wind influences continue eastward to the summits of the Coast Range, and through the valleys of the Skeena and Stikine Rivers (Fig. 7.3).

7.2.1 Open Pacific Coast

Western Vancouver Island, including Pacific Rim National Park reserve, is influenced by a Maritime West Coast temperate climate (modified Köppen-Geiger Cfb). Daily mean temperatures at Tofino (49.1°N 125.9°W) in January and July are 3 °C and 14 °C, respectively (Environment and Climate Change Canada 2020; Fig. 7.4). The very low range in temperature, characteristic of a maritime west coast temperate climate, indicates the limited seasonal variability throughout the region.

Slightly cooler conditions prevail along the northwest coast (Köppen-Geiger Cfb, transitional to Cfc near Prince Rupert, 54.3°N, 130.3°W; Fig. 7.5). January daily mean temperatures are 2 °C at Prince Rupert, and July daily mean values approximate 12 °C. Comparable values for Gwaii Haanas National Park are 4 °C and 11 °C. The temperature range is the lowest in Canada, and indicates

Fig. 7.2 The Salish Sea encompasses all coastal marine waters extending from Vancouver Island west to the open Pacific Ocean. Left: the sea includes more open water in the northeast, fed by fjords such as Bute Inlet; the Fraser Delta, source of substantial sediment input; the Gulf Islands, including Gulf Islands National Park Reserve; Puget Sound (USA); and the open strait west of Victoria separating Vancouver Island from Washington (USA). (NASA). Right: View from space looking west across the Salish Sea (International Space Station, NASA)

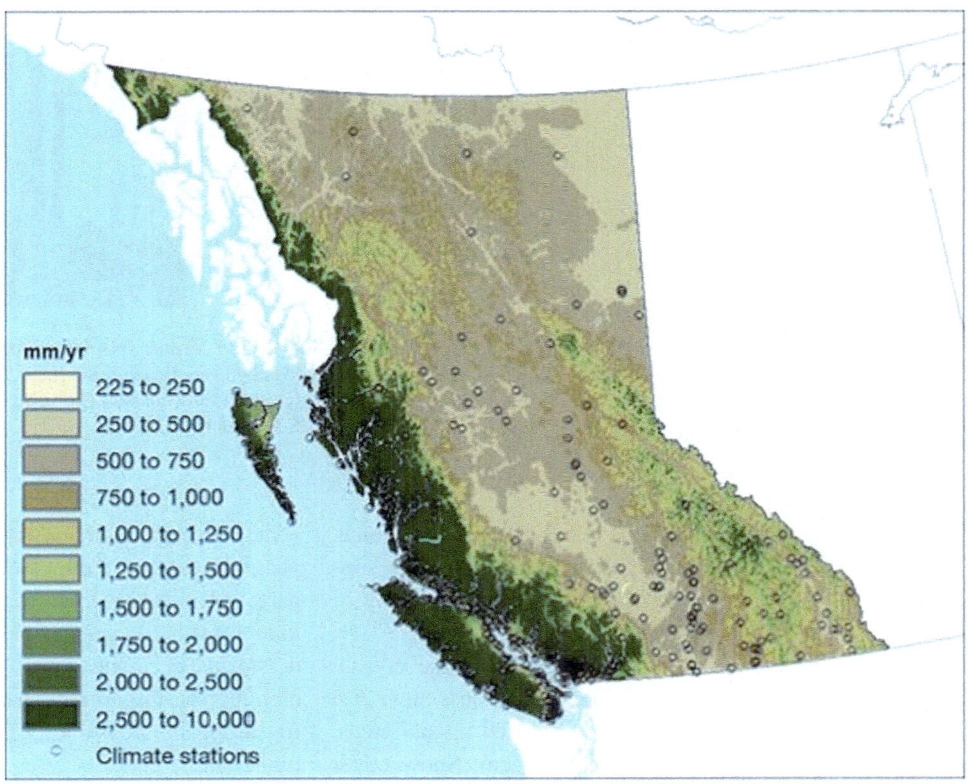

Fig. 7.3 Precipitation in British Columbia (Environment and Climate Change Canada 2020)

the effect of climatic moderation by the waters of Hecate Strait on Prince Rupert, and by the open Pacific Ocean on Gwaii Haanas. The mild climate allows some vegetation species to flourish, but the low summer temperatures and lack of sunlight hinder the establishment of many deciduous species.

Winds blowing from the Pacific Ocean to the land bring moisture-laden air to the west flanks of the ranges, creating large amounts of orographic precipitation on the windward side. Some west-facing slopes on Haida Gwaii and Vancouver Island receive more than 4000 mm of precipitation per year, comparable to totals for monsoonal areas and considerably

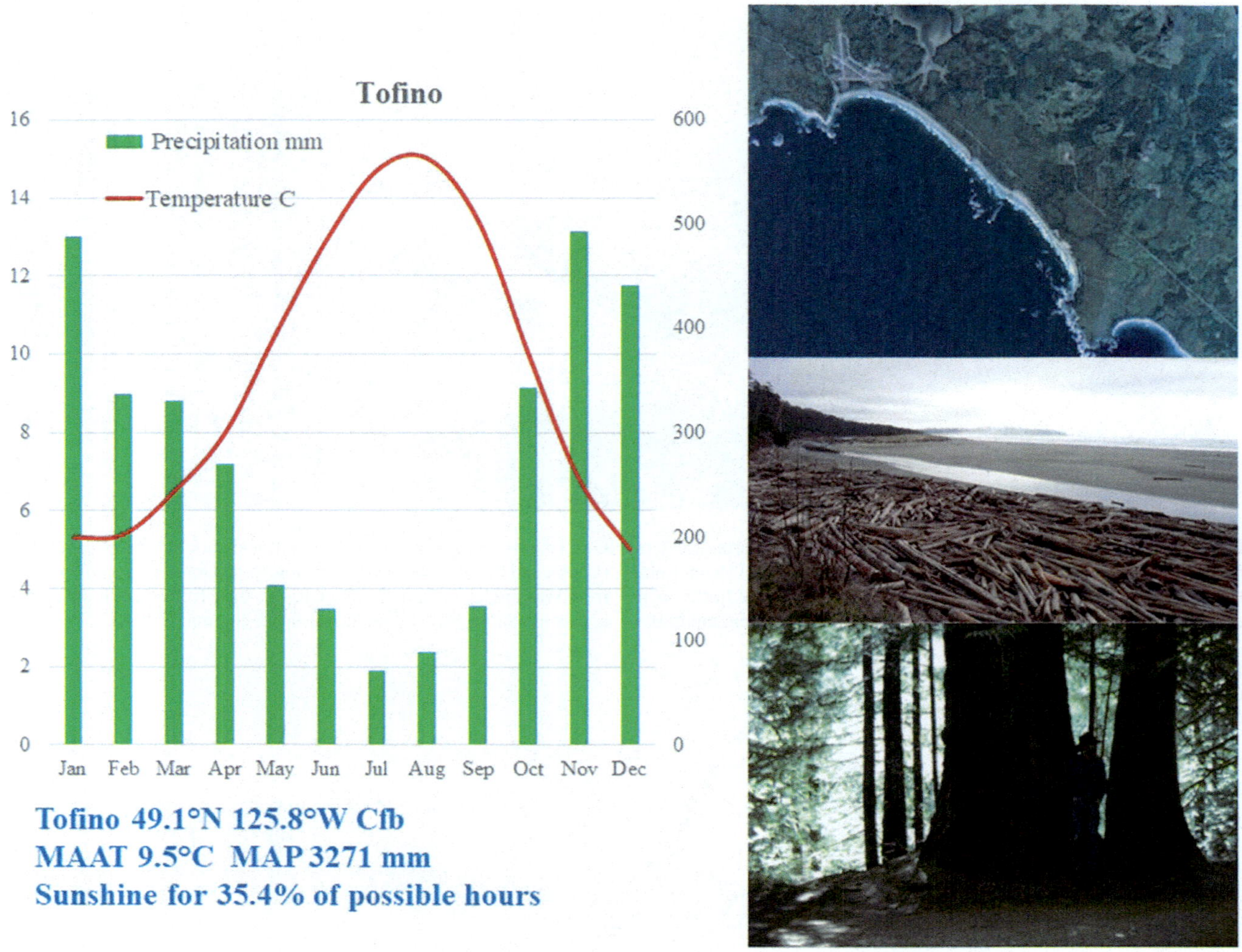

Fig. 7.4 Tofino is located adjacent to Pacific Rim National Park and has a Köppen-Geiger Cfb climate with abundant rainfall. The driest period in summer corresponds with maximum tourist visitation. (*Data* Canadian Climate Normals, Environment and Climate Change Canada 2020). Upper right: The Long Beach segment of Pacific Rim is located southeast of Tofino Airport. (NASA). Centre right: Accumulation of Douglas Fir logs on Long Beach following a storm. Lower right: large *Tsuga plicata* (western red cedar) at Cathedral Grove in the interior benefit from the Cfb climate

more than is received in rain forests in Brazil and New Guinea. Prince Rupert, which receives more than 2600 mm of precipitation in an average year, advertises itself as "The City of Rainbows".

Snowfall reaches 120–140 cm per year on the coastal plains and the lower flanks of the ranges, less than 5% of total precipitation. Within densely forested upland areas, however, the snowpack may exceed 100 cm. Snow cover persists for 50–90 days in the most densely forested areas, but exposed regions seldom retain snow for more than a few days each winter. Snow is more common on east-facing than on west-facing slopes.

The moisture regime is extremely perhumid, and the soils remain saturated in many locations throughout the year. Prince Rupert averages less than 1250 h of sunshine per year, the lowest total of any major town in Canada, and parts of Haida Gwaii (particularly Gwaii Haanas) receive less than 1100 h per year. Violent weather is rare, with 4 thunderstorms occurring in an average year. Instead of violent cloudbursts, the area is characterized by steady drizzle and fog along the coast, with 55 days/year when navigation is hindered.

Prevailing winds shift throughout the year. During the winter, the winds blow from the southwest, bringing relatively warm moisture-laden air from the northward-flowing Alaska Current. Winter winds thus act to warm the area. In the summer, the winds blow from the northwest, bringing cool air from the Gulf of Alaska, and thus cooling the land.

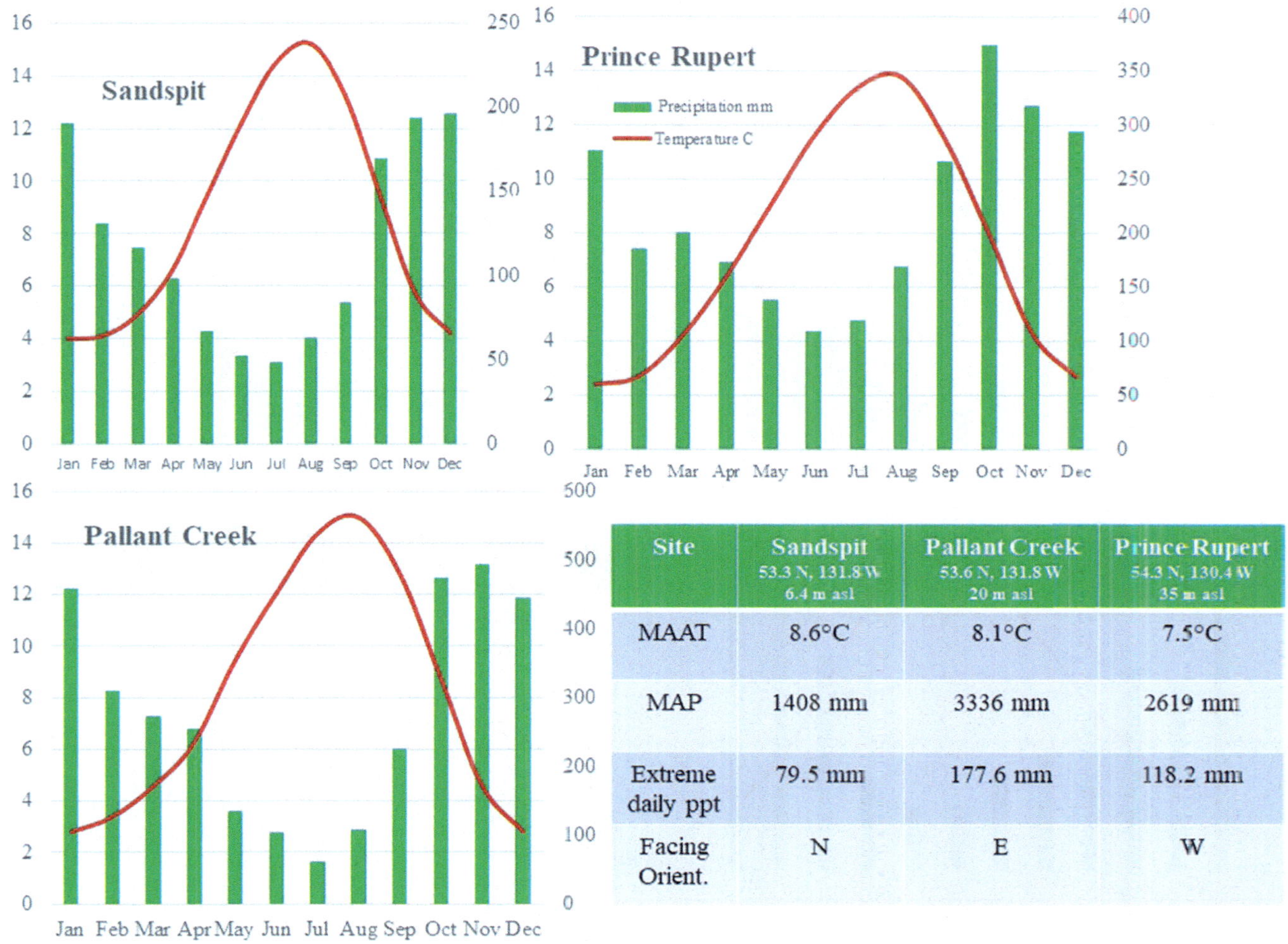

Site	Sandspit 53.3 N, 131.8 W 6.4 m asl	Pallant Creek 53.6 N, 131.8 W 20 m asl	Prince Rupert 54.3 N, 130.4 W 35 m asl
MAAT	8.6°C	8.1°C	7.5°C
MAP	1408 mm	3336 mm	2619 mm
Extreme daily ppt	79.5 mm	177.6 mm	118.2 mm
Facing Orient.	N	E	W

Fig. 7.5 Comparison of climates at Sandspit (northern Haida Gwaii, Cfb), Pallant Creek (southern Haida Gwaii, Cfb), and Prince Rupert (Cfc). (*Data* Canadian Climate Normals, Environment and Climate Change Canada 2020)

7.2.2 Salish Sea Shorelines

Along the Salish Sea shorelines, aspect is a critical factor (Fig. 7.6). Prevailing winds in the summer blow from the southwest, and thus the southwestern-facing slopes are warmer and much wetter than the northeastern slopes. During the winter, winds are more variable, and originate from both the southwest and southeast. Southeast winds blow across the Salish Sea, bringing precipitation from Puget Sound to the southeastern sides of the Gulf Islands and to Vancouver. As a result, mean annual precipitation on Salt Spring Island (9 km across, high point 698 m asl) varies from 800 mm on the dry northeastern side of to 1350 mm on the wet southwestern side. Victoria (48.4°N, 123.4°W; 883 mm annual precipitation) and Nanaimo (49.2°N, 123.9°W; 1165 mm), on the leeward side of the island, have Köppen-Geiger Csb (dry summer) climates (Environment and Climate Change Canada 2020).

Vancouver (49.4°N, 123.1°W), facing west along the Salish Sea and flanked by the Coast Range, may be classified as either Cfb or Csb (1189 mm annual precipitation). The driest month (July) has 20% of the precipitation of the wettest month (December), in the transition separating dry seasonal from year-round moist climates. Vancouver's summer is drier than Tofino's or Prince Rupert's, but is it dry enough to be classed as Csb? Visitors from coastal Norway might find Vancouver quite dry, while a visitor from Calgary could feel damp. Climate zone boundaries are not sharply defined lines.

Limited summer precipitation, at the height of the tourism, recreational, and agricultural seasons, has started to create difficulties, even in relatively wet Tofino. Increasing population of Salt Spring and other Gulf Islands has resulted in significant over-use of the groundwater supply. Water table levels have dropped, and saltwater intrusion in coastal wells has become evident since 2012 (Larocque et al. 2015).

Site	Victoria 48.6 N, 123.4 W 19.5 m asl	Saltspring 48.9 N, 123.5 W 45.7 m asl	Vancouver 49.2 N, 123.2 W 4.3 m asl
MAAT	10.0°C	10.7°C	10.4°C
MAP	883 mm	987 mm	1189 mm
Prevail. Wind	W	W	E
Frost-free days	211	200	237

Fig. 7.6 Comparison of climates at Victoria, Saltspring Island, and Vancouver. (*Data* Canadian Climate Normals, Environment and Climate Change Canada 2020)

7.2.3 Interior

Interior sites, isolated from the warm westerly winds of the Pacific, such as Mount Edziza (58°N, 131°W) and Tweedsmuir Provincial Parks (52°N, 126°W), show substantial differences in climate (Fig. 7.7). On east-facing slopes, the climate is mid-boreal (Köppen-Geiger Dfb) below 1500 m asl and highland (Köppen-Geiger H) above 1500 m asl. In the eastern part of Mount Edziza, January daily mean temperatures range between −16 °C and −22 °C for valley sites along the Iskut River. July daily mean temperatures range between 11 °C and 14 °C. Precipitation totals vary between 600 mm (leeside valley bottom sites) and 1100 mm (west-facing slopes along the Iskut River, and volcanic summits). Snowfall exceeds 400 cm (approximately 40% of total precipitation) on the summit areas of Mount Edziza and Ice Peak, but represents about 25% of precipitation in the drier areas. Overall, humid moisture regimes are characteristic. Scattered alpine permafrost is present above 1800 m asl. In contrast, the western areas along the Stikine River, influenced by the strong prevailing westerly winds from the Pacific Ocean, are classified as southern boreal (Dfa) regimes.

7.3 Bedrock

Geologically, the Pacific Maritime region lies within the Insular Structural Belt (Vancouver Island and Haida Gwaii, along with narrow coastal plains) and the Coast Structural Belt (Cascade Mountains; Coast Ranges, including the Pacific, Kitimat, and Boundary Ranges and Stikine Plateau, with numerous subdivisions; Gabrielse and Yorath 1991, Massey et al. 2005; Figs. 7.8, 7.9, 7.10). Earthquake activity characterizes the offshore Pacific continental shelf and adjacent islands. Recently and currently active volcanoes occur in the Cascade Mountains and Stikine Plateau (Hickson and Edwards 2001).

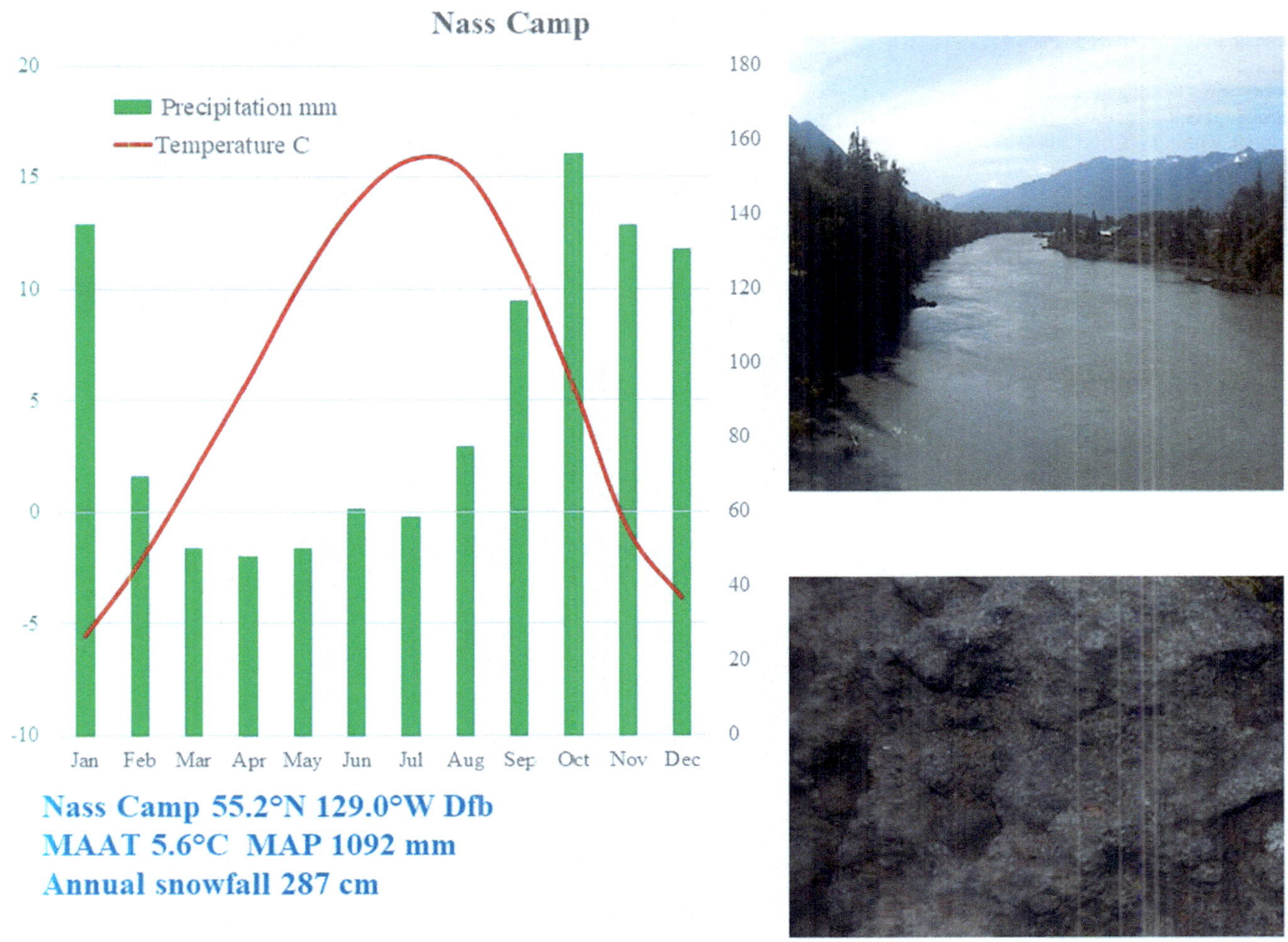

Fig. 7.7 Nass Camp, north of Nisga'a Memorial Lava Bed Provincial Park, has a Dfb climate. (*Data* Canadian Climate Normals, Environment and Climate Change Canada 2020). Upper right: Nass River at Nass Camp. (Wikimedia/Keith Freeman). Lower right: volcanic lava is present at several locations through the region

7.3.1 Insular Structural Belt

The spine of Vancouver Island is made up of the Vancouver Island Range, a complex of northwest–southeast trending rolling to rounded ridges to 1200 m asl developed on Mesozoic granite and granodiorite batholiths (Fig. 7.9).The batholiths formed as British Columbia's land mass was assembled by systematic accretion of terrain on to the western limit of North America during the Jurassic and Cretaceous (Haggart et al. 2006). Tectonic motion has continued to the present, and minor earthquakes occur periodically.

Metamorphic rocks, including schists, metaquartzites, and marbles, are associated with the margins of the batholiths. Altitudes are too low to support active glaciers, with the exception of three small remnants surrounding Red Pillar southwest of Comox. Evidence of former glaciation is present, including cirques, glacial troughs, and depositional features.

Along the west coast, Jurassic reef limestones and pillow basalts, representing the periodic formation and destruction of tropical volcanic islands and coral atolls 170 million years ago, are present at Long Beach. The rock units have undergone complex and repeated deformation, and faults of many orientations are common.

The Broken Islands are composed of several igneous plutons (quartz diorite, magmatite), metamorphic amphibolite, and gneiss. Ages range from 620 to 270 million years old, spanning the Palaeozoic.

The discontinuous Estevan Coastal Plain flanks the western coast of the island, extending from the shore (e.g. Long Beach) to 5–15 km inland. The plain slopes gently seaward, with a total relief of 30–50 m. Much of the coastal plain is underlain by Cenozoic rocks ranging in age from 11 to 65 million years. Fossilized Eocene and Oligocene marine molluscs and crabs can be seen along the West Coast Trail.

The oldest rocks in eastern Vancouver Island are Triassic volcanic rocks and Jurassic granite and granodiorite plutons. These igneous and metamorphic rocks are overlain by Upper Cretaceous sandstones and coals of the Nanaimo Group. Coal

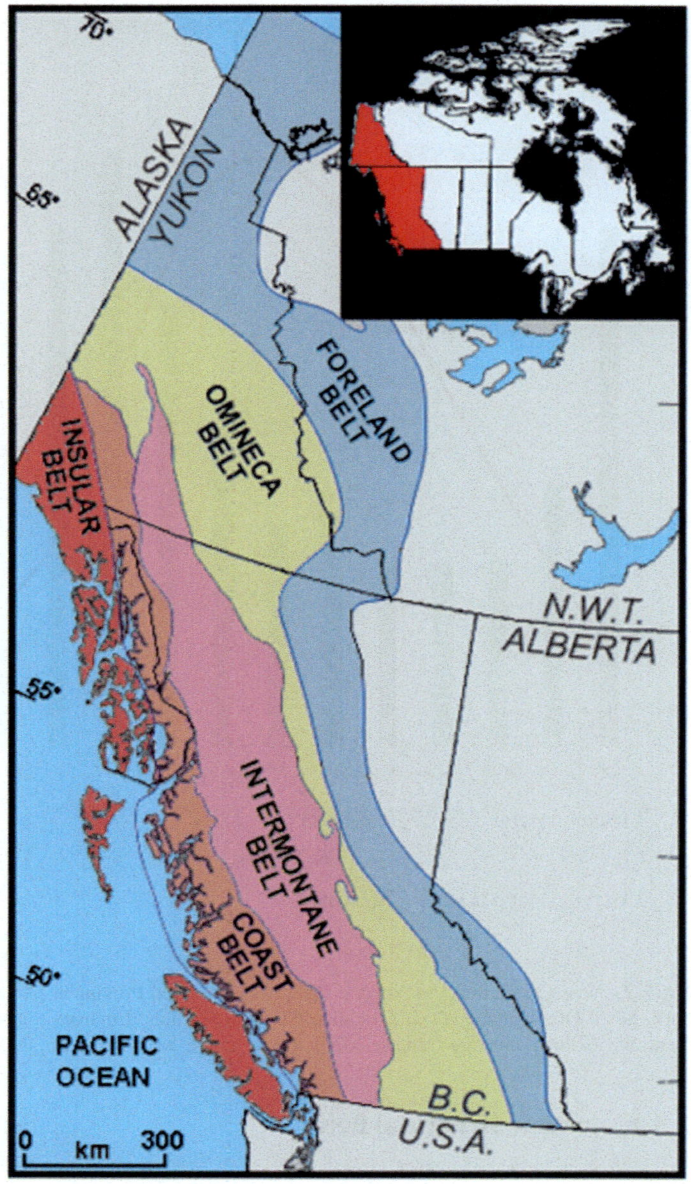

Fig. 7.8 Structural Belts of the Canadian Cordillera. The Pacific Maritime ecoregion includes areas of the Coast and Insular structural belts. Southwestern Yukon, within the Coast and Insular Belts, is discussed in Chap. 6. (Wikimedia)

was mined at Nanaimo during the early 1900s. The sedimentary rocks have been folded during the Late Cretaceous-Early Palaeogene Orogenic events into a series of west-northwest—east-southeast trending folds.

The Gulf Islands are aligned as a series of ridges parallel to the general trend of deformation and topography throughout southern Vancouver Island. The southwestern slopes of the islands are gently sloping to rolling-rugged, with shoals and coastal plains separated by steep, locally vertical cliffs. The northeastern sides of the islands vary from steep cliffs to rolling topography. Access from the water is restricted in many areas, and all of the major harbours (except Vesuvius) are located along the northeastern and eastern shores.

On Haida Gwaii, the Haida Gwaii Ranges (including the San Christoval Mountains (Moresby) and the Cameron Range (Graham)) are also formed of Mesozoic granitic batholiths (Anderson and Reichenbach 1989). These are relatively low, rolling mountains, 1150 m relief, with rounded summits and relatively gentle western slopes. Isolated peaks and plateaux are common. The ranges are cut by deep fjord valleys such as Cumshewa Inlet, locally termed "canals", developed along faults aligned northeast-southwest and north–south. Glacial features including cirques and glacial troughs are present, but do not dominate the landscape.

The Hecate Lowland and Hecate Strait separate Haida Gwaii from the mainland Kitimat Ranges, also composed of Mesozoic granite and granodiorite (Fig. 7.10). Ultramafic intrusions of Cretaceous age are also present. The igneous rocks are readily modified in the perhumid climate, producing the rounded summits characteristic of the Coast Ranges, from

Fig. 7.9 The Insular Ranges form the islands of Vancouver and Haida Gwaii. The term "Insular" refers to the islands, in contrast to the Coast Range Structural belt to the east. Note the differences in snow cover between the higher Coast Range and lower mountains on Haida Gwaii and Vancouver Island. (NASA). Lower left inset: winter view of northern Vancouver Island and adjacent Mainland Coast (NASA). Upper right inset: Nigei Island, off northern Vancouver Island. Centre right inset: Coast Range, Fjordland conservatory area

the Boundary Ranges along the Alaska-BC border south to the Pacific Ranges of Tweedsmuir Provincial Park.

The Mesozoic rocks underlying Skaga Island, Haida Gwaii, and the Hecate Lowland formed during the time that the tectonic Farallon Plate was gradually subducted under North America (Goes 2013, Hawley and Allen 2019). Currently, only two small remnants of the ancient Farallon Plate exist, the Explorer Microplate in the north and the Juan de Fuca Microplate in the south. The microplates are separated from the Pacific Plate to the west by the still-operating Explorer spreading ridge system. To the east, the microplates are being subducted under North America along the Cascadia Subduction Zone. To the north of the Explorer Microplate, the North American and Pacific Plates are in direct contact along the Haida Gwaii Transform Fault.

7.3.1.1 Earthquakes and Tsunami

The continuing spreading, combined with sliding and subduction along the Haida Gwaii Fault and Cascadia Subduction Zone, causes repeated earthquakes in the region (Figs. 7.11, 7.12 and 7.13). The largest reliably measured earthquake in Canada, with a M_W magnitude of 8.1 on the Richter scale, occurred southwest of Haida Gwaii on 22 August 1949 (Natural Resources Canada 2022). Damage to buildings was recorded at Terrace and Prince Rupert, and an oil tank at Cumshewa Inlet on Moresby Island collapsed. No human injuries were recorded in the relatively low population area of Haida Gwaii, although observers noted that cows had been knocked off their feet.

On 27 October 2012, a magnitude M_W 7.8 earthquake along the Haida Gwaii Fault affected hot springs in Gwaii

Fig. 7.10 The Coast Ranges are composed of granitic batholiths, with active cirque glaciers at the summits (upper left) and inactive cirques at lower elevations (lower left). Logging has been conducted at lower elevations in many locations. Upper right: Shannon Falls, Squamish, tumbles 335 m across Coast Range granite. Lower right: *Pseudotsuga menziesii* (Douglas Fir) dominates much of the shoreline of the Coast Ranges

Haanas National Park. Infrastructure damage was negligible. Although there were only a few minor physical injuries, significant psychological trauma resulted from both the earthquake and subsequent aftershocks (Bird et al. 2016).

All localities along the southern Salish Sea are potentially vulnerable to earthquakes. Liquefaction is a distinctive style of sediment failure associated with earthquake disturbance of coastal, deltaic, and fluvial sediments. Unlike slides, debris flows, and avalanches, liquefaction frequently occurs on flat, low-lying terrain.

In metropolitan Vancouver, the liquefaction hazardous areas include floodplain and delta sediments of the Fraser River, and infilled port areas along Burrard Inlet and False Creek. Areas underlain by bedrock or glacial sediments are not hazardous. In Victoria, infilled stream valleys and the port facilities of Esquimalt are susceptible to liquefaction.

Thorough assessment of earthquake vulnerability combines the physical factors of liquefaction, amplification of surface waves, slope instability, and earthquake intensity, with socio-economic and demographic sensitivity data. Assessment of the vulnerability of neighbourhoods requires consideration of Social (income; demographics; unemployment; education; home ownership; language) and Systemic (road network driving times, ability to evacuate) factors, in addition to geological data. Combining these in a geographic Information Systems model allows determination of relative vulnerability for each neighbourhood (Walker et al. 2014).

In British Columbia, as elsewhere along the Pacific Coast of North America, "Shake-Out" drills have become standard in preparation efforts (Great BC Shakeout n.d.). The drill is conducted at a set, pre-announced time. Participation is optional, but strongly encouraged through public-service messages, social media, and efforts by major employers. Under these circumstances, participation is more likely by some than by others, depending on social media connections,

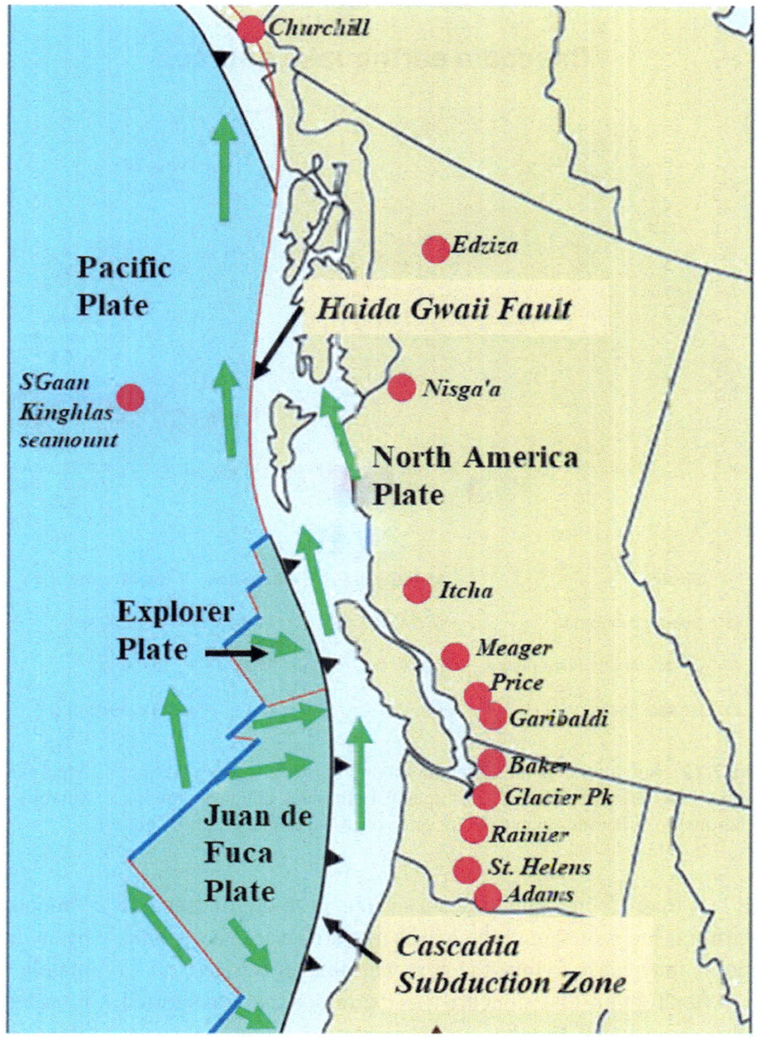

Fig. 7.11 Tectonic regime, British Columbia. Red dots: volcanoes. Green arrows: directions of plate movement. Red lines: strike-slip (transcurrent) faults. Blue lines: spreading ridge segments flanking the Explorer and Juan de Fuca Plates. Black lines: subduction zones, with teeth indicating direction of relative movement

available time to prepare, significance attributed to preparation, and place of employment. As with fire drills, elementary schools are more likely to participate than universities are.

In California, comparison of the total population with those who participated in a 2008 "Shake-out" drill revealed that female, white, older, and higher-income people were more likely to participate (Adams et al. 2017). Participation rates were lowest for Hispanic/Latino and African American people. Service businesses, small businesses, and self-employed people were less likely to participate than were schools and government offices. Broad-brush campaigns may miss several of the most sensitive population groups. Most participants only did the two-minute "drop, cover, hold still" part of the drill, rather than being involved in the other components such as attending disaster planning meetings, participating in simulations or games, or conducting personal assessments.

Participation is influenced by recent events. Following the disastrous Canterbury earthquake, the 2012 New Zealand "Shake-out" drill saw enhanced participation. Schools which were presented with several optional activities selected those that applied to their communities, rather than only doing the drill. Students were encouraged to extend emergency preparedness activities to their homes.

All British Columbia coastal communities are subject to tsunami (Fig. 7.14). The 1964 Anchorage earthquake produced a tsunami that caused significant damage at Port Alberni. Port Alberni is exposed to tsunami generated by far-field earthquakes (such as the December 2004 Banda Aceh and the 2011 Tohoku-Oki events), tsunami generated by near-field earthquakes (such as those near Haida Gwaii or along the Juan de Fuca Plate subduction zone), and those generated by slope failures (e.g. in Alaskan fjords). Vancouver, sheltered by Vancouver Island, is exposed only to near-field earthquake tsunami.

Several earthquakes since 2000 have produced tsunami that washed ashore on British Columbia, including Banda Aceh and Tohoku-Oki, but the maximum wave heights were

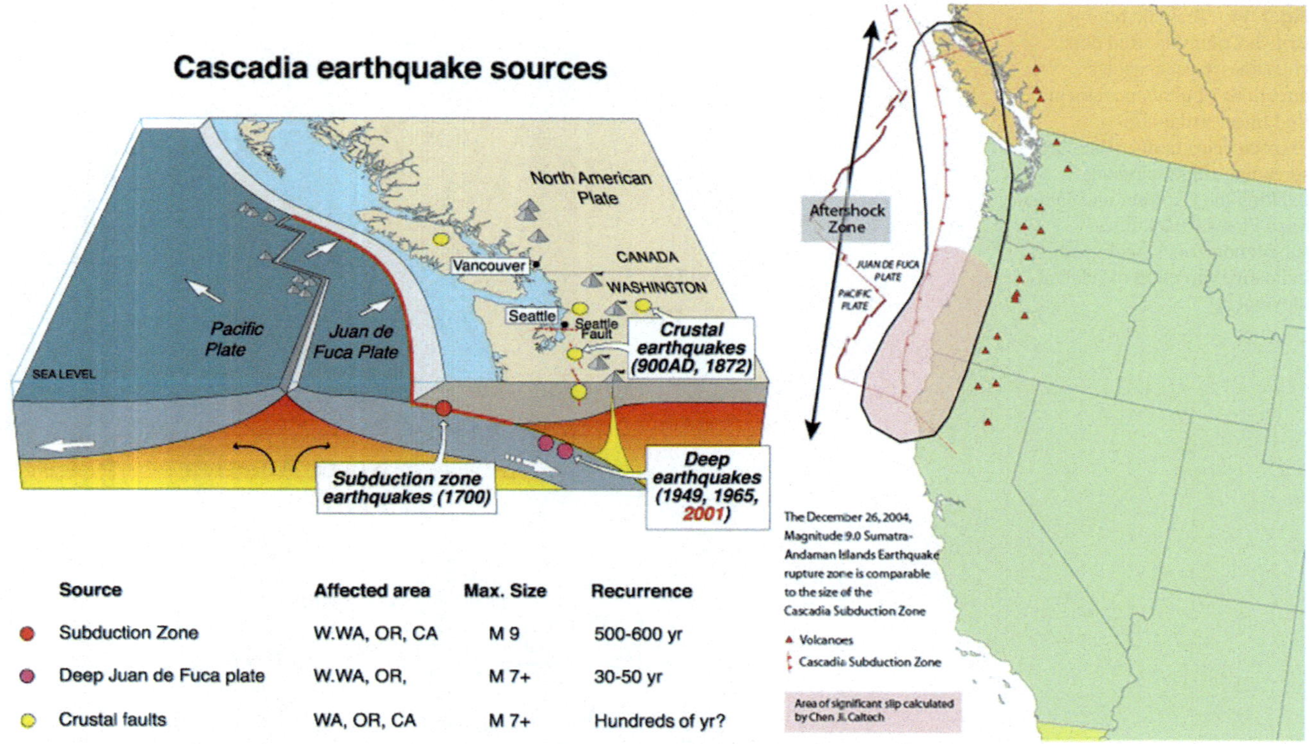

Fig. 7.12 Subduction in the Cascadia subduction zone was responsible for the 1700 ce major Cascadia earthquake. Left: block diagram showing relationship of Pacific, Juan de Fuca, and North American Plates. Right: Areal extent of the Cascadia Subduction Zone. (United States Geological Survey)

all less than 15 cm. This poses a serious problem: if tsunami warnings are issued, and the waves that arrive are very small, people may cease to take the next warning seriously.

After Tohoku-Oki, concern was expressed that transported debris could foul beaches, or contain radioactive materials. Computer modelling predicted transit across the Pacific, driven along the North Pacific gyre, with the main plume reaching British Columbia in three years. In reality, the plume moved faster than was predicted, with pieces of debris following highly divergent tracks. Large items were concentrated on Haida Gwaii and southwestern Vancouver Island (e.g. Hipfner et al. 2018). The debris arrived in a succession of waves in mid-2012, early-mid 2013, mid-2014, and mid-2015.

Distinctive sand deposits produced by tsunami are preserved along western Vancouver Island and other sites in Washington, Oregon, and California (Fig. 7.15). Although no written records exist, the oral histories of the Haida, Skeena, Salish, and other indigenous peoples record destructive waves impacting communities prior to 1800 (e.g. American Museum of Natural History 2018).

At Port Alberni, investigations (Clague and Bobrowsky 1994) revealed 7 tsunami sand layers beneath the known 1964 deposits, dating from about 3500 years ago. Investigations at other sites indicated that the coastline from southern Vancouver Island to northern California had been struck by at least 9 tsunami in the past 3,500 years, or approximately one every 400 years (Peterson et al. 2013). The record revealed that a particularly large event had occurred about 300 years ago, but there was no indication as to whether the event recorded a local (near-field) or cross-Pacific (far-field) tsunami, or what the magnitude of the earthquake responsible was.

In Japan, written records of tsunami impacts encompass several hundred years. Most tsunami had been correlated to near-field events, from the Japanese perspective: devastating tsunami matched to significant earthquake damage onshore, as occurred during Tohoku-Oki. Other smaller events were attributed to relatively weak far-field tsunami. However, the written records did reveal that a damaging tsunami had struck Japan on about 26 January, 1700 (Satake et al. 2003). As there was no indication of damage caused by a local Japanese earthquake, the powerful event was designated as the "Orphan" tsunami, attributed to an unknown, far-field earthquake. Combined analysis of the cultural Japanese and geological North American records indicated that the Orphan tsunami's parent was an earthquake in the Cascadia subduction zone, which struck on 26 January 1700 with an estimated M_S magnitude of 9. Computer modelling confirmed that a magnitude 9 earthquake off the coastline of Vancouver

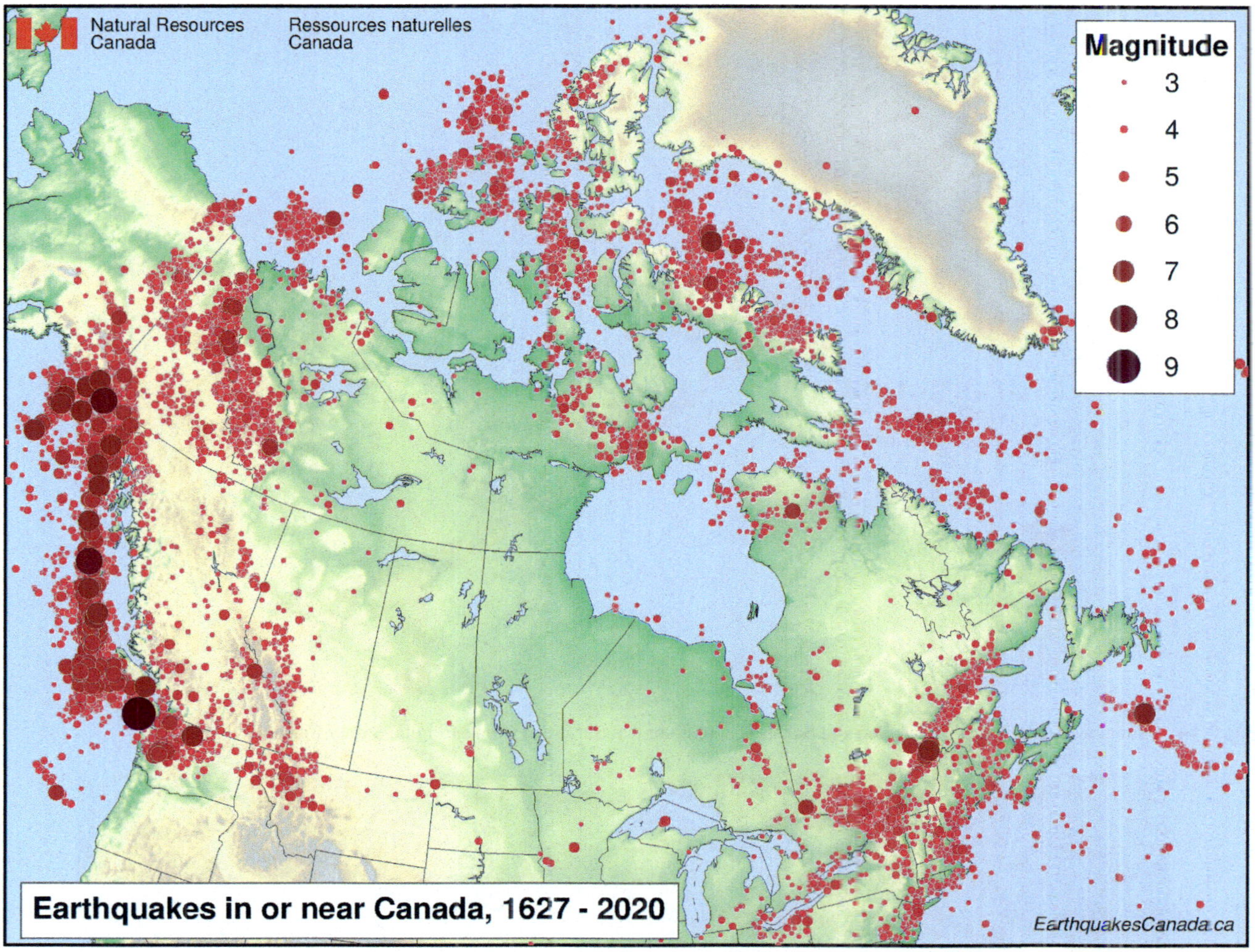

Fig. 7.13 Earthquake record for Canada, 1627–2020. Note concentration of large events along the Pacific margin, including the Cascadia earthquake of 1700 (estimated magnitude 9). Earthquakes Canada, Natural Resources Canada

Island would generate a tsunami capable of causing the damage recorded across the Pacific Ocean from the Cascadia subduction zone to Japan (Fig. 7.12).

Investigation of the Orphan Tsunami was of more than academic interest: it demonstrated that a magnitude 9 earthquake had recently occurred, producing a near-field tsunami that impacted sites throughout the Pacific coastline of Vancouver Island, with run-up heights locally in excess of 15 m. An earthquake occurring at 1000 m water depth in the Cascadia subduction zone would produce a tsunami travelling at 350 km h^{-1}, requiring 40–60 min to reach Port Alberni.

Earthquake-generated tsunami are compared based on the magnitude of the originating earthquake, the height of the incident wave, and run-up heights, although standards are no consistent. In particular, confusion arises if incident wave height and run-up height are not clearly differentiated. Local circumstances are significant influences on the amount of infrastructure damage and human casualties. Several intensity scales, measuring impacts on infrastructure, environments, and human communities, have been proposed, the most commonly used being the Integrated Tsunami Intensity Scale (ITIS-2012; Lekkas et al. 2013). The ITIS-2012 divisions differentiate between human perceptions and physical effects. Criteria are combined to make an overall assessment, designated by Roman numerals I (no perceptions, effects, or impacts) to XII (most severe).

Emphasis on infrastructure and building damage makes intensity assessment of the Orphan tsunami, and the consequent impacts on coastal British Columbia, difficult. Based on comparison with the environmental effects produced by Tohoku-Oki in a topographic and built environment resembling southwestern British Columbia, assuming an ITIS-2012 intensity of X to XI (moderately to very severe) is reasonable. Vancouver and Victoria are at significant hazard for earthquake damage, both physical and psychological.

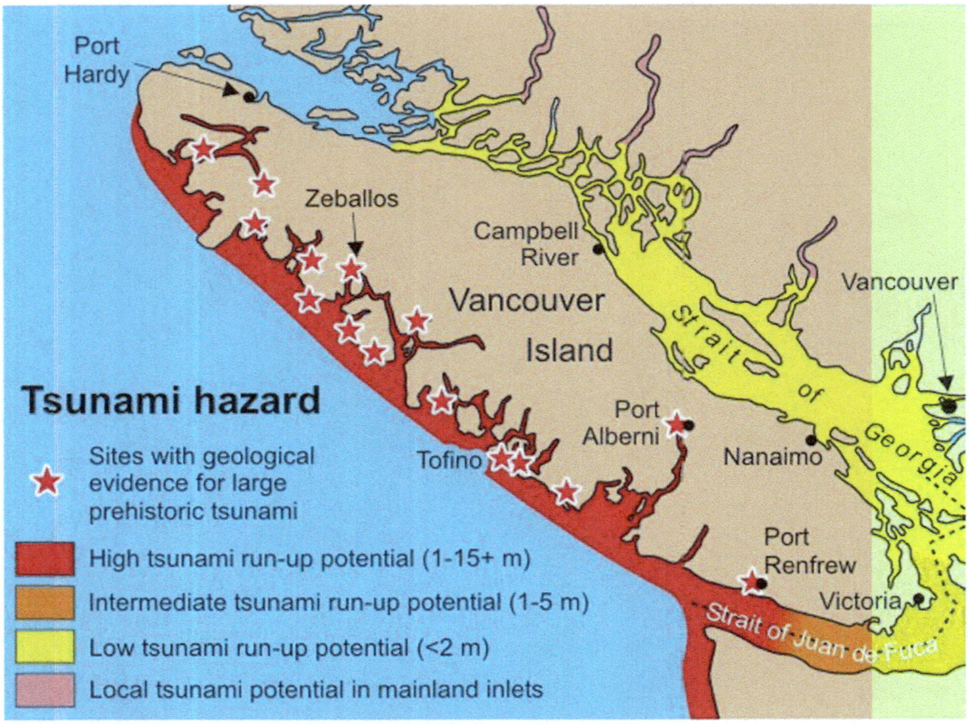

Fig. 7.14 Tsunami hazard, Vancouver Island and Salish Sea. (Geological Survey of Canada/Public Safety Canada)

Fig. 7.15 Tsunami hazard assessment for Vancouver Island is based in part on the presence of Tsunami Laid Sediments (TLS) produced by former larger earthquakes. (Courtesy of Peter Bobrowsky, Geological Survey of Canada)

7.3.2 Coast Structural Belt

The Coast Structural belt includes the Coast Ranges, including the Pacific, Kitimat, and Boundary Ranges formed of Mesozoic granitic batholiths similar to those of the Vancouver Island and Haida Gwaii ranges, the sedimentary and volcanic Stikine Plateau north of Prince Rupert, and the volcanic Cascade Mountains east of Vancouver and the Lower Mainland (Gabrielse and Yorath 1991; Massey et al. 2005; Fig. 7.16).

7.3.2.1 Mount Edziza

Mount Edziza Provincial Park, 360 km northwest of Prince Rupert, was established in northwestern British Columbia to preserve 1000 km^2 of the volcanic landscape of the Stikine Plateau (Fig. 7.17). The park is accessible by road from the Cassiar-Stewart Highway (#37), providing an alternative to the Alaska Highway as a route north to Yukon. The park is cut by the Telegraph Trail, the trace of the now-abandoned telegraph line from British Columbia to Yukon. Built immediately after the height of the Klondike Gold Rush, and abandoned in 1936, the trail is now largely overgrown and is impossible to vehicles. The community of Telegraph Creek, where the former trail crossed the Stikine River, is one of the few reminders of this early effort in communications technology. The terrain has been occupied by the Talhtan First Nation and their ancestors for the past 12,000 years, and the obsidian outcrops have been utilized to make numerous artifacts.

The Mount Edziza Volcanic Complex consists of the remains of four volcanoes, aligned along a south–north axis (Souther 1992). Topography is controlled by the type of lava involved in each volcano, the age of each, and the amount of dissection which the lava flows have undergone. The volcanoes decrease in age systematically from south to north, and the prevailing westerly winds cause the west-facing slopes to receive more rainfall than the eastern slopes. As a result, the eastern and western sides of the Edziza Volcanic Complex are topographically distinct.

The volcanoes formed as a result of five cycles of volcanic activity, extending from 11 million to 2000 years ago. Each

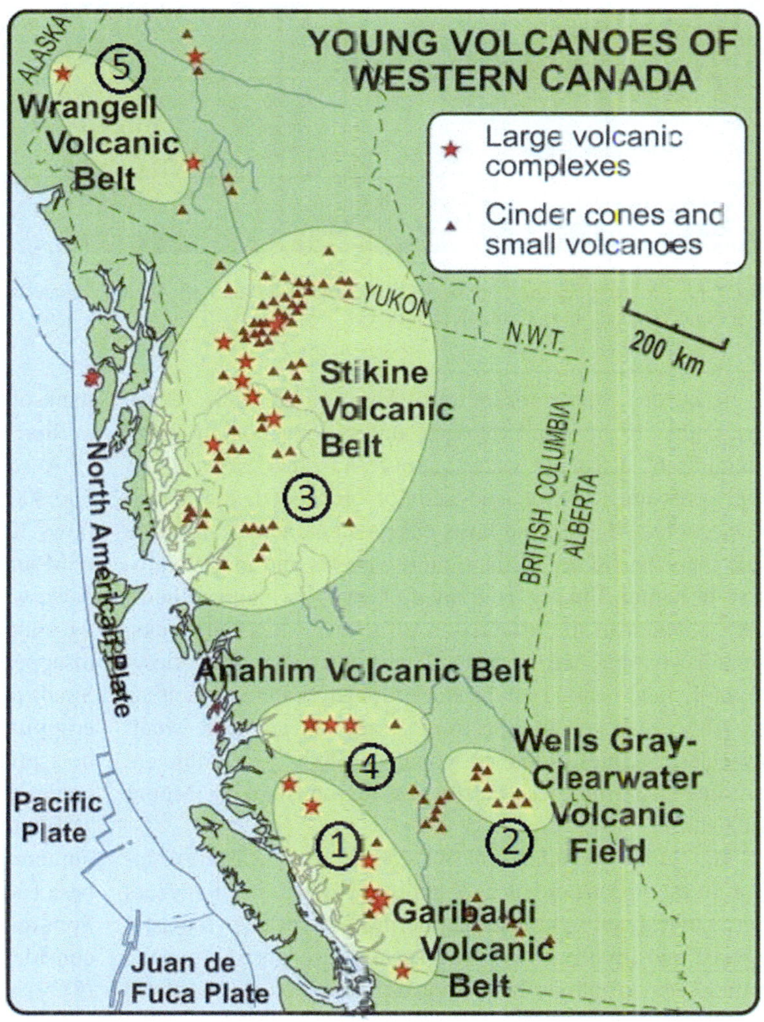

Fig. 7.16 Volcanoes of western Canada. (Geological Survey of Canada)

Fig. 7.17 Volcanic features, northern Coast Structural Belt. Left: Mount Edziza (NASA). Upper right: Triplex Cones and lava flow. Lower right: Eve Cone. (United States Geological Survey)

cycle began with the eruption of basic basalt lavas. These lavas tend to be non-explosive, unless they contain large amounts of dissolved water vapour and other gases under high pressure. As volcanic activity proceeded, the volcanoes cooled slightly, and lavas changed character to become intermediate, richer in silica, more viscous and more explosive in nature. Finally, continued changes in magma chemistry resulted in the production of acidic, silica-rich rocks, such as rhyolite and trachyte. These lavas were very slow-flowing, and tended to produce domal features, such as the rhyolite dome at Spectrum Ridge. In addition, small pyroclastic cones, some as young as 2000 years old, are present in the northern part of the region, flanking Mount Edziza.

The initial volcanic event of the Quaternary was the eruption of the Spectrum volcano, approximately 2 million years ago. Large quantities of basalt, intermediate rocks, pyroclastic debris, and rhyolite were successively discharged from the vent. Later eruptions built volcanic cones on the northern flank of the Spectrum Range. A period of approximately 1 million years of volcanic inactivity followed.

A second major burst of volcanic activity 1 million–500,000 years ago formed Ice Peak and the associated lava flows. The most recent major burst, resulting in the creation of Mount Edziza, occurred about 80,000 years ago, and coincided with the expansion of glaciers across the region. After the major eruptions ceased, the Cordilleran Inlandsis receded, disappearing from Mount Edziza about 14,000 years ago. Small pyroclastic cones and fissure volcanoes continued to erupt until 2000 years ago, and the presence of three active hot springs and hydrothermal fumaroles indicates that the potential for volcanic activity has not ceased.

Fissure volcanoes also produce basaltic lava, issuing from numerous fissures, forming overlapping lobes and plains of basalt across broad expanses of terrain. Although fissure eruptions are generally not violent, they may nevertheless present considerable hazards. Discharges of toxic carbon dioxide (CO_2), sulphur dioxide (SO_2), and hydrogen sulphide (H_2S)

gas commonly accompany fissure eruptions. These gases are denser than air, and tend to hug the surface, asphyxiating any unfortunate life they overwhelm.

7.3.2.2 Nisga'a Memorial Lava Bed Provincial Park

In the 1700s, Tseax Cone (100 km northeast of Prince Rupert) erupted, producing basaltic lava, tephra, and gas discharges (Williams-Jones et al. 2020; Fig. 7.18). Lava flowed into the Tseax River, damming it to form Lava Lake, and then progressed 22 km downstream into the Nass River valley. Oral traditional history of the Nisga'a people relates that poisonous gases and smoke descended from Tseax into two villages over several days. Residents had time to dig pits for shelter, indicating that the gases moved relatively slowly. However, toxicity was high, and at least 2,000 Nisga'a people were killed by gas discharge. Tseax Cone is now part of Nisga'a Memorial Lava Bed Provincial Park.

7.3.2.3 Stratovolcanoes

Stratovolcanoes (also called composite volcanoes) consist of interbedded layers of lava and pyroclastic deposits. Pyroclastic debris, produced during some eruptions, tends to be concentrated around the vent of the volcano. The lava flows radially outwards from the vent, and extends farther away from the source of the eruption. The volcano thus develops a curved slope, steeper near the summit than at the base, because the pyroclastic debris tends to stand at steeper angles than does the slowly cooling lava.

Stratovolcanoes are conspicuous features in landscapes, with heights to 3 km and maximum diameters to 25 km. The flanks of stratovolcanoes are partially covered by younger sediments. Although beautiful and often personified by their human neighbours, stratovolcanoes are marked by explosive, difficult to predict eruptions. Most loss of human life is associated with stratovolcanoes.

Many stratovolcanoes have complex life histories, with multiple eruptions over 100,000 years or more. A stratovolcano may produce mostly lava in one eruption, mostly pyroclastic debris in another, and mixed assemblages in a third. A composite volcano may lie dormant for hundreds or thousands of years before erupting explosively. Dormant and recently active Canadian stratovolcanoes include Cayley, Edziza, Garibaldi, Price, and Meager in British Columbia, and Volcano Mountain in Yukon (Stasiuk et al. 2003).

Eruptive behaviour and intervals between eruptions vary widely. Mount Baker, 30 kms south of Chilliwack, last erupted in 1880, and is considered to be extremely active. None of

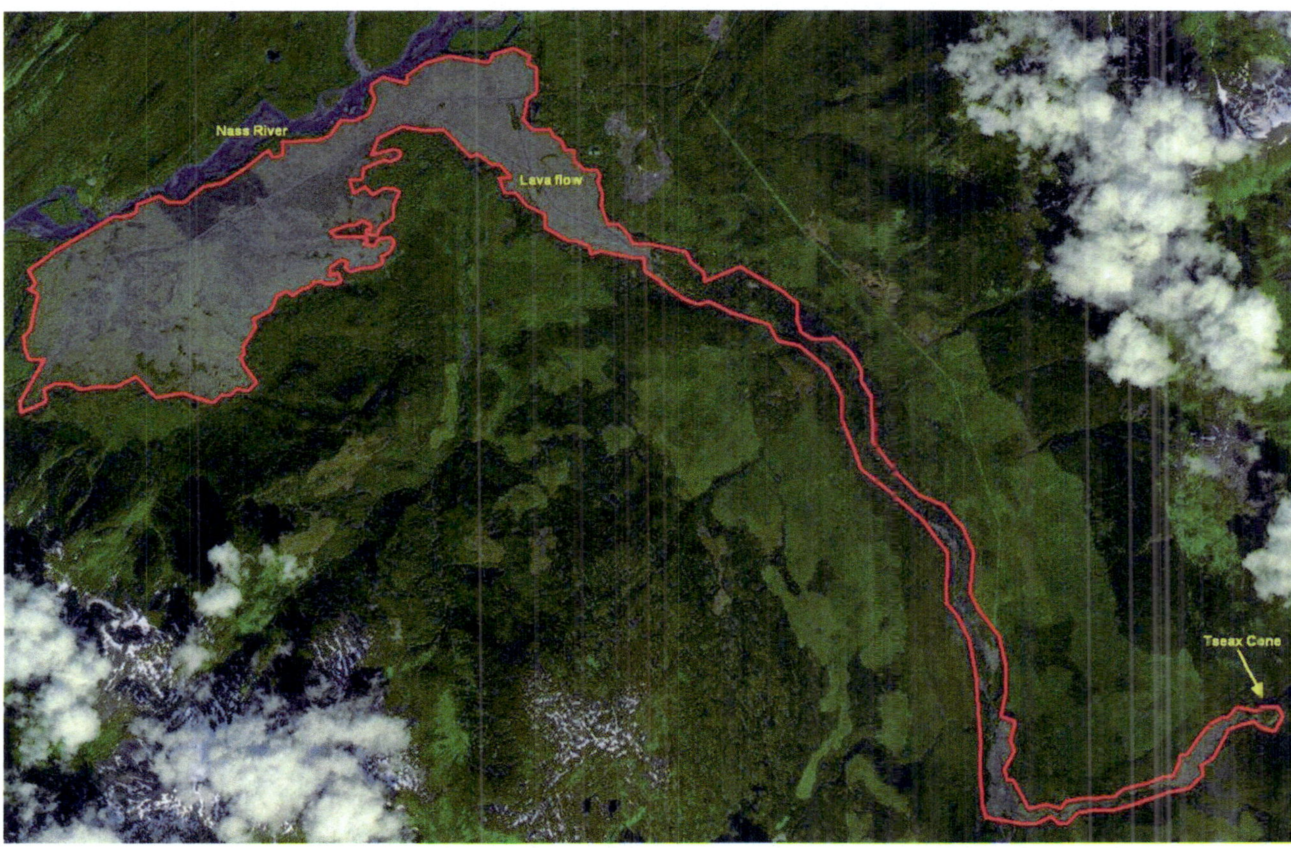

Fig. 7.18 Tseax Cone and lava flow. (NASA)

the stratovolcanoes in Mount Garibaldi Provincial Park has erupted in the past 10,000 years, but there is no evidence that they are extinct.

Stratovolcano eruptions produce tephra, ash and dust-sized glass fragments ejected in solid form. Each eruption produces tephra with distinctive geochemical and physical characteristics, so that multiple eruptions of the same volcano can be differentiated and dated. Eruptions of Cascade and St. Elias volcanoes over the past 15,000 years produced tephra layers preserved in lake sediments across western Canada (e.g. Lakeman et al. 2008, Jensen et al. 2019).

Tephra transport dynamics are influenced by weather, particle size, shape and density, and the style of eruption. Explosive eruptions produce plumes reaching the tropopause, incorporating tephra into atmospheric circulation and transporting it thousands of kilometres. Plume height is related to the volume of material displaced, but not to magma temperature, velocity of upward movement, or gas content. Understanding of transport mechanisms allows more accurate predictions of the spread and density of tephra plumes, leading to better assessment of the hazard to aviation and terrestrial organisms.

Between 833 and 850 ad, an explosive hydrovolcanic eruption of Mt. Churchill, in the St. Elias Range of Alaska, spread the White River Tephra across southern Yukon and northern British Columbia (Lerbekmo 2008; Fig. 7.19). More than 20 cubic kilometres of tephra was deposited, locally exceeding 1 m in thickness, smothering vegetation and starving herbivores across 200,000 km^2 of southern Yukon. Oral tradition of the Dene First Nation indicates that the residents were compelled to temporarily abandon southern Yukon, moving south into adjacent British Columbia. Shards of White River Tephra are preserved as thin layers and cryptotephra in Greenland and northern Europe, indicating transport over 7000 km.

Jet aircraft travel from Vancouver is subject to potential interruption from tephra eruptions. Canadian stratovolcanoes erupt between 1100 °C and 1500 °C. Tephra is generally erupted in molten form, and cools rapidly in the atmosphere to form glass shards at temperatures between 600° and 1000 °C. Jet aircraft are sensitive because the internal temperatures of engines are high enough to partially or completely melt the solid tephra (Clarkson et al. 2016). Gases discharged from jet turbines exceed 1000 °C, converting the solid shards into sticky droplets of molten glass. The addition of liquid glass to the air intake disrupts the fuel/air mix necessary for effective jet engine performance. As the tephra is sucked into the hot engine, it congeals to form films of fine glass over pitot tubes, nozzle guide vanes, ducts, air compressors, and turbine blades. Failure of any of these components results in

Fig. 7.19 Exposures of the White River Tephra in southwestern Yukon can exceed 1 m in thickness

Fig. 7.20 Volcanic landscapes, southern British Columbia. Upper: Mount Meager. Lower: Mount Garibaldi. (Smithsonian Institution, Lee Siebert)

compressor overload, causing the engine to shut down. The aircraft will stall, rapidly losing altitude. Unless the force of the descent and the decrease in engine temperature are sufficient to dislodge the coating and allow the engines to be restarted, a crash is inevitable. Lower levels of exposure will not result in crashes, but will abrade and compromise engine performance over time.

Mount Meager (Fig. 7.20) is located 150 km north of Vancouver in the Coast Range Mountains. During the most recent eruption, 2360 years ago, volcanic debris flows (lahars) reached the Pemberton area in the Lillooet River valley (Simpson et al. 2006).

Mass movements on Mount Meager remain a major hazard. On 6 August 2010, a rockslide, transitional to a high-velocity debris flow (sturzstrom), was generated from the collapse of the secondary peak of Mount Meager (Guthrie et al. 2012). The debris temporarily blocked both the Lillooet River and Meager Creek, necessitating a precautionary nighttime evacuation of 1500 residents of Pemberton, 65 km downstream. Although no human lives were lost, the economic impact of this mass movement is estimated at approximately $10 million. A similarly volatile geomorphic situation exists at nearby Mount Cayley, in the Garibaldi Volcanic Belt (Evans et al., 2001).

The Coquihalla Highway links coastal and interior British Columbia, representing an alternative route to the Fraser Canyon (Figs. 7.21 and 7.22). The highway is subject to damage from river flooding, closure by avalanches, and slope failures in metamorphosed greenstone bedrock.

7.4 Quaternary

7.4.1 Glaciation

Quaternary glaciation involved ice from two sources. On several occasions, Cordilleran ice expanded from interior British Columbia to reach the Pacific Coast along the major valleys, deepening fjords, and flowed westward onto Vancouver Island and Haida Gwaii (Clague and Ward 2011, Slaymaker 2017). At Pacific Rim, glaciers as much as 300 m thick covered the entire park region between 20,000 and 14,000 years ago. Although glacial ice reached the continental shelf in some locations (Shaw et al. 2020), recent evidence suggests that the northwestern tip of Vancouver Island may not have been ice-covered during the last glaciation, providing a biological refugium 20,000 years ago (Hebda et al. 2022). Retreat of the glaciers began about 18,000 years ago, with rapid deglaciation once climate warming began. Gwaii Haanas became ice-free by 17,000 years ago, the last glaciers receded from the Hecate Lowland about 13,500 years ago, and the Gulf Islands were deglaciated approximately 13,000 years ago.

Local mountain glaciers developed in the Cascades, Coast Ranges, and Vancouver Island Range (e.g. Hawkins et al. 2021, Harris and Russell 2022; Fig. 7.23). The mountains on Haida Gwaii are too low to have supported large glaciers during the Quaternary period.

Subglacial eruptions result in a variety of volcanic forms, including linear tindar ridges flanking fissures, and flat-topped, conical, and complex tuya volcanoes. Tuyas differ from steeply sloping volcanoes with thin alpine glaciers on the flanks, because the thickness of the glacial ice influences the style of eruption, the subsequent flat-topped tuya geomorphology, and the hazard to humans. Although Canada has no active subglacial basaltic volcanoes, several tuyas that developed under glacial ice between 10,000 and 740,000 years ago in British Columbia are now exposed (Friele and Clague 2009, Edwards et al. 2011, Harris and Russell 2022).

Glaciers in the Cascades and Coast Ranges are particularly sensitive to climate warming. All are currently suffering from negative mass balance, with accelerated rates of retreat, upslope movement of equilibrium lines, and shrinking accumulation zones. In some cases in the Coast and Vancouver Island Ranges, the entire glacier lies in the ablation zone, and there is no area of net supply. This condition is fatal, leading to a glacier's rapid disappearance from the landscape. In the decade 2011–2020, the rate of ablation increased by sevenfold over the period 1984–2010 for glaciers across British Columbia, and up to 32-fold for glaciers in the Vancouver Island Range (Bevington and Menoumos 2022).

7.4.2 Relative Sea Level Histories

Relative sea level histories following deglaciation are complex, and vary substantially along the Pacific coast (e.g. Shugar et al. 2014, Barrie et al. 2021, Fedje et al. 2021, Letham et al. 2021). Ongoing research is linked to investigations of the archaeological record. On Haida Gwaii, isostatic depression allowed marine waters to flood areas up to 200 m asl, compared to 100 m asl in the Gulf Islands and 60 m along the west coast of Vancouver Island. Almost all of the coastal plain was flooded after deglaciation, resulting in the formation of terraces and deposition of nearshore sediment locally reworked into coastal dunes.

This rapid transgression was followed by an equally rapid regression, which dropped sea levels below the present coastline. Throughout Hecate Strait, relative sea level may have been as much as 150 m below the present coastline 14,000 years ago. Although this drop was insufficient to allow faunal transit across the strait, it would have greatly increased the area of the coastal plain, allowing faunal migration along the coast from the unglaciated Aleutian Islands of coastal Alaska to newly deglaciated British Columbia, and south to Washington and Oregon. A growing body of archaeological evidence suggests that early *Homo sapiens* followed this route to populate western North America. A 13,400 year old submerged fishing weir has been found in Juan Perez Sound, Haida Gwaii, and other submerged archaeological sites have been documented offshore of Haida Gwaii and in the Salish Sea. Relative sea level then rose sharply, reaching 20 m above its present position 9500 years ago in northern Haida Gwaii, only to recede throughout the early and mid-Holocene.

At present, relative sea level is rising along most of the Pacific coast, at rates approximating 0.4–1.3 mm/year (Government of British Columbia 2016). In the Gulf Islands, the slowly rising sea level, coupled with mean tidal ranges of 2.3–2.7 m, has resulted in the development of coastal erosional features including wave-cut notches and isolated small sea stacks. Slow relative sea level fall has been recorded at Tofino.

In the Fraser Delta, Richmond and Delta are currently undergoing natural subsidence due to sediment compaction.

Fig. 7.21 The Coquihalla Highway links coastal and interior British Columbia. Upper: Zopkios Ridge. Lower: Site of 1963 sturzstrom east of Hope, British Columbia, along Highway #3. The sturzstrom originated in weak serpentine clays in metamorphosed greenstone bedrock. Exposed bedrock (buff) shows signs of abrasion and scouring. Four lives were lost

Fig. 7.22 Fraser Canyon. The Fraser Canyon route was the major link between coastal and interior British Columbia prior to construction of the Coquihalla Highway

Loading resulting from construction, including the weight of the dykes designed to contain the Fraser River and prevent flooding, is also contributing. Subsidence locally reaches 8 mm/y (Mazzotti et al. 2009). Coupled with ongoing sea level rise, this could result in submergence of 2 m along the Salish Sea by 2200 ce. Relative sea level rise of 1.5 m would put much of Delta and Richmond, Vancouver International Airport, and Pitt Meadows, in addition to port facilities along Burrard Inlet and False Creek, under water (Kulp and Strauss 2019).

7.4.3 Fraser River

The Fraser (Halqemeylem: Sto:lo) River is the largest discharging into the Pacific Ocean in Canada (Fig. 7.24). The mean annual discharge exceeds 3500 m^3 s^{-1}, but has exceeded 20,000 m^3 s^{-1} during flood events (Brice et al. 2021). The river carries 20 million tonnes of sediment annually, most of which is deposited in the lower reaches west of Hope and in the delta. The geomorphology varies from braided to wandering in most reaches, in addition to straight bedrock-confined segments in reaches such as Hell's Gate (Fig. 7.22). Although not navigable upstream of Hope, the river valley formed the most convenient route to the British Columbia interior, and was used successively for the Cariboo Trail, CPR, Canadian Northern Railway (now CN), and the Trans-Canada Highway. In addition to being a major salmon river, deeper pools support populations of white sturgeon (*Acipenser transmontanus*).

Flooding of the Fraser and its tributaries can occur as a result of spring freshet, glacial melting, or excessive summer–autumn rainfall. Snow melt at various altitudes produces peak discharges from May through July. Although much interannual variability occurs, peak discharges currently are 7–14 days earlier than during the 1980s (Shrestha et al 2012).

A large drainage basin which receives significant rainfall or abundant snowfall over a long period can accumulate sufficient water to generate flooding over broad areas, lasting for weeks, as has occurred along the lower Fraser from Hope westward. This style of prolonged, regionally extensive flooding is generally associated with river reaches that have broad floodplains with extremely low gradients, or form deltas. Along the lower Fraser, river channelization and development of riparian terrain for agriculture play significant roles. The combined effects of climate change, extreme weather events, anthropogenic modification, and growing

Fig. 7.23 Montane glaciation was present throughout the Coast and Insular Structural Belts during the Quaternary. Upper left: Cirques indicate glaciation throughout the Coast Ranges. Elongated, sloping cirque dominated by frost erosion at the backwall, Kitimat Ranges. Upper right: Crevassing and frontal calving, Bear Glacier. (Wikimedia/Not Home). Lower Left: Salmon Glacier, near Stewart. (Lynn Kolund, United States Forestry Service). Lower right: Montane glaciation developed fjords along the Pacific Coast of the Coast Ranges. (NASA)

human population have resulted in elevated vulnerability to flooding.

In November 2021, widespread flooding triggered by extreme rainfall linked to climate change (Gillett et al. 2022) occurred along the Fraser's tributaries, progressing downslope to the valley. In addition to damage to highways (notably the Coquihalla Route, #5) and infrastructure inland, extensive flood damage occurred along the lower Fraser. Five human lives were lost. At Sumas Prairie, near Abbotsford, land which had been drained for agricultural production was reclaimed by the river. Breached dykes necessitated the evacuation of more than 1100 homes, businesses, and farms. Unfortunately, more than 600,000 farm animals could not be evacuated, and subsequently drowned. Disruptions to agriculture, road and rail transport to and from Vancouver, and damage to infrastructure totalled at least $10 billion (2022 Canadian).

7.5 Vegetation

7.5.1 Southern Pacific Coast Assemblages

The defining trees of the Coastal Southern Pacific Cordilleran ecoclimatic region are large, warmth-loving conifers (Demarchi 2011; Ministry of Forests, Lands and Natural Resource Operations 2014). Along the west coast of Vancouver Island, the perhumid moisture regime allows the development of a dense coniferous forest (Fig. 7.26). *Abies amabilis* (lovely fir; Pacific Silver fir) requires a minimum of 2800 mm of precipitation annually for its growth (Klinkenberg 2020). It is found only on western slopes of the coastal ranges, facing the open Pacific Ocean. Other conifers include *Pseudotsuga menziesii menziesii* (coastal Douglas-fir), *Tsuga heterophylla* (western hemlock), *Abies grandis* (grand fir;

Fig. 7.24 The Fraser (Halqemeylem: Sto:lo) River is the largest discharging into the Pacific Ocean in Canada. The mean annual discharge exceeds 3500 $m^3 s^{-1}$, but has exceeded 20,000 $m^3\ s^{-1}$ during flood events. The river carries 20 million tonnes of sediment annually (NASA)

Halkomelem: *T'a'xwulhp*), *Picea sitchensis* (Sitka spruce), and *Thuja plicata* (western redcedar, provincial tree of British Columbia; Fig. 7.25, 7.26). Soil types include dystric and sombric brunisols and humo-ferric podzols.

Thuja plicata (Halkomelem: *xepá:y)* grow rapidly in the moist environment, with 600 year old trees reaching 50 m in height and 5 m in diameter (Hosie 1979; Menary 2020; Fig. 7.27). In 2022, the Cheewhat Giant, in Pacific Rim, was Canada's largest tree, at 449 m^3 trunk volume. Some of the tallest trees on Vancouver Island are located in Cathedral Grove along the Port Alberni Highway #4. One specimen reached 71 m in height before being killed by vandals in 1972.

The root system extends above ground, allowing cedar to survive flooding and deposition of flood sediment. Shallow roots extract moisture from constantly wet surfaces, indicated by ferns in most cases. A 40-m tall tree may have roots extending less than 5 m into the ground. Although its thin bark makes it highly susceptible to fire, under natural conditions fire frequency in the perhumid moisture regime is typically in excess of 300 years.

Indigenous peoples preferred the wood of xepá:y for building construction, and large logs were used to shape canoes (Stewart 1995). Cedar bark was used for clothing and footwear. Western red cedar also finds its way into beehives, and many of the 14 billion pencils manufactured each year.

The oil of western red cedars older than 100 years contains the monoterpene thujaplicin. Thujaplicin's antifungal and antibacterial properties make it useful for treating stomach disorders, toothache, arthritis, fungal infections (athlete's foot), respiratory infections, and ringworm (Moerman 1998). Western red cedar pulp has recently found extensive employment in N-95 masks. Thujaplicin also repels fleas and other biting insects. Thujaplicin gives western red cedar its distinctive aroma.

Closely spaced trees are not exposed to strong winds. However, removal of some trees, particularly by logging, exposes others to wind toppling. Logging has long been a major industry on Vancouver Island, and one attended by considerable controversy. Formerly, the practice was to clear-cut areas of coniferous vegetation, creating open spaces in the forest canopy, ironically called "black holes". Selective cutting is now practiced in some regions, but efforts to log old-growth forest (more than ±100 years) are always met with protests. Efforts to return to traditional practices of cedar management (Zahn et al. 2018) are underway in many areas. Tree growth in the Pacific Cordilleran forest is rapid, but the new forests lack the biodiversity of the old (Hoffman et al. 2021, Gilhen-Baker et al. 2022), and small mammals and birds seldom survive the transition period during reforestation.

Pseudotsuga menziesii menziesii (coastal Douglas-fir; Halkomelem: *lá:yelhp*) ranges throughout southern British

Fig. 7.25 Southern Pacific Coast forests are dominated by conifers such as *Pseudotsuga menziesii* (Douglas Fir) and *Tsuga heterophylla* (western hemlock), extending to the Pacific shoreline in calm water locations

Columbia. Preferring dystric brunisiols and humo-ferroc podzols, Douglas-fir extends a deep taproot, making it less susceptible to windthrow. It is also more fire-resistant than *Thuja plicata*, but is not tolerant of flooding. Most browsers avoid Douglas-fir, making it more likely to spread in areas where herbivore populations have been reduced. In areas outside the densest forests, Douglas-fir is the dominant tree species of the Coastal Southern Pacific Cordillera. The largest known example in Canada, the 1000-year-old Red Creek Fir adjacent to Pacific Rim, has a diameter of 4.2 m, trunk diameter of 3.1 m, and a height of 73.8 m (Menary 2020). Historical records indicate trees to 120 m height.

With its very high strength-to-weight ratio, Douglas-fir is extensively used for architectural structural purposes, including laminated and composite beams. It represents about one-third of the total timber volume harvested from BC coastal forests.

7.5.1.1 Deciduous Species

Deciduous trees are subordinate in most areas. The major deciduous types are *Quercus garryana* (Garry oak), *Populus trichocarpa* (black cottonwood), *Acer macrophyllum* (big-leaved maple), *Alnus rubra* (red alder), and *Cornus nuttalli* (Pacific dogwood). These species have very large leaves, designed to catch the small amounts of sunlight which filter through the dense canopy of 40 m high conifers. The deciduous trees generally are found within 2 km of the shore, or near artificial clearings, as the inland conifer forests are too dense and have canopies sufficiently continuous to block the sunlight and prevent the deciduous trees from establishing themselves.

Quercus garryana (Garry oak; Halkomelem: *p'xwulhp*; Fig. 7.28) is the only oak native to western Canada (Hosie 1979, Catherall et al. 2018). It lives on the Gulf Islands, adjacent Vancouver Island, and in an isolated stand in the Fraser Valley. As it prefers warmer and drier climates (especially during summer), the current trend of climate change in southwestern British Columbia would generally favour the expansion of the species and associated ecosystem, outcompeting moisture-loving conifers. However, the intrusion of non-native species, coupled with habitat destruction, represent severely limiting factors. The oaks reproduce both vegetatively and through dispersal of acorns by Steller's jay (*Cyanocitta stelleri*, provincial bird of BC). Spread of Garry oak would be favoured by fire, and previous agricultural activities of the indigenous Halkomelem and other Coast Salish peoples probably played a significant role in expanding its range until the mid-1800s (Pellatt and Gedalof 2014). Subsequently, Garry Oak has been replaced by Douglas-fir.

Acer macrophyllum (big-leaved maple; Senćoŧen: Tŧá'ełć; Halkomelem: q'umun'ulhp) may reach 35 m in open areas, but more commonly is 10–15 m tall in coastal BC. Individual

Fig. 7.26 Southern Pacific Coast species include *Tsuga heterophylla* (western hemlock; left), *Taxus* (yew; upper right); *Pinus monticola* (western white pine, lower centre), and *Pseudotusga menziesii* (Douglas fir, lower right). (*Pinus monticola*: Wikimedia/Twisty)

leaves can be 60 cm in diameter. Similarly to Garry Oak, big-leaved maple was favoured by periodic Halkomelem cultural burning and clearance. Traditional uses included tea (from bark), sap, containers, cooking pot covers, and seasoning (from leaves), wooden bowls, and paddles.

Cornus nuttalli (Pacific dogwood; Halkomelem: kwit-xulhp) is the provincial floral emblem of British Columbia, although its range does not extend to northern Vancouver Island, north of Knight Inlet, or inland beyond the Coast Range Mountains. It is commonly found in clearings adjacent to Douglas-fir stands. "Dogwood" comes from the Sanskrit word "dag", the sharpened wooden pegs of the European species *C. sanguinea*, used in ship construction. The Halkomelem and other indigenous groups used kwit-xulhp wood to make arrows, bows, and clothing needles.

Other species include *Malus fusca* (Pacific Crab apple; Halkomelem: qwa'up-ulhp), *Acer circinatum* (vine maple; Halkomelem: síits'əɬp), *Sambucus cerulea* (blue elderberry), *Sambucus callicarpa* (red elderberry; Halkomelem: th'iwuq'ulhp), *Rhamnus purshiana* (cascara), *Prunus emarginata* (bitter cherry), *Prunus virginiana demissa* (western choke cherry), *Rhododendron macrophyllum* (Pacific rhododendron), and *Toxicodendron diversiloba* (poison oak). Wooded bog areas support Labrador tea (*Rhododendron groenlandicum*), bog laurel (*Kalmia angustifolia*), and evergreen huckleberry (*Vaccinium membranaceum*).

Along moist coastal areas in the Gulf Islands and adjacent southern Vancouver Island, a common species is *Arbutus menziesii* (arbutus, also called madrone; Senćoŧen: kekeiłć; Halkomelem*: qaanlhp;* Fig. 7.29). This species grows to 15 m in height, and is marked by conspicuous, smooth bright red bark. Arbutus is an evergreen ericad, related to the blueberry, partridge-berry, and Labrador Tea. The Nanaimo Lowland and adjacent islands is the only natural Canadian habitat for arbutus.

Other plants associated with the arbutus community include *Berberis* (Oregon Grape), the mint *Prunella vulgaris* (heal-all), and several moss species including *Rhacomitrium canescens, Stereocaulon tomentosum, Dicranum howellii*, and *Selaginella wallacei* (spikemoss). Disturbed areas support *Ulex* (gorse), *Urtica* (thistle), and *Ilex* (holly). The Vancouver Island town of Duncan is the country's largest producer of commercial holly for Christmas decorations.

The mild climate allows many non-native species to flourish in gardens, and many have escaped to the wild. Californian species such as redwood and sequoia, and more

Fig. 7.27 *Thuja plicata* (western red cedar) is a signature and culturally important tree of the moist Southern Pacific Coast forest. Lower left *Sphyrapicus ruber* (Red-Breasted Sapsucker) utilizing western red cedar trunk

exotic species such as Japanese maple and eucalyptus, can be found growing untended on the Gulf Islands. Some isolated areas appear to support crops of *Cannabis sitiva* (marijuana), which grows rapidly in this climate. Many imported species have become invasive, particularly Scotch Broom (*Cystisus scoparius*), and the native floral assemblage is undergoing alteration as a result.

7.5.1.2 Coastal and Altitudinal Influences

The driest, most exposed coastal slopes support *Pinus monticola* and scrubby, krummholz *Pinus contorta* (shore pine; Fig. 7.30). This growth habit accounts for the species name "contorta". In other, less-exposed environments, *Pinus contorta* grows as a straight-trunked tree. Lodgepole pine in the Rocky Mountain foothills, noted for its straight trunk, is also a representative of species *Pinus contorta*. Regosols and brunisols dominate in these areas. Other species present include wild beach rye (*Elymus spp.*), Nootka rose (*Rosa nutkana*), verbena species (*Verbena*), salal, black twinberry, marram grass (*Ammophila brevigulata*), and kinnikinick (*Arctostaphylous rubra*). Beaches subject to wave disturbance support scattered sandwort (*Arenaria*), beach pea (*Japonicus glaber*), and giant vetch (*Vicia gigantia*).

Elevation controls the distribution of some conifer species. At altitudes above 250 m, *Callitropsis nootkatensis* (yellow cypress, yellow cedar) is an important component of the assemblage. Semi-exposed coastal areas, especially those with sandy soil, are dominated by *Pinus monticola* (western white pine), *Picea sitchensis* (Sitka spruce), and *Pinus contorta* (shore pine). The tallest trees are 25 m high and 300 years old. *Picea sitchensis* tends to live longer than either of the pine species. Understory vegetation includes salal (*Gaultheria shallon*), black twinberry (*Lonicera involucrata*), and shrubby Pacific crabapple (*Malus fusca*).

7.5.2 Northern Pacific Coast Forest

The Northern Pacific Coast Forest grows on Gwaii Haanas, Naikoon, and the Hecate Lowland (Demarchi 2011; Ministry of Forests, Lands and Natural Resource Operations 2014). It is dominated by *Tsuga heterophylla* (western hemlock), *Abies amabilis* (amabilis fir, where annual precipitation exceeds 2800 mm), *Thuja plicata* (western red cedar), *Picea sitchensis* (Sitka Spruce; Haica kiid), and *Callitropsis nootkatensis* (yellow cedar or Nootka cypress; Fig. 7.31). The tallest trees, generally Sitka spruce in near-coastal areas and western red cedars inland, exceed 60 m. Poor soils, often developed on sand, support *Pinus contorta* (shore pine) and *Picea sitchensis*.

Fig. 7.28 Characteristic deciduous species of the Southern Pacific Coast. Upper: *Acer macrophyllum* (bigleaf maple). Lower: *Quercus garryana* (Garry oak)

The coolness of the summer climate precludes the establishment of some southern conifers. *Pseudotsuga menziesii* (Douglas Fir) is very uncommon in Gwaii Haanas, although it does grow in the drier areas of the Hecate Lowland, particularly on east-facing slopes. *Abies grandis* (grand fir) is not present anywhere in the region. The damp conditions and saturated soils prevent the establishment of *Pinus monticola* (western white pine) on mainland sites, and although the terrain surrounding Masset would be suitable, pollen cannot spread to the islands due to the prevailing westerly winds. Recently, parasitic mistletoe (*Viscum album*) has reduced the population of western hemlock, its preferred host.

Callitropsis nootkatensis (Haida: *sgahláan*; Fig. 7.32) wood is flexible and soft, ideal for carving and shaping into bowls, paddles, bows, and beans for housing. It also floats well, making it suitable for canoes. Sgahláan were harvested by carving a channel in the trunk, surrounding the cavity with wet branches of western hemlock, and then setting a fire. The pre-1800 distribution of sgahláan adjacent to communities, and the forest assemblage in general, was influenced by anthropogenic fire in this perhumid environment (Schang et al. 2022).

Significant die-offs of *Callitropsis nootkatensis* have plagued northern coastal British Columbia in recent decades (Comeau et al. 2021). The cause has been traced to freezing of near-surface root systems during winters with reduced snow cover. Ongoing climate change has increased the number of freeze–thaw cycles, particular at higher altitudes, while reducing snowfall and pack retention. Locally, logging and road construction have also funnelled winds upslope, resulting in further snow removal.

Areas of saturated soils with low nutrient contents support wetland assemblages with stunted and contorted Sitka spruce. Understory species in these areas include *Arctostaphylous uva-ursi* (bearberry), *Arctostapylous alpina* (alpine bear-

Fig. 7.29 *Arbutus menziesii* (arbutus), within family Ericaceae, is noted for its red trunk with branching habit. In Canada, it is confined to sheltered, drier locations along the fringes of the southern Salish Sea

berry), *Calamagrostis purpurasens* (reed-bentgrass), *Astragalus williamsii* (milk-vetch), *Potentilla* (cinquefoil), *Pyrola* (wintergreen), *Linnaea borealis* (twinflower), *Carex* (sedge), and *Artemisia* (sage).

Deciduous trees are subordinate components of the assemblage, and are confined to exposed coastal situations and cleared areas (today, commonly as a result of logging; formerly reflecting community development). The dominant deciduous species are *Acer macrophyllum* (bigleaf maple, on sheltered, generally east-facing sites or riverbanks), *Alnus rubra* (red alder), *Alnus viridis sinuata* (Sitka Alder, generally on west-facing slopes), *Alnus viridis fruticosa* (Siberian Alder, throughout the area despite its "subarctic" name), and *Populus trichocarpa* (black cottonwood, generally on east-facing slopes). Understory vegetation is limited due to the completeness of the forest canopy. The limited influx of organic matter to the soils retards their development, and many are relatively poor in organic matter and humus. The dominant soil types are sombric, dystric, and eutric brunisols, with regosols developed in exposed localities.

7.5.3 Northern Inland Assemblage

Further inland in northern British Columbia, the eastern slopes, and areas at higher elevations, support a vegetation assemblage adapted to colder and drier conditions (Fig. 7.33). Major species include *Pinus contorta* (lodgepole pine, in its interior straight-trunked habit) *Abies lasiocarpa* (Alpine fir), *Picea engelmannii* (Engelmann Spruce), *Picea glauca* (white spruce), *Tsuga mertensiana* (mountain hemlock), *Betula papyrifera* (white birch), *Salix* (willow; in shrub and tree forms), and *Populus tremuloides* (trembling aspen). *Picea mariana* (black spruce) only occurs along the Iskut River and further east in the Skeena Mountains (Demarchi 2011; Ministry of Forests, Lands and Natural Resource Operations 2014). Brunisolic and regosolic soils are dominant below 1800 m asl. Static cryosols are present at higher elevations.

Understory vegetation includes *Amelanchier* (saskatoon), *Rhododendron albiflorum* (white-flowered rhododendron), *Menziesii ferruginea* (false azalea), *Claytonia sarmen-*

Fig. 7.30 *Pinus contorta* (shore pine). The shoreline habitat produces contorted trunks under wind and sea spray stress, leading to the specific name

tosa (Alaska springbeauty), *Epilobium latifolium* (broad-leaved fireweed), *Epilobium angustifolium* (narrow-leaved fireweed), *Trientalis arctica* (northern starflower), *Polemonium* (sky-pilot), *Aconitum delphiniifolium* (mountain monkshood), and *Listeria borealis* (northern twayblade). Above treeline, between 1800 and 2000 m, an alpine tundra assemblage is present, including *Silene acaulis* (moss campion), *Draba alpina* (alpine draba), *Corydalis pauciflora* (blue corydalis), *Dryas,* and *Saxifraga* (see Marr et al. 2012).

7.6 Fauna

7.6.1 Interior Sites

Faunal assemblages in interior sites such as Mount Edziza, Tweedsmuir Provincial Park, and Huchsduwachsdu Nuyem Jees/Kitlope Heritage Conservancy include species from both the Pacific Maritime and the Interior Cordilleran Ecoclimatic regions (Banfield 1981, Naughton 2012). Inhabitants include *Odecolieus hemonius sitkensis* (mule deer), *Neogale vison* (mink), *Martes americana* (pine marten), *Martes pennanti* (fisher), *Mustela richardsonii* (ermine), *Gulo gulo* (wolverine), *Lynx canadensis* (lynx), *Alces alces* (moose), *Lontra Canadensis pacifica* (Pacific Coast Otter), *Lepus americanus macfarlani* (Pacific Varying Hare), *Marmota monax petrensis* (Stikine groundhog), *Marmota caligata* (hoary marmot), *Tammiascirus hudsonicus* (red squirrel), *Glaucomys sabrinus* (flying squirrel), *Castor canadensis* (beaver), *Ondrata zibethicus* (muskrat), *Erethizon dorsatum* (porcupine), *Canis latrans* (coyote), *Canis lupus* (wolf), *Vulpes vulpes* (red fox), *Oreamnos americanis* (mountain goat), *Ursus arctos horribilis* (grizzly bear), and *Ursus americanus* (black bear).

Fig. 7.31 Northern Pacific Coast forest assemblages include *Pseudotsuga menziesii* (Douglas Fir), with *Callitropsis nootkatensis* (Nootka cypress) in sheltered locations and *Picea sitchensis* (Sitka spruce) in open or rocky substrate locations

Fig. 7.32 *Callitropsis nootkatensis* (Nootka cypress)

Fig. 7.33 Inland vegetation assemblages, northern Coast Range region east of Prince Rupert. Species include *Pinus contorta* (lodgepole pine) *Abies lasiocarpa* (Alpine fir), *Picea engelmannii* (Engelmann Spruce), *Picea glauca* (white spruce), *Tsuga mertensiana* (mountain hemlock), *Betula papyrifera* (white birch), and *Populus tremuloides* (trembling aspen)

Mount Edziza lies along the border of the ranges of *Ovis canadensis* (bighorn sheep) to the south, and *Ovis dalli* (Dall's sheep) to the north. Although reduced in numbers and confined to higher altitudes compared to their distribution ca. 1850 (Wild Sheep Working Group 2014), bighorn sheep are expanding their range northward at present. Boreal species, such as mule deer, wolverine, and lynx are more common in the eastern areas.

Interchange with interior BC populations occurred in the Skeena Valley and along the Hecate Lowland. Here, the fauna present include *Odocoileus hemonius sitkensis* (mule deer), *Canis latrans* (coyote), *Canis lupus* (wolf), *Felis concolor* (cougar), *Lynx canadensis* (lynx), *Ursus americanus* (black bear), and *Ursus arctos horribilis* (grizzly bear). Although most boreal species are present in this northern Pacific Cordillera, *Alces alces* (moose) are not. At montane elevations, *Ovis canadensis* (bighorn sheep) are present. Species populations are generally low, reflecting an absence of prey for most predatory species and the unsuitability of the dense coniferous forest for some herbivores. A conspicuous exception to this are bears, thriving on the rich harvest of Pacific salmon of the Great Bear Rainforest region (e.g. Mowat and Heard 2006).

Two subspecies of the black bear exist: the common *Ursus americanus altifrontalis*; and *Ursus americanus kermodei*, a subspecies with light grey to white fur that is otherwise almost identical to the common black bear. White-furred kermode bears (Spirit bears; Tsimshianic*: moksgm'ol*) evolved on Princess Royal Island along the eastern shore of Hecate Strait, and have spread to a number of adjacent islands and the mainland south and east of Prince Rupert, as far inland as Terrace. Kermode bears show a variety of fur colours, with white-furred animals representing 10–20% of the population (Hedrick et al. 2012). The subspecies probably developed its white coat during the most recent glaciation, and has been able to breed without interaction with the more common black bears due to the isolated nature of the terrain. Terrestrial fauna in southern coastal British Columbia represent descendants of northward-migrating animals, constrained by mountainous topography inland.

7.6.2 Island Fauna

Terrestrial faunal assemblages of the islands are very limited. The warm water temperatures keep the Salish Sea, Johnstone Channel, and Hecate Strait ice-free throughout the year, precluding faunal migration from the British Columbia mainland. The dense forests also afforded little opportunity for nourishment for small animals, and hence limited large predators as well. Consequently, mainland species such as *Mephitis mephitis* (skunk), *Lynx canadensis* (lynx), and *Ursus arctos horribilis* (grizzly bear) are absent.

On Vancouver Island, only 18 indigenous species of mammals are present (Banfield 1981). They include *Cervus canadensis rooseveltii* (Roosevelt Wapiti), *Ursus americanus* (black bear), *Felis concolor* (cougar), *Canis lupus crassodon* (Vancouver Island wolf), *Lontra canadensis vancouverensis* (Vancouver Island river otter), *Martes americana* (marten), and *Neogale vison* (mink), all of which are excellent swimmers. The only large species present which is not known for its swimming skill is *Odocoileus hemonius* (mule deer). Although some mule deer were introduced to Vancouver Island and have spread, it is uncertain as to whether all mule deer on the island are descendants of the introduced individuals.

The isolation of Vancouver Island has led to the development of several island species and subspecies, such as the Vancouver Island river otter. Most of these animals have restricted habitats, due to the limited area, and many have probably always been present in small numbers. With increased logging throughout the island, many of these Vancouver Island residents have suffered severely. The rarest of these species is *Marmota vancouverensis* (Vancouver Island marmot), a close relative of the yellow-bellied marmot and the groundhog. From a low of 30 individuals in 2004, the current population of Vancouver Island marmots is estimated at about 500 individuals (see Marmot Recovery Foundation, n.d.).

Noteworthy terrestrial fauna in Gwaii Haanas include *Ursus americanus carlottae*, a local subspecies of Black Bear, noted for its large, broad forehead. *Procyon lotor vancouverensis* (Vancouver Island raccoon) is also present, and it is uncertain whether this subspecies is native to Haida Gwaii or has been introduced. Other introduced rodents include beaver (*Castor canadensis*), red squirrel (*Tamiasciurus hudsonicus*), and several species of rats and mice, none of which is appreciated by those interested in preserving the native vegetation assemblages.

The only native ungulate subspecies, *Rangifer tarandus dawsoni* (dwarf Dawson Caribou), formerly confined to Graham Island, has been extinct since 1910 (Byun et al. 2002). The introduction of Sitka blacktail deer (*Odocoileus hemonius sitkensis*), also a dwarf ungulate, was successfully accomplished between 1900 and 1916, although the local forest vegetation has paid a price. Wapiti (*Cervus canadensis*) have also been introduced to Graham Island.

In the Gulf Islands, settlement has resulted in the elimination of many species, and the small areas of the islands provide insufficient ranges for most large mammals. Distinctive bird species include the Rufous hummingbird (*Selasphorus rufus*), orange-crowned warbler (*Vermivora celata*), Wilson's warbler (*Wilsonia pusilla*), and yellow-billed cuckoo (*Coccyzus americanus*). Fallow deer (*Cervus dama*) were introduced to James Island in 1902, and have spread to several adjacent islands, reaching nuisance invasive status.

There are a number of insect, amphibian, and reptile species with very limited Canadian distribution (BC Ministry of the Environment n.d). Reptiles are represented by four species of garter snakes (*Thamnophilis*). Sharp-tailed snake (*Contia tenuis*) and gopher snake (*Pituophis catenifer catenifer*) are both considered as endangered on the COSEWIC listing. These snake species prefer open forests and woodlands, especially along the fringes of forests. They are more common in the deciduous fringes (Garry oak, bigleaf maple) than in the dense conifer-dominated areas. Warmth-loving, the snakes prefer drier, south-facing slopes. The major threats include habitat loss and invasive plant colonization, particularly of Scotch broom. Human persecution of snake populations has occurred as well.

Amphibians include Pacific treefrog (*Hyla regilla*), red-legged frog (*Rana aurora*), rough-skinned newt (*Taricha granulosa*), Olympic salamander (*Rhyacotritum olympicus*), northwestern salamander (*Ambystoma gracile*), and clouded salamander (*Aneides ferreus*). The amphibians are generally confined to the dense coniferous forests, and are threatened with habitat loss resulting from logging.

Several bird species, in particular the northern spotted owl (*Strix occidentalis*) inhabited "old-growth" conifer forests on Vancouver Island and the lower mainland. The spotted owl cannot survive outside of the old-growth forest, as it is dependent on the hollow trunks of older or dead Douglas-firs for its dens. The population in British Columbia is definitely less than 100 individuals, and may be less than 10 (Northern Spotted Owl Breeding Program, n.d.). In an attempt to compensate both the owl and other species such as the bald eagle, logging companies have begun the practice of erecting telephone poles or dead hollow trees in recently logged areas, to provide alternate nesting sites. Other bird species resident in Pacific Rim National Park Reserve (Godfrey 1986, Bird 2017) include Red-throated loon (*Gavia stellata*), scoter (*Oidemia spp.*), bufflehead (*Bucephala albeola*), Brandt's cormorant (*Phalacrocorax penicillatus*), Black oystercatcher (*Haematopus bachmani*), Steller's jay (*Cyanocitti stelleri*), Pileated woodpecker (*Dryocopus pileatus*), chestnut-backed chickadee (*Parus rufescens*), golden-crowned kinglet (*Regulus satrapa*), red crossbill (*Loxia curvirostra*), song sparrow (*Melospiza melodia*), and dark-eyed junco (*Junco oreganus*). Trumpeter swans (*Olor buccinator*) winter in the park.

7.6.3 Pacific Salmon

Canadian landings of all species of Pacific salmon (*Oncorhynchus*) represent 15–20% of the world total. The Pacific salmon, unlike the Atlantic (*Salmo salar*), spawns only once in its lifetime. The species of Pacific salmon include *O. gorbuscha* (pink; Halkomelem: haan), *O. keta* (chum, dog salmon; Halkomelem: k^{w}'úlux), *O. kisutch* (coho; Halkomelem: caw'in), *O. nerka* (sockeye; Halkomelem*: suk-kegh,* red fish), and *O. tshawytscha* (tyee, chinook, spring; Halkomelem: sc'uq^{w}ay'; Fisheries and Oceans Canada n.d.; see Pincott 2018; Fig. 7.34). Although all of these species are harvested commercially, the sockeye, coho, and tyee are most prized. A large tyee may exceed 12 kg.

Pink salmon were initially considered to be the least desirable, due to their relatively small size and low fat and oil contents. Since 1980, pink salmon has become increasingly important, and the catch shares held by coho and sockeye salmon have been declining (Grant et al. 2019; Pacific Salmon Foundation n.d., Flores et al. 2021, Wilson et al. 2021). The decline is due to several factors. In addition to logging, local overfishing and poaching, chemical pollution, and construction of hydroelectric dams on the BC mainland have all had an impact. The removal of sea otters (*Enhydra lutris)* from the offshore kelp beds has also had an adverse impact, allowing destruction of the kelp beds by over-grazing sea urchins.

During El Niño years, the warmer waters off the coast of southern Vancouver Island, an important feeding territory for growing Coho salmon (as well as the other species), are invaded by Pacific Jack mackerel (*Trachurus symmetricus*), a warm-water species from southern Baja California (Beamish and Neville 2021). Disputes concerning allocation of stocks of Coho and the other salmon species among BC user groups, and between Canada and the United States, continue.

Species distribution is greatly affected by local conditions, with adjacent rivers and coastal waters supporting very different populations (Temple 2005, Ferriss et al. 2014). On Haida Gwaii, the streams flowing into Juan Perez Sound are dominated by pink salmon, with lesser chum populations and almost no other species. Dixon Entrance, along the northern margin of Graham Island, is also dominated by pink salmon, but contains coho, sockeye, and chum. The Skeena River has healthy populations of chinook and chum salmon, in addition to the dominant pink, but lacks coho and sockeye. Burke Channel, in the southern part of the Hecate Lowland, is dominated by sockeye salmon, with lesser amounts of pink, coho, chum, and chinook.

One local effect which has become evident recently is the change in habitat induced by logging, both as a result of increased sediment and wood fibre input into the streams and the physical transport of thousands of floating logs in the rivers. The species are affected to different degrees by logging activity. For example, *Oncorhynchus keta* (chum salmon) is uniformly adversely affected by logging, but some data suggests that *O. kisutch* (coho) is unaffected or only marginally affected (Tschaplinski and Hartman 1983). Logging appears to increase the overall yearly variability of the catches of almost all of the salmon species.

Declining salmon stocks throughout the Pacific Northwest have resulted in tensions between the various users of the resource, including sport fishers, commercial offshore vessels, and native groups (Price et al. 2017; Grant et al. 2019). Tensions have also grown between the US and Canadian governments, due to the trans-border nature of the stocks. Salmon spawning in the headwaters of the Columbia River (in Canada) must traverse the river (and its dams) to reach the sea; many of the salmon offshore of the Washington coast spawn

Oncorhynchus gorbuscha (pink; haan)

Oncorhynchus keta (chum, dog, kʷ'úlux)

Oncorhynchus kisutch (coho; caw'in)

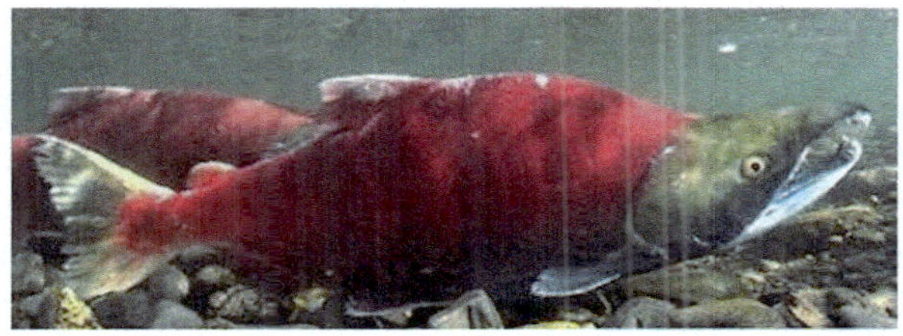
Oncorhynchus nerka (sockeye, red, suk-kegh)

Oncorhynchus tshawytscha (tyee, chinook, spring, sc'uqʷay')

Fig. 7.34 Pacific salmon species. (Illustration credits: pink, sockeye, and chinook (tyee): US Fish and Wildlife Service; chum, United States Geological Survey; coho, National Park Service (US))

in the tributaries of the Fraser River. Disputes over allocation of salmon quotas have escalated into threats to interrupt or hinder coastal traffic and communications along the Pacific seaboard.

7.7 Marine Environments

7.7.1 Currents. Tides, and Waves

The dominant ocean surface current is the Alaska Current, which flows northward from Oregon to Alaska, bringing relatively warm water along open Pacific shorelines. Approximately 60% of the North Pacific Current flow moves northward, with the remainder turning south along the California coast (Freeland 2006). During a typical year, water temperatures do not vary greatly, with mean temperatures of 8 °C in January (the coldest month) and 13 °C in August (the warmest month) offshore of Vancouver Island (Thomson 1981, Okey and Dallimore 2001). Ongoing climate change is gradually warming the current, with a predicted change of 1.8 °C to sea surface temperature by 2050. Typically, the Alaska Current is 100 km wide, with a discharge of 10^6 m^3s^{-1}. At 2–3 km/h peak velocity it is a relatively slow current. Onshore winds create downwelling conditions over most of the current, resulting in warmer temperatures in coastal locations. Fog is occasionally a problem for coastal navigation, with sufficient fog to preclude navigation occurring on 20 days in a typical year at Haida Gwaii and Prince Rupert.

The configuration of the Continental Shelf and the presence of Vancouver Island and Haida Gwaii reduce the influence of freshwater discharge on the offshore current, with the exceptions of the southern Salish Sea (Fraser River) and Dixon Entrance (Skeena and Nass rivers). In these areas, the downwelling dominant along most of the coastline is countered by estuarine and fluvial input, resulting in enhanced productivity.

ENSO (El Niño) years see some modification to this pattern, with the Alaska Current weakening due to partial interruption of the feeding Kuroshio Current from the central

Pacific. However, water temperatures remain relatively warm, as southwesterly winds bring warm air from the pool of warm water west of Baja California, created by the disruption of the Humboldt Current-Equatorial Current system in the South Pacific. The coastal zones are thus characterized by slower northward flow of the Alaska Current, but lesser changes in temperature.. The effects of El Niño events are less pronounced along the Hecate Strait shoreline than to the south, and also are less noticeable than along the coastline of Alaska west of Prince William Sound. Temperatures at shoreline sites vary little with El Niño.

In the North Pacific, the tidal amphidromic point is located west–southwest of San Diego. The tide progresses northwestward along the British Columbia Pacific coast, moving approximately 740 km/h in the deep water offshore (Dohler 2007). Because the Continental Shelf along the open Pacific coast is relatively narrow (50–70 km), there is little difference in high-tide times along the open coast of Vancouver Island, and only 30 min difference between high tide times at Cape Scott and southern Haida Gwaii (Fig. 7.35).

Cliffs and narrow coastal plains characterize the coastline at Gwaii Haanas (Bevington et al. 2017). Sand spits and dunes are found at Naikoon and Masset on Graham Island. Hecate Strait has a tidal range of 2.4 -3.7 m, and the tidal range exceeds 4 m locally in the Skeena estuary.

The Pacific Rim coastline is wave-dominated, marked by lengthy sandy beaches and tombolos in lower energy areas (such as Long Beach; Fig. 7.36) and erosional sea stacks and sea cliffs in higher energy regions (such as the West Coast Trail segment). Waves arriving at Pacific Rim are driven across the Pacific by westerly winds, producing long-wavelength swells which construct offshore bars. Multiple breakers are common, but seldom reach sufficient amplitudes for good surfing. Storms are less common here than along most of the Pacific shore.

7.7.2 Offshore Bathymetry

The Continental Shelf-Slope break, located 50–70 km seaward of the open Pacific shorelines, is very sharp (Barrie et al. 2014). Submarine canyons are present seaward of the Juan de Fuca Strait, Hecate Strait, and Dixon Entrance (Fig. 7.37). However, the contrast in relief is limited compared to the Laurentian Channel of the North Atlantic coastline, due to the differences in postglacial relative sea level history.

The Pacific Ocean waters generally exceed 2000 m in depth, and locally the EEZ extends more than 300 km beyond the slope break. Bottom current flow is northwest offshore of Vancouver Island, and northward along Haida Gwaii. Contour currents dominate over turbidity currents, and the outer shelf, slope, and offshore regions are generally sediment-starved.

The Pacific Ocean waters north of Vancouver Island are designated as Queen Charlotte Strait, Queen Charlotte Sound, Hecate Strait (K̲andaliigwii), and Dixon Entrance (Crawford and Thomson 1991). The former three are areas of continental shelf (maximum depth 250 m), with shallow banks (Middle Bank, Goose Island Bank) separated by troughs (Moresby, Mitchell's (or Reed), Goose Island (or Sea Otter)) feeding the Queen Charlotte Fan. Microtidal conditions prevail along the coastline. The Hecate Strait and Queen Charlotte Sound Glass Sponge Reefs Marine Protected Area, established in 2017, conserves four locations with 9,000 year old reefs (Thornborough and Dunham 2019).

Dixon Entrance flanks the north coast of Haida Gwaii, and has a maximum depth of 450 m. The Skeena is the major river flowing to the ocean along the Hecate Lowland shore, discharging through a braided channel system at Prince Rupert. The narrow, fjordal Skeena estuary and river valley are notable for landslides and mudflows, which frequently hinder traffic along the Yellowhead Highway (#16) and the Canadian National Railway (Clague 1984, Blais-Stevens et al. 2018). The Skeena valley lies along the proposed Northern Gateway petroleum pipeline. On 8 June 2002, the Zymoetz River slope failure moved 4 km downslope to fracture a natural gas pipeline east of Terrace. Repair of the natural gas pipeline and lost access to timber cost approximately $45 million (2022 equivalent; McDougall et al., 2006).

Another source of freshwater for Dixon Entrance is the Nass River, entering Portland Canal (*K'alii Xk'alaan,*) along the Alaska border. "Canal" refers to the Spanish term for "channel", rather than designating an artificial waterway.

7.7.2.1 Seamounts

Plate tectonic activity has produced several volcanic vents and seamounts, which form unique ecological settings.. The Endeavour Hydrothermal Vents Marine Protected Area occurs at ~ 2250 m depth, along the Explorer spreading ridge (Fisheries and Oceans Canada 2021). Chemosynthesis, whereby organic compounds are derived from chemical synthesis of hydrogen sulphide and other inorganic molecules, without sunlight, results in unique biological communities. One species of sulphide-hosted microbe can survive to 121 °C. Twelve species only live within the Endeavour MPA, including the sea spider (*Sericosura venticola*).

Volcanic seamounts are also present offshore of Vancouver Island and Haida Gwaii. Seamounts appear to be the surface expression of hot plumes of partially molten rock rising from the lower mantle, not necessarily associated with Plate Tectonic boundaries. Plumes rising under the ocean floor result in magma eruptions as basalt lava, often capped by obsidian, tephra, and larger volcaniclastic fragments formed as hot magma meets cold ocean water. Submarine volcanic

Fig. 7.35 Mesotidal conditions prevail along much of the British Columbia coastline. Upper: Sandy tidal flat with boulders (some added for shoreline protection) and bedrock outcrops, Bella Bella. Lower: Sandy tidal flat at Nanaimo

eruptions are influenced by seawater interaction with magma, producing substantial quantities of volcanic gases.

SGaan Kinghlas (Bowie) seamount is located 180 km west of Haida Gwaii (Herzer 1971, Dower and Fee 1991). It rises from 3000 m depth to within 24 m of the surface of the Pacific Ocean, greater relief than many terrestrial mountains in BC. As wave heights (and trough depths) in the region can exceed 27 m, SGaan Kinghlas is considered a hazard to navigation.

Fig. 7.36 Coastline, Pacific Rim National Park. Upper left: conifer forests extend to the shoreline at Long Beach and other locations with gentle slopes and sediment beaches. Lower left: Long Beach during a period of relative storm inactivity displays low-energy fine sand with well-developed gentle ripples, an ideal place for a walk. Upper right: after a storm, displaced logged conifer trunks form concentrations of large woody debris. Lower right: storm damage to Parks Canada infrastructure near Tofino

During the postglacial lower relative sea level episode, SGaan Kinghlas was exposed above the ocean waters, and was well-known to the Haida residents. Composed of basalt capped by sand- and ash-sized tephra, SGaan Kinghlas last erupted approximately 18,000 years ago. A submarine volcanic eruption could pose a hazard to navigation, similarly to Kick'em Jenny near Grenada in the Caribbean.

The combination of shallow water and volcanic nutrients provide unique ecological conditions at SGaan Kinghlas, leading to its designation as a Marine Protected Area co-managed by the Haida nation and Fisheries and Oceans Canada. Among the organisms present are longhorn decorator crab *(Chorilia longipes),* brittle stars (Ophiacantha), Giant Pacific octopus (*Enteroctopus dofleini*), Pacific Ocean perch (*Sebastes alutus*, rockfish), Pacific halibut (*Hippoglossus stenolepis*), and sablefish (*Anoplopoma fimbria*). Fishing, particularly for sablefish, is not permitted, and marine traffic is restricted for both ecological and navigational reasons.

7.7.3 Salish Sea

The Salish Sea includes the waters of Juan de Fuca Strait, Haro Strait and the other passageways through the San Juan and Gulf Islands archipelagos, and the Strait of Georgia (Freelan 2009; Tucker and Rose-Redwood 2015; Fig. 7.38). This relatively shallow (maximum depth 400 m) sea is dominated by tidal influx through Juan de Fuca Strait (west) and Johnstone Strait (north), along with freshwater discharge from the Fraser River. The Salish Sea drainage basin is defined by the height of land on Vancouver Island, and by the drainage divide on the BC and Washington mainland, including the Fraser River south of Hell's Gate.

Juan de Fuca Strait has a maximum width of 18 km and a maximum depth of 250 m across the Continental Shelf, shallowing to 120 m depth at the Victoria-Green Point Moraine (Thomson 1981). The moraine marks the maximum seaward extent of the glacially eroded segment of the outlet valley

Fig. 7.37 Bathymetry of the Pacific offshore of Canada. Seamounts are indicated with red dots. (Base map: NOAA)

for the Fraser. The numerous submarine glacial depositional landforms present south of the Gulf Islands are susceptible to slope failure, including that generated by earthquakes, and reworking by currents. The main part of the strait, south of Vancouver Island, is not a fjord.

Shallow coastal waters slow the tidal progression. The tide requires 2 to 4 h to progress from the southwestern tip of Vancouver Island to Haro Strait (Thomson 1981; Dohler 2007). Tides are semi-diurnal (more diurnal and weaker at Victoria), with influx at 4 km/h forming a modified salt-water wedge at depth. At Victoria and Colwood (Esquimalt harbour), the tide becomes more diurnal in character, with only one pronounced high and low water each day for about 20 days a month. Limited tidal activity at Esquimalt harbour was one factor contributing to the establishment of the naval installation. The mean tidal range varies from 1.8 m at Victoria to 2.4 m at the southern entrance to Haro Strait (Fig. 7.39). Tidal influx is displaced by the Coriolis Effect to the east, resulting in higher and more rapidly moving tidal waters in Puget Sound and Boundary Bay.

Summer water temperatures are typically 11 °C–14 °C, with the coolest water associated with the strongest tidal influx. Due to general climate warming, sea surface temperatures are predicted to increase by 2–3 °C by 2095 (Khangaonkar et al. 2019). Typical salinity ranges from 29–32 psu.

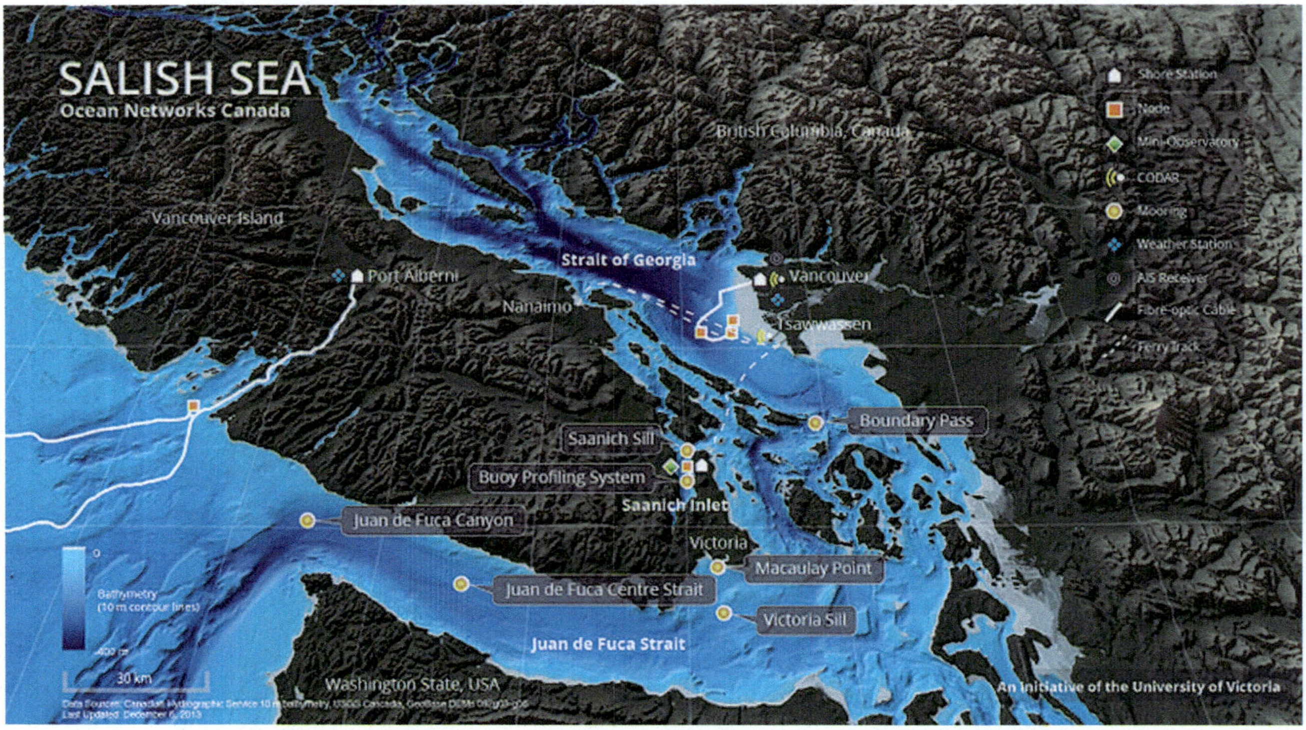

Fig. 7.38 Salish Sea oceanography. Ocean Networks Canada, hosted by the University of Victoria, uses ships, UAVs, gliders, buoys, coastal installations, and subsea instruments to collect marine data

The San Juan Islands archipelago, flanking Haro Strait and Boundary Pass to the east, was the subject of the 1859 boundary dispute between USA settlers and Hudson's Bay Company personnel (Neering 2011). This "Pig War" cost the life of one pig ("It's up to you to keep your potatoes out of my pig"), but resulted in considerable military deployments by the USA and Great Britain. The dispute was settled in 1872 by Wilhelm I of Germany, in favour of the USA.

The islands to the north and west of Haro Strait and Boundary Pass are referred to as the Gulf [of Georgia] Islands. Gulf Islands National Park Reserve was established in 2003. It currently consists of 62 km^2, including parts of 15 islands, several exposed rocks and reefs, and adjacent submerged areas. Consultation is ongoing, with the goal of establishing Gulf Islands as a full National Park.

Haro Strait has a maximum depth of 200 m, and is the main inlet for tidal waters into the interior of the Salish Sea (Thomson 1981; Fig. 7.39). Salinity varies from 22–28 psu. Other channels through the Gulf Islands are shallow and relatively narrow. Active Pass, separating Galiano and Mayne Islands as a 5 km long, 450 m wide channel with two 90° bends, is part of the main ferry route between the mainland and Vancouver Island. Tidal currents in Active Pass frequently exceed 15 km/h.

The open waters to the north, washing Vancouver and Nanaimo, are relatively shallow, warm in summer (~21 °C offshore, ~25 °C at Vancouver), and hyposaline offshore (23–27 psu in summer). Tides at Nanaimo and Vancouver (Burrard Inlet) are low mesotidal (2.1–2.3 m range) and mixed mainly semi-diurnal, with greater variation in Burrard Inlet. Wave action at Vancouver is restricted by the limited fetch afforded to the prevailing southwesterly winds, and is almost insignificant at Nanaimo.

With a mean discharge of 3500 m^3 s^{-1}, increasing by a factor of ten during late spring and early summer, the Fraser River has a major influence on water quality in the northern Salish Sea. A plume of freshwater reduces mean salinity offshore to 12–16 psu, with water temperatures varying by as much as 10 °C from the shallow, isolated Boundary Bay to the southeast. The freshwater plume generally trends southwest, augmenting ebb tidal flow through Haro Strait.

Tidal influx to the northern Salish Sea also comes from Queen Charlotte and Johnstone Straits. Separating northern Vancouver Island from the mainland, Johnstone Strait is 50 km long and 2.5 to 5 km wide, with a maximum depth of 400 m. The strait is marked by semi-diurnal, mesotidal (3.0–3.5 m) conditions, and strong tidal currents, with velocities to 7.7 m s^{-1}, and discharge to 3.2 × 10^5 m^3 s^{-1} (Sutherland et al. 2007). As the tide progresses inland (west to east), tidal rapids (e.g. in Cordero Channel) and whirlpools form, obstructing navigation. Offshore of Campbell River, Ripple Rock was destroyed in 1958 to remove one whirlpool, in what was possibly the "world's largest non-atomic blast" (Museum at Campbell River n.d.).

Fig. 7.39 Haro Strait has a maximum depth of 200 m, and is the main inlet for tidal waters into the interior of the Salish Sea

7.7.3.1 Bute Inlet

The narrowness of Johnstone Strait led to suggestions that the Canadian Pacific Railway could be constructed from Bute Inlet across to Vancouver Island. This undertaking would have required construction of seven high bridges across the tidal channels, in addition to the effort required to build along Bute Inlet and the Homathko River. For a combination of engineering and political reasons, the Fraser Canyon route was selected for the CPR in 1881.

Bute Inlet is the largest and best-studied of several fjords along the Salish Sea, Johnstone Strait, and Queen Charlotte Sound coasts (Jackson et al. 2021). At 80 km length, 4 km width, and 660 m maximum depth, flanked by peaks to 2700 m asl, Bute Inlet is a classic fjord developed in a cubic parabolic glacially eroded trough. The deep waters of the fjord are isolated from marine influx by an erosional sill 220 m below the surface. Combined with the freshwater influx from the Homathko River, the result is sharp stratification. Throughout most of the year, a 20 m thick surface fresh layer overlies a sharp halocline and thermocline to 50 m depth, below which salinity and temperature remain essentially constant at 30–31 psu and 8 °C–8.5 °C. Changes in the deepwater regime over time are largely driven by slow interchange with surface waters. Water temperatures increased by 1.2–1.3 °C over 1950–2020 (Jackson et al. 2021).

Steep fjord and trough valley walls are subject to slope failure. Failure of steep bedrock slopes can result from a release of lateral pressure. An active alpine glacier exerts stress on the slopes flanking the glacial trough, as well as providing lateral support. When the glacier melts, the slopes are relieved of stress. This debuttressing allows the rock to expand, forming joints and weakening the rock. the steep walls of a fjord extending to sea level may be subject to the same style of mass movements as are steep cliffs along glaciated alpine valleys.

In this environment, near-field tsunami generated by slope failure into lakes are of significant concern. A failure-driven tsunami at Chehalis Lake, BC, produced run-up heights in excess of 10 m, overwhelming campgrounds around the lake (Roberts et al. 2013). Fortunately, the tsunami occurred on 4 December 2007, when the campgrounds were unoccupied.

On 28 November 2020, a slope failure released 18 million m^3 of rock 1,000 m downslope into a glacial lake on Elliot Creek, at the head of Bute Inlet (Geertsema et al. 2022). The 50 million tonnes of rock generated a 100 m high tsunami in the lake. Subsequently, much of the rock and water arrived at the bottom of Bute Inlet, perturbing the water stratification and bottom sedimentation. Similar failures have occurred at Kitimat and other fjords. Ongoing glacial retreat means that increased debuttressing and local tsunami generation can be anticipated in the future.

Fig. 7.40 *Enhydra lutris* (sea otter). Removal of the sea otter has had a severe effect on the ecological balance of the Pacific Coast. See text for discussion

7.7.4 Marine Fauna

7.7.4.1 Sea Otter

Enhydra lutris (sea otter; Fig. 7.40) formerly inhabited coastal waters (< 40 m depth) throughout the region (Silverstein et al. 1995). Mature otters are between 14 and 50 kg, and 1–1.5 m long. Adaptations to the marine environment include small nostrils and ears which can be closed, flat webbed hind feet, short front paws with ridged padding and retractable claws, the ability to drink seawater, and reduced bone density (osteosclerosis), facilitating floating on the back. Sea otters use flat pebbles for cracking mollusc shells, making them possibly the only mammals other than primates to consistently use tools.

Before widespread commercial exploitation for fur began in 1741, there were between 150,000 and 300,000 sea otters in Pacific coastal waters. In the period 1741–1867, more than 800,000 were killed by Russian harvesters in the Aleutian Islands. Subsequently in the latter nineteenth century, the species was hunted to extinction in Canada. After imposition of an international hunting ban in 1911, small populations of sea otters survived in Alaska and along the California coast.

Removal of the sea otter has had a severe effect on the ecological balance of the Pacific Coast. Sea otters eat molluscs and urchins, and thus serve to keep these populations under control. Removal of the otters initiated a rise in the populations of most marine molluscs, including abalone (*Haliotis kamtschatkana*). A productive abalone fishery quickly developed, as new markets for abalone opened across North America and the Asian market became more lucrative. Simultaneously, the red sea urchin (*Stronglyocentrolus franciscanus*) population began to expand. The red sea urchins began to compete with the abalone for food, and both started to overgraze the kelp forests along the shores.

Overharvesting of abalone became a severe problem in the early 1980s, when landings collapsed. Between 1977 and 1984, abundance of northern abalone declined by more than 75%. Since 1952, the zone of maximum exploitation shifted along the length of the British Columbia coast, with the northern areas dominating since 1975. However, harvest for all northern abalone fisheries in BC has remained closed since December 1990, including indigenous communities.

The decline in abalone and the absence of sea otters has permitted the red urchins to expand, decimating the kelp beds and creating "urchin barrens", where only urchins are present on the ocean floor. The removal of the kelp exposes the shorelines to higher wave energies, depriving the Pacific salmon of a major sanctuary and productive food source area. As kelp beds are destroyed by urchin overgrazing, salmon numbers suffer accordingly. The initial removal of the sea otter, therefore, has resulted in a series of repercussions throughout the ecosystem, ultimately impacting on the economic stability of the west coast fishing industry.

Between 1969 and 1972, 89 sea otters were re-introduced to the Bunsby Islands, northwest of Pacific Rim National Park (Nichol et al., 2015). Numbers declined initially, but the population in BC coastal waters is currently increasing, estimated at 8000 animals. The original otter colony has split into separate populations, a common response to limited marine resources. Restoration of the otters has allowed some recovery of kelp beds, increasing salmon habitat, while simultaneously decreasing populations of some shellfish. Although the net benefits outweigh the costs, communities which are dependent upon shellfish harvesting have suffered disproportionate effects. Threats remain: the 1989 Exxon Valdez oil spill in Prince William Sound, Alaska, killed half that area's population.

7.7.4.2 Marine Mammals

Other marine mammals of British Columbia coastal waters include California sea lion (*Zalophus californicus;* Fig. 7.40), Steller's sea lion (*Eumetopias jubatus*), harbour seal (*Phoca vitulina*), northern fur seal (*Callorhinus ursinus*), and northern elephant seal (*Mirounga angustirostris*; Ford 2014). As many as 20,000 Grey whales (*Eschrichtius robustus*) migrate through British Columbia waters between mid-March and mid-April. Other cetaceans include orca (*Orcinus orca*), humpback (*Megaptera novaengliae*), sperm (*Physeter catodon*), minke (*Balaenoptera acutorostrata*), and fin whales (*Balaenoptera physalus*) and several species of dolphins.

Marine mammals of most species have been increasing in numbers in coastal British Columbia. Steller's sea lions numbered more than 32,000 in 2013, when the last census was taken (Olesiuk 2018). The estimated population of Pacific harbour seals is in excess of 100,000. Approximately 300 orcas live in the nearshore waters of Gwaii Haanas. A marine area of 3,400 km^2 north of Haida Gwaii is protected as "critical habitat" for orca and other cetaceans.

7.7.4.3 Orca

Orcinus orca (killer whale) is an iconic species in British Columbia (Fig. 7.40). There are four distinct populations in BC waters, varying in habitat, diet, possibly social structure, and conservation status (Barrett-Lennard and Heise 2007, BC Cetacean Sightings Network n.d.).

Resident orca remain in pods incorporating several family units, termed matrilines, involving a complex matriarchal society. Pods are based in a central marine territory, and travel as a group. Their diet is largely chinook salmon taken offshore, with other salmon species forming much of the remainder. Along the coast of BC, the "southern" resident community is centred in the southern Salish Sea. The combination of reliance on diminished salmon stocks, increased toxin levels in coastal waters, disturbance from vessel traffic, and reduced numbers makes this population critically endangered.

The "northern" resident orca community is centred in Johnstone Strait and Queen Charlotte Sound. Although more numerous than the southern resident community, the northern community faces a similar array of threats, and is listed as threatened by COSEWIC. Approximately 150 of these orca frequent the Robson Bight (Michael Bigg) Ecological Reserve.

The third community is migratory, and is referred to as Bigg's orcas (after Michael Bigg, formerly an orca specialist with Fisheries and Oceans Canada, who pioneered the technique of identification through photography of whale fins and flukes). Bigg's orcas primarily consume seals, sea lions, dolphins, and humpback whale and grey whale calves, hunting mainly in coves rather than offshore. They migrate along the Pacific coastline of North America. The available evidence indicates that the population of Bigg's whales is increasing, and more than 1000 unique whales have been identified. Matrilines are present, but generally are less extensive than in resident whale populations. Resident and Bigg's orcas do not socialize or interact with each other.

The fourth population involves offshore orcas. These orcas live over the outer Continental Shelf and Slope, although they occasionally venture into coastal waters, where they can be sighted and photo-identified. Little is known about the offshore orcas, and their population is estimated at approximately 250.

7.7.5 Seabirds

Birds utilize the Gwaii Haanas area extensively. With an estimated annual bird population of 750,000, and 1.5 million breeding seabirds, Haida Gwaii is one of the more important birding areas of the Pacific coast (Parks Canada 2021) Species include the black oystercatcher (*Haematopus bachmani*), Leach's fork-tailed petrel (*Oceanodroma leucorhoa*); rhinoceros auklet (*Cerorhinca monocerata*); semipalmated plover (*Charadrius semopalmatus*), least sandpiper (*Erolia minutilla*), bald eagle (*Haliaeetus leucocephalus*), ancient murrelet (*Synthliboramphus antiquus*), tufted puffin (*Lunda cirrhata*), horned puffin (*Fratercula corniculata*), Cassin's auklet (*Ptychoramphus aleuticus*), and pelagic cormorant (*Phalacrocorax pelagicus*). Parks Canada has undertaken a vigourous programme to eradicate rats, major predators of seabird eggs and chicks (Brandes 2021).

7.8 Coasts, Harbours, and Ports

Approximately 80% of British Columbia's human population lives in the coastal zone of the Pacific Maritime Ecozone. The Lower Mainland, extending from tidewater along the Fraser River to Hope, was home to more than 3 million people in 2022, about 60% of BCs total population. Vancouver Island accounted for almost 900,000 people, with ~ 400,000 in the Victoria CMA. Prince Rupert, the only city outside the Salish Sea drainage basin, had a population of 13,442 in 2021 (Fig. 7.41), and Kitimat had 8236. The Lower Mainland, including Vancouver, and Victoria are experiencing population growth, whereas Prince Rupert's and Kitimat's populations are essentially stable.

Interaction with the Pacific and its 27,000 km shoreline has been essential for all Pacific Maritime Ecozone residents for the past 15,000 years. Marine and coastal terrestrial resources, including cedar and salmon, shape the economies and cultures of the region. Declines in salmon populations are felt immediately. Many residents feel emotional connections to the great conifers, conflicting with others' desires to use them economically.

Fig. 7.41 Pacific marine mammals. Upper: *Orcinus orca* (orca); Lower *Zalophus californianus* (sea lion)

Vancouver's emergence as the dominant port (Fig. 7.42) was not entirely inevitable. In the 1840s, the original Fort Vancouver near the mouth of the Columbia River was regarded as the pre-eminent outlet for east–west trade. Subsequently, Victoria and Esquimalt developed as the preferred site for governmental and naval facilities. New Westminster, along the Fraser River, appeared to be the most likely for port development on the Lower Mainland. When discussion of the CPR began in 1871, Esquimalt, Bute Inlet, Douglas Channel (Kitimat), Portland Canal, Bella Coola, and the Skeena Estuary were all considered as potential tidewater termini, in addition to several other fjords.

The choice of Burrard Inlet was based in part on military considerations, in part because the CPR was able to acquire land for port facilities, in part because there was no substantial pre-existing settler population (with land speculators), in part for political reasons within British Columbia, and only in part for logistical and engineering considerations. Vancouver differed from Québec, Montreal, and Saint John in that a river location was not a dominant factor in port selection, but resembled Halifax and St. John's in its establishment around the deepest and most sheltered harbour. CPR traffic was based on shipping goods (including wheat) from eastern locations, and trans-Pacific imports, rather than transporting local BC lumber and other products. The result was that the Fraser Delta remained relatively undeveloped commercially, and the river community of New Westminster languished, while the Burrard Inlet waterfront grew. Subsequent development by the Canadian Northern Railway on the north side of Burrard Inlet followed the same pattern. The Grand Trunk Pacific also developed port facilities at Prince Rupert with the intention of promoting long-distance rather than local commerce, making little use of the Skeena River (Fig. 7.43).

Kitimat was constructed by the Aluminium Company of Canada (ALCAN) as part of the development of the aluminium fabrication plant and associated hydroelectric dam at Kenamo, including reversal of the Nechako River and flooding for reservoir construction. The choice of site was based on the requirements for a sheltered deepwater port

Fig. 7.42 Prince Rupert, the only city outside the Salish Sea drainage basin, had a population of 13,442 in 2021 (Wikimedia/Flickr/Sam Beebe)

without large amounts of competing vessel traffic, and the necessity for a captive source of electricity. Kitimat has also been suggested as the terminus for Liquefied Natural Gas and petroleum pipelines, as has Prince Rupert, but substantial opposition exists among many residents of the northern Pacific coastal regions.

British Columbia waters have been closed to hydrocarbon exploration and development since 1972, by both provincial and federal action. Although geological potential exists in offshore sedimentary basins, thee is very little appetite or public support for exploration. Opposition to pipeline development in the north and expansion of the existing hydrocarbon pipelines to Vancouver is strong, as is concern for the potential impacts of petroleum spills in coastal waters.

Responsibility for containment of an oil spill rests with the Western Canada Marine Response Corporation (WCMRC), an industry-funded body encompassing all large vessels (400 t) and oil-handling facilities (WCMRC n.d.). WCMRC currently has response bases at Vancouver, Duncan, Kitimat, and Prince Rupert, with more planned, along with caches of response equipment at other sites. The ideal response time in the event of an oil spill within the marine waters of British Columbia for a relatively small oil spill (<150 t) is 5 h, with 72 h for complete deployment to contain spills up to approximately 26,000 t.

In the event of a tanker-based spill in the Salish Sea, current circulation indicates that the Vancouver Island coastline south of Nanaimo and the Gulf Islands are more exposed than are the Fraser Delta, Burrard Inlet, and Lower Mainland. Enhanced Fraser River discharge during Spring would also tend to drive floating oil westward. Another concern is that the routes currently used by tankers arriving and departing Vancouver parallel salmon migration routes, particularly pink salmon (*Oncorhynchus gorbuscha*).

Fig. 7.43 Metropolitan Vancouver, showing the major port of Burrard Inlet in the north. Vancouver International Airport (Sea Island), Richmond, and Delta to the south are undergoing active subsidence. The terminals for Vancouver Island ferries are located at Horseshoe Bay in the north and Tsawwassen in the south. (NASA, Sentinel-2 image)

References

Adams RM, Karlin B, Eisenman DP, Blakley J, Glik D (2017) Who Participates in the Great ShakeOut? Why audience segmentation is the future of disaster preparedness campaigns. Int J Environ Res Public Health 14:1407. https://doi.org/10.3390/ijerph14111407

American Museum of Natural History (2018). Ghosts of Tsunamis past. American Museum of Natural History. Accessed 14 July 2022.

Anderson, R G Reichenbach I (1989) A note on the geochronometry of Late Jurassic and Tertiary plutonism in the Queen Charlotte Islands, British Columbia. Geol Surv Can Paper 89–1H:105–112, https://doi.org/10.4095/127417

Barrett-Lennard LG Heise K (2007) The natural history and ecology of killer whales: Foraging specialization in a generalist predator. In: Estes JA, Brownell RL, DeMaster DP, Doak DF, Williams TM (2007) Whales, whaling and ocean ecosystems. University of California Press, Berkeley. BC Cetacean Sightings Network (n.d.). Killer wahles. https://wildwhales.org/speciesid/whales/killer-whale/ Accessed 18 July 2022

Barrie JV, Hetherington R, MacLeod R (2014) Pacific margin, Canada shelf physiography: a complex history of glaciation, tectonism, oceanography and sea-level change. In: Chiocci, F. L. & Chivas, A. R. (eds) Continental Shelves of the World: Their Evolution During the Last Glacio-Eustatic Cycle. Geol Society, London, Memoirs, 41:303–313, https://doi.org/10.1144/M41.22

Barrie JV, Greene HG, Conway, KW Brothers DS (2021) Late Quaternary Sea level, isostatic response, and sediment dispersal along the Queen Charlotte Fault: Geosphere 17:375–388

BC Ministry of Forest, Lands and Natural Resource Operations (2014) Veg Res Inv

Bevington AR Clague JJ Millard T Walker IJ Geertsema M (2017) The diverse landscapes of Haida Gwaii. In Slaymaker O (ed.), Landscapes and landforms of Western Canada. Springer, pp 291–302

Bevington AR, Menomous B (2022) Accelerated change in the glaciated environments of western Canada revealed through trend analysis of optical satellite imagery. Remote Sens Environ 270:112862. https://doi.org/10.1016/j.rse.2021.112862

Bird L et al. (2016) The October 2012 magnitude (MW) 7.8 earthquake offshore Haida Gwaii, Canada. Int Seismol Centre. https://doi.org/10.5281/zenodo.999242 Accessed 18 July 2022

Bird DM (2017) Birds of Canada. DK Canada

Blais-Stevens A, Behnia P, Castagner A. (2018) Historical landslides that have resulted in fatalities in Canada (1771–2018); Geological Survey of Canada, Open File 8392, map poster

Brandes H (2021) A seabird haven. https://www.heidebrandes.com/master-blog/a-seabird-haven-on-the-canadian-islands-of-haida-gwaii-the-work-to-save-the-ancient-murrelet-goes-on. Accessed 18 July 2022

Brewer CE, Bleich VC, Foster JA, Hosch-Hebdon T, McWhirter DE, Rominger EM, Wagner MW, Wiedmann BP (2014) Bighorn Sheep: Conservation Challenges and Management Strategies for the 21st Century. Wild Sheep Working Group, Western Association of Fish and Wildlife Agencies, Cheyenne, Wyoming, USA

Brice RL, Coulthard B, Homfeld I, Dye L, Anchukaitis K (2021) Paleohydrological context for recent floods and droughts in the Fraser River Basin, British Columbia, Canada. Environ Res Lett 16:124074

British Columbia Ministry of Environment (n.d). The reptiles of British Columbia. www.bcreptiles.ca. Accessed 18 July 2022

British Columbia Ministry of Environment (n.d.). Guide to Amphibians of Vancouver Island. https://www2.gov.bc.ca/assets/gov/environment/plants-animals-and-ecosystems/wildlife-wildlife-habitat/amphibians-reptiles-and-turtles/guide_to_amphibians_of_vancouver_island_gulf_islands.pdf. Accessed 18 July 2022

Byun SA, Koop BF, Reimchen TE (2002) Evolution of the Dawson caribou (*Rangifer tarandus* dawsoni). Can J Zool 80:956–960

Catherall EE, Janes JK, Josefsson CA, Gorrell JC (2018) Spatial and temporal assessments of genetic structure in an endangered Garry oak ecosystem on Vancouver Island. Botany, 96:257–265

Clague JJ, Bobrowsky PT (1994) Evidence for a large earthquake and tsunami 100–400 years ago on western Vancouver Island, British Columbia. Quatern Res 41:176–184

Clague JJ (1984) Quaternary geology and geomorphology, Smithers-Terrace-Prince Rupert area, British Columbia. Geol Surv Can Mem 413, p 71

Clarkson RJ, Majewicz EJE, Mack P (2016) A re-evaluation of the 2010 quantitative understanding of the effects volcanic ash has on gas turbine engines. J Aerosp Eng 230:2274–2291

Comeau VM, Daniels LD, Zeglen S (2021) Climate induced yellow-cedar decline on the island archipelago of Haida Gwaii. Ecosphere. 12(3). https://doi.org/10.1002/ecs2.3427

Demarchi DA (2011) The British Columbia Ecoregion Classification 3rd Edition. Ecosystem Information Section, BC Ministry of Environment, Victoria, BC. http://www.env.gov.bc.ca/ecology/ecoregions/index.html

Dohler G (2007). Tides in Canadian waters. Canadian Hydrographic Service, Fisheries and Oceans Canada

Edwards BR, Russell JK, Simpson K (2011) Volcanology and petrology of Mathews Tuya, northern British Columbia, Canada: glaciovolcanic constraints on interpretations of the 0.730 Ma Cordilleran paleoclimate. Bulletin of Volcanology 73;479–496

Environment and Climate Change Canada 2020. Canadian Climate Normals 1981–2010 https://climate.weather.gc.ca/climate_normals/index_e.html. Accessed 18 July 2022

Evans SG, Hungr O, Clague JJ (2001) Dynamics of the 1984 rock avalanche and associated distal debris flow on Mount Cayley, British Columbia, Canada: implications for landslide hazard assessment on dissected volcanoes. Eng Geol 61:29–51

Fedje D, Mackie Q, McLaren D, Wigen B, Southon J (2021) Karst caves in Haida Gwaii: archaeology and paleontology at the Pleistocene-Holocene transition. Quat Sci Rev 272:107221

Ferriss BE, Trudel M, Beckman BR (2014) Regional and inter-annual trends in marine growth of juvenile salmon in coastal pelagic ecosystems of British Columbia, Canada. Mar Ecol Prog Ser 503:247–261

Fisheries and Oceans Canada (n.d.) Pacific Salmon. https://www.pac.dfo-mpo.gc.ca/pacific-smon-pacifique/index-eng.html Accessed 18 July 2022

Fisheries and Oceans Canada 2019. State of the Canadian Pacific salmon: Responses to changing climate and habitats. https://www.dfo-mpo.gc.ca/species-especes/publications/salmon-saumon/state-etat-2019/ebook/index-eng.html. Accessed 18 July 2022

Flores A-M, Davies MM, Kushneryk K, Lawn P, Helms S, Thomson HM, Nelson KR, Burns CW, Roias S, Gerwing TG (2021) Declines of Juvenile Coastal Cutthroat Trout and Coho Salmon over Fifteen Years in a Salmon-Bearing Stream in the Salish Sea. Western North American Naturalist 81(4):550–557

Ford JKB (2014) Marine Mammals of British Columbia. Royal Museum of British Columbia

Freelan S (2009) The salish sea map. http://staff.wwu.edu/stefan/salish_sea.html. Accessed 18 July 2022

Freeland HJ (2006) What proportion of the North Pacific Current finds its way into the Gulf of Alaska? Atmosphere-Ocean, 44:321–330

Gabrielse, H Yorath, C J 1991. Geology of the Cordilleran Orogen in Canada. Geological Survey of Canada, Geology of Canada Series 4. https://doi.org/10.4095/134069

Geertsema M, Menounos B, Bullard G, Carrivick JL, Clague JJ, Dai C, et al. (2022) The 28 November 2020 Landslide, Tsunami, and Outburst Flood—A Hazard Cascade Associated with Rapid Deglaciation at Elliot Creek, British Columbia, Canada. Geophys Res Lett 49:6. https://doi.org/10.1029/2021GL096716

Gilhen-Baker M, Roviello V, Beresford-Kroeger D, Roviello GN (2022) Old growth forests and large old trees as critical organisms connecting ecosystems and human health. A review. Environ Chem Lett 20(2):1529–1538. https://doi.org/10.1007/s10311-021-01372-y

Gillett NP et al. (2022) Human influence on the 2021 British Columbia floods. Weather and Climate Extremes 36:100441. https://doi.org/10.1016/j.wace.2022.100441

Goes S (2013) Western North America's jigsaw. Nature 496(7443):25–27

Government of British Columbia (2016) Indicators of Climate Change for British Columbia.ww2.gov.bc.ca/assets/gov/environment/research-monitoring-and-reporting/reporting/envreportbc/archivedreports/climate-change/climatechangeindicators-13sept2016_final.pdf

Grant SCH, MacDonald BL, Winston ML (2019) State of the Canadian Pacific salmon: responses to changing climate and habitats. Can Tech Rep Fish Aquat Sci 3332

Great BC Shake-out (n.d). Great BC Shake-out https://www.shakeoutbc.ca/. Accessed 18 July 2022

Guthrie RH, Friele P, Allstadt K, Roberts N, Evans SG, Delaney KB, Roche D, Clague JJ, Jakob M (2012). The 6 August 2010 Mount Meager rock slide-debris flow, Coast Mountains, British Columbia:

characteristics, dynamics, and implications for hazard and risk assessment. Nat Hazards and Earth Sys Sci 12:1277–1294
Haggart J, Enkin RJ, Monger JWH (2006) Paleogeography of the North American Cordillera. Geological Association of Canada, Special Publication 46
Hawley WB, Allen RM (2019) The fragmented death of the Farallon plate. Geophys Res Lett 46:7386–7394
Hawkins AC, Menounos B, Goehring BM, Osborn GD, Clague JJ, Jensen B (2021) Tandem dating methods constrain late Holocene glacier advances, southern Coast Mountains, British Columbia, Quaternary Sci Rev 274:107282,
Hebda CFG, McLaren D, Mackie Q, Fedje D, Pedersen MW, Willerslev E, Brown KJ, Hebda RJ (2022) Late Pleistocene palaeoenvironments and a possible glacial refugium on northern Vancouver Island, Canada: Evidence for the viability of early human settlement on the northwest coast of North America. Quat Sci Rev 279, Article 107388. https://doi.org/10.1016/j.quascirev.2022.107388
Hedrick et al. (2012) Population genetics of the white-phased "spirit" black bear of British Columbia. Evolution 66:305–313
Hickson CJ, Edwards BR (2001) Volcanoes and volcanic hazards. Bull Geol Surv Canada 548:145–181
Hipfner JM et al. (2018) Beach-cast debris surveys on Triangle Island, British Columbia, Canada indicate the timing of arrival of 2011 Tōhoku tsunami debris in North America. Marine Poll Bull 136:407–413
Hoffman KM, Davis EL, Wickham SB, Schang K, Johnson A, Larking T, Lauriault PN, Quynh Le N, Swerdfager E, Trant AJ (2021). Conservation of Earth's biodiversity is embedded in Indigenous fire stewardship. Proc Natl Acad Sci U S A 118(32):e2105073118
Hosie RC (1979) Native trees of Canada. Canada Department of Forestry, Don Mills, Ontario
Jackson JM, Bianucci L, Hannah CG, Carmack EC, Barrette J (2021) Deep waters in British Columbia mainland fjords show rapid warming and deoxygenation from 1951 to 2020. Geophysical Research Letters, 48(3): e2020GL091094. https://doi.org/10.1029/2020GL091094
Jensen BJ, Beaudoin, AB Clynne MA, Harvey J, Vallance JW (2019) A re-examination of the three most prominent Holocene tephra deposits in western Canada: Bridge River, Mount St. Helens Yn and Mazama. Quat Int 500:83–95
Jones J, Keller P, van der Flier-Keller E (2021) Review of official responsibility for the Salish Sea marine environment. Ocean Coast Manag 211:105748. https://doi.org/10.1016/j.ocecoaman.2021
Khangaonkar T, Nugraha A, Xu W, Balaguru K (2019) Salish Sea response to global climate change, sea level rise, and future nutrient loads. Journal of Geophysical Research: Oceans, 124. https://doi.org/10.1029/2018JC014670
Klinkenberg B (2020) E-Flora BC: electronic atlas of the plants of British Columbia. Department of Geography, University of British Columbia. Accessed 18 July 2022
Kulp SA Strauss BH (2019) New elevation data triple estimates of global vulnerability to sea-level rise and coastal flooding. Nat Commun 10, Article number:4844. https://doi.org/10.1038/s41467-019-12808-z
Lakeman TR, Clague JJ, Menounos B, Osborn GD, Jensen BJL, Froese DG (2008) Holocene tephras in lake cores from northern British Columbia, Canada. Can J Earth Sci 45:935–947
Larocque I, Allen DM, Kirste D 2015. The hydrology of Salt Spring island. Department of Earth Sciences, Simon Fraser University. https://www.sfu.ca/personal/dallen/SFU. Accessed 13 July 2022
Lekkas EL, Andreadakis E, Kostaki I, Kapourami E (2013) A Proposal for a new integrated tsunami intensity scale (ITIS-2012). Bull Seismol Soc Am 103:1493–1502
Lerbekmo JF (2008) The white river ash: largest holocene Plinian tephra. Can J Earth Sci 45:693–700
Letham B, Lepofsky D, Greening S (2021) A post-glacial relative sea level curve for the central Douglas Channel area, British Columbia, Canada. Quaternary Science Reviews, 263:106991. https://doi.org/10.1016/j.quascirev.2021.106991
Marmot Recovery Foundation (n.d.). https://marmots.org/. Accessed 18 July 2022
Marr KL Hebda RJ Mackenzie WH (2012) New alpine plant records for British Columbia and a previously unrecognized biogeographical element in western North America. Botany https://doi.org/10.1139/b2012-009
Massey NWD, MacIntyre DG, Desjardins PJ, Cooney RT (2005) Geology of British Columbia. BC Geological Survey, Map 2005–5
Mazzotti S, Lambert A, Van der Kooij M, Mainville A (2009) Impact of anthropogenic subsidence on relative sea-level rise in the Fraser River delta. Geology, 37(9):771–774
Menary D (2020) Great trees of Canada. Lulu.com
McDougall S, Boultbee N, Hungr O, Stead D, Schwab J (2006) The Zymoetz River landslide, British Columbia, Canada; description and dynamic analysis of a rock slide debris flow. Landslides 3:195
Moerman DE (1998) Thuja plicata. Moerman, D E, Native American ethnobotany. Timber Press, Portland, Oregon, pp 558–561
Mowat G, Heard DC (2006) Major components of grizzly bear diet across North America. Can J Zool 84:473–489
Museum of Campbell River (n.d.). The ripple rock. https://crmuseum.ca/learn/ripple-rock/. Accessed 18 July 2022
Natural Resources Canada 2022. Earthquakes Canada. https://earthquakescanada.nrcan.gc.ca/index-en.php. Accessed 18 July 2022
Naughton D (2012) The Natural history of Canadian Mammals. Canadian Museum of Nature
Neering, R 2011. The pig war: the last Canada–US Border Conflict. Heritage House, Victoria
Nichol LM, Watson JC, Abernethy R Rechsteiner E, Towers J (2015) Trends in the abundance and distribution of sea otters (Enhydra lutris) in British Columbia updated with 2013 survey results. DFO Can Sci Advis Sec Res Doc 039
Northern Spotted Owl Breeding Program (n.d). https://www.nsobreedingprogram.com/. Accessed 18 July 2022
Olesiuk, P.F 2018) Recent trends in abundance of steller sea lions (*Eumetopias jubatus*) in British Columbia. Fisheries and Oceans Canada, Scientific Advisory Secretariat, Research Document 2018/006. v + p 67
Parks Canada 2021. Birds of Gwani Haanas. https://www.pc.gc.ca/en/pn-np/bc/gwaiihaanas/activ/experiences/oiseaux-birds. Accessed 18 July 2022
Peterson CD, Clague JJ, Carver GA, Cruikshak KM (2013) Recurrence intervals of major paleotsunamis as calibrated by historic tsunami deposits in three localities: Port Alberni, Cannon Beach, and Crescent City, along the Cascadia margin, Canada and USA. Nat Hazards 68:321–336
Pincott E (2018) Contact and change in Central Salish words for salmon. Papers for the International Conference on Salish and Neighbouring Languages 53, University of British Columbia Working Papers in Linguistics 47, Marianne Huijsmans, Roger Lo, Daniel Reisinger, and Oksana Tkachman (eds.), 181–195
Price MH, English KK, Rosenberger AG, MacDuffee M, Reynolds JD (2017) Canada's Wild Salmon Policy: an assessment of conservation progress in British Columbia. Can J Fish Aquat Sci 74(10):1507–1518
Roberts NJ, McKillop RJ, Lawrence MS, Psutka JF, Clague JJ, Brideau ME, Ward BC (2013) Impacts of the 2007 landslide-generated tsunami in Chehalis Lake, Canada, in Landslide Science and Practice, pp 133–140, eds Margottini,C., Canuti, P.& Sassa, K., Springer
Satake K, Wang K, Atwater BF (2003) Fault slip and seismic moment of the 1700 Cascadia earthquake inferred from Japanese tsunami descriptions. J Geophys Res: Solid Earth 108(B11):2535. https://doi.org/10.1029/2003JB002521
Schang K, Cox K Trant AJ (2022) Habitation Sites Influence Tree Community Assemblages in the Great Bear Rainforest, British

Columbia, Canada. Front Ecol Evol 9:791047. https://doi.org/10.3389/fevo.2021.791047
Shaw J, Barrie JV, Conway KW, Lintern DG, Kung R (2020) Glaciation of the northern British Columbia continental shelf: the geomorphic evidence derived from multibeam bathymetric data. Boreas 49:17–37
Shrestha RR, Schnorbus MA, Werner AT, Berland AJ (2012) Modelling spatial and temporal variability of hydrologic impacts of climate change in the Fraser River basin, British Columbia, Canada. Hydrol Process. https://doi.org/10.10.1002/hyp.9283
Shugar DH, Walker IJ, Lian OB, Eamer JBRR, Neudorf C, McLaren D, Fedje D (2014) Postglacial sea-level change along the Pacific coast of North America. Quaternary Science Reviews, 97:170–192. https://doi.org/10.1016/j.quascirev.2014.05.022
Silverstein A, Silverstein V, Silverstein RA (1995) The Sea Otter. Millbrook Press, Brookfield CT
Simpson KA, Stasiuk M, Shimamura K, Clague JJ, Friele, P (2006) Evidence for catastrophic volcanic debris flows in Pemberton Valley, British Columbia. Can J Earth Sci 43:679–689
Slaymaker O (2017) Landscapes and Landforms of Western Canada. Springer
Stasiuk MV, Hickson CJ, Mulder T (2003) The vulnerability of Canada to volcanic hazards. Nat Hazards 28:563–589
Stewart H (1995) Cedar: tree of life to the Northwest Coast Indians. Douglas and McIntyre, Vancouver, BC
Sutherland G, Foreman MG Garrett C (2007) Tidal current energy assessment for Johnstone Strait, Vancouver Island. Proc Inst Mech Eng, Part A, 221:147–157
Temple N (2005) Salmon in the Great Bear Rainforest. Raincoast Conservation Society, Victoria, BC
Thomson RE (1981) Oceanography of the British Columbia coast. Can Spec Publ Fish Aquat Sci 56: 291p. Fisheries and Oceans Canada
Thornborough K, Dunham J (2019) Development of risk-based indicators for the Hecate Strait/Queen Charlotte Sound Glass Sponge Reefs Marine Protected Area. DFO Can Sci Advis Sec Res Doc. 2019/004
Trimbach DJ, Gaydos JK, Biedenweg K (2021) Salish sea survey: geographic literacy enhancing natural resource management. Soc Nat Resour 34:1147–1154
Tucker B, Rose-Redwood R (2015) Decolonizing the map? Toponymic politics and the rescaling of the Salish Sea. Can Geogr 59:194–206
VanBlaricom GR (2001) Sea otters. Voyageur Press, Stillwater, MN
Walker BB, Taylor-Noonan C, Tabbernor A, McKinnon T'B, Bal H, Bradley D, Shuurman N, Clague JJ (2014) A multi-criteria evaluation model of earthquake vulnerability in Victoria, British Columbia. Nat Hazards 74:1209–1222
Williams-Jones G, Barendregt RW, Russell JK, Le Moigne Y, Enkin RJ, Gall R (2020) The age of the Tseax volcanic eruption, British Columbia, Canada. Can J Earth Sci 57:1238–1253
Wilson KL, Bailey CJ, Davies TD, Moore JW (2021) Marine and freshwater regime changes impact a community of migratory Pacific salmonids in decline. Glob Change Biol 15895
Zahn MJ, Palmer, MI, Turner NJ (2018) Everything We Do, It's Cedar: First Nation and Ecologically-Based Forester Land Management Philosophies in Coastal British Columbia. Journal of Ethnobiology 38:314–332. https //doi.org/10.2993/0278-0771-38.2.314

Taiga Ecozones

8

Abstract

The Taiga Shield and Plains represent the northerly continuations of the Boreal Shield and Plains ecozones. Several environmental aspects of the Hudson Plains, along the southern shorelines of Hudson and James Bays, are transitional between the Taiga and Arctic regions. The taiga region is marked by discontinuous permafrost and conifer-dominated assemblages. Taiga Shield and Plains ecozones have modified Köppen-Geiger Dfc (northern boreal) climates, with strong seasonal variation. Mean monthly temperatures and mean annual precipitation vary along east-west and south-north gradients, with locations in the north and west having lesser precipitation and lower winter temperatures. Along the transition between Köppen-Geiger Dfc and ET (Arctic Tundra) climate zones, temperatures and precipitation decrease northward. Geologically, the Taiga Plains represent the northern extension of the Interior Plateau, underlain by Phanerozoic, Palaeozoic, and Mesozoic sedimentary rocks. The Taiga Shield includes parts of the Bear and Tu Nedhe (Slave) Provinces, parts of the Churchill (Kivalliq, Nunavik, Labrador), Superior (Nunavik), and Grenville (southern Labrador) Provinces, and the Makkovik Province of the Canadian Shield. The Hudson Plains, and the region overlain by Hudson Bay, consists of Palaeozoic and Mesozoic sedimentary units. All parts of the Taiga Shield, Taiga Plains, and Hudson Plains were glaciated during the Quaternary. Ice advanced from two major centres of the Laurentide Inlandsis, centred on Kivalliq and western Labrador. Taiga forests are dominated by black spruce (Picea mariana), with associated white spruce (Picea glauca). Southerly areas include balsam fir (Abies balsamea) and tamarack (Larix laricina). Taiga fauna represent communities intermediate between the boreal regimes to the south and the arctic tundra to the north. Typical mammal species of taiga regions include caribou (Rangifer tarandus caribou), wolverine (Gulo gulo), timber wolf (Canis lupus) black bear (Ursus americanus), red fox (Vulpes vulpes), lynx (Lynx canadensis), snowshoe hare (Lepus americanus), red squirrel (Tamiasciurus hudsonicus), beaver (Castor canadensis), and muskrat (Ondatra zibethica). Marine environments discussed include brackish Lake Melville, and the sea-ice dominated Labrador Sea and Hudson Bay. The Taiga Shield, Taiga Plains, and Hudson Plains have a combined area of 2.3 million km2, with a human population density of 0.04 / km2. Of the combined population of approximately 92,000, the three largest communities of Yellowknife (20,340), Labrador City-Wabush (9376), and Happy Valley-Goose Bay (8040) account for approximately 40%. Approximately 60% of the total population is indigenous.

Keyword

Taiga Shield · Taiga plains · Hudson plains · Hudson bay · Labrador sea

8.1 Introduction

This chapter looks at the Taiga Shield, Taiga Plains, and Hudson Plains ecozones, along with the marine environments of the Labrador Sea and Hudson Bay. The Taiga Shield and Plains represent the northerly continuations of the Boreal Shield and Plains ecozones discussed in Chap. 3. Several environmental aspects of the Hudson Plains, along the southern shorelines of Hudson and James Bays, are transitional between the Taiga and Arctic regions (Fig. 8.1).

The taiga (northern boreal) region is marked by discontinuous permafrost and conifer-dominated assemblages. The boundary between the taiga and arctic regions is commonly, although a little inaccurately, referred to as the "tree line". Climate, vegetation, and permafrost boundaries do not universally coincide. The transitions are generally not sharp, but commonly occupy an irregular belt which locally may be several tens of kilometres in width. Large wetland areas

N. Catto and G. R. Vincent, *Canada: Land and People*, World Regional Geography Book Series,
https://doi.org/10.1007/978-3-032-04391-7_8

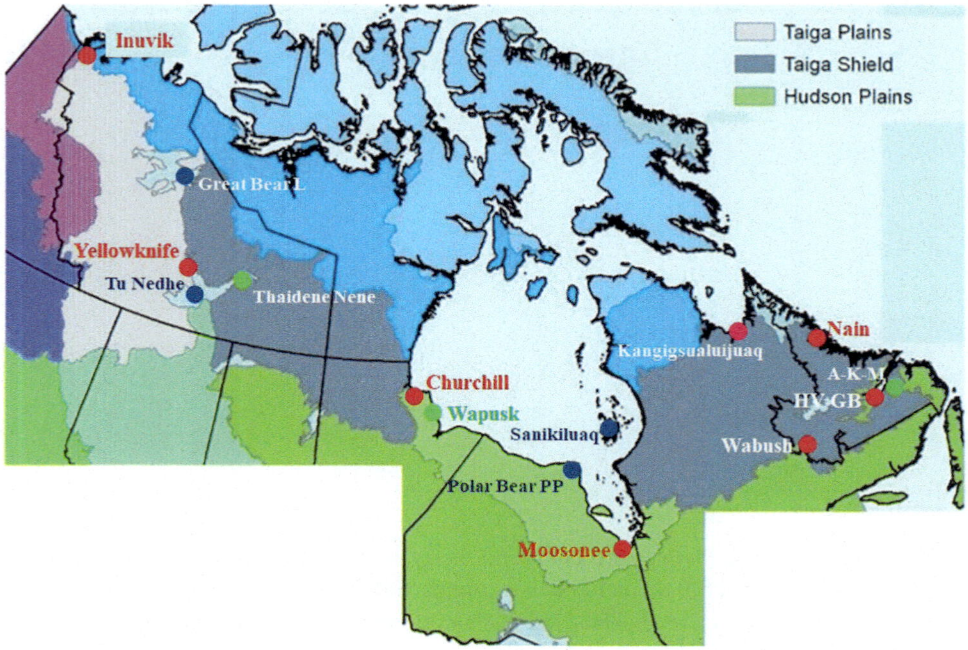

Fig. 8.1 Taiga Plains, Taiga Shield, and Hudson Plains Ecoregions. Green dots: national parks; Blue dots: other locations discussed in text; red dots: communities. A-K-M: Akami Uapishk^{U}-KakKasuak-Mealy Mountains National Park; HV-GB: Happy Valley-Goose Bay. (Base map: Environment and Climate Change Canada and Agriculture and Agri-foods Canada)

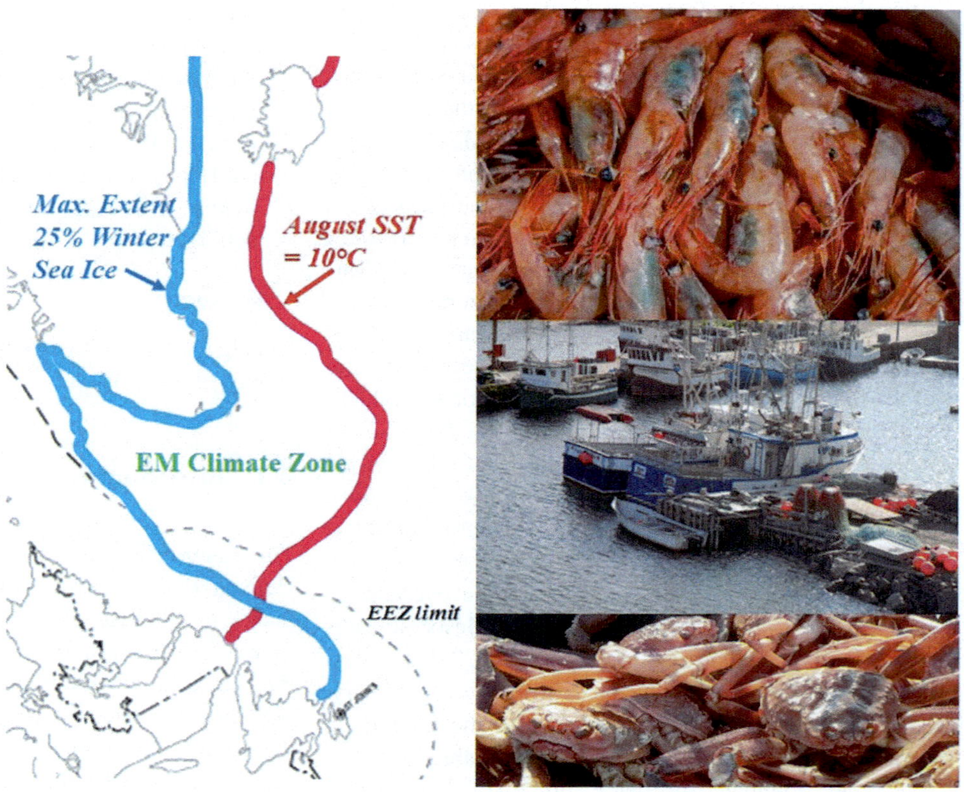

Fig. 8.2 The Labrador Sea has a Köppen-Geiger polar marine EM climate. Left: the EM climate zone (Ballinger et al. 2013) is defined as the region between the mean 10° August sea surface temperature isotherm (south) and the mean position with at least 25% winter ice coverage (north) Right: this region of the Labrador Sea has important fisheries for *Pandalus borealis* (Northern Shrimp) and *Chionoecetes opilio* Snow Crab for Newfoundland-based vessels

are common, especially in the Hudson Plains (Abraham and Keddy 2005).

Lacking the perceived uniqueness of the Arctic, and the familiarity of the more populated Boreal regions, the Taiga regions are the least known for many Canadians. A common perception is that these regions are uniform, without significant variation. "Southerners" commonly do not distinguish between the northern taiga and arctic areas, but they are vastly different in all respects. Diversity within the taiga ecozones is as prevalent as in the other ecozones of Canada.

Recent years have seen the emergence of a stronger Indigenous presence in the use of place names for terrain features in Canada. This long overdue acknowledgment of the reality of Canadian Indigenous culture can complicate place name recognition for English, French, and international readers. We have included a partial list of equivalences between English and Indigenous names for terrain features and communities (Table 1.2). In this chapter, we will use both English and Indigenous names. Many terrain features and smaller communities have only designated Indigenous names.

8.2 Climate

Taiga Shield and Plains ecozones have modified Köppen-Geiger Dfc (northern boreal) climates, with strong seasonal variation (Beck et al. 2006). Mean monthly temperatures and mean annual precipitation vary along east–west and south–north gradients, with locations in the north and west having lesser precipitation and lower winter temperatures (Phillips 1990). Coastal locations, such as Happy Valley-Goose Bay and Makkovik, see more precipitation and higher winter temperatures, although seasonal variations are evident.

The offshore region of the Labrador Sea has an EM (polar maritime) climate (Ballinger et al. 2013; Fig. 8.2). EM zones are defined as the area bordered by the 10 °C summer isotherm (measured as sea surface temperature, SST), and the maximum latitude of winter sea ice concentrations with a minimum of 25%. EM regions have cool summer temperatures, but very limited sea ice cover in winter. Winter average monthly temperatures do not fall below −7 °C. Precipitation occurs throughout the year, with totals greater than in tundra regions (typically 500–800 mm/y). Relative humidity is high throughout the year as well, and more than 50% of days are "wet" (receiving at least trace amounts of moisture).

Offshore of Makkovik, the temperature of the Labrador Sea varies from −2 °C in January to 6 °C in July, while on land the corresponding mean values are −15 and 11 °C. The overall range of land temperatures, from hottest summer to coldest winter, is in excess of 60 °C; the corresponding value for the adjacent ocean is less than 10 °C. As a result, the water forms a cool reservoir during the summer months, which acts to generate onshore breezes and decreases the temperature of the coastline (Catto 2020). In areas where westerly winds from the interior counter the effects of onshore breezes (as at Makkovik), the oceanic effect is confined to a thin ribbon of shoreline (<10 km distant from the open ocean). However, in areas where westerly winds are less dominant, or where the sites are more exposed, oceanic modification of temperature, winds, and associated precipitation extends further inland.

8.2.1 Taiga Shield

Labrador City-Wabush (53°N, 67°W), in western Labrador, has a typical interior Dfc climate, sometimes described as having only two seasons (Environment and Climate Change Canada 2020; Fig. 8.3). Winter is definitely noteworthy, with

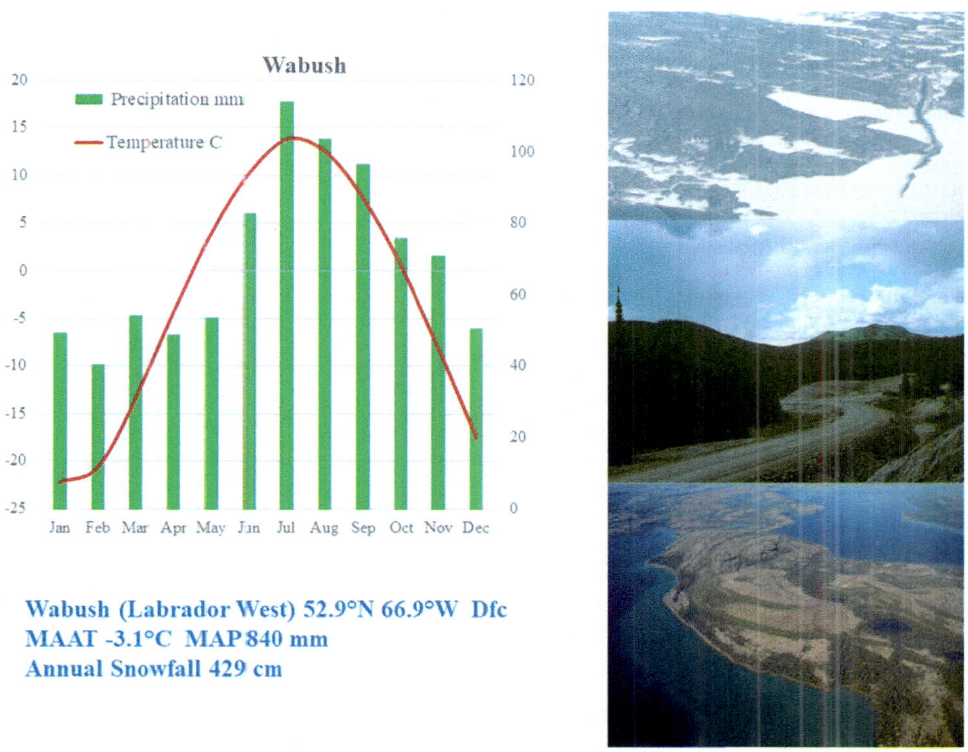

Fig. 8.3 Wabush, located in western Labrador, has a Köppen-Geiger Dfc (northern boreal/taiga) climate. *Data* Canadian Climate Normals, Environment and Climate Change Canada (2020). Upper right: Esker, Molson Lake, east of Wabush; Centre right: Trans-Labrador Highway; Lower right: Canadian shield topography

a January daily mean temperature of −22.2 °C, and temperatures fall below −40 °C on several occasions during a typical winter. The prevailing west-northwesterly winds blow from the interior of the continent, creating dry, cold winters marked by calm days and minimal humidity. In an average winter, Labrador City-Wabush has more than 60 days without any recorded wind. Strong winds are particularly rare in the winter, although the light snow can readily be propelled to create white-out or blizzard conditions.

Summers are cool, with a July daily mean temperature of 13.8 °C in Labrador City-Wabush. Prevailing winds are west-southwesterly during the summer months. Three to four high summer days occur during most years, and the frost-free season is approximately 90 days. In a typical year, sunshine occurs during 35–40% of daylight hours, exceeding 1700 h annually.

The mean annual precipitation is 840 mm, of which approximately 40% of the precipitation occurs as snow. Freezing rain is relatively uncommon. Snow cover persists up to 210 days in wooded areas.

In comparison, the Lake Melville region (including Happy Valley-Goose Bay, 53°N 60°W; Fig. 8.4) shows the effects of topography and marine waters on the northern boreal Dfc climate. The daily mean temperature in January is − 16 °C, with a corresponding July value of 14 °C Environment and Climate Change Canada 2020). The topography funnels southwesterly winds along the axis of Lake Melville, and also limits the influence of northwesterly winds. Thermal inversion conditions can develop in the Happy Valley-Goose Bay area, resulting in warmer temperatures (particularly in the summer) for lakeside sites. Temperatures decline with elevation, and the summits of the Akami-Uapishk[U]-KakKasuak-Mealy Mountains (~1100 m asl; a National Park with three official names) are generally 7°–8 °C colder than the sea level communities. Temperatures can fall below −30 °C in winter, and rise above 30 °C in the summer. The area receives 65–70 frost-free days annually, and a mean of 1620 h of sunshine.

Annual precipitation varies from 850 mm at Happy Valley-Goose Bay, the lowest value in the region, to more than 1000 mm at the summits of the mountains. Precipitation also increases northeastward along Lake Melville. Snowfall comprises approximately 30% of the precipitation at lakeside sites, rising to 40% in the mountains. Summer droughts are common, and forest fires ignited by lightning occur (Newfoundland and Labrador 2020).

The prevailing wind at Goose Bay Airport is west-southwesterly, directed along the axis of the valley. At other sites, swirling winds are induced by topography, and some locations are dominated by easterly or east-southeasterly winds. The summits of the Akami-UapishkU—KakKasuak-Mealy Mountains are influenced by easterly and westerly winds for approximately equal amounts of time each year. Calm days are infrequent, with a mean of only 22 occurring at Goose Bay Airport, and even lower totals in the mountains and in the eastern parts of Hamilton Inlet. Winds travelling from central North America bring acidic precipitation to Happy Valley-Goose Bay on occasion, but unfortunately bring other chemical contaminants with much greater frequency.

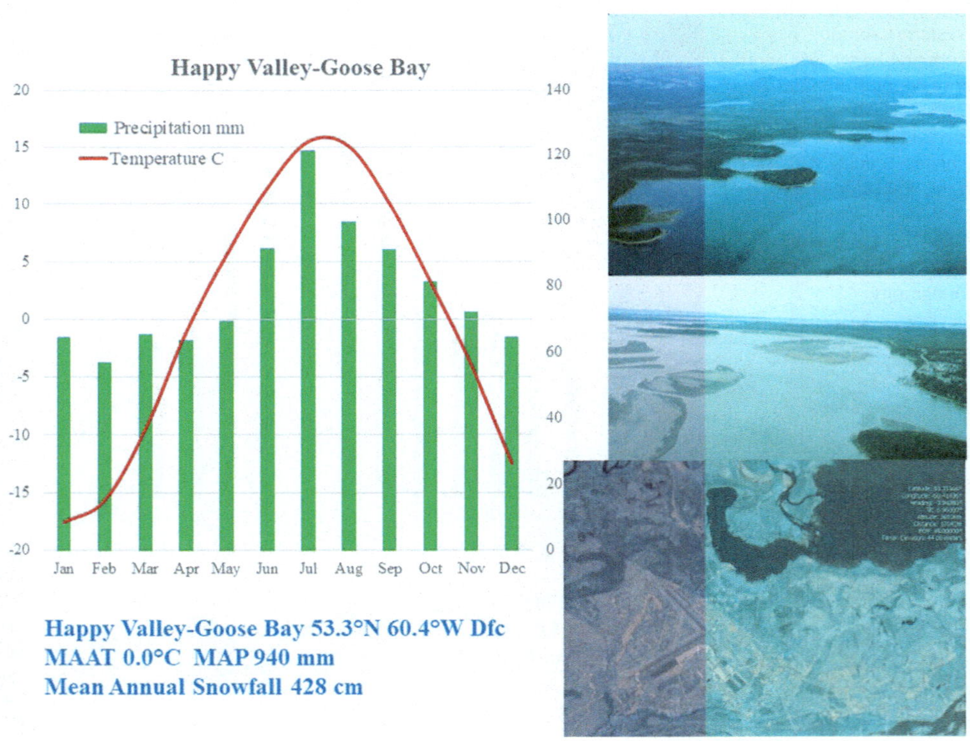

Fig. 8.4 Happy Valley-Goose Bay has a Köppen-Geiger Dfc (northern boreal/taiga) climate. *Data* Canadian Climate Normals, Environment and Climate Change Canada (2020). Upper right: Lake Melville; Centre right: Churchill River; Lower right: Goose Bay is located north of the airport and military base. Community of Goose Bay is directly east of the airport; Happy Valley is located along the Churchill River to the east. (ISS/NASA image)

Kangigsualuijuag (Fort George; "very large bay"), the home town of Governor-General Mary Simon, is located in northern Nunavik, on the eastern shore of Ungava kangiqluk (Bay). The northernmost location of taiga vegetation in Nunavik, the climate is Northern Boreal (Köppen-Geiger Dfc), with mean temperatures ranging from 14 (July) to −17 °C (January and February; Fig. 8.5). Sea surface temperature in Ungava kangiqluk varies from −2 °C (winter) to 3 °C (August). Kangigsualuijuag receives approximately 1500 h of sunshine per year.

The coastal location, combined with the topographic setting in a glacial cubic parabolic trough, results in enhanced precipitation. Approximately half of the 600 mm of precipitation is snow, leading to accumulations on the slopes. In 1999, an avalanche struck the school in Kangigsualuijuag during a New Year's Day party, killing 9 and injuring 25.

At Yellowknife NT (62°N 114°W), January and July mean temperatures are −25.6 °C and 17 °C, respectively (Environment and Climate Change Canada 2020; Fig. 8.6). Transitions between winter and summer are sharp, with little "spring" or "autumn" weather. Record highs exceeding 30 °C have

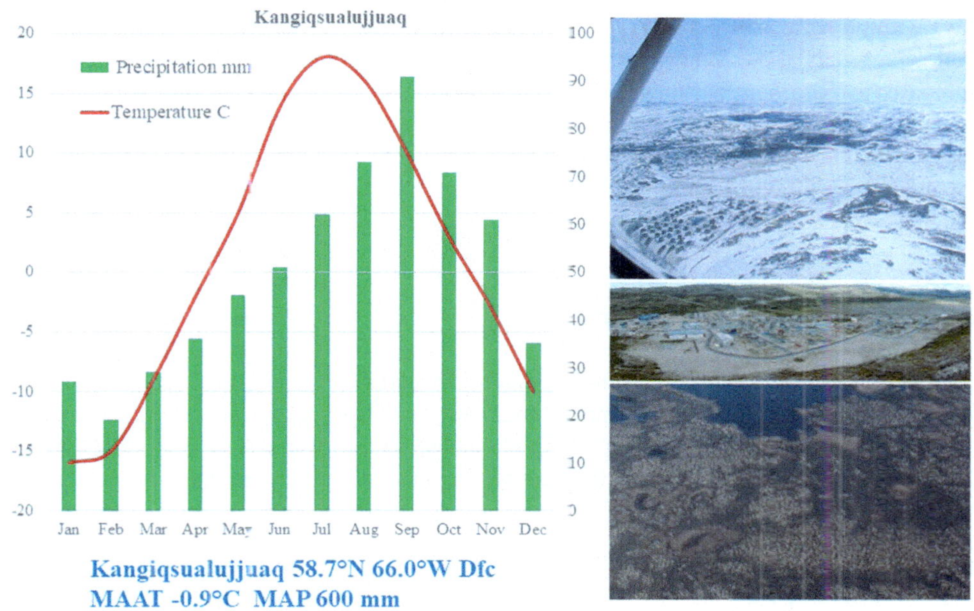

Fig. 8.5 Kangiqsualuijuaq, Nunavik, has a Köppen-Geiger Dfc climate. *Data* Canadian Climate Normals, Environment and Climate Change Canada (2020). Upper right: overview (Wikimedia/Nicholas Perrault); Centre Right community view (Wikimedia/Lester Kovac); Lower right. Taiga Shield terrain, Nunavik

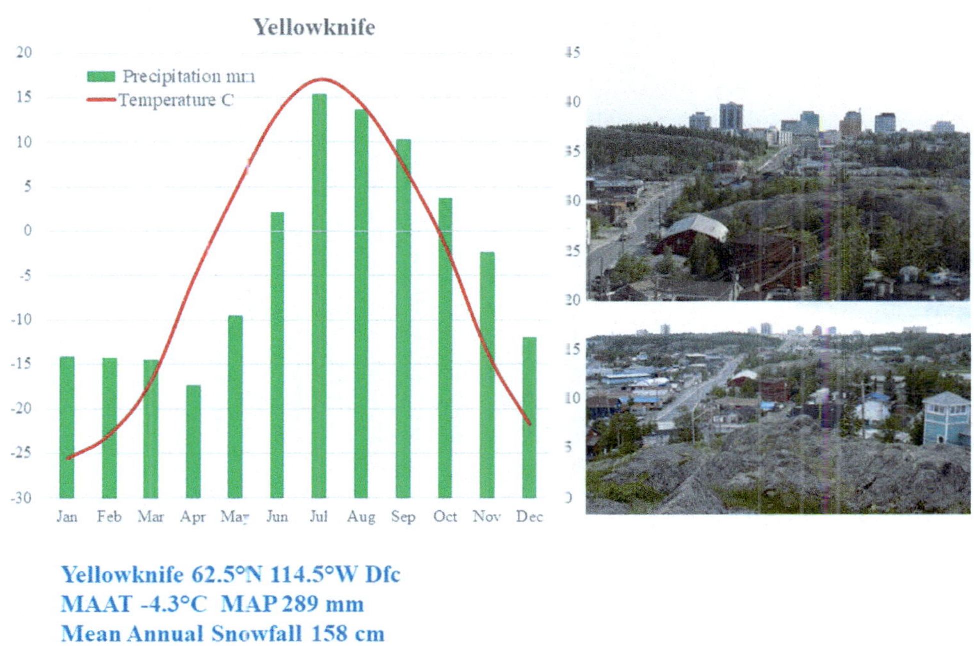

Fig. 8.6 Yellowknife has a Köppen-Geiger Dfc climate. *Data* Canadian Climate Normals, Environment and Climate Change Canada (2020). Right: Downtown Yellowknife has abundant outcrops of Tu Nedhe Canadian Shield bedrock. (Wikimedia/Trevor MacInnis)

been recorded in June, July, and August. Record lows below −40 °C have been recorded in all months from November through April, and temperatures below 0 °C in all months except July. Compared to other Canadian communities with 20,000 or more residents, Yellowknife has the coldest winters, most sunshine hours (2257 y^{-1}), and longest duration of snow cover (192 days y^{-1}) (Phillips 1990).

In Yellowknife, the mean annual precipitation is 289 mm, with about 60% occurring as snow. Snowflakes to be low density and dominantly columnar in shape, in contrast to the higher-density and dendritic forms characteristic of warmer boreal climates (Snowcrystals 2013). In combination with generally light winter winds, this reduces the number of days with blowing snow, or blizzard or whiteout conditions. Yellowknife has about 8 days with blowing snow annually, compared to more than 35 in taiga locations in Labrador. Snow cover duration, annual maximum snow depth, and duration of snow cover have all decreased at Yellowknife since 1950, as for most Taiga Shield locations (Environment and Climate Change Canada 2020).

8.2.2 Taiga–Tundra Transition

Along the transition between Köppen-Geiger Dfc and ET (Arctic Tundra) climate zones, temperatures and precipitation decrease northward (Rouse 1993). The boundary "line" is currently shifting northward as climate warming proceeds (Smith 2011; French and Slaymaker 2012; McLaughlin and Packaler 2021), and some communities such as Moosonee (Ontario) which were formerly "arctic" are now firmly boreal (Fig. 8.7). Along Hudson Bay, Wapusk National Park has a tundra (ET) climate in coastal areas, transitional to a high northern boreal/taiga (Dfc) regime approximately 30 km inland. At Churchill, January daily mean temperatures are −26 °C, with nighttime lows commonly below −35 °C (Environment and Climate Change Canada 2020). Temperatures above −10 °C are extremely rare in winter. The daily mean temperature in July is 10 °C. The region commonly does not experience any high summer days. Fog is very rare on Hudson Bay, and hence sunshine at Wapusk is seldom obstructed in the atmosphere. The northern latitude, however, limits the hours of bright sunshine to 1700 per year.

Mean annual precipitation is 600 mm, of which about 30% (180 cm) falls as snow. The snowpack reaches a modal depth of 70 cm in the taiga (marginally less in the coastal areas), and persists for approximately 220 days per year. The moisture regime is humid. Prevailing winds vary throughout the year. Summer winds blow from the southwest across Manitoba, whereas winter winds blow from the west-northwest and northwest, dominantly from Kivalliq. Both systems, however, result in dry conditions in Wapusk. The humid moisture regime reflects the short evapotranspiration season, the low summer temperatures, and the effects of slowly degrading permafrost, rather than being the result of large quantities of precipitation.

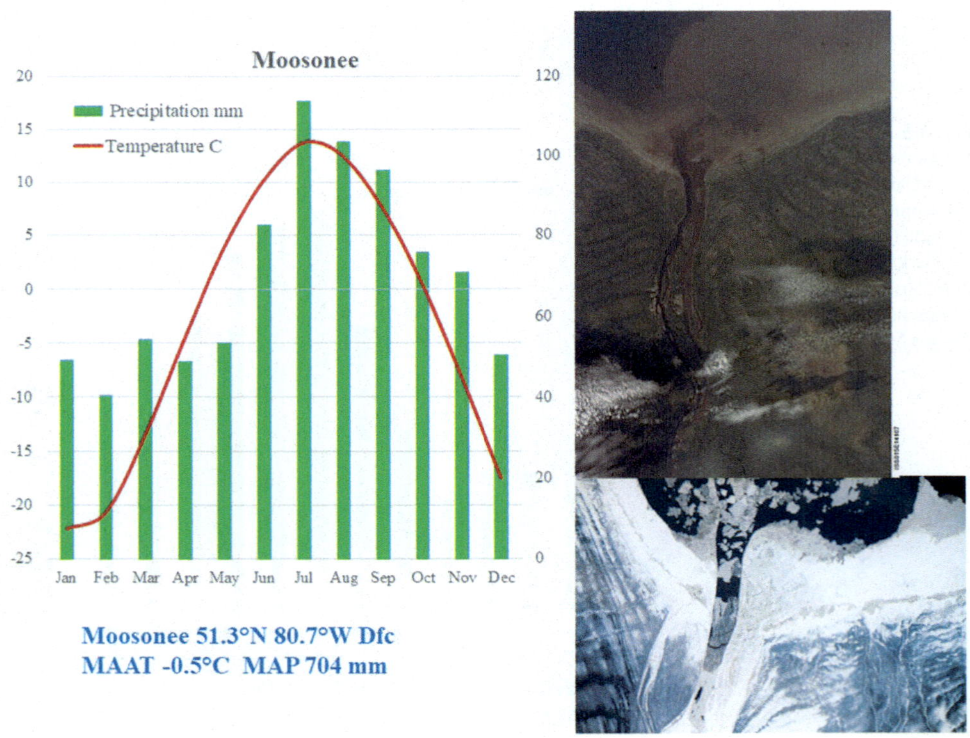

Fig. 8.7 Moosonee, Hudson Plains ecoregion, has a Köppen-Geiger Dfc climate. *Data* Canadian Climate Normals, Environment and Climate Change Canada (2020). Right: Summer and winter images of the mouth of Moose River in James Bay. (ISS/NASA images)

Climate warming has occurred since the establishment of Churchill as a port in the 1930s, as indicated by temperature readings, northward spread of some plant species, and the length of the shipping season through Hudson Bay and Hudson Strait (Halsey et al. 1997; Andrews et al. 2018; Dredge and Dyke 2020; Environment and Climate Change Canada 2020). The pattern of climate warming has been ongoing since the 1850s. At the height of the "Little Ice Age" (ca. 1500–1850 ce), Hudson and James Bays experienced a truly arctic climate. Food resources were scarce, and palynological, plant macrofossil, and lacustrine sediment data suggest that the tree line was more than 50 km south of its present position in some areas (e.g. Kettles et al. 2000; McLaughlin and Webster 2013; Hobbs et al. 2017). When Henry Hudson sailed into the bay in 1610, he probably had to endure more pack ice, colder temperatures, and a greater shortage of food than any visitor has since 1850. In contrast to the taiga regime that greets the visitor to Moosonee today, Hudson and his crew were confronted with conditions similar to those currently prevalent on Qikiqtaaluk. As Hudson was unable to discover the Northwest Passage from James Bay to the Orient, his crew decided to afford him the opportunity of unlimited, independent exploration without their assistance (Chadwick 2007).

8.3 Permafrost

Permafrost terrain in Canada is divided into zones reflecting the areal extent and location, which are characterized by distinctive landform assemblages and landscapes (Heginbottom et al. 1995; French 2007; Fig. 8.8). In the Contin-

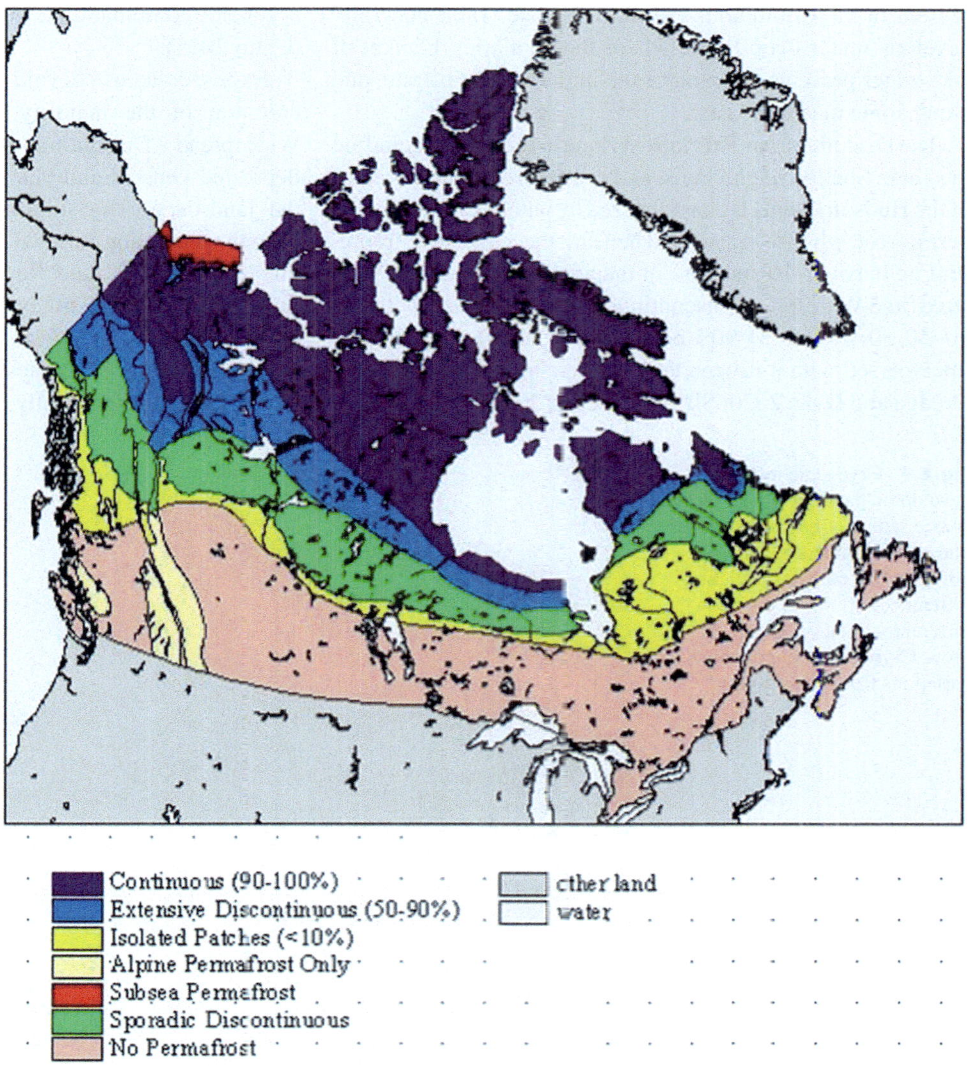

Fig. 8.8 Permafrost zones in Canada. (Natural Resources Canada)

uous Permafrost Zone, permafrost conditions are generally (90–100%) present beneath the seasonally thawed active layer. The Continuous Permafrost zone is characterized by tundra (Köppen-Geiger ET) climates, with a Mean Annual Air Temperature (MAAT) of no greater than −7 °C. The Continuous zone is also marked by limited snow cover, with thickness typically less than 30 cm and total annual snowfall less than 100 cm.

Seasonal thawing does occur within the Continuous Permafrost Zone. The active layer is the uppermost part of the sediment profile subject to seasonal freezing and thawing. The short summers permit plant growth in the thin, seasonally thawed active layer, with plant roots extending downward only to the base of the thawed material. Thicker, active layers develop in the spring and summer in areas of thick sediment cover, in regions where snow accumulations insulate the soil, and in areas where vegetation cover is more extensive.

Isolated patches of unfrozen terrain, called talik, are also present in the Continuous Permafrost Zone. Talik generally develops under deep lakes, where the insulating blanket of lake water prevents freezing of the underlying substrate, and along some marine coasts.

The Discontinuous Permafrost Zone to the south, including the Taiga Shield, Taiga Plains, and southern and interior areas of the Hudson Plains, is characterized by patchy distribution of permafrost, with frozen ground beneath the active layer representing between 10 and 90% of the area. The Sporadic, Scattered, and Widespread Discontinuous subzones have between 10–30, 30–50, and 50–90% permafrost coverage, respectively, interspersed with unfrozen talik (Abraham and Keddy 2005; Dredge and Dyke 2020; Slaymaker et al. 2020a, b).

Within the Discontinuous Permafrost Zone, the MAAT varies from a maximum of −0.5 °C along the southern fringe to −7 °C at the transition to the Continuous Zone. Total annual snowfall is greater, but usually does not exceed 300 cm. Thick active layers occur where MAAT is greater than −3 °C.

In undisturbed landscapes, the patchwork of permafrost and unfrozen talik reflects local differences in vegetation cover, as well as the underlying sediment or bedrock, the topography, and the effects of fire (Glaser et al. 2004; Abraham and Keddy 2005; Dyke and Sladen 2010; Ackley et al. 2021; Li et al. 2021). Seasonal freeze-thaw cryogenic mounds are developed on exposed areas of sediment, where vegetation cover is minimal or has been disturbed by fire, and where snow is periodically removed by wind (Fig. 8.9). Frost creep and gelifluction occur where silt and fine sand lie on slopes steeper than approximately 2°. Mudflows and debris flows can be activated on steeper slopes with thick sediment cover. Exposed plateaux develop gravitational-cryogenic combination features, such as stone circles and nets (Catto 2015).

Exposed areas devoid of vegetation cover are not common in the majority of the Scattered and southern Widespread Discontinuous zones. Repeated glaciation deposited veneers and blankets of sediment across much of the landscape. The summer climate facilitated vegetation growth, producing mineral Cryosols and Brunisols over the glacial sediments, and Regosols over the bedrock, in turn providing better substrates for enhanced forest development. Accumulation of peat in fens, marshes, and bogs proceeded under the summer climates as well, forming thick blankets which could be gradually colonized along the fringes by

Fig. 8.9 Cryogenic mounds develop on flat to gently sloping coarse sediment-covered surfaces under frost-influenced Dfc and northern Dfb climates. Height differences of 10 s of cm allow different vegetation species to grow. Cryogenic mounds show variations from year-to-year

black spruce. Evolving vegetation assemblages responded to forest fires. Throughout much of the Scattered and southern Widespread Discontinuous zones of Canada, the vegetation assemblages and the permafrost-talik assemblages have developed together since deglaciation, creating distinctive combinations of wetlands, vegetation-covered, peat-dominated and sediment-dominated landforms (National Wetlands Working Group 1988; Warner and Rubec 1997; Abraham and Keddy 2005, Dredge and Dyke 2020).

8.3.1 Discontinuous Permafrost Landscape Features

8.3.1.1 Peat Plateaux

Areas with thickest, most extensive blankets of peat associated with topographically low terrain can be marked by extensive peat plateaux (Sannel and Kuhry 2009; Samson et al. 2010; Fig. 8.10). A typical peat plateau is a flat plain, several hundred square metres in area, locally flanked by marginal ramparts of peat 10 s of centimetres high. Configurations vary, but many have oval to teardrop shapes. Peat plateaux tend to occur in groups or fields. The 1–3 m thick blanket of peat overlies mineral sediment or bedrock. In the Hudson Plains and eastern Kivalliq they are underlain by marine Tyrrell Sea silt (Dredge 1992; Dredge and Dyke 2020).

Peat plateaux develop in discontinuous permafrost areas with MAAT of −2 to −5 °C. Maintenance of the plateau requires that the lower part of the peat remains frozen throughout the year. Summer climates allow peat to slowly accumulate, less than 1 mm per year, and winter freezing and stable permafrost at depth prevent organic decomposition. Successive layers of peat underlying the plateau surface commonly show evidence of successive vegetation assemblages indicating gradual development of progressively less water-saturated conditions. Wetland sedge peat (*Carex*, cottongrass) at the base is overlain by moss-dominated peat (*Sphagnum*, with *Cladonia* lichen), capped by heathland vegetation or forest duff from black spruce. The sequential development results from gradual freezing of the peat under permafrost conditions as the seasonal freezing front migrates

Fig. 8.10 Landscape features in northern Discontinuous Permafrost terrain. Upper left: Peat Plateau vegetated with *Cladonia* lichen. Upper right: Palsa are low irregular features, colonized by *Picea mariana* (black spruce). Lower left: Lithalsa, overlain by heath assemblage. Lower right: *Eriophorum* (cottongrass) is a common sedge species in wet locations in Discontinuous Permafrost landscapes

downward. However, the layers of peat are generally not disturbed by cryoturbation.

Fire plays a role in peat plateaux development. In areas where permafrost is under pressure from warming summer temperatures, fire can result in the melting of the plateau core, producing a shallow, unstable wetland. In many cases, however, a late-summer fire (when vegetation is generally dry) can act to rejuvenate the peat plateau, by exposing the surface to renewed frost penetration in winter, along with reduced snow cover. Burning the vegetation, along with the uppermost peat in the active layer allows deeper frost penetration.

The margins of the peat plateaux are areas of reduced permafrost thickness or unfrozen talik. Snow trapped by the treed plateau feeds the surrounding wetland areas in summer, causing accumulation of water and melting along the fringes. Trees fringing the plateaux lean towards the melting areas, creating drunken forests. Because drunken forest development results from gradual subsidence of the fringes rather than heave of the plateau centre, it is most apparent in areas where melting of permafrost is underway.

Discontinuous permafrost, peat plateaux, and vegetation assemblages develop together. Undisturbed forest communities over peat plateau will act to preserve the underlying permafrost. Gradual climate warming will cause the permafrost to melt, but if the process occurs relatively slowly, the differences in relief will continue to support distinctive vegetation assemblages. Southerly examples of relict peat plateaux, with pine and tamarack growing on low flat areas flanked by spruce in depressions, are present on taliks and under boreal climates, where the former permafrost has melted slowly.

More rapid increases in temperature result in permafrost melting, causing collapse of the plateau and leading to fenland development. Conversion of peat plateaux to wetland fens has occurred within the last 200 years across the Hudson Plains and southern Taiga Shield and Taiga Plains.

8.3.1.2 Palsa

Cryoturbation and local frost heaving of sediment under taiga forest cover result in the formation of mounds of sediment, with an ice-rich core. Landforms formed primarily of peat are termed palsa; those formed primarily of clastic sediments (silt, sand, or glacial sediment) are referred to as lithalsa (Calmels and Allard 2008; Seppälä 2011; Wolfe et al. 2014). Palsa and lithalsa are confined to the southernmost part of the Widespread Discontinuous zone, the Scattered Discontinuous zone, and along the fringes of Alpine Permafrost areas. MAAT suitable for formation range from +1 to −4 °C.

Palsa are low circular or oval irregular domes 10–30 m in diameter, standing 1–7 m above the surrounding terrain. Inside the palsa, a blanket of frost-heaved and cryoturbated peat 1–4 m thick overlies a mound of cryoturbated clastic mineral sediment. Unlike a peat plateau, frost-heaving is necessary for the development and maintenance of a palsa. The upper surface of a palsa is usually colonized by black spruce (*Picea mariana*), often with *Cladonia* lichen (caribou moss) and heaths such as Labrador Tea (*Rhododendron groenlandicum*) and blueberry species (*Vaccinium sp.*).

Maintenance of a palsa requires limited snow cover, allowing winter frost penetration. In the southern taiga and boreal forests, warmer climates are generally accompanied by increased snow depth and persistence, which will gradually lead to palsa degradation. Feedback relationships exist between vegetation cover, snow accumulation, and warmer winter temperatures. Increased winter temperatures under taiga climates result in more snowfall, which is more effectively trapped by the vegetation, causing wetter conditions to persist in the summer. This leads to less effective insulation of the underlying permafrost by the wet peat, facilitating growth of shrubs and trees, which trap more snow during the ensuing winter.

Palsa are also greatly influenced by forest fires. In contrast to peat plateaux, the effect on wildfire on palsa is usually to accentuate degradation, by exposing the frozen core to seasonal melting and initiating collapse of the flanks. Because the palsa is initiated by differential frost heave over a prolonged time, rather than seasonal penetration of the freezing front, disruption of the surface vegetation results in collapse. Warmer summer temperatures can result in increased fire frequency due to lightning strikes, also impacting vegetation assemblages. Locally, degradation of palsa near communities has occurred, producing wetlands. This is especially evident in areas where little human care was taken to avoid taiga and peat fires.

Lithalsa resemble palsa in their dimensions, but are composed entirely of clastic mineral sediment. The internal structures of lithalsa suggest that vertical frost-heave, lateral frost-thrusting, and cryoturbation all are involved in formation. Some lithalsa appear to have been initiated as palsa, but subsequently lost their original peat blanket, leaving the frost-heaved mineral core exposed. Once exposed, these lithalsa are highly susceptible to thermokarst degradation, completely collapsing to form pitted terrain or small ponds.

8.3.2 Southern Scattered Permafrost Features

The most southerly areas of the Scattered Permafrost zone contain only isolated permafrost features, although frost-influenced landforms such as cryogenic mounds, felsenmeer, and gelifluction features are common. Peat plateaux and palsa are absent, and most peat landforms resemble those of the seasonal frost-influenced boreal areas to the south.

Aapa (Fig. 8.11), also referred to as ladder fens, string fens, or ribbed fens, are organic landforms consisting of alternating elongate peat-covered ridges (strings, ribs) and depressed areas (flarks). Aapa are characteristic of northern boreal and taiga climates, both within and south of the discontinuous permafrost zone (Laitinen et al. 2005; Catto 2015; Slaymaker et al. 2020a, b). Actively growing aapa occur in many regions where permafrost and taiga forests are not present, and they are thus not considered to be a diagnostic permafrost indicator. They generally require perhumid to humid moisture regimes, and MAAT of +6 to −4 °C. Even when present in the Scattered Permafrost zone, aapa are not underlain by permafrost.

Aapa develop where reduced summer evaporation combined with seasonal frost table (tjäle) development allows accumulation of organic matter in a wetland environment. The moisture regime is a dominant factor: in perhumid coastal areas, aapa fens are common, whereas humid interior areas support aapa bogs, with fewer aapa fens. Aapa are extremely rare in drier interior areas of the southern taiga.

Aapa ridges or strings of peat are 1–5 m wide, either gently curved or straight, standing 10–40 cm above the general level of the fen, and commonly support black spruce, heath-land vegetation (e.g. Labrador Tea, blueberries (*Vaccinium*)), and *Cladonia* lichen (caribou moss). The flarks, depressed areas, support sedges (*Carex*, cottongrass (*Eriophorum*)), or standing water. The overall gradients of the aapa are very low, generally less than 1°, but the surface of each successive flark is less than 10 cm below its upslope neighbour. Aapa grade laterally into intricately patterned net fens, with net-like arrangements of low ridges separating depressions with standing or slowly flowing water.

The genesis of aapa is uncertain. Among numerous possible origins, the most plausible involve slow downslope creep (<10 cm per year) of partially frozen organic-rich sediments driven by groundwater hydraulic pressures, combined with differential heaving and possibly subsoil buildup of pressurized methane gas, produced from decaying vegetation.

Fig. 8.11 Aapa form a variety of patterns varying from net-like to complicated arrangements of strings and flarks

8.4 Bedrock

8.4.1 Taiga Plains

Geologically, the Taiga Plains represent the northern extension of the Interior Plateau, underlain by Phanerozoic, Palaeozoic, and Mesozoic sedimentary rocks. The Taiga Shield includes parts of the Bear and Tu Nedhé (Slave) Provinces, parts of the Churchill (Kivalliq, Nunavik, Labrador), Superior (Nunavik), and Grenville (southern Labrador) Provinces, and the Makkovik Province of the Canadian Shield. The Hudson Plains, and the region overlain by Hudson Bay, consists of Palaeozoic and Mesozoic sedimentary units (Wheeler et al. 1996; Natural Resources Canada 2009).

The Taiga Plains are predominantly low-relief terrain, dominated by Deh Cho ("Big River", Mackenzie) and its tributaries. Phanerozoic metamorphosed sedimentary rocks are exposed in the drainage basin between Tu Nedhé (Great Slave Lake) and Łíídlįį Kųę (Fort Simpson). Other regions, extending from northeastern British Columbia to the tundra and treeline in Northwest Territories and northeastern Yukon, are underlain by flat-lying to gently dipping sandstones, shales, and lignite coals, ranging in age from Early Cretaceous to Neogene. These sedimentary rocks are soft and are readily eroded.

The Peel Plateau, flanking the Peel (Teetł'it Gwinjik) River and extending east to Tsiigèhnjik (Arctic Red River) and Deh Cho, is a broad, gently eastward-sloping surface ranging from 650 m asl elevation in the west to 100–150 m asl elevation in the east (Smith et al. 2004; Fig. 8.12). The surface of the plateau is generally even, with relief less than 100 m. The Peel River and its major tributaries, however, have cut deep canyons into the Cretaceous Arctic Red Formation shale and siltstone. The Peel River is entrenched to 300 m depth throughout its lower reaches below its confluence with the

Fig. 8.12 Peel River (Teetł'it Gwinjik) and its tributaries drain the Peel River Plateau along the western margin of the Taiga Plains. Upper: Peel River (left) and Caribou River (right) flow through sedimentary bedrock with spruce-dominated taiga forest. Lower: Glacial ice thrust ridges of weak sedimentary bedrock constrain the Peel River's northward flow

Fig. 8.13 Taiga Shield bedrock, Thaidene Nene. Left: Roche moutonées indicate glacial flow from northeast to southwest across the Tu Nedhe structural province. (Wikimedia/Michael Iannicelli). Right: Glacial erosion of the resistant Shield bedrock is displayed on the Simpson Islands. (National Air Photo Library)

Wind River, and its position cross-cutting the plateau at almost 90° azimuth to the regional slope is suggestive of an origin as a glaci-marginal meltwater channel (Catto 2015). Under the Yukon Land Claims Agreement, the 68,000 km^2 part of the watershed in Yukon is designated as the Peel Watershed Planning Region.

8.4.1.1 Tu Nedhé

Tu Nedhé straddles the boundary between the Taiga Plans and Taiga Shield, although it is underlain by Canadian Shield rock (Wolfe et al. 2017; Fig. 8.13). Yellowknife is located along this physiographic boundary. The boundary is marked by North Arm and the Macdonald Fault, which forms the straight northeast–southwest shoreline of the lake, including Thaidene Nene (East Arm) southwest of Łutselk'e (Snowdrift). Relief along the fault escarpment reaches 40 m. The terrain south of the lake has very low relief, generally less than 100 m.

Tu Nedhé is Canada's deepest surveyed lake, at 614 m (Schertzer 2000), and is the fourth largest in area, at 28,568 km^2 (Fig. 8.14). The lake developed due to erosion along the Macdonald Fault system, and along the margin of the Canadian Shield in the area now occupied by North Arm. Tu Nedhé, like most of Canada's large lakes, is primarily a product of preglacial erosion. Glaciation resulted in the deepening of some of the embayments of Thaidene Nene, but had a minimal effect on North Arm.

Fig. 8.14 Tu Nedhe. The Canadian Shield terrain to the north and east of the lake, and the Taiga Plains to the south and west underlain by sedimentary bedrock, show contrasting patterns of lakes in response to the bedrock geology. Thaidene Nene National Park Reserve is located along the eastern arm of Tu Nedhe. The Athabasca Sand Dunes and William River Delta are visible along the southern shore (see Chap. 3). Orange dots and whitish plumes mark active forest fires. (NASA)

The East Arm Hills bedrock forms the intricate pattern of embayments and islands in Thaidene Nene (Bostock 2014). These rocks are a complex of metamorphosed sedimentary rocks, granites, gneisses, gabbros, volcanic basalts and diabases, and mylonites. The presence of the mylonites, highly metamorphosed rocks composed of disaggregated fragments of pre-existing rock units eroded by faulting and subsequently re-combined in different structural arrangements, indicate that the Macdonald Fault and other similarly aligned faults in the East Arm Hills were active approximately 1900 Ma. These faults are tectonically inactive at the present time. Stromatolites, columnar and branching structures composed of fossilized blue-green algae, are also preserved in the rocks of the East Arm Hills (Hoffman 1989).

8.4.2 Western Taiga Shield

8.4.2.1 Bear Province

The Bear Province of the Canadian Shield is located west of Coronation Gulf, and includes the terrain from Kugluktuk southwest to Great Bear Lake. Bear province rocks are Proterozoic metasediments and metamorphosed basic volcanic units, and range in age from 1400 to 2200 Ma (Wheeler et al. 1996; Natural Resources Canada 2009).

Great Bear Lake, with a surface water elevation of 156 m asl (virtually identical to Tu Nedhé), and a surface water area of 31,328 km^2, is the largest lake entirely contained within Canada (Johnson 1975). The lake consists of five deep arms (Dease, McTavish, McVicar, Keith, and Smith Arm) connected in a central basin. The arms were developed along weaker bedrock units of the Tu Nedhé geological province. Great Bear Lake's arms are controlled by the structural geology of the region, and are not primarily the products of glacial erosion. Glaciation did modify the arms to a small extent, but the bedrock topography of the terrain formed prior to the Quaternary.

8.4.2.2 Kivalliq: Tu Nedhe and Churchill Provinces

The Tu Nedhe province of the Canadian Shield includes granite, gneiss, migmatite, metamorphosed volcanic rocks, and metamorphosed sedimentary rocks and schists (Yellowknife Supergroup), formed during the Archean Eon and deformed ca. 2400 Ma. Younger Proterozoic basalt sills, metamorphosed volcanic units, and dolomites are also present.

Kimberlite pipes have intruded the Precambrian rocks. Kimberlite, a basic rock containing olivine and phlogopite mica, also can contain diamonds. The BHP (Ekati) diamond mines, located on Lac De Gras southeast of Kugluktuk, produce diamonds from a sequence of 45–62 Ma kimberlite pipes (Creaser et al., 2004).

The Taiga Shield part of the Kivalliq region of Nunavut south of the treeline is underlain by Churchill province granites, gneisses, and volcanic rocks. Latest Proterozoic sandstones, some of which contain iron and uranium, are exposed in the Thelon Wildlife Sanctuary northwest of Dubawnt Lake.

8.4.3 Hudson Plains

The Hudson Plains involve two bedrock geological units. The western shore and Hudson Bay are underlain by Churchill and Superior province metamorphic and igneous rocks at

Fig. 8.15 Akimiski Island shows a typical Hudson Plains landscape, with Palaeozoic bedrock overlain by Tyrrell Sea deposits (note raised beaches along the southern side). Permafrost ablation is contributing to sediment load in rivers reaching the coast. (NASA)

depth (Mints et al. 2021). In most regions, these Canadian Shield rocks are overlain by Palaeozoic sedimentary rocks. The oldest rocks in the Hudson Plains are Superior Province granites and gneisses of the Sutton Hills, a southeast–northwest arcuate band rising to 290 m asl south of Cape Henrietta Maria (Nehinaw: *Mooshawow*, barren land). The prominence of the Sutton Hills is an indication of the generally subdued topography.

The younger rocks are limestones and shales of Early and Middle Silurian and Devonian age, extending across Hudson and James Bays (Ontario Geological Survey 1991; Fig. 8.15). A variety of marine fossils are present, including graptolites (in the shales) and corals, brachiopods, and bryozoans (in the limestones). The youngest Cretaceous rocks are exposed west of Moosonee. The bedrock is covered with thick Quaternary deposits at most locations, and it has little influence on the topography.

The Ordovician and Silurian rocks continue beneath the waters of Hudson Bay, and are also exposed on Shugliaq (Southampton) Island in the north (Zhang 2010). In the centre

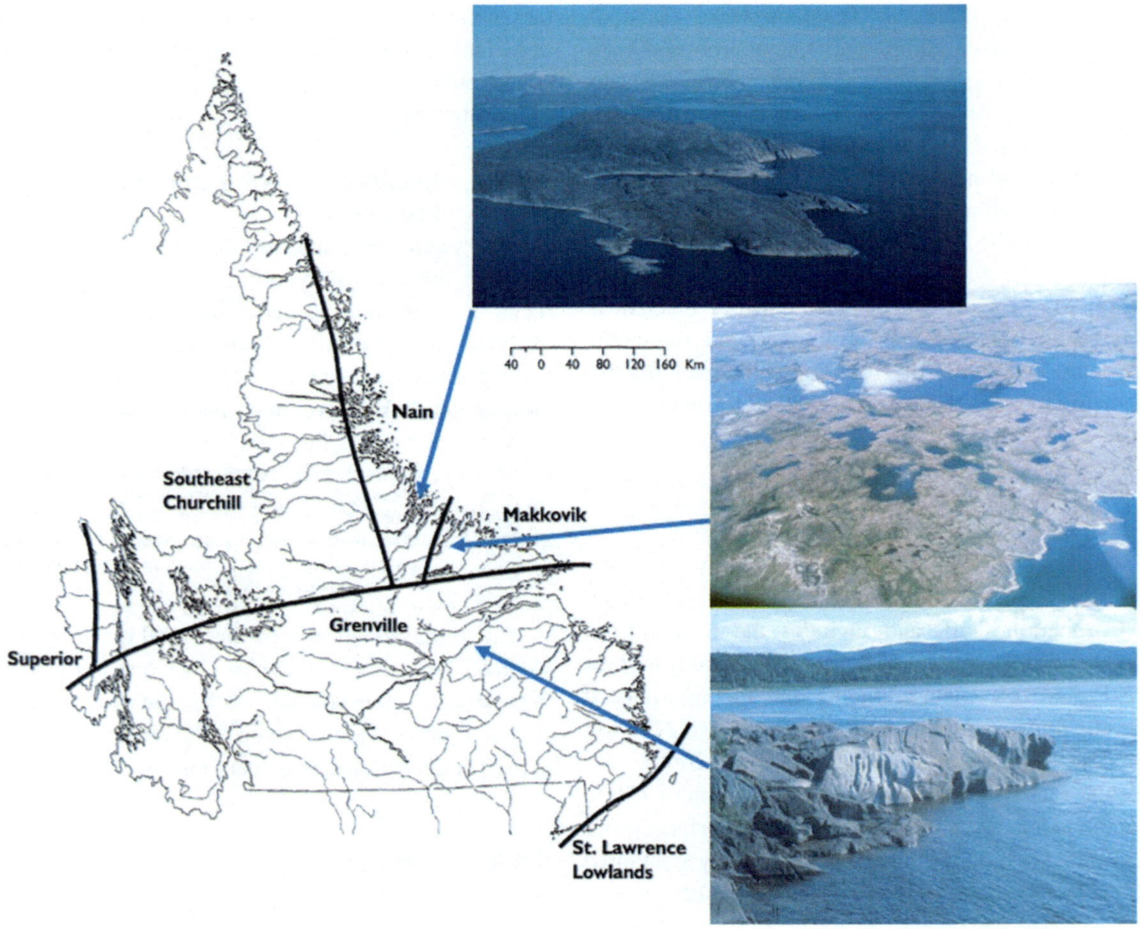

Fig. 8.16 Geological structural Provinces, Labrador

of the bay, Devonian sedimentary rocks overlie the Silurian deposits. The presence of this relatively young, soft rock in the centre of Hudson Bay indicates that glacial erosion was minimal in the area.

8.4.4 Eastern Taiga Shield: Nunavik–Labrador

The Nunavik-Labrador region contains contains parts of four geologic provinces of the Canadian Shield (Wardle et al. 1997; Valley et al. 2011; Hocq 1994a, b; Hocq and Verpaelst 1994; Fig. 8.16). Churchill province rocks form an arcuate belt extending north from near Waskaganish along southeastern James Bay (Wînipekw) to the Inukjuak (Inoucdjouac) Hills along the eastern shore of Hudson Bay, eastward through the Povungnituk Hills (Cape Smith Fold Belt) of the tundra of northern Nunavik, and southward from Kuujjuaq on Ungava kangiqluk to western Labrador. The rocks in western Labrador are referred to as the Labrador Trough, a NNW-SSE trending belt of deformed metasedimentary, metavolcanic, volcanic, metamorphic, and igneous rocks, in addition to the iron-bearing rocks at Kawawachikamach (Schefferville), Labrador City, and Wabush. The intensely folded Churchill province metasediments are exposed on Sanikiluaq (Belcher) and Arviliit (Ottawa) Islands (Fig. 8.17). Resistant dolomites form the intricate landscapes of Sanikiluaq, one of Earth's truly unique archipelagoes (Flaherty 1918).

Rocks of the interior areas of the Taiga Shield of Nunavik and westernmost Labrador are assigned to the Superior Province of the Canadian Shield (Hocq 1994a). These granites and gneisses are old, many in excess of 2 billion years in age. They resemble the Grenville Province rocks to the south in lithology, but have undergone relatively fewer orogenic and metamorphic events.

Fig. 8.17 False-colour satellite image of Sanikiluaq (Belcher Islands) highlights differential erosion of bedrock units. Red and purple tones indicate vegetated terrain. (NASA)

The most conspicuous topography of the Taiga Shield in Labrador is associated with the Grenville province, formed by orogenic activity in several events between 2000 and 950 Ma (Hocq 1994b; Wardle et al. 1997; Slaymaker et al. 2020a, b). Rock types include igneous acidic rocks (granites and related types) and acidic, highly metamorphosed rocks (gneiss and schist) derived from the metamorphism of pre-existing sedimentary and igneous rocks, and metasedimentary schists, argillites, quartzites, and gneisses.

South of Lake Melville, the Akami Uapishk^{U}-KakKasuak-Mealy Mountains are composed of granites, gneisses, and anorthosites, all resistant to glacial and fluvial erosion (Catto 2020). Short high-gradient streams and steep-sided resistant plateaux mark the landscape, with maximum elevations of 1190 m asl. Proterozoic igneous intrusive activity *ca.* 1645–1625 Ma produced mafic and ultramafic intrusive suites, with high concentrations of iron, magnesium, and nickel (although not in economical quantities for mining in this region). Rock types include anorthosite (almost 100% labradorite plagioclase feldspar), charnockite (granitic rock with abundant iron-rich orthopyroxene), and monzonite (similar to anorthosite, but containing both potassium feldspar and plagioclase feldspar). These lithologies can support only limited vegetation assemblages, as many plant species cannot cope with the ultrabasic nickel-rich mineralogy. Anorthosites are confined to areas of old granitic crust, and have not formed on Earth since the early Cambrian.

The Makkovik province occupies a wedge in coastal Labrador, flanked by the Grenville province to the south and by Arctic Cordilleran landscapes of the Nain province to the west (see Chap. 9). Rock units range from Archaean to Palaeo-Proterozoic, and consist of a variety of granitic plutons, metamorphosed sediments, and mylonite shear zones (Wardle et al. 1997).

8.5 Quaternary

All parts of the Taiga Shield, Taiga Plains, and Hudson Plains were glaciated during the Quaternary. Ice advanced from two major centres of the Laurentide Inlandsis, centred on Kivalliq and western Labrador.

8.5.1 Kivalliq and Taiga Plains

West of Hudson Bay, Kivalliq and the Taiga Plains were glaciated by ice flowing westward and southwestward (Dyke 2004, Sharpe et al. 2017, Slaymaker 2017, Storrar and Livingstone 2017; Wolfe et al. 2017; Stokes 2020). Glacial flutings, erosional bedrock features, and eskers

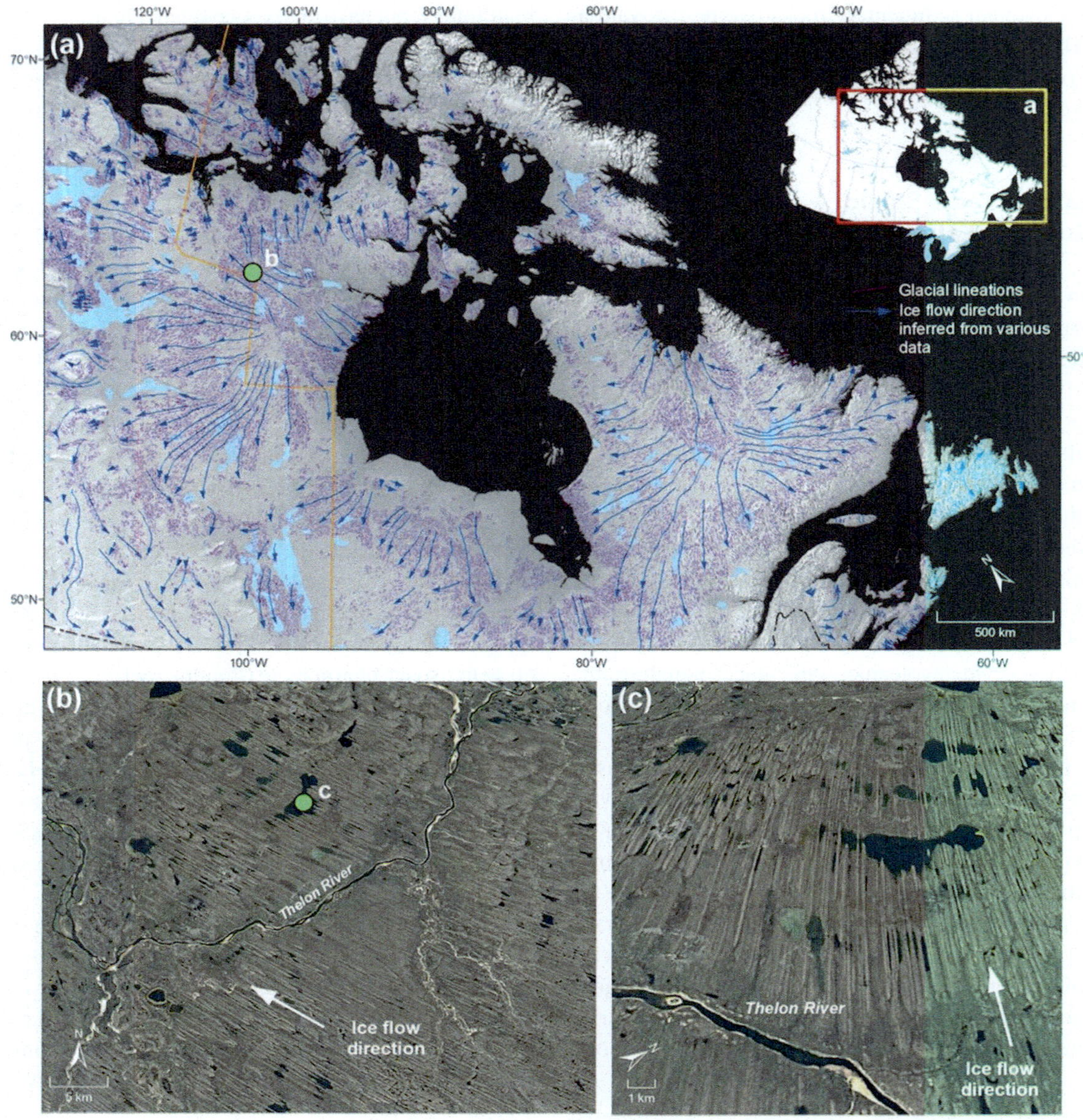

Fig. 8.18 Streamlined bedrock-dominated features are common in glaciated Shield regions. **a** Distributions and orientation of streamlined features. *Source* Glacial lineations and ice flow direction from Shaw et al. (2010). **b** Landsat image showing examples of mega-scale (>5 km long) glacial lineations with long axis aligned with ice flow direction; part of the Dubawnt Lake Ice Stream, northwest of Dubawnt Lake in the vicinity of Thelon River. *Source* Google Earth: Landsat/Copernicus; centred on about 64°15′N, 102°06′W. **c** Close-up, oblique view of features looking northwest across the field of mega-scale glacial lineations. *Source* Google Earth: Landsat/Copernicus; centred on about 64°19′N, 102°09′W. (Slaymaker and Catto 2020, ch 2)

indicate the ice flow direction (Fig. 8.18). As the ice flowed across the southern Taiga Plains, the pre-existing Tu Nedhe basin topography diverted movement to the southwest, into Alberta and northeastern British Columbia, and to the northwest along the Deh Cho (Mackenzie) valley. Ice covered the Taiga Plains, reaching the eastern edge of the Cordillera along the margins of the valley north to the Peel Plateau, and expanding north to the Beaufort Sea coast (Catto 1996; Catto et al. 1996; Dyke et al. 2003; Stokes et al. 2015).

Deglaciation began about 17,000 years ago. As the ice retreated northeastwards, the tributary valleys in the Deh Cho watershed opened, creating diversions from previous drainage patterns and impounding glacial lakes. In eastern Yukon, the Snake River was diverted from its preglacial northeastern course as a direct tributary of Deh Cho by glacial advance. Forced to flow westward, the Snake became a tributary of the Peel River.

Continued glacial retreat resulted in the impoundment of Glacial Lake McConnell, extending from Lake Athabasca to the western tip of Great Bear Lake (Wolfe et al. 2017, see Chap. 3). The Thaidene Nene area of Tu Nedhé only was submerged by Lake McConnell for a very short period, as the valley opened almost simultaneously, allowing the lake to drain towards the northwest. By about 10,000 years ago, the configuration of the Tu Nedhé shoreline and the topography of Thaidene Nene resembled those of the present.

8.5.2 Hudson Plains

To the south, the western Hudson Plains region also was glaciated by ice originating from Kivalliq (Dredge 1992; Dyke et al. 2003). Although the area was covered by several glaciers during the Quaternary, the surface features present are related to the most recent glacial event. Glaciers moved across the terrain from the northwest and northeast, depositing several tills and eroding the Sutton Inlier. Many events of glacial advance and retreat can be distinguished. Final glacial retreat from the area occurred about 11,000 years ago.

8.5.2.1 Relative Sea-Level Changes

Following deglaciation, removal of the 2 km thick ice sheet left the newly exposed terrain glaci-isostatically depressed. The Tyrrell Sea (named for J.B. Tyrrell, Geological Survey of Canada) flooded all of the terrain in coastal Kivalliq, Manitoba, and northern Ontario to approximately 180–210 m asl, with inundation increasing northward (Fulton 1995; Simon et al. 2014; Catto et al. 2020; Dredge and Dyke 2020; Fig. 8.19). In Kivalliq, marine waters extended up to 250 km west of the modern Hudson Bay shoreline at Ukkusiksalik National Park. The marine clays deposited by the Tyrrell Sea formed a thick blanket over all of the pre-existing Quaternary deposits and bedrock. Withdrawal of the marine waters began shortly after the Tyrrell Sea reached its maximum extent about 8000 years ago, producing flights of raised beaches marking successive shoreline abandonment. In southern Kivalliq and inland from Wapusk National Park, more than 100 former shoreline positions can be seen.

Isostatic recovery is still ongoing in the region, at a rate of approximately 30 cm/100 y, and Hudson Bay can be considered as a remnant of the Tyrrell Sea. Thus, the Hudson Platform is the fastest growing physiographic region of Canada, in terms of adding new terrain. As Hudson Bay has a mean depth of 93 m, continued isostatic

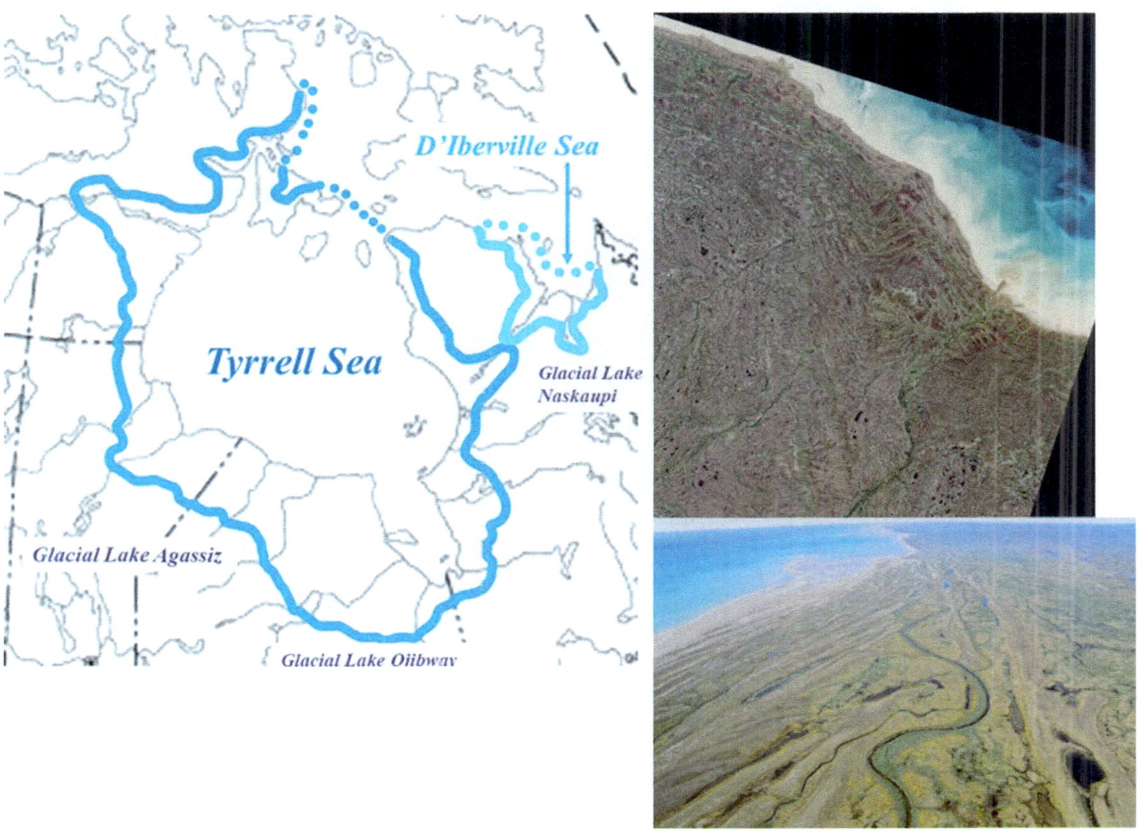

Fig. 8.19 Tyrrell Sea. Left: Areal extent of the Tyrrell Sea around Hudson Bay (Wînipekw), and the d'Iberville Sea around Ungava kangiqluk. Right: beaches of the Tyrrell Sea southwest of Hudson Bay in the Severn River area (upper) and Ridges Creek (lower), Ontario

rebound will cause the bay to shrink significantly through the next 5,000 years. Modern coastal features such as Cape Churchill will become raised beaches as relative sea level continues to drop.

When considering geomorphology, coastal hazards, or port construction and operation, it is the relative sea level with respect to the terrain (or a harbour) that is important, rather than the absolute volume of marine water in the ocean. Changes in sea level are measured relative to a point, such as a wharf or cliff. From a relative sea level perspective, rapidly rising land coupled with a slower increase in ocean water volume, as is currently occurring in Hudson Bay, will result in a fall in relative sea level. Churchill is located near the site of the Hudson's Bay Company factory at Fort Prince of Wales. Although marginally better than "Port Nelson" (the abortive attempt to build at the mouth of the Nelson River, near York Factory; Catto et al. 2020), it was a difficult choice for a port, due to shifting tidal flats and large offshore bars, necessitating the construction of an artificial island offshore. Declining relative sea level will pose difficulties for port operations in Churchill.

8.5.2.2 River Landscapes

Declining relative sea level also influences river landscapes inland. The Nelson River serves as the outlet for Lake Winnipeg, and thus carries the entire drainage of the Red, Assiniboine, Saskatchewan, and Bow Rivers (Déry et al 2005). The Nelson and other major rivers, including the Winisk, Attawapiskat, Albany, and Moose, are braided, wandering, and anabranching streams, marked by high volumes of spring discharge and greatly reduced summer flow. The streams are dominated by sand, although finer sediments are provided by reworking the marine deposits of the Tyrrell Sea. The river valleys are not deeply entrenched. The designation of the Winisk as a "Wild" river refers to the lack of settlement, not its challenge to canoeists.

The anabranching reach of the Attawapiskat River at Attawapiskat, Ontario, carries sand and silt in suspension, with gravel as bedload (Catto 2015; Fig. 8.20). Large woody debris is transported downstream by freshet and river ice. The islands are erosional forms, dissected by avulsion. The low surrounding relief allows avulsion channels to depart from the existing river valley, gradually expanding the area

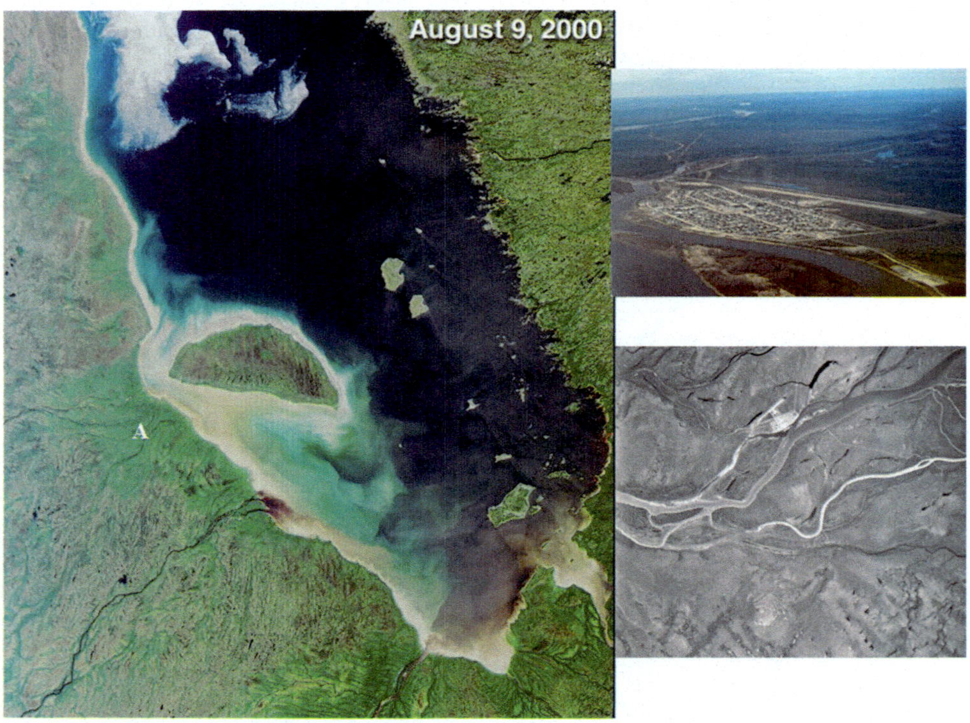

Fig. 8.20 The lower reaches of the Attawapiskat River, and other streams flowing into western James Bay, are anabranching streams subject to flooding. Left: Flow from the Attawapiskat River ("A") and other streams southeastward into James Bay indicated by sediment plumes, August 2000. (NASA). Upper right: community of Attawapiskat (Government of Ontario). Lower right: The reach carries sand and silt in suspension, with gravel as bedload. Large woody debris is transported downstream by freshet and river ice. The islands are composed of cohesive glacimarine Tyrrell Sea clay and glacial till, and are erosional forms, dissected by avulsion. The low surrounding relief allows avulsion channels to depart from the existing river valley, gradually expanding the area directly influenced by the Attawapiskat River (National Air Photo Library)

directly influenced by the Attawapiskat River. In combination with ablating permafrost, the result is increased susceptibility to flooding resulting from freshet and dynamic ice jamming.

In Canada, many Indigenous communities endure flooding and contamination of water supplies on a frequent basis. In several cases, notably in the Hudson Plain, the locations of these communities were not selected by the residents, but by people from outside, without consultation. Flood damage does not result from lack of traditional knowledge.

Flooding of the land of the Kashechewan First Nation of northern Ontario induced by snowmelt, freshet, and ice jamming along the Albany River now occurs earlier in the spring than it did 50 years ago (Khalafzai et al. 2021). The same pattern is evident for all the major rivers draining into James and Hudson Bays across northern Ontario from the south and west, impacting the communities of Fort Severn, Peawanuck, Attawapiskat, and Kashechewan. Although the peak discharge of the Albany River has not changed markedly, the combination of poorly designed and constructed infrastructure, community layout, and disruption due to ice road maintenance downstream has resulted in increased flood hazard in the community, along with water contamination. Unfortunately, in common with the experience of many Indigenous communities, no consultation or application of traditional knowledge of flooding was undertaken.

8.5.3 Taiga Shield, Nunavik, and Labrador

East of Hudson Bay, the Taiga Shield of Nunavik and Labrador was glaciated by the Laurentide Inlandsis, originating in the terrain of western Labrador (Klassen et al. 1992; Dyke et al 2003; Catto et al. 2020). Part of the source area now lies beneath the Smallwood Reservoir. The Laurentide Inlandsis expanded radially from the source area. Several phases of activity, involving shifting ice divides and accumulation areas, are recorded by striations, glacial flutings, and other landforms around Labrador City (Figs. 8.21 and 8.22). As much as 2.5 km of ice covered Lake Melville at the height of the last glaciation, about 20,000 years ago. Western Labrador was among the last areas of North America to be deglaciated, approximately 6800 years ago.

In most areas, glaciation had relatively little influence on the topography of the Taiga Shield. Resistant rocks limited glacial erosion, removing less than 10 m of pre-existing rock from most areas during the entire Quaternary period. In these areas, the drainage pattern reflects the faults, folds, and differential resistance of the bedrock units. In a few areas, thicker accumulations of glacial sediments are present, forming till blankets, drumlins, or DeGeer moraines.

Fig. 8.21 Drumlinized terrain in western Labrador indicates glacial flow towards the east from the Laurentide Inlandsis accumulation centre in western Labrador. (George Blackwood photo)

Fig. 8.22 Eskers in western Labrador were produced by water flow in subglacial tunnels. (George Blackwood photo)

When a glacier reaches deep water, the ice toe is decoupled from the underlying surface, and it begins to float. DeGeer moraines developed in the deep water of eastern Hudson Bay. They are narrow (10–20 m), low (less than 5 m), arcuate ridges, oriented convex in the direction of ice flow, and drape the underlying topography (Catto 2015). The successive DeGeer moraines formed as a result of sequential glacial retreat. As the ice receded, the previous DeGeer moraine was submerged and preserved.

Glacifluvial eskers and kames are also present. Formation of a large kame terrace or delta complex, such as that at Muskrat Falls, Churchill River, Labrador, requires a significant amount of time to allow sediment to accumulate. As with eskers, these accumulations of sand and gravel are important aggregate resources for construction. The relatively flat-topped surfaces, well-drained due to the coarse sediment, deposits form suitable sites for construction of airports, roads, and communities, and also are valuable sources of sand and gravel. In many areas, including Muskrat Falls, glacifluvial kame landforms indicate former ice-front positions.

The Akami Uapishk^{U}-KakKasuak-Mealy Mountains did not support independent ice caps (Fig. 8.23). The eastern part of the mountains is marked by the highest elevations, with local areas of permafrost (generally above 900 m asl). More than 20 cirques, carved by local glaciers, are present in the headwaters of the Etagulet and English Rivers, with floor elevations at 600 m or greater. Cirques facing northwards have linear or undulatory slopes, whereas tarns and reigels are common in cirques developed on the southern and eastern flanks of the Mountains. No active glaciers are present.

Fig. 8.23 Taiga forest, northern flank of Akami Uapishk^{U}-KakKasuak-Mealy Mountains National Park

Parc national Kuururjuaq is a 4,460 km^2 co-managed provincial park extending from Ungava kangiqluk to Mont D'Iberville/Mount Caubvick (1646 m) on the Labrador/Québec border. The 166 km Korok River runs north through Kuururjuaq to Ungava kangiqluk. Parc National Ulittaniujalik is a 5,293 km^2 park, created in 2016 in partnership with the Inuit communities of Kangiqsualujjuaq and Kuujjuaq, and the Naskaupi community of Kawawachikamach. Ulittaniujalik translates as "the place where there are shorelines", referring to the beaches of Glacial Lake Naskaupi. Ulittaniujalik is the result of collaboration between local communities, scientists, the Kativik Regional Government (KRG), Makivik Corporation and the Québec government. Most of the park's staff comes from nearby Kangiqsualujjuaq and speak three languages.

8.5.3.1 Relative Sea-Level Changes

Following deglaciation about 12,000 years ago, Lake Melville and the lower reaches of the Churchill River were flooded by marine waters (Catto 2020; Fig. 8.24). The sea reached elevations of 130–135 m at the southwestern end of the lake, and ice formed the edge of the embayment. As the glacier receded, isostatic rebound elevated the land surface, limiting penetration of marine waters along the Churchill River valley. The marine waters eventually extended inland to Gull Island Rapids. As isostatic rebound continued, relative sea level fell. Former marine beaches southwest of Happy Valley-Goose Bay were reworked by southwest winds to form parabolic dunes, and other marine sediments were reworked by the Churchill and Goose Rivers (Fig. 8.25). At present, sea level continues to fall in Lake Melville, at a rate estimated at approximately 1 mm each year.

Inland from the modern shoreline of Ungava kangiqluk, the postglacial d'Iberville Sea extended south and west along the major river valleys. Developed after Ungava kangiqluk was deglaciated after 8000 bce, allowing the drainage of Glacial Lake Naskaupi, the d'Iberville Sea was laterally contiguous with the eastern shoreline of the Tyrrell Sea. At present, glaci-isostatic rebound continues along the shore of Ungava kangiqluk.

Fig. 8.24 Landscapes adjacent to Lake Melville show the influence of progressive decline in relative sea level since deglaciation. Upper left: Elevated strandlines are marked by alternating *Picea* (dry sediment, predominantly sand) and *Cladonia* lichen (pale yellow terrain, fibric peat); Upper right: Moderate-viscosity debris flow in brackish-water silty clay was triggered by a combination of precipitation, undercutting by the Churchill River, and insect infestation of spruce. The resulting landform and disordered diamicton deposit indicates loss of internal cohesion during debris flow; Lower left: Brackish-water silty clay with dropstone; Lower right: Glacimarine terrace forms a suitable location for the community of North West River

Fig. 8.25 Churchill River. Upper left: Transverse gravel ribs formed by seasonal ice push. Lower left: channel bar forms and flats indicate a mixed gravel-sand sediment load. Right: West of Happy Valley-Goose Bay, the landscape includes parabolic dunes developed under a drier mid-Holocene climate, slope failures in brackish-water silty clay (see Fig. 8.23), and longitudinal bars in the gravel-dominated braided Churchill River. (NL Air Photo)

8.6 Vegetation

Taiga forests are dominated by black spruce (*Picea mariana*), with associated white spruce (*Picea glauca*; Rowe 1972; Bonneya et al. 2018; Baldwin et al. 2019; Montesano et al 2020; Castilla et al. 2022; CNVC n.d.; Fig. 8.26). Southerly areas include balsam fir (*Abies balsamea*) and tamarack (*Larix laricina*). Jack pine (*Pinus banksiana*) occurs along the transition to mid-boreal assemblages in Nunavik, northern Ontario, and Manitoba, replaced by lodgepole pine (*Pinus contorta*) and *Pinus contorta x banksiana* hybrids further west, confined to the best-drained areas of gravelly soil (Lusebrink et al. 2013). Pines are currently expanding their ranges northward.

Black spruce (*Picea mariana*) is the most common and widespread tree in taiga forests (Hosie 1979; Ryan 1995; Earle 2018; CNVC n.d.). Its range covers 60% of Canada, encompassing all but the warmest areas south of treeline, and growing on land in all provinces and territories. Growth habits vary from upright, 20 m high trees in the southern taiga, to prostrate specimens at treeline.

Needles are four-sided, almost square in cross-section, compared to the flat needles of balsam fir. The grooves in the needles have whitish stomata, giving the needles a blue-green appearance (compared to the lighter greens of white spruce and balsam fir). The darker green needles and the dense, conical configuration of the relatively short branches gives the tree a darker appearance than the white spruce, hence the name "black". Feather moss commonly grows on black spruce.

Black spruce prefers organic-rich or peaty, wet to moist, acidic soil. It can reproduce by seed (pollen), or vegetatively by "layering". Lower branches take root when they are buried by litter. In northerly taiga areas, the trees vegetatively reproduce in most years, but in the southern taiga and mid-boreal regions, seed reproduction almost always occurs. The long fibres in black spruce wood are ideal for the manufacture of pulp designed for fine paper products. Black spruce grows rapidly, and thus has come to dominate many taiga and boreal forest areas. It is also the preferred species for making spruce tea (to combat scurvy) and spruce beer. Black spruce is the provincial tree of Newfoundland and Labrador.

White spruce (*Picea glauca*) is the tallest spruce species in the taiga, up to 25 m in height in favourable locations (Hosie 1979; Ryan 1995; Earle 2018; CNVC n.d.). Needles are four-sided in cross-section, and are lighter green than those of black spruce. The more open configuration of the branches and the lighter tone of the needles led to the term "white".

White spruce prefers dry to moist mineral soils, slightly to moderately acidic, and does not grow well on peat or water-saturated soils. It survives in more exposed coastal locations than does black spruce, as it is more salt-tolerant. It is more resistant to windthrow than are balsam fir or black spruce. In marine coastal areas, it survives further north than does black spruce, to 58°N in Labrador, whereas black spruce does not extend north of 55°N. White spruce is used for pulp-and-paper manufacture. It is more useful for lumber than is black spruce, as it tends to grow straighter trunks, with fewer knots. The straight, tall trunks make it preferred over black spruce for

Fig. 8.26 Taiga forests are dominated by *Picea mariana* (black spruce), with associated *Picea glauca* (white spruce). Deciduous shrubs include *Alnus viridis crispa* (green alder) and *Salix sp.* (willow)

planting as an ornamental. However, its strong aroma makes it unsuitable for use in spruce tea or spruce beer.

The narrow-leaved deciduous species *Populus tremuloides* (trembling aspen), *Populus balsamifera* (balsam poplar), *Alnus incana rugosa* (hoary alder), and *Betula papyrifera* (white birch) form lesser components of southern taiga assemblages (Rowe 1972; CNVC n.d.). These species are at or near their northern limits. Rarer participants can include *Acer spicatum* (mountain maple) and *Sorbus decora* (showy mountain ash; Ryan 1995). *Alnus viridis crispa* (green alder) and *Salix sp.* (willow) shrubs are common. In northern and elevated locations, shrubs present along the treeline include *Betula glandulosa* (dwarf birch; Northwest Territories n.d.) and *Salix arctica* (willow) (Walker et al. 2005).

Typical understory species of the taiga include *Rhododendron decumbens* (northern Labrador Tea), *Rhododendron groenlandicum* (southern Labrador Tea), *Vaccinium uliginosum* (bilberry), *Arctostaphylous uva-ursi* (bearberry), *Empetrum nigrum* (crowberry), *Epilobium latifolium* (broad-leaved fireweed), *Eriophorum* (cottongrass), *Rubus chamaemorus* (bakeapple), *Carex* (sedge), *Cassiope tetragona* (arctic white heather), *Cladonia rangifera* lichen (caribou moss), *Dryas integrifolia* (mountain avens), *Pyrola* (wintergreen), *Saxifraga* (saxifrage), *Silene* (moss campion), and *Vaccinium vitis-idaea* (partridgeberry) (Fig. 8.27). Along the shores of Hudson Bay, continuing sea level decline and climate warming are altering the distribution of this assemblage, as these species occupy newly emerged terrain and arctic species abandon southern areas to the taiga forest (Fig. 8.28).

An example of an altitudinal transition in vegetation occurs in the Lake Melville- Akami Uapishk[U]-KakKasuak-Mealy Mountains area (Catto 2020). A taiga/mid-boreal forest assemblage is present on the Lake Melville Plain and in the lower elevations of the Churchill Plateau (below approximately 400 m asl). Podzols (lesser) and brunisols (dominant) are present.

Above 400 m, a true taiga forest replaces the mixed assemblage. Balsam fir, aspen, mountain maple, chuckley-pear, pin cherry, and dogberry disappear, showy dogberry is replaced by northern showy dogberry (*Sorbus decora groenlandica*), and white birch is uncommon. *Alnus viridis crispa* (green alder), and the dwarf birches *Betula michauxii* (Newfoundland), *Betula borealis* (northern), *Betula glandulosa* (tundra), *Betula pumila* (swamp), and *Betula minor* (dwarf white) are present.

The summits of the mountains (above 900 m asl) support a tundra vegetation assemblage. Tree species are absent, and the largest plants are scrubby dwarf birches, willows, and green alders. Northern species, such as *Rhododendron decumbens* (northern Labrador Tea) replace their southern equivalents.

Fig. 8.27 Taiga Forest understory. Upper left: *Rhododendron groenlandicum* (southern Labrador tea). Upper right: *Vaccinium vitis-idaea* (partridgeberry). Lower left: *Epilobium latifolium* (broad-leaved fireweed). Lower right: *Vaccinium uliginosum* (bilberry)

Fig. 8.28 Taiga Forest vegetation transitional zones show mixtures of boreal and taiga flora and fauna. Inset: *Betula glandulosa* (dwarf birch) foliage

8.7 Soils

Soil development is minimal in most taiga regions (Soil Classification Working Group 1998). In the more southerly areas with the densest vegetation and thickest accumulations of Quaternary sediments, such as western Labrador, soils present include ferro-humic and humo-ferric podzols, with the latter having less humic matter and more iron than the former. Brunisols, regosols, gleysols, and organic soils are also frequently encountered. All soils are acidic, and the buffering capacity of the bedrock is low. The potential for acidic precipitation resulting from the iron ore operations exists, and local examples of acid rain and acidified terrain have been noted. To date, the impact of acid precipitation has not become an environmental issue in western Labrador.

In more northerly regions, and those without thick Quaternary sediments, soils are poorly developed brunisols, regosols, and static and turbic cryosols. Organic fibrisols are associated with fen, bog, and permafrost organic landforms.

8.8 Fauna

Taiga fauna represent communities intermediate between the boreal regimes to the south and the arctic tundra to the north (Banfield 1981; Naughton 2012). Locally, topographic isolation resulting from altitude, modern water bodies such as Hudson Bay, and lingering effects of previous Quaternary events, along with ongoing climate change, also influence faunal distribution. For many boreal species, such as raccoon and skunk, the lack of snow cover coupled with the cold winter temperatures are the limiting factors which prevent them from inhabiting the taiga. Others, such as white-tailed deer and porcupine, find the scarcity of deciduous trees to be a crucial negative factor.

Typical mammal species of taiga regions include caribou (*Rangifer tarandus caribou*), wolverine (*Gulo gulo*), timber wolf (*Canis lupus*), black bear (*Ursus americanus*), red fox (*Vulpes vulpes*), lynx (*Lynx canadensis*), snowshoe hare (*Lepus americanus*), red squirrel (*Tamiasciurus hudsonicus*), beaver (*Castor canadensis*), and muskrat (*Ondatra zibethica*). As elsewhere south of the treeline, Coyote (*Canis latrans*) has succeeded in colonizing taiga in the past 50 years, locally benefitting from the removal of the competing wolverine and wolf. With ongoing climate warming, skunk (*Mephitis mephitis*) is also steadily expanding its range (CABI 2022).

Moose (*Alces alces*) is resident in the southern taiga, although it is expanding its range northward in response to both climate change and decrease in woodland caribou (Chubbs and Schafer 1997; Rempel et al. 2021). For solitary animals such as moose and wolf, eskers form convenient trackways across taiga landscapes.

8.8.1 Caribou

The taiga regions are migratory homes to both barren-ground (*Rangifer tarandus groenlandicus*) and woodland caribou (*Rangifer tarandus caribou*) (Bergerud 2000, COSEWIC 2011; Northwest Territories 2022). North of the taiga, Tuktut

Nogait (Inuvialuktun: young caribou) National Park was designated in 1998, in large part to protect the calving grounds of the Bluenose barren-ground caribou herd (Parks Canada 2007; Davison 2015). The Bluenose caribou migrate northeastward from taiga areas of Dease and Smith Arms of Great Bear Lake to the open tundra of Tuktut Nogait. The topographic obstacles of Dolphin and Union Strait to the north and Great Bear Lake to the south (in addition to the denser southern taiga forest) preclude longer migrations.

Barren ground caribou are blocked by Hudson Strait from reaching northern Labrador. *Rangifer tarandus caribou* (woodland caribou) form several herds in northern Labrador and adjacent Québec, of which the George River herd is the largest and best-studied (Bergerud 2000; COSEWIC 2017). The species, *tuktu* in Inuktitut and *atok* in Innu-Aimun, was and is vitally important to both peoples as a food source and for hides. The caribou are also regarded as a key diagnostic species, indicative of the health of the faunal community and the environment as a whole.

The George River Herd is the only migratory group of caribou in Labrador (Fig. 8.29). The population was very tentatively estimated at 5,500 animals in 2018, down from 385,000 in 2001 (NL 2020). Population estimates are difficult due to the migratory nature of the caribou. The George River Herd migrates throughout northern and central Labrador and Nunavik. Calving is the most sensitive period in the caribou's life cycle. The principal calving grounds of the Herd are on the upland tundra plateau along the Quebec-Labrador border. Caribou show great fidelity to calving areas but not to other seasonal ranges. Calving areas appear to be chosen primarily because they offer reduced predation risks, especially from wolves. Quantity and quality of food resources are secondary considerations, only affecting caribou movements and distribution within a wider area chosen primarily for its safety from predators. Food reserves in the calving range in northwestern Labrador have been in decline since the 1960s.

Caribou migrate to their summer range, the tundra plateau east of George River and the north coast of Labrador, after calving is complete in late June. Coastal areas in northern Labrador provide important summer habitat, as cool on-shore breezes deter biting insects. In August, the large herds of caribou break up into smaller groups. Southward migration begins in October, along with mating. By late November, the caribou have reached their winter ranges in the southwestern taiga forests, often feeding in windswept areas with minimal snow cover. The major predator is wolf (*Canis lupus*). Black bears occasionally eat caribou, and lynx and wolverine will prey on calves. Males lose their antlers in the late fall, while females retain their antlers until the calving season. Females thus have an advantage in the competition for food sources during the difficult winter season. In caribou society, females are dominant in the winter and spring, whereas males dominate in the summer and autumn.

The effect of development on caribou migration is a particular concern (Rempel et al. 2021). Woodland caribou are reluctant to approach a road in forested terrain, possibly because it represents a greater contrast to the surrounding boreal environment. Roads in northern taiga or tundra areas evoke less apprehension. Caribou prefer to cross where the road is straight and level, since this provides the best visibility.

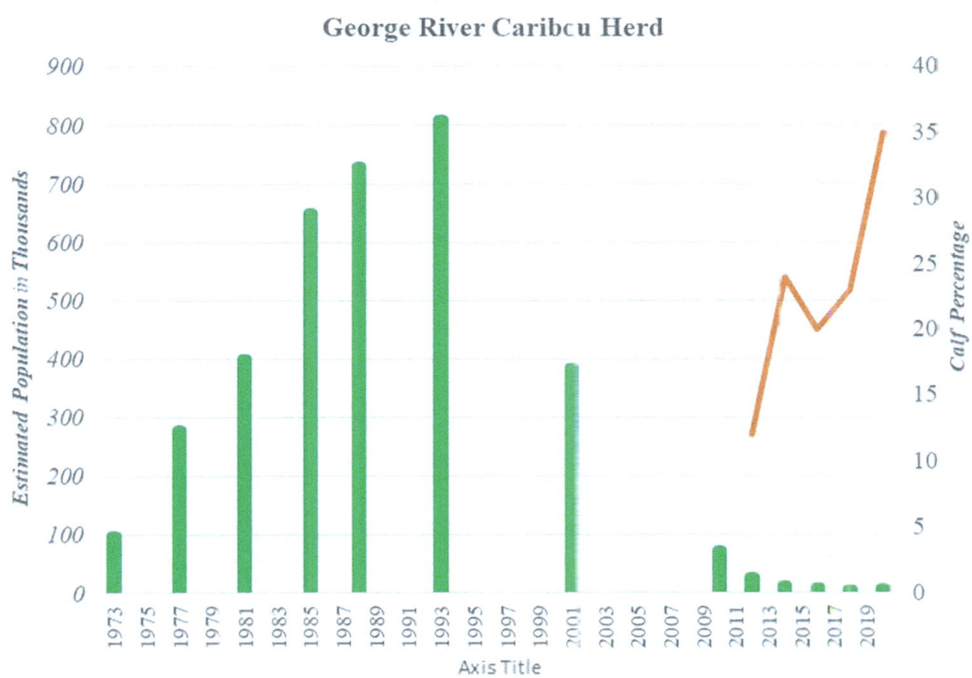

Fig. 8.29 George River Caribou Herd population has varied since 1973, but the sharp downward trend since 1993 is evident. The increasing percentage of calves since 2011 is encouraging for the prospect of future herd recovery. *Data* Government of Newfoundland and Labrador

In tundra areas, caribou shun roads which are elevated above the surrounding terrain on embankments, or those flanked by high snowbanks.

Three Labrador herds are sedentary (non-migratory), living in individual forested areas: Mealy Mountains (2500 animals), Lac Joseph (1500), and Red Wine Mountains (less than 100). These caribou are genetically distinct but physically similar to the George River population.

8.8.2 Predators

Wolves in both tundra and taiga regions represent the same species, *Canis lupus*. "Timber" wolves have brown to black fur, enabling them to blend into the treed environment, whereas "arctic" wolves have white fur for camouflage against the snow. The two populations are identical in other respects, however, and will interbreed freely.

The decline in wolverine (*Gulo luscus*) numbers is regarded as an extremely serious matter by indigenous peoples across the taiga. In traditional Innu thought, *Kuekuatsheu* was considered to have created the Earth, along with *Otshisk* (*Ondrata zibethicus*, muskrat) (Wolverine Foundation 2022). *Kuekuatsheu* is a central figure, responsible for forming much of the landscape and having numerous dealings with animals and people. Removal of this culturally important species from the faunal assemblage is thus regarded as an indication of failing environmental health, an opinion shared by many biologists.

Wolverines are solitary animals, with large home ranges (Naughton 2012; Wolverine Foundation 2022). Each wolverine occupies a home range of 100–200 km^2, and individuals have been tracked for distances in excess of 80 km, particularly when following migrating caribou herds. Although the wolverine is largely a scavenger, it is quite capable of defending its food from wolves and black bears. A large adult male wolverine has a mass of 15 kg, and is capable of overpowering a human.

Wolverines construct dens by burrowing in deep snow. Tunnels are dug 1–3 m below the surface of the snow, and may extend for more than 50 m. These "snow caves" provide a well-insulated environment for bearing young. In some cases, the mother wolverine will drag spruce boughs into the den for the young to lie upon. The dens are usually constructed in early to mid-March, with the birth of 1–3 kits commonly following shortly thereafter. Kits are weaned at approximately 10 weeks of age. Wolverines will immediately abandon any den if disturbed by humans.

Human hunting and trapping is the major cause of mortality among wolverines. Secondary causes include starvation and injuries from other predators (particularly wolves) during fights over kills. However, the current decline in the wolverine population is almost entirely related to human pressure throughout their range.

8.8.3 Bears

Climate change is also influencing the distribution of the bear species *Ursus americanus* (black), *Ursus arctos horribilis* (grizzly), and *Ursus maritimus* (polar) (Clark et al. 2019). In the Taiga Plains and interior Taiga Shield of Kivalliq, black and grizzly bears are expanding their ranges northward. In coastal Labrador, polar bears (*nanuq*) are more frequently encountered further south. Individuals are transported south on ice floes and pans, and sightings are common south to the Strait of Belle Isle, and occasionally in coastal Newfoundland as far south as St. John's. The difficulties posed by polar bears near human settlements, however, together with their preference for arctic environments, have prevented establishment of the species along the coastline south of 57°N. The Torngat Mountains forms a topographic barrier, and polar bear absence from the interior has allowed the black bear (Innu-aimun *mask*) to expand its range north of the tree line. Expansion reflects increasingly successful competition with polar bears. As climate warming affects the distribution of sea ice in the Labrador Sea, conditions have become progressively less suitable for *nanuq* to inhabit northernmost Labrador, similarly to the situation in southern Hudson Bay.

The most visible population of polar bears in Canada occupies the shoreline of Hudson Bay at Churchill, Wapusk (Nehinaw*: white bear*) National Park, and Polar Bear Provincial Park (Obbard and Walton 2004; Laforge et al. 2017; Ontario 2022; Fig. 8.30). Polar bears are able to swim the waters of Hudson Bay without difficulty, and thus have occupied the entire southern shore. In the Churchill region, individual bears have achieved nuisance status, and have required wildlife officers to remove or destroy them.

Fig. 8.30 Polar Bears have become a focus of tourism and scientific research at Churchill and Wapusk National Park. (Wikimedia/Ian Stewart)

Adult male bears are up to 450 kg, and females reach 275 kg (Banfield 1981, Naughton 2012, Nunavut Environment 2022). An adult male can run at 50 km/h for short bursts, can maintain a speed of 30 km/h indefinitely, has no problem with hills or steep slopes, and is capable of swimming across Hudson Bay. Females appear to be a little faster. As the *Homo sapiens* world record for the 100-m dash is 35 km/h, people cannot outrun polar bears. Although polar bears prefer ringed seal (*Phoca hispida*) and bearded seal (*Erignathus barbatus*), they will eat any type of meat (Michaux et al. 2021).

Bears den in snow hollows or scooped out excavations in marine clay in November. The cubs are born in December, and family groups appear in late March to begin hunting live meat. The bears follow the pack ice front seaward throughout the spring and early summer. When the pack ice dissipates, many bears come ashore into summer retreats, such as Wapusk. Here, they exert themselves as little as possible, although they will not pass up any opportunity to get an easy meal.

In Wapusk and Churchill, wardens and residents go to considerable length to keep tourists off the menu (see Clark 2003; Schmidt et al. 2022). Camping is strictly regulated in Wapusk, for the protection of "southerners". Local residents avoid camping along the coastline, especially near icebergs (favourite haunts of the bears). Dogs are useful, but inexperienced dogs have the distressing tendency to bring the bears back to camp. Tents should be set up in straight lines, rather than circles, to avoid giving any bears the impression that they are surrounded or trapped. Cooking should be done in the open, away from tents.

Despite its name, Polar Bear Provincial Park does not have a large population of polar bears, as they are less common in the more southern areas. The advancing taiga forest climate has not acted as a major deterrent for the bears as yet, as the countering influence of increased garbage production by humans has kept pace.

8.8.4 Small Mammals, Nunavik, and Labrador

The presence of Hudson Bay has isolated small mammal populations in Nunavik and Labrador from Kivalliq and the Taiga Plains. Species present in Labrador include *Condylura cristata* (star-nosed mole), *Lepus americanus* (snowshoe hare; oaposh), *Tamiascurus hudsonicus regalis* (red squirrel; annikotash), *Glaucomys sabrinus makkovikensis* (Labrador flying squirrel; oponikotsash), *Castor canadensis canadensis* (beaver; anisk), *Ondrata zibethicus aquilonius* (muskrat; otshisk), *Sorex cinereus* (shrew; tsinitoapokoshish); *Mustela erminea richardsonii* (ermine; socoshish), *Mustela nivalis* (least weasel), *Mustela vison lowii* (Labrador mink; tchékash), *and Lutra canadensis chimo* (Ungava Otter; nkok). The mixture of subspecies confined to Labrador and Nunavik with those more widely distributed across the continent indicates faunal isolation to some extent. Climate variations throughout the Holocene are responsible for some of this faunal distribution.

Glaucomys sabrinus makkovikensis (flying squirrel; oponikotsash; Fig. 8.31) typically has light brown fur on its back and pale grey fur on its stomach (Wells-Gosling and Heaney 1984). A patagium (fleshy, fur-covered membrane) connects the wrist to the ankle. When extended, this enables the squirrel to glide up to 90 m. Labrador flying squirrels appear to be more effective fliers than more southerly residents, possibly in response to the greater distance between trees in the northern part of their range. The tail is furred, flattened, rounded at the end, and is almost as long as the

Fig. 8.31 *Glaucomys sabrinus* (Northern Flying Squirrel). (upper: Bob Cherry/National Parks Service) Lower: current range. *Data* International Union for Conservation of Nature

remainder of the body. In summer, the species is nocturnal, but in Labrador flying squirrels are active throughout much of the winter daylight hours. Fungi and lichens are its major food sources, with lesser amounts of seeds, spruce and balsam, fir buds, and sap.

One litter is born each year, with 2–3 young. Although the male is not involved in caring for the young, northern flying squirrels sometimes share nests and may live in groups of up to 8 adults and juveniles. Individuals of *Glaucomys sabrinus* form single-sex groups for warmth during the winter. Although female northern flying squirrels are territorial, males are not. Northern flying squirrels chirp, and when distressed they make low, quacking sounds. In western Labrador, the main predators of northern flying squirrels are owls, hawks, marten, fishers, mink, and ermine.

Mink (*Neogale vison*; tchékash) are larger than ermine (typically 1 kg), and unlike the ermine are aquatic (Banfield 1981). Labrador mink eat fish, frogs, insects, shrews, voles, and deer mice. Preferred habitats are always adjacent to watercourses. Mink are usually nocturnal, but are sometimes active around dawn and dusk. They usually hunt alone, and do not hibernate, frequently swimming under ice in rivers and ponds.

8.8.5 Freshwater Fish

The distribution of fish species which cannot migrate across saline waters east and west of Hudson Bay, such as northern pike *(Esox lucius)*, reflects previous deglacial linkages of drainage systems such as Labrador's Churchill River and Kivalliq streams to southerly systems.

Freshwater species include *Salvelinus fontinalis* (brook trout), *Esox lucius* (northern pike), *Salvelinus namaycush* (lake trout), and *Coregonus clupiformis* (whitefish) (Scott and Crossman 1998). Whitefish (Nehinaw: *atihkamêk*, Gwich'in: *luk dagaii*, Nunatsiavut Inuit: *killalugak*) is an important human food species in the southern taiga, but does not occur in tundra or northernmost taiga areas. Whitefish range from 20 to 50 cm in length and 2–12 kg in mass. Whitefish cannot move readily through rapids or shallow waters, and require cold conditions. They spawn on gravelly or stony substrates in late October and early November, and require water temperatures of less than 8 °C for successful reproduction (the mortality of the eggs reaches 99% at a water temperature of 10 °C).

Salvelinus alpinus (Arctic Char or Charr; Inuit *ikaluk*; Gwich'in *ddhah luk*) is an important human food source (e.g. Beaumier et al. 2015; Pannikkar and Lemmond 2020; Fig. 8.32). Char colonized taiga regions from the coastlines inland (Salisbury et al. 2019), and thus are not found inland of obstructions or rapids on river systems. Although considered an "arctic" species, char co-habit waters with brook trout and lake trout wherever their ranges overlap (Scott and Crossman 1998). Competition between the species exists, however. Char spawn and hatch earlier in the season than do brook or lake trout, and thus young hatchling char can prey upon the unhatched eggs of the other salmonids. This is one factor that has kept the trout species from expanding their ranges to the Labrador Sea coast south of Groswater Bay. The second factor is the differing paths of migration to eastern Labrador, with char moving from the marine coastline and trout moving from the southwest.

Fig. 8.32 *Salvelinus alpinus* (Arctic char), fish and fillet. (Wikimedia/Saibling)

Throughout Labrador and Nunavik, *Salmo salar* (Atlantic salmon) has been a major food source for raptors, mammals, and humans since shortly after deglaciation. Culturally and economically significant salmon populations are present in all of the major rivers of Labrador, notably those entering the Labrador Sea from Postville south to the Strait of Belle Isle. Conflicts over salmon management, particularly the relative roles of the subsistence fisheries of the local communities and the sports fishery of commercial lodges, have occurred.

In Labrador, salmon spawn in late August through early October, preferring fine gravel in riffle areas or shallow pools. Deep pools are preferred feeding sites, and hence are the areas where salmon are generally caught, but spawning fish require faster water and coarser sediment. The gestation period is approximately 110 days, and hatching usually coincides with water temperatures of 4 °C or greater. The young fish remain in the gravel after hatching for three to six weeks, emerging in late June-early July. Maturation from hatchling to seabound parr requires 4–5 years in Labrador. The salmon return, as grilse, after 2–3 years.

Acipenser oxyrhynchus (Atlantic Sturgeon) is confined to Lake Melville. This large species (individuals may exceed 50 kg) is primarily a nearshore marine and estuarine fish. It prefers brackish, quiet, slow-flowing waters, with muddy to fine sand bottoms, and water temperatures above 4 °C. The presence of Atlantic sturgeon in Lake Melville, along with tomcod *(Microgadus tomcod)* and eel *(Anguilla rostrata)*, indicates that the Lake Melville estuary is currently a healthy environment, receiving adequate quantities of fresh water from the Churchill River throughout the year and being well-aerated in the spring and summer. All three species are migratory, moving in response to changes in water temperature and salinity, and possibly in response to monthly variations in tides (particularly eels).

8.9 Marine Environments

8.9.1 Lake Melville

Lake Melville is brackish (5–15 psu). The 1900 m^3/s mean discharge of the Churchill River at Happy Valley-Goose Bay reflects upstream regulation, with freshet reduced, and seasonal low flow increased. Turbid underflows during freshet bring cold, dense, sediment-laden waters to Lake Melville.

Limited wave activity and dominant southwesterly winds influence the shoreline. Tidal platforms and marshes extend to elevations of 2 m asl, and inactive features are associated with higher strandlines to ±10 m asl (Catto 2020).

8.9.2 Labrador Sea

8.9.2.1 Bathymetry

The bathymetry of the Labrador Sea consists of a number of troughs, separating shallower banks. Some banks are linked by saddles, representing areas of moderately deep water (more than 200 m below sea level) that are not as linear or well-defined as the deeper troughs (Bell 1989).

The northernmost part of the coastline, from Killiniq (Cape Chidley) south to Saglek Bay, is flanked by the shallow Saglek Bank, generally less than 200 m below sea level. The Saglek Bank is cut by a trough aligned WNW-ESE, extending seaward from Nachvak Fjord. The southern margin of the Saglek Bank is defined by a trough that extends from Saglek Bay to junction with a second trough extending seaward from Hebron Fjord.

To the south of the Hebron Fjord Trough is Okak Bank, centred around 58°N, 60°30′W. Okak Bank is also less than 200 m below sea level. To the south, Okak Bank is bordered by the deeper Okak Bay trough, reaching to below 300 m.

The Nain Bank borders the southern margin of the Okak Bay Trough. Locally, the shallowest parts of the Nain Bank are less than 100 m below sea level. South of Cape Kiglapit, the Nain Bank is flanked to the west by the deep (>600 m) NW–SE aligned Labrador Marginal Trough, which parallels the coastline southeastward to Hawke Saddle. The trough is deeply indented, showing a pinnate drainage pattern on the landward side. The shallowest area of the Nain Bank is located directly to the east of Kikiktaksoak Island. The Nain Bank is bordered to the southeast by the Hopedale Saddle, with a maximum depth in excess of 600 m below sea level, and minimum depths of less than 400 m.

The Labrador Marginal Trough reaches its greatest depth northeast of Cape Makkovik, more than 800 m below sea level. To the south are the Makkovik Bank and Harrison Bank. The shallowest part of Makkovik Bank, north of Cape Harrison, is less than 100 m below sea level. The Makkovik Bank is separated from the Harrison Bank to the southeast by a trough more than 300 m deep, extending northeastwards from Cape Harrison. The shallowest part of the Harrison Bank is less than 200 m below sea level.

The Cartwright Saddle extends northeastward from the Labrador Marginal Trough, to the south of Harrison Bank. Depths at the landward head of the marginal trough are in excess of 600 m. The water shallows over the Cartwright Saddle to less than 400 m. The saddle extends NNE from Sandwich Bay. This alignment, indicating that Sandwich Bay and not the Narrows of Lake Melville formed the main channel between the shoreline and the open Labrador Sea, indicates that the Churchill River-Lake Melville axis has not always been the dominant drainage route for water from western and central Labrador.

To the southeast of the Cartwright Saddle, Hamilton Bank (minimum depth <200 m), Hawke Saddle (maximum depth in excess of 500 m), and Belle Isle Bank (minimum depth <200 m) extend southward to 52°N latitude. The Labrador Marginal Trough shallows to the southeast. The Bardore Marginal Trough is located 10–20 km offshore from north of Hawke Harbour to south of Alexis Bay. It is aligned parallel to shore and has a maximum depth in excess of 300 m southeast of Alexis Bay, representing a seaward submarine canyon extension of the Alexis River. The absence of digitate margins along the narrow (generally <5 km wide) Bradore Trough suggests that it developed primarily as a submarine canyon.

The Strait of Belle Isle is a northeastward extension of the Mecatina Trough and Esquiman Channel. It reaches a maximum depth in excess of 100 m off Point Amour and northeast of 56°W. The seaward extension of the trough is more than 200 m deep to the north of Belle Isle.

8.9.2.2 Surface Currents

Cold currents flow southwestward from the Barents Sea along the Kalaallit Nunaat (Greenland) coast, forming the northern loop of the North Atlantic gyre system (Figs. 8.33 and 8.34). The East Greenland Current brings cold water along the coast through Denmark Strait between Iceland and Kalaallit Nunaat (The older terms "East Greenland Current" and "West Greenland Current" have not yet been replaced by "Kalaallit Nunaat Currents" in the oceanology literature as of 2022). This water circulates along the Kalaallit Nunaat coast to Cape Farewell, and then flows northwestward along the Kalaallit Nunaat coast as the West Greenland Current. Augmented by cold,

Fig. 8.33 Deployment of a Sediment Trap during Shipboard research on the Labrador Sea. (Evan Edinger)

fresh water from the glaciers, the flow returns southward along the Labrador coast as the Labrador Current (Farmer 1981; Talley and McCartney 1982).

The Labrador Current transports approximately 6 million m^3/s of water southwards at a velocity of 1–2 km/h (measured offshore of Belle Isle), making it among the slowest currents in the ocean. The peak velocity is usually reached in late summer. Typically in late August, the surface velocity offshore of Nain is 10 cm/s (0.4 km/h). Influxes from the streams have relatively minimal impacts on surface current velocities, but the addition of water from a southwestward-flowing arm of the West Greenland Current offshore of Nain and Hopedale increases surface velocity to 20 cm/s (0.8 km/h) offshore of Groswater Bay (Kangerliorsoak). Velocities in the winter months are approximately half the summer values.

Part of the reason for this low velocity is the presence of the low pressure system situated over Baffin Bay and Davis Strait. As the Labrador Current is attempting to flow southwards away from the low pressure region, it is working against the pressure gradient. The increase in pressure from north to south is matched by an increase in sea level. Although the modal difference in sea level between Nain and St. John's is less than 5 cm, and therefore cannot be seen or felt by mariners, this gradient is sufficient to reduce the volume and velocity of the Labrador Current. Thus, travellers moving northward against the Labrador Current are moving down the pressure and sea level gradients to northern Labrador.

During some seasons when the low pressure system is particularly strong, weak northward flow of the Labrador Current marks the area of Davis Strait from Iqaluit northwards. This results in the transport of relatively warmer waters northwards into Davis Strait, hindering ice formation there. It also precludes southward transport of colder Davis Strait and Baffin Bay water southward along the Labrador Coast. In areas to the south of Nain, this delays seasonal ice formation in the autumn, and hastens the breakup of seasonal ice in the spring. In years when the northward reversal does not occur, or is weakened, sea ice persists for longer periods along the Labrador coastline from Cartwright northwards, hindering coastal navigation during the normal late spring-early summer season.

Even along the southern segment of the Labrador Sea, the effects of the current reversal further north are felt. Although the Labrador Current does not reverse itself offshore of communities to the south of Nain, strong southerly winds will provide sufficient energy through frictional transfer to the upper layers of the water mass to drive water northwards. The result is to bring periodic influxes of water from south to north along the coastline. Although the normal result of this is to increase surface temperatures (at least temporarily) as warm water is brought from south to north, decreased temperatures may result where colder waters are transported northwards. This can occur where rivers (Eagle River, Paradise River) or estuaries (Lake Melville) discharge quantities of colder water into the Labrador Sea following spring breakup. A stronger effect is observed along the southeastern coastline, where cold waters from the St. Lawrence River exit the Strait of Belle Isle in the spring and may be transported northwards towards Black Tickle.

The occasional weakness of the Labrador Current has produced anomalous results for those investigating the offshore waters. During the environmental investigations

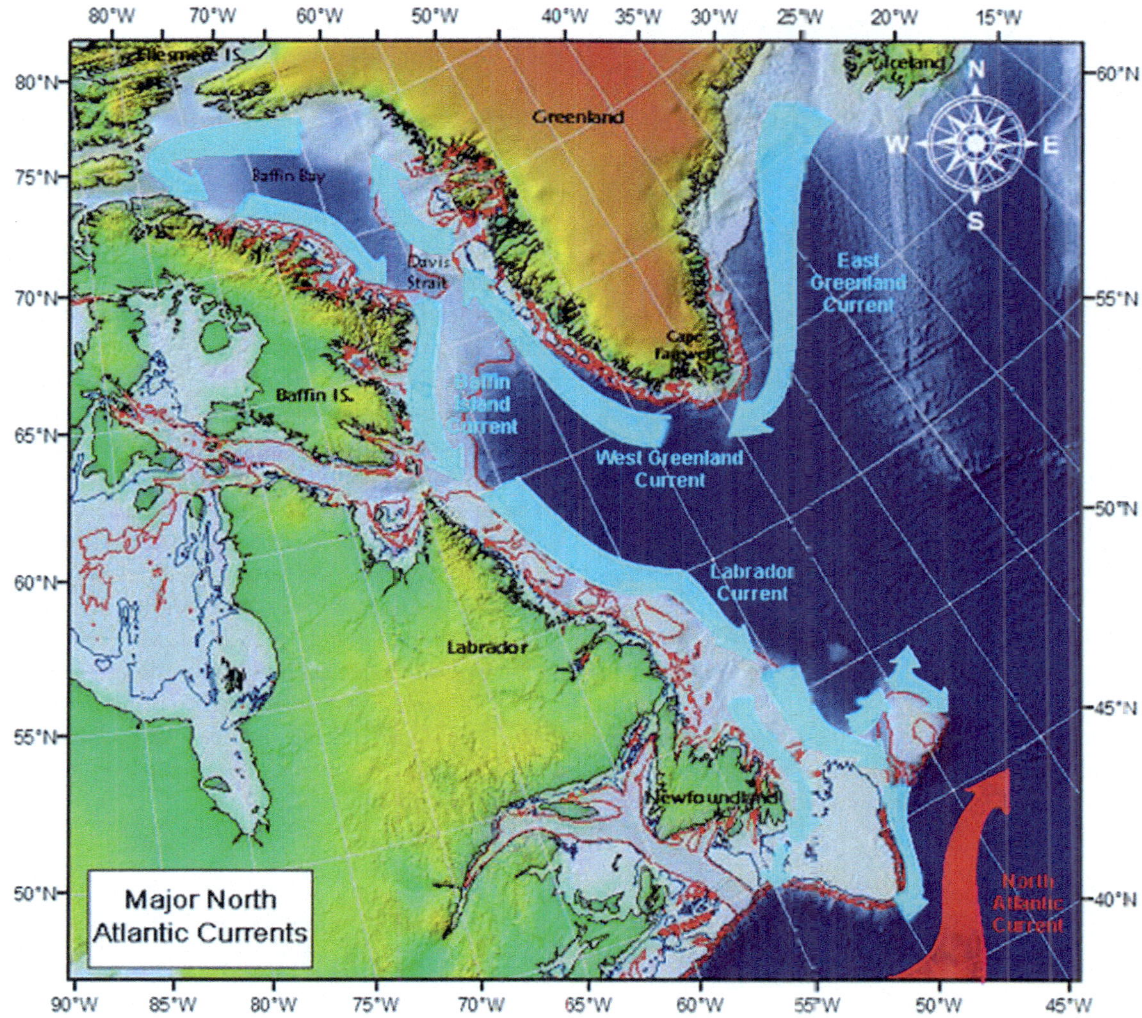

Fig. 8.34 Labrador Current. (United States Coast Guard)

related to the "White Rose" petroleum development off Newfoundland, the consultants indicated that they were unable to find the Labrador Current, although they knew that it should be present. The short-term study conducted for this project coincided with a period of northward flow in northernmost Labrador and Nunavut, and minimal flow along the remainder of the current's length, accounting for the Labrador Current's temporary absence. This behaviour must also be considered in any development of offshore gas reserves beneath the Labrador Sea.

Along the coastline of Labrador, interaction between the Labrador Current and inputs of seasonally warmer water from estuaries and rivers results in increases of biological productivity. Due to the seasonal nature of both the Labrador Current and the freshwater inputs from land, however, the periods of biological productivity are shorter than over the Grand Banks further south. In addition, the presence of seasonal ice cover also influences upwelling and mixing rates, even during spring breakup on land when freshet discharge is occurring.

8.9.2.3 Sea Ice

Sea ice is ubiquitous along the entire Labrador coast (Fig. 8.35). Ice reaches its maximum thickness and maximum seaward extent in early February, more than 1000 km seaward of Nain and Hebron and completely filling Hudson Strait (Andrews et al. 2018) For the terrestrial animal life, as well as human community residents, this expanse of sea ice provides additional terrain for movement. In mid- to late March, recession of sea ice in the north corresponds with expansion to the south and southeast. By early May, ice typically extends up to 600 km offshore of Makkovik (Markham 1980; Environment Canada 2025; Canadian Ice Service 2020).

Sea ice generally leaves the Strait of Belle Isle in late May or early June (Catto 2020). Open leads (*polyna; ashkui*) are often present along the shoreline south of Black Tickle.

Fig. 8.35 Sea Ice, Labrador Sea. (Earth Observatory, NASA)

Outflow of the spring freshet from the Churchill River and Lake Melville results in the formation of a tongue of sea ice extending southeastward from the main ice limit into the open Labrador Sea. Sea ice typically leaves the Killiniq area in early July, opening Hudson Strait for navigation. Ice expands southwards to close Hudson Strait to navigation in September.

Since 2000, sea ice coverage in the southern Labrador Sea has generally been below the historical average for the past 60 years, although individual anomalous years with extensive or restricted ice are common (Canadian Ice Service 2022). Reduced sea ice cover restricts travel between coastal communities and opportunities for harvesting marine resources, and can lead to unsafe conditions resulting in loss of life. Enhanced locally based monitoring undertaken through the SMART ICE (Sea-ice Monitoring And Real-Time Information for Coastal Environments) and other initiatives blends community and scientific observations to assist Inuit community residents in coping with shifting ice patterns resulting from weather events influenced by climate change (Forbes 2011; Riedlsperger et al 2017; Bell 2018; Wilson et al. 2020).

In coastal regions, sea ice cover reduces the climate effects of oceanic modification are reduced during the winter months. Sea ice does not entirely prevent exchange of heat from the ocean to the atmosphere, but by preventing winds from interacting with surface waters, it greatly reduces the efficiency and rate of heat transfer. Thus, in areas with thick and persistent ice cover, conditions essentially revert to continental climates during the winter months.

Fig. 8.36 Icebergs characterize the coastline of the Labrador Sea from Killiniq to Newfoundland

8.9.2.4 Icebergs

Icebergs are present throughout the year in the open Labrador Sea, and drift southward during the summer (Fig. 8.36). Almost all are confined to the area of the Labrador Marginal Trough. Since 1990, between 1800 and 2000 icebergs reach Killiniq annually. Icebergs are produced by calving along Baffin Bay and Davis Strait, predominantly from Kalaallit Nunaat.

Icebergs begin to ablate rapidly once they move southward from Hudson Strait. During the peak iceberg transit season (typically mid- July), the sea surface temperature off Hopedale is 6 °C. An iceberg 40 m high above the water line and 100 m long (approximate volume 2,000,000 m^3, mass approximately 10 million tonnes) will completely ablate in 24 days in 2 °C water, and requires only 15 days in 4 °C water. Large icebergs make up approximately 20% of the iceberg population off Killiniq, about 10% off Makkovik, and about 5–6% at the mouth of the Strait of Belle Isle. Smaller fragments ("growlers") increase in proportion southwards as the larger icebergs disintegrate, but an average "growler" will endure for less than 5 days in 4 °C water. Less than 10% of the icebergs which enter Labrador waters offshore of Killiniq succeed in running the gauntlet to 52°N.

The number of icebergs reaching a particular point is only indirectly related to climate conditions. As the larger icebergs break into smaller fragments en route, in many instances a greater number of icebergs at St. Anthony, Twillingate, or St. John's results from the progressive breakup of a single large mass, rather than an increase in the initial number. Advances and recessions of tidewater glaciers in Kalaallit Nunaat, leading to calving beyond the grounding line, are more closely related to the total volume of berg ice produced.

8.9.2.5 Tides

In Atlantic Canada, the semi-diurnal tide progresses southwestward (Dohler 2007). The northwest–southeast alignment of the Labrador coastline means that the tidal front strikes virtually the entire length of the shore simultaneously. Thus, there is little difference in the times of high tides along all open segments of the coastline from Killiniq to Battle Harbour.

Once the tide enters a restricted embayment, its progression is slowed as the water begins to interact frictionally with the bay floor and the coastline. In Groswater Bay, the tidal front requires approximately 2 h to travel from the mouth to Rigolet. The rate of travel is further slowed in Hamilton Inlet, and in the shallow waters of Lake Melville. Almost seven hours are required for the tidal front to progress through Groswater Bay, Hamilton Inlet, and Lake Melville to Happy Valley-Goose Bay. As a result, the time of high tide at Goose Bay and North West River almost coincides with low tide along the mouth of Groswater Bay.

Similar effects are observed in other inlets along the coastline. The tidal front requires approximately 1 h to travel inland from the Bay of Islands to Postville, about 1 h to reach the head of Voisey's Bay from its mouth, about 1.5 h to reach Port Hope Simpson from Spear Harbour, and about 2.5 h to reach Paradise River from Huntingdon Island.

Most of Labrador's coastline is classed as microtidal. Small areas of mesotidal conditions have developed where tides are funnelled into progressively narrowing embayments, such as the heads of Kaipokok Bay and Ugioktok Bay, at Cartwright, and in scattered localities along Lake Melville and the Strait of Belle Isle.

8.9.2.6 Coastal Labrador

The Labrador coastline is dominated by rocky shores, including fjord cliffs, fjard granite and gneiss with skerries, and rock platforms resulting from frost action combined with seasonal ice shove (Catto 2020; Fig. 8.37). Gravel pocket beaches and bayhead bars develop in coves where faulting and jointing ae modified by frost and wave action. Tidal platforms are present along the shores of Lake Melville, in addition to the large delta of the Churchill River. Sandy beaches are less common, although examples exist along the north shore of the Strait of Belle Isle (e.g. Blanc-Sablon), at Iron Strand in Torngat Mountains National Park (see Chap. 9), and at Porcupine Strand.

Tidally influenced bedrock platforms mark the mesotidal coastline of Cartwright-Sandwich Bay. During the winter, the platforms are scoured by drifting sea ice, pushed onshore by tides until freeze-up occurs, with renewal in spring. The impact of seasonal ice activity is directly related to tidal range, with greater effects evident in mesotidal areas. Sea ice must be driven aground, either by waves or tides, in order to be effective as an erosional agent.

Bedrock platforms are exposed during early to mid-spring (June). Small amounts of sediment cover are removed during autumn storm events. Above high tide level, the development of rock platforms is primarily of result of frost action when the rock is exposed.

Porcupine Strand, along the shore of Trumore Bay in Akami-Uapishk^{U}-KakKasuak-Mealy Mountains National Park Reserve, may have been the original "Wunderstrand" site mentioned in the Norse sagas (Fig. 8.38). Dominated by southeastwards littoral drift, the white medium-grained sand beaches extend over 55 km. Bedrock knobs divide the strand into segments, and interrupt littoral drift during some wind conditions.

8.9.3 Marine Labrador Sea Birds

More than 110 bird species currently are classified as residents of Labrador. Bird species have varied relatively little under human pressure, with only two—the Great Auk (*Pinguinus impennis*) and the Labrador Duck (*Camptorhynchus labradorius*)—becoming extinct through overhunting (Godfrey 1986; Bird 2017).

Oceanodroma leucorhoa (Leach's storm-tailed petrel) lives in dense colonies on coastal islands. The most northerly, and the largest colony, is located on the Gannet Islands. St. Peter's Island also houses a large colony. Unlike other marine coastal species, the petrel constructs burrows for its nests. During nesting season, many of the petrels are thus underground. Feeding is nocturnal, on molluscs and crustaceans. Petrels come ashore only during the breeding and nesting season. The burrowing habit of the petrel leaves it very vulnerable to predation from burrowing mammals, which explains why petrel colonies are invariably located on islands isolated from the mainland by strong currents.

Eider (*Sommateria mollissima)* is a coastal species, living in large colonies on low-lying, rocky shorelines. This habitat, more accessible to hunters than the cliff environments preferred by some other marine species, has proven to be vulnerable to overhunting. Although eider populations in Labrador have not suffered as severely as those further south, undisciplined hunting could create problems for the species.

Among the numerous bird species frequenting coastal and near-coastal rivers is the harlequin duck (*Histrionicus histrionicus;* in Innu-Aimun, *Nost paostokoeshish*). Harlequin ducks breed along rocky shorelines in intertidal or surf zone areas. Although the Harlequin duck has succeeded in occupying the ecological niche vacated by the extinct Labrador duck (*Camptorhynchus labradorius*), it is under threat outside Labrador due to tourist disturbance and habitat loss.

A key species that has largely disappeared from Labrador is the Brant (*Branta bernicla;* Innu-aimen *apishtiss*). *Apishtiss* is culturally important to the Innu nation. The initial cause of the demise of the Brant appears to be primarily natural.

Prior to 1930, Brant wintered in the Mid-Atlantic US states and in southern New England, feeding primarily in salt marshes dominated by eelgrass (*Zostera marina*), and migrated northward after spring breakup to lakes in the

Fig. 8.37 Coastal Labrador. Upper left: Coastline of the Makkovik structural province is dominated by glaciated metamorphic bedrock, restricted coastal heath barrens assemblages and exposed rock, and deep indentations along geological faults and joints. Upper right: Sediment ridges formed by seasonal thrusting of sea ice are common along gently sloping coastlines. Lower: In summer and autumn (left), the coastline is defined by the water's edge as depicted on a typical topographic map (false-colour Landsat Image, Canadian Space Agency). In winter, the edge of land-fast ice marks the border between "terrestrial" and "marine", both physically and for human community resource utilization and travel. (NASA image)

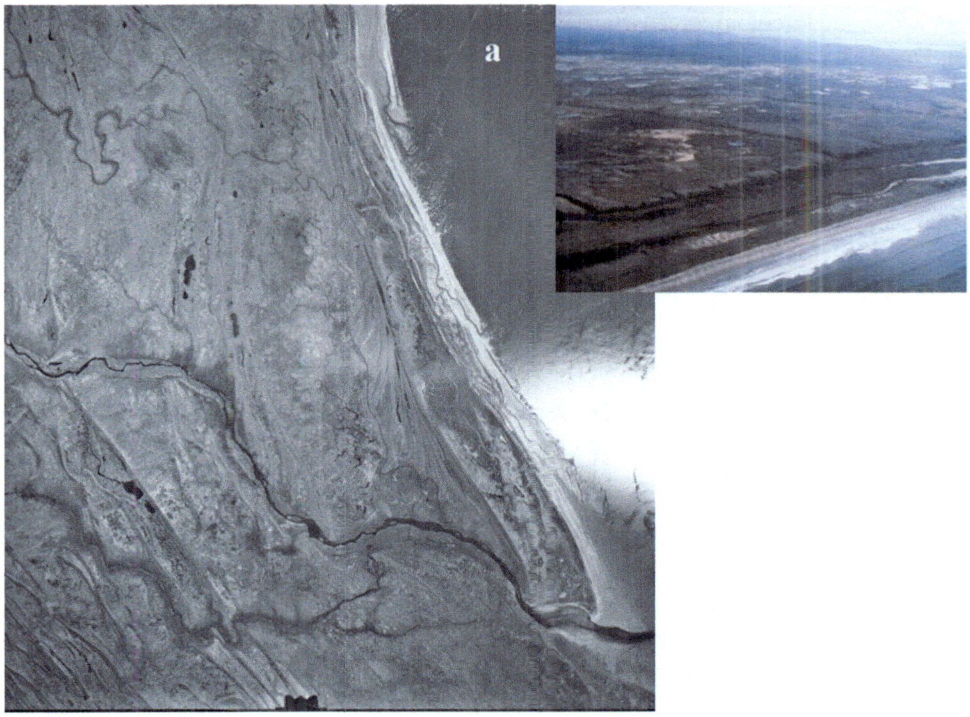

Fig. 8.38 Porcupine Strand, Trumore Bay. Left: Multiple marine strandlines indicate progressive fall in relative sea level from ~150 m asl. (National Air Photo Library) Right: The shoreline is dominated by a dissipative regime and southeastward littoral drift. Ephemeral foredunes develop during summers. (Photo courtesy of Joyce MacPherson)

tundra and northern taiga (Godfrey 1986). The birds flew along interior routes northward from the eelgrass marshes at Kamouraska and Sept-Iles, utilizing the many shallow lakes of western Labrador as staging areas. Eelgrass vegetation communities are found in sandy, relatively sheltered lowshore locations. *Zostera* is not as tolerant to freshwater influxes as is saltmarsh vegetation, so eelgrass areas tend to be associated with mesotidal flats that are not adjacent to major rivers. Eelgrass beds are important feeding and resting areas for salmon, trout, eels, and other fish, as well as for Brants.

Unfortunately, the Brant was almost entirely dependent on eelgrass (see Ganter 2000). Between 1930 and 1935, virtually all of the eelgrass in eastern North America was destroyed by the fungus *Labyrinthula* ("wasting disease"). Ironically, the pathogen was spread throughout the eelgrass range by migrating birds. As the Brant is dependent upon eelgrass, removal of the species in only five years effectively doomed the Brant population.

Although the Brant did not become extinct, and has recovered in some areas of North America, the migration route through western Labrador has not been re-established. Today, the Brant is a rare visitor to Nain and Jack Lane's Bay. The reasons for the failure of Apishtiss to re-establish its migration pattern are not certain, as some of the eelgrass beds have recovered from the fungal pathogen.

Gull species in Labrador include a disjunct population of *Larus delawarensis* (Ring-billed gull), which is confined to Lake Melville. Other gulls include the ubiquitous *Larus argentatus* (herring gull); *Larus hyperboreus* (Glaucous gull), in coastal areas marginal to the area north of Hopedale; and *Larus marinus* (Great black-backed gull) throughout the coastal zone. The black-legged kittiwake (*Rissa tridactyla*) breeds on the Outer Gannet Islands, along with *Morus bassanus* (northern gannet; Fig. 8.39).

Three species of terns inhabit Labrador. *Sterna caspia* (Caspian tern) is confined to the Backway area of Lake Melville. *Sterna hirundo* (common tern) can be found south of the Churchill River. The Arctic tern (*Sterna paradisaea*) is found throughout Labrador, and prefers sandy beaches and dunes as nesting sites. Although it is primarily a marine

Fig. 8.39 *Morus bassanus* (northern gannet)

species, a single colony breeds in the Labrador interior at Attikamagen Lake. Other common coastal birds include the thick-billed murre (*Uria lomvia*); the razorbill (*Alca torda*), along the coastline to Nain; and the Atlantic puffin (*Fratercula arctica*), in coastal clifftop sites south from Nain.

8.9.4 Marine Mammals

Marine mammals in the Labrador Sea include walrus, several species of whales, and seals (Mansfield 1967; Lien et al. 1985; Reeves 1988; Proctor and Lynch 2005). Walrus (*Odobenus rosmarus;* see Chap. 9) is extremely rare in Labrador waters and have always been more common north of Okak than to the south.

Commercial whaling in Labrador waters began with the Basque factory operations at Red Bay in the early 1500s (see Chap. 10) and continued into the mid-twentieth century. Today, whaling is not practised in Labrador. Narwhal, beluga (Fig. 8.40), orca, humpback, and bowhead whales, present in Labrador Sea waters, are discussed in Chap. 9. The North Atlantic Right Whale is discussed in Chap. 2. Population updates for all marine mammal species are provided by Fisheries and Oceans Canada.

Hyperoodon ampullatus, the Northern Bottlenose whale, is an arctic toothed whale species. It is not found in coastal areas, and prefers water depths of at least 1000 m (500 fathoms). Weighing up to 4 tonnes, and with a maximum length of 9 m, it feeds primarily on squid. The Northern Bottlenose was a preferred target of whalers off northern Labrador, due to its habit of travelling in pods of up to 20 members and the tendency of pod members to remain with a striken individual. This allowed whalers to successively kill each member of the group.

Fig. 8.40 *Delphinapterus leucas* (beluga). (NOAA)

The Atlantic White-sided dolphin (*Lagenorhynchus acutus*) and the White-beaked dolphin (*Lagenorhynchus albirostris*) are both common off Labrador. Large males of both species have lengths of 3 m and masses of 200 kg. The dolphins live in social groups of 10–20 individuals, cooperating to hunt smelt, herring, and capelin. They are pelagic species, and do not venture into estuaries. About 5000 White-beaked dolphins and perhaps 2000 Atlantic White-sided dolphins can be found in the Labrador Sea.

The Pilot whale (*Globicephala melas*), also called the Pothead whale, is a toothed whale that feeds primarily on squid in waters along the edge of the Continental Shelf. The Pilot whale is highly social, and lives in extremely vocal pods. Adult males reach 7 m length and 3 tonnes, and may live for 50 years or more. Currently, there are approximately 14,000 Pilot whales in the Labrador Sea.

Phocoena phocoena (harbour porpoise) is the smallest of Labrador's toothed whales. Adult males reach 1.8 m in length, 60 kg in mass, and typically live 10–12 years. The Harbour Porpoise attacks schooling fish, including herring, silver hake, pollock, capelin, and whiting. It lives in subarctic and boreal coastal waters and estuaries.

Baleen whales lack teeth, and strain plankton, invertebrates and small fish from the ocean waters through the baleen in their mouths. *Balaenoptera physalis*, the Fin whale, is the second largest of the baleen species in the Labrador Sea, and among the most common with an estimated 8000 migrating through the sea annually. A mature male Fin whale can reach 7 m in length and weigh 45 tonnes, and may live for 100 years. Fin whales eat small fish, krill, and squid. They are pelagic, but will occupy water depths as shallow as 150 m.

The most common of the baleen whales is the Minke or pammiuligak, *Balaenoptera actuorostrata.* More than 200,000 Minke whales are present worldwide, with an estimated 30,000 in the northwestern North Atlantic. Minke whales are coastal, and are seldom encountered more than 100 km from the shoreline. They are particularly fond of estuarine waters. Minke whales feed on krill, squid, and small fish. A mature male Minke is typically 8 m in length and weighs 8 tonnes.

Balaenoptera borealis, the Sei Whale, requires 900 kg of copepods and krill as food each day. There are approximately 1000 Sei Whales in the Labrador Sea during the summer months, skimming copepods and krill from ocean waters far from shore. Thus, they are rarely seen in the company of minke whales. The Sei Whale migrates to subtropical waters in late summer. A mature male may weigh 30 tonnes and have a length of 18 m.

The largest of the baleen whales, and the largest mammal on Earth, is the Blue Whale, *Balaenoptera musculus.* The Blue Whale is an endangered species, as a result of whaling activities. An estimated 500 blue whales use the Labrador

Sea yearly. Blue whales are highly migratory, feeding almost exclusively on krill in open waters with depths in excess of 500 m. Sightings of blue whales are very rare in coastal Labrador. Young blue whales grow up to 90 kg each day, and a mature adult male can reach 28 m long and have a mass of 180 tonnes, equivalent to more than 400 adult male polar bears.

Another preferred species of whalers was the Sperm whale, *Physeter macrocephalus*. The head of an adult male Sperm whale has up to 2000 L of oil, contained in the spermacetti organ which gives the whale its common English name. The function of the spermacetti organ is uncertain. Suggested uses include absorption of nitrogen gas, facilitating deep and rapid dives; focussing sounds underwater; and acting to control buoyancy as the waxy oil is alternately hardened and melted. An adult male can be 20 m long and have a mass of 50 tonnes, equal to 15 large elephants or 120 adult male polar bears. Females are shorter (up to 12 m) and considerably lighter (maximum 18 tonnes). Sperm whales may live in excess of 77 years. Their preferred diet is squid, in common with other toothed whales in the Labrador Sea. They are a pelagic species, inhabiting ocean areas with minimum depths of 1000 m. The Sperm whale has recovered since commercial whaling has been discontinued, and the estimated numbers today worldwide are in excess of 100,000.

8.9.4.1 Seals

Sealing was an important industry in the nineteenth and early to mid-twentieth century, both for trade with Europe and for local consumption of meat and making clothing. Sealing efforts, were necessary for the survival of coastal residents until the 1940s. Subsequent economic changes have rendered the seal hunt less of an economic necessity. Changing attitudes towards sealing in Europe and North America have severely limited or eliminated the markets for many seal products. The role of seals in the decline of commercial marine fish stocks is strongly debated and somewhat uncertain, although seals can and will eat any swimming species smaller than themselves.

The most common seal in Labrador waters is *Phoca groenlandica* (Chaulk et al. 2013; Fig. 8.41). Mature seals are referred to as saddleback seals in the scientific literature, harp seals by many Labradorians, and beaters by sealers. The Inuktitut terms for mature harp seals are xaygolik and kaigulik, and the Innu-Aimun term is pitshuatshuk. Young harp seals are termed xaygoliatsuk by Inuit, and are also called bedlames or whitecoats. The spring hunt of whitecoat seals, now discontinued, was a source of international controversy for many years.

The northwest Atlantic population of harp seals whelps throughout the area from the southeastern Labrador Sea to the Gulf of St. Lawrence. Harp seals migrate northward in spring, following the margin of the pack ice "front", and travelling north of the Arctic Circle in the summer months. They return to southernmost Labrador waters in December. Whelping on pack ice off the southeast coast of Labrador occurs in late February to early March, followed successively by mating and moulting. Adult male harp seals reach 2 m in length, and weigh up to 250 kg. Dark patches on their backs and sides form horseshoe-shaped "harps".

Fig. 8.41 *Phoca groenlandica* (harp seal)

Population updates for all marine mammal species are provided by Fisheries and Oceans Canada. In 2019, the estimated harp seal population offshore of Atlantic Canada was approximately 7.6 million individuals. Sculpins, Arctic cod, Atlantic (northern) cod, and capelin accounted for most of the prey consumed by mature Labrador harp seals. The young pups ate turbot, capelin, sand lance, and Arctic cod.

Crystophora cristata is known as the hooded seal, bluddernose seal, or netsivak. Young seals are called bluebacks. Adult males can reach 3 m in length and 300 kg mass. The hooded seal takes its name from the hood-like sack on top of its muzzle. Hooded seals are found on pack ice offshore of Labrador in late March and early April in most years.

Harbour Seal is the most common name for *Phoca vitulina*. Other names employed include bay seal, common seal, ranger, and spotted seal; xasigiatsiak or kassigiak (Inuktitut), and innatshuk kasigiak (Innu-Aimun). Very mature individuals are called dotter (or dodder), or xasigiak. A mature male harbour seal can be 1.5 m long and weigh 100 kg. Harbour seals are non-migratory, preferring open water areas with abuyndant supplies of crab, northern cod, herring, sculpin, capelin, and small flatfish. They whelp in coastal sites.

The grey seal, *Halichoerus grypus*, is termed netsiak in Inuktitut. Mature males reach 2.4 m and 250 kg. The grey seal prefers estuarine areas, especially Hamilton Inlet. More common in southern Labrador, they do not like open coastal areas.

The ringed seal (*Phoca hispida*), also known as jar, aetsik or natsik (Inuktitut), and pupunatshuk (Innu-Aimun), tends to inhabit coastal areas north of Groswater Bay. They require stable landfast ice for reproductive purposes.

Erignathus barbatus is the scientific name for the bearded seal, also called lassie, singing seal, square flipper, udjuk (Inuktitut), and uapistitun (Innu-Aimun). The Inuktitut term for younger bearded seals is udjuasuk. The largest of the arctic seals, male bearded seals may reach 2.5 m and 200 kg. During the spring, the bearded seal is found on coastal pack ice, most commonly to the north of Hopedale. Occasionally, bearded seals range south to the Strait of Belle Isle. The estimated population off Labrador is approximately 200,000. Bearded seals are primarily bottom feeders, using their snouts and flippers to capture benthic fish species including sculpins, flatfish, cod, eelpouts, clams, crabs, and shrimp.

8.9.5 Hudson Bay (Wînipekw)

Including James Bay, Hudson Bay has a surface water area of 1,230,000 km^2, a mean depth of 93 m, and a maximum depth of 270 m (Ingram and Prinsenberg 1998; Macdonald and Kuzyk 2011). The bathymetry consists of the central Hudson Bay Basin, flanked to the east by the NNE-SSW Midbay Bank and Winisk Trough (Fig. 8.42). Offshore ridges extend SW from Mansel Island and N-S from Sanikiluaq. Most of James Bay is less than 40 m deep. The absence of submarine canyons seaward of the mouths of the major rivers

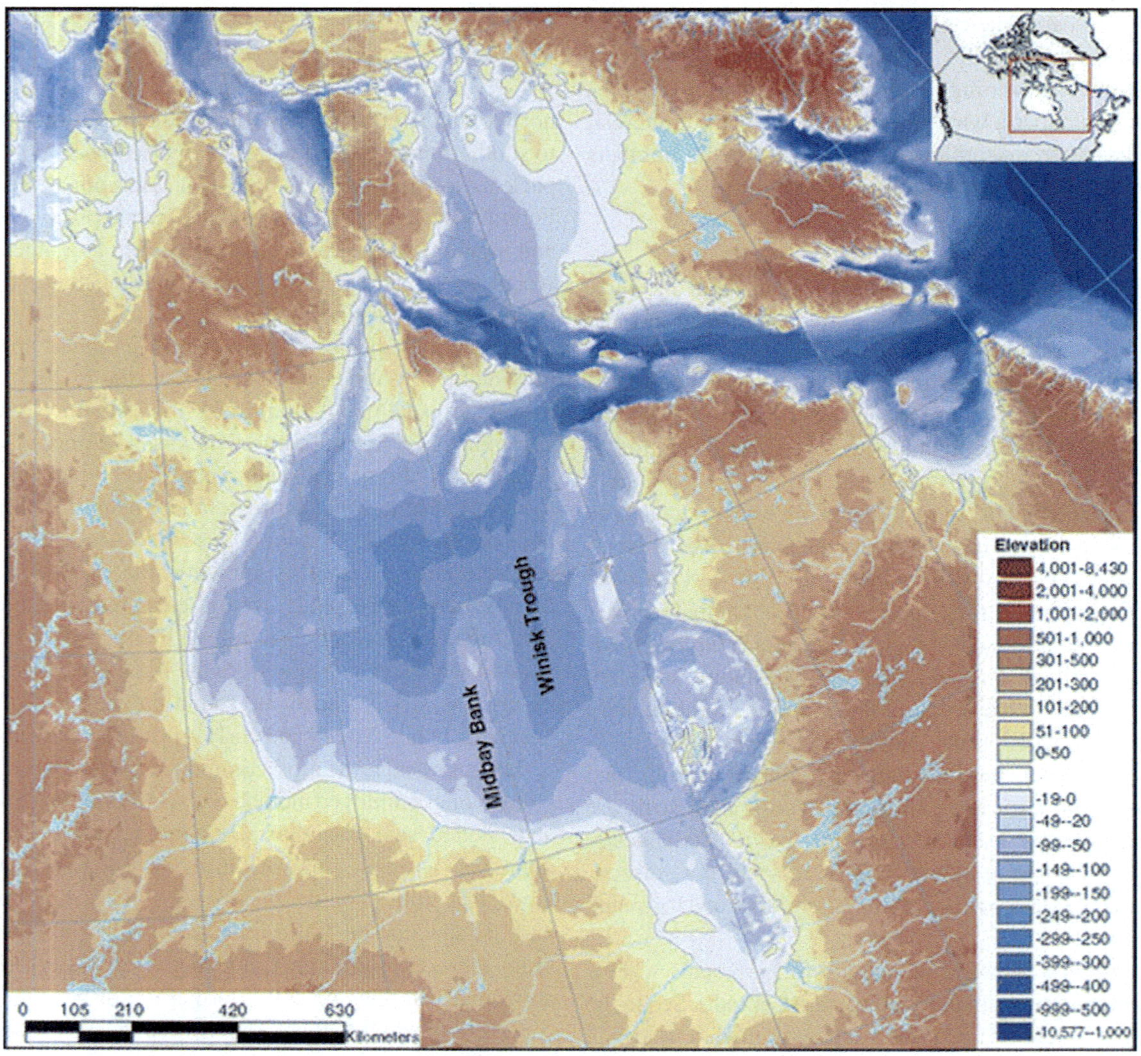

Fig. 8.42 Hudson Bay bathymetry. (Springer, The Ocea-Sea Ice-Atmosphere System, ch. in "A Little Less Arctic")

Fig. 8.43 Rupert Bay, summer circulation. (Landsat 8, NASA)

indicates that lower sea level has not been significant in the bay's Quaternary history. The bathymetry also indicates that glacial erosion has not been a significant factor.

Water circulation is counter-clockwise. A strong coastal conduit along the eastern side is fed by currents from the major streams and Kivalliq shoreline. Surface currents in the central basin are relatively weak. Outflow through Hudson Strait is augmented by a current along the northern shore, fed by water from Foxe Basin (Ilagiit nunagivaktangit) and Fury and Hecla Strait.

The freshwater stream influx, dilution resulting from seasonal ice melt, and limited subsurface influx through Hudson Strait result in low salinity throughout the bay (Fig. 8.43). Offshore James Bay salinity in summer is typically 33.0–33.5 psu, declining to 30 psu in northern Hudson Bay. Brackish water (maximum 15–25 psu) characterizes the coastal areas. Summer surface temperatures reach 8–9 °C in the south, with a strong pycnocline offshore at 15–20 m depth. Semidiurnal microtidal conditions prevail, with ranges of 1.1–1.3 m along the western coast, to less than 0.2 m along the eastern shore and at Sanikiluaq.

Freeze-up begins in December, and ice cover typically lasts until May–June (Fig. 8.44). Current action breaks the ice cover along the eastern side initially, producing large tabular floes. The central Hudson Basin and the southwestern coastlines are the last areas to lose ice cover each year. Ice is thin, typically less than 2 m thick in the northern bay, and does not persist throughout the summer. Since 1971, the length of the freeze-up season has shortened by up to a month in the major river mouths, compared to 7–10 days along the coastline in general.

Hudson Bay supports populations of Arctic cod (*Boreogadus saida*). Capelin (*Mallotus villosus*) spawn on coarse sediment tidal flats and beaches along the Nelson River estuary, Sanikiluaq, and Arviliit. The shallow waters prevent deepwater fish such as turbot (*Reinhardtius hippoglossoides*)

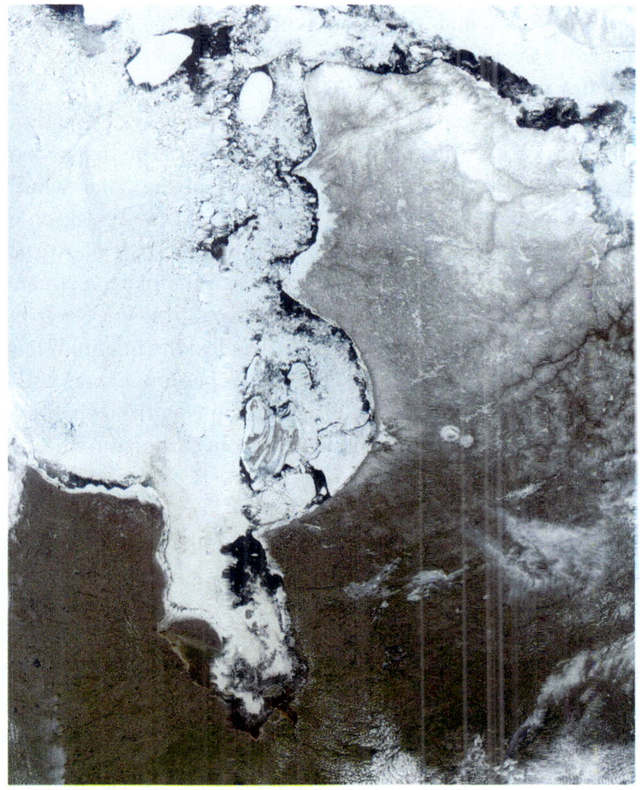

Fig. 8.44 Sea ice, Hudson Bay. (Earth Observatory, NASA)

and redfish (*Sebastes sp.*) from entering through Hudson Strait. Brackish coastal areas in the south support lake trout (*Salvelinus namaycush*) and lake whitefish (*Coregonus clupeaformis*). Arctic charr (*Salvelinus alpinus*) are present, locally with brook trout (*Salvelinus fontinalis*) and Atlantic salmon (*Salmo salar*).

8.10 Human Communities

The Taiga Shield, Taiga Plains, and Hudson Plains have a combined area of 2.3 million km^2, with a human population density of 0.04/km^2. Of the combined population of approximately 92,000, the three largest communities of Yellowknife (20,340), Labrador City-Wabush (9376), and Happy Valley-Goose Bay (8040) account for approximately 40%. Approximately 60% of the total population is indigenous.

Yellowknife (Tlicho: Sǫ̀ǫmbak'è) is the capital of Northwest Territories, located on the North Arm of Tu Nedhé. Initially founded as a gold mining community in 1934, Yellowknife has undergone transitions to government service centre, transportation hub, service centre for diamond mining to the northeast, and tourism destination. The last gold mine closed in 2004, but the legacy of toxic arsenic-bearing waste and environmental pollution will remain as a significant threat for decades to centuries (see Keeling and Sandlos n.d.; Sandlos and Keeling 2016, 2021).

Yellowknife and coastal Labrador are primary destinations for viewing the Aurora Borealis (Fig. 8.45). The displays of the Aurora Borealis result as electrons in the solar wind's plasma interact with Earth's Magnetic Field. The position of the Magnetic Pole controls alignment of the belt of Aurora activity. Charged particles move through the plasma stream, increasing kinetic energy of the electrons. As free electrons collide with those bound in gas molecules, the quantum leap converts kinetic energy to visible light as photons are released. Green tones result from electron collisions with oxygen, red from nitrogen, and blues and violets from carbon dioxide and other gas molecules.

The combination of the iron ore resources at Kawawachikamach (Schefferville) Québec), Labrador City, and Wabush, and the 5,428,500 kW power station at Churchill Falls, have resulted in economic activity in western Labrador. The post-World War II economic development of western Labrador began with the exploitation of iron ore at Kawawachikamach (1950–1982), and the construction of the Québec North Shore and Labrador Railway linking the mine site to the coastline at Sept-Iles. As was the practice at the time, the railway construction was undertaken without a prior Environmental Impact Assessment, and without much in the way of preliminary geological assessment. Numerous engineering problems were encountered in building the railway, particularly directly north of Sept-Iles where the line ran across sensitive marine clay that was prone to slumping and flow when disturbed. In addition, local patches of permafrost and muskeg posed engineering challenges.

After iron mining began in 1962, Labrador City became the region's largest community in 1968. The open-pit mine sites are located to the north of Labrador City, west of Wabush Lake. In 1965, an additional mine was developed at Wabush, located approximately 2 km south of Labrador City. The name comes from the Innu-Aimun *waboz*, snowshoe hare. The regional airport is located northeast of Wabush.

Happy Valley-Goose Bay, formed from the official amalgamation of the Goose Bay military base (constructed in 1941) and environs and the residential community of Happy Valley (constructed in 1943) in 1974. The site of Goose Bay was selected for the purpose of airfield construction. Consequently, the core of the community developed on a sandy marine terrace at 30 m asl, formerly covered with small aeolian parabolic dunes and sand sheets and cut by minor fluvial channels, adjacent to the two NE-SW (primary) and NW–SE (subordinate) runways. The foreslopes of the marine terrace isolated the site from the surrounding terrain. This simultaneously provided good drainage (preferred by the military engineers) and good security (preferred by the military guards). The only construction problems involved backfilling of the shallow river channels that cut across the site. Goose Bay served as a key staging point for aircraft traveling across the North Atlantic, and was also a base for aerial anti-submarine warfare.

Today, the major employer is the military base. Other employment centres around the tourism industry (particularly outfitters); mining developments, notably the Voisey's Bay nickel mine; ongoing hydroelectric development of the lower

Fig. 8.45 Aurora borealis over Labrador. (International Space Station/NASA image)

Churchill River; construction and shipping associated with the Trans-Labrador Highway, and lesser manufacturing and agricultural initiatives.

Forestry operations in taiga areas are subject to several biological constraints. Taiga forests are patchy by nature, and the isolated small patches of trees are poor resources in terms of wood volume. Removal of the stands would adversely impact the landscape. Low wood volumes necessitate clear-cutting, as there are seldom enough trees to pick and choose from. The isolated stands require comparatively long lengths of road, as measured in wood volume per km of road required. Growth rates in taiga environments are extremely slow, limiting the possibility of regeneration. Many new trees are established by vegetative reproduction, meaning that new trees are dependent on the presence of older ones on individual sites. In a black spruce taiga forest, vegetative reproduction tends to dominate during poorer (colder, wetter) summers. These factors in combination greatly limit the potential for commercial forestry, resulting in the preservation of almost all of the taiga forests.

Most indigenous communities in the taiga regions are located along riverbanks, or lacustrine or marine coastlines. In the Hudson Plains, the Wemindji communities within Eeyou Istchee display complex multi-layered relationships with all facets of the physical environment (Mulrennan 2020; Wren et al. 2020).

References

Abraham KF, Keddy CJ (2005) The Hudson Bay Lowland. In: Fraser LH, Keddy PA (eds) The World's Largest Wetlands: ecology and conservation. Cambridge University Press, New York, pp 118–148

Ackley C, Tank SE, Haynes KM, Rezanezhad F, McCarter C, Quinton WL (2021) Coupled hydrological and geochemical impacts of wildfire in peatland-dominated regions of discontinuous permafrost. Sci Total Environ 782:146841

Andrews J, Babb D, Barber DG (2018) Climate change and sea ice: shipping in Hudson Bay, Hudson Strait, and Foxe Basin (1980–2016). Elementa: Sci Anthr 6:19. https://doi.org/10.1525/elementa.281

Baldwin K, Allen L, Basquill S, Chapman K, Downing D, Flynn N, MacKenzie W, Major M, Meades W, Meidinger D, Morneau C, Saucier J-P, Thorpe J, Uhlig P (2019) Vegetation Zones of Canada: a Biogeoclimatic Perspective. Scale 1:5,000,000. Natural Resources Canada, Canadian Forestry Service, Sault Ste. Marie, ON

Ballinger TJ, Schmidlin TW, Steinhoff DF (2013) The polar marine climate revisited. J Clim 26:3935–3952

Banfield AWF (1981) The mammals of Canada. National Museum of Natural Sciences, National Museums of Canada. 438 p.

Beaumier MC, Ford JD, Tagalik S (2015) The food security of Inuit women in Arviat, Nunavut: the role of socio-economic factors and climate change. The Polar Record 51:550–559

Beck C, Grieser J, Kottek M, Rubel F, Rudolf B (2006) Characterizing global climate change by means of Köppen climate classification. http://koeppen-geiger.vu-wien.ac.at/pdfs/beck_et_al_ksb_2006.pdf

Bell JS (1989) Labrador Sea. Geological Survey of Canada, East Coast Basin Atlas Series 1989, 112 pages (53 sheets). https://doi.org/10.4095/127152

Bell T (2018) SmartICE: an agent for transformational change of coastal monitoring in Inuit Nunangat. J Ocean Technol 13:132–133

Bergerud AT (2000) Caribou. In: Demarais S Krausman PR (eds) Ecology and management of large mammals in North America. Prentice-Hall, New Jersey, pp 658–693

Bird DM (2017) Birds of Canada. DK publishing, London

Bonneya MT, Danbya RK, Treitz PM (2018) Landscape variability of vegetation change across the forest to tundra transition of central Canada. Remote Sens Environ 217:18–29

Bostock HS (2014) Physiographic Regions of Canada. Geological Survey of Canada Map 1254A, 2nd edn, scale 1:5,000,000. https://www.nrcan.gc.ca/maps-tools-and-publications/maps/geology-and-geosciences/16876. Accessed 28 July 2022

CABI (2022) Mephitis mephitis. https://www.cabi.org/isc/datasheet/75675

Calmels F, Allard M (2008) A structural interpretation of the palsa/lithalsa growth mechanism through the use of CT scanning. Earth Surf Proc Land 33:209–225

Canadian Ice Service (2020) Canadian Ice Service archive. https://www.canada.ca/en/environment-climatechange/services/ice-forecasts-observations/latest-conditions/archive-overview.html

Castilla G, Hall RJ, Skakun R, Filiatrault M, Beaudoin A, Gartrell M, Smith L, Groenewegen K, Hopkinson C, van der Sluijs J (2022) The multisource vegetation inventory (MVI): a satellite-based forest inventory for the Northwest Territories Taiga Plains. Remote Sens 14:1108. https://doi.org/10.3390/rs14051108

Catto NR (1996) Richardson Mountains, Yukon-Northwest territories: the Northern Portal of the Postulated Ice-Free Corridor. Quatern Int 32:3–20

Catto N (2015) Geomorphology: landscapes we live in. Oxford University Press, Don Mills, Ontario

Catto NR (2020) Labrador and the Northern Peninsula of Newfoundland: foreshore and flanking geomorphology. In: Slaymaker O, Catto NR (eds) Landscapes and Landforms of Eastern Canada. Springer International Publishers, Cham, Switzerland, pp 179–210

Catto NR, Liverman DGE, Bobrowsky PT, Rutter NW (1996) Laurentide, Cordilleran, and Montane Glaciation in the western Peace River-Grande Prairie region, Alberta and British Columbia, Canada. Quatern Int 32:21–32

Catto N, Slaymaker O, Kovanen DJ (2020) Quaternary Glacial, Glaciamarine, and Glacilacustrine History. In: Slaymaker O, Catto NR (eds) Landscapes and Landforms of Eastern Canada. Springer International Publishers, Cham, Switzerland, pp 49–78

Chadwick I (2007) Henry Hudson. http://www.ianchadwick.com/hudson/hudson_05.htm. Accessed 26 July 2022

Chaulk KG, Michelin D, Williams M, Wolfrey T (2013) Community-based observations of marine mammal occurrences in Groswater Bay, Labrador. Can Field-Naturalist 127(1):31–37

Chubbs TE, Schaefer JA (1997) Population growth of Moose, *Alces alces*, in Labrador. Can Field-Naturalist 111:238–242

Clark DA (2003) Polar bear-human interactions in Canadian national parks, 1986–2000. Ursus 14:65–71

Clark DA, Brook R, Oliphant-Reskanski C, Laforge MP, Olson K, Rivet D (2019) Novel range overlap of three ursids in the Canadian subarctic. Arctic Science 5:62–70

CNVC (Canadian National Vegetation Classification) (2016). http://cnvc-cnvc.ca. Accessed 25 July 2022.

COSEWIC (2011) Designatable units for Caribou (Rangifer tarandus) in Canada. Committee on the Status of Endangered Wildlife in Canada, Ottawa, p 88

COSEWIC (2017) COSEWIC assessment and status report on the Caribou (*Rangifer tarandus*) Eastern Migratory population, Torngat Mountains population in Canada. Ottawa. https://www.registrelepsararegistry.gc.ca/default.asp?lang=En&n=CAA68B17-1

Creaser RA et al (2004) Macrocrystal phlogopite Rb-Sr dates for the Ekati property kimberlites, Slave Province, Canada: evidence for multiple intrusive episodes in the Paleocene and Eocene. In: 8th International Kimberlite conference selected papers 1, pp 399–414

Davison T (2015) Technical Report on the Cape Bathurst, Bluenose-West, and Bluenose-East Barren-Ground Caribou Herds. Govt of NWT. https://wrrb.ca/sites/default/files/BEBWCB_Tech_Report_Feb2015Final2.pdf

Day MPD (2006) Hypothermia. Nursing 36(12). www.nursing2006.com

Déry SJ, Stieglitz M, McKenna EC, Wood EF (2005) Characteristics and trends of river discharge into Hudson, James, and Ungava Bays, 1964–2000. J Clim 18:2540–2557

Dohler G (2007) Tides in Canadian waters. Canadian Hydrographic Service, Fisheries and Oceans Canada

Dredge LA (1992) Glacial and environmental geology of northeastern Manitoba. Geological Survey of Canada, Memoir 432

Dredge LA., Dyke LD (2020) Landscapes and landforms of the Hudson Bay Lowlands. In Landscapes and landforms of eastern Canada, Edited by O. Slaymaker, N. R. Catto. World Geomorphological Landscapes. Cham, Springer. pp 211–227

Dyke AS (2004) An outline of North American deglaciation with emphasis on central and northern Canada. In: Ehlers J, Gibbard PL (eds) Quaternary Glaciations—Extent and Chronology, Part II: North America. Elsevier, Amsterdam, pp 371–424

Dyke LD, Sladen WE (2010) Permafrost and Peatland evolution in the Northern Hudson Bay Lowland, Manitoba. Arctic 63:429–441

Dyke AS, Moore A, Robertson L (2003) Deglaciation of Canada. Geological Survey of Canada Open File 1574

Earle CJ (2018) Picea mariana. The Gymnosperm Database. https://www.conifers.org/pi/Picea_mariana.php

Environment Canada (2025) Sea Ice in Canada. https://www.canada.ca/en/environment-climate-change/services/environmental-indicators/sea-ice.html

Environment and Climate Change Canada (2020) Canadian Climate Normals 1981–2010. https://climate.weather.gc.ca/climate_normals/index_e.html

Ernst R Bleeker W (2010) Large igneous provinces (LIPs), giant dyke swarms, and mantle plumes: significance for breakup events within Canada and adjacent regions from 2.5 Ga to the Present. Can J Earth Sci 47. https://doi.org/10.1139/E10-025

Farmer G (1981) The Cold Ocean Environment of Newfoundland. In: MacPherson AG, MacPherson JB (eds) The natural environment of Newfoundland, Past and Present. Memorial University of Newfoundland, St. John's, pp 56–82

Flaherty RJ (1918) The Belcher Islands of Hudson Bay: their discovery and exploration. Geogr Rev 5(6):433–458

Forbes DL (2011) State of the arctic coast 2010—Scientific review and outlook. In: International arctic science committee, land-ocean interactions in the coastal zone, arctic monitoring and assessment programme, international permafrost association. Helmholtz-Zentrum, Geesthacht, Germany, 178 p. http://arcticcoasts.org

French HM (2007) The Periglacial environment, 3rd edn. Wiley, Chichester

French HM, Slaymaker O (2012) Changing cold environments: a Canadian perspective. Wiley, Chichester, UK

Fulton RJ (1995) Surficial materials of Canada. Geological Survey of Canada Map 1880A

Ganter B (2000) Seagrass (*Zostera* spp.) as food for brent geese (*Branta bernicla*): an overview. Helgol Mar Res 54:63–70

Glaser PH, Hansen BCS, Siegel DI, Reeve AS, Morin PJ (2004) Rates, pathways and drivers for peatland development in the Hudson Bay Lowlands, northern Ontario, Canada. J Ecol 92:1036–1053

Godfrey WE (1986) The birds of Canada. National Museum of Natural Sciences, National Museums of Canada, 595 p.

Halsey L, Vitt D, Zoltai S (1997) Climatic and physiographic controls on wetland type and distribution in Manitoba, Canada. Wetlands 17:243–262

Handford C, Thomas O, Imray CHE (2017) Frostbite. Emerg Med Clin North Am 35:281–299

Heginbottom JA, Dubreuil MA, Harker PA (1995) Canada—Permafrost. National Atlas of Canada, 5th edn, National Atlas Information Service, Natural Resources Canada, MCR 4177

Hobbs WO, Edlund MB, Umbanhowar CE Jr, Camill P, Lynch JA, Geiss C, Stefanova V (2017) Holocene evolution of lakes in the forest-tundra biome of northern Manitoba, Canada. Quatern Sci Rev 159:116–138

Hocq M, (1994a) La Province du Supérieur. Dans: Géologie du Québec. Ministère des Ressources naturelles en collaboration avec Les publications du Québec; MM 94-01, pp. 7–20

Hocq M (1994b) La province de Grenville. Géologie du Québec. C. Dubé, Marquis Montmagny: 75–94

Hocq M, Verpaelst P (1994) Les sous-provinces de l'Abitibi et du Pontiac C. Dubé (Eds) Géologie du Québec Les Publications du Québec MM 94–01 21–37

Hoffman PF (1989) Pethei reef complex (1.9 Ga), Great Slave Lake, N.W.T. In: Geldsetzer HH, James NP, Tebbutt GE (eds) Reefs: Canada and adjacent areas: Canadian society of petroleum geologists memoir, vol 13, pp 38–48

Hosie RC (1979) Native Trees of Canada. Canada Department of Forestry, Don Mills, Ontario

Ingram RG, Prinsenberg S (1998) Coastal Oceanography of Hudson Bay and Surrounding Eastern Canadian Arctic Waters. In: Robinson AR, Brink KH (eds) The Sea. Volume 11: Coastal Segment. Chapter 29

Johnson L (1975) Physical and chemical characteristics of Great Bear Lake. J Fish Res Board Can 32:1971–1987

Keeling A, Sandlos J (n.d.) Toxic Legacies Project. http://www.toxiclegacies.com/author/arnkeeling/. Accessed 27 July 2022

Kettles IM, Garneau M, Jette H (2000) Macrofossil, pollen, and geochemical records of peatlands in the Kinosheo Lake and Detour Lake areas, northern Ontario. Geol Surv Can Bull 545:1–24

Khalafzai AK, McGee TK, Parlee B (2021) Spring flooding and recurring evacuations of Kashechewan First Nation, northern Ontario, Canada. Int J Dis Risk Reduct 63:102443. https://doi.org/10.1016/j.ijdrr.2021.102443

Klassen RA, Paradis S, Bolduc AM, Thomas RD (1992) Glacial landforms and deposits, Labrador (Newfoundland) and eastern Québec. Geological Survey of Canada, Map 1814A

Laforge MP, Clark DA, Schmidt AL, Lankshear JL, Kowalchuk S, Brook RK (2017) Temporal aspects of polar bear (Ursus maritimus) occurrences at field camps in Wapusk National Park, Canada. Polar Biol 40:1661–1670

Laitinen J, Rehell S, Huttunen A (2005) Vegetation-related hydrotopographic and hydrologic classification for aapa mire (Hirvisuo, Finland). Ann Botanicae Fennici 42:107–121

Langlais K, Bhiry N, Lavoie M (2021) Holocene dynamics of an inland palsa peatland at Wiyâshâkimî Lake (Nunavik, Canada). Écoscience 28:269–282

Li X-Y, Jin H-J, Wang H-W, Marchenko SS, Shan W, Luo D-L, He R-X, Spektor V, Huang Y-D, Li X-Y, Jia N (2021) Influences of forest fires on the permafrost environment: a review. Adv Clim Chang Res 12:48–65

Lien J, Fawcett L, Staniforth S (1985) Wet and fat: whales and seals of Newfoundland and Labrador. Breakwater Books, St John's

Lusebrink I, Erbilgin N, Evenden ML (2013) The lodgepole × jack pine hybrid zone in Alberta, Canada: a stepping stone for the mountain pine beetle on its journey East across the boreal forest? J Chem Ecol 39:1209–1220

Macdonald RW, Kuzyk ZZA (2011) The Hudson Bay system. J Mar Syst 88:337–488

Mansfield AW (1967) Seals of Arctic and Eastern Canada. Fisheries Research Board of Canada

Markham WE (1980) Ice Atlas, Eastern Canadian Seaboard. Environment Canada

McLaughlin J, Webster K (2013) Effects of a changing climate on peatlands in permafrost zones: a literature review and application to Ontario's Far North. Climate Change Research Report 34, Ontario Ministry of Natural Resources

McLaughlin JW, Packalen MS (2021) Peat carbon vulnerability to projected climate warming in the Hudson Bay Lowlands, Canada: a decision support tool for land use planning in peatland dominated landscapes. Front Earth Sci 9:650662. https://doi.org/10.3389/feart.2021.650662

Michaux J, Dyck M, Boag P, Lougheed S, van Coeverden de Groot P (2021) New insights on polar bear (*Ursus maritimus*) diet from faeces based on next-generation sequencing technology. Arctic 74:87–99

Mints MV, Dokukina K, Afonina TB (2021) Precambrian lithosphere beneath Hudson Bay: a new geological model based on the Hudson Bay Lithospheric Experiment (HuBLE), Canadian Shield. Tectonophysics 799. https://doi.org/10.1016/j.tecto.2020.228701

Montesano PM, Neigh CSR, Macander M, Feng M, Noojipady P (2020) The bioclimatic extent and pattern of the cold edge of the boreal forest: the circumpolar taiga-tundra ecotone. Environ Res Lett 15:105019. https://doi.org/10.1088/1748-9326/abb2c7

Mulrennan ME (2020) Do landscapes listen? Wemindji Eeyou knowledge, adaptation, and agency in the context of coastal landscape change. In: Slaymaker O, Catto NR (eds) Landscapes and Landforms of Eastern Canada. Springer International Publishers, Cham, Switzerland, pp 543–556

National Wetlands Working Group (1988) Wetlands of Canada. Ecological Land Classification Series # 24 Environment Canada, Ottawa and Polyscience Publications, Montreal

Natural Resources Canada (2009) Geological Provinces of Canada. https://www.nrcan.gc.ca/maps-tools-and-publications/maps/geology-and-geosciences/16876. Accessed 12 July 2022

Naughton D (2012) The natural history of Canadian mammals. Canadian Museum of Nature

Newfoundland and Labrador Fisheries, Forestry and Agriculture Forestry and Wildlife. https://www.gov.nl.ca/ffa/files/forestry-forest-fires-pdf-fire-stats-2021.pdf

Newfoundland and Labrador (2020) Fisheries forestry agriculture. https://www.gov.nl.ca/ffa/files/2020-George-River-Caribou-Census-Results_.mp4

Northwest Territories (2022) Barren-Ground Caribou. https://www.enr.gov.nt.ca/en/services/barren-ground-caribou/bluenose-east-bluenose-west-and-cape-bathurst-herds

Northwest Territories n.d. Dwarf Birch—Betula glandulosa. https://www.enr.gov.nt.ca/sites/enr/files/fact_sheets/dwarf_birch.pdf

Nunavut Environment (2022) Polar Bears in Canada. https://www.polarbearscanada.ca/en

Obbard ME, Walton LR (2004) The importance of Polar Bear Provincial Park to the Southern Hudson Bay polar bear population in the context of future climate change. In: Parks research forum of Ontario proceedings, pp 105–116. https://polarbearscience.files.wordpress.com/2013/07/obbard-and-walton-2004_prfo-proceedings-p105-116.pdf

Ontario Geological Survey (1991) Bedrock Geology of Ontario, northern sheet. Ontario Geological Survey, Map 2541, scale 1:1 000 000

Ontario (2022) Polar bears. https://www.ontario.ca/page/polar-bears

Paal P, Gordon L, Strapazzon G, Brodmann Maeder M, Putzer G, Walpoth B, Wanscher M, Brown D, Holzer M, Broessner G Brugger H (2016) Accidental hypothermia–an update. Scand J Trauma Resusc Emerg Med 24:111

Parks Canada (2007) Tuktut Nogait National Park Management Plan. https://www.pc.gc.ca/en/pn-np/nt/tuktutnogait/gestion-management/plan1

Panikkar B, Lemmond B (2020) Being on land and sea in troubled times: climate change and food sovereignty in Nunavut. Land 19:508–518

Phillips D (1990) The climates of Canada. Environment Canada

Proctor NS, Lynch PJ (2005) A field guide to North Atlantic wildlife: marine mammals, seabirds, fish, and other sea life. Yale University Press, New Haven

Reeves RR (1988) Cetaceans of Canada. Fisheries and Oceans Canada

Rempel RS, Carlson M, Rodgers AR, Shuter JL, Farrell CE, Cairns D, Stelfox B, Hunt LM, Mackereth RW, Jackson JM (2021) Modeling cumulative effects of climate and development on moose, wolf, and caribou populations. J Wildl Manag 85:1355–1376

Riedlsperger R, Goldhar C, Sheldon T, Bell T (2017) Meaning and means of 'sustainability': an example from the Inuit Settlement Region of Nunatsiavut, northern Labrador. In: Fondahl G, Wilson G (eds) Northern sustainabilities: understanding and addressing change in the circumpolar world. Springer International Publishing, Cham, Switzerland, pp 317–336

Rouse WR (1993) Northern climates. In: French HM, Slaymaker O (eds), Canada's cold environments, pp 65–92

Rowe JS (1972) Forest regions of Canada. In: Canadian forest service, Department of the Environment Publication 1300, Ottawa. 172 p. + map

Ryan AG (1995) Native Trees and Shrubs of Newfoundland and Labrador. Park Interpretation Publication 14, Parks and Natural Areas Division, Newfoundland and Labrador

Salisbury SJ, McCracken GR, Keefe D, Perry R, Ruzzante DE (2019) Extensive secondary contact among three glacial lineages of Arctic Char (*Salvelinus alpinus*) in Labrador and Newfoundland. Ecol Evol 9:2031–2045

Samson DO, Bhiry N, Lavoie M (2010) Late-Holocene palaeoecology of a polygonal peatland on the south shore of Hudson Strait, northern Québec, Canada. Holocene 20(4):525–536

Sandlos J, Keeling A (2016) Toxic legacies, slow violence, and environmental injustice at Giant Mine, Northwest Territories. Northern Rev 42:7–21

Sandlos J, Keeling A (2021) Mining country: a history of Canada's mines and miners. James Lorimer and Company, Toronto, p 192

Sannel ABK, Kuhry P (2009) Holocene peat growth and decay dynamics in sub-arctic peat plateaus, west-central Canada. Boreas 38:13–24

Schertzer WM (2000) Digital bathymetry of Great Slave Lake. NWRI Contribution No 00-257, National Water Research Institute, Burlington, ON, Canada

Schmidt AL, Loring P, Clark DA (2022) Local experts' observations, interpretations, and responses to human-polar bear interactions in Churchill, Manitoba. Arctic 75:257–271

Scott WB, Crossman EJ, (1998) Freshwater Fishes of Canada. Galt House, Oakville. Canada. Fisheries Research Board of Canada. Bulletin 184

Seppälä M (2011) Synthesis of studies of palsa formation underlining the importance of local environmental and physical characteristics. Quatern Res 75:366–370

Sharpe DR, Kjarsgaard BA, Knight RD, Russell HAJ, Kerr DE (2017) Glacial dispersal and flow history, East Arm area of Great Slave Lake, NWT, Canada. Quatern Sci Rev 165:49–72

Simon KM, James TS, Forbes DL, Telka AM, Dyke AS, Henton JA (2014) A relative sea-level history for Arviat, Nunavut, and implications for Laurentide Ice Sheet thickness west of Hudson Bay. Quatern Res 82:185–197

Slaymaker O (2017) Landscapes and landforms of Western Canada. Springer International Publishers, Cham, Switzerland

Slaymaker O, Catto NR (2020) Landscapes and landforms of Eastern Canada. Springer International Publishers, Cham, Switzerland

Slaymaker O Catto N Kovanen DJ (2020a) Eastern Canadian landscapes as a function of structure, relief and process: the big picture. In: Slaymaker O, Catto NR (eds) Landscapes and Landforms of Eastern

Canada. Springer International Publishers, Cham, Switzerland, pp 3–48
Slaymaker O Catto N Kovanen DJ (2020b) Holocene geomorphic processes in Eastern Canada. In: Slaymaker O, Catto NR (eds) Landscapes and landforms of Eastern Canada. Springer International Publishers, Cham, Switzerland, pp 79–117
Smith CAS, Meikle JC, Roots CF (2004) Ecoregions of the Yukon territory: biophysical properties of Yukon landscapes. In: Agriculture and Agri-Food Canada, PARC Technical Bulletin No. 04-01, Summerland, British Columbia, 313 p.
Smith S (2011) Trends in permafrost conditions and ecology in northern Canada. Canadian Biodiversity: ecosystem Status and Trends 2010. Technical Thematic Report No. 9. Canadian Councils of Resource Ministers. Ottawa
Snowcrystals.com (2013) A Snowflake Primer: the basic facts about Snowflakes and Snowcrystals. Accessed 27 July 2022
Soil Classification Working Group (1998) The Canadian system of soil classification, 3rd edn. Publication, Agriculture and Agri-Food Canada, p 1646
Stares J, Kosatsky T (2015) Hypothermia as a cause of death in British Columbia 1998–2012, a descriptive assessment. Can Med Assoc J Open 3(4):E352–E358
Stokes CR, Tarasov L, Blomdin R, Cronin TM, Fisher TG, Gyllencreutz R, Hättestrand C, Heyman J, Hindmarsh RCA, Hughes ALC, Jakobsson M, Kirchner N, Livingstone SJ, Margold M, Murton JB, Noormets R, Peltier WR, Peteet DM, Piper DJW, Preusser F, Renssen R, Roberts DH, Roche DM, Saint-Ange F, Stroeven AP, Teller JT (2015) On the reconstruction of palaeo-ice sheets: recent advances and future challenges. Quatern Sci Rev 125:15–49
Stokes CR (2020) Glacial landform assemblages of mainland Nunavut, west of Hudson Bay, and their palaeo-glaciological significance. In: Slaymaker O, Catto NR (eds) Landscapes and landforms of Eastern Canada. Springer International Publishers, Cham, Switzerland, pp 121–144
Storrar RD, Livingstone SJ (2017) Glacial geomorphology of the northern Kivalliq region, Nunavut, Canada, with an emphasis on meltwater drainage systems. J Maps 13:153–164
Talley LD, McCartney MS (1982) Distribution and circulation of Labrador Sea water. J Phys Oceanogr 12:1189–1205
Valley P, Bulle F, Skipton D (2011) Geology of the Southeastern Churchill Province, western Labrador. Newfoundland and Labrador Department of Natural Resources Geological Survey, Current Research (2011), Report 11-1, pp 295–311
Walker DA, Raynolds MK, Daniëls FJA, Einarsson E, Elvebakk A, Gould WA, Katenin AE, Kholod SS, Markon CJ, Melnikov ES, Moskalenko NG, Talbot SS, Yurtsev BA (2005) The circumpolar Arctic vegetation map. J Veg Sci 16:267–282
Wardle RJ, Gower CF, Ryan B, Nunn GAG, James DT, Kerr A (1997) Geological Map of Labrador; 1:1 million scale. Government of Newfoundland and Labrador, Department of Mines and Energy, Geological Survey, Map 97-07
Warner BG Rubec CD (1997) The Canadian wetland classification system. Wetlands Research Centre, University of Waterloo, Waterloo, 68 p.
Wells-Gosling N, Heaney LR (1984) Glaucomys sabrinus. Mammalian Species 229:1–8 (American Society of Mammalogists)
Wheeler JO, Hoffman PF, Card KD, Davidson A, Sanford BV, Okulitch AV, Roest WR (1996) Geological map of Canada. Geological Survey of Canada, Map D1860A
Wilson KJ, Bell T, Arreak A, Koonoo B, Angnatsiak D, Ljubicic GJ (2020) Changing the role of non-indigenous research partners in practice to support Inuit self-determination in research. Arctic Sci. https://doi.org/10.1139/as-2019-0021
Wolfe SA., Kerr DE, Morse PD (2017) Slave Geological Province: An Archetype of Glaciated Shield Terrain, In: Landscapes and Landforms of Western Canada, edited by: Slaymaker O, Springer, Switzerland, pp 77–86
Wolfe SA, Stevens CW, Gaanderse AJ, Oldenborger GA (2014) Lithalsa distribution, morphology and landscape associations in the Great Slave Lowland, Northwest Territories, Canada. Geomorphology 204:302–313
Wolverine Foundation (2022) Information about Kuekuatsheu. https://kuekuatsheu.ca/informationwolverines.html
Wren CD, Costopoulos A, Hawley M (2020) Settlement choice under conditions of rapid shoreline displacement in Wemindji Cree Territory, subarctic Quebec. Quatern Int 549:191–196
Zhang S (2010) Upper Ordovician stratigraphy and oil shales on Southampton Island, field trip guidebook. Geological Survey of Canada, Open File 6668, 48 p. https://doi.org/10.4095/285559

9 Arctic Ecozones

Abstract

The Arctic ecozones of Canada include all tundra and glacier-covered terrain north of the treeline, internal Arctic waters, the northernmost North Atlantic Ocean washing Qikiqtaaluk (Baffin Island) and Umigmak Nuna (Ellesmere Island), and the Beaufort Sea coast of the Arctic Ocean. Subdivisions include the Southern Arctic, Arctic Cordillera, Arctic Coastal Plain, the Northern and Western Arctic, islands in the Inuvialuit Settlement region, and waterways threading through the archipelagos. Arctic regions in Canada have Tundra ET climates, with H (highland) climates above 1500 m asl and EF climates on active glaciers below 1500 m asl. In addition to latitudinal and altitudinal gradients, significant differences exist from west to east across the Arctic. The Arctic ecozones of Canada overlie several geological provinces. The Interior Platform, Bear Province, and Tu Nedhe (Slave) Province extend north to the shoreline of Amundsen Gulf. The Arctic Cordilleran regions of the Torngat Mountains, eastern Qikiqtaaluk, Tatlurutit, and Umigmak Nuna supported alpine glaciation during the Quaternary. Modern glaciers exist in all these regions. Continental glaciation covered Nunavik, Qikiqtaaluk, Kivalliq, the Western Arctic, and the Northern Arctic (Innuitian Ice Sheet). Arctic vegetation communities can be divided into three major assemblages, influenced by combinations of latitude, altitude, and snow cover. Erect dwarf shrub Low Arctic assemblages characterize areas adjacent to the treeline ecotone. Prostrate dwarf-shrub Mid-Arctic communities occupy a west-east belt. Cushion-forb High Arctic assemblages occupy areas north of M' Clure Strait-Viscount Melville Sound-Tallurutiup Imanga. Distinctive communities also occur in restricted areas with deep snowpacks, over broken rock felsenmeer, and in salt marshes. Musk ox, Barren-ground, and Peary caribou are iconic species of the Canadian Arctic. Polar bears are present in most areas below about 1200 m asl. The waters of Baffin Bay and Davis Strait are fed by the West Greenland Current from the North Atlantic, by southward flow from Nares Strait, and by eastward flow through Ausuituu Tariunga (Jones Sound). Tallurutiup Imanga (Lancaster Sound) between Qikiqtaaluk and Tatlurutit is marked by interchange of North Atlantic and interior Arctic Archipelago waters. The Arctic Ocean is fed by relatively warm, saline North Atlantic waters from the Norwegian (North Atlantic) Current, entering through Fram Strait and the Barents Sea; by freshwater discharges from the Yenisey, Lena, Ob, and other Siberian and Nordic Rivers; by freshwater from the Mackenzie (Deh Cho) and other North American rivers; and by cool Pacific Ocean waters through the Bering Strait. Arctic regions have experienced the greatest amount of climate change in the recent past and are predicted to see the greatest changes in the future.

Keyword

Arctic ecozones · Qikiqtaaluk · Beaufort sea · Arctic cordillera · Arctic coastal plain

9.1 Introduction

This chapter discusses the Arctic ecozones of Canada. Included are all tundra and glacier-covered terrain north of the treeline, internal Arctic waters, the northernmost North Atlantic Ocean washing Qikiqtaaluk (Baffin Island) and Umigmak Nuna (Ellesmere Island), and the Beaufort Sea coast of the Arctic Ocean. Subdivisions include the Southern Arctic (mainland NWT, central Kivalliq, and Nunavik); Arctic Cordillera (Torngat Mountains, eastern Qikiqtaaluk, Tatlurutit (Devon Island), and Umigmak Nuna, and central Axel Heiberg): Arctic Coastal Plain (southern and central Beaufort Sea coast); the Northern and Western Arctic (Qikiqtaaluk, northeastern Kivalliq, Kitikmeot, islands in the Inuvialuit Settlement region, and waterways threading through the archipelagos; Fig. 9.1).

N. Catto and G. R. Vincent, *Canada: Land and People*, World Regional Geography Book Series,
https://doi.org/10.1007/978-3-032-04391-7_9

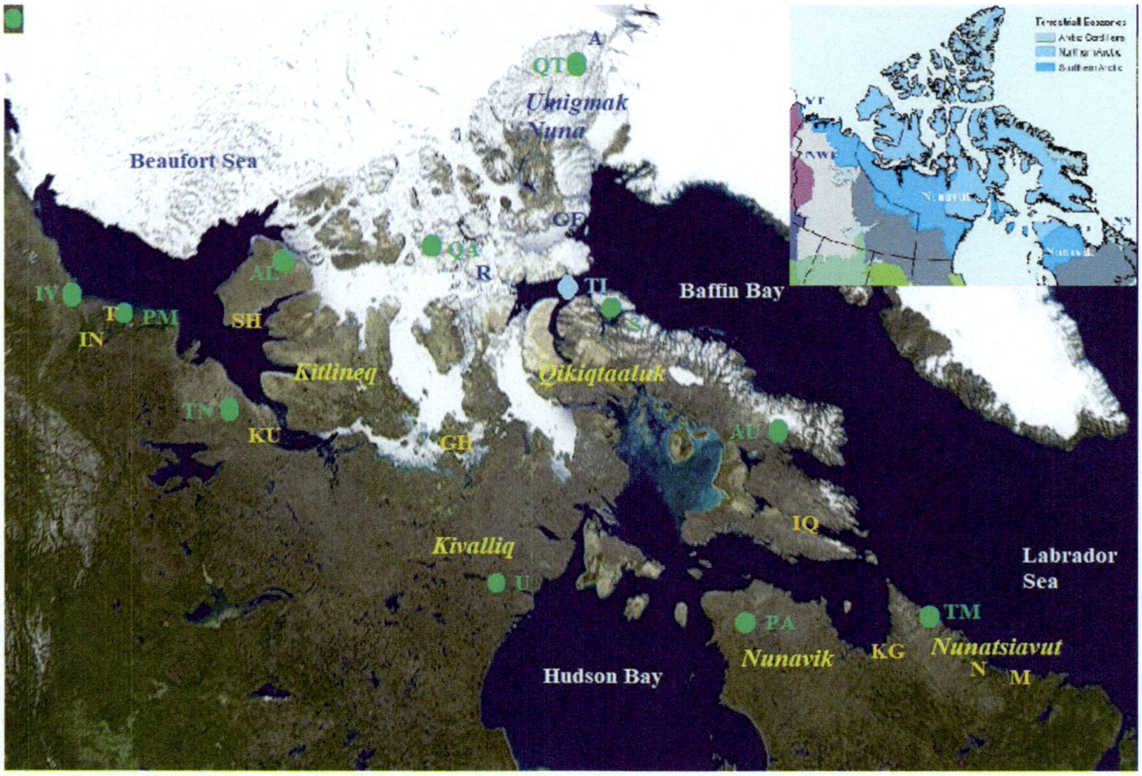

Fig. 9.1 Arctic Cordillera, Northern Arctic, and Southern Arctic ecoregions, showing locations discussed in text. National Parks: AL: Aulavik, AU: Auyuittuq, IV: Ivvavik, PM: Pingo National Landmark, QA: Qausuittuq, QT: Quttinirpaaq, S: Sirmilik, TM: Torngat Mountains, TN: Tuktut Nogait, U: Ukkusiksalik. PA: Pingualuit astrobleme. TI: Tallurutiup Imanga. Communities: A: Alert, GF: Grise Fjord, GH: Gjoa Haven, IN: Inuvik, IQ: Iqaluit, KG: Kuujjuaq, M: Makkovik. N: Nain, R: Resolute, SH: Sachs Harbour, T: Tuktoyaktuk

Recent years have seen the emergence of a stronger Indigenous presence in the use of place names for terrain features and communities, particularly in Arctic Canada. This long overdue acknowledgement of the reality of Canadian Indigenous culture can complicate place name recognition for English, French, and international readers, as well as for aspiring geographers who memorized many English colonial names for Arctic features in their youth.

In this chapter, we will use both English and Indigenous names. English Canadian names applied to topographical features and communities are listed in Table 1.2 (Chap. 1). Many terrain features and smaller communities have only designated Indigenous names.

9.2 Climate

Arctic regions in Canada have Tundra ET (modified Köppen-Geiger) climates, with H (highland) climates above 1500 m asl and EF climates on active glaciers below 1500 m asl (Phillips 1990; Beck et al. 2006; French and Slaymaker 2012; Druckenmiller et al. 2021). In addition to latitudinal and altitudinal gradients, significant differences exist from west to east across the Arctic (Maxwell 1980–82, 1981; Serreze and Barry 2005). Dry conditions dominate in the western Arctic, influenced by the high-pressure system generally prevalent over the Beaufort Sea and central Arctic Ocean. In the eastern Arctic, low pressure is more common, leading to onshore maritime winds and enhanced precipitation (especially during summer).

Eastern Qikiqtaaluk (Baffin), including Iqaluit (63.7°N, 68.6°W) and Auyuittuq National Park (67°N, 65°W), is influenced by a tundra climate (modified Köppen-Geiger ET; with glacial climates in the Penny Ice Cap region; Fig. 9.2). January daily mean temperatures in Iqaluit and Auyuittuq are −26.9 °C and −32 °C, respectively (Environment and Climate Change Canada 2020), and fjords remain ice-covered and inaccessible to shipping until late-June-mid-July (Hatcher and Forbes 2015). The July daily mean temperature at Iqaliut is 8.2 °C, but most parts of Auyuittuq have daily mean temperatures below 0 °C throughout the year.

During the summer, maritime arctic winds from both the northwest and southeast are common, but there are no Polar Easterlies. The southeasterly winds bring moisture from Davis Strait, and hence southeast-facing slopes

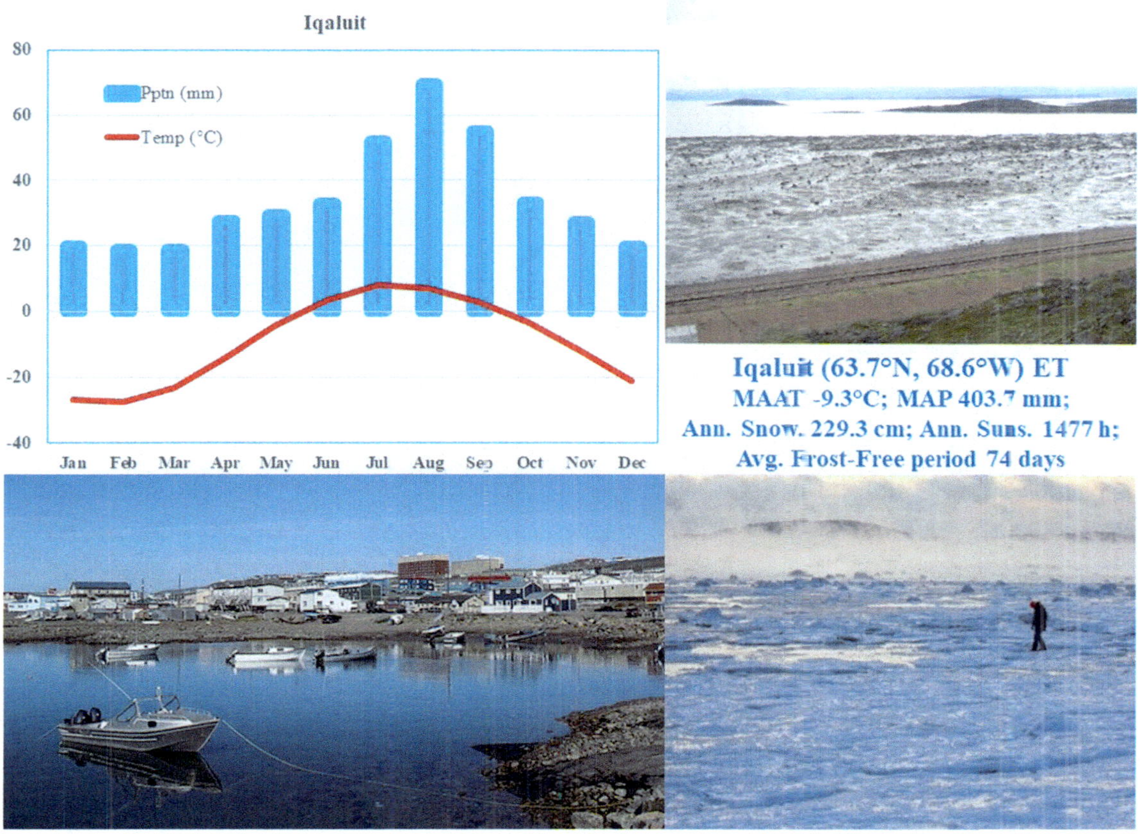

Fig. 9.2 Iqaluit. Climate data: Canadian Climate Normals, Environment and Climate Change Canada. Photos courtesy Don Forbes

(such as Iqaluit) tend to receive precipitation at this time (Maxwell 1980–82, 1981). Local river flooding can result (Kjikjerkovska 2016).

In contrast, winds from the northwest have travelled across the spine of Qikiqtaaluk, and are relatively dry. Local variations are common, however, as the winds are funnelled by the deep valleys and fjordal systems, particularly along Akshayuk Pass. During the winter months, the winds come dominantly from the southern quadrant (southwest to southeast). These winds, however, blow across the frozen waters of Cumberland Sound and the coastal areas of Davis Strait, and thus carry little moisture. As in most arctic regions, the winter months are the driest, and snowfall tends to occur from late summer to late autumn and from late spring to early summer.

Mean annual precipitation varies from approximately 600 mm at Pangnirtung, on southward-facing slopes, to 404 mm at Iqaluit and less than 200 mm at northwest-facing sites (Environment and Climate Change Canada 2020). Aspect plays a critical role in precipitation totals and style. Snowfall forms a major component of the precipitation, with much coming during the spring and summer. At Iqaluit, approximately 60% of the annual precipitation is snow, compared to 80% in the interior of Auyuittuq.

Snow cover persists for a minimum of 260 days in a typical year at Pangnirtung (at sea level). The presence of glaciers at higher interior elevations indicates year-round snow cover in these areas. At 700 m elevation, winter snow cover is usually established by late August. Maximum snow depth at Pangnirtung is usually about 80 cm. Interior areas commonly have snow depths of 150 cm during the winter, representing more than half of the total seasonal snowfall. The accumulation of snow is vital for the maintenance of the glaciers of the region, but ongoing climate warming has resulted in the recession of most.

Gjoa Haven (Uqsuqtuuq, 68.6°N, 95.9°W; within the Kitikmeot region) is colder (−33.8 °C January mean) and markedly drier (191 mm y^{-1}) than Qikiqtaaluk (Environment and Climate Change Canada 2020; Fig. 9.3). Winds are dominantly north and northwest, blowing from land and across frequently frozen straits, providing little moisture (Maxwell 1980–82, 1981). Topographic effects are relatively unimportant. Snow represents about 60% of total precipitation, with both mean depth (15 cm in February) and density (150 g/ml) being relatively low. Snow cover duration and depth have both decreased since 1950, with duration 20–25 days less now than in 1998.

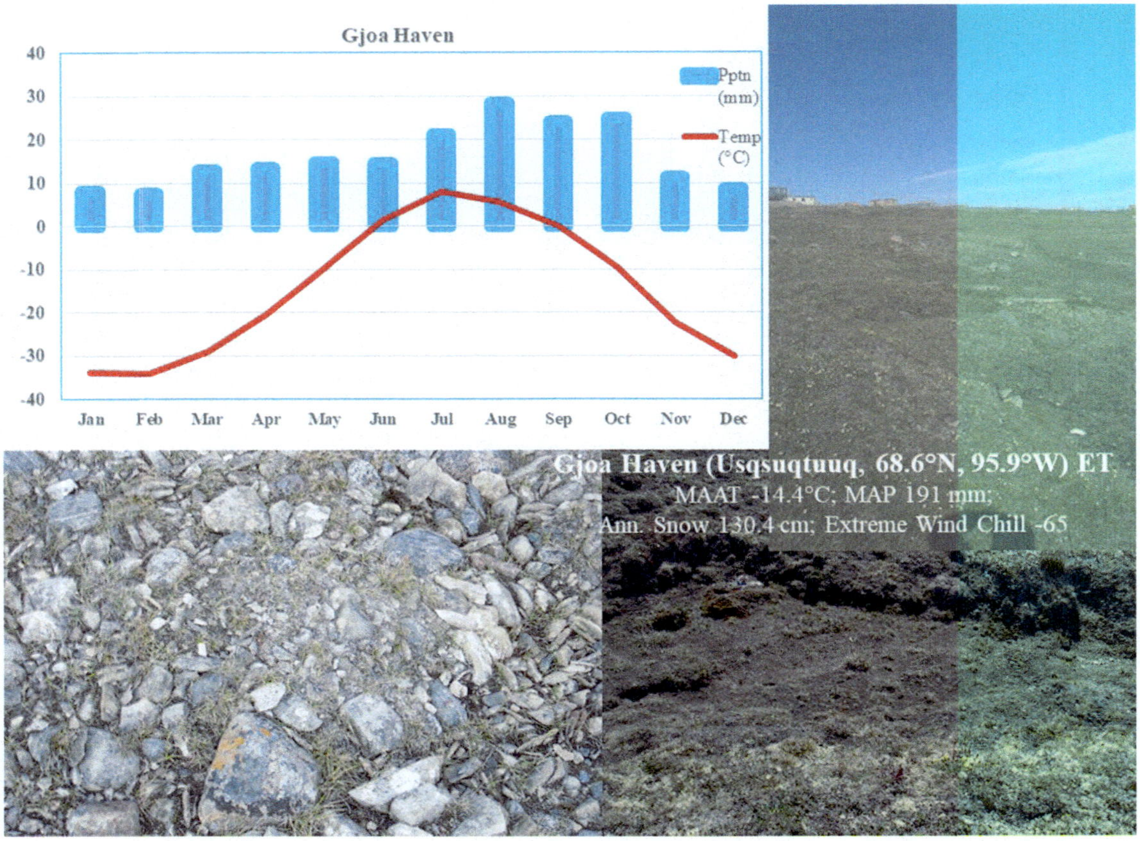

Fig. 9.3 Gjoa Haven. Climate data: Canadian Climate Normals, Environment and Climate Change Canada. Photos illustrate local permafrost stone circles and erosional features

Inuvik (68.4°N, 133.8°W) is under the influence of a taiga climate (modified Köppen-Geiger Dfc), with three months of mean temperatures above 10 °C (Fig. 9.4). Daily mean temperatures in January and July are −26.9 °C and 14.1 °C, respectively (Environment and Climate Change Canada 2020). Marked warming has occurred in recent years (Government of Northwest Territories Environment 2015; Druckenmiller et al. 2021), which is resulting in permafrost ablation and northward migration of vegetation and fauna. The wide temperature range indicates a strong continental influence, which is abnormal in arctic regions. The area benefits from 20 to 25 high summer days per year and receives 1900 h of sunshine. At present, the frost-free period is typically 66 days.

Deh Cho (Mackenzie River) is usually frozen from early November to late May (Fig. 9.5). Formerly, the freeze-up was of great economic significance, as virtually all supplies were brought to Inuvik by river barge from Łíídlı̨ı̨ Kų́ę́ (Fort Simpson). The 1979 completion of the Dempster Highwayto Dawson Cityalmost completely killed commercial river freighting. In addition, air freight has grown greatly in importance since 1980.

The northernmost point of the public road system of Canada is located approximately 10 kms north of Inuvik, along the eastern edge of the Mackenzie Delta. A winter road extends to Tuktoyaktuk, on the Beaufort Sea. These two communities are the only points in the Canadian Arctic accessible by road. Even with road access, the role of Deh Cho remains central to landscape and people.

Mean annual precipitation in the Inuvik region is 250 mm, of which 50–60% falls as snow. The modal snow pack depth is 50 cm, and snow cover typically persists for 210–230 days, decreasing by about 20 days since 1990. Technically, the area is considered to have a dry subhumid moisture regime, but the effects of slope orientation and vegetation cover are more important than the absolute amount of precipitation. The prevailing winds are easterly.

Drier conditions mark more northerly sites (Maxwell 1980–82, 1981), including Sachs Harbour (Banks Island, 72°N, 125.3°W; 128 mm/y, January daily mean −29 °C) and Qausuittuq National Park (Bathurst Island, 76°N, 100°W; 150 mm y^{-1}, January daily mean −30 °C; Environment and Climate Change Canada 2020). At Sachs Harbour, 55–60%

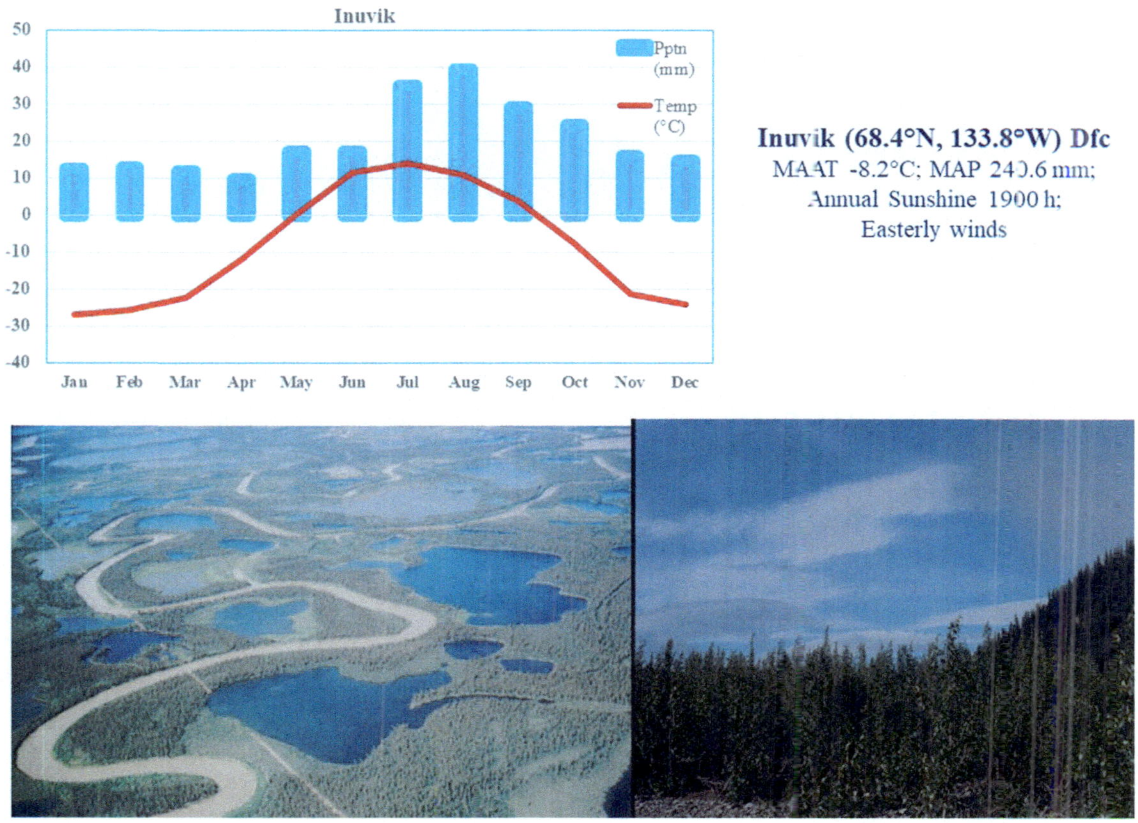

Fig. 9.4 Inuvik. Climate Data: Canadian Climate Normals, Environment and Climate Change Canada. Inuvik is located along the eastern margin of the Mackenzie Delta (lower left) and at the northern taiga/tundra transition (lower right)

Fig. 9.5 Static Ice Jam, Deh Cho (Mackenzie River), formed by sharp drop in surface water temperature (Natural Resources Canada)

Fig. 9.6 Sachs Harbour. Climate Data: Canadian Climate Normals, Environment and Climate Change Canada. Sachs Harbour was established as a permanent community on southern Banks Island, with buildings aligned in a grid pattern. Lower right: *Ovibos moschatus* (musk ox) is a key food and cultural resource for the community (Kathleen Parewick photo)

falls as snow (Fig. 9.6). The heaviest snowfalls occur during September and October, when snowfall totals 40–50% of the yearly average. Snowpack depths seldom exceed 30 cm. August is the wettest month, with 15–20 mm of rain and 40–45 cm of snow. Winters are extremely dry, with Sachs Harbour receiving only 10 mm of precipitation (10% of the annual total) during the four months from January through April. Prevailing winds are westerly and northwesterly in winter, and southeasterly in summer.

At Qausuittuq National Park, approximately 70% of the precipitation falls as snow, mostly during the late spring, summer, and early fall (Edlund and Alt 1989; Fig. 9.7). Although the island supports no active glaciers, due to its low elevation, snow cover persists for 300 days annually. Snow depths, however, seldom exceed 25 cm, and individual snowfall events bringing more than 5 cm are very rare. Thunderstorms are curiosities, with records since 1950 suggesting an average of one every two years (Maxwell 1980–82; Environment and Climate Change Canada 2020).

The prevailing winds vary seasonally, with relatively weak northwesterly arctic winds during the winter months and stronger northerly and north-northeasterly winds during summer. The southern part of Bathurst Island, and the settlement of Resolute (Qausuittuq, "*the place with no dawn*") on adjacent Cornwallis Island, are notorious for their strong, often bitingly cold "summer" winds. The summer winds blow across seasonally open waters of the Sverdrup Basin to the north, picking up small amounts of moisture. Winter winds from the Beaufort Sea traverse frozen pack ice, and thus acquire virtually no moisture. Seasonal shifts bring 24 h of darkness to Bathurst Island during the winter, and 1600 h of sunshine to the island during a typical summer.

As the northernmost part of Canada, Umigmak Nuna (Ellesmere Island) might be anticipated to be the coldest region. However, although Alert (82.5°N, 62.4°W) does have the coldest mean daily July temperature at 3 °C (Environment and Climate Change Canada 2020), its daily mean January temperature of −39 °C is marginally warmer than values for

Fig. 9.7 Resolute (Qausuittuq). Climate Data: Canadian Climate Normals, Environment and Climate Change Canada. Upper Left: Aerial view (Wikimedia/Dave Brosha). Upper right: location of Resolute ("R") on Cornwallis Island. On this false-colour image, areas of tundra vegetation cover appear as pale red. (NASA). Lower right: *Papaver radicatum* (Arctic Poppy)

some localities in the northern taiga of Yukon and Kivalliq with Dfd climates (Fig. 9.8). Temperature fluctuations during the winter are characteristically less than in the taiga regions, but this is cold comfort to those stationed at Alert. In Quttinirpaaq National Park, Lake Hazen (81.8°N, 71°W), sheltered in the fault valley of the Hazen Plateau, is a thermal oasis, with temperatures as much as 10 C° warmer than on adjacent highlands (France 1993; Clark 2007; Köck et al. 2012). The lake remains ice-free between early July and late September. In mid-May, the ice cover typically exceeds 2 m in thickness.

The low temperatures and frozen water surfaces limit precipitation (Maxwell 1980–82, 1981; Serreze and Barry 2005; Environment and Climate Change Canada 2020). Northern interior Umigmak Nuna receives 60–90 mm of precipitation annually, making it the driest region of Canada (and comparable to most major deserts of the world). The highest summits of the Agassiz Ice Cap area receive up to 250 mm y^{-1}. The low humidity effectively precludes thunderstorm activity, and no thunderstorms have ever been recorded in the Lake Hazen area. Approximately 30% of the precipitation in the Lake Hazen valley (20–30 cm) falls as snow, during many small events in the spring, summer, and autumn. Umigmak Nuna thus receives less snow throughout a typical year than most Canadian localities receive during a single winter storm. Cold conditions, however, enable the snow to last 11 months each year in the Lake Hazen lowland. Mountainous areas receive up to 150 cm of snow annually. Winter is the driest season throughout northern Umigmak Nuna.

Winds are generally westerly to northerly in the Hazen Plateau, travelling from the British Empire Range towards the valley. Aspect is not critical as a vegetation or climatic control in this region, however. The long days of summer, when Sun stands above the horizon for 24 h, give the area large amounts of sunshine. Between mid-May and late September, Lake Hazen receives 2150 h of sunshine.

Torngat Mountains National Park lies along the transition between taiga and tundra climates (Catto 2020). Coastal areas such as Nachvak Fjord have a rigourous northern boreal climate (modified Köppen-Geiger Dfc), modified by the influence of the cold Labrador Current. Daily mean temperatures at Nachvak Fjord and Saglek Bay approximate −22 °C in January and 10 °C in July. Mean annual precipitation is

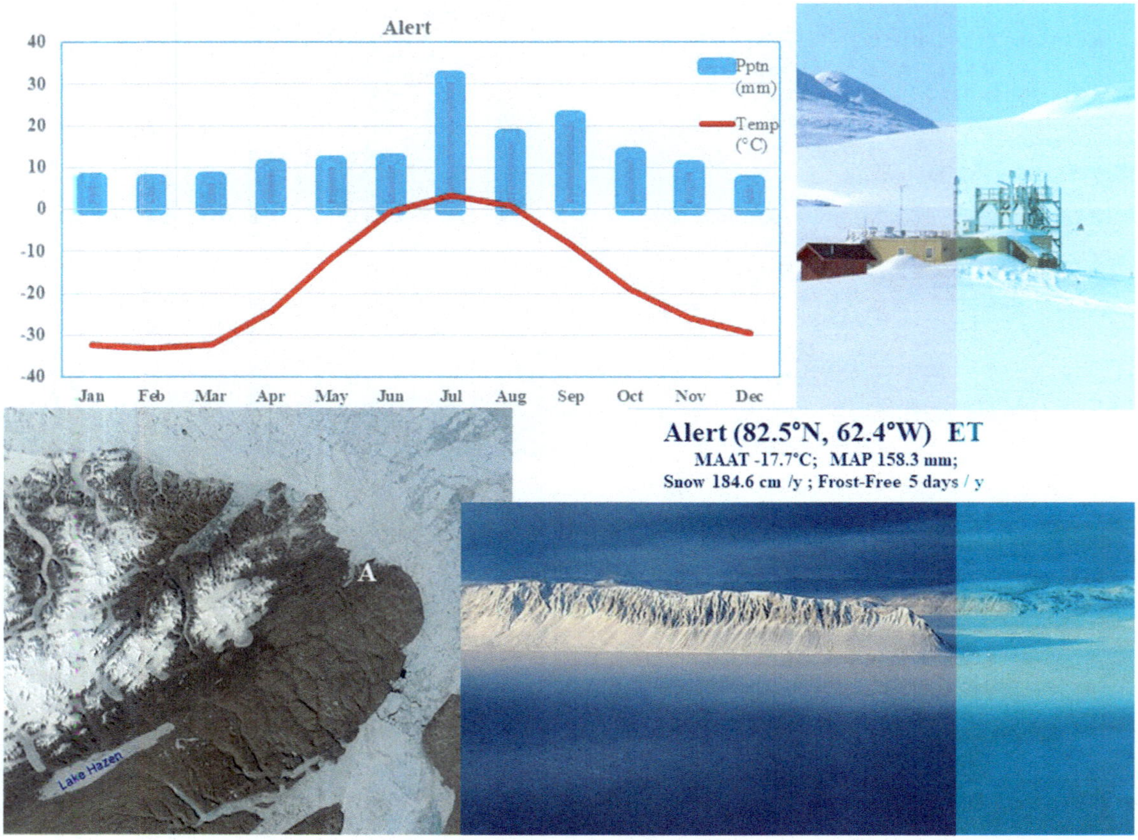

Fig. 9.8 Alert. Climate Data: Canadian Climate Normals, Environment and Climate Change Canada. Upper right: Global Atmosphere Watch Observatory. Lower left: Location of Alert ("A") and Lake Hazen in Quttinirpaaq National Park on northern Umigmak Nuna. (NASA). Lower right: Eureka Sound (Wikimedia/Michael Studinger/NASA)

Fig. 9.9 Helga River outlet valley, Torngat Mountains

500 mm, of which 55–60% falls as snow. Snow cover persists for approximately 260 days per year. The area receives 3 thunderstorms in a typical year, and 1400 h of bright sunshine (Fig. 9.9).

The relative dryness of the area is due to the prevailing winds, which are northwesterly during July (crossing Nunavik) and come from the west-southwest in January (crossing the width of the continent). The summer winds carry most of the moisture to the region, from Hudson Strait; winter winds are extremely dry. As a result, there is no significant rain shadow effect on the east-facing slope.

Inland regions, higher summits above ±900 m, and the terrain north of Seven Islands Bay to Killiniq (Cape Chidley) are under the influence of a true arctic tundra (ET) climate. The area of the few summits above 1500 m is too small to warrant classification of these areas as "highland" climatic zones. A typical year has no high summer days.

In contrast to most areas of the Arctic, Northern Labrador saw decreased temperatures during the period 1980–1995. This led to suggestions that enhanced flow of the Labrador Current in association with the North Atlantic Oscillation could produce anomalous climate change marked by cooling (Hurrell et al. 2003; Pocklington et al. 1994; Morgan and Pocklington 1997), with possible analogies to the Younger Dryas glacial readvance (e.g. Labreyrie and Elliot 1999; see Chap. 4). However, the pattern changed to warming in the late 1990s, leading to glacial recession, reduced sea ice, and significant ecological changes (e.g. Way et al. 2015; Barrand et al. 2017; Davis et al. 2021). Labrador is currently ongoing substantial warming and precipitation change, resulting in significant problems for its residents, requiring multiple adaptations (e,g. Bell et al. 2008; Slaymaker et al. 2020b; Harper et al. 2021; Sawatzky et al. 2021).

Fig. 9.10 Coronation Gulf. The Bear province upland of the Canadian Shield extends eastward to the Coronation Hills and the Napaktulik area south of Kugluktuk. Numerous small islets aligned southwest-northeast are conspicuous in Coronation Gulf. These islands mark the alignment of resistant strata of metamorphosed bedrock of the Tu Nedhe geological province of the Canadian Shield. (NASA image)

9.3 Bedrock

The Arctic ecozones of Canada overlie several geological provinces (De Kemp et al. 2006; St-Onge et al. 2015; Harrison et al. 2016). The Interior Platform (see Chap. 5), Bear Province, and Tu Nedhe (Slave) Province (see Chaps. 3 and 8) extend north to the shoreline of Amundsen Gulf. Metasedimentary rocks associated with the Bear Province also form the Shaler Mountains across northwestern Kitlineq (Victoria Island).

9.3.1 Canadian Shield

9.3.1.1 Bear Province

The Bear Province of the Canadian Shield extends from Coronation Gulf southwest to Great Bear Lake (Fig. 9.10). Proterozoic metasediments and metamorphosed basic volcanic units from 1400 to 2200 Ma are present (Wheeler et al. 1996; Natural Resources Canada 2009).

Volcanic greenstone belts are present and have been exploited for copper by the Kilusiktormiut at Kugluktuk since at least 1000 ce. Quartz sandstones and quartzites form resistant units. Younger diabase dykes and sills (~1000 Ma) have intruded the older rocks, forming resistant ridges and ledges. The numerous small islands in Coronation Gulf are exposed pieces of the diabase sills (Ernst and Bleeker 2010). Bear Province rocks are also present throughout most of Tuktut

Nogait National Park, forming the core of the Melville Hills west of Bluenose Lake. The Hornaday River follows the western margin of the Bear Province units.

The central plateau area of Tuktut Nogait is drained by streams flowing in four directions. Most of the drainage flows eastward, into the Rae River and Coronation Gulf. A second component flows westward via the Hornaday and Horton Rivers, across the Horton Plain of the Interior Plateau (Okulitch 1992). West of Tuktut Nogait, the Hornaday River turns north and flows into Amundsen Gulf at Paulatuk on Darnley Bay. Drainage also flows northward through Bluenose Lake to Amundsen Gulf, and southward to Dease Arm of Great Bear Lake. The rivers which cut through or flank the Melville Hills (Hornaday, Horton, Bluenose) have carved deeply incised valleys and canyons. In the plateau area, the rivers are meandering with thermokarst lake beads.

9.3.1.2 Churchill Geological Province

Churchill Province rocks are exposed in Kivalliq, form the peninsula of Kingngailap Nunanga (Boothia) and Kuganajuup Qikiqtang (Somerset Island), and are present in Sanirajak (Melville Peninsula), northern Nunavik, and throughout much of Qikiqtaaluk (St-Onge et al. 2020).

Eastern Qikiqtaaluk is part of the Qikiqtaaluk subunit of the Churchill geological province. The granites and gneisses of Qikiqtaaluk were originally formed during the Archaean Eon, *ca.* 3 billion years ago, and were then extensively deformed and further metamorphosed during the Hudsonian orogeny at the conclusion of the Palaeo-Proterozoic, *ca.* 1.8 billion years ago (Fig. 9.11). This area is one of the most intensively deformed and modified zones of the Canadian Shield (Johns and Young 2006; Scott and deKemp 1998).

The Cumberland Peninsula is marked by extremely rugged relief, with sharp-crested mountains and deep fjords aligned south-southwest to north-northeast (Dyke et al. 1982; Andrews 1989). Sivanitirutinguak stands at 2015 m above sea level. Qaisualuk has a near-vertical cliff more than 1100 m high, and a total relief of 1675 m. Qulitasaniakvik Falls has a vertical descent of 660 m in several steps. Although higher peaks are present in the Victoria and Albert Range of Umingmak Nuna, the Cumberland Peninsula exhibits the sharpest contrasts in elevation within the Canadian Arctic. Most of the major peaks were at one time named for person-

Fig. 9.11 Auyuittuq National Park lies within the Churchill geological province on Qikiqtaaluk. Left: Qaisualuk has a near-vertical cliff more than 1100 m high, and a total relief of 1675 m (Mel Reasoner). Upper right: pegmatite dykes in Archaean rock (Wikimedia/Mike Beauregard). Lower right: Flat-topped summits of the Sivanitirutinguak Range are remnants of a Palaeogene erosional plane

alities or places taken from Norse mythology, but these names are being replaced with the original Inuit terms (Parks Canada 2022b; see Table 1.2).

The Sivanitirutinguak Range stretches along the entire length of the Cumberland Peninsula, from the eastern margins at Cape Dyer and Hoare Bay to the northwestern boundary of Auyuittuq NP, at Nudlung Fjord. Most of the major chains in the range trend northwest-southeast, and are cut by faults at 90° which have been widened by erosion into valleys and fjords. As a result, the topography has a rather rectilinear appearance when viewed from the air.

Some of the summits of the Sivanitirutinguak Range are flat-topped or marked by very gently northeastward-sloping plateaux. These summits are remnants of an erosional plane, which represented the surface of the land during the early part of the Palaeogene, *ca.* 50 million years ago. At this time, Qikiqtaaluk had been separated from Kalaallit Nunaat (Greenland) for about 5–10 million years, as Davis Strait was widening due to plate tectonic movement.

Mid-ocean ridge basalts were produced along both spreading ridges, and as the ridges continued to widen these basalts moved away from the centres of the rift valleys. On easternmost Qikiqtaaluk, mid-ocean ridge basalts formed during the Palaeocene, *ca.* 60 million years ago, are present at Cape Dyer and Cape Searle (Burden and Langille 1990; Dafoe and Bingham 2020). Gradually, the eastern rift system between Scandinavia and Kalaallit Nunaat spread progressively more efficiently and faster than did the western system (Funck et al. 2012; Oakey and Chalmers 2012). The western system ceased to spread approximately 40 million years ago, and Qikiqtaaluk and Kalaallit Nunaat have remained in the same relative position since that time. The basic topography and landscape of Cumberland Peninsula was established approximately 10 million years ago, and glaciation has resulted in only minor fjord widening and deepening since that time.

Severe earthquakes (Richter magnitude >7) periodically occur off Qikiqtaaluk and Bylot Island (Earthquakes Canada n.d.). These seismic events indicate that tectonic activity did not cease entirely after sea-floor spreading between Kallaalit Nunaat and Qikiqtaaluk stopped in the Eocene.

9.3.1.3 Nain Geological Province

Torngat Mountains National Park lies within the Saglek Block of the Nain geological province of the Canadian Shield (Wardle et al. 1997), dominated by Eoarchean to Paleoarchean gneisses (3800 to 3600 Ma). The Saglek Block is cut by numerous NNW-SSE faults. Fjords extend up to 15 kms inland from the open coast, and main valleys dissect the ranges, extending across the drainage divide and Labrador/Québec boundary. Relief along the coastline is approximately 1200 m, and individual peaks rise above 1500 m (Fig. 9.12). The fjords generally cut across the structural grain of the fault systems.

Younger rock units within the Saglek Block include the Ramah sandstones, shales, and chert beds (approximately 2000 Ma). The chert units of the Ramah Group have been used for the manufacture of artefacts, traced along the coastlines of Labrador and northern Newfoundland (Anstey and Renouf 2011). Layered intrusions of basic gabbro, anorthosite, and other igneous rocks of the Nain Plutonic Suite (1322–1305 Ma) are exposed in the Kiglapiat Mountains and in scattered outcrops southwestward to the head of Voisey's Bay. Anorthosite, composed of labradorite feldspar, is used as

Fig. 9.12 Torngat Mountains National Park lies within the Nain geological province. Inset: Labradorite crystal in anorthosite

Fig. 9.13 Pingualuit astrobleme, Nunavik, formed 1.4 Ma (NASA)

ornamental building stone and in jewellery production. The Voisey's Bay nickel mine exploits the layered intrusions of the Nain Plutonic Suite.

9.3.1.4 Superior Geological Province

The western and northern Arctic ecozones of Nunavik lie within the Superior Geological Province (Hocq 1994), with Archean granites and gneisses (2700–2650 Ma). Most rocks were further metamorphosed or modified during the later orogenic events of the Proterozoic. The physiographic Larch Plateau (Bostock 2014) consists of planar to gently rolling (50 m relief) surfaces developed on the granites and gneisses The drainage divide between Hudson Bay and Ungava kangiqluk (Bay) cuts across the terrain irregularly from north to south, without a conspicuous height of land.

The most prominent geomorphic feature in the Larch Plateau is the 1.4 million year old Pingualuit astrobleme (61°17′N, 73°40′W; Grieve 2001; Guyard et al. 2011; Fig. 9.13). Discovered during World War II by pilots flying military missions, the astrobleme was formerly known as the Chubb, Ungava, and Nouveau-Québec crater. Pingualuit was designated as a parc national in 2004 and is co-managed by the Kativik Regional Government.

Pingualuit means "where the land rises", a reference to the raised rim around the edge of the crater. The Pingualuit astrobleme is almost perfectly circular, 3.4 km in diameter, and approximately 410 m deep. The lake in the centre of the crater is 246 m deep. The impact responsible for forming Pingualuit was approximately 150 megatonnes, roughly equal to 8,500 times that of the atomic bomb dropped on Hiroshima in 1945.

Fig. 9.14 The Povungnituk Hills structure, extending west-southwest from northern Ungava kangiqluk to Cape Smith along Hudson Bay, is highlighted by differences in snow cover between valleys and ridges. Note the ice-free area around Pingualuit. (NASA image)

The northern flank of the Sugluk Plateau in northernmost Nunavik has been modified by glacial erosion, resulting in steep cliffs locally exceeding 600 m height. The interior of the Sugluk plateau consists of planar to gently rolling surfaces.

The Povungnituk Hills (Fig. 9.14) form part of a broad arc of deformed metasedimentary, metavolcanic, volcanic, metamorphic, and igneous rocks that extends northwards through Nunavik, west-southwest to Cape Smith along Hudson Bay, through Povungnituk and the Inoucdjouac Hills, and southward through Hudson Bay, exposed on the Sanikiluaq (Belcher) Islands (see Chap. 8). In eastern Nunavik, the rocks

Fig. 9.15 Coastal geomorphology of the Povugnituk Hills along Hudson Bay. Upper: Low-lying rocky coast and skerries between Cape Dufferin and Puvirnituq (59°06′N, 78°15′W). Lower: Coastline at Akulivik (60°48′N, 78°14′W). (Boisson and Allard 2020) *{Open access, used with permission}*

of the Povungnituk Hills were formed between 2100 and 1840 Ma, and subsequently deformed during the collision between the Superior and Nain provinces (Hocq 1994; Wardle et al. 1997; Fig. 9.15).

9.3.2 Arctic Platform

The Arctic Platform is underlain by flat-lying to gently deformed sedimentary rocks of Palaeozoic to Mesozoic ages (Trettin 1991; St-Onge et al. 2015; Lamoureux and Rudy 2020). In Sirmilik National Park (73°N, 81.1°W), most of the Borden Peninsula is composed of Ordovician and Silurian carbonates, shales, and sandstones. Locally, the red sandstones have been modified and eroded to form isolated spires, or hoodoos (Parks Canada 2022g). Similar rocks crop out on the Brodeur Peninsula to the west, and throughout Lancaster Plateau. The youngest rocks in the area are Cretaceous and Paleocene-Eocene sedimentary units, which are preserved as isolated patches that escaped erosion during the Neogene prior to Quaternary glaciation. These rocks crop out along the south coast of Bylot Island, on the coastal fringes of Nalluata imanga and Tasiujan (Currie et al. 2020).

Fig. 9.16 On Banks Island (west), the Arctic Platform is underlain by flat-lying to gently deformed Palaeozoic to Mesozoic sedimentary rocks. Aulavik National Park is located in northern Banks Island. Duck Hawk Bluffs (Fig. 9.27) are marked by pale brown sediment exposure along the southwest coast. Neo-Proterozoic sedimentary rocks crop out on Kitlineq (Victoria I) to the east (NASA)

In the Arctic Platform of Banks Island, bedrock units decrease in age systematically towards the northwest (Fig. 9.16). The oldest unit is Neo-Proterozoic dolomites, cherts, mudstones, algal stromatolites, sandstones, and siltstones. Devonian units include siltstone, shale, and coral-algal biohermal reef rocks, exposed at Mercy Bay (Copper and Edinger 2009).

Lower Cretaceous sandstones and siltstones of the Isachsen Formation flank the Devonian rocks and have locally been eroded into badlands topography. Other Cretaceous units include claystones, glauconite marine sandstones, and bituminous shales, bentonites, and volcanic tuffs. The Cretaceous units occupy the eastern and southern third of the island.

The central third of Banks Island is occupied by a north–south belt of Paleocene and Eocene sandstones, shales, siltstones, and lignite coals of the Eureka Sound Formation, including bony fish and crocodilian fossils (Eberle et al. 2014). These rocks are soft and have been modified to form the gently sloping westernmost part of the Arctic Platform, grading westward into the Arctic Coastal Plain.

9.3.3 Arctic Coastal Plain

The Arctic Coastal Plain is marked by low relief (<200 m), and slopes gently to the Beaufort Sea (Trettin 1991; Bostock 2014; St-Onge et al. 2015). At Ivvavik National Park, Qikiqtaryuk (Herschel Island), and Tuktoyaktuk, it is underlain by flat-lying to gently dipping sandstones, shales, and lignite coals, ranging in age from Early Cretaceous to Neogene (Burn

Fig. 9.17 Qikiqtaryuk (Herschel Island) formed by ice thrusting of friable bedrock from southeast to northwest along the Yukon Beaufort Sea coast. Today, active erosion of the overlying Quaternary deposits threatens archaeological and cultural heritage sites (False-colour image, NASA)

2012, 2017b, Parks Canada 2022c; Fig. 9.17). These sedimentary rocks are soft and are readily eroded. The bedrock of the Mackenzie Delta is covered with a thick succession of Quaternary glacial and fluvial deposits in all areas, with the exception of Rocky Hill east of Inuvik.

Banks Island's west coast also lies within the Arctic Coastal Plain. This terrain is an almost featureless plain, extending approximately 50 km inland from the Beaufort Sea and rising from sea level to 100 m asl. Braided rivers form networks up to 10 km wide of shallow channels, and thermokarst lakes and beaded streams are common. All of the terrain is underlain by continuous permafrost. In most areas of the Arctic Coastal Plain, the bedrock has no influence on the topography of the Beaufort Sea coast (e.g. Lantuit and Pollard 2008; Obu et al. 2017).

9.3.4 Innuitian Orogen

The Innuitian Orogen geological province occupies the most northerly islands of the Canadian Arctic (Trettin 1991; De Kemp et al. 2006; Harrison et al. 2016; Fig. 9.18). Three main divisions can be recognized: a north–south trending Cornwallis fold belt of deformed Early Devonian and older rocks, extending from the Grantland Mountains of northern Umigmak Nuna (Ellesmere) to western Devon Island and Bathurst Island; a younger Late Devonian Parry Islands fold belt, aligned ESE-WNW through Bathurst and Melville Islands; and the Sverdrup Basin infill of Carboniferous to Mesozoic rocks, in the elbow created by the fold belts, opening towards the northwest.

The Grantland Mountains are composed of a succession of Cambrian to Devonian sedimentary rocks, both clastic sandstones and shales and carbonate limestones and dolomites (Piepjohn and von Gosen 2017). Coal seams are also present. These rocks, deformed into north–south oriented folds, were deposited under tropical to semi-tropical conditions, when Umingmak Nuna lay adjacent to the Equator in the Northern Hemisphere. The highest mountain is Barbeau Peak, rising to 2616 m asl. Most of the area is covered by alpine glaciers and ice fields of the Barbeau Glacier Complex, with ice thicknesses in excess of 900 m.

The northwestern coast is cut by the deep Greely, Tanquary, and Archer Fjords, developed along southwest-northeast trending fault valleys (Trettin 1991). These valleys extend across the width of northern Umigmak Nuna. Lake Hazen is developed along the low-lying Hazen Plateau, an extension of the fault valley occupied by Tanquary Fjord. Although the fjords and valleys have been accentuated by glaciation, their original position and basic morphology is controlled by the orientation of the pre-Quaternary fault systems (England 1996; England et al. 2000). Cliffs along Archer Fjord are more than 700 m high.

South of Lake Hazen and Tanquary Fjord, the mountains of Judge Daly Promontory and the Victoria and Albert Range, buried beneath the Agassiz Ice Cap, isolate the Hazen Plateau from central Umigmak Nuna (van Wychen et al. 2020). Relief in this area approximates 1800 m. Northern Umigmak Nuna is separated from Kalaallit Nunaat by two narrow straits, Robeson Channel and Kennedy Channel, with maximum widths of 30 and 50 km respectively.

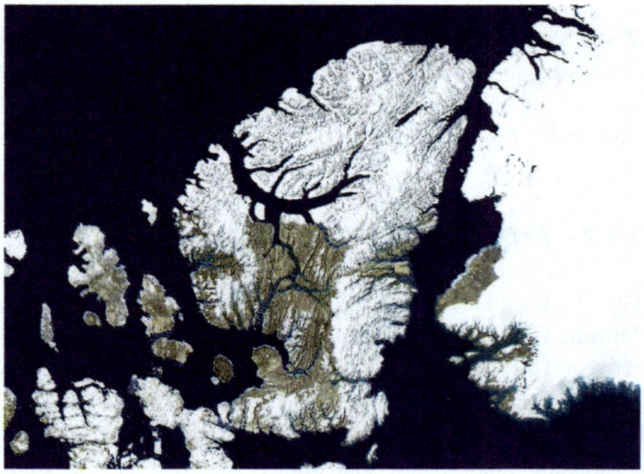

Fig. 9.18 The Innuitian Orogen geological province occupies the most northerly islands of the Canadian Arctic. This image shows the north–south trending Cornwallis fold belt in the Grantland Mountains of northern and eastern Umigmak Nuna (Ellesmere), and the Sverdrup Basin infill of Carboniferous to Mesozoic rocks opening towards the northwest (see Fig. 9.21) (NASA)

Bathurst Island lies within the Parry Plateau, within the Innuitian region. The maximum relief on the island is 412 m. The terrain consists of a series of sharply defined ridges and valleys, locally rolling to undulating. In the eastern part of the island, the ridges and valleys are aligned north–south, reflecting the folded terrain of the Cornwallis fold belt (Fig. 9.19).

The terrain in the area where the fold belts intersect, in the eastern third of Bathurst Island, is marked by irregular rolling hills and hummocks (Kerr 1974; Harrison and de Frietas 1998; Harrison 2018). Faulting and folding has shaped the coastline into a series of deep, linear embayments, aligned both parallel and at right angles to the fold systems.

Fig. 9.19 The Late Devonian Parry Islands fold belt is aligned ESE-WNW through Qausuittuq (Bathurst) and Ilulliq (Melville) Islands. The eastern third of Qausuittuq, where the older Cornwallis and younger Parry Islands fold belts intersect, is marked by irregular rolling hills and hummocks (NASA)

In the western, central, and northern parts of the island, the ridges and valleys are more sharply defined than in the east, and are aligned west-southwest–east-northeast. These ridges and valleys were created during the Parry Islands event of Late Devonian age, and reflect both folding and faulting (Fig. 9.20). Polar Bear Pass extends along one of these fold-created valleys, connecting Bracebridge and Goodsir Inlets. The wildlife sanctuary includes two elongated lakes and numerous small and shallow thermokarst ponds, and is flanked by rolling ridges to 240 m elevation.

The tectonic history of Bathurst Island is the dominant factor which is evident in the physiography. The Early Devonian Cornwallis fold belt, which affected the easternmost parts of the island, buckled pre-existing Ordovician, Silurian, and Devonian limestones, dolomites, shales, and gypsum beds into a series of tightly spaced north–south ridges and valleys (Trettin 1991; Harrison 2018). The folds are upright, and plunge both to the north and south. During the folding, thrust and normal faults were also generated, accentuating the ridge-and-valley pattern. This folding event was associated with the uplift of the Boothia Arch region to the south, creating a spine of disturbed terrain from Bathurst Island through Somerset Island to the Boothia Peninsula (Okulitch et al. 1986).

During the Late Devonian, similar rock units in the central and western parts of Bathurst Island were deformed into west-southwest–east-northeast folds. These anticlines and

Fig. 9.20 Resistant bedrock ridge resulting from folding in the Parry Islands Fold Belt, May Inlet, Qausuittuq (Bathurst Island)

synclines are also upright and doubly plunging. The Parry Islands fold belt stretches west to Melville Island (Harrison et al. 2014a, b). This deformation was also accompanied by thrust and normal faulting.

The youngest rocks in the area occupy the northwesternmost points of Bathurst Island, and Cameron Island to the northwest. These Mesozoic sandstones and shales were draped over the folded rocks of the Parry Islands fold belt when the area subsided, and form the southernmost edges of the geological Sverdrup Basin (Beauchamp et al. 1998). The Mesozoic sediments are the only non-folded or faulted rocks on Bathurst Island.

The Sverdrup Basin is infilled by younger sedimentary rocks (Galloway et al. 2021; Fig. 9.21). On Ellef Ringnes and Melville Islands, alternating beds of gypsum-anhydrite and limestone have been deformed into domes, which form distinctive landforms (Harrison 1995). The arid ET climate allows preservation of the highly soluble gypsum and anhydrite on the surface.

9.4 Quaternary

The Arctic Cordilleran regions of the Torngat Mountains, eastern Qikiqtaaluk, Tatlurutit, and Umigmak Nuna supported alpine glaciation during the Quaternary. Modern glaciers exist in all these regions. Continental glaciation covered Nunavik, Qikiqtaaluk, Kivalliq, the Western Arctic, and the Northern Arctic (Innuitian Ice Sheet).

9.4.1 Torngat Mountains: Alpine Glaciation

The Torngat Mountains preserve evidence of four alpine glacial advances (Clark 1988; Klassen et al. 1992; Fig. 9.22). The most recent, Saglek glaciation, is considered to represent the Last Glacial Maximum advance, beginning about 30,000 years ago. The Saglek glaciers extended to the coast along the major fjords, leaving the intervening coastlines ice-free. Mountain glaciers were topographically confined. The limited extent of the glaciers prevented significant fjord erosion north of Saglek Bay. Fjords are plugged on the seaward side by accumulations of glacial sediment. Deglaciation began 14,000 years ago, and the last glacial ice had left the coastline no later than 11,000 years ago.

The postglacial marine limit declines from south to north, primarily in response to glaci-isostatic recovery as the Laurentide Inlandsis in central Labrador retreated. The marine limit is approximately 90 m near Hopedale, 50 m at Saglek Fjord, and 17 m at Killiniq (Klassen et al. 1992). There is no definitive evidence concerning ongoing relative sea level change in Torngat Mountains National Park, but it is not considered to be significant.

At Hopedale, sea level is falling slowly as a result of glaci-isostatic rebound, exposing new coastal plains and scattered islets and skerries offshore. Kretschmer Island is now linked to the mainland by a tombolo exposed as a result of the gradual decline in sea level. To the south, a larger tombolo forms the north shore of Tooktoosner Bay.

Fig. 9.21 Sverdrup Basin. Left: Ellef Ringnes Island. Enhanced false-colour image shows bedrock deformation and evaporite dome. (NASA). Upper right: Hoodoo in Isachsen Formation sandstone, Ellef Ringnes (Natural Resources Canada). Lower right: Beaufort Formation sediments, Prince Patrick Island (Natural Resources Canada)

Fig. 9.22 Cubic parabolic glacier troughs terminate in fjords throughout Torngat Mountains National Park

Fig. 9.23 Iron Strand is characterized by density-sorted concentrations of garnet and ilmenite, with associated quartz and feldspar. Beach placer deposits are covered with patchy quartz-dominated aeolian sand. Net sediment transport is from south (background) to north

In 2022, about 70 active glaciers are present in the Torngat Mountains. Most are less than 1 km long and are retreating (Way et al. 2015; Barrand et al. 2017). Rock glaciers are also present on the flanks of some of the higher summits.

Beach deposits have developed from reworking of older glacially transported material. Garnet, ilmenite, magnetite, and sillimanite reworked from gneisses form beach placer deposits along Iron Strand and Seven Islands Bay (Fig. 9.23). At Iron Strand, promontories interfere with waves from the northeast. To the south of Helga River, shore-parallel transport is southeastward. To the north of Helga River, net littoral transport is northerly. Both segments of the beach are swash-aligned and shore-parallel. Iron Strand is a relatively low energy system, protected from winter storm events by seasonal ice, with abundant sediment flux (Catto 2020).

9.4.2 Nunavik: Continental Glaciation

Scattered distributions of glacial sediments and computer-generated ice sheet modelling suggest that extensive glacial ice cover existed throughout Nunavik for at least 150,000 years until the Holocene (Stokes et al. 2015). Ice advanced northward from the Laurentide centre in western Labrador, covering Nunavik and interacting with the Torngat glaciers to the east and the Foxe Basin ice to the north (Klassen et al. 1992). Large ice streams developed in Hudson Strait as the combined glacial mass flowed eastward.

As the glaciers withdrew, eastern Nunavik remained isostatically depressed. River systems drained to the west, following the isostatically controlled topographic slope, and were impounded against the retreating ice in Glacial Lake Naskaupi in Ulittaniujalik. Glacial Lake Naskaupi was short-lived, lasting for a few hundred years ca. 8,300 years ago (Dubé-Loubert et al. 2018). By 8,000 years ago, glacial ice had completely withdrawn from northern Nunavik. The George River drained north to Ungava kangiqluk, into the glaci-isostatic d'Iberville Sea (Allard et al. 1989). Retreat continued rapidly throughout the early Holocene, and all of Nunavik was deglaciated by 7,000 years ago.

9.4.3 Umigmak Nuna: Alpine Glaciation

Northern Umigmak Nuna was glaciated during the Quaternary (Hodgson 1989; England 1996; England et al. 2000). Glaciers advanced from the Barbeau and Agassiz Ice Caps, and also from Kalaallit Nunaat. The area was almost completely covered during Quaternary glaciations, with the exception of a few nunataks on Judge Daly Promontory. Deglaciation is still ongoing in the area. Lake Hazen's north shore was deglaciated approximately 2,000 years ago, and ongoing climatic changes continue to affect the glaciers of the area (van Wychen et al. 2020). The current landscape of northern Umigmak Nuna thus resembles that which existed shortly following deglaciation in many areas of Canada (Fig. 9.24).

The sediment record at the bottom of Hazen Lake indicates regional climate changes (Clark 2007; Köck et al. 2012; Lehnherr et al. 2018). The "Little Ice Age" reached its maximum from 1700 to 1850 AD, marked by extensive ice cover over the lake. Subsequently, climate warming is indicated by changes in sediment, geochemistry, and diatoms preserved in the sediments. Recently, the diatom genus *Campylodiscus* has appeared in the lake and bottom sediments, indicating ongoing climate warming.

Fig. 9.24 Glacial landscape of Tanquary Fjord region, southwestern Quttinirpaaq National Park, Umigmak Nuna. (NASA Earth Observatory)

9.4.4 Qikiqtaaluk Glaciations

Qikiqtaaluk was glaciated several times during the Quaternary (Dyke et al. 1982; Andrews 1989; Wolfe 1996; Briner et al. 2009; Margreth et al. 2016; Birch et al. 2017). The oldest event, the Qikiqtaaluk (Baffin) Glaciation, was also the most extensive, with ice flowing from Foxe Basin (Ilagiit nunagivaktangit) towards the north and northeast. A major glacial tongue occupied Tallurutiup Imanga (Lancaster Sound), flowing eastward as it was fed by outlet glaciers from Admiralty Inlet and Prince Regent Inlet (west of the Brodeur Peninsula). Following glaciation, a climate similar to today's prevailed, resulting in glacial retreat. This phase, referred to as the Salmon River Interglacial, could be about 120,000 years old.

A subsequent glacial advance from the Foxe Basin centre of the Laurentide Inlandsis expanded east, south, and west to interact with ice from western Labrador, Nunavik, and Kivalliq. Foxe Basin ice moved northward from Qikiqtaaluk onto Bylot Island, along Admiralty Inlet, and north to Tallurutiup Imanga. Local glaciers developed simultaneously on the Byam Martin Mountains. The ages of these glaciations are uncertain, but they could be 80,000 years old. Associated marine deposits, up to 90 m asl, formed around the margins of the glaciated terrain. Following each glaciation, sea levels rose, depositing marine clays and marine fossils in the coastal areas.

The most recent glacial event to affect the region involved both local glaciers expanding down the fjords from the Penny Ice Cap and the Sivanitirutinguak Range (Fig. 9.25), and continental glaciation originating from Foxe Basin (Dyke et al. 2002; Dowdeswell et al. 2016b). The Foxe Basin ice overran almost all of Qikiqtaaluk (including mainland Sirmilik National Park), forming drumlins and bringing limestone erratics to the granitic and gneissic terrains. The timing of the glacial advance was later than in the mainland regions to the south, with the maximum late glacial extent being reached about 12,000 years ago.

Sirmilik was glaciated by Qikiqtaaluk glaciers, but the local glaciers on Bylot Island appear to have advanced only minimally if at all (Dyke 2004, 2008; Bennett et al. 2014; Dowdeswell et al. 2016a; Furze et al. 2018). A large tongue of eastward-flowing ice in Tallurutiup Imanga extended inland to 140 m asl, and offshore on the Continental Shelf (Li et al. 2011; MacLean et al. 2017). Another event involved ice flowing northwards from Qikiqtaaluk across Tasiuja to Bylot Island. Isostatic depression associated with these glaciations resulted in raised marine deposits to 100 m asl at Milne Inlet, 80 m at Cape Hatt, 60 m at Pond Inlet, and 35–45 m on southern Bylot Island (Nalluata imanga). Raised marine shorelines are not present on northern Bylot Island.

In eastern Qikiqtaaluk, rapid retreat of the glaciers began when Hudson Strait was deglaciated. The influx of water through Hudson Strait, lapping against the core of the Foxe Basin glaciers, caused the Foxe ice to rapidly disintegrate. By 8,500 years ago, the entire Foxe Basin was free of glacial ice (Utting et al. 2016), and the extremities of the former dome-shaped Foxe continental glacier were left as beheaded remnants on the Melville Peninsula, Southampton Island, and western Qikiqtaaluk. The Penny Ice Cap was now free of the influence of Foxe Basin ice, and thus was able to expand and contract unimpeded. The warm interval in the mid-Holocene caused the Penny ice cap to slowly contract, deglaciating the flanks of Pangnirtung Pass *ca.* 6,000 years ago. Similar rapid retreat resulted from the deglaciation of Tallurutiup Imanga north of Sirmilik.

Fig. 9.25 Alpine glacial features, Qikiqtaaluk. Upper left: Truncated spurs formed as glacial erosion along the main valley eroded bedrock ridges extending at right angles to the direction of main valley flow, leaving the spurs at the end of aretes along the tributary glacier. Upper right: Small terminal moraine formed by tributary glacier, Auyuittuq National Park. This terminal moraine is under attack by the braided river in the trunk valley. Lower Left: Pangnirtung Fjord, formed by alpine glacial flow. Lower right: Cubic parabolic glacial trough, Auyuittuq National Park. (Courtesy Mel Reasoner)

Deglaciation and isostatic depression initially allowed marine sediments to be deposited around the margins of the fjords, to elevations of 70 m asl. Subsequently, continued rebound caused relative sea level to fall to 45–51 m below the modern level approximately 8500 years ago (Cowan et al. 2021). At present, relative sea level rise and increased wave action are occurring at Iqaluit and elsewhere along Qikiqtaaluk (Hatcher et al. 2014; Hatcher and Forbes 2015).

Many of the alpine glaciers in Auyittuq and Sirmilik readvanced during the latter stages of the Holocene. This glacial event coincided with general climate cooling throughout the North Atlantic ("Little Ice Age"), climaxing between 1000 and 1850 ce. Moraines marking the "Little Ice Age" maximum are preserved at several places in Auyuittuq, including Akulutarusiq Lake on the west side of Weasel River, as well as Windy Lake and Summit Lakes. Ongoing retreat is evident (Sharp et al. 2014; van Wychen et al. 2020).

On Bylot Island, the "Little Ice Age" maximum positions were reached within the last 150 years (Fig. 9.26). Although recession has been ongoing since, all of the terrain above 1000 m, with the exception of the steepest cliffs, is ice-covered (van Wychen et al, 2020, Parks Canada 2022g). Glaciers cover more than 50% of Bylot Island. Active accumulation of glacial ice occurs above ~700 m asl, and several glaciers extend to tidewater.

9.4.5 Western Arctic

In the western Arctic, ice from Kivalliq (west of Husdon Bay) and Kingngailap Nunanga (Boothia Peninsula) expanded west to cover Kitlineq (Victoria Island), Banks Island, and the margins of the Beaufort Sea (Dyke 1983, 1984; MacLean et al. 2010; Margold et al. 2015; Storrar and Livingstone 2017). Banks Island was glaciated repeatedly by ice from the M'Clintock Channel area (east of Kitlineq; Vincent 1982, 1983; Evans et al. 2014). Ice moved in two streams, following Amundsen Gulf in the south, and Viscount Melville Sound and M'Clure Strait in the north, and onlapped onto Banks Island (Figs. 9.27 and 9.28). During the most extensive glaciations, all of Banks Island was ice-covered. Some lesser glaciations affected only the areas fringing the coasts. Each glaciation was preceded by marine inundation, due to isostatic depression, and was followed by another marine event. Locally, glacial lakes were temporarily impounded in the Thomsen River valley and other valleys during each phase of glacial retreat.

Following each glaciation, parts of Banks Island were submerged by marine waters. The most extensive of these events formed the Big Sea, flooding all terrain below approximately 220 m asl, following the extensive Thomsen Glaciation. In contrast, the marine incursions following the

Fig. 9.26 Many of the alpine glaciers in Sirmilik National Park readvanced during the latter stages of the Holocene. On Bylot Island, all of the terrain above 1000 m, with the exception of the steepest cliffs, is ice-covered, and glaciers cover more than 50% of the terrain (NASA image)

Fig. 9.27 Laurentide glaciation, Banks Island. Upper: Most of Banks Island is covered with a thin blanket of Quaternary sediments, displaying permafrost features without conspicuous glacial landforms. Extents of ice coverage are inferred from distribution of erratics from bedrock units to the east. Lower: Southern Banks Island was glaciated repeatedly by ice moving through Amundsen Gulf from the M'Clintock Channel area, producing the thick sediment successions at Duck Hawk Bluffs west of Sachs Harbour (see Vincent 1982, 1983; Evans et al. 2014)

M'Clure Stade only flooded the fringes of the island, with water reaching 120 m asl on the eastern and southern shores (East Coast Sea), 30 m asl on the north (Investigator Sea), and 20 m asl on the west (Meek Point Sea). All of Banks Island was free of glacial ice at the conclusion of the last glaciation.

Approximately 25,000 years ago, ice extended along the Beaufort Sea coast west of Qikiqtaryuk (Herschel Island; Rampton 1982). Subsequently, ice retreated to the southeast, punctuated by minor readvances and stillstands, until the entire area was deglaciated no later than 13,000 years ago. There are no raised marine sediments of Late Glacial or Holocene age along the southeastern Beaufort Sea coast.

The Beaufort Sea has been steadily rising for at least the past 30,000 years, and possibly since 120,000 years ago, drowning the permafrost which is currently degrading beneath it (Hill et al. 1993; Heginbottom et al. 1995; Allard and Pollard 2013; Paulla et al. 2022). This Subsea Permafrost Zone developed during the last glacial maximum period, when sea level was lower and the southern parts of the sea floor were exposed, beyond the limit of glaciation. As sea level subsequently rose, the flooded permafrost became unstable. The insulating effect of the water keeps temperatures above 0 °C on the sea floor. The permafrost melts through the input of geothermal heat flux from below, and from the input of heat and salt water from the sea, causing subsidence of approximately 1 cm per year throughout the southern Beaufort Sea. The subsea permafrost is thus thawing at present, and will continue to do so inexorably.

9.4.6 Northern Arctic

Quaternary glaciation in the area north of M'Clure Strait and Viscount Melville Sound has left fewer traces on the landscape (French 2017b; Lamoureux and Rudy 2020). Most areas show little evidence of glaciation apart from scattered erratic boulders across the surface, but all appear to have been covered by glacial ice at some time during the past 20,000 years. Collectively, these features are assigned to the Innuitian Ice Sheet Complex (England et al. 2006, 2009; Nixon et al. 2014; Fig. 9.29). Cold-based ice expanded across Ellef Ringnes Island and Ilulliq (Melville Island; Lamoureux and Rudy 2020), reaching the Beaufort Sea coastline.

On Bathurst Island, deglaciation occurred 9,000–10,000 years ago, and was followed by marine incursion into the isostatically depressed coastal areas to approximately 90 m above present sea level (Fig. 9.30). Gravel beaches and erosional marine benches mark the elevated positions (Parks Canada 2022e).

9.4.7 Continuous Permafrost Zone Features

With the exception of thin strips of shoreline the Arctic regions are underlain by continuous permafrost extending from centimetres below the surface to 100 or 1000 s of metres depth (Heginbottom et al. 1995; Government of Canada 2008; Smith 2011; Natural Resources Canada 2012; Woo 2012; Slaymaker et al. 2020a; Fig. 9.31). Although permafrost develops in all geological solids, creation of distinctive features and landforms depends upon deformation induced by

Fig. 9.28 Beaufort Sea coastline of Banks Island. Upper left: Duck Hawk Bluffs contains a thick succession of Quaternary sediments. Slope failures result from exposure of segregated ice lenses. Upper right: ongoing failure due to segregated ice lens ablation near Sachs Harbour. Within two hours after the photograph was taken, the bluff where the person is kneeling had failed. Lower left: thermokarst ponds and ice-wedge polygons are subject to erosion by wave action, with sediments transported by littoral drift. Lower right: Cape Kellett, an elongated spit with a hooked tip

freezing and thawing. As a result, most permafrost features in the Continuous Permafrost Zone are associated with silt, the most readily deformed material and one which provides the appropriate balance between permeability and retention of water along freezing fronts (Catto et al. 2020). Most features are associated with coastal regimes in the southern areas of the Continuous Permafrost Zone, marked by varying sediment temperatures and winds facilitating periodic removal of insulating snow blankets. Diagnostic features include large ice wedges, patterned ground, thermokarst lakes, thaw flows, and pingoes.

9.4.7.1 Ice Wedge Features

Under extremely cold and relatively dry conditions, thermal contraction can cause exposed terrain to crack vertically (French 2007; Burn and Kokelj 2009; French and Slaymaker 2012; Catto et al. 2020). Although the temperatures required to freeze the porewater in the sediment are relatively high (−0.5 °C for coarse sand, −2 °C for clay), substantially colder temperatures are required to cause the ground to split vertically. In the Arctic Coastal Plain, cracking occurs when air temperatures are −25 to −30 °C at maximum, with ground surface temperatures of −13 to −15 °C. Sharp drops in temperature are conducive to crack formation. The presence of thermal vertical cracks is thus indicative of extreme periglacial conditions in the northernmost Widespread Discontinuous or southernmost Continuous zones, and formation requires minimal snow and vegetation cover.

Cracking can be initiated either from the surface downwards, or from the permafrost table upwards. Once a crack forms, it represents a plane of weakness that can subsequently snap during the next cold event, so repetitive failure at the same crack site is common. However, failure does not necessarily occur during each cold event: typically, less than half the cracks fail on an annual basis. The developing cracks are influenced by the micro-relief of their surroundings, and in some instance by the growth of vegetation.

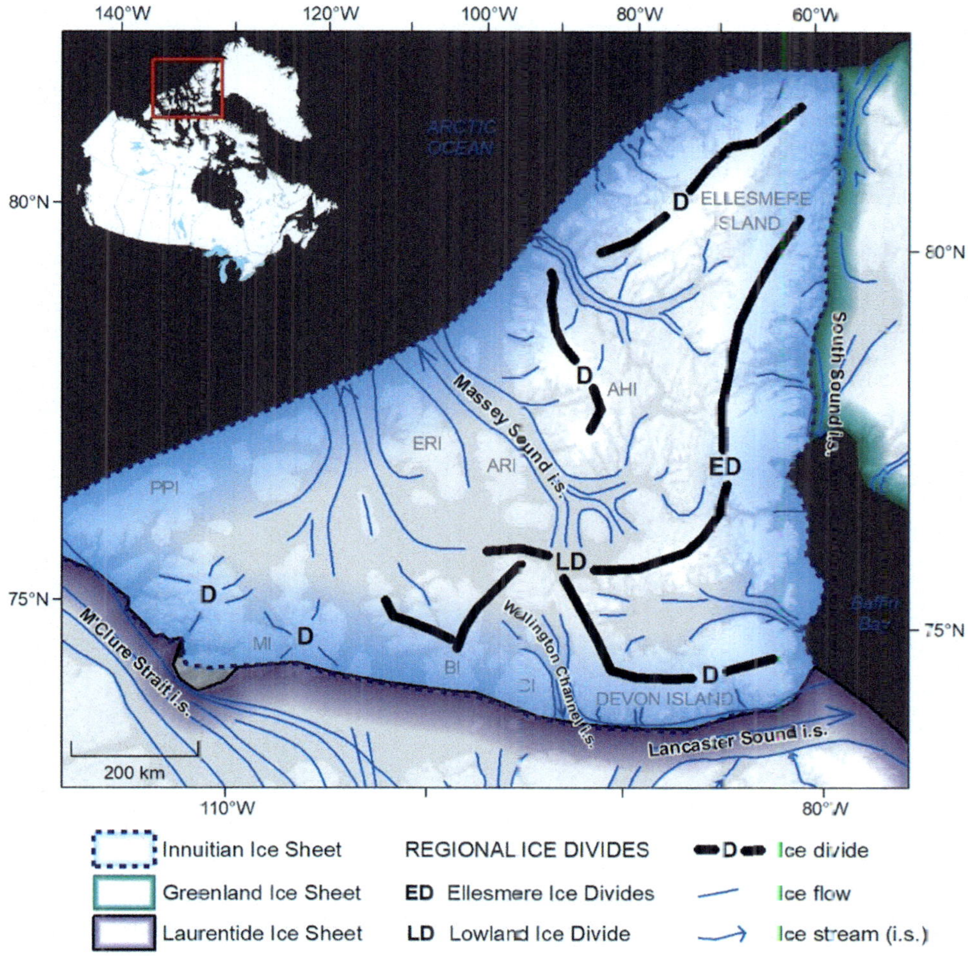

Fig. 9.29 Reconstruction of the Innuitian Ice Sheet over parts of the Canadian High during the Last Glacial Maximum, showing ice divides, and ice streams (modified from England et al. 2006, 2009; Stokes et al. 2015). AHI—Axel Heiberg Island, ARI—Amund Ringnes Island, BI—Bathurst Island, CI—Cornwallis Island, ERI—Ellef Ringnes Island, MI—Melville Island, PPI—Prince Patrick Island (Slaymaker and Catto ch 2)

If it survives as an open fissure into the summer, cracks in relatively wetter environments can be filled with liquid water, in addition to melted snow and ice pellets, all of which are subject to freezing during the following winter. Freezing produces a slender ice wedge. If the ice wedge survives the next summer, due to an insulating blanket of sediment, a peat mat, or detritus, it may gradually grow as additional interstitial water migrates and freezes along the margins of the wedge, or renewed cracking allows an influx from the surface. Repetitive freezing forces the surrounding host sediment apart by shearing, gradually broadening the wedge. The sediment along the wedge boundaries commonly has small normal faults, indicating that displacement has occurred.

Thermal cracks and ice wedges do not develop as isolated examples. Over time, cracking results in a polygonal pattern of intersecting wedges, which forms a network of ice-wedge polygons (Fig. 9.32).

Ice-wedge polygons (also called tundra polygons or patterned ground) vary in size and shape. Typical features are metres to 10 s of metres across, but some may exceed 100 m width. Geometry varies from rectangular or square, to irregular polygons, to hexagons. Adjacent polygons in the same area can exhibit significantly different forms, reflecting combinations of sediment, pre-existing topography, snowdrift accumulation, active layer thickness, interstitial ice content, and vegetation.

High-centred ice-wedge polygons have dry centres, without standing water or wetland vegetation. Summer melting of the uppermost part of the bordering wedges produces trenches of standing or slowly flowing water around the margins. Flowing water can contribute to overdeepening of the trenches, which together with movement of the water-saturated sediment flanks, further accentuates the shallow dome in the centre.

Fig. 9.30 Arctic Platform landscapes. Upper: regressive raised beaches, formed during glaci-isostatic uplift, Bathurst Island. Lower: Braided large river, Sabine Peninsula, Melville Island. The river is incised into late Holocene prograded fluvial deposits that formed in the shallow coastal setting. Subsequent glaci-isostatic emergence has resulted in base level lowering and incision. (Courtesy Scott Lamoureux and Ashley Rudy)

Low-centred ice-wedge polygons have wet central areas, which may have shallow pools of standing water or wetland vegetation. The surrounding trenches developed at the exposed tops of the ice wedges are flanked on either side by narrow rampart ridges of elevated sediment. Although the total difference in height between the water surface of the trench and the crest of the flanking rampart is minimal, characteristically less than 30 cm, the height of the rampart is sufficient to isolate the centre of the polygon from the trench, impounding the shallow pond. The ramparts appear to develop as sediment is moved by cryoturbation and thrusting from the centre of the polygon to the periphery, not by heaving along the ice-wedge trenches.

9.4.7.2 Thermokarst Features

Thermokarst is the subsidence of the ground surface in permafrost regions due to differential melting of ice or ice-rich sediments (French 2007, 2017a; French and Slaymaker 2012; Catto et al. 2020). Exposure and subsequent melting can result from a variety of natural causes. Physical removal of overlying or flanking material, by wave erosion, river cutting, slope failure, or wind action can expose ice-rich sediment. Removal of vegetation cover, particularly by fire, is another cause. Climate warming and increased precipitation are also factors in many areas (Smith 2011; Kokelj et al. 2015; Bush and Lemmen 2019; Slaymaker et al. 2020a).

In many cases, human activities also can cause or accelerate thermokarst development. Removal of vegetation by logging, burning, or through vehicle and foot traffic; cutting through ice-rich sediment banks during road construction; inducing erosion by diverting drainage; and melting the surface from inadequately insulated buildings, pipelines, or sewage lagoons have all resulted in local thermokarst development.

Shallow thermokarst lakes (also called thaw lakes) develop from local melting of ice-rich silty or sandy sediments, induced by combinations of thermal and mechanical erosion (Fig. 9.33). Typically, multiple generations of lakes occupy the landscape, including actively developing features and drained basins. The Mackenzie Delta is marked by numerous thermokarst lakes (Burn and Kokelj 2009; Burn 2017a). A thermokarst lake expands until erosion breaches a bank, causing drainage. Disturbance along the lakeshore due to vehicle, ATV, or foot traffic also accelerates erosion. Undercutting and drainage can also be triggered by erosion if the lake is adjacent to a river channel or an eroding marine coastline.

Exposure of ice-rich sediment and segregated ice lenses triggers slope failures. Active Layer Detachment Slides (ALDS) result from sliding of a cohesive sediment layer over permafrost, sometimes exceeding 5 m/h (Rudy et al. 2017). ALDS occur when friction between the cohesive layer and

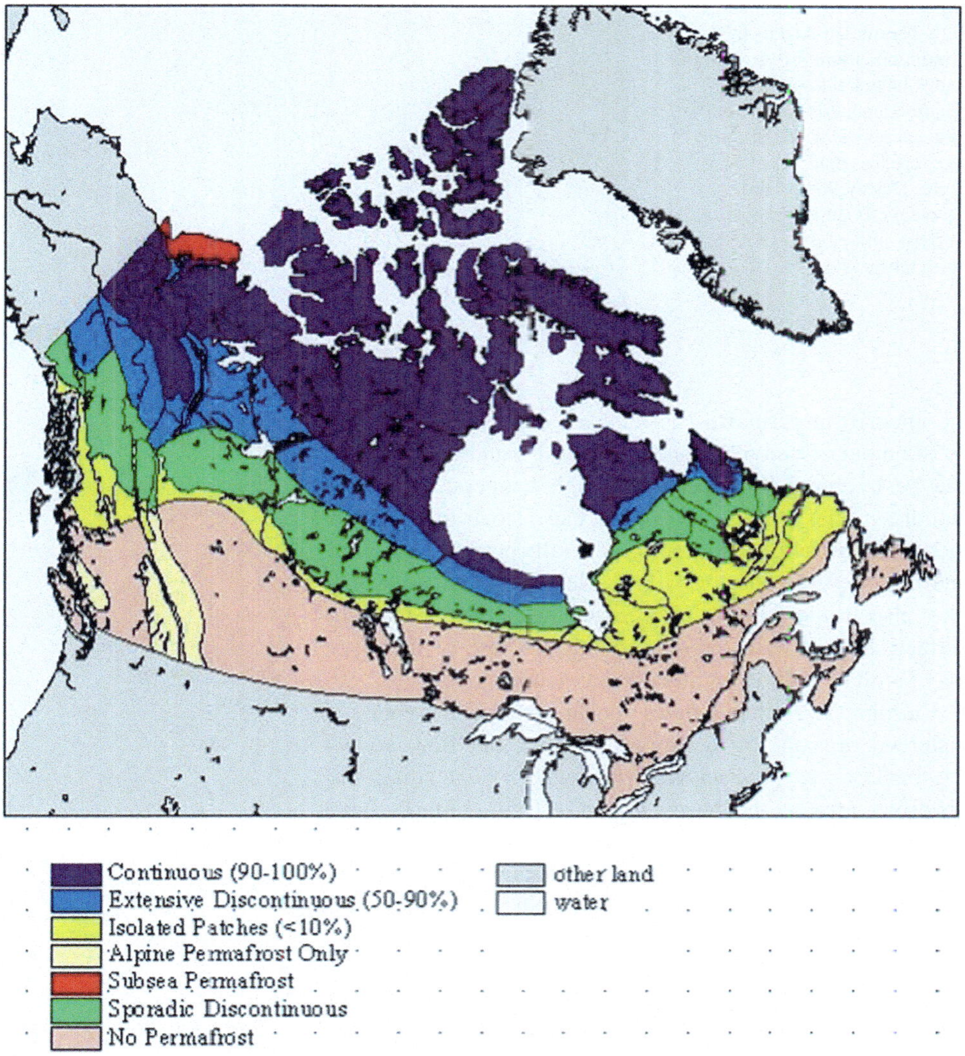

Fig. 9.31 Permafrost zones in Canada (Natural Resources Canada)

Fig. 9.32 High-centred ice-wedge polygons are bordered by ice wedges, which melt on the surface throughout the summer. Sea Otter River area, Banks Island

Fig. 9.33 Thermokarst lakes. Left: Thermokarst lakes in Continuous Permafrost, western Banks Island. Circular, square-edged, and compound forms are present. All of these lakes are less than 1.5 m deep. Right: Dewey Soper Bird Sanctuary, Foxe Basin coast, western Qikiqtaaluk. (NASA Earth Observatory)

the underlying permafrost is reduced, either by melting or by the input of rainwater, snowmelt, or groundwater. ALDS can be triggered by a period of high temperatures under cloudless skies. Moderate rainfall can also trigger ALDS movement. Heavier precipitation generally produces gullying and mudflow as cohesion is lost. Melting of interstitial ice may also trigger sliding of a frozen layer over a thawing surface. Higher water pressure along the base of the layer may facilitate motion on moderate to gentle slopes.

Retrogressive thaw flows occur where thawing of disturbed area at the head or crown of the flow allows the bank to erode backward (retrogress), extending the disturbed area at the headward end. These failures typically develop and flow rapidly, causing substantial erosion. Thawing of ice-rich sediment in the backwall can exceed 20 mm per hour. Thaw failures occur in silt or fine sand units with high ice content, either as interstitial or segregated ice lenses, along linear or gently curved rotational shear planes. Backwall slopes are steep, with a minimum angle of 20°, although most are in excess of 60°. Retrogression rates can reach 16 m per year.

Retrogressive thaw flow activity is particularly significant along marine coastal bluffs where ice-rich silts are exposed, rendering the shoreline susceptible to erosion (Lantuit and Pollard 2008; Ramage et al. 2017; Cunliffe et al. 2019; Fig. 9.34). Thawing is accentuated by thermal abrasion due to wave action, as well as melting induced by air temperature. Inundation of ice-rich sediments by marine water also accentuates melting. Higher sea surface temperatures result in increased thawing and coastal retreat. In cases where wave action removes sediment from the toes of coastal bluffs, new exposures of fresh, ice-rich sediment and steeper slopes promote enhanced thaw and retrogression. However, shorelines with less effective wave action may gradually accumulate sediment, prograding the shoreline, flattening the slope, and reducing the likelihood of large-scale retrogressive failure.

Coastal erosion resulting from permafrost thawing, reduced sea ice cover, and enhanced wave and storm activity, is a significant problem across all areas of the Canadian Arctic where sediments flank the shore (Forbes 2011; Allard and Pollard 2013; Forbes and Hansom 2011; Hatcher and Forbes 2015; Ford et al. 2021; MacLeod and Dallimore 2021). Sea level rise compounds the problem along the Beaufort Sea coast, notably at Qikiqtaryuk (Shaw et al. 1999; Burn 2012, 2017b; Radosavljevic et al. 2016).

Fig. 9.34 Upper: Coastal wave action along the Tuktoyaktuk Peninsula exposed this segregated ice lens, resulting in enhanced erosion. Lower: Coastal erosion along the Beaufort Sea breached this thermokarst lake basin, causing the lake to drain. Newly developed small thermokarst lakes have formed in the basin

Accelerating erosion at Tuktoyaktuk has necessitated relocation of infrastructure and discussion of remedial and adaptive measures (Catto and Parewick 2011). The highest point of the Tuktoyaktuk Peninsula is less than 25 m above sea level. Underlain by continuous permafrost, the peninsula contains numerous thermokarst lakes and other permafrost features. Sea level along the Beaufort Sea coast is currently rising at approximately 3.5 mm/y, submerging subsea permafrost.

The combination of rising sea level, increased storm and wave activity, and ice-rich coastal sediments expose the Tuktoyaktuk coastline to erosion. Average rates of recession approximate 1 m/y, with maximum rates reaching 10 m/y (Forbes 2011).

Ongoing climate change is predicted to result in flooding, increased thermal and mechanical erosion, beach migration, thermokarst lake breaching, and destabilization of sediments in the coastal zone at Tuktoyaktuk. Coastal protection measures including concrete slabs, rip-rap, sandbags, and beach nourishment are ineffective under current storm conditions. Continued shoreline retreat will impact many buildings within the next 20 years. Some areas of Tuktoyaktuk are predicted to suffer complete inundation during future storm events. Efforts are underway to relate residents to less exposed housing.

9.4.7.3 Pingoes

"Pingo" is derived from the original Inuvialuktun word describing a large, conical mound of unconsolidated clastic sediment. The Tuktoyaktuk (69.5°N, 133°W) area has more than 2,000 pingoes developed in silty lacustrine sediments, although most are less than 10 m high (Mackay 1979; Burn 2017a; Kunz and Kneisel 2021). Eight of the largest pingoes, including Ibyuk (49 m high, 300 m diameter), are protected in the federally designated Pingo Canadian Landmark" administered by Parks Canada (2022d; Fig. 9.35). Ibyuk has been used by the Inuvialuit residents of the Tuktoyaktuk area as a lookout for migrating caribou and marine life, and as an aid to navigation.

Pingo formation requires continuous permafrost, with mean annual air temperature no higher than −6 °C. Pingoes also are associated with aeolian, lake, or exposed marine silt or fine sand. Growth must be unrestricted by vegetation, or animal (including human) disturbance.

Pingo development results when a plug of unstratified ice formed below the surface is intruded upward as a solid plug. At Tuktoyaktuk, pingoes develop in recently drained shallow lake basins. As the exposed lake silt freezes, a plug of pure (segregated) ice develops, confined by a thin cap of frozen sediment, resting on a layer of liquid water. As freezing encroaches from the sides and top, the liquid water is compressed into a lens under high hydrostatic pressure. If hydrostatic pressure exceeds the confining pressure due to the weight of overlying sediment, the plug of segregated ice Is forced upwards, acting as a plunger in a cylinder. The intrusive ice plug deforms the thin sediment layer, creating a hydrostatic pingo. Compression cracks form at the base, with tension and dilation cracks forming a star-shaped pattern at the crown.

Although development of a pingo can be initiated upon lake drainage, growth of a 10 m high pingo requires centuries. Ibyuk reflects climate conditions that existed several thousands of years ago.

The intrusive ice lens will continue to rise until either the capping sediment layer fractures under tension, exposing the intrusive ice plug or failure at the base allows the groundwater

Fig. 9.35 Ibyuk Pingo, Tuktoyaktuk. The surface of the pingo is marked by large tension cracks. Some leakage of the high-pressure water lens at the base has occurred, accentuating the growth of tundra vegetation. High-centre ice-wedge polygons and thermokarst ponds are also visible (Courtesy Nat Rutter)

to leak to the surface, causing the water pressure to drop. A pingo cannot grow infinitely. Once the pingo ceases to grow, collapse results as the ice plug melts and the unsupported sediment cap fails. Pingoes can also cease growth as a result of lateral erosion along the coastline or by streams cutting into the flanks.

9.5 Vegetation Assemblages

Arctic vegetation communities can be divided into three major assemblages, influenced by combinations of latitude, altitude, and snow cover (Bliss 1997; Walker et al. 2005). Erect dwarf-shrub Low Arctic assemblages characterize areas adjacent to the treeline ecotone, including coastal Ivvalik NP, Tuktoyaktuk, Tuktut Nogait NP, Kangiqtiniq, Kuujjuaq, Torngat Mountains NP, Nunavik, Iqaluit, and the lower areas of Auyittuq NP. Prostrate dwarf-shrub mid-arctic communities occupy a west–east belt, including Aulavik NP, Usqsuqtuuq, Kuganajuup Qikiqtang, Sirmilik NP, and the higher areas of Auyittuq NP. Cushion-forb High Arctic assemblages occupy areas north of M'Clure Strait-Viscount Melville Sound–Tallurutiup Imanga, including Ilulliq, Qausuittuq, Tatlurutit, and Umingmak Nuna. Distinctive communities also occur in restricted areas with deep snowpacks, over broken rock felsenmeer, and in salt marshes.

Fig. 9.36 Erect dwarf-shrub low arctic species. Upper: *Salix uva-ursi* (bearberry willow). Lower: *Eriophorum angustifolium* (cottongrass)

9.5.1 Erect Dwarf-Shrub Low Arctic

The treeline is marked by *Picea mariana* (black spruce) in NWT and Kivalliq, and *Picea glauca* (white spruce) in Nunavik, the Torngat Mountains, and coastal Yukon. Scattered shrubs of *Salix arctica* (arctic willow), *Betula glandulosa* (tundra birch), *Betula borealis* (northern dwarf birch), *Betula pumila* (swamp birch), *Betula minor* (dwarf white birch), and *Alnus viridis crispa* (green alder) mark the Erect Dwarf-Shrub community (Porsild and Cody 1980; Ryan 1995; Smith et al. 2004; Baldwin et al. 2019; Montesano et al. 2020; Fig. 9.36). *Salix* and *Betula glandulosa* commonly do not exceed 60 cm in height, but may survive more than 100 years. *Salix* leaves were important supplements to traditional diets, as they are rich in Vitamin C. Although *Salix* and *Betula glandulosa* are common in the Canadian low- and mid-arctic, *Betula nana nana* (arctic birch) is confined to Qikiqtaaluk in Canada.

Other constituents of the southernmost tundra assemblage include *Empetrum nigrum* (crowberry), *Oxyria* (sorrel), *Polygonum* (knotweed), *Chamerion latifolium* (broad-leaved fireweed), *Pyrola* (wintergreen, generally on the most favourable sites), *Rhododendron decumbens* (Labrador tea), *Arctostaphylos* (bearberry), *Potentilla* (cinquefoil), *Eriophorum* (cottongrass), *Vaccinium vitis-idaea* (bog cranberry), *Cassiope tetragona* (arctic white heather), *Silene* (moss campion), *Diapensia lapponica* (confined to gravelly zones), *Dryas integrifolia* (mountain avens), and *Saxifraga oppositifolia* (Purple saxifrage, floral emblem of Nunavut; Fig. 9.37). *Azalea loiseleuria* and *Rhododendron lapponicum* (Lapland rosebay) do not occur north of southern Qikiqtaaluk. Many of these plants are edible, and some are of great medicinal value, especially sorrel, which prevents scurvy and was known to European visitors as "arctic salad". Turbic and static cyrosols

Fig. 9.37 *Saxifraga oppositifolia* (purple saxifrage)

Fig. 9.38 Arctic shoreline vegetation. Upper left: *Rhodiola rosea* (golden root). Upper right: *Dryas integrifolia* (dryas). Lower left: *Saxifraga oppositifolia* (purple saxifrage). Lower right: *Papaver nudicale* (arctic poppy)

and regosols dominate the soil assemblage (Soil Classification Working Group 1998).

The shore communities contain salt-tolerant plants which can endure repetitive freeze-thaw cycles, and prefer gravelly substrates (Fig. 9.38). The assemblage includes *Diapensia lapponica, Armeria maritima* (thrift), *Arenaria* (sandwort), and *Cochlearia* (scurvy grass). The bright orange lichen *Caloplaca* is present in areas where birds nest and provide ample nourishment through their excrement to the lichens (Parks Canada 2022b). Soils are poorly developed regosols.

9.5.2 Prostrate Dwarf-Shrub Mid-Arctic

Prostrate Dwarf-Shrub communities are dominated by *Salix arctica* (Fig. 9.39), with *Betula nana exilis* (arctic birch) and *Betula glandulosa* only present in the most favourable, southerly sites (Edlund 1983; Larter et al. 2009; Fontaine and Mallory 2011; Parks Canada 2022a). Common plants include *Cochlearia groenlandica* (scurvygrass), *Empetrum nigrum* (crowberry), *Oxyria* (sorrel), *Polygonum* (knotweed), *Rhododendron decumbens* (Labrador tea), *Arctostaphylos* (bearberry), *Potentilla* (cinquefoil), *Eriophorum* (cottongrass), *Parrya arctica, Puccinellia* (goosegrass), and *Vaccinium vitis-idaea* (bog cranberry; Porsild 1964). Although the herbaceous species assemblages do not differ greatly from those of the Erect Dwarf-Shrub region, the percentage of ground cover is greatly reduced, with vegetation frequently confined to the most favourable areas.

Fig. 9.39 Prostrate dwarf-shrub communities include *Salix arctica*

In more rigourous climate areas, *Salix* disappears (Bliss and Svoboda 1984; Bliss et al. 1984; Edlund and Alt 1989). However, *Salix arctica* survives in the thermal oasis flanking Hazen Lake on Umingmak Nuna (Parks Canada 2022f).

Fell-field communities develop in areas where frost heaving has disturbed the surface of the land, producing areas of shattered and thrusted blocks of rock ("felsenmeer") and areas of cryogenic hummocks where more sediment was initially present (Fig. 9.40). Few plants are present, the most noticeable being *Silene* (moss campion) and the bright yellow arctic poppy (*Papaver nudicale*). Scattered lichens

Fig. 9.40 Fell-field communities develop in areas where frost heaving has disturbed the surface of the land, producing felsenmeer

and bryophytes (mosses) are also present. Soils are poorly developed, and are classified as regosols and turbic cryosols (Tedrow 1974; Soil Classification Working Group 1998).

The snow-patch communities are the most areally restricted assemblages. These communities require deep accumulations of snow for their development, but also require lesser amounts of frost heave disturbance and thicker soils. The dominant plant types are *Saxifraga* (saxifrage) and *Ranunculus* (buttercup). Soils in these regions are static cryosols.

9.5.3 Cushion-Forb High Arctic

Cushion-forb vegetation assemblages characterize High Arctic settings such as Qausuittuq (Porsild 1964; Sheardj and Geale 1983; Parks Canada 2022e). The assemblages resemble those of mid-arctic regions, differing primarily in the size of individual plants and the proportion of vegetation cover of the surface. Typically, less than half of the exposed sediment is covered by vegetation (Fig. 9.41).

In these environments, *Silene acaulis* (moss campion) forms cushion-like mats 10–30 cm in diameter, with individual stems extending 2–5 cm above the surface (Porsild 1964). Stems tend to have similar heights above the surface, so that no one stem is more exposed. *Phippsia algida, Cassiope tetragona* (arctic heather), *Oxyria* (sorrel), *Potentilla* (cinquefoil), and *Empetrum nigrum* (curlewberry) also adopt cushion-like forms in these habitats. Locally, *Papaver cornwallisensis* (Cornwallis poppy), *Papaver radicum* (arctic poppy), *Eriophorum* (cottongrass), *Saxifraga* (saxifrage), *Arctanthemum integrifolium* (entire-leaved daisy), and lichens such as *Arctoparmelia centriguga lapponicum* are present. Shoreline communities contain salt-tolerant genera such as *Arenaria humifusa* (sandwort).

In High Arctic regions, freeze-thaw cycles are not common, and all of the terrain inland from the coast has developed continuous permafrost. Small areas of turbic cryosols, where seasonal freezing and thawing of the uppermost 5 cm of sediment ("active layer") occurs, are developed in low-lying areas, primarily adjacent to the major rivers and the deep coastal embayments. Small areas of patterned ground and

Fig. 9.41 Cushion-Forb vegetation communities occupy High Arctic settings (upper left), marked by periglacial processes and semi-permanent snowpacks (lower left). Upper right: *Silene acaulis* (moss campion). Lower right: *Cassiope tetragona* (arctic heather; National Parks Service)

gelifluction features are associated with these turbic cryosol areas.

The short growing season (typically less than one month) prevents most vegetation from establishing itself in northern Umigmak Nuna (Lévesque1997; Desjardins et al. 2021). Most areas are devoid of any vegetation cover. The few plants which do survive in sheltered areas include lichens, bryophytes, and *Silene acaulis*. Soil development is minimal, with regosols dominating and static cryosols present on the thicker sedimentary successions.

9.6 Fauna

9.6.1 Musk Ox

Musk ox (*Ovibos moschatus;* umimmak) is an iconic Arctic Species (Banfield 1981; Naughton 2012; Fig. 9.42). Aulavik National Park was intended to preserve terrain suitable for musk ox (Ferguson 1991, Parks Canada 2010, 2022a). *Ovibos moschatus* inhabits the tundra zone of the Beaufort coastal plain.

Throughout Banks Island, musk ox populations oscillate substantially, in common with those on other Canadian Arctic islands (Nagy et al. 1996; Larter and Nagy 2001; Davison et al. 2013). The species appeared to be nearing extinction in 1915, with the populations on both Banks Island and Kitlineq estimated at almost zero. Although overhunting was blamed at that time (and has been cited on several occasions since), recent analysis has suggested that the simultaneous decline throughout all of the arctic islands can more plausibly be attributed to climatic fluctuations. Anomalous snowfalls are particular threats to musk ox, as they are unable to clear the snow for grazing.

In 1930, the population of Kitlineq was estimated at 30 musk ox. Numbers rose slowly, to approximately 900 in 1961. This number, however, has proven to be a sufficient critical mass for breeding, as the population swelled to 12,000 in 1980 and possibly as high as 30,000 in 1991. Simultaneously, growth brought the Banks Island population to more than 40,000 animals, with more than 10,000 living within Aulavik (Davison et al. 2013). Fluctuations in musk ox populations not related to human hunting have been documented throughout the Arctic Islands.

In the last ten years, the musk ox population on both islands has dropped again: some estimates suggest that the musk ox population in 2022 is currently <60% of those living in the mid-1990s. The Banks Island population, however, appears to fluctuate between a low of ~11,000 and a high of ~70,000 animals (e.g. Davison and Baryluk 2021).

Musk ox numbers on Bathurst Island have remained relatively low since the early 1900s, in contrast to the changes registered on Kitlineq and Banks Island (Ferguson 1987). From a population estimated at 100 in 1958, numbers

Fig. 9.42 *Ovibos moschatus* (Musk ox). Upper: Male Stand-off. (National Parks Service, USA). Lower left: Typical habitat, eastern Banks Island. Lower right: *Ovibos moschatus* is a key food and cultural resource for Sachs Harbour residents (Kathleen Parewick photo)

increased to ±1100 in 1961. By March 1973, the population had stabilized at approximately 670 animals, but a hard winter in 1974 caused the population to drop to <200. Since 1974, the population has gradually risen to 600, remaining stable since 1990.

Ovibos moschatus travels in small herds with 6–12 individuals, although larger groups may form temporarily. Male musk ox stand 150 cm high at the shoulder, and have masses up to 400 kg. The males are formidable opponents for wolves, and even polar bears could not successfully tackle an alert male. When threatened, a herd of musk ox will form a defensive circle, with the males facing outward and the females and calves positioned in the centre. The males tend to remain stationary until the threat has passed, although some young males will occasionally charge directly at the threat. The charging musk ox almost always travel in straight lines. This defensive posture works well against single predators, and against packs of wolves that cannot break the circle of bulls, but it was disastrous when applied against humans armed with rifles. The musk ox will not break the defensive circle, even when one animal is wounded or killed, and hunters thus were able to systematically shoot all of the animals in a circle. Most early Arctic European explorers relied heavily on the musk ox for food and fur.

Although musk ox can traverse rivers, they cannot swim long distances. Hudson Strait precludes migration of *Ovibos moschatus* to northern Labrador and Nunavik. *Ovibos moschatus* was formerly present in the Lake Hazen area of Quttinirpaaq National Park (Parks Canada 2022f). The species was utilized extensively by the polar explorers Greely and Peary as a source of food. The population appears to have collapsed during the period between 1910 and 1920, as occurred elsewhere in the Arctic Islands. Unlike most other regions, however, muskoxen have been unable to re-colonize northern Umigmak Nuna, although they are present in the south (Anderson and Kingsley 2015).

Ovibos moschatus also are absent from Qikiqtaaluk (Barr 1991). Inuit oral history suggests that the species was present in the fjord valleys of southern Qikiqtaaluk until *ca* 1450 ce. This date coincides with the timing of a general climatic cooling in the northern North Atlantic, which precipitated glacial readvances in European and North American mountains and hastened the disintegration of the Norse colonies on adjacent Kalaallit Nunaat. Musk ox disappearance from Qikiqtaaluk may thus be linked to renewed glacial (Little Ice Age) activity.

9.6.2 Caribou

Barren-ground (*Rangifer tarandus groenlandicus;* tuktu) and Peary caribou (*Rangifer tarandus pearyi*) also are iconic species of the Canadian Arctic (Banfield 1981; COSEWIC 2004a, b; Fauchald et al. 2017). Barren-ground caribou occupy terrain from the northern taiga to Viscount Melville Sound, from the Beaufort Sea coast of Ivvavik to Sanirajak and Qikiqtaaluk. In Ivvavik, the tundra supports barren-ground caribou, whereas woodland caribou live in the taiga areas, particularly in Vuntut (Parks Canada 2022c; Fig. 9.43).

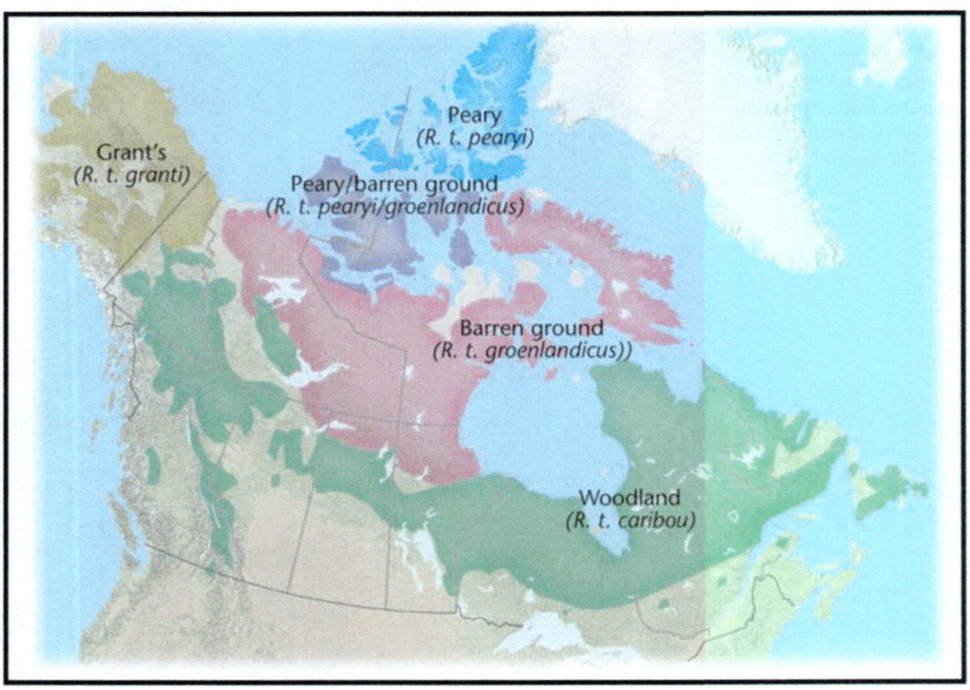

Fig. 9.43 Caribou species in Canada (COSEWIC et al. 2011b) (Environment and Climate Change Canada, by permission)

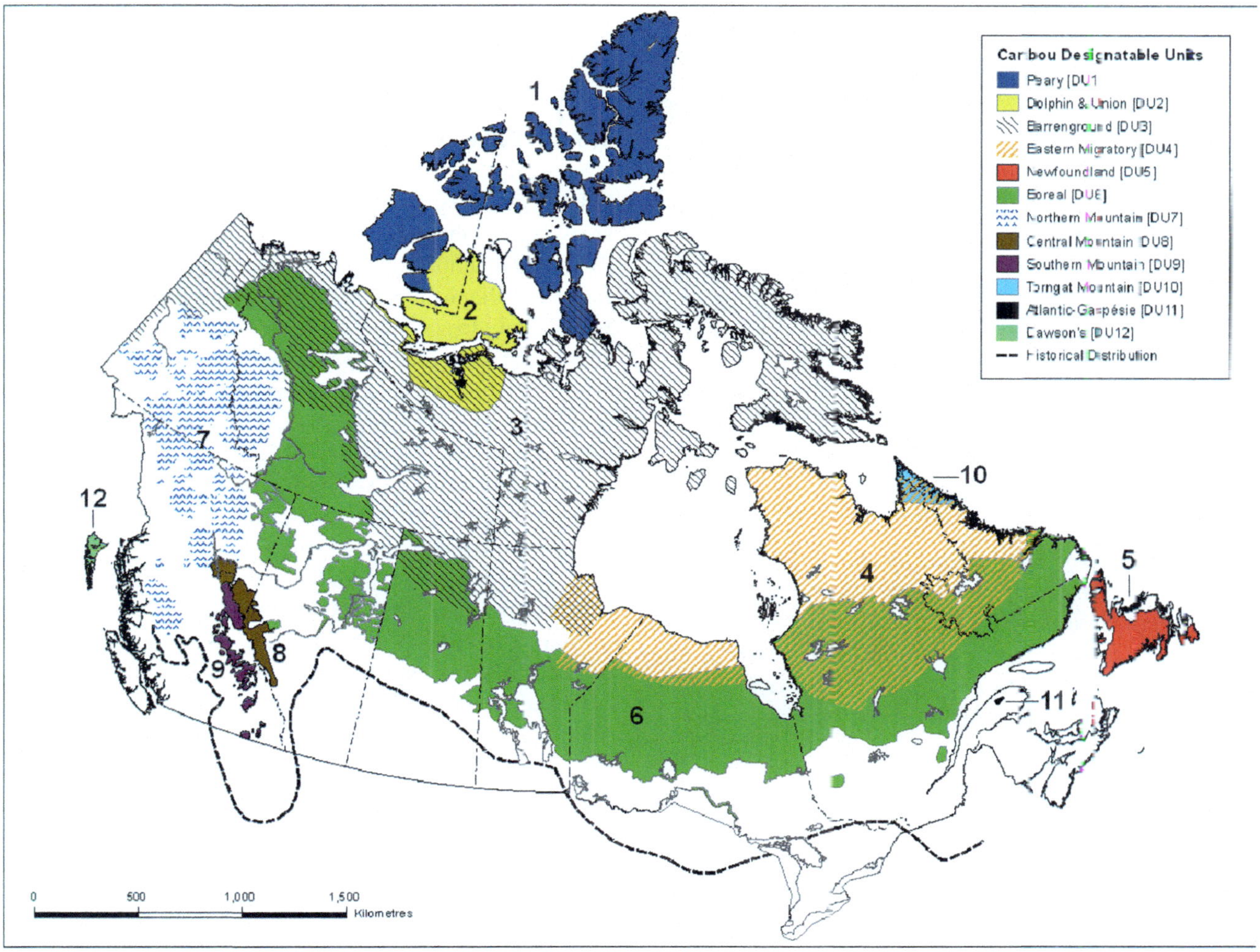

Fig. 9.44 Caribou Designatable Units list caribou populations (herds) by region and genetic affinity (COSEWIC et al. 2011). (Environment and Climate Change Canada, by permission)

Caribou of both species migrate freely throughout the Beaufort coastal region on a seasonal basis.

The Dolphin and Union herd is a genetically distinct population, resulting from interbreeding of barren-ground and Peary caribou (COSEWIC 2004a, b; Fig. 9.44). Named for Dolphin and Union Strait, the herd migrates over the terrain of Kitlineq and western mainland Nunavut from Kugluktuk to Kingauk (Bathurst Inlet), moving south during the winter. East of Tuktut Nogait NP, their migratory range overlaps with that of the Bluenose East Herd. In 2015, the Dolphin and Union herd was estimated to have between 18,000 and 14,000 animals, a substantial decline from the 1997 estimates of 28–35,000 (Leclerc 2015). However, caribou herd populations can fluctuate substantially from strictly natural causes, and limited pre-existing data make conclusions as to long-term trends difficult.

On Qikiqtaaluk, the rugged topography and harsh interior climates confine barren-ground caribou to areas below 1200 m asl. Based on population distributions, limited genetic data, and extensive Inuit qaujimajatuqangit (traditional ecological knowledge), there are between 3 and 5 distinct herds (Government of Nunavut 2018). However, the total caribou population has fallen sharply since 1990, and there are currently less than 5,000 animals throughout Qikiqtaaluk.

Peary caribou (Fig. 9.45) is a distinctive subspecies of the Canadian Arctic Archipelago (Nagy et al. 1996; Larter and Nagy 2001; COSEWIC 2015). Populations live on Kingngailap Nunanga (Boothia), Kuganajuup Qikiqtang (Somerset Island), Banks Island, Kingailik (Prince of Wales Island), Umingmak Nuna (Ellesmere), and Qausuittuq. Peary caribou are much smaller than other species, with mature males averaging 110 kg and 1.7 m from nose to tail. Unlike barren-ground caribou, Peary caribou do not undertake long seasonal migrations. The total population in Canada is about 13,000 individuals. The available data suggest that, although protected by the federal Species-at-Risk-Act (SARA), Peary caribou numbers have declined since 1980, and the species is currently listed as "threatened" in Canada (COSEWIC 2015).

Fig. 9.45 *Rangifer tarandus pearyi* (Peary caribou). (NOAA, L. David Mech)

The total population has declined by about 37% since 1980, although the Qausuittuq population has shown periods of growth over that time (Parks Canada 2022e). On Umingmak Nuna, Peary caribou numbers are increasing slowly, after almost being extirpated by Peary and colleagues in the early twentieth century. Deeper snow cover and icing events in winter, preventing the caribou from reaching the ground cover vegetation food source, are the major natural threats.

9.6.3 Polar Bear

Polar bears (*Ursus maritimus*, Nanuk) are present in most areas below about 1200 m asl (Banfield 1981). Distinctive subpopulations live in the coastal areas of Davis Strait, Baffin Bay, Ilagiit nunagivaktangit, the northern Beaufort Sea, and several inter-archipelago areas, in addition to the better-known and more-studied populations of Hudson Bay (Jensen et al. 2020; Fig. 9.46). On Qausuittuq, the Tallurutiup Imanga (Lancaster Sound) population uses Polar Bear Pass as a convenient overland shortcut between the western Arctic and Baffin Bay, in preference to swimming around the island (Parks Canada 2022e). The bears travel through the pass steadily, often without disturbing the surrounding fauna.

At present, the total population of polar bears in Canadian waters is estimated at 16,000 (Government of Nunavut 2021). In sharp contrast to the declining number and health of the southernmost populations in Hudson Bay (see Chap. 8) and the Yukon Beaufort Sea coast (Bromaghin et al. 2021), the

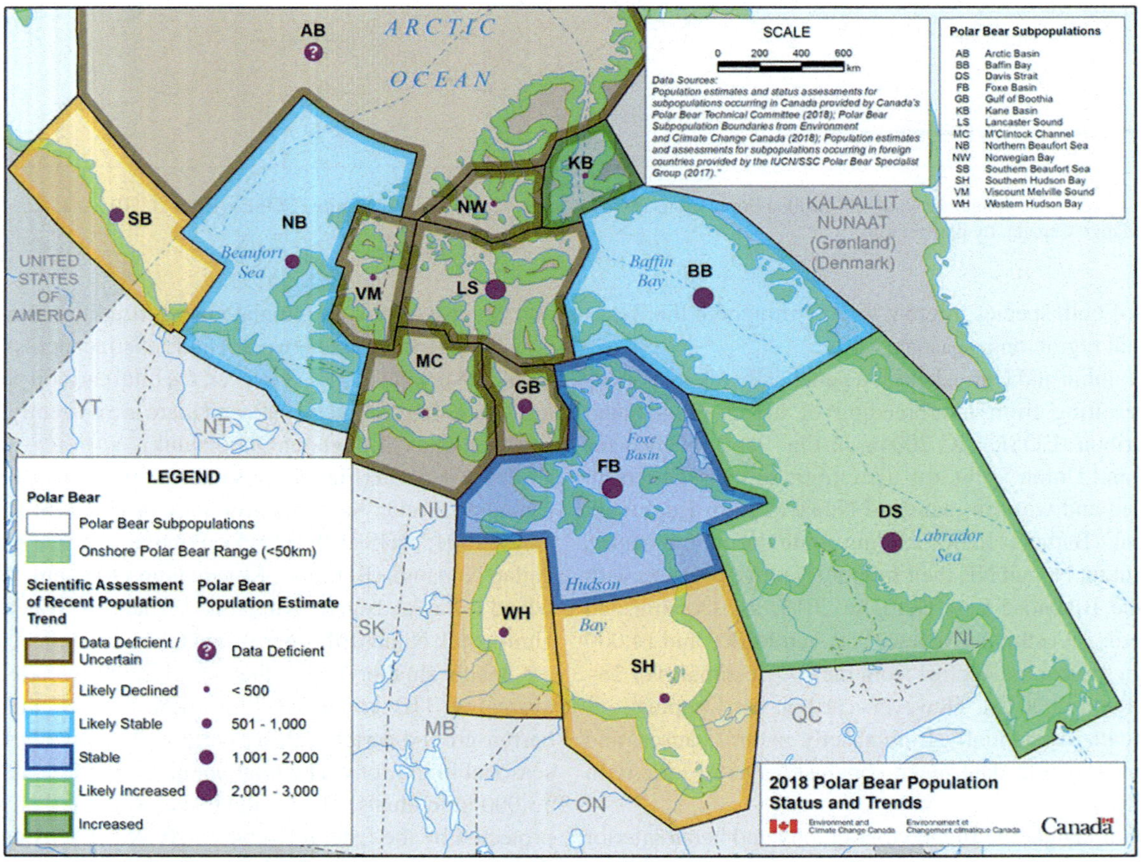

Fig. 9.46 Polar Bear Population Status and Trends, 2018. See text for discussion. (Environment and Climate Change Canada, by permission)

available evidence indicates increasing populations in Davis Strait (coastal Nunavik, Torngats, and Qikiqtaaluk), and stable or likely stable populations in Ilagiit nunagivaktangit, Baffin Bay, and the Beaufort Sea coast of NWT. Additional data is required to assess trends for the Tallurutiup Imanga population, and those of the other inter-archipelago regions (Government of Nunavut 2021). Ongoing climate change is currently improving habitat for the more northerly populations.

Inuit qaujimajatuqangit provides guidance to local residents concerning the health of polar bear populations. Polar bear harvesting is strictly regulated, and confined to the indigenous populations of NWT and Nunavut.

9.6.4 Other Terrestrial Species

Other Arctic carnivores and scavengers are wolf (*Canis lupus;* amaruq), Arctic Fox (*Vulpes lagopus;* Tiriganiaq or KaKuttâsuk), Ermine (*Mustela richardsoniii;* tigiak), and wolverine (*Gulo gulo;* kapviq). Arctic foxes prey primarily upon small rodents, and the arctic wolves attack caribou and arctic hares (*Lepus arcticus*).

Lepus arcticus (arctic hare; Ukaliq) is a common herbivore throughout the area, extending to Quttinirpaaq National Park (Parks Canada 2022f). With mass to 5 kg, arctic hares are generally larger than arctic foxes. Establishment of weather and military installations, such as Alert, have resulted in population increases as the animals scavenge. At Alert, the scarcity of vegetation in the area has influenced the arctic hare's lifestyle, and many have become cannibalistic.

Two other common herbivores, Collared Lemming (*Dicrostonyx torquatus;* ivinngaq) and Brown Lemming (*Lemmus trimucronatus*), are common prey species for most of the region's predators (Banfield 1981; Fig. 9.47). Lemmings undergo periodic population explosions, with numbers commonly tripling within a breeding season. When the carrying capacity of the land is exceeded, the lemming population crashes dramatically, and the subsequent year is marked by very few lemmings and many hungry predators. The explosions and crashes are not simultaneous throughout the arctic lemming's vast range, but depend upon local weather conditions and the availability of food. In the Tuktoyaktuk area, populations appear to have expanded and contracted on approximately 3–5 year cycles within the past 60 years.

As food supplies within one region begin to dwindle, lemmings will often attempt to migrate *en masse* in order to find nourishment. They are powerful swimmers, and will cross small ponds without difficulty (although they cannot survive for long in sea water). These migrations have given rise to the myth that lemmings periodically commit suicide as the food supply runs low. In reality, lemmings do not appear to exhibit this degree of social responsibility.

Fig. 9.47 *Lemmus trimucronatus* (brown or collared lemming). (Wikimedia/Robert Macleod)

Arctic char (*Salvelinis alpinus;* ikaluk) is present in many freshwater bodies. Hazen Lake in Quttinirpaaq contains a small population of arctic char, which entered the lake when sea levels were elevated following deglaciation and have since become isolated (Babaluk et al. 2001). The char can survive the winters, as the headwaters of Hazen Lake do not completely freeze over.

9.7 Marine Environments

9.7.1 Northern North Atlantic

The northernmost North Atlantic Ocean, washing Qikiqtaaluk (Baffin), Tatlurutit (Devon), and Umigmak Nuna (Ellsemere), can be divided into several components. The waters of Baffin Bay (*Saknirutiak Imanga*) and Davis Strait are fed by the West Greenland Current from the North Atlantic, by southward flow from Nares Strait, and by eastward flow through Ausuituu Tariunga (Jones Sound). Tallurutiup Imanga (Lancaster Sound) between Qikiqtaaluk and Tatlurutit is marked by interchange of North Atlantic and interior Arctic archipelago waters (Figs. 9.48 and 9.49).

Nares Strait, including Robeson Channel, Hall Basin, Kennedy Channel, Kane Basin, and Smith Sound, separates Umigmak Nuna from Kalaallit Nunaat, with width varying from 30 to 150 km. Southward current flow, dominated by the Beaufort Gyre, moves seasonal ice southward and deflects the West Greenland Current, closing this potential southern door to the Arctic Ocean.

Hans Island (80.8°N, 66.5°W; 1.3 km^2, devoid of vegetation) is located in a 35-km-wide reach of Kennedy Channel. Although the remainder of the maritime boundary between Canada (or behalf of Nunavut) and Denmark (on behalf of Kalaallit Nunaat) has been settled, Hans Island remained in

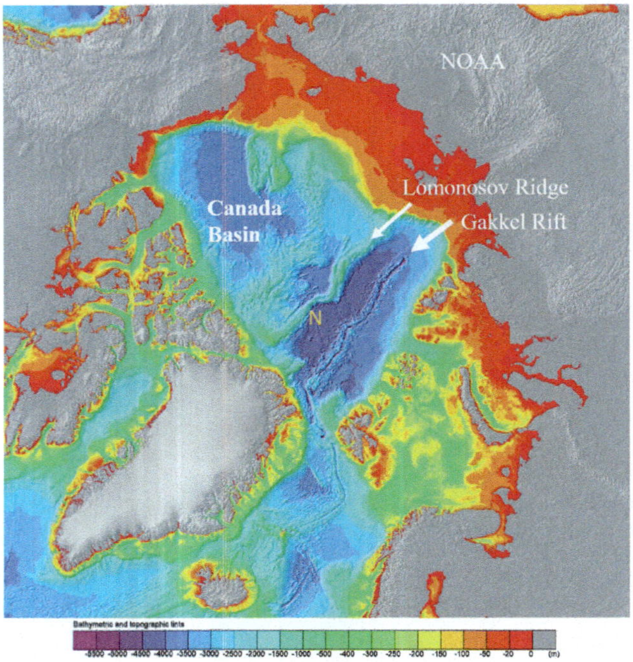

Fig. 9.48 Arctic Ocean bathymetry. Gakkel Rift marks the tectonic boundary between the North American and Eurasian Plates. Lomonosov Ridge is considered as part of the Continental Shelf. N: North Pole (NOAA)

dispute until June 2022, when it was divided in half. During the dispute, both sides planted flags and left bottles of liquor for the other on Hans Island.

9.7.1.1 Tallurutiup Imanga

Tallurutiup Imanga (Lancaster Sound) separates Qikiqtaaluk and Bylot Islands from Devon Island (Tatlurutit) to the north, and is the eastern entrance to the shipping route through the Arctic islands (Kochanowicz et al. 2021; Rattue 2021; Halliday et al. 2022). Tallurutiup Imanga National Marine Conservation Area (108,000 km^2) encompasses critical marine habitat (Parks Canada 2019). It is home to a large stock of Arctic cod (*Arctogadus glacialis*), in addition to numerous other fish species. Marine mammals inhabiting coastal waters include ringed and beaded seals; walrus; beluga, orca, bowhead, and humpback whales; and the faunal emblem of the Northwest Territories, the narwhal (*Monodon monoceros*, Qilaluga)(Mansfield 1967; Reeves 1988).

Approximately 45,000 walrus (*Odobenus rosmarus;* aaveq) live throughout Arctic Canada (COSEWIC 2017; Fig. 9.50)). A mature male may be as much as 3 m long and weigh 900 kg. Walruses live up to 25 years. Their preferred habitat is moving pack ice over shallow water feeding grounds. The large tusks are used for digging invertebrates from the ocean floor. A mature male will eat up to 3000 clams each day.

Fig. 9.49 Arctic Ocean currents (Woods Hole Oceanographic Institution)

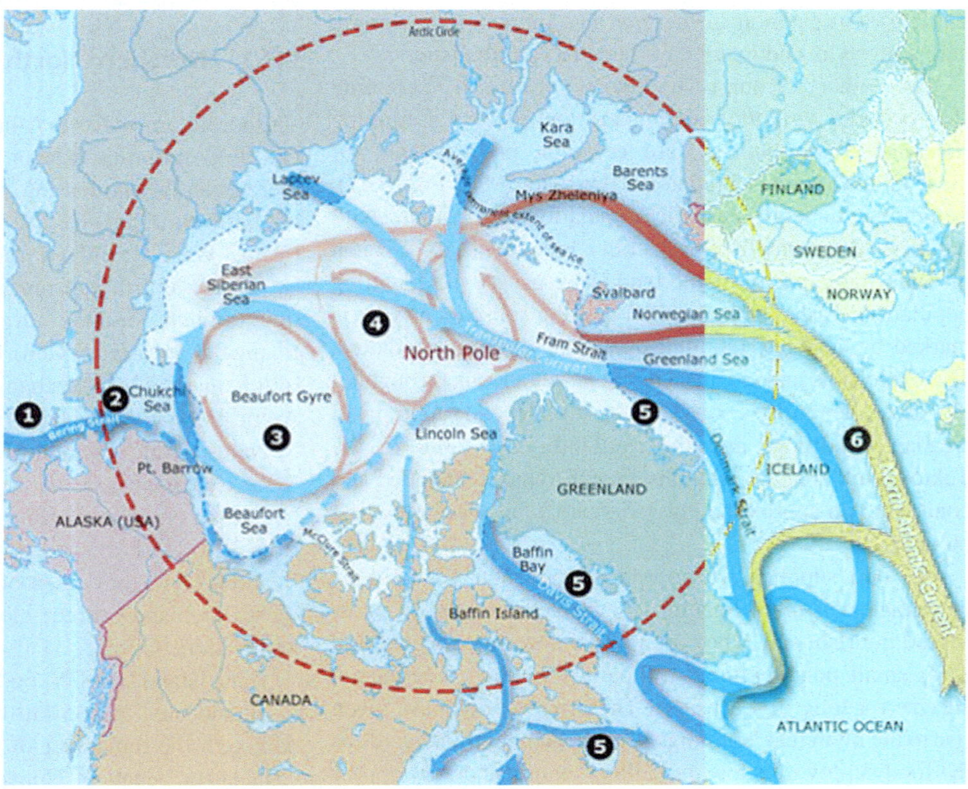

Fig. 9.50 *Odobenus rosmarus* (walrus) live throughout Arctic Canada, occasionally venturing south along the Labrador Sea

Fig. 9.51 Arctic cetaceans. Upper: *Balaena mysticetus* (bowhead whale; NOAA). Lower: *Monodon monoceros* (narwhal; Kristin Laidre, NOAA)

Most of the world's 70,000 narwhal live in or travel through Tallurutiup Imanga (COSEWIC 2004a, b, Parks Canada 2019, Louis et al. 2020). "Narwhal" comes from the Old Norse term meaning "corpse whale", referring to its pale white skin. It is a close relative of the beluga. The Narwhal's conspicuous "horn" is an overextended canine tooth. All mature males have at least one extended canine, and some have a second, shorter extended tooth. Mature females may have a short extended tooth, but most lack this feature. Among the suggested uses of the tusk, sexual and physical display seems to be the most reasonable proposed. A large male may reach 6 m and 1.5 tonnes, and can live for 40 years. Narwhals eat squid, skate, flounder, halibut, turbot, and cod. Orcas and walruses are predators.

Beluga (*Delphinapterus leucas;* qilalugaq; Fig. 9.51) is similar in dimensions to the Narwhal, and the two species appear to be closely related evolutionarily (Reeves 1988). The name comes from the Russian word for "white", and belugas were among the most common whales in subarctic and boreal waters throughout the Northern Hemisphere prior to the mid-twentieth century. Although hunted by whalers for its 40% blubber content, beluga was seldom the preferred target. Belugas are found in coastal embayments and estuaries. They prefer open pack ice, especially areas with open leads (polyna, ashkui). Belugas use their long fibrous dorsal ridge to break through thin ice to create a breathing hole. Belugas are opportunistic feeders, frequently taking grilse and spawning salmon. They are preyed upon by orcas, walruses, and polar bears.

Sightings are relatively common, perhaps because the white beluga is very conspicuous. However, many populations of belugas in eastern Canadian waters have been in decline since the late 1700s, especially after 1900. Rates of decline have varied throughout the different population groups. The population in Kangiqtualuk (Cumberland Sound) declined rapidly in the early 1900s, and may number less than 1000 individuals (Marcoux and Hammill 2016). Tallurutiup Imanga is home to about 25,000 beluga, approximately 20% of the total population in Canadian waters.

Orcinus orca (aarluk) are common, primarily in shallow coastal environments. Orcas are opportunistic feeders, eating fish, squid, seabirds, small seals and occasionally porpoises, dolphins, and immature whales. Males reach 8 m in length and 9 tonnes in mass. The species is notable for its longevity, with some females living for more than 100 years.

The humpback whale (*Megaptera novaeangliae;* keporkak), a long-distance migrator, was a preferred whaling target, and is currently listed as of Special Concern (COSEWIC 2011a). They inhabit coastal waters, feeding on krill, squid, and capelin. An adult male humpback may have a mass of 20 tonnes, a length of 15 m, and can live for 80 years. Humpbacks communicate by singing, primarily as a competitive sport between males. The tail flukes are also used in sexual competition by the rival males. The configurations of the tail flukes are unique to each individual humpback.

An indication of Quaternary climate fluctuations is provided by fossils of the Bowhead whale (*Balaena mysticetus;* arvik). The summer range of the bowhead is controlled by its requirement to breathe, which means that it can only frequent ice-free channels (Dyke et al. 1996). ^{14}C dated bowhead whale remains can thus be used as proxy data to determine which channels were open at particular times. Between 12000 and 9500 years ago, relatively arid conditions led to ice-free channels. During this period, bowheads spread across all of the central Arctic. A climate shift to colder, wetter conditions between 9500 and 5500 years ago led to the development of pervasive summer sea ice in the central arctic, forcing the bowheads out.

Renewed climate warming between 5500 and 3000 years ago allowed bowheads to extend their range from Qikiqtaaluk into the central Arctic as the extent of summer sea ice diminished. Most of the range expansion occurred from east to west, as ice remained as an obstacle in M'Clure Strait and M'Clintock Channel. Subsequently, climate conditions worsened, once again dividing the bowheads into the separate populations that exist today. The greatest restrictions in the summer range occurred *ca.* 1870–1900 AD, co-incident with "Little Ice Age" advances on Bylot Island. During the twentieth century, progressive climate warming has allowed bowheads to expand the ranges again.

9.7.1.2 North Water Polynya (*Sarvarjuaq)*

North Water Polynya (*Sarvarjuaq)* extends offshore of Ausuituu Tariunga and Tallurutiup Imanga (Barber et al. 2001; Tang et al. 2004; Oceans North Conservation Society 2018; Fig. 9.52). Exceeding 85,000 km^2 at its greatest extent, the polynya remains as open water throughout the year, with higher salinity (32–34 ppt) and a warmer climate than the surrounding areas. The polynya forms a resting and feeding area for cetaceans, walrus, fish, and seabirds. Wind and tidal activity drive the relatively warm West Greenland current onshore. Seasonal upwelling of colder, less dense currents from Ausuituu Tariunga and Tallurutiup Imanga, combined with overturning of the warmer, more saline West Greenland waters, enhance microplankton productivity. North Water is the most biologically productive area of the northern North Atlantic north of the Grand Banks. The resources of North Water are critical to the lifestyles of the traditional Inuit communities of Ikpiarjuk (Arctic Bay) and Mittimatalik (Pond Inlet), as well as Ausuiktuq (Grise Fjord).

The lack of fauna kept Inuit from occupying Umigmak Nuna, until a group from Inukjuak (Port Harrison, Inoucdjouac), Québec was semi-forcibly resettled in Grise Fjord (Ausuiktuq) in the south by the federal government during the 1950s (Cavell 2021). The Inuit were offered inducements to settle in Grise Fjord, but had a limited number of options and cannot be considered to have made the move freely; as they were moved from the southernmost Canadian Arctic to the northernmost area, they were also unfamiliar with the environment. The establishment of this community on southern Umigmak Nuna was essentially a political manoeuvre, designed to establish Canadian sovereignty. The federal government has now acknowledged its responsibility and its obligations in this regard, and has issued a formal apology to those involved and their descendants.

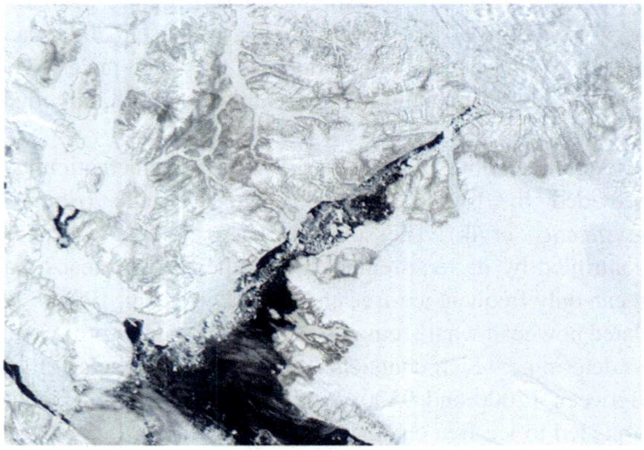

Fig. 9.52 North Water Polynya (Sarvarjuaq), Spring 2019 (NOAA)

The entire North Water polynya has been proposed to be designated as a Marine (or Indigenous) Protected Area. At present (2022), the Nirjutiqavvik National Wildlife Area in Ausuituu Tariunga (Cobourg Island, Lady Ann Strait, 1,650 km^2) conserves habitat for approximately 385,000 thick-billed murres, black-legged kittiwakes, and other marine birds (Environment Canada 2019a). The Nirjutiqavvik polynya remains open throughout the year, expanding during the summer to coalesce with North Water. Nirjutiqavvik is co-managed by the Ausuiktuq Co-management Committee. Inuit residents can harvest wildlife for economic, social and cultural purposes. In addition, Prince Leopold Island Migratory Bird Sanctuary conserves 31,000 ha in Tallurutiup Imanga (Environment Canada 2019b).

9.7.1.3 Davis Strait

Baffin Bay and Davis Strait developed as sea-floor spreading during the Cretaceous and Paleocene pushed Qikiqtaaluk and Kalaallit Nunaat apart. Maximum water depth is 2136 m in the Baffin Basin. Qikiqtaaluk coastal waters are dominated by the southward-flowing Labrador Current, transporting cold water and icebergs from Kalaallit Nunaat. Macrotidal and hypertidal regimes characterize many of the embayments along Cumberland Sound (Kangiqtualuk). At Iqaluit, the mean spring tidal range is 11.3 m (Dale et al. 2002), with tidal currents reaching 3.6 m/s, posing a challenge for port development. Seasonal ice shove and storm surge activity are ongoing hazards in Iqaluit and other coastal communities (Hatcher et al. 2014). Climate change has impacted the waters of Baffin Bay and Davis Strait, both in terms of warmer water temperatures and Arctic Amplification as a result of diminished sea ice cover.

Colonies of cold-water corals are present at depths of 200–400 m along the continental slope. The Davis Strait Conservation Area (17,298 km^2) in the Hatton Basin conserves cold-water gorgonian coral colonies, along with black coral, stony coral, hydrocoral, sea pens and sponges, including prohibition of bottom trawling fisheries (Fisheries and Oceans Canada 2019). Fishing for pelagic species, notably Greenland halibut and shrimps, is permitted.

9.7.2 Arctic Ocean

The Arctic Ocean is fed by relatively warm, saline North Atlantic waters from the Norwegian (North Atlantic) Current, entering through Fram Strait and the Barents Sea; by freshwater discharges from the Yenisey, Lena, Ob, and other Siberian and Nordic rivers; by freshwater from the Mackenzie (Deh Cho) and other North American rivers; and by cool Pacific Ocean waters through the Bering Strait, in order of relative importance (Timmermans and Marshall 2020). Surface water circulation in Canadian waters and the offshore Canada Basin is dominated by the clockwise Beaufort Gyre, moving southwest from Umingmak Nuna to the Yukon coast. Return surface flow is dominated by the Transpolar Current, flowing south through Fram Strait to eastern Kalaallit Nunaat, Iceland, and the North Atlantic, with lesser volumes passing eastern Umingmak Nuna through Kane Basin to Davis Strait (Timmermans and Marshall 2020).

9.7.2.1 Beaufort Sea

Salinity and temperature differences among these distinctive water masses result in density stratification in the Beaufort Sea (Timmermans and Marshall 2020; Farmer et al. 2021). Relatively fresh (32–28 ppt, decreasing from north to south) and cold (−1 to −3 °C) waters extending to 100–200 m depth characterize the Beaufort Gyre. Fresher, lower-density water tends to float and is concentrated in the centre of the Gyre, just as floating debris or plant matter forms patches in the Sargasso Sea and Pacific Ocean. Occasionally, driftwood derived from Siberian rivers is transported by the gyre as ice-rafts, to be deposited on the shore (Dyke et al. 1997). The freshwater concentrations facilitate formation of sea ice, if surface water temperatures remain suitable (Fig. 9.53). Saline but warmer (0–3 °C) Atlantic waters are present to the sea floor. Pacific waters are more prevalent at depth in the Chukchi and East Siberian Seas, but extend to the Lincoln Sea and the coastline of northern Umingmak Nuna. The relatively shallow Lincoln Sea (maximum depth 300 m) is commonly ice-covered throughout the year, with substantially greater thicknesses of ice than over the North Pole or the deepwater Canada Basin.

Increased and warmer freshwater influx would result in substantial changes to sea ice production and marine circulation (Timmermans and Marshall 2020; Feng et al. 2021). At present, rivers from northern Norway to Chukotka supply about 85% of the freshwater influx (~2500 km^3/y). Of the

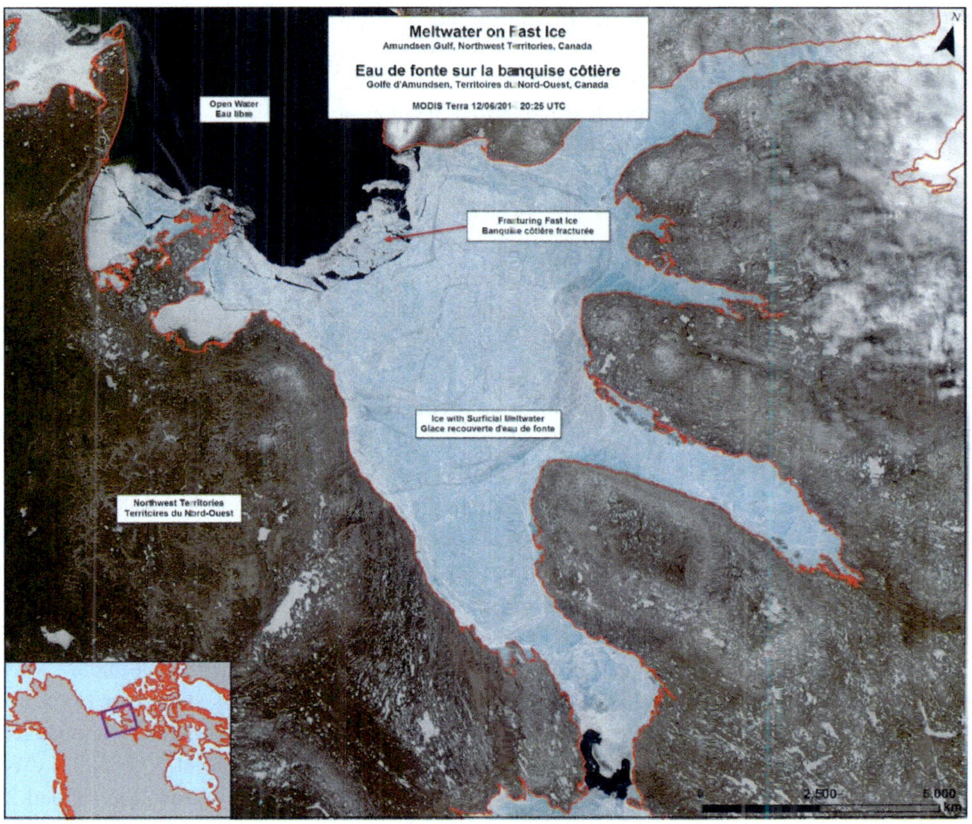

Fig. 9.53 Sea Ice with Surficial Meltwater cover, Amundsen Gulf (Image: MODIS Terra, NASA; annotations: Canadian Ice Service, Environment and Climate Change Canada)

North American contribution, including northern Kalaallit Nunaat, the Arctic Islands, and the Beaufort Sea coast, more than 75% is contributed by the Mackenzie River. Whereas the treeline ecotone and the 10 °C July isotherm largely coincide in northern Canada, there are substantial areas with permafrost north of treeline but south of the 10 °C July isotherm in Russia, notably in the Yenisey, Ob, and Pechora drainage basins. As a result, climate warming leading to permafrost ablation is increasing the discharges of these systems, as well as producing warmer river water and longer seasonal flows. In contrast, climate warming in the headwaters of the Mackenzie Basin in boreal and prairie Alberta may result in decreased discharge in future.

Increased freshwater influx, coupled with decreased sea ice production, would augment flow into the Transpolar Current, potentially affecting the production of North Atlantic Deep Water off Kalaallit Nunaat, and hence Atlantic Meridional Overturning and thermohaline circulation (Böning et al. 2016). It would also result in northeastward displacement of the Beaufort Gyre, reducing the influx of Arctic Ocean water into the Canadian archipelagos.

Predicted changes in mean annual air temperature along the Beaufort Sea coast could exceed 10 C° by 2100 ce. Much of this change is linked to Arctic Amplification resulting from diminished sea ice cover (Notz and Stroeve 2016; Serreze and Meier 2019). The thickest sea ice is driven by the Beaufort gyre, with minimum volumes and thickness in August and September, and maximums in April and May.

During the summer, open ocean waters store insolation, which gets converted into long wave thermal infrared radiation as heat. During the autumn, the heat, slows the rate at which new ice can grow across the surface. During the winter, only relatively thin single-year ice can grow, which is not as effective at trapping the heat within the marine water below. As thermal long wave infrared radiation is radiated from the ocean, it warms the overlying air, producing Arctic amplification.

Increased summer energy input from insolation delays the growth of single-year ice in autumn, resulting in only thin single-year ice in the subsequent winter. The result is greater lengthening of the ice-free season in the autumn in the spring. Although the total amount of time that ice-free is decreasing, the decrease is most evident in the autumn months.

Ecological communities in the Beaufort Sea include 150 Plankton species, more than 700 marine invertebrate species, Arctic cod (*Arctogadus glacialis*), polar cod (*Boreogadus saida),* Pacific fish species including chum((*Oncorhynchus keta*) and herring (*Clupea pallasii*), and species from adjacent freshwater including Arctic char (*Salvelinus alpinus*), whitefish (Coregonus clupeaformis, and inconnu (*Stenodus leucichthys*) (Suprenand et al. 2018). There are approximately 40,000 genetically distinct beluga, along with narwhal, walrus, bearded seal (*Erignatus barbatus*), ringed seal (*Pusa hispida*), and polar bear. The Anguniaqvia Niqiqyuam and Tarium Nirvutait Marine Protected Areas are located in the Amundsen Gulf and Mackenzie Delta respectively.

Fig. 9.54 Mackenzie Delta. Fluvial activity dominates, along with seasonal ice. Suspended load of clay, silt, and very fine sand gives the river a characteristically milky brown appearance. (NASA)

9.7.2.2 Mackenzie Delta

The Mackenzie Delta is dominated by fluvial activity, along with seasonal ice, as it enters the microtidal Beaufort Sea (Burn 2017a; Fig. 9.54). Wave action results in detachment and erosion of blocks of permafrozen sediment. Suspended load of clay, silt, and very fine sand gives the river a characteristically milky brown appearance. Medium-coarse sand is carried as bedload. In the delta region, relief is less than 20 m in almost all areas.

9.7.2.3 Beaufort Sea Wedge Boundary Dispute

The Yukon-Alaska border, as defined by the 1825 Treaty of St. Petersburg between Great Britain and the Russian Empire, is 141°W. Between the coastline and the EEZ 200 nautical mile (370 km) limit, the boundary was left undefined. Canada, as successor to Great Britain (1880), has taken the position that the maritime boundary should follow 141°W seaward. The USA, as successor to the Russian Empire (1867), maintains that the boundary should follow a projection at 90° from the shoreline at 69.6°N, 141°W. The configuration of the coast results in a wedge-shaped 21,000 km^2 area between the two positions, widening to about 110 km along the EEZ (Russell 1992; Bekker et al. 2018; Acikgonul and Lucas 2019).

The dispute has been given new life with the potential for hydrocarbon exploitation, the ecological and ethical considerations involved, and indigenous rights. The Inuvialuit Settlement Region, as agreed between the Inuvialuit and Canada, extends to 141°W seaward. Although the wedge contains an estimated 6 billion barrels of oil, along with substantial amounts of natural gas, environmental concerns form a powerful brake on potential exploitation. Canada currently does not permit exploration or licencing in Arctic waters. The USA leased eight plots of subsea terrain for exploration in the disputed area in 2004, which resulted in a diplomatic protest from Canada.

9.7.3 Inter-Archipelago Arctic Waters

The numerous channels threading through the archipelagoes between the Beaufort Sea and northernmost North Atlantic Ocean have undergone a variety of designations over the years. Traditionally, they are referred to collectively as the "Northwest Passage". The Government of Canada has consistently referred to them as "Internal Waters", and hence not open to free usage by shipping as an international waterway (unlike the Bosporus). This interpretation is not accepted by some other countries, including the United States. Here, we'll use the "neutral" term "inter-archipelago waters" to describe these channels.

A traditional topographic map dividing areas into "land" and "water" gives a misleading view of the system. The Beaufort Gyre transports large volumes of ice southwestward, consistently blocking M'Clure Strait north of Banks Island, northern Prince of Wales Strait between Banks Island and Kitlineq, and Fitzwilliam Strait, between Melville and Prince Patrick Islands: even during the warmest summers, ice cover exceeds 80% at the western ends. M'Clure Strait, to 600 m deep, formed through a combination of tectonic processes and glacial erosion. Prince of Wales Sound, along the eastern margin of Banks Island, is relatively shallow and narrow, and does not serve as an obstacle sufficient to isolate Banks Island from to the southeast (Oceans North Conservation Society 2018; Figs. 9.55 and 9.56).

The narrower Byam, Byam Martin, and Belcher Channels, and Penny Strait, typically exhibit 50–60% July ice cover. Open marine salinities vary between 25 and 32 psu during the warmest months. Water flows relatively slowly through these channels into Hassel Sound and Norwegian Bay. From there, water moves gradually eastward north of Tatlurutit into Ausuituu Tariunga, and southward into Viscount Melville Sound.

Along the southern shore of the Beaufort Sea, the combination of the Beaufort Gyre and influx from Deh Cho divert flow to the west. Water in Amundsen Gulf, Dolphin and Union Strait, and Coronation Gulf is warmed and freshened by freshwater input from Kivalliq (July 3–5 °C, 24–27 psu), with ice cover limited to 20% or less. This channel serves as a barrier to terrestrial faunal migration during the summer months, contributing to genetically distinct populations in Kitlineq, Tuktut Nogait, and western Kivalliq. Further east, Bathurst Inlet receives less marine influx, resulting in colder, less saline waters and more extensive summer ice cover, typically exceeding 40% in July.

Fig. 9.55 Western section of the Inter-archipelago Arctic waters, showing typical extents of summer ice cover. See text for discussion. (NASA)

Queen Maud Gulf, Gjoa Haven, and Rae Strait also are characterized by low salinity (19–21 psu in Rae Strait in July) and cold summer waters, but the combination of prevailing wind and topography result in limited summer ice cover (5–25%). This combination produces a longer navigation season, eventually allowing Roald Amundsen to make the first marine transit of the route around eastern Qikiqtaq (King William Island) via Gjoa Haven in 1903–1906.

Water flow through Parry Channel proceeds from west to east through Viscount Melville Sound, M'Clintock Channel, and Barrow Strait (controlling sill depth 140 m). The dominant west-to-east flow pattern obstructed navigators approaching from Tallurutiup Imanga. Temperature, salinity, and ice cover all show west-to-east gradients, with the eastern Barrow Strait being the warmest (1–2 °C), most saline (29–32 psu), and least ice-covered (0–10%) during July. Additional flows into Parry Channel occur from Peel Sound and the deep Prince Regent Inlet (south) and Wellington Channel (north). Augmented by the strong flow from Prince Regent Inlet, Barrow Channel feeds water to Tallurutiup Imanga and receives North Atlantic water in return.

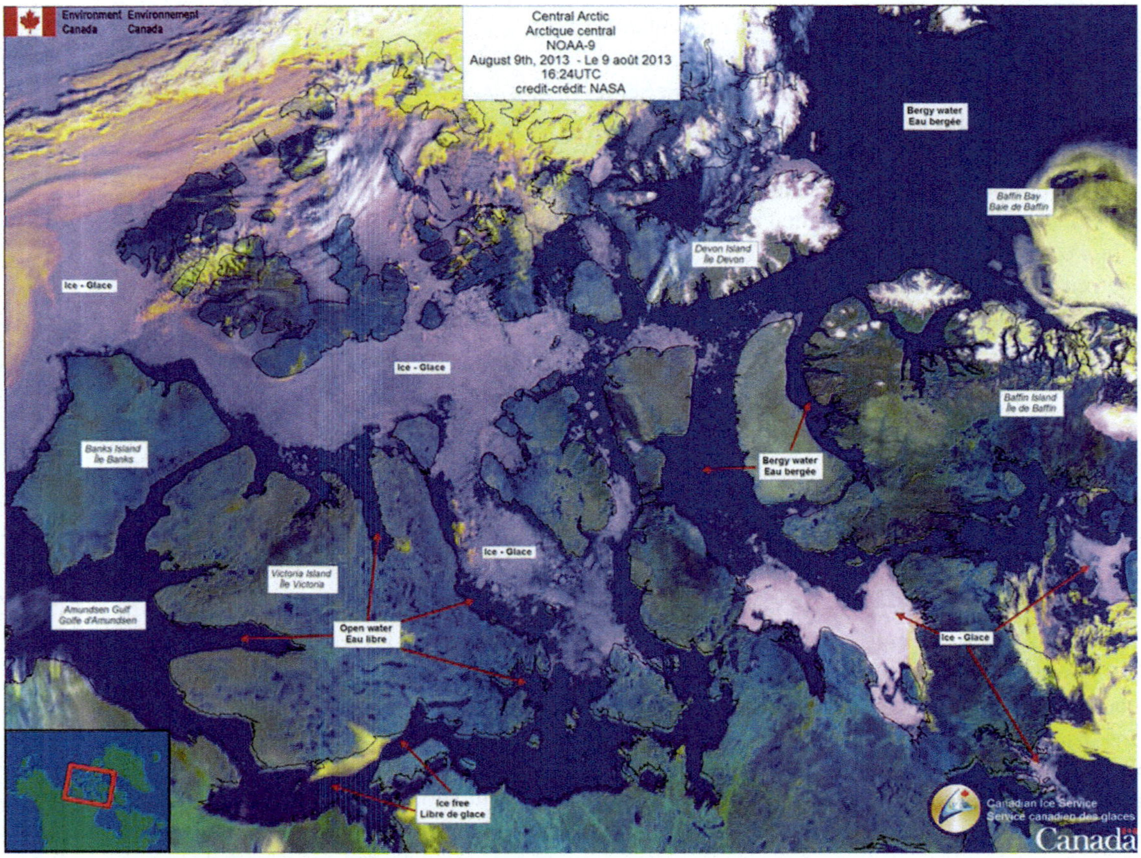

Fig. 9.56 Sea Ice cover, Central Arctic, August 2013. Note the open water of the Northwest Passage (Image: MODIS Terra, NASA; annotations: Canadian Ice Service, Environment and Climate Change Canada)

The Gulf of Boothia flows southward into Committee Bay (Nattiligaarjuk), Fury and Hecla Strait (controlling depth 50 m), and Ilagiit nunagivaktangit. For early navigators, the perhaps aptly named Committee Bay proved to be a dead end, with the coldest July waters (−1 to −2 °C) south of Ellef Ringnes Island and 80–90% summer ice cover (Fig. 9.57).

9.8 Northern Communities and Climate Change

Arctic regions have experienced the greatest amount of climate change in the recent past, and are predicted to see the greatest changes in the future (Bush and Lemmen 2019). These changes impact communities in a variety of ways, including distribution and abundance of faunal resources and northward and altitudinal movement of vegetation. Summer precipitation may induce local flooding, leading to infrastructure damage and potential contamination of water supplies. Changes in sea ice formation affect transportation by snowmobile and access to marine resources. The impetus for ice-breaking to allow lengthened shipping seasons may run counter to the best interests of local communities (Bishop et al. 2022; Huntington et al. 2022). Ablating permafrost may disrupt infrastructure, lead to local pollution problems, and require relocation of homes (Gibson et al. 2021; Zufelt 2021). Shoreline erosion triggered by permafrost ablation, reduced sea ice cover, and enhanced wave action poses difficulties for coastal communities, compounded by rising relative sea level along the Beaufort Sea coast (Forbes 2011).

Physical, demographic, socio-economic, and political factors also influence northern communities' ability to adapt to climate change and variation. Communities in Nunavut, Nunavik, Nunatsiavut (Inuit-governed north Labrador), and the Inuvialuit Settlement Region of Northwest Territories are all coastal (Fig. 9.58). Regions of cultural and economic significance, such as Ivvavik, Aulavik, and Qausuittuq National Parks, Qikiqtaryuk, Ungava kangiqluk, and northern Nunatsiavut are also coastal and subject to change.

Canadian arctic communities have small populations. Iqaluit (7429, Statistic Canada 2021 Census) is the largest, augmented by its status as capital of Nunavut. Kuujjuag, Igloolik, Inukjuak, Pangnirtung, and Gjoa Haven, significant centres for governance and logistics, all have popu-

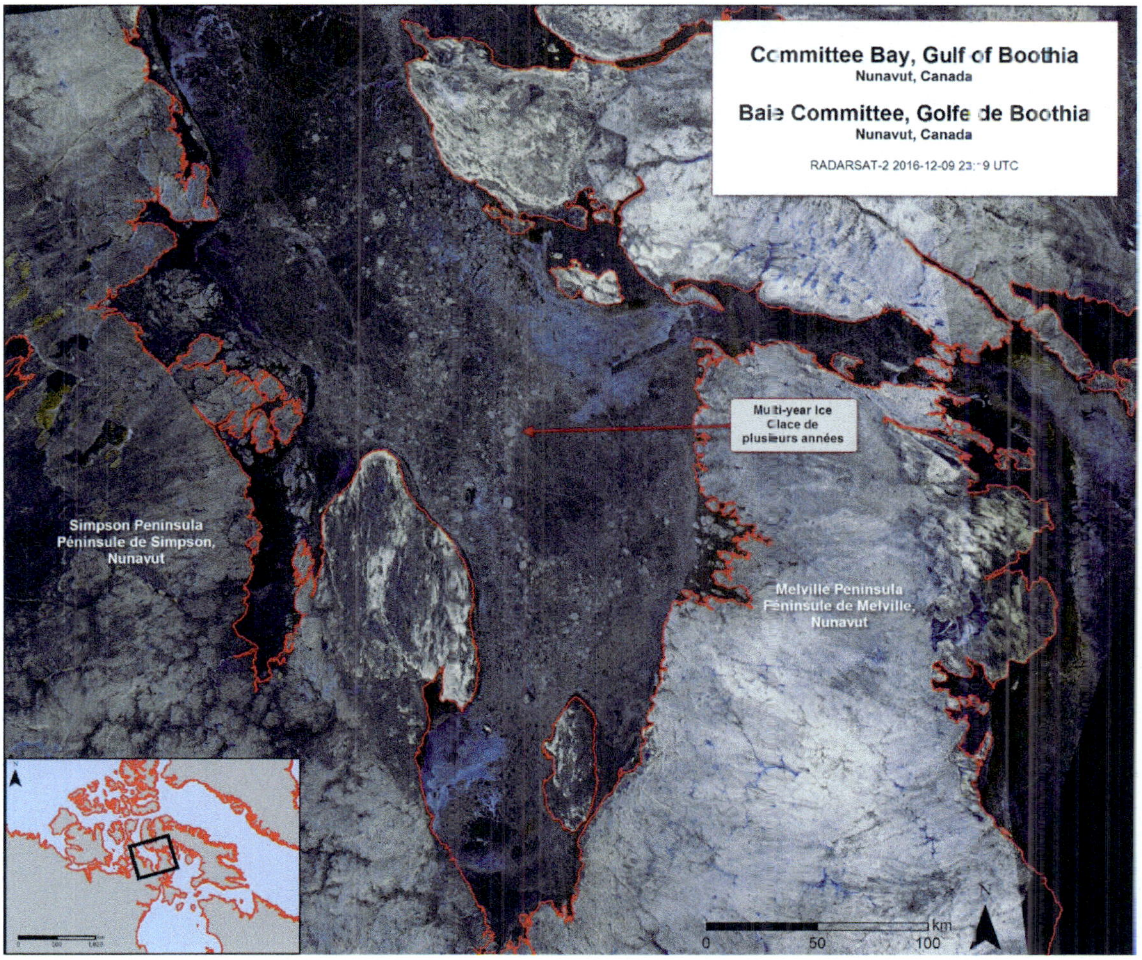

Fig. 9.57 Sea Ice cover, Committee Bay, Gulf of Boothia, 2016. Note the extensive multi-year ice, precluding navigation (Image: Radarast-2; annotations: Canadian Ice Service, Environment and Climate Change Canada)

lations between 1000 and 3000. The logistical and transportation centres of Tuktoyaktuk, Nain (seat of Nunatsiavut government), Natuashish (largest Innu community), Resolute (Qausuittuq), and Sachs Harbour (Ikahuak) all have less than 1000 inhabitants. Census population figures for northern communities can be deceptive, depending on when the census is taken seasonally (with respect to the traditional round of activities) and which community boundaries are considered. For Nain, the registered decline of 24.7% population between 2016 and 2021, potentially impacting federal government funding, was attributed to this effect. Registered population changes varies from negative (e.g. Taloyoak, −9.2%) to strongly positive (e.g. Igloolik, 17.5%). For Nunavut as a whole, the population increased by 2.5% between 2016 and 2021.

Although people may spend considerable time outside a community (in the "country"), official residence outside a community is uncommon. All but 18 of Nunavut's 36,858 residents were designated within a community in 2021.

Northern communities are overwhelmingly indigenous demographically. Iqaluit is the only community with a substantial non-indigenous population (approximately 40%). In Nain, Natuashish, Gjoa Haven (Uqsuqtuaq), Tuktoyaktuk, and Kuujjuag, more than 90% of the population are indigenous. In smaller communities, the only non-indigenous people would be associated with governmental or administrative occupations, and many of these would be transitory residents.

Most northern communities were planned as population centres by southern-based governments (Fig. 9.59). Many were established in the period 1950–1970, in attempts to concentrate indigenous populations for a variety of reasons, both positive (e.g. provision of services such as health and education) and negative (such as exertion of political control, e.g. Cavell (2021)). Communities were built in compact areas, with grid or curvilinear road patterns and equally spaced houses, often with limited regard for the topography or prevailing winds. The resulting population densities can

Fig. 9.58 Inuit regions and Communities (Statistics Canada)

be surprisingly high. Iqaluit (144 people/km^2), Pangnirtung (188.5/km^2), and Kugaaruk (204/km^2) have population densities comparable to towns in southern Canada. Pangnirtung, Kugarruk, and other communities are built on narrow coastal plains, adjacent to terrain suitable for an airstrip, with limited room for expansion or relocation. Others such as Tuktoyaktuk (Forbes 2011) are influenced by ablating permafrost, existing infrastructure, traditional land uses, and sediment unsuitable for construction.

Sachs Harbour is located in the Inuvialuit Settlement Region of Northwest Territories. Established as a trading post in 1929, Sachs Harbourdid not grow in population until the 1960s, when federal government efforts encouraged establishment of permanent Inuvialuit communities. The community is predominantly Inuvialuit, with the only non-Inuvialuit residents being teachers. All residents are involved in traditional activities.

The community is laid out in a grid pattern, with the main road paralleling the shoreline atop a sediment bluff. Buildings are exposed to slope failure due to wave undercutting and thermal ablation of permafrozen sediment. Rising sea level (3 mm/y), decreased sea ice, increased effectiveness of storm events, and increased precipitation are all contributing to erosion of the bluff (Belliveau 2007).

Gjoa Haven (47.3/km^2), the only permanent community on King William Island, was also laid out with grid and curvilinear road patterns in a succession of developments beginning in 1957. Its location along the Northwest Passage provides opportunities for economic growth from shipping and tourism, but also carries risk of potential damage from petroleum spillage (Papadimitriou 2007). Gjoa Haven is currently undergoing glaci-isostatic emergence, with sea level falling at approximately 1 mm/y. The community is built on a succession of raised gravel and sand beaches, forming

Fig. 9.59 Arctic Communities were largely developed using grid street patterns designed for southern landscapes. Upper left: Sachs Harbour. Upper right: Iqaluit (Don Forbes photo) Lower left: Gjoa Haven (National Air Photo Library. Lower right: Nain

a continuous series of low terraces from the highest point to the shoreline. There are no wave-cut bluffs in excess of 1 m in height. Continuous permafrost underlies the community, but segregated ice lenses are rare. Thermokarst features are present only in areas of anthropogenic disturbance, and are very shallow. Surface gullying is evident along the river valleys, and on sandy slopes (Papadimitriou 2007).

Gjoa Haven is an Inuit community that, while conscious of subtle and ongoing environmental changes, initially has not encountered debilitating local consequences (Catto and Parewick 2008). The community is isolated but well-resourced and managed. Its steady population growth means that erosion does present some challenges to administrators tasked with identifying suitable areas for additional housing and other new development.

Alternatives to this development pattern are possible. Hopedale (273.4/km^2) and Makkovik (123.7/km^2) in Nunatsiavut differ in settlement history and community layout from the Arctic sites. Hopedale was chosen as the site of a Moravian mission in 1752. The community faces southward, receiving the maximum amount of solar insolation during the summer months, and is located on the northern side of Hopedale Harbour, in the area least affected by tidal flats and seasonal ice shove. In addition to these advantages, Anniowaktorusek Island provides some shelter from south-easterly winds. The site also sits on the broadest and driest marine terrace. The agglomerated pattern, providing access to the harbour, indicates some local involvement in community design.

Makkovik was initially established in 1838, and the name chosen for the community, by Torsten Andersen a Norwegian-born employee of the Hudson's Bay Company. Until 1957, the community's residents were non-indigenous "settlers", largely descended from Andersen and summer fish harvesters from Newfoundland. The settler community formed a typical agglomerated settlement pattern along the northwestern coastline of Makkovik Harbour.

In the late 1950s, the federal government drive to resettle Inuit from Hebron led to the establishment of a Inuit community at Makkovik, marked by uniform government-constructed housing and an "orderly" road pattern, south of the original community. The "Hebron End" was thus culturally and physically distinctive from the original community. Problems were soon experienced as the two populations initially failed to mesh, and in the mid-1960s the situation essentially involved two adjacent communities with relatively little interaction. However, the situation gradually changed, partially as a result of integration between the communities, and also because many ex-Hebron residents subsequently relocated northwards to Nain. Today, the former differentiation between the settler and Inuit communities has largely disappeared although the different settlement patterns remain.

Arctic communities are thus faced with natural geological and environmental hazards compounded by climate change, within a socio-economic and political framework largely imposed from outside (Archer et al. 2017). Residents have a close connection to their landscapes. Most are either directly engaged in primary resource occupations or in tourism. Many families have lived in the same communities for generations, albeit perhaps in different physical locations. In the process of living in and looking at their environments, they have observed much. In Arctic Canada, where all communities are coastal and dominantly ethnically Inuit, knowledge of the land remains widespread, particularly among community elders. Residents have acquired a large volume of environmental knowledge, although sometimes unrecognized, uncategorized, and unquantified.

Successful adaptation will require both physical assessments and appropriate communication involving communities with an intimate cultural understanding (Catto and Parewick 2008, 2011; Ford et al. 2014). An effective plan for adaptation needs substantial input from all research, administrative, socio-economic, and cultural communities, which will lead to more appropriate and valuable analyses of risk, sensitivity, and vulnerability.

References

Acikgonul YE, Lucas ER (2019) Developments in maritime delimitation law over the last decade: emerging principles in modern case law. Can Yearb Int Law 57:156–222. https://doi.org/10.1017/cyl.2020.9

Allard M, Fournier A, Gahé E, Seguin MK (1989) Le Quaternaire de la côte sud-est de la baie d'Ungava, Québec nordique. Géographie physique et Quaternaire 43:325–336

Allard M, Pollard W (2013) Permafrost and climate change in Northern Coastal Canada. In: ArcticNet annual research compendium (2013–14). http://www.arcticnet.ulaval.ca/pdf/compendium2012-13/4.4_permafrost.pdf

Anderson M, Kingsley MCS (2015) Distribution and abundance of Peary caribou (*Rangifer tarandus pearyii*) and muskoxen (*Ovibos moschatus*) on southern Ellesmere Island. Nunavut Department of Environment, Wildlife Research Section, Status Report, Igloolik, NU

Andrews JT (1989) Quaternary geology of the northeastern Canadian Shield. In: Fulton RJ (ed) Quaternary geology of Canada and Greenland. The Geology of Canada, no 1. Supply and Services Canada, Ottawa, pp 276–317

Anstey RJ Renouf MAP (2011) Down the Labrador: Ramah Chert Use at Phillip's Garden, Port au Choix. In: The cultural landscapes of Port au Choix Precontact Hunter-Gatherers of Northwestern Newfoundland, Memorial University of Newfoundland, pp 189–207. https://doi.org/10.1007/978-1-4419-8324-4_10

Archer L, Ford JD, Pearce T, Kowal S, Gough WA, Allurut M (2017) Longitudinal assessment of climate vulnerability: a case study from the Canadian Arctic. Sustain Sci 12:15–29

Babaluk JA, Wissink HR, Troke BG, Clark DA, Johnson JD (2001) Summer movements of radio-tagged arctic charr (*Salvelinus alpinus*) in Lake Hazen, Nunavut, Canada. Arctic 54:418–424

Baldwin KA, Basquill L, Chapman S, Downing K, Flynn D, MacKenzie N, Major W, Meades M, Meidinger W, Morneau D, Saucier C, Thorpe J-P, Uhlig JP (2019) Vegetation Zones of Canada: a Biogeoclimatic Perspective. In: Scale 1:5,000,000. Natural Resources Canada, Canadian Forestry Service, Sault Ste. Marie, ON

Banfield AWF (1981) The Mammals of Canada. University of Toronto Press

Barber D, Marsden R, Minnett P, Ingram G, Fortier L (2001) Physical processes within the North Water (NOW) polynya. Atmos Ocean 39:163–166

Barr W (1991) Back from the brink: the road to muskox conservation in the Northwest Territories. The Arctic Institute of North America of the University of Calgary, Alberta, Canada. Kamotik Series No. 3. 127pp.

Barrand NE, Way RG, Bell T, Sharp MJ (2017) Recent changes in area and thickness of Torngat Mountain glaciers (northern Labrador, Canada). Cryosphere 11:157–168

Beauchamp B, Thériault P, Henderson CM, Pinard S, Lin R (1998) Carboniferous to Triassic formations of the Sverdrup Basin. Geol Surv Can Bull 526:195–233

Beck C, Grieser J, Kottek M, Rubel F, Rudolf B (2006) Characterizing Global Climate Change by Means of Köppen Climate Classification. http://koeppen-geiger.vu-wien.ac.at/pdfs/beck_et_al_ksb_2006.pdf

Bekker P, Olswang CMN, van de Poll R (2018) Resolving the undelimited arctic boundary between Canada and the United States in the Beaufort Sea by using a law-and-science approach. In: OTC-29178-MS offshore technology conference. Accessed 10 Aug 2022

Bell T, Jacobs JD, Munier A, Leblanc P, Trant A (2008) Climate change and renewable resources in Labrador: looking toward 2050. In: Proceedings and report of a conference held at North West River, Labrador. Labrador Highlands Research Group, Memorial University of Newfoundland

Belliveau K (2007) Coastal geomorphology of Southwest Banks Island, NWT: historical and recent shoreline changes and implications for the future. Unpublished M.Sc. thesis, Environmental Science, Memorial University, St. John's NL, Canada

Bennett R, Campbell CC, Furze FA (2014) The shallow stratigraphy and geohazards of the Northeast Baffin Shelf and Lancaster Sound. Bull Can Pet Geol 62:217–231

Birch L, Cronin T, Tziperman E (2017) Glacial Inception on Baffin Island: The role of insolation, meteorology, and topography. J Clim 30:4047–4064

Bishop B, Owen J, Wilson L, Eccles T, Chircop A, Fanning L (2022) How icebreaking governance interacts with Inuit rights and livelihoods in Nunavut: a policy review. Mar Policy 137:104957. https://doi.org/10.1016/j.marpol.2022.104957

Bliss LC (1997) Arctic ecosystems of North America. In: Wielgolaski FE (ed) Ecosystems of the world 3, Polar and Alpine tundra. Elsevier, Amsterdam, pp 551–683

Bliss LC, Svoboda J (1984) Plant communities and plant production in the western Queen Elizabeth Islands. Holarct Ecol 7:325–344

Bliss LC, Svoboda J, Bliss DI (1984) Polar deserts, their plant cover and plant production in the Canadian High Arctic. Holarct Ecol 7:305–324

Bostock HS (2014) Physiographic Regions of Canada. Geological Survey of Canada Map 1254A (2nd edition), scale 1:5,000,000. https://www.nrcan.gc.ca/maps-tools-and-publications/maps/geology-and-geosciences/16876. Accessed 12 July 2022

Böning CW, Behrens E, Biastoch A, Getzlaff K, Bamber JL (2016) Emerging impact of Greenland meltwater on deepwater formation in the North Atlantic Ocean. Nat Geosci 9. https://doi.org/10.1038/ngeo2740

Briner JP, Davis PT, Miler GH (2009) Latest Pleistocene and Holocene glaciation of Baffin Island, Arctic Canada: key patterns and chronologies. Quatern Sci Rev 28:2075–2087

Bromaghin JF, Douglas DC, Durner GM, Simac KS, Atwood TC (2021) Survival and abundance of polar bears in Alaska's Beaufort Sea, 2001–2016. Ecol Evol 11:14250–14267

Burden ET, Langille AB (1990) Stratigraphy and sedimentology of Cretaceous and Paleocene strata in half-grabens on the southeast coast of Baffin Island, Northwest Territories. Bull Can Soc Pet Geol 38:185–196

Burn CR (2012) Herschel Island Qikiqtaryuk: a natural and cultural history of Yukon's Arctic island. Wildlife Management Advisory Council (North Slope), Whitehorse

Burn CR (2017a) Mackenzie Delta: Canada's Principal Arctic delta. In: Slaymaker O (ed) Landscapes and landforms of Western Canada. Springer International Publishers, Cham, Switzerland, pp 321–334

Burn CR (2017b) Herschel Island (Qikiqtaryuk), Yukon's Arctic Island. In: Slaymaker O (ed) Landscapes and Landforms of Western Canada. Springer International Publishers, Cham, Switzerland, pp 335–348

Burn CR, Kokelj SV (2009) The environment and permafrost of the Mackenzie Delta area. Permafrost Periglac Process 20:83–105

Bush E, Lemmen DS (2019) Canada's changing climate report. Government of Canada, Ottawa, p 444

Catto NR (2020) Labrador and the Northern Peninsula of Newfoundland: foreshore and flanking geomorphology. In: Slaymaker O, Catto NR (eds) Landscapes and landforms of Eastern Canada. Springer International Publishers, Cham, Switzerland, pp 179–210

Catto NR, Parewick K (2008) Hazard and vulnerability assessment and adaptive planning: mutual and multi-lateral community-researcher communication, arctic and Atlantic Canada. In: Liverman D (ed) Communicating geoscience. Geological Society of London, vol 305, pp 123–140

Catto N Parewick K (2011) Vulnerability, adaptation, adaptive capacity and resilience. In: Forbes DL (ed) State of the Arctic Coast 2010—Scientific review and outlook. International Arctic Science Committee, Land-Ocean Interactions in the Coastal Zone, Arctic Monitoring and Assessment Programme, International Permafrost Association. Helmholtz-Zentrum, Geesthacht, Germany, pp 104–114

Catto NR, Slaymaker O, Kovanen DJ (2020) Human-modified landscapes. In: Slaymaker O, Catto NR (eds) Landscapes and landforms of Eastern Canada. Springer International Publishers, Cham, Switzerland, pp 475–508

Cavell J (2021) "Consolidation and control of all Eskimo income": the motive for the 1953 high arctic relocation. J Can Stud 55:118–151

Clark D (2007) Assessing the health of the Lake Hazen ecosystem, Ellesmere Island, Northwest Territories. In: Environment Canada, ecological monitoring and assessment network. Report prepared for Parks Canada, Nunavut Field Unit, Iqaluit, Nunavut

Clark PU (1988) Glacial geology of the Torngat Mountains, Labrador. Can J Earth Sci 25:1184–1198

Copper P, Edinger E (2009) Distribution, geometry and palaeogeography of the Frasnian (Late Devonian) reef complexes of Banks Island, NWT, western arctic, Canada. In: Geological society of London Special Publication, vol 314, pp 109–124

COSEWIC (2004a) COSEWIC assessment and update status report on the narwhal *Monodon monoceros* in Canada. Committee on the Status of Endangered Wildlife in Canada, Ottawa, vii + 50 pp.

COSEWIC (2004b) COSEWIC assessment and update status report on the Peary caribou *Rangifer tarandus pearyi* and the barren-ground caribou *Rangifer tarandus groenlandicus* (Dolphin and Union population) in Canada. Committee on the Status of Endangered Wildlife in Canada, Ottawa

COSEWIC (2011) COSEWIC assessment and status report on the Humpback Whale *Megaptera novaeangliae* in Canada. Committee on the Status of Endangered Wildlife in Canada, Ottawa

COSEWIC (2011b). Designatable Units for Caribou (Rangifer tarandus) in Canada. Committee on the Status of Endangered Wildlife in Canada

COSEWIC (2015) COSEWIC assessment and status report on the Peary Caribou *Rangifer tarandus pearyi* in Canada. Committee on the Status of Endangered Wildlife in Canada, Ottawa

COSEWIC (2017) COSEWIC assessment and status report on the Atlantic Walrus *Odobenus rosmarus rosmarus*, High Arctic population, Central-Low Arctic population and Nova Scotia-Newfoundland-Gulf of St. Lawrence population in Canada. Committee on the Status of Endangered Wildlife in Canada, Ottawa

Cowan B, Carter J Forbes DL, Bell T (2021) Postglacial sea-level lowstand on Cumberland Peninsula, Baffin Island, Nunavut. Can J Earth Sci 58. https://doi.org/10.1139/cjes-2021-0026

Cunliffe AM, Tanski G, Radosavljevic B, Palmer WF, Sachs T, Lantuit H, Kerby JT, Myers-Smith HL (2019) Rapid retreat of permafrost coastline observed with aerial drone photogrammetry. Cryosphere 13:1513–1528

Currie LD, Brent TA, Turner EC (2020) Offshore bedrock geology of Eclipse Sound and Pond Inlet: connecting the structure and stratigraphy of Bylot and northern Baffin islands. Can J Earth Sci 57:1254–1267

Dafoe LT, Bingham-Koslowski N (2020) Baffin Island and the Labrador-Baffin Seaway geological synthesis. Geological Survey of Canada, Bulletin 608

Dale JE, Leech S, McCann SB, Samuelson G (2002) Sedimentary characteristics, biological zonation and physical processes of the tidal flats of Iqaluit, Nunavut. In: Hewitt K, Byrne ML, English M, Young G (eds) Landscapes in transition: landform assemblages and transformations in cold regions. Kluwer, Dordrecht. pp 205–234

Davis EL, Trant AJ, Way RG, Hermanutz L, Whitaker D (2021) Rapid ecosystem change at the Southern Limit of the Canadian Arctic, Torngat Mountains National Park. Remote Sens 13:2085. https://doi.org/10.3390/rs13112085

Davison T Baryluk S (2021) Aerial Survey of Muskoxen (*Ovibos moschatus*) and Peary Caribou (*Rangifer tarandus pearyi*) on Banks Island, July and August 2019. Environment and Natural Resources Government of the Northwest Territories Manuscript Report 295. https://www.enr.gov.nt.ca/sites/enr/files/resources/295_manuscript.pdf. Accessed 8 Aug 2022

Davison T, Pongracz J, Williams J (2013) Population survey of Peary caribou (*Rangifer tarandus pearyi*) and muskoxen (*Ovibos moschatus*) on Banks Island, Northwest Territories, July 2010. Rangifer 33(21) 135–140

De Kemp EA, Gilbert C, James DT (2006) Geological Provinces in Nunavut. Geological Survey of Canada Map D1860A

Desjardins É, Lai S, Payette S, Vézina F, Tam A, Berteaux D (2021) Vascular plant communities in the polar desert of Alert (Ellesmere Island, Canada) establishment of a baseline reference for the 21st century. Écoscience 28:243–267

Dowdeswell EK, Todd BJ, Dowdeswell JA (2016a) Crag-and-tail features: convergent ice flow through Eclipse Sound, Baffin Island, Arctic Canada. In: Dowdeswell JA, Canals M, Jakobsson M. Todd BJ, Dowdeswell EK, Hogan KA (eds) Atlas of Submarine glacial landforms: modern, quaternary and ancient. Geological society of London Memoir, vol 46, pp 55–56

Dowdeswell EK, Todd BJ, Dowdeswell JA (2016b) Submarine medial moraines and convergent ice flow, Scott Inlet, Baffin Island, Arctic Canada. In: Dowdeswell JA, Canals M, Jakobsson M, Todd BJ, Dowdeswell EK, Hogan KA (eds) Atlas of Submarine glacial landforms: modern, quaternary and ancient. Geological society of London Memoir, vol 46, pp 193–194

Druckenmiller ML, Moon T, Thoman R (2021) The arctic. Bull Am Meteorol Soc 102:S263–S315. https://doi.org/10.1175/BAMS-D-21-0086.1

Dubé-Loubert H, Roy M, Schaefer JM, Clark PU (2018) ^{10}Be dating of former glacial Lake Naskaupi (Québec-Labrador) and timing of its discharges during the last deglaciation. Quatern Sci Rev 37. https://doi.org/10.1016/j.quascirev.2018.05.008

Dyke AS (1983) Quaternary geology of Somerset Island, District of Franklin, NWT In: Geological survey of Canada Memoir, vol 404

Dyke AS (1984) Quaternary geology of Boothia Peninsula and northern District of Keewatin, central Canadian Arctic. In: Geological survey of Canada Memoir, vol 407

Dyke AS (2004) An outline of North American deglaciation with emphasis on central and northern Canada. In: Ehlers J, Gibbard PL (eds) Quaternary Glaciations—Extent and chronology, Part II: North America. Elsevier, Amsterdam, pp 371–424

Dyke AS (2008) The Steensby Inlet Ice Stream in the context of the deglaciation of Northern Baffin Island, Eastern Arctic Canada. Earth Surf Proc Land 33:573–592

Dyke AS, Hooper J, Savelle JM (1996) A history of sea ice in the Canadian Arctic Archipelago based on postglacial remains of the bowhead whale (*Balaena mysticetus*). Arctic 49:235–255

Dyke AS, England J, Reimnitz E, Jetté H (1997) Changes in driftwood delivery to the Canadian Arctic Archipelago: the hypothesis of postglacial oscillations of the Transpolar Drift. Arctic 50:1–16

Dyke AS, Andrews JT, Miller GH (1982) Quaternary history of Cumberland Peninsula, Baffin Island, District of Franklin. In: Geological Survey of Canada Memoir, vol 403, Ottawa

Dyke AS, Andrews JT, Clark PU, England JH, Miller GH, Shaw J, Veillette JJ (2002) The Laurentide and Innuitian ice sheets during the Last Glacial Maximum. Quatern Sci Rev 21:9–31

Ernst RE, Bleeker W, (2010) Large igneous provinces (LIPs), giant dyke swarms, and mantle plumes: Significance for breakup events within Canada and adjacent regions from 2.5 Ga to present. Canadian Journal of Earth Sciences, 47, 695–739

Eberle JJ, Gottfried MD, Hutchison JH, Brochu CA (2014) First record of Eocene Bony Fishes and Crocodyliforms from Canada's Western Arctic. Plos One 9(5):e96079. https://doi.org/10.1371/journal.pone.0096079

Edlund SA (1983) Reconnaissance vegetation studies on western Victoria Island, Canadian Arctic Archipelago. In: Geological Survey of Canada, Paper 83–1B, pp 75–81

Edlund SA, Alt BT (1989) Regional congruence of vegetation and summer climate patterns in the Queen Elizabeth Islands, Northwest Territories, Canada. Arctic 42:3–23

England J (1996) Glacier dynamics and paleoclimatic change during the last glaciation of eastern Ellesmere Island, Canada. Can J Earth Sci 33:779–799

England JH, Furze MFA, Doupe JP (2009) Revision of the NW Laurentide Ice Sheet: implications for paleoclimate, the northwest extremity of Beringia, and Arctic Ocean sedimentation. Quatern Sci Rev 28:1573–1596

England J, Smith IR, Evans DJA (2000) The last glaciation of east-central Ellesmere Island, Nunavut: ice dynamics, deglacial chronology, and sea level change. Can J Earth Sci 37:1355–1371

England J, Atkinson N, Bednarski J, Dyke AS, Hodgson DA, O'Cofaigh C (2006) The Innuitian Ice Sheet: configuration, dynamics and chronology. Quatern Sci Rev 25:689–703

Environment Canada (2019a) Nirjutiqavvik National Wildlife Area. https://www.canada.ca/en/environment-climate-change/services/national-wildlife-areas/locations/nirjutiqavvik.html

Environment Canada (2019b) Prince Leopold Island Migratory Bird Sanctuary. https://www.canada.ca/en/environment-climate-change/services/migratory-bird-sanctuaries/locations/prince-leopold-island.html

Environment and Climate Change Canada (2020) Canadian Climate Normals 1981-2010. https://climate.weather.gc.ca/climate_normals/index_e.html

Evans DJA, England J, La Farge C, Coulthard RD, Lakeman TR, Vaughan JM (2014) Quaternary geology of the Duck Hawk Bluffs, southwest Banks Island, Arctic Canada: a re-investigation of a critical terrestrial type locality for glacial and interglacial events bordering the Arctic Ocean. Quatern Sci Rev 31:82–123

Farmer JR, Sigman DM, Granger J et al (2021) Arctic Ocean stratification set by sea level and freshwater inputs since the last ice age. Nat Geosci 14:684–689. https://doi.org/10.1038/s41561-021-00789-y

Fauchald P, Park T, Tommervik H, Myneni R, Hausner VH (2017) Arctic greening from warming promotes declines in caribou populations. Sci Adv 3:e1601365

Feng D, Gleason CJ, Lin P et al (2021) Recent changes to Arctic river discharge. Nat Commun 12:6917. https://doi.org/10.1038/s41467-021-27228-1

Ferguson MAD (1987) Status of Peary caribou and muskox populations on Bathurst Island, N.W.T., August 1981. Arctic 40:131–137

Ferguson RS (1991) Detection and classification of muskox habitat on Banks Island, Northwest Territories, Canada, using Landsat thematic mapper data. Arctic 44(Supplement 1):66–74

Fisheries and Oceans Canada (2019) Davis strait conservation area. https://www.dfo-mpo.gc.ca/oceans/oecm-amcepz/refuges/davisstrait-detroitdavis-eng.html

Fontaine AJ, Mallory ML (2011) Detection and classification of land cover classes of Southampton Island, Nunavut, using LANDSAT ETM+ data. In: Canadian Wildlife Service Occasional Paper No. 119. Environment Canada, Ottawa

Forbes DL (2011) State of the Arctic Coast 2010—Scientific review and outlook. International Arctic Science Committee, Land-Ocean Interactions in the Coastal Zone, Arctic Monitoring and Assessment Programme, International Permafrost Association. Helmholtz-Zentrum, Geesthacht, Germany

Forbes DL, Hansom JD (2011) Polar coasts. In: Wolanski E, McLusky DS (eds) Treatise on Estuarine and coastal science, vol 3. Academic Press, Waltham, pp 245–283

Ford JD, McDowell G, Jones J (2014) The state of climate change adaptation in the Arctic. Environ Res Lett 9:104005

Ford JD, Pearce T, Canosa IV, Harper S (2021) The rapidly changing Arctic and its societal implications. Wiley Interdiscip Rev: Climate Change 12(6):e735. https://doi.org/10.1002/wcc.735

France RL (1993) The Lake Hazen trough: a late winter oasis in a polar desert. Biol Cons 63(2):149–151

French HM (2007) The Periglacial environment, 3rd edn. Wiley, Chichester

French HM (2017a) The Banks Island Tundra. In: Slaymaker O (ed) Landscapes and landforms of Western Canada. Springer International Publishers, Cham, Switzerland, pp 97–108

French M (2017b) The Polar Desert of Prince Patrick Island. In: Slaymaker O (ed) Landscapes and landforms of Western Canada. Springer International Publishers, Cham, Switzerland, pp 177–186

French HM, Slaymaker O (2012) Changing Cold environments: a Canadian perspective. Wiley, Chichester, UK

Funck T, Gohl K, Damm V, Heyde I (2012) Tectonic evolution of southern Baffin Bay and Davis Strait: results from a seismic refraction transect between Canada and Greenland. J Geophys Res Solid Earth 117. https://doi.org/10.1029/2011JB009110

Furze MFA, Pienkowski AJ, McNeely MA, Bennett R, Cage AG (2018) Deglaciation and ice shelf development at the northeast margin of the Laurentide Ice Sheet during the Younger Dryas chronozone. Boreas 47:271–296

Galloway BJ, Dewing K, Beauchamp B, Matthews W (2021) Upper Paleozoic stratigraphy and detrital zircon geochronology along the northwest margin of the Sverdrup Basin, Arctic Canada: insight into the paleogeographic and tectonic evolution of Crockerland. Can J Earth Sci 58:164–187. https://doi.org/10.1139/cjes-2019-0226

Gibson CM, Brinkman T Cold H, Brown D Turetsky M (2021) Identifying increasing risks of hazards for northern land-users caused by permafrost thaw: integrating scientific and community-based research approaches Environ Res Lett 16:064047. https://doi.org/10.1088/1748-9326/abfc79

Government of Canada (2008) Canadian Permafrost thickness, Geological Survey of Canada. Natural Resources Canada; Earth Sciences Sector; Canada Centre for Mapping and Earth Observation. http://geogratis.gc.ca/api/en/nrcan-rncan/ess-sst/092b663d-198b-5c8d-9665-fa3f5970a14f.html. Accessed 30 Jan 2015

Government of Northwest Territories, Environment (2015) Climate observations. https://www.enr.gov.nt.ca/sites/enr/files/page_3_nwt-climate-observations_06-13-2015_vf_1_0.pdf

Government of Nunavut, Environment (2018) Baffin Island Caribou Management Plan Working together to ensure Baffin Island caribou harvest is sustainable GN DOE

Government of Nunavut, Environment (2021) Polar Bears in Canada. https://www.polarbearscanada.ca/en/polar-bears-canada/canadas-polar-bear-subpopulations

Grieve RAF (2001) Impact cratering on Earth. In: Brooks GR (ed) A synthesis of geological Hazards in Canada. Geological Survey of Canada Bulletin, vol 548, pp 207–22

Guyard H, St. Onge G, Pienitz R, Francus P (2011) New insights into Late Pleistocene glacial and postglacial history of northernmost Ungava (Canada) from Pinualuit Crater Lake sediments. Quat Sci Rev 30:3892–3907

Halliday WD, Dawson J, Yurkowski DJ, Doniol- Valcroze T, Ferguson SH, Gjerdrum C, Hussey NE, Kochanowicz Z, Mallory ML, Marcoux M, Watt CA, Wong SNP (2022) Vessel risks to marine wildlife in the Tallurutiup Imanga National Marine conservation area and the eastern entrance to the Northwest Passage. Environ Sci Policy 127:181–195

Harper SL, Dorough DS, MacDonald JP, Cunsolo A, King N (2021) Climate change and Inuit health: research does not match risks posed. One Earth 4:1656–1660. https://doi.org/10.1016/j.oneear.2021.11.017

Harrison JC (1995) Melville Island's salt-based fold belt, Arctic Canada. Geol Surv Canada Bull 472

Harrison JC, de Freitas TA (1998) Bedrock Geology, Bathurst Island Group (NTS 68G, 68H, 69A, 69B; small Parts of 78H and 79A), District of Franklin (Nunavut), Northwest Territories. Geol Surv Canada Open File 3577. https://doi.org/10.4095/209408

Harrison JC, Lynds T, Ford A (2014a) Geology, tectonic assemblage map of the Byam Martin Channel area, Melville and surrounding Islands, Nunavut-Northwest Territories. Geol Surv Canada Can Geosci Map 32

Harrison JC, Thorsteinsson R, Mayr U, Lynds T, Ford A, Turner EC (2014b) Geology, tectonic assemblage map of Wellington Channel, Cornwallis and Western Devon Islands, Nunavut. Geological Survey of Canada, Canadian Geoscience Map 33

Harrison JC Lynds T, Ford A, Rainbird RH (2016) Simplified tectonic assemblage map of the Canadian Arctic Islands, NWT-Nunavut. Canadian Geoscience Map 80

Harrison JC (2018) Intersecting fold belts in the Bathurst Island region, Nunavut. J Geodyn 118:82–105

Hatcher SV, Forbes DL, Manson GK (2014) Coastal geoscience field work near Iqaluit, Nunavut, 2009–2011. In: Geological Survey of Canada, Open File 7653, 65 p. https://doi.org/10.4095/295528

Hatcher SV, Forbes DL (2015) Exposure to coastal hazards in a rapidly expanding Northern Urban Centre, Iqaluit, Nunavut. Arctic 68:453–471

Heginbottom JA, Dubreuil MA, Harker PA (1995) Canada—Permafrost. National Atlas of Canada, 5th edn. National Atlas Information Service, Natural Resources Canada, MCR 4177

Hill PR, Héquette A, Ruz M-H (1993) Holocene sea-level history of the Canadian Beaufort shelf. Can J Earth Sci 30:103–108

Hocq M (1994) La Province du Supérieur. Geologie du Quebec. Ministère des Ressources naturelles. Québec MM 94-01, pp 7–20

Hodgson DA (1989) Quaternary geology of the Queen Elizabeth Islands. In: Fulton RJ (ed) Quaternary geology of Canada and Greenland. The Geology of Canada, no 1. Supply and Services Canada, Ottawa, pp 443–478

Huntington HP, Zagorsky A, Kaltenborn BP, Shin HC, Dawson J, Lukin M, Dahl PE, Guo P, Thomas DN (2022) Societal implications of a changing Arctic Ocean. Ambio 51:298–306. https://doi.org/10.1007/s13280-021-01601-2

Hurrell JW, Kushnir Y, Ottersen G, Visbeck M (2003) An overview of the North Atlantic oscillation. In: Hurrell JW (ed) The North Atlantic Oscillation: climatic significance and environmental impact geophysical monograph, vol 134. American Geophysical Union, pp 1–35

Jensen EL, Tschritter C, de Groot PVC, Hayward KM, Branigan M, Dyck M, Clemente-Carvalho RBG, Lougheed SC (2020) Canadian polar bear population structure using genome-wide markers. Ecol Evol 10:3706–3714

Johns SM, Young MD (2006) Bedrock geology and economic potential of the Archean Mary River group, northern Baffin Island, Nunavut. In Geological Survey of Canada, current research, p C5. https://doi.org/10.4095/222954

Kerr JW (1974) Geology of Bathurst Island Group and Byam Martin Island, Arctic Canada (Operation Bathurst Island). In: Geological Survey of Canada Memoir, vol 378

Klassen RA, Paradis S, Bolduc AM, Thomas RD (1992) Glacial landforms and deposits Labrador (Newfoundland) and eastern Québec. In: Geological Survey of Canada Map, vol 1814A

Kjikjerkovska E (2016) Long-term hydroclimatic change and interannual variability in water sources, Apex River (Iqaluit), Baffin Island, Nunavut. M.Sc. thesis. Queen's University, Kingston, Ontario. 107 p.

Kochanowicz Z, Dawson J, Halliday WD, Sawada M, Copland L, Carter NA, Nicoll A, Ferguson SH, Heide-Jørgensen MP, Marcoux M, Watt C, Yurkowski DJ (2021) Using western science and Inuit knowledge to model ship-source noise exposure for cetaceans (marine mammals) in Tallurutiup Imanga (Lancaster Sound). Nunavut Can Marine Policy 130:104557. https://doi.org/10.1016/j.marpol.2021.104557

Köck G, Muir D, Yang F, Wang X, Talbot C, Ganter N, Moser D (2012) Bathymetry and Sediment Geochemistry of Lake Hazen (Quttinirpaaq National Park, Ellesmere Island, Nunavut). Arctic 65:56–66

Kokelj SV, Tunnicliffe J, Lacelle D, Lantz TC, Chin KS, Fraser R (2015) Increased precipitation drives mega slump development and destabilization of ice-rich terrain, northwestern Canada. Glob Planet Change 129:56–68

Kunz J, Kneisel C (2021) Three-dimensional investigation of an open- and a closed-system Pingo in northwestern Canada. Permafrost Periglac Process 32:541–557

Labreyrie L, Elliot M (1999) Abrupt climate changes—Causes and consequences. In: Abrantes F, Mix A (eds) Reconstructing ocean history. Kluwer Academic, Dordrecht, pp 73–81

Lamoureux SF Rudy ACA (2020) Melville, Bathurst, and Cornwallis Islands: low to moderate relief Innuitia. In: Slaymaker O, Catto NR (eds) Landscapes and landforms of Eastern Canada. Springer International Publishers, Cham, Switzerland, pp 315–332

Lantuit H, Pollard WH (2008) Fifty years of coastal erosion and retrogressive thaw slump activity on Herschel Island, southern Beaufort Sea, Yukon Territory, Canada. Geomorphology 95:84–102

Larter NC, Nagy JA (2001) Variation between snow conditions at Peary caribou and muskox feeding sites and elsewhere in foraging habitats on Banks Island in the Canadian High Arctic. Arct Antarct Alp Res 33:123–130

Larter NC, Raillard M, Epp H, Nagy JA (2009) Vegetation mapping of Banks Island with particular reference to Aulavik National Park. File Report No. 138 Government of the Northwest Territories, Department of Environment and Natural Resources

Leclerc L-M (2015) Population Estimate and 2015–2017 Demographics Population Indicators of the Dolphin and Union Caribou Herd (*Rangifer tarandusgroenlandicusx pearyi).* Nunavut Department of Environment. Accessed 11 Aug 2022

Lehnherr I, St. Louis VL, Sharp M et al (2018) The world's largest High Arctic lake responds rapidly to climate warming. Nat Commun 9:1290. https://doi.org/10.1038/s41467-018-03685-z

Lévesque E (1997) Plant Distribution and Colonization in extreme Polar deserts, Ellesmere Island, Canada. Unpublished Ph.D. thesis, Department of Botany, University of Toronto. Accessed 10 Aug 2022

Louis M et al (2020) Influence of past climate change on phylogeography and demographic history of narwhals, *Monodon monoceros*. Proc R Soc Ser B 287:20192964. https://doi.org/10.1098/rspb.2019.2964

Li G, Piper DJW, Campbell DC (2011) The Quaternary Lancaster Sound trough-mouth fan, NW Baffin Bay. J Quat Sci 26:511–522

Mackay JR (1979) Pingos of the Tuktoyaktuk Peninsula Area, Northwest Territories. Géog Phys Quatern 33:3–61

MacLean B, Blasco S, Bennett R, England J, Rainey W, Hughes-Clarke J, Beaudoin J (2010) Ice keel seabed features in marine channels of the central Canadian Arctic Archipelago: evidence for former ice streams and iceberg scouring. Quatern Sci Rev 29:2280–2301

MacLean B, Blasco S, Bennett R, Lakeman T, Pienkowski AJ, Furze MFA, Hughes Clarke J, Patton E (2017) Seafloor features delineate Late Wisconsinan ice stream configurations in eastern Parry Channel, Canadian Arctic Archipelago. Quatern Sci Rev 160:67–84

MacLeod RF, Dallimore SR (2021) Assessment of storm surge history as recorded by driftwood in the Mackenzie Delta and Tuktoyaktuk Coastlands, Arctic Canada. Front Earth Sci 9:698660. https://doi.org/10.3389/feart.2021.698660

Mansfield AW (1967) Seals of Arctic and Eastern Canada. Fisheries Research Board of Canada

Marcoux M, Hammill MO (2016) Model estimates of Cumberland Sound beluga (*Delphinapterus leucas*) population size and total allowable removals. Fisheries and Oceans Canada, Scientific Research Document 2016/077

Margold M, Stokes CR, Clark CD, Kleman J (2015) Ice streams in the Laurentide Ice Sheet: a new mapping inventory. J Maps 11:380–395

Margreth A, Gosse JC, Dyke AS (2016) Quantification of subaerial and episodic subglacial erosion rates on high latitude upland plateaus: Cumberland Peninsula Baffin Island, Arctic Canada. Quat Sci Rev 133:108–129

Maxwell JB (1980–82) The climate of the Canadian Arctic islands and adjacent waters. Environment Canada, Atmospheric Environment Service. Climatol Stud 30

Maxwell JB (1981) Climatic Regions of the Canadian Arctic Islands. Arctic 34:225–240

Montesano PM, Neigh CSR, Macander M, Feng M, Noojipady P (2020) The bioclimatic extent and pattern of the cold edge of the boreal forest: the circumpolar taiga-tundra ecotone. Environ Res Lett 15:105019. https://doi.org/10.1088/1748-9326/abb2c7

Morgan MR, Pocklington R (1997) The chilling aspects of global warming in Atlantic Canada. In: Shaw RW (ed) Climate change and climate variability in Atlantic Canada. Environment Canada, Atlantic region, occasional paper 9, pp 184–194

Nagy JA, Larter NC, Fraser VP (1996) Population demography of Peary caribou and muskox on Banks Island, NWT, 1982–1992. Rangifer (9):213–222

Natural Resources Canada (2009) Geological Provinces of Canada. https://www.nrcan.gc.ca/maps-tools-andpublications/maps/geology-and-geosciences/16876 Accessed 12 July 2022

Natural Resources Canada (2012) Atlas of Canada, Northern Geodatabase. https://open.canada.ca/data/en/dataset/702ebdea-39ff-50e4-ab5f-de1150d16b7a. Accessed 10 July 2022

Naughton D (2012) Natural History of Canadian Mammals. Canadian Museum of Nature

Nixon FC, England JH, Lajeunesse P, Hanson MA (2014) Deciphering patterns of postglacial sea level at the junction of the Laurentide and Innuitian Ice Sheets, western Canadian High Arctic. Quat Sci Rev 91:165–183

Notz D, Stroeve J (2016) Observed Arctic Sea-ice loss directly follows anthropogenic CO_2 emission. Science 354:747–750

Oakey GN, Chalmers JA (2012) A new model for the Paleogene motion of Greenland relative to North America: plate reconstructions of the Davis Strait and Nares Strait regions between Canada and Greenland. J Geophys Res 117:B10401

Obu J, Lantuit H, Grosse G, Günther F, Sachs T, Helm V, Fritz M (2017) Coastal erosion and mass wasting along the Canadian Beaufort Sea based on annual airborne LiDAR elevation data. Geomorphology 293:331–346

Oceans North Conservation Society (2018) Canada's Arctic Marine Atlas. Ottawa, Ontario: Oceans North Conservation Society, World Wildlife Fund Canada, and Ducks Unlimited Canada. Accessed 11 Aug 2022

Okulitch AV (1992) Geological Atlas, Horton River, Franklin–Mackenzie. Geological Survey of Canada Map NR-9/10/11/12-G

Okulitch AV, Packard JJ, Zolnai AI (1986) Evolution of the Boothia Uplift, Arctic Canada. Can J Earth Sci 23:350–358

Papadimitriou S (2007) Climate-related hazards and Risks to Community infrastructure, Gjoa Haven, Nunavut. Unpublished M.Sc. thesis. In: Environmental science. Memorial University, St. John's NL, Canada

Parks Canada (2010) State of the Park Report 2010: Aulavik NP of Canada. parkscanada.gc.ca. ISBN 978-1-100-16507-3

Parks Canada (2019) Tallurutip Imanga. https://www.pc.gc.ca/en/amnc-nmca/cnamnc-cnnmca/tallurutiup-imanga. Accessed 10 Aug 2022

Parks Canada (2022a) Aulavik National Park. https://www.pc.gc.ca/en/pn-np/nt/aulavik. Accessed 10 Aug 2022

Parks Canada (2022b) Auyuittuq National Park. https://www.pc.gc.ca/en/pn-np/nu/auyuittuq. Accessed 10 Aug 2022

Parks Canada (2022c) Ivvavik National Park. https://www.pc.gc.ca/en/pn-np/yt/ivvavik. Accessed 10 Aug 2022

Parks Canada (2022d) Pingo National Landmark. https://www.pc.gc.ca/en/lhn-nhs/nt/pingo. Accessed 10 August 2022

Parks Canada (2022e) Qausutttuq National Park. https://www.pc.gc.ca/en/pn-np/nu/qausuittuq. Accessed 10 Aug 2022

Parks Canada (2022f) Quttinirpaaq National Park. https://www.pc.gc.ca/en/pn-np/nu/quttinirpaaq. Accessed 10 Aug 2022

Parks Canada (2022g) Sirmilik National Park https://www.pc.gc.ca/en/pn-np/nu/sirmilik. Accessed 10 Aug 2022

Paulla CK, Dallimore SR, Jin YK, Caress DW, Lundsten E, Gwiazda R, Anderson K, Hughes-Clarke J, Youngblut S, Melling H (2022) Rapid seafloor changes associated with the degradation of Arctic submarine permafrost. PNAS 119:e2119105119. https://doi.org/10.1073/pnas.2119105119

Phillips D (1990) The climates of Canada. Environment Canada, 176 p.

Piepjohn K, von Gosen W (2017) Structural transect through Ellesmere Island (Canadian Arctic): superimposed Palaeozoic Ellesmerian and Cenozoic Eurekan deformation. In: Pease V, Coakley B (eds) Circum-arctic lithosphere evolution. Geological society, vol 460, London, Special Publications, pp 33–56. https://doi.org/10.1144/SP460.5

Pocklington R, Morgan R, Drinkwater K (1994) Why we should not expect 'greenhouse warming' to be a significant factor in the Eastern Canadian coastal zone in the near future. In: Wells PG, Ricketts PJ (eds) Coastal Zone Canada 1994, Cooperation in the Coastal Zone, Bedford Institute of Oceanography, vol 4, pp 1824–1830

Porsild AE (1964) Illustrated flora of the Canadian Arctic Archipelago, vol 146. Nature Museum of Canada Bulletin, Ottawa

Porsild AE, Cody WJ (1980) The vascular plants of continental Northwest Territories, Canada. National Museum of Canada, Ottawa

Radosavljevic B, Lantuit H, Pollard W, Overduin P, Couture N, Sachs T, Helm V, Fritz M (2016) Erosion and Flooding—Threats to Coastal Infrastructure in the Arctic: a case study from Herschel Island, Yukon Territory, Canada. Estuaries Coasts 39:900–915

Ramage JL, Irrgang AM, Herzschuh U, Morgenstern A, Couture N, Lantuit H (2017) Terrain controls on the occurrence of coastal retrogressive thaw slumps along the Yukon Coast, Canada. J Geophys Res Earth Surf 122:1619–1634

Rampton VN (1982) Quaternary geology of the Yukon Coastal Plain. Geol Surv Canada Bull 317

Rattue KG (2021) Environmental protection in the Canadian High Arctic: Tallurutiup Imanga/Lancaster Sound as a case study for reconceptualising governance. Unpublished Ph.D. Thesis, University of Oxford. Accessed 2 Aug 2022

Reeves RR (1988) Cetaceans of Canada. Fisheries and Oceans Canada

Rudy ACA, Lamoureux SF, Treitz P, Van Ewijk K, Bonnaventure PP, Budkewitsch P (2017) Terrain controls and landscape-scale susceptibility modelling of active-layer Detachments, Sabine Peninsula, Melville Island, Nunavut. Permafrost Periglacial Process 28:79–91

Russell D (1992) International ocean boundary issues and management arrangements. In: VanderZwaag D (ed) Canadian ocean law and policy. Butterworths, Toronto, pp 463–507

Ryan AG (1995) Native Trees and Shrubs of Newfoundland and Labrador. Park Interpretation Publication 14, Parks and Natural Areas Division, Newfoundland and Labrador

Sawatzky S, Cunsolo A, Shiwak I, Flowers C, Jones-Bitton A, Gillis D, Middleton J, Wood M, Rigolet Inuit Community Government, Harper SL (2021) "It depends…": Inuit-led identification and interpretation of land-based observations for climate change adaptation in Nunatsiavut, Labrador. Reg Environ Change 21:54. https://doi.org/10.1007/s10113-021-01772-4

Scott DJ, deKemp EA (1998) Bedrock geology compilation, northern Baffin Island and northern Melville Peninsula, Northwest Territories. Geological Survey of Canada, Open File 3633

Serreze C, Barry RG (2005) The Arctic climate system. Cambridge University Press, New York

Serreze MC, Meier WN (2019) The Arctic's sea ice cover: trends, variability, predictability, and comparisons to the Antarctic. Ann N Y Acad Sci 1436:36–53

Sharp M, Burgess DO, Cawkwell F, Copland L, Davis JA, Dowdeswell EK, Dowdeswell JA, Gardner A, Mair D, Wang L, Williamson, S, Wolken GJ, Wyatt F (2014) Remote sensing of recent glacier changes in the Canadian Arctic. In: Kargel JS, Leonard GJ, Bishop MP, Kääb A, Raup BH (eds) Global land ice measurements from space, Ch. 9. Praxis-Springer, pp 205–228. https://doi.org/10.1007/978-3-540-142

Shaw J, Taylor RB, Forbes DL, Solomon S, Ruz M-H (1999) Sensitivity of the coasts of Canada to sea-level rise, vol 505. Geological Survey of Canada, Bulletin

Sheardj W, Geale DW (1983) Vegetation studies at Polar Bear Pass, Bathurst Island, N.W.T. I. Classification of plant communities. Can J Botany 61:1618–1636

Slaymaker O, Catto NR, Kovanen DJ (2020a) Holocene geomorphic processes in Eastern Canada. In: Slaymaker O, Catto NR (eds) Landscapes and landforms of Eastern Canada. Springer International Publishers, Cham, Switzerland, pp 79–119

Slaymaker O, Catto NR, Kovanen DJ (2020b) Protecting geodiversity in eastern Canada. In: Slaymaker O, Catto NR (eds) Landscapes and landforms of Eastern Canada. Springer International Publishers, Cham, Switzerland, pp 557–582

Smith CAS, Meikle JC, Roots CF (2004) Ecoregions of the Yukon Territory: biophysical properties of Yukon landscapes. In: Agriculture and Agri-Food Canada, PARC Technical Bulletin No. 04-01, Summerland, British Columbia, 313 p

Smith S (2011) Trends in permafrost conditions and ecology in northern Canada. In: Canadian biodiversity: ecosystem status and trends 2010. Technical Thematic Report No. 9. Canadian Councils of Resource Ministers, Ottawa, ON, 22 p

Soil Classification Working Group (1998) The Canadian system of soil classification. 3rd edn. Publication, Agriculture and Agri-Food Canada, p 1646

St-Onge MR, Harrison JC, Paul D, Tella S, Brent TA, Jauer CD, Maclean BC (2015) Tectonic map of Arctic Canada. In: Geological Survey of Canada, Canadian Geoscience Map 187. https://doi.org/10.4095/295945

St-Onge MR, Scott DJ, Rayner N, Sanborn-Barrie M, Skipton D, Saumur B et al (2020) Archean and Paleoproterozoic cratonic rocks of Baffin Island. In: Dafoe LT, Bingham-Koslowski N (eds) Baffin Island and the Labrador-Baffin Seaway geological synthesis. Geological Survey of Canada, Bulletin, vol 608

Stokes CR, Tarasov L, Blomdin R, Cronin TM, Fisher TG, Gyllencreutz R, Hättestrand C, Heyman J, Hindmarsh RCA, Hughes ALC, Jakobsson M, Kirchner N, Livingstone SJ, Margold M, Murton JB, Noormets R, Peltier WR, Peteet DM, Piper DJW, Preusser F, Renssen R, Roberts DH, Roche DM, Saint-Ange F, Stroeven AP, Teller JT (2015) On the reconstruction of palaeo-ice sheets: recent advances and future challenges. Quatern Sci Rev 125:15–49

Storrar RD, Livingstone SJ (2017) Glacial geomorphology of the northern Kivalliq region, Nunavut, Canada, with an emphasis on meltwater drainage systems. J Maps 13:153–164

Suprenand PM, Ainsworth CH, Hoover C (2018) Ecosystem model of the entire Beaufort Sea Marine ecosystem: a temporal tool for assessing food-web structure and marine animal populations from 1970 to 2014. Marine Science Faculty Publications 261. https://scholarcommons.usf.edu/msc_facpub/261. Accessed 11 August 2022.

Tang CCL Ross CK Yao T Petrie B DeTracey BM Dunlap E (2004) The circulation, water masses and sea-ice of Baffin Bay. Bedford Institute of Oceanography Fisheries and Oceans Canada. Report F56–020

Tedrow JCF (1974) Soils of the high arctic landscape. In: Smiley TL, Zumberge JH (eds) Polar deserts and modern man. University of Arizona Press, Arizona, pp 63–69

Timmermans M-L, Marshall J (2020) Understanding Arctic Ocean circulation: a review of ocean dynamics in a changing climate. J Geophys Res: Oceans 125:e2018JC014378. https://doi.org/10.1029/2018JC014378

Trettin HP (1991) Geology of the Innuitian Orogen and Arctic Platform of Canada and Greenland. Geology of Canada, no. 3. Geological Survey of Canada, Ottawa, pp 29–56

Utting DJ, Gosse JC, Kelley S, Vickers K, Ward EC, Trommelen MT (2016) Advance, deglacial, and sea level chronology for Foxe Peninsula, Baffin Island, Nunavut. Boreas 45:439–454

Van Wychen W, Copland L Burgess D (2020) Ice Masses of the eastern Canadian Arctic Archipelago. In: Slaymaker O, Catto NR (eds) Landscapes and Landforms of Eastern Canada. Springer International Publishers, Cham, Switzerland, pp 297–314

Vincent J-S (1982) The quaternary history of banks Island, N.W.T., Canada. Géographie physique et Quaternaire 36:209–232

Vincent JS (1983) La géologie du Quaternaire et la géomorphologie de l'île de Banks, Arctique canadien. Commun Géol Can Mémoire 405

Walker DA, Raynolds MK, Daniëls FJA, Einarsson E, Elvebakk A, Gould WA, Katenin AE, Kholod SS, Markon CJ, Melnikov ES, Moskalenko NG, Talbot SS, Yurtsev BA (2005) The Circumpolar Arctic vegetation map. J Veg Sci 16:267–282

Wardle RJ, Gower CF, Ryan B, Nunn GAG, James DT, Kerr A (1997) Geological Map of Labrador; 1:1 million scale. Government of Newfoundland and Labrador, Department of Mines and Energy, Geological Survey, Map 97-07

Way RG, Bell T, Barrand NE (2015) Glacier change from the early Little Ice Age to 2005 in the Torngat Mountains, northern Labrador, Canada. Geomorphology 246:558–569

Wheeler JO, Hoffman PF, Card KD, Davidson A, Sanford BV, Okulitch AV, Roest WR (1996) Geological Map of Canada. Geol Survey of Canada, Map D1860A. https://ftp.maps.canada.ca/pub/nrcan_rncan/raster/atlas_4_ed/eng/environment/land/025_26.jpg Accessed 12 July 2022

Wolfe AP (1996) Wisconsinan Refugial landscapes, eastern Baffin Island, Northwest Territories. Can Geogr 40:81–87

Woo M (2012) Permafrost hydrology. Springer International Publishing, Berlin, p 516

Zufelt J (2021) Permafrost 2021 merging permafrost science and cold regions engineering. In: Regional conference on permafrost 2021 and the 19th international conference on cold regions engineering October 24–29, 2021. American Society of Civil Engineers

10 National Parks and Conserved Areas

Abstract

Canada is perceived as dominated by vast, unaltered wilderness by many people. Statistically, the population density of 3.70 people km^{-2} spread across 9,984, 570 km^2 suggests open, untrammelled space. In reality, approximately 75% of the population live in the 41 largest Census Metropolitan Areas. "Natural" areas of more than 500 km^2 no longer exist in most of southern Canada. Although humans have occupied Canada for shorter periods than in many areas of Africa and Eurasia, pre-1600 CE modifications of landscapes have been significant locally and regionally, particularly from selective burning. Traditional uses by residents since 1600 CE have resulted in significant changes in beaches and coastal terrain. National Parks are protected areas designated under the Canada National Parks Act, owned by the Government of Canada. The parks are administered by Parks Canada, a Crown agency under the Ministry of Environment and Climate Change Canada. The goal of the national parks system is to set aside lands representing Canada's 43 terrestrial natural regions (including subdivisions) as described in the National Parks System Plan. Concepts and practices in Canadian parks developed through a combination of USA and UK influences. The first were created as pleasure parks for the benefit and enjoyment of the people, not to conserve ecology. Parks Commissioner James B Harkin, serving 1911-1935, saw exposure to "nature" as a societal cure, stressed the economic aspects of tourism, and believed in "nature as medicine". Harkin was not a natural scientist and tended to hire people who similarly lacked ecological training or experience. Harkin's vision continued to guide national park establishment and management until the early 1970s. After 1975, the revised mission of Parks Canada was primarily to protect the ecological integrity of the land, and secondarily to allow the public to explore, learn about, and enjoy Canada's natural spaces. Current challenges to the protected area system in Canada include insufficient protection, visitation pressures, parks as isolated ecological islands, stresses to ecological integrity, and the roles of money and politics. Analysis of Canada's history with parks and protected area reveals a chequered pattern of changing motivations, assumptions, and intentions; community engagement and centralized bureaucratic impositions; and overall successes and shortcomings. Establishment and maintenance of parks, and all decisions involved with them, reflect the socio-economic, political, and cultural contexts of various times. Although parks may be thought of as elements of the natural environment, they are also a form of infrastructure, societal creations.

Keywords

Landscapes · National Parks · Park management · Protected areas · Current challenges

10.1 Landscapes, Wilderness, and Attitudes

Canada is perceived as dominated by vast, unaltered wilderness by many people. Statistically, the population density of 3.70 people km^{-2} spread across 9984, 670 km^2 suggests open, untrammelled space. In reality, approximately 75% of the population live in the 41 largest Census Metropolitan Areas, and only NWT and Nunavut have more urban than rural residents (Fig. 10.1). "Natural" areas of more than 500 km^2 no longer exist in most of southern Canada.

Although some Canadian geographers identify "cultural landscapes", where human imprints are dominant, as distinct from "natural landscapes", others have recently discarded this distinction. In southern Canada, human imprints are more subtle and pervasive than may first appear. Landscapes throughout much of Atlantic Canada, the Great Lakes–St. Lawrence Lowland, the Prairies, southern Cordillera, coastal Pacific maritime, and even much of the Boreal and southern

N. Catto and G. R. Vincent, *Canada: Land and People*, World Regional Geography Book Series
https://doi.org/10.1007/978-3-032-04391-7_10

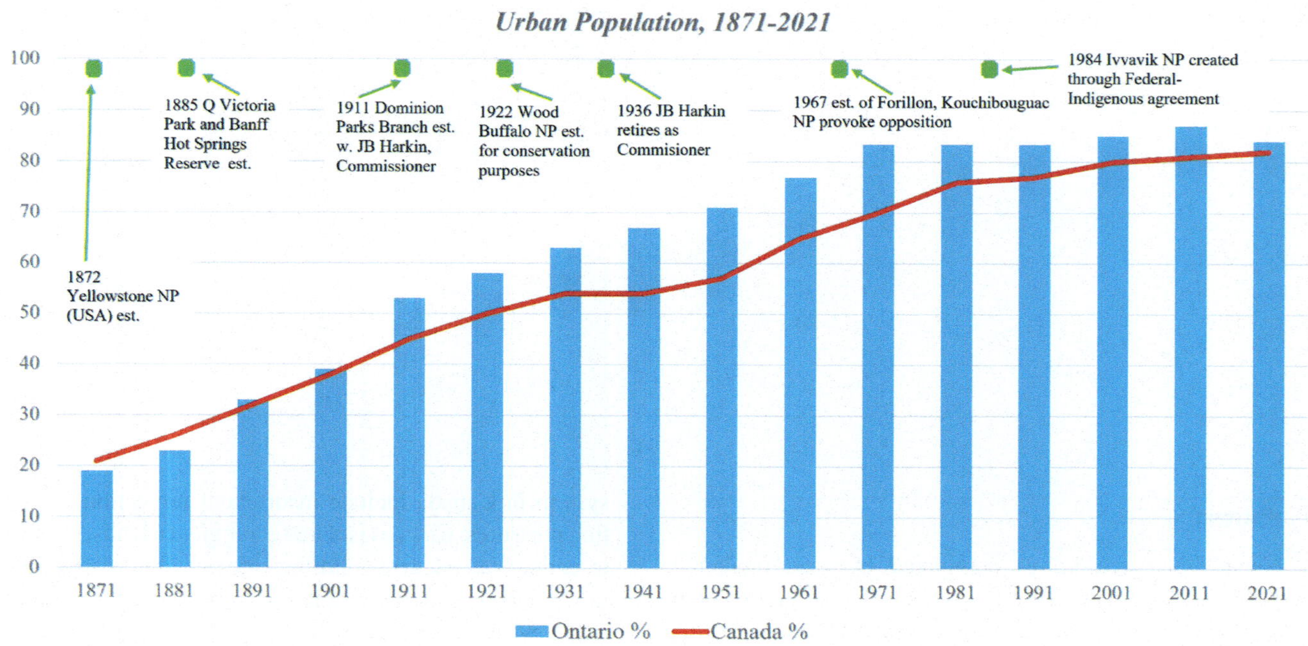

Fig. 10.1 Urban Population, 1871–2021. Urban populations for Canada and Ontario (the most urbanized province) are shown along with events in park management discussed in the text. Population data from Statistics Canada, Census of Canada

Taiga Shield regions have been altered by humans so extensively that the concept of a "natural landscape" or "untouched wilderness" is meaningless: all landscapes should be considered in part as cultural (Higgins et al. 2012; Förster et al. 2013; Youdelis et al. 2020). This approach is exemplified in the definition of landscape as an environment "as perceived by people" (European Landscape Convention; Council of Europe, 2000).

Although humans have occupied Canada for shorter periods than in many areas of Africa and Eurasia, pre-1600 ce modifications of landscapes have been significant locally and regionally, particularly from selective burning (Ryan et al. 2013). Traditional uses by residents since 1600 ce have resulted in significant changes in beaches and coastal terrain (e.g. Catto and Catto 2012).

Human-environmental relations have marked many communities and residents of Canada for hundreds of years. The frequent personification of rivers and individual mountains indicates the connections felt by people with terrain. The sense of place and human connectivity with landscape has led to the desire for conservation of landscapes. However, environmental perceptions in Canada have also varied over both place and time, colouring both human-environmental relationships and strategies for preserving or conserving land and water.

Exposure and attitudes to weather events reflect one aspect of human–environment relations. Winter-related injuries and deaths can be divided into occupational, socio-economic, and recreationally related events. In Canada, increasing proportions of casualties from snow, cold, and avalanches involve people engaged in recreational activities, rather than those whose occupations require outdoor work or those who lack economic resources to avoid cold.

All areas of Canada are exposed to at least some snowfall (and all its consequences), although the amount of snow cover and its duration vary significantly across the country. Sensitivity, the potential impact of a hazard, depends upon human, community, or societal conditions. Familiarity with the hazard, or measures taken to reduce the impact, can result in a reduction of sensitivity.

St. John's, NL, receives more than 3 m of snow annually, with as much as 90 cm in one day (Environment Canada 2020a). Based on experience, residents have developed coping strategies, involving shovels, snowploughs, snow tyres, and snowmobile suits. Weather forecasting has improved to allow "snow days" to be announced one to two nights in advance, letting people prepare and make alternate arrangements for work, school, and social commitments. Although exposure to snowfall is high, sensitivity is low. Even during the week-long closure of the city resulting from the mid-latitude cyclone of mid-January 2020, experience and coping reflected the residents' adaptive capacity (see Chap. 13).

Early European settlers saw snow as a threat, rather than as a recreational opportunity or source of groundwater recharge. Throughout the period 1605–1850, most settlers came from regions in Europe with lesser winter snowfalls. Non-adaptability was exemplified by Champlain's first less-than-successful colony at Port-Royal (1605). Snow remained a perceived threat through European settlement of Ontario, where residents tended to "hive-up" during the winter months, and in western Canada, where the CPR actively discouraged discussion of snow and promotion of Winter Carnivals in the 1880s. Winnipeg's first winter festival was not held until 1922. Today's perception of snow as a valued resource for groundwater recharge, reduction of frost penetration, impact on surface albedo, and winter recreation and tourism, together with concern over declining snowfall in mountain parks, would shock the average nineteenth-century Canadian of European descent. Similarly, a pre-Confederation seafarer from Newfoundland would be amazed to hear twenty-first-century residents express concern about the scarcity of icebergs. Fog appears romantic to the visitor, but dangerous to the fish harvester. The transition from weather as exclusive threat to a more nuanced view varies with time, place, community, and individual.

Changing attitudes with time and place can be recognized for other aspects of the natural environment as well. Logging, clearing (or "reclaiming") the land was seen as a mark of civilization throughout the nineteenth century. The idea of conserving a landscape which was not markedly "scenic"—not mountainous, with no agricultural value; or rugged coastal, with no possibility for port development—did not take hold until the mid-twentieth century.

In 1921, slightly less than half of the Canadian population lived in cities (Fig. 10.1). Montréal (618,505) was the largest, with five others having populations exceeding 100,000. In 2022, the urban population exceeded 31 million (82%), with 34 large urban centres having populations in excess of 100,000. In addition, the growth in urbanized area increasingly separated urban from non-urban landscapes.

Approximately 50% of Canada's Canada Land Inventory Class 1 farmland is located within a 50 km radius of downtown Toronto, and much of this has been urbanized or suburbanized (Catto et al. 2020). In 1921, the distance from downtown Toronto to the outer edge of urban development was approximately 5 km: in 2024, it was 45 km. Urban dwellers in 1921 had much more direct connections to diverse environments and landscapes than at present. Humans' relationships with their environments, including whether humans perceive themselves as part of nature or separate from it (c.f. Levin and Unsworth 2013), influence attitudes towards parks.

The desire for conservation, seeing the landscape and ecology as inherently valuable, developed initially in urban-based academic circles with people living physically distant from the areas they wished to conserve. Sometimes, those thoughts did not fully account for people, both indigenous and non-indigenous, who had been living in and with the landscapes for some time. The concept of "wilderness"—uninhabited by humans, or essentially undisturbed by human activity, with its natural ("original") ecology—assumes that any previous or present occupants have not modified the landscape, and that the ecology has remained unaltered throughout time. As our understandings of climate and consequent ecological changes throughout the past 20,000 years, combined with human activity over much of that timespan in Canada, have grown, the concepts of a "natural landscape" and conservation in general require revision.

10.2 National Parks

National Parks are protected areas designated under the Canada National Parks Act (2000), owned by the Government of Canada. The parks are administered by Parks Canada, a Crown agency under the Ministry of Environment and Climate Change Canada (ECCC). The goal of the national parks system is to set aside lands representing Canada's 43 terrestrial natural regions (including subdivisions) as described in the National Parks System Plan. Parks Canada also manages Pingo National Landmark, three National Marine Conservation Areas, and one National Marine Conservation Area Reserve. National historic sites also are administered by Parks Canada.

As of April 2024, there were 49 federal National Parks and Park Reserves, representing 33 of the natural regions (Fig. 10.2, Table 10.1). The Manitoba Lowlands and the Interior Dry Plateau (southern British Columbia) regions have areas which have been formally proposed as National Park Reserves. Of the 8 remaining areas, all except the Shugliaq (Southampton Island) Plain have designated provincial or co-managed parks. Shugliaq has two ECCC Migratory Bird Reserves and is flanked by a designated "Area of Interest" for future establishment of a Marine Protected Area, administered by Fisheries and Oceans Canada.

The process of establishing a National Park in Canada involves initial designation of the terrain as a reserve, subsequently followed by official proclamation as a National Park in print by the Government of Canada ("gazetting"). Reserve status indicates that some unresolved issues remain, which may include land ownership, compensation for former owners or for relinquished mineral rights, precise delineation of boundaries, indigenous title or co-management, and access to terrain for indigenous and other local residents (e.g. for hunting, wild plant harvesting, or other traditional uses). Significant time may elapse between initial Reserve designation and formal gazetting. Gros Morne was established as a Reserve in 1973, and as a UNESCO World Heritage site in 1987, but was gazetted as an official National Park only in 2005. Pacific Rim and Nahanni were designated as Reserves in 1970 and 1974, respectively, but remained ungazetted as of April 2024. Ten National Park Reserves existed as of April

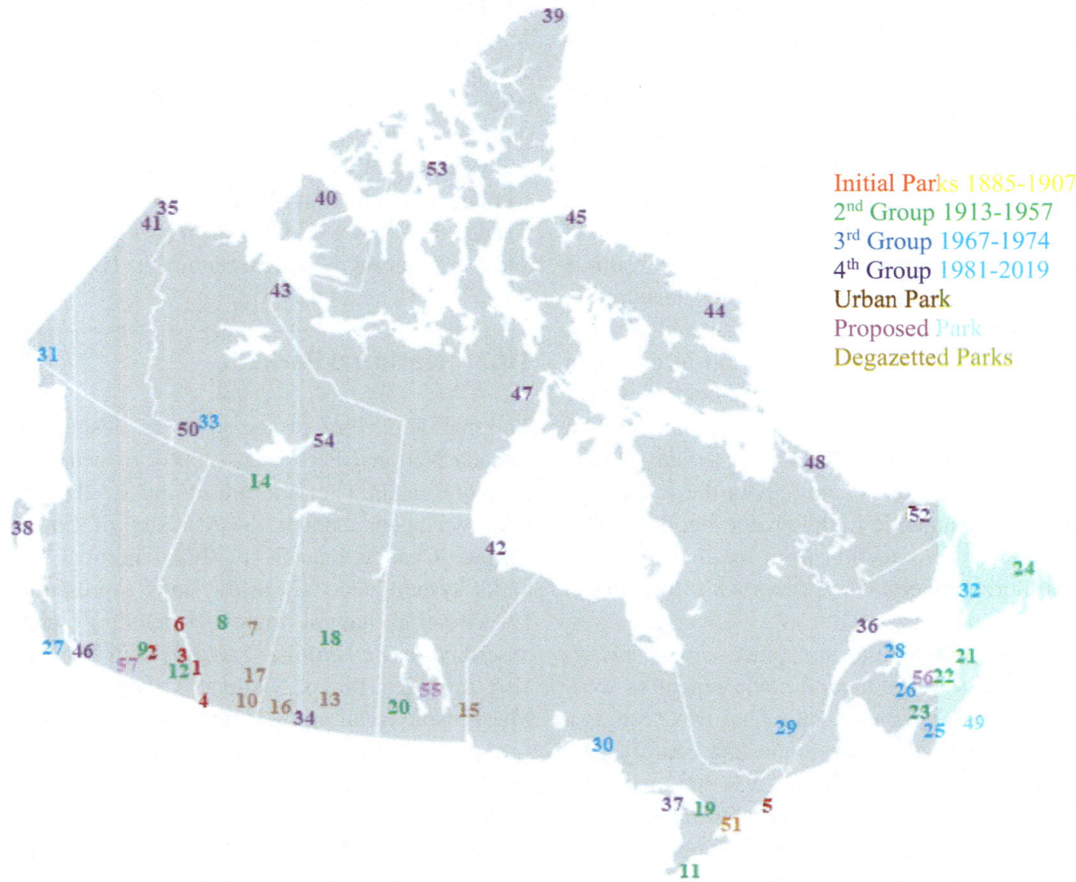

1 Banff	16 Menissawok	30 Pukaskwa	44 Auyuittuq
2 Glacier	17 Wawaskesy	31 Kluane	45 Sirmilik
3 Yoho	18 Prince Albert	32 Gros Morne	46 Gulf Islands
4 Waterton Lakes	19 Georgian Bay Islands	33 Nahanni	47 Ukkusiksalik
5 Thousand Islands	20 Riding Mountain	34 Grasslands	48 Torngat Mountains
6 Jasper	21 Cape Breton Highlands	35 Ivvavik	49 Sable Island
7 Buffalo	22 Prince Edward Island	36 Mingan Archipelago	50 Nááts'ihch'oh
8 Elk Island	23 Fundy	37 Bruce Peninsula	51 Rouge
9 Mountt Revelstoke	24 Terra Nova	38 Gwaii Haanas	52 Akami-Uapishk^{u}-KakKasuak-Mealy Mts
10 Nemiskam	25 Kejimkujik	39 Quttinirpaaq	53 Qausuittuq
11 Point Pelee	26 Kouchibouguac	40 Aulavik	54 Thaidene Nëné
12 Kootenay	27 Pacific Rim	41 Vuntut	55 "Manitoba Lowlands"
13 Vidal Point	28 Forillon	42 Wapusk	56 Pituamkek
14 Wood Buffalo	29 La Mauricie	43 Tuktut Nogait	57 South Okanagan—Similkan
15 Brereton Lake			

Fig. 10.2 National Parks in Canada, 2024. Parks are listed in numerical order of establishment. National Marine Conservation Areas, National Landmarks, and National Historic Sites are not illustrated. Parks are divided into phases of establishment, as discussed in text

2024. Most visitors will notice no difference in experiences between Reserves and gazetted National Parks.

Removal of some or all protection ("degazetting") can occur. Whiteshell (Brereton Lake, MB) and Katepwa (Vidal) Point (SK) parks were transferred to their respective provincial jurisdictions in 1930, remaining as provincial parks. Menissawok (SK), Nemiskam (AB), and Wawaskesy (AB), all initially established to protect threatened pronghorn (*Antilocapra americana*), were degazetted in 1947 after populations rebounded (Lothian 1987). The terrain of Wawaskesy was also used as Canadian Military Forces Base (CFB) Suffield, beginning in World War II, and is currently part of the Suffield National Wildlife Refuge. Buffalo National Park (AB), established in 1908, was also degazetted in 1947 after incorporation in CFB Wainwright.

Table 10.1 National Parks, National Landmark, and National Marine Conservation Areas. Sites are listed in order of establishment. Colour coding follows Fig. 10.2

Park	Est	Status[a]	Natural region	Ecozone	UNESCO affiliation[b]	Recr. Facil.[c]	Adjacent communities[d]
Banff	1885		Rocky Mountains	Montane Cordillera	WHS: CRM Parks	G, S, H	Banff I; Lake Louise I
Glacier	1886		Columbia Mts	Montane Cordillera			Donald S
Yoho	1886		Rocky Mountains	Montane Cordillera	WHS: CRM Parks		Field I; Golden S
Waterton Lakes	1895		Rocky Mountains	Montane Cordillera	WHS: CRM Parks; Internat. Peace Park	G	Waterton Park Village I
Thousand Islands	1904		Gt Lks-St. Law.; Precambrian	Boreal Shield; Mixedwood Plains	BR: Frontenac Arch	G	Ivy Lea S
Jasper	1907		Rocky Mountains	Montane Cordillera	WHS: CRM Parks	G, S, H	Jasper I
Buffalo	1908	dgz 1947	S Boreal Plains & Plateaux	Prairies			Wainwright
Elk Island	1913		S Boreal Plains & Plateaux	Boreal Plains			Edmonton
Mount Revelstoke	1914		Columbia Mts	Montane Cordillera			Revelstoke S
Nemiskam	1914	dgz 1947	Prairie Grasslds	Prairies			Nemiskam
Point Pelee	1918		St Lawrence Lowlands	Mixedwood Plains	Ramsar Convention wetland		Leamington S
Kootenay	1920		Rocky Mountains	Montane Cordillera	WHS CRM Parks	H	Radium Hot Springs S
Vidal Point (Katepwa SK Prov Park)	1921	dgz 1930	Prairie Grasslands	Prairies		G	Katepwa Beach I
Wood Buffalo	1922		Northern Boreal Plains	Boreal Plains; Taiga Plains	WHS: Wood Buffalo		Fort Smith S
Brereton Lake (Whiteshell MB Prov Park)	1922	dgz 1930	Precambrian	Boreal Shield		G	Whiteshell I; Rennie S
Menissawok	1922	dgz 1930	Prairie Grasslands	Prairies			Maple Creek
Wawaskesy	1922	dgz 1947	Prairie Grasslands	Prairies			Medicine Hat
Prince Albert	1927		S Boreal Plains & Plateaux	Boreal Plains		G	Waskesiu I
Georgian Bay Islands	1930		Gt Lks-St. Law.; Precambrian	Boreal Shield	BR: Georgian Bay Littoral		Honey Harbour S
Riding Mountain	1933		S Boreal Plains & Plateaux	Boreal Plains		G	Wasagaming I
Cape Breton Highlands	1936		Maritime Acadian Highlds	Atlantic Maritime		G	Cheticamp S; Ingonish E
Prince Edward Island	1937		Maritime Plain	Atlantic Maritime		G	Cavendish S; Stanhope S; Brackley Beach S
Fundy	1948		Maritime Acadian Highlds	Atlantic Maritime	BR: Fundy	G	Alma S
Terra Nova	1957		E Nfld Atlantic	Boreal Shield		G	Charlottetown E, Port Blandford S
Kejimkujik	1967		Atlantic Coastal Uplands	Atlantic Maritime	BR: Southwest Nova		Maitland Bridge S

(continued)

Table 10.1 (continued)

Park	Est	Status[a]	Natural region	Ecozone	UNESCO affiliation[b]	Recr. Facil.[c]	Adjacent communities[d]
Kouchibouguac	1969		Maritime Plain	Atlantic Maritime			Kouchibouguac S
Pacific Rim	1970	Reserve	Pacific Coast Mts	Pacific Maritime			Tofino S
Forillon	1970		Laurentian Boreal Highlands	Atlantic Uplands			Gaspé S; Cap-aux-Os E
La Mauricie	1970		Gt Lks-St. Law.; Precambrian	Boreal Shield			St-Mathieu-du-Parc S
Pukaskwa	1971		Central Boreal Uplands	Boreal Shield			Heron Bay S
Kluane	1972	gaz 1993	Northern Coast Mountains	Pacific Maritime	WHS: Kluane/ Wrangell–St. Elias/ Glacier Bay/ Tatshenshini-Alsek		Haines Junction
Gros Morne	1973	gaz 2005	W Newfoundland Uplands	Boreal Shield	WHS: Gros Morne	G	Norris Hr E, Rocky Hr E, Sallys Cove E, St. Paul's E, Trout River E
Nahanni	1974	Reserve	Mackenzie Mts	Taiga Cordillera	WHS: Nahanni	*	Łíídlı̨ı̨ Kų́ę́
Grasslands	1981		Prairie Grasslds	Prairies			Val Marie S
Ivvavik	1984		Northern Yukon	Southern Arctic; Tundra Cordillera		*	Old Crow
Mingan Archipelago	1984	Reserve	St Lawrence Lowlands	Gulf of St. Lawrence			Havre-St-Pierre S
Bruce Peninsula	1987		St Lawrence Lowlands	Mixedwood Plains	BR: Niagara Escarpment		Tobermory S
Gwaii Haanas	1988	Reserve	Pacific Coast Mountains	Pacific Maritime	WHS: SG̱ang Gwaay Llanagaay	*	Skidegate S
Quttinirpaaq	1988		Eastern High Arctic	Arctic Cordillera		*	Alert/Resolute
Aulavik	1992		Western Arctic Lowlands	Western Arctic		*	Sachs Harbour
Vuntut	1993		Northern Yukon	Tundra Cordillera		*	Old Crow
Wapusk	1996		Hudson-James Lowlands	Hudson Plains			Churchill
Tuktut Nogait	1998		Tundra Hills	Southern Arctic		*	Paulatuk
Auyuittuq	2001		Northern Davis	Northern Arctic		*	Iqaluit
Sirmilik	2001		Eastern Arctic Lowlands	Arctic Cordillera; Northern Arctic		*	Mittimatalik
Gulf Islands	2003	Reserve	Salish Sea Lowlands	Pacific Maritime			Sidney
Ukkusiksalik	2003		Central Tundra	Northern Arctic		*	Qamani'tuaq
Torngat Mountains	2008		North Labrador Mountains	Arctic Cordillera		*	Nain
Sable Island	2013	Reserve	Scotian Shelf	Scotian Shelf		*	Halifax
Nááts'ihch'oh	2014	Reserve	N Interior Plateaux & Mts	Taiga Cordillera		*	Łíídlı̨ı̨ Kų́ę́

(continued)

Table 10.1 (continued)

Park	Est	Status[a]	Natural region	Ecozone	UNESCO affiliation[b]	Recr. Facil.[c]	Adjacent communities[d]
Rouge	2015		St Lawrence Lowlands	Mixedwood Plains			Toronto
Akami-Uapishk^{u}-KakKasuak-Mealy Mts	2015	Reserve	East Coast Boreal	Taiga Shield			Happy Valley-Goose Bay
Qausuittuq	2015		Western High Arctic	Western Arctic		*	Resolute
Thaidene Nënë	2019	Reserve	NW Boreal Uplands	Taiga Shield		*	Yellowknife
"Manitoba Lowlands" (unofficial name)	–	Proposed	Manitoba Lowlands	Boreal Plains			Dauphin
Pituamkek	–	Proposed	Maritime Plain	Atlantic Maritime			Lennox Island
South Okanagan-Similkameen	–	Proposed	Interior Dry Plateau	Cordillera Plateaux			Osoyoos S
Pingo National Landmark	1984		Mackenzie Delta	Southern Arctic			Tuktoyaktuk
Fathom Five	1987	NMCA	St Lawrence Lowlands	Great Lakes			Tobermory S
Saguenay-St. Lawrence	1998	NMCA	Gt Lks-St. Law.; Precambrian	Gulf of St. Lawrence			Tadoussac
Gwaii Haanas	2010	NMCA	Pacific Coast	Northern Shelf		*	Skidegate S
Lake superior	2015	NMCA	Precambrian	Great Lakes			Thunder Bay
Tallurutiup Imanga	2021	NMCA	Tallurutiup Imanga	Eastern Arctic		*	Ikpiarjuk
Iles-de-la-Madeleine	–	Proposed NMCA	Gulf of St Lawrence	Gulf of St. Lawrence			Havre-Aubert
Salish Sea	–	Proposed NMCA	Salish Sea	Salish Sea			Victoria

[a]Status: dgz, degazetted (date); gaz, gazetted (date) after establishment; reserve, not gazetted as of March 2024; NMCA, National Marine Conservation Area
[b]UNESCO Affiliation: WHS, World Heritage Site; CRM, Canadian Rocky Mountains; BR, Biosphere Reserve
[c]Recreational Facilities: G, golf course; H- hot springs; S, ski hill
[d]Adjacent communities: I, internal, within park boundary; E, enclave, surrounded by park; S, service community
*no road access

Discussions concerning the appropriate balance between preserving or conserving ecological integrity and recreational use, other permitted uses, operational concerns, relationships with adjacent human communities, the role of educational programmes, and budgetary constraints have marked the history of provincial and national parks alike. Between 1995 and 1999, Newfoundland and Labrador degazetted 56 provincial parks and "scenic attractions" due to financial constraint, turning them over to community groups and private owners.

10.3 Other Protected Areas

Gatineau Park is located north of the Ottawa River, west of Gatineau, Québec. Gatineau is not considered part of the national park system of Canada, as it is administered by the National Capital Commission (NCC) in the same manner as the museums, national gallery, and art centre located in the national capital region (Gagnon and Filion 2004, NCC n.d.). Gatineau is considered primarily as a recreational park by the NCC, rather than being regarded exclusively as a conservation area. Private residences pre-dating the park's establishment are permitted within the park boundaries. Gatineau is also noted for containing numerous lakes, including Meech Lake, the location of the "retreat" of the Prime Minister of Canada.

Rouge River, established in 2015 to include 40 km^2 of riparian and ravine terrain in eastern metropolitan Toronto, is currently the only National Urban Park (Table 10.1). Plans exist to establish others in Saskatoon (Meewasin Valley), Halifax, Edmonton, Montreal, Winnipeg, Windsor (ON), and Colwood (BC), focused around rivers and coastal areas (Parks Canada 2019a). Although most municipal parks were established for recreation, some also include conservation areas.

In the Canadian Federal system, provinces and territories have the authorization to establish parks. Beginning with Queen Victoria Park (Niagara Falls, 1885) and Algonquin Provincial Park (1893) in Ontario, and Laurentides and Mont-Tremblant (1895) in Québec, all Provinces and Territories have also designated areas as parks, with protection, permitted uses, and terminology varying. Sizes vary from picnic areas to the 2.36 million hectare Polar Bear Provincial Park (ON). As of April 2024, Ontario had 329 provincial parks, in addition to 292 conservation areas. Many are small, intended to provide an opportunity for recreation, or less commonly to conserve a particular local landscape.

In Québec, the designation "*Parc National*" refers to the 27 parks established and administrated by the Government of Québec through the *Société des établissements de plein air du Québec* (Sépaq), not by the federal Government of Canada. Laurentides, Mont-Tremblant, Miguasha and Percé (Gaspé), and Saguenay Fjord are examples. Kuururjuaq, Pingualuit crater, and Ulittaniujalik in Nunavik are jointly administered with the Kativik Regional Government. Forillon, La Mauricie, and Mingan Archipelago are the only parks in Québec administered exclusively by the Government of Canada.

Canada ratified the UNESCO Convention Concerning the Protection of the World Cultural and Natural Heritage (World Heritage Convention; UNESCO 2011a, 2011b) in 1976. Parks Canada is the lead agency for implementation in Canada (Slaymaker et al. 2017, 2020b). Initially, the Convention divided World Heritage sites into exclusive natural and cultural categories. The effects of human population growth, recent climate and environmental change, recognition of human modification of landscapes and environments over the past 1000 or more years, and growing appreciation of the inter-relationships between people and land have increasingly made this distinction both moot and inappropriate. All landscapes are cultural, and landforms are integral components of places of cultural or historical significance. This changing appreciation also continues to influence park establishment, maintenance, conservation, and perceptions (Kareiva et al. 2007, Ellison 2013, Förster et al. 2013, Chylińska 2019, Youdelis et al. 2020, Slaymaker et al. 2021, Council of Europe 2022).

10.4 National Parks and Parks Canada: Establishment and Evolution

Definitions of "park" in the Canadian Oxford Dictionary (Barber 1999) include:

- a piece of land, usually with lawns, gardens, etc., in a town or city, maintained at public expense for recreational use;
- a large area of government land kept in its natural state for recreational use, wildlife conservation, etc.;
- a large enclosed area of land, etc., either public or private, used to accommodate wild animals in captivity;
- a large enclosed piece of ground, usually with woodland and pasture, attached to a stately home.

The common use of the same word to describe both an area for the protection of wildlife or natural habitats and an area intended for human enjoyment and recreation illuminates the issues that have surrounded national and provincial park establishment and management in Canada since 1885.

10.4.1 Initial Influences

As in several matters, concepts and practices in Canada developed through a combination of USA and UK influences. The works of George P. Marsh (1864, 1885, see Curtis et al. 1982, Garvey 2009), John Muir (Worster 2008; Lasky 2014; McGuckin 2015), Nathaniel P. Langford (1905), and other USA conservationists led to the establishment of Yellowstone National Park in 1872 (Haines 1977). Langford's vision of Yellowstone as a "pleasure ground for the benefit and enjoyment of the people" was influenced by his observations of sites in UK and elsewhere in Europe that combined recreational, aesthetic, and perceived natural conservation. In addition, Langford regarded the commercial development of the area surrounding Niagara Falls as an example to be repudiated. Yellowstone's establishment was not based purely on ecological grounds, and the geological and geomorphic features were the primary natural aspects deemed worthy of conservation. Subsequent developments in USA National Parks, including the forced exclusion of indigenous peoples, the importance of railway and automobile tourism, and the construction of recreational infrastructure, found echoes in Canada.

In the nineteenth century UK, landscape preservation and conservation proceeded differently. Influenced by the earlier "picturesque" movement driven by visual artists and poets (Gilpin 1786, Andrews 1990, Bate 1991), emphasis was placed on landscape aesthetics rather than on either ecology or recreation. Although most of the areas had been substantially modified by forest clearance and agriculture over the previous 700 years, landscapes were romanticized as "natural". Human settlements continued, along with activities such as hunting. No government measures were taken to expropriate land or to restrict activities other than those which would alter the aesthetics: traditional uses were strongly encouraged. When the first UK national parks were established after World War II, the essential criteria were "great natural beauty, a high value for open-air recreation, and substantial continuous extent" (Hobhouse 1947). This view of landscape, including human settlements and land uses as integral parts, thus differs substantially from the sharp cultural/natural separation envisaged by both the early USA conservationists and the initial UNESCO World Heritage Convention.

10.4.2 Initiation in Canada

The initial efforts to establish parks in Canada were influenced by these cross-currents. Queen Victoria Park was established at Niagara Falls in 1885 to preclude the commercial development decried by Langford. Named for the reigning monarch, the park was (and is) a manicured space, designed to set a natural feature in a pleasant aesthetic environment as perceived by nineteenth-century designers. Admission was free, unless the visitor requested a tour guide. The park provided a space for recreation, allowing people to step outside the industrial environment: a pleasure ground (Ontario 2015).

10.4.2.1 Banff

In the same year, Banff Hot Springs Reserve was established, as John A. Macdonald's response to a dispute among competing entrepreneurs as to who should be allowed to exploit the site. The CPR was not one of the competing entrepreneurs, but preferred government ownership and controlled access for tourism purposes and because it envisaged an aesthetically pleasing establishment (by nineteenth century criteria). The Rocky Mountains Park Act (1887) formally established the site as Rocky Mountains National Park (renamed Banff National Park in 1930) (Lothian 1976, 1977; McNamee and Kraulis 1994; Luxton 2008; Campbell 2011; MacEachern 2011).

The presence of the CPR was critical. Banff, and the subsequent southern Cordilleran parks (Glacier, 1886, CPR), Yoho (1886, CPR), Waterton Lakes (1895, Great Northern Railway and CPR), Jasper (1907, Canadian Northern and Grand Trunk Pacific), the former Buffalo (1908, Grand Trunk Pacific), Mount Revelstoke (1914, CPR), and Kootenay (1920, CPR), along with Elk Island (1913, Grand Trunk Pacific) were established along railway lines with tourism in mind, rather than for the preservation of diagnostic or characteristic landscapes.

Banff Park was created as a pleasure park for the benefit and enjoyment of the people, using the language of Langford and the description applied to Yellowstone. As at Niagara Falls, the intent was to protect the hot springs from inappropriate development, not to conserve ecology. Development deemed "appropriate" included the original CPR Banff Springs Hotel, opened in 1888, along with bathing facilities, observation platforms and trails, and infrastructure necessary to support the transcontinental railway (Bella 1987; Lothian 1987; Luxton 2008). From the original 26 km^2, the park was expanded to 11,400 km^2 in 1902, contracted to 4663 km^2 in 1911, and subsequently re-expanded to its current 6641 km^2 (Lothian 1976; Luxton 2008). Much of the adjacent area is conserved in Kananaskis Country and other provincially administered Alberta parklands.

Initially, logging, coal mining, cattle grazing, and hunting of bighorn sheep and wapiti were all permitted within Banff (Lothian 1987). The park combined the aspects of "scenery" and economic use with recreation (Bella 1987), melding concepts from USA and UK. Banff and the other southern Cordilleran parks were explicitly intended to be "islands of civilization in a sea of wilderness". "Wilderness" in this context meant undeveloped, not ecologically undisturbed. "Civilization" implied economic and recreational activities involving development.

The Rocky Mountains Park Act (1887) also prohibited human settlement or occupation of any of Banff's area (not including the CPR townsite of Banff, or other installations), and authorized the removal of "trespassers". This provision was aimed at three groups: agricultural settlers, unapproved commercial entrepreneurs, and the indigenous Îyârhe (Stoney) Nakoda people, signatories to Treaty 7 covering the land in 1877. In the first of several expulsions from national and provincial parks, the Nakoda were compelled to leave Banff between 1890 and 1920 (Binnema et al. 2006, Finegan 2018).

10.4.2.2 Other Initial Parks

Similar patterns characterized the establishment and early development of Glacier, Yoho, Waterton Lakes, and Jasper National Parks (Lothian 1976; McNamee and Kraulis 1994). Glacier was designed to showcase the Illecillewaet ("swift water") Glacier, but unfortunately the ice has receded substantially since 1886 and is no longer visible from the railway or Trans-Canada Highway through Rogers Pass. "Yoho" is translated from the Kootenai language as "it is magnificent" or "place of tremendous beauty" or "wow". Waterton Lakes was established in 1895 after it was determined that the area had no agricultural or ranching potential, although petroleum exploration and development were encouraged. The Great Northern Railway constructed the Prince of Wales Hotel on the shore of Waterton Lake in 1927 (Glacier Park Collection 2022). The lake's position straddling the Canada-USA border allowed the GNR to ferry passengers north to view the scenery and drink alcohol during Prohibition in the USA. Jasper, established in 1907, was the last of the national parks initially intended in the "scenic preserve" mould.

In eastern Canada, the absence of mountainous scenery along transportation routes subject to tourism marketing limited park establishment. Until World War 1, the only national park east of Alberta was St. Lawrence Islands (1904, now Thousand Islands). Unlike the Cordilleran parks, it was a private initiative, rather than an act of policy: the initial waterfront property was donated by the local Mallory family for "park purposes" (Parks Canada 2020a). Nine islands owned by the federal government for navigational and defence infrastructure were then added, and recreational pleasure-boating facilities constructed. In 2024, the 6 km^2 park consists of 21 islands, many smaller islets, and 3 km^2 on the mainland.

10.4.2.3 Algonquin Provincial Park

The large provincially managed parks of Ontario and Québec also differed from the Cordilleran model. Algonquin Provincial Park, established in 1893, is the oldest provincial park in Canada (Ontario 2015). Its 7653 km^2 area and its proximity to Toronto and Ottawa make Algonquin one of the best known and most popular Canadian provincial parks. Algonquin was named a National Historic Site in Canada in 1992, based on its role in the development of park management; its pioneering visitor interpretation programmes, and its role in inspiring artists such as Tom Thomson, who in turn inspired Canadians' appreciation of natural landscapes. Today, images designed to portray the Canadian wild for commercial or overseas tourist purposes frequently are taken from Algonquin. Efforts to assess or depict the degree of environmental damage as a result of human activity (for example, from acid rain), also are commonly focused on Algonquin.

Ironically, the environment of Algonquin at the time of its creation and in the years since is significantly different from its pre-1850 state. Algonquin initially was established to provide protection ***for*** logging, not from logging. Access was provided by the Ottawa, Arnprior, and Parry Sound (later Canada Atlantic) Railway, financed by Ottawa lumber magnate JR Booth (Bell 1991; Brown 1994). Jack pines, the symbol of the area as painted by Thomson, were an insignificant component of the forest prior to the removal of more commercially desirable species prior to 1910. The indigenous Algonkian and Ojibwe populations were removed, and human settlement or occupation prohibited (except for loggers, railway personnel, and subsequently tourist lodges and private cottage owners). Commercial logging continues in some areas of Algonquin today. Leases for existing private cottages have been extended to 2038, but new construction is not permitted (Friends of Algonquin Park 2022).

The park was proposed primarily as an economic decision to promote logging and the 1890s equivalent of "adventure" tourism—canoe-based camping, hunting, and fishing. Mineral resources were assessed throughout the park region, and development was rejected on strictly economic grounds. The chief proponent, James Dickson, was sincere and a dedicated naturalist and outdoorsman (Dickson 1886, Kawartha Virtual Museum n.d.). He was also highly realistic and suggested Algonquin as a preservation target because of its location and political visibility, rather than as a sampling of "wilderness". In this sense, the establishment of Algonquin served as a prototype for park development as revenue and employment generator in economically depressed areas.

As a metaphor for an unspoiled Canadian wilderness, Algonquin is somewhat inaccurate. Despite that, however, it remains by far the largest piece of essentially undeveloped land in southern Ontario, and hence retains a high political and psychological visibility. Understanding of the environment of Algonquin, and the ongoing changes to it, enables Canadians in other parts of the country to put environmentally related statements by Ontarians into context.

10.5 Economics, Scenery, and "Nature as Medicine"

In 1911, Canada's seven national parks were essentially independent, with serendipitous management decisions and no guiding overall philosophy. The establishment of the Dominion Parks Branch (now Parks Canada) of the Ministry of the Interior (subsequently Natural Resources Canada) and the appointment of James B Harkin as Commissioner brought sweeping changes. Over the next 25 years, beginning with Elk Island (1913) and Mount Revelstoke (1914), Harkin oversaw the expansion of the park system, with 14 new parks (5 of which were subsequently degazetted or transferred to provincial jurisdiction). Firmly stamped with his own philosophy. Harkin's vision continued to guide national park establishment and management until the early 1970s (Lothian 1976, Bella 1987, Foster 1998, Hart 2010, Campbell 2011).

Born in the agricultural Ottawa Valley community of Vankleek Hill, James Bernard Harkin trained as a journalist before entering the federal public service. As an ardent but amateur conservationist, he was influenced by the ideas of Muir, Langford, Henry David Thoreau, and his contemporary Gifford Pinchot (Miller 2001), as well as by the UK "picturesque" movement, and the urban park designs of Frederick Law Olmsted (Hall 1995) and his contemporary Frederick Todd (Pollock-Ellwand 2017). As with many of his contemporaries, Hardin was greatly influenced by the impacts on societies and human psychologies wrought by World War I and saw exposure to "nature" as a societal cure.

Harkin stressed the economic aspects of tourism in park establishment and development (Bella 1987). Building on railway tourism, automobile visitation was encouraged through road construction, and strongly marketed to domestic and international audiences. Parks were sold as scenery. Areas with no previous history of tourism, notably Prince Edward Island, became international attractions. Two consequences were that park scenic landscapes came to be seen as symbolic of Canada as a whole, and that areas without suitably dramatic scenery (such as prairie and boreal forest Shield landscapes) were not prioritized for park establishment.

The aspects of economic development and diversification in rural areas were also important (MacEachern 2001). Used initially in Algonquin (1893) and Buffalo (1908), the model was applied in Kootenay (1920), Prince Albert (1927), Riding Mountain (1933), and Cape Breton (1936), and subsequently in Prince Edward Island (1937), Fundy (1948), Terra Nova (1957), Kejimkujik (1967), Kouchibouguac (1969), Forillon and La Maurice (1970), Pukaskwa (1971), and Gros Morne (1973). These parks were seen as economic drivers in areas where logging had become uneconomic, or where small-scale agriculture provided limited incomes. Unfortunately, a consequence was that pre-existing residents of these new park areas, both indigenous and non-indigenous, were summarily expropriated or deported, and traditional land uses prohibited. Similar actions occurred as provincial parks were established, notably at Quetico (1910) in northwestern Ontario (Killan 1993). Initially received with resignation, the practice met with increasing resistance from residents, particularly at Kouchibouguac, Forillon, and Gros Morne (Herd and Heintzman 2011; Rudin 2011).

The second philosophical pillar can be described as humanitarian, societal, or moralistic. Harkin believed in "nature as medicine", perhaps mystically: exposure to the outdoors, even in an automobile, was good for the human body, mind, and spirit, physically and mentally necessary, and morally uplifting (Hart 2010). Just as the UK "pleasure grounds" and urban parks such as Niagara Falls, Halifax's Public Gardens, Mont-Royal (Montréal), High Park (Toronto), or Stanley Park (Vancouver) were intended to do, visiting the national parks would allow urban workers to refresh and recharge, returning as healthier, happier, more productive, and even more moral and patriotic members of society. This philosophy led to the establishment of national parks as close as possible to cities by automobile, with the provision that the areas should be suitably "scenic": Kootenay for Vancouver, Prince Albert for Saskatoon, Vidal Point (now Katepwa Provincial Park) for Regina, Riding Mountain and Brereton Lake (now Whiteshell Provincial Park) for Winnipeg, Cape Breton Highlands and Prince Edward Island for Halifax, and subsequently Fundy for Fredericton, Moncton, and Saint John; and Terra Nova for St. John's. Toronto, Ottawa, Montreal, and Québec were considered to be adequately served by the adjacent large provincial parks, such as Algonquin and Mont-Tremblant.

A second effect was the priority given to recreational infrastructure. Kootenay (1920) was built around the bathhouse and spa at Radium Hot Springs, and was connected to Banff by the Banff-Windermere Highway, blasted through the canyons of the Vermilion River and Sinclair Creek. Recreational infrastructure, including ski resorts, golf courses, and pleasure boat facilities, along with cleared and manicured areas, were important elements of all the parks established in this era, including Georgian Bay Islands (1929), and were added to existing parks such as Banff, Jasper, and Thousand Islands.

10.6 Limited Role of Ecology

Harkin was not a natural scientist and tended to hire people who similarly lacked ecological training or experience (Hart 2010). Park wardens were former trappers, "mountain men", or reformed poachers, charged with enforcement and not ecological assessment or conservation (Burns 2000). Wardens were seen as law enforcement, preserving the parks from local human activity. Policy decisions allowing logging or development were not within their purview. Ecology was invoked for only 5 of the parks established during Harkin's tenure. Menissawok, Nemiskam, and Wawaskesy were single-species preserves for pronghorn, and were degazetted when the species was considered to be out of danger of extinction

Point Pelee (1918) evolved from a migratory bird sanctuary and was designated due to lobbying from Harkin's friends in the birder community (Green 1976). Private cabin owners (including some birders) were permitted to remain and practice recreational activities (Parks Canada 2016). Duck hunting was permitted in the park until 1969. The Caldwell First Nation Chippewa people lived on Point Pelee for at least 6000 years before European colonization. In 1790, a treaty officially ceded a large tract of land, which included Point Pelee, to the Crown. The Caldwell First Nation was not a signatory of that treaty. After Point Pelee was gazetted, Caldwell First Nation residents were forcibly removed from the land (Leclair 1988).

Wood Buffalo (1922) was established as a conservation area for the subspecies *Bison bison athabascae*, mirroring the establishment of Buffalo (1908) for the plains *Bison bison bison*. Distant from railways and roads, Wood Buffalo represented a departure from the philosophy of economic diversification and providing recreational opportunities for urban dwellers. Although the stated intent was to preserve the northern subspecies, the introduction of almost 7000 plains bison from Buffalo National Park on several occasions (most recently after Buffalo was commandeered by CFB Wainwright) and other locations led to interbreeding, with the result that "pure" *Bison bison athabascae* no longer exist in the region (McKennirey and Richardson 1985, Gates and Jung 2013, Environment Canada 2018a, b). In addition, the newcomers brought bovine tuberculosis and brucellosis to the Wood Buffalo population. At other times, bison from Wood Buffalo were transferred to Elk Island and other spots Ironically, the hybridized bison in Elk Island are genetically closer to *Bison bison athabascae* than are those in Wood

Buffalo. The widespread transfer of bison was not compatible with basic ecological principles, reflecting Harkin's lack of knowledge and managerial approach.

As a conservation area, hunting was not permitted in Wood Buffalo, including by the region's indigenous Dene (Athabascan) and Mikisew Nehinaw (Cree) populations (McCormack 1992). This prohibition was extended to all aspects of traditional land use, including plant harvesting and the use of salts for medicinal purposes from Wood Buffalo's saline flats. Although the indigenous populations were not expelled, unlike the situation in the other parks, severe restrictions were placed on their ability to interact with the landscape and be refreshed and recharged.

10.7 Reassessment

The period 1967–75 saw questioning of the basic philosophical tenets which had prevailed since Harkin's appointment in 1911. Factors involved included:

- Expansion of the system, with 9 new parks and reserves created between 1967 and 1974;
- Opposition by local residents facing removal or restrictions, particularly at Kouchibouguac and Forillon;
- Public reaction to the resulting adverse publicity;
- Conflicts between economic and ecological approaches, particularly at Kejimkujik, Pacific Rim, and Nahanni;
- Increased emphasis on ecological integrity by parks management and staff;
- Increased interest in ecology by prospective wardens, coupled with public expectations of the role;
- Opposition to development of new recreational infrastructure, including golf courses and ski areas;
- Increased desire for local residents' involvement, including more decentralized management and recruitment of local personnel;
- Realization that public support would be necessary for any new parks developed in more southern areas, to broaden the ecological representation; and
- Realization that establishment of new parks in northern areas would require full cooperation with indigenous peoples, based on respectful relationships.

The revised mission of Parks Canada was primarily to protect the ecological integrity of the land, and secondarily to allow the public to explore, learn about, and enjoy Canada's natural spaces. This not only represented a departure from the previous economic and recreational motives, but emphasized ecology (or biology) above all other aspects of landscapes. Conflicts still existed between those primarily interested in the preservation of particular species, seemingly necessitating both "purely ecological" approaches and removal of all potential human threats, and those more willing to accept traditional human use of landscapes.

The summarial expropriations which had caused anger at Kouchibouguac and Forillon (Rubin 2011) were replaced with a series of negotiations, land purchases, and traditional usage agreements. This slowed reserve creation and gazetting at Pacific Rim, Kluane, Gros Morne, Nahanni, Grasslands, Mingan Archipelago, and Gwaii Haanas, but eventually resulted in better Parks Canada-community relations. Traditional land uses were more widely (although not universally) accepted. Enclaves involving communities flanked by parkland were permitted to remain at Pacific Rim, Kluane, and Gros Morne. Gazetting of Kluane, Gros Morne, and Grasslands was delayed until all matters were settled: as of April 2024, Pacific Rim, Nahanni, Mingan, and Gwaii Haanas remain as Reserves. Emphasizing the co-operative aspects of working with communities and respecting traditional knowledge are important facets of current practice.

Bruce Peninsula (1987), the most recent national park in rural southern Canada, was formed from land previously set aside as Cyprus Lake Provincial Park and Fathom Five National Marine Park in Georgian Bay. Negotiations have continued at Gulf Islands reserve since 2003 (Parks Canada 2018a, b). Sable Island reserve, without permanent human habitation, was designated in 2013 (Parks Canada 2019b).

10.7.1 Co-management of Northern Parks

In 1984, Ivvavik became the first national park created through an indigenous-federal land settlement, the Inuvialuit Final Agreement (NWT Executive and Indigenous Affairs Parks Canada n.d.). This set the pattern for further national parks in Nunavut, NWT, and Yukon, as well as Nunavik. Vuntut (1993), Tuktut Nogait (1998), Ukkusiksalik (2003), Torngat Mountains (2008), Kuururjuaq (Nunavik, Québec 2009); Nááts'ihch'oh (Reserve, 2014), Akami-Uapishk-Kakkasuak-Mealy Mountains (Reserve, 2015), Qausuittuq (2015); Ulittaniujalik (Nunavik Québec 2016), and Thaidene Nëné (Reserve, 2019) are managed or co-managed by Indigenous residents. Parcs national Kuururjuaq and Ulittaniujalik are Nunavik parks developed and managed by Inuit residents, in conjunction with the Government of Québec (Slaymaker et al. 2020a, b). Most of Kuururjuaq park's staff comes from nearby Kangiqsualujjuaq. Aulavik (1992), Wapusk (1996), Auyuittuq (2001), and Sirmilik (2001) also have substantial indigenous input in operations and management.

The uniqueness of the Quttinirpaaq landscape on Umigmak Nuna (Ellesmere Island) suggested conservation. Park establishment in 1988 was not unanimously accepted, however, as some scientists familiar with the area argued that designation of a National Park would attract excessive visitation (Parks Canada 2022b). Today, visitation is tightly controlled by Parks Canada, and in 2019 was limited to 18 people (Parks Canada 2021).

10.8 UNESCO World Heritage Sites

UNESCO has designated 20 World Cultural and Natural Heritage sites in Canada, including 7 in national parks. Provincially protected sites include Dinosaur, Head-Smashed-In, and the archaeological site of Aisinai'pi (Writing-on-Stone), Alberta (see Chap. 5), Miguasha (QC), Joggins (NS), and Mistaken Point (NL) (see Chap. 2). Pimachiowin Aki (ON, see Chap. 3), one of the few UNESCO sites officially designated for both cultural and natural values, was officially designated in 2018.

Five other sites were designated primarily for cultural criteria (Slaymaker et al. 2020a, b). The Rideau Canal in Ottawa is a National Historic Site (see Chap. 4) L'Anse aux Meadows (NL) National Historic Site contains the excavated remains of a Viking settlement ca. 1015 ce, the earliest evidence of Europeans in North America. Partially reconstructed timber-framed turf structures include three dwellings, one forge, and four workshops. With increased storm activity evident since 2000, L'Anse aux Meadows could be vulnerable to storm surge, even under relatively stable relative sea level (Catto 2011).

Red Bay National Historic Site (NL) includes the Basque whaling station, established ca. 1530 ce. The remains of three Basque whaling galleons and four small chalupas used in the capture of whales are present. Between 1550 and 1600, at least 20,000 bowhead whales (*Balaena mysticetus*) were killed for oil production (McLeod et al. 2008). A combination of European political developments and increased accessibility of whales off Svalbard led to the disuse of Red Bay ca. 1620 ce.

Québec City was founded by Samuel de Champlain in 1608. It is the only North American city to have preserved ramparts, bastions, gates, and defensive works. The Upper Town, built on Cap Diamant, has remained the religious and administrative centre of Québec. The Lower Town is built around Place Royale and the harbour.

Grand-Pré National Historic Site (NS), along the Minas Basin coastline, illustrates the adaptations made by the Acadian settlers. The site is a memorial to the Acadian way of life, as well as to the Grand Derangement deportation, beginning in 1755.

Lunenburg (NS), built on a drumlin, is the best surviving example of a planned British colonial settlement in North America. Established in 1753, it has retained its original grid pattern layout and wooden architecture.

10.9 Challenges to Protected Areas

Dearden and Dempsey (2004) identified five challenges to the protected area system in Canada: (a) large ecoregions with no or insufficient protection; (b) exponential increase in visitation; (c) parks are isolated ecological islands; (d) pursuit of ecological integrity has significant logistical, financial, enforcement, and scientific challenges; and (e) budgetary and political pressures at federal and provincial levels. A review of the current situation (2024) highlights both change and continuity.

10.9.1 Insufficient Protection

The International Union for the Conservation of Nature defines a protected area as a clearly defined geographical space, recognized, dedicated, and managed, through legal or other effective means, to achieve the long-term conservation of nature (IUCN 2022). Canada's National Parks generally qualify as IUCN Category II, based on area and the stated commitment to conserving functioning ecosystems. Some smaller national parks are classed as Category III (Natural Monument or Feature). The marine parts of Pacific Rim and Kouchibouguac are rated as IUCN Category VI (protected area with sustainable use of natural resources), as fishing is permitted.

As of April 2024, Canada had designated 12.6% of its terrestrial (land and freshwater) and 9.1% of its marine territory in 9900 protected areas, compared to 7.9% and 0.52% respectively in 2004 (Environment and Climate Change Canada 2022). All but one of the 43 major terrestrial ecoregions, the Shugliaq (Southampton Island) Plain, have national, provincial, or co-managed parks, or planned areas for national park reserves. Although this situation appears promising, more detailed assessment reveals shortcomings.

Canada has 825 IUCN Category 1a (Strict Nature Reserve) and 303 Category 1b (Wilderness Area) protected areas (IUCN 2022). Nature Reserves are generally small, intended for the conservation of particular species, especially birds, and are closed to most human visitation. Wilderness areas are larger, and generally non-motorized human traffic is permitted.

Buffalograss Ecological Reserve (SK) is a Category 1a area, 0.15 km^2 along the Souris River. Its intent is to preserve one of the few remaining areas with undisturbed buffalograss (*Bouteloua dactyloides*), classified as a threatened species in Canada by the Committee on the Status of Endangered Wildlife in Canada (COSEWIC 2011).

The 1070 km^2 Avalon Wilderness Reserve (NL) is a Category 1b region. Although some areas had undergone low-intensity logging by residents, in 1964 it was established to preserve a boreal forest community in a relatively untouched inland region of the Avalon Peninsula, distant from coastal communities (Newfoundland and Labrador 2022). Newfoundland and Labrador Wilderness Reserves are open to non-motorized access by humans, including fishing and hunting of certain species. In reality, all-terrain vehicle and snowmobile traffic is common, and there is no enforcement. All-terrain vehicle damage to some fens along the eastern margin of the Wilderness Area is evident.

Most ecoregions, including all those south of the treeline/ Arctic-Subarctic transition and east of the Cordillera, have less than 12% of their terrain afforded some degree of protection, generally not to national park level (IUCN Category II). Only the Arctic Cordillera, Tundra Cordillera, and Pacific Maritime ecoregions have more than 20% of their terrain afforded some degree of protection. Most of the protected marine areas are in the deeper waters of the Pacific, Arctic, and northern North Atlantic. Coastal and nearshore waters in Atlantic Canada and Hudson Bay have less than 5% of their areas afforded some protection. Many terrestrial and marine areas are in IUCN Categories IV (habitat or species management area) or V (protected landscape or seascape), lower and less comprehensive levels of protection.

10.9.2 Visitation Pressures

Increased visitation was evident in all parks prior to the COVID-induced restrictions of 2020–21 (Parks Canada 2023, Table 10.2). Banff National Park provides Parks Canada with an instantly recognizable, well-defined image. This popularity has a significant price, however. The 4,130,081 paying visitors who swarmed through Banff in the 2022–23 tourist season (1.7 times the total at Jasper, the next most popular national park, and not including those who drive through along the Trans-Canada Highway) have a large impact on the natural environment. In 2020, COVID restrictions closed Banff for two months and prevented virtually all international travel, but the park still saw 3.2 million visitors.

Development pressures are strong within Banff. Together with winter skiing facilities, hotels, summer campgrounds, and tourist amenities, Banff townsite's permanent population of 8905 (2021) also requires services and housing. In addition, the park's position straddling the major transportation artery from Calgary to Vancouver results in large amounts of through traffic. Traffic volumes, with over 8,000,000 people per year, have grown to the point where the Trans-Canada Highway through the park was converted to a four-lane expressway. Prior to the construction of overpasses (6) and underpasses (38) for animals, significant (and ever-growing) numbers of wildlife-vehicle collisions occurred during the summer months (Clevenger et al. 2009).

Visitation numbers for the southern parks in particular are underestimates. Many people transit parks on through highways, and other use the park services without registering or paying (especially during off-seasons). Pacific Rim, Saguenay-St. Lawrence, Glacier-Mount Revelstoke, Prince Edward Island, and Yoho all saw more than 650,000 visitors in 2022–23 (Parks Canada 2023).

The effects of visitation are magnified in parks with smaller areas, or where much of the activity is concentrated in a small zone, usually paralleling a highway. Banff's visitors are spread unevenly across 6641 km^2, and access to the backcountry areas is tightly controlled by permit, as is the case in the other Cordilleran parks and Pacific Rim. Average visitation in the 750 km^2 belt flanking the Trans-Canada Highway was approximately 5500 people/km^2 in 2022–23. Overall, 9 national parks saw more than 1000 visitors/km^2 in 2022–23 (Parks Canada 2023; Table 10.2), with the highest totals registered at Prince Edward Island (20,920/km^2).and Point Pelee (32,808/km^2).

10.9.2.1 Prince Edward Island National Park

Prince Edward Island National Park was established in 1937 and has subsequently become one of Canada's major tourist attractions, both for its natural attributes and because of its association with "Anne of Green Gables". The park encompasses two areas. The older part includes 27 km^2 of beaches and adjacent shoreline areas in the vicinities of Cavendish, Rustico, and Stanhope. With the growth of tourism, additional areas have undergone tourist development, including Cabot Beach Provincial Park, on the northeastern margin of Malpeque Bay. The newer part includes 18 km^2 of terrain on the Greenwich Peninsula, north of St. Peters. This area has remained undeveloped, with minimal tourist infrastructure and no vehicle access, and is the closest approximation to "backcountry". In the older part of the park, visitation exceeds 28,000 people/km^2 (Parks Canada 2023).

Decreased snowfall and limited ice cover over the southern Gulf of St. Lawrence, in combination with sea level rise, have resulted in increased wave erosion and narrowing of beaches. Along the barrier island protecting Cascumpec Bay to the west of PEI National Park, erosion between 1974 and 2004 caused coastal retreat of 115 m, a rate of 3.8 m/y (Conroy 2007). Climate variations also result in increased degradation of coastal dunes. Stronger winter winds sweeping across exposed dunes, some eroded due to human pressure during the previous summers, have resulted in enhanced erosion and lowering and flattening of dune profiles. Increased foot traffic and vehicle access to beaches and backing dunes have accelerated erosion of coastal tourist sites.

Dunes at Cabot Beach Provincial Park have receded more than 3 m over the period 1990–2020, predominantly due to erosion generated by human activity (Catto 2002, Catto

Table 10.2 Visitation in national parks and reserves, 2022–23 fiscal year. Data from Parks Canada

National park/reserve	Visitation 2022–23[a]	Area (km^2)	Persons /km^2
Banff	4,130,081	6,641	621.9
Jasper	2,415,463	11,228	215.1
Pacific Rim	1,163,362	510	2,281.1
Saguenay-St. Lawrence	1,070,634	1,245	859.9
Mt Revelstoke & Glacier	775,190	1,611	481.2
Yoho	663,878	1,313	505.6
Kootenay	574,126	1,406	408.3
Prince Edward Island[b]	564,843	27	20,920.1
Point Pelee	492,122	15	32,808.1
Waterton Lakes	475,842	505	942.3
Bruce Peninsula	460,435	125	3,683.5
Elk Island	409,891	194	2,112.8
Riding Mountain	336,560	2,968	113.4
Fundy	303,575	206	1,473.7
Prince Albert	287,372	3,875	74.3
La Mauricie	278,923	536	520.4
Cape Breton Highlands	277,203	948	292.4
Kouchibouguac	243,489	239	1,018.8
Forillon	164,130	240	683.8
Gros Morne[c]	163,152	1,805	90.4
Thousand Islands	109,284	24	4,553.5
Kejimkujik	75,284	404	186.3
Kluane	47,098	5,900	8.0
Terra Nova	45,191	399	113.3
Mingan Archipelago	44,017	151	291.5
Gulf Islands	28,434	37	768.5
Georgian Bay Islands	27,482	14	1,963.0
Grasslands	19,656	730	26.9
Pukaskwa	17,285	1,878	9.2
Gwaii Haanas	2,651	1,474	1.8
Wood Buffalo	2,399	44,972	0.053
Nahanni	1,104	3,000	0.37
Sable Island	806	30	26.9
Auyuittuq	256	19,089	0.01
Ivvavik & Vuntut	179	9,750	0.02
Nááts'ihch'oh	109	4,850	0.02
Wapusk	82	11,475	0.007
Aulavik	34	12,200	0.003
Tuktut Nogait	12	18,181	0.0007
Quttinirpaaq	7	37,775	0.0002
Sirmilik	6	22,200	0.0003

[a]2022–23 Fiscal year, based on officially recorded visitation
[b]Data not available for 2022–23; Value is mean for 2019–2022
[c]Data not available for 2022–23; Value is mean for 2019–2021
No data available for: Akami-Uapishk[u]-KakKasuak-Mealy Mts; Thaidene Nëné; Torngat Mts; Ukkusiksalik

et al. 2020, Catto and Catto 2009). The dunes at Cavendish, Rustico, and Stanhope in the national park are also showing signs of erosion. Roads constructed by Parks Canada through dune areas have also accelerated erosion, and several must be periodically cleared of sand to keep them passable to vehicles.

Recently, efforts have begun to stabilize the dunes, by planting vegetation such as Marram grass on their surfaces and attempting to restrict human foot traffic to boardwalks. These measures can work: erosion of dunes at Lakeside, near Greenwich, stopped entirely after a walkway was constructed across the dunes (Catto 2015, Catto et al. 2020). Interestingly, the walkway was made from recycled plastic soft-drink bottles, a product that PEI has banned in an attempt to reduce litter. Restrictions on visitor access to the newer Greenwich unit, notably limited vehicle access, are designed to simultaneously promote tourism and conserve the ecological integrity of the terrain. Plans to designate the adjacent relatively undisturbed Pītaweikek area as a National Park Reserve, co-managed by the Mi'kmaq people and Parks Canada, have benefitted from experience gained through previous errors.

Tourists outnumber islanders 8:1 during the summer months. In addition to environmental impacts, there is a danger that tourist influxes of this magnitude can result in erosion of the local culture and distortion of socio-economic patterns: that the island will fall into a "tourist trap" of its own making.

Tourism is a vital economic industry in PEI (Tourism PEI 2022). It is a vital component in future economic sustainability of most coastal communities, and in particular is seen as a key component in the economic revitalization of formerly fishery-dependent communities. Intensive marketing by the provincial government promotes tourism. Lengthening of the tourist season into the winter, enhanced tourist use of natural areas, and geomorphic stresses induced by climate change in combination have affected sites throughout PEI and could have an impact on the long-term sustainability of both natural areas and tourism.

10.9.2.2 Point Pelee National Park

Point Pelee is the southernmost National Park in Canada, and at 15 km^2 is one of the smallest. The small size of the park and the number of visitors places extreme strains upon the area, necessitating restrictions on access (Parks Canada 2016).

These concerns reached a head in 1963, when 781,000 visitors used the park. Environmental degradation resulting from excessive visitation was so severe during the late 1950s and early 1960s that a Royal Commission recommended decommissioning the park. Rather than abandoning the terrain, however, restrictions on tourist access were imposed. The campground within the park was closed permanently, and camping is no longer permitted anywhere in the park. Private automobiles were first restricted in area, then banned during the peak tourist seasons, and now are prohibited altogether. Public transportation, initially by diesel bus, is now exclusively by solar-powered train or by foot. Private cottages, which had been established prior to the formation of the park, were gradually purchased. Although no expropriation was practised, a policy was instituted whereby cottages could not be sold (except to the park) and could not be willed or deeded (except to direct family relatives living with the deceased). These measures have resulted in virtually all of the formerly private lands becoming parts of the park. Considering the tiny area involved, it would be difficult to conceive of a solution that could allow private development and park preservation to co-exist.

In the northern parks, especially those without road access, visitation is limited and tightly controlled by Parks Canada. Twelve of the parks and park reserves in taiga and Arctic regions saw less than 1000 outside visitors in 2022–23, spread across large areas (Parks Canada 2023).

In 1995, Aulavik National Park ("place where people travel") was established in northern Banks Island, encompassing 12,200 km^2 (17% of the island). The management plan indicates that the park will remain predominantly (or entirely) as an undeveloped area, and that access will be limited and confined to aircraft and guided paddling tours along the Thomsen River (Parks Canada 2010a, 2022a). The plan states that, "visitors must be self-sufficient".

10.9.3 Isolated Ecological Islands

The concept of an "ecological island" suggests that areas surrounded by anthropogenically modified terrain can remain less disturbed. The park serves as a refugium for species, although it is an island washed by the waves of human activity. This concept looks at the stresses from outside park boundaries. In addition to broader environmental impacts including climate change, land use practices in the surrounding area have a direct impact on a park's flora and fauna.

An Intact Forest Landscape can be defined as an area which contains forest and non-forest ecosystems minimally influenced by human economic activity (IFL 2021). Intact forest landscapes must have minimum areas of 500 km^2 (50,000 ha), with minimal width of 10 km. Human settlements, infrastructure, agriculture, logging, and mining terrains are conventionally excluded. Areas with evidence of low intensity and older disturbances are treated as subject to "background" influence and are eligible for inclusion. Here, we'll modify the definition to include terrain not naturally occupied by forest, including grasslands and tundra.

Fifteen Canadian national parks have areas less than 500 km^2. Several of these are further divided into physically disconnected units, or by highways and infrastructure.

10.9.3.1 Forillon

Forillon has a relatively small area of 245 km^2, including conserved offshore marine areas, which does not provide most animal species with ample territory for movement. The majority of visitors come during July and August (at least 80%), and most who come stay for less than 1 day (75%) and come by private vehicle (>95%). Thus, the visitation pressure during the summer months is extreme, and traffic jams on roads dead-ending at coastal cliffs are common occurrences during the summer. The balance between providing visitor services and conservation of ecological integrity is precarious (Parks Canada 2010b).

Forillon is bordered on the landward side by Route 197, which serves as a bypass road. It was hoped that this route would remove through vehicle traffic from the park and hence support ecological integrity. Unfortunately, construction of the highway isolated the Forillon Peninsula from the hinterland of Gaspé, preventing animal migration and hemming the fauna into the small peninsula. There is virtually no interchange or interbreeding of populations separated by Route 197, with consequent losses in genetic diversity and faunal population viability.

Measures have recently been put in place to limit visitor impact. All freshwater fishing was banned in park lakes in 1995, and traditional marine fisheries were restricted in the coastal zone. The park management plan identifies forms of recreation to be encouraged and also identifies forms which will not be promoted. Mountain biking, for example, is described as "generally detrimental to the natural environment and is not thought to contribute in any specific way to the authentic experience" of a visit to a national park.

10.9.3.2 Elk Island and Fundy

Elk Island's 194 km^2 is insufficient to allow grazing animals or most predators the territorial minimums they need for survival. Elk Island is the only Canadian National Park that is completely surrounded by a fence. As a result, species reintroduced into the park (in particular the hybridized *B. bison athabascae x B.b.bison*) cannot be free-ranging and must be fed and confined at various times throughout the year.

Much of Fundy National Park (206 km^2) and the surrounding areas have been logged extensively (Cooper and Clay 1994). With the shift from timber harvesting to pulp-and-paper operations, and with the forestry industry preference for monoculture stands of identical characteristics, reforestation efforts since 1960 have concentrated on developing uniform stands of fast-growing conifers (particularly *Picea* and *Abies*) for market. This practice has led to a loss of habitat for many of the animal species in the region, including the park. Reductions in the population of *Martes americana* (Pine Marten) have attracted much attention, while comparable changes in *Glaucomys sabrinus* (northern flying squirrel) numbers have not been as extensively publicized. Unfortunately, the 206 km^2 area of Fundy National Park is too small to provide many of the forest species with the amount of territory which they need to maintain successful populations.

10.9.3.3 Wood Buffalo

Adverse effects are not confined to the smaller parks. Wood Buffalo (44,972 km^2) is threatened by the expansion of pulp-and-paper logging and processing to the south and west of the park. Water influx to the Peace-Athabasca Delta, a Ramsar Wetland of International Significance, and an important breeding area for White Pelicans (*Pelecanus erythrorhynchos*). Whooping cranes (*Grus americana*; see Chap. 3), and 212 other bird species, 45 species of mammals, and 20 fish species, are under pressure from the industrial oil sands development upstream along the Athabasca River, including contamination and water usage in refining. Upstream along the Peace River, hydroelectric dams and use of water for agriculture have also affected discharge to the delta.

Concern has arisen among the ranching communities to the west about the presence of tuberculin, brucellosis, and other viruses within the bison herd in Wood Buffalo. Tuberculin is not immediately fatal to bison (or domestic cattle), although in the long term, these diseases do appear to weaken the herd and lessen reproductive success. However, contaminated meat cannot be sold commercially, and any cattle so contaminated must be destroyed. The potential for transfer of the viruses to domestic free-ranging cattle has prompted the ranching industry to request a large-scale cull (as much as 80%) of the bison herd.

The combination of environmental threats has prompted a review of Wood Buffalo's status as a World Heritage Site by UNESCO (2021). In addition, the Mikisew Cree First Nation has encouraged the formation of the Delta Research Institute in Fort Chipewyan in efforts to integrate ecological research with traditional indigenous knowledge.

10.9.4 Ecological Integrity

Efforts to conserve and enhance ecological integrity face multiple challenges. In many of the southern parks, ecological change since the 1600 s has been substantial. The growing realization that pristine, primeval pre-1600 conditions, unaffected by human activity, did not exist in many areas raises the question: What ecological condition is meant when contemplating "restoration"? Possible efforts to rehabilitate natural habitats under current conditions may substantially differ from those that would be involved in recreating an earlier ecology. "Restoration" is not possible where the "original, unaltered" state is unknown due to the absence of contemporary data for whichever date is selected as representatively "original". Introduced pathogens have eliminated or greatly restricted species (e.g. chestnut, *Castanea dentata*; elm, *Ulmus americana*). Others have been rendered extinct (passenger pigeon, timber rattlesnake). In southern areas, only a few mammal species (white-tail deer, raccoon, coyote) have increased in numbers under recent human pressure. Introduced species, such as moose (*Alces alces*) in Terra Nova, Gros Morne, and Cape Breton Highlands, and autumn olive (*Elaeagnus umbellata*) in Point Pelee, and rapidly increasing indigenous species, such as herring gulls (*Larus argentatus*) in Mingan, also pose challenges. Range expansions and contractions induced or facilitated by climate change have affected species in all parks.

10.9.4.1 Logistics and Politics

These scientific challenges are accentuated by logistical and political constraints. Canada's governmental system, with powers divided between federal and provincial ministries and authorities, can hamper efforts to develop unified policies aimed at ecological conservation. Natural resources, particularly non-renewable resources, are primarily a provincial responsibility. Agriculture is divided between the jurisdictions. Crown-indigenous relations are largely the purview of the federal government, although the provinces are also involved. The provinces are responsible for the management of public lands belonging to the province, property rights, energy, and "matters of a merely local or private nature". Issues which affect the ecological integrity of national parks, including logging, mining, dam construction, and wildlife management outside park boundaries, are largely governed by the provinces.

Environmental matters, including water, were not mentioned in the 1867 division of responsibilities in the British North America Act. The Dominion Parks Branch was initially placed in the Department of the Interior, also responsible for immigration, indigenous matters, and natural resource exploitation, all of which had higher priority than did federal land management. Subsequently, national parks were administered through Environment Canada (and the Department of Canadian Heritage from 1994 to 2003).

Enforcement of ecological integrity within parks rests with the wardens. Strictness of enforcement varies, resulting in some anomalies. In all national parks, fossil collecting without a scientific permit is prohibited. Yoho and Kootenay park wardens frequently inspect the packs of hikers returning from the Burgess shale beds for contraband. In May 2022, a person received a $20,000 fine and five months of house arrest for possession of 45 Burgess Shale fossils (Canadian Press 2022). However, when fossil-bearing outcrops along Highway 16 in Jasper National Park were blasted for road construction, geologists from the University of Alberta were not permitted to take specimens: wardens required the fossils to be replaced in the new roadbed, where they were promptly paved over. Although all park resources—plants, animals, rocks, fossils—are theoretically considered equally important and must remain within the park, differences in practice are common.

10.9.4.2 Traditional Relationships with Landscapes

Enforcement of strict "no-take" policies runs afoul of traditional harvesting practised by both indigenous and non-indigenous people. Some of the fiercest battles over national park establishment and management have been triggered by attempts at rigid enforcement. Rigid enforcement frequently comes from a lack of perception by outsiders, a belief that one approach fits all circumstances, and a lack of understanding of local human-environmental relationships. In some senses, it remains linked to the 1911 view of a national park.

Residents of the small communities adjacent to national parks frequently have intimate connections to the landscape. Most residents of boreal, taiga, tundra, or rural coastal regions are directly engaged in some combination of primary resource occupations, tourism, or traditional land uses. Many families have lived in the same communities and environment for generations, have observed much, and have acquired immense environmental knowledge.

In coastal regions, features having deeply personal significance for fish harvesters, mariners, and residents may not be appreciated by visitors or tourists. Many first-time viewers see shores without the emotions that familiarity brings, producing both more positive and more negative perceptions of aesthetic worth than are held by local residents. The same phenomenon can be observed when considering any landscape in Canada—prairie, boreal forest, tundra—from the perspectives of different persons. These psychological differences in attitude necessitate consideration of many diverse viewpoints in issues where human perceptions are involved, including parks.

Parks Canada personnel potentially have advantages in communication. They can discuss matters in non-technical fashion with people who already have acquired ecological information. Local residents see the consequences of environmental changes directly, without the technological buffers of urban North America, and have time series of traditional knowledge and oral history for comparison. However, for Parks Canada, going from ecological assessment to meaningful and effective co-management requires an intimate understanding of the people and communities involved. The approach must shift from the concept that clear presentation of data will inevitably result in an "ecologically educated" community that will respond in a manner deemed appropriate. Efforts that are strictly confined to avoiding terminology, without considering the socio-economic and cultural perspectives of the audience in their environment, will be unsuccessful. Fortunately, the increasing involvement of local communities in co-management, coupled with realization that there are multiple ways to understand and appreciate landscapes, is starting to improve matters.

10.9.5 Money and Politics

In democratic societies, national parks reflect political priorities and societal will. Wright et al. (2019) reported that 93% of Canadians surveyed agreed or strongly agreed that protected areas are necessary, with no statistically significant differences by gender, education, income, level of use of protected areas, or region. Similar results have been noted in other surveys since 2000.

Protection of wildlife and scenic beauty were ranked as most important. Leisure opportunities, stress relief, cultural significance, water supply, and economic benefits were ranked as progressively less important, although significant. Inadequate understanding of value, insufficient funding, low government priority, lack of public support, and competing demands for space were considered the greatest challenges to increasing protected areas. Interestingly, respondents substantially over-estimated the percentage of protected land in Canada. Parks Canada's tracking surveys in recent years have produced similar results, in addition to revealing that less than 40% of those surveyed could identify the agency responsible for Canada's National Parks and Historic Sites.

This general level of expressed public support contrasts with fiscal reality. Federal government budget allocations to Parks Canada have declined consistently since 2005, under both Conservative and Liberal administrations. A decrease of 40% is projected between 2019 and 2025 (e.g. Parks Canada 2022c). In most provinces, the majority of funding for provincial parks comes from user fees, partnership arrangements, and private donations. With the exception of directly adjacent, contained, and enclave communities, park issues seldom receive attention at election time.

10.10 Conclusion

Analysis of Canada's history with parks and protected area reveals a chequered pattern of changing motivations, assumptions, and intentions; community engagement and centralized bureaucratic impositions; and overall successes and shortcomings. Establishment and maintenance of parks, and all decisions involved with them, reflect the socio-economic, political, and cultural contexts of various times. Although parks may be thought of as elements of the natural environment, they are also a form of infrastructure, societal creations.

Much of the recent discussion has focused on biodiversity, rather than other elements of landscapes (such as geodiversity, geology, and geomorphology). This focus reflects a variety of causes: the number of biologists involved in protected area administration and management; a tendency to focus on single, charismatic species; a tendency to generalize from one site to a broader region ("polar bears are in trouble along southern Hudson Bay; therefore, they face extinction everywhere"); and a lack of understanding or acceptance of the ecological/conservation measures practised by indigenous or local trappers, hunters, and fish harvesters.

Future evolution will depend upon building relationships among the different communities, including conservationists, ecologists, geomorphologists, park users, and those living adjacent to on within parks and protected areas. Local indigenous and non-indigenous residents commonly have a great appreciation of the linkages among geo-, bio-, and cultural diversity, and respect for their knowledge needs to continue to develop in conjunction with scientific research and conservationist ethics. Understanding and appreciation of landscapes is not confined to one group, community, or culture, and there are multiple ways of knowing.

References

Andrews M (1990) The search for the picturesque: landscapes, aesthetics and tourism in Britain, 1760–1800. Scholar Press, Aldershot

Barber K ed. (1999) Canadian Oxford Dictionary 3rd edition. Oxford University Press

Bate J (1991) Romantic ecology: Wordsworth and the environmental tradition. Routledge, London

Bell A (1991) A way to the west. Barrie, Ont.: privately published

Bella L (1987) Parks for Profit. Harvest House

Binnema T, Niemi M (2006) Let the Line be Drawn Now': wilderness, conservation, and the exclusion of aboriginal people from Banff National Park in Canada. Environ Hist 11:724–750

Brown R (1994) The Ottawa Arnprior and Parry Sound: John Booth's ghost town trail. Broadview Press, Ghost Railways of Ontario. Peterborough, pp 191–201

Burns RJ (2000) Guardians of the wild: a history of the warden service of Canada's national parks. University of Calgary Press

Campbell CE (2011) A century of Parks Canada 1911–2011. University of Calgary Press

Environment Canada (2018) Recovery Strategy for the Wood Bison (*Bison bison athabascae*) in Canada. Species at Risk Act Recovery Strategy Series. Environment and Climate Change Canada. Ottawa. viii +, pp 59

Canadian Press (2022) Parks Canada recovers 45 fossils stolen from Burgess Shale, levies $20,000 fine May 12, 2022. https://ckpgtoday.ca/2022/05/12/parks-canada-recovers-45-fossils-stolen-from-burgess-shale-levies-20000-fine. Accessed 14 Aug 2022

Catto NR (2002) Anthropogenic pressures on coastal dunes, southwest Newfoundland. Canad. Geogr. 46:17–32

Catto NR (2015) Geomorphology: landscapes we live in. Oxford University Press, Don Mills, Ontario

Catto NR, Catto G (2012) Landscape response to human impact in coastal Newfoundland, Canada: 29,000 km of "untouched" coast-

line. Landscape Archaeology Conference, LAC 2012. Topoi, Special 3:225–227
Catto NR, Slaymaker O, Kovanen DJ (2020) Human-modified landscapes. In: Slaymaker O, Catto NR (eds) Landscapes and landforms of Eastern Canada. Springer International Publishers, Cham, Switzerland, pp 475–508
Catto NR, Catto G (2009) Geomorphology, sedimentology, and management issues, Hog Island (Pemamgiag) Sandhills, PEI. Report to the Mi'kmaq Confederacy of Prince Edward Island, p 99
Catto NR (2011) Coastal erosion in newfoundland. Newfoundland and Labrador, Ministry of Environment and Conservation, pp 144
Chylińska D (2019) The role of the picturesque in geotourism and iconic geotourist landscapes. Geoheritage 11:531–543
Clevenger AP, Ford AT, Sawaya MA (2009) Banff wildlife crossings project: Integrating science and education in restoring population connectivity across transportation corridors. Final report to Parks Canada Agency, Radium Hot Springs, British Columbia, Canada. 165pp. Accessed 8 Aug 2022
Conroy D (2007) Habitat enhancement in Cascumpec and Tracadie Bays, PEI; M.E.S. thesis, Environmental Science, Memorial University of Newfoundland, St. John's
Cooper L, Clay D (1994) An historical review of logging and river driving in Fundy National Park. Research Notes of Fundy National Park. No. FUN94-05. Parks Canada, p 73
COSEWIC (2011) COSEWIC assessment and status report on the Buffalograss Bouteloua dactyloides in Canada. Committee on the Status of Endangered Wildlife in Canada, Ottawa
Council of Europe (2022) European Landscape Convention. https://www.coe.int/en/web/landscape. Accessed 14 Aug 2022
Curtis J, Curtis W, Lieberman F (1982) The World of George Perkins marsh. Countryman Press, Woodstock, VT
Dearden P, Dempsey J (2004) Protected areas in Canada: decade of change. Can Geogr 48:225–239
Dickson J (1886) Camping in the Muskoka Region. Reprint, Ontario Department of Lands and Forests, (Toronto) 1959
Ellison EM (2013) The suffocating embrace of landscape and the picturesque conditioning of ecology. Landsc J Des, Plan, Manag Land 32:79–94
Environment and Climate Change Canada (2022) Canadian Environmental Sustainability Indicators: Canada's conserved areas. www.canada.ca/en/environment-climate-change/services/environmental-indicators/conservedareas.html. Accessed 14 Aug 2022
Environment Canada (2020) Canadian Climate Normals 1981–2010. https://climate.weather.gc.ca/climate_normals/index_e.html. Accessed 18 July 2022
Finegan C (2018) Reflection, acknowledgement, and justice: a framework for indigenous-protected area reconciliation. Int Indigen Policy J 9(3). https://doi.org/10.18584/iipj.2018.9.3.3
Förster F, Großmann R, Hinz M, Iwe K, Kinkel H, Larsen A, Lungershausen U, Matarese C Meurer P, Nelle O, Robin V, Teichmann M (2013) Towards mutual understanding within interdisciplinary palaeo environmental research: An exemplary analysis of the term landscape. Quat Int 312:4–11
Foster J (1998) Working for Wildlife: The Beginning of Preservation in Canada (Second ed.). University of Toronto Press
Friends of Algonquin Park (2022) Cottages and commercial leases. https://www.algonquinpark.on.ca/visit/park_management/leases.php Accessed 14 Aug 2022
Gagnon S Filion M (2004) The creation and early development of Gatineau park. National Capital Commission. http://www.canadascapital.gc.ca/data/2/rec_docs/1663_gatineau_study_e.pdf
Garvey TG (2009) The civic intent of George Perkins Marsh's Anthrocentric environmentalism. N Engl Q 82:80–111
Gates CC, Jung TS (2013) COSEWIC assessment and status report on the plains bison (*Bison bison bison*) and the wood bison (*Bison bison athabascae*) in Canada. Committee on the Status of Endangered Wildlife in Canada. Ottawa. xv + 109
Gilpin W 1786. Observations, relative chiefly to picturesque beauty, made in the year 1772, on several parts of England; particularly the mountains, and lakes of Cumberland, and Westmoreland. London: R. Blamire,
Glacier Park Collection (2022) 90 years: Inside the Opening of the Prince of Wales. https://www.glacierparkcollection.com/lodging/prince-of-wales-hotel/stories/inside-the-opening-of-the-prince-of-wales/ Accessed 14 August 2022
Green J (1976) The federal government and migratory birds: the beginning of a protective policy. Can Hist Assoc 11:207–227
Haines AL (1977) The yellowstone story: a history of Our First National Park. Colorado Associated University Press
Hall L (1995) Olmsted's America: An "Unpractical" Man and His Vision of Civilization. Bullfinch Press, Boston, MA
Hart EJ (2010) JB Harkin: father of Canada's National Parks. University of Alberta, Edmonton
Herd M, Heintzman P (2011) The impact of park development on the lives of local inhabitants within Gros Morne National Park. Proceedings of the 2010 Northeastern Recreation Research Symposium GTR-NRS-P-94 197–203
Higgins S, Mahon M, McDonagh J (2012) Interdisciplinary interpretations and applications of the concept of scale in landscape research. J Environ Manag 113:137–145
Hobhouse A (1947) Report of the National Parks Committee (England and Wales) London, H.M. Stationery Office
IFL (2021) Intact forest landscapes. https://intactforests.org/. Accessed 14 Aug 2022
IUCN 2022. Protected areas. https://www.iucn.org/our-work/protected-areas-and-land-use. Accessed 14 Aug 2022
Kareiva P, Watts S, McDonald R, Boucher T (2007) Domesticated nature: shaping landscapes and ecosystems for human welfare. Science 316:1866–1869
Kawartha Virtual Museum (n.d.) James Dickson. https://maryboro.ca/virtual/james-dickson-founds-algonquin-park/. Accessed 14 Aug 2022
Killan G (1993) Protected places: a history of Ontario's provincial parks system. Dundurn, Toronto
Langford NP (1905) The discovery of Yellowstone Park—diary of the washburn expedition to the Yellowstone and Firehole Rivers in the Year 1870. Frank Jay Haynes, St Paul, MN
Lasky K (2014) John Muir: America's first environmentalist. Candlewick Press, Somerville MA
Leclair L (1988) The Caldwell People of Point Pelee and Pelee Island: a brief history and survey of documents, department of Indian affairs and northern development
Levin WE, Unsworth SJ (2013) Do humans belong with nature? The influence of personal vs. abstract contexts on human-nature categorization at different stages of development. J Environ Psychol 33:9–13
Lothian WF (1976) A history of Canada's National Parks. Volume I. Environment Canada Supply and Services Canada
Lothian WF (1977) A history of Canada's National Parks. Volume II. Environment Canada Supply and Services Canada
Lothian WF (1979) A history of Canada's National Parks. Volume III. Environment Canada Supply and Services Canada
Lothian WF (1987) A history of Canada's National Parks. Volume IV. Environment Canada Supply and Services Canada
Luxton EG (2008) Banff : Canada's first national park : a history and a memory of Rocky Mountain Park (2nd ed.). Summerthought Pub., Banff
MacEachern A (2001) Natural selections: National Parks in Atlantic Canada, 1935–1970. McGill-Queen's University Press
MacEachern A (2011) M.B. Williams and the Early Years of Parks Canada. In Campbell, CE (ed.). A century of Parks Canada 1911–2011. Univ. of Calgary Press

Marsh GP (1864) Man and nature; or, physical geography as modified by human action. C. Scribner's sons, New York

Marsh GP (1885) The earth as modified by human action. C. Scribner's sons, New York

McCormack PA (1992) The political economy of bison management in Wood Buffalo National Park. Arctic 45:367–380

McGuckin T (2015) John Muir: Father of the National Parks. Lulu Press

McKennirey M, Richardson B (1985) The great buffalo saga. National Film Board/Parks Canada video. https://www.nfb.ca/film/great_buffalo_saga/

McLeod BA, Brown MW, Moore MJ, Stevens W, Barkham SH. Barkham M, White BN (2008) Bowhead whales, and not right whales, were the primary target of 16th- to 17th-century Basque Whalers in the Western North Atlantic. Arctic 61:61–75

McNamee T, Kraulis JA (1994) The national parks of Canada. Key Porter Books, Toronto

Miller C (2001) Gifford pinchot and the making of modern environmentalism. Island Press

NCC (n.d.) Gatineau Park. https://ncc-ccn.gc.ca/places/gatineau-park. Accessed 14 Aug 2022

NL Environment and Climate Change (2022) Avalon Wilderness Reserve. https://www.gov.nl.ca/ecc/natural-areas/wer/r-aw/. Accessed 14 Aug 2022

NWT Executive and Indigenous Affairs. Inuvialuit Final Agreement. www.gov.nt.ca/EIA/Priorities/Concluding and Implementing Land Claim and Self-Government Agreements. Accessed 14 Aug 2022

Ontario (2015) Tourism in Ontario: government and the tourism industry—growth of the park system. Archives of Ontario. Queen's Printer for Ontario. Accessed 14 Aug 2022

Parks Canada (2010a) State of the Park Report 2010: Aulavik NP of Canada. parkscanada.gc.ca

Parks Canada (2010b) Forillon National Park of Canada Management Plan. ForillonManagmentPlan2010.pdf https://www.pc.gc.ca/en/pn-np/qc/forillon/gestion#plan. Accessed 14 Aug 2022

Parks Canada (2016) Multi-species action plan for Point Pelee National Park of Canada and Niagara National Historic Sites of Canada. *Species at Risk Act* Action Plan Series. Parks Canada Agency, Ottawa. iv + 39 pp

Parks Canada (2018) Gulf Islands National Park Reserve. Park Management. https://www.pc.gc.ca/en/pn-np/bc/gulf/plan/c. Accessed 14 Aug 2022

Parks Canada (2019a) Rouge River Urban National Park. https://www.pc.gc.ca/en/pn-np/on/rouge/info. Accessed 14 Aug 2022

Parks Canada (2019b) Sable Island National Park Reserve of Canada Management Plan. https://www.pc.gc.ca/en/pn-np/ns/sable/info/plan/gestion-management-2019 Accessed 14 Aug 2022

Parks Canada (2020) Thousand Islands National Park: Culture. https://www.pc.gc.ca/en/pn-np/on/1000/culture. Accessed 14 Aug 2022

Parks Canada (2022a) Aulavik National Park. https://www.pc.gc.ca/en/pn-np/nt/aulavik. Accessed 10 Aug 2022

Parks Canada (2022b) Quttinirpaaq National Park. https://www.pc.gc.ca/en/pn-np/nu/quttinirpaaq. Accessed 10 Aug 2022

Parks Canada (2022c) Parks Canada 2022–23 Departmental Plan. https://www.pc.gc.ca/en/agence-agency/bib-lib/plans/dp/dp2022-23/index. Accessed 14 Aug 2022

Parks Canada (2023) Parks Canada attendance 2022–3. https://www.pc.gc.ca/en/docs/pc/attend. Accessed 16 March 2024

Parks Canada (n.d.) National Urban Parks. https://www.pc.gc.ca/en/pun-nup. Accessed 14 Aug 2022

Pollock-Ellwand N (2017) The prolific interpreter of the Olmsted vision: Frederick G. Todd Canada's first landscape architect. Plan Perspect 34(4):1–24

Rudin R (2011) Kouchibouguac: representations of a park in Acadian popular culture. In: Campbell C ed. A century of Parks Canada, University of Calgary Press, pp 205–33

Ryan KC, Knapp EE, Varner JM (2013) Prescribed fire in North American forests and woodlands: history, current practice, and challenges. Front Ecol Environ 11:e15–e24. https://doi.org/10.1890/120329

Slaymaker O, Catto NR, Kovanen D (2017) Geomorphological landscapes and Geoconservation. In: Slaymaker O (ed) Landscapes and landforms of Western Canada. Springer International Publishers, Cham, Switzerland, pp 413–435

Slaymaker O, Catto NR, Kovanen DJ (2020a) Holocene geomorphic processes in Eastern Canada. In: Slaymaker O, Catto NR (eds) Landscapes and Landforms of Eastern Canada. Springer International Publishers, Cham, Switzerland, pp 79–119

Slaymaker O, Catto NR, Kovanen DJ (2020b) Protecting geodiversity in Eastern Canada. In: Slaymaker O, Catto NR (eds) Landscapes and landforms of Eastern Canada. Springer International Publishers, Cham, Switzerland, pp 557–582

Slaymaker O, Spencer T, Embleton-Hamann C (2021) Recasting geomorphology as a landscape science. Geomorphology 384:107723. https://doi.org/10.1016/j.geomorph.2021.107723

Tourism PEI (2022) Industry performance and trends. https://www.tourismpei.com/sites/default/files/2022-08/202205TourismIndicatorsRptFin.pdf. Accessed 14 Aug 2022

UNESCO (2011a) Operational guidelines for the implementation of the World Heritage convention. UNESCO, Paris. Accessed 14 Aug 2022

UNESCO (2011b) Preparing world heritage nominations, 2nd ed. UNESCO, Paris. Accessed 14 Aug 2022

UNESCO (2021) Decision 44 COM 7B.190 Wood Buffalo National Park (Canada) (N 256). https://whc.unesco.org/en/decisions/7906. Accessed 14 Aug 2022

Worster D (2008) A passion for nature: the life of John Muir. Oxford University Press

Wright PA, Moghimehfar F, Woodley A (2019) Canadians' perspectives on how much space nature needs. FACETS 11 April 2019. https://doi.org/10.1139/facets-2018-0030

Youdelis M, Nakoochee R, O'Neil C, Lunstrum E, Roth R (2020) "Wilderness" revisited: is Canadian park management moving beyond the "wilderness" ethic? Can Geogr 64:232–249. https://doi.org/10.1111/cag.12500

Migration, Immigration, Ethnicity 11

Abstract

Over 23% of the total population of Canada in 2021 were immigrants, defined as people who were or had been a landed immigrant in their lifetime, the highest percentage since 1921. The migratory increase in population has exceeded the rate of natural increase since 1996 and is currently responsible for more than 90% of population growth, contributing an estimated 80% of the growth in labour force between 2016 and 2021. In the city of Toronto in 2021, 46.6% of the population were not born within Canada, including almost 7% who had arrived within the previous 5 years. Ca. 1500 CE, the estimated total human population of the terrain currently known as Canada was between 200,000 and 2,000,000, with 500,000 accepted as a most likely approximation. Indigenous people represented 5% of the population (1.8 million) in 2021, including 70,545 Inuk and 624,250 Métis. Initial colonial efforts involved French communities of Nouvelle France (Québec) and Acadie (Maritime Canada, including Plaisance (Placentia) NL), and English communities in Newfoundland and Nova Scotia. Following the conquest of Nouvelle France, further immigration from France effectively ceased until the mid-1800s. Britain had an open unrestricted immigration policy until 1905, and the same absence of controls applied to the British North American colonies. Local colonial administrations gradually assumed more authority over settlement, but control over numbers of immigrants only began in the 1850s. Net emigration characterized most years between 1860 and 1900. Federal government policy after 1896 changed to strongly encourage immigration, and organized campaigns (1898-1905) targeted rural residents of Central and Eastern Europe. After years of racial discrimination, in 1962 the Federal government removed ethnicity as a criterion for consideration for admission to Canada, with emphasis being placed on education, skills, and potential economic contribution. Recognition of ethnic origin is based on self-identification on the Census. The 2021 responses indicate an increasing variety of self-identifications with more than 450 ethnic or cultural origins cited. In total, "Canadian" (5.7 million, compared to 11.2 million in 2016), "English" (5.3; 6.2), "Irish" (4.4; 4.7), "Scottish" (4.4; 4.8), and "French" (4.1; 4.7) were the most commonly cited (including those citing multiple origins), in a population of 38 million. In 2021, about 18% of Canadians were bilingual in English and French. Emphasis on English and French, while valid in terms of the historical ethnic mixture and the role of Québec, does not account for the increasing number of Canadians who speak more than one language other than French or English. Multiculturalism has been official policy in Canada since the early 1970s. Multiculturalism involves questions of both personal and collective identity, language, nationality or ethnicity, religion, cultural practices, and sense of place. Although challenges remain and success is not universal, nor universally acknowledged, support for multiculturalism largely remains strong.

Keyword

Migration · Immigration · Ethnic diversity · Historical patterns · Multiculturalism

This chapter outlines the historical context and present (January 2025) conditions involving migration and immigration to Canada, and the current ethnic composition of the country. Unless otherwise indicated, statistical data cited throughout were obtained from the Census of Canada, available on the Statistics Canada website (Statistics Canada 2021–2024). Information from each census is released over several years following the enumeration.

N. Catto and G. R. Vincent, *Canada: Land and People*, World Regional Geography Book Series,
https://doi.org/10.1007/978-3-032-04391-7_11

11.1 Introduction

Over 23% of the total population were immigrants in 2021, defined as people who were or had been a landed immigrant in their lifetime, the highest percentage since 1921 (Demosim Team 2017, 2022a; Statistics Canada 2021–2024). Barring a substantial change in government policy, immigrants will represent almost 30% of the population by 2036.

The migratory increase in population has exceeded the rate of natural increase since 1996 and is currently responsible for more than 90% of population growth (Demosim Team 2022a). Immigration is the primary factor in continuing population increase (Fig. 11.1). It also contributed an estimated 80% of the growth in labour force between 2016 and 2021. About 92% of immigrants were 54 or younger (Demosim Team 2022a).

In the city of Toronto in 2021, 46.6% of the population were not born within Canada, including almost 7% who had arrived within the previous 5 years. Vancouver (41.8%) and Calgary (31.5%) also had immigrant percentages substantially above the national average. Québec (6.7%), Saint John NB (6.6%), and St. John's NL (4.6%) all had percentages substantially below the national average, as well as below the 6.9% average for centres outside the 41 large Census Metropolitan Areas with at least 100,000 residents (Fig. 11.2).

For people who had obtained landed immigrant or permanent resident status within 5 years of the 2021 Census, India (18.6%), Philippines (11.4%), China (8.9%), Syria (4.8%), and Nigeria (3.0%) were the most common countries of birth. Including Pakistan and Iran, the six Asian countries account for 48.3% of the total, compared with 2.0% from France and 1.7% from UK. Overall, more than 50% of immigrants in 2021 were born within the continent of Asia, compared with 22.5% from Europe and 10% from Africa. About 70% of people with Asian ethnic or cultural origins are first-generation residents, compared with about 25% of the total population. Two-thirds of those reporting European origins (including UK) are third-generation or longer residents (Fig. 11.3).

11.2 Indigenous Peoples

All of Canada, with the exception of northwestern Yukon and isolated nunataks in southern Alberta, was glaciated

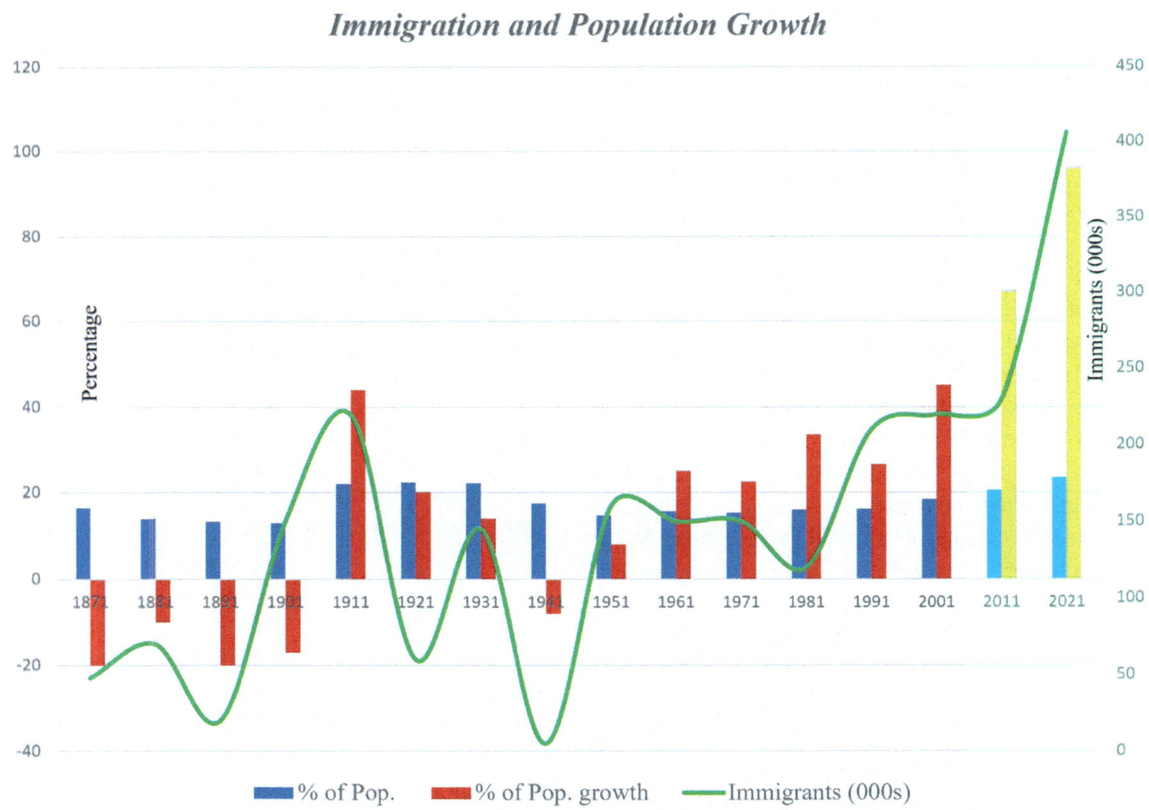

Fig. 11.1 Immigration and Population Growth 1871–2021, as recorded in decadal censuses. 1911 marked the only census where immigration accounted for more than 40% of population growth until 2001. The 2021 Census indicated the more than 90% of population growth in Canada was due to immigration. Negative population growth percentages indicate periods where emigration exceeded immigration (Statistics Canada, Census of Canada data)

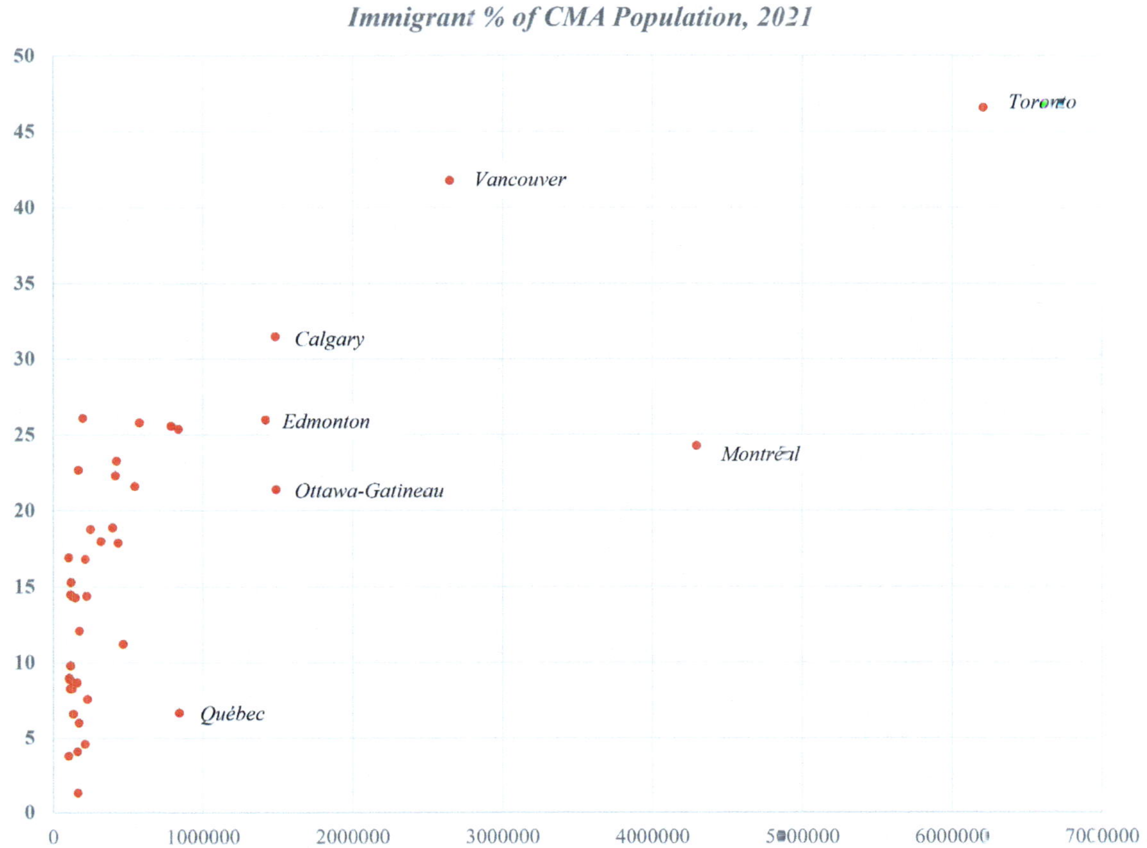

Fig. 11.2 Percentage of population of CMA represented by immigrants, 2021 Census of Canada. The largest CMA by total population are labelled

during the Quaternary, precluding animal or human occupation (see Chap. 1). Although much effort has been expended in research and conjecture, no definitively human occupation sites outside of northwestern Yukon (Beringia) predating the Last Glacial Maximum (*ca.* 20,000 years ago) have been documented as of 2024. Southeastward movement in interior North America was blocked by ice and inhospitable terrain (Catto 1996). The archaeological evidence indicates that early *Homo sapiens* followed the Pacific coastline southward from Kamchatka and Alaska, using suitable occupation sites now submerged by sea level change, as the first stages in populating North America. Estimated dates of the initial occupation of the terrain which is now Haida Gwaii vary from 14,000 to 18,000 years ago (see Chap. 7). Subsequently, Phylogenetic analysis and ^{14}C dated sites indicate that people spread southward along the western margin of the Cordillera, and then eastward and northward to occupy the terrain now known as Canada (Mendez et al. 2013). Human migration followed the retreating glacial ice and impounded glacial lakes. People reached southern Alberta ca. 14–15,000 years ago, and southern Ontario and Nova Scotia *ca.* 11,000 years ago. By *ca.* 6500 years ago, the last continental glaciers had ablated and conditions were suitable for human occupation throughout almost all of Canada.

Phylogenetic analysis and ^{14}C dated sites suggest that migration across the Arctic occurred in waves from *ca.* 5000 years ago to 1500 ce (Raghavan et al. 2014; Kanatami 2020). Initial movement was from the Beaufort Sea coastline northeast through the Arctic Archipelago to Kalaallit Nunaat, followed by southward migration along the Atlantic coast. The Sivullirmiut ("first people" or "Tunnit" to the Inuit; "pre-Dorset" and "Dorset" to archaeologists) migrated ca. 4000 years ago to occupy the entire Atlantic coastline to the Baie Verte Peninsula (NL) The most southerly site, the Fleur-de-Lys soapstone quarries, was exploited *ca.* 500 bce–500 ce (Fig. 11.4).

Subsequently, the Taissumanialungmiut ("Thule" to archaeologists, from Thule, Kalaallit Nunaat) people moved southward from Qikiqtaaluk *ca.* 1350 ce (Inuit Tapiriit Kanatami, 2020). The Taissumanialungmiut are the direct ancestors of today's Inuit ("the people"). The Taissumanialungmiut are notable for the development of techniques and tools (such as the toggle harpoon) that could be used to successfully harvest whales and seals offshore, in open marine waters. The Taissumanialungmiut thus were successful at adapting to the gradually cooling climate of the northern North Atlantic, at the same time that the Viking colonies were failing to respond and were drifting towards extinction

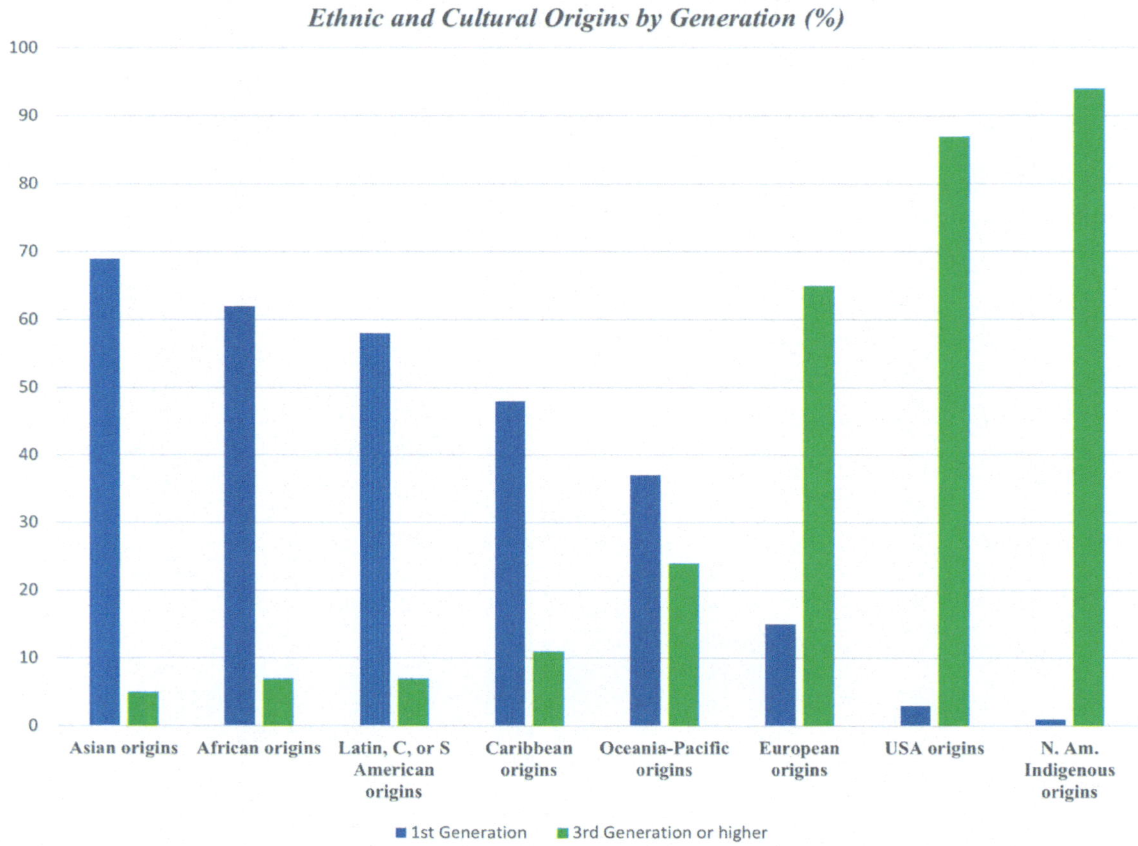

Fig. 11.3 Ethnic and cultural origins by generations. First-generation people are defined as recent immigrants, not born in Canada. Third-generation people are those with at least one grandparent born in Canada. Ethnic and cultural origins are based on self-identification

The style of living in relation to the environment, including development of kayaks and larger open umiak skin boats for offshore hunting, which was successfully practised by the Taissumanialungmiut differed little from that of the Inuit until recent times.

Fig. 11.4 Soapstone quarry exploited by Sivullirmiut ca. 500bce–500 ce, Fleur-de-Lys, NL

Subsequent waves of Arctic residents genetically related to the Taissumanialungmiut, the *Inuvialuit*, occupied the Beaufort Sea coast and Banks Island from the west (Kanatami 2020). Genetic interchange occurred with Dene (Tlicho, Dehcho, and Denésoliné) in the west, and with Innu from Nitassinan (interior Labrador and southern Nunavik) and Nîhithaw from the southwest coast of Hudson Bay.

Ca. 1000 ce, the broad linguistic groups in Canada included the Na-Dene in the western Taiga, Boreal Shield, and northern Cordillera south of the tundra; the Algonquian in the Boreal Shield, northern Prairies and Rocky Mountain foothills, and Maritime Canada; the Haudenosaunee of the St. Lawrence–Great Lakes Mixedwood Plains; the Lakotan of the eastern Midgrass Prairie, the Salishan language groups of the southern Cordillera and Pacific Maritime ecoregions; the Wakashan of the southern Pacific Maritime ecoregion; the Tsimshianic of the northern Pacific mainland coast and adjacent interior; the Haida of Haida Gwaii; and the Inuit language groups of the Arctic and northern Taiga (Goddard 1999; Oxford 2019). Limited information (Marshall 2001–2009) prevents classification of the Beothuk language of the residents of Newfoundland.

11.2.1 Indigenous Peoples After 1500 Ce

Ca. 1500 ce, the estimated total human population of the terrain currently known as Canada was between 200,000 and 2,000,000, with 500,000 accepted as a most likely approximation (Thornton 2000). Socio-political organization varied from the sophisticated Haudenosaunee Confederacy (est. 1134 ce) to the apparently completely independent and isolated Beothuk hunter-gatherer groups (Crown-Indigenous Relations and Northern Affairs Canada, 2017). Rivalries between and within linguistic groups existed, including competition for resources. Historically, contract with European new arrivals occurred initially from the east, with Leif Eirikson at L'Anse-aux-Meadows NL. *ca.* 1015 ce being followed by others along the Atlantic seaboard 1500–1600 ce, from the south and southeast *ca.* 1620–1750 ce, and finally from the Pacific coast *ca.* 1780–1820 ce (Harris 1987).

Estimates of the demographic catastrophe that followed range from 25 to 90% decline of the Indigenous population from each group's initial contact (see Thornton 2000). Suggested causes in Canada include imported diseases (smallpox, measles, influenza), forced removal and expulsion, imprisonment on colonial legal grounds, and unfair competition for resources. Although outright military attacks and attempts at biological warfare were rare (although unfortunately reality), loss of life in eastern Canada did occur as a result of conflicts due to European-Indigenous alliances and trade rivalries (see Chap. 1). Earlier writers tended to emphasize disease and overt armed conflict, while disregarding the psychological effects of the invasion of newcomers and disruption of lifestyles and cultures. Today, the processes and effects of colonialism are widely accepted as responsible for the demoralization, loss of purpose and sense of self, and emotional trauma felt by many Indigenous people (Crown-Indigenous Relations and Northern Affairs Canada 2017). Whether the eighteenth-century actions should be considered as planned and organized genocide, or whether they represented blind obtuseness, lack of insight and empathy, and stupidity, is debatable. They certainly lacked the intent behind the planned and purposeful genocidal establishment of residential schools in the nineteenth century.

The 1871 Census indicated an "aboriginal" population of 102,358 in Canada (Saku 1999). The estimates for Manitoba, British Columbia, "Rupert's Land" (Northwest Territories), and "Labrador and the Arctic Watersheds" (not including the part of Labrador administered by Newfoundland) were highly arbitrary, with only New Brunswick, Nova Scotia, Ontario, Québec, and PEI residents being accurately counted (PEI was included despite not joining Canada until 1873). A very rough approximation suggested ~45% Algonquian (using the linguistic groups above), ~40% Na-Dene and Pacific coastal peoples combined, ~10% Haudenosaunee and related peoples, and 4,000 Inuit. The Indigenous people were further classified by lifestyle as "living chiefly by fishing" (~23%), by "prairie hunting in camps" (~17%), as "families in the woods" (~43%), and in "villages in settled districts" (~17%, more than 70% of those enumerated in the five accurately counted provinces). These data, although probable underestimates and showing evident colonial bias, do provide an indication of the demographic impact between 1610 and 1871.

11.3 Initial Colonial Settlements

11.3.1 Francophone Communities

Initial colonial efforts involved French communities of Nouvelle-France (Québec) and Acadie (Maritime Canada, including Plaisance (Placentia) NL), and English communities in Newfoundland and Nova Scotia (Harris 1987). The four communities differed substantially in demographics and immigration policy.

In Nouvelle-France (Table 11.1; Fig. 11.5), immigration was encouraged through a series of commercial transactions between Crown and entrepreneurs, often with indifferent success (Moogk 1989; Gossage and Little 2015). Many residents were military personnel and government officials, who tended to leave after their terms of duty ended. Of the ~15,000 people who came from France to Québec prior to 1759, perhaps 10,000 returned to France after a few years or died childless.

However, upon prompting from the chief administrator, Intendent Jean Talon, Louis XIV sponsored the emigration of 832–852 young women to Québec as the *filles du roi* between 1663 and 1673, with the intent of spurring population growth through marriage (Landry 1992). Most were from Paris or Normandy, and two-thirds were 25 or younger (to age 12). All were volunteers, had to be judged capable of the hard work involved in farming, and had to present a letter of good conduct, generally from the parish priest. All had the option to decline the marriage upon arrival and return to France, which about 100 did, in addition to 300 who declined before departure. About 737 of the *filles du roi* married in Nouvelle-France, giving birth to almost 700 children by 1670. By 1672, the population had more than doubled to 6700 since 1663, and the programme was cancelled as having accomplished its goal (Landry 1992). On average, each woman had 6.4 children. The *filles du roi* accounted for more than half of all women arriving in Nouvelle-France between 1610 and 1759, and a much larger percentage of those who remained and raised families. Approximately, two-thirds of French-Canadians and 20 million people overall can trace their genealogy to a *fille du roi*. The concentration of mothers from one region of France effectively established the dialect spoken in Québec. By 1759, the population was approximately 80,000 (Gossage and Little 2015).

Table 11.1 Communities in Nouvelle-France, 1600–1717

Community	Lat N	Long W	Founded	Geographical notes
Tadoussac	48.15	69.74	1600	Saguenay–St. Lawrence confluence. Initially, seasonally-occupied fur trading post, forming competition with Québec
Québec	46.81	71.22	1608	Chosen for defence against marine attack, as St. Lawrence confined by steep cliffs
Charlesbourg	46.87	71.27	1626	Charles–St. Lawrence confluence. Agricultural. Today, part of Québec
Beauport	46.85	71.19	1634	Montmorency–St. Lawrence confluence. Agricultural Today, part of Québec
Trois-Rivières	46.36	72.54	1634	St-Maurice–St. Lawrence confluence. Agriculture, fur trading, lumbering
Lévis	46.76	71.24	1636	South bank of St. Lawrence opposite Québec. Initially agricultural
Montreal	45.51	73.55	1642	Lachine Rapids formed the head of navigation on the St. Lawrence
Sorel	46.03	73.11	1642	Richelieu-St. Lawrence confluence. Defence against Haudenosaunee descending Richelieu
L' Assomption	45.81	73.43	1647	L' Assomption–St. Lawrence confluence. Agricultural. Today, suburb of Montreal
Sault-Ste-Marie	46.55	84.36	1668	Falls separating Superior and Huron. Religious mission and trading post
Kamouraska	47.34	69.51	~1670	Kamouraska–St. Lawrence confluence. Eel fishery and agricultural
Verchères	45.77	73.36	~1670	South bank of St. Lawrence downstream from Montreal. Agricultural and defence against Haudenosaunee. Today, suburb of Montreal
Repentigny	45.74	73.38	~1670	North bank of St. Lawrence downstream from Montreal. Agricultural. Today, suburb of Montreal
Varennes	45.68	73.44	~1672	South bank of St. Lawrence downstream from Montreal. Agricultural. Today, suburb of Montreal
Ft. Frontenac	44.24	76.47	1673	Outlet of Lake Ontario. Defence against Haudenosaunee and English. Today, Kingston
Montmagny	46.98	70.55	1674	Rivière du Sud–St. Lawrence confluence. Agricultural
Chambly	45.45	73.28	1680	Head of navigation on the Richelieu. Defence against Haudenosaunee
La Malbaie	47.65	70.15	1688	La Malbaie–St. Lawrence confluence. Agricultural, fur trade, fishery
Rimouski	48.45	68.53	1696	Rimouski–St. Lawrence confluence. Agricultural and fishery
Ft. Kaministiquia	48.38	89.25	1717	Trading post. Today, Thunder Bay

Fig. 11.5 Francophone communities. Upper left: reconstruction of "habitation" at Port-Royal NS. Upper right: Québec was selected as a defensible position on high cliffs along the St. Lawrence. Lower left: Plaisance (now Placentia NL) was the centre of French colonial administration in Acadie until 1713. Lower right: remains of Fort Beauséjour NB

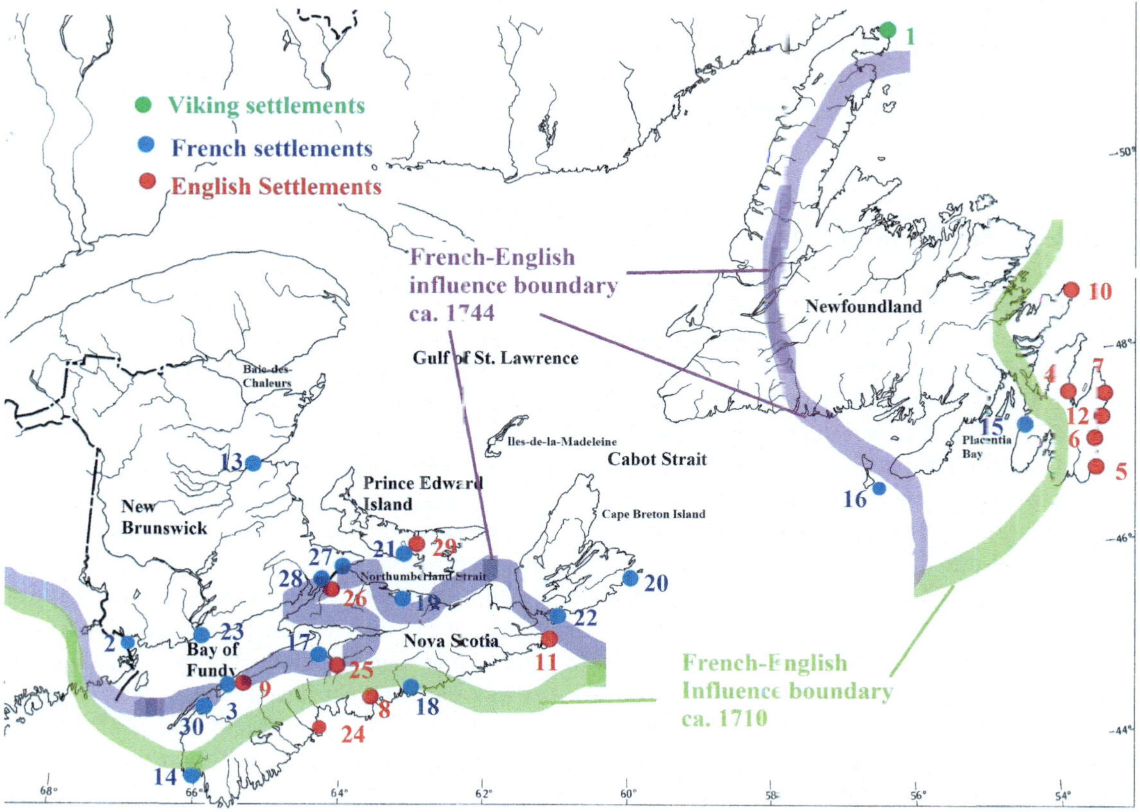

Fig. 11.6 Atlantic Communities. Viking (L'anse-aux-Meadows, 1015), selected French, and selected English settlements and fortifications identified by number and date of founding. French–English influence boundaries are approximate, crossed by much unofficial trading and travel. The Mi'qmaq community of Skmaqn was subsequently sequentially occupied by Port-la-Joye and Fort Amherst. 1. L'Anse-aux-Meadows 1015; 2. Ile-Ste-Croix 1604; 3. Port-Royal 1605; 4. Cupids 1610; 5. Renews 1617; 6. Ferryland 1620; 7. St. John's 1622; 8. Halifax 1625–1632; 9. Fort Anne/Annapolis Royal 1629; 10. Bonavista 1630; 11. Canso 1632; 12. Bay Bulls 1638; 13. Miramichi 1648; 14. Tebcuque 1650; 15. Plaisance/Placentia 1660; 16. St. Pierre 1665; 17. Grand-Pré 1680; 18. Chezzetcouk/Chezzetcook 1701; 19. Tantamagouche 1718; 20. Louisbourg 1719; 21. Skmaqn/Port-la-Joye/Fort Amherst 1720; 22. Nérichac/Isle Madame 1720; 23. Menagouèche/Fort Saint-Jean/Saint John 1748; 24. Lunenburg 1749; 25. Fort Edward 1750; 26. Fort Lawrence 1750; 27. Fort Gaspareaux 1751; 28. Fort Beauséjour 1752; 29. Charlottetown 1765; 30. Saulnierville 1785

Acadie did not benefit from a royally sanctioned immigration programme. Initial settlement was hampered by clashes between rival colonial entrepreneurs Menou and LaTour, English raids from Virginia and New England, and disputes over whether the settlements should focus on fishing or farming. In 1671, Acadia's total population was about 400, rising primarily through natural increase to about 13,000 in 1755. Indigenous (Mi'kmaq and Maliseet)-settler conflicts were minimal, without the irritants provided by a lucrative fur trade and military forces. Unlike Nouvelle-France, most Acadians came from the Loire Valley and Bordeaux, with the result that Acadian French was (and is) differentiated from Québecois French.

Also in contrast to Nouvelle-France, Acadia was treated as an administrative and military backwater. The primary exception was Plaisance (Placentia) NL, founded ca. 1660 as a military base, rather than a fishing port or colonial venture (Fig. 11.5). The inhabitants, however, persisted in conducting themselves in non-military ways, including fishing and commercial occupations, and even traded with the "enemy" in St. John's and New England. Trade was extensive between 1676 and 1710, only temporarily disrupted by war in 1696–7.

In the 1713 Treaty of Utrecht, France conceded the Atlantic shoreline of Nova Scotia and Plaisance to Britain (Fig. 11.6) and moved its military personnel to the under-construction base at Louisbourg, Cape Breton Island (Harris 1987). Louisbourg remained administratively, conceptually, and psychologically separate from Acadia under its capture by Britain in 1758, with almost all of the garrison returning to France. The Acadian residents were on good terms with the Nova Scotian colonists and had no military capacity or interest. Nevertheless, they were declared a threat and forcibly expelled by the British commander Lawrence, beginning in 1755. About two-thirds (~10,000) were deported, most to the Caribbean and Louisiana, but some as far as the Falkland Islands. Others fled the original settlement regions of the Annapolis Valley, Saint John River mouth, and Chignecto Isthmus for more remote areas including the Acadian Peninsula (NB), western PEI, Cheticamp and southern Cape Breton Island, and the Port-au-Port Peninsula (NL).

11.3.2 Anglophone Communities

The English colonies also were distinct (Reid 2016). Nova Scotia was initially seized from the Acadian and Indigenous residents *ca.* 1615. Although returned to France in 1632, small colonies remained at Lunenburg, Annapolis Royal, and the future site of Halifax, primarily consisting of Londoners, New Englanders, and German Protestants fleeing the Thirty Years War, but no Scots. Population growth was slow, even after the assignment of Atlantic Nova Scotia to Britain in 1713 (Fig. 11.6). The 1749 establishment of Halifax (named for George Dunk, a.k.a Lord Halifax) was a military response to the perceived threat of Louisbourg. Halifax, as its rival Louisbourg, was conceived in isolation to its surroundings, and grew up as an entity independent of Acadia. As far as Acadians were concerned, both fortresses could have been located in mid-Atlantic or on Jupiter.

For several years after its possession was confirmed by the 1763 Treaty of Paris, Britain did little towards settling Acadia. Interest was finally sparked by the necessity for re-locating the Scottish population displaced by the agricultural clearances in the Highlands, the difficulties resulting from local problems in Ireland, the problems of over-population in the Hanoverian and Hessian areas of Germany, and the requirement to resettle and compensate those who had remained loyal to George III throughout the USA War of Independence (United Empire Loyalists, UEL) and now were unwanted in their home communities (Harris 1987; Reid 2016). The UEL were concentrated in southwestern Nova Scotia (Shelburne, Yarmouth), in Charlotte County, NB (named for George's wife), and along the Saint John River (discussed below).

Newfoundland's initial settlement was distinguished by being unwanted by many. Desirous of retaining control over the lucrative cod fishery, using the waters for Royal Navy training, and hindering the desertion of sailors to civilian communities, the British government did not support permanent settlement of the coast (Rowe 1980; Cadigan 2009). In direct contrast to the *filles du roi* initiative, women were explicitly prohibited to sail on the fishing vessels from England until the 1670s (Handcock 1989). In practice, this was an unenforceable regulation, as the numerous embayments and archipelagos afforded many opportunities for those who chose to stay throughout the year. The "livyers" (those who "live here") gradually began to settle "winter locations", in sheltered embayments away from view of the open coastline. Some local researchers suggest that European occupation of parts of the Conception Bay coastal areas (Harbour Grace-Brigus) and the St. John's area began in the 1580s.

Cupids (from a misspelling of Cuper's (Cooper's) Cove) was the first official English settlement in what is now Canada, by John Guy (1610). Others followed at Renews and Ferryland, along the Southern Shore, 1617–1620 (Rowe 1980; Cadigan 2009). Renews was *Mayflower*'s first North American port of call. The choice of the Ferryland-Renews area was based on previous knowledge gleaned from Basque and French fish harvesters (Fig. 11.7). Local place names, such as Aquaforte (Ria de Aguea, fresh water), Fermeuse (Ria Formoso, beautiful river), and Ferryland (Forilhao, reef) have Basque roots. Other names with religious significance, such as Cabo de Pena (Cape Penitence) and Baie des Trépassés (Bay of Departed Souls) were transformed to Cape Pine and Trepassey, respectively.

The settlements which developed after 1670 were fishing communities, with residents from Devonshire and Dorsetshire (West Country), southern Ireland (Waterford, Kerry, and Galway), and the Channel Islands ("Jerseymen", including Huguenots) (Rowe 1980; McManus 1991; Cadigan 2009). Communities grew by natural increase, with many being either completely ethnically homogeneous or with distinct sections separated by origin. Newfoundland was the most populous English-speaking colony in British North America until *ca.* 1780. By 1816, its population had reached 52,000, compared with 81,000 in Nova Scotia, about 35,000 in New Brunswick, and 95,000 in Upper Canada (Ontario).

At Bay Bulls, Irish Catholic settlement began in the region in the 1770s and continued through the 1850s. As with many religiously split communities, the town was physically divided into distinctive Catholic and Protestant areas. A 1-km wide buffer zone ("no-man's land?") was established on the south side of the harbour, separating Protestants (who occupied the flatter land with the best fish processing sites) from the Catholics. Over time, encroachments on the formerly vacant terrain occurred, coupled with gradual emigration of the Protestant population (to the "Boston states") and higher birth rate among Catholics. By 1900, the entire community was dominantly Catholic.

Together, these four colonial regions represent the earliest French and English settlements of Canada, setting patterns which would persist until the mid-1800s (and beyond). The ratio of French: English settlement over time indicates the rapid growth of Nouvelle-France, followed by population growth in Newfoundland and Nova Scotia, coupled with the expulsion of the Acadians and limiting of immigration from France after 1755. The English population of Canada only equalled the French population *ca.* 1841, following immigration into Upper Canada (now Ontario; Fig. 11.8).

Fig. 11.7 Avalon Peninsula of Newfoundland communities. Top left: Excavation of Ferryland archaeological site, occupied ca. 1620–40. Lower left: Site of Cupids, occupied 1610. Top right: "Livyers" occupied sites such as Frogmarsh (near Brigus), where they could not be readily seen by marine vessels. Middle right: aligned stones are interpreted as sites used by Basque fish harvesters for drying cod; mistaken point. Lower right: unofficial marker erected by residents of Renews

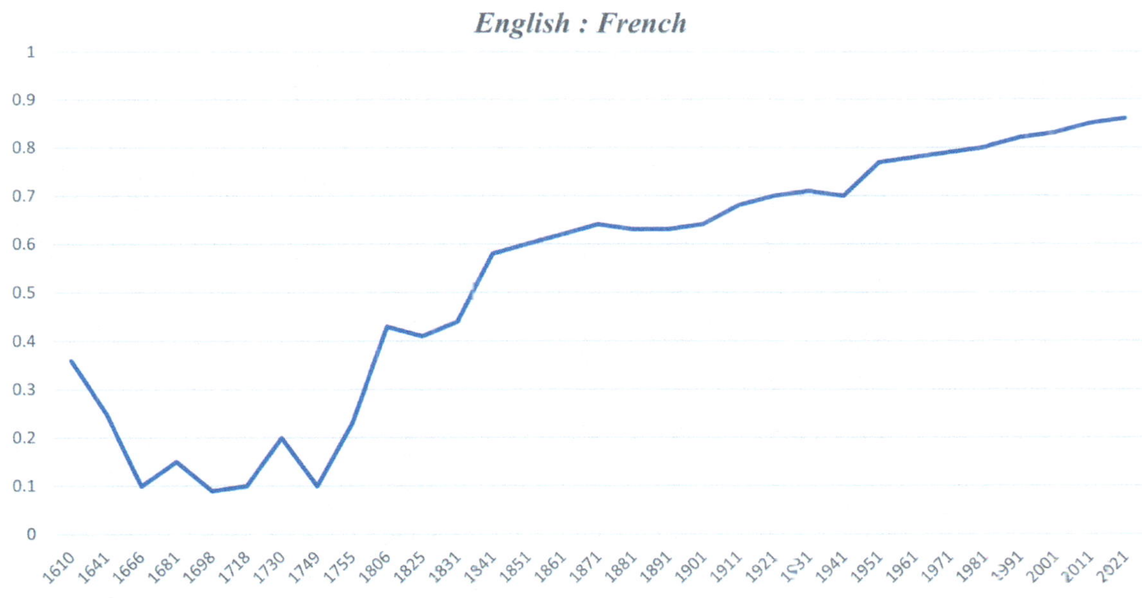

Fig. 11.8 Ratio of English: French speakers in territory currently known as Canada, 1610–2021. Data from censuses and reports compiled and maintained by Statistics Canada. See text for discussion

11.4 Immigration 1763–1867

Following the conquest of Nouvelle-France, further immigration from France effectively ceased until the mid-1800s. Britain had an open unrestricted immigration policy until 1905, and the same absence of controls applied to the British North American colonies. Immigration proceeded in a somewhat haphazard fashion, marked by private initiatives in colonization and substantial regional differences. Few if any controls applied to people migrating across BNA-USA borders until 1815. Until 1828, the British government did not regulate carriage of potential immigrants across the Atlantic. Local colonial administrations gradually assumed more authority over settlement, but control over numbers of immigrants only began in the 1850s.

11.4.1 United Empire Loyalists

Following the conclusion of the USA War of Independence, the British were faced with the problem of accommodating numerous civilians, supporters, military personnel, and indigenous people who had supported their wartime efforts and consequently felt unwanted or unwelcome in USA. Collectively, these people were termed United Empire Loyalists (UEL) by Guy Carleton (Lord Dorchester), the British Governor-General. Beginning in 1784, the British assisted ~50,000 UEL to resettle in British North America in the initial phase (Humber 1991; Conrad and Finkel 2006). About 30,000 others subsequently migrated across the undefined land borders into eastern Québec (the "Eastern Townships", Estrie) and across the St. Lawrence and Niagara Rivers into southwestern Ontario in a series of phases from 1790 to 1812 (Fig. 11.9).

Carleton and his successors considered the UEL from a mixed perspective of agricultural colonization and defence. Settlement sites such as Sandwich (today, Windsor ON), Newark (Niagara-on-the-Lake), Dick's Creek (St. Catharines), York (Toronto), Hardscrabble (Cobourg), Hallowell (Picton), and Elizabethtown (Brockville) guarded river crossings and lake shorelines. Land along the Grand River was allocated to Haudenosaunee under the leadership of Thayendenegea (Joseph Brant), with the expectation that they would help to defend against USA attacks (realized in 1812). The initial wave of ~10,000 ethnically English, German (many soldiers from Hesse and Hanover), Dutch (New York state), and Black (including military personnel such as Richard Pierpoint, early settler at Dick's Creek, along with 500 slaves; Walker 1992; Johnson and Aladejebi 2022) arrived between 1785 and 1795 (Conrad and Finkel 2006). For better civilian administration and military operations, Upper Canada was declared a separate colony in 1791. At the same time, slavery was abolished (Humber 1991). Fuelled by later migration from USA (~30,000) along with natural increase, Upper Canada's population increased to ~100,000 by 1812.

In contrast to Upper Canada, the UEL communities in Québec did not have military origins. Initially, the colonial administration under Frederick Haldimand discouraged UEL settlement in Québec, with the goals of reserving the land for Francophone residents and preserving an uninhabited buffer zone with New York, New Hampshire, and the independent republic of Vermont. UEL housed in refugee camps at Sorel and Chambly, along the Richelieu River, were moved to the newly established Haldimand County along Lake Erie. After Vermont became a state (1791), the policy was abandoned, and later-arriving UEL (~20,000, with ~300 Black slaves; Walker 1992) allowed to buy farmland and establish communities including Cowansville, Granby, Lennoxville,

Fig. 11.9 Successive groups of United Empire Loyalists formed conspicuous immigrant communities in eastern Canada. Upper left and lower right: urban and rural UEL residences are commonly grand local stone buildings, indicating social position. Adolphustown-Glenora, southern Ontario. Upper right: Plaque commemorating UEL arrival in Saint John NB, one of many historical markers. Lower left: Idealized monument to UEL arrivals, Hamilton, Ontario

Magog, and Sherbrooke. Others were involved in lumbering, including Philemon Wright, founder of Wrightsville (now Gatineau, Québec).

Approximately, 40,000 of the initial UEL were settled in Atlantic Canada (Rees 2000). Major settlements included Antigonish, Lunenburg, and Shelburne (NS); St. Andrews and other communities in Charlotte County, and Doaktown, Fredericton, Woodstock, and Perth-Andover along the Saint John River valley (NB); and Bedeque (PEI). As much of Nova Scotia and PEI had previously been allocated for British settlement by the King and Government, UEL were primarily concentrated in the newly separated province of New Brunswick (Conrad and Finkel 2006). Ethnically, English, German (Braunschweig (Brunswick) Saxony, Hesse, and Hanover), and Scottish settlers dominated, along with ~2000 Black settlers (including military personnel and 1200 slaves; Walker 1992). Unfortunately, prejudice remained among the UEL (Whitehead 2013, Johnson and Aladejebi 2022). The first recorded racism-based riot in Canada occurred in 1784, when English UEL of Shelburne attacked Black UEL of Birchtown in a dispute over land allocation (Table 11.2). Many Black UEL throughout Atlantic Canada subsequently relocated to Sierra Leone.

11.4.2 Maritime Canada 1763–1867

In addition to the UEL, immigration from Britain and Ireland populated Maritime Canada. Much of Nova Scotia had been allocated to friends of the British court throughout the 1700s, and could now be sold for profit or traded to pay debts to jewellers, tailors, and others, which also acted as a subtle incentive to encourage development. Direct immigration from Britain was relatively limited before 1815. Most immigrants came from impoverished regions—Highland Scots, Irish, West County English, Cumbrians (e.g. Ommer 1986; Reid 2016; McGowan 2023). Scottish settlement began with the first "Clearances" (*Fuadaich nan Gàidheal*) in 1762, during which Highland residents were forcibly displaced to make way for sheep. Settlement was concentrated in Cape Breton Island's central Bras d'Or valley and along the "Gaelic shore" north of Sydney, and in the Pictou and Antigonish areas of Nova Scotia. Acadian settlements persisted along St. Mary's Bay in the southwest.

After 1815, immigration involved primarily individuals, rather than organized colonial groups. Economic depression, rural depopulation (sometimes forced), reduced military establishments, and urban poverty triggered emigration from England, Scotland, and urban Ireland. Immigrants arrived as individuals and joined established communities, leaving fewer traces (e.g. Campey 2010). In contrast to Ontario and Québec, but similarly to PEI and Newfoundland, Irish immigration had largely ceased by 1840.

Settlement proceeded much more slowly on PEI than in mainland Nova Scotia, due in part to the exactions and speculative activities of absentee landlords, and in part due to the absence of economic opportunities other than farming (Driscoll 1988; Whitcomb 2010). Organized efforts began with Samuel Holland's survey in the 1760s, establishing Charlottetown and other communities and dividing PEI into 8000 ha lots. George III allocated these lots to court friends and military personnel but did not require the new owners to take up residence on PEI. Settlers ("tenants") paid rent to the absentee landlords and had no possibility of eventually gaining ownership of their land. Exchanging tenant status on farmland in Britain for tenant status on farmland in PEI was not attractive to many prospective settlers, and the lack of requirement for development encouraged land speculation. Tenancy was not completely abolished until after Confederation (Driscoll 1988; Whitcomb 2010).

Settlers were primarily Scottish islanders (Argyll, Bute, Skye, Na h-Eileanan an Iar (Outer Hebrides), rather than the Mainland Scots of Ontario and Nova Scotia), and Irish (Kilkenny, Tipperary, Wexford, Waterford), arriving largely between 1770 and 1810 (Fig. 11.10). Many subsequently left PEI after discovering the facts of tenancy. Subsequent immigration included west county English (settlements in Prince County), and Irish from County Monaghan (1830s, the last major influx). By the mid-1850s, immigration from Britain to PEI was essentially finished. Subsequently, out-migration dominated through the later 1800s, with the exception of family-based immigration from Lebanon and Syria in the 1880s.

11.4.3 Newfoundland 1763–1867

The Island of Newfoundland's limited suitability for agriculture, the lack of economic opportunities outside of the fishery and marine trade, Catholic-Protestant tensions, and the restrictive practices of the dominant merchants of St. John's and other communities precluded large-scale immigration throughout the period. Immigration from West Country England (Dorset, Devon, Somerset), southern Ireland (Waterford, Wexford, Galway; ~35,000), and the Channel Island of Jersey (including Huguenots) was primarily connected with the fishery. As the Beothuk had retreated to the interior prior to 1780, becoming extinct in 1821 (Pastore 1992; Marshall 2001–2009), the coastline was unoccupied. Throughout 1800–1830, the growing population dependent on the small-boat muscle-propelled cod fishery was able to expand to settle unoccupied coves. However, after 1830, most of the good fishing grounds were claimed, and immigration from Ireland largely ceased (Rowe 1980; Cadigan 2009).

Scottish immigration involved merchants and entrepreneurs from Lanark and Renfrewshire, using their

Table 11.2 Examples of restrictive immigration policies and racial discrimination, Canada 1784–1947. See text for further discussion. Official discrimination and exclusion policies ended in 197, and ethnic origin was removed as a consideration for immigration eligibility in 1962

Year	Targeted group	Comments
1784	UEL Blacks	Riot in Birchtown NS, first recorded racism-based riot in Canada. Many victims subsequently chose to relocate to Sierra Leone
1825–1845	Irish	~200,000 arrivals. Irish formed more than 60% of total immigration to Canada 1825–1860. Irish blamed for cholera outbreak in Montreal 1832; accusations that British govt. encouraged Irish immigration to reduce influence of francophone population in Lower Canada. Catholic Irish (majority) discriminated against in Upper Canada
1845–1852	Irish, fleeing Potato Famine	Ships infected with typhus interned at Grosse Ile. Dr. GM Douglas had 250 beds to serve >100,000 arrivals in 1847. ~400,000 arrivals total. More than 50% subsequently emigrated to USA. Ships also interned at Partridge I., NB
1863	Blacks (from USA)	Oil Springs (ON) riot. Irish protest Blacks' willingness to work for lower wages. All Black people leave Oil Springs after riot. One person convicted and sentenced to prison
1869	Paupers; disabled; women	People excluded included: lunatic, idiotic, deaf, dumb, blind, infirm, over age 60, widows with children, women with children but no husband
1872	Undesirable persons	Act legalized interment and deportation, with limited legal procedure
1873	Chinese	BC Chinese Tax Act (provincial) imposed extra taxes on Chinese people and businesses
1879	Blacks	Refusal and hindrance of Blacks migrating from USA (fleeing Ku Klux Klan)
1885	Chinese	Chinese Immigration Act (federal). $50 Head Tax imposed
1886	Chinese	Vancouver anti-Chinese riots
1900	Chinese	$100 Head Tax imposed
1903	Chinese	$500 Head Tax imposed. Chinese people categorized as "unhealthy… unfit for full citizenship…obnoxious to a free community and dangerous to the state."
1906	Chinese	$300 Head Tax imposed by Newfoundland. Repealed 1949 upon entry into Confederation
1906	Italians, Spaniards, Portuguese, Russians, Poles	Restrictive Immigration Act prohibited "any class of immigrants whenever it was considered necessary or expedient". Ethnic groups were not explicitly named, but policies of exclusion developed. People from southern Italy were particularly targeted
1906–1911	Blacks	Unofficial discrimination by agents began 1906 (e.g. not replying to correspondence). "Continuous journey" regulation applied to Caribbean immigrants 1908. Blacks essentially excluded as immigrants 1911. Repealed 1947
1907	Asians	Anti-Asian riots in Vancouver and elsewhere in BC
1908	Japanese	Hayashi-Lemieux "Gentlemen's Agreement" restricts Japanese immigration to 400 males and domestic servants per year, with immediate family members. Development of *shasshin kekkon* "picture bride" strategy as response
1908	Sikhs, Punjabis, Bengalis, Fijians, Caribbeans, Latin Americans, others	"Continuous journey" regulation bars transit from ports without regularly-scheduled passenger shipping service to Canada. *Komagata Maru* refused permission to land Sikhs 1914. Right to vote granted and restrictions repealed 1947
1910	Italians, Portuguese, Spaniards, Greeks, Blacks, Jewish	Immigration Act gives government more authority to deny entry or deport after landing. Charitable institutions not permitted to sponsor immigrants. Exclusion for people deemed "unsuited to the climate of Canada". Ethnic groups not explicitly named, but policies of exclusion developed
1918	Greeks	Anti-Greek riots in Toronto
1919	Germans, Austrians, Hungarians, Turks	Additional restrictions added as amendments to bar "enemy aliens" from World War I
1922	British favoured	Empire Settlement Act promotes immigration of English-Scottish-Welsh juveniles, farm labourers, domestic servants etc. Irish excluded. Largely unsuccessful
1923	Chinese	Chinese Immigration Act (Exclusion Act). Essential prohibition of Chinese immigration. Repealed 1947
1927–47	Jewish	Restrictive criteria imposed 1927. Christie Pits (Toronto) antisemitic, pro-Nazi riot 1933. Canada accepted only 5000 Jews in total 1933–39. Restrictions repealed 1947
1931–45	Almost everybody	… except USA (white) and British citizens with sufficient capital resources. Theoretically repealed with change of government 1935; in practice, remained through 1945
1941–47	Japanese	Internment in Rocky Mountain camps 1942. Deportation to Japan ("repatriation") 1946. Repealed 1947 after protests

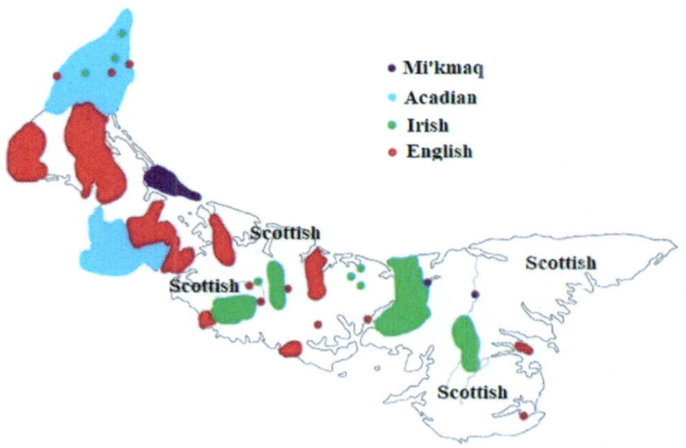

Fig. 11.10 Distribution of ethnic communities in PEI, based on the 2106 and 2021 Censuses. Scottish settlers dominated in King's (east) and Queen's (central) counties. Prince County (west) has significant Acadian and English populations. Boundaries are approximate and should not be taken to indicate ethnically exclusive areas. Lennox Island (Malpeque Bay) is home to the Mi'kmaq Lennox Island First Nation

expertise to establish fishery-related businesses in St. John's, Carbonear, and Harbour Grace. Scottish communities in western Newfoundland (St. Andrews, Codroy) date from 1840–1870, with small numbers of migrants primarily from Cape Breton and PEI driven by scarcity of farmland, disgust with tenancy, and fears that impending Confederation would result in tax increases. Acadian communities were established on the Port-au-Port Peninsula.

After 1850, immigration was minimal in Newfoundland (Rowe 1980; Cadigan 2009). Emigration ("out-migration") dominated due to lack of opportunity in the traditional small-boat fishery and no potential for other forms of employment or socio-economic advancement. High rates of natural increase, with many families having more than 6 children, only partially compensated for out-migration to Ontario, western Canada, USA ("Boston States"), and port cities around the world. The last Census of Newfoundland (including Labrador) in 1945 revealed that the population was more than 96% British by ethnic origin (Newfoundland 1945; Fig. 11.11). As late as the mid-1990s, a person whose surname was not listed in Seary's "Family Names of the Island of Newfoundland" (1977) would be met by an inquiry of where they were from and why they were here.

11.4.4 Canada 1815–1867

From 1791 through 1841, Lower and Upper Canada were politically separate, with separate immigration policies. They were merged in 1841 to form a single province of Canada, which existed until Confederation in 1867.

After 1815, immigration and emigration with USA continued. Concerns about the political loyalty of the former USA residents, some of whom had collaborated with or actively campaigned with USA troops, affected the governments of Upper Canada (to a great degree) and Lower Canada

Fig. 11.11 Ethnic Origins of the Population of Newfoundland and Labrador, based on the 1945 Newfoundland Census. "Others" includes Chinese and other European ethnicities. "French" represents predominantly Acadians in the port-au-Port Peninsula. Indigenous populations were likely to be underestimated, especially for Labrador. Ethnic origins are based on identification by census takers

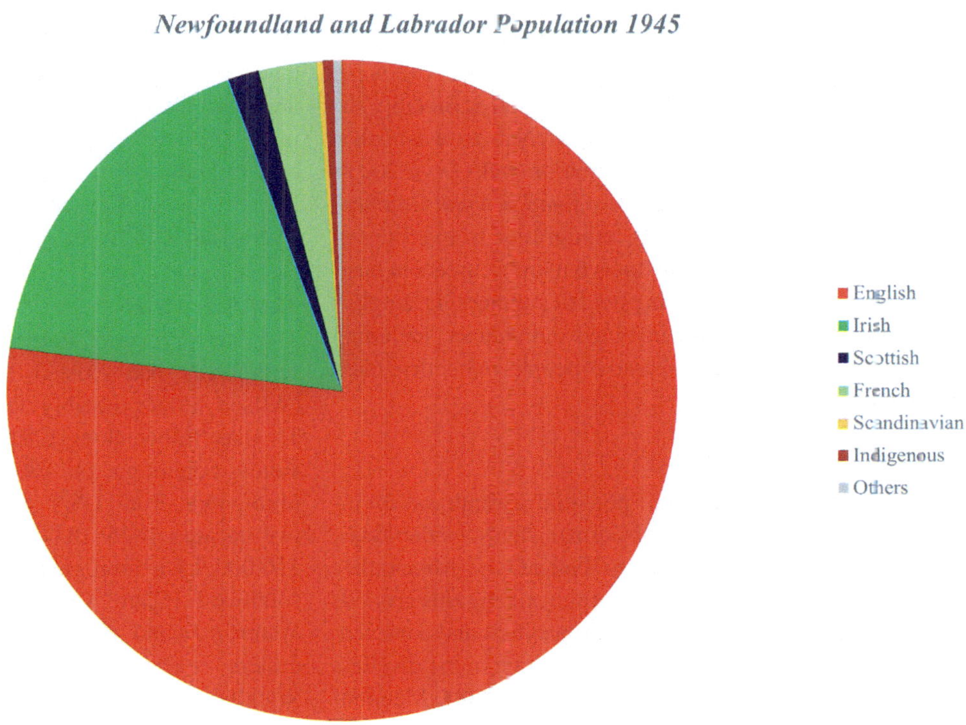

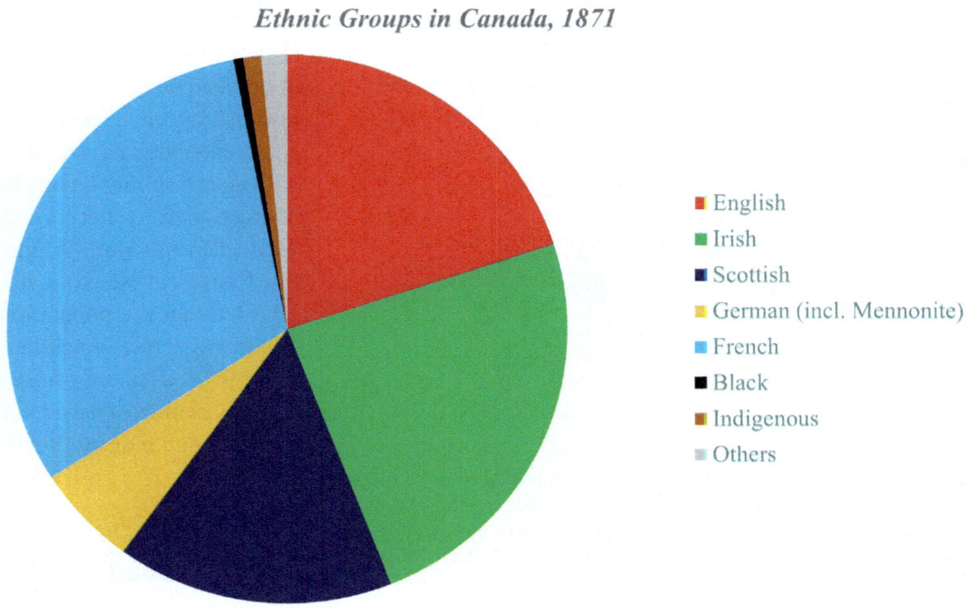

Fig. 11.12 Ethnic Groups in Canada, based on 1871 Census of Canada. "German" includes UEL and Mennonites, rather than representing country of origin. "Others" includes Chinese and other European ethnicities. Ethnic origins are based on identification by census takers

(to a lesser degree). This led the governments to consider other potential sources of immigrants.

Upper Canada's preferred candidates were English and Scottish. An earlier 1804 experiment with Lord Selkirk (discussed further below) at Baldoon along the St. Clair River was partially successful, suggesting that additional groups of Scots could be enticed to come. Following the Baldoon model, the focus was on organized colonies with a designated leader (or landowner), relying on clan loyalties and family ties along with religion. Selkirk's colonial ventures in Ontario, PEI, and Manitoba were all religiously segregated, with separate Catholic and Presbyterian communities (Bumsted 2008). Although the initiative met with some initial success, there were conspicuous failures, notably the 1825 attempt of Archibald "Laird" McNab to establish a feudal dictatorship at Arnprior (Cameron and Gwyn 2003). More significantly, the completion of *Fuadaich nan Gàidheal* changed the character of Scottish immigration from groups of impoverished crofters to individuals pursuing careers in business, finance, and the developing field of engineering. English immigration also primarily involved individuals, including many professional and administrative people, rather than mass migration (Fig. 11.12).

11.4.4.1 Mennonites

Mennonites formed a group of potential agricultural immigrants. Originating from several German states, Netherlands, Switzerland, Alsace, and western Poland, the Mennonites were subject to discrimination in Europe on religious grounds as Anabaptists, on military grounds as pacifists and conscientious objectors, and on social grounds as commune dwellers (Epp 1974). Between 1800 and 1850, about 50,000 Mennonites arrived in southwestern Upper Canada, both directly from Europe and from Pennsylvania, the majority after 1830 (Epp 1974). As with "Jewish", "Mennonite" can refer either to ethnic origin or to religious affiliation. Censuses of the time generally identified the Mennonites as "German", regardless of place of origin (Fig. 11.12).

Almost immediately, Mennonites began dividing into strongly traditional agriculturalists rejecting all modern technology ("Old Order"), and those with more diverse lifestyles, including urban dwellers. Mennonite influences were strong in the development of Elmira, Kitchener (formerly Berlin), Leamington, London, and Waterloo (Bloomfield 1995). When land availability in southwestern Ontario became a problem, new agricultural Mennonite communities were established in the Red River Valley of Manitoba and the Clay Belt of Northern Ontario (Epp 1974). Today, of the ~180,000 Mennonites in Canada, only about 2% follow the ultra-traditional "Old Order" or Amish lifestyle.

11.4.4.2 Irish

The other group of potential immigrants were Irish. Irish settlements had existed in southern Ontario previously (Wilson 1989), notably the colonies established by Col. Thomas Talbot along Lake Erie and in the Thames River valley beginning in 1803, primarily with ethnic Irish immigrants from USA (Conrad and Finkel 2006). In the two decades 1825–1844, Upper Canada attracted 120,000–150,000 Irish immigrants, primarily from Munster in the southwest (Clare, Cork, Kerry, Limerick, Tipperary, and Waterford) and the southern Leinster counties along the eastern coast (Carlow, Kildare, Kilkenny, Wexford, and Wicklow). Most immigrants were Catholic. As well, Lower

Canada also accepted Catholic Irish immigrants through church initiatives (Wilson 1989). An estimated 80,000–100,000 Irish immigrants arrived in Lower Canada before 1845.

Tensions developed almost immediately (Table 11.2). Fears that the British government was dumping Anglophone Irish immigrants into Francophone Lower Canada were only partially alleviated by the newcomers' Catholicism, and their tendency to become rapidly involved with existing communities and societies. An 1832 cholera outbreak in Montreal, which killed 9000 residents, was blamed on the "dirty" Irish by many (see Sendzik 1997). Although most immigrants came from rural areas, the absence of unclaimed farmland meant that most settled in Montreal and Québec. Unlike the agricultural *Fuadaich nan Gàidheal* Scottish and the Mennonites, the Irish formed conspicuous ethnically distinct communities of manual labourers. In addition, as all immigrants destined for Upper Canada initially landed at Québec, the negative impressions of a steady stream of rootless newcomers grew. Perhaps half of the Irish immigrants from this period eventually left for USA.

In Anglican and Presbyterian Upper Canada, the arrival of thousands of Catholics sparked consternation among many (Wilson 1989). Rapid expansion of the Grand Orange Lodge of British North America was one consequence. Initially, the land shortage problem was not acute, and more Irish migrants were able to find agricultural employment. Until 1845, the newcomers were perceived as part of a population growth spurt, as the population increased to ~800,000 from ~100,000 in 1812. However, among the previously established residents and the English colonial authorities, the stereotypes of the "dirty, lazy, drunken" Irish capable only of manual labour, along with the Scottish merchants and engineers and the English administrators and professionals, has taken firm hold. The attitude was exemplified by Francis Bond Head, appointed Governor of Upper Canada in 1835 after a stint establishing the Poor Law Workhouses in England subsequently discussed by Charles Dickens.

Matters became much more critical in 1845, when Ireland accidentally imported the potato blight, *Phytophthora infestans*, from New York. More than 3 million rural residents had become entirely dependent on potatoes for food. The blight killed at least a third of the potato crop in 1845, and about 75% in 1846 (Ó Gráda 2006). Initially, the Potato Famine (*Drochshaol)* was triggered by the loss of potatoes to blight, but the loss of seed potatoes proved to be more devastating in the longer term.

It is difficult to attribute deaths specifically to the Potato Famine, as other diseases such as cholera, diphtheria, tuberculosis, typhoid, and typhus were rampant (Ó Gráda 2006). The best estimates suggest approximately 1 million Irish, about 12% of the population, died from the direct (starvation) or indirect (disease) effects of the Famine, approximately one-third of the rate of mortality during the Black Death of 1349–1350. Rates varied widely, ranging from 25–30% of the population of Connacht to virtually no one in county Carlow. The urban centres of Dublin, Cork, and Belfast saw population increases, resulting from internal migration.

Emigration was extensive from the worst-hit areas of Connaught (Mayo, Galway), Munster (Cork, Limerick, Tipperary), and southern Ulster (Monaghan), representing different populations from those of Waterford and Leinster which dominated previous migrants. In many cases, families designated boys, teenagers, and young adult males to go, with the hope that they would send remittances once established in North America. Some of these "unaccompanied minors" were incorrectly assumed to be orphaned or abandoned upon arrival at the quarantine entry points of Grosse Ile (Québec) (O'Gallagher 1984) and Partridge Island (NB).

Due to Britain's open immigration policy, no restriction on migrant transit was possible. Immigrants could avoid USA restrictions on shipboard arrivals by sailing to Canada first, with a view to migrating southwards later. At least 50% subsequently emigrated to USA. Passage to Canada was much less expensive than to USA, particularly as many of the ships were involved in the lumber trade, and would otherwise have sailed in ballast. Immigrants were unexpected paying customers, but the ships were not designed for passenger transport (Mackay 1990). Some of the immigrants were suffering from contagious diseases, particularly typhus, when they boarded. Medical facilities at Grosse Ile and Partridge Island were completely inadequate, with predictable results (O'Gallagher 1984). At Grosse Ile, Dr. GM Douglas had 250 beds to serve the more than 100,000 arrivals in 1847; he subsequently committed suicide

At least 400,000 people passed through Grosse Isle and Partridge Island between 1845 and 1852 (Wilson 1989). Perhaps 200,000–250,000 left for USA within a short time. The remainder settled in Upper and Lower Canada (Mackay 1990). Unaccompanied Irish children were adopted by francophone families and educated in French, but were permitted to retain their own names. Irish people formed more than 60% of total immigration to Canada between 1825 and 1860. The 1871 Census of Canada revealed that almost a quarter of the population was ethnically Irish (Fig. 11.12).

11.4.4.3 Blacks

Blacks also came from USA between 1850 and 1865, escaping slavery through the Underground Railroad. Although Upper Canada had abolished slavery in 1791, prejudice proved hard to eliminate. In Oil Springs, a riot in 1863 resulted from Irish workers protesting Blacks' willingness to work for lower wages. Following the riot, one person was convicted and imprisoned, but all Black people left Oil Springs. After 1865, perhaps three-quarters of the Black immigrants to southern Ontario returned to USA (Johnson and Aladejebi 2022).

11.4.5 Rupert's Land

Prior to 1869, the Hudson's Bay Company (HBC) actively discouraged immigration to the Prairies, in the belief that agriculture and urban communities would adversely affect the fur trade. Métis descendants of fur traders, trappers, and hunters lived throughout the region, but did not practice agriculture, provided pemmican from hunted *Bison* to the traders, and were active participants in the fur trade.

The only exception was Lord Selkirk's Red River colony of Scottish settlers (Bumsted 2008). Initially blocked by HBC, Selkirk circumvented the policy by acquiring HBC stock, becoming a part owner, and lifting the ban for his colony only in 1812. The colony was positioned along the river south of Lake Winnipeg, because the chernozemic soil was better there, but also because the new colony blocked the trade route used by HBC's competitor, the North West Company (NWC) to ship furs via the Great Lakes to Montreal. Selkirk's colony also dispossessed the existing Métis community.

Due to poor planning, the colony's food supply failed in 1814, which caused the colony administrator to issue a ban on all food sales, including pemmican. Not surprisingly, the Métis objected, culminating in the deaths of 22 people in the clash at Seven Oaks (1816), including the administrator. Convinced that the NWC were behind the attack, Selkirk forced a trial of NWC personnel, in addition to attacking the Fort William NWC post. After their acquittal, Selkirk forced another trial, lost again, sued the NWC, lost the countersuit, was convicted for his attack on Fort William, lost several appeals, and died bankrupt in 1820 (Bumsted 2008). The following year, the NWC and HBC merged, and the Scots and Métis got along well thereafter. Ca. 1856, the population of Manitoba included approximately 7000 European settlers (Beaujot and McQullan 1982).

11.4.6 British Columbia

Aside from a few HBC personnel and occasional naval visits, British Columbia's entire non-indigenous population came from the south along the Pacific coastline. Prior to 1857, the estimated non-indigenous population was less than 300 (Barman 1991). In that year, placer deposits along the Thompson River triggered a gold rush, followed in quick succession by finds and rushes along the Fraser River and in the Cariboo country (1862). Victoria's population increased from ~100 to ~30,000 in four weeks in 1857, with additional towns in the Fraser Valley and Cariboo. Settlement came almost exclusively from California, estimated as about one-third Anglo-American; one-third Spanish, French, German, and other European ethnicities; and one-third Chinese. A single party of ~60 "Overlanders" from eastern Canada reached the Cariboo in 1862.

Many of the European and Anglo-American immigrants were gold seekers who had come from eastern Canada, chasing gold across North America and Australia and back to interior British Columbia. The Chinese immigrants (~7000 between 1857 and 1870) came predominantly from Guangdong province in southern China. Initially, immigrants travelled through Guangzhou or Hong Kong to California and then north. Direct immigration to BC from China began in the mid-1860s.

As with all gold rushes, the boom was temporary (Barman 1991). The Cariboo Road was completed to Barkerville in 1865, just as the gold supply was dwindling. A new strike in Wyoming two years later sent the gold seekers southeast. By 1867, BC had a population of 40,000–50,000, of whom 75–80% were indigenous.

11.5 Immigration 1867–1962

Net emigration characterized most years between 1860 and 1900 (Canadian Museum of History n.d.; Whitaker 1991; Kelley and Trebilcock 1998; Edmonston and Fong 2011; Bumsted and Bumsted 2016; Fig. 11.13). Recruitment efforts were minimal until 1880, and a combination of economic depression (1873–79) and a shortage of good agricultural land resulted in substantial migration of both residents and newer arrivals to USA. Efforts by the Federal government and CPR to encourage homesteading along the railway were hampered by the difficulties encountered by new arrivals in farming the Shortgrass Prairie. Declining population growth in Britain, the preference of many Irish immigrants to settle in USA rather than British colonial Canada (see Mannion 2022), and the inability of many immigrants from urban Britain to find employment in the limited Canadian manufacturing sector also limited net migration (see Gentilcore 1993; Bumsted and Bumsted 2016).

Federal government policy after 1896 changed to strongly encourage immigration (Canadian Museum of History n.d.; Burnett 1988; Whitaker 1991; Gentilcore 1993; Kelley and Trebilcock 1998; Raska 2020; Fig. 11.14). In contrast to earlier efforts, organized campaigns (1898–1905) targeted rural residents of Central and Eastern Europe. Advertisements translated into a variety of languages stressed free farms for all, available money and supports, glorious possibilities, abundant railway connections for transporting grain to market, vast coal fields at convenient distances, grants offered to facilitate settlement, and the Prairie climate, the healthiest in the world. Clifford Sifton, the federal minister responsible, identified the target audience as "peasants in sheepskin coats", experienced farmers familiar with chernozemic soils and steppe climates who were interested in owning their own land and escaping political, religious, and socio-economic oppression (Whitaker 1991). Advertising was conducted with the permis-

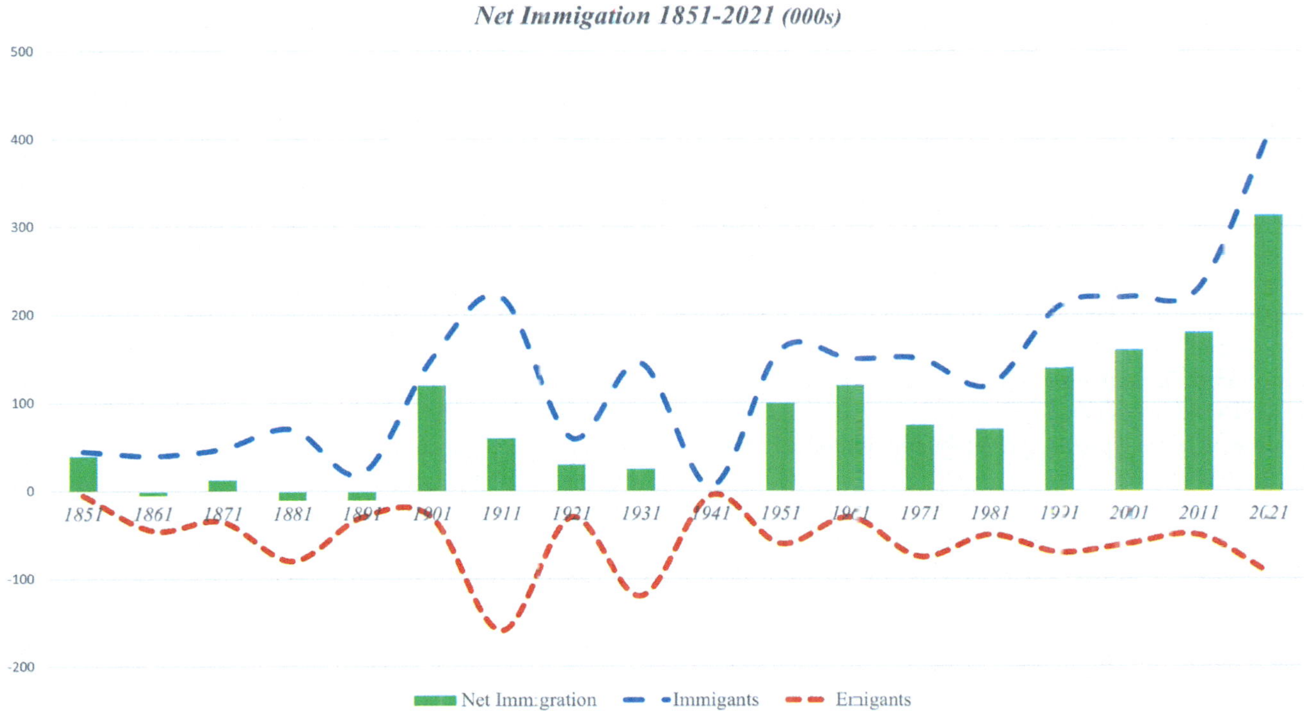

Fig. 11.13 Net Immigration 1851–2021. Net immigration data indicates totals for the 10 years prior to the census date. Immigrant and emigrant totals are based on 5-year running averages. Data prior to 1871 refers to totals for all areas presently within Canada, except Newfoundland and Labrador (Data from Statistics Canada Census of Canada, and Edmonston and Fong 2011)

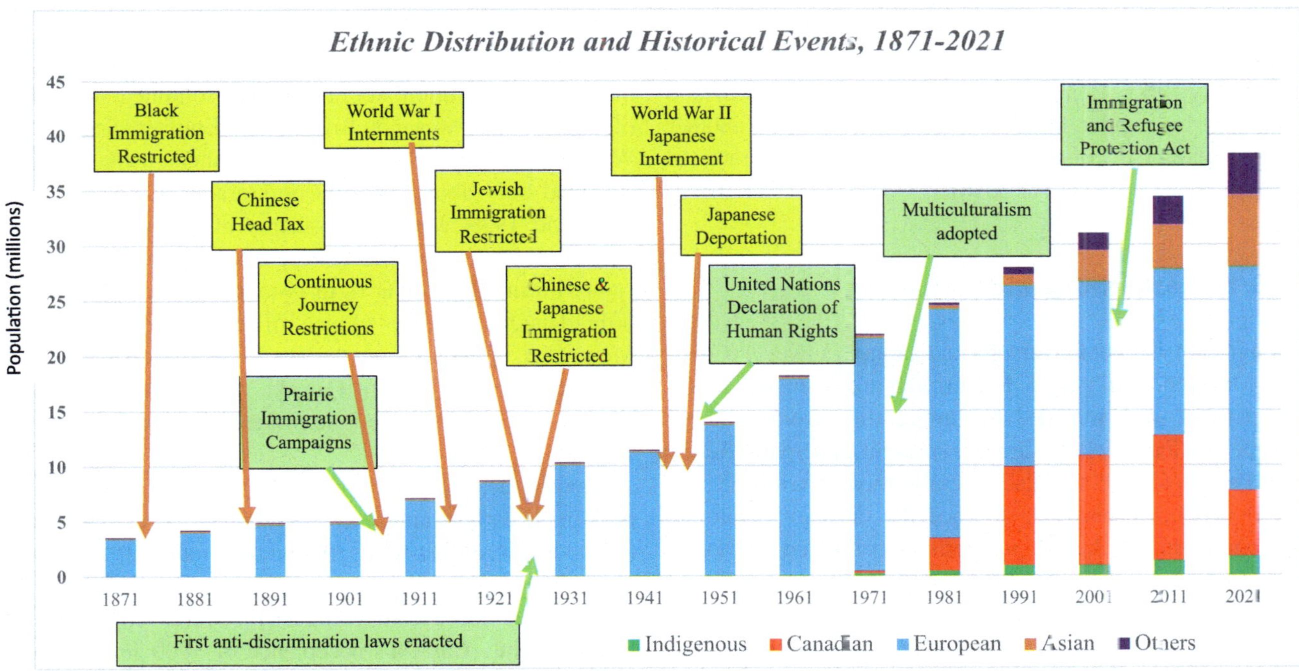

Fig. 11.14 Timeline of Ethnic Distributions and Immigration-related Historical Events, 1871–2021. Yellow boxes represent discriminatory or exclusionary policies; green boxes represent events favouring broader immigration. Ethnic origins prior to 1971 are based on identification by census takers; from 1971, based on self-identification by respondents. See text for discussion of "Canadian" ethnicity. "Others" refers to African, Central and South American, Black, and Oceania-Pacific origins. Indigenous totals prior to 1961 are too low to be readily visible (Historical timeline data from Canadian Museum of Immigration and sources cited in text; ethnic data from Censuses of Canada)

sion (or indifference) of the governments of Austria-Hungary (Cisleithania) and the Russian Empire, in part to encourage politically, ethnically, or religiously undesirable people to emigrate (Canadian Museum of History n.d.; Bumsted and Bumsted 2016). The campaigns successfully enticed immigration from Galicia (Polish and Ukrainian) and Moravia (Czech) in Cisleithania, and Ukraine, Belarus, and Moldova (Russian Empire), as well as adjacent Bulgaria and Rumania (Isajiw and Makuch 1990; Kukushkin 2007; Hinther and Mochoruk 2011). Ongoing pogroms led to Jewish immigration, initially to Montreal and Toronto. Although emphasis was placed on individual migration, communal groups such as the Doukhobors (Transcaucasia) were permitted to come, albeit with some reluctance from federal officials (Kelley and Trebilcock 1998).

The policy of encouragement of immigrants from eastern Europe did not survive the replacement of Sifton by Frank Oliver in 1906. Progressively more restrictive regulations in 1906, 1910, and 1919 were aimed at limiting immigration from Eastern and Mediterranean Europe, in addition to Black and Asian immigration (discussed below; Table 11.2). Most Ukrainian immigrants to Canada arrived after World War II (see Hinther and Mochoruk 2011).

Many prairie immigrants during this period were from USA, as farmland was made available, and much of the USA farmland had been occupied. In many family histories, parents left Canada (Ontario) in the 1870s or 1880s, and then their descendants would return to Canada (prairies) in the early 1900s. Remittance men, immigrants from Britain supported by regular family contributions, formed another socio-economically distinctive group, not always appreciated by local residents (Zuehlke 2016).

11.5.1 Restrictions and Prejudices

Immigration was not universally encouraged, however (Canadian Museum of History n.d.; Bumsted and Bumsted 2016). Restrictions were imposed on immigration from southern Europe, particularly Italy, Greece, Portugal, and Spain (Kelley and Trebilcock 1998). Black immigration (Caribbean and USA) had been discouraged since the onset of economic depression in 1873, and restrictions continued thereafter (Milan and Tran 2004). By 1908, would-be immigrants were required to come to Canada directly from their original country by "continuous journey", rather than purchasing tickets that entailed a stopover in a third country. Travel itineraries involving rail travel across Europe to a suitable port with a direct connection to Canada were not affected, but as there were no direct passenger sailing routes from India or Japan, emigration from these countries was largely prohibited (Pottie-Sherman and Wilkes 2019). However, Japanese women who married Japanese men living in Canada by proxy ("picture brides") were permitted entry until 1928.

Imposition of Head taxes on all prospective Chinese immigrants began in 1885 (after completion of the CPR) (Canadian Museum of History n.d.). By 1903, the Head Tax stood at a prohibitive $500 (2024 eq. ~ $17,000), about two years' salary for a typical full-time worker. Newfoundland followed suit in 1906 with a $300 Head tax, designed to stop the estimated 300 Chinese immigrants from overwhelming the colony (Cadigan 2009). In 1923, virtually all immigration from China was prohibited, a restriction which persisted until 1947 (Kelley and Trebilcock 1998; Table 11.2).

Immigration declined from its peak in 1906 (Canadian Museum of History n.d.; Gentilcore 1993; Edmonston and Fong 2011), as the supply of agriculturally suitable land became limited, government recruiting efforts were curtailed, especially after 1910, and country-of-origin governments raised restrictions (Whitaker 1991). The outbreak of World War I ended this phase of enhanced immigration, with numbers not surpassed until the mid-1990s (Fig. 11.14). Ironically, numerous immigrants from the Austro-Hungarian Empire were detained as "enemy aliens" during World War I, although most had chosen to leave the empire behind. Sharp restrictions on Bengali, Black, Chinese, Japanese, Pakistani, Punjabi, and other non-European groups continued until after World War II, even while anti-discrimination laws were being passed to apply to Canadian residents (Kelley and Trebilcock 1998). The Great Depression and World War II effectively curtailed immigration until the late 1940s (Canadian Museum of History n.d.). Under Prime Minister RB Bennett, the 1931 Immigration Act effectively prohibited acceptance of anyone who was not British (i.e. English, Scottish, or Welsh) or a white USA citizen with financial means. Canada accepted only about 140,000 immigrants in total between 1933 and 1939. Japanese residents of Canada numbering 22,000, including citizens, were involuntarily interned in camps during World War II. In 1945, the Federal government offered the Japanese the "choice" of moving east of the Rocky Mountains, or deportation ("repatriation") to Japan, although virtually all were born in Canada. About 3,000 were ultimately deported before public protest reversed the policy in 1948.

Jewish immigration to Canada began in the 1700s. After a peak in the early twentieth century, reduced overall immigration and antisemitism led to sharp restrictions in 1927 (Davies 1992; Tulchinsky 2008). An antisemitic riot at Christie Pits baseball stadium (Toronto, August 1933) was led by the local Swastika club. Canada's acceptance of 5000 Jewish immigrants between 1933 and 1948 contrasts adversely with USA (95,000 prior to 1940), Great Britain (40,000), China (18,000), and Mexico and South America (75,000 in total). Refugees fleeing Franco, Mussolini, or Stalin faced similar intolerance from Canadian immigration officials. Antisemitism was rife in Canada, manifested by a senior bureaucrat's 1945 statement on Jewish immigration that "none is too many" (Abella and Troper 1983; Davies 1992).

After 1948, changed international circumstances, changes in Canadian government and civil service, changed economics, and changed public attitudes all led to support for substantial increases in immigration (Canadian Museum of History n.d.; Bumsted and Bumsted 2016; Richmond 2019; Raska 2020; Fig. 11.14). Canada's involvement in the United Nations and support for the Universal Declaration of Human Rights led to the repeal of the worst anti-immigration laws (although attitudes were slower to change, especially for Black and South Asian immigrants; see Atak et al. 2023). The emphasis on family reunification favoured potential British, French, Italian, Portuguese, and other European immigrants, who were more likely to have family members who had previously immigrated to Canada (Richmond 2019). Refugee aid programmes favoured Jewish Holocaust survivors and eastern European people impacted by conflict (e.g. 37,500 Hungarians after 1956) (Immigration Refugees and Citizenship Canada 2023), with lesser attention to people afflicted in Africa, Asia, and South America (e.g. Raska 2015). In 1961, about 95% of immigrants arrived from European countries or USA.

Canada has the fourth largest Jewish community in the world, exceeded in number by those in Israel, USA, and France (Tulchinsky 2008; Weinfeld et al. 2012). The 2021 Census recorded 335,295 people, predominantly Ashkenazi Jews.

11.6 Immigration After 1962

In 1962, the Federal government removed ethnicity as a criterion for consideration for admission to Canada, with emphasis being placed on education, skills, and potential economic contribution (Canadian Museum of History n.d.; Hawkins 1988; Richmond 2019). Potential immigrants are scored on a points system allocating credit for these attributes, as well as other factors. Although emphasis on education and potential economic contributions could favour applicants from countries over those whose homes lacked opportunities to acquire skills or training, the revisions reduced ethnic bias. Significant discretion remains with officers of Immigration, Refugees, and Citizenship Canada (IRCC), at the points of application, entry, and within Canada.

Although family reunification and sponsorship by Canadian citizens remained, along with refugee admissions, the revised emphasis in combination with adoption of multiculturalism policies led to increased immigrant diversity (Cameron 2004; Kymlicka 2010; Edmonston and Fong 2011; Boyd and Alboim 2012; Boyd 2015). Fluctuations since 1971 reflect economic conditions in Canada, refugee relief (Immigration Refugees and Citizenship Canada 2023), programmes targeted to skills- and occupation-based recruitment, and decisions as to the appropriate number of applicants to admit (Triadafilopoulos 2021). Limits of 200–235,000 admitted applicants through 1991–2015 (with allowance for refugee influxes) were replaced by higher limits in recent years, reaching 440,000 in 2023.

11.6.1 Asylum Claimants

Canada allows requests for asylum upon arrival at a Canadian port of entry, or online. Technically, people are considered as asylum claimants until they are accepted into Canada, at which time they are classified as refugees, although in practice this distinction is commonly overlooked. Between 2000 and 2021, numbers of these documented asylum claimants recorded by IRCC and border security officers (Public Safety Canada) varied between 10,365 (2013) and 64,030 (2019). After 2021 (24,910), a sharp spike saw 91,735 claimants in 2022, rising to 144,035 in 2023 (Fig. 11.15). A substantial part of this spike can be attributed to online applications, allowing review and decision without requiring travel to a Port of Entry. The largest numbers of asylum-seekers came from Mexico, Haiti, Türkiye, Colombia, Iran, and India. Rejection rates (~35% overall) are difficult to calculate, as adjudication of many claims by the Immigration and Refugee Board of Canada (independent from IRCC) can require more than 1 year. The claims adjudicated in 2022 (many dating from previous years) show high rates of acceptance of claimants from Afghanistan, Egypt, El Salvador, India, Iran, Nigeria, and Ukraine, among others (Immigration and Refugee Board of Canada 2023a). Low rates of acceptance categorize applications from Türkiye, Syria, Mexico, Haiti, Democratic Republic of the Congo, Colombia, and Bahamas, among others.

Undocumented (or "illegal") migration, particularly across the USA-Québec border south of Montreal, was a political concern in 2017–2018. Public Safety Canada estimates suggest that approximately 28,000 people were involved, primarily originating from Central America and Mexico (Immigration and Refugee Board of Canada 2023b).

11.6.2 Emigration

Emigration remained at a constant level of 40–45,000 between 1971 and 2016 (Statistics Canada 2021–2024). Most emigrants went to USA, with specific job positions, spousal unification, and general economic opportunity the principal reasons. Throughout this period until today, Canadian politicians and commentators have expressed concerns over "brain drain" of highly educated people to USA, citing a variety of causes including differences in taxation and licensing requirements.

In the period 2022–2023, emigration increased to more than 90,000 per year, maintaining a proportion of ~25% of immigration (440,000 in 2023, compared with 200,000–235,000 between 1991 and 2015). Immigrants from Taiwan, Hong Kong, France, India, and USA are statistically more

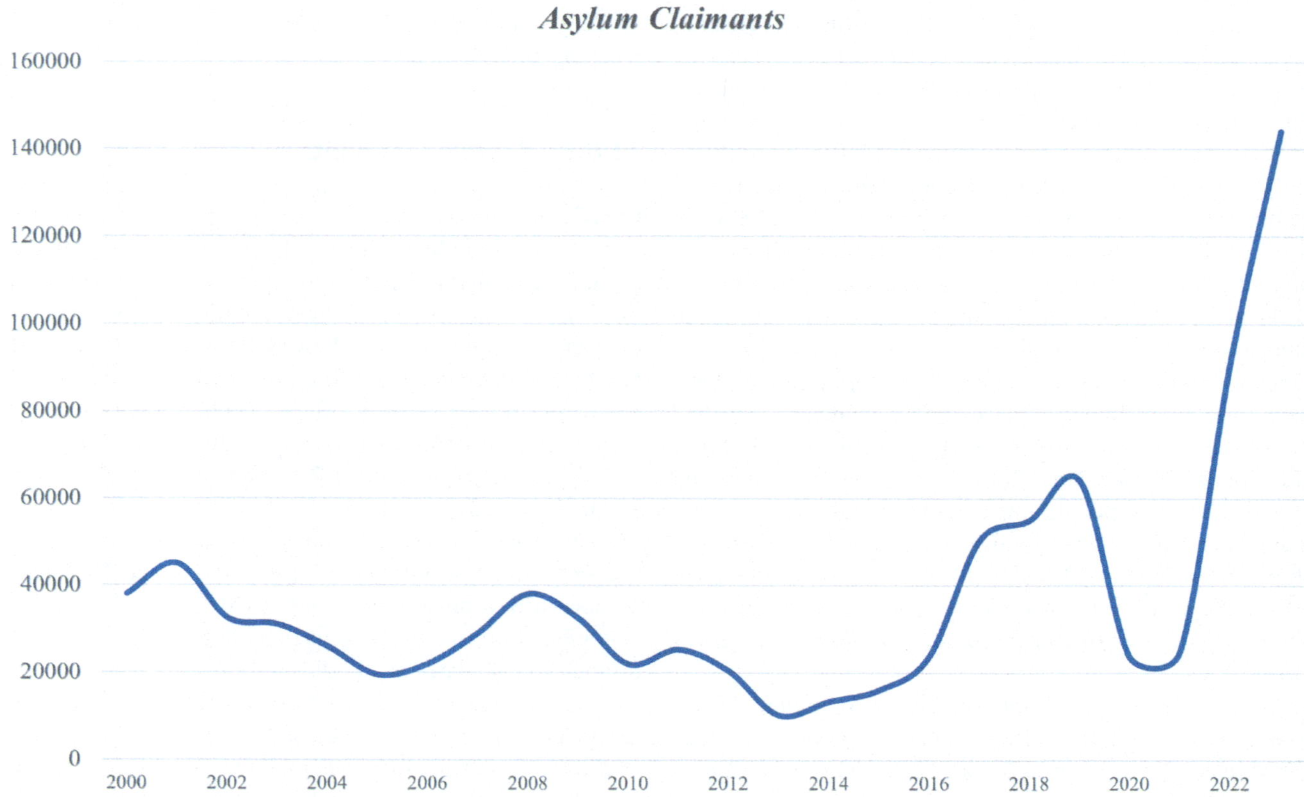

Fig. 11.15 Asylum Claimants by year, 2000–2023. Numbers include those making formal requests for asylum. Data from Public Safety Canada, IRCC, and Immigration and Refugee Board of Canada)

likely to subsequently emigrate. Some evidence suggests that Canada may have been an intermediate destination for people who ultimately wish to immigrate to USA, particularly for those from Taiwan and Hong Kong. The imposition of a visa requirement for visitors from Mexico (removed 2017; under discussion for possible re-imposition in 2024) was intended in part to allay USA government concerns that Canada was being used as an intermediate destination by Latin American migrants. Reverse migration, with people returning to their original country, appears to be more common for those from France, India, and USA.

11.6.3 Immigration Policy

Statistics Canada classifies admitted immigrants as economic (including skills-based and education-based admissions), those involved in business programmes (involving committed capital investment), those sponsored by family members, refugees, and those nominated by provincial and territorial governments (except Québec and Nunavut). The Provincial Nominee Programme is designed to help areas other than urban Alberta and BC, Ontario, and Québec (the main destinations; Fig. 11.16) attract immigrants. Provinces nominate applicants based on their perceived need for specific occupations or skill-sets, time-based commitments to work in specific locations, and on their general desire to attract immigrants based on overall demographic concerns. With otherwise ageing populations, the Atlantic Provinces have been eager to take advantage. Applicants are still assessed by Immigration, Refugees, and Citizenship Canada (IRCC), but nomination by a province virtually guarantees acceptance. Although provincial nominations have influenced immigrant destinations, with more than fourfold increases in percentages in PEI and Saskatchewan between 2006 and 2021, Ontario (44%), Québec (15.3%), BC (14.9%), and Alberta (14.5%) remain the most popular choices (Statistics Canada 2021–2024, Demosim Team 2022a).

Under Canada-Québec agreements (Béchard 2018), Québec assumes responsibility for all applicants. Potential immigrants must be approved by the Ministère de l'immigration, de la Francisation et de l'Inclusion (MIFI). Approved applicants then file a request for admission and a resident visa with IRCC, but can only be rejected on the bases of severe health issues or a criminal record. Québec admits immigrants for family reunification and as refugees, but all economic immigrants now require knowledge of French for admission (Termote 2023).

Economic immigrants currently represent about 60% of those admitted in the past 10 years (Figs. 11.17 and 11.18).

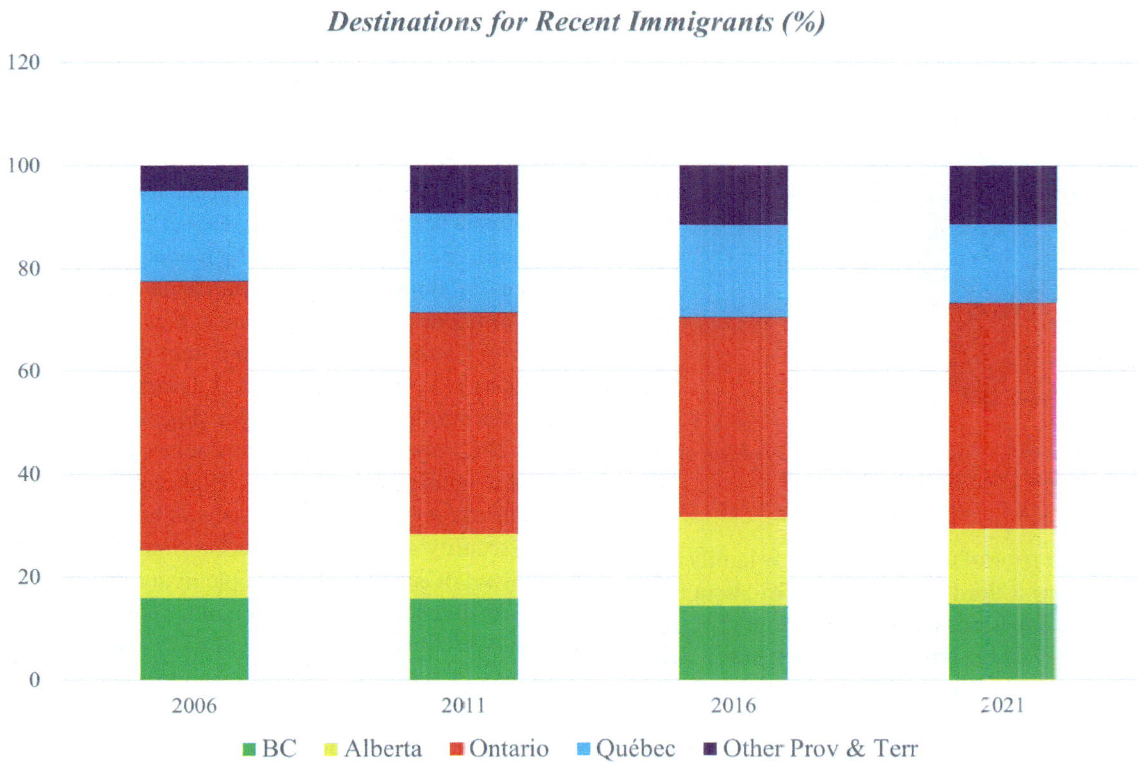

Fig. 11.16 Initial destinations for recent immigrants by Province, 2006–2021. Increasing percentages of those heading to other provinces and territories indicates the effect of Provincial Nominee programmes (Statistics Canada data)

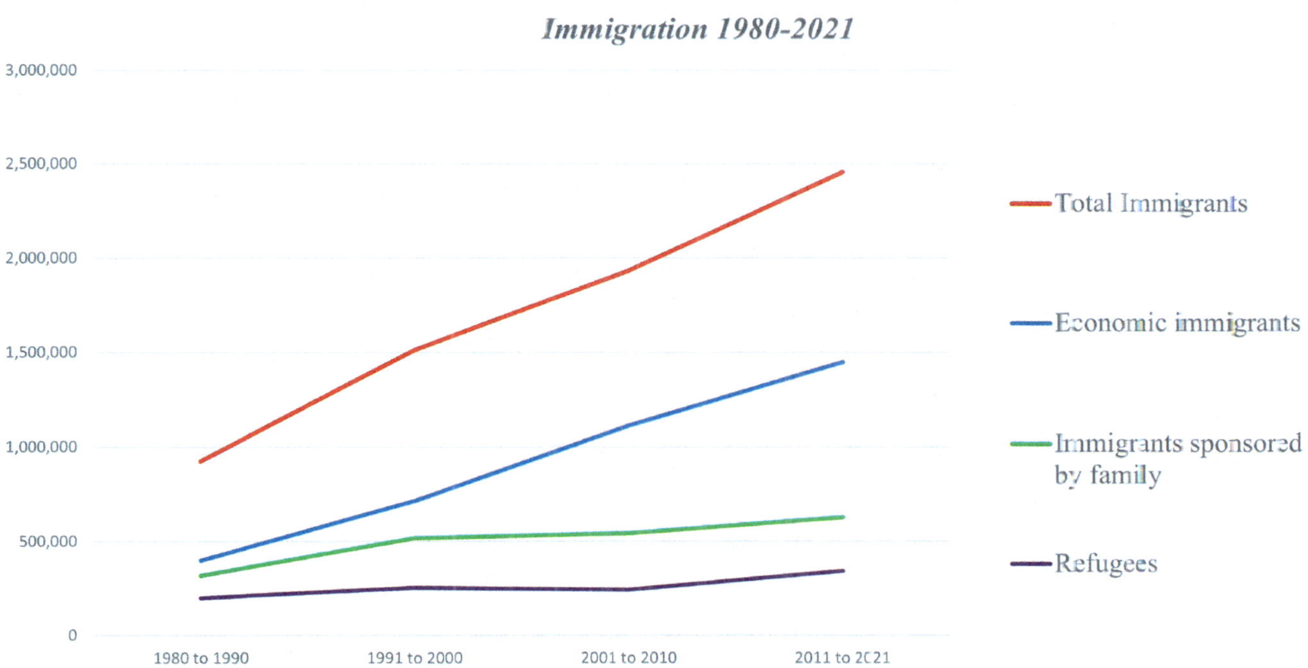

Fig. 11.17 Immigration by classification of immigrant, 1980–2021. Totals are for 10-year periods as indicated. Note the growth of both numbers and relative percentages of economic immigrants (Statistics Canada data)

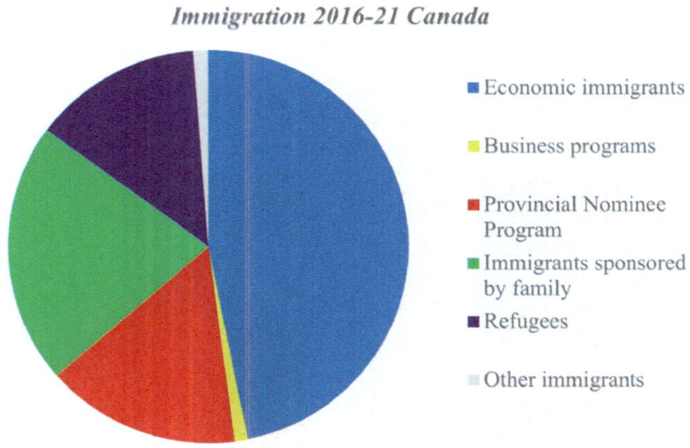

Fig. 11.18 Immigration to Canada 2016–2021, based on programme. See text for discussion (Statistics Canada data)

The relative proportion of those sponsored by family members has dropped from a third in 1980–1990 to about a quarter today. Refugee admissions have declined from 22% in 1980–1990 to about 14% of the total today, although some refugees are admitted as provincially sponsored nominees (Fig. 11.19).

People admitted as landed immigrants are eligible to apply for citizenship after 5 years residence, including filing tax returns. Those between ages 18 and 54 must also demonstrate adequate knowledge of either English or French, and normally must pass a citizenship test (Immigration Refugees and Citizenship Canada 2021; also see Joshee and Derwing 2005). Less than half of immigrant permanent residents became citizens within 10 years of arrival, and one-sixth left the country, usually after residence of less than seven years (Hou and Picot 2019).

Typically, new arrivals are not evenly distributed across all economic sectors (Demosim Team 2022a; Aziz et al. 2023). It is informative to look at which occupations do draw high shares of immigrants in employment sectors (Fig. 11.20). Government-funded health care is regarded as a key component of Canadian society. More than one in four healthcare workers is a recent immigrant, is a migrant, providing health care for residents of the country (Fig. 11.21). Between 2015 and 2022, about a thousand physicians arrived in Canada as landed immigrants each year. Higher percentages of immigrants are involved in fields such as hospitality, retail trade, and security and cleaning. Recent immigrants are providing a key set of services to all people living in Canada.

The diversity of mother tongues spoken by recent immigrants is illustrated in Fig. 11.22 and Table 11.3 (see Demosim Team 2017, 2022a). English, the largest category and including people from many countries, represents about a quarter of the total. Mandarin, Tagalog, Arabic, Punjabi, Yue, French, and Spanish each account for 5–6% of the population, with each including more than 400,000 people. In total, 43 distinct languages were mother tongues of a minimum of 20,000 immigrants.

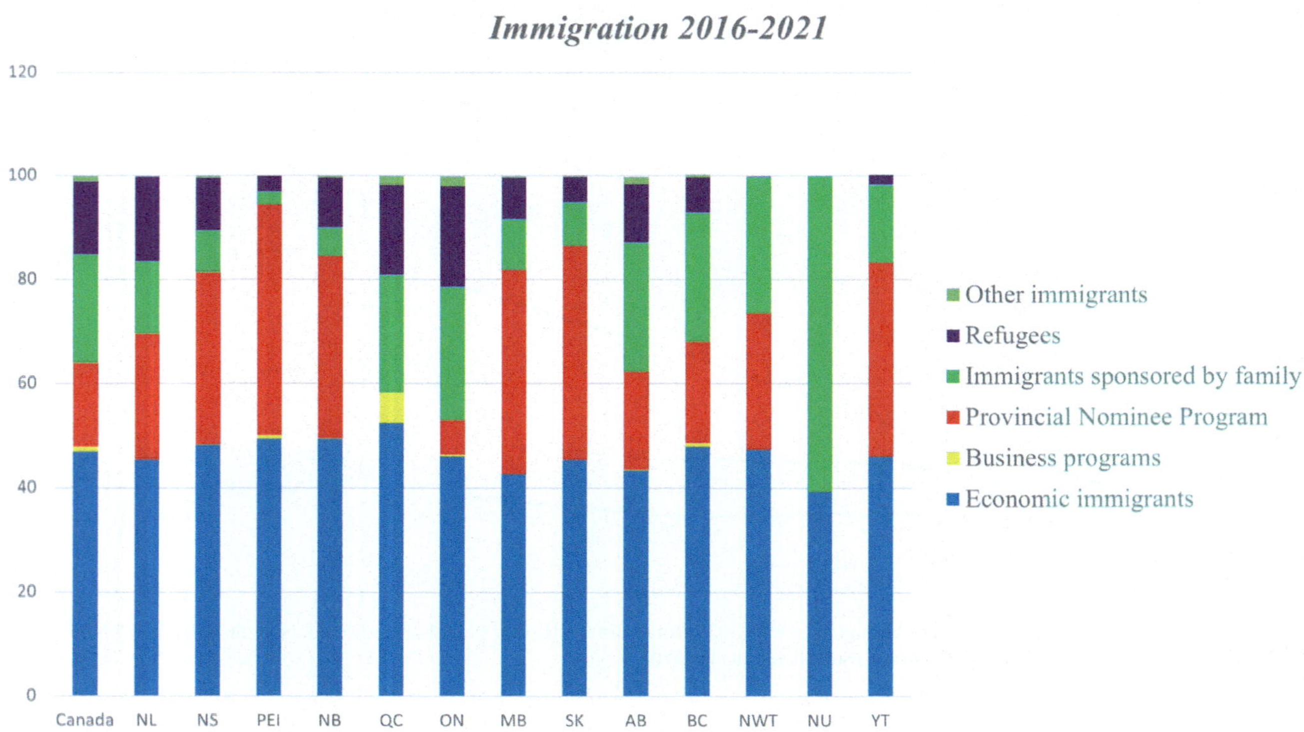

Fig. 11.19 Immigration to Canada 2016–2021, divided by province of initial destination and programme. Québec and Nunavut do not participate in Nominee Programmes. Business programmes available vary by province (Statistics Canada data)

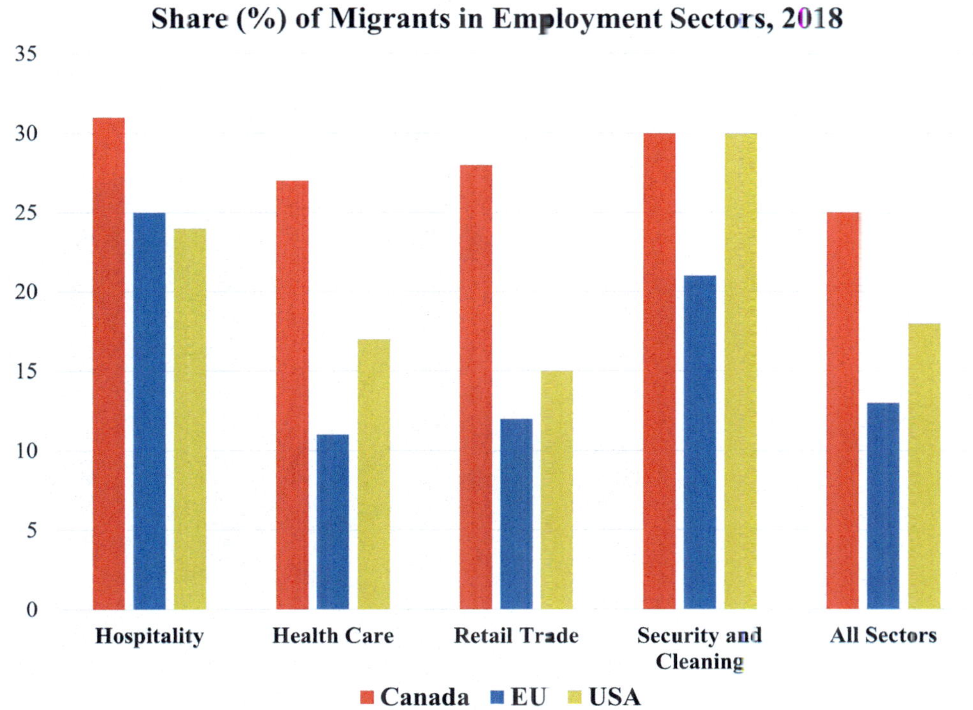

Fig. 11.20 Percentage shares of migrants in selected employment sectors, 2018 (Data: government agencies responsible for immigration and statistical compilation in Canada, EU, and USA)

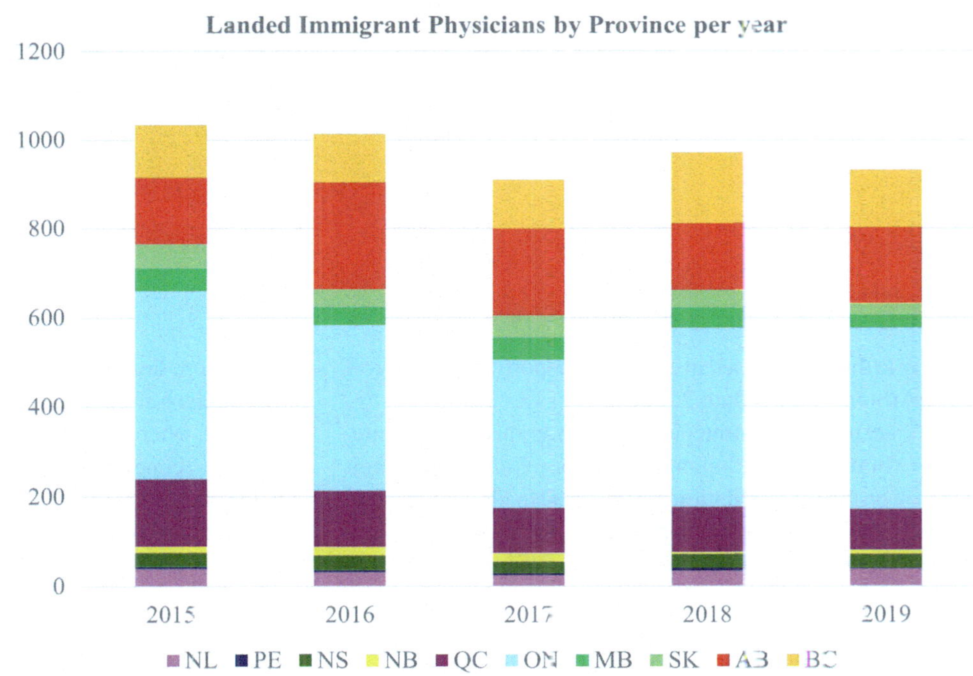

Fig. 11.21 Landed immigrant physicians by province, 2015–2019 (Data: Immigration, Refugees, and Citizenship Canada)

11.6.4 Destinations and Acculturation

The primary factors influencing the destination for most immigrants are economic, and social factors including safety, schooling for children, absence of discrimination, and quality of life. Language training and local community services were generally less important (Akbari 2020).

The former emphasis on family connections as a driver for immigration (Boyd and Alboim 2012; Boyd 2015) has been replaced with economic concerns, leading to shifts in who applies to immigrate. Family connections are secondary for the majority of immigrants and can prove insufficient to retain new arrivals. From regions such as Latin America, immigration has shifted from refugees seeking political asylum and

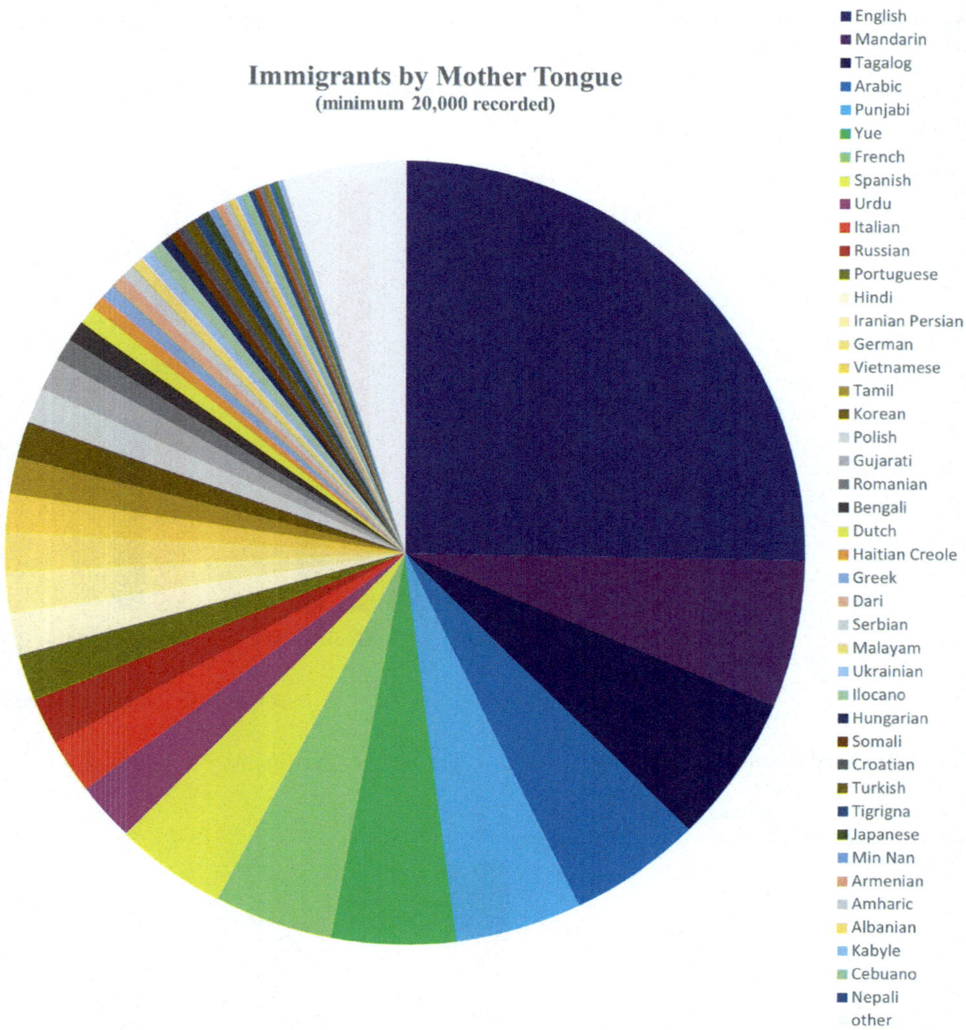

Fig. 11.22 Immigrants by mother tongue, with minimum of 20,000 respondents, based on self-identification (Census of Canada 2021)

stability to those interested in economic opportunity (Vengroff 2013; Armony 2015; Mata 2021).

Four acculturation strategies for immigrants include integration (high sense of belonging to both Canada and their place of origin), assimilation (high for Canada and low for place of origin), separation (low for Canada and high for place of origin), and marginalization (low for both). All immigrant groups, historical and current, show evidence of individuals' adoption of each strategy (e.g. Shuva 2021), but most show tendencies towards integration and assimilation over time.

As multiculturalism policy developed in the 1960s–1970s, the concept of Canada as a cultural mosaic took hold (Kelley and Trebilcock 1998; Li 2000; Kymlicka 2010; Boyd and Alboim 2012; Boyd 2015). This postulate stressed integration, with the idea that self-identification as hyphenated (e.g. Belgian-Canadian) did not imply lower commitment to Canadian residence or society. Frequently, the mosaic concept was compared favourably with "melting pot" assimilation models, in which immigrants are gradually subsumed into an overall, more homogenous society. However, some residents and politicians argued that hyphenated status was indicative of separation, whereby immigrants enjoyed (or exploited) Canadian socio-economic conditions, without developing a sense of belonging.

Aside from those exhibiting racism or xenophobia (see Henry et al. 1998; Breton 2005; Baker et al. 2015), or those uncomfortable with societal change, concerns about separation may reflect misunderstandings of the changed nature of immigration. From the nineteenth century to the 1980s, immigrants to urban areas frequently initially settled in ethnically distinct communities (e.g. Zucchi 2007; Sanzone 2015). Many came with limited economic resources and limited formal education. Over time, they got their bearings, they started to advance economically, socially, and educationally speaking, and then they tended to move from the initial ethnic community to another area of the city. In some neighbourhoods, that could create flux and a high poverty rate, because the people who are coming in have relatively few resources

Table 11.3 Mother tongue for largest immigrant groups by number, Census of Canada 2021. Note the increase in Arabic, Hindi, Gujarati, and Korean speakers in 2016–21, in comparison to the total number of immigrants recorded in the 2021 Census

Mother tongue	Immigrants 2016–2021[a] Number (ranking)	Total Immigrants recorded in 2021 Census of Canada Number (ranking)
English	324,555 (1)	2,244,280 (1)
Arabic	136,580 (2)	479,040 (4)
Tagalog	112,160 (3)	541,000 (3)
Mandarin	104,795 (4)	541,540 (2)
French	86,960 (5)	426,010 (7)
Punjabi	85,835 (6)	451,650 (5)
Spanish	56,540 (7)	402,010 (8)
Hindi	52,320 (8)	155,420 (13)
Urdu	35,305 (9)	216,930 (9)
Gujarati	27,790 (10)	118,370 (20)
Russian	25,440 (11)	175,100 (11)
Portuguese	23,145 (12)	170,640 (12)
Iranian Persian	22,220 (13)	150,500 (14)
Korean	20,590 (14)	133,625 (18)

[a] minimum 20,000 people recorded in 2021 Census of Canada

to begin with, increasing the overall total in terms of people who are in economic need.

That model, which sees the neighbourhood as a gateway or arrival site, with successive occupation of immigrant groups, did apply to some communities at that time (Zucchi 2007). Longer-term residents may base their perceptions of immigration on personal family history, which recorded such transitions over time. As an example, the Spadina-Kensington Market area of Toronto has seen successive periods where Irish, German, Jewish, Italian, Portuguese, Chinese, Vietnamese, and Chinese immigrants have formed neighbourhoods (e,g. Sanzone 2015). Family and community histories may influence perception of current immigration and immigrants (e.g. Paquet and Lawlor 2022). However, the model has significant inaccuracies, both in the assumption that arrivals came in discrete waves and that all immigrants were financially impoverished upon arrival. Neighbourhood differentiation occurred on economic grounds, as well as ethnicity.

The shift in dominant motives for immigration from family migration and fleeing from oppression ("push factors") to economic and educational opportunity ("pull factors") has produced a change in immigration patterns (e.g. Rice and Quan 2023). The older model of sequential ethnic neighbourhoods is increasingly not applicable, although examples still exist, particularly for groups who are fleeing from oppression or poverty today.

Berry and Hou (2016, 2017) found that integration and assimilation produced the highest scores for life satisfaction, while separation and marginalization strategies produced significantly lower scores. For mental health, integration and separation had higher scores than assimilation and marginalization. In life satisfaction and mental health, immigrants had significantly higher scores than non-immigrants. A majority of immigrants display integration and assimilation strategies (Hou et al. 2016), with conditions in the source country as a critical influence. In a study of Sri Lankan Tamil Canadians, Amarasingham et al. (2016) described Canadian multiculturalism as the product of daily practices of tacit nation-building.

Remittances, money that migrants send home, are a major factor when people first arrive in Canada and sometimes considerably thereafter (Elcioglu and Shams 2023; MacIsaac 2023). In many developing countries, remittances are a vital part of the economies. Canadian residents tend to be generous in remittances (e.g. Syed 2021). Although COVID-19 made a major impact in many areas related to immigration (Hamilton et al. 2022), remittances persisted at very high levels throughout the pandemic.

11.6.5 Perceptions

Canadians in general have a fairly accurate perception about the total number of immigrants, but generally overestimate the number of refugees and asylum-seekers (Paquet and Lawlor 2022; also see Kocher 2023). Many underestimate the education, qualifications, and financial resources of immigrants as a group, and overestimate the importance of family connections in decisions by both immigrants and IRCC officials. Mental images are stronger influences than are immigration statistics.

These views represent a mixture of perceived memories from earlier phases of immigration to Canada, coupled with images of border crossings elsewhere. Media, particularly social media, play a large role in shaping perceptions, with perceptions of threat and border insecurity influenced by coverage of migrants along the Mexico-USA border.

11.6.6 Retention Issues

Governmental perspectives have shifted from recruitment to retention, especially in communities outside the major CMA of southern Ontario and Vancouver (Fang et al. 2018; also see Paquet 2019). Commonly cited reasons for leaving the initial province, or Canada in general, were lack of employment opportunities, underemployment, or low salaries; unsatisfactory health care; high taxation; and incidents of discrimination (Akbari 2020; Lu and Hou 2020; Aziz et al. 2023).

Recruitment efforts, including provincial nomination programmes and university competition for international students, are focused on people with transferrable skills and potential economic contributions. Restrictions on professional and skilled trades employment, or underemployment which does not give the immigrant the opportunity to employ their education or training, are significant obstacles (e.g. Lo et al. 2017). Workplace discrimination and non-recognition or devaluation of foreign education and experience are more important than ethnic differences in influencing career satisfaction (Lu and Hou 2020; Nnamdi Okafor and Kalu 2023; see Rice and Quan 2023).

Regulated occupations require a license or certificate, usually issued by provincial governments or regulatory authorities. Approximately, 20% of occupations in Canada are regulated, including certified general accountant, engineer, lawyer, nurse, physician, and veterinarian. Skilled trades (carpenter, electrician, plumber, welder) are also regulated in most provinces. For example, Manitoba lists 31 regulated professions, including all those related to health (e.g. physicians, dentists, nurses, midwives, naturopaths, opticians, pharmacists), education (teachers, child-care workers), engineers, social workers, and veterinarians. For most of the more than 55 designated skilled trades, prospective workers can undertake an apprenticeship or can write the trade's qualification examination. For nine trades, including electrician, crane operator, pipefitter, esthetician, and hairstylist, compulsory certification is required. Some other provinces require long-haul truckers to pass certification.

As professional organizations (e.g. Colleges of Physicians and Surgeons) regulate professional certification, they are largely insulated from political pressure, and judge applicants based on the similarity of their credentials to those of the provincial organization. Skilled trades are regulated by provincial organizations, often involving the input of colleges and unions. Applicants for trades are frequently required to complete either apprenticeships under provincial supervision, or demonstrate that they have accumulated sufficient supervised work experience. For example, a prospective plumber in New Brunswick requires 7200 h of apprenticeship or 10,800 h of work experience; for a cook, the hours required are 5400 and 8100, respectively. As a further complication, some trade qualifications from one province may not be recognized in another, even for Canadian citizens.

These complications, a facet of the division of responsibility between federal government and provinces, result in delay and frustration for professional and skilled trades immigrants. Many are underemployed in the interim; some eventually give up the effort or opt to go elsewhere. The problem is most acute for professional applicants, who may be asked to repeat all their schooling at a Canadian institution.

Social conditions, quality of life concerns, and discrimination are also important in immigrants' decisions to stay or depart. Differences exist by community. Regional differences are also evident, with immigrants to Ontario and the Atlantic Provinces more likely to have a strong sense of belonging to Canada than for immigrants in British Columbia (Stick et al. 2023). Relative affordability, quality of life, and access to nature are involved in immigrants' decisions to arrive or stay in smaller centres, with social attachments of lesser importance (Kelly and Nguyen 2023). Typically in a smaller urban centre, relatively few organizations dominate immigrant employment and reception (Crea-Arsenio et al. 2023). Immigrant entrepreneurs must overcome obstacles of recognition and acceptance, in addition to the challenges present in business (Graham and Pottie-Sherman 2020). Because federal policies dominate, even the largest urban centres have very limited responsibilities regarding settlement, integration, and language education (Ellyson et al. 2016). Although participation in winter sports appears to have beneficial efforts on economic integration for some immigrants, these were not evident for visible minority immigrants (Japardze and Kaplan 2023).

Unsatisfactory health care is a factor in immigrant decisions to leave the original point of arrival for another province or country. Provincial public health programmes and available services are often budgetarily limited and understaffed. Sociocultural differences may hinder effective provision of healthcare services, especially those involving mental health (Emerson et al. 2021; Gyan et al. 2023). Earlier immigrants tend to be satisfied economically, but health concerns lower their satisfaction with quality of life in general as they age (Nangia and Gingrich 2020).

Federal policy currently encourages Francophone immigration to combat demographic changes in francophone communities outside Québec (Houle et al. 2014). In provinces outside Québec, communities with Francophone residents were more likely to successfully retain Francophone immigrants (Haan et al. 2017).

Approximately, 60% of recent immigrants have become citizens. The overall citizenship rate has declined since 2006, notably for immigrants with low family income, poor official language skills, and lower levels of education, with the decline largest among Chinese immigrants (Hou and Picot 2019).

11.7 Ethnicity

In contrast to modern immigration statistics, recognition of ethnic origin is based on self-identification on the census. Changes in category definitions over time and in the options available to respondents can make direct comparisons difficult. Prior to 1981, "ethnic origin" referred to the respondent's father's ethnicity exclusively. Self-identification may not correspond with genetic or genealogical data. In a study of Gaspésie residents, Vézina et al. (2014) noted that although 9 of 10 respondents indicated a single ethnic origin (commonly based on their fathers'), almost all had multiple ethnicities in their genealogy and genetics.

Prior to 1971, respondents could not indicate "Canadian" as an ethnicity. That change has resulted in more people considering themselves as Canadians, rather than identifying themselves on the basis of where their ancestors may have come from. By 2006, more respondents in Québec, Atlantic Canada, and eastern and northern Ontario identified as Canadian than as any other ethnicity, with Indigenous ethnicities most commonly identified throughout the Arctic and Taiga. "English" (as distinct from Irish, Scots, or Welsh) was the most commonly identified ethnicity in Alberta, southern BC, and southwestern Manitoba, with German and Ukrainian categorizing some census divisions in Manitoba, Saskatchewan, and Alberta. All of these identifications indicate pluralities, not majorities. Many respondents cited multiple ethnic or cultural origins.

By 2021, however, the picture had changed: Canadian remained the most commonly cited ethnicity in Québec, but English led in Newfoundland, southern and northwestern Ontario, southern Manitoba and Alberta, and throughout most of British Columbia (Statistics Canada 2021–2024). German was the most commonly cited option in southern Saskatchewan, with Ukrainian forming a belt along the Midgrass Prairie route of the former Canadian Northern Railway (constructed 1903–1910) in Manitoba and eastern Saskatchewan. In northeastern Ontario, French was the most commonly cited option. In the Maritimes, various census divisions reported Acadian, Irish, and Scottish as the plurality choice. In all regions, however, the 2021 responses indicate an increasing variety of self-identifications, with more than 450 ethnic or cultural origins cited. In total, "Canadian" (5.7 million, compared to 11.2 million in 2016), "English" (5.3; 6.2), "Irish" (4.4; 4.7), "Scottish" (4.4; 4.8), and "French" (4.1; 4.7) were the most commonly cited (including those citing multiple origins), in a population of 38 million.

11.7.1 Bilingual Status

In 2021, about 13% of Canadians were bilingual in English and French (Fig. 11.23). Over the past 60 years, percentages have increased in Québec (45%) and New Brunswick, the

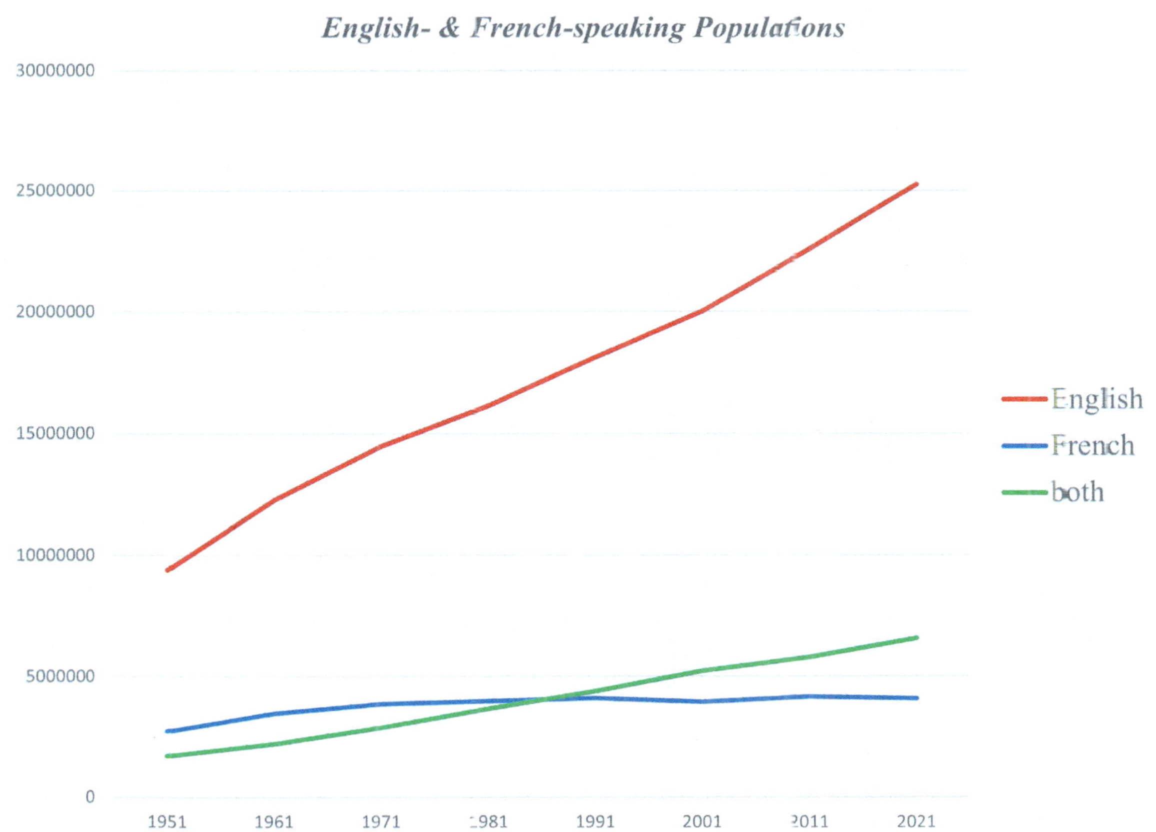

Fig. 11.23 English and French-speaking populations in Canada, 1951–2021. Languages prior to 1971 are based on identification by census takers; from 1971, based on self-identification by respondents (Censuses of Canada)

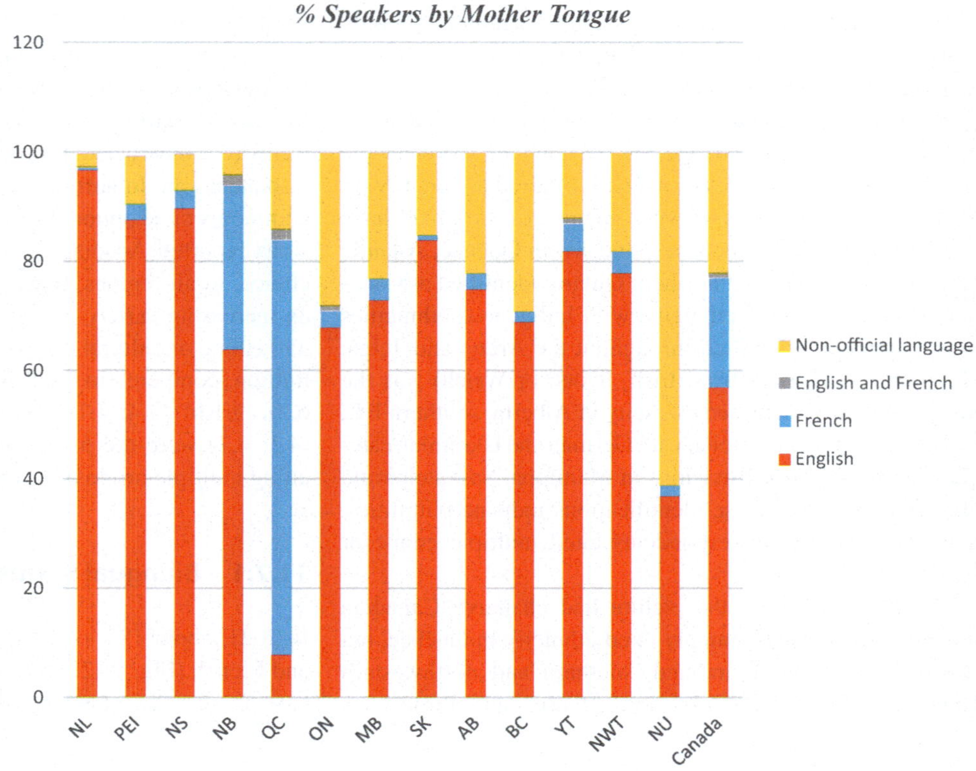

Fig. 11.24 Speakers by Mother Tongue for Provinces and Territories, 2021. "Non-official" languages include all except English and French. Census of Canada, based on self-identification)

only officially bilingual province (34%). In the remainder of Canada, less than 10% of Canadians are functionally bilingual. The number of people speaking both languages exceeds the French-speaking population (Houle and Corbeil 2017). The shift reflects increased immigration from countries where French is not commonly spoken, in addition to the perceived advantages in learning English for employment in the private sector, post-secondary education, and the influence of globalization and social media.

Emphasis on English and French, while valid in terms of the historical ethnic mixture and the role of Québec, does not account for the increasing number of Canadians who speak more than one language other than French or English (Houle and Corbeil 2017; Fig. 11.24). The Official Languages Act (1969) declared French and English as the two official languages of Canada, giving Canadians the right to service in either from federal institutions and courts. All others, including Indigenous languages, are classified as "non-official". The distinction is useful in urban locations such as Toronto and Edmonton (with more than 100 languages spoken by students in public schools), Vancouver (more than 170 spoken in homes), and Montreal (where more than 20% can speak at least three languages). It is less useful for Nunavut (where the population is dominantly Indigenous) or Newfoundland and Labrador (where virtually all of the non-official language speakers are Indigenous people living in Labrador).

11.7.2 Religions

A slight majority (53.3%) identified as Christian in 2021, compared with 77.1% in 2001.

The other religions most commonly identified were Islam (4.9%), Hindu (2.3%), Sikh (2.1%), Buddhist (1.0%), and Jewish (0.9%). Numerous other religions were reported as well. Immigration accounts for much of the increase for non-Christian religions, as immigrants represent the majority of Buddhists, Hindus, Muslims, and Sikhs.

The second-largest category was 12.6 million people who reported no religious affiliation, 34.6%, representing a substantial increase from 2011 (23.9%). This change mirrors other surveys suggesting that about half of the population do not consider religious or spiritual beliefs to be important in their lives, including about a fifth of those reporting a religious affiliation. Admitting a lack of religious affiliation has become much more socially acceptable in the past decade. Of the immigrants admitted from 2011 to 2021, 21.5% reported no religious affiliation.

11.7.3 Indigenous Peoples

Indigenous people represented 5% of the population (1.8 million) in 2021, including 70,545 Inuk and 624,250 Métis (Statistics Canada 2021–2024; Demosim Team 2022b).

Self-identified groups with the largest populations were Néhinaw (Cree), Anishinaabe (including Algonquin and Ojibwe), Haudenosaunee, Mi'kmaq, Dene, Salish, and Siksika ("Blackfoot"). Slightly more than 20% have knowledge of an Indigenous language, and about 15% speak an Indigenous language in their home, including more than half of the Inuk population.

Less than 30% of First Nations people live on a reserve. Proportions among provinces vary from slightly more than half (MB, SK) to about 95% (NL). More than half the off-reserve population lives in urban areas, with more than 110,000 in Winnipeg CMA, 100,000 in Montreal, 95,000 in Edmonton, 70,000 in Vancouver, and 60,000 in Toronto. Indigenous people represent more than 10% of the CMA populations of Edmonton, Saskatoon, Sudbury, Thunder Bay, and Winnipeg.

Approximately, 30% hold a post-secondary qualification, roughly half of the percentage for all Canadians. The unemployment rate for those voluntarily seeking employment is about 15%, compared to ~11% nationally. The median total income from all sources (age 15 and older) was $34,800, approximately 15% below the national figure. Almost 20% of Indigenous people are considered to have low income, compared to 11% of all Canadians.

Lifestyle statistics tell a similarly repetitive and depressing story. Among many examples, although some improvement has occurred, a significant life expectancy gap remains. Overall life expectancy is estimated as 9 years less than that of the general population, and 11 years for Inuk. Between 55 and 60% of the Inuk population are without a regular healthcare provider, compared with 20% of First Nations people living off-reserve and about 15% of the general population. The suicide rate among First Nations people is three times higher than the rate among non-Indigenous people (2016 data), with higher rates for people living on reserve in general. For Inuk, the suicide rate is 9 times that of the general population. More than one quarter of homicide victims in 2022 was Indigenous. About 25% of those incarcerated in federal and provincial custody are Indigenous.

Recognition of the issues and the need for Truth and Reconciliation is generally high among the non-Indigenous population (Crown-Indigenous Relations and Northern Affairs Canada 2017), although action could be swifter. The increasing visibility of Indigenous people and cultures, Indigenous entrepreneurship and management, and presence in academic institutions are all positive signs. Much remains to be accomplished.

11.7.4 Visible Minorities

Statistics Canada (2021–2024) uses respondents' self-reported data to identify visible minorities (Fig. 11.25).

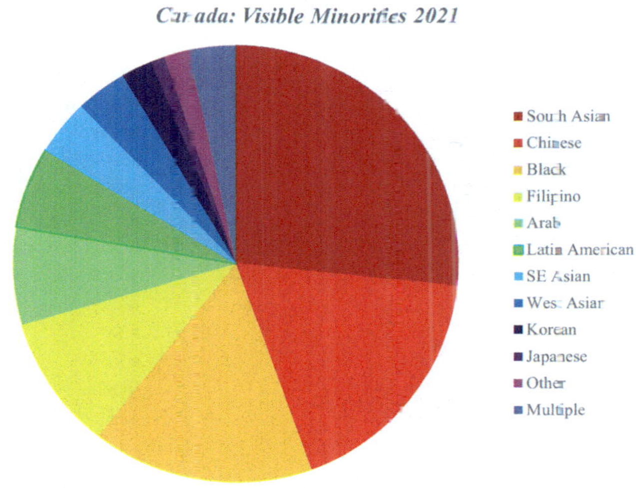

Fig. 11.25 Visible minorities in Canada, 2021. Ethnicity initially based on self-identification by respondents, grouped into categories by Statistics Canada. Indigenous people are explicitly not considered to be a visible minority (Census of Canada 2021)

Visible minorities are classified by ethnicity, rather than national origin. Ten larger population groups are distinguished (Figure). Visible minority populations with smaller numbers are grouped as "not included elsewhere" (other), as well as those reporting multiple visible minority origins. By definition, Indigenous people are not included as a visible minority. In total, about 23% of the population reported data indicating a visible minority. Although this percentage is similar to that for recent immigrants, these population groups are not directly related. Many members of visible minorities have long been resident in Canada, or are of second, third, or higher generations. "South Asians" (including people of Bengali, Indian, Pakistani, and Sri Lankan ethnicities, among others) represent the largest visible minority group in Canada. "West Asians" include people of Afghani, Armenian, Iranians, Syrian, Turkish, Uzbeki, and other ethnic origins from Asia west of Pakistan and south of Russia, including the Caucasus, but not including those who self-identify as Arab.

Differences among cities reflect both recent immigration and long-standing visible minority communities. In Toronto CMA, visible minorities represented more than 55% of the population in 2021 (Statistics Canada 2021–2024; Fig. 11.26). The Chinese population of Toronto grew largely after 1975, fuelled by emigration from Hong Kong. Substantial Vietnamese emigration occurred between 1972 and 1980. Most Filipinos arrived after 2000, at the same time as most South Asians (Tolentino 2023). Over half of the emigrants from India are destined for Toronto.

Black people are predominantly of Caribbean origin and/or Nova Scotian origin, including many third-generation Canadians whose ancestors settled in Toronto in the late 1800s,

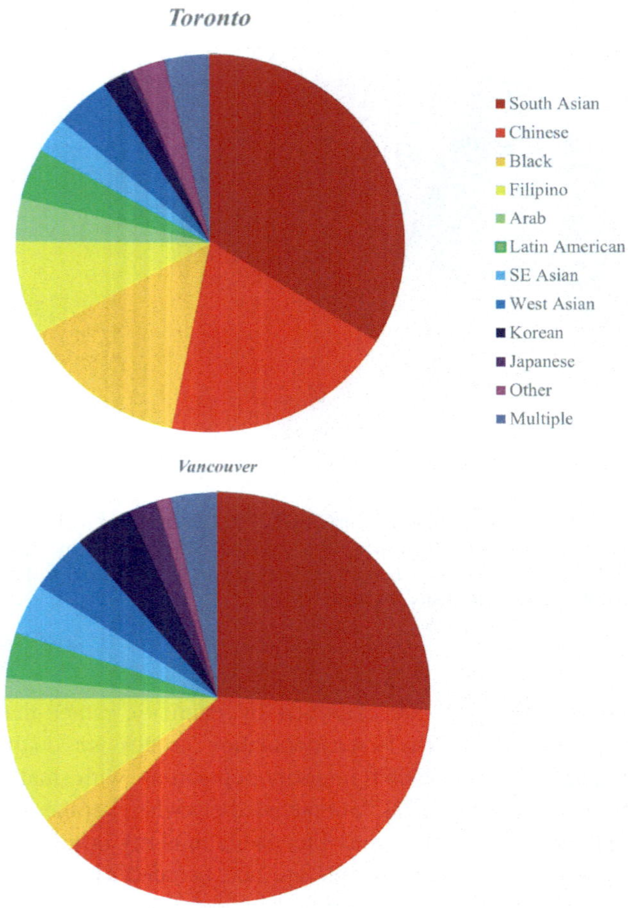

Fig. 11.26 Visible minorities in Toronto (upper) and Vancouver (lower), 2021. Ethnicity initially based on self-identification by respondents, grouped into categories by Statistics Canada. Indigenous people are explicitly not considered to be a visible minority (Census of Canada 2021)

although recent immigration from eastern Africa and Nigeria have become significant (Milan and Tran 2004). Substantial differences exist in socio-economic conditions: Barbadians have the highest median income among all Black Canadian groups, higher than for Toronto CMA as a whole (Ornstein 2006), whereas more than half of Eritrean and Somali residents are considered low income.

Vancouver's Chinese population grew rapidly after 1882, resulting in a series of repressive measures to limit immigration, culminating in almost total prohibition between 1923 and 1947. However, natural increase and movement from rural BC and elsewhere in Western Canada resulted in continual population growth (Fig. 11.26). Immigration from Hong Kong (1990–1997) and Taiwan was eclipsed by that from China by 2000, although ex-Hongkongers tended to have higher professional qualifications and greater incomes. Mandarin speakers now outnumber Cantonese speakers, indicative of the shift from initial immigration through Guangzhou in the nineteenth century to Shanghai and Beijing today. Mandarin, Punjabi, and Cantonese are the second-, third-, and fourth-most commonly spoken languages in Vancouver CMA (Statistics Canada 2021–2024).

Immigration from South Asia, initially dominated by Sikhs, was actively discouraged until the 1950s (Barrier and Dusenberry 1989; Johnston 2005; Pottie-Sherman and Wilkes 2019). Subsequently, Punjabi immigration dominated until the late 1990s, along with Bangladesh, Nepal, Pakistan, Sri Lanka, and people of south Asian ancestry from Fiji. Vietnamese and Filipino communities grew after 1975 and 2000, respectively. The Black community in Vancouver CMA is relatively small, but growing from recent immigration, notably from francophone countries in western and central Africa.

Other CMA show distinctive visible minority distributions (Fig. 11.27). The relatively smaller percentages of South Asian, Chinese, and Filipino people in Halifax, Montreal, and Ottawa-Gatineau partially reflect immigrants' preferences for Toronto, Vancouver, and Calgary (see Coloma et al. 2012; Tolentino 2023). Calgary's populations also reflect eastward migration from Vancouver over time. Historical patterns of Black immigration from USA (Halifax; Saney 1998; Clairmont and Magill 1999) and the Caribbean and western Africa (Montreal and Ottawa-Gatineau) have been supplemented by recent immigration, with more than 100,000 of Haitian descent living in Montreal (Jadotte 1977). Lebanese and Syrian immigrants and refugees, many francophone, form substantial Arab populations in Montreal and Ottawa-Gatineau (Houle et al. 2014). Latin American residents come from 21 different countries, with Mexico, Colombia, and El Salvador the largest sources of recent immigration (Vengroff 2013; Mata 2021).

St. John's visible minority populations reflect recent "Gateway City" government efforts at recruitment in the face of an ageing domestic population (see Graham and Pottie-Sherman 2020; Pottie-Sherman and Graham 2020). High proportions of all visible minority groups are recruited professionals (physicians, engineers, IT and other technical specialists), professors at Memorial University, or their family members. Consequently, the demographics of these populations differ from those in the larger CMA across Canada.

Although difference in opportunities, racial profiling, and well-documented discrimination undoubtedly affects visible minorities, the varieties of experiences both within and among the groups make quantification difficult and overall conclusions uncertain. Despite Canada's reputation as generally welcoming and overall support for immigration, diversity, and inclusion (Immigration Refugees and Citizenship Canada 2023), it is clear that people of visible minorities continue to face socio-economic barriers and challenges (O'Neill et al. 2019; Crossman et al. 2021).

Visible minority groups generally showed higher poverty in the first (recent immigrants) and second (children of immi-

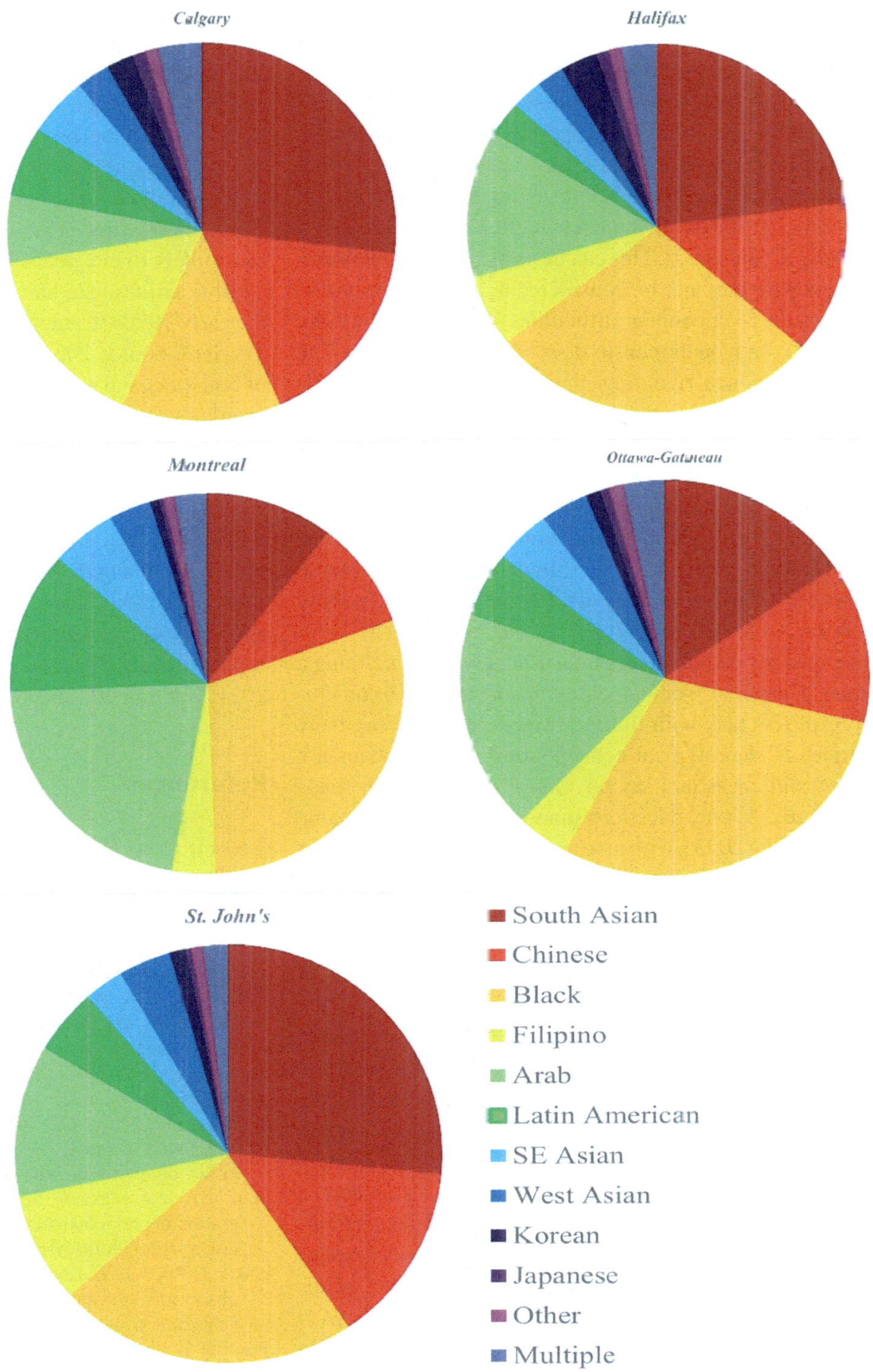

Fig. 11.27 Visible minorities in selected CMA, 2021. Ethnicity initially based on self-identification by respondents, grouped into categories by Statistics Canada. Indigenous people are explicitly not considered to be a visible minority (Census of Canada 2021)

grants) generation than the general population, with the exception of Filipinos (Schimmele et al. 2023). While differences disappeared in the third generation for South Asian, Chinese, Japanese, and Koreans, they remained for the Arab, Black, Latin American, and West Asian groups. As another example, although the percentage of Black Canadians with a Bachelor's degree or higher was comparable to the general population, the rate for African-origin Blacks was three times that of Canadian-origin Blacks, a distinction particularly marked by gender (Wall and Wood 2023). While Black women are more likely to have post-secondary qualifications than are Canadian women in general, the reverse is true for Black men.

11.7.5 Social Identity?

Differences in opportunities and discrimination can also be faced by those who could be considered as "invisible minorities". Persons with similar facial features often differ substantially in their other attributes. Classification of people into broad groups such as Black, South Asian, or White obscures or conceals significant differences in ethnicity, religion, and culture (e.g. Cénat 2022). This can be particularly problematic for groups who may not be considered as a visible minority, but nevertheless encounter difficulties (e.g. Se'ver 1999). Self-identification on the census does not necessarily translate into lived experience. Equating ethnicity with a particular religion, or the reverse, can also be problematic for people who self-identify as an invisible minority. On the 2021 Census, the number of people self-identifying as Jewish by religion was more than 50,000 greater than those identifying as Jewish by ethnicity. Although invisible minorities are difficult or impossible to quantify, personal attitudes and experiences do contribute to the overall picture of ethnic and cultural diversity (Kymlicka 2010; Demosim Team 2022a).

Given the myriad of factors indicating diversity, defining a common "Canadian" social identity would seem to be an ultimately fruitless task with limited value. Nevertheless, many have tried. A stereotypical concept suggests Canadians are noted for politeness and saying "sorry" frequently, perhaps too frequently. Survey data does support this concept to some extent (McHugh 2023), with nine apologies per day per person surveyed. This survey also noted that over four in five respondents agreed that Canadians are polite. The most polite city was Markham (ON), which also had the greatest ethnic diversity in Canada in 2021, including an immigrant population of 45%.

A more rigourous assessment is the General Social Survey—Social Identity periodically conducted by Statistics Canada (2021a, b). The 2020 assessment (conducted during the COVID-19 pandemic) noted that the healthcare system received the highest score for pride in a Canadian achievement. The functioning of democracy in Canada also received high marks. Asked whether they felt proud of Canada's treatment of all groups in society, 64% of visible minority respondents said yes, compared with 44% of others. Approval scores ranged from 52% from Black respondents, to 77% from West Asian respondents. Men in general expressed more pride in society than did women.

Other recent public opinion surveys have produced varied results, with higher levels of pride in multiculturalism, the armed forces, hockey, and the Canadian Flag, and lower scores for the state of democracy (reflecting political opinions) and health care (reflecting difficulties in accessibility and available primary healthcare providers). All of these suggested characteristics can be challenged: hockey currently ranks behind swimming, soccer, running, cycling and golf in participation rates for different age groups, and increasing ethnic and cultural diversity has broadened both the range of and interest in available activities. Costs for equipment and ice time and perceived potential for injury are other factors. The Canadian Flag has become less respected by some in response to its use in political demonstrations, and others associate it as a symbol of government's failure to Indigenous peoples. Those expressing pride in the Canadian Armed Forces are most likely to cite peacekeeping and disaster response.

Multiculturalism has been official policy in Canada since the early 1970s (Cameron 2004; Brosseau and Dewing 2018; Triadafilopoulos 2021). Multiculturalism involves questions of both personal and collective identity, language, nationality or ethnicity, religion, cultural practices, and sense of place. What is it about a particular community, town, or country that allows people to look at others coming in from elsewhere, or various people who have developed cultures in different ways to blend together, to make an overall society? Many Canadians believe that the idea that Canada is multicultural as a key facet of life. Although challenges remain and success is not universal, nor universally acknowledged, support for multiculturalism largely remains strong (Adams 2021; Banting 2022).

References

Abella I, Troper H (1983) None is too many: Canada and the Jews of Europe, 1933–1948. University of Toronto Press

Adams M (2021) 50 years of multiculturalism: it's as Canadian as maple syrup. Can Geogr. https://canadiangeographic.ca/

Akbari AH (2020) Immigration in Nova Scotia: who comes, who stays, who leaves and why? Nova Scotia Office of Immigration

Amarasingam A, Naganthan G, Hyndman J (2016) Canadian multiculturalism as banal nationalism: understanding everyday meanings among Sri Lankan Tamils in Toronto. Can Ethn Stud 48:119–141

Armony V (2015) Settling North of the U.S. Border: Canada's latinos and the particular case of Québec. Lasaforum 46(4):20–23

Atak I, Linley-Moore C, Kim J (2023) Canada's implementation of the UN global compacts on migration and refugees: advancing foreign policy objectives and the status quo? Refug Surv Q 42:453–474

Aziz N, Aziz A, Mahar G (2023) Role of provincial migration and immigration in provincial trade of Canada. J Risk Fin Manage 16. https://doi.org/10.3390/jrfm16070328

Baker J, Price J, Walsh K (2015) Unwelcoming communities: youth observations of racism in St. John's, Newfoundland and Labrador, Canada. J Youth Stud 19:103–116

Banting K (2022) Multiculturalism policy in Canada: conflicted and resilient. In: Lindquist et al (ed) Policy success in Canada. Oxford University Press, pp 183–205

Barman J (1991) The West beyond the west: a history of British Columbia. University of Toronto Press, Toronto

Barrier NG, Dusenbury VA (1989) The Sikh diaspora: migration and the experience beyond Punjab. South Asia Books, Columbia, MO

Beaujot R, McQuillan K (1982) Growth and dualism: the demographic development of Canadian society. Gage, Toronto

Béchard J (2018) Immigration: the Canada-Québec accord. Legal and Social Affairs Division, Library of Parliament. Publ. 2011-89-E, revised 2018

Berry J, Hou F (2016) Immigrant acculturation and wellbeing in Canada. Can Psychol 57:254–264

Berry J, Hou F (2017) Acculturation, discrimination, and well-being among second generation of immigrants in Canada. Relations 61:29–39

Bloomfield E (1995) Waterloo township through two centuries. Waterloo Historical Society, Waterloo

Boyd M (2015) Ethnicity and race in Canada: past and present. In: Sáenz R, Embrick D, Rodríguez N (eds) the international handbook of the demography of race and ethnicity. International handbooks of population, vol 4. Springer, Dordrecht. https://doi.org/10.1007/978-90-481-8891-8_2

Boyd M, Alboim A (2012) Managing international migration: the canadian case. Rodríguez-García D (ed) Managing immigration and diversity in Canada: a transatlantic dialogue in the new age of migration. McGill-Queen's University Press, Queen's Policy Studies Series, Montreal and Kingston, pp 123–150

Breton R (2005) Ethnic relations in Canada: institutional dynamics. McGill-Queens

Brosseau L, Dewing M (2018) Canadian multiculturalism. Library of Parliament, 2009-20-E revised

Bumsted JM (2008) Lord Selkirk: a life. University of Manitoba Press, Winnipeg, Manitoba

Bumsted JM, Bumsted MC (2016) A history of the Canadian peoples, 5th ed. Oxford University Press Canada Toronto

Burnett D (1988) Coming Canadians. McClelland & Stewart. Toronto

Cadigan ST (2009) Newfoundland and Labrador: a history. University of Toronto Press

Cameron A, Gwyn J (2003) McNab, Archibald, 17th Chief of Clan Macnab. In: Dictionary of Canadian biography, vol 8. University of Toronto/Université Laval

Cameron E (2004) Multiculturalism and immigration in Canada: an introductory reader. Canadian Scholars' Press, Toronto

Campey LH (2010) Planters, paupers, and pioneers—English settlers in Atlantic Canada. Dundurn, Toronto

Canadian Museum of History (n.d.) Historical overview of immigration to Canada. https://www.historymuseum.ca/cmc/exhibitions/tresors/immigration/imf0300e.html

Catto NR (1996) Richardson Mountains, Yukon-Northwest Territories: the Northern Portal of the Postulated Ice-Free Corridor. Quatern Int 32:3–20

Cénat JM (2022) Who is black? the urgency of accurately defining the black population when conducting health research in Canada. Can Med Assoc J194:E948–9. https://doi.org/10.1503/cmaj.220274

Clairmont DH, Magill DW (1999) Africville: the life and death of a Canadian black community. Canadian Scholars' Press, Toronto

Coloma RS, McElhinny B, Tungohan E, Catungal JPC, Davidson LM (2012) Filipinos in Canada: disturbing invisibility. University of Toronto Press

Conrad M, Finkel A (2006) History of the Canadian peoples: beginnings to 1867. Pearson Longman, Toronto

Crea-Aresenio M, Newbold KB, Baumann A, Walton-Roberts M (2023) A few big players: systems approach to immigrant employment in a mid-sized city. Can Geogr 67:513–524

Crossman E, Hou F, Picot G (2021) Are the gaps in labour market outcomes between immigrants and their Canadian born counterparts starting to close? Stat Can Econ Soc Rep Cat 36-28-0001 1(4)

Crown-Indigenous Relations and Northern Affairs Canada (2017) First nations in Canada. https://www.rcaanc-cirnac.gc.ca/eng/1307460755710/1536862806124

Davies AT (1992) Antisemitism in Canada: history and interpretation. Wilfrid Laurier University Press, Waterloo

Demosim Team (2017) Immigration and diversity: population projections for Canada and its regions, 2011 to 2036. Statistics Canada

Demosim Team (2022a) Population projections on immigration and diversity for Canada and its regions, 2016 to 2041: overview of projection assumptions and scenarios. Statistics Canada

Demosim Team (2022b) Projections of the indigenous populations and households in Canada, 2016 to 2041: overview of data sources, methods, assumptions and scenarios. Statistics Canada

Driscoll F (1988) History and politics of prince Edward island. Can Parliamentary Rev Summer 1988. http://www.revparl.ca/11/2/11n2_88e_Driscoll.pdf

Elcioglu EF, Shams T (2023) Brokering immigrant transnationalism: remittances, family reunification, and private refugee sponsorship in neoliberal Canada. Curr Sociol 1–9. https://doi.org/10.1177/001139212311556

Ellyson C, Andrew C, Clément R (2016) Language planning and education of adult immigrants in Canada: contrasting the provinces of Quebec and British Columbia, and the cities of Montreal and Vancouver. Lond Rev Educ 14(2). https://doi.org/10.18546/LRE.14.2.10

Edmonston B, Fong E (2011) The changing Canadian population. McGill-Queen's University Press

Emerson SD, Ritland L, Guhn M (2021) A scoping review of associations between ethno-cultural context and mental health in Canada. Can J Commun Mental Health 40(2). 7870/cjcmh-2021-010

Epp FH (1974) Mennonites in Canada 1786–1920. Macmillan Toronto

Fang T, Sapeha H, Neil K (2018) Integration and retention of refugees in smaller communities. Int Migr 56(6):83–99

Gentilcore RL (1993) Historical Atlas of Canada: Volume II: the land transformed. University of Toronto

Goddard I (1999) Native languages and language families of North America (map). University of Nebraska Press, Lincoln

Gossage P, Little JI (2015) Une histoire du Québec: Entre tradition et modernité. Hurtubise, Montréal

Graham N, Pottie-Sherman Y (2020) The experiences of immigrant entrepreneurs in a medium-sized Canadian City: the case of St. John's, Newfoundland and Labrador, Canada. Can Geograph 1–13. https://doi.org/10.1111/cag.12627

Gyan C, Chowdhury E, Yeboah AS (2023) Adapting to a new home: resettlement and mental health service experiences of immigrant and refugee youth in Montreal. Humanit Soc Sci Commun 2023. https://doi.org/10.1057/s41599-023-01572-7

Haan M, Arbuckle J, Prokopenko E (2017) Individual and community-level determinants of retention of anglophone and francophone immigrants across Canada. Can Stud Popul 44:59–76

Hamilton LK, Esses VM, Walton-Roberts M (2022) Borders, boundaries, and the impact of COVID-19 on immigration to Canada. Stud Soc Justice 16(1)

Handcock G (1989) Soe longe as there comes Noe women: origins of English settlement in Newfoundland. Newfoundland History Series, #6, Breakwater Press, St. John's, Newfoundland, Canada

Harris RC (1987) Historical Atlas of Canada: Volume I: From the beginning to 1800. University of Toronto

Hawkins F (1988) Canada and immigration: public policy and public concern. McGill-Queen's

Henry F, Tator C, Mattis W, Rees T (1998) The colour of democracy: racism in Canadian society. Nelson Thomson, Scarborough. ON

Hinther RL, Mochoruk J (2011) Re-imagining Ukrainian-Canadians: history, politics, and identity. University of Toronto Press

Hou F, Picot G (2019) Trends in the citizenship rate among new immigrants to Canada. Economic insights. Stat Can Catalogue no. 11-626-X

Hou F, Schellenberg G, Berry J (2016) Patterns and determinants of immigrants' sense of belonging to Canada and their source Country. Anal Stud Branch Stat Can

Houle R, Corbeil J-P (2017) Language projections for Canada, 2011 to 2036. Stat Can

Houle R, Pereira D, Corbeil JP (2014) Statistical portrait of the french-speaking immigrant population outside Quebec (1991–2011). Citizenship and Immigration Canada, Ottawa
Humber CJ (1991) Allegiance: the Ontario story. Heirloom Publishing, Mississauga
Immigration and Refugee Board of Canada (2023a) Claims by country of alleged persecution—2022. https://www.irb-cisr.gc.ca/en/statistics/protection/Pages/RPDStat2022.aspx
Immigration and Refugee Board of Canada (2023b) Irregular border crossing statistics. https://www.irb-cisr.gc.ca/en/statistics/Pages/Irregular-border-crosser-statistics.aspx
Immigration Refugees and Citizenship Canada (2021) Discover Canada: the rights and responsibilities of citizenship. https://www.canada.ca/en/immigration-refugees-citizenship/corporate/publications-manuals/discover-canada.html
Immigration Refugees and Citizenship Canada (2023) Canada: a history of refuge. https://www.canada.ca/en/immigration-refugees-citizenship/services/refugees/history.html
Inuit Tapiriit Kanatami (2020) Our 5000 year heritage. https://www.itk.ca/5000-years-inuit-history-heritage/
Isajiw W, Makuch A (1990) Ukrainians in Canada. In: Pawliczko AL (ed) Ukraine and Ukrainians throughout the world. University of Toronto Press, Toronto, p 333. ISBN 0-8020-0595-0
Jadotte H (1977) Haitian immigration to Quebec. J Black Stud 7(4):485–500
Japardze I, Kaplan U (2023) Cultural and economic integration of immigrants in Canada: "Do you play Hockey?" Labour Market Res 57:13. https://doi.org/10.1186/s12651-023-00340-z
Johnson MA, Aladejebi F (2022) Unsettling the Great White North: Black Canadian history. University of Toronto Press
Johnston H (2005) Sikhs in Canada. Encyclopedia of diasporas. Springer 1075–1083. https://link.springer.com/referencework/https://doi.org/10.1007/978-0-387-29904-4#
Joshee R, Derwing TH (2005) the unmaking of citizenship education for adult immigrants in Canada. JIM1/RIMI 6(1):61–80
Kelley N, Trebilcock M (1998) The making of the Mosaic: a history of Canadian immigration policy. University of Toronto Press, Toronto
Kelly M, Nguyen M (2023) Choosing to stay: understanding immigrant retention in four non-metropolitan counties in Southern Ontario. J Int Migr Integr. https://doi.org/10.1007/s12134-023-01034-8
Kukushkin V (2007) From peasants to labourers: Ukrainian and Belarusian immigration from the Russian empire to Canada. McGill-Queen's University Press, Montreal
Kymlicka W (2010) Ethic, linguistic, and multicultural diversity of Canada. In: Courtney J, Courtney JC, Smith D (eds) Oxford handbook of Canadian politics, pp 301–320
Landry Y (1992) Les Filles du roi au XVIIe siècle. Bibliothèque Québécoise, Montréal
Li PS (2000) Cultural diversity in Canada: the social construction of racial differences. Department of Justice, Canada
Lo L, Li W, Yu W (2017) Highly-skilled migration from China and India to Canada and the United States. Int Migr. https://doi.org/10.1111/imig.12388
Lu Y, Hou F (2020) Immigration system, labor market structures, and overeducation of high-skilled immigrants in the United States and Canada. Int Migr Rev 54:1072–1103
MacIsaac S (2023) Remittance modality: unpacking Canadian money transfer mechanism choices. Int Migr Rev 1–27. https://doi.org/10.1177/01979183231181564
Mackay D (1990) Flight from famine: the coming of the Irish to Canada. Mclelland and Stewart, Toronto
Mannion P (2022) "I'm as Good an Irishman as You": the ancient order of hibernians and the construction of Irish ethnicity in Canada and the United States, 1908–1918. J Am Ethn Hist 41(2)
Marshall I (2001–2009) The Beothuk. Breakwater Books, St. John's
Mata F (2021) The immigrant waves from Latin America to Canada: a look at immigration and census statistics. In: International conference women, gender and intersectionality in the Lusophone World, June 29 to July 2, 2022, Ponta Delgada, Azores, Portugal
McGowan MG (2023) Overview: Irish migration and settlement in Canada. Embassy of Ireland, Ottawa
McHugh R (2023) The rudest cities in Canada, revealed. https://preply.com/en/blog/rudest-cities-in-canada. Accessed 07 Feb 2024
McManus G (1991) Atlas of Newfoundland and Labrador. Breakwater Books, St, John's
Mendez FL et al (2013) An African American paternal lineage adds an extremely ancient root to the human y chromosome phylogenetic tree. Am J Hum Genet 92:454–459
Milan A, Tran K (2004) Blacks in Canada: a long history. Statistics Canada, 11-008. https://www150.statcan.gc.ca/n1/pub/11-008-x/2003004/article/6802-eng.pdf
Moogk PN (1989) Reluctant exiles: emigrants from France in Canada before 1760. William Mary Quart 46:463–505
Nangia P, Gingrich K (2020) Quality of life of Immigrants in Canada. Can Ethn Stud J 52. https://doi.org/10.1353/ces.2020.0012
Newfoundland (1945) Census of Newfoundland [and Labrador], vol 1, p 105. https://collections.mun.ca/digital/collection/cns_tools/id/122013
Nnamdi Okafor O, Kalu K (2023) Integration challenges, immigrant characteristics and career satisfaction for immigrants in the field of accounting and finance: an empirical evidence from Canada. Crit Perspect Account. https://doi.org/10.1016/j.cpa.2023.102602
O'Gallagher M (1984) Grosse Île: gateway to Canada, 1832–1937. Carraig Books, Ste-Foy, Québec
Ó Gráda C (2006) Ireland's great famine: interdisciplinary perspectives. Dublin Press
Ommer RE (1986) Primitive accumulation and the Scottish 'Clann' in the old world and the new. J Hist Geogr 12:121–141
O'Neill T, Makaju U, Shrestha R, Tamrakar N, Shrestha BM, Sthapit BM (2019) The Nepali-Canadian living standards survey: newcomer incorporation in the Greater Toronto area. Can Ethnic Stud J 51:69–88
Ornstein M (2006) Ethno-racial groups in Toronto, 1971–2001: a demographic and socio-economic profile. Institute for Social Research, York University. http://www.isr.yorku.ca/
Oxford W (2019) Indigenous languages in Canada. Canadian Language Museum, Toronto. anguagemuseum.ca/wp-content/uploads/2022/12/indigenous_languages_in_canada_0.pdf
Paquet M (2019) Province building and the federalization of immigration in Canada. University of Toronto Press, Toronto
Paquet M, Lawlor A (2022) Numbers and images: representations of immigration and public attitudes about immigration in Canada. Can J Polit Sci 55:827–851
Pastore RT (1992) Shanawdithit's people: the archaeology of the Beothuks. Breakwater Books, St, John's
Pottie-Sherman Y, Graham N (2020) Live, work, and stay? Geographies of Immigrant receptivity in Atlantic Canada's aspiring gateways. Geograph Rev. https://doi.org/10.1080/00167428.2020.1804301
Pottie-Sherman Y, Wilkes R (2019) Anti-immigrant sentiment in Canada. In: Anti-immigrant sentiments, actions, and policies: the North American Region and the European Union, Universidad Nacional Autónoma de México, pp 275–289
Raghavan M et al (2014) The genetic prehistory of the New World Arctic. Science 345. https://doi.org/10.1126/science.1255832
Raska J (2015) Forgotten experiment: Canada's resettlement of palestinian refugees, 1955–1956. Histoire Sociale/Soc Hist 48:445–473
Raska J (2020) Starting points: researching Canadian immigration and ethnic history. Can Mus Immigr Pier 21. https://pier21.ca/blog/jan-raska/starting-points-researching-canadian-immigration-and-ethnic-history

Rees R (2000) Land of the loyalists: their struggle to shape the maritimes. Nimbus, Halifax

Reid JG (2016) Immigration to Atlantic Canada: historical reflections. J Roy Nova Scotia Hist Soc 19:38–53

Rice FE, Quan T (2023) Beyond "economic immigration": understanding the role of labor market and lifestyle expectations in technology sector newcomer experiences in Canada. J Int Migr Integr 24(Suppl 6):S1167–S1188 p

Richmond AH (2019) Post-war immigrants in Canada. University of Toronto Press

Rocher F (2023) In Canada's French-speaking Quebec, immigration sparks anxieties about language and identity. Migration Policy Institute Online Journal. https://www.migrationpolicy.org/article/immigration-quebec-language

Rowe F (1980) History of newfoundland and Labrador. McGraw-Hill Ryerson, Toronto

Saku JC (1999) Aboriginal census data in Canada: a research note. Can J Nativ Stud 19:365–379

Saney I (1998) Canada: the black nova Scotian odyssey: a chronology. Race & Class 40(1):78–91

Sanzone D (2015) From mass immigration to professional workers. a portrait of the present Italian "Comunità" in Ontario, Canada. Italian Canadiana 26–29:31–52

Seary ET (1977) Family names of the Island of Newfoundland. McGill-Queen's University Press

Sendzik W (1997) The 1832 Cholera epidemic: a study in state formation. McGill University, Montréal, Dept of History

Se'ver A 1999. What's in a name?: Subtle jeopardies Turkish immigrants experience. Bizim Anadolu 6 (70–72): 2. https://tspace.library.utoronto.ca/handle/1807/17548

Schimmele C, Hou F, Stick M (2023) Poverty among racialized groups across generations. Statistics Canada 36-28-001

Shuva NZ (2021) Internet, social media, and settlement: a study on Bangladeshi immigrants in Canada. Can J Inf Libr Sci 43:291–315

Statistics Canada (2021) General social survey—social identity. https://www23.statcan.gc.ca/

Statistics Canada 2021–2024 (2021) Census of Canada. https://www12.statcan.gc.ca/census-recensement/2021

Stick M, Schimmele C, Karpinski M, Cissokho S (2023) Immigrants' sense of belonging to Canada by province of residence. Statistics Canada, Economic and Social Reports 36-28-0001

Syed IU (2021) Remittance flows from healthcare workers in Toronto, Canada. Sustainability 13:9536. https://doi.org/10.3390/su13179536

Termote M (2023) L'immigration et la pérennité du français au Québec. Ministère de l'Immigration, de la Francisation et de l'Intégration, Montréal. https://cdn-contenu.quebec.ca/cdn-contenu/adm/min/immigration/publications-adm/rapport/RapportMIFI_mai2022_MarcTermote.pdf

Thornton R (2000) Population history of Native North Americans. In: Haines MR, Steckel RH (eds) A population history of North America. Cambridge University Press 13. ISBN 978-0-521-49666-7

Tolentino JF (2023) Navigating waters: experiences of Filipino Canadian identity making in the diaspora. PhD thesis, Wilfrid Laurier University, Waterloo

Triadafilopoulos T (2021) The foundations, limits, and consequences of immigration exceptionalism in Canada. Am Rev Can Stud 51:3–17. https://doi.org/10.1080/02722011.2021.1923150

Tulchinsky GJJ (2008) Canada's Jews: a people's journey. University of Toronto Press

Vengroff R (2013) Immigration policy at the subnational level in North America: Quebec and Georgia in comparative perspective. Norte América 8(2013):141–170

Vézina H, Tremblay M, Lavoie È-M, Labuda D (2014) Concordance between reported ethnic origins and ancestral origins of Gaspé Peninsula Residents. Population 69:7–27 (Stringer, L., transl.)

Walker JWSG (1992) The Black Loyalists. University of Toronto Press

Wall K, Wood S (2023) Education and earnings of Canadian-born Black populations. Statistics Canada https://www150.statcan.gc.ca/n1/pub/75-006-x/2023001/article/00009-eng.htm

Weinfeld M, Schnoor RF, Koffman DS (2012) Overview of Canadian Jewry. The American Jewish Year Book, pp 109–112. 55-93

Whitaker R (1991) Canadian immigration policy since confederation. Canadian Historical Association, Ottawa

Whitcomb E (2010) A short history of prince Edward island. From Sea To Sea Enterprises, Ottawa

Whitehead RH (2013) Black loyalists: southern settlers of nova scotia's first free black communities. Nimbus Publishing, Halifax

Wilson DA (1989) The Irish in Canada. Canadian Historical Association, Ottawa

Zucchi J (2007) A history of ethnic enclaves in Canada. Canadian Historical, Ottawa

Zuehlke M (2016) Scoundrels, dreamers, and second Sones: British remittance men in the Canadian west. Harbour Publishers, La Vergne

12 Aspects of Urban Geography

Abstract

The majority of Canadians live in urban communities, particularly in Census Metropolitan Areas (CMA), the largest urban centres. CMA are divided into five zones, based on commuting times by automobile during off-peak hours. Downtowns are defined as the core of the historic central municipality within the CMA, based on employment density and input from municipal officials and professional planners. Larger CMA may include secondary downtowns, associated with other cities within the CMA. In the largest CMA, downtown residents represent a small proportion of the total population. Population growth rate between 2016 and 2021 does not show a correlation with overall CMA population. In 2021, Statistics Canada recognized 141 Census Agglomerations (CA), with total and core populations of at least 10,000. Of the 141 CA, 26 saw declines in population between 2016 and 2021, generally directly related to downtowns in mining, petroleum extraction, pulp-and-paper and forestry, and steel manufacturing industries. Provincial capitals in most instances were established in the province's primate city, with little dispute. Automobile use represents about three-quarters of total commuter traffic. CA and smaller CMA have higher rates of automobile commuting, due to the limited public transit. Factors influencing automobile use include population density, availability of public transit, fuel price, and roadway congestion. Increasing commuting expenses, particularly gasoline, have some influence on travel behaviour. Suburban residents tend to consider changing vehicle types more readily than adopting public transit. Canada's six largest CMA have rail rapid transit systems. Active transport, including walking and cycling, accounts for slightly more than 6% of all commutes in Canada. Heat is regarded as a significant hazard in Canada's urban centres. Exposure to urban heat and limited vegetation cover are coupled with inequities in other environmental aspects in urban areas, with negative consequences for physical and mental health. High geographic variability characterizes environment conditions within CMAs. Increased income inequality is primarily limited to the three largest CMA. Governance and taxation regimes mean that the tools theoretically able to address rising inequality are not within the control of municipalities.

Keyword

Census Metropolitan Areas · Census agglomerations · Provincial capitals · Urban transportation · Urban heat · Inequality

The majority of Canadians live in urban communities. This chapter focuses on the Census Metropolitan Areas (CMA), the largest urban centres, with emphasis on population, growth, urban transit and commuting, environmental stresses, urban vegetation, and inequality in urban systems. Other aspects of Canadian urban geography are considered in Chap. 13, as well as Chaps. 11 (Immigration, Migration, and Ethnicity), 14 (Transportation), 15 (Urban Vegetation), 17 (Energy), and 19 (Winter).

12.1 Census Metropolitan Areas and Census Agglomerations

Statistics Canada (2021b, 2022a) defines a Census Metropolitan Area (CMA) to have a population of at least 100,000, of whom at least 50,000 must live in the core (Table 12.1). Other municipalities within a CMA must have a high degree of integration with the core, as determined by commuting traffic to and from work. A Census Agglomeration (CA) has a total and core population of at least 10,000, but less than 100,000. Both CMA and CA are based on labour market data, rather than on political boundaries or built-up areas. Population centres within a CMA or CA have minimum

N. Catto and G. R. Vincent, *Canada: Land and People*, World Regional Geography Book Series,
https://doi.org/10.1007/978-3-032-04391-7_12

Table 12.1 Census Metropolitan Areas (CMA) Populations (2021) and Population Growth Rates (2016–21). Statistics Canada (2022a), Census of Canada

	Population 2021	Growth 2016–21 (%)
Toronto ON	6,202,225	4.63
Montreal QC	4,291,732	4.57
Vancouver BC	2,642,825	7.28
Ottawa-Gatineau ON-QC	1,488,307	8.51
Calgary AB	1,481,806	6.41
Edmonton AB	1,418,118	7.32
Québec QC	839,311	4.08
Winnipeg MB	834,678	6.59
Hamilton ON	785,184	5.04
Kitchener-Cambridge-Waterloo ON	575,847	9.92
London ON	543,551	10.02
Halifax NS	465,703	9.08
St. Catharines-Niagara ON	433,604	6.78
Windsor ON	422,630	6.00
Oshawa ON	415,311	9.34
Victoria BC	397,237	8.01
Saskatoon SK	317,480	7.59
Regina SK	249,217	5.29
Sherbrooke QC	227,398	7.21
Kelowna BC	222,162	13.99
Barrie ON	212,856	8.02
St. John's NL	212,579	2.00
Abbotsford-Mission BC	195,726	8.42
Kingston ON	172,726	7.06
Sudbury (Greater) ON	170,605	2.80
Guelph ON	165,588	8.95
Saguenay QC	161,567	0.03
Trois-Rivières QC	161,489	3.49
Moncton NB	157,717	8.91
Brantford ON	144,162	7.42
Saint John NB	130,613	3.50
Peterborough ON	128,624	5.67
Lethbridge AB	123,847	5.50
Thunder Bay ON	123,258	1.35
Nanaimo BC	115,459	10.03
Kamloops BC	114,142	9.95
Chilliwack BC	113,767	12.07
Belleville-Quinte West ON	111,184	7.53
Fredericton NB	108,610	5.76
Chatham-Kent ON	104,316	2.23
Drummondville QC	101,610	4.59
Red Deer AB	100,844	0.42

populations of 1,000. Areas within a CMA or CA which are not population centres are defined as rural areas. A CMA can include several distinct population centres, along with one or more cores and surrounding rural areas. Toronto CMA extends more than 75 km from downtown and contains several other population centres and rural areas.

Rural areas form more than 60% of the land surface in all CMA except Red Deer (37%). A majority of the CMA (24) have at least 90% rural land, including Ottawa-Gatineau, Edmonton, Winnipeg, Halifax, Saskatoon, and Regina (Figs. 12.1 and 12.2). In some cases, this reflects political extension of urban boundaries by amalgamation (e.g. Halifax Regional Municipality), annexation, or establishment of "green belts" to restrict suburban development. CMA with 80–90% rural land include Calgary, Québec, London, and St. Catharines-Niagara. Of the 20 largest CMA, only Toronto, Montreal, Vancouver, Hamilton, Kitchener-Cambridge-Waterloo, and Victoria have less than 75% rural land area within their boundaries. Combined, 87% of the land area in the CMA is classified as rural.

"Rural" in this context refers to terrain outside the boundaries of a population centre (Statistics Canada 2021b, 2022a). It does not imply undeveloped or unoccupied land, or land used for agriculture, parkland, or forestry purposes. Rural areas may include widely spaced "large lot" suburban and exurban development, designated green belts with existing development conserved, or recreationally zoned terrain. Residents of these rural areas may have a variety of occupations.

12.1.1 CMA Zones

CMA are divided into five zones, based on commuting times by automobile during off-peak hours. Downtowns are defined as the core of the historic central municipality within the CMA, based on employment density and input from municipal officials and professional planners. Larger CMA may include secondary downtowns, associated with other cities within the CMA. The Toronto CMA includes the cities of Mississauga (717,961), Brampton (656,480), Markham (338,503), Vaughan (323,103), and Oakville (213,759). The Montreal CMA includes Laval (438,366), Longueil (254,483), St-Jerome (100,859), St-Jean-sur-Richelieu (88,083), and Chateauguay (75,891). Vancouver CMA includes Surrey (568,322), Burnaby (249,125), Richmond (209,937), Coquitlam (148,625), and Delta (108,455).

Broader unofficial urban areas are commonly recognized. The Greater Toronto Area (GTA) includes the Toronto CMA and the city of Burlington to the southwest. Some authors extend the GTA to the east, including the Oshawa CMA, and to the southwest, including the Hamilton CMA (which

Fig. 12.1 Population of CMA (2021) compared to percentage of land area within large urban population centre. Most CMA, especially those with population less than 500,000, have less than 15% of their land area within a large urban centre. See text for discussion. Statistics Canada (2022a), Census of Canada

includes Burlington). The "Golden Horseshoe" along the southwest shore of Lake Ontario adds St. Catharines-Niagara CMA. Commuter studies can include Barrie CMA (north) and Guelph and Kitchener-Waterloo-Cambridge CMAs (west). Urban and suburban developments flank Highway 401 for 170 km between Newcastle and Cambridge. The BC Lower Mainland extends east from Vancouver CMA along the Fraser River through Abbotsford-Mission CMA, Chilliwack, and Hope.

The second zone of a CMA is the urban fringe, located within a 10-min drive from downtown. Together, the downtown and urban fringe represent the central built-up area of the CMA, the large urban population centre. Suburbs are designated based on driving times as near, 10–20 min from downtown; intermediate, 20–30 min from downtown; and distant, 30 min or more from downtown.

In the largest CMA, downtown residents represent a small proportion of the total population (Fig. 12.3, from 2.6% in Montreal to 5.1% in Ottawa-Gatineau; Statistics Canada 2022a). Victoria, Abbotsford-Mission, Peterborough, Kamloops, Chilliwack, Belleville-Quinte West, and Fredericton are the only CMA with more than 10% of residents living in the downtown. Overall, downtown residents represented about 3.5% of the total Canadian population. Although downtown populations in most CMA grew more rapidly than urban populations as a whole, downtowns are still largely places to work, rather than places to live. The implication that downtown service businesses are largely dependent on commuter rather than residents' traffic became evident during the COVID-19 restrictions and remains apparent in 2024 (Statistics Canada 2024a, b). Declines in downtown populations in Montreal and Vancouver over 2020–2021 (Statistics Canada 2022a) may also reflect peoples' decisions to move to less urban settings in response to COVID-19.

Downtown population densities range from 1491 km^{-2} (Lethbridge) and 1795 km^{-2} (Moncton) to 8,367 km^{-2} (Montreal), 16,608 km^{-2} (Toronto), and 18,837 km^{-2} in Vancouver, with a mean value of 5,835 km^{-2} for all CMA downtowns (Statistics Canada 2022a). Approximate values for built-up areas (downtown, urban fringe, and near suburbs) range from 20 to 55% of those for downtowns (e.g. ~4400 km^{-2} for Toronto, 4800 km^{-2} for Montreal, 5300 km^{-2} for Vancouver, and 1600 km^{-2} for Calgary). Comparable values for large USA cities (United States Census 2020) include New

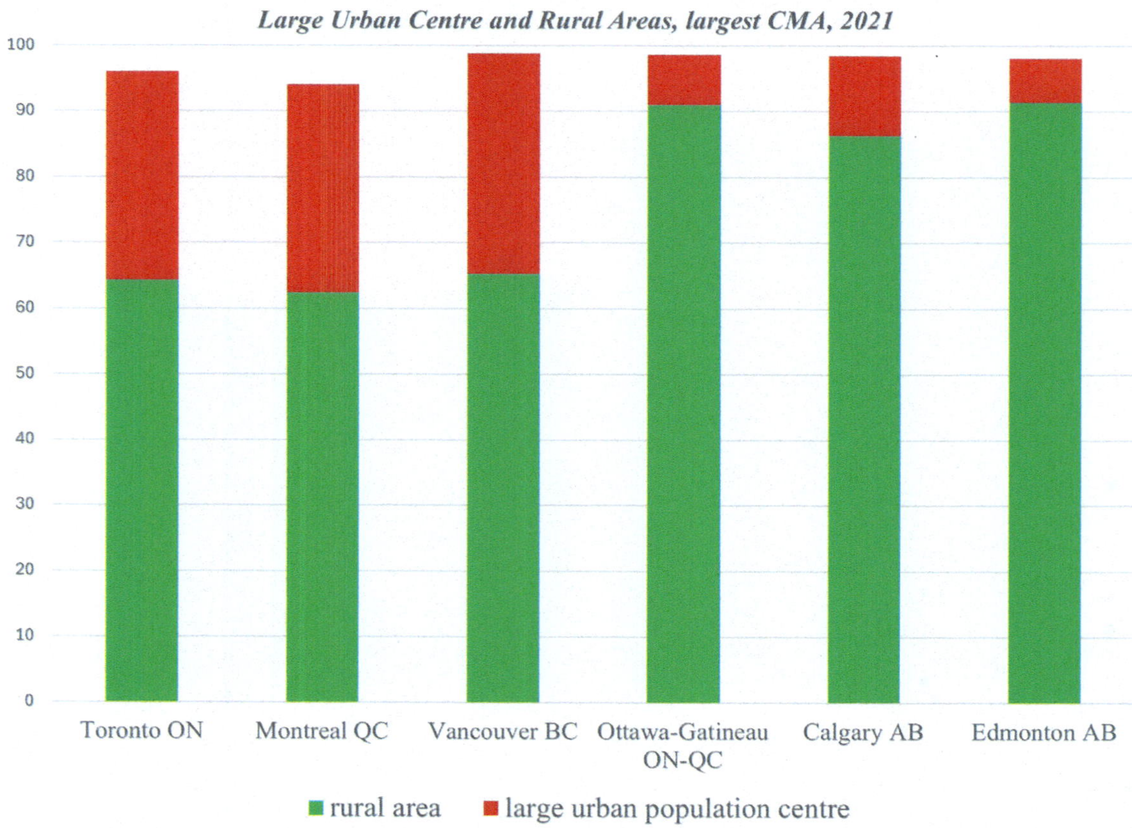

Fig. 12.2 Comparison of land area occupied by large urban centre (Downtown, Urban Fringe and Near Suburb) and Rural Areas, largest CMA by population (2021). Totals less than 100% indicate areas of intermediate suburbs and smaller population centres surrounded by rural areas. See text for discussion. Data from Statistics Canada (2022a), Census of Canada

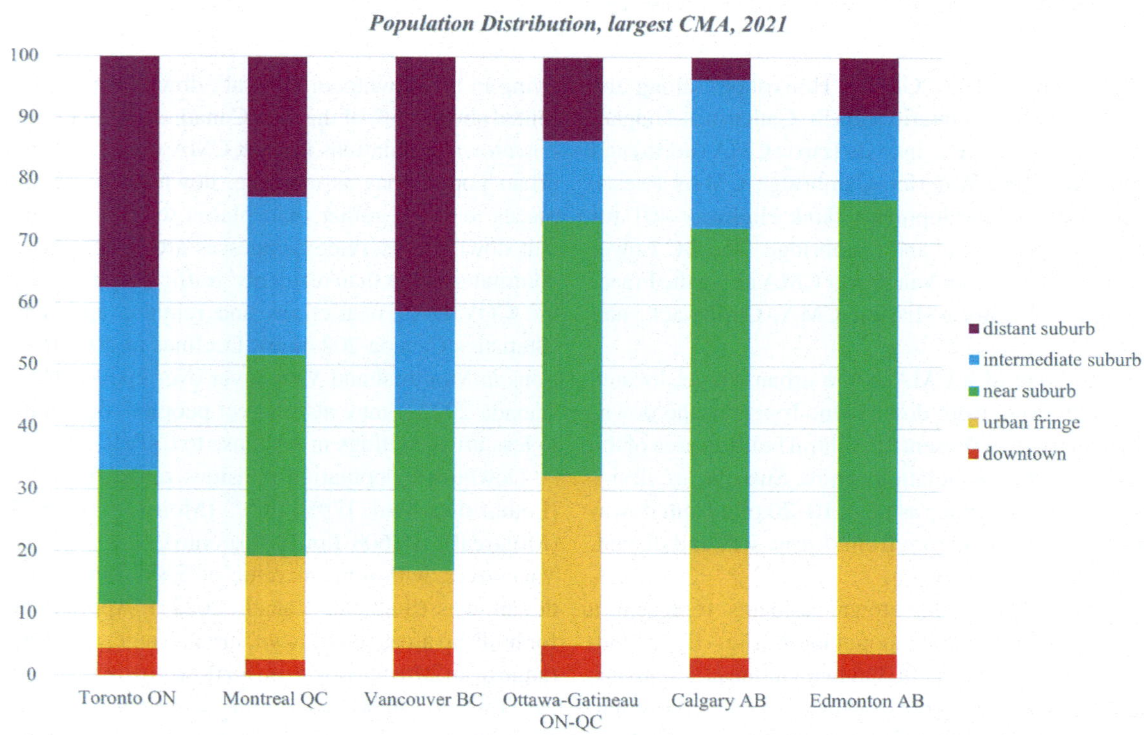

Fig. 12.3 Population distribution by location for the 6 largest CMA by population, 2021. See text for discussion. Data from Statistics Canada (2022a), Census of Canada

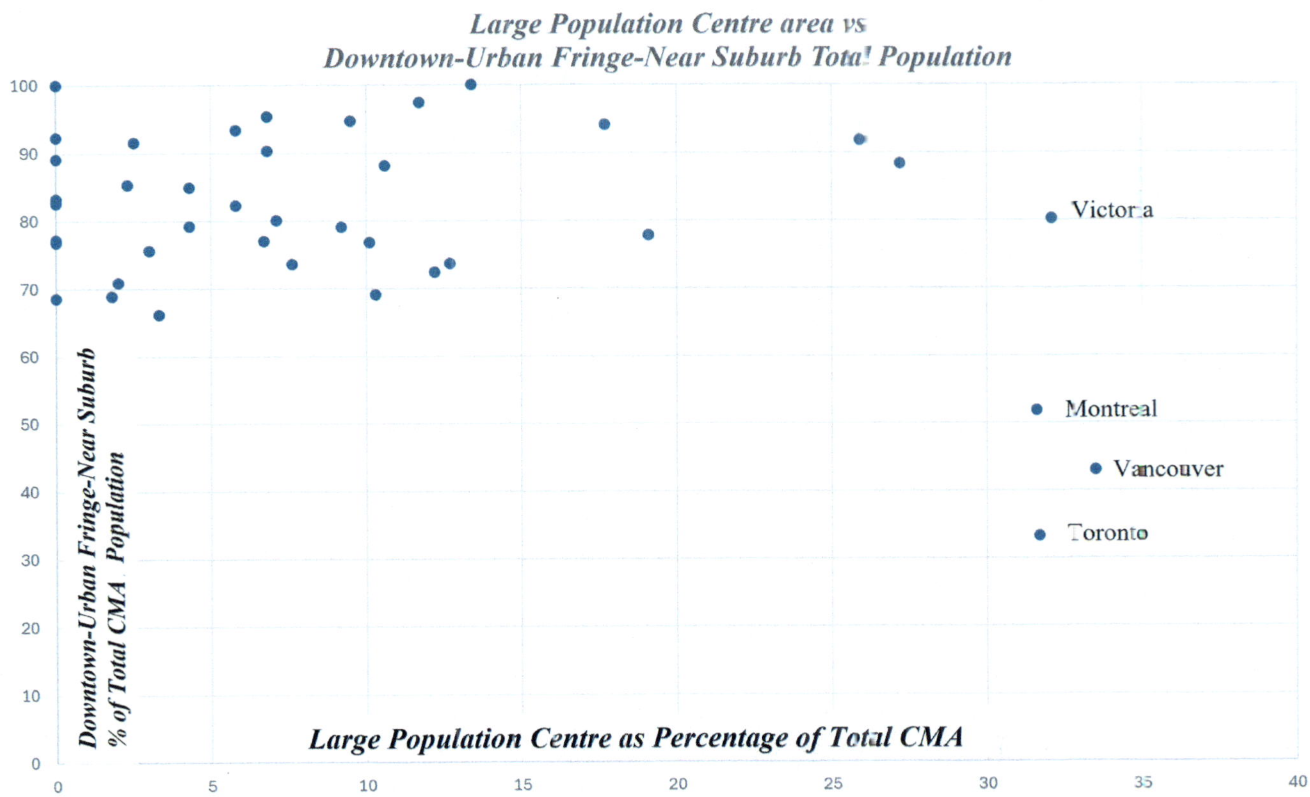

Fig. 12.4 Large Population Centre as percentage of total CMA area compared with percentage of population living in the combined Downtown-Urban fringe-Near Suburb areas. See text for discussion. Data from Statistics Canada (2022a), Census of Canada

York (11,319 km^{-2}), San Francisco (7,193 km^{-2}), Chicago (4,656 km^{-2}), Los Angeles (3,206 km^{-2}), and Houston (1,389 km^{-2}).

The three largest CMA have 30–35% of their land in the large urban population centre (downtown, urban fringe, and inner suburbs; Fig. 12.4). The smaller CMA have less than 15% of the land area in the centre, with the exception of Victoria (~32%) and St. John's (~19%). A similar pattern is evident in comparing the Large Population Centre areas with the percentage of population living in the large urban population centre (Figure). The CMA with populations of 100,00–200,000 combine lesser percentages of land in the urban centre (and thus greater percentages of rural areas) with population concentrations in the centre (and lesser percentages in intermediate and distant suburbs).

Intermediate and distant suburban residents represent greater percentages in Toronto, Montreal, and Vancouver than in the other large CMA (Fig. 12.3). In Montreal and Vancouver, topographic constraints (mountains and rivers, respectively) restrict suburban location. Agricultural and conservation restrictions ("green belts") adjacent to the cities encourage development of more distant suburbs in Toronto and Vancouver, and to a lesser extent in the Ontario area of Ottawa-Gatineau. Differences in municipal zoning policies in the communities within the CMA also have considerable influence.

Automobile travel time is also a significant factor. Although the designations of distance are based on travel time in off-peak hours, the increasing traffic on Highway 401 (Toronto), Autoroutes 10, 15, and 40 and the St. Lawrence bridges and tunnel (Montreal), and Highway 1 (Vancouver) are making the distinction between peak and off-peak hours less relevant (e.g. CAA 2017; Allen et al. 2022). Significantly lower travel times in Calgary and Edmonton, along with lower land prices and fewer building restrictions (in general) mean that fewer distant suburbs develop. In 2016–2021, growth in the largest three CMAs was most rapid in the distant suburbs, whereas the intermediate suburbs increased most rapidly in Calgary and Edmonton (Statistics Canada 2022a).

The classification's assumptions are that all travel is by automobile, without commuter rail or rapid public transit; that people travel primarily from suburbs to downtown for work, rather than to other locations; and that commuting occurs regularly, without working from home. In 2024, about 20% of the Canadian population is working from home. Decisions to live at a distance from downtown may not reflect commuting commitments. This is also apparent with increased migration from southern Ontario and Québec during COVID-19 restrictions to the perceived safer environments of Atlantic Canada, with suitable internet connections for remote work.

The seven CMA in the Prairie Provinces all show low percentages of population in intermediate and distant suburbs

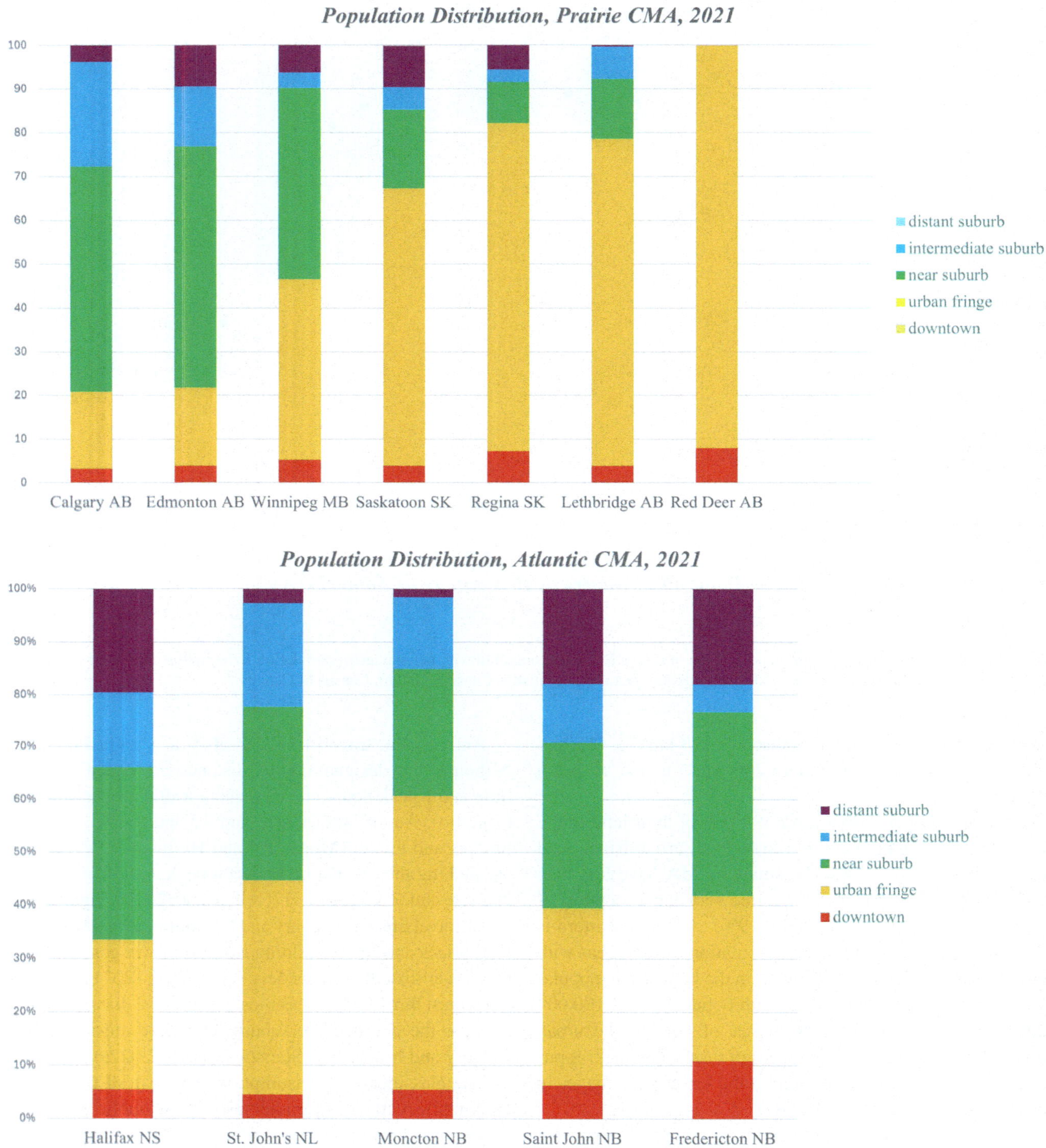

Fig. 12.5 **a** Population distribution by location for the Prairie CMA, 2021. **b** Population distribution by location for the Atlantic CMA, 2021. See text for discussion. Data from Statistics Canada (2022a), Census of Canada

(Fig. 12.5a). Low traffic volumes and lesser populations allow residents to live in the urban fringes, with 15% or less living in intermediate and distant suburbs (none in Red Deer, the smallest CMA). Rapid transit and commuter rail are not present in Winnipeg, Lethbridge, and Red Deer, and the Light Rail Systems in Calgary and Edmonton are limited and do not extend to intermediate suburban locations.

The five Atlantic region CMA show different patterns (Fig. 12.5b). Downtown residents represent 4.6% (St. John's) to 10.9% (Fredericton). Although some gentrification has

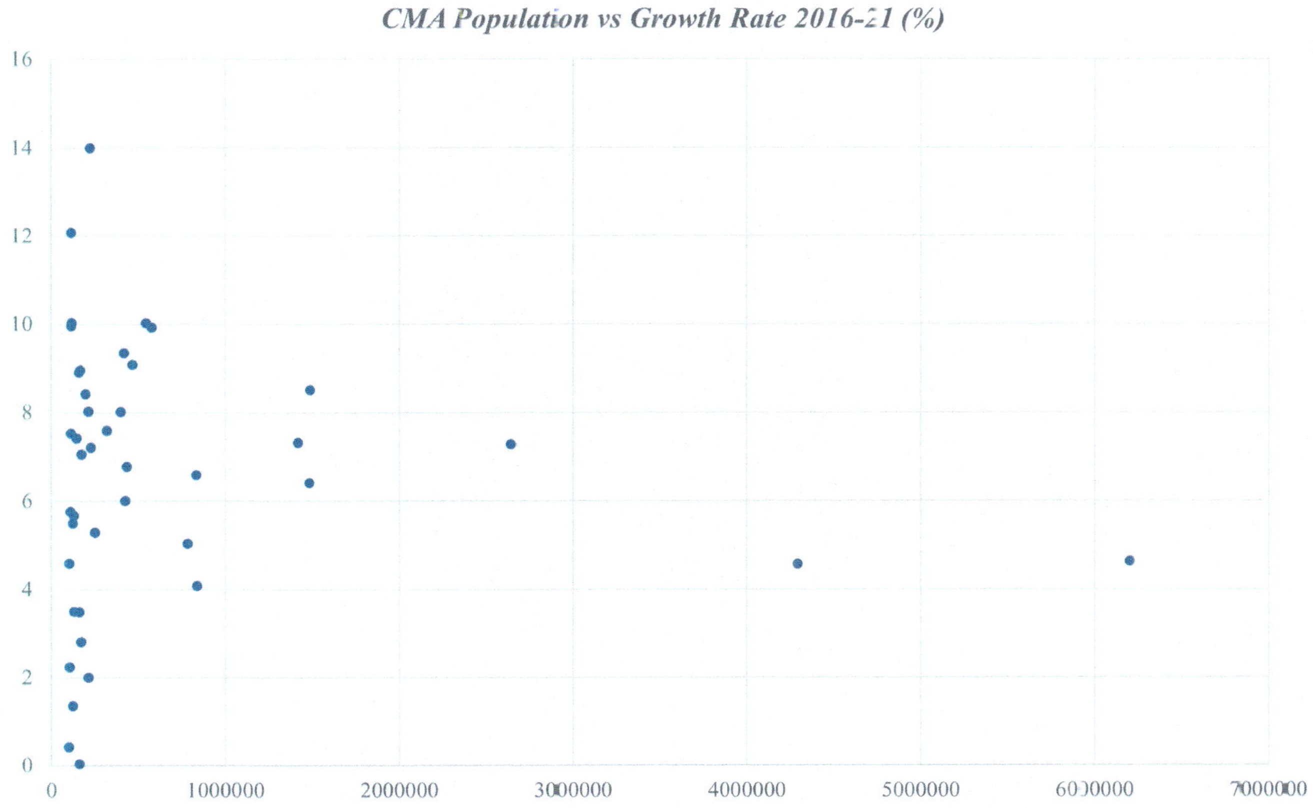

Fig. 12.6 CMA Population (2021) versus Population Growth Rate (2016–2021). See text for discussion. Data from Statistics Canada, Census of Canada

occurred, notably in Halifax and Fredericton, higher proportions of lower household incomes and employment, and higher levels of residential instability characterize the residential areas adjacent to the downtown cores. The importance of government, health facilities, and higher education as employers, several of which are not located in the historic downtowns, change traffic patterns, real estate values, and residential choices to a much greater extent than in the larger CMA with more large employers. Near and intermediate suburbs have higher percentages of residents in the CMAs due to short travel times coupled with zoning and taxation differences. The lesser populations of St. John's and the New Brunswick CMAs preclude construction of rapid transit.

12.2 Population Growth and Employment

Population growth rate between 2016 and 2021 does not show a correlation with overall CMA population (Fig. 12.6). The highest growth rates occurred in the British Columbia CMA of Kelowna (14.0%), Chilliwack (12.1%), and Nanaimo (10.0%), along with London ON (10.0%). Ottawa-Gatineau (8.5%) had the highest growth rate among the CMA with at least 1 million residents. CMA in southwestern Ontario generally had higher growth rates than the mean (~6.5%), while CMA in Prairie Canada and Québec generally grew more slowly. Regional and local economic factors had greater effects on growth rates than population size. The lowest growth rates marked St. John's (2.0%), Thunder Bay (1.4%), Red Deer (0.45), and Saguenay (0.03%).

Of the \$2 × 10^{12} total GDP generated in Canada in 2021, the CMA accounted for more than 75% (Statistics Canada 2023b). Toronto (21%), Montreal (11%), Vancouver (8%), and Calgary (5%) represented 45% of the total, with Ottawa-Gatineau, Edmonton, Québec, Winnipeg, Hamilton, and Kitchener-Cambridge-Waterloo accounting for another 16% (Fig. 12.7, Table 12.2). In all regions except Atlantic Canada (with relatively small CMA), CMA represent the dominant economic factor (Fig. 12.8). Hydrocarbon production increases the total for non-CMA communities in Alberta and Saskatchewan. Toronto CMA accounts for more than half of the total GDP produced in Ontario.

Nationally, trade, health care and social assistance, manufacturing, educational services, accommodation and food, public administration, and professional, scientific, and technical services represented the largest employment sectors (Statistics Canada 2021a; Fig. 12.9). The majority of those

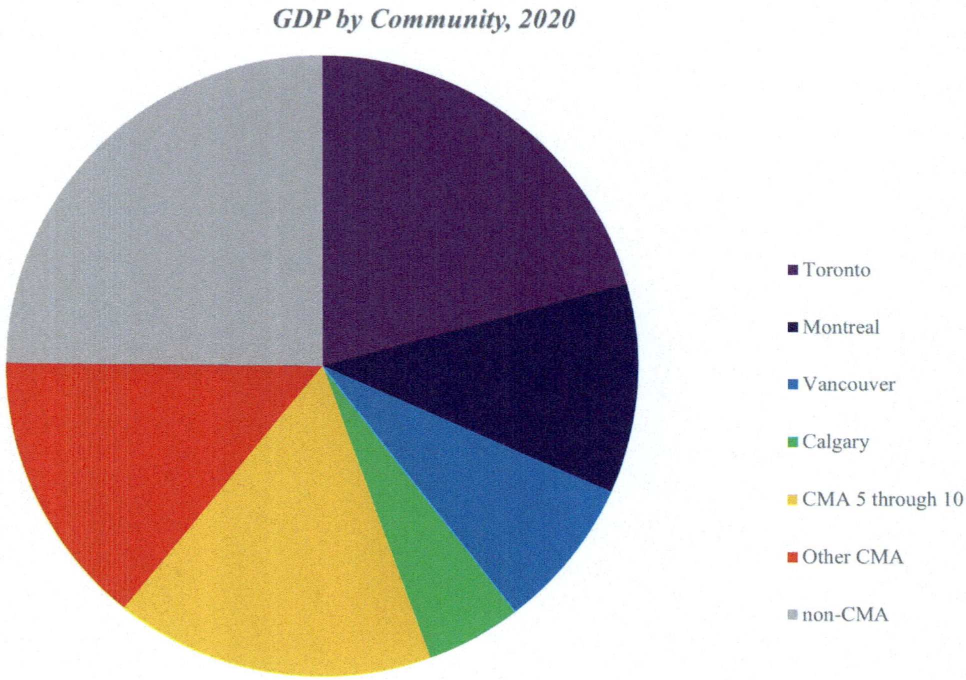

Fig. 12.7 GDP percentages by community type, 2020. Data from Statistics Canada (2023b)

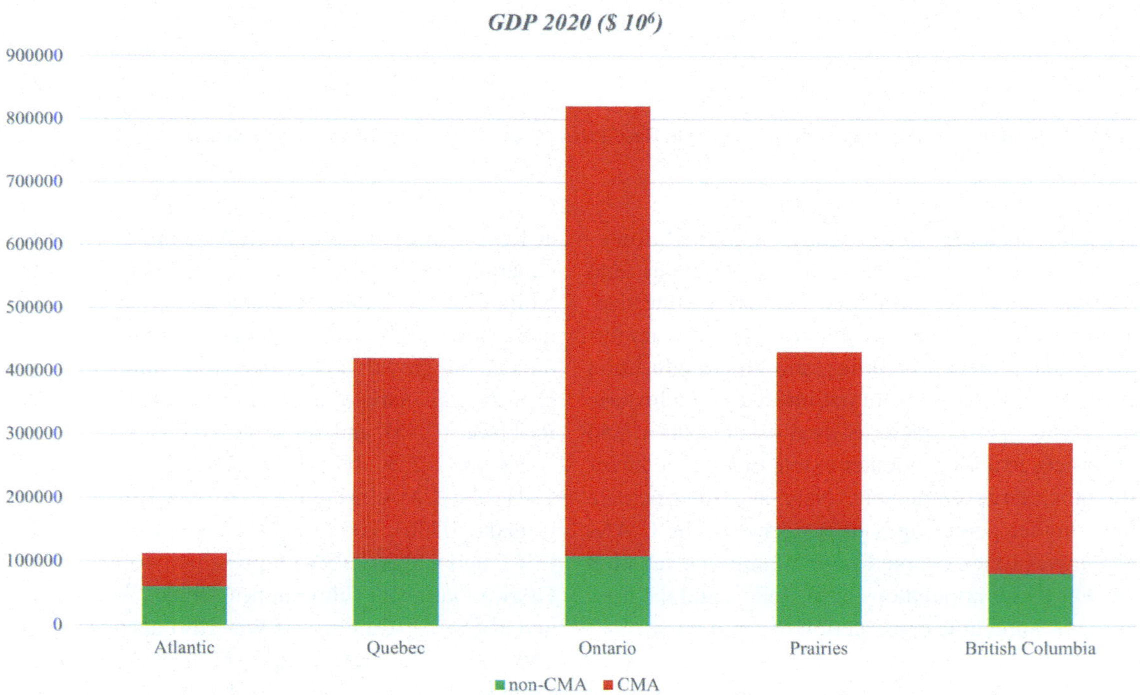

Fig. 12.8 Share of GDP (2020, $\$10^6$ Canadian) by community type. Data from Statistics Canada (2023b)

practitioners are located in CMAs. The primary industries of agriculture, mining and petroleum extraction, forestry, and fisheries represent less than 3% of total employment.

Comparison of numbers of those employed in economic sectors among the CMA reveals some patterns. In the manufacturing and utilities sector, fluctuating in employment numbers reflect economic conditions, along with the effects of COVID-19 restrictions in 2019–20 (Fig. 12.10). As health is a provincial responsibility, the timing and severity of restrictions varied among the CMA. The nature of manufacturing also played a role, with the electronic and information technology industries in Kitchener-Cambridge-

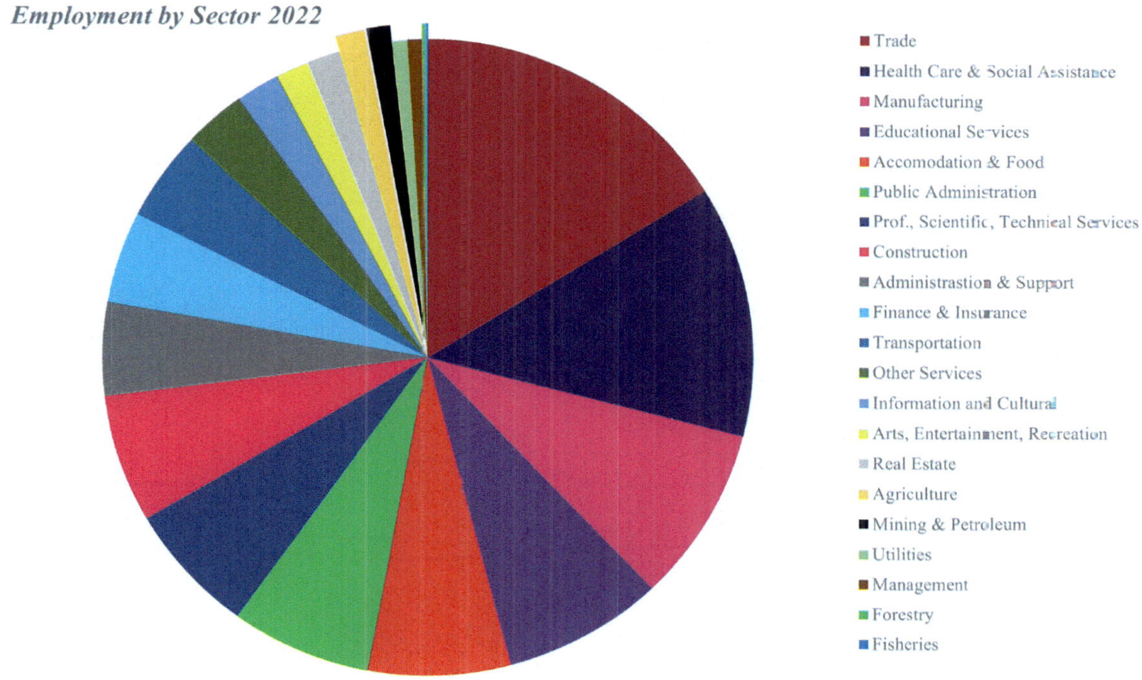

Fig. 12.9 Employment by sector in Canada, 2022. Statistics Canada

Table 12.2 GDP ($\$10^6$ Canadian) for the 10 largest CMA by population, 2021. Statistics Canada (2023b)

CMA	GDP $\$10^6$
Toronto ON	430,935
Montreal QC	228,707
Vancouver BC	163,772
Calgary AB	102,663
Ottawa-Gatineau ON-QC	89,914
Edmonton AB	87,464
Québec QC	47,935
Winnipeg MB	45,005
Hamilton ON	37,037
Kitchener-Cambridge-Waterloo ON	32,987
Total	1,266,419

Waterloo (2022) less impacted than the more general industrial bases in Toronto and Montreal. The data highlight the relative importance of the high-technology manufacturing sector in Kitchener-Cambridge Waterloo (total 2021 population 575,847) compared to Calgary (1,481,806), Vancouver (2,642,825) and Ottawa-Gatineau (1,488,307, not shown in Fig. 12.10).

Toronto CMA, with the largest Stock Exchange (TSX) and head offices of the "big five" largest banks, has traditionally been considered as the business and financial centre of Canada (Dinnie 2011). Financial operations are centred on Bay Street in the downtown (Fig. 12.11). The Royal Bank of Canada (RBC), initially founded in Halifax, and the Bank of Montreal (BMO) relocated their head offices from Montreal to Toronto in 1976 and 1977, respectively. The Bank of Nova Scotia (Scotiabank) established its headquarters in Toronto in 1900, joining Toronto-Dominion (TD Canada Trust) and Canadian Imperial Bank of Commerce (CIBC). Relocation of several company headquarters from Montreal after the election of the Parti Québecois government in 1976 (see Levine 1991) strengthened the economic dominance of Toronto. Economically, Montreal recovered after 2000, with a focus on digital and biotechnology industries. Calgary (petroleum and natural gas corporate offices), Vancouver (technology), and Ottawa-Gatineau (public service and technology) also have significant business, financial, and administrative employment.

Natural and applied sciences-related employment (within professional, scientific, and technical services) forms significant components of the workforce in the largest CMA (Statistics Canada 2021a; Fig. 12.12). Growth since 2011 is particularly evident in Toronto, Montreal, and Vancouver, with little downturn in 2019–20. In all the largest CMA, employment in this sector exceeds that in manufacturing and utilities, reflecting the ongoing transition from industrial to information and high-technology enterprises.

Arts, culture, recreation, and sport form significant components of the employment picture, although downturns resulting from COIVD-19 are evident in Toronto, Vancouver, and Ottawa-Gatineau (Statistics Canada 2021a; Fig. 12.13). Alberta imposed lesser restrictions and for a shorter period.

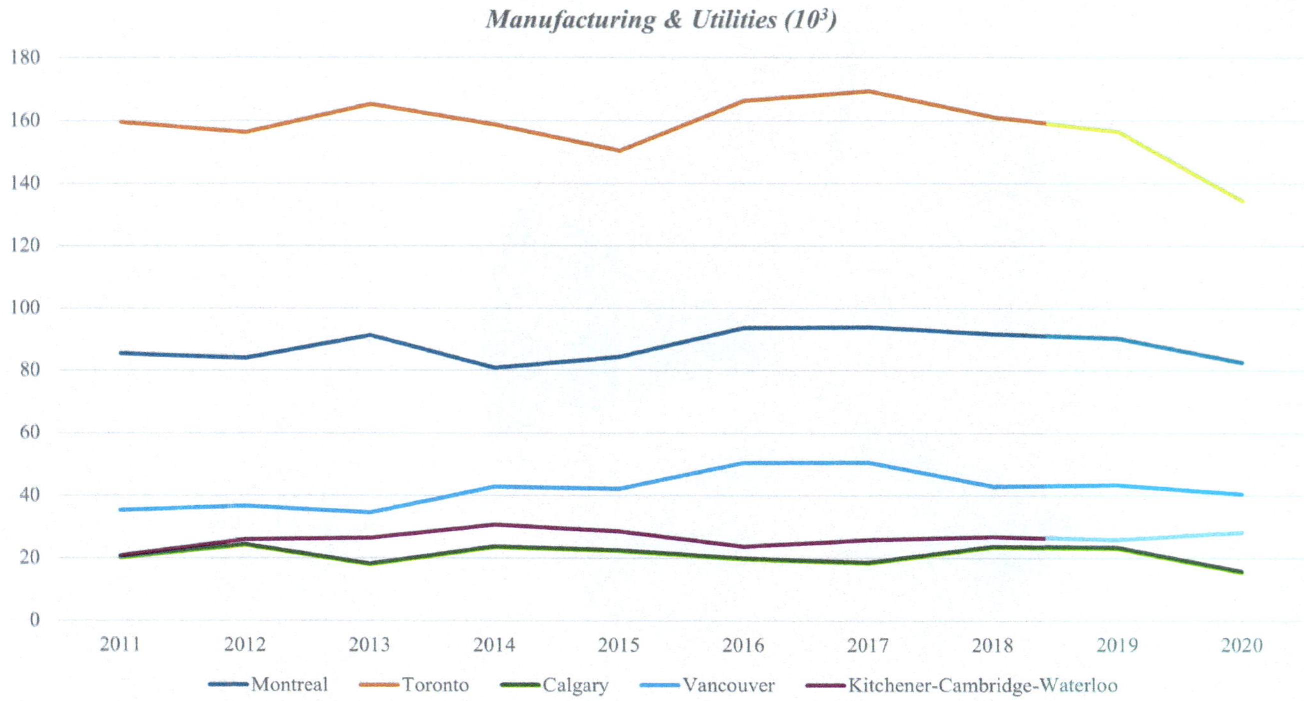

Fig. 12.10 Employment in manufacturing and utilities (10^3). Data from Statistics Canada (2021a)

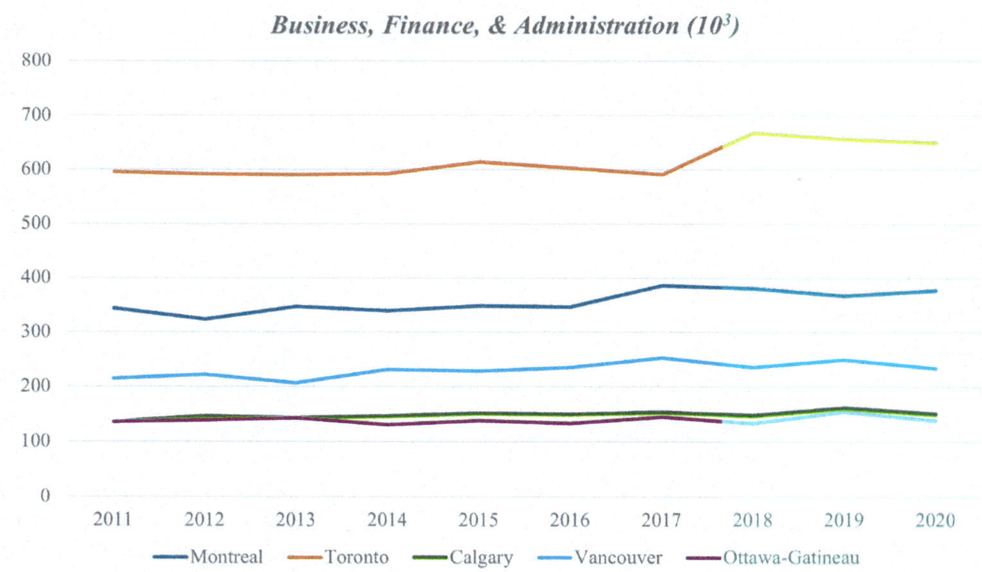

Fig. 12.11 Employment in business, finance, and administration in largest CMA (10^3). Data from Statistics Canada (2021a)

In general, the largest CMA have greater proportions of residents employed in financial and administrative services, professional and technical services, and arts, culture, recreation, and sport than do the other CMA. Health care and social assistance and public administration form smaller relative sectors in the ten largest CMA (with the exception of Ottawa-Gatineau).

The five largest CMA all saw overall increases in the numbers of people employed from 2011 through 2019 (Statistics Canada 2021a; Fig. 12.14). However, with the exception of Toronto, the rate of growth in overall population in each has exceeded the rate of employment growth (Fig. 12.15). The discrepancies are most marked in Ottawa-Gatineau (12.4% increase in population compared with 5.5% increase in employment) and Calgary (29.4% increase in population compared with 13.2% increase in employment). Some restrictions on public service hiring in Ottawa-Gatineau, and the economic downtown in the petroleum sector in Calgary, account for some of these differences.For the CMA as a group, employment decreased in four CMA over 2011–2020

Fig. 12.12 Employment in natural and applied sciences in largest CMA (10^3). Data from Statistics Canada (2021a)

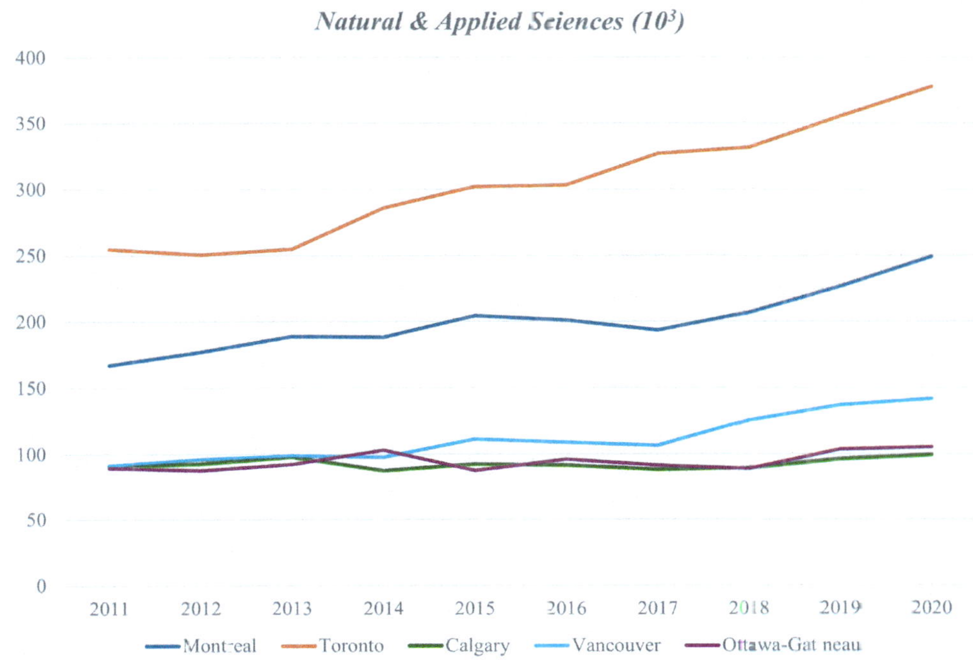

Fig. 12.13 Employment in arts, culture, recreation, and sport in largest CMA (10^3). Data from Statistics Canada (2021a)

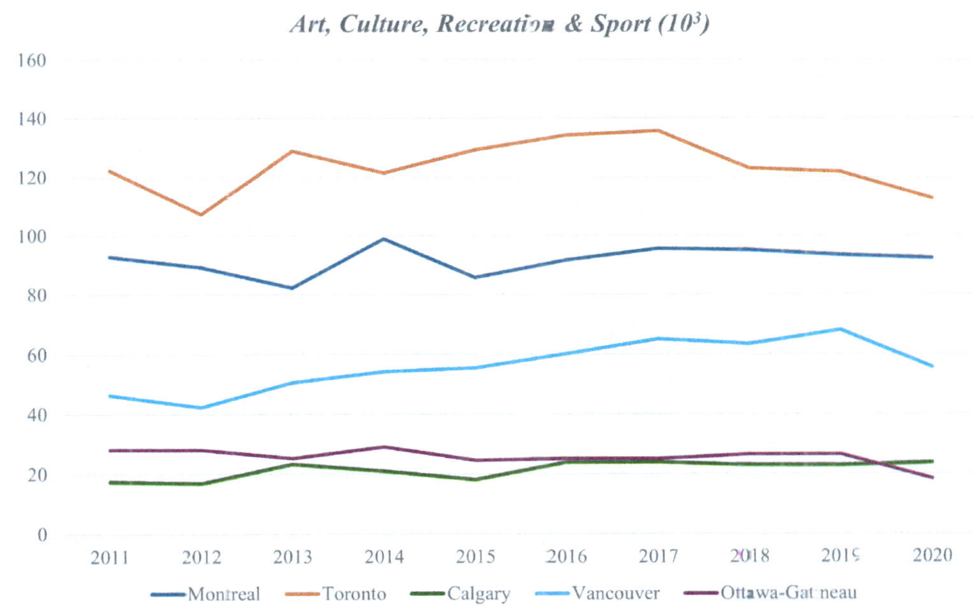

(Fig. 12.15). The discrepancy between increasing population and decreasing employment was most evident in Peterborough (closure of General Electric and other industrial facilities), Greater Sudbury (downturn in mining), St. Catharines-Niagara (closure of General Motors and other industrial facilities), and Saint John (downturn in heavy industry). In addition to Calgary, population growth outpaced employment growth in Edmonton (similar to Calgary), and Kelowna (with increasing percentage of older and retired residents). Ten CMA, including Saskatoon, showed employment change rates equal or greater than population growth rates. The relationship between employment and population growth rates does not directly indicate unemployment numbers, as other demographic factors also are involved. Declines in employment during 2019–2020, reflecting COVID-19 restrictions, have affected the overall rates. Data currently (April 2024) being processed by Statistics Canada may demonstrate the longer-term implications.

12.3 Census Agglomerations

In 2021, Statistics Canada (2021b, 2022a) recognized 141 Census Agglomerations (CA), with total and core populations

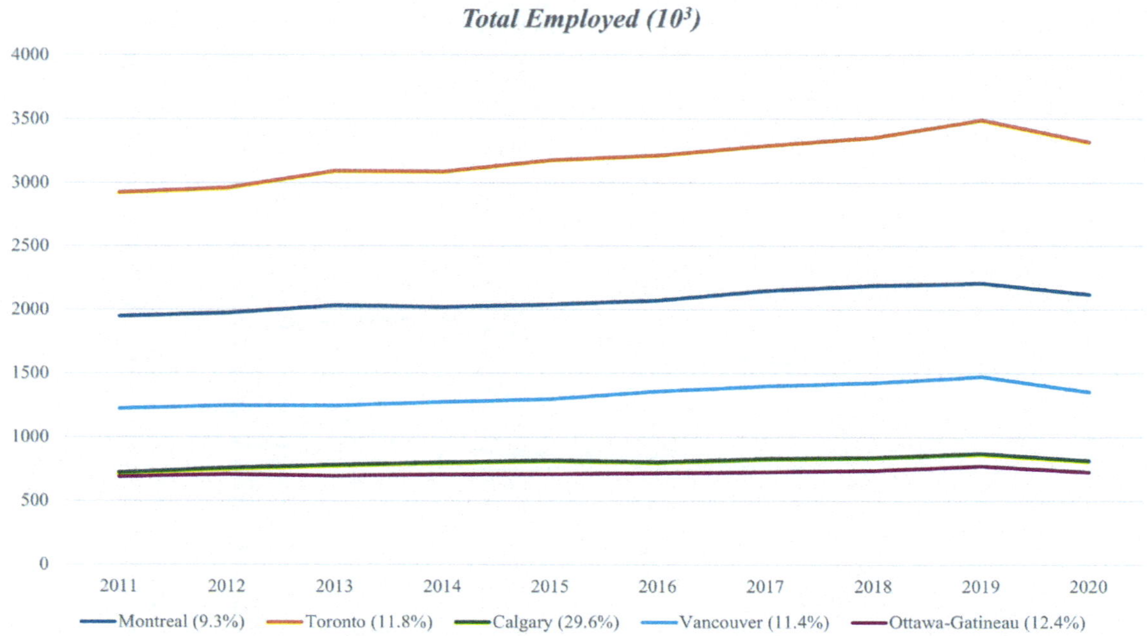

Fig. 12.14 Total employment in largest CMA (10^3). Data from Statistics Canada (2021a)

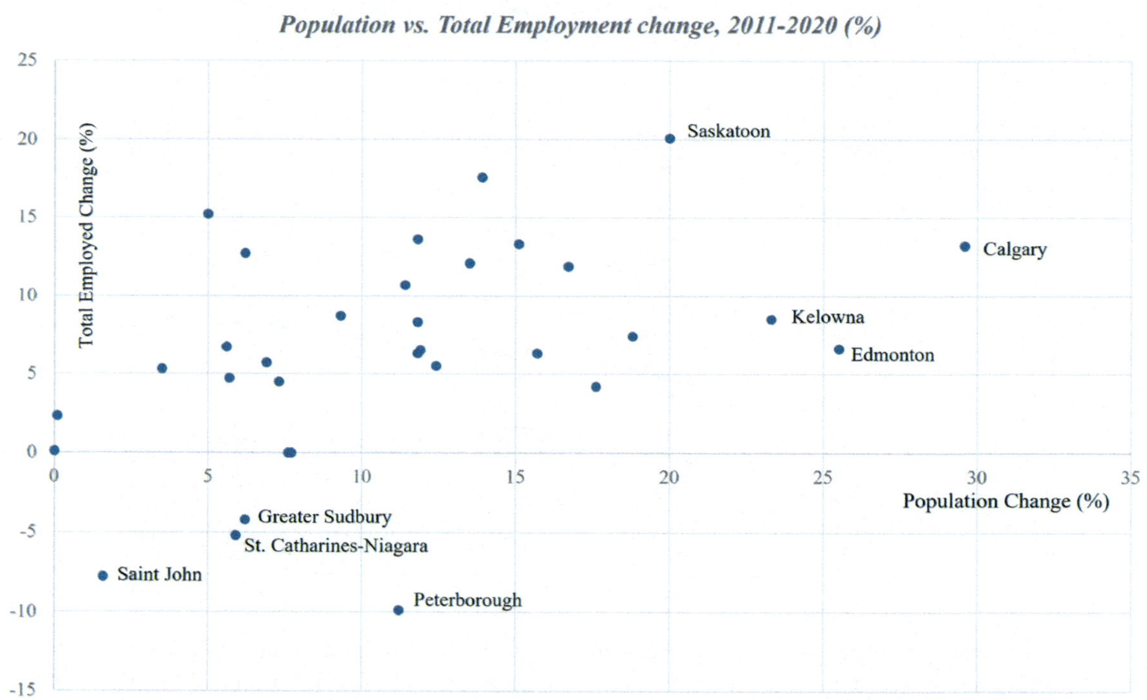

Fig. 12.15 CMA Population (2021) compared with percentage change in employment, 2011–2020. See text for discussion. Data from Statistics Canada (2021a, 2022a), Census of Canada

of at least 10,000. The CA of Charlottetown PEI (78,858, the 48th largest in Canada in 2021), Whitehorse YK (31,913), and Yellowknife NWT (20,340) are the largest urban centres in their respective jurisdictions.

Employment in Charlottetown is dominated by the governmental, healthcare, and education sectors (IRCC 2019; Statistics Canada 2021a). Chemical manufacture, biotechnology, machining, and agricultural-related business are also present.

Of the 141 CA, 26 saw population growth of 7.0% or greater during 2016–2021 (Table 12.3). Several benefitted from a combination of tourism and recreational attractions and relative proximity to CMA, providing advantages for

Table 12.3 Census Agglomerations (CA) showing the largest increase in population, 2016–2021. Statistics Canada (2022a), Census of Canada

CA	2021 Population	% Increase 2016-2021	Factors
Squamish BC	24,232	21.8	Tourism, residential
Wasaga Beach ON	24,862	20.3	Tourism, residential
Tillsonburg ON	18,615	17.3	Industry, residential
Canmore AB	15,990	14.3	Tourism, residential
Collingwood ON	24,811	13.9	Tourism, residential
Woodstock ON	46,705	13.6	Industry
Whitehorse YK	31,913	13.1	Capital
Cowansville QC	15,234	11.6	Residential
Steinbach MB	17,806	11.1	Local population growth
Sainte-Agathe-des-Monts QC	19,892	10.5	Tourism, residential
Centre Wellington ON	31,093	10.3	Residential
Salmon Arm BC	19,705	10.1	Tourism, residential
Charlottetown PEI	78,858	9.8	Capital
Lachute QC	14,100	9.6	Residential
Vernon BC	67,086	9.4	Residential, tourism
Courtenay BC	63,282	9.2	Residential, tourism
Essa ON	22,970	9.0	Residential
Penticton BC	47,380	8.8	Residential, tourism
Midland ON	27,894	8.0	Residential, tourism
Sylvan Lake AB	16,514	7.9	Residential
Summerside PEI	18,157	7.9	Residential
Winkler MB	32,655	7.8	Local population growth
Campbell River BC	40,704	7.5	Tourism, residential
Parksville BC	31,054	7.4	Residential, tourism
Ingersoll ON	13,693	7.3	Industry
Orillia ON	33,411	7.2	Residential
Duncan BC	47,582	7.0	Residential, tourism

commuting or work-from-home residents. Squamish, adjacent to the Whistler ski resort and connected to Vancouver CMA by the Sea-to-Sky Expressway, saw the most rapid growth among CA. Other examples include the southern Georgian Bay (Wasaga Beach, Collingwood, Midland), the Okanagan Valley (Vernon, Penticton), Vancouver Island (Courtenay, Campbell River, Parksville, Duncan), Canmore (80 km west of Calgary), Sainte-Agathe-des-Monts (80 km northwest of Montreal), and Salmon Arm CA (transitioning from a forestry-based economy). Residential growth associated with an adjacent CMA is evident in Cowansville (connected to Montréal by Autoroute 10), Lachute (Autoroute 50), Centre Wellington (adjacent to Guelph and Kitchener-Cambridge-Waterloo CMA), Essa (nestled between Toronto and Barrie CMA), Orillia (directly north of Barrie CMA), and Sylvan Lake (connected to Edmonton CMA by the Yellowhead Highway #16). Summerside, the second-largest urban centre in PEI, owes its population growth to increased immigration and inter-provincial migration

The more traditional factors of industrial activity and local population growth are not as apparent. The southwestern Ontario CA of Woodstock (Toyota assembly plant opened 2015), Tillsonburg, and Ingersoll all have significant automobile-related establishments. The Manitoba communities of Steinbach and Winkler are dominated by German-Mennonite communities, with relatively high birth rates and low median ages. The demographics of Steinbach and Winkler differ substantially from the other CAs with high growth rates (Vineberg 2010).

Of the 141 CA, 26 saw declines in population between 2016 and 2021 (Statistics Canada 2022a; Table 12.4). In all but two cases (Sainte-Marie and Alma), the declines can be directly related to downtowns in mining, petroleum extraction, pulp-and-paper and forestry, and steel manufacturing industries. Prince Rupert serves as exporting port for several of these commodities. "One-industry" towns are always susceptible to downtowns in natural resource-based operations. Downturns in petroleum extraction (see Chap. 17) and pulp-and-paper manufacture (see Chap. 18) are related to broader international factors. While the downturn for the pulp-and-paper industry has been endemic for some time, leading to steady population decreases after contraction or closure, the petroleum communities are marked by a boom-and-bust phenomenon, with rapid increases in population before 2016 followed by sharp decreases. In contrast to the pulp-and-paper and mining communities, the oil-and-gas CA are relatively adjacent to CMA (Calgary, Edmonton, Regina, Saskatoon).

Table 12.4 Census Agglomerations (CA) showing decreases in population, 2016–2021. Statistics Canada (2022a), Census of Canada

Census aggregation	Change 2016–2021 (%)	Principal industry
Campbellton NB	−9.2	Mining
Estevan SK	−6.0	Petroleum
Dawson Creek BC	−5.4	Petroleum
Thompson MB	−4.7	Mining
Corner Brook NL	−3.9	Pulp and paper
Baie-Comeau QC	−3.8	Pulp and paper
North Battleford SK	−3.4	Petroleum
Sainte-Marie QC	−3.2	–
Sept-Iles QC	−2.8	Mining
Dolbeau-Mistassini QC	−2.8	Pulp and paper
Grand Falls-Windsor NL	−2.2	Pulp and paper
Matane QC	−2.2	Pulp and paper
Sault Ste Marie ON	−1.8	Steel
Timmins ON	−1.5	Mining
Alma QC	−1.4	–
Kenora ON	−0.85	Mining, forestry
Lloydminster AB-SK	−0.62	Petroleum
Moose Jaw SK	−0.52	Petroleum
Wetaskiwin AB	−0.48	Petroleum
Cape Breton NS	−0.41	Steel, mining
New Glasgow NS	−0.26	Mining
Portage la Prairie MB	−0.26	Petroleum
Medicine Hat AB	−0.19	Petroleum
Prince Rupert BC	−0.15	Transportation
Quesnel BC	−0.14	Pulp and paper
Rouyn-Noranda QC	−0.05	Mining

Mining communities are particularly susceptible to partial or total abandonment. In Canada, the long list includes Wabana NL (peak population 14,000; 2021 population 1,815), Tilt Cove NL (1,370; 4), Gagnon QC (4,000; 0), Cobalt ON (10,000; 1,000), Flin Flon MB (10,500; 4,940), Uranium City SK (2,500; 91), Pine Point NWT (1,900; 0), Keno City YK (2,000, 20), Cassiar BC (1,500; 0), and Kitsault BC (1,200; 1). Elliot Lake ON, a rare exception (25,000; 11,372) managed to transition from a uranium mining to a retirement community (Elliot Lake Standard 2024), with a population growth rate of 5.9% between 2016 and 2021.

Cape Breton CA, the largest by population in 2021, incorporates the communities of Sydney, North Sydney, Glace Bay, and others. Industrial Cape Breton developed around the subsea coal mines and the steel mills of Dominion Coal and Steel (DOSCO; later DOFASCO). Iron was obtained from the company-owned properties in Wabana. DOSCO attempted to operate as a fully integrated concern. As the major markets for finished steel products were in the industrial heartland of the Great Lakes, only semi-finished steel was produced in Sydney: the semi-finished products were shipped to Hamilton, Ontario, for final processing. The company operated in a somewhat paternal fashion, a practice which ended when the Nova Scotian operations were sold to Hawker-Siddeley of Great Britain in 1957. Shortly thereafter, the first coal mine closures began, followed by closure of the Wabana iron mines (1966) and the Sydney mills (October 1967).

The Development Corporation of Cape Breton (DEVCO), a federal government crown corporation, took over the mills temporarily, and in December 1967 transferred them to Sydney Steel Corporation (SYSCO). As SYSCO fell further behind in the international steel competition, the need for centralization and specialization was felt. SYSCO responded in 1983 by shutting down all of its operations, except for the mill producing rails. When the federal government terminated subsidies in 1998, the last mill closed. Cape Breton was left with a greatly diminished economic base, along with substantial cleanup costs (see MacKinnon 2016, 2020).

12.4 Provincial and Territorial Capitals

Provincial capitals in most instances were established in the province's primate city, with little dispute. Halifax, St. John's, Toronto, and Winnipeg were the largest commercial and population centres in their respective provinces. Charlottetown and Fredericton were established as capitals, the latter designated in part to counter the economic influence of Saint John. Québec and Victoria were the commercial and population centres in their respective provinces when designated, and remained their status partly due to political pressure. Regina was established as the territorial capital of the Northwest Territories in 1883, due to the political influence of Lieutenant-Governor Edgar Dewdney and the commercial influence of the Canadian Pacific Railway. There were no other major settlements along the CPR line at the time, and Battleford (the capital 1877–1883) had been bypassed. When Saskatchewan became a province in 1905, Regina remained as the capital despite some minor complaint from Saskatoon.

Only in Alberta was there controversy over the capital location. The major contenders were Calgary (along the CPR, supported by the Conservative politician RB Bennett) and Edmonton (along the Grand Trunk Pacific and Canadian Northern railways, supported by the Liberal politician Frank Oliver and the Lieutenant-Governor GHV Bulyea). Banff, Medicine Hat, Red Deer, and other centres also raised claims (Kilpatrick 1980). The federal Liberal government of Wilfred Laurier, supportive of the Grand Trunk Pacific, backed Edmonton, which was chosen in 1906. The Calgary-Edmonton rivalry, in all aspects, remains as the strongest intercity competition in Canada.

Territorial capitals were designated by the federal government. After Yukon was separated from NWT in 1898, Dawson City (1898 population ~ 17,000 (NWMP 1898), 1901 population 9,142 (Statistics Canada 2022d) was designated as the capital. The decline in gold production was matched by a decline in population, with only 4,157 people registered throughout Yukon in the 1921 Census (although the Indigenous population was probably underestimated). Construction of the Alaska Highway through southern Yukon led to the transfer of the capital to Whitehorse in 1953.

After Alberta and Saskatchewan became provinces in 1905, Ottawa was designated as the official capital of NWT. Fort Smith served as the administrative capital from 1911 until 1967. After two years of hearings by the Carrothers Commission (1967), Yellowknife was selected as the new capital, based on its access to road transportation, its commercial significance as a gold-mining community, and the expressed preference of NWT residents. Yellowknife was the first capital chosen in part in response to residents' input, rather than as a result of political manoeuvring (as were Regina and Edmonton). Fort Smith continued to house much of the territorial civil service until the 1993 construction of the Legislative Building in Yellowknife.

Prior to establishment of Nunavut as a separate territory, a non-binding plebiscite of residents chose Iqaluit (60.23%) over Rankin Inlet (Kangiqtiniq) as the desired capital (Nunatsiaq News 1995). Iqaluit became the capital when Nunavut was proclaimed on 1 Apil 1999.

12.5 Universities

Government-sponsored universities were initially located in capital cities, beginning with Université Laval (1663, the oldest institution of higher education in Canada, Québec) and continuing with the University of New Brunswick (1785), Dalhousie University (1818, Halifax), University of Toronto (1827), University of Manitoba (1877), University of Alberta (1906), Memorial University of Newfoundland (1925), and University of Prince Edward Island (1969). The University of Saskatchewan (1907) and University of British Columbia (1908) were established in Saskatoon and Vancouver, respectively. Other universities were initially established by religious and secular groups, beginning with King's College (initially Windsor NS 1789, subsequently Halifax). Branch campuses of the provincial universities also were established in metropolitan areas, many of which subsequently became independent institutions (e.g. Calgary, Regina, and Université de Montreal (from Laval)). In 2024, there were more than 80 publicly funded universities in Canada.

12.6 Commuting

For both environmental and human health, reductions in commuter automobile traffic would be beneficial. Automobile use represents about three-quarters of total commuter traffic in Canada (Figure; Statistics Canada 2017, 2022c; Figs. 12 16,

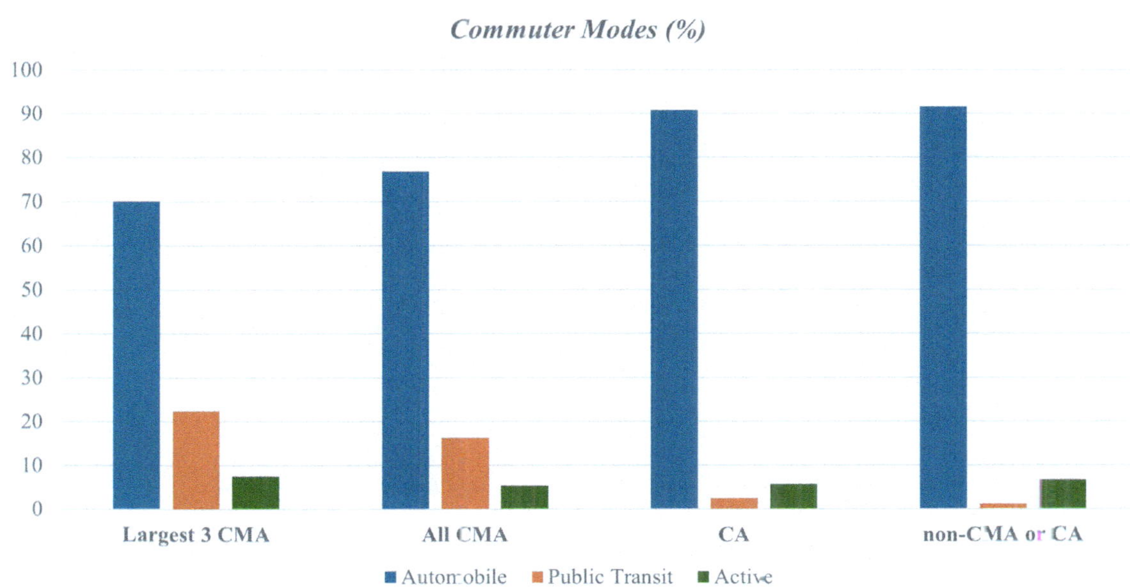

Fig. 12.16 Preferred modes of commuter travel. See text for discussion Data from Statistics Canada (2017)

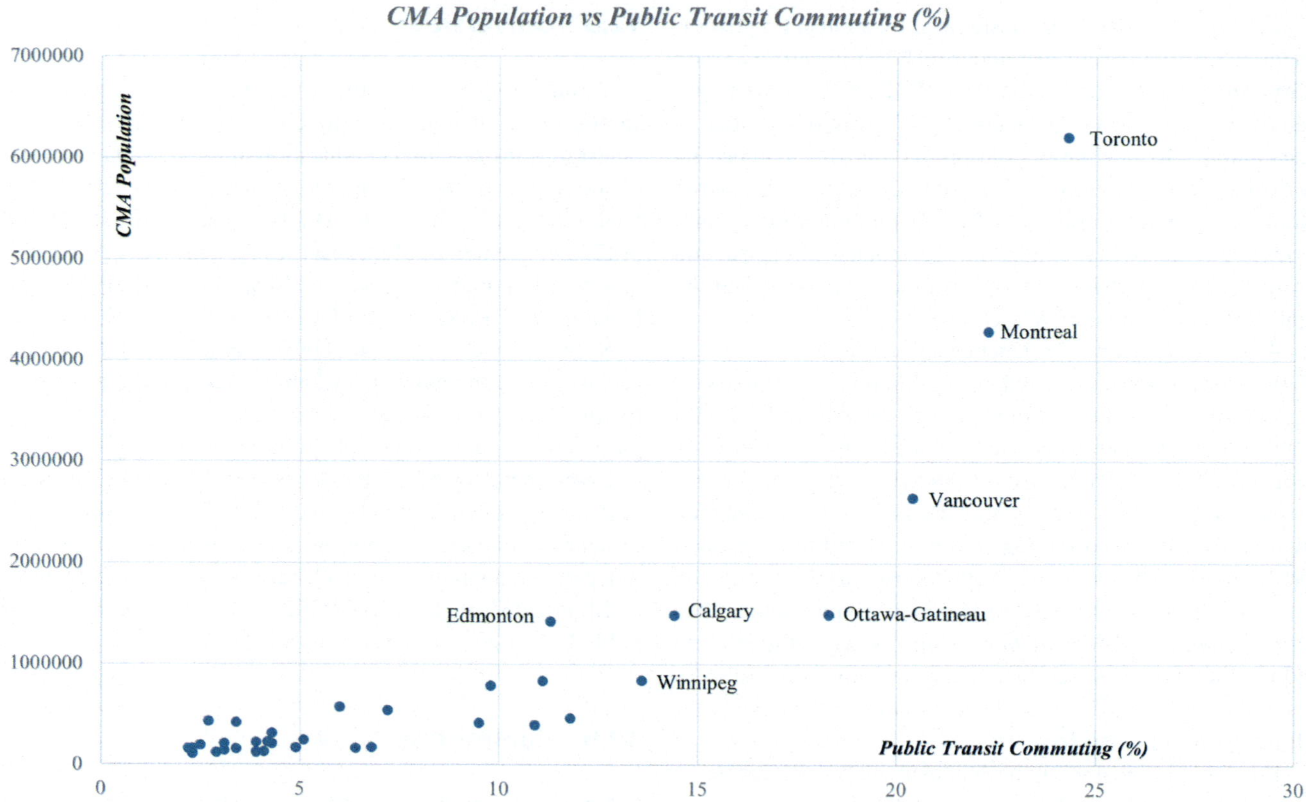

Fig. 12.17 CMA Population (2021) compared with percentage of Commuters using Public Transit (2017). Points for the 6 largest CMA are indicated Data from Statistics Canada (2017, 2022a)

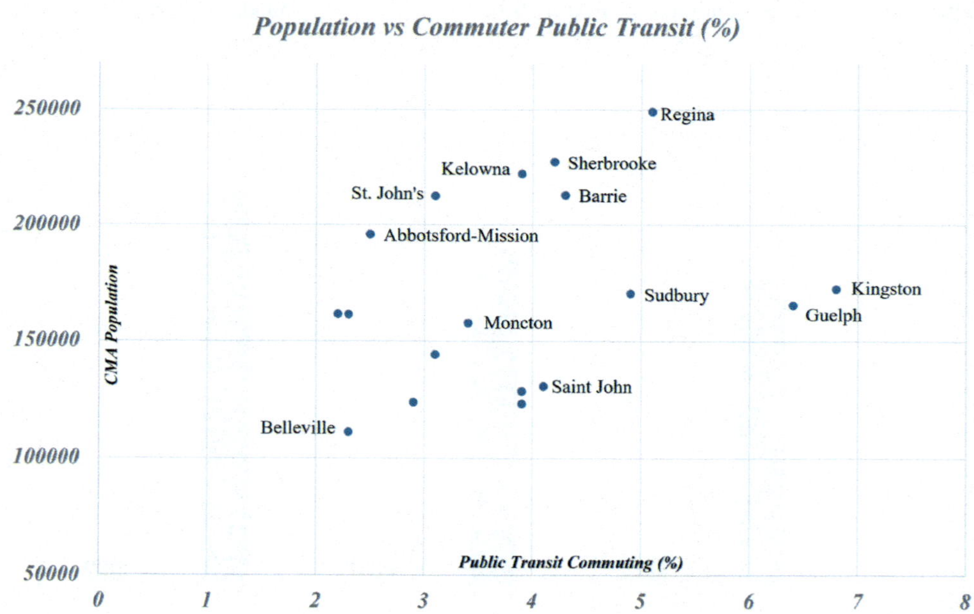

Fig. 12.18 Population for CMA with 100,000–250,000 residents (2021) compared with percentage of Commuters using Public Transit (2017). Points for the selected CMA are indicated. See text for discussion. Data from Statistics Canada (2017, 2022a)

12.17 and 12.18). CA and smaller CMA have higher rates of automobile commuting, due to the limited public transit in those communities. Areas outside CA generally have no or very limited public transport options, particularly where mid- and long-distance intercity bus service has been discontinued (see Chap. 19). Possession or access to an automobile (car-sharing, carpooling, taxi) is almost a necessity in the smaller urban centres, posing significant obstacles to people who cannot or choose not to drive. In the CMA, carpooling, ride-sharing, and taxi use accounted for about 10% of commutes

(Statistics Canada 2017). The active modes of walking and cycling as commuting choices vary little among community size, with slightly higher rates in the largest CMA attributed to proximity of home and workplace. The largest CMA have rail rapid transit systems and dedicated bus lanes and routes (discussed further below).

For all modes combined, daily commuting times average about 30 min in each direction (Allen et al. 2022; Statistics Canada 2023c). Toronto commuters in 2022 spent almost two hours in transit (both ways), with delays in both automobile and public transit increasing, compared to about 90 min in Montreal and 100 min in Vancouver. About 1 in 6 Canadians spend 45 min or longer commuting each way on workdays, in contrast to the almost 90% who would prefer a commute of 15 min or less (IWG 2021).

COVID-19 restrictions and health concerns led to reductions in commuting in 2020–2021, with an estimated 2.5–3 million fewer commuters (Savage and Turcotte 2020). Most of the decreases resulted from the reduced use of public transit, especially in the largest CMA. Expressed preference for personal vehicle use rose from 42 to 70% during COVID-19 (Traffic Injury Research Foundation 2020), while those who preferred public transit fell from 41 to 4.4%. In 2023, about 20% of the population was working from home (decreased from 40% during 2020; Statistics Canada 2024a), and public transit ridership generally is about 75–80% of 2016–18 levels (APTA 2024; Statistics Canada 2024b). Overall commuting numbers have exceeded pre-pandemic values.

12.6.1 Private Automobiles

Typical per capita energy expenditure from private vehicle transport in CMA (see Newman and Kenworthy 2021) varies from 30,000 MJ y^{-1} (Montreal) to about 60,000 MJ y^{-1} (smaller low-density CMA with limited public transit). Gasoline consumption per capita varies from about 3.2 l d^{-1} (Montreal and Toronto) to about 4.5 l d^{-1} (smaller CMA in Atlantic and Prairie Canada; Canada Energy Regulator 2024). Reduction in vehicle use in urban areas resulted in a 4–5% decrease in GHG emissions in 2020 (USEPA 2024). Electric (not including hybrid) vehicles accounted for 2.3 of 1000 new passenger automobile and light truck registrations in 2021 (Statistics Canada 2022b, c), with the highest rates among CMA in Vancouver (6.5/1000) and Sherbrooke (5.2/1000). Together, hybrid and zero emission vehicles represent about 10% of the new vehicles registered in the three largest CMA, and about 3% of passenger vehicles on urban roads.

Declines in NO_x and particulate emissions reflected decreased diesel use. Transportation accounts for more than 50% of Canada's 1.7 Mt NO_x emissions (ECCC 2020).

Ozone (O_3) emissions remained stable or increased in many areas, including the southern Great Lakes. Reduced NO_x causes a reduction in photochemical O_3 production in less-polluted areas (Bozem et al. 2017). In urban centres with initially high NO_x and volatile organic compound concentrations, increased near-surface O_3 results as titration by NO decreases (e.g. Collivignarelli et al. 2020; Shi and Brasseur 2020; Siciliano et al. 2020; Venter et al. 2020). O_3 concentrations exceeding 40 ppb are common at several urban stations in southwest Ontario, with concentrations of 35–40 ppb in urban centres in the Great Lakes–St. Lawrence Lowland and those adjacent to areas of petroleum refining (Health Canada 2021). Weekly variations in near-surface O_3 levels resulted from decreased urban traffic during weekends in southern Ontario (Leung et al. 2021).

O_3 at ground level is a significant pollutant and health hazard. Infection of airways and lungs can result in asthma, emphysema, and chronic bronchitis, particularly affecting elderly people, children, and those with existing medical conditions. Health Canada (2016, 2021) estimated that acute O_3 exposure contributed to about 2,800 deaths per year, with chronic exposure responsible for an additional 1,300 deaths. O_3 concentrations > 62 ppb are considered hazardous to human health, and Health Canada (2016) recommends that residential concentrations over 8 h exposure should not exceed 20 ppb.

Factors influencing automobile use include population density, availability of public transit, fuel price, and roadway congestion (Choi 2018; Saeidizand et al. 2022). Increasing commuting expenses, particularly gasoline, have some influence on travel behaviour (Young et al. 2016). Suburban residents tend to consider changing vehicle types more readily than adopting public transit (Papaioannou et al. 2020). Suburban residents are more likely to more to another suburb than to the urban fringe or core. Personal demographics including age, income, gender and political and environmental attitudes also play important roles (Axisa et al. 2012). In Montreal CMA males with high income and full-time work consume more gasoline per capita (Manaugh et al. 2010). Vancouver is notable for a high rate of vehicle ownership and relatively inexpensive gasoline among global urban centres (Saeidizand et al. 2022), although taxes make gasoline in Vancouver more expensive than any other CMA.

Committing modes have been shown to be related to ethnicity, impacting both immigrants and those born in Canada (Dilmaghani 2022). Black millennial commuters in particular are less likely to use private automobiles for commuting. "Extreme commuting" rates (one-way time in excess of 60–90 min, regardless of mode) were estimated at 11.5% for low-income households, 13.5% for recent immigrants, 13.4% among non-white Canadians, 14.7% for Latin

American Canadians and 18.6% for Black Canadians prior to the COVID-19 pandemic (Allen et al. 2022). Housing and employment discrimination appear to be primary factors. In suburban areas, however, home ownership appears to be the dominant factor (Moos and Mendez 2015).

Freeway construction in urban areas has concentrated on bypass routes, rather than routes into or through downtowns. Freeways constructed into downtowns are present in only a few CMA (Toronto, Montreal, Ottawa-Gatineau, Québec, Hamilton, Windsor, St. Catharines-Niagara, Saint John). In Toronto (Gardiner Expressway) and Ottawa (Queensway, Highway 417), the routes followed former railway corridors. Other downtown freeways connect bridges and service marine and lake ports. Successful opposition to Vancouver's downtown freeway (1967; Places that Matter 2016) and Toronto's proposed Spadina Expressway (1971; Read the Plaque 2024) marked the practical end to downtown freeway construction: today, the latter route is occupied by a subway line.

Freeways within many of the smaller CMA are bypass routes or sections of longer highways designed to connect larger centres (e.g. Moncton, Trois-Rivières, Kingston, Red Deer, Kamloops). Ring or loop roads carry traffic around the cores of Vancouver, Calgary, Edmonton, Winnipeg, Kitchener-Waterloo, Montréal, and Halifax, among other CMA. Freeways initially built as bypasses around the core (e.g. highway 1 in Vancouver, highway 401 in Toronto, Boulevard métropolitain (Autoroute 40) in Montréal) are now surrounded by dense urban development. The government of Ontario has constructed a new bypass Highway 407 around Toronto, which has served as a locus of suburban development.

Toronto CMA has more kilometres of metropolitan and core freeway, and more lane width per person than comparable USA cities (Townsend and Ellis-Young 2018). Although Canadian CMA in general have higher population densities and lower metropolitan core freeway kilometres than USA cities, the differences are mostly related to size and growth history than to differences in urban planning and design. The 18-lane Highway 401 through Toronto is currently the widest freeway in North America.

Transportation is a provincial responsibility. Federal financial contributions to road projects in municipalities vary to a maximum of 40% of costs (compared to 50% for public transit projects). The remainder comes from provincial sources. Toll roads and bridges are relatively uncommon in Canada (Highway 407 is an exception). Some projects are funded exclusively by provinces, particularly Ontario and Alberta. The 13 February 2024 announcement that the Federal Government will not fund future major urban road projects (Montreal Gazette 2024) represents a substantial departure from previous practice, and has drawn sharp reaction from other levels of government.

12.6.2 Rapid Rail Transit

Canada's six largest CMA have rail rapid transit systems (Table 12.5). Construction of the Tramway de Québec, scheduled to begin in early 2024, was abruptly paused by political decision in November 2023.

12.6.2.1 Subways

The Toronto Transit Commission's (TTC) subway system consists of the U-shaped north-south-north Yonge-University-Spadina line; the west-east Bloor-Danforth line; and the west-east Sheppard line. Daily weekday ridership has varied from more than 1.4 million in 2019 to approximately 350,000 during the height of the COVID-19 restrictions, and in early 2024 was estimated at 1.2 million (TTC 2023; Munro 2023; APTA 2024, Statistics Canada 2024b). Train frequency during rush hours varies from 3–5 min. Accessibility for the physically challenged remains an issue at about 20% of the stations, but is scheduled to be completely addressed by 2025. Rolling stock is manufactured at Bombardier's Thunder Bay plant, and the tracks have a unique "Toronto gauge" (1495 mm) which dates to the horse-drawn urban railways of the 1860s (Stamp 1989). Much of the funding comes from the Government of Ontario (Metrolinx), and as a result the government has considerable say in all aspects of the operation.

The Métro de Montréal's rubber-tyred subway system consists of the east–west Green and Blue lines, the north–south Yellow Line tunnelling under the St. Lawrence to Longueil, and the north–south-north U-shaped Orange Line. Weekday ridership is currently over 1 million, approximately 80% of the pre-pandemic level (APTA 2024). During rush hours, the system (especially the Green Line) is extremely crowded. Compared to the CMA population of 4.3 million (2021), Montréal has the highest frequency of transit use per capita of CMA, second only to New York City in North America. Train frequency during rush hours varies from 2–5 min. In 2024, only 27 of the 73 stations are fully accessible, and full conversion is not scheduled to be completed until 2038. Rolling stock is manufactured by Bombardier, and the tracks have the North American Standard 1435 mm gauge. Trains are not air-conditioned. About a quarter of direct funding comes from the Government of Québec.

Globally, subways generally do not result in enhanced urban population growth (Gonzalez-Navarro and Turner 2018). The Toronto and Montréal CMA and urban cores show general, but not accelerated, population growth since construction of the various segments of subway. Increases in subway extent result in lesser increases in ridership. Increased decentralization, noted in some cities (Gonzalez-Navarro and Turner 2018) did not occur as a result of subway construction in Canada, as the subways were confined to the urban core and not linked to extensive park-and-ride systems.

Table 12.5 Rapid transit rail and busway systems for CMA. Ridership estimates are daily weekday means for Q4 2023, based on figures from APTA (2024) and websites of transit authorities. See text for discussion

Community	System name	Technology	Initial opening	Ridership (000 s)	Length (km)	Length u.c.(km)
Toronto (GTA)	GO transit	Commuter rail	CNR/CPR	171	526.1	–
	Subway	Subway	1954	1200	70.5	37.8
		Light rail	1985*	–	–	
		Light metro	u.c.**	–	–	
	Streetcar	Streetcar	1892	300	83	–
	Viva (York RM)	Busway	2005	9	38	–
	Mississauga transitway	Busway	2014	~30	18	–
Montréal CMA	Exo	Commuter rail	CNR/CPR	30	204.5	–
	Métro	Rubber-tyre métro	1966	1046	69.2	6
	Réseau express métropolitain	Light métro	2023	190pr	16.5	50.4
Edmonton	Edmonton LRT	Light rail	1978	85	37.4	14
Calgary	C-Train	Light rail	1981	264	59.9	20
	MAX	Busway	2018	8	56	–
Vancouver CMA	West coast express	Commuter rail	1995	5.6	69	–
	Skytrain	Light metro	1985	446	79.6	5.7
Ottawa-Gatineau CMA	Ottawa transitway	Busway	1983	220	58.9	–
	O-Train	Light rail	2001	75	12.5	48.5
	Gatineau rapidbus	Busway	2013	7	12	–
Winnipeg	Winnipeg RT	Busway	2012	17	11.2	–
Kitchener-Cambridge-Waterloo	ION	Light rail	2019	25	19	–
Hamilton	B-Line	Light rail	u.c	–	–	14
Québec	Tramway de Québec	Light rail	n.d	–	–	19pr

Estimated daily weekday mean for Q4, 2023. pr = projected by Exo.
*Scarborough Line 1985–2003 ** u.c. = under construction.

12.6.2.2 Streetcars

Toronto has the only streetcar system in Canada. The TTC lines have survived many attempts at elimination since 1892 and are regarded as iconic by many residents (Stamp 1989; Cal 2007). Streetcars operate on 9 lines in the downtown, encompassing 83 km, with the east–west lines along King Street and Queen Street the most used. Weekday ridership has recovered since the COVID-19 restrictions and was at 300,000 in late 2023 (Munro 2023; TTC 2023). The cars operate on the Toronto Gauge (1495 mm), and most share the streets with automobile traffic. Rolling stock is manufactured by Bombardier and is electrically powered.

12.6.2.3 Commuter Rail

Commuter rail systems serve Toronto (GO (Government of Ontario; Metrolinx) Transit), Montréal (Exo), and Vancouver (West Coast Express). For the most part, these systems utilize existing routes established by CNR and CPR for regular rail traffic. Metrolinx currently owns about 80% of its 526 km used trackage, giving it control over scheduling and maintenance. GO lines currently extend from Toronto's downtown Union Station to Oshawa, Barrie, Kitchener-Waterloo (via Guelph), St. Catharines-Niagara Falls (via Hamilton), Markham, Milton, and Stouffville, covering much of the GTA and the "Golden Horseshoe". Proposals exist to extend service to Bolton (northwest) and Bowmanville and Peterborough (east), currently served by GO buses to Oshawa. Metrolinx is currently undertaking an extensive programme of electrification, track enhancement, and station modification. The goal is to institute 15–30 min service at peak hours to all stations in the system by 2025. The trains use the North American Standard 1435 mm gauge. Approximately, weekday ridership for all routes was 171,000 in late 2023.

GO expansion intended to increase ridership on the regional rail system. Service quantity, population density, fuel price, and unemployment rate were significantly associated with commuter rail ridership (Shantz et al. 2022). Employment density and seasonal variation also were significant.

Exo operates 5 lines (and additional bus routes) to Mont-St-Hilarie (E), Candiac (SE), Vaudreuil (W), St-Jérôme (N), and Mascouche (NE), using CNR and CPR routes. Unlike GO, Exo is completely integrated with Montreal's Métro and other public transit facilities. Weekday ridership declined drastically from an estimated 77,000 in 2017–2018 to about 30,000 in late 2023 (APTA 2024), attributed to a combination of COVID-19 restriction effects and replacement of the most travelled Deux Montagnes line with Réseau express métropolitain service (discussed below). The trains use the North American Standard 1435 mm gauge.

The West Coast Express connects Abbotsford-Mission and Vancouver CMA with peak hour diesel-electric trains using standard-gauge track along the CPR route. Weekday ridership dropped by 70% during the COVID-19 restrictions, and in late 2023 sat at 5,600 (APTA 2024). Travel time of 75 min from Mission to downtown Vancouver is generally faster than driving, but the West Coast Express has not proven popular with Abbotsford-Mission CMA commuters.

12.6.2.4 Light Rail Transit

Light rail transit (LRT) has become the rail rapid transit system of choice in Canada. LRT systems exist in Edmonton, Calgary, Ottawa, and Kitchener-Cambridge-Waterloo, and are under construction in Hamilton and Toronto.

Edmonton LRT has three lines totalling 37.4 km (with another 14 km under construction). The lines are electrified, standard gauge, and primarily at grade. Tunnel sections under the University of Alberta (through Cretaceous bedrock) and through downtown (glacigenic sediment, posing significant geotechnical challenges; Ellwood et al. 2011), and two dedicated bridges across the North Saskatchewan River are the most significant engineering elements. Weekday ridership was approximately 85,000 in late 2023 (APTA 2024), with the university generating much of the traffic. All stations are accessible.

Calgary LRT was constructed after Edmonton's, but posed no significant engineering challenges. The system was designed by Oliver Bowen, a Black engineer descended from two of Calgary's first Black settlers (Spotlight Productions 2017). The two lines are electrified, standard gauge, and primarily at grade with short elevated and tunnelled sections. Despite initial dire predictions of failure in petroleum-based Calgary, the LRT has proven the most successful of Canadian services, with 264,000 weekday riders in late 2023 (APTA 2024). Much of the 14.4% public transit share of Calgary commuting is attributable to the C-Train. The decline from the peak ridership in 2015–16 is largely attributed to reduced employment in downtown Calgary. Approximately, half of the commuters working in downtown Calgary used the C-Train.

LRT development in Calgary increases property values for adjacent apartments, but not for other housing styles (Choi et al. 2022). Property value increases are linked to pedestrian accessibility and welcoming. In comparison, commuter rail development appears to have a more general effect on residential property values (e.g. around Montreal, Dubé et al. 2013).

In contrast, Ottawa's O-Train has undergone a series of challenges. Initial operation of the north–south Trillium Line, using CPR standard-gauge track, began in 2001 as a pilot project. Rather than being the product of the route planning practised in Calgary, the line west of the downtown was commissioned because the tracks were there: aside from Carleton University, the potential for increased ridership was low. After years of political controversy, an east–west Confederation Line was constructed to link the downtown with eastern Ottawa and the Federal Government complex at Tunney's Pasture, opening in 2019. Construction of the downtown tunnelled segment along Queen and Rideau Streets encountered significant geotechnical challenges, including a major collapse in June 2016. The Ottawa Light Rail Transit Commission (2022) public inquiry report detailed significant problems with construction, maintenance, and operation of the Confederation Line. Additional difficulties have occurred since the initial opening. Meanwhile, the Trillium Line closed for repairs in 2020, and is scheduled for reopening late in 2024. The system carried 12,500 passengers per weekday in late 2023.

Kitchener-Waterloo's ION system opened in 2019. It currently carries about 25,000 riders per weekday, using standard-gauge CNR tracks. The system was politically controversial before, during, and after construction. Although considered to be a success by the municipal government, ION created inequalities in transport mobility. Olsson and Thomas (2024) argued that ION did not improve travel times or access to transit for existing users, and that construction had negative effects on neighbourhoods. This study considered ION to be an example of political promotion of new infrastructure, termed "transit boosterism". In April 2024, the long-term consequences and utility of the ION LRT remain to be determined.

Toronto's experience with LRT development is politically chequered. The Scarborough LRT opened in 1985. Although it was commonly used, a combination of equipment failures and a desire to upgrade the service led to the line's long-anticipated closure in 2023. Suggested replacements in 2010 included express bus service, an enhanced LRT, or extension of the Bloor-Danforth subway line. Although the TTC and Toronto City Council favoured LRT development, Ontario Premier Rob Ford preferred subways, and intervened in the Scarborough, Sheppard, and Eglinton line discussions to express his viewpoint. As urban transit funding largely came from the Government of Ontario, Ford's views had effect. Following several controversies and governmental administrations, the result was the extension of the Bloor-Danforth subway, construction of the Sheppard Avenue subway (discussed above), and a commitment to construct an

LRT line with at-grade and tunnelled sections along Eglinton Avenue, with completion scheduled for late 2024.

12.6.2.5 Light Metro

Light metro systems have capacity intermediate between LRT and subways. Canada's first light metro system was Skytrain in Vancouver CMA (1985). The 3 lines include 79.6 km (5.7 km under construction) of elevated and tunnelled sections. Estimated weekday ridership was 446,000 in late 2023 (APTA 2024). The major engineering feature is the Skybridge across the Fraser River. Lines include the original NW–SE route connecting downtown, New Westminster, and Surrey; a south-north route from the airport to downtown; and a W-E route connecting downtown to Burnaby, Simon Fraser University, and Coquitlam. Peak hour service has trains every 2–6 min. TransLink ridership in the Vancouver CMA is approximately one-third on SkyTrain, 64% by bus, about 1% on the SeaBus ferry across Burrard Inlet from downtown to North Vancouver, and about 0.4% on the West Coast Express.

Réseau express Métropolitain, a light metro line connecting Brossard with downtown Montreal, opened in 2023. Extensions to Deux Montagnes (along the former Exo line), Sainte-Anne-de-Bellevue, and the international airport, are planned for late 2024–2027. Projected weekday ridership is 190,000.

The light metro Ontario Line is currently under construction in Toronto. A north–south diagonal line designed to connect the Eglinton LRT, downtown Toronto, and the Lake Ontario waterfront, it is to be built as a combination elevated, at-grade, and tunnelled standard-gauge line. The 2019 announcement of routing, construction design, and operation under Metrolinx were substantially influenced by Ontario Premier Doug Ford, brother of the late Rob Ford.

12.6.3 Buses

Dedicated busway services are provided in York Regional Municipality, Mississauga, Calgary, Ottawa-Gatineau, and Winnipeg. The Ottawa Transitway is the oldest (1983) and most used (220,000 riders per weekday in late 2023, APTA 2024). The Transitway system extends west to Kanata and Stittsville, south to Barrhaven and Leitrim, east to Orleans ad Blackburn Hamlet, and connects with Gatineau north of the Ottawa River.

In Winnipeg, the new BRT has proven popular. Use of the BRT depends more upon the individual stop locations rather than the system as a whole (Baker and Linovski 2022). The greatest influence on ridership is felt on the connecting routes.

Bus ridership declined or remained stable in most CMA between 2010 and 2019 (e.g. Boisjoly et al. 2018; Diab et al. 2020, 2021). Vehicle revenue kilometres and automobile ownership are the main factors affecting bus use. While some studies suggest that increased ride-sourcing services and cycle use are associated with higher levels of transit ridership (Boisjoly et al. 2018), others show mixed results, with increases in larger CMA and decreases in smaller ones (Diab et al. 2020). Frequently, ridership also depends on the service characteristics of the individual routes, rather than the overall system (Diab et al. 2021; Baker and Linovski 2022).

Reduction of GHG emissions is a primary motivation behind increasing bus usage. Adoption of electric buses in Canada, both for public transit and school use, has been limited by concerns with cold-weather range and reliability, initial adoption costs, and budgetary constraints under decreasing or stagnant ridership (Mohamed et al. 2019). The majority of emissions result from on-board operation. Use of LNG propulsion systems could reduce emissions by ~8% (Pourahmadiyan et al. 2021), but adoption costs remain an obstacle, along with local shortages in LNG supply. On-request and small feeder bus systems have been instituted in some smaller CMA (e.g. Bürstlein et al. 2021; Sanaullah et al. 2021).

COVID-19 restrictions and commuters' health concerns resulted in large drops in public transit uses in 2020. Recovery since 2021 has been variable, typically ranging between 75–80% of pre-pandemic ridership (APTA 2024; Statistics Canada 2024b). In the largest CMA, differences in recovery by mode are generally minor. In Ottawa, C-Train ridership in late 2023 was at or below 50% of pre-pandemic level, compared with 70% for bus usage. Service reliability and interruption issues, along with the higher level of working-form-home and hybrid work schedules among the federal public service, can account for the difference. Ottawa's public transit mix differs from the other large CMA in its emphasis on dedicated express busways and limited success with rail (Fig. 12.19).

Bus rider demographics have been subject to study, notably in smaller CMA where rail systems are not financially feasible. Millennials are generally less likely to have a valid driver's license, and those with licenses tend to drive fewer kilometres than do other urban residents. Hence, millennials are more likely users of public transit, a pattern which has persisted as they age, despite changes in occupation and family status (Newbold and Scott 2018). Single-person households also are less likely to drive, and more likely to use public transit or active modes (Young and Lachapelle 2017). Although their overall trip frequency does not differ from the urban population in general, they do undertake more entertainment and social-related trips. Higher levels of environmental consciousness and concern also results in more willingness to use public transit and accept delays and minor inconveniences (Eldeeb and Mohamed 2022).

Lower-income residents are generally assumed to be more likely to use public transit, essentially forming dependent populations. However, this simple picture is complicated in CMA where low-income residents may own or share vehi-

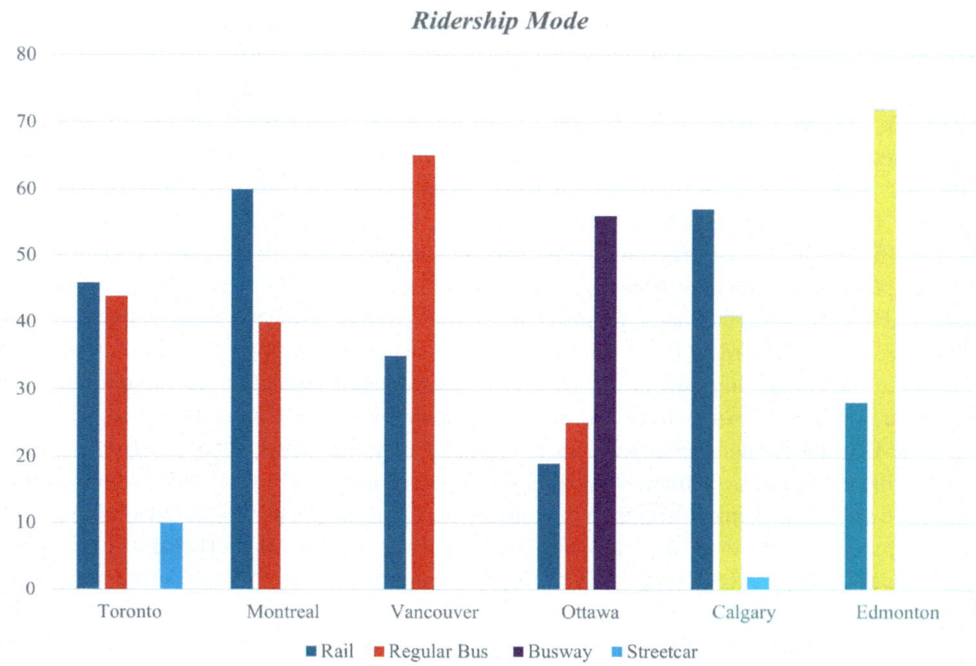

Fig. 12.19 Preferred public transit ridership modes for 6 largest CMA. Ridership estimates derived from daily weekday means for Q4 2023, based on figures from APTA (2024) and websites of transit authorities. Note that all modes are not available in any CMA. See text for discussion

cles. In the Greater Toronto-Hamilton region, low-income households with vehicles have the most elastic relationship between transit accessibility and use, switching to transit in response to improved service (Yousefzadeh Barrir et al. 2021). In situations with more limited public transit, as in most CA and smaller CMA usage declines as car ownership increases. Usage is linked to build environment attributes, transit service, and socio-economic aspects (Diab et al. 2020).

Transit ridership and shifting from automobile use varies significantly within the smaller CMA. In Kingston, a spatially small CMA with the dominant centrally located transit destination of Queen's University (more than 42,000 students, staff, and faculty, compared to the CMA population of 172,726), transit use doubled to 7.8% between 2013 and 2016 (Collins and Macfarlane 2018). Females, those with lower household income, living 5–10 km from the university campus, those without a campus parking permit, and those willing to spend at least 21 min en route were most likely to shift from automobile to bus.

In St. John's, a spatially large CMA with four significant suburbs with limited Metrobus bus service and multiple large employers, rider demographics and reasons for use differ. Although Metrobus spent considerable effort targeting Memorial University and college students, a pre-pandemic survey (Metrobus 2019) revealed that students only represented 20% of the ridership. Respondents' reasons for bus use focused on access to vehicles and ability to drive (Fig. 12.20). Environmental concerns were not cited by respondents.

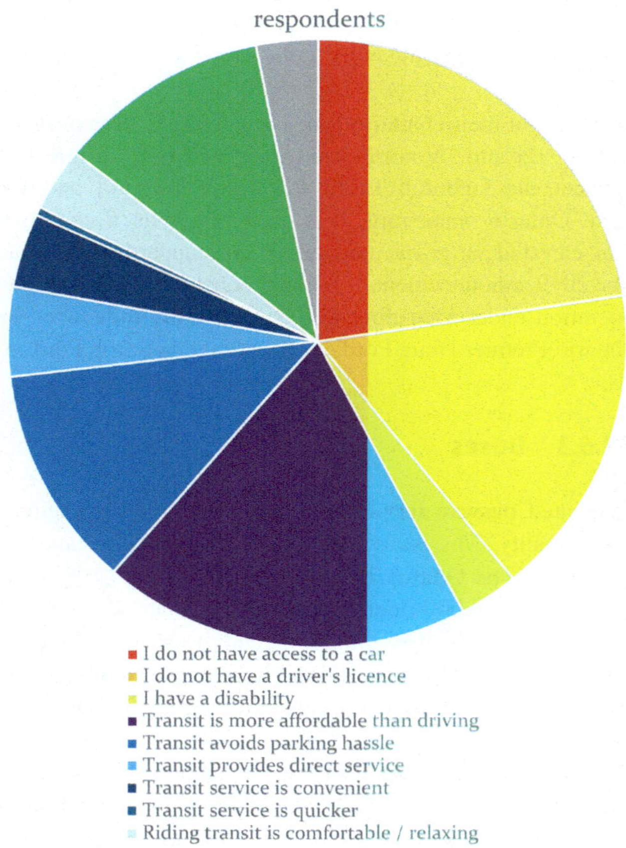

Fig. 12.20 Reasons cited by respondents for using Metrobus in St. John's, 2019 (Metrobus 2019)

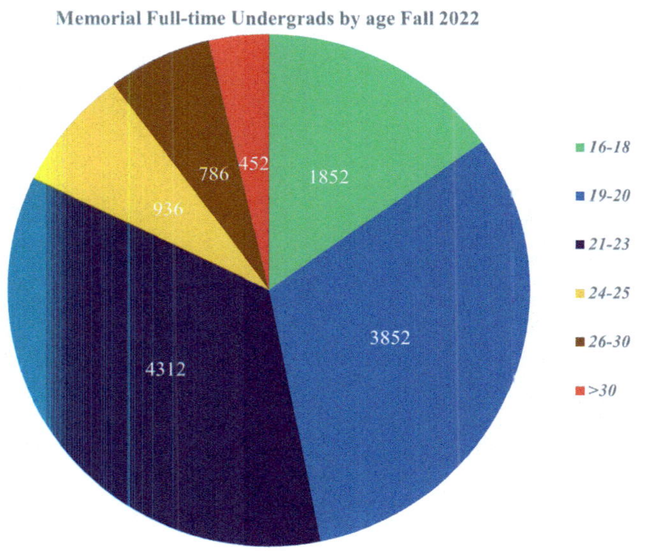

Fig. 12.21 Memorial University of NL Full-time undergraduates by age, Fall 2022. Age distributions should be considered in discussions of transit pass programmes. Data analysed by author from Memorial University records, Spring 2023

Programmes offering discounted bus passes in association with universities have had varying degrees of reception from students (Butler and Sweet 2020; Lachapelle et al. 2022). As the proposals generally involve charging all students the same fee, collected by the university, they favour students who can use transit and disadvantage those who live either more distally from the university or within walking distance, as well as those who already own or have access to vehicles for other reasons. The shift from market-based equity towards equity of opportunity has been embraced by some students, but rejected by others. Memorial University student's firm rejection of a universal bus pass proposal (CBC 2019) reflects the evolving demographics of student populations (Fig. 12.21).

12.6.4 Commuting Using Active Modes

Active transport, including walking and cycling, accounts for slightly more than 6% of all commutes in Canada (Fig. 12.16; Statistics Canada 2017). Victoria (16.9%), Kingston (9.5%), Vancouver, and Halifax had the highest rates in 2021. The lowest rates (less than 4%) characterize Oshawa and Abbotsford-Mission, two communities with substantial commuting to the adjacent large CMA. Active commuting percentages do not show a relationship with annual snowfall or other weather conditions (Fig. 12.22). There is somewhat of an inverse relationship between snowfall and cycling, although it is statistically weak (Fig. 12.23). Many CMA have challenging winter conditions, including snow clearing (see Chap. 13), making cycling largely a seasonal activity. Cyclists typically represent approximately 20% of active commuters, varying from 40% (Victoria) to 4% (Saint John, St. John's).

Although motivations for utilitarian cycling differ (e.g. Galway et al. 2023), a general desire among municipal planners to encourage cycling has led to study of motivations and policies. Cultural and demographic factors, particularly age, has led to increased utilitarian cycling (Assunçao-Denis and Tomalty 2019). Local measures, policies favouring cycling, and construction of cycling lanes and other infrastructure are generally effective. In some communities, the activities and advocacy of cycling groups have been very influential. Designation of cycling lanes on streets has led to opposition from automobile drivers and residential parkers in some cities, resulting in removal or scaling-back (). Most cycling paths also serve as pedestrian routes, and purely dedicated paths are rare and disconnected. Active transit options are generally favoured for schoolchildren, but more intrusive policy options have lesser support (Molner et al. 2023).

12.7 Heat as Urban Environmental Hazard

Heat is regarded as a significant hazard in Canada's urban centres (Health Canada 2011, 2012; Stewart et al. 2017). In human hazard terms, an extreme heat event (EHE) is a period of unusually high atmospheric-related heat stress that may have adverse health consequences for the affected population. Although a heat event represents an extended period of temperatures above the normal value, there is no generally accepted definition of for how long and how high above normal should qualify. Environment and Climate Change Canada currently defines an extreme heat event as involving three consecutive days with maximum temperature of at least 32 °C (see Health Canada 2011). Individual, community, or society responses to heat (extreme, excessive, or expected) strongly depend on past experience, preparation, and preferences.

Attributing economic damage and human health consequences to extreme heat events is difficult. Much damage and human health impacts relates from the combination of heat with other environmental stressors, notably wildland fire smoke and air pollution. Increases in levels of suspended particulate matter and air pollutants are correlated to increases in rates of myocardial infarction and pulmonary infection cases. Increased temperature and relative humidity, combined with air pollution, can increase the frequency of smog days in urban areas and cause serious health problems, both morbidity and mortality (Steib et al. 2000, 2023; Burnett et al. 2004, 2018; Crouse et al. 2012, 2017; Health Canada 2016, 2021).

Because heat has a greater physiological effect on elderly people and those in poor health, an argument can be made that those who unfortunately succumb during an extreme heat

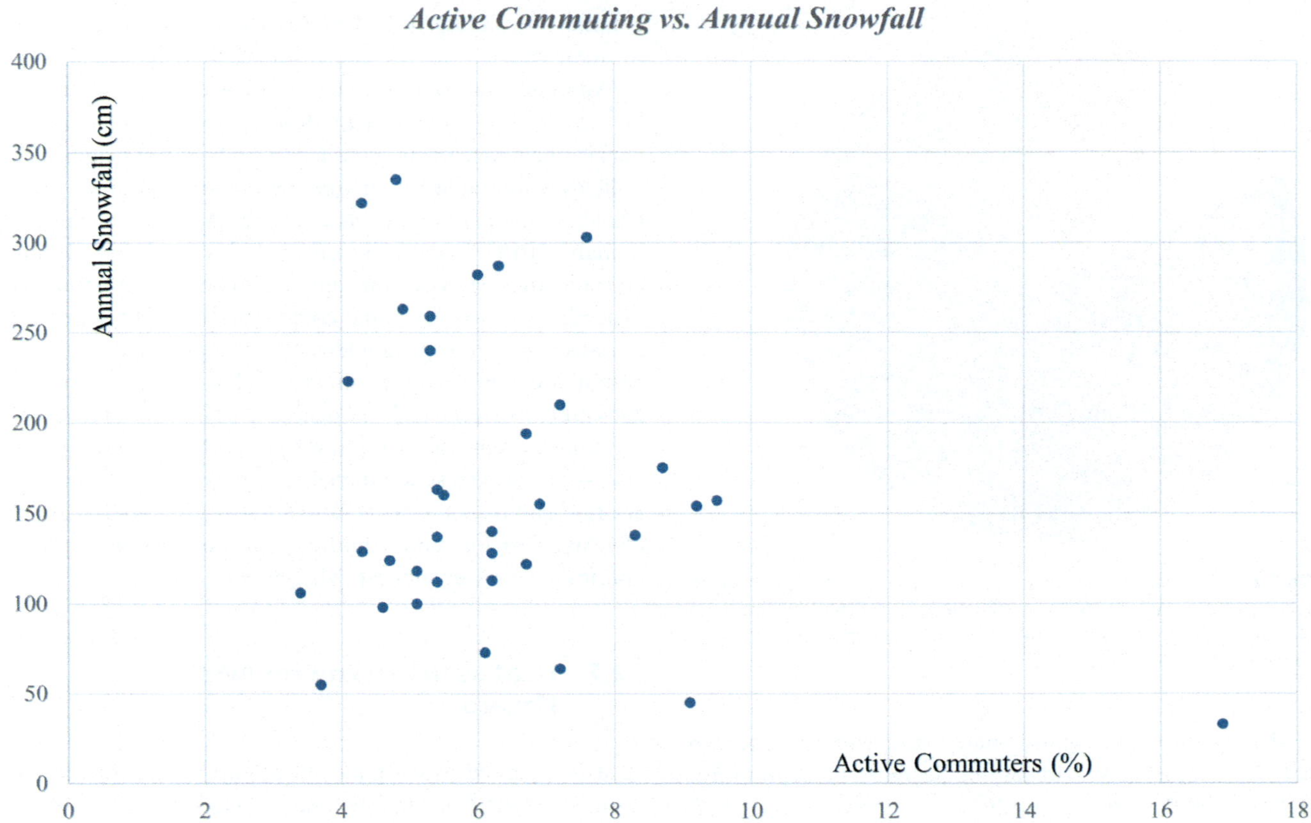

Fig. 12.22 Annual snowfall versus percentage of commuters using active transportation methods (walking and cycling). Snowfall from Canadian Climate Normals (Environment and Climate Change Canada; see Chaps. 1 and 13). Commuting totals from Statistics Canada (2017)

Fig. 12.23 Annual snowfall versus percentage of commuters cycling. Snowfall from Canadian Climate Normals (Environment and Climate Change Canada; see Chaps. 1 and 13). Commuting totals from Statistics Canada (2017)

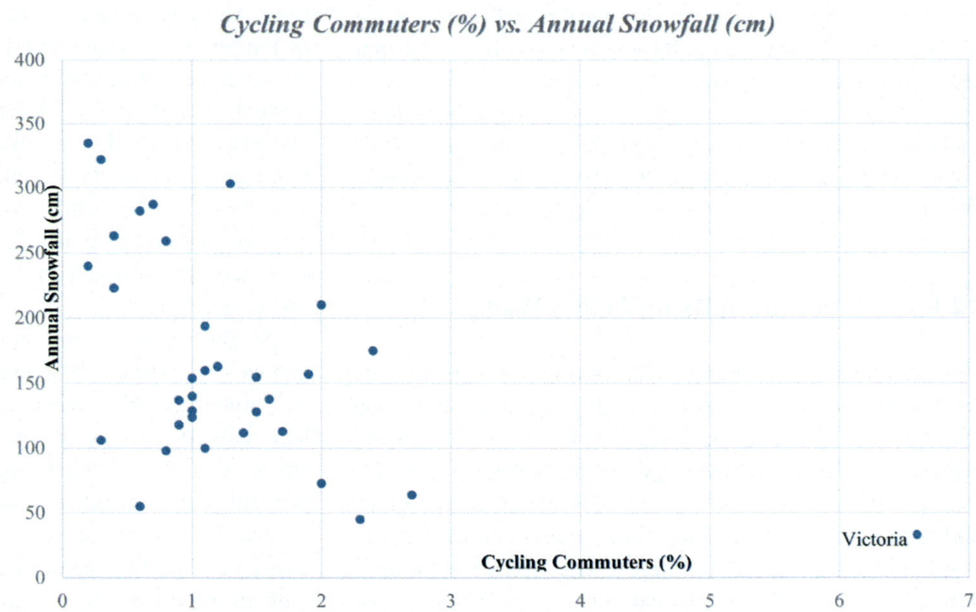

event would not have survived for a long subsequent period, unlike many victims of tsunami or tornadoes. "Heat" is often not listed as the cause of death by physicians, making subsequent estimates of mortality difficult. Use of different criteria by coroners and medical examiners can influence calculated mortality rates.

Although extreme heat events until recently have been of lesser concern in Canada, Toronto Public Health (City of Toronto 2022) currently estimates that extreme heat contributes to 120 deaths annually in Toronto. The 2018 EHE in Montreal may have contributed to 86 deaths (Lamothe et al. 2019; Lebel et al. 2019). The British Columbia Heat Dome of 2021 was responsible for 619 deaths (Government of Canada 2021; British Columbia Coroners' Service 2022). Increasing morbidity and lost productivity, combined with the stress on emergency response and medical personnel and the implications of power failures during hot summers, have heightened concern surrounding extreme heat events. Most of the preparation and studies have been conducted in the larger CMA, notably in Montreal, Toronto, Vancouver, and Winnipeg (Pengelly et al. 2007; Friedel 2011; Stewart et al. 2017; Price et al. 2018; Lamothe et al. 2019; Lebel et al. 2019; Health Canada 2021, City of Toronto 2022).

In addition to meteorological and climatological factors, land surface conditions and human land use are also significant. Changes in land use that reduce the amount of available water (e.g. by replacing forests with urban areas) result in decreased evaporation, lesser storage of energy as latent heat, and increased surface temperatures (Rinner and Hussain 2011; Stewart and Oke 2012; Phelan et al. 2015; Roman et al. 2018; Chen et al. 2020; Czekailo et al. 2020, 2021; Knight et al. 2021). Reductions in soil moisture have similar effects, whereas high soil moisture conditions largely prevent extreme heating. Replacement of soil by asphalt and concrete thus contributes to extreme heat events. Feedback relationships with suitable meteorological conditions and oscillations increase these effects.

The frequency and duration of extreme heat events, measured by temperature, have increased globally. In Canada, increased numbers of extreme heat events have been recorded since 1953, as measured by both daytime and nighttime Humidex values (Mekis et al. 2015; ECCC 2020). Nighttime increases are more common, as also noted in many locations. Projections based on climate modelling suggest that the number and intensity of extreme heat events will continue to increase in future.

12.7.1 Heat and Humans

Apparent or equivalent temperatures are intended to express what humans may feel physically (biometeorology) or psychologically. Analogous to wind chill, the equivalent temperature is not a measurement in degrees Celsius, but is intended to represent an equivalent "feeling" by people of an apparent temperature, allowing for differences in wind speed, humidity, and other factors.

The Humidex is a common measure of the combined effects of temperature and humidity (ECCC 2019). Designed initially for Canadian weather forecasting, the Humidex compares temperatures with relative humidity values to estimate equivalent temperatures. A measured temperature of 30 °C combined with a relative humidity of 50% would produce a Humidex value of 36. The same temperature with a relative humidity of 90% gives a Humidex of 45: Environment Canada advises people to avoid exertion under these conditions.

However, a Humidex value does not necessarily imply either suitable or dangerous conditions. The same equivalent temperature affects people differently, as can become evident when comparing the heat tolerances of two humans sharing the same classroom, car, or house. An equivalent temperature of 45 may pose health risks for some, but not for others. An equivalent temperature of 22 ("no discomfort") may be considered very uncomfortable by some.

Heat indexes are based on weather parameters recorded outdoors. Applying the results to humans suggests that they are also living outdoors, and that their skin is directly exposed to the environmental parameters—i.e., that they are not wearing clothing. Neither of these assumptions is valid in most cases.

For groups of people who spend time working outdoors, wearing similar clothing (e.g. police officers, postal delivery works), comparative heat stress indexes can be generated and used to assess employee health, such as the Relative Strain Index (Lee and Henschel 1966). Study of the results can lead to changes in uniform design to reduce heat effects. In addition, detailed studies can be used to determine which particular body areas (head, arms) show the greatest effects of extreme heat. This has resulted in the development of thermal comfort indices for specific groups, as well as the more generalized Universal Thermal Climate Index

As most urban-dwelling Canadians spend the majority of their time indoors, particularly children and unemployed seniors, assessment of thermal conditions within buildings is important to determine overall heat stress in a population. Individual buildings and rooms (factories, office towers, university lecture theatres) can be assessed, and criteria developed to warn occupants when heat stress is likely. For individual dwellings, however, systematic assessment is not practical, and would not account for individual physiological differences. Especially in buildings without air conditioning or with poor internal air circulation, the heat stress may be substantially in excess of that estimated using outdoor atmospheric measurements. Issuing a general statement such as the Humidex may be the most effective way to get individuals to consider possible adverse effects of heat, allowing for their personal circumstances. However, for detailed assessment, a more comprehensive measure such as the Universal Thermal Climate Index (Bröde et al. 2012) is more reliable.

12.7.2 Urban Heat Island

Within a city, temperatures are generally higher than those in surrounding rural areas. The Urban Heat Island phenomenon was initially recognized early in the twentieth century, and has been investigated in many cities since 1970 (e.g. Oke 1973, 1982; Böhm 1998; Rinner and Hussain 2011; Stewart and Oke 2012; Gago et al. 2013; Liao et al. 2021).

The overall magnitude of the urban heat island effect for a particular city depends on its population, with increases having an exponential effect (Kim and Brown 2021). In North America, effects are negligible for urban populations less than 100,000. Wind speed and prevalence, surrounding topography, nature and amount of energy consumption, and style of urban construction also play critical roles. Urban sprawl tends to produce stronger heat island effects in North American cities. Changes in energy consumption and urban landscape have resulted in notable increases in urban heat island intensity.

Toronto is a highly developed city economically, with high automobile usage, and relatively light winds, with a population of approximately 4 million. In downtown Toronto, the present urban heat island effect produces a 2–4 °C increase in temperature compared to surrounding rural areas. Seasonally, the greatest differences occur in late summer and in winter, with the minimum differences in May. Differences of 1–2 °C are evident between downtown Toronto and the temperatures recorded at Pearson International Airport, the principal weather station of Environment Canada for metropolitan Toronto (Moshin and Gough 2013). Substantial variations exist throughout Toronto, resulting from local land use and infrastructure. Local microclimates could permit urban agriculture of cold-sensitive crops in the city, in addition to influencing urban vegetation.

In contrast, St. John's NL, with a lesser population and stronger winds, has an urban heat island effect of approximately 1 °C. Temperatures at St. John's International (Torbay) Airport are approximately 1 °C cooler than temperatures at Memorial University of Newfoundland.

12.7.3 Health Effects

Humans subjected to extreme heat undergo three stages of reaction (Health Canada 2011, 2012; CCOHS 2024). Initial exposure may produce heat rash, nausea, muscle spasms, cramps, or fainting. Usually, patients recover if they are able to drink water and rest in a cool location. It is important to respond to these early warning signs.

Heat exhaustion occurs when fluid loss through excessive perspiration causes the body's cooling system to fail. Skin is moist and cool to the touch, but internal body temperature can rise above the normal 38 °C. The person may suffer nausea and vomiting, shallow breathing or panting, or blurred vision. Patients must be moved to cool areas, provided with water, and fanned or sprayed with water as well. Tight or excess clothing should be loosened or removed. As heat exhaustion can have persistent health effects if untreated, summoning medical help is advisable.

Heat stroke is the most severe effect. A person suffering heat stroke has expended all their internal water and salt reserves, and stops perspiring. Without any mechanism to remove excess heat, the internal temperature rises above 41 °C. Disorientation, racing or irregular pulse, headache, irrational behaviour or speech, and dry, red skin are symptomatic. Convulsions may occur, and the patient may suffer blackouts or lose consciousness. Heat stroke is a medical emergency, requiring immediate attention from paramedics or physicians. While awaiting medical assistance, patients must be moved to cool areas, tight or excess clothing loosened or removed, and fanned or sprayed with water. The patient should be provided with water if conscious, but do not administer water to an unconscious person.

Mortality due to extreme heat or heat stroke is predominantly associated with cardiovascular, respiratory, and cerebrovascular causes (Bustinza et al. 2013; Lamothe et al. 2019; Steen et al. 2019; Meade et al. 2020; Grigorieva and Lukyanets 2021; Henderson et al. 2022). Cardiovascular disease has been recognized as a significant factor in many heat-related mortality studies, although a connection with increased morbidity is not always apparent. Hypertension is also a significant factor and susceptibility can be increased by taking drugs to control it. Underlying respiratory conditions are worsened by extreme heat, especially when accompanied by elevated particulate levels or pollution. Cerebrovascular illness, ischaemic strokes, kidney failure, and pulmonary conditions are also affected by extreme heat events.

12.7.4 Susceptible and Vulnerable Urban Populations

Differences in response to extreme heat events and medical consequences between females and males have been noted. Increased mortality and morbidity among females aged 75 and older may reflect the greater proportion of older women in demographic populations (see Steen et al. 2019; Arsad et al. 2022). Other gender differences can be related to differences in lifestyle and occupation.

Although anyone exposed to prolonged extreme heat (including outdoor workers) is susceptible, elderly people are the most frequent victims (Lebel et al. 2019). Physiologically, elderly people in general perspire less have reduced sensations of thirst and therefore consume less water and are more subject to dehydration. They are more likely to suffer visual, cognitive, or hearing impairments, have challenges with mobility or agility, and are less physically fit. Elderly people are also more likely to live alone or in social isolation.

In addition, populations of elderly people may be concentrated in particular areas, such as urban heat islands or socio-economically disadvantaged neighbourhoods, leading to increased mortality rates. In southern Ontario, mortality among elderly people in individual cities is correlated to high levels of urbanization and high costs of living (Smoyer et al. 2000). Heat-related mortality among all age groups, but especially the elderly, tends to be higher in the downtown core, usually the centre of the urban heat island. Urban growth and sprawl have resulted in more demographic differences between adjacent communities in Canada, accentuating the effects of extreme heat events.

Children are also considered to be susceptible to extreme heat events. Children's increased sensitivity to heat is due to both physiological and psychological factors. Children produce more body heat during physical activity than do adults. If the air temperature is higher than skin temperature, their smaller bodies absorb heat faster. Young children also perspire less, and their cardiac systems are not capable of rapidly increasing blood flow. In Montreal, Auger et al. (2015) noted increased incidence of sudden infant death syndrome during hot days. Reduced birth weight during extreme heat events also occurs.

Young children have less experience and less ability to monitor their body state, take effective remedial action, or communicate that they are in trouble. They may also be susceptible to related conditions, such as excessive ozone, air pollutants, or other causes of asthma and lung ailments. About 13% of Canadians under age 12 have been diagnosed with asthma (Health Canada 2021).

Typically, children do not control the number of hours that they spend outdoors or in hot environments. In schools, adjustment of recess times to avoid prolonged hot periods and installation of air conditioning are possible measures. Closures are not as effective if children are merely sent home to equally hot conditions.

Workers in outdoor occupations, particularly construction, road maintenance, agriculture, trash collection, mail carriers, and patrolling police officers are particularly susceptible to heat exhaustion and stroke. Independent workers, self-employed persons, and those working for smaller firms or in smaller job teams can have greater susceptibility. Accidents resulting from inattentiveness or stress induced by heat are an additional problem. Health-and-safety campaigns aimed at occupations frequently link heat with ultraviolet exposure and sun safety.

People who live alone are at greater hazard, because they are alone, and sometimes because of the circumstances that result in that condition. Heat is considered by physicians to exacerbate mental health challenges, notably schizophrenia. Increased admissions for mental health issues and for effects of Alzheimer's disease during extreme heat events have been documented (Vanečkova and Bambrick 2013; Lee et al. 2023). Taking antidepressant or antipsychotic drugs can interfere with the body's cooling and salt regulation.

People who are confined to their bed or home are dependent upon caregivers for assistance with daily living. Those with sensory or cognitive impairments also are more susceptible. In many cases, physical issues are compounded by social isolation or financial challenges.

Socially disadvantaged people, including those with low incomes or who are homeless, face greater stress from extreme heat events. They are less likely to have access to air conditioning or other cool locations. Individuals with alcohol or drug addictions may also be more susceptible, although they may not consider themselves as part of an at-risk group.

Recent immigrants and tourists are potentially susceptible, due to difficulties in language communication, or lack of knowledge of local alert procedures health and social service programmes. Access to warnings broadcast by communications media, formerly considered a problem, has become less significant as social media use spreads.

Combinations of factors are generally responsible for extreme heat event mortality. Underlying medical conditions (cardiovascular or hypertension), age, living alone, living in disadvantageous socio-economic or personal circumstances, air pollution, and exposure to an urban heat island with limited green space increase sensitivity and vulnerability to heat events: most people who unfortunately succumb have more than one of these disadvantages (Lebel et al. 2019).

Perception of knowledge and hazard represented by extreme heat events varies substantially among people and between general populations and researchers and healthcare professionals. Official presentations which concentrate on factual information, on the assumption that people are generally unaware of extreme heat events, may be ineffective.

As many people have experience with heat, in contrast to earthquakes or volcanic eruptions, their experiential knowledge is based on personal contexts. Knowledge and response tend to focus on individuals' direct surroundings, and information presented on a broader city- or country-wide basis may be discounted. To many people, especially in colder climates, hot weather is a positive event, and they may lack any perception of heat as a hazard. Some people regard heat as an integral element of their environment or sense of place, choosing to live in hot locations. Others may consider that weather events cannot be changed, and thus the consequences must be endured in a fatalistic way.

Socio-economic factors condition response to heat events. Attitudes towards air conditioning are a good example. In urban Canada, air conditioning use by residents is widespread, except by those living in communities which are climatically colder, and those who are economically disadvantaged. Events later in summer may not be perceived as severe as those early in the season. Mortality rates tend to be higher for early season events in Canada. Acclimatization occurs both seasonally and over longer periods.

12.7.5 Building Design

Building design or retrofitting can aid in reducing the impact of extreme heat events. The most common approaches are the installation of solar shading, improvements in ventilation, external wall insulation, reflective coverings ("cool" walls and roofs), and green (vegetation-covered) roofs (e.g. Zinzi and Agnoli 2012; Buckland-Nicks et al. 2016; Hosseini et al. 2017). Additional internal wall insulation is less effective. As heat is exchanged between adjacent buildings, the effects of reduction of thermal energy output are cumulative.

Cool roofs and facades are designed to reflect insolation, precluding absorption of solar energy and subsequent conversion to thermal infrared radiation. Cool roofs may be constructed of non-absorbing materials, coated, or may be painted white rather than darker colours. In Canada, where the desire for summer cooling must be balanced against the requirement for winter heating, cool roofs and facades are generally not used. However, retrofitting of older buildings in Montreal has proven effective.

Green roofs have planted vegetation, as gardens, in boxes, or as continuous covers. Although any plant suitable for the particular climate could be used, the preferred species are low-maintenance plants with shallow roots, such as the boreal succulent *Sedum* (stonecrop). As a biological assemblage, any green roof will respond to microclimatic conditions, including shading, wind patterns, and moisture influenced by adjacent buildings. The species assemblage will change over time if untended.

In addition to their aesthetic appeal, green roofs are intended both to provide energy efficiency in the building and to influence the surrounding urban heat island. Depending upon the plant species used, evapotranspiration may occur largely during the day (for most flowering plants) or largely at night (for succulents such as *Sedum*). Production of extra moisture will result in absorption of some insolation and thermal infrared radiation as latent heat, as the moisture is vapourized. Consequently, the time of the maximum cooling effect will vary by plant species: for *Sedum*, maximum cooling will occur at night, when the urban heat island effect is relatively strongest.

12.7.6 Drowning

Annually, approximately 460 drowning fatalities occur in Canada (Drowning Prevention Research Centre 2022). Accidental outdoor drowning (not associated with occupations such as fishing and shipping) tends to increase during heat events, particularly associated with high relative humidity. Swimming deaths in open water especially increase on summer weekends. In Ontario, where most victims were male (80–90%, mean age 40 years), temperatures exceeding 30 °C resulted in a 69% increase in the probability of outdoor drowning (Fralick et al. 20103). Outdoor drowning rates tend to be higher for rural residents than for urban dwellers. Differences in ethnicity among drowning victims have also been noted in Canada. Rates of personal floatation device use vary with age, gender, and ethnic origin, with particularly low usages in northern Canadian indigenous communities.

Across Canada, male fatalities outnumber female by 5:1, with the highest death rates for ages 20–34 and older than 65 (Drowning Prevention Research Centre 2022). Younger adults tended to be involved in recreational activities, and >50% of the victims had consumed alcohol. Older adults tended to drown in urban settings (such as pools and tubs), and other medical conditions (particularly cardiac) were often involved. The lowest rate is for children between ages 5 and 14, who are also most likely to wear a personal flotation device. The rate for children under age 5 has declined in the past 20 years, with many fatalities resulting from falling into unattended swimming pools or while bathing.

12.8 Urban Vegetation

Vegetation within a city contributes to the removal of air pollutants, stores CO_2, provides habitat for wildlife, and exchanges water with the atmosphere (Gober et al. 2013; McGovern and Pasher 2016; Nowak et al. 2018; Cheng et al. 2021). Trees further act to reduce noise transmission and provide shade, particularly during the hotter months. In addition to having aesthetic value, urban vegetation is a potential contributor to reducing the urban heat island effect, and also provides land surface characteristics that modify excessive heat buildup (Duick et al. 2014; Nesbitt et al. 2017; Ko 2018; Aminipouri et al. 2019).

Urban tree cover density is measured by the percentage of land surface area shielded from direct insolation in summer (Conway and Bourne 2013; Shahtahmassebi et al. 2021). Greater tree cover density is correlated with reduced average and daytime summer temperatures at local scales. Rates of morbidity are lower in neighbourhoods with greater tree cover. Amounts of cooling beneath and in the vicinity of tree cover varied from less than 1 °C to more than 6 °C, depending on the tree species, wind speeds, humidity, and overall climate region. Differential effects on ultraviolet, visible, and short-wave infrared radiation have also been noted. In drier or semi-arid climates, tree spacing is more significant than in more humid areas, reflecting both the rates of evapotranspiration under differing conditions and the difference in canopy habit for different species. Tree species vary in their effectiveness at modulating urban heat islands, based on growth rate, geometry of leaf cover, canopy habit, and ability to tolerate urban conditions.

Most urban residents prefer to see trees in their landscapes and are generally supportive of planting and maintenance efforts (Steenberg et al. 2013; Ordóñez et al. 2017; Butt et al. 2021; Milbourne 2021; Wirtz et al. 2021). Municipalities with urban forestry plans tend to emphasize the planting of native tree species, although under constraints such as availability of stock, cost, and survivability in urban environments. Emphasis varies from municipality-wide to neighbourhood-focused, and from government-driven to resident-initiated. Climate change represents another factor complicating species viability and planting choices. Losses in cover have occurred due to introduced pest species such as Asian Longhorned Beetle, which attacks maples (*Acer sp.*), and Emerald Ash Borer, which attacks white ash (*Fraxinus americana*).

Urban forests are not evenly distributed throughout Canadian cities (e.g. Crouse et al. 2017; Nesbitt et al. 2018; Cheng et al. 2021; Ng et al. 2021; Pinault et al. 2021). In many cases, parks and other public green spaces are concentrated in more economically affluent neighbourhoods. Analysis of high-resolution satellite images and unmanned aerial vehicle data reveals correlations between income level and vegetation cover, combined with inverse relationships between vegetation cover and temperature: economically disadvantaged neighbourhoods tend to be hotter and less green. In some cases, racial and ethnic differences between tree-rich and tree-poor neighbourhoods also are present.

Trees planted on private lands are generally chosen by the residents. Cost, species availability, available space, growth rate, local climate (wind, snow cover, freezing rain), personal taste, neighbours' preferences all factor into these cultural decisions (Shakeel and Conway 2014; Elsadek et al. 2019). Municipal and other initiatives to encourage planting of some species (typically native) and discourage others (typically non-native or invasive, such as Norway Maple (*Acer platanoides*) typically have limited or no effect on residents' decisions (Almas and Conway 2016). As the choice reflects at least some elements of personal culture and values, residential trees are seldom planted specifically to reduce urban heat island effects.

Augmentation of the urban forest is a priority in many Canadian cities (Ordóñez and Duinker 2013; Crouse et al. 2017; Cheng et al. 2021; Wirtz et al. 2021). The effects may not be uniform: affluent residents are more likely to plant private trees, and the resulting increases in real estate value, combined with gentrification, may accentuate disparities. Affluent neighbourhoods are also more likely to be aware of municipal or non-profit initiatives for urban greening. Lot size is also a factor: larger areas, whether public or private, tend to have more and healthier trees.

12.9 Inequities for Canadian Urban Populations

Exposure to urban heat and limited vegetation cover are coupled with inequities in other environmental aspects in urban areas, with negative consequences for physical and mental health. High geographic variability characterizes environment conditions within CMAs (Davis et al. 2022).

Increased exposure to air pollution ($PM_{2.5}$, NO_2, SO_2, and O_3) affects marginalized low-income (Toronto), immigrant (Montreal), and Indigenous (Vancouver) residents (Giang and Castellani 2020). Census tracts with a higher prevalence of low income and Indigenous residents had significantly higher air pollution attributable mortality (Stieb et al. 2023). Inequality in air pollution attributable mortality tended to be driven more by baseline mortality inequalities than exposure inequalities.

In Toronto, premature mortality was correlated with poor cancer screenings, pollution, lack of tree canopy, increased uptake of student nutrition programmes, high walkability index, and presence of visible minorities (Auwor and Melles 2019). In Vancouver, inequality ratios increased for almost all-cause-specific mortalities from 1990 through 2016, notably for HIV/AIDS and sexually transmitted infections, maternal and neonatal disorders, and neoplasms (Yu et al. 2021).

Premature mortality rates declined in all socio-economic groups between 1991 and 2016 in Canada as a whole, except for women without a high school diploma (Shahidi et al. 2020). However, absolute income-related inequalities narrowed among men and widened among women, and widened for those with less education. Socio-economic inequalities involved in premature mortality and avoidable mortality widened. Income-related inequalities in some hospitalization indicators, including cardiac arrest, stroke, injury and self-injury, and admissions related to substance abuse, have persisted in Canada's major cities (Nichols et al. 2018).

Economically marginalized communities faced greater risks from COVID-19 (Mishra et al. 2022) In Toronto, the Gini coefficient of cumulative cases by population size was 0.41, reflecting household income, visible minority status, recent immigration housing, multigenerational households, and designated essential workers.

For youths, income inequality was associated with an increased hazard of deaths of despair, drug overdose, and all-cause deaths, but not significantly associated with suicide deaths (Benny et al. 2022). Students living in census divisions within the second and third quintiles of income inequality were significantly more likely to engage in monthly or weekly binge drinking (Lowe et al. 2023). There was no significant association between higher income inequality and current monthly alcohol consumption. Mental health outcomes are significantly worse for lower-income urban Canadians, along with large and increasing inequalities and differences in mental disorder prevalence rates (Missiuna et al. 2021).

Age, female gender, less education, lower income, functional impairment, and chronic diseases were significantly related to social isolation and loneliness (Menec et al. 2019). Living in an urban core area with a high proportion of low-income older adults was related to increased odds of social isolation.

12.9.1 Income Inequalities

Across Canada, the Gini coefficient for 2021 increased to 0.353 (Statistics Canada 2023a, b, c) following a drop in 2020 attributed in part to COVID-19 restrictions. Toronto (0.407), Calgary (0.386), Vancouver (0.385), Windsor (0.356) and the Ontario part of Ottawa-Gatineau (0.350) had the highest rates of income inequality among CMA. The three CMAs with the lowest Gini coefficients were Saguenay (0.284), Québec (0.293) and Drummondville (0.303). Interestingly, Saguenay and Vancouver were ranked as the most and least happy CMAs respectively by their residents (Helliwell et al. 2019).

Increased income inequality is primarily limited to the three largest CMA (Fong 2017), which together with Windsor had the only values above the national mean. Income mobility has generally declined in recent years, together with a rising share of income to the most wealthy (IMF 2021). Increased housing costs in the largest CMA have led to combinations of neighbourhood gentrification and concentrations of low-income residents elsewhere (Siddiq et al. 2023). In Toronto, Montreal and Vancouver, many neighbourhoods with a high concentration of immigrants have significantly higher levels of mortgage debt than other neighbourhoods (Simone and Walks 2019). The largest CMA are marked by income polarization and increasing spatial fragmentation (Ilic and Sawada 2021).

Outside the three largest CMA, income inequality in Canada has been relatively stable over the past 25 years (Burkinshaw et al. 2022). The exceptions are communities with persistent and substantial declines in population, labour force outcomes, working-age share of the population, and industrial diversity (Weaver et al. 2023). Increased income inequality was felt mainly by younger people and those with low incomes, defined as ~$26,000 for a single person (Canada Revenue Agency), or $18,000–$26,000, based on area (Statistics Canada). The low-income population, approximately 14% of the total, include higher proportions of visible minorities, recent immigrants in urban areas, activity-restricted persons, lone parents (with declining rates), and Urban Indigenous persons (with declining rates). Low-income rates for seniors are declining steadily, reflecting pensions, taxation policies, societal safety-net programmes, and the general ageing demographic of those formerly holding high-paying positions. Income status transitional changes affect approximately 30% of the number from year-to-year.

Increases in market income for lower-income families do not fully offset decreases in government transfers (Statistics Canada 2023a). Of the CMA, Québec (10.4%), Victoria (11.5%), Saguenay (11.8%) and Guelph (11.8%) had the lowest percentages of low-income persons in 2021, with Windsor (19.3%), Toronto (18.6%) and Winnipeg (17.2%) have the highest rates.

Governance and taxation regimes mean that the tools theoretically able to address rising inequality are not within the control of municipalities (Fong 2017). Federal and provincial governments set policies and programmes within their mandates. Municipalities are left to pick up the pieces and attempt to help the people.

References

Allen J, Palm M, Tiznado-Aitken I, Farber S (2022) Inequalities of extreme commuting across Canada. Travel Behav Soc 29:42–52

Almas AD, Conway TM (2016) The role of native species in urban forest planning and practice: a case study of Carolinian Canada. Urban Forest Urban Green 17:54–62

Aminipouri M, Knudby AJ, Krayenhoff ES, Zickfeld K, Middel A (2019) Modelling the impact of increased street cover on mean radiant temperature across Vancouver's local climate zones. Urban Forest Urban Green 39:9–17

APTA (2024) American public transportation association. Public Transp Ridership Rep Fourth Quart. https://www.apta.com/wp-content/uploads/2023-Q4-Ridership-APTA.pdf

Arsad FS, Hod R, Ahmad N, Ismail R, Mohamed N, Baharom M, Osman Y, Radi MFM, Tangang F (2022) The impact of heatwaves on mortality and morbidity and the associated vulnerability factors: a systematic review. Int J Environ Res Publ Health 19(23):16356. https://doi.org/10.3390/ijerph192316356

Assunçao-Denis M-É, Tomalty R (2019) Increasing cycling for transportation in Canadian communities: understanding what works. Transp Res Part A 123:288–304

Auger N, Fraser WD, Smargiassi A, Kosatsky T (2015) Ambient heat and sudden infant death: a case-crossover study spanning 30 years in Montreal, Canada. Environ Health Perspect 123(7):712–716

Awuor L, Melles S (2019) The influence of environmental and health indicators on premature mortality: an empirical analysis of the City of Toronto's 140 neighborhoods. Health Place 58. https://doi.org/10.1016/j.healthplace.2019.102155

Axisa JJ, Scott DM, Newbold KB (2012) Factors influencing commute distance: a case study of Toronto's commuter shed. J Transp Geogr 24:123–129

Baker DM, Linovski O (2022) The impact of a single bus rapid transit corridor on transit ridership: the winnipeg example. Transp Res Rec 2676:94–109

Benny C, Smith BT, Hyshka E, Senthilselvan A, Veugelers PJ, Pabayo R (2022) Investigating the association between income inequality in youth and deaths of despair in Canada: a population-based cohort study from 2006 to 2019. Epidemiol Commun Health. https://doi.org/10.1136/jech-2022-219630

Böhm R (1998) Urban bias in temperature time series—a case study for the city of Vienna, Austria. Clim Change 38:113–128

Boisjoly G, Grisé E, Maguire M, Veillette M-P, Deboosere R, Berrebi E, El-Geneidy A (2018) Invest in the ride: A 14 year longitudinal analysis of the determinants of public transport ridership in 25 North American cities. Transp Res Part A 116:434–445

Bozem H, Pozzer A, Harder H, Martinez M, Williams J, Lelieveld J, Fischer H (2017) The influence of deep convection on HCHO and H_2O_2 in the upper troposphere over Europe. Atmos Chem Phys 17:11835–11848

British Columbia Coroners' Service (2022) Extreme heat and human mortality: a review of heat-related deaths in B.C. in Summer 2021. https://www2.gov.bc.ca/assets/gov/birth-adoption-death-marriage-and-divorce/deaths/coroners-service/death-review-panel/extreme_heat_death_review_panel_report.pdf

Bröde P, Fiala D, Błażejczyk K et al (2012) Deriving the operational procedure for the Universal Thermal Climate Index (UTCI). Int J Biometeorol 56:481–494

Buckland-Nicks M, Heim A, Lundholm J (2016) Spatial environmental heterogeneity affects plant growth and thermal performance on a green roof. Sci Total Environ 553:20–31

Burkinshaw S, Terajima Y, Wilkins CA (2022) Income inequality in Canada. Bank of Canada Staff Discussion Paper 2022–16. https://doi.org/10.34989/sdp-2022-16

Burnett RT, Chen H, Szyszkowicz M et al (2018) Global estimates of mortality associated with long-term exposure to outdoor fine particulate matter. Proc Natl Acad Sci 115(38):9592–9597

Burnett RT, Stieb D, Brook JR, Cakmak S, Dales R, Raizenne M, Vincent R, Dann T (2004) Associations between short-term changes in nitrogen dioxide and mortality in Canadian cities. Arch Environ Health 59(5):228–236

Bürstlein J, López D, Farooq B (2021) Exploring first-mile on-demand transit solutions for North American suburbia: a case study of Markham, Canada. Transp Res Part A 261–283

Bustinza R, Lebel G, Gosselin P, Bélanger D, Chebana F (2013) Health impacts of the July 2010 heat wave in Québec, Canada. BMC Public Health 13(1):56–56

Butler A, Sweet M (2020) No free rides: winners and losers of the proposed Toronto transit commission U-Pass program. Transp Policy 96:15–28

Butt S, Smith SM, Moola F, Conway TM (2021) The relationship between knowledge and community engagement in local urban forest governance: a case study examining the role of resident association members in Mississauga, Canada. Urban Forest Urban Green 60. https://doi.org/10.1016/j.ufug.2021.127054

CAA (2017) Canadian national bottlenecks study. Canadian Automobile Association https://www.caa.ca/app/uploads/2021/01/16170_Canadian_National_Bottlenecks_Study_EN_1_4_17.pdf

Cal C (2007) Streetcars for Toronto—35th anniversary https://web.archive.org/web/20101128071525/http://spacingtoronto.ca/2007/12/01/streetcars-for-toronto-35th-anniversary/

Canada Energy Regulator (2024) Provincial and territorial energy profiles—Canada https://www.cer-rec.gc.ca/en/data-analysis/energy-markets/provincial-territorial-energy-profiles/provincial-territorial-energy-profiles-canada.html

Carrothers Commission (1967) Archives. Prince of Wales Northern Heritage Centre

CBC (2019) U-pass vote results. cbc.ca/news/canada/newfoundland-labrador/upass-vote-results-1.5047243#:~:text=A%20mandatory%20fee%20increase%20that,pass%20at%20a%20reduced%20price

CCOHS (2024) Heat-related illness. Canadian Council for Occupational Health and Saftety. https://www.ccohs.ca/products/posters/heat_illness

Chen F, Lu S, Zhao P, van de Voorde T, Xu W (2020) Urban land cover mapping from airborne hyperspectral imagery using a fast jointly sparse spectral mixture analysis method. Can J Remote Sens 46:330–343

Cheng Z, Nitoslawski S, van den Bosch CK, Sheppard S, Nesbitt L, Girling C (2021) Alignment of municipal climate change and urban forestry policies a Canadian perspective. Environ Sci Policy 122:14–24

Choi K (2018) The influence of the built environment on household vehicle travel by the urban typology in Calgary, Canada. Cities 75:101–110

Choi K, Park HJ, Uribe FA (2022) The impact of light rail transit station area development on residential property values in Calgary, Canada: focus on land use diversity and activity opportunities. Case Stud Transp Policy 10. https://doi.org/10.1016/j.cstp.2022.11.008

City of Toronto (2022) Heat relief strategy. https://www.toronto.ca/wp-content/uploads/2022/05/8f66-Heat-Relief-Strategy-2022.pdf

Collins PA, MacFarlane R (2018) Evaluating the determinants of switching to public transit in an automobile-oriented mid-sized Canadian city: a longitudinal analysis. Transp Res Part A 118:682–695

Collivignarelli MC, Abbà A, Bertanza G, Pedrazzani R (2020) Lockdown for CoViD-2019 in Milan: what are the effects on air quality? Sci Total Environ 732:139280

Conway TM, Bourne KS (2013) A comparison of neighborhood characteristics related to canopy cover, stem density and species richness in an urban forest. Landsc Urban Plan 113:10–18

Crouse DL, Peters PA, van Donkelaar A, Goldberg MS, Villeneuve PJ, Brion O, Khan S, Atari DO, Jerrett M, Pope CA, Brauer M, Brook JR, Martin RV, Stieb D, Burnett RT (2012) Risk of nonaccidental and cardiovascular mortality in relation to long-term exposure to low concentrations of fine particulate matter: a Canadian national-level cohort study. Environ Health Perspect 120:708–714

Crouse DL, Pinault L, Balram A, Hystad P, Peters PA, Chen H, von Donkelaar A, Martin RV, Ménard R, Robichaud A, Villeneuve PJ (2017) Urban greenness and mortality in Canada's largest cities: a national cohort study. Lancet Planet Health 1:e289–e297

Czekajlo A, Coops NC, Wulder MA, Hermosilla T, Lu Y, White JC, van den Bosch M (2020) The urban greenness score: a satellite-based metric for multi-decadal characterization of urban land dynamics. Int J Appl Earth Obs Geoinf 93. https://doi.org/10.1016/j.jag.2020.102210

Czekajlo A, Coops NC, Wulder MA, Hermosilla T, White JC, van den Bosch M (2021) Mapping dynamic peri-urban land use transitions across Canada using Landsat time series: spatial and temporal trends and associations with socio-demographic factors. Comput Environ Urban Syst 88. https://doi.org/10.1016/j.compenvurbsys.2021.101653

Davis Z, de Groh M, Rainham DG (2022) The Canadian Environmental Quality Index (Can-EQI): development and calculation of an index to assess spatial variation of environmental quality in Canada's 30 largest cities. Environ Int 170. https://doi.org/10.1016/j.envint.2022.107633

Diab E, DeWeese J, Chaloux N, El-Geneidy A (2021) Adjusting the service? understanding the factors affecting bus ridership over time at the route level in Montréal, Canada. Transportation 48:2765–2786

Diab E, Kasraian D, Miller EJ, Shalaby A (2020) The rise and fall of transit ridership across Canada: understanding the determinants. Transp Policy 96:101–112

Dilmaghani M (2022) Ethnoracial disparities in commute duration and mode among Canadian-born millennials. Travel Behav Soc 29:266–278

Dinnie K (2011) City branding: theory and cases. Palgrave Macmillan

Doick KJ, Peace A, Hutchings TR (2014) The role of one large greenspace in mitigating London's nocturnal heat island. Sci Total Environ 493:662–671

Drowning Prevention Research Centre (2022) Canadian drowning prevention plan. https://www.lifesaving.bc.ca/

Dubé J, Thériault M, DesRosiers F (2013) Commuter rail accessibility and house values: the case of the Montreal South Shore, Canada, 1992–2009. Transp Res Part A 54:49–66

ECCC (2019) Humidex. https://www.canada.ca/en/environment-climate-change/services/seasonal-weather-hazards/warm-season-weather-hazards.html#toc7

ECCC (2020) Canada's air pollutant emissions inventory report. Cat. En81-30E-PDF. www.canada.ca/en/environment-climate-change/services/air-pollution/publications/emissions-inventoryreport-2020.html

Eldeeb G, Mohamed M (2022) How do psychological factors sway customers' preferences towards transit service attributes? Case Stud Transp Policy 10:905–914

Elliot Lake Standard (2024) Elliot Lake has the country's third highest concentration of people 65 and older. https://www.elliotlakestandard.ca/news/elliot-lake-has-the-countrys-third-highest-concentration-of-people-65-and-older-2

Ellwood D, Martin CD, Dafni J (2011) Ground characteristics for the design of the Edmonton North LRT Twin Tunnels. In: 2011 Pan-Am CGS geotechnical conference, Paper 302. https://www.issmge.org/uploads/publications/84/52/GEO11Paper302.pdf

Elsadek M, Sun M, Sugiyama R, Fujii E (2019) Cross-cultural comparison of physiological and psychological responses to different garden styles. Urban Forest Urban Green 38:74–83

Fong F (2017) Income inequality in Canada: the Urban Gap. Chartered Professional Accountants of Canada, cpacanada.ca

Fralick M, Denny CJ, Redelmeier DA (2013) Drowning and the influence of hot weather. PLoS ONE 8(8):e71689. https://doi.org/10.1371/journal.pone.0071689

Friedel N (2011) Winnipeg: the importance of region-specific triggers. Institute for Catastropihc Risk Reduction. https://www.iclr.org/wp-content/uploads/PDFS/01_Winnipeg.pdf

Gago EJ, Roldan J, Pacheco-Torres R, Ordónez J (2013) The city and urban heat islands: a review of strategies to mitigate adverse effects. Renew Sustain Energy Rev 25:749–758

Galway LP, Deck E, Carastathis J, Sanderson R (2023) Exploring social-ecological influences on commuter cycling in a midsize northern city: a qualitative study in Thunder Bay, Canada. J Transp Geogr 92:102995

Giang A, Castellani K (2020) Cumulative air pollution indicators highlight unique patterns of injustice in urban Canada. Environ Res Lett 15:124063

Gober P, Larson KL, Quay R, Polsky C, Chang H, Shandas V (2013) Why land planners and water managers don't talk to one another and why they should! Soc Nat Res Int J 26(3):356–364

Gonzalez-Navarro M, Turner MA (2018) Subways and urban growth: evidence from earth. J Urban Econ 108:85–106

Government of Canada (2021) Surviving the heat: the impacts of the 2021 western heat dome in Canada. https://science.gc.ca/site/science/en/blogs/science-health/surviving-heat-impacts-2021-western-heat-dome-canada

Grigorieva E, Lukyanets A (2021) Combined effect of hot weather and outdoor air pollution on respiratory health: literature review. Atmosphere 12:790. https://doi.org/10.3390/atmos12060790

Health Canada (2011) Communicating the health risks of extreme heat events: toolkit for public health and emergency management officials. Health Canada, Ottawa, p 60

Health Canada (2012) Heat alert and response systems to protect health: best practices guidebook. Health Canada, Ottawa, p 92

Health Canada (2016) Residential indoor air quality guideline: ozone. https://www.canada.ca/en/health-canada/services/publications/healthy-living/residential-indoor-air-quality-guideline-ozone.html

Health Canada (2021) Health impacts of air pollution in Canada: estimates of morbidity and premature mortality outcomes—2021 Report. https://www.canada.ca/en/health-canada/services/publications/healthy-living/health-impacts-air-pollution-2021.html

Helliwell JF, Shiplett H, Barrington-Leigh CP (2019) How happy are your neighbours? variation in life satisfaction among 1200 Canadian neighbourhoods and communities. PLoS ONE 14(1):e0210091. https://doi.org/10.1371/journal.pone.0210091

Henderson SB, McLean KE, Lee MJ, Kosatsky T (2022) Analysis of community deaths during the catastrophic 2021 heat dome. Environ Epidemiol 6(1):e189. https://doi.org/10.1097/ee9.0000000000000189

Hossieni M, Lee B, Vakilinia S (2017) Energy performance of cool roofs under the impact of actual weather data. Energy Build 145:284–292

Ilic L, Sawada M (2021) The temporal evolution of income polarization in Canada's largest CMAs. PLoS ONE 16(6):e0251430. https://doi.org/10.1371/journal.pone.0251430

IMF (2021) income inequality: where do we stand? Canada: selected issues. International Monetary Fund Country Report 21/55

IRCC (2019) Economic profile series: Charlottetown, prince Edward island. Immigration Refugees and Citizenship Canada. https://lipdata.ca/wp-content/uploads/sites/28/2020/02/Charlottetown-Economic-Profile-2019.pdf

IWG (2021) The 15-minute commute. International Workplace Group. https://assets.iwgplc.com/image/upload/v1620812137/IWG/MediaCentre/xsndewyg2co5d2va6dgq.pdf

Kilpatrick AB (1980) A lesson in boosterism: the contest for the Alberta provincial capital, 1904–1906. Urban Hist Rev 8:47–109

Kim SW, Brown RD (2021) Urban Heat Island (UHI) intensity and magnitude estimations: a systematic literature review. Sci Total Environ 779. https://doi.org/10.1016/j.scitotenv.2021.146389

Knight T, Price S, Bowler D, Hookway A, King S, Konno K, Richter RL (2021) How effective is 'greening' of urban areas in reducing human exposure to ground-level ozone concentrations, UV exposure and the 'urban heat island effect'? an updated systematic review. Environ Evid 10. https://doi.org/10.1186/s13750-021-00226-y

Ko Y (2018) Trees and vegetation for residential energy conservation: a critical review for evidence-based urban greening in North America. Urban Forest Urban Green 34:318–336

Lachapelle U, Manaugh K, Hamelin-Pratte S (2022) Providing discounted transit passes to younger university students: are there effects on public transit, car and active transportation trips to university? Case Stud Transp Policy 10:811–820

Lamothe F, Roy M, Racine-Hamel S-H (2019) Epidemiological investigation—heat wave in the summer of 2018 in Montreal. Regional Public Health Department of the CIUSSS du Centre-Sud-del'Île-de-Montréal. https://santemontreal.qc.ca/fileadmin/fichiers/professionnels/DRSP/Directeur/Rapports/Resume_EnqueteChaleurMtl_2018_Anglais.pdf

Lebel G, Dubé M, Bustinza R (2019) Surveillance des impacts des vagues de chaleur extrême sur la santé au Québec à l'été 2018, INSPQ. https://www.inspq.qc.ca/bise/surveillance-des-impacts-des-vagues-de-chaleur-extreme-sur-la-sante-au-quebec-l-ete-2018#ref

Lee DH, Henschel A (1966) Effects of physiological and clinical factors on response to heat. Ann NY Acad Sci 134:743–749

Lee MJ, McLean KE, Kuo M, Richardson GRA, Henderson SB (2023) Chronic diseases associated with mortality in British Columbia, Canada during the 2021 western north America extreme heat event. GeoHealth. https://doi.org/10.1029/2022GH000729

Leung KHY, Arnillas CA, Cheng VYS, Gough WA, Arhonditis GB (2021) Seasonality patterns and distinctive signature of latitude and population on ozone concentrations in Southern Ontario, Canada. Atmos Environ 246. https://doi.org/10.1016/j.atmosenv.2020.118077

Levine MV (1991) The reconquest of Montreal: language policy and social change in a Bilingual City. Temple University Press, Philadelphia

Liao W, Hing T, Heo Y (2021) The effect of spatial heterogeneity in urban morphology on surface urban heat islands. Energy Build 244. https://doi.org/10.1016/j.enbuild.2021.111027

Lowe SAJ, Basnet S, Leatherdale ST, Patte KA, Pabayo R (2023) Inequality's on tap: a longitudinal study of area-level income inequality and alcohol consumption among Canadian adolescents. J Adolesc Health 73:1093–1100

MacKinnon L (2016) Deindustrialization on the periphery: an oral history of Sydney steel, 1945–2001. PhD thesis, History. Concordia

MacKinnon L (2020) Closing sysco: industrial decline in Atlantic Canada's steel city. University of Toronto Press

Manaugh K, Miranda-Moreno LF, El-Geneidy AM (2010) The effect of neighbourhood characteristics, accessibility, home-work location, and demographics on commuting distances. Transportation 37:627–646

McGovern M, Pasher J (2016) Canadian urban tree canopy cover and carbon sequestration status and change 1990–2012. Urban Forest Urban Green 20:227–232

Meade RD, Akerman AP, Notley SR, McGinn R, Poirier P, Gosselin P, Kenny GP (2020) Physiological factors characterizing heat-vulnerable older adults: a narrative review. Environ Int 144:105909. https://doi.org/10.1016/j.envint.2020.105909

Mekis E, Vincent LA, Shephard MW, Zhang X (2015) Observed trends in severe weather conditions based on humidity, wind chill and heavy rainfall events in Canada for 1953–2012. Atmos Ocean 53:383–397

Menec VH, Newall NE, Mackenzie CS, Shooshtari S, Nowicki S (2019) Examining individual and geographic factors associated with social isolation and loneliness using Canadian Longitudinal Study on Aging (CLSA) data. PLoS ONE 14(2):e0211143. https://doi.org/10.1371/journal.pone.0211143

Metrobus (2019) St. John's transit review. https://www.metrobus.com/PDF/StJohnsTransitReview2019.pdf

Milbourne P (2021) Growing public spaces in the city: community gardening and the making of new urban environments of publicness. Urban Stud 58:2901–2919

Mishra S, Ma H, Moloney G, Yiu KCY, Darvin D et al (2022) Increasing concentration of COVID-19 by socioeconomic determinants and geography in Toronto, Canada: an observational study. Ann Epidemiol 65:84–92

Missiuna S, Plante C, Pahwa P, Muhajarine N, Neudorf C (2021) Trends in mental health inequalities in urban Canada. Can J Public Health 112:629–637

Mohamed M, Ferguson M, Kanaroglou P (2019) What hinders adoption of the electric bus in Canadian transit? perspectives of transit providers. Transp Res Part D 64:134–149

Molner B, Curtin KD, Kongats K, Brown JA, Thomson M, Atkey K, Nykiforuk CIJ (2023) Policy influencer support for active transportation policy in two Canadian provinces: implications for advocacy. J Transp Health 28:101537

Montreal Gazette (2024) Electric vehicles not a panacea for climate change: Steven Guilbeault. https://montrealgazette.com/news/local-news/electric-vehicles-not-a-panacea-for-climate-change-steven-guilbeault

Moos M, Mendez P (2015) Suburban ways of living and the geography of income: how homeownership, single-family dwellings and automobile use define the metropolitan social space. Urban Stud 52:1864–1882

Moshin T, Gough WA (2013) Characterization and estimation of urban heat island at Toronto: impact of the choice of rural sites. Theoret Appl Climatol 108:105–117

Munro S (2023) TTC ridership update 2023. https://stevemunro.ca/2023/11/13/ttc-ridership-update-october-2023/

Nesbitt L, Hotte N, Barron S, Cowan J, Sheppard SRJ (2017) The social and economic value of cultural ecosystem services provided by urban forests in North America: a review and suggestions for future research. Urban Forest Urban Green 25:103–111

Nesbitt L, Meitner MJ, Sheppard SRJ, Girling C (2018) The dimensions of urban green equity: a framework for analysis. Urban Forest Urban Green 34:240–248

Newbold KB, Scott DM (2018) Insights into public transit use by Millennials The Canadian experience. Travel Behav Soc 11:62–68

Newman P, Kenworthy J (2021) Gasoline consumption and cities revisited: what have we learned? Curr Urban Stud 9:532–553

Ng A, Gergel SE, Eskelson BNI (2021) Inequality and allergenic cover of urban greenspaces surrounding public elementary schools in Vancouver. Brit Columbia, Can Ecosyst People 17:178–190

Nichols M, Chuang J, Gronwold S, Hynes G, Harvey J, Plante C, Neudorf C, Allen S (2018) Measuring trends in health inequalities across urban cities in Canada: a focus on health system outcomes. Int J Popul Data Sci 3. https://doi.org/10.23889/ijpds.v3i4.971

Nowak DJ, Hirabayashi S, Doyle M, McGovern M, Pasher J (2018) Air pollution removal by urban forests in Canada and its effect on air quality and human health. Urban Forest Urban Green 29:40–48

Nunatsiaq News (1995) December 11: Nunavut's choice. https://archive.ph/20070630074212/http://www.nunanet.com/~nunat/pages/nunavut1.html

NWMP Census of Dawson City (1898) Canadian research knowledge network. Government of Canada

Oke TR (1973) City size and the urban heat island. Atmos Environ 7:769–779

Oke TR (1982) The energetic basis of the urban heat island. Q J R Meteorol Soc 108 1–24

Olsson L, Thomas R (2024) Mobility justice or transit boosterism? the use of rail transit as an urban transformation strategy in Kitchener, Canada, and Malmö. Mobilities, Sweden. https://doi.org/10.1080/17450101.2024.2304844

Ordónez C, Beckley T, Duinker PN, Sinclair AJ (2017) Public values associated with urban forests: synthesis of findings and lessons learned from emerging methods and cross-cultural case studies. Urban Forest Urban Green 25:74–84

Ordónez C, Duinker PN (2013) An analysis of urban forest management plans in Canada: implications for urban forest management. Landsc Urban Plan 116:36–47

Ottawa Light Rail Transit Commission (2022) Report of the Ottawa light rail transit public inquiry. https://www.archives.gov.on.ca/en/e_records/OLRTPI/documents/final-report/index.html

Papaioannou EM, Hawkins J, Nurul Habib KM (2020) A study of car and home ownership decisions in the face of increasing commuting expenses (CHOICE) in the Greater Toronto Area (GTA). Case Stud Transp Policy 8:971–983

Pengelly LD, Campbell ME, Cheng CS, Fu C, Gingrich SE, Macfarlane R (2007) Anatomy of heat waves and mortality in Toronto. Lessons for public health protection. Can J Public Health 98:364–368

Phelan PE, Kaloush K, Miner M, Golden J, Phelan B, Silva H, Taylor RA (2015) Urban heat island: mechanisms, implications, and possible remedies. Annu Rev Environ Resour 40:285–307

Pinault L, Christidis T, Toyib O, Crouse DL (2021) Ethnocultural and socioeconomic disparities in exposure to residential greenness within urban Canada. Health Rep 32(5). https://doi.org/10.25318/82-003-x202100500001-eng

Places That Matter (2016) Georgia street viaducts. https://placesthatmatter.ca/location/georgia-st-viaduct

Pourahmadiyan A, Ahmadi P, Kjeang E (2021) Dynamic simulation and life cycle greenhouse gas impact assessment of CNG, LNG, and diesel-powered transit buses in British Columbia, Canada. Transp Res Part D 92:102724

Price K, Benmarhnia T, Gaudet J, Kaiser D, Sadoine ML, Perron S, Smargiassi A (2018) The Montreal heat response plan: evaluation of its implementation towards healthcare professionals and vulnerable populations. Can J Public Health 109(1):108–116

Read the Plaque (2024) Spadina expressway. https://readtheplaque.com/plaque/spadina-expressway

Rinner C, Hussain M (2011) Toronto's urban heat island—exploring the relationship between land use and surface temperature. Remote Sens 3:1251–1265

Roman LA, Pearsall H, Eisenmanc TS, M. Conway TM, Fahey RT, Landry S Vogt J, van Doorn NS, Grove JM, Locke DH, Bardekjian AC, Battles JJ Cadenasso ML, van den Bosch CCK, Avolio M, Berland O, Jenerette DG, Mincey SK, Patakir DE, Staudhammer S (2018) Human and biophysical legacies shape contemporary urban forests: a literature synthesis. Urban Forest Urban Green 31:157–168.

Saeidizand P, Fransen K, Boussauw K (2022) Revisiting car dependency: a worldwide analysis of car travel in global metropolitan areas. Cities 120:103467

Sanaullah I, Alsaleh N, Djavadian S, Farooq B (2021) Spatio-temporal analysis of on-demand transit: a case study of Belleville, Canada. Transp Res Part A 145:284–301

Savage K, Turcotte M (2020) Commuting to work during COVID-19. Statistics Canada. https://publications.gc.ca/site/eng/9.890247/publication.html?wbdisable=true

Shahidi FV, Parnia A, Siddiqi A (2020) Trends in socioeconomic inequalities in premature and avoidable mortality in Canada, 1991–2016. Can Med Assoc J 192:E1114–E1128. https://doi.org/10.1503/cmaj.191723

Shahtahmassebi AR, Li C, Fan Y, Wu Y, Lin Y, Gan M, Wang K, Malik A, Blackburn GA (2021) Remote sensing of urban green spaces: a review. Urban Forest Urban Green 57. https://doi.org/10.1016/j.ufug.2020.126946

Shakeel T, Conway TM (2014) Individual households and their trees: fine-scale characteristics shaping urban forests. Urban Forest Urban Green 14:136–144

Shantz A, Casello J, Woudsma C, Guerra E (2022) Understanding factors associated with commuter rail ridership: a demand elasticity study of the GO transit rail network. Transp Res Rec 2676:131–143

Shi X, Brasseur GP (2020) The response in air quality to the reduction of Chinese economic activities during the COVID-19 outbreak. Geophys Res Lett 47. https://doi.org/10.1029/2020GL088070

Siciliano B, Dantas G, da Silva C, Arbilla G (2020) Increased ozone levels during the COVID-19 lockdown: analysis for the city of Rio de Janeiro, Brazil. Sci Total Environ 737. https://doi.org/10.1016/j.scitotenv.2020.139765

Siddiq FK, Klymentieva H, Lee TJC (2023) Applying the Lorenz Curve and Gini coefficient to measure the population distribution. Int Adv Econ Res 29:177–192

Simone D, Walks A (2019) Immigration, race, mortgage lending, and the geography of debt in Canada's global cities. Geoforum 98:286–299

Smoyer K, Rainham D, Hewko J (2000) Heat-stress-related mortality in five cities in Southern Ontario: 1980–1996. Int J Biometeorol 44:190–197

Spotlight Productions (2017) Secret Alberta: the former life of amber valley. https://www.youtube.com/watch?v=OxXJubB2cPQ

Stamp RM (1989) Riding the radials. The Boston Mills Press, Toronto's Suburban Electric Streetcar Lines

Statistics Canada (2017) Commuters using sustainable transportation in census metropolitan areas. https://www12.statcan.gc.ca/census-recensement/2016/as-sa/98-200-x/2016029/98-200-x2016029-eng.cfm

Statistics Canada (2021a) Employment by census metropolitan areas and occupation, annual (x 1,000). https://www150.statcan.gc.ca/t1/tbl1/en/tv.action

Statistics Canada (2021b) Census Metropolitan Area (CMA) and census agglomeration (CA). Dictionary, Census of Population 2021. https://www12.statcan.gc.ca/census-recensement/2021/ref/dict/az/Definition-eng.cfm?ID=geo009.

Statistics Canada (2022a) Population counts for census metropolitan areas, census agglomerations, population centres, and rural areas. https://www150.statcan.gc.ca/t1/tbl1/en/tv.action?pid=9810000601

Statistics Canada (2022b) Market snapshot: record high electric vehicle sales in Canada. https://www.cer-rec.gc.ca/en/data-analysis/energy-markets/market-snapshots/2022/market-snapshot-record-high-electric-vehicle-sales-canada.html

Statistics Canada (2022c) Automotive statistics. https://www.statcan.gc.ca/en/topics-start/automotive

Statistics Canada (2022d) North to Yukon... the rush is on again. https://www.statcan.gc.ca/o1/en/plus/691-north-yukon-rush-again

Statistics Canada (2023a) Main highlights on Income of families and individuals: Subprovincial data from the T1 Family File. https://www150.statcan.gc.ca/n1/daily-quotidien/230809/dq230809b-eng.pdf

Statistics Canada (2023b) Gross domestic product (GDP) at basic prices, by census metropolitan area. https://www150.statcan.gc.ca/t1/tbl1/en/tv.action

Statistics Canada (2023c) Commuting to work by car and public transit grows in 2023. https://www150.statcan.gc.ca/n1/daily-quotidien/230822/dq230822b-eng.htm

Statistics Canada (2024a) Working from home in Canada. https://www150.statcan.gc.ca/n1/daily-quotidien/240118/dq240118c-eng.htm

Statistics Canada (2024b) Urban public transit. https://www150.statcan.gc.ca/n1/daily-quotidien/240320/dq240320d-eng.htm

Steen Y, Ntarladima A-M, Grobbee R, Karssenberg D, Vaartjes I (2019) Sex differences in mortality after heat waves: are elderly women at higher risk? Int Arch Occup Environ Health 92:37–48

Steenberg JWN, Duinker PN, Charles JD (2013) The neighbourhood approach to urban forest management: the case of Halifax, Canada. Landsc Urban Plan 117:135–144

Stewart ID, Oke TR (2012) Local climate zones for urban temperature studies. Bull Am Meteorol Soc 1879–1900

Stewart RE, Betancourt D, Davies JB, Harford D, Klein Y, Lanigan R, Mortsch L, O'Connell E, Tang K, Whitfield PH (2017) A multi-perspective examination of heat waves affecting Metro Vancouver: now into the future. Nat Hazards. https://doi.org/10.1007/s11069-017-2793-7

Stieb DM, Beveridge RC, Brook JR, Smith-Doiron M, Burnett RT, Dales RE, Beaulieu S, Judek S, Mamedov A (2000) Air pollution, aeroallergens and cardiorespiratory emergency department visits in Saint John, Canada. J Expo Anal Environ Epidemiol 10:461–477

Stieb DM, Smith-ron M, Quick M, Christidis T, Xi G, Miles RM et al (2023) Inequality in the distribution of air pollution attributable mortality within Canadian cities. GeoHealth 7:e2023GH000816. https://doi.org/10.1029/2023GH000816

Townsend C, Ellis-Young M (2018) Urban population density and freeways in North America: a re-assessment. J Transp Geogr 72:75–83

Traffic Injury Research Foundation (2020) The Impact of the COVID-19 pandemic on travel behaviour and Road Safety. https://tirf.ca/news/tirf-reports-on-the-impact-of-the-covid-19-pandemic-on-travel-behaviour-road-safety/

TTC (2023) CEOs report, October 2023. Toronto Transit commission. https://www.ttc.ca/transparency-and-accountability/CEOs-Report.

United States Census (2020) US census bureau, Washington USA

USEPA (2024) Sources of greenhouse gas emissions. United States Environmental Protection Agency, https://www.epa.gov/ghgemissions/sources-greenhouse-gas-emissions

Vaneckova P, Bambrick H (2013) Cause-specific hospital admissions on hot days in Sydney, Australia. PloS One 8(2):e55459. https://doi.org/10.1371/journal.pone.005545

Venter ZS, Aunan K, Chowdhury S, Lelieveld J (2020) COVID-19 lockdowns cause global air pollution declines. Earth Atmos Planet Sci 117:18984–18990

Vineberg R (2010) Immigration and demographic change in Canada and Manitoba. Rural Development Institute, Brandon University. https://www.brandonu.ca/rdi/files/2011/06/Immigration-and-Demographic-Change-in-Manitoba-report-2010.pdf

Waterloo (2022) Tech and digital media. https://www.regionofwaterloo.ca/en/ng-business/techanddigitalmedia.aspx

Weaver D, Krawchenko T, Markey S (2023) The index of economic disparity: measuring trends in economic disparity across Canadian census subdivisions and rural and urban communities. Can Geogr 68:142–158

Wirtz Z, Hagerman S, Hauer RJ, Konijnendijk CC (2021) What makes urban forest governance successful? A study among Canadian experts. Urban For Urban Greening 58. https://doi.org/10.1016/j.ufug.2020.126901

Young M, Lachapelle U (2017) Transportation behaviours of the growing Canadian single-person Households. Transp Policy 57:41–50

Young M, Tanguay GA, Lachapelle U (2016) Transportation costs and urban sprawl in Canadian metropolitan areas. Res Transp Econ 60:25–34

Yousefzadeh Barri E, Farber S, Kramer A, Jahanshahi H, Allen J, Beyazit E (2021) Can transit investments in low-income neighbourhoods increase transit use? exploring the nexus of income, car-ownership, and transit accessibility in Toronto. Transp Res Part D 95:102849

Yu J, Dwyer-Lindgren L, Bennett J, Ezzati M, Gustafson P, Tran M, Brauer M (2021) A spatiotemporal analysis of inequalities in life expectancy and 20 causes of mortality in sub-neighbourhoods of Metro Vancouver, British Columbia, Canada, 1990–2016. Health Place 72. https://doi.org/10.1016/j.healthplace.2021.102692

Zinzi M, Agnoli S (2012) Cool and green roofs. An energy and comfort comparison between passive cooling and mitigation urban heat island techniques for residential buildings in the Mediterranean region. Energy Build 55:66–76

Urban Geography of Canada II

13

Abstract

To study Canadian cities is to delve into the very heart of the nation's economic, social, and cultural evolution. As epicentres of innovation, diversity, and economic dynamism, cities like Toronto, Vancouver, Montreal, and Calgary hold a pivotal place in shaping Canada's global competitiveness. Canadian cities are complex, shaped by their geographical, social, economic, and cultural diversity. Their development is marked by a unique combination of historical factors such as European colonization, mass immigration, and industrialization. The history of urbanization in Canada is marked by several key phases. The nineteenth century saw a rapid acceleration in urbanization, driven by the Industrial Revolution and massive immigration, particularly from Europe. In addition to Montreal and Québec, other cities including Toronto and Vancouver developed around industry and burgeoning railway infrastructure. In the twentieth century, urbanization intensified alongside industrialization with the modernization of infrastructure, while suburbanization gained momentum after World War II, spurred by the rise of the automobile and the expansion of suburban areas. Canada experienced a growing cultural diversity with the arrival of immigrants from various regions as the country underwent a significant economic transition shaped by global forces with local ramifications. As the driving force behind urbanization, industrialization spurred the expansion of Canada's manufacturing industries in major cities such as Montreal, Toronto, and Hamilton. Industrialization also spurred the development of new infrastructure, including railways, ports, and roads, which facilitated the movement of goods and labour. As Canada's traditional manufacturing economy began to decline from the 1970s due to international competition and technological changes, the post-war era was also marked by the modernization of industry, with the rise of high-tech industries, financial services, and new economic sectors. In the final decades of the twentieth century and the early twenty-first century, Canadian cities underwent a transformation driven by globalization, continued population growth, and economic diversification. Nonetheless, although the role of traditional industries has diminished, the connections between industrialization and urbanization continue to shape Canadian cities today. Suburban sprawl is arguably one of the most defining characteristics of modern Canadian cities, with sprawling low-density residential neighbourhoods, shopping malls, and extensive highway use. However, in recent years, sustainable urban planning policies have emerged to curb this sprawl and encourage urban densification, particularly through transit-oriented development. The COVID-19 pandemic has suddenly and profoundly reshaped the trajectory of Canadian cities, catalysing changes that will influence urban evolution for years to come. One of the most profound changes was the rise of telecommuting, with many workers opting to work from home, reducing the need for daily commutes. The Canadian urban system is organized into a network of hierarchical cities and urban regions, with complex interactions between them depending on their size, economic function, and geographical influence who play key roles in their respective contexts. Canadian cities offer a high quality of life and lead the way in sustainability and multiculturalism. However, they must also navigate complex challenges, particularly in terms of housing affordability, urban sprawl, social inequality, and ageing infrastructure, to remain competitive and inclusive in the global landscape.

Keywords

Cities · Innovation · History of urbanization · Economic diversification · Inequality

N. Catto and G. R. Vincent, *Canada: Land and People*, World Regional Geography Book Series,
https://doi.org/10.1007/978-3-032-04391-7_13

13.1 Introduction

Canada is a predominantly urban nation. In 2021, nearly 82% of Canadians resided in a "population centre" (Statistics Canada 2021, 2022). Even more striking is that nearly half (47.4%) of the population lived in one of the six "Census Metropolitan Areas" (CMA) with over a million inhabitants. An equal share, 9.7%, was distributed between two population size groups: the five CMAs with populations ranging from 500,000 to 999,999 and the eleven CMAs with 200,000 to 499,999 residents. The remaining 16.7% of Canadians lived in nineteen CMAs with populations between 100,000 and 199,999 (Fig. 13.1, 13.2). However, these figures obscure the fact that the Atlantic provinces exhibit significantly lower rates of urbanization, ranging from 51% in New Brunswick to 60% in Newfoundland and Labrador. Only Nunavut is lower, with an urbanization rate of 43.2%.

To study Canadian cities is to delve into the very heart of the nation's economic, social, and cultural evolution. As epicentres of innovation, diversity, and economic dynamism, cities like Toronto, Vancouver, Montreal, and Calgary hold a pivotal place in shaping Canada's global competitiveness. These urban landscapes gather a tapestry of diverse populations, embodying immigration and multiculturalism, while simultaneously driving economic engines in fields such as finance, technology, and the cultural industry. Moreover, Canadian cities of all sizes stand as arenas where pressing issues of sustainability, infrastructure management, and social inequality are keenly felt, offering rich ground for examining urban planning and public policy (Walks 2001, 2013; Bourne

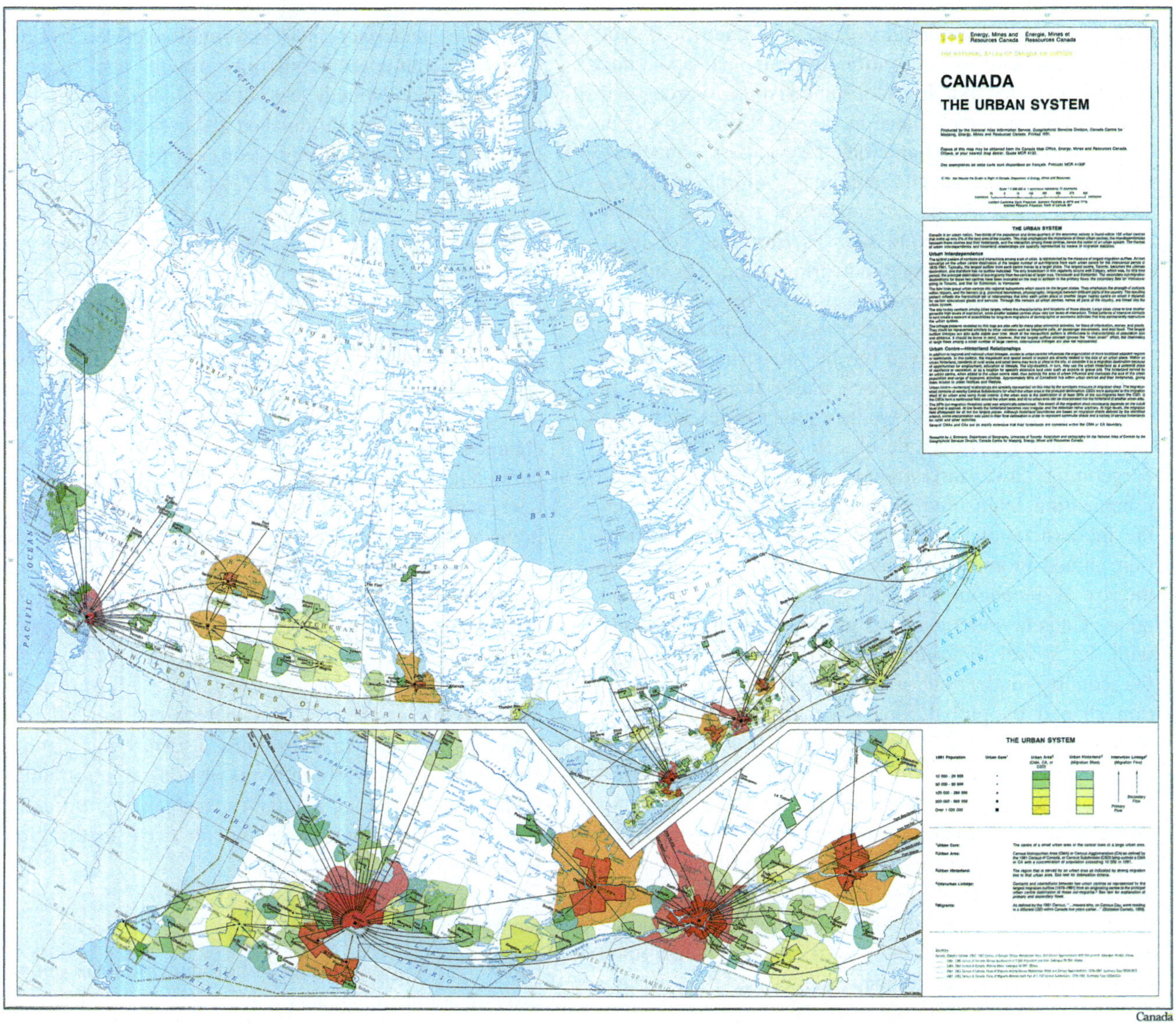

Fig. 13.1 Canada: the urban system. (Energy, Mines, and Resources Canada)

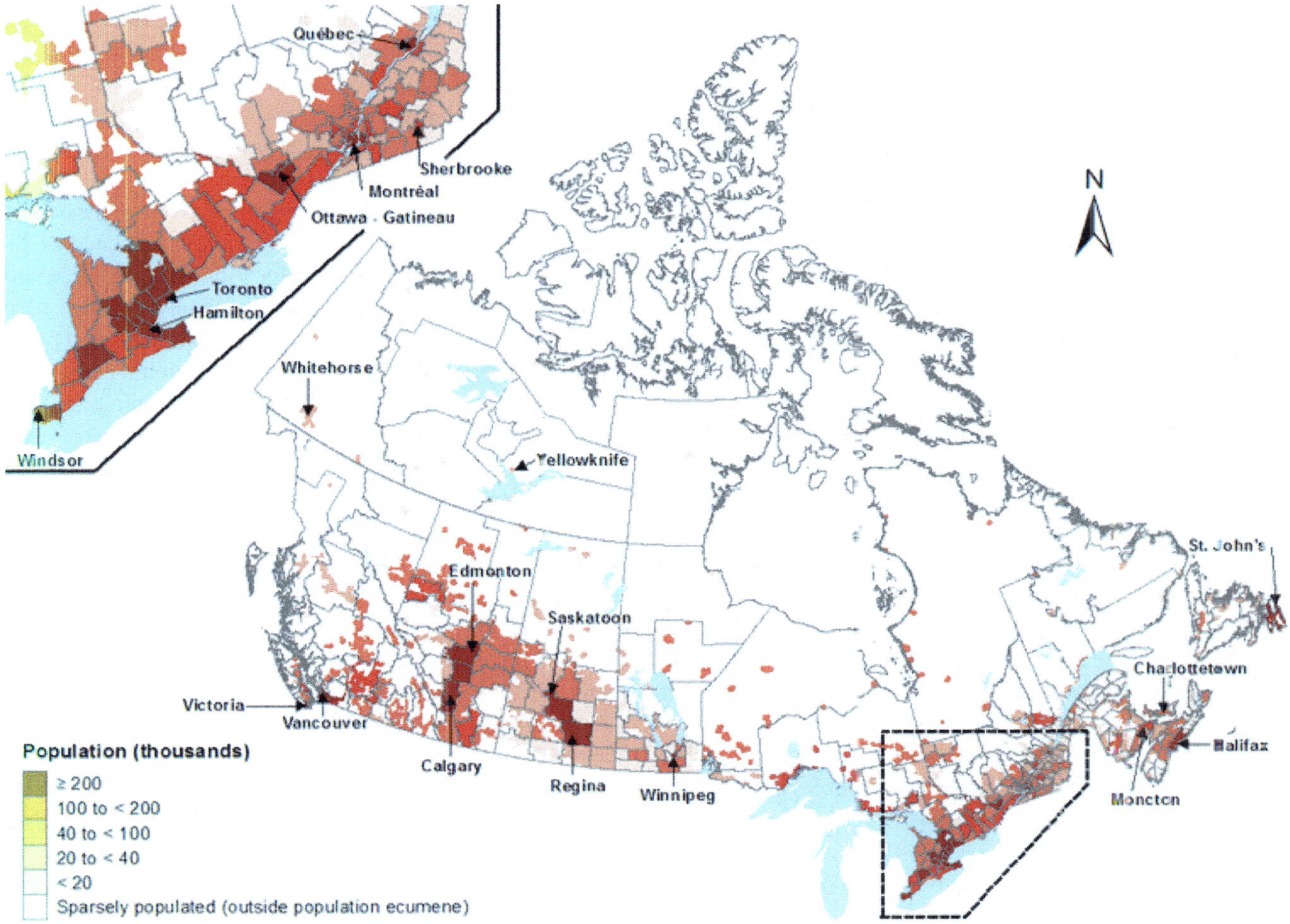

Fig. 13.2 Population density. (Statistics Canada)

and Simmons 2003a, 2003b). Understanding the challenges and opportunities these cities present is vital to crafting an inclusive and resilient future for the entire nation.

Canadian cities are complex, shaped by their geographical, social, economic, and cultural diversity. Their development is marked by a unique combination of historical factors such as European colonization, mass immigration, and industrialization. Geographically, they span vast and varied territories, from coastal areas such as Vancouver to inland cities such as Winnipeg, each facing distinct challenges related to climate, infrastructure, and natural resources. Socially, Canadian cities are defined by profound multiculturalism, born from a long history of immigration and coexistence with Indigenous peoples, creating intricate dynamics in terms of social cohesion and diversity management. Economically, they serve as engines of innovation while grappling with growing inequalities and housing crises. This complexity is further heightened by environmental challenges, particularly those linked to climate change and sustainable urban planning. Analyzing these cities, therefore, requires a flexible approach to fully grasp the interplay between their many dimensions.

The urban population, which stood at 37% in 1901, surged to 50% between 1921 and 1931, and swiftly climbed to 75% by the 1960s (Statistics Canada 2021). Within the span of a century that saw Canada shift from an industrial to a post-industrial society, Canadian cities now play a pivotal role as hubs of trade, immigration, culture, and innovation in a modern globalized world. Their connections to global networks have profoundly reshaped their economic development, urban structure, and the animation of their cultural and demographic landscapes.

Globalization has transformed Canadian cities into vibrant hubs at the heart of global economic and cultural networks. However, this process brings with it challenges such as rising inequalities, gentrification (the incoming of wealthier households in a neighbourhood; Ley 1996; Lees et al. 2010), and the pressing need to develop sustainable infrastructure. To remain competitive in this globalized context, Canadian cities must continue to attract talent, secure investments, and innovate in their responses to economic, social, and environmental challenges.

13.2 History of Urbanization of Canada

The history of urbanization in Canada is marked by several key phases. During the precolonial period, Indigenous peoples lived in semi-nomadic communities, without cities in the modern sense. With French colonization, urban centres such as Québec and Montreal began to emerge (Trudel 1979). The nineteenth century saw a rapid acceleration in urbanization, driven by the Industrial Revolution and massive immigration, particularly from Europe. In addition to Montreal and Quebec, other cities including Toronto and Vancouver developed around industry and burgeoning railway infrastructure.

In the twentieth century, urbanization intensified alongside industrialization with the modernization of infrastructure, while suburbanization gained momentum after World War II, spurred by the rise of the automobile and the expansion of suburban areas (Harris 2004; see Filion and McSpurren 2007). Canada experienced a growing cultural diversity with the arrival of immigrants from various regions as the country underwent a significant economic transition shaped by global forces with local ramifications (Simmons and Simmons 1974, Artibise and Stelter 1981, Artibise 1986, 1988a, 1988c; Fig. 13.3a, b).

By the twenty-first century, roughly 80% of the population is urban. Cities, increasingly integrated into a global system (Sassen 1991; Olds 2001; Walks 2001), continue to expand, yet face challenges such as densification, gentrification, affordable housing, and environmental sustainability. Public transit initiatives aim to better manage urban sprawl (Canadian Urban Institute 2019).

13.2.1 Colonial Cities

The first colonial cities in Canada began to develop in the seventeenth century, primarily along rivers and coastlines, due to their strategic importance for trade, exploration, and defence. These early Canadian cities emerged within the context of European colonization and the growth of the fur trade. Their locations were determined by strategic geographic factors such as access to waterways, trade routes, and natural resources (e.g. Harris 2008, Fougères and MacLeod 2018).

As the fur trade was a major economic driver in the development of these cities, particularly Québec, Montreal, and Trois-Rivières, most of the early colonial cities in Canada were situated along rivers or coastlines, facilitating transportation and trade. Locations were often chosen for defensive reasons as well, especially in the context of colonial rivalries, such as those between the French and the British (Trudel 1979). Access to natural resources (timber, fish, fur) was equally vital for the survival and economic growth of these colonies. These early towns were small, with rudimentary infrastructure, often planned in a centralized manner around an administrative and religious core.

Tadoussac (founded 1600) is regarded as one of the earliest French trading posts. Its position at the confluence of the Saguenay River made it an ideal location for commerce with the Indigenous Peoples. Although officially established in 1783 by Loyalists, the area of Saint John was explored by the French as early as 1604. Its position at the mouth of the Saint John River was particularly strategic for maritime trade. Port Royal, established in 1605, is recognized as the first French colony in North America. Despite facing repeated British attacks, it remains a significant symbol of French presence in the Atlantic region and the birthplace of Acadia.

Founded in 1608 on the St-Lawrence River, Québec stands as the first permanent colonial city of New France. It served as a strategic political, military, and economic centre for the fur trade and the defence against the British. Further west on the same river, Trois-Rivières founded in 1634, became a crucial transit point for the fur trade between Europeans and Indigenous Peoples. However, owing to its strategic location along the Saint Lawrence River, Montreal, founded in 1642, swiftly evolved into a significant commercial hub, further augmented by the advent of more efficient modes of transportation.

In response to British aggressions, France established Louisbourg on the Atlantic shore of Cape Breton Island in 1713. This fortress served as a strategic bastion for commerce and fishing, playing a pivotal role in French defence until its eventual capture by the British. Subsequently, the British preferred and established Halifax in 1749, in response to French influence. Halifax emerged as a strategic and virtually invincible military port, as well as a major commercial centre in the North Atlantic.

Despite some economic progress, Canada remained predominantly a natural resource exporter with minimal industrial development until the mid-nineteenth century. While the fur trade, timber, fisheries and few minerals sustained much of the economy, large-scale industrialization did not take root until advancements in transportation and the establishment of railways began to stimulate domestic manufacturing and urban growth. The rapid pace of industrialization that would follow had a profound and transformative effect on Canadian cities, driving their expansion and spurring the development of entirely new urban centres. Factories and railways attracted a large influx of workers, leading to significant population growth, while cities such as Montreal, Toronto, and Hamilton emerged as industrial powerhouses. This urbanization reshaped the social and economic fabric of the country, fostering the rise of distinct working-class neighbourhoods, expanding infrastructure, and creating a demand

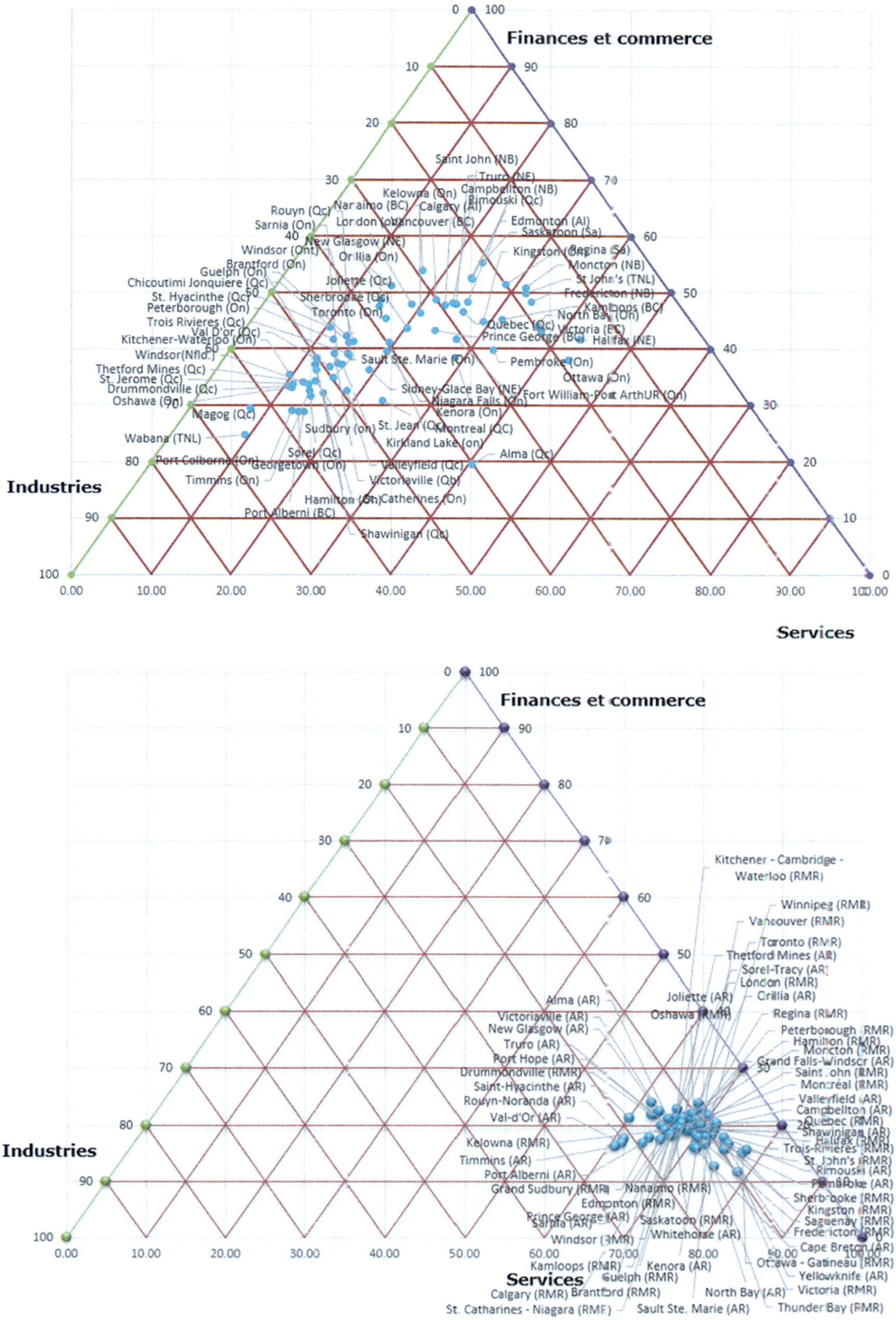

Fig. 13.3 Economic roles fulfilled by Canadian Cities. A: 1961; B: 2021. (Guy Vincent; A: adapted from Simmons and Simmons 1974)

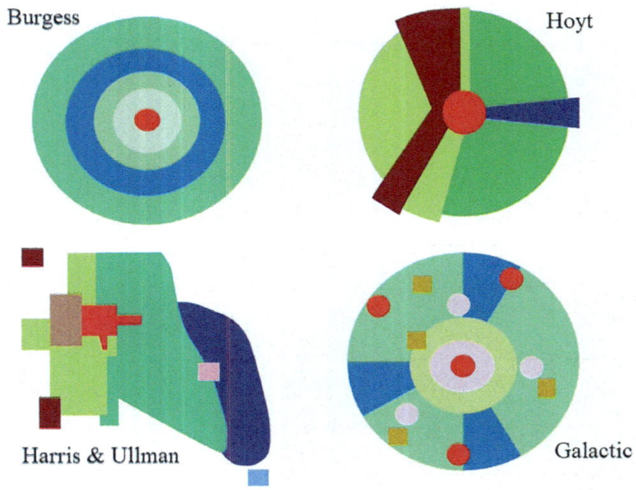

Fig. 13.4 Models of city structure. Top left: The Burgess model features a Central Business District core (red), surrounded by concentric rings of lower-income housing (pale green), middle-income (green), higher-income (blue), and mid-high income commuter housing (green). Top right: The Hoyt sectoral model features a Central Business District core (red), flanked by sectors of lower-income housing (pale green), middle-income (green), higher-income (blue), and industrial development (brown). Bottom left: The Harris and Ullman model features a Central Business District core (red), flanked by sectors of lower-income housing (pale green), middle-income (green), and light industrial development (light brown). Heavy industrial (brown) and higher-income housing (blue) are distal to the core. Secondary business cores (pink) and distal suburbs (light blue) are also present. Bottom right: The Galactic City model features a Central Core (red), multiple secondary cores (pink), sectors of lower-income (pale green), middle-income (green), and higher-income housing (blue), and multiple industrial parks with light industry (light brown). See text for discussion

for housing, services, and new industries that would define Canada's urban landscape for decades.

The next steps of Canadian urban development can be interpreted through the frameworks of classical land use models, particularly those developed by the Chicago School. These models, such as Burgess's Concentric Zone Model, Hoyt's Sector Model, and Harris and Ullman's Multiple Nuclei Model, offer insights into how economic forces, transportation networks, and social stratification shape cities (Fig. 13.4). In Canada, the application of these models can help understand the ongoing trends of suburbanization, densification, and polycentric growth, particularly in larger metropolitan areas, where distinct zones of business, residential, and industrial areas evolve in response to both local and global dynamics.

13.3 Classical Theoretical Models of Land Use

The classical models of land use are urban theories that describe the spatial distribution of human activities within urban environments, based on economic, social, and geographical factors. To fully understand the evolution of urban areas, it is essential to present these influential models, which laid the foundation for modern urban studies.

13.3.1 Concentric Zone Model (Burgess 1925)

Ernest Burgess introduced the idea that cities expand in concentric rings from a central point, typically the Central Business District (CBD). Each zone in Burgess's Concentric Zone Model represents a distinct pattern of land use, reflecting socio-economic gradients. At the core lies the Central Business District (CBD), a hub of commercial and administrative activity. Surrounding it is the "zone in transition", marked by industrial activities and deteriorating housing, often occupied by recent immigrants or low-income residents. As one moves outward, successive rings depict progressively more desirable residential areas, from working-class neighbourhoods to middle- and upper-class suburban zones, where housing quality and socio-economic status improve with distance from the city centre. Regional growth manifests as the creation of new businesses in the central business district (CBD), which triggers a ripple effect through neighbouring zones. This phenomenon of "invasion and succession" occurs as industries and populations expand outward, invading the next contiguous zone. Consequently, higher-income residents move outward to less congested zones, while lower-income groups settle into the vacated areas closer to the CBD, a pattern known as the "filtering process".

The housing filtering process refers to the gradual movement of housing through different economic classes over time. As newer, higher-quality housing is built, wealthier households tend to move into these new homes. This creates a process where older housing stock, now considered less desirable, "filters down" to lower-income households. The housing filtering process can happen for several reasons, including physical depreciation of buildings, changes in neighbourhood characteristics, or economic shifts (Bourne 1981, CMHC, Davidoff and Somerville (2025)).

In this model, affluent residents are generally the first to occupy new housing. Over time, as these buildings age and deteriorate, or as newer housing becomes available, higher-income residents move to newer or better-maintained areas. The older housing then becomes affordable to middle-income residents, and the cycle continues, with lower-income households eventually occupying the oldest and most worn-out housing.

This process has significant implications for urban planning, affordable housing, and social inequality. While filtering can provide affordable housing options for lower-income groups, it can also lead to a concentration of poverty in aging neighbourhoods that may lack investment and services. Urban planners often monitor the filtering process to manage housing policy and ensure sustainable development across all income groups.

The Concentric Zone Model developed by Ernest Burgess in 1925 emerged in the context of rapid urbanization and industrialization in the early twentieth century, particularly in the city of Chicago. Burgess, a sociologist at the University of Chicago, was part of the "Chicago School of Sociology", a group of scholars who focused on understanding the dynamics of urban environments. At the time, Chicago was a booming industrial city, experiencing a massive influx of immigrants and migrants, and undergoing significant economic, social, and spatial transformation.

This period of rapid industrial growth, combined with the development of railroads and streetcars, contributed to stark divisions in land use and social class, creating distinct zones within cities. Burgess observed that the city's expansion followed a pattern of concentric circles, where different social groups and economic activities were spatially arranged in distinct zones based on their proximity to the Central Business District (CBD).

Burgess's Concentric Zone Model reflects this dynamic, with the idea that cities grow outward from a central point, with distinct zones radiating from the centre. His model offered a way to understand how factors such as socio-economic status, land use, and industrial growth influenced urban form, and it became foundational in urban sociology and planning.

Burgess's model was not just descriptive but also emphasized how economic and social forces shaped urban geography, leading to the patterns of residential segregation and urban sprawl still observed today. The model was a product of its time, capturing the spatial dynamics of early 20th-century industrial cities such as Chicago.

13.3.2 Sector Model (Hoyt 1939)

Hoyt's sector model introduces the idea of urban growth along transportation routes, suggesting that cities expand not only in concentric rings but also in elongated sectors along major roads or railways. This model also accounts for the influence of topography, where higher-income groups often settle on elevated land, away from the crowded and industrial areas below. A compelling example of this dynamic is James Ames' "The City Below the Hill" (1897), which contrasts the working-class St-Henri neighbourhood in Montreal, located in the lower areas near industrial zones, with Westmount, an affluent suburb situated on the slopes of Mount Royal. The stark difference in social class between these two areas, divided by both physical geography and the railway tracks, underscores how topography and transportation corridors shape urban land use, as wealthier residents seek distance from the city's bustling and industrial districts.

Hoyt's Sector Model can be seen as a natural progression from Burgess's Concentric Zone Model. While Burgess conceptualized cities as expanding in concentric circles outward from the central business district (CBD), Hoyt introduced the idea that urban growth is more dynamic, developing along key transportation routes and influenced by natural geography. This addition of movement, shaped by circulation axes like railways and major roads, allowed for a more nuanced understanding of urban development, where higher-income residential areas often grow in specific sectors, away from industrial zones. Hoyt's model built on Burgess's foundation but accounted for more complexity in urban growth patterns. For instance, in 1926, Harland Bartholomew was commissioned to develop a city plan for Vancouver. His approach, inspired by the concentric zone model, emphasized the need for wide streets and a comprehensive layout that would accommodate future growth. However, Bartholomew's focus on transportation infrastructure, particularly major roads, gradually aligned Vancouver's development more with Hoyt's sector model, where urban growth followed transportation routes. Bartholomew's influence extended well beyond Vancouver; his firm created urban plans for over 500 cities and counties across the U.S., profoundly shaping American urban form (Kollatz 2017). He is often regarded as a major figure in the development of the modern American city, playing a pivotal role in shaping city planning principles.

Homer Hoyt's model suggested that urban growth follows sectors rather than concentric rings. These sectors radiate out from the CBD along transportation routes. Residential zones, industrial areas, and commercial districts expand outward in wedges, depending on factors like income or proximity to major roads. This model is a logical evolution of the Burgess model as it highlights the influence of transportation and social stratification on urban land use. Calgary's land use eloquently illustrates this model (Artibise 1988b).

Homer Hoyt developed the Sector Model (1939) as a refinement of the earlier Concentric Zone Model by Ernest Burgess. Hoyt, an economist and land-use planner, aimed to provide a more detailed explanation of urban growth, particularly in response to the role of transportation infrastructure and how it influences city development. Using a wide array of statistical data, his model was based on the study of land values and urban patterns in Chicago, much like Burgess, and 141 more cities, but offered a different perspective by focusing on how cities expand along transportation corridors, like railways or major roads.

The Sector Model proposed that urban growth occurs in wedges or sectors, radiating outward from the city centre, rather than in uniform concentric rings. Certain land uses, especially residential areas, develop along particular axes, typically following transportation routes. High-income residential areas, for instance, are likely to extend in specific directions, away from industrial zones, benefiting from better transportation and scenic landscapes. Lower-income housing tends to be concentrated near industrial areas, often downwind due to pollution concerns.

Hoyt's model was particularly relevant during the first half of the twentieth century, a time when the automobile and rail systems had an increasing influence on urban sprawl and neighbourhood development. The model has since been used to explain how socio-economic classes distribute within cities, focusing on the roles of transportation and geographical features in shaping urban form. It was particularly influential for understanding American and Western urban patterns during the mid-twentieth century.

13.3.3 Multiple Nuclei Model (Harris and Ullman 1945)

This model recognizes that large cities often grow around multiple centres, each serving a distinct function (e.g. residential, industrial, commercial). Instead of a single CBD, cities evolve with various specialized hubs or "nuclei", reflecting the complexity of urban activities. This model is particularly relevant for modern cities where multiple business districts and industrial zones coexist across different regions.

The model was developed in response to the limitations of earlier urban models like the Concentric Zone Model by Burgess and the Sector Model by Hoyt. While Burgess and Hoyt explained urban growth around a single city centre, Harris and Ullman recognized that as cities grew larger and more complex, they did not revolve around just one central business district (CBD). Instead, they proposed that cities develop multiple "nuclei" or centres of activity.

Their model reflects the urban realities of mid-twentieth century American cities, where suburbanization, increasing automobile use, and diverse economic activities were leading to polycentric city structures. In these cities, different sectors or neighbourhoods evolved around specific functions: a commercial centre, an industrial zone, or a residential area. These distinct nodes of development often emerged around specialized land uses, such as manufacturing districts, shopping centres, or university campuses, creating a multi-centred urban structure.

The multiple nuclei model underscores that modern cities grow organically, responding to economic activities, social patterns, and geographical factors. For instance, industrial centres tend to develop on the outskirts where land is cheaper and space is more available, while residential areas for the wealthy may cluster in different parts of the city, often avoiding proximity to industrial zones.

Harris and Ullman's model emphasizes that a city's growth is shaped by various forces that create distinct, separate centres of activity rather than expanding uniformly from a single core. This was particularly reflective of the postwar urban development seen in the U.S., but it has been applied globally to understand complex city structures.

13.3.4 Peripheral Model or Galactic City Model, 1960s) (Harris 1978, 1997; Lewis 2013; Vance 1964; City of Montreal 2021, Among Others)

As cities expanded during the mid-twentieth century, this model emerged to explain the development of suburban areas. Here, peripheral zones became increasingly important, with residential and commercial growth spreading outwards from the core, often linked by highways and transportation networks. Unlike earlier models, the traditional downtown area loses its dominance, as suburban nodes become centres of activity.

The Galactic City Model, also known as the Peripheral Model, was developed in the 1960s to describe the emerging spatial organization of modern cities in the postindustrial era, particularly in North America (Pierce 1983). It reflects the shift away from the traditional centralized city structure seen in earlier models like the Concentric Zone or Sector models, towards a more decentralized, polycentric urban form driven by the rise of the automobile and suburbanization.

This model depicts cities as sprawling, decentralized entities with a strong focus on the periphery. Unlike earlier models, which positioned the central business district (CBD) as the core around which all urban development occurred, the Galactic City Model recognizes that cities are increasingly characterized by a diffuse network of nodes, or "urban realms" connected by highways and ring roads. These nodes—such as suburban shopping malls, business parks, and residential suburbs—function as semi-autonomous centres of activity that reduce the primacy of the downtown area.

The model captures the dynamics of the post-World War II era, where suburbanization, car dependency, and the expansion of highways led to the growth of edge cities (Garreau 1981) and peripheral commercial districts. The rise of these decentralized zones reflects the way that cities expanded outwards, with the urban core often losing its traditional dominance over the metropolitan area. This model applies particularly well to cities that experienced rapid suburban growth, like Los Angeles, Toronto, or Atlanta, where suburban nodes developed to meet the needs of growing populations living further from the downtown.

In summary, the Galactic City Model explains the increasingly polycentric nature of urban development, highlighting how transportation infrastructure and suburbanization redefined the layout and functioning of metropolitan areas in the late twentieth century.

13.3.5 Classical Land Use Models as Continuum of Change

These models provide crucial insight into the historical evolution of cities, but contemporary factors such as globalization, technological advances, and sustainability challenges have added new layers of complexity to traditional urban forms. Understanding these classical models is key to appreciating the patterns and drivers of modern urbanization.

But it is trivial to understand that they are evolutive models and not separate from one another as the example of Bartholomew's Vancouver plan illustrates. Comparing the zones of classical land use models to elastic bands offers a useful metaphor for understanding the fluidity and adaptability of urban spaces. In both Burgess's Concentric Zone Model and Hoyt's Sector Model, the zones or sectors represent different land uses that are subject to pressures that cause them to shift and stretch.

In the Concentric Zone Model, each ring (zone) represents a specific land use, from the central business district (CBD) outward to residential zones. Much like elastic bands, these zones are not fixed. They stretch and contract depending on factors like economic growth, migration, or changes in urban policy. For example, the "transition zone" (often characterized by industrial activity and lower-income housing) can expand outward as industries invade adjacent residential areas, or contract if gentrification occurs and wealthier populations reclaim the space.

Similarly, in the Sector Model, urban growth occurs along sectors, typically aligned with transportation routes. These sectors can be seen as elastic bands stretched along highways or railways. Higher-income groups may push sectors of wealthier housing outward, while industrial or commercial uses may also expand along these routes, exerting pressure on adjacent zones. The metaphor of elastic bands here reflects how these sectors can stretch in response to the city's needs, reflecting dynamic changes in the urban fabric, until they break into small pieces thus forming specialized nodes throughout the metropolitan areas.

The Multiple Nuclei Model of Harris and Ullman, and the Galactic City Model, extend this concept further by imagining a city that is no longer dominated by a single centre but by multiple hubs. Each hub acts like a node where different land use types interact, and the elastic bands connecting these zones may be even more fluid, allowing for a more decentralized, sprawling urban pattern.

In all these models, the flexibility of zones, much like elastic bands, reflects the pressures of growth, socioeconomic change, and infrastructural development, adapting to the ever-shifting demands of the urban environment.

13.4 Industrialization and Urbanization in Canada

The intertwining of urbanization and industrialization in Canada is both profound and pivotal, shaping the nation's economic and social landscape since the late nineteenth century. Industrialization, marked by the shift from an agrarian economy to one rooted in manufacturing and the exploitation of natural resources, became the catalyst for rapid urban growth, forging the rise of cities across the country.

As the driving force behind urbanization, industrialization spurred the expansion of Canada's manufacturing industries in major cities such as Montreal, Toronto, and Hamilton. These urban centers, strategically located near the Great Lakes and key rivers, were ideally positioned for the transportation of raw materials and finished goods. The burgeoning industrial economy generated employment opportunities in the production sectors, drawing waves of workers from rural areas into the urban fabric. However, while industrialization prompted the growth of large cities and manufacturing hubs, these very cities have had to reinvent themselves over time to adapt to globalization and the new economic and environmental challenges.

Montreal emerged as a significant industrial hub, particularly in textiles food production, and the transportation industries (Linteau 1990). The development of the railway system further accelerated its urban sprawl, linking the city to broader national and international markets. Similarly, Toronto, initially known as a commercial centre experienced swift industrial growth, with its manufacturing sector expanding rapidly in textiles, steel, and chemicals, marking its transformation into a key player in Canada's industrial narrative.

13.4.1 Urbanization and the Specialization of Industrial Cities

Industrialization also spurred the development of new infrastructure, including railways, ports, and roads, which facilitated the movement of goods and labour. This enabled industrial cities to expand rapidly, becoming magnets for rural populations. The construction of such infrastructure, in turn, created jobs, drawing even more people to urban centres and further accelerating their growth. Urbanization driven by industrialization led to the specialization of certain cities based on their dominant industries.

Cities such as Hamilton in Ontario became specialized in heavy industry, particularly steel production. Hamilton emerged as one of the country's leading industrial cities, home to Dofasco and Stelco, solidifying its role as a manufacturing and metal processing hub.

In other regions, the exploitation of natural resources gave rise to industrial cities. Sudbury, Ontario, for instance, flourished due to its mining industry, primarily nickel extraction. This industrial growth attracted workers and stimulated the development of urban infrastructure necessary to support a growing population.

Cities including Vancouver, Winnipeg and Halifax played key roles as transportation and trade centres, facilitating the export of natural resources (such as timber, wheat, minerals, and oil) and manufactured goods. These cities thrived as crossroads in global trade and maritime networks.

13.4.2 Social Effects of Urbanization and Industrialization

Industrialization and urbanization profoundly transformed Canadian society, bringing about changes in working conditions, social dynamics, and population distribution. Industrialization in Canada, particularly after Confederation in 1867, led to a radical transformation of cities. Urban centres became major industrial hubs. Industrialization gave rise to a new urban working class, often composed of European and rural immigrants seeking employment in factories and mines. This internal and external migration to urban centres resulted in a significant population increase in cities. This model of development was characterized by industrial concentration along ports and railway lines and stratified neighbourhood development. The influx of workers, often immigrants, fuelled the rapid expansion of cities. Social classes became segmented, with workers often living near factories in poor conditions, often overcrowded with poor sanitation, while the middle and upper classes settled in more distant, affluent and exclusive neighbourhoods, Montreal's Westmount, Toronto's Rosedale, Winnipeg's Tuxedo Junction and Vancouver's Shaughnessy are such examples.

13.4.3 Shift to a Diversified Urban Economy and the Rise of Suburbia

In the twentieth century, after World War II, urban expansion took on a new dimension. Canada witnessed a profound transformation in its urban landscape, marked by the rapid expansion of suburbia. As returning veterans sought homes and families grew during the baby boom, the suburban ideal took root, fuelled by the rise of the automobile and the construction of highways that connected these newly minted residential enclaves to the urban core. Suburbia became synonymous with the promise of space, home ownership, and a middle-class lifestyle—features that lured city dwellers to the quiet edges of towns.

Canada also opened its doors to numerous waves of immigrants, particularly from Europe, and from the 1970s onward, from Asia, Latin America, and other parts of the world (see Chap. 11). The country's major cities became multicultural hubs, contributing to both the cultural diversity and the economic dynamism of urban life. Urbanization accelerated, and Canadian cities began to expand more diffusely toward the outskirts. The construction of roads and highways, intended to improve access to city centres, as well as the availability of affordable mortgages, had the unintended consequence of encouraging the development of residential neighbourhoods farther afield and drawing families away from crowded urban centres. Residential areas were separated from commercial and industrial zones, creating urban environments heavily reliant on cars.

The simultaneous population surge of the baby boom and the consequent increase in housing, economic growth, and the rise of the widespread use of automobiles allowed people to move away from city centres and settle in residential suburbs to fuel urban sprawl. Cities like Vancouver, Calgary, and Edmonton experienced rapid growth, as the construction of major public infrastructures such as highways and bridges facilitated the creation of suburbs that, in turn, necessitated investments in health, education and public works infrastructures. Large urban areas saw the birth of sprawling suburban communities, with single-family homes dotting vast stretches of land, neatly arranged along wide streets and cul-de-sacs.

Over time, these suburbs evolved into more than just bedroom communities. Suburban poles—new commercial and business hubs on the city's periphery—began to emerge, creating a decentralization of economic activity. Places such as Mississauga near Toronto or Burnaby near Vancouver, or Laval near Montréal transformed into thriving centres in their own right, offering shopping malls, corporate offices, and industrial parks. This rise of suburban poles redefined the geography of Canadian cities, creating a multi-nodal urban structure where the once clear boundary between city and suburb blurred. These poles became magnets of economic opportunity, shaping the country's urban identity and paving the way for a new era of suburban modernity. Some examples such as the following may appear surprising!

13.4.3.1 "Suburbia, Bridge and… Beer!"

Although its construction predates World War II, the Lion's Gate Bridge is a good example of infrastructure that facilitated the spread of postwar suburban Vancouver.

The Lion's Gate Bridge, a Vancouver landmark, was brought to life by the British Pacific Properties, a real estate development company owned by the Guinness family, famed for their brewing empire (Places that Matter 2016,

Changing Vancouver–Then and Now 2020, British Pacific Properties 2024, Canada's Historic Places 2024, March 2024). The project began in 1937, with construction completed by 1938, primarily to unlock the residential potential of West Vancouver. Recognizing that a direct link to downtown Vancouver would elevate the value of their North Shore land holdings, the Guinness family strategically invested in this ambitious infrastructure (National Post 2015).

Designed and constructed by the engineering firm Monsarrat and Pratley, with the collaboration of multiple contractors, the bridge was privately funded by the Guinness family. Upon its completion, the bridge was later sold to the City of Vancouver and the provincial government in 1955. Its official opening on November 14, 1938, marked the beginning of a new era for Vancouver's connectivity and suburban expansion.

Stretching majestically across Burrard Inlet, the Lion's Gate Bridge connected the North Shore to the heart of the city, transforming previously isolated areas like West and North Vancouver into burgeoning residential and commercial hubs. Before its construction, access to the North Shore was limited primarily to ferries, hindering significant growth.

As car ownership surged in the post-World War II years, the bridge became a critical artery, facilitating the expansion of suburban neighbourhoods. Residents could now live farther from the city centre while easily commuting to Vancouver's urban core. This development paralleled the wider suburbanization trend seen across North America during the mid-twentieth century, where highways and bridges paved the way for suburban communities.

In recent years, the Lion's Gate Bridge continues to be essential to the region's transportation network. However, increased traffic congestion underscores the challenges of sustainable suburban growth, as Vancouver balances development with the need for environmental stewardship and infrastructure modernization.

13.4.4 The Era of Canadian "Metropolization"

This era also marked the emergence of Canadian "megacities," with the accelerated growth of metropolitan regions, particularly Toronto, which surpassed Montreal to become the economic hub of the country. However, this expansion brought significant challenges, such as managing public services, social segregation, and sustainable urban planning in the face of sprawling cities (e.g. Donald 2005, Hodge and Gordon 2014). Over time, urban centres that were initially focused on industrialization evolved into diversified economies, where the service, finance, technology, and education sectors gained increasing prominence. However, the resurgence of city centres was not an immediate process. For a time, urban cores languished as households flocked to the emerging suburban landscapes, lured by the promise of more space and modern amenities. Manufacturing sectors, confined by the limitations of older urban infrastructure, struggled to expand or integrate more efficient machinery, while retail businesses, facing the exodus of their once-loyal urban clientele, turned to suburban shopping malls in "search of the lost customers". Many city centres slipped into a gradual decline, with once-thriving downtowns witnessing the slow drain of vitality and economic energy.

The Canadian Ministry of State for Urban Affairs was established in 1971 as a federal department tasked with addressing the growing challenges associated with urbanization in Canada. Its creation was spurred by the rapid postwar urban expansion and the rising complexity of issues in Canadian cities, such as housing, transportation, infrastructure development, and the deteriorating condition of urban centres (Spicer 2011). The ministry sought to coordinate federal efforts with provincial and municipal governments to ensure coherent urban planning and to support cities in handling population growth, suburban sprawl, and the emerging need for sustainable development. The ministry was disbanded in 1979, and urban policy initiatives were absorbed into other governmental departments or left to provincial and municipal authorities.

Although short-lived, the Ministry of Urban Affairs contributed to the national conversation about the importance of urban planning and the federal government's role in supporting sustainable and liveable cities. Its legacy is seen in later federal initiatives focusing on housing, infrastructure, and environmental sustainability.

McLemore et al. (1975) addressed the conditions of ten Canadian metropolitan areas. It primarily focused on the transformation of Canadian cities, particularly in the 1960s and early 1970s, and outlined the socio-economic shifts in urban centres, driven by policies that focused on housing, transportation, and industrial development. These policies led to both the decline and revitalization of inner-city areas. The report also highlights how inner-city changes were influenced by the broader trends of deindustrialization, suburbanization, and shifts in public policy. They developed a relatively simple typology that identified four distinct types of neighbourhoods that characterized the inner cities they studied: neighbourhoods that were declining, stable, revitalizing and undergoing massive redevelopment. Though straightforward, their findings indicated that distinct processes required tailored tools and approaches, highlighting the need for a nuanced, flexible methodology.

Their study emphasized the evolving roles of urban spaces, transitioning from industrial centres to hubs of service-oriented economies, leading to a transformation in both the physical landscape and the demographic makeup of inner cities. Moreover, public policy, though not solely responsible for urban decline, played a pivotal role in both the deterioration and recovery of these areas.

13.4.4.1 Transition to a Service-Based Economy

As Canada's traditional manufacturing economy began to decline from the 1970s due to international competition and technological changes, the post-war era was also marked by the modernization of industry, with the rise of high-tech industries, financial services, and new economic sectors (see Chap. 18). Many cities diversified towards the service sector. Toronto, for instance, transformed into a major world class financial hub, while Montreal invested in creative industries and information technologies (Conseil des Arts du Canada 2019). This development solidified the position of certain cities as industrial and economic hubs, particularly in the provinces of Ontario and Québec.

13.4.4.2 Impact on Urban Geography

This diversification led to the expansion, transformation and reorganization that has profoundly changed the structure of Canadian cities, their geographic configuration, and their role in the national and global economy. The downtown areas of many major cities were redeveloped to accommodate skyscrapers, shopping centres, and office complexes. In Montreal, Expo 67 spurred a wave of development, while Toronto, benefiting from the economic growth of the financial sector, saw the rise of its modern business district. Former industrial districts were often converted into residential or commercial areas. Urban revitalization projects, such as the redevelopment of Montreal's Old Port and Vancouver's Granville Island, emerged to modernize disused industrial zones while accommodating the needs of growing urban populations. Simultaneously, suburbia has continued to spread.

13.4.5 Place Ville Marie: A Contemporary Cathedral and Montreal's Underground City

Place Ville Marie (PVM) is one of Montreal's most iconic skyscrapers and a defining landmark in the city's downtown core, symbolizing the beginning of Canada's architectural "race to the skies". Since its completion in 1962, Place Ville Marie has stood as a defining structure in Montreal's skyline, not just as a commercial hub but as a symbolic "cathedral" of modernity and has played a pivotal role in the modernization of Montreal and the development of the Underground City (RÉSO).

The idea for Place Ville Marie emerged in the late 1950s, as Montreal sought to transform itself into a modern financial and commercial hub. The project was developed by New-York city promoter William Zeckendorf and backed by the Royal Bank of Canada, which wanted to establish a new headquarters.

Designed by I.M. Pei, its sleek, modernist and towering presence with a distinctive cruciform (cross-shaped) design, grand esplanade, and central location reflect the city's transformation in the mid-twentieth century. Just as historic cathedrals once stood as the heart of a city's social and spiritual life, Place Ville Marie has served as a contemporary epicentre of business, culture, and urban connectivity. This unique shape maximized natural light and offered panoramic views of the city. At 188 m (617 feet) and 47 stories tall, it was the tallest building in Canada at the time and remained the tallest in Montreal until the completion of 1000 de La Gauchetière in 1992.

When it opened in 1962, Place Ville Marie became a symbol of Montreal's transformation into a global financial centre. It attracted major corporations, banks, and law firms, cementing its status as the city's business heart.

13.4.5.1 Architectural and Symbolic Parallels

In the medieval era, cathedrals were designed to be awe-inspiring structures, drawing people inward for communal gathering, contemplation, and worship. Place Ville Marie, with its cruciform (cross-shaped) design, evokes a similar sense of purpose—not for religious devotion, but for the pursuit of progress, commerce, and innovation.

The central tower, soaring 188 m (617 feet) above Montreal, mirrors the vertical ambition of gothic cathedrals, which reached toward the heavens. Its beacon light, visible at night, acts as a modern-day equivalent of a church spire, guiding residents and visitors alike.

Below the tower, the Esplanade PVM serves as an open gathering space, much like the plazas in front of traditional cathedrals. It has hosted celebrations, protests, and public events, reinforcing its role as a civic centre where people converge to share in collective experiences.

Much like a cathedral attracted merchants, pilgrims, and thinkers, Place Ville Marie has been a magnet for business leaders, intellectuals, and everyday Montrealers. The underground shopping complex—the first piece of the Underground City (RÉSO)—transformed the way people interacted with the urban space, providing a sheltered, interconnected hub that remains a key part of Montreal life.

The building's role in the economy is also akin to how cathedrals once fuelled commerce in medieval cities. Housing major corporations, banks, and cultural institutions, PVM has long been a symbol of economic faith in the city's future.

While traditional cathedrals were places of prayer and reflection, modern society often finds its rituals and sense of belonging in different spaces. The daily rhythm of office workers ascending its elevators, the public gatherings on its esplanade, and the cultural events hosted within its walls all point to a reimagined sense of community.

Through its architectural grandeur, symbolic cross-shaped structure, and role as a gathering place for commerce and culture, Place Ville Marie stands as a contemporary cathedral—a testament to Montreal's aspirations, resilience, and identity in the modern world.

13.4.6 Montreal's Underground Network

At the same time, PVM was a catalyst for the Underground City (RÉSO). Its construction included a shopping concourse beneath the tower, marking the beginning of Montreal's vast underground pedestrian network. This innovation allowed people to travel comfortably between buildings, metro stations, and commercial spaces without exposure to the city's harsh winters.

RÉSO is one of the largest underground pedestrian networks in the world. Spanning over 32 kms (20 miles), this extensive system connects metro stations, shopping centres, office towers, hotels, cultural institutions, and more. It serves as a vital part of daily life for thousands of Montrealers, throughout the year.

The Underground City dates back with the construction of Place Ville Marie, a major office and retail complex. The project included an underground shopping mall, which became the first section of what would later expand into a vast interconnected network.

In 1966, the opening of Montreal's Metro system accelerated underground expansion. Many metro stations were integrated into underground walkways, allowing pedestrians to move through downtown without exposure to extreme weather. By the 1980s and 1990s, new shopping centres, office buildings, and cultural sites were connected, forming the extensive network that exists today.

Montreal's Underground City is a vast, interconnected network that offers a unique urban experience, blending convenience, commerce, and culture. One of its defining characteristics is its extensive interconnectivity, linking over 1,700 shops, 200 restaurants, office buildings, hotels, and cultural landmarks. Pedestrians can move seamlessly between major locations such as Place des Arts, the Bell Centre, Complexe Desjardins, and the Montreal Convention Centre without stepping outside.

Another key feature of RÉSO is its climate-controlled environment, making it an essential part of daily life, especially during Montreal's harsh winters and hot summers. The underground passageways provide a comfortable alternative to outdoor streets, allowing people to commute, shop, and dine, and even run while avoiding extreme weather conditions. Incidentally RÉSO is not only a convenient way to navigate the city during winter but also a venue for exciting running events. One of the most notable is the 5KM BougeBouge Montreal Underground, an annual race that takes participants through the vast network of tunnels, shopping centers, and metro connections beneath downtown Montreal.

Accessibility is also a major strength of the Underground City. With multiple entry points integrated into metro stations, office towers, and shopping centres, it is designed for ease of navigation. Its connection to the Montreal Metro system ensures smooth transit for thousands of daily users, reinforcing its role as a key part of the city's infrastructure.

Beyond its practical use, RÉSO is also a cultural and social hub, hosting art exhibits, concerts, and seasonal events throughout the year. The underground space is not just a transit route but a lively area where visitors can experience Montreal's vibrant culture. Public art installations and architectural features add to the aesthetic appeal of the network, transforming it into a unique urban landscape.

Finally, the Underground City plays a significant role in Montreal's economy, supporting businesses and attracting both locals and tourists. The retail spaces, dining establishments, and entertainment venues contribute to the city's commercial vitality, making RÉSO not just a passageway but a thriving economic zone.

With its seamless connectivity, weather-proof environment, cultural significance, and economic impact, Montreal's Underground City stands as a remarkable feat of urban planning, enhancing both the functionality and experience of the city's downtown core.

13.4.7 Underground (and Aboveground) Networks in Canadian Cities

Consequently, several major Canadian cities have developed extensive underground pedestrian networks, allowing residents and visitors to navigate urban centres without exposure to extreme weather conditions. These systems, designed for convenience and efficiency, connect commercial buildings, transit hubs, shopping centres, and cultural institutions, playing a crucial role in the daily lives of many Canadians.

Toronto's PATH system is another remarkable underground network, stretching 30 kms (19 miles) beneath the city's downtown core. Originally developed in the early twentieth century and significantly expanded in the 1970s, PATH is recognized as the largest continuous underground shopping complex in the world, with over 1,200 stores and services. It provides direct access to Union Station, Scotiabank Arena, the CN Tower, and Eaton Centre, making it a vital hub for both commuters and tourists.

In contrast to the underground networks of Montreal and Toronto, Calgary's + 15 Skywalk System takes a different approach. Instead of tunnels, it features elevated pedestrian bridges that connect buildings 15 feet above ground. Covering 18 kms (11 miles), the system was developed in the 1970s to facilitate pedestrian movement during harsh conditions. It links key locations such as Calgary Tower, the CORE Shopping Centre, and various office towers, contributing to the city's walkability.

Edmonton also has an underground system known as the Pedway, which extends 13 kms (8 miles) and combines underground pathways, above-ground walkways, and enclosed bridges. Established in the 1970s, the Pedway connects important destinations such as Edmonton City Centre Mall, Churchill Square, and Rogers Place, serving as an essential part of the downtown transportation network.

Unlike other major cities, Vancouver has a limited underground walkway system, primarily due to its milder climate. The existing underground connections are found mainly between Pacific Centre Mall, Vancouver City Centre, and Waterfront Station, offering a more modest but still useful pedestrian network.

These underground systems provide numerous benefits, including protection from extreme weather, economic stimulation through retail and tourism, improved transit connectivity, and enhanced urban accessibility. While Montreal and Toronto boast the largest and most extensive networks, cities like Calgary and Edmonton have adapted their systems to meet local needs. As urban centres continue to grow, these underground and elevated networks remain essential components of Canada's modern cityscapes, shaping the way people navigate and experience urban life.

On a smaller scale, both the Halifax Pedway (1.5 km) and Winnipeg Walkway (2 km) System serve as essential winter-friendly pedestrian networks, improving walkability in their respective downtown areas. While smaller than the systems in Montreal or Toronto, they enhance urban connectivity, support local businesses, and improve accessibility for residents and visitors.

Since the 1960s, Canada's Central Business Districts (CBDs) have undergone significant growth, driven by economic expansion, urbanization, and infrastructure development. The rise of modern skyscrapers in the 1960s–1980s marked a new era, with major projects like Place Ville Marie (1962) and Toronto-Dominion Centre (1967) transforming Montreal and Toronto into financial hubs. The oil boom fuelled the rapid development of Calgary and Edmonton's downtowns, while expanding public transit systems improved connectivity in all major cities.

By the 1990s and 2000s, Canadian CBDs evolved beyond financial centres, incorporating mixed-use developments that blended commercial, residential, and entertainment spaces. Toronto solidified its position as Canada's financial capital, while Montreal embraced tech and cultural industries. Cities like Vancouver attracted international investment, and Alberta's urban centres diversified their economies beyond oil.

In the 2010s and beyond, sustainability and technology became key drivers of change. The rise of LEED-certified skyscrapers, public transit expansions, and global tech investments reshaped urban cores. The COVID-19 pandemic accelerated shifts toward hybrid work, prompting cities to rethink office spaces and explore residential conversions. Today, Toronto remains Canada's leading business hub, while Montreal, Vancouver, Calgary, and Edmonton continue adapting to new economic realities, emphasizing innovation, sustainability, and urban liveability.

13.5 Towards Postindustrial Cities

In the final decades of the twentieth century and the early twenty-first century, Canadian cities underwent a transformation driven by globalization, continued population growth, and economic diversification (Hutton 2009, Bunting and Filion 2015, Vinodrai 2015; Figs. 13.5, 13.6). Nonetheless, although the role of traditional industries has diminished, the connections between industrialization and urbanization continue to shape Canadian cities today.

13.5.1 Industrial Cities and Deindustrialization

The shift from industrial to service-based economies has redefined Canadian cities. Deindustrialization brought about the decline of traditional manufacturing jobs, while tertiarization has fostered new opportunities in the service sectors. However, this transformation has also introduced new challenges, particularly in terms of social inequality and urban affordability (e.g. Vinodrai 2015). The ongoing adaptation of these cities continues to shape Canada's urban future. Industrial cities such as Hamilton and Windsor as well as large metropolitan regions that contained heavily industrialized districts such as Montreal, Toronto and Vancouver have faced the challenges of deindustrialization and significant economic transitions, grappling with factory closures and the relocation of production elsewhere in Canada or countries with cheaper labour. Former industrial zones became derelict, leading to increased poverty and a reduction in the urban tax base. Cities had to diversify their economies, often focusing on devel-

Fig. 13.5 Ongoing high-rise construction, a common scene in postindustrial Canadian downtowns. (Guy Vincent photo 2007)

Fig. 13.6 From the top: view of the urban system of Toronto from the CN Tower. (Guy Vincent photo)

oping new industries, such as technology or education, and revitalizing downtown areas. Efforts to revitalize these cities and districts, often focused on infrastructure modernization, sustainable development, and attracting new industries (such as technology, specialized services and renewable energy), have become essential. The workforce profile has undergone a profound transformation over the decades, marked by a significant rise in educational attainment and an increasing presence of women, who have played a pivotal role in driving the tertiarization of the Canadian economy. This shift reflects broader social and economic changes, as both education and gender diversity have become cornerstones of modern labour dynamics, steering the country away from its industrial roots toward a service-oriented economic landscape.

As a result, urban centres have undergone significant densification and modernization, marked by the emergence of residential towers, expansive commercial complexes, and modern office buildings. This transformation is particularly evident in Toronto, Montreal, and Vancouver, which have evolved into key service hubs, focusing on finance, technology, and

Fig. 13.7 Place Ville-Marie, Montreal, one of the first of Canada's recent skyscraper boom. (Guy Vincent photo 2023)

cultural industries, rather than their historical roles as industrial centres. Medium size metropolitan areas such as Ottawa, Calgary, Edmonton, Winnipeg and Halifax have witnessed a remarkable renewal of their downtown landscapes into "new downtown canyons" as they have shifted from being industrial hubs to become financial, technological, and cultural centres of services and emerged as global hubs for immigration and business, fostering greater ethnic and cultural diversity (Figs. 13.7, 13.8).

Tertiarization, the expansion of the service sector, has become the dominant force in shaping the modern Canadian urban economy. This transformation involves the rise of jobs in areas like finance, education, healthcare, and information technology, moving away from reliance on heavy industries. Service industries often cluster in urban cores, leading to gentrification, new housing developments, and enhanced infrastructure. While the service economy has created high-paying jobs, especially in tech and finance, it has also led to greater income inequality, as many service jobs (e.g. retail, hospitality) offer lower wages and less job security.

Gentrification, the process of urban transformation marked by rising property values, the displacement of lower-income residents, and the influx of wealthier populations, is deeply connected to global industrial restructuring. This phenomenon is not just a local issue—it is driven by economic shifts, globalization, and changing labour markets that have reshaped cities worldwide.

Gentrification in Canada is closely linked to global industrial restructuring, particularly as cities transition from manufacturing-based economies to service-oriented and postindustrial frameworks. This shift has led to the transformation of urban neighbourhoods, to often leading to socio-economic shifts through the displacement of long-standing, lower-income residents by more affluent newcomers.

In large cities like Toronto, Montreal, and Vancouver, as well as smaller and more industrially focused cities, the decline of traditional manufacturing industries, has resulted in widespread deindustrialization. This shift has left numerous urban areas with abandoned industrial sites and declining populations. Concurrently, the rise of service sectors, including finance, technology, and creative industries, has increased the demand for centrally located, high-quality residential and commercial spaces. This demand has spurred the redevelopment of former industrial zones and working-class communities into upscale neighbourhoods, a hallmark of gentrification.

Chris Hamnett's research, among others, highlights that gentrification is now firmly established as a major phenomenon in Western cities, deeply embedded in both academic and popular discourse. Hamnett notes that this process is closely associated with postindustrialism and the restructuring of occupational and industrial sectors within global cities.

The concept of uneven development offers a theoretical framework for understanding these dynamics. This perspective posits that capitalism inherently produces spatial and temporal unevenness, leading to disparities in development both within and between regions (N. Smith). Such uneven development often results in the marginalization of certain urban areas, that Smith identified as Rent gaps (N. Smith), making them prime targets for gentrification as capital seeks new avenues for investment.

Therefore, Gentrification is not just about wealthier residents moving into urban neighbourhoods—it is the result of global economic restructuring, financial speculation, and policy-driven urban redevelopment. As cities continue to evolve in the post-industrial economy, the challenge remains: how to balance economic growth with social equity and ensure cities remain accessible to all residents.

While tertiarization has fuelled economic growth in major urban centres, it has also exacerbated issues like housing affordability and the displacement of lower-income residents. For example, the gentrification of former industrial neighbourhoods in Toronto, Montreal and Vancouver has pushed out long-time residents, making urban centres increasingly unaffordable for middle- and lower-income families. Recent waves of condominiums high-rises have surged throughout

Fig. 13.8 Vancouver: Classic past reflected in glittering present. (Guy Vincent photo 2011)

the once industrial downtown cores and waterfronts resulting in strikingly different skylines.

Meanwhile suburbanization in Canada has persisted, and with the growing decentralization of business operations from urban centres to suburban business parks, the nation's urban landscape increasingly adheres to a polycentric model (Poitras 2015, 2018). This is particularly evident in large metropolitan areas such as Toronto, Vancouver, and Montreal, where peripheral zones have evolved into bustling urban hubs that became satellite cities in their own right boasting their own employment hubs, commercial areas, and services that can be viewed as "Suburban Business Districts" (or SBD as opposed to CBD). As businesses relocate and new commercial and residential developments sprout beyond traditional city cores, multiple centres of activity begin to emerge, each fostering its own economic and social ecosystems often times adopting New Urbanism principles. Former suburbs are now striving to densify through transit-oriented development projects, aiming to reduce car dependency by focusing on public transportation hubs. This diffusion of urbanization, while alleviating pressures on central business districts, has reshaped the geography of Canadian cities into a web of interconnected but distinct nodes of growth.

13.6 Urban Sprawl and Automobile Dependency

Suburban sprawl is arguably one of the most defining characteristics of modern Canadian cities with sprawling low-density residential neighbourhoods, shopping malls, and extensive highway use (Canadian Urban Institute 2019). However, in recent years, sustainable urban planning policies have emerged to curb this sprawl and encourage urban densification, particularly through Transit-Oriented Development (TOD).

Transit-Oriented Development promotes urban development centred around public transportation hubs, such as subway stations, tramways, or railway stations, to limit sprawl and reduce car dependency. In Toronto, Montreal and more recently Vancouver, this model is gaining traction with the development of densely populated neighbourhoods around subway lines or new rapid transit projects (for instance, Vancouver's SkyTrain or Montreal's REM). These efforts aim to create more sustainable, connected, and liveable urban spaces, reducing the need for long commutes and making public transit a more viable option for residents.

Québec City is advancing urban densification projects along key transportation corridors started in the early 2000s, aimed at promoting sustainability and reducing suburban sprawl (Ville de Québec 2011, 2020). These projects focus on areas like the Sainte-Foy corridor and Laurier Boulevard, introducing high-rise and mixed-use developments near public transport hubs. The city's new (and contested) tramway project is central to these efforts, fostering transit-oriented developments (TODs) to encourage public transportation and reduce car dependency (Ville de Québec 2023).

However, challenges remain, such as zoning laws, public opposition, historic preservation, and infrastructure limitations. Balancing modernization with preserving Québec City's historical character is particularly complex due to its UNESCO World Heritage status. Despite these hurdles, the city remains committed to integrating urban growth with transit development for sustainable expansion.

13.7 Impact of COVID Pandemic

The COVID-19 pandemic has suddenly and profoundly reshaped the trajectory of several Canadian cities, catalyzing changes that will influence urban evolution for years to come. One of the most profound changes was the rise of telecommuting, with many workers opting to work from home, reducing the need for daily commutes. Remote work has driven decentralization even further away, reducing the dominance of city cores as workers have moved to suburban and rural areas in search of space and flexibility. This shift has amplified demand for housing in these regions, stretching affordability and transforming the housing market. Smaller urban centres became more attractive.

Public transit systems, once a lifeline for commuters, saw drastic declines in ridership, pushing cities to rethink mobility infrastructure, favouring biking, walking, and other alternatives. The pandemic also highlighted the value of public spaces, leading urban planners to prioritize green spaces and human-centric urban designs. At the same time, it exposed deep socioeconomic inequalities, sparking calls for more equitable policies in housing, healthcare, and public services.

Ultimately, COVID-19 has prompted a renewed focus on sustainability, pushing cities toward greener, more resilient futures. The pandemic has laid bare vulnerabilities, but it has also created opportunities for reinvention, as Canadian cities adapt to a new urban reality.

13.7.1 Towards a Reversing Trend?

In the wake of the COVID-19 pandemic, Canadian cities have begun exploring the conversion of vacant office spaces into residential units to address both the decline in demand for office spaces and the ongoing housing crisis. This trend has emerged in urban centres such as Toronto, Vancouver, and Montreal, where the shift to remote work has left many office buildings underutilized.

This initiative seeks to revitalize downtown cores by transforming idle office buildings into much-needed housing, particularly affordable units. The conversion process, however, comes with challenges, including structural complexities and zoning restrictions that often complicate the adaptation of office layouts for residential purposes. Additionally, developers must consider the need for amenities such as natural lighting, ventilation, and parking, which differ significantly between office and residential spaces.

In cities like Calgary, there have been government-led initiatives to incentivize these conversions, with the city allocating funds to repurpose office buildings into mixed-use developments. This is part of broader efforts to create more vibrant, sustainable urban environments that integrate housing, workspaces, and public amenities. The long-term success of these projects hinges on adapting urban planning policies and fostering collaboration between governments and private developers.

The trend toward office-to-residential conversions highlights a key post-pandemic urban development strategy aimed at both reducing vacant office space and addressing urban housing shortages (Canadian Urban Institute 2023).

However, whether these trends will persist or if urban patterns will eventually revert to their pre-pandemic norms remains uncertain. The evolving dynamics of work, housing, and public space utilization are still in flux, and the long-term trajectory will likely depend on factors such as economic recovery, technological integration, and evolving social behaviours.

13.8 Canada as a Haven in a World of Turmoil

Canada's current immigration trends (as of 2024) are putting significant pressure on cities, particularly in terms of housing and infrastructure. In 2023, Canada's population surged by over 1.2 million people, driven largely by immigration, which contributed to the country's fastest growth rate since 1957.

While immigration is essential for addressing labour shortages and economic growth, it has also exacerbated issues like housing shortages, especially in major urban areas.

The rapid increase in newcomers has led to growing concerns over the country's ability to absorb such numbers without compromising the standard of living. Economists have warned that this influx could strain critical services, including healthcare, education, and housing. Canada's housing supply is particularly under pressure, with the Canada Mortgage and Housing Corporation estimating a shortfall of 3.5 million homes needed by 2030 to restore affordability. High immigration levels, especially the influx of temporary residents like international students and foreign workers, have also been linked to soaring rental and property prices, making it harder for Canadians to enter the housing market.

In response, the federal government started to adjust its immigration policies in late 2024. For instance, it is capping the number of international student permits and reducing temporary resident numbers to ease pressures on urban infrastructure. Despite these efforts, managing the balance between welcoming new immigrants and ensuring sustainable urban growth remains a complex challenge.

13.9 The Canadian Urban System

Canada is a highly urbanized country, with about 80% of the population living in urban areas. The majority of this population is concentrated in the Québec-Windsor Corridor, passing through Ottawa and Hamilton. This region is home to nearly 60% of Canada's population and constitutes the country's economic heartland. Nevertheless, even by being outside the major urban corridors, intermediate and smaller cities and towns that are scattered across Canada play a key role in sustaining vital industries and providing raw materials. These include regional hubs and communities tied to natural resource extraction (forestry, mining, oil) and agriculture.

The Canadian urban system is organized into a network of hierarchical cities and urban regions, with complex interactions between them depending on their size, economic function, and geographical influence who play key roles in their respective contexts. It is dominated by several large metropolitan areas, followed by medium and small cities, organized into a hierarchy that reflects their size and economic importance. It relies on well-developed transportation networks consisting of a web of highways and railways, as well as international airports and ports, which connect cities to each other and to international markets.

Toronto, Montreal, and Vancouver are global metropolises that serve as vital economic, cultural, and international hubs. They house corporate headquarters, financial institutions, and major transportation infrastructures. These cities exert substantial global influence. Toronto, with a population of about 6.7 million in the Greater Toronto Area (GTA), is standing as a financial powerhouse (Fig. 13.9), Montreal, with about 4.3 million residents in its metropolitan area, as a centre of culture and innovation especially in the French world (Fig. 13.10) and Vancouver, with around 2.6 million residents in Metro Vancouver, as a key gateway for trade and technology in the Pacific region (Figs. 13.12, 13.13, 13.14). Their vibrant economies, diverse populations, and cultural significance make them central players on the world stage, driving growth and connecting Canada to global networks.

Regional centres play crucial roles in their respective regions by providing services and often serving as transportation hubs or industrial centres. Calgary (Fig. 13.15) and Edmonton, each with a little over 1 million inhabitants, are respectively a major energy hub, and a centre for petrochemical industries and distribution. Ottawa-Gatineau is the national capital and a centre for politics and technology with a population of about 1.4 million residents. Québec City, with around 830,000 residents, is a historical and cultural centre of francophone Canada and an important player in tourism and technology.

Medium and small cities form a significant part of the urban system, acting as local centres for regional economic activities and services. They include cities like Halifax (Fig 13.16), Winnipeg (Fig. 13.17), London, Saskatoon, Regina, and Victoria, often linked to specific sectors like agriculture, natural resources, or niche industries. Cities that are part of larger CMAs (Census Metropolitan Areas) benefit from proximity to major urban hubs, while "standalone" cities often rely on local industries and services for economic stability.

In addition to its major urban centres, Canada is home to a vast array of smaller cities and rural communities. Although these areas play a less central role in the national economy, they play an important role in regional development as they are vital to specific industries, particularly agriculture, natural resources (such as forestry and mining), and tourism. These intermediate and rural communities form an essential backbone to the country, supporting both regional economies and Canada's natural resource sectors (Figs. 13.18, 13.19).

13.10 Types of Urban Fabrics

Canadian cities showcase a rich diversity of urban fabrics, shaped by various phases of development, cultural influences, and economic and social priorities. This urban diversity spans from low-density suburban areas to dense downtown cores featuring contemporary "cathedrals"—skyscrapers that house offices and residential spaces. The fabric of these cities includes bustling central business districts, suburban commercial hubs, historical neighbourhoods, old and newly developed suburban areas, and redeveloped industrial spaces, all

Fig. 13.9 Toronto: **a** Postindustrial waterfront, reflecting growth and transition. (Waterfront Toronto) **b** One consequence is an explosion in housing prices, continuing from 2014 through 2024. (Guy Vincent photo 2014)

of which reflect the unique context of each city's economic, social, and geographic evolution. These varied urban forms make Canadian cities an invaluable laboratory for studying urbanism and the evolution of cityscapes.

The cities of Eastern and Western Canada present contrasting urban models, shaped by their history, density, transportation networks, and geographical constraints. Eastern cities, being older and more densely built, are marked by compact historical centres and more developed public transit infrastructures. In contrast, Western cities, which are newer, tend to have more expansive urban layouts, a greater reliance on cars, and are still in the process of developing densification strategies, although exceptions including Vancouver, "partly due to its natural surroundings,"categorized as basic and non-basic jobs, showcase more concentrated urban growth.

13.10.1 Economic Specialization

The Canadian urban system is marked by a division of labour, with certain cities specializing in specific sectors. In urban economics, a city's economic base is often divided into two main categories: basic and non-basic jobs.

Basic Jobs (or "town forming") generate income by bringing new capital into the city from outside sources. They are tied to industries or sectors that produce goods and services for consumption outside the city's boundaries. Common examples include manufacturing, tourism, and large-scale service industries like finance or tech that serve global or national markets. For instance, if a car factory in a city produces cars sold across the country or internationally, the jobs related to that factory are considered basic.

Non-Basic Jobs (or "town serving") supply the local population and economy directly by circulating money within the region. They do not bring in money from outside the city but are essential for supporting the local economy by providing services to residents. Examples of non-basic jobs include local retail, healthcare, education, and hospitality. A local grocery store or a schoolteacher would be considered non-basic.

The relationship between basic and non-basic jobs can also be understood through the economic base multiplier. This concept explains how every basic job can lead to the creation of additional non-basic jobs. Basic jobs, tied to the company's production of goods or services for external markets, often stimulate the local economy by creating non-basic jobs, which serve the needs of the local population and support the city's internal economic ecosystem. For example, a factory (basic) job might create demand for local services like restaurants, retail, construction and housing, all of which are non-basic. Often, the arrival of a new company or industry in a city is celebrated for generating a range of direct and indirect employment opportunities—categorized as basic and non-basic jobs, respectively. Conversely, a loss of basic jobs will also trigger a loss in non-basic, or indirect, jobs.

In Canadian cities, large economic sectors like finance, professional services, technology in Toronto, aerospace, arts, artificial intelligence in Montreal, or International trade, technology, film industry in Vancouver function as basic industries, exporting services and generating income from outside sources. The same can be said for medium size metropolitan areas such as Calgary's energy (oil and gas)

Fig. 13.10 Montreal's former industrial and port areas have undergone substantial transition and redevelopment in the postindustrial era **a** Now inactive Lachine Canal, a remnant of the industrial era. (Guy Vincent photo, 2008) **b** Post-industrial redevelopment of Old Montreal port. (Guy Vincent photo, 2006)

Fig. 13.11 House in the long-established exclusive Westmount neighbourhood of Montreal, divided in two for easier sale. (Guy Vincent photo 2000)

sector, Winnipeg's transportation and logistics, agri-food sectors, and Halifax's maritime industries, government and military services. Smaller CMAs have developed strong economic ties with the features of their regions: Saskatoon by focusing on agriculture and innovation; Moncton, as a transportation and logistic hub of the Maritimes; St. John's as an energy and fisheries core. These industries, in turn, support a variety of non-basic jobs in sectors like hospitality, retail, and local government. Understanding the balance between basic and non-basic jobs helps planners, economists, and policymakers assess the sustainability and growth potential of urban economies. The economic specializations of these cities contribute to the cohesion of the national urban system, with intercity exchanges of goods, services, and people.

13.11 Globalization Effects on Industrial and Resource Towns

Over the past 60 years, the economic landscape of Canadian cities has undergone a profound transformation, as clearly demonstrated by trinary graphs depicting their evolving economic roles (Fig. 13.3a, b). A key driver of this shift has been globalization which has profoundly impacted Canada's industrial and resource towns (Bradbury 1988), as well as larger cities industrial districts, reshaping their economies, employment structures, and communities. Global economic integration has redefined their economic foundations, reshaping employment structures and altering the fabric of local communities. These towns, historically reliant on industries such as manufacturing or resource extraction have had to adapt to the forces of international competition, technological advancements, and shifts in global demand. This has pushed many cities to diversify their economies toward services, technology, and knowledge-based industries. Some of the key effects are economic shifts, job polarization and outsourcing, population decline, a need for diversification to revitalize its economic base, and being subjected to global fluctuations.

13.11.1 Economic Shift and Decline

Globalization has exposed Canadian industrial and resource towns to intense global competition, particularly from countries with lower production costs. As industries in sectors such as manufacturing and natural resources have become increasingly globalized, many Canadian towns, especially those in the manufacturing heartlands of Ontario and Quebec, have experienced job losses due to the relocation of factories to countries with cheaper labour. For instance, industries like automobile manufacturing have moved production to Mexico or the U.S. South, leading to the decline of towns like Windsor and Oshawa, which had become heavily reliant on auto plants.

Resource towns (mining, oil, and forestry) have suffered from volatility due to fluctuating commodity prices driven by international demand. Communities in Alberta and Saskatchewan have experienced boom-bust cycles in response to global oil prices. These fluctuations can lead to periods of prosperity, followed by rapid downturns, as seen in Fort McMurray, a hub for the oil sands industry. When oil prices

Fig. 13.12 Vancouver. **a** Older mansion in Shaughnessy neighbourhood (Guy Vincent photo, 2003) **b** Postindustrial landscape, Granville Island (Guy Vincent photo 2011)

Fig. 13.13 Downtown Vancouver **a** View in the 1970s (Vancouver City Archives) **b** Recent view from suburban area, looking north (Guy Vincent photo)

Fig. 13.14 False creek industrial landscape ca. 1936 (**a**) and 1941 (**b**). Today, this area has been gentrified and extensively re-developed. (Vancouver City Archives)

Fig. 13.15 These two houses in downtown Calgary, threatened by gentrification-driven re-development in 2007, had disappeared two years later. (Guy Vincent photo)

Fig. 13.16 Halifax from Fort Charlotte on George Island. (Guy Vincent photo 2024)

a

b

Fig. 13.17 a East railway yard at The Forks, confluence of Assiniboine and Red River, Winnipeg, 1970. (University of Manitoba Archives). b Today, this area has been developed with recreational and entertainment facilities. (Tourism Winnipeg/Mike Peters)

Fig. 13.18 Hudson's Bay Company building, Qu'Appelle SK, now repurposed for small business enterprise. Qu'Appelle was founded as a trading and farming community. (Guy Vincent photo 2007)

Fig. 13.19 Manyberries, Alberta, welcome sign. With a population of 84 (2022) and declining, Manyberries is a place at the base of Canadian Urban System (Guy Vincent photo 2016)

collapsed in 2014–2015, the town faced layoffs, reduced investments, and economic uncertainty. Even a large city like Calgary has suffered tremendously during and since the COVID19 epidemic.

13.11.2 Job Polarization and Outsourcing

Globalization has also contributed to job polarization, where middle-skill jobs, particularly in manufacturing, have been outsourced or automated. While high-skill jobs in industries such as technology and finance have grown in urban centres like Toronto and Vancouver, many blue-collar jobs in industrial towns have vanished. This has increased economic disparities between larger cities and smaller towns, leading to social and economic challenges in industrial regions.

While globalization has opened up international markets for Canadian commodities, it has also led to job losses due to technological advancements in extraction and production processes in resource-based towns. Mining towns in northern Canada, for instance, have seen reduced employment due to automation and more efficient production methods.

13.11.3 Migration and Population Decline

The decline in traditional industries due to globalization has often led to outmigration from smaller industrial and resource towns, especially among younger populations seeking better opportunities in larger cities. This demographic shift has left many of these towns with aging populations and shrinking tax bases, further limiting their ability to invest in infrastructure and services.

For example, Sudbury in Ontario, a historically mining-dependent town, has seen significant outmigration as mining jobs declined. Similar trends are evident in other towns tied to the forestry sector in British Columbia, where automation and global competition have led to the closure of sawmills and pulp mills, driving residents to seek employment elsewhere.

13.11.4 Efforts to Diversify and Revitalize

In response to these challenges, many industrial and resource towns have sought to diversify their economies. Efforts include transitioning to tourism, retirement communities, service industries, or renewable energy. For instance, the town of Corner Brook in Newfoundland, once a hub for the pulp and paper industry, has pivoted towards tourism and outdoor recreation to attract new visitors and investment.

Some towns, such as Sault Ste. Marie, have invested in green energy projects to create new jobs and reduce dependence on traditional manufacturing. Others, such as Sudbury, have shifted focus towards research and development in mining technologies to remain competitive in the global market.

13.11.5 Environmental and Social Concerns

The globalization of resource extraction industries has raised concerns about the environmental impact on Canadian resource towns. With increasing global demand for resources like oil, gas, and minerals, extraction activities have intensified, leading to environmental degradation in certain regions. Communities dependent on fossil fuels and mining face not only economic risks from global price fluctuations but also environmental damage, which affects local ecosystems and public health.

In addition to environmental concerns, the social fabric of many resource towns has been impacted. The boom-bust nature of industries like oil extraction in Fort McMurray has created social challenges, including housing shortages during boom periods and a lack of social services during downturns. As if this was not enough, the city of Fort McMurray was hit by devastating wildfires in 2016. This economic instability has also led to issues like mental health struggles and community dislocation.

13.12 Canadian Cities' Strengths and Challenges

Many Canadian cities, such as Vancouver and Toronto, have garnered global recognition for their efforts in sustainable urban planning. Vancouver, for instance, has ambitious goals to become the "greenest city in the world", emphasizing eco-friendly building practices, renewable energy, and extensive public transportation networks. Similarly, Toronto's Greenbelt helps limit urban sprawl, preserve natural spaces, and encourage densification within the city. However, such mandates are more difficult to implement in smaller urban regions where public transportation ridership and financing remain an important concern.

Toronto, Montreal, and Vancouver are renowned for their diversity and multiculturalism (see Chap. 11). Toronto, often called "the most multicultural city in the world", embraces an official policy of multiculturalism, which allows it to attract talent and investment from a globally diverse population. This diversity is not just cultural but also reflected in economic activities, such as Toronto's status as a financial powerhouse or Montreal's vibrant arts and creative industries. Smaller urban regions are increasingly becoming ethnically more diverse as well.

Canadian cities consistently rank highly in global liveability indices, offering robust healthcare systems, high-quality education and relatively low crime rates. Cities like Calgary, Toronto, and Vancouver frequently top these lists, emphasizing the strength of Canadian urban infrastructure, social systems, and overall well-being.

On the other hand, one of the most significant challenges for Canadian cities, particularly Toronto and Vancouver, but increasingly all over Canada, is the affordability of housing. The cost of homes and rent has surged dramatically in recent decades, driven by limited housing supply in times of high immigration rates, high demand, and increasing foreign investment. Efforts to create affordable housing options have struggled to keep pace with the growing demand, leading to issues of gentrification and displacement in certain neighbourhoods.

Although some cities such as Vancouver have successfully managed to curb urban sprawl, others, particularly in Alberta (e.g. Calgary and Edmonton), continue to grapple with the effects of suburbanization. The sprawling nature of these cities has led to a heavy reliance on automobiles, contributing to traffic congestion, air pollution, and inefficiencies in public transit systems. Investment in transportation infrastructure has not always kept up with population growth, leading to ongoing challenges with congestion and accessibility in key urban areas.

Rapid urbanization in major Canadian cities has also exacerbated social and spatial inequalities. The rising cost of living in urban centres has marginalized lower-income groups, pushing them to the outskirts of cities where access to essential services and employment opportunities may be more limited. Gentrification in central city neighbourhoods has further heightened the visibility of income disparity, raising questions about inclusivity and the social fabric of these cities.

Canadian cities face challenges in maintaining and upgrading their aging infrastructure that have accompanied the Postwar Baby Boom, particularly in older cities like Montreal and Toronto. Outdated public transit systems, deteriorating roads, and water supply issues (i.e. Montreal and Calgary water main breaks, summer 2024) often necessitate significant public investment, but limited budgets can slow down necessary improvements. The need for infrastructure investment becomes even more pressing given the pressure of growing populations and the demand for modern amenities and services.

13.12.1 Prospects and Future Directions

To address these challenges, Canadian cities are increasingly focusing on smart growth, sustainable policies, and inclusive urban design. Initiatives like intensification of public transit networks, mixed-use development, and green building practices are all part of an effort to create more resilient, liveable cities. The shift towards density, rather than outward sprawl, is becoming a hallmark of Canadian urban planning, as cities attempt to balance growth with environmental stewardship and social equity.

Canadian cities are increasingly vulnerable to climate risks such as floods, heatwaves, storms, and wildfires. With the growing impact of climate change, these challenges are intensifying, underscoring the need for effective adaptation policies to protect urban infrastructure, communities, and ecosystems, a step many cities have already begun (e.g. Toronto City Planning Division 2019; City of Vancouver 2020a, 2020b; City of Montreal 2021; Institut urbain du Canada 2021). Reducing greenhouse gas (GHG) emissions is one of Canada's top priorities in combating climate change as of 2024 (Government of Canada 2021, 2023). The national goal, as outlined in the Paris Agreement, is to reduce emissions by 40–45% below 2005 levels by 2030 and achieve carbon neutrality by 2050.

In summary, Canadian cities offer a high quality of life and lead the way in sustainability and multiculturalism. However, they must also navigate complex challenges, particularly in terms of housing affordability, urban sprawl, social inequality, and aging infrastructure, to remain competitive and inclusive in the global landscape.

References

Ames Sir HB (1897) The City Below the Hill: a sociological study of a portion of the city of Montreal, Canada, Montreal : Bishop Engraving and Printing. Also available on the Internet Archive website

Artibise AFJ (1988a) Canada's urban landscape. Can Encycl

Artibise AFJ (1988b) Western Canada: an urban geography. Univ Ott Press

Artibise AFJ (1988c) Canada's urban transformation: 1945–1975. Univ Ott Press

Artibise AFJ, Stelter GA (1981) Guide to Canadian Urban Studies. University of British Columbia Press, Vancouver

Bourne LS, (1981) The geography of housing. London: Edward Arnold. Also available on Internet Archive website

Bourne LS, Simmons J (2003a) The urban system and social polarization in Canada. Univ Tor Press

Bourne LS, Simmons J (2003b) New fault lines? Recent trends in the Canadian urban system and their implications for planning and public policy. Can J Urban Res 12:22–47

Bradbury JH (1988) Living with Boom and Bust Cycles: New Towns on the Resource Frontier in Canada, 1945–1986. Can Geogr 32:270–282

British Pacific Properties (2024) A proud legacy. https://britishproperties.com/about-us/legacy/

Bunting T, Filion P (2010) Canadian cities in transition: new directions in the twenty-First century. Oxford University Press, Toronto

Burgess EW (1925) The growth of the city: an introduction to a research project. In Park RE, Burgess EW, McKenzie RD, eds, The City. University of Chicago Press

Canada's Historic Places (2024) Lions gate bridge national historic site of Canada. https://www.historicplaces.ca/en/rep-reg/place-lieu.aspx?id=11711

Canadian Urban Institute (2019) Urban Sprawl and Sustainability in Canadian Cities. https://www.iciss.ca/2019/09/urban-sprawl-in-canadas-cities/

Canadian Urban Institute (2023) The case for conversion: Understanding opportunities for conversions of office space to housing in Canadian downtowns. https://canurb.org/fr/publications/the-case-for-conversions

Changing Vancouver–Then and Now (2020) Guinness Tower. https://changingvancouver.wordpress.com/2020/09/14/guinness-tower-west-hastings-street/

City of Montréal (2011) 1966: Galactic Metropolis. https://www.cca.qc.ca/en/articles/issues/2/what-the-future-looked-like/32738/1966-galactic-metropolis

City of Montréal (2021) Plan, Climat Montréal. https://montreal.ca/articles/plan-climat-montreal-objectif-carboneutralite-dici-2050-7613

City of Vancouver (2020a) Climate emergency action plan. https://vancouver.ca/green-vancouver/vancouvers-climate-emergency.aspx

City of Vancouver (2020b) Greenest City Action Plan. https://vancouver.ca/green-vancouver/greenest-city-action-plan.aspx

CMHC (Canadian Housing and Mortgage Corporation), Understanding Filtering: A long-Term Strategy to New Supply and Housing Affordability

Conseil des Arts du Canada (2019) Les industries créatives au Canada

Davidoff T, Somerville T (2025) Filtering in the Canadian Housing Sector: An Empirical Evaluation A Report for the Canadian Mortgage and Housing Corporation, University of British Columbia's Sauder School of Business, on behalf of CMHC. https://assets.cmhc-schl.gc.ca/sf/project/archive/research_6/filtering_en.pdf

Donald B (2005) The politics of local economic development in Canada's global city: Toronto. Space and Polity 9:261–281

Filion P, McSpurren K (2007) Smart Growth and Development Reality: The Difficult Co-ordination of Land Use and Transport Objectives. Urban Studies 44:501–523

Fougères D (2018) MacLeod R The History of a North American City. McGill-Queen's University Press, Montreal

Government of Canada (2021) Canada's climate action plan: a healthy environment and a healthy economy. https://www.canada.ca/en/services/environment/weather/climatechange/climate-plan/climate-plan-overview/healthy-environment-healthy-economy.html

Government of Canada (2023) Canada's national adaptation strategy. https://www.canada.ca/en/services/environment/weather/climatechange/climate-plan/national-adaptation-strategy/full-strategy.html

Hamnett C (2003) Gentrification, postindustrialism, and industrial and occupational restructuring in global cities. In Bridge G, Watson S, eds, Gentrification. Chapter 28, https://doi.org/10.1002/9780470693414.ch28

Harris CD, Ullman EL (1945) The Nature of Cities. Annals of the American Academy of Political and Social Science 242:7–17

Harris R (2004) Creeping conformity: how canada became suburban, 1900–1960. University of Toronto Press

Harris RC (1978) The growth of urban Canada. Historical Atlas of Canada

Harris RC (1997) The urban experience in Canada: 1850–1920. McGill-Queen's University Press

Harris RC (2008) The founding of montréal. Urban History Review 17(3):200–215

Hodge G, Gordon D (2014) Planning canadian communities. Nelson Education, Toronto

Hoyt H (1939) The structure and growth of residential neighborhoods in American cities. Federal Housing Administration, Washington, USA

Hutton TA (2009) The New Economy of the Inner City: Restructuring, Regeneration, and Dislocation in the Twenty-First-Century Metropolis. Routledge, London

Institut urbain du Canada (2021) Stratégies d'adaptation climatique pour les villes canadiennes. https://canurb.org/fr/publications/

Kollatz H (2017) A man with a plan—'Dean of city planners' Harland Bartholomew left his mark on Richmond. Richmond Magazine. https://richmondmagazine.com/news/richmond-history/city-planner-harland-bartholomew/

Lees L, Slater T, Wyly E (2010) The gentrification reader. Routledge, New York

Lewis R (2013) The industrial suburbs of toronto and chicago, 1850–1950. Urban History Review 42(1):4–19

Ley D (1996) The new middle class and the remaking of the central city. Oxford University Press

Linteau P-A (1990) Histoire de Montréal depuis la Confédération. Boréal, Québec

Marsh JH (2024) Lion's gate bridge, Canadian encyclopedia. https://www.thecanadianencyclopedia.ca/en/article/lions-gate-bridge

McLemore, R., Aass C, Keilhofer P (1975) The changing Canadian Inner City. Urban paper A.75.3. Ottawa: Ministry of State for Urban Affair.

National Post (2015) British investors who bought B.C. land 85 years ago sitting on thousands of acres worth millions. https://nationalpost.com/news/canada

Olds K (2001) Globalization and urban change: capital, culture, and pacific rim Mega-Projects. Oxford University Press

Peirce FL (1983) The galactic metropolis. In: Platt R, Macinko G (eds) Beyond the urban fringe. University of Minnesota Press, Minneapolis, pp 23–49

Places that matter (2016) Lion's gate bridge. Vancouver Heritage Foundation. https://placesthatmatter.ca/location/lions-gate-bridge

Poitras C (2015) La ville en mouvement. Les formes urbaines et architecturales du système automobile, 1920–1960 In: Atlas historique du Québec—Le fait urbain. Centre interuniversitaire d'études québécoises (CIEQ), Québec, 1–16

Poitras C (2018) Quand la banlieue était le futur. Zeitschrift Für Kanada-Studien 38(2):8–24

Sassen S (1991) The Global City: New York, London. Princeton University Press, Tokyo

Simmons JW, Simmons R (1974) Urban Canada. Copp Clark Publishing, Toronto

Smith N (2011) Uneven development redux. In Smith N, ed. Uneven Development 25 years on

New Political Economy 16, 261–265. https://doi.org/10.1080/13563467.2011.542804

Spicer Z (2011) The rise and fall of the ministry of state for Urban affairs: a Re-Evaluation. Canadian Political Review 5:117–126

Statistics Canada (2021) Census of population. https://www12.statcan.gc.ca/census-recensement/2021/

Statistics Canada (2022) Canada's large urban centres continue to grow and spread. https://www150.statcan.gc.ca/n1/daily-quotidien/220209/dq220209b-eng.htm

Toronto City Planning Division (2019) Transform TO: Climate Action Strategy. https://www.toronto.ca/services-payments/water-environment/environmentally-friendly-city-initiatives/transformto/

Trudel M (1979) Histoire de la Nouvelle-France. Fides, Montréal.

Vance JE (1964) Geography and Urban evolution in the San Francisco Bay Area. University of California, Berkeley, Institute of Government

Ville de Québec (2011) Plan de mobilité durable—Ville de Québec. https://www.ville.quebec.qc.ca/apropos/planification-orientations/transport/docs/PlanMobiliteDurable.pdf

Ville de Québec (2020) Plan de développement du centre-ville de Québec. https://www.quebec.ca/gouvernement/ministere/economie/publications

Ville de Québec (2023) Le tramway de Québec. https://tramwaydequebec.info/projet/amenagement/

Vinodrai T (2015) Economic change in Canadian cities: Innovation, creativity and the knowledge-based economy. In: Canadian Cities in Transition: New Directions in the Twenty-First Century. 5th ed. Oxford University Press, Toronto

Walks A (2001) The social ecology of the post-fordist/global city? economic restructuring and socio-spatial polarisation in the Toronto Urban Region. Urban Studies 38:407–447

Walks A (2013) Income inequality and polarization in Canadian cities. University of Toronto. https://tspace.library.utoronto.ca/handle/1807/94383

Transportation

14

Abstract

Use of rivers and canals for transport in Canada is severely limited by winter climates. Many of the largest rivers are braided or wandering (Catto 2014), limiting steamboat traffic in shallow-draught, shifting channels. The history and development of the St. Lawrence Seaway-Great Lakes canal system is discussed at length in Chapter 4 of this volume. Marine ports, including those along the Great Lakes—St. Lawrence Seaway, are a federal responsibility. Seventeen of the largest, including Vancouver, Montreal, and Halifax, are managed by Canada Port Authorities (CPA). The only commercial Arctic port is located at Churchill (MB), on western Hudson Bay. Canadian railways can be examined as the major modes of land transport between 1850 and 1950. The railway often dictated its own route, rather than being constrained to follow a predetermined course. Important historical lines include the Intercolonial (Maritime Canada); Grand Trunk (Great Lakes-St. Lawrence lowlands); and the Canadian Pacific (CPR), Canadian Northern, and Grand Trunk Pacific (transcontinental). Indebtedness forced the Canadian Northern into liquidation, with the federal government taking over the railway. In 1923, the former Canadian Northern was amalgamated with the Grand Trunk, Grand Trunk Pacific, and Intercolonial Railway to form Canadian National Railways (CNR). CNR (now CN) was privatized in 1995, and CPR merged with the Kansas City Southern railroad in 2023. Passenger rail service outside the Québec-Montréal-Toronto corridor is minimal. Road construction and maintenance in the provinces is a provincial responsibility, commonly downloaded to municipalities for local streets. Canada's total road network has exceeded 1 billion kilometres. Road construction in the Boreal, Taiga, and northern Cordillera of Canada is linked to military requirements, resource extraction, and government administration, in addition to providing access to and from northern communities. Air travel is regulated by the federal government. Airports throughout Canada, formerly administered by the federal government, were largely privatized in the 1990s. In southern Canada, air freight represents less than 3 % of the total carried domestically. Toronto Pearson International airport handles approximately 30% of all domestic passenger traffic, 45% of transborder, and almost half of international travel.

Keywords

Rivers and canals · Marine ports · Railways · Roads and highways · Air travel

The earliest communities, both indigenous and European, were established to utilize nearby natural resources, to regulate and guard trade routes, or for military or political purposes. Settlements developed where combinations of transport by water and reasonable agricultural conditions (often on fine-grained glacial, glacifluvial, glacilacustrine, or glacimarine sediments) existed (Catto et al. 2020). Until about 1840 in Canada, all towns in the interior were largely dependent on human or animal power for energy. Communities interested in trade had to be near a river or coast to take advantage of transportation. Goods had to be moved over portages around waterfalls or rapids. Falling water also provided one of the few energy alternatives to muscle.

14.1 Rivers and Canals

Use of rivers and canals for transport in Canada is severely limited by winter climates. Many of the largest rivers are braided or wandering (Catto 2014), limiting steamboat traffic in shallow-draught, shifting channels. The Rideau Canal (Watson 2007) and Trent-Severn Waterway (Angus 1988), although intended at least in part for commercial traffic when

N. Catto and G. R. Vincent, *Canada: Land and People*, World Regional Geography Book Series,
https://doi.org/10.1007/978-3-032-04391-7_14

constructed, did not develop as commercial routes. Smaller canals in eastern Canada, including the Shubenacadie and St. Peters (NS), Carillon and Chambly (QC), and Desjardins, Murray, Port Carling, Port Sandfield, and Wolfe Island (ON), also were not commercially successful (Catto et al. 2020). The history and development of the St. Lawrence Seaway–Great Lakes canal system is discussed at length in Chap. 4 (also see Andrey et al. 2014).

River (York Boat) transport was used by the Hudson's Bay Company to link their inland trading posts with the port at York Factory, in competition with the North West Company's canoe-based system along the Ottawa River through Lake Superior (HBC Heritage 2016). Steamboat traffic served Winnipeg along the meandering Red River between 1859 and 1910 (Berton 1970; Riverkeepers 2015), and Dawson City along the braided Yukon River from 1892 to 1955 (Anderson 1983). Commercial vessels plied the Athabasca and lower Peace Rivers in the early to mid-twentieth century (Affleck 2000). Barge traffic along Deh Cho (Mackenzie R) remains significant from mid-May to late October (NWT 2024).

Log driving along rivers was an integral part of early lumbering and cultural identities in eastern Canada (McCarthy 1991, Boréalis 2020, Simard 2022; see also Garceau 1957). Techniques and tools were brought to North America by Fennoscandian immigrants, and spread across eastern Canada with the advancing *Pinus strobus* lumbering trade (see Chap. 4). Communities developed at choke points created by rapids or constrictions, including Badger (NL, Exploits R), Shawinigan (Québec, Riv. St-Maurice) and Ottawa-Gatineau (Chaudière Falls). Although the drives represented the most effective way to move logs prior to railway construction, they were relatively inefficient, with typically 15–20% of logs lost *en route* and others extensively damaged and rendered unusable during transit. Driving generally ceased shortly after the arrival of rail or road in a logging area. Floating of large rafts along the Fraser River persists in British Columbia.

14.2 Marine Ports

Constitutionally, marine ports, including those along the Great Lakes–St. Lawrence Seaway, are a federal responsibility (Brooks 2007). Seventeen of the largest are managed by Canada Port Authorities (CPA), charged with reaching financial self-sufficiency while maintaining safety, security, and environmental protection during operations (Ashrafi et al. 2014, Transport Canada 2020a, 2020b, ACPA 2024). Currently, the 17 CPA handle 60–65% of the total marine tonnage (Table 14.1).

14.2.1 Vancouver, VFPA

With increasing trans-Pacific trade, the Port of Vancouver (Vancouver Fraser Port Authority, VFPA) has become Canada's largest, handling 140–150 million t of cargo, handling more than $300 billion worth of goods and generating $11–$13 billion GDP annually (Stevens 1999; Transport Canada 2020b, 2023; VFPA 2024). Facilities in 29 terminals covering 1500 ha handle grain, coal, minerals, forest products, and other bulk cargo; petroleum and chemicals; automobiles; intermodal containers; and cruise ships and their passengers (Figs. 14.1, 14.2; Table 14.1).

Port facilities are concentrated in four areas (VFPA 2024). Initially, traffic used the Fraser River for access to the interior, with the port located at New Westminster, on tidewater 30 km from the seaward margin of the delta. Development was restricted by tidal currents, fluvial sediment accumulation, and river-induced flooding over Annacis and Tilbury Islands and the narrow upper tidal flats. As New Westminster grew, serving as the capital of British Columbia 1858–1866, port facilities were concentrated on the south bank of the Fraser (now part of Surrey), facing the developing downtown. Navigational difficulties, combined with political factors, led the Royal Navy to locate its base at Esquimalt adjacent to Victoria on Vancouver Island (CFB Esquimalt 2024). This, along with the subsequent CPR decision to develop Burrard Inlet as its marine terminus, greatly hindered New Westminster's career as a port. The VFPA (2024) currently manages facilities in Surrey, as well as maintaining dykes, dredging, and environmental monitoring of the Fraser River from the delta margin upstream to Pitt Meadows, Port Coquitlam, Maple Ridge, and Langley. Lumber, pulp, steel, and intermodal containers dominate the freight mix. Annacis Island, now incorporated in the city of Delta, is the site of the Annacis Auto Terminal, importing vehicles manufactured in Asia.

False Creek, named because it is not connected to the Fraser, flanks the south side of the city of Vancouver. After a career as an industrial centre, today it serves as a harbour for private recreational vessels and cruise ships (Alexander 2000).

Burrard Inlet was chosen as the Pacific terminus of the CPR in 1881 (Berton 1971). Extending 22 km inland from its mouth at the Lion's Gate Bridge to Port Moody, Burrard Inlet was preferred as a port site due to its deep bathymetry and lack of large fluvial influx (reducing complications from sedimentation, terrestrial flooding, and fluvial-tidal interactions). As elsewhere, land speculation and political considerations also played significant roles in the CPR decision to locate the initial port in the western part of the south shore, adjacent to the modern downtown of Vancouver. The north shore (City

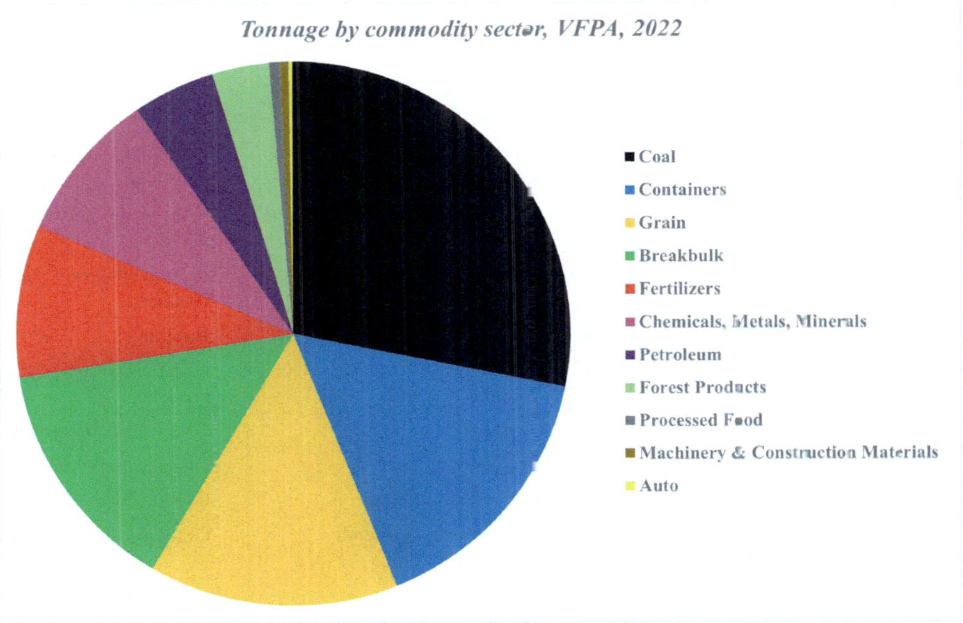

Fig. 14.1 Tonnage by commodity sector, Vancouver Fraser Port Authority, 2022. (Transport Canada 2020b, 2023, VFPA 2024)

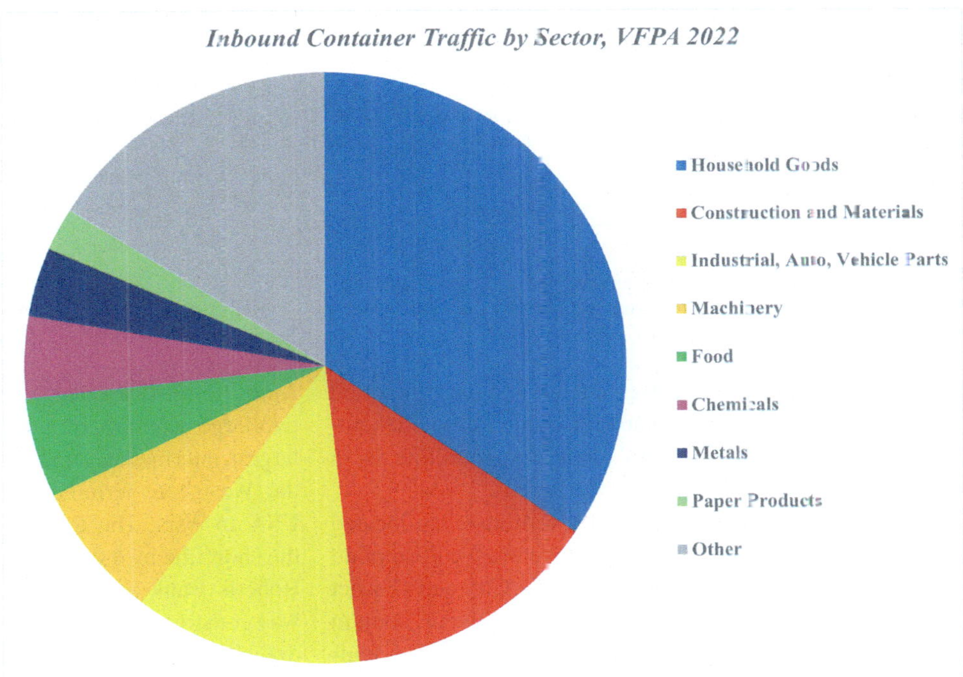

Fig. 14.2 Inbound container traffic by sector, Vancouver Fraser Port Authority, 2022. (Transport Canada 2020b, 2023; VFPA 2024)

and District of North Vancouver) was subsequently developed by the CNR (initially, Canadian Northern Railway, see below). The restriction imposed by steeply sloping shorelines was partially overcome by infilling along the south and north shorelines. Some areas of "made ground" are potentially susceptible to liquefaction in the event of an earthquake, and are also exposed to ongoing sea level rise.

On the south shore of Burrard Inlet, the Burnaby Westridge marine terminal is connected to the contentious Trans-Mountain Pipeline, transporting petroleum from Alberta, in addition to two other petroleum terminals at Burnaby. In total, VFPA manages a minimum of 500,000 t of transported petroleum annually (VFPA 2024). Pollution management is the shared responsibility of the VFPA (within port waters) and the industry-funded Western Canada Marine Response Corporation (2024) (seaward).

Although much of the port activity is located elsewhere, the facilities within the City of Vancouver remain significant, supporting $10.4 billion in economic activity and more than 52,000 jobs in 2023 (VFPA 2024). Eight major termi-

Table 14.1 Traffic at the 17 Canada port authorities, 2017–2023. (Transport Canada 2020a, 2020b; ACPA 2024; and websites for each port authority)

Port authority	Tonnage (Mt y^{-1})	Terminals	Incoming goods	Outgoing goods	Cruise ships** (y^{-1})
Vancouver-Fraser	140–150	29	Automobiles containers sugar chemicals bulk goods	Grain; canola coal; minerals forest products petroleum; chemicals containers	>300
Montreal	35–40	23	Containers sugar salt fertilizer	Grain; canola minerals containers bulk goods	45–50
Québec	30–35	14	Containers chemicals steel; minerals petroleum	Containers; grain minerals bulk goods forest products	120–135
Sept-Iles QC	25–33	1	Bulk goods	Iron ore	–
Saint John NB	25–30	4	Containers petroleum; LNG bulk goods chemicals; foods	Grain; minerals gypsum products forest products	70–75
Prince Rupert	25–30	3	containers	Grain; coal forest products	40–50
Thunder Bay	8–10	1	Bulk goods	Grain; canola; minerals	–
Hamilton ON*	7–11	1	Chemicals iron ore	Grain; soybeans steel; bulk goods	–
Halifax	5–6	5	Containers petroleum automobiles	Grain gypsum products containers	170–180
Windsor ON	4–5	1	Bulk goods	grain; salt; aggregate	–
Nanaimo BC	3.8–5	1	Petroleum bulk goods	forest products	10–15
Trois-Rivières	3.5–5	1	Bulk goods	Bulk goods	–
Toronto	2	1	Bulk goods; sugar; cement	Bulk goods	–
Port Alberni BC	1–2	1	Bulk goods	Forest products; seafood	1–3
St. John's NL	1–2	1	Containers bulk goods automobiles	Regional trans-shipment	30–35
Belledune NB	1–2	4	–	Forest products minerals; armour stone	–
Oshawa ON*	1	1	Chemicals	Bulk goods	–
Saguenay QC	0.5–0.6	1	Containers	Forest products	–

* Hamilton and Oshawa are operated as a single Canada port authority
** total number of visits, including multiple visits by the same vessel

nals handle exported grain, pulses, canola, and beef; imported sugar; intermodal containers; and cruise ship traffic. Restoration of former industrial areas along the shoreline is in progress.

On the north side of Burrard Inlet, terminals are located at Port Moody (petroleum, potash, ethylene glycol, sulphur, and refined canola oil exports), and in the City and District of North Vancouver (caustic soda, ethylene glycol, sodium chlorate, lumber, wood chip and pellet, coal, and agricultural exports; intermodal container, dry bulk, phosphate rock, steel, and refined sea salt imports). The VFPA drydock is also located in the City of North Vancouver. The eight North Vancouver terminals focus on bulk exports and imports, and have a more industrial character than the ports in the central area of the city of Vancouver.

The fourth area includes newer port facilities along the open Salish Sea in Richmond and Delta, extending from Vancouver International Airport south to Tsawwassen at 49°N. The Richmond Auto Terminal is a major import site for vehicles manufactured in Asia, generating almost $1 billion in economic activity annually. In addition to the Annacis Auto Terminal located along the Fraser River, Delta hosts two large marine terminals at Roberts Bank: Deltaport (the largest intermodal container port in the VFPA system), and the Westshore terminal, exporting coal from Canada and USA to Asia. The Roberts Bank complex is connected to the shoreline by a 4-km-long causeway extending across the Roberts Bank tidal flat into the Salish Sea. The causeway was constructed in 1970 with the goals of facilitating deep-draught marine vessel docking and environmental protection of the flanking Roberts Bank Wildlife Management Area (designated 2011; British Columbia 2021), important wintering habitat for dunlins (*Calidris alpina*), trumpeter swans (*Cygnus buccinator*), black-bellied plovers (*Pluvialis squatarola*), and other marine bird species. To the south, another 3 km-long causeway connects the Tsawwassen Ferry Terminal, serving passenger and freight traffic to Vancouver Island.

Natural hazards to the VFPA facilities include earthquakes, tsunami, and relative sea level rise (see Chap. 7, Souza Almeida et al. 2023). More than 90% of the potential economic cost of relative sea level rise throughout Canada is associated with vulnerability of the VFPA infrastructure and associated activity (Andrey et al. 2014; Kulp and Strauss 2019). Disruptions to the Fraser Canyon and Coquihalla highway and rail routes, severing connections to the remainder of Canada, also have severe economic impacts, along with any disruptions affecting port facilities (British Columbia 2023, also see Pratson 2023).

14.2.2 Other Pacific Ports

Prince Rupert is the other major port on the Pacific coast (Transport Canada 2020b, 2023, Port of Prince Rupert 2022), although it typically handles only 20% of the traffic of the VFPA, despite being 2 days shipping time closer to Shanghai (Table 14.1). Intended by the Grand Trunk Pacific Railway as a rival to Vancouver, Prince Rupert was relatively slow to develop throughout the twentieth century, handicapped by the single-track railway prone to interruptions from weather and slope failures (Clague 1984; Blais-Stevens et al. 2018). The Prince Rupert Port Authority (PRPA; Port of Prince Rupert 2022) manages three terminals exporting grain (primarily to Central and South America, northern Africa, and the Arabian Peninsula), lumber and wood pellets (primarily to Europe), and metallurgical coal for steel manufacture (primarily to China, Japan, and South Korea). Imports are dominantly intermodal containers. Prince Rupert is the terminal for ferry services to northern Vancouver Island, Haida Gwaii, and southern Alaska, and has grown in importance as a port-of-call for cruise ships.

The harbour has the physical advantages of deeper water (35 m), and is easier to navigate than either Burrard Inlet or the Fraser River (Port of Prince Rupert 2022). However, the large imbalance between container imports and exports, and the export of grain and lumber to European ports, reduce the theoretical advantage in faster trans-Pacific transport.

Port authorities in Nanaimo and Port Alberni, Vancouver Island, handle primarily local and lumber-related commercial traffic, cruise ships, and recreational vessels (Transport Canada 2020b). The port at Kitimat handles traffic associated with aluminium production.

14.2.3 St. Lawrence CPA

Montreal is the largest port by tonnage port in Eastern Canada, handling 35–40 million t y^{-1}, 15–20% of the national total (Transport Canada 2023a, b; Table 14.1). The port handles direct imports and exports, as well as trans-shipped outbound containers from the St. Lawrence Seaway, and domestic traffic to and from Atlantic Canada, with traffic of 2000 commercial vessels annually (Transport Canada 2020b, 2023; Abu Aisha et al. 2022; Montreal Port Authority 2023). The port and city (Dagenais 2017) grew at the foot of the Lachine Rapids, the inland limit for marine vessel navigation prior to construction of the St. Lawrence Seaway (see Chap. 4).

The port authority manages 23 terminals and other facilities, along the St. Lawrence River, extending from Cité du Havre (Bickerdale Basin) and the old port along the north bank in the southwest to the newly developed container port at Contrecoeur along the south bank, 40 km downstream (Montreal Port Authority 2023). Facilities along the north bank of the St. Lawrence include a complex of dry bulk terminals handling road salt, sugar, ores, fertilizer, and grain; container terminals handling rail- and truck-transported freight; and a complex of 6 liquid bulk terminals handing petroleum products, petrochemicals, and fertilizers, concentrated along 5 km of waterfront. Construction of the Louis-Hippolyte Lafontaine Tunnel between Montreal and Longueuil facilitates shipping through the main (western) channel of the St. Lawrence. The area of the old port (last operative 1980) and Cité du Havre have largely been repurposed for other urban uses, with ship traffic diverted to the southern side of the river through the Canal de Rive Sud and St-Lambert locks. Recreational marinas are located at Longueuil and Boucherville, along the south bank.

The ports of Québec and Sept-Iles handle comparable volumes of traffic (Transport Canada 2020b, 2023). Québec is open to marine traffic throughout the year (with Canadian Coast Guard ice-breaking services), handling approximately 30 million t and 1300–1500 vessels annually (Port de Québec 2024). The 14 terminals are located at Anse-au-Foulon (3), Pte-à-Carcy (1), Estuarie (3), and Beauport (5) on the north (Québec) bank, and Valero (1) and Davie-Québec (1) on the south (Lévis) bank. The Québec shore terminals handle chemicals, minerals, steel, grain, forest products, and dry bulk goods. The Valero terminal serves the Jean Gaulin Valero petroleum refinery (formerly Ultramar). A major shipyard is located at Davie-Québec.

Sept-Iles serves as the marine outlet for pelletized iron ore mined in Labrador West. Port construction dates from the Iron Ore Company of Canada's mining activity, with the mines connected to Sept-Iles by the Québec North Shore and Labrador railway (Rodon et al. 2022). In addition to the Port Authority at the smaller Saguenay marine port, other ports along the St. Lawrence Estuary are owned and operated as public ports, including Baie-Comeau, Gros-Cacouna (northeast of Rivière-du-Loup), Gaspé, Matane, Port-Cartier, and Rimouski. A Port Authority also operates the facilities at Trois-Rivières, midway between Québec and Montreal (Transport Canada 2020b, 2023).

14.2.4 Great Lakes CPA

Federal Port Authorities also manage facilities at the Great Lakes ports of Toronto, Hamilton, Windsor, Thunder Bay, and

Oshawa (jointly with Hamilton) (Transport Canada 2020b; Table 14.1). Extensive modification has been required to construct all of these ports. In Toronto, infilling since 1840 (Front Street) has moved the Lake Ontario shoreline 700–900 m southward (Baker et al, 1998), and infilling of Hamilton Harbour has reduced the area by 40% since 1820 (Karrow and Morin 1998). Building port facilities over the Kaministiquia River delta at Thunder Bay required pilings extending 50 m to Precambrian bedrock (Karrow 1998).

Changes in traffic modality, particularly container shipping, and the growth of trucking have greatly impacted the Great Lakes ports. Thunder Bay, once the largest port for grain shipments in Canada, now handles less than 5% of the tonnage of shipping passing through VPFA annually (Transport Canada 2020b, 2023; Fig. 14.3). The ports along Georgian Bay, which once trans-shipped grain across southern Ontario to Toronto and Ottawa, are all inactive and largely abandoned or repurposed (e.g. Holland-Stergar 2020). Grain for trans-Atlantic export largely goes to Montreal and points east. Of the Great Lakes Port Authorities, Thunder Bay handles the most traffic, and Toronto the least, about 20% of the tonnage passing through Thunder Bay (Transport Canada 2020b, 2023).

14.2.5 Atlantic Canada CPA

In Atlantic Canada, Port Authorities operate the facilities in Halifax, Saint John, St. John's, and Belledune (NB) (Transport Canada 2020b; Table 14.1). Smaller facilities are present in other harbours, including Charlottetown and Sydney. Numerous Small Craft Harbours serve fish harvesters and processors.

Halifax was initially conceived as a Royal Navy base, as part of the overall military strategy for British North America (see Chapter xxx). The grid pattern evident in downtown Halifax, below the military fortification of Citadel Hill, initially was a product of British military surveyors. Grid patterns mark other port towns, including Saint John and Charlottetown, which had a significant British military presence at their founding or were laid out by retired British military engineers (e.g. Rider 2009). St. John's, without an initial dominant military presence, has a street pattern that developed in accordance with the dominant economic activity, fishing and mining.

The ports added space for docks, warehouses, and rail lines by infilling the margins of harbours with sediment and rubble. The South End Terminal, Seaport, (former) Naval Dockyard, and Fairview Cove Terminal in Halifax are all built on harbour infill (Lewis et al. 1998).

Halifax has a deep, sheltered, ice-free harbour with a microtidal range. The developed port encompasses approximately 10 km^2. The facilities (Halifax Port Authority 2024) include the South End Container Terminal, along the western side of the harbour south of downtown; Ocean Terminals (grain, north of the container terminal), Richmond Supply Terminals (supply base for offshore petroleum activities, north of downtown along the west shore of The Narrows), Fairview Container Terminal (Bedford Basin), and the Sheet Harbour Industrial Port (90 km east of Halifax Harbour). Separate facilities not administered by the Halifax Port Authority serve the Wrights Cove Gypsum Terminal (north-

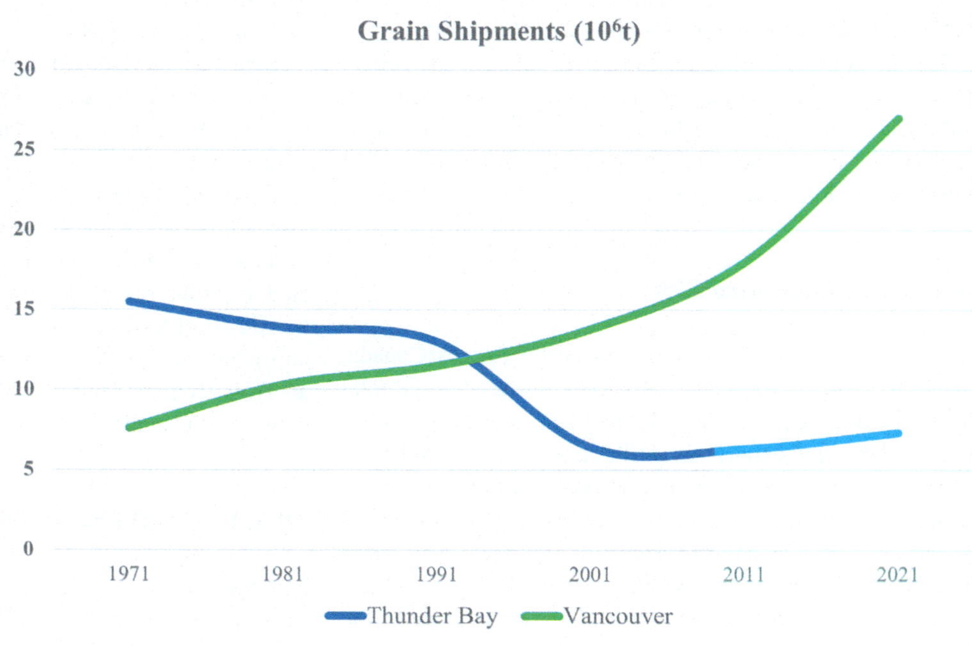

Fig. 14.3 Grain shipments (10^6 t) through Thunder Bay and VFPA, 1971–2021, indicating the shift from Transatlantic to Transpacific grain movements. Transport Canada (2020b, 2023)

east side of Bedford Basin), the Dartmouth Refinery and Woodside support base for offshore petroleum operations (east shore of the harbour, Dartmouth), and the Autoport (Eastern Passage).

Although Halifax is the largest port in Atlantic Canada, and among the 5 largest in Canada, its annual traffic of less than 5 million t is a fraction of that routed through Vancouver or Montreal (Transport Canada 2023a, b). Lesser tonnage reflects the increasing dominance of trans-Pacific over trans-Atlantic trade, and competition from other North American Atlantic ports (c.f. Ansary 1993). Much of the waterfront activity and interest today focuses on the maritime history of the trans-Atlantic and Caribbean "triangular" trade, military and convoy activity during the two World Wars, and Halifax's role as a major port for immigration to Canada from Europe prior to 1971 (see Chapter mmm). Nevertheless, the port serves as a trans-shipment point for goods travelling from the west to points eastward and northward, as well as a cruise port of call. Halifax is the home base for Canadian Fleet Atlantic of the Royal Canadian Navy and of the Halifax Shipyard, along the western side of the harbour north of the downtown.

Compared to Halifax, Saint John harbour has the physical disadvantages of hypertidal range, narrow nearshore rock-bound areas with steep slopes, three separate peninsulas, and significant fluvial influx from the Saint John (Wolastoq) River (McGahan 1982). Although Saint John is sometimes perceived to be overshadowed by Halifax, its port handles substantially more traffic and tonnage, 25–30 million t y^{-1} (Transport Canada 2023a, b; Port Saint John 2024). Petroleum and refined products, liquefied natural gas, forestry products, dry bulk, foodstuffs, liquid cargo (including molasses), potash, gypsum, and other minerals, container, and other cargo pass through the port, with its connections to more than 500 others. Saint John's rejuvenation as a port can be attributed to the increased value of cargoes.

Port facilities are located in Saint John West, Barack Point (potash terminal, central peninsula), Courtenay Bay (east embayment of the harbour, including the petroleum and gypsum terminals, privately managed), and Mispec Point (LNG, also privately managed). Facilities occupy 5 km of shoreline, with 21 berths. The historic port adjacent to the downtown on the central peninsula has been redeveloped, with all the facilities repurposed.

St. John's insular location and occasional (although rare) restrictions from drift ice and icebergs have limited its attractiveness for marine traffic. The port developed primarily from trans-shipment for the cod fishery, with small vessels off-loading for subsequent trans-Atlantic transport. As the city was not intended to be a naval base, the physical layout and street patterns were established in response to the needs of the fish harvesters and merchants. Infilling has decreased the water area by approximately 30% since initial European arrival (*ca.* 1510; Catto 2008). Today, a small container facility is located at the southwest end of the harbour, almost exclusively for imports.

14.2.6 Churchill

The only commercial Arctic port is located at Churchill (M3), on western Hudson Bay. Other Arctic communities are served by barge traffic during the short summer navigation season. After much agitation from Prairie farmers looking for an alternative Atlantic outlet for grain (Regehr 1976), the Hudson Bay Railway and Port of Churchill were constructed in 1929, shortly before the onset of the Great Depression (Stevens 1962; Larson et al. 2019). The loss of the monopoly held by the Canadian Wheat Board in 2012 allowed shippers to find less expensive alternatives. Today, both railway and port are owned and operated by an indigenous-led consortium. Sea ice restricts navigation to mid-July through early October. Grain shipments account for more than 90% of traffic when the port and railway are operational (Larson et al. 2019). Between 2014 and 2023, the port was able to ship grain in only 3 summers (2014, 2019–2020). The peak of 621,000 t was reached in 2006. Ongoing shifts in the grain trade to Vancouver and Prince Rupert suggest that Churchill will not be an economically viable port in the near future, unless subsidized, regardless of sea ice extents in Hudson Bay.

Originally, Port Nelson, at the mouth of the Nelson River, was to be the site of Canada's Hudson Bay port. However, this macrotidal estuary is dominated by constantly shifting mudbanks, propelled by sediment from marine and fluvial input, in addition to reworking by seasonal ice shove (Catto 2014). Attempts to dredge the channels and construct elaborate infrastructure failed. Consequently, the Hudson Bay Railway was diverted at Amery MB to extend 240 km north to Churchill (Stevens 1962; Catto et al. 2020).

14.3 Rail

14.3.1 Historical Geographical Context

Throughout the 1840s, it became evident to many residents of eastern Canada that future prosperity depended upon modernization, on transportation, and on communication. In the climate of the mid-nineteenth century, those things could only be gained through railway construction.

In a geographical sense, a railway can be examined in terms of its position as the major mode of land transport between 1850 and 1950 (Table 14.2). In North America, this timing meant that the railway often dictated its own route, rather than being constrained to follow a predetermined course. Railways can be considered as transportation links, and also be placed

Table 14.2 Historical railways in Canada. See text for further discussion

Railway	Dates of construction	Mainline terminals	Mainline length (km)	Total max. length* (km)	Ownership	Subsequent fate and comments
Intercolonial (IC)	NS: 1853–67 IC Truro-Rivière-du-Loup:1867–1876	Halifax NS; Rivière-du-Loup QC (subsq. Lévis QC)	898	1200	Govt of Nova Scotia 1853–1867; Govt of Canada	Merged with canadian national railways (CNR) 1923
Prince Edward Island	1871–1875	Tignish PEI; Borden PEI; Elmira PEI	235	446	Govt of PEI 1871–1873; Govt of Canada	Merged with canadian national railways 1923; abandoned 1989
Newfoundland	1881–1898	St. John's NL; Port-aux-Basques NL	882	1458	Private 1881–1923 (w. govt. support); Govt of Nfld, 1923–1949; CNR	Merged with canadian national railways 1949; abandoned 1989
Grand Trunk (GTR)	1852–1859	Portland (ME, USA); Lévis QC; Sarnia ON; Chicago (IL, USA)	1744	9800	Private (w. federal Govt. support)	Declared bankrupt and merged with canadian national railways, 1923
Canadian Pacific (CPR)	1874–1885	Montreal QC; Vancouver BC; subsq. Saint John NB ()	4650	32,000	Govt of Canada 1874–1880; Private 1881 (initially w. govt. support)	Merged with Kansas City Southern 2023 (CPKC)
Canadian Northern (CNoR)	1896–1915	Montreal QC; Toronto ON; North Vancouver BC	3500	15,100	Private (w. provincial and federal govt. support)	Declared bankrupt and merged with canadian national railways, 1923
Grand Trunk Pacific (GTP)	1905–1914	Winnipeg MB; Fort William ON; Prince Rupert BC	2004	3000	Private (GTR) (w. federal govt. support)	Declared bankrupt and merged with canadian national railways, 1923
National Transcontinental (NTR)	1905–1917	Moncton NB; Winnipeg MB	1800	1800	Govt of Canada; operational partnership w. GTR	Declared bankrupt and merged with canadian national railways, 1923
Hudson Bay (HBR)	1904–1929	The Pas MB; Churchill MB	864	1300	CNR; private	Operated by Indigenous-led consortium from 2018
Quebec North Shore & Labrador (QNSLR)	1951–1954	Sept-Iles QC; Schefferville -Kawawachikamach QC; Labrador City NL	578	690	Private	Operated to transport iron ore to Sept-Iles
White Pass & Yukon (WPYR)	1898–1900	Skagway (AK, USA); Whitehorse (YT)	177	177	Private	Seasonal tourist excursions only
Pacific Great Eastern (PGE)	1913–1971	North Vancouver BC; Fort Nelson BC	1570	2320	Private; Govt of BC	Renamed BC Rail 1972; Leased to CN 2004
Temiskaming & Northern Ontario (TNO)	1902–1932	North Bay ON; Moosonee ON	595	1086	Govt of Ontario	Renamed Ontario Northland Railway 1946
Mackenzie Northern (MNR)	1961–1964	Grimshaw AB; Hay River NWT	969	1050	Private; CN/CP; CN	Purchased by CN 2006

in a societal context. A 19th-century railway can be viewed as a megaproject, with similar potentials for either success or failure.

A trunk line railway designed to move diverse goods and people from point to point differs from a railway tied to a single commodity or a single community. In some cases, such as the Quebec North Shore & Labrador line, the railway can be considered simply as an extension of the mines at Labrador City, Wabush, and Schefferville (Rodon et al. 2022; Table 14.1). In order to make a geographical success of a railway, one must be sure what the basic geographical purpose is, and that it makes sense.

14.3.1.1 Atlantic Canada

Each of the eastern provinces of Canada invested heavily in railway construction between 1850 and 1900, with varying degrees of foresight and varying degrees of success, but with ramifications that continue today. In Nova Scotia, railway construction began as a publicly funded enterprise (Stevens 1960–1962). Joseph Howe regarded operation as a public service, and was not terribly concerned with profit. Innovations, such as allowing farmers to attach their carriages and sleighs to flat cars in piggyback fashion, were to be encouraged, even if they brought in minimal revenue and ate up the workers' time and energy. Freight rates were kept low for individuals, as were passenger rates; discounts for bulk freight were minimal. The Nova Scotia Railway was intended to encourage people to use the lines, and to become railway-minded.

The system was designed to service Nova Scotia, although Howe did talk about the possibility of linkage to Québec (and even to the Pacific Ocean). Halifax, as the best harbour with the best infrastructure, was to be the focus, with lines radiating outwards and inland. Although the relative costs of rail and water shipping varied from time to time and place to place, water transport dominated rail transport in cost terms.

By 1870, the Nova Scotia Railway operated three lines: from Halifax to Windsor (eventually to Annapolis), to service the Annapolis and Avon River valleys, and to bring the produce and gypsum from these areas to the best harbour; from Halifax to Truro, to bypass the difficult tidal estuary and Minas Basin, and to allow gold to be shipped from Waverley and the Rawdon Hills to the port; and from Truro to Pictou, to allow coal to be shipped through Halifax's superior harbour rather than New Glasgow's difficult one, and to provide coal to Halifax (Stevens 1960–1962; Stephens 1972).

Later ventures were less successful. Political interference had not been a major factor in the early stages, but politicians began to interfere with construction routes, hiring of personnel, and management decisions (Stevens 1960–1962). Lines suffered from politically inspired diversions, adding awkward curves and taking them across barren, difficult terrain (such as the aptly named "Folly Mountain", north of Truro). Political manoeuvring was evident in many of the later efforts. Lines built in direct competition with water traffic—Yarmouth to Halifax, Yarmouth to Digby, Halifax to Country Harbour, and Port Hawkesbury to St. Peter's on Cape Breton's southern coast, proved unprofitable (Stevens 1960–1962; Woods 1992).

In New Brunswick, the only substantial port was Saint John, served by river traffic. As New Brunswick wished to export goods through Saint John, rather than shipping them to Halifax, it was not very interested in constructing a line to junction with the Nova Scotia Railway at Truro. Building an Atlantic outlet for the inland regions of Canada, ice-bound by the winter freezing of the St. Lawrence, required constructing a line overland from a marine port to Québec City or Montreal across or around the Appalachians. As this was an expensive proposition, New Brunswick politicians sought to get Britain to pay for the effort.

The British government was willing to finance the Intercolonial Railway through its colonies of New Brunswick and Canada, but it insisted on having the line built to its specifications (Stevens 1960–1962; Legget 1973). The line had to be built to the same 5′ 6″ gauge as the Grand Trunk Railway between Québec and Toronto, so that trains could move freely along the line without trans-shipment being necessary.

The second specification demanded by Britain was tied to military defence (Underwood 2005). Britain had fought one war with the United States in 1812–1815, and engaged in a boundary dispute between New Brunswick and Maine over the Aroostook Valley. The British government was eager to link New Brunswick and Canada, but it also looked upon the line from a military perspective. Any line had to be militarily defensible, as far away from the Maine border as possible, and certainly not on the south or west bank of the Saint John River. The best place for the track, from Britain's view, was along the coast of the Gulf of St. Lawrence. From New Brunswick's view, that was the worst place for the line: away from the settled communities of the Saint John Valley, away from the port of Saint John, and serving only the Acadians of the St. Lawrence coast. That line would undoubtedly be extended to join the Nova Scotia system, and would thus serve primarily to develop Halifax as a winter port for Montreal, while providing little economic growth in the most populous regions of New Brunswick. After Confederation, the Intercolonial Line was built, absorbed the Nova Scotia Railway, and subsequently became part of the Canadian National system (Cruikshank 1986, Forbes 1994).

The Prince Edward Island Railway story is one of the truly incredible stories of railway geography (Stevens 1960–1962; Lavallee 1972). At the time of its construction, some islanders opposed to the effort denounced it as a nefarious plot orchestrated by the British government to bankrupt the province (which had a healthy financial surplus) and drive it into Confederation (Bolger 1961; Baldwin 2009). As some of these opponents subsequently went on to become staunch advocates of Confederation, as the government of Canada showed a limited interest in Prince Edward Island, and as the British government was determined to push P.E.I. into union bankrupt or solvent, these accusations have to be treated as unproved. Nevertheless, it is remarkable how quickly the railway did go bankrupt. Less than 2 years after construction started, the line was insolvent, and Prince Edward Island was in debt and in Canada--to the relief of John A. Macdonald, who saw this development primarily in terms of the elimination of a potential smuggling problem such as that represented by St.-Pierre-et-Miquelon.

As an island geared to agriculture, the best justification for a railway would be to get produce to market, suggesting a radial pattern of short lines from Charlottetown. Prince Edward Island decided to build their railway as a government enterprise through the length of the island, parallel to both coasts. The bill approved by the island legislature specified that the main line would have a length of "about 120 miles".

Although the initial terminals of Alberton and Georgetown are 138 km (86 miles) apart by road, the line as constructed extended 235 km (147 miles; Table 14.2). The extra length was due to the necessity to service all those who had contributed to the line's success by voting in the legislature, by electing suitable individuals, by smoothing out legal problems, and by performing other such tasks. The track meandered gracefully throughout the island, passing through all appropriate properties and past all appropriate doors. As the line passed north of Hillsborough Bay, it did not serve Charlottetown, necessitating the construction of a branch line along the western margin of the bay, one of several.

Prince Edward Islanders had no previous experience in building railways, which led to problems in construction. Stations were built to political rather than geographical criteria: both Charlottetown and Summerside initially had stations that had no access to Northumberland Strait. Although the railway serviced everybody, nobody wanted to be inconvenienced by it or have their access to fields restricted. This meant that level crossings had to be installed at an average of more than 4 per kilometre. Farmers generally assumed that they and their animals had the right-of-way at level crossings, so a trip on the island railway involved a rather leisurely pace.

When Prince Edward Island entered Confederation in 1873, the primary concern of its politicians was to pass responsibility for the railway's completion, operation, and deficit financing to the federal government (Bolger 1961). This objective was successfully accomplished, and the railway continued to lose money for its new owner. Comparison of operating expenses with revenues reveals that the PEI Railway managed to lose \$1.44 for every dollar of revenue during its lifespan (Stevens 1960–1962). This state of affairs continued until the line was closed in 1989 and the tracks were sold for scrap (Fig. 14.4).

Newfoundland also determined to build a railway. Initially intended to connect the copper mines at the port of Hall's Bay to the port of St. Johns, construction began in 1881, with the company entering bankruptcy in 1884 after completing 92 km of track (Lavallee 1972; Cuff 2001; Harding 2008). Subsequently, RG Reid took over construction, completing the new 882-km mainline to Port-aux-Basques in 1898, bypassing Hall's Bay and crossing the high, windswept Gaff Topsails (Table 14.2). For economy, the line was a single-track

Fig. 14.4 Abandoned railbed of the Prince Edward Island Railway is now used as a hiking, cycling, and snowmobile trail. (Author's photo)

narrow gauge (3′ 6″) system. The Newfoundland government also contracted for the construction of several branch lines, connecting Bonavista, Carbonear, Trepasssey, and Placentia to the main line (Cuff 2001; Harding 2008). After the railway went bankrupt again in 1923, the Newfoundland government took operational control, which continued until the Commission of Government after bankruptcy of the government in 1934. Following its only profitable years during World War II, the Newfoundland Railway was incorporated into the CNR in 1949, but failed to be economically viable. After completion of the cross-island Trans-Canada Highway, passenger service was discontinued on the remaining main line in 1969 (Harding 2008), and freight service in 1988 (Cuff 2001). The rails were removed shortly thereafter. Today, the route serves as the "T'Railway" hiking trail.

14.3.1.2 Grand Trunk Railway

The Grand Trunk Railway (GTR) was a 19th-century megaproject designed to facilitate traffic along the corridor paralleling the St. Lawrence River and Lake Ontario, extending westward across southern Ontario to the St. Clair River and eastward across Québec, connecting the colony of Canada to the Atlantic Ocean at Portland, Maine (Currie 1957). Construction began in 1852, with completion of the main line in 1856 (Fig. 14.5, Table 14.2). As a result of poor decisions, the initial capitalization of the GTR was inadequate.

Construction across the relatively flat terrain of southern Ontario and Québec was relatively easy technically by 19th-century standards, but costs were underestimated, the work was not always performed well, and labour shortages restricted progress (Currie 1957; Stevens 1960–1962; Legget 1973; Lavallée 1982; Carlos and Lewis 1995). The design, based partially on engineering standards geared to British climate conditions, had to be modified over time. Incorporation of the Atlantic and St. Lawrence connection to Portland required the GTR to use the 5′ 6″ gauge, to insulate Portland from traffic over standard gauge (4′ 8.5″) USA railroads through Boston and New York. Eventually, economic realities required the GTR to convert all its lines to standard gauge by 1885. In addition, the Atlantic and St. Lawrence was poorly constructed and the acquisition over-valued. The GTR overestimated potential revenues, under-estimated competition from lake and river traffic, and was under political pressure to construct branch lines and maintain low freight rates and passenger fares. This pressure contributed to the rapid construction of the expensive Victoria Bridge across the St. Lawrence at Montreal (1859), and the branch line along the south shore of the St. Lawrence to Rivière-du-Loup, designed to link with the Intercolonial from NB and NS (Lavallée 1982; Cruikshank 1986). Construction of this line was invoked in a clause of the British North America Act establishing Confederation in 1867. The GTR suffered the first of several financial crises in 1860, four years after opening (Currie 1957).

Conflicting, somewhat mutually incompatible objectives were another problem. British investors (the vast majority) were encouraged to regard the GTR as an Imperial project, necessary for defence (Currie 1957; Smith 2013). The British

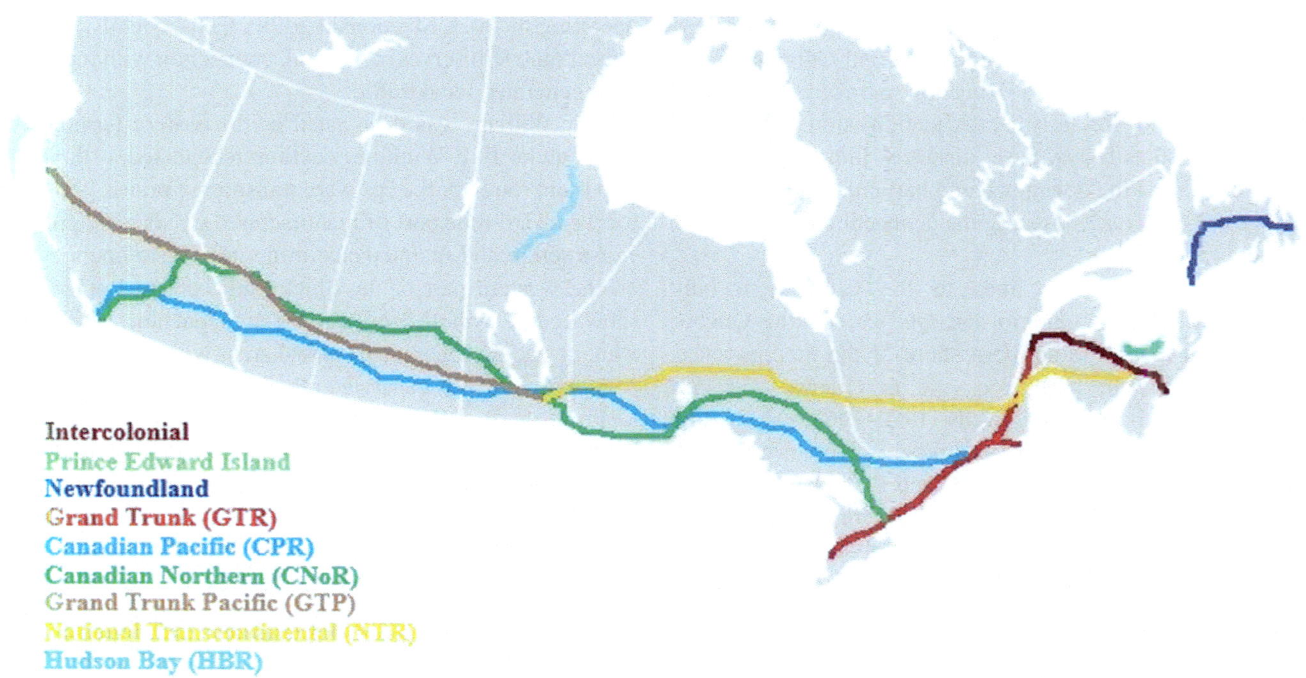

Fig. 14.5 Major historical railways in Canada. See Table 14.2 and text for further discussion

government certainly argued the case for the GTR for troop transport, as had been done when locating the Intercolonial Railway through New Brunswick. Under the circumstances, many investors believed that the Government of Canada would always supply the necessary funding to the GTR as loans and bond support, not allowing the company to fail. Federal, provincial, and municipal politicians regarded the GTR as a source of employment, a patronage opportunity, a public service for their particular communities, and a large corporation to be kept in line. GTR officials recognized that local freight and passenger traffic did not generate sufficient revenue to cover operating expenses, and preferred to concentrate on longer hauls, particularly from Chicago across Canada to eastern USA port cities.

The GTR shaped the geography of areas that it served, particularly in southwestern Ontario and eastern Québec, where river and road transport had been limited. West of Toronto, the cities of Guelph, Berlin (renamed Kitchener in 1916), Stratford, and Sarnia developed along the line. Railway and urban construction on glacial till and glacilacustrine clay plains was straightforward (Catto et al. 2020). Town plans were geometric, although not confined to 90° angles, reflecting centralized planning by the GTR rather than agglomeration. The railway and station were typically situated either downtown (Guelph) or within a short distance (500 m in Stratford). Richmond and Sherbrooke, developed from existing towns along the R. St-Francois, show similar patterns although modified by the valley topography and pre-existing construction.

Along the St. Lawrence and Lake Ontario shore, the GTR largely connected existing towns. Stations were located some distance from the shoreline and pre-existing downtowns (e.g. 800 m at Prescott, 1 km at Brockville and Cobourg, 2 km at Belleville, and 2.5 km at Kingston, the former capital of Canada). As with similarly located stations along the PEI and Newfoundland railways, these distances limited connection of the GTR with the downtown and harbour, hindering interchange of goods and loosening the connection between town residents and railway.

Numerous connector lines to the GTR were built throughout southern Ontario and Québec, linking Québec to Montreal along the northern shore of the St. Lawrence, Montreal, Prescott, and Brockville to Ottawa, and Port Hope to Peterborough (Stevens 1960–1962, McIlwraith 2006). The GTR purchased competing or connector lines throughout southwestern Ontario, such as the Great Western (Windsor-Niagara Falls), Northern (Toronto-North Bay), and Midland (Belleville-Port Hope-Peterborough-Midland), especially after 1881 to outbid the CPR for possession (Currie 1957; Berton 1971). By 1900, the GTR was the dominant railway in the St. Lawrence–Great Lakes plains.

14.3.1.3 Transcontinental Railways

Construction of a transcontinental railway was a condition of BCs adherence to Confederation. After a slow start marred by scandal, construction began in 1874 under government contracts, accelerating after 1881 with the formation of the private (although heavily subsidized by government loans) Canadian Pacific Railway (CPR) company (Berton 1970, 1971; Lamb 1977; Murray 2011). As with the Grand Trunk, although the CPR was nominally a private economic venture, the work had strong political and societal overtones linked to colonization and imperialism (McKenzie 1996). Significant fiscal challenges for the CPR throughout the construction period continued until substantial immigration began in the late 1890s (Lamb 1977).

The CPR line connected in the east with lines from Ottawa (now disused) and Toronto at North Bay (Table 14.2, Fig. 14.5). Westward, the initially surveyed route across the Boreal Shield would have taken the CPR substantially to the north of Lake Superior. As revised after 1881, the new route crossed the Shield diagonally from Sudbury to Heron Bay, following the Lake Superior shore west to Nipigon and Fort William (now part of Thunder Bay). This change combined relatively straightforward engineering and construction on the eastern segment with challenging work across the ridges and valleys of the Superior Geological Province west of Heron Bay (see Chap. 3), necessitating numerous tunnels and rock cuts (Berton 1971; Legget 1973; Murray 2011). The shoreline afforded no large harbours suitable for shipping east of Thunder Bay, limiting lake transport and commercial development, and known mineral resources at the time were limited to the Silver Islet mine near Thunder Bay (which ceased operating in 1884, Macdonald 2008). Consequently, the line served (and still serves) as a trunk connection with relatively little generated local traffic.

Fort William was designated as the eastern terminus of the Thunder Bay–Winnipeg section, recognized as the most important route for the eastward transport of prairie grain and for westward transport of manufactured goods. The greatest construction and maintenance difficulties were encountered on the western part of the Shield, adjacent to Rat Portage (now Kenora), where deep peat bogs and fenlands were significant obstacles. This line traversed areas with limited mineral resources, and also generated relatively little local traffic.

From the Red River, where population pressure forced a southward deviation from Selkirk to the growing Winnipeg (Bellan 1962; Berton 1971), the line traversed the southern part of the Midgrass Plains and the Shortgrass Plains through Portage-la-Prairie to Calgary. This line was located substantially to the south of the existing centres of population (Fort Carlton, Battleford, and Edmonton), the established Carlton Trail (foot and horse traffic), and the initially surveyed route.

The southern line allowed the CPR to control town formation and development at Brandon, Regina, Moose Jaw, Medicine Hat, Calgary, and other communities, and also served to deter potential competing lines extended northwest from the USA. Relocation substantially shifted the axis of immigrant population growth and agricultural settlement southward from the North Saskatchewan Valley, and resulted in the political and commercial development of Regina as the new capital of the NWT (and subsequently of Saskatchewan).

Another effect was to delay agricultural settlement in the dry Shortgrass Prairie west of Swift Current. Climate variation in the region resulted in anomalously wet conditions favourable to agriculture immediately prior to construction, turning to the more normal drier climate in the later 1880s, coinciding with efforts to encourage agricultural immigrants. Alternating wet and dry conditions have persisted in the Shortgrass Prairie (see Chap. 5). The requirement to construct sidings at 13 km intervals across the prairies, based on the distance a horse could pull a full wagonload of grain in one day, served as loci for agricultural community formation, with construction of grain elevators and branch lines.

Indigenous relations with railway construction crews were frequently tense, but never erupted into overt violence. Individual leaders' reactions varied from acceptance and negotiation coupled with resignation (notably Isapo-Muxika (Crowfoot) of the Siksika) (Dempsey 2003) to resistance (Payipwât (Piapot) of the Nēhiyawi-pwātak) (Wishart, 2007). However, the CPR and resulting immigration and settlement were a major factor contributing to the expansion of the Residential School system, which proved devastating for indigenous peoples throughout.

West of Calgary, the line followed the Bow River to Banff, crossed the Rocky Mountains continental divide at Kicking Horse Pass, and traversed the Cordillera to Kamloops through Rogers and Eagle Passes. This route posed greater engineering and construction challenges than the originally surveyed route through Yellowhead Pass and the North Thompson River valley presented (Legget 1973). The presence of hot springs at Banff led to the establishment of Canadia's first national park, followed by others along the route. National parks were established to promote tourism along railways, contributing substantially to the international image of Canada (see Chap. 10). The section through the Selkirk and Monashee Mountains was the last to be completed in 1885.

Kamloops was connected to the Pacific Ocean at Vancouver through the Thompson and Fraser River valleys. This segment of the CPR also posed substantial engineering and construction difficulties (Berton 1971; Murray 2011). Chinese workers played a major role during construction, despite encountering racist attitudes (see Chapter mmm).

Following completion of the CPR main line, increasing population in western Canada led to pressure to construct additional railways. Unfortunately, replicating experience in eastern Canada, economic and logistical concerns were sometimes not the motivation for construction. Through a combination of political pressures and corporate decisions, three major transcontinental railways were constructed in Canada between 1896 and 1916, which turned out to be one and a half too many.

The Canadian Northern Railway (CNoR) originated in eastern Manitoba (Regehr 1976). The product of the promotors and contractors William Mackenzie and Donald Mann, the CNoR developed gradually from the coalescence of local, rapidly built "colonization" railways, constructed in some cases at the request of agricultural communities. Rather than a trunk with large branches, the early CNoR more resembled a bush, with shoots extending in various directions. The CNoR tended towards inexpensive basic construction, use of second-hand rolling stock, and relatively low freight rates, focusing on local business (Table 14.2, Fig. 14.5). Construction of a route from Winnipeg to Thunder Bay around the southern shore of Lake of the Woods in Minnesota through Rainy River and along the old Dawson portage road provided farmers with an alternative to the CPR. By 1906, the CNoR system had reached northwest to Edmonton, primarily through progressive expansion of branch lines, and the company dominated railway trade across the prairies north of the South Saskatchewan River through the settled agriculturally fertile belt of the Midgrass Prairie and Aspen Parkland (Regehr 1976; Cameron 1992; Murray 2011).

At this critical juncture, Mackenzie and Mann made an unfortunate and ultimately fatal decision for the CNoR. Faced with the presence of the federally supported Grand Trunk Pacific Railway (GTP) and unable to come to an agreement for cooperation with it (discussed below), Mackenzie and Mann determined to simultaneously build the CNoR west from Edmonton to Burrard Inlet, and east from Thunder Bay across the Boreal Shield to Ottawa and Toronto, as well as acquiring segments of railway in Ontario, Québec, and Nova Scotia (Regehr 1976; Woods 1992). This drastic over-expansion proved too much for the CNoR financial resources, especially after the outbreak of World War I curtailed immigration and availability of labour and railway materials (Stevens 1960–1962; MacKay 1986). The eastern section, following the route originally surveyed for the CPR across the Boreal Shield north of Lake Superior, was not economically viable. The western segment through Yellowhead Pass and along the Thompson and Fraser valleys to Port Mann on Burrard Inlet, was well-conceived economically and subsequently formed a viable route. However, indebtedness forced the CNoR into liquidation, with the federal government taking over the railway. In 1923, the former CNoR was amalgamated with the GTP, Grand Trunk, and Intercolonial Railway to form Canadian National Railways (CNR) (Perl 1994).

The Grand Trunk Railway (GTR) had declined the opportunity to build the original CPR (Berton 1970). At the beginning of the twentieth century, under new management led by CM Hays, the GTR decided to construct a trunk line from Winnipeg to the Pacific Ocean (Stevens 1960–1962; Lavallée 1982; Cameron 1992; Murray 2011). The GTP was intended as a trunk line, carrying goods between Asia and eastern Canada through the to-be-constructed port of Prince Rupert (of Rupert's Land). Emphasis on local prairie traffic, even grain, was relatively limited. The line was to be built to the highest prevailing standards for high-speed express traffic, bypassing existing population centres on the Prairies (except Edmonton), through the Yellowhead Pass (directly paralleling 210 km of the CNoR), and west across difficult construction country through the Cordillera across Tsimshian land (McDonald n.d.) to the mouth of the Skeena River at Prince Rupert (Stevens 1960–1962; Legget 1973). New communities to service the GTP were established at Melville, Prince George, Smithers, and Terrace (Leonard 1996). Prince Rupert was conceived using the latest ideas in town planning, including curved streets rather than a grid pattern and open vistas, as a suitable grand Pacific Ocean terminus for international travellers.

The GTR required a connection from the GTP eastern terminus at Winnipeg to its lines in eastern Canada (Currie 1957, Stevens 1960–1962, Cameron 1992, Murray 2011). In 1903, GTR management led by Hays negotiated an agreement with Prime Minister Wilfred Laurier, under which the federal government would construct a National Transcontinental Railway (NTR) from Winnipeg to eastern Canada as a joint venture with the GTR. Negotiations between the GTR and CNoR to avoid duplicate construction and cooperate (with the GTR working in eastern Canada and the CNoR in western Canada) failed, with participants and observers suggesting a variety of reasons (Regehr 1976). Ultimately, financial resources were insufficient for construction of the GTP, combined with management disarray following Hays' death on *RMS Titanic* in 1912, the impact of World War I, and difficulties for the NTR. Failure of the GTP caused the collapse of the GTR, and both were incorporated in the CNR in 1923, without financial compensation to the GTR shareholders (Currie 1957, Stevens 1960–1962, Perl 1994, Smith 2013).

The Laurier government's NTR project had its eastern terminus in Moncton. The line ran diagonally through sparsely populated central New Brunswick to Edmunston, where it connected with the line to Québec City, crossing the St. Lawrence over the Québec Bridge. From there, the NTR proceeded directly westward to Winnipeg via La Tuque, Cochrane, and Nakina, across the Boreal Shield, bypassing the St. Lawrence Valley, Montreal, Ottawa, Toronto, Sudbury, and Thunder Bay. Construction of a branch line to Thunder Bay was delayed, proving frustration from the GTR (Stevens 1960–1962). Although strenuous efforts were made to promote agricultural colonization in the Abitibi area of Québec and the Cochrane area of Ontario, taking advantage of the clayey soil resulting from Glacial Lake Agassiz (see Chap. 3), settlement was slow and profits absent. As with the GTR and CPR, political considerations were central for the NTR, but financial viability was not achieved. The NTR was handed over to CNR as a bankrupt and generally unwelcome gift in 1923.

14.3.2 Rail Today

After 1923, CNR was faced with surplus trackage, parallel lines in northern Ontario lacking integration, inherited debts, public expectations of maintained service, hopes raised in newly established communities, increasing competition from road transport, and requests for additional construction of branch lines, including a connection to an additional marine outlet for grain through Hudson Bay at Churchill (Stevens 1960–1962, Perl 1994; Fig. 14.6; Table 19.2). A connection between the former CNoR and NTR lines through Longlac increased east–west freighting efficiency. CNR gradually coped with these challenges and the Great Depression to maintain service, providing effective competition and complementary service to CPR (Legget 1973, see Bonsor 1995). After World War II, continued growth in truck freighting and private automobiles surpassed the railways' abilities (Roberts et al. 2020). CNR acquired the Newfoundland Railway in 1949, incurring further expenses. Both railways closed branch lines, reduced passenger service, and focused on particular commodities (Bonsor 1995). CNR (now CN) was privatized in 1995 (see Boardman et al. 2013), and CPR (now CPKC) merged with the Kansas City Southern railroad in 2023.

Throughout the 1960s and 1970s, both railways continued to operate decreasing passenger service, incurring substantial financial losses. As a private corporation, CPR required subsidies of up to 80% of losses to retain services, but CNR also required substantial injections of financial support as well. Various marketing schemes failed to arrest the persistent decline in passenger traffic. In 1977, CNR passenger services were assumed by a new Crown (government) Corporation, Via Rail Canada, which also assumed the few remaining CPR passenger routes in 1978 (Nelligan 1982, Canada 1992).

Via Rail remains as a government corporation in 2024, with decisions on service levels substantially dependent on political considerations, and operations requiring substantial government money (Canada 1992, Milke 2014; Via Rail Canada 2018). Passenger numbers have stabilized at approximately 5 million/y, compared to more than 60 million rail passenger trips in 1944 (Fig. 14.7). More than 75% of the travel occurs along the Québec-Montréal-Toronto corridor.

Fig. 14.6 Operational railway km in Canada, 1871–2021, compared with population. Figures include Newfoundland and Labrador. (Statistics Canada, Census of Canada)

For every $1 collected from passenger fares, the federal government contributes more than $2 in subsidy (Milke 2014).

Passenger rail service outside the Québec-Montréal-Toronto corridor is minimal. Vancouver is served by local commuter routes, an AMTRAK connection to Seattle, and by thrice-weekly service to Edmonton, Saskatoon, Winnipeg, and Toronto. Calgary and Regina have no passenger service. Twice- or thrice-weekly trains connect Montreal and Halifax, but there is no service to Fredericton, Saint John, Prince Edward Island, or Newfoundland. Local service exists to Senneterre and Jonquière (Québec) from Montreal, to Churchill (MB) from Winnipeg, and from Jasper (AB) to Prince Rupert (BC). Services on other lines not connected to the Via passenger rail network connect Sept-Iles (QC) and Labrador City (NL), Sudbury and White River (ON), Cochrane and Moosonee (ON), and Carcross (YT) to Skagway (Alaska, USA).

Periodic calls for the establishment of high-speed rail service along the Québec-Montréal-Toronto corridor (with connections to Ottawa), and less frequently Edmonton-Calgary and Calgary-Vancouver, have not resulted in construction. Politicians (as well as some geographers interested in transportation and climate change) have pushed for improved passenger rail service, but the costs of infrastructure renewal and new construction and projected passenger numbers are both significant obstacles (Katz-Rosene 2017). As Via Rail currently uses existing CN and CPKC routes, built for slower-moving freight trains, construction of high-speed links would require completely new lines. At present, freight trains have priority over passenger trains on the single-track lines in use outside the Québec-Montréal-Toronto corridor, with the resulting delays impacting passenger service schedules (Via Rail Canada 2018).

Rail traffic in Canada in 2024 generates approximately $10 billion y^{-1}, 95% of which comes from freight (Railway Association of Canada 2021; Statista 2022a; Statistics Canada 2023c; Fig. 14.8). Of the total Canadian rail transport industry revenues, CN accounts for over 50% and CPKC for approximately 35%. The railways moved 330 million t in 2022, about 70% of intercity and 50% of exported freight. Minerals, including coal (~30%), agricultural goods (~11%), chemicals, including hydrocarbons (~11%), lumber and paper products (~6%), and automobiles (~5%) are the most commonly transported commodities (Statistics Canada 2023c; see Macciotta et al 2018). Intermodal container traffic constitutes 35–40% of the total freight transported.

14.4 Road

Road construction and maintenance in the provinces is a provincial responsibility, commonly downloaded to municipalities for local streets. Routes between cities are gener-

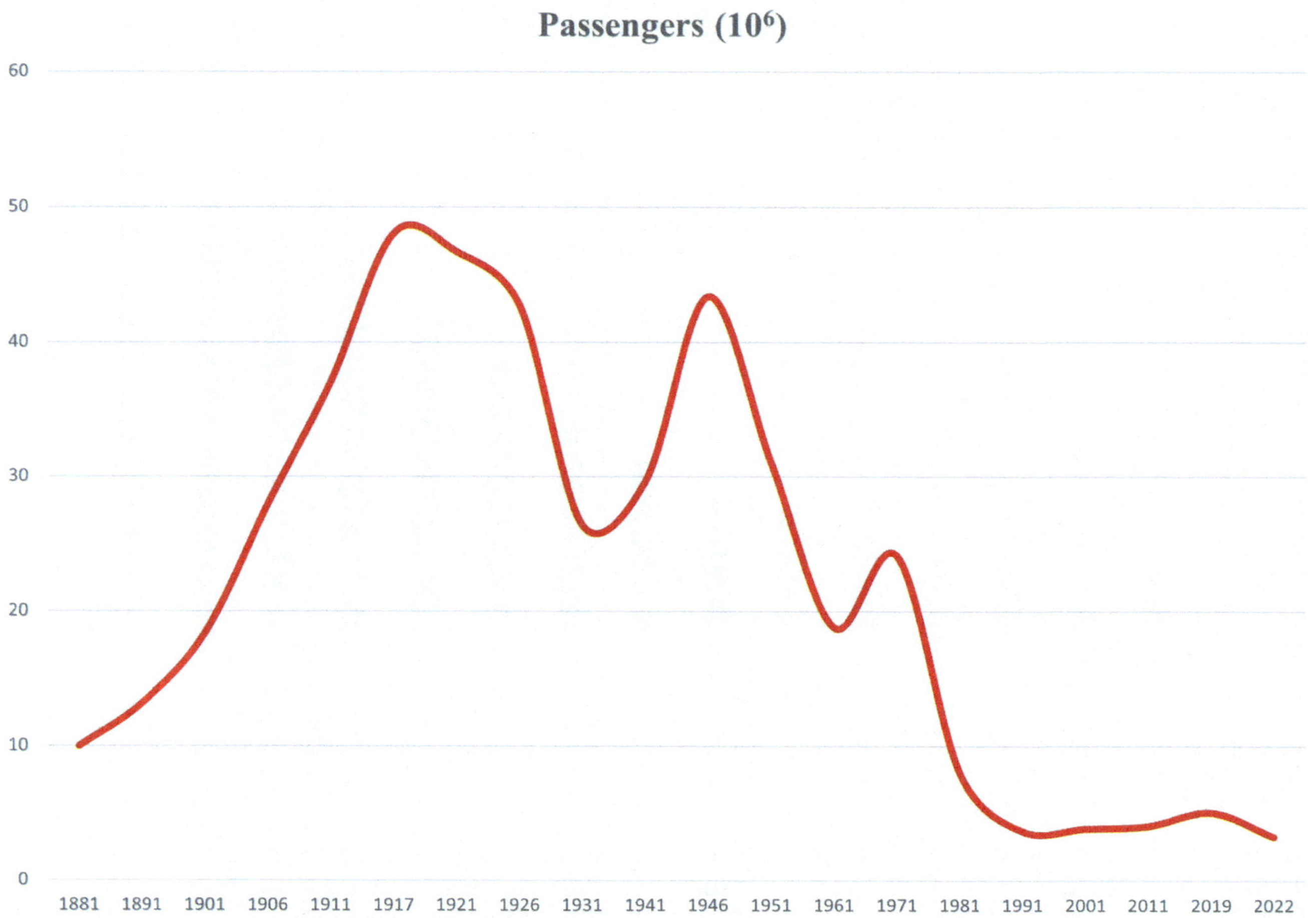

Fig. 14.7 Passenger traffic (millions) on Canadian railways, 1881–2022. Figures include Newfoundland and Labrador. (Via Rail Canada 2018, and annual reports)

ally maintained by provinces, with the exception of Ontario, where 4976 km of the provincial highway system (~30% of the former total) was downloaded to municipalities in 1997–98 (Bevers 2023). Although the federal government has identified almost 40,000 km of routes as the "National Highway System" (~3.7% of public roads in Canada), the highways are provincially controlled, and there is no equivalent of the USA Interstate Highway or the UK National Highway systems (Council of Ministers Responsible for Transportation and Highway Safety 2019). Canada's total road network has exceeded 1 billion kilometres (Fan and Heminthavong 2022).

The initial roads followed existing Indigenous travel routes, generally paralleling rivers or bypassing falls and rapids around portages. In eastern Canada, road improvement was undertaken by military groups, government-organized work levees, or private interests. Roads were generally trampled earth or gravel, with stone paths uncommon. Corduroy roads made by lashing logs together in rafts at right angles to the transit route, were used to traverse wetlands and mudholes, but were difficult to maintain. Overland travel was often easier in winter, over frozen rivers and wetland terrain, without accompanying mosquitoes. The shortcoming of the roads was a factor encouraging the development and use of railways. In Prairie Canada, the few long-distance trails were rapidly superseded as railway construction progressed. Road development generally had to wait until the advent of the automobile (Roberts et al. 2020).

The Trans-Canada Highway (TCH) main route extends for 7500 km, eastward from Victoria (BC) through Vancouver, Calgary, Regina, Winnipeg, Ottawa, Montreal, Fredericton, North Sydney (NS), and St. John's, traversing 9 provinces. The main route from Lake Superior to Vancouver parallels the original CPR main line. Ferry services from North Sydney (NS) to Channel-Port aux Basques (NL), 178 km apart, are operated by Marine Atlantic. The provincially owned BC Ferries connects Horseshoe Bay with Nanaimo on Vancouver Island. A separate TCH route connects Prince Edward Island with New Brunswick (via the Confederation Bridge/Epekwitk Crossing) and Nova Scotia (via ferry). Several alternate routes are designated across western Québec and central and northern Ontario. The TCH and its branches

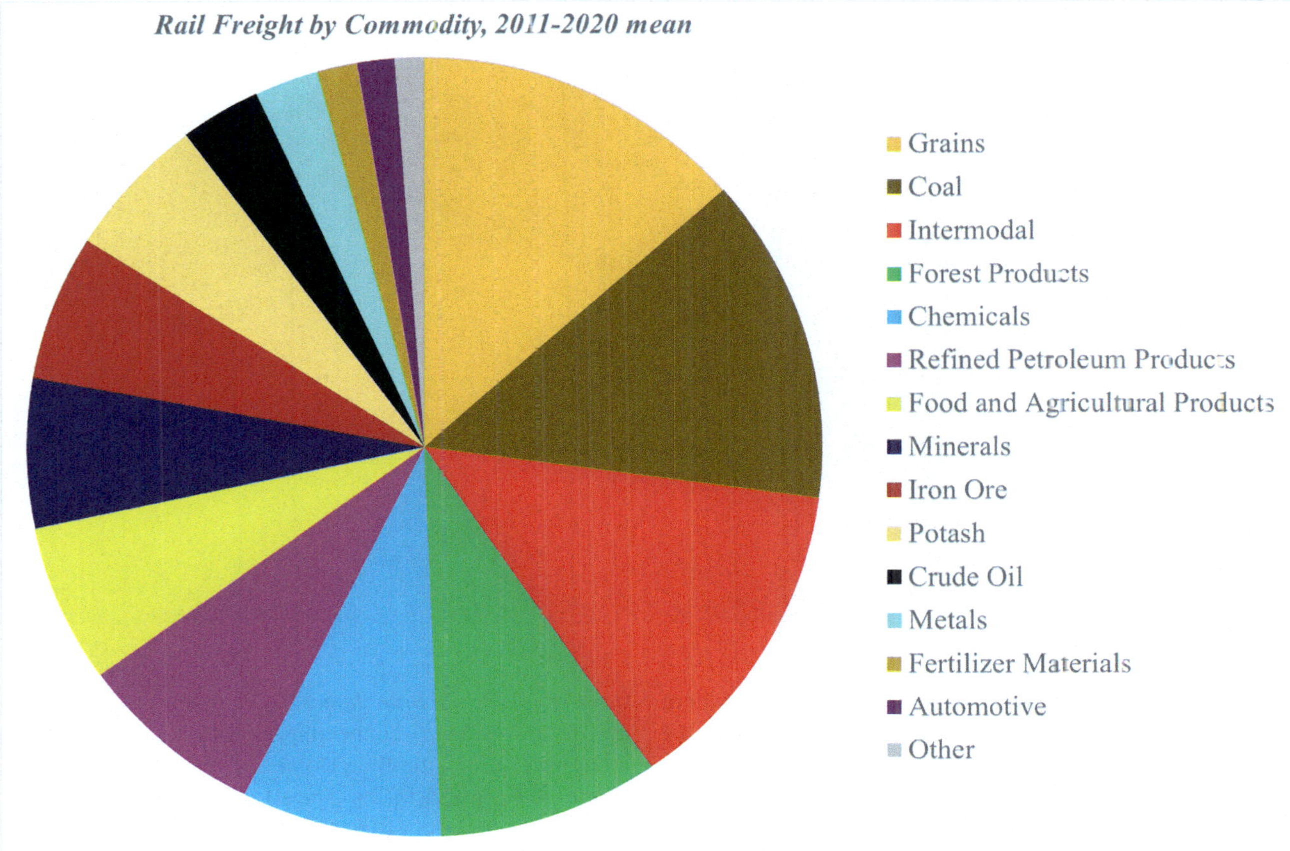

Fig. 14.8 Rail freight traffic by sector, 2011–2020 mean proportions. (Railway Association of Canada 2021, Statista 2022a, Statistics Canada 2023c)

do not extend through Toronto or Halifax. The Yellowhead Highway forms an alternate northern route across western Canada through Winnipeg, Saskatoon, Edmonton, Yellowhead Pass, and Prince Rupert, paralleling the original GTP railway route.

Construction of the TCH was undertaken by each province to provincial standards and schedules. The last sections of the main route to be completed were along the Lake Superior shoreline in 1960 (although the Northern Ontario branch route via Cochrane and Hearst was completed in 1943), across Rogers Pass (BC) in 1962 (bypassing the older gravel road following the Big Bend of the Columbia River), and across central Newfoundland in 1965 (at "Pearson's Peak", a small hill west of Grand Falls-Windsor). The Thunder Bay–Atikokan–Fort Frances branch route in northwestern Ontario, and the section of the Yellowhead Highway along the upper Fraser River east of Prince George BC, were the last to be completed.

Much of the TCH main route consists of two-lane, non-controlled access highways, particularly across Newfoundland, central and northern Ontario, and Cordilleran BC. Lower traffic volumes and higher construction costs have mitigated against political promises to build controlled-access freeways. The branch routes through PEI, western Québec, Ontario, and the Yellowhead Highway are also largely two-lane, non-controlled access highways. At Nipigon, the bridge across the Nipigon River forms the only road connection linking eastern and western Canada. Other choke points along the TCH include the Wreckhouse area of western NL, the Canso Causeway linking Cape Breton Island and mainland NS, the Kenora-Whiteshell segment in westernmost Ontario, and the mountain passes in BC. Freeways link the major cities in the Great Lakes–St. Lawrence Lowland, New Brunswick, and Nova Scotia.

14.4.1 Northern Roads

Road construction in the Boreal, Taiga, and northern Cordillera of Canada is linked to military requirements, resource extraction, and government administration, in addition to (or in some cases, of greater importance than) providing access to and from northern communities. In northwestern Canada, major construction projects include the

Fig. 14.9 Curved wooden bridge over the Kiskatinaw River, BC, constructed as part of the original Alaska Highway in 1942. The bridge is open to foot traffic only. (Author's photo)

Alaska Highway, built across northern BC and southern Yukon by US military personnel in the 1940s (Fig. 14.9); the Canol Road, now partly abandoned, paralleling the World War II oil pipeline from Norman Wells NWT to Ross River YT; the Robert Campbell Highway, connecting Watson Lake, the mining communities of Faro and Mayo, and Carmacks YT; the Nahanni Range Road, to the now-abandoned mining community of Tungsten; the Stewart–Cassiar Highway, across northwestern BC; the Dempster Highway, connecting Dawson YT and Inuvik NWT; the Inuvik-Tuktoyaktuk Highway, the only road reaching Arctic Ocean tidewater in Canada; the Mackenzie Highway, extending from Grimshaw and High Level (AB) along Deh Cho north from Yellowknife; and the Liard Highway along the Liard River in southern NWT and northeastern BC. Roads connecting resource-based Boreal communities to southern infrastructure in the Prairie Provinces are present in Alberta (Fort McMurray, petroleum), Saskatchewan (La Ronge, mining and forestry), and Manitoba (Flin Flon and Thompson, copper, and nickel). Although Nunavut has approximately 850 km of roads, there are no connections to the road network elsewhere in Canada.

In eastern Canada, northern roads include the Ignace-Pickle Lake-Windigo Lake route (Highway 599 and Northern Ontario Resource Trail; Ontario 2020), serving mineral resource activities including those in the Ring of Fire, associated with REE used in electric batteries; the Vermilion Bay–Red Lake–Nungesser route (Highways 105, 125, and Northern Ontario Resource Trail), serving gold and other mining; the Cochrane-Kattawagami River road (Highway 652, gold mining); the *Route de la Baie James* connecting the La Grande Hydroelectric reservoir to Matagami and Val d'Or in Québec; the *Route Transtaïga* extending east from *Route de la Baie James* to Brisay and Caniapiscau; the *Route du Nord*, connecting Chibougamau with *Route de la Baie James*; the Baie Comeau-Fermont route (Québec Route 389), linked to the Trans-Labrador Highway across Labrador to Happy Valley-Goose Bay; the southern extension of the Trans-Labrador Highway to Cartwright and Red Bay (NL); and the uncompleted (as of 2024) Québec North Shore highway (Route 138) between Sept-Iles, Natashquan, and Blanc-Sablon. As of 2024, no all-weather land routes link Nunavik or Nunatsiavut to southern Canada.

In northern Canada, overland travel in winter is often easier than in summer, as long as the traveller is adequately prepared. Snowmobiles are essential components of northern lives, much as automobiles in more southerly climes. Transportation of bulky or heavy goods requires the construction and maintenance of snow and ice winter roads (Barrette 2015, 2018).

A minimum snow density of 400 kg/m^3 is required to support smaller vehicles, with at least 500 kg/m^3 required for commercial truck travel (NWT Dept of Transportation 2007). In many cases, snow must be mechanically compacted, particularly in Taiga regions where freshly fallen snow frequently is less than 100 kg/m^3. In Boreal Manitoba and northwestern Ontario, snow density commonly does not exceed 150 kg/m^3 even after several weeks of natural accumulation. Snow hardness is assessed using the Rammsonde equation as equal to 3.826 (snow density)–1618. Minimum Rammsonde values of 450 are required for commercial snow and ice road traffic. Testing involves the use of both mechanical probes to assess density, and geophysical techniques to measure resistivity (linked to snow density) (Barrette 2015, 2018).

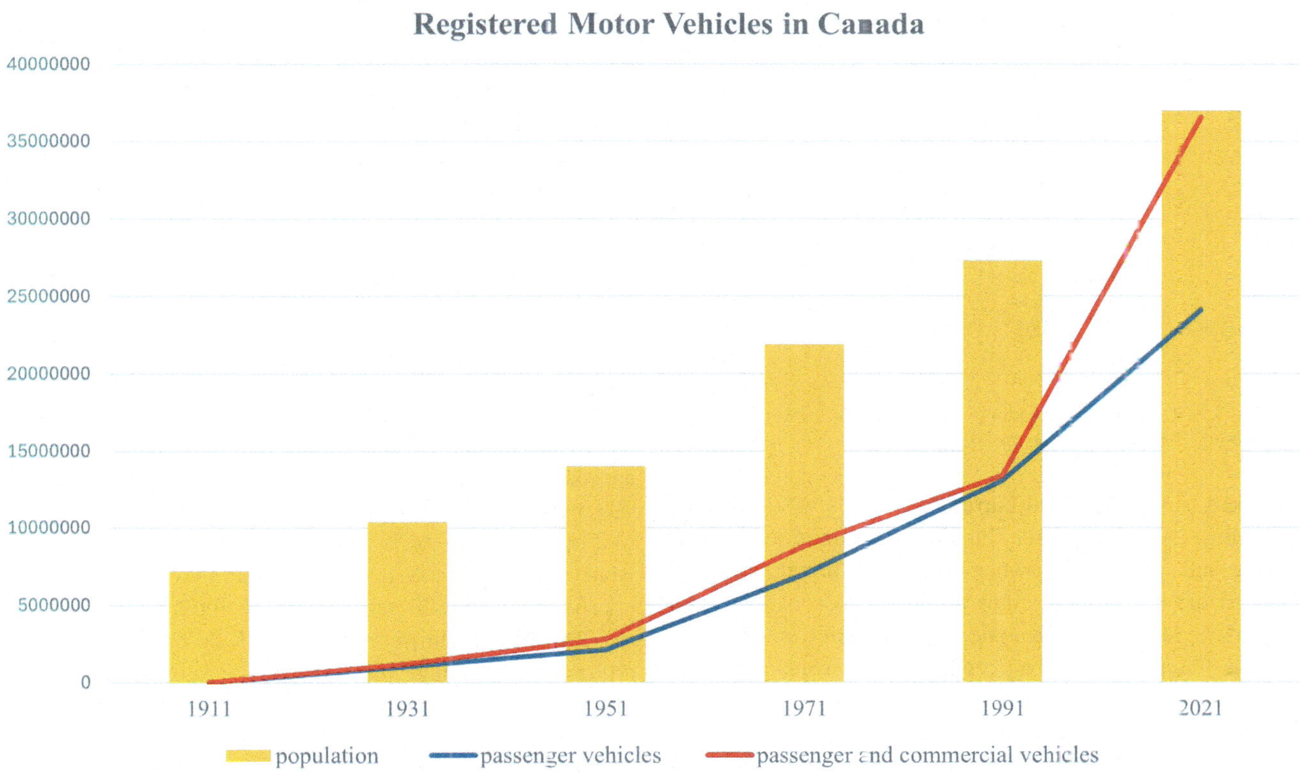

Fig. 14.10 Motor vehicle registrations in Canada compared with population growth, 1911–2021. Figures include Newfoundland and Labrador. (Statistics Canada 2023a, and Census of Canada)

Ice-penetrating radar and physical augering are used to assess ice thickness over lakes and rivers (NWT Dept of Transportation 2007; Hayley and Proskin 2008). A minimum thickness of 40 cm of clear blue ice (without evidence of regelation, zones of trapped air bubbles, or debris) is required to support a 1-t pickup truck, rising to 120–150 cm for a fully loaded commercial 18-wheeler semi. Although the winter routes are colloquially referred to as "ice roads", in practice river and lake crossings are kept to the absolute minimum necessary. Monitoring and road openings and closures are the responsibility of provincial, territorial and/or municipal government employees, and truckers need permission from monitors to undertake any journey.

Although driving to supply communities can make interesting television, most ice roads service mining and petroleum operations. Considering the preparation, maintenance, monitoring, driver salaries and availability, and provision of rescue services, the break-even time for most winter roads is between 40 and 50 days/winter. Late freeze-ups and early thaws, coupled with increasingly frequent mid-winter warm events due to climate change, mean that the operation of many winter roads is becoming problematic (Barrette 2015, 2018).

14.4.2 Vehicle Registrations and Drivers

Registrations of both personal passenger and commercial vehicles have steadily increased throughout the past hundred years (Fig. 14.10). From 1971 through 2021, the ratio of people to registered passenger vehicles dropped from 3.1 to 1.5 (Statistics Canada 2023a). Provincially, current rates vary from 1.3 (SK) to 1.9 (PEI), with slightly lower rates in the largest urban centres. Through the same period, average household size declined from 3.5 to 2.9, indicating a significant increase in the number of multi-car households. The average Canadian household owned 1.5 cars in 2021 (Statistics Canada 2023a).

Electric (including hybrid) vehicles accounted for 3.0% of passenger automobile and light truck registrations in 2022 (Statistics Canada 2023a). Although concern over GHG emissions is abundant, higher initial costs, shortages of charging stations (especially in rural areas), range concerns, and battery performance issues are significant factors limiting adoption of electric vehicles for personal use.

The percentage of Canadians of driving age (16–17 for restricted licences, 17–19 for full licences for passenger vehicles, varying by province) owning vehicles has steadily

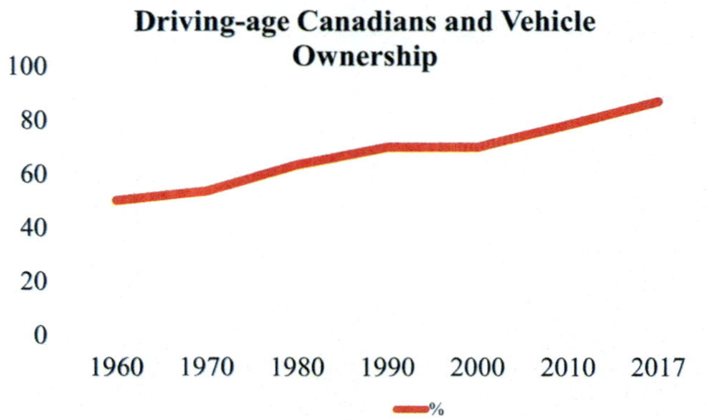

Fig. 14.11 Driving-age Canadians and Vehicle Ownership (by percentage), 1960–2017. (Tests.ca 2024)

increased since 1960, and currently exceeds 80% (Tests.ca 2024; Fig. 14.11). Younger Canadians continue to embrace driving and automobile ownership. Male: female ratios of licenced drivers show a slight preponderance of males for ages 65 or less, although males 75 and older are twice as likely as females to retain a licence. Driving is the main form of transportation for those between 65 and 84, and for males 85–89. The average private driver covers more than 15,000 km annually, and about 12 million Canadians commute to work by automobile each year. Clearly, despite concerns with fuel prices, environmental consequences, and the health disadvantages of sedentary lifestyles, automobile driving remains a central component of Canadian life, necessitating continued investment in streets and highways.

Elementary and high school students living at least 1 km from school generally have the option of using a school bus. Safe arrival at school is approximately 80 times more likely than when travelling by automobile, with only a single school bus passenger fatality between 2009 and 2018 (Tests.ca 2024). Nevertheless, many parents prefer using their automobiles to transport their children.

Numbers of collisions, injuries, and fatalities have all seen steady declines on Canadian roads. Collisions involving injuries dropped from 153,832 in 2002 to 77,933 in 2021, despite increases in vehicle and driver numbers (Transport Canada 2023a, b; Tests.ca 2024). The fatality rate (collisions/10000 vehicles registered) dropped from 1.57 to 0.67 over the period. About two-thirds of fatalities include drivers and passengers, together with pedestrians (15–20%), cyclists (2–3%), and motorcyclists (10–15%). Drivers and passengers who were not wearing seat belts, legally required throughout Canada with ~95% compliance, accounted for 30% of fatalities.

Although weather conditions are involved in ~25% of accidents with attributed causes, fatalities peak in the warmer months (Transport Canada 2023a, b). Speeding and distracted driving are major causes: 25–30% of fatalities are attributable to speeding. Health care costs and lost productivity resulting from automobile accidents exceed $10 billion annually (Canadian Association of Chiefs of Police 2023). Unfortunately, aggressive driving remains a significant problem. Collisions (including automobile-infrastructure collisions) and insurance claims are typically highest in December and January (Tests.ca 2024).

14.4.3 Wildlife-Vehicle Collisions

Wildlife-vehicle collisions are significant problems across Canada. In British Columbia, at least 9,800 wildlife-vehicle collisions occur each year, resulting in the deaths of 4 humans, significant injury to 380 others, and the deaths of 9,300 animals (Vanlaar et al 2012). On Newfoundland, more than 700 moose-vehicle collisions occur annually, costing the lives of 2 humans and ~ 685 moose. Damage to vehicles resulting from collisions with deer or moose in Canada averaged approximately $5600 per incident, with total costs in excess of $200 million (Huijser et al. 2009). Moose collisions generally result in damage in excess of the value of the vehicle involved.

Accidents are most likely to occur in the evening, either because deer and moose are more active at these times, or because humans and animals are less able to see each other at those times (Vanlaar et al. 2014, Laliberté and St-Laurent 2020, and references therein). Deer in isolated, protected, or non-hunting areas show more daytime activity than those in human-influenced areas, suggesting adaptation to human pressure. Typically, collision rates increase at the beginning of the plant growing season, suggesting a connection with animal movement in search of browse. Snow cover, calving, use of roadside salt, abandoned yearlings, and rutting all have been linked to increased collisions. Moose preferentially avoid roadsides when biting insects were most abundant. Not surprisingly, collision rates tend to be higher on roads with high traffic volumes, either throughout the year or seasonally. Roads through forested areas or with curves, requiring lower speed driving, have lower collision rates. Studies undertaken on a regional scale tend to reveal intuitively obvious relationships between collision sites and suitable habitat, but more intensive locally focused ("road-by-road") work turns up apparent anomalies and exceptions.

14.4.4 Intercity Bus Travel

Long-distance intercity bus travel has become increasingly limited since 2010, with many companies curtailing or ceasing operations due to financial losses (Statistics Canada 2023b). COVID-19 impacts resulted in a 50% cut in revenue compared to the preceding three years. As the people impacted are most likely to be living in smaller communities, partic-

Fig. 14.12 Truck freight traffic by commodity sector, 2020. Compare with Fig. 14.8. (Statistics Canada 2021, Transportation Association of Canada 2021, Transport Canada 2021)

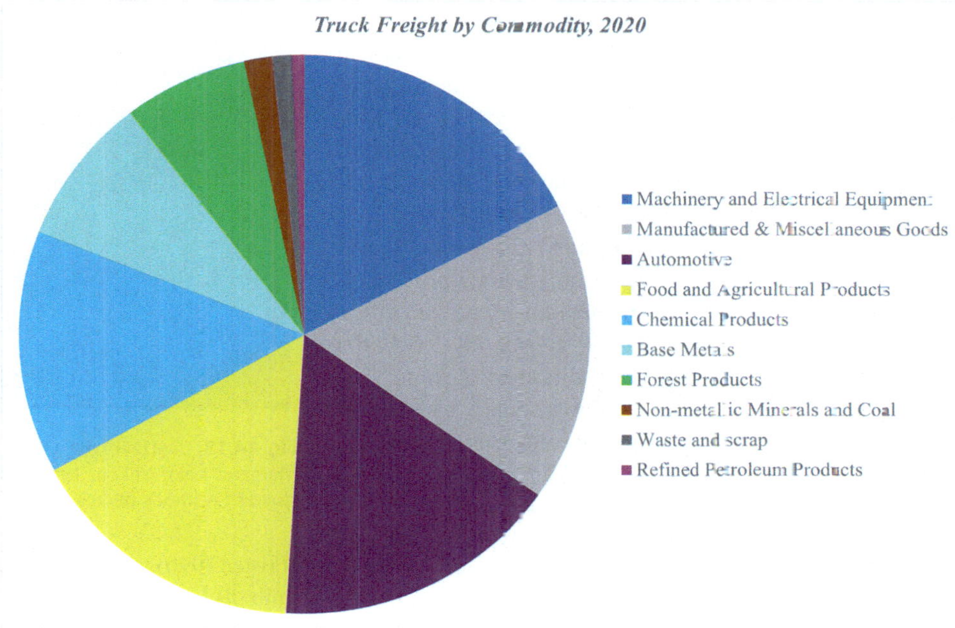

ularly indigenous communities, and are least likely to be able to avail of automobile transportation, the decline of bus services has become a societal and political concern (House of Commons 2023).

14.4.5 Trucking

Trucking employs approximately 400,000 people in Canada, including 300,000 drivers. The sector accounts for approximately 3.5% of the national GDP, generating roughly $40 billion annually (Statistics Canada 2021, Transportation Association of Canada 2021, Transport Canada 2021, Fan and Heminthavong 2022, Statista 2022b). Truck transportation moves 35–40 billion t-km in 60–65 million shipments annually (Fig. 19.12). Trucking accounts for 40–50% of merchandise trade and 75–80% of domestic freight transportation, including moving over 90% of consumer products and foodstuffs. Trucking represents 60% of overall Canada–USA trade traffic, more than 80% by value, including electrical machinery, automobiles and parts, agricultural products, and chemical and petrochemical products. Automobile manufacture in Canada and the adjacent USA frequently involves multiple trips across the border, carrying parts and components in both directions at Sarnia and Windsor. Considering all aspects (driver salaries and benefits, fuel, maintenance, insurance, and administrative costs), long-distance trucking costs approximately $2 per km travelled.

Trucking, including driver training and licencing, company licencing, enforcement of weight limits and vehicle maintenance and safety requirements, permitted goods, emissions and other environmental requirements, and operational hours are enforced by the provinces. Local municipal bylaws may also apply. Cross-border (USA-Mexico) trade, including health standards for drivers (such as vaccination status) is regulated by the federal government. Some provinces permit Long Combination Vehicles (LCV, B-trains) with two trailers attached to a single tractor.

For several years, industry representatives and clients have been concerned about a possible shortage of drivers, with estimates of about 20,000 (compared to the current total of 300,000) common (Ontario Trucking Association 2023). Almost 97% of drivers are male (Crizzle et al. 2017, Statista 2022b). More than 30% are recent immigrants, including an estimated 18% from southern Asian countries, forming a majority of drivers in the Vancouver CMA (Ontario Trucking Association 2018). Truck driving was designated as an Express Entry occupation for potential immigrants by Immigration Canada in 2023. The average salary for full-time truckers is about 20% less than the national average for fully employed people, with female truckers receiving about 10% less than males. Drivers on long-haul routes (minimum 160 km, but typically much longer) are generally paid more than those trucking locally. Most drivers are not unionized and act as self-employed contractual workers.

Concerns have been raised concerning driver health (Crizzle et al. 2017). Long-haul drivers typically convey 18-wheeler semi-trucks (12 t empty, to a legal maximum of 40 t loaded), driving a legal maximum of 13 h/day (15 h/day north of 60°N), with mandatory 8-h breaks between shifts, over distances of 500 km or more (Canada 2024). Studies indicate that drivers suffer from hypertension, cardiovascular disorders, and diabetes linked to occupational stress, poor diet, obesity, sleep apnea, and general fatigue. Long-haul truckers also tend to be older and have higher smoking rates than the general population (Crizzle et al. 2017). All these factors increase the hazard of being involved in collisions. Inadequacies in driver training, especially for winter conditions and Cordilleran highways, have also been flagged as an issue (Malkin et al. 2021). Training standards, methods, evaluations, and follow-up vary considerably among provinces.

Together, heavy (>4.5 t) and medium (two-axle) trucks travel almost 30 billion vehicle-km, consuming 9.5 billion litres of diesel fuel annually, in addition to gasoline used by some medium and light trucks (Environment and Climate Change Canada 2022). Diesel burning emissions are generally lower in CO_2 (170–200 g per km travelled, compared to 180–240 for standard gasoline), but are higher in particulates (0.4–0.8 g km^{-1}, compared to <0.1), and much higher in nitrogen oxides (0.45–0.50 g km^{-1}, compared to 0.10–0.15) (Environment and Climate Change Canada 2023a, b). As N_2O is 200–350 times more effective than CO_2 at adsorbing thermal infrared wavelengths, the use of diesel is an ongoing environmental concern. Efforts have been made to introduce butane, biomethane, bioethanol, and electric-powered vehicles, but practical heavy trucks for long-haul travel have not yet been developed. Biodiesel emits lesser amounts of CO_2, but much higher amounts of nitrogen oxides (up to 0.65 g km^{-1}) (Environment and Climate Change Canada 2023a, b). Long combination vehicles can reduce fuel consumption by up to 13.4% per t-km, and overall transportation costs by up to 30%.

14.5 Air

Air travel, not anticipated in the 1867 BNA Act, is regulated by the federal government. Trans-Canada Airlines (TCA) was established as a subsidiary of CNR in 1937, and was given a monopoly on trans-Canada and international air travel by Prime Minister King. Restrictions on competing airlines, including the CPR subsidiary Canadian/CP Air, continued through the 1970s. The airline was renamed Air Canada in 1965, became an independent Crown Corporation in 1978, and was privatized in 1989 (Pigott 2014). Meanwhile, the private airline industry underwent a series of reorganizations, including the corporate disappearance of CP Air, resulting in the formation of several regional and international carriers, including WestJet.

Perhaps the initial governments were not paying attention, as the first piloted balloon flight in Canada occurred at Saint John in 1840 (Louis Anselm Lauriat), and the first passengers were carried by balloon in 1856 at Montreal (pilot: Eugène Godard). AG Bell's Silver Dart flew at Baddeck (NS) in 1909. Returning World War I aviators popularized aviation in Canada, especially for isolated and northern communities

Fig. 14.13 DeHavilland Otter, 8th aircraft built, delivering fuel to Margaret Lake Yukon, 1982. This aircraft has since been decommissioned. (Author's photo)

(Foster 1990), using a variety of surplus aircraft and purpose-built flying boats, aircraft with attached floats, and small "bushplanes", including the C-64 Norseman (1935), which had oversized tyres and could use rough, short runways. Later bushplanes included the De Havilland DHC-2 Beaver (1946), the DHC-3 Otter (1952; Fig. 14.13), and the larger DHC-6 Twin Otter (1965). Today, bushplanes remain in service across Boreal, Taiga, and Cordilleran Canada, along with jet-fuelled helicopters.

Bush flying originated for mineral exploration, terrain reconnaissance and aerial mapping, forest fire spotting, search-and-rescue operations, and as air ambulances; use for regular transportation and freighting came later (Foster 1990). Pioneer pilots such as Stuart Graham, Wilfrid "Wop" May, and CH "Punch" Dickins became local heroes. As a method of connecting northern communities to the south with tedious overland travel, bush flying was a significant advance for most northern residents, and is commemorated in many communities. Unfortunately, as some bushplanes were also used to transport Indigenous children involuntarily to residential schools, these monuments reveal a chequered legacy (CBC 2021).

Aviation in northern Canada remains extremely important, as the only method to effectively connect roadless communities year-round. Indigenous-owned airlines link communities in Nunavik, Nunatsiavut, Nitassinan, Nunavut, Northwest Territories, northern Ontario, and Yukon, as well as several in southern Canada, using jet, turboprop, and propellor-driven aircraft. Northwest Territories has more than 60 airports, landing strips, water aerodromes, and heliports. Aujuittuq (Grise Fiord) at 76.4°N, 82.9°W is the most northerly airport in Canada with regularly scheduled passenger and freight service. Charter and military aircraft service Alert (82.5°N, 62.3°W) on northern Umingmak Nuna (Ellesmere Island).

Icing of aircraft, both on the ground and in flight, is a significant winter hazard throughout almost all of Canada

(Catto et al. 2006). Most incidents result from ice buildup on the ground or during take-off, as aircraft fly through near-surface air strata of different temperatures. Icing is involved in approximately 10% of fatal aircraft crashes, including the March 1989 crash of a Fokker F-28 at Dryden, Ontario, with 24 fatalities. Many severe icing incidents are associated with the passage of weather fronts or coastal topography, producing uplifting of moist surface air.

Ice fog consists entirely of ice crystals and forms at temperatures of −15 °C or less (Gultepe and Heymsfeld 2016). In northern Canada, ice fog is common, occurring up to 25% of the time, and frequently results in low visibility at airports. Reduced visibility and aircraft icing due to ice fog causes high accident rates, perhaps more than 30 times the average across Canada. At present, forecasting ice fog occurrences is complicated by limited monitoring and remote sensing data, both at ground level and from satellites, and from ongoing challenges in understanding the physics of formation.

In southern Canada, airfield construction began in 1932, in many areas serving as make-work projects during the Great Depression. The goal was to construct airstrips spaced 150 km apart (or less). Civil aviation was delayed by the Great Depression and World War II, and the first non-stop Halifax-Vancouver flight occurred only in 1949. Until the 1970s, federal regulations significantly impacted permitted carriers, routes, schedules, and passenger fares.

14.5.1 Air Traffic

In southern Canada, air freight represents less than 3% of the total carried domestically (690 kt in 2022) and to the USA (252 kt) (Statistics Canada 2024). International freight (498 kt), not including courier services, also represents a relatively small proportion. Goods commonly transported by air include pharmaceuticals, precision machinery, high-value manufactures, and high-value, highly perishable goods. Canada exported $3.26 billion worth of lobster in 2021 to USA, China, UK, and EU, primarily as live shipments by air.

Passenger air travel has steadily recovered since the COVID-19 restrictions of 2020–21, reaching approximately 90% of pre-pandemic levels by late 2023 (Statistics Canada 2023d). Total traffic in 2022 saw 117 million embarkations and debarkations at Canadian airports, with about 60% carried on Canadian-based airlines. Traffic was 168.7 billion passenger-km in 2022 (Fig. 14.14). With a mean journey length exceeding 2670 km (approximately the air-line distance between Calgary and Toronto), trips tend to be long-distance. Flights connecting Montreal and Toronto (approximately 500 km) carry an average of 30,000 passengers weekly, comparable to the number using the Via rail service and about 15% of those making the trip by automobile.

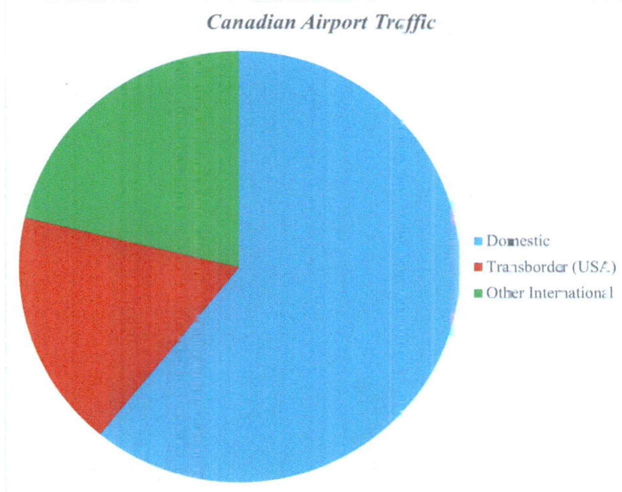

Fig. 14.14 Canadian airport passenger traffic, 2020–2023. (Statistics Canada 2023d, 2024)

The five busiest airports by passenger numbers account for 60–65% of traffic between Canadian destinations, 90–95% of transborder traffic to USA, and 95–97% of other international traffic (Statistics Canada 2023d; Fig. 14.15). Federal government licencing and landing rates restrict transborder and international travel. Toronto Pearson International airport handles approximately 30% of all domestic traffic, 45% of transborder, and almost half of international travel. Toronto's dominance reflects federal government policy, increased aircraft range, and transatlantic routing, in addition to consumer demand, despite the frustration of Atlantic Canadian residents who must fly considerable distances west before crossing the Atlantic to the east.

Airports throughout Canada, formerly administered by the federal government, were largely privatized in the 1990s. This policy has resulted in some increase in operational efficiency at the larger airports, although coupled with an increase in landing fees deemed necessary to pay for infrastructure maintenance and upgrades. For smaller airports, divestiture has resulted in significant fiscal challenges, particularly for those less likely to attract funding from industry or other governmental sources (Davis 2012). Partial or total loss of passenger and freight services has occurred at several locations in Atlantic Canada.

The airports at Gander, Happy Valley-Goose Bay, and Stephenville were built initially for military purposes, and evolved to serve Trans-Atlantic traffic when aircraft had shorter ranges insufficient to fly directly between Europe and mainland North America. These airports, together with St. John's and Halifax, served as safe harbours for flights diverted on 11 September 2001. Gander (pop. 10,000 in 2001) received 38 aircraft and 6595 passengers and crew (DeFede 2011). Gander and Happy Valley-Goose Bay remain open for mechanical, emergency, and weather-related diversions,

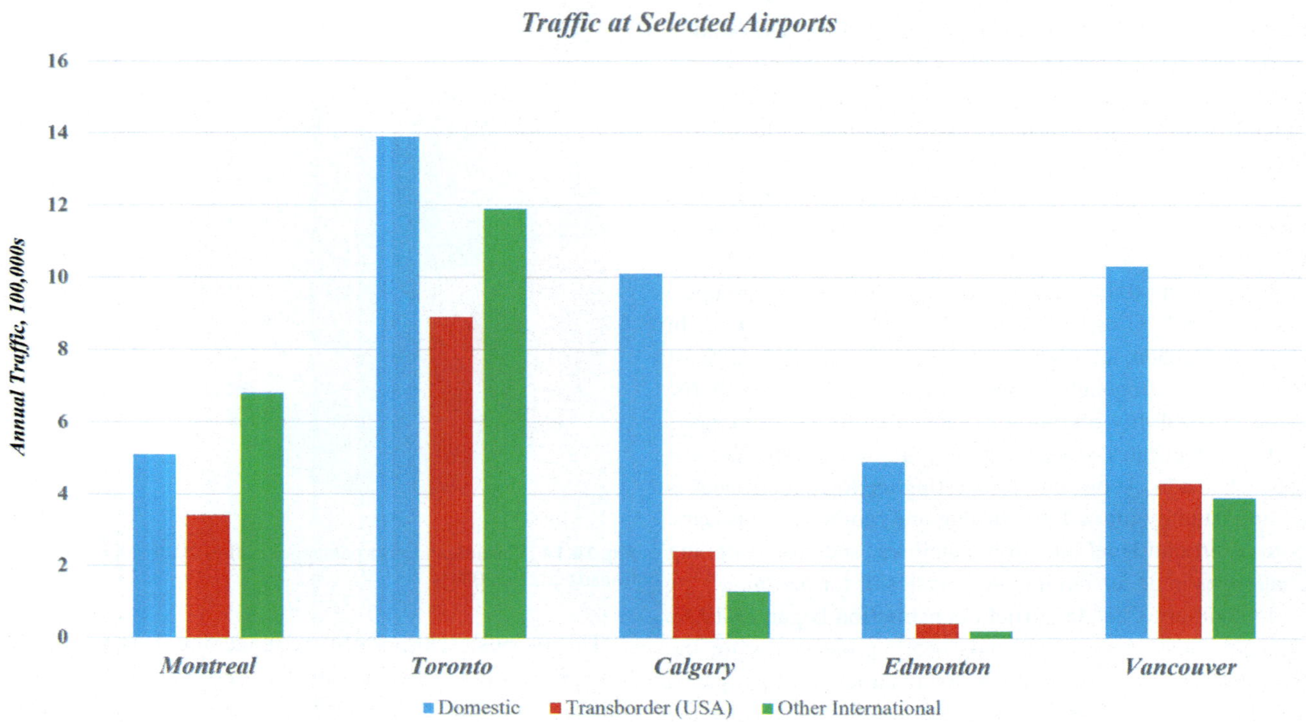

Fig. 14.15 Traffic at selected Canadian airports. (Statistics Canada 2023d, 2024)

as well as for regional flights. Gander Control is the primary air traffic control station in North America for aircraft flying over the North Atlantic.

The federal government has also privatized air traffic control and air navigation to the private company NavCanada (2024). Airport security is also handled by private firms.

References

Abu Aisha T, Ouhimmou M, Paquet M, Montecinos JA (2022) Developing the seaport container terminal layout to enhance efficiency of the intermodal transportation system and port operations—case of the Port of Montreal. Marit Policy Manag 49:181–198

Affleck EL (2000) Steamboating on the lower Peace River. BC Historical News 33(1). http://www.bchistory.ca/awards/article/Affleck.pdf

Alexander D (2000) The best so far: Vancouver's remarkable approach to the Southeast False Creek redevelopment is a big step towards sustainable redevelopment planning for urban sites. Altern J 26(3):10–15

Anderson BC (1983) Lifeline to the Yukon: A History of Yukon River Navigation. Superior Publishing Co.

Andrey J, Kertland P, Warren FJ (2014) Water and transportation infrastructure. In Canada in a Changing Climate: Sector Perspectives on Impacts and Adaptation, (ed) Warren FJ, Lemmen DS Government of Canada, Ottawa, ON, p 233–252

Angus JT (1988) A respectable ditch: a history of the Trent-Severn Waterway 1833–1920. Queen's University Press, Kingston

Ansary HJ (1993) North american ports and the internationalization of large markets. https://onlinepubs.trb.org/Onlinepubs/trcircular/459/459-002.pdf

APCA (2024) Association of canadian port authorities. https://acpa-aapc.ca/

Ashrafi M, Acciaro M, Walker TR, Magnan GM, Adams M (2014) Corporate sustainability in Canadian and US maritime ports. J Clean Prod 220:386–397

Baker CL, Lahti LR, Roumbanis DC (1998) Urban Geology of Toronto and Surrounding Area. In Karrow PF, White OL (eds), Urban geology of Canadian Cities. Geol Assoc Can Spec Paper 42, 323–352

Baldwin D (2009) Prince Edward Island: an illustrated history. Nimbus, Halifax.

Barrette PD (2015) Overview of ice roads in Canada: Design, usage and climate change adaptation. OCRE-TR-2015–011. National Research Council of Canada, https://nrc-publications.canada.ca/eng/view/object/?id=5984226f-bee8-48fe-a138-5a23c800f435

Barrette PD (2018) The Canadian winter road infrastructure and climate change adaptation: prospective solutions through R&D. Technical Report, National Research Council of Canada. Ocean, Coastal and River Engineering, 2018–04–30 https://doi.org/10.4224/40000402

Bellan RC (1962) Rails Across the Red—Selkirk or winnipeg. manitoba historical society transactions series 3, 1961–62 season https://www.mhs.mb.ca/docs/transactions/3/railsacrossthered.shtml

Berton P (1970) The National Dream. McLelland and Stewart, Toronto

Berton P (1971) The Last Spike. McLelland and Stewart, Toronto

Bevers C (2023) Jan. 1 1998: The day Ontario's Highways died. https://www.thekingshighway.ca/jan11998.html#:~:text

Blais-Stevens A, Behnia P, Castagner A (2018) Historical landslides that have resulted in fatalities in Canada (1771–2018). Geological Survey of Canada, M183-2-8392-eng.pdf

Boardman AE, Laurin C, Moore MA, Vining AR (2013) Efficiency, profitability, and welfare gains from the Canadian National Railway privatization. Res Transp Bus Manag 6:19–30

Bolger F (1961) Prince edward island and confederation. Can Cathol Hist Assoc Rep 28:25–30

Bonsor N (1995) Competition, regulation and efficiency in the Canadian railway and highway industries. In Paldi F (ed.), Essays in Canadian surface transportation. Vancouver: Fraser Institute

Boréalis (2020) Restoring balance: the ups and downs of log driving. https://www.borealis3r.ca/en/exposition/drave-dedrave-ramener-lequilibre/

Brooks MR (2007) Port devolution and governance in Canada. Res Transp Econ 17:237–257

British Columbia (2021) Roberts bank wildlife management area. https://www2.gov.bc.ca/gov/content/environment/plants-animals-ecosystems/wildlife/wildlife-habitats/conservation-lands/wma/wmas-list/roberts-bank

British Columbia (2023) BC Highway Flood Recovery Projects—Highway 5—Coquihalla. https://www2.gov.bc.ca/gov/content/transportation-projects/bc-highway-flood-recovery/2021-flood-road-recovery-projects-highway-5

Cameron D (1992) The people's railway: A history of Canadian National. Vancouver: Douglas & McIntyre

Canada (1992) Royal commission on national passenger transportation. Z1–1989–1–1992-eng.pdf.

Canada (2024) Commercial vehicle drivers hours of service regulations SOR/2005–313. Justice Laws website, https://laws-lois.justice.gc.ca/eng/regulations/SOR-2005-313/fulltext.html

Canadian Association of Chiefs of Police (2023) Natl Facts Stats. https://cacp.ca

Carlos AM, Lewis FD (1995) The Creative financing of an unprofitable enterprise: the Grand Trunk Railway of Canada 1853–1881. Explor Econ Hist 32:273–301

Catto NR (2008) Urban and environmental geology of St. John's and environs. GAC-MAC Field trip Guidebook

Catto NR (2014) Geomorphology: Landscapes we live in. Oxford Toronto

Catto NR Hyndman D Hyndman D (2006) Winter Hazards. In: Catto NR Hyndman D Hyndman D, Natural Hazards and Disasters, Nelson, 343–362

Catto NR, Slaymaker O, Kovanen DJ (2020) Human-modified Landscapes. In: Slaymaker O, Catto NR (eds) Landscapes and landforms of eastern Canada. Springer International Publishers, Cham, Switzerland, pp 475–507

CBC (2021) Bush plane monument in Thompson, Man., a painful reminder for residential school survivors. https://www.cbc.ca/news/indigenous/thompson-plane-monument-residential-schools-1.6068398

CFB Esquimalt (2024) History of naden at CFB Esquimalt. https://navalandmilitarymuseum.org/museum/about-the-museum/the-cfb-esquimalt-military-base

Clague JJ (1984) Quaternary geology and geomorphology, Smithers-Terrace-Prince Rupert area, British Columbia. Geol Surv Can, Mem 413

Council of Ministers Responsible for Transportation and Highway Safety (2019) Canada's national highway system, annual report 2017. https://www.comt.ca/Reports/NHS%20Annual%202017.pdf

Crizzle AM, Bigelow P, Adams D, Gooderham S, Myers AM, Thiffault P (2017) Health and wellness of long-haul truck and bus drivers: A systematic literature review and directions for future research. J Transp Health 7:90–109

Cruikshank K (1986) The people's railway: the intercolonial railway and the canadian public enterprise experience. Acadiensis 16:78–100

Cuff R (2001) Railway. Memorial University of Newfoundland, Newfoundland and Labrador Heritage

Currie AW (1957) The grand trunk railway of Canada. University of Toronto Press

Dagenais M (2017) Montreal. An environmental history. UBC Press, City of water

Davis MD (2012) Federal transportation divestiture of remote ports and airports in Atlantic Canada: An introductory analysis. Res Transp Bus Manag 4:61–68

DeFede J (2011) The day the world came to town. Everand

Dempsey HA (2003) Isapo-Muxika. dictionary of canadian biography, 11, University of Toronto/Université Laval

Environment and Climate Change Canada (2022) National inventory report 1990–2021: greenhouse gas sources and sinks in Canada. https://publications.gc.ca/collections/collection_2021/eccc/En81-4-1-2019-eng.pdf

Environment and Climate Change Canada (2023) Emission factors and reference values https://www.canada.ca/en/environment-climate-change/services/climate-change

Fan D Heminthavong K (2022) Road transportation: heavyweight of the Canadian economy. Library of Parliament 2022–04-E. https://lop.parl.ca/staticfiles/PublicWebsite/Home/ResearchPublications/HillStudies/PDF/2022-04-E.pdf

Forbes ER (1994) The intercolonial railway and the decline of the maritime provinces revisited. Acadiensis 24:3–26

Foster JA (1990) The Bush Pilots. McClelland & Stewart Inc., Toronto

Garceau R (1957) Log Drive. National Film Board of Canada

Gultepe I, Heymsfeld AJ (2016) Introduction: Ice fog, ice clouds, and remote sensing. Pure Appl Geophys 173:2977–2982

Halifax Port Authority (2024) Port of Halifax. https://www.portofhalifax.ca/

Harding, L (2008) The newfoundland railway, 1898–1969. A History. McFarland

Hayley D Proskin S (2008) Managing the safety of ice covers used for transportation in an environment of climate warming. In: Proceedings of the 4th Canadian Conference on Geohazards: From Causes to Management

HBC Heritage (2016) York Boat. https://www.hbcheritage.ca/things/technology/the-york-boat

Huijser, M.P, Duffield JW, Clevenger AP, Ament RJ, McGowen PT (2009) Cost-benefit analyses of mitigation measures aimed at reducing collisions with large ungulates in North America; a decision support tool. Ecol Soc 14(2): 15. http://www.ecologyandsociety.org/vol14/issue2/art15/

Holland-Stergar P (2020) Depot Harbour: the rise and fall of an Ontario Grain Port. Unpublished PhD, Dept of History, Western University, https://ir.lib.uwo.ca/etd/6863

House of commons (2023) Improving bus connectivity in Canada. https://www.ourcommons.ca/content/Committee/441/TRAN/Reports/RP12444313/441_TRAN_Rpt13_PDF/441_TRAN_Rpt13-e.pdf

Karrow PF (1998) Urban geology note: thunder bay, Ontario. In Karrow PF, White OL, (eds.), Urban Geology of Canadian Cities. Geological Association of Canada Special Paper 42, 191–196

Karrow PF, Morin FJ (1998) Urban Geology Note: Hamilton, Ontario. In Karrow PF, White OL, (eds.), Urban Geology of Canadian Cities. Geological Association of Canada Special Paper 42, 279–286

Katz-Rosene R (2017) To build or not to build? Competing narratives of high-speed rail development in Canada. Mobilities 12:758–777

Kulp SA, Strauss BH (2019) New elevation data triple estimates of global vulnerability to sea-level rise and coastal flooding. Nat Commun 10(1):4844. https://doi.org/10.1038/s41467-019-12808-z

Lavallee O (1972) Narrow gauge railways of Canada. Montreal: Railfare

Leonard F (1996) A thousand blunders: the grand trunk pacific railway and northern british columbia. Univ Br C Press, Vancouver

Laliberté J St-Laurent M-H (2020) In the wrong place at the wrong time: Moose and deer movement patterns influence wildlife-vehicle collision risk. Accid Anal Prev 135, 105365. https://doi.org/10.1016/j.aap.2019. 105365

Lamb WK (1977) History of the Canadian pacific railway. Macmillan Publishing, New York

Larson PD, Lin Y, Ng AKY (2019) Hudson bay railway and the port of churchill: a balancing act. In: 54th Annual Meeting of the Canadian Transportation Research Forum, 26–29, 2014. https://ctrf.ca/wp-content/uploads/2022/08/2019192Larson.pdf

Lavallée O (1982) The grand trunk railway of canada: an overview. Railroad History 147:12–18

Legget, RF (1973) Railways of Canada. Douglas & McIntyre

Lewis CFM, Taylor BB, Stea RR, Fader GBJ, Horne RJ, MacNeill SG, Moore JG (1998) Earth science and engineering: urban development in the metropolitan halifax region. In Karrow PF, White OL, (eds.), Urban Geology of Canadian Cities. Geological Association of Canada Special Paper 42, 409–444

Macciotta R, Robitaille S, Hendry M, Martin CD (2018) Hazard ranking for railway transport of dangerous goods in Canada. Case Stud Transp Policy 6:43–50

MacDonald B (2008) Excavating for a mine: Silver Islet, 1868–2008: 140th anniversary. Porphyry Press, Thunder Bay, Ontario

MacKay D (1986) The asian dream: the pacific rim and Canada's National Railway. Douglas & McIntyre

Malkin J, Crizzle AM, Zello G, Bigelow P, Shubair M (2021) Long-haul truck driver training does not meet driver needs in Canada. Saf Health Work 12:35–41

McCarthy S (1991) The last drive. Canadian Geographic, December 1991, 18–31

Mcdonald JA n.d. Bleeding day and night: the construction of the Grand Trunk Pacific Railway across Tsimshian reserve lands. Royal Ontario Museum 33–68

McGahan E. (1982) The Port of Saint John: volume one, from confederation to nationalization. Saint John: National Harbours Board

McIlwraith TF (2006) Ontario railways before 1880: a scrap-iron bonanza. Prog Ind Ecol 3:344–360

McKenzie KJ (1996) Damned if you do, Damned if you don't: An option valve to evaluating the subsidy of the CPR Mainline. Canadian Journal of Economics: 255

Milke M (2014) Subsidies in Canada: a $684 billion price tag. Fraser Institute. https://www.fraserinstitute.org/sites/default/files/government-subsidies-in-canada-a-684-billion-price-tag.pdf

Montreal Port Authority (2023) https://www.port-montreal.com/en/

Murray T (2011) Rails across Canada: the history of canadian pacific and canadian national railways., Voyageur Press.

NavCanada (2024) What we do. https://www.navcanada.ca/en/corporate/about-us/what-we-do.aspx

Nelligan T (1982) Via Rail Canada: The first five years. PJT Publishing

NWT Dept of Transportation (2007) A field guide to ice construction safety. a263493.

NWT (2024) Marine transport services. https://www.inf.gov.nt.ca/en/MTS

Ontario (2020) Connecting the North: a draft transportation plan for Northern Ontario. Government of Ontario

Ontario trucking association (2018) Mag: Immigrants more than half of Truck Drivers in Major Cities. https://ontruck.org/mag-immigrants-more-than-half-of-truck-drivers-in-major-cities/

Ontario Trucking Association (2023) PWC report: truck driver shortage a risk to Canada's economy. https://ontruck.org/pwc-report-truck-driver-shortage-a-risk-to-canadas-economy

Perl A (1994) Canada's Railway nationalization: Economic necessity or political initiative? Can J Polit Sci 27:23–52

Pigott P (2014) Air Canada: the History. Dundurn, Toronto

Port de Québec (2024) https://www.portquebec.ca/en/

Port of Prince Rupert (2022) Annual report. https://2022.rupertport.com/

Port Saint John (2024) https://www.sjport.com/

Pratson LF (2023) Assessing impacts to maritime shipping from marine chokepoint closures. Commun Transp Res 3:100083. https://doi.org/10.1016/j.commtr.2022.100083

Proskin S, Groznic E, Hayley D, Mathison F, McGregor R, Neth V (2011) Guidelines for the construction and operation of winter roads. Transportation Association of Canada

Railway Association of Canada (2021) Rail Trends. www.railcan.ca

Rider PE (2009) Charlottetown: a history. Canadian Museum of Civilization, Ottawa-Gatineau

Roberts C, Meadowcroft J, Layzell DB (2020) The rise of the automobile: lessons from historical canadian transportation transitions. Transit Accel Rep 2(4):1–35

Rodon T, Keeling A, Boutet J-S (2022) Schefferville revisited: The rise and fall (and rise again) of iron mining in Québec-Labrador. Extr Ind Soc 12:101008. https://doi.org/10.1016/j.exis.2021.101008

Regehr TD (1976) The Canadian northern railway. MacMillan, Toronto

Riverkeepers (2015) History of steamboats on the Red River. https://riverkeepers.org/wp-content/uploads/2016/06/Steamboat_fact_sheet_Jan-2015-JC.pdf

Simard M (2022) Log-drivers: superheroes of times gone by. Tourisme Charlevoux, https://www.tourisme-charlevoix.com/en/blog/log-drivers-superheroes/

Smith A (2013) Patriotism, self-interest and the 'empire effect': britishness and british decisions to invest in Canada, 1867–1914. J ImpAl Commonw Hist 41:59–80

Souza Almeida L, Rodriges L, Goerlandt F, Ancona-Segovia J, Pelot R, Bristow D, Chang S (2023) Datasets of disrupted transportation networks on Canada's West Coast in a plausible M9.0 Cascadia Subduction Zone earthquake scenario. Data in Brief 46 108762 https://doi.org/10.1016/j.dib.2022.108762

Statista (2022a) Rail industry in Canada

Statista (2022b) Trucking industry in Canada

Statistics Canada (2021) Trucking financial statistics. https://www150.statcan.gc.ca/t1/tbl1/en/tv.action?pid=2310029101

Statistics Canada (2023a) Automotive statistics. https://www.statcan.gc.ca/en/topics-start/automotive

Statistics Canada (2023b) Passenger bus and urban transit 2021. https://www150.statcan.gc.ca/n1/daily-quotidien/230602/dq230602b-eng.htm

Statistics Canada (2023c) Rail transportation in Canada 2021. https://www150.statcan.gc.ca/n1/pub/11-627-m/11-627-m2023008-eng.htm

Statistics Canada (2023d) Air passenger traffic at Canadian airports. https://www150.statcan.gc.ca/t1/tbl1/en/tv.action?pid=2310025301

Statistics Canada (2024) Air cargo traffic at Canadian airports. https://www150.statcan.gc.ca/t1/tbl1/en/tv.action?pid=231002540

Stephens DE (1972) Iron roads of nova scotia. Lancelot Press, Windsor NS

Stevens GR (1960–1962) Canadian national railways. Toronto: Clarke, Irwin, and Company Limited

Stevens H (1999) Canada: the port of Vancouver. In: InstAl Position Seapt. Springer, Dordrecht: Ch. 7 189–205

Tests.ca (2024) The ultimate list of Canada Driving Statistics for 2023. https://tests.ca/driving-statistics/

Transportation Association of Canada (2021) Understanding goods management in Canada: Trends and Best Practices. www.tac-atc.ca

Transport Canada (2020a) Canadian port authorities. https://tc.canada.ca/en/corporate-services/policies/canadian-port-authorities

Transport Canada (2020b) List of Canadian port authorities. https://tc.canada.ca/en/marine-transportation/ports-harbours-anchorages/list-canada-port-authorities

Transport Canada (2021) Transportation in Canada 2020 overview report. https://tc.canada.ca/sites/default/files/2021-07/transportation_canada_2020_overview_report.pdf

Transport Canada (2023) Canadian Motor Vehicle Traffic Collision Statistics: 2021 https://tc.canada.ca/en/road-transportation/statistics-data/canadian-motor-vehicle-traffic-collision-statistics-2021

Transport Canada (2023) Key traffic and Volume Statistics. transportation-canada-annual-reports/transportation-canada-2022/key-traffic-volume-statistics

Underwood J (2005) Built for War: Canada's intercolonial railway. Railfare Books

Vanlaar WGM, Barrett H, Hing MM, Brown SW, Robertson RD (2014) Canadian wildlife-vehicle collisions: An examination of knowledge and behavior for collision prevention. J Safety Res 68:181–186

Vanlaar WGM, Gunson KE, Brown SW, Robertson RD (2012) Wildlife-Vehicle collisions in Canada. Traffic Inj Res Found, https://tirf.ca/wp-content/uploads/2017/01/WildlifeVehicle_Collision_Deliverable1_Eng_6.pdf

VFPA (2024) Vancouver fraser port authority. https://www.portvancouver.com

Via Rail Canada (2018). 2018 Public meeting, questions and answers. https://www.viarail.ca/sites/all/files/media/pdfs

Watson KW (2007) The Rideau route: exploring the pre-canal waterway. Privately published, Elgin, Ontario

Western Canada Marine Response Corporation 2024. WCMRC, https://wcmrc.com/

Wishart DJ (2007) Payipwat. University of Nebraska Press. p, Encyclopedia of the Great Plains Indians, p 150

Woods SE (1992) Cinders and saltwater: the story of atlantic Canada's railways. Nimbus, Halifax

Forestry and Wildland Fires 15

Abstract

Forested area accounted for 362 Mha in 2022, about 38% of Canada, 14% of the global naturally treed areas, and 9% of forested land. About 77% lies within the boreal forest regions of the Boreal Shield and Plains, Taiga Shield and Plains, and interior and northern Cordillera. The total wood volume of 4.5 x 108 m3 has declined by about 0.01% since 1990, primarily due to fire losses. Canada's forests are largely provincially owned Crown Land (75.4%), with 13% under territorial control, 1.7% federally controlled, and 2.1% managed by Indigenous communities and organizations. Only 6.7% is in private hands, mostly as small woodlots. The choice of harvesting strategy depends upon the species mix, age structure, intended use, management and forester-training effort required, and economics of production. Forestry products contributed $34.3 billion GDP in 2021, including pulp-and-paper manufacture ($9.7 billion), lumber ($5.3 billion), and wood product manufacture ($19.8 billion). Pulp-and-paper mills are present in all provinces, except PEI. Domestic production of softwood lumber in 2021 was 56 million m3, together with 8.9 million m3 of structural panels (plywood and oriented strandboard), and 873500 m3 of hardwood lumber. BC, Québec, and Alberta are the largest producers, typically accounting for 80-85% of production. Canada and USA have been engaged in a trade dispute over Canadian exports of softwood lumber, primarily used for residential construction. Initiated in 1982, the latest phase of dispute began in 2015 and remained unresolved as of 2025. Net emissions (CO_2 equivalent) from managed forested areas have varied from 50 to 250 Mt annually since 2010, reflecting the prevalence of wildland fires and insect disturbances.

Keywords

Forests · Forest management · Pulp and paper · Softwood lumber · GHG emissions · Wildland fires

15.1 Forests

A global estimate suggests there are about 3 trillion trees worldwide, 420 trees for each human. About 740 billion trees are in the boreal forest. Canada has approximately 380 billion trees, about 10,000 per person. Forested area accounted for 362 Mha in 2022, about 38% of Canada, 14% of the global naturally treed areas, and 9% of forested land (Natural Resources Canada 2022). About 77% lies within the boreal forest regions of the Boreal Shield and Plains, Taiga Shield and Plains, and interior and northern Cordillera. Other forested areas include the Pacific Maritime, Atlantic Appalachian, and (in many areas, formerly) the Mixedwood Plains, currently 19% forest-covered. Additional area with some tree cover (high taiga and Cordillera above timber line) represents 50 Mha, not included as forest. The tree line is defined as the altitude limit of upright woody-stemmed vegetation. Timberline is defined as the altitude and the limit of practically usable timber, which effectively equates to the limit of forest cover.

Of the 348 Mha of forest, about 18 Mha represents human-planted trees (afforestation), with the remainder naturally regenerating (Natural Resources Canada 2022). The term "afforestation" is preferred to "reforestation", as the newly planted forest will of necessity not have the same species assemblage and ecology as the former.

Canadian forests show a variety of stand ages, with most between 41 and 120 years (Fig. 15.1, Natural Resources Canada 2022). Broadleaf and mixed forests show similar patterns in stand age classes (Fig. 15.2). Coniferous forests in boreal regions generally have younger stand assemblages than the conifer-dominated Pacific Maritime forests.

Deforestation, although frequently upsetting, is uncommon, with about 0.5% of Canadian forests undergoing deforestation due to industrial causes since 1990 (Dyk et al. 2015). Industrial sector deforestation currently stands at about 34 kHa per year, with hydrocarbon infrastructure (~50%), agriculture (~40%), and urban and suburban devel-

N. Catto and G. R. Vincent, *Canada: Land and People*, World Regional Geography Book Series,
https://doi.org/10.1007/978-3-032-04391-7_15

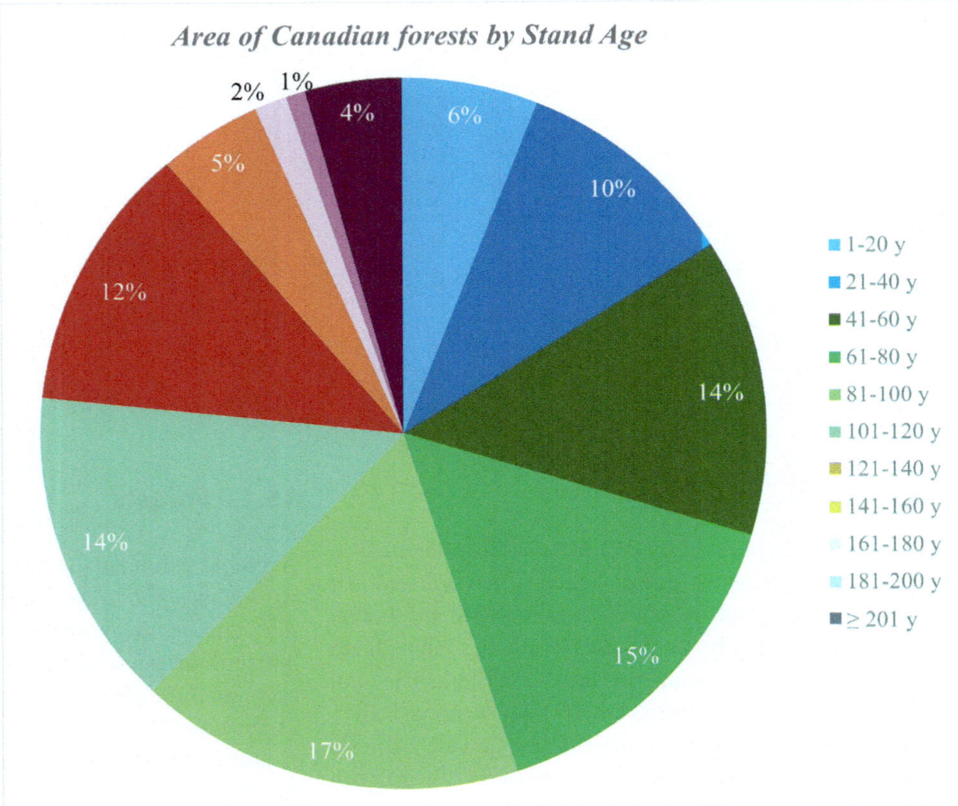

Fig. 15.1 Estimated area of Canadian Forests by Stand Age, 2020–2021. Natural Resources Canada (2022)

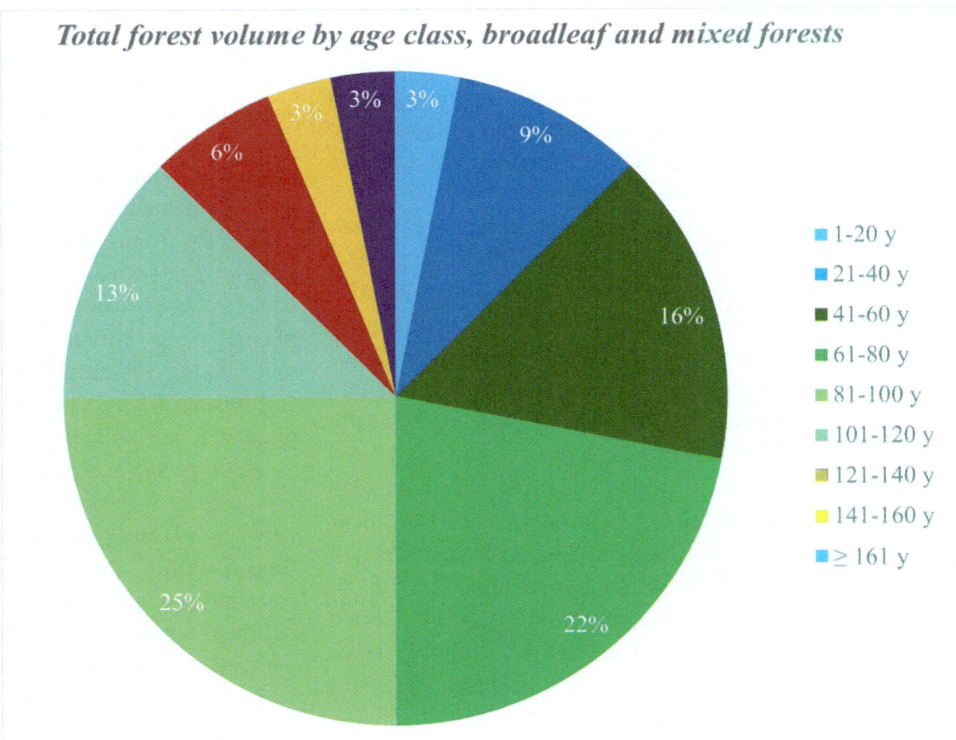

Fig. 15.2 Estimated total forest volume by age class, broadleaf, and mixed forests, 2020–2021. Natural Resources Canada (2022)

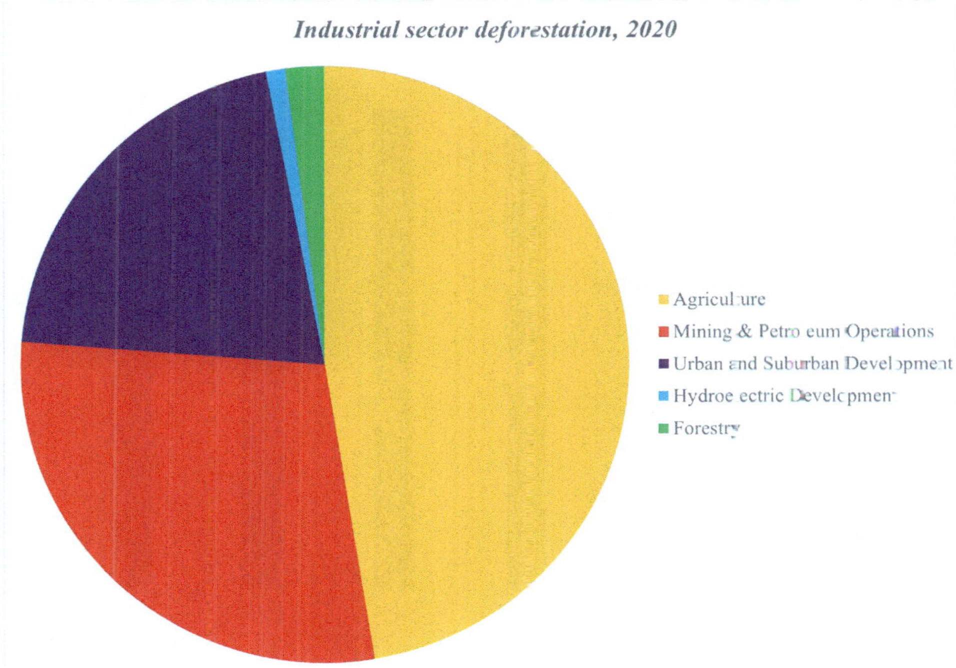

Fig. 15.3 Industrial sector-induced deforestation by percentage, 2020. Natural Resources Canada (2022)

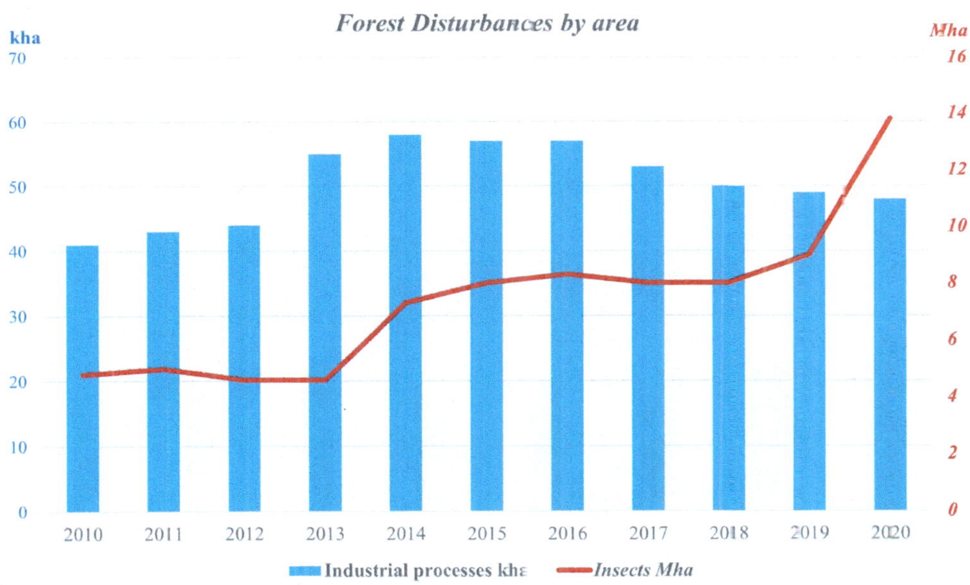

Fig. 15.4 Forest disturbances by area, 2010–2020. Industrial processes: kha, left-hand scale; Insect disturbances Mha, right-hand scale. National Forestry Database (2022b, 2022c)

opment (~10%) the largest factors (Natural Resources Canada 2022; Figs. 15.3, 15.4). Forestry operations are responsible for less than 5% of industrial deforestation, and harvest less than 0.5% of the forested area each year (National Forestry Database 2022c). Insect defoliation accounted for 18.8 Mha of deforestation in 2020 (National Forestry Database 2022b), with wildland fires accounting for 4.3 Mha (National Forestry Database 2022a, discussed below). The total wood volume of 4.5×10^8 m^3 has declined by about 0.01% since 1990, primarily due to fire losses (Natural Resources Canada 2022: Figs. 15.5, 15.6). Climate change is anticipated to increase fire frequency and intensity in the boreal forest, as well as potentially contributing to insect and fungi outbreaks and invasions (Gauthier et al. 2015, Boucher et al. 2018, Hanes et al. 2019, Williamson et al. 2019, Natural Resources Canada 2020b, Environment and Climate Change Canada 2022).

Major native insects responsible for losses include the eastern spruce budworm (*Choristoneura fumiferana*, Atlantic Canada), hemlock looper (*Lambdina fiscellaria*, Atlantic Canada), jack pine budworm (*C. pinus*, Boreal Shield), forest tent caterpillar (*Malacosoma disstria*, Prairie Parkland and Boreal Plains), and mountain pine beetle (*Dendroctonus ponderosae*, Cordillera and western Boreal Plains). After a sharp spike in 2008–2010, mountain pine beetle disturbances

Fig. 15.5 Total wood volume (10^9 m^3), 1990–2020. Natural Resources Canada (2022)

had declined by >95% by 2016. In contrast, eastern spruce budworm had disturbed more than 6 Mha by 2018 (Natural Resources Canada 2022). Invasive species are increasing in variety and number, particularly in boreal regions, as a result of human disturbance, inadvertent transportation, evolving habitats, and climate change.

15.2 Governance and Harvesting

Canada's forests are largely provincially owned Crown Land (75.4%), with 13% under territorial control, 1.7% federally controlled, and 2.1% managed by Indigenous commu-

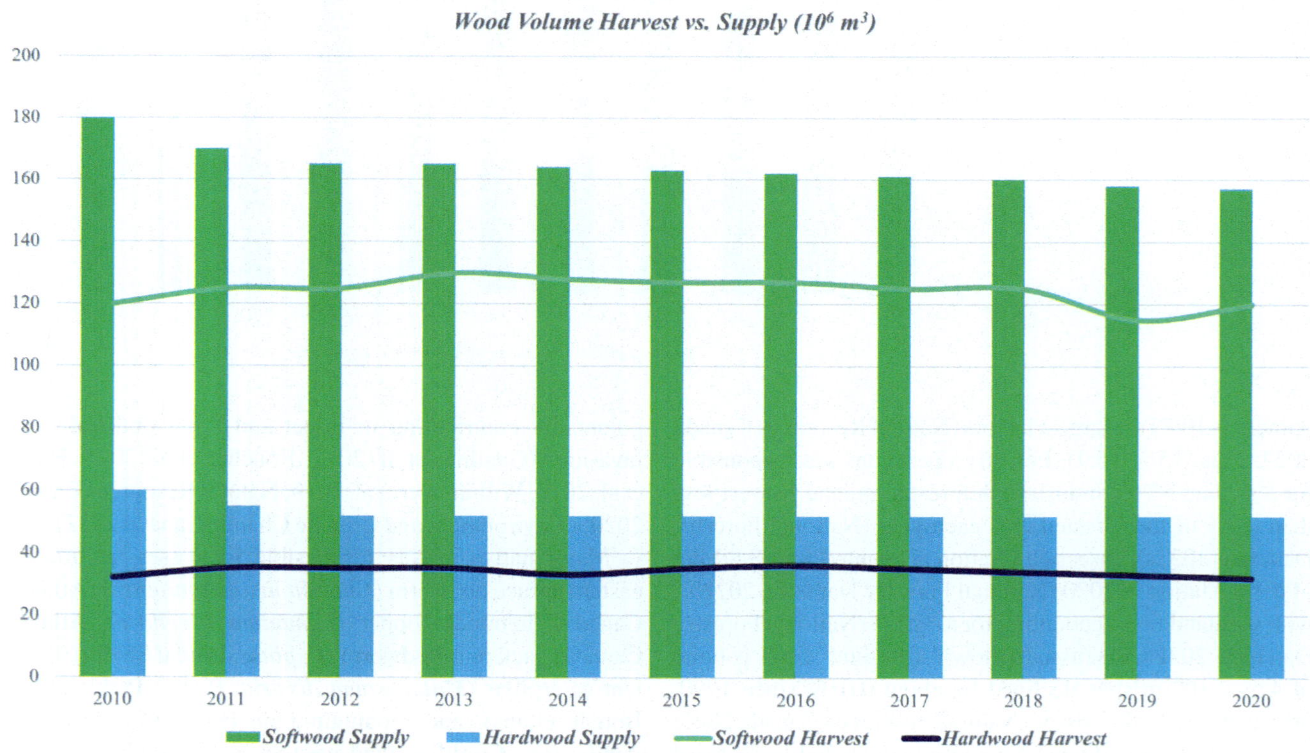

Fig. 15.6 Wood volume harvest compared to wood supply for hardwoods and softwoods, 2010–2020. Data from Natural Resources Canada (2022)

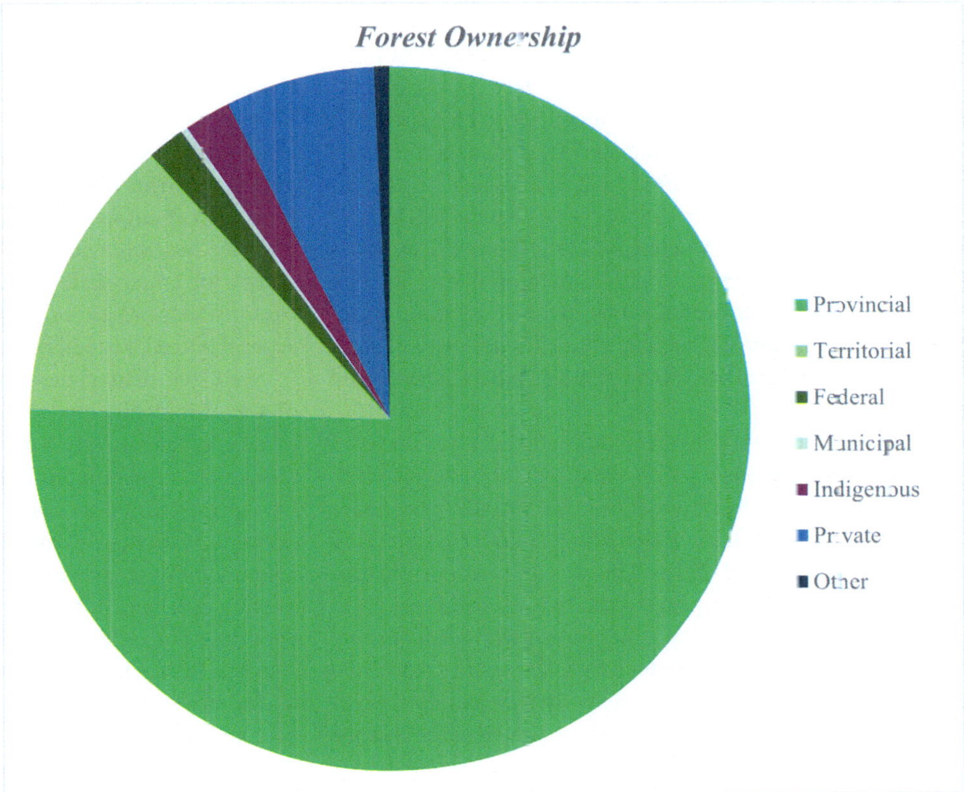

Fig. 15.7 Forest ownership, 2021. Natural Resources Canada (2022)

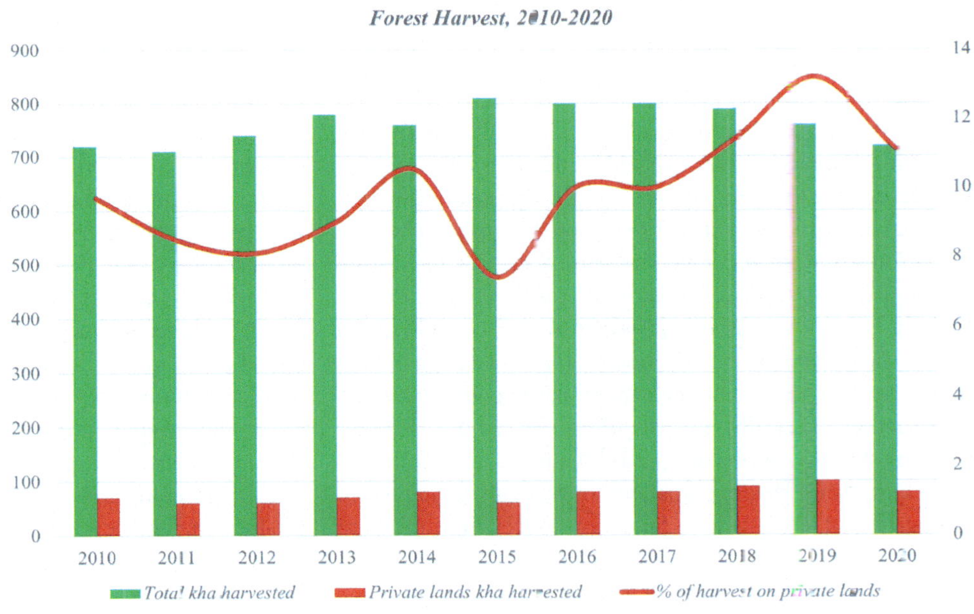

Fig. 15.8 Forest harvest, 2010–2020. Privately owned lands are harvested at greater rates than are other forests (compare with Fig. 15.7). Data from Natural Resources Canada (2022)

nities and organizations (Fig. 15.7). Only 6.7% is in private hands, mostly as small woodlots (Natural Resources Canada 2022; Fig. 15.8). Private companies lease Crown Land for their operations, subject to government regulations and paying stumpage fees. This factor is a major contributor to the ongoing Canada–USA softwood lumber trade dispute (discussed below).

Forest certification, somewhat analogous to fair trade or to organic certification, is one method that can be used to assess harvesting methods and forest management (Natural Resources Canada 2020a; Certification Canada 2023a, 2023b). Three organizations, the Forestry Stewardship Council, which has the most stringent standards, and the Canadian Standards Association and Sustainable Forestry

Initiative offer certification. Forest operators can apply for certification under any or all three. The three third-party organizations had certified 168 Mha in 2020.

The choice of harvesting strategy depends upon the species mix, age structure, intended use, management and forester-training effort required, and economics of production (Matthews 1989; Nova Scotia Department of Labour 1994; Sauvageau 1995; Macdonald 1999; Natural Resources Canada 2022). In addition to public opinion and regulatory restrictions, consumer preference is involved in some areas. Methods used in Canada include selection harvesting, shelterwood cutting, seed tree management, patch-cutting, and clear-cutting.

Selection harvesting involves the detailed assessment of individual trees or small groups, and then choosing with to harvest. In the Pacific Maritime conifer forest, individual larger diameter western red cedar (*Tsuga plicata*) higher in value may be chosen (Gregory et al. 2018). This was the common practice of indigenous harvesters prior to ca. 1850. Commonly, individual trees are cut in a very specific manner without disturbing others, which means fairly detailed pre-harvesting study and expensive harvesting processes. Selective harvesting is practised for hardwood sugar maple (*Acer saccharum*), red oak (*Quercus rubra*), and yellow birch (*Betula alleghaniensis*), when they are harvested in Atlantic Appalachian and Mixedwood Plains forests in the Maritimes, Ontario, and Québec. These trees are frequently harvested for customized furniture, requiring specific wood attributes. Woodlots may be used for other purposes such as maple syrup production.

Selection harvesting is designed to create a varied age structure. Removal of only the oldest trees, or only middle-aged trees, would result in an ecologically unsound forest. The goal of a varied age structure is more attainable when the initial forest has a varied age structure. Selection harvesting is generally unsuitable for shade-intolerant trees (e.g. *Pinus*), in many boreal assemblages (Bose et al. 2014), and for old growth stands with extensive canopies. It works well in mixed-wood forests with a few large canopy hardwoods, spreading trees such as beech (*Fagus grandifolia*) or red oak. Priority selection starts by cutting diseased or damaged trees, followed by mature trees shading younger ones, with a maximum cutting rate of 25% or less. Selection harvesting maintains wildlife habitat, allows browsing organisms to continue to thrive, and has a lesser visual and aesthetic impact.

Shelter wood cutting removes stands in successive cutting episode. All or most of the trees in one patch are cut, leaving adjacent patches undisturbed. Saplings are planted in the cut areas. The intent is to produce even-aged stands, encouraging regeneration. Selected large specimens are left to act as sheltering ("nurse") trees, protecting young saplings from wind, freezing rain, and other extreme weather. Suitable shelter trees are large, broad-leaved deciduous (oak, beech, maple) or spreading conifers (white pine). They are robust and wind resistant. As the young trees grow, the shelter trees will be harvested, so they are also selected for eventual harvesting characteristics.

Afforestation of the cut patches allows the use of a mix of different species for replanting, encouraging biodiversity. Shelter wood harvesting thus can be used in previously logged forests, in attempts to (at least partially) reconstruct an original tree assemblage by choosing seedling species. It also allows for larger-scale use than does selective harvesting in circumstances where wood qualities are important, with lesser ecological and aesthetic disruption.

Seed tree management uses uncut seed trees as sources of forest rejuvenation, rather than for harvest. The goal is to select "genetically superior" specimens, robust, wind-resistant, and maximum pollen producers, in order to allow for regrowth after harvesting. The number of seed trees left depends on the species and the areal extent involved. Conifers tend to produce more pollen per tree than do deciduous species, and pine pollen is more effectively spread by wind. Seed tree management tends to produce even-aged prodigy stands, and may lead to decreased genetic diversity. In contrast to shelterwood practice, the ecological effects and aesthetic perceptions can resemble those of clear-cuts or large fires.

Clear cutting can be defined to have several meanings. In a clear- or patch-cut, all trees are removed, regardless of species. Patch-cutting involves complete clearance of smaller areas (patches or strips), with patch dimensions or strip widths typically 3–5 times mean tree heights. Shade-intolerant trees (such as pines, birches, and poplars) are favoured by larger patches, whereas smaller dimensions are used for shade-tolerants (such as spruce and fir). Each patch is managed as a distinctive unit, producing a mosaic of tree ages over time, but with each patch having a stand of uniform height and age. In order to extract the maximum value, both economically and ecologically, patch-cutting requires more pre-harvest analysis and detailed record keeping. Cutting in strips can create a wind tunnel effect, producing damage to flanking trees.

Clear-cutting involves removal of all vegetation. In many cases, the ground is tilled, and new seedlings are planted simultaneously. The goal then is to replace old stands of trees, particularly conifers used for framestock lumber or pulp production, with new and vigorous stands. Despite its appearance, clear-cutting can have some advantages (Kimmis 1991; Bose et al. 2014; Hanberry 2021). It reduces susceptibility to fire, particularly in boreal coniferous forests, and insects. It also allows rapid salvage of damaged trees. Clear-cutting can be suitable for very large-scale harvesting, designed to produce the largest possible quantity of wood within a given period of time. However, although it does produce open areas for some species such as berries, or fauna such as deer, it does result in the greatest degree of ecological change. It does not look particularly attractive and it can result in erosion on steep slopes.

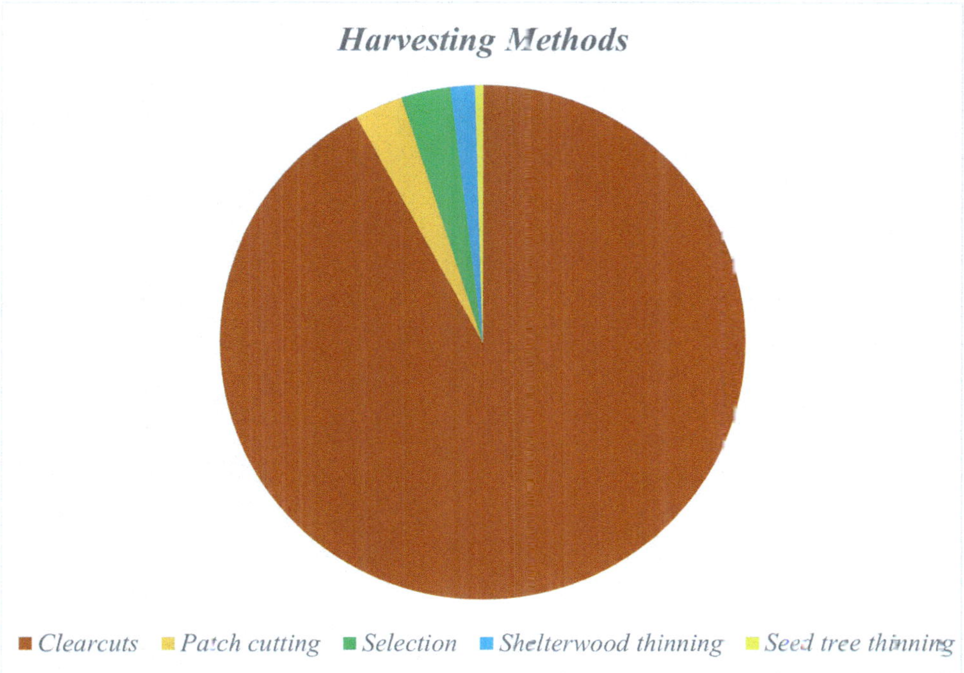

Fig. 15.9 Harvesting methods 2021. Date from Natural Resources Canada (2022)

Clear-cutting is currently used in 90–95% of forestry operations (Natural Resources Canada 2022; Fig. 15.9). Although there are situations where clear cut is the right forestry management choice, the problem is that it is used in too many situations, not just the most suitable ones. There are differing forests, different fire histories, and different human intervention histories. Clear-cutting stirs differing philosophical, theoretical, and practical considerations. For many people, it tends to be considered as a purely exploitative solution.

15.3 Forestry Products

Forestry products contributed $34.8 billion GDP in 2021 (Fig. 15.10), including pulp and paper manufacture ($9.7 billion), lumber ($5.3 billion) and wood product manufacture ($19.8 billion). Direct (Fig. 15.11) and indirect employment was 345,825 persons. The balance of trade is highly favourable to Canada, dominated by manufactured wood products and paper. Primary wood represents less than 5% of total exports by value (Statistics Canada 2022a, 2022b; Fig. 15.12).

15.3.1 Pulp and Paper

Pulp and paper mills are present in all provinces, except PEI (Environment and Climate Change Canada 2022). Facilities include chemical pulp mills (27%), mechanical pulp mills (8%), and those using pulp produced either on-site or elsewhere in paper manufacture (65%). A mill producing specialty paper or other products may utilize pulp from multiple sources.

Mechanical production involves grinding and attrition by friction to gradually separate the wood into fibres (Environment and Climate Change Canada 2022). The physical composition is relatively unaffected, and most of the lignin binding the cellulose is retained. Mechanical processing results in greater pulp yield, stiffness, and relatively low strength. Mechanical pulps typically are used for newsprint, increasingly limited in demand, as well as for books and commercial flyers and posters.

Chemical processing involves partially dissolving the wood in acidic or alkaline liquid, destroying lignin and hemicellulose to extract cellulose (Environment and Climate Change Canada 2022). Cellulose fibres are left intact, producing stronger, more flexible paper, but with lesser pulp yield. Use of sodium hydroxide and sodium sulphide for kraft processing (from the German word for "strong") is common, but produces a darker product. To make printing paper, facial tissue, or toilet paper, most chemically produced pulp is bleached. Unbleached darker pulp is used for packaging. Paper is made by compressing and drying pulp, using a steam heated can dryer.

Stocks for pulp and paper processing are classified as "softwood" (conifers) and "hardwood" (angiosperms). Processing uses primarily black spruce (*P. mariana*), white spruce (*Picea glauca*), jack pine (*Pinus banksiana*), lodgepole pine (*Pinus contorta*), balsam fir (*Abies balsamea*), tamarack (*Larix*

Fig. 15.10 Forestry sector contributions to GDP, 2021. Natural Resources Canada (2022), Statistics Canada Census of Canada 2021

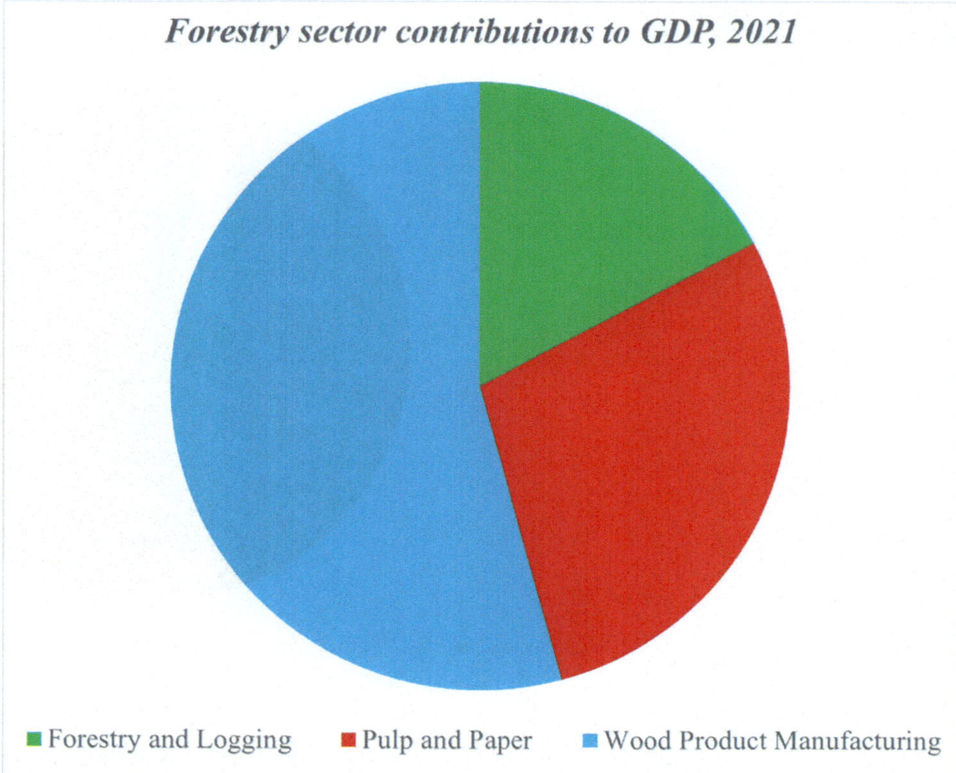

Fig. 15.11 Forestry industry direct employment, 2021. Natural Resources Canada (2022), Statistics Canada Census of Canada 2021

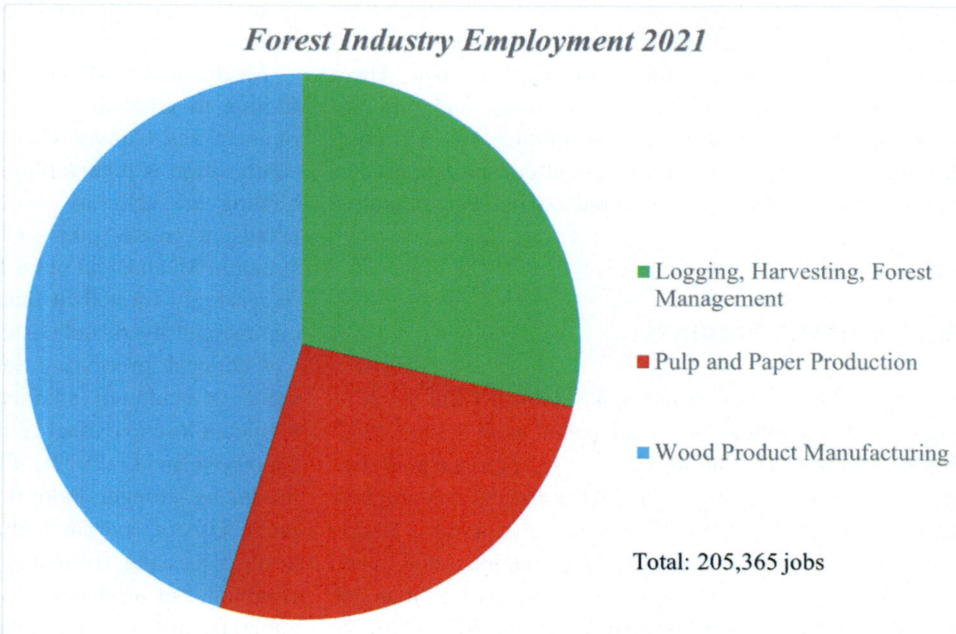

laricina), and aspen (*Populus tremuloides*). Although aspen is an angiosperm and therefore technically a "hardwood", its wood is softer than most conifers. In the Boreal Shield and Atlantic regions, black spruce and jack pine are most commonly used, whereas aspen use is more typical in the Boreal Plains and northern Prairies.

In Canada, long-fibre (5–7 mm cells) black spruce and jack pine are preferred for book paper, coated paper, and maps, whereas short-fibre (2–3 mm cells) aspen can be used for high-quality printing paper (ABB 2023). Speciality papers are made from mixtures of pulp from different species, allowing combinations of strength, flexibility, and opaqueness. Toilet paper

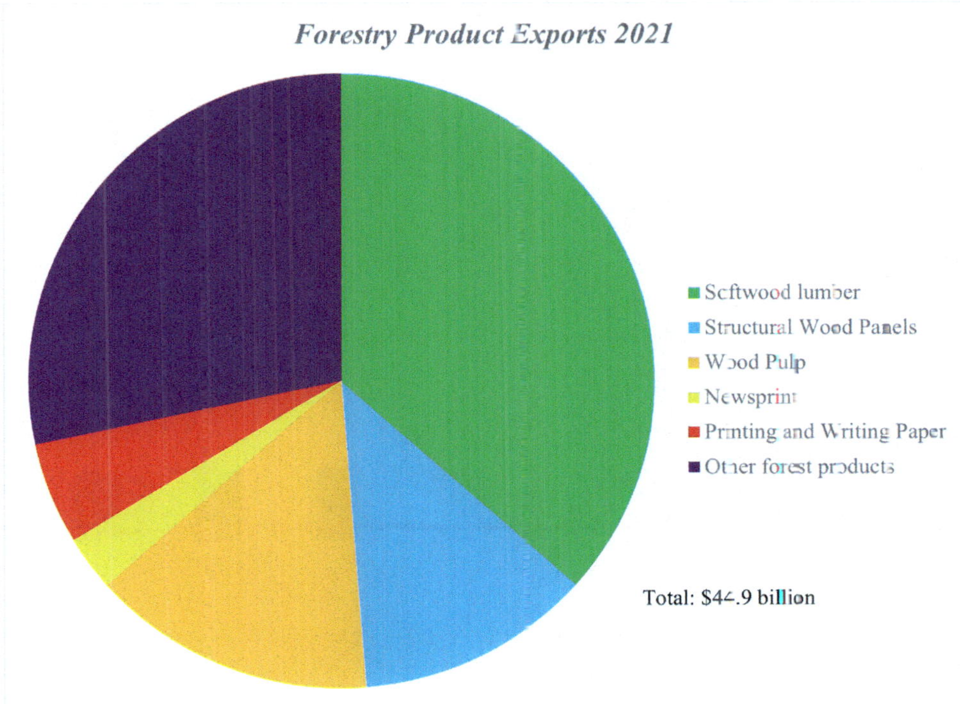

Fig. 15.12 Forestry product exports by percentage, 2021. Natural Resources Canada (2022), Statistics Canada Census of Canada 2021

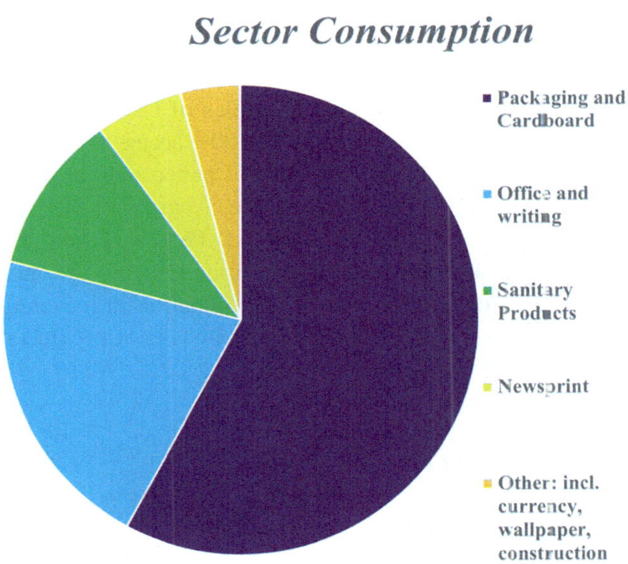

Fig. 15.13 Sector consumption of paper production, by percentage. Natural Resources Canada (2022), Statistics Canada Census of Canada 2021

manufacture in Canada dominantly uses spruce, fir, and pine, due both to cost considerations and relative shortages of hardwoods such as maple and oak. Softer toilet papers have higher percentages of hardwood fibres (see Figs. 15.13, 15.14).

Product diversification, termed biotransformation, involves use of the individual wood components, including cellulose, hemicellulose, and lignin, extracted by chemical processing. Examples include disinfecting wipes and rayon towels (cellulose), makeup and toothpaste (carboxymethyl cellulose), medications (microcrystalline cellulose), guitar picks, nail polish, and table tennis balls (nitrocellulose), paints (hydroxyethyl cellulose), and LCD screens (cellulose triacetate) (ABB 2023). By 2050, these and other advanced bio-based products may represent 50% of forest consumer goods (House of Commons 2018, Environment and Climate Change Canada 2022, Natural Resources Canada 2022). Such options are generally not available for mechanical processing mills affected by the apparently irreversible decline in demand for newsprint, more than 80% since 2000 (Bogdanski 2014). Canada's newsprint production declined by 70% during that time (Natural Resources Canada 2022).

Despite predictions of paperless offices and universities, 1980–2020 saw a global increase of 400% in total paper consumption. Annual Canadian use of paper for all purposes, including the majority use for packaging and cardboard, is about 160 kg per person, 4 to 5 harvestable trees. Printing and writing paper consumption amounted to 937,245 t in 2021 (Natural Resources Canada 2023), with education as the largest sectoral user. Production of 1 kg of paper requires 300–350 l of water.

Toilet paper (bathroom tissue) is a commonly used bleach kraft paper product. Canadian usage is approximately 22 kg per capita (85–95 rolls, or about 13,000 sheets) (Cottonelle 2023; Statista 2024). Annual production requires about 2 million trees (at 450 kg/tree). Approximately 60% of the content of an average roll of toilet paper is recycled material from other uses, although this varies considerably by brand.

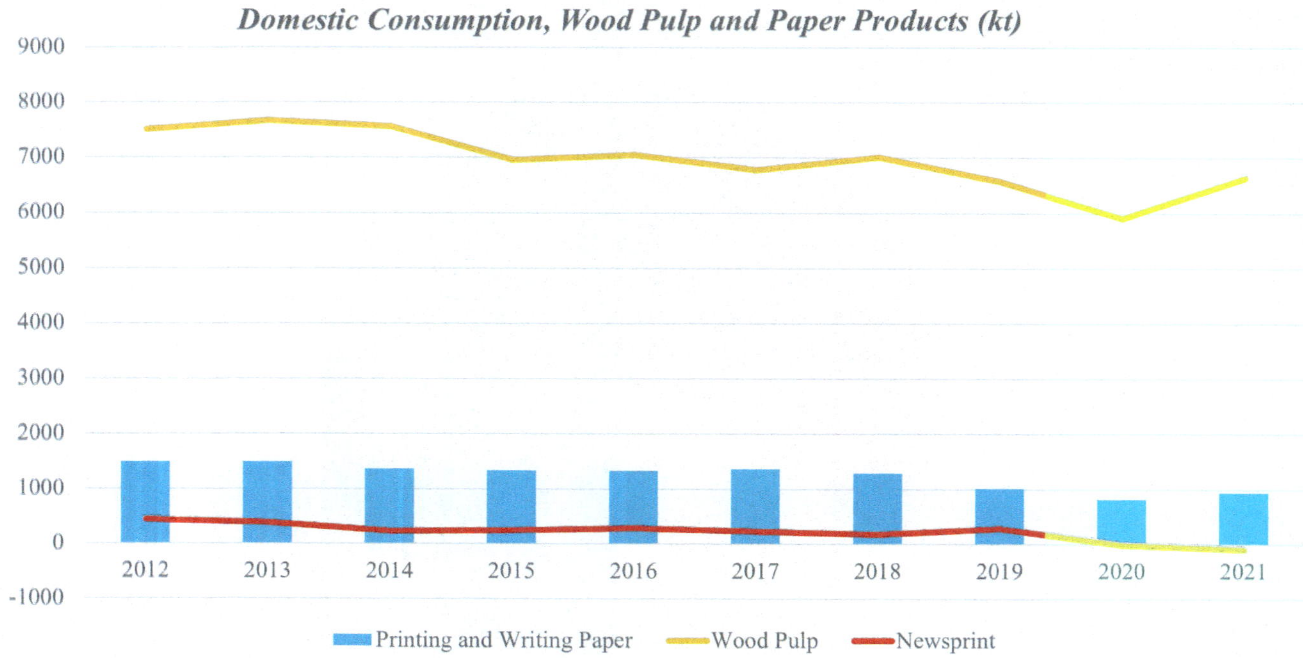

Fig. 15.14 Domestic consumption of wood pulp and paper products (kt), 2012–2021. Negative domestic consumption of newsprint reflects exports. Natural Resources Canada (2022)

Paper manufacture is energy-intensive, representing 8% of industrial demand (Environment and Climate Change Canada 2020). However, due to a combination of operational changes and reduced production (particularly from newsprint plants), energy use has declined by almost 50% since 2000.

Pulp and paper production releases a variety of pollutants into the atmosphere and water bodies (Environment and Climate Change Canada 2022). Only about 4% of the potential contaminants are recovered, primarily phosphorous, manganese, aluminium, and heavy metals. Criteria air contaminants (CAC) released include carbon monoxide (51%), nitrogen oxides (17%), sulphur dioxide (13%), particulate matter (10%), methanol and other volatile organic compounds (8%) methanol, and ammonia (1%). A bleach kraft pulp mill emits airborne hydrogen sulphide, dimethyl sulphide and disulphide, and other volatile sulphur compounds. Mills also release reduced sulphur, phosphorus, and nitrates (Environment and Climate Change Canada 2022).

Depending on species and climate, trees used in paper manufacture require 20–50 years to grow to marketable size (Bogdanski 2014). Afforestation is commonly conducted, often sequentially, on Crown lands leased for production.

Paper and cardboard represent 15–20% of the waste in a typical urban landfill, a substantial proportion of the 720 kg y^{-1} disposed of per capita. Recycling programmes have resulted in the diversion of more than 3 million t of paper and cardboard annually (Environment and Climate Change Canada 2022). However, overall and paper-recycling rates remain low, and many paper-based products (such as single-use beverage containers containing plastics) are not recyclable. International restrictions on acceptance of recycled materials have impacted the economics of Canadian paper recycling programmes. With almost 100% access to services, Canadians recycle more than 80% of paper and cardboard (6 million t y^{-1}), compared to the average of 28% for all materials (~50% for recyclables). Perceived time and effort required and storage space limitations are the most commonly cited reasons for non-participation in recycling. Waste reduction and management are joint federal, provincial, and municipal responsibilities.

15.3.2 Lumber

15.3.2.1 Softwood

Domestic production of softwood lumber in 2021 was 56 million m^3, together with 8.9 million m^3 of structural panels (plywood and oriented strandboard), and 873,500 m^3 of hardwood lumber (Natural Resources Canada 2022, Capuano 2023; see Figs. 15.15, 15.16). BC, Québec, and Alberta are the largest producers, typically accounting for 80–85% of production.

Spruce (*Picea sp.*), pine (*Pinus sp.*), and Balsam fir (*Abies balsamea*) represent the majority of dimension lumber production, used in house construction. A 200 m^2 house requires about 30 m^3 of lumber. Eastern Canadian operations process primarily white spruce (Picea glauca), balsam fir, and red pine (*Pinus resinosa*). Their strength, low density,

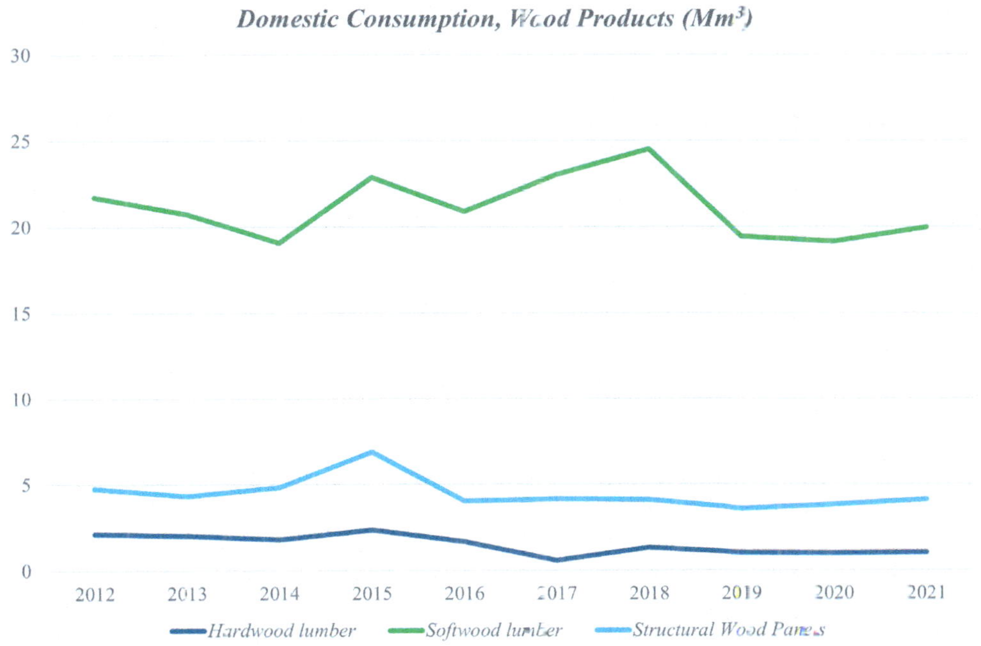

Fig. 15.15 Domestic consumption of wood products (Mm^3). Natural Resources Canada (2022)

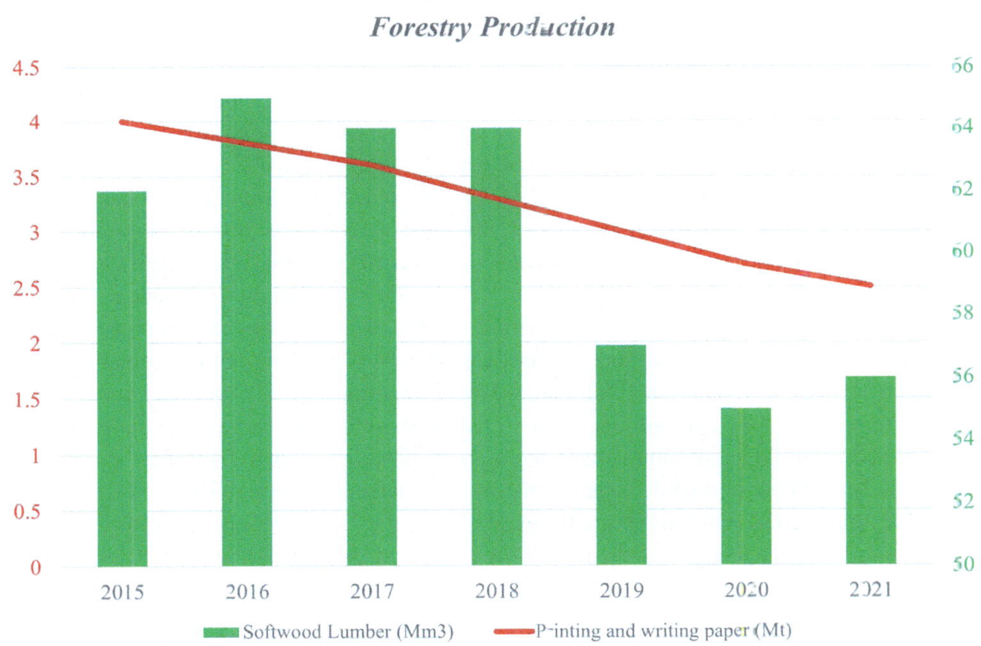

Fig. 15.16 Production of softwood lumber (Mm^3, green columns, right-hand scale) compared with printing and writing paper (red line, Mt, left-hand scale). Data from Natural Resources Canada (2022)

durability, ease of working, and relatively low cost make them popular choices. In general construction work, supplies of these woods are frequently combined as spruce-pine-fir (SPF). Afforestation of SPF assemblages is increasingly common, with all species harvested and processed together. Eastern white cedar (*Thuja occidentalis*) and eastern hemlock (*Tsuga canadensis*), considered "northern species" in the construction industry, are also used for dimension lumber. Due to its relative scarcity and cost, eastern white pine (*Pinus strobus*) is more commonly used for wood product manufacturing.

In the interior Cordillera of BC, Douglas fir (*Pseudotsuga menziesii*) is commonly processed for dimension lumber. Douglas fir lumber typically has a straight, knot-free grain, is relatively lightweight, flexible, easy to work and nail, and stains well. It is the major species of softwood lumber exported. Western larch (*Larix occidentalis*) is used together with Douglas fir in general construction work, and its wood has similar physical characteristics. Western hemlock (*Tsuga heterophylla*) is also frequently grouped with Douglas fir in both general construction usage and afforestation plantings. It is also used in decks, panels, and trusses.

Western white pine (*Pinus monticola*), lodgepole pine (*P. contorta*), and Ponderosa pine (*P. ponderosa*) are also used. Pinus have high strength and are easy to work, paint, and stain, but they are also soft and prone to damage during construction. Much of the pine used in Canadian building is imported southern yellow (*P. echinata*) and loblolly pine (*P. taeda*). Interior BC accounts for about 90% of total softwood production in the province.

Western red cedar (*Thuja plicata*) is considered as a Northern species and is a valuable commercial tree in British Columbia (Gregory et al. 2018). It is decay-resistant, making it suitable for exterior applications such as siding, fences, and decks. However, it is relatively heavy and dense, and is generally expensive. Western red cedar is the characteristic tree of the southern Pacific Maritime ecozone (see Chap. 7). Western red cedar is slow-growing, with older trees subject to fungal decay. Groves exhibiting fungal decay may be culled in salvage operations, with higher harvesting costs and lower returns (Renzie and Han 2008).

Plywood, oriented strandboard, fibreboard, and particleboard are composite materials made of wood veneers, bands, or fragments glued together in various orientations. In Canada, most is made from spruce, pine, and fir (eastern Canada), or Douglas fir and hemlock (western Canada). Plywood is used for roofing and subfloors. Oriented strandboard and high-density fibreboard can also be used for vertically aligned studs and low-load beams. Particleboard, made from wood chips, is a lighter-weight and less expensive alternative to plywood, but should not be used where load-bearing is a concern.

Canada and USA have been engaged in a trade dispute over Canadian exports of softwood lumber, primarily used for residential construction. Initiated in 1982, marked by alternating periods of tension and compromise, court challenges and appeals, bellicose statements and patient negotiations (Zhang 2010), the latest phase of dispute began in 2015 and remained unresolved as of 2024 (Global Affairs Canada 2024).

Although approximately two-thirds of USA consumption uses lumber produced domestically, USA lumber companies argue that Canadian softwood lumber, primarily from Cordilleran BC, is unfairly subsidized by the Government of Canada. As it is alleged that the prices charged are lower than the true costs of production in Canada, countervailing duties have been imposed by USA Department of Commerce as an antidumping measure.

Lumber companies lease rights to log on Crown Land, with ownership remaining with the provincial government. Stumpage fees charged by government are set administratively, based on the amount and value of timber to be harvested. Stumpage fees vary between provinces and by location within a province. The USA duties vary by province and by individual company, reflecting the differences in stumpage fees. This practice contrasts with the USA situation, where land used for logging is privately owned and stumpage fees reflect the competitive free market. While USA companies lobby their federal government to impose countervailing duties to remove the perceived advantage, Canadian companies lobby individual provincial governments over fees and the federal government to obtain a general international tariff agreement.

The dispute is therefore complicated by political positions in both countries and in individual provinces. The Canada-USA difference in public attitudes to government involvement and public ownership, manifest over discussions of health care, is also deeply involved. The matter is further complicated by increasing USA export of domestic lumber production to third countries, and by the increasing purchase of USA land and softwood producers by private companies based in Canada.

Both western red and eastern white cedar, along with Nootka cypress (*Callitropsis nootkatensis,* also called yellow cedar) are also used to manufacture shingles and shakes for roofing and siding. Shingles are machine-cut and milled, whereas shakes are split using hand tools. Shakes have irregular shapes and are generally used to give a rustic appearance to roofing. Typically 40–70 cm long with a tapered cross-section, shingles and shakes last from 30 to 50 years under most Canadian climates.

Shingle and shake production from western red cedar uses 200 to 300-year-old trees. Eighty per cent of the North American western red cedar population is located in British Columbia. Current levels of shake and shingle production may be sustainable for about 100 years, but serious concerns about both sustainability and the use of old-growth trees have arisen. Nootka cypress grows more rapidly, and populations are larger, but consumer preference is still strongly for red cedar. Export of shingles and shakes to USA accounts for ~90% of BC production. In contrast to softwood lumber, from 2021 all cedar shingles and shakes are exempt from USA Department of Commerce antidumping and countervailing duties (Global Affairs Canada 2024).

Eastern white cedar is used for shingles in Québec and New Brunswick. Eastern white cedar trunk diameters are smaller and knots are more common, making them less suitable for shakes. Combined, eastern Canada represents 10–12% of total Canadian production, primarily shipped to northeastern USA.

Mass timber production involves gluing multiple veneers, flakes, and thin planks of dimension wood to construct large beams (Natural Resources Canada 2021). The intent is to replace large natural timbers, increasingly hard to find and expensive, in bearing load construction. Types include glue-laminated, cross-laminated, and nail-laminated timber. Mass timber production uses coniferous softwoods.

15.3.2.2 Hardwood

Hardwood species used for lumber include maple (*Acer sp.*), oaks (*Quercus sp.*), yellow birch (*Betula alleghaniensis*), and black cherry (*Prunus serotina*), as well as aspen (*Populus tremuloides*). Hardwoods are used for flooring, cabinet doors, and veneers, where aesthetics are desirable. Birch is primarily used in millwork, flooring, and cabinetry (often in the form of plywood). Aspen is used in oriented strandboard (fibreboard), plywood, structural composite lumber, and cross-laminated timber.

Aspen is also used to make chopsticks. Chopstick manufacture uses about 50 million trees annually. Trees must be knot-free and even grained, which cannot be determined until the tree is cut. About 85% of the aspens cut are wasted in the process.

15.3.3 Wood Product Manufacturing

Wood product manufacturing adds value and increases employment. There has been a general tendency for the wood product manufacturing sector in particular to increase in value over recent years.

A typical profile of a Canadian furniture manufacturer shows Canadian ownership (97% of the 7,000 companies) of a micro- to small-sized employer (Blair 2024). Only about 1% of companies employ 100 workers or more. About 95% of production is exported, primarily to USA. Furniture sales are major revenue generators, approximately $24 billion. Living room furniture represents the largest component ($5.2 billion in 2021), followed by bedroom ($3.6 billion) and kitchen and dining room furniture ($2.2 billion).

Both hardwoods and softwoods are used in furniture manufacture. In eastern Canada, oak (*Quercus sp.*) is available from the Mixedwood Plains and Nova Scotia, and is durable and attractive to consumers, but it is also dense, difficult to work with, slow-growing, and expensive. Sugar maple (*Acer saccharum*), also used extensively in eastern Canada, is somewhat durable, and is faster-growing, more widespread, and less expensive than oak. Less dense than oak, it is prone to cracking and requires more maintenance both during working and after consumer purchase. Bigleaf maple (*Acer macrophyllum*) is used in BC, but is considered less suitable for furniture manufacture due to its small trunk dimension, softness, and tendency to receive damage. Yellow birch (*Betula alleghanesis*) is sturdy, easy to paint or stain, and relatively inexpensive, and is frequently used to make kitchen tables and chairs. Cherry (*Prunus sp.*) is also used. Less expensive furniture usually uses softwood, including pine, western hemlock, Douglas fir, and Nootka cypress.

Hardwoods are also used for musical instruments. Maple is popular for making violins and wind instruments. Stringed instruments use yellow birch, cherry, and ironwood (*Carpinus caroliniana).* High-end guitars commonly use old-growth softwoods, particularly Sitka spruce (*Picea sitchensis*), a practice criticized on ecological grounds (Gibson and Warren 2021). Large-diameter logs, cut at 90° to the growth rings, are required to provide the front face, back, and sides.

For the ~25% of Canadians who choose burial over cremation, walnut and cherry are the most expensive casket choices. Other options include maples, pines, and poplars, in decreasing order of expense.

15.4 GHG Emissions

Net emissions (CO_2 equivalent) from managed forested areas have varied from 50 to 250 Mt annually since 2010 (Environment and Climate Change Canada 2020; Fig. 15.17). The variations reflect the prevalence of wildland fires and insect

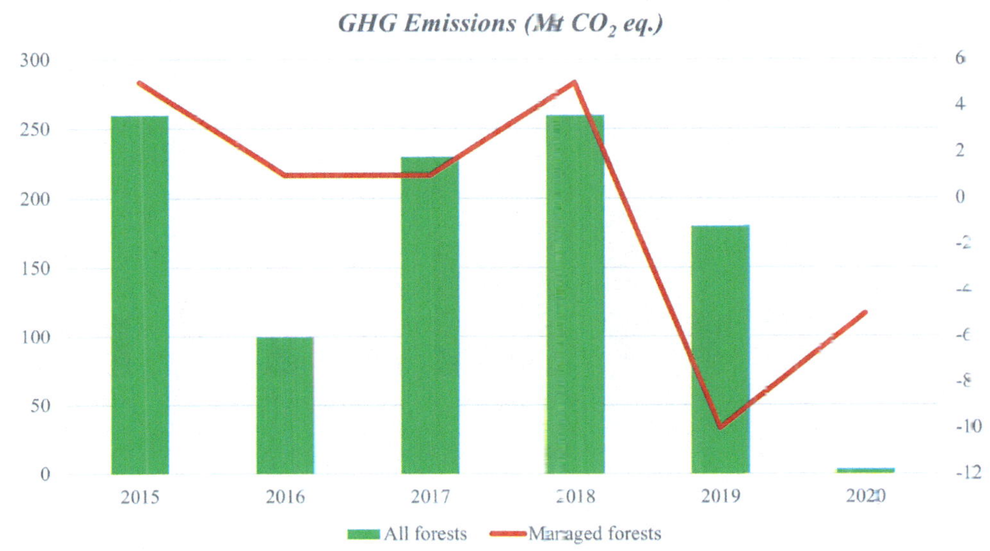

Fig. 15.17 Net GHG Emissions (Mt CO_2 eq.) from managed forests (red line, left hand scale) and all forests (green columns, right hand scale), 2015–2020. Forests frequently serve as CO_2 sinks. Environment and Climate Change Canada (2020)

disturbances in specific years (Natural Resources Canada 2022). Human activities, including harvesting, afforestation, recycling, waste management, and burning of cleared vegetation (slash), result in a net sink of 5–10 Mt (CO_2 equivalent) annually. The increasing use of wood products in preference to energy-intensive concrete and metal in construction is suggested as a climate mitigation strategy. 1 m^3 of lumber, if burned, would emit CO_2 equivalent to combustion of 9,000 l of gasoline.

15.5 Wildland Fires

On average, between 2006 and 2022 more than 6700 wildland fires occurred in Canada each year, burning an average of 2.5 Mha (Natural Resources Canada 2022; Fig. 15.18). There is considerable variation from year to year. Most are relatively small: 97% of fires burn 200 ha or less, but the fires larger than 200 ha are responsible for more than 97% of the total area burned (Canadian Inter-Agency Forest Fire Centre 2024). A single large fire in a particular year, such as the Horse River (Fort McMurray) fire in May 2016, may skew the statistics: this single fire burned 589,000 hectares, more than 35% of the total area burned in Canada that year (Canadian Inter-Agency Forest Fire Centre 2017; Elmes et al., 2018).

Most wildfire in Canada occurs in the Boreal Forest and Cordillera (Coogan et al. 2020). Numerous communities are situated within forested areas, leading to potential exposure to fires, particularly where urban development meets wildland vegetation along the Wildland-Urban Interface (WUI). In Canada, the fire season generally runs from April to October, with most wildland fires occurring in May to September.

Wildland fires may be started by natural or human causes. Lightning is the most common natural cause (Catto et al. 2006; Coogan et al. 2020). Humans start fires unintentionally in the course of recreation (e.g. camping), industrial activity, failure of infrastructure, transportation (e.g. sparks from locomotives), accidents, or carelessness. A small percentage of wildland fires are started deliberately by humans. Arson is legally defined in Canada to require negligence, disregard for human life, or intent to damage property. Conviction for arson requires that the accused knew the probable consequences of starting the fire. Very few fires are deliberately started by humans suffering from pyromania, for emotional gratification or to induce euphoria, rather than for personal motives or financial gain.

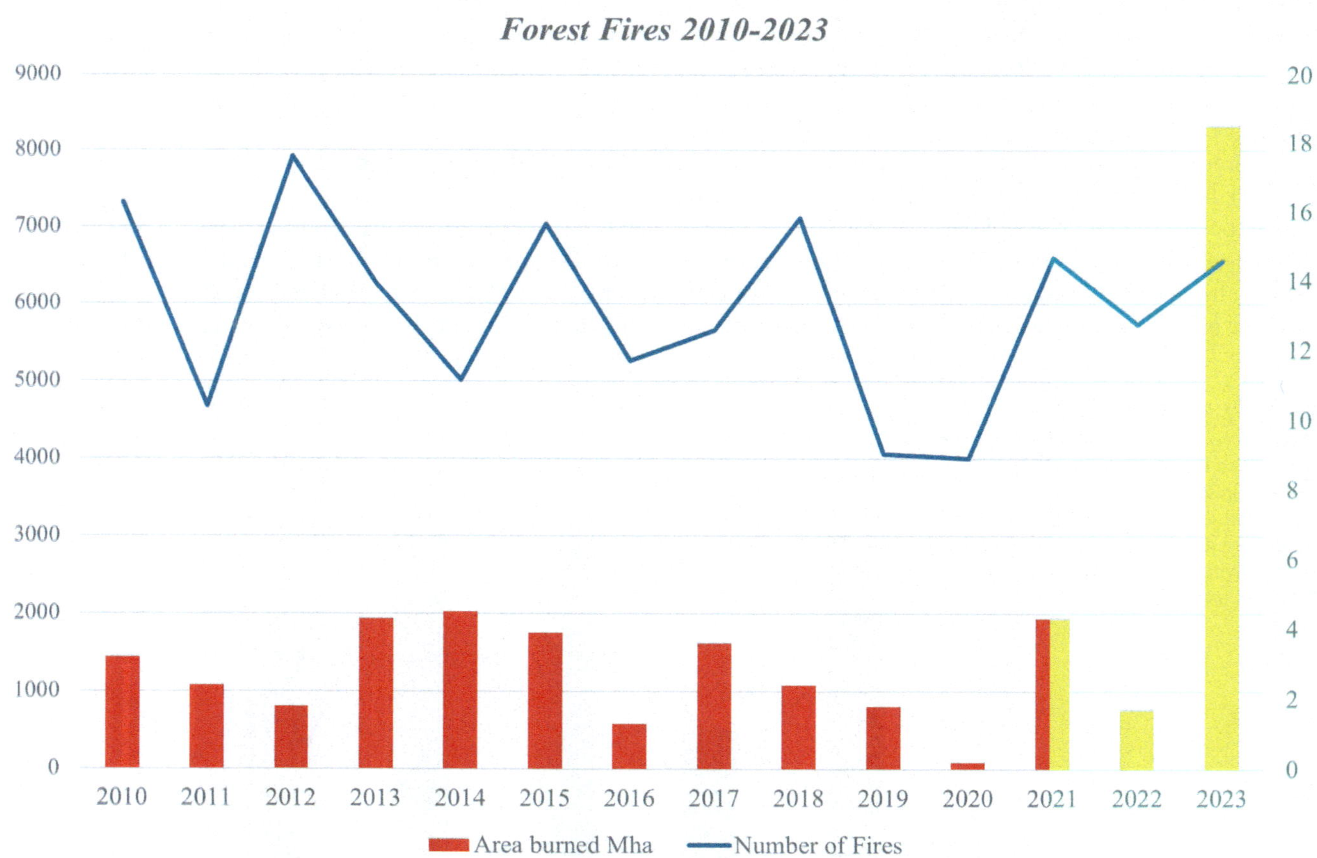

Fig. 15.18 Number of wildland fires and area burned (Mha), 2010–2023. Note the lack of direct correlation. Canadian Inter-Agency Forest Fire Centre (2024)

Throughout Canada, the percentages of recent wildland fires caused by lightning or humans are nearly equally divided. Lightning causes most large wildland fires in northern Canada, typically accounting for more than 80% of the area burned annually, whereas most human caused fires occur in the more populated and southern parts of the country (Canadian Inter-Agency Forest Fire Centre 2024). Annual variations occur in response to differences in weather: abundant thunderstorms may generate more lightning, while discouraging human recreational use of forests. Socio-economic factors and educational campaigns may also influence both human numbers and care in wildlands.

Regional variation is common: 97% of wildland fires in Nova Scotia are human caused, compared to only 14% of wildland fires in the Northwest Territories. Differences reflect combinations of human use of wildlands, industrial and transportation activities, fire weather and climate, species assemblages in forests, and distance from the nearest community (Fig. 15.19).

Burning produces smoke. Black smoke indicates incomplete combustion, with fragments and microscopic pieces of charcoal included in the convective mix (Urbanski 2013). Smoke carried by convection to heights of 1000 m or more can be incorporated into atmospheric circulation, spreading hundreds to thousands of kilometres from the initial fire (e.g. Hu et al. 2019).

When a wildfire impacts a settlement, the burning buildings, cars, and possessions add petroleum-based components, heavy metals, and other toxins to smoke (Sparks and Wagner 2021). Combustion of polystyrene, insulation, vinyl siding, and nylon carpet, along with particleboard and fibreboard treated with melamine, contribute significant emissions.

In Canada, approximately 30% of fine particulate emissions in the atmosphere originate from wildland fires (Environment and Climate Change Canada 2021). Wildfire smoke increases hospitalization, use of respiratory medication, coughing, wheezing, and eye irritation (Johnston et al. 2012). Respiratory hospital admissions may double during wildfire smoke events. Exposure to fine particulate matter concentrated in smoke may also impact cardiovascular health, leading to stress for people with heart problems. Infants, the elderly, pregnant women, and people with chronic health problems such as asthma and chronic obstructive pulmonary disease are particularly sensitive. Women with asthma may be more severely affected than are men. In North America, smoke allergies and sensitivities could affect one-third of households.

Firefighters typically must cope with sleep deprivation, physical and psychological stresses, and isolation as they engage fire and smoke. Exposure to heat and carbon

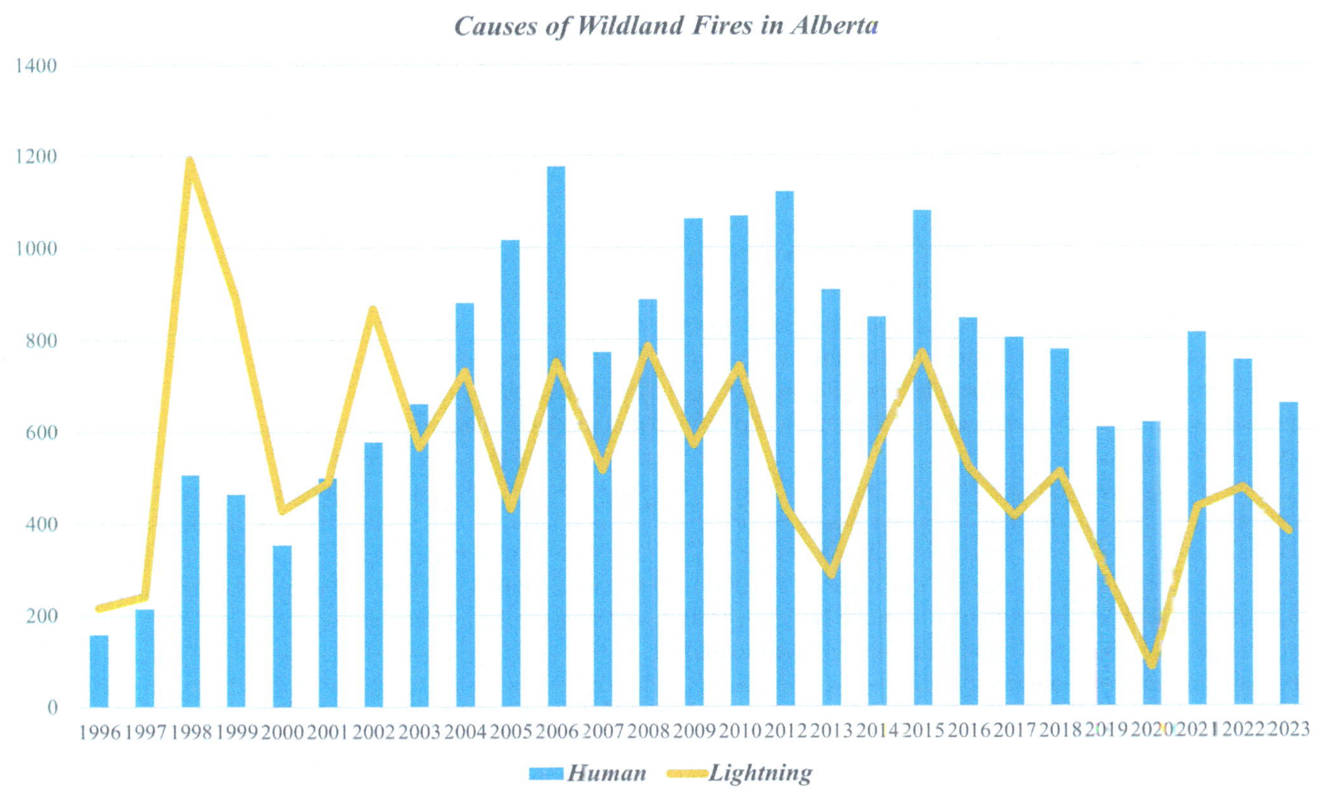

Fig. 15.19 Causes of wildland fires in Alberta, 1996–2023. Reliable data for 2011 are unavailable. 1998 represented a strong El Niño year. Canadian Inter-Agency Forest Fire Centre (2024)

monoxide increases the physical effort demanded of firefighters, although their cognitive functions may not be significantly impaired (Thompson et al. 2024). While in action, firefighters are more commonly injured by slipping or falling, and by accidents related to machinery operation, than by the effects of heat and smoke, although the long-term chronic effects of smoke inhalation have been less studied (Firefighter Life Safety Research Centre 2008, Tyakoff et al. 2015, Ramsden et al. 2018). Firefighter injuries tend to be more severe near the end of the fire season than at the beginning, suggesting that fatigue and burnout may be important factors (Britton et al. 2013).

Wildland fires also emit substantial amounts of CO_2, and thus both contribute to climate change and are accentuated by it (Environment and Climate Change Canada, 2020). In a typical year, 2 Petagrams of carbon are released into the atmosphere by wildland fires globally, with emissions from fires in boreal and temperate forests typically accounting for about 15%. Typically, burning conifers produce more gas emissions (CO, CO_2, CH_4) than do deciduous trees (e.g. Kipping et al. 2022). Emissions in both 2021 and 2023 from the large wildland fires in Canada greatly surpassed typical values (Wang et al. 2023; Zheng et al. 2023).

Wildland fires also result in significant mercury emissions (McLagan et al. 2021). In Canada, Hg emissions from forest fires (annual average 3.5 t) are typically 30% of anthropogenic mercury emissions, but may exceed anthropogenic emissions during severe fire years (Environment and Climate Change Canada 2022c). Atmospheric circulation results in net transport of Hg from land to ocean, and from low latitudes to high latitudes, leading to increased concentrations in Taiga and Arctic environments.

The two main responses to wildland fires are suppression and monitoring (Catto et al. 2006). Suppression involves extinguishing fires as soon as possible. Total suppression has been practised, at least attempted, over most of North America since World War II. Human-caused fires were causing considerable property damage, and most of the fires in eastern North America were caused by people. Evolving technology, especially air tanker ("water-bomber") design, made total suppression more feasible.

When a fire is detected, it may be monitored and not suppressed, as long as human lives and assets valued by humans (such as settlements) are not at hazard, saving time, energy, and money. It also recognizes that there are benefits to letting some fires burn for ecological reasons. However, only a very small proportion of wildland fires are allowed to burn without being suppressed immediately.

Problems arise with this monitoring approach when the assets that are at risk are less clearly defined than towns or people: should habitat for endangered species be allowed to burn? Should national parks be considered as natural areas ("burn"), or as valuable socio-economic assets for human recreation ("suppress")? Are forests valuable timber resources ("suppress") or natural ecosystems ("burn")? Who owns a particular forest and gets to decide–governments, society, indigenous communities?

Provincial and territorial government wildfire management agencies across Canada are engaged in wildfire prevention, preparedness, and suppression. Prevention includes education programmes to reduce human-caused fires, which are aimed at the public, industry, and other forest users. Provincial, territorial, and federal agencies share firefighting resources such as aircraft, personnel, equipment, and other items through the Canadian Interagency Forest Fire Centre (CIFFC). During the fire season, CIFFC provides daily reports on the fire situation across the country to provincial and territorial government agencies, and to the public (Canadian Inter-Agency Forest Fire Centre 2024).

The federal government is responsible for wildfire management in Indigenous communities, military bases, and national parks. The federal government has agreements with provinces to suppress fires on Indigenous settlements and on military bases. Parks Canada is responsible for wildfire management in National Parks.

In wildfire studies, megafires are defined on the basis of overall socio-economic and ecological impact, rather than considering only the number of hectares burned (Stephens et al. 2014). As an extension of the original fire triangle (fuel, heat, oxygen), the megafire triangle recognizes humans as a fundamental factor, along with fuels and climate. Fuel availability and development are influenced by vegetation pattern (including density of tree cover, height, topographic influence), amount (all types, including fine fuel, duff, deadfall), and disturbance history (including previous fires, insect infestations, and storm events). The climate factor considers longer-term events in addition to daily, seasonal, and monthly weather, including climate oscillations and the ramifications of ongoing climate change. The human factor includes consideration of land use (past and present), and land management policies, both directly related to fire and those that are indirectly related (e.g. zoning, building codes, provision of fire services, road construction). Landscape configurations conducive to megafire disasters, including fire exclusion from the landscape (total suppression), development of wildland-urban interfaces, and weather factors, are also considered. A megafire results from human, biological, and meteorological factors operating in combination, and has significant physical and socio-economic impacts.

References

ABB (2023) Characteristics of wood and papermaking fibers. ABB Group, https://new.abb.com/pulp-paper/abb-in-pulp-and-paper/articles/characteristics-of-wood-and-papermaking-fibers

Blair N (2024) Furniture industry statistics. https://madeinca.ca/furniture-industry-canada-statistics/

Bose AK, Harvey BD, Brais S, Beaudet M, Leduc A (2014) Constraints to partial cutting in the boreal forest of Canada in the context of natural disturbance-based management: a review. For: Int J For Res, 87, 11–28

Boucher D, Boulanger Y, Aubin I, Bernier PY, Beaudoin A, Guindon L, Gauthier S (2018) Current and projected cumulative impacts of fire, drought, and insects on timber volumes across Canada. Ecol Appl 28:1245–1259

Bogdanski BEC (2014) The rise and fall of the Canadian pulp and paper sector. Forestry Chronicle 90:785–793

Britton C, Lynch CF, Ramirez M, Torner J, Buresh C, Peek-Ash C (2013) Epidemiology of injuries to wildland firefighters. Am J Emerg Med 31:339–345

Canadian Inter-Agency Forest Fire Centre (2017) Canada Report 2016. https://www.ciffc.ca/sites/default/files/2019-03/2016_canada_report_2017_05_15_v02.pdf

Canadian Inter-Agency Forest Fire Centre (2024) Publications. http://www.ciffc.ca/

Capuano T (2023) Sawmill industry in Canada: 15 years in review. Statistics Canada. 11-621-M

Catto NR Hyndman D Hyndman D (2006) Wildland Fires. In: Catto NR Hyndman D Hyndman D, Natural Hazards and Disasters, Nelson, 363–384

Certification Canada (2023a) Canadian statistics. http://certificationcanada.org/en/statistics/canadian-statistics/

Certification Canada (2023b) Provincial statistics. http://certificationcanada.org/en/statistics/provincial-statistics/

Coogan SCP, Daniels LD, Boychuk D, Burton PJ, Flannigan MD, Gauthier S, Kafka V, Parks JS, Wotton BW (2020) Fifty years of wildland fire science in Canada. Can J for Res 51:283–302

Cottonelle (2023) How much toilet paper do we use? https://www.cottonelle.com/en-ca/tips-advice/toilet-paper-101/how-much-toilet-paper-do-we-use

Dyk, A., Leckie, D., et al. (2015) Canada's national deforestation monitoring system: system description. Victoria, British Columbia: Natural Resources Canada–Canadian Forest Service, Pacific Forestry Centre. http://cfs.nrcan.gc.ca/publications?id=36042

Elmes MC, Thompson DK, Sherwood JH, Price JS (2018) Hydrometeorological conditions preceding wildfire, and the subsequent burning of a fen watershed in Fort McMurray, Alberta, Canada. Nat Hazard 18:157–170

Environment and Climate Change Canada (2020) National inventory report 1990–2018: greenhouse gas sources and sinks in Canada. https://unfccc.int/documents/224829

Environment and Climate Change Canada (2021) Canadian environmental sustainability indicators: Emissions of harmful substances to air. www.canada.ca/en/environment-climate-change/services/environmental-indicators/emissionsharmful-substances-air.html

Environment and Climate Change Canada (2022a) Canadian environmental sustainability indicators: sustainability of timber harvest.: https://www.canada.ca/en/environment-climate-change/services/environmental-indicators/sustainability-timber-harvest.html.

Environment and Climate Change Canada (2022b) National pollutant release inventory: pulp and paper sector. https://www.canada.ca/en/environment-climate-change/services/national-pollutant-release-inventory/tools-resources-data/pulp-and-paper.html

Environment and Climate Change Canada (2022c) National pollutant release inventory: mercury. https://www.canada.ca/en/environment-climate-change/services/national-pollutant-release-inventory/tools-resources-data/mercury.html

Environment and Climate Change Canada (2023) Canadian Environmental Sustainability Indicators. https://www.canada.ca/en/environment-climate-change/services/environmental-indicators.html

Firefighter Life Safety Research Center (2008) Firefighter Fatalities and Injuries: the role of heat stress and PPE. University of Illinois Fire Service Institute, Urbana-Champaign

Gauthier S, Bernier P, Kuuluvainen T, Shvidenko AZ, Schepaschenko DG (2015) Boreal forest health and global change. Sci 349:819–822

Gibson C Warren A (2021) The guitar industry's hidden environmental problem. https://phys.org/news/2021-05-guitar-industry-hidden-environmental-problem.html

Global Affairs Canada (2024) Softwood Lumber dispute. https://www.international.gc.ca/controls-controles/softwood-bois_oeuvre/index.aspx?lang=eng

Gregory C, McBeath A, Filipescu C (2018) An Economic Assessment of the western redcedar industry in British Columbia. Canadian Forest Service, Natural Resources Canada Information Report FI-X-017

Hanberry BB (2021) Forest disturbance types and current analogs for historical disturbance-independent forests. Land 10:136. https://doi.org/10.3390/land10020136

Hanes C, Wang X, Jain P, Parisien M-P, Little JM, Flannigan MD (2019) Fire-regime changes in Canada over the last century. Can J for Res 49:256–269

House of Commons (2018) Value-added products in Canada's Forest Sector: Cultivating Innovation for a competitive bioeconomy. Report of the Standing committee on Natural Resources

Hu Q et al (2019) Long-range-transported Canadian smoke plumes in the lower stratosphere over northern France. Atmos Chem Phys 19:1173–1193

Johnston FH, Henderson SB, Chen Y, Randerson JB, Marlier M, DeFries RS, Kinney P, Bowman DMJS, Brauer M (2012) Assessing global morality attributable to smoke from landscape fires. Environ Health Perspect 120:695–701

Kimmins JP (1991) Clearcutting: responsible forest management or environmentally-destructive forest exploitation. Can Pulp Pap Assoc, Montréal

Kipping L, Gossner MM, Koschorreck M, Muszynski S, Maurer F, Weisser WW, Jehmlich N, Noll M, (2022) Emission of CO_2 and CH_4 from 13 deadwood tree species is linked to tree species identity and management intensity in forest and grassland habitats. Glob Biogeochem Cycles, 36, e2021GB007143. https://doi.org/10.1029/2021GB007143

MacDonald AJ (1999) Harvesting systems and Equipment in British Columbia. BC ministry of forests. For Eng Res Inst Can Handb HB-12

Matthews JD (1989) Silvicultural systems. Clarendon Press, Oxford, p 284

McLagan DS, Stupple GW, Darlington A, Hayden K, Steffen A (2021) Where there is smoke there is mercury: Assessing boreal forest fire mercury emissions using aircraft and highlighting uncertainties associated with upscaling emissions estimates. Atmos Chem Phys 21:5635–5653

National Forestry Database (2022a) Forest fires, Table 3.1.2, Area burned by jurisdiction, cause class, response category and protection zone. http://nfdp.ccfm.org/en/data/fires.php

National Forestry Database (2022b) Forest insects, Table 4.1, Area of moderate to severe defoliation (including beetle-killed trees) by insects. http://nfdp.ccfm.org/en/data/insects.php

National Forestry Database (2022c) Harvest, Table 5.2, Area harvested by jurisdiction, tenure, management and harvesting method. http://nfdp.ccfm.org/en/data/harvest.php

Natural Resources Canada (2020a) Sustainable Forest Management. https://www.nrcan.gc.ca/our-naturalresources/forests-forestry/sustainable-forest-management/forestcertification-canada/17474

Natural Resources Canada (2020b) Changing climate, changing forest zones. https://www.nrcan.gc.ca/climate-change/impacts-adaptations/climate-change-impacts-forests/impacts/changing-climate-changing-forest-zones/13093

Natural Resources Canada (2021) The state of mass timber in Canada 2021. Nat Resour Can, Report Fo1–20E

Natural Resources Canada (2022) The state of Canada's forests 2022. Fo1–6E. cfs.nrcan.gc.ca/stateoftheforests

Natural Resources Canada (2023) Domestic Consumption. https://cfs.nrcan.gc.ca/statsprofile/consumption

Nova Scotia Department of Labour 1994. The Forest Professional—a code of practice for the Stewards of tomorrow's forests. N S Dep Labour Occup Health Saf Div. pp 54

Ramsden R, Smith J, Turcotte K, Garis L, Kunz K, Maxim P, Thomas L, Pike I (2018) Determinants of injury and death in canadian firefighters. Univ Fraser Val

Renzie C Han H-S (2008) Harvesting productivity and cost of clearcut and partial cut in interior British Columbia, Canada. J For Sci 24, 1–14

Sauvageau F (1995) Silvicultural Terms in Canada. Can For Serv, Report Fo42–170/1995

Sparks TL, Wagner J (2021) Composition of particulate matter during a wildfire smoke episode in an urban area. Aerosol Sci Technol 55:734–747

Statista (2024) Toilet Paper-Canada. https://www.statista.com/outlook/cmo/tissue-hygiene-paper/toilet-paper/canada

Statistics Canada (2022a) Lumber production, shipments, and stocks by species, monthly (x 1,000). https://www150.statcan.gc.ca/t1/tbl1/en/tv.action?pid=1610001701

Statistics Canada (2022b) Gross domestic product (GDP) at basic prices by industry (x 1,000,000). https://www150.statcan.gc.ca/t1/tbl1/en/tv.action?pid=3610040101

Stephens SL, Burrows N, Buyantuyev A, Gray RA, Keane RE, Kubian R, Liu S, Seijo F, Shu L, Tolhurst KG, van Wagtendonk J (2014) Temperate and boreal forest mega-fires: characteristics and challenges. Front Ecol Environ 12:115–122

Thompson C, Ferrie L, Pearson SJ, Highlands B, Matthews MJ (2024) Do extreme temperatures affect cognition? A short review of the impact of acute heat stress on cognitive performance of firefighters. Front Psychol 14:1270898. https://doi.org/10.3389/fpsyg.2023.1270898

Tyakoff A, Garis L, Thomas L (2015) Firefighter Injuries in British Columbia: an examination of frequency, severity, locational, and temporal aspects. University of the Fraser Valley

Urbanski S (2013) Combustion efficiency and emission factors for wildfire-season fires in mixed conifer forests of the northern Rocky Mountains, US. Atmos Chem Phys 13:7241–7262

Wang Z et al (2023) Severe global environmental issues caused by Canada's record-breaking wildfires in 2023. Adv Atmos Sci. https://doi.org/10.1007/s00376-023-3241-0

Williamson T, Johnston M, Nelson HW, Edwards JE (2019) Adapting to climate change in Canadian forest management: Past, present and future. For Chron 95:76–90

Zhang, D (2010) The Softwood Lumber War: Politics, Economics, and the Long U.S.-Canadian Trade Dispute. Routledge

Zheng B et al (2023) Record-high CO_2 emissions from boreal fires in 2021. Science 379:912–917

Agriculture and Food

16

Abstract

Primary agriculture represents about 1.4% of the Canadian workforce and 1.6% of Canada's GDP. Canadian agricultural exports, with the exception of supply managed dairy and poultry, are substantial. Imports are substantial, particularly of fruits and vegetables. Agriculture is a shared federal-provincial responsibility. The median age of farm operators was 58.0 y in 2021, compared to an overall Canadian population median of 41.6 y. More than 60% were age 55 or older, with larger percentages in Atlantic Canada and southern British Columbia. Regional shortages of available (or willing) labour have led to the large-scale use of foreign migrant farm workers. Major grain types include wheat, oat, barley, rye, and corn. Canola is the major oilseed. Vegetable production is dominated by tomatoes, carrots, and onions. Other economically significant vegetables include cabbage, sweet corn, broccoli, peppers, lettuce, cauliflower, and asparagus. Climate makes vegetable and fruit importation a seasonal reality, with USA and Mexico being the primary sources. Potato production is increasingly split among provinces, with Alberta, PEI, and Manitoba each holding slightly more than 20%, and New Brunswick and Québec 10-15% each. Soft fruits grown in Canada include cherries, peaches, pears, plums, nectarines, and apricots. Apples are the second most harvested fruit by hectares, and the largest by mass. Beef production varies from 1.5 to 2.0 Mt annually. Cattle numbers vary substantially between and within years. The average herd size is about 70 head, with 60% of operations having 47 or fewer, and only 1% having more than 500. Production is concentrated in Alberta and Saskatchewan, although all provinces have active beef operations. The dairy industry is the main economic agricultural activity in Québec, Ontario, British Columbia, Nova Scotia, and Newfoundland and Labrador and is also significant in PEI and NB. Dairy supply management, instituted in the 1960s, directly links domestic supply with domestic demand. Supply management applies to all dairy products, including milk, butter, and cheese. Dairy supply management continues to be a significant problem for Canada-USA trade negotiations, as well as others. Pork production is the fourth largest farming industry in Canada. Production is concentrated in southeastern Manitoba, southern Québec, and southern Ontario. Poultry and egg production operates under a supply management framework similar to that for dairy. Honeybees are key components for agriculture and food, for honey production but especially for their role as pollinators. Dietary preferences and eating habits are changing. Approximately 10% of the population self-identifies as vegetarian or vegan, with 2% as subject to celiac disease, and larger percentages restricting consumption of meat, gluten, sugar, carbohydrates, or other food types. These changes will inevitably impact food production destined for domestic consumption.

Keywords

Agriculture · Farmer demographics · Agricultural production · Supply management · Dietary preferences

This chapter outlines the present conditions involving agriculture in Canada, including implications connected to food consumption and climate change. Unless otherwise indicated, statistical data cited throughout were obtained from the Census of Canadian Agriculture (Agriculture and Agri-Foods Canada (AAFC), and Statistics Canada). Information from each census is released over several years following the enumeration.

16.1 Introduction: Farms in Canada

Primary agriculture, including work that is performed within the boundaries of a farm, nursery, or greenhouse, repre-

N. Catto and G. R. Vincent, *Canada: Land and People*, World Regional Geography Book Series,
https://doi.org/10.1007/978-3-032-04391-7_16

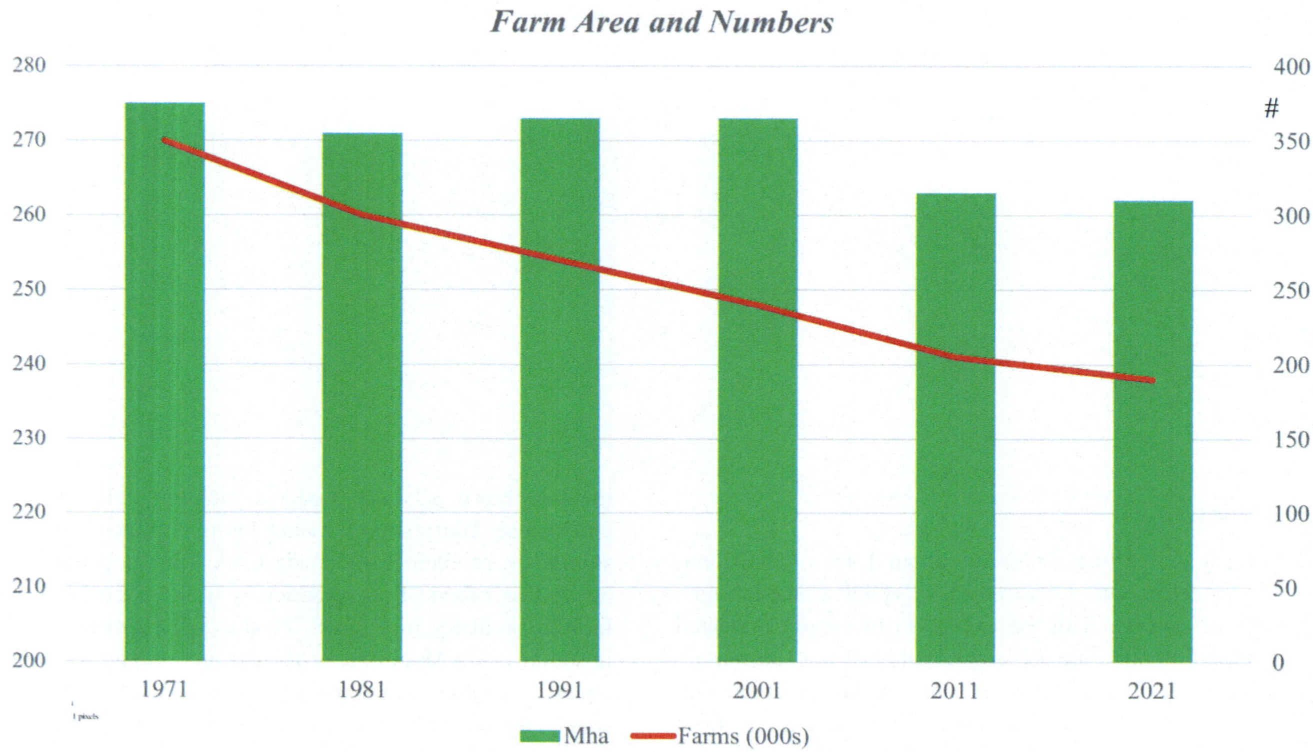

Fig. 16.1 Farm area (Mha) and number of farms in Canada, 1971–2021. Decline in area has been outpaced by decreasing farm numbers, reflecting the tendency for larger farm size over time. Census of Canadian Agriculture

sents about 1.4% of the Canadian workforce and 1.6% of Canada's GDP (2022; Agriculture and Agri-Food Canada 2020a, 2022a, Statistics Canada 2022–23). Approximately 190,000 farms occupy 62 Mha (Figs. 16.1 and 16.2). Farm consolidation, as well as farm operators owning more than one piece of property, has been a trend for several decades. About 20% of operators own multiple properties. Larger farms are increasingly dominant, with 50% of sales attributable to the largest 5% by area, with 67% to the largest 10% (Mangan 2015, Mussell 2021; Statistics Canada 2022–23). The average farm size has doubled since 1970.

Canadian agricultural exports, with the exception of supply-managed dairy and poultry, are substantial. Imports are also substantial, particularly of fruits and vegetables (Dyer et al. 2011). The leading agricultural trading partners are the USA and China. Domestically, including beverages, cannabis, and tobacco, agricultural purchases represent almost $200 billion annually.

Most Canadian farms are high-technology, energy-intensive operations involving machinery, including aircraft and UAVs (unmanned aerial vehicles, aka. "drones") and the use of chemical fertilizers and pesticides (Phillips et al. 2019; Lychuk et al. 2021; Wen et al. 2022). UAV are used for reconnaissance, soil moisture assessment, detection of erosion problems, crop health monitoring, locating livestock, and targeting specific areas for pesticide application (e.g. Rejeb et al. 2022). Canadian grain, oilseed, vegetable, and fruit farmers own or lease ~6 tractors per farm, along with 300,000 combines, forage harvesters, hay-balers, and swathers (Statistics Canada 2022–23).

Usage of fertilizers has increased by more than 60% since 2010 and currently is ~4.6 Mt, including synthetics (Malaj et al. 2020; Environment and Climate Change Canada 2021). P_2O_5 is imported, whereas nitrogen-based fertilizers and K_2O are produced domestically. Canada is the largest producer and exporter of potash (KCl), used to manufacture K_2O (Natural Resources Canada 2022). About ~60% of total fertilizer production is exported (Statistics Canada 2022–23). The federal government is attempting to reduce GHG emissions from agriculture, particularly N_2O (Fig. 16.3), by 30% by 2030 through voluntary action by farmers (Agriculture and Agri-Foods Canada 2023e).

About 90% of crop farms use at least one herbicide, generally glyphosate to control invasive weeds (Fig. 16.4). Fungicides and insecticides are used by ~30% and ~25%, respectively. Pesticide licencing, use, and monitoring fall within provincial jurisdiction, governed by Health Canada's Pest Management Regulatory Agency. Farmers are required to take courses and undergo monitoring to be licenced to use pesticides, many of which (such as malathion) are banned for most non-agricultural uses. In 2024, more than 40 municipalities have severe restrictions or bans on glyphosate use.

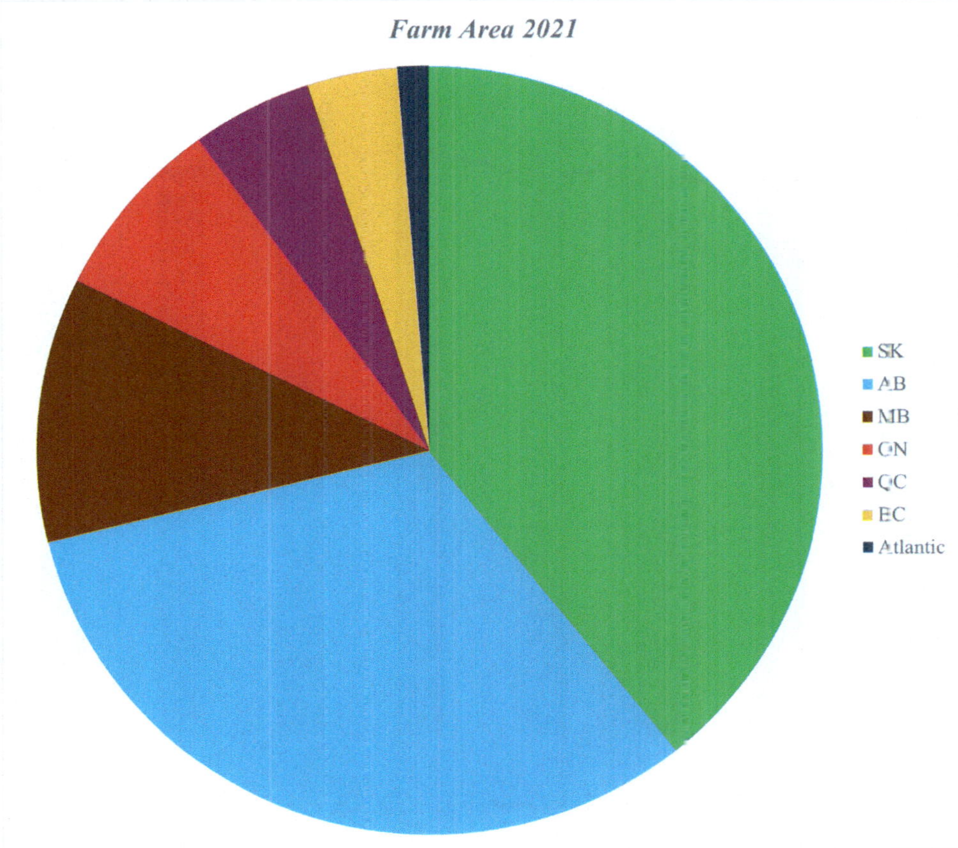

Fig. 16.2 Farm area by percentage within provinces, 2021 Census of Canadian Agriculture

16.1.1 Organic Farming

Organic farming involves more than 4200 farms, primarily in Québec (~30% of the total), Saskatchewan, and BC (Bialais 2020; Statistics Canada 2022–23). Although the number of organic farms has increased while total farm numbers decreased, organic farms still represent less than 2.5% of the total. Organic farms are more likely to be co-operated by men and women, and organic farmers are generally younger (~10% age 35 or younger). Organic production is most concentrated in fruits and tree nuts (8%), vegetable, and greenhouses and nursery operations. Designation as organic production under the AAFC Organic Product Regulations requires certification by an International Organization for Standardization, rather than being directly set by AAFC. For international or interprovincial trade, the use of "organically grown" or "organically produced" requires certification. For goods produced and sold in the same province, that province is responsible for certification and regulation. Organic production avoids the use of chemical fertilizers and pesticides, but the enhanced use of manure has the potential for local *Escherichia coli* contamination, and to increase N_2O emissions.

16.1.2 Food Insecurity and Northern Farming

Food insecurity is endemic in Arctic, taiga, and northern boreal communities, marked by high prices, limited selection, and inferior quality of fresh produce (Burnett et al. 2017, Leblanc-Laurendeau 2020, Taylor et al. 2023). Pricing reflects transport and limited demand from small populations, although suspicions of profit-taking are prevalent. Federal subsidy programmes for marketers and northern residents, such as Nutrition North Canada (Government of Canada 2024), have not been as effective as desirable (e.g. Galloway 2017; Galloway and Li 2023). For example, the Government of NL (Newfoundland and Labrador 2023) estimates that the average weekly cost of a standard food basket is almost twice as high in Nain as in St. John's. Nain, at the end of the marine transport route along the Labrador coast, has the highest cost of any community in Canada.

Northern communities are actively promoting local agriculture, including indoors, in an attempt to reduce food insecurity and promote local economic development and self-reliance (e.g. City of Yellowknife 2019). However, energy

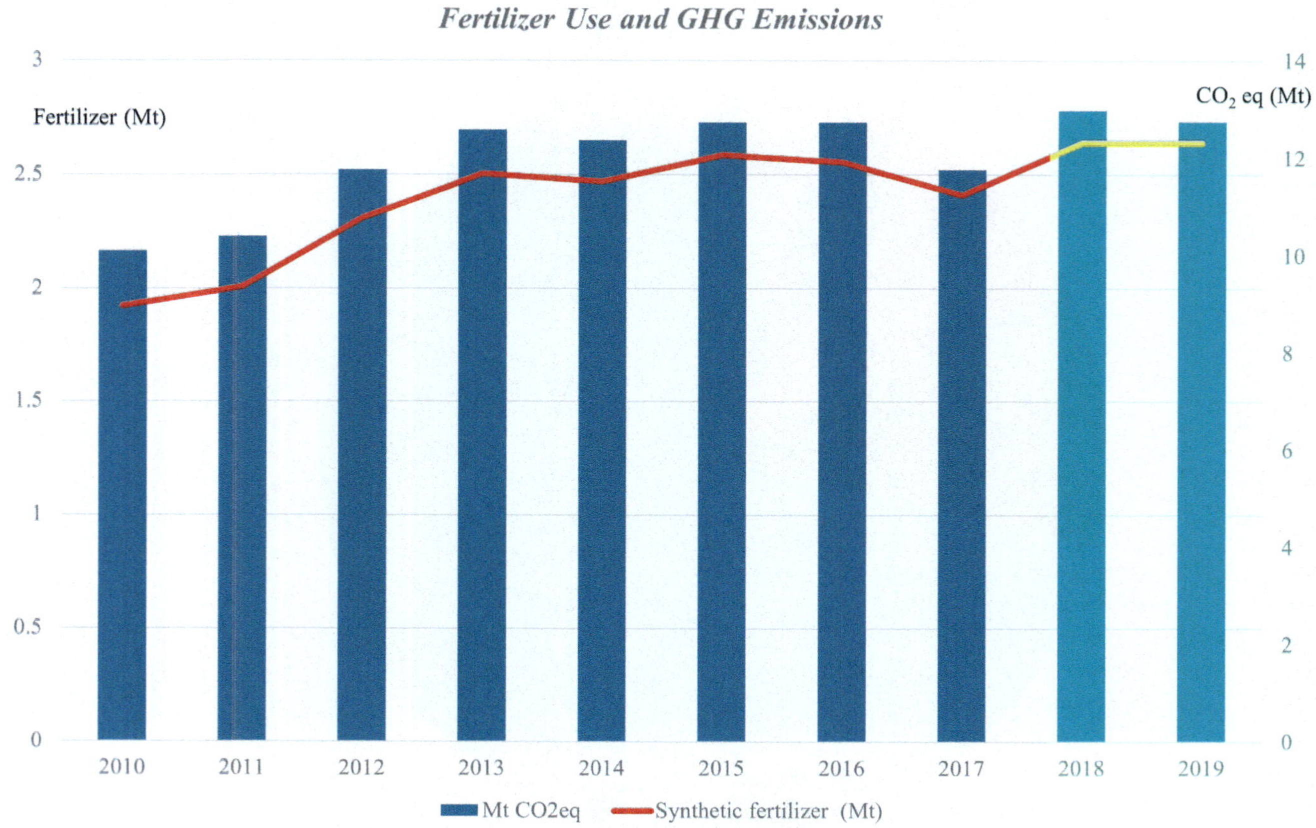

Fig. 16.3 Fertilizer use and GHG emissions, 2010–2019. Application of synthetic nitrogen-based fertilizers is correlative with emissions (Environment and Climate Change Canada 2021)

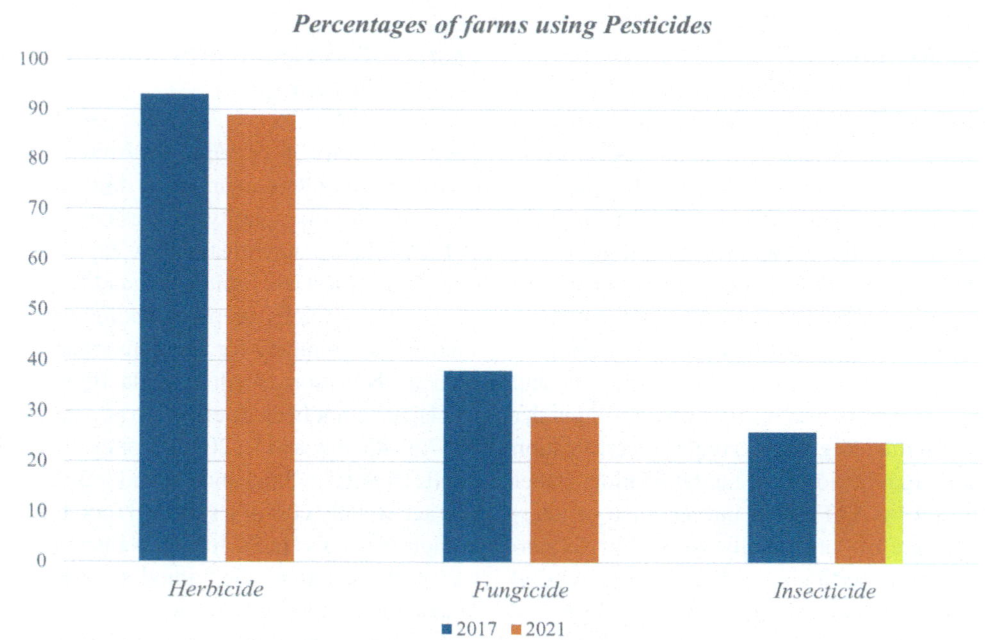

Fig. 16.4 Pesticide use on farms, 2017–2021. Census of Canadian Agriculture

use and taiga and boreal forest clearance have been identified as issues (KC et al. 2021, Seguin et al. 2021, Klöffel et al. 2022). Policies should consider environmental effects as well as local socio-economic concerns.

16.2 Governance

As in many matters, agriculture is a shared federal-provincial responsibility (see Koc et al. 2008). The federal ministries of Agriculture and Agri-Foods Canada (AAFC) and Health Canada are responsible for agricultural research, technology assessment, drought monitoring, financial assistance programmes, international trade, food standards, food safety, packaging and labelling, and advising on nutritional health (including preparation of the Canada Food Guide for consumers; Health Canada 2019a, b, 2021). Supply management of dairy and poultry products is a federal responsibility. Environment and Climate Change Canada is also involved through weather forecasting, climatological research, and programmes to monitor and reduce GHG emissions (Environment and Climate Change Canada 2021). Provinces are responsible for intra-provincial trade and food safety issues, land use regulations, designation of land for agricultural reserves, business taxation, sales tax on food and agricultural goods, public health, postsecondary agricultural education programmes, pesticide use, and most issues involving environmental protection or pollution (although pollution of designated waterways involves Fisheries and Oceans Canada). In many cases, joint federal-provincial initiatives are funded 60%:40%, with the provinces responsible for delivery and monitoring.

Municipalities have responsibilities as designated by the province involved. Municipalities are responsible for zoning, property tax, waste management and disposal, and provision of services (particularly water supply and sewage disposal). Even in rural areas, many farms lie within municipal boundaries. Urban municipalities may conduct food inspections and initiate nutritional health programmes. Through zoning, urban municipalities can determine warehouse, wholesale, and retail food operations, and can use zoning to promote (or discourage) various forms of urban agriculture. Most urban municipalities restrict or prohibit keeping of farm animals within residential areas.

16.3 Farmer Demographics

Compared to Canada's population as a whole, farmers are older (e.g. Vasseur and Catto 2008; Luby et al. 2020; Statistics Canada 2022–23). The median age of farm operators was 58.0 y in 2021, compared to an overall Canadian population median of 41.6 y (Fig. 16.5). More than 60% were age 55 or older, with larger percentages in Atlantic Canada and southern

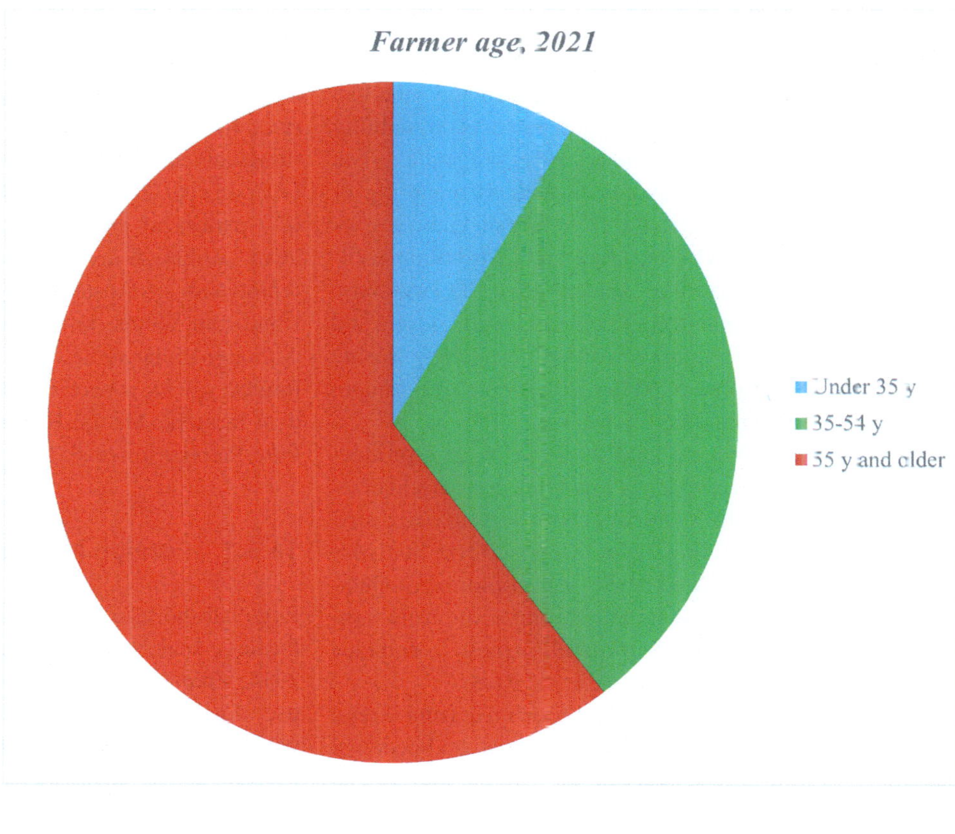

Fig. 16.5 Farmer age, 2021. The median age of farm operators was 58.0 y in 2021 (Census of Canadian Agriculture; Statistics Canada 2022–23)

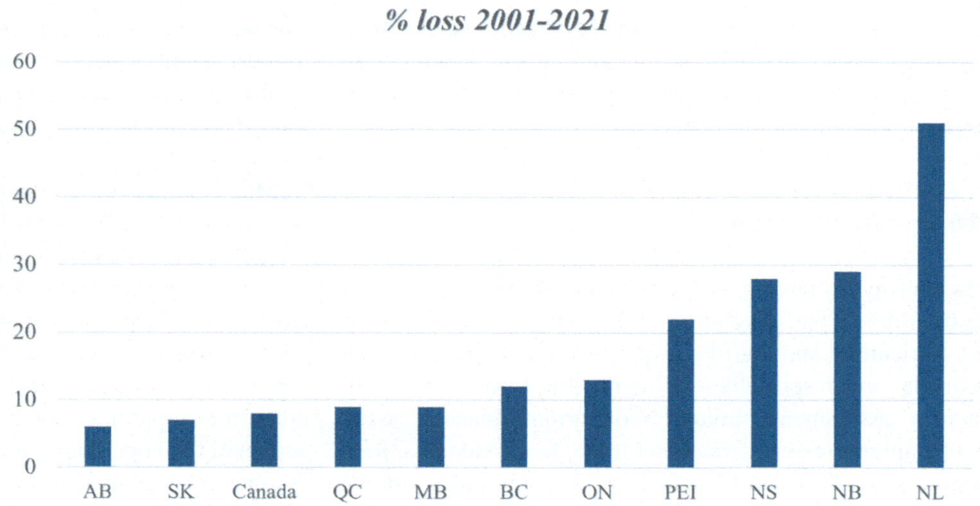

Fig. 16.6 Percentage of farm area lost by province between 2001 and 2021. High rates of loss in Atlantic Canada reflect both abandonment by older retiring farmers and suburban development. Urbanization is the main cause of loss in Ontario and BC. The overall Canadian value reflects the larger hectare totals in Alberta and Saskatchewan. Census of Canadian Agriculture

British Columbia. Young operators (35 or younger) represent less than 9%. The total of 262,455 farm operators continues the declining trend over the past 25 years. Ageing agricultural communities result in loss of potential farmland for production, due to both the reduced activity and sale of land upon retirement (Fig. 16.6). More than 30% of farm operators were women. Only about 12% of farm operators have established a succession plan, with plans more common for grain and oilseed farmers.

Attention has been focused on challenges faced by younger people interested in becoming farm operators, as well as those currently farming (Nasielski et al. 2024). The most significant obstacles are the limited availability and high cost of farmland, along with the difficulty in obtaining credit financing for machinery and livestock. Expanding urban and suburban development increases land costs and restricts availability (Fig. 16.6). In some instances, newer suburban and exurban residents object to farming practices such as manuring or livestock grazing. Land distant from urban centres may also be distant from food processing and distribution facilities. Younger farmers are less able to grow the large crop volumes necessary to obtain contracts from warehousers and supermarket chains. Obtaining workers, especially during peak activity, can also be a challenge. Despite government initiatives, these impediments are growing.

New and young farmers tend to adopt organic production, agro-ecological approaches involving less mechanization and chemical fertilizer use, and community-based distribution (Bialais 2020; Laforge et al 2021a, 2021b; Statistics Canada 2022–23). Compared to the overall farming population, they are more likely to be female and come from non-farming or urban backgrounds, with willingness to embrace non-traditional approaches (Laforge et al. 2018).

Farmers of all commodities are under increasing stress, stemming from financial, operational, societal, and personal challenges. In rural areas, isolation is common and mental health supports may be unavailable. Many Canadian farmers still do not have access to high-speed internet services. Farming, especially that involving livestock, is a 24-h-day proposition involving working from home. Rates of anxiety, stress, and depression substantially in excess of the general population are common (Jones-Bitton et al. 2020). COVID-19 also adversely impacted mental health of farm-dwellers more severely than for the general population, particularly for women and those who had reported previous mental health challenges (Thompson et al. 2022).

Firearms, commonly available tools for wildlife control, unfortunately can be used for suicide. A linkage between adverse weather conditions, particularly drought, stress, decreased mental health, and increased suicide rates among male farmers is a common perception. In Australia, suicide rates for rural men are significantly higher than those for rural women or urban residents (e.g. Arnautovska et al. 2014; O'Brien et al. 2014). However, there is no statistical evidence of a pattern of increasing farming suicides during drought years. Although drought certainly contributes to stress, the remoteness of Australian farm life seems to be the most significant factor, rather than financial stress or weather. Although comparable Canadian statistical studies to date are limited, suicide is perceived as a significant problem for Canadian farmers (Mental Health Commission of Canada 2022; Purc-Stephenson et al. 2023).

16.3.1 Foreign Migrant Farm Workers

Regional shortages of available (or willing) labour have led to the large-scale use of foreign migrant farm workers (Beaulieu 2014; Zhang et al. 2021; Statistics Canada 2022–23). The use of temporary migrant workers allows operators to cope with the short period of high demand during the harvest. Mechanization is impractical in many fruit and vegetable operations which employ most of the workers, in contrast to the large numbers of migrant grain harvesters and cattle-drivers of the early twentieth century. Wages are low (20–25% below comparable non-agricultural occupations) and physical workload is high. Operations in rural areas lack an available domestic workforce, both numerically and due to older demographics.

The two Immigration and Citizenship Canada (federal) programmes involved are the temporary foreign worker programme, which requires prospective employers to determine that no Canadian workers are available, and the International Mobility Program, which encourages hiring from specific countries. Whereas work permits issued through the International Mobility Program are generally flexible with respect to both employer and exact job duties, those issued through the temporary foreign worker programme are employer- and task-specific. These more restrictive work permits in practice give employers more leverage and control over foreign migrant workers, with the potential for abuse (e.g. Landry et al. 2021, Soroka 2022). The maximum duration of employment is 8 months, but is usually substantially less for agricultural workers, commonly only for the harvest period.

From approximately 6% in 2005, foreign migrant workers currently represent about 20% of the employed agricultural workforce, about 70,000 people, with more than 20,000 working in southwestern Ontario (Zhang et al. 2021, Statistics Canada 2023a). Almost half are from Mexico (Soroka 2022), with substantial totals from Jamaica, Guatemala, Thailand, Philippines, and Péru, with other people from Trinidad and Tobago, Barbados, Nicaragua, Vietnam, and USA. Caribbean workers are more likely to be hired through the International Mobility Program, particularly those from Commonwealth-affiliated countries. Under some circumstances, time spent by workers under the International Mobility Program may be counted towards permanent residency and Canadian citizenship (if desired), but this is not available to temporary foreign workers. More than half are employed in food and beverage processing and manufacturing, with the next largest groups in greenhouses, nurseries, and floriculture operations, vegetable and melon farming, and fruit and tree nut farming (Fig. 16.7; Zhang et al. 2021, Statistics Canada 2023a).

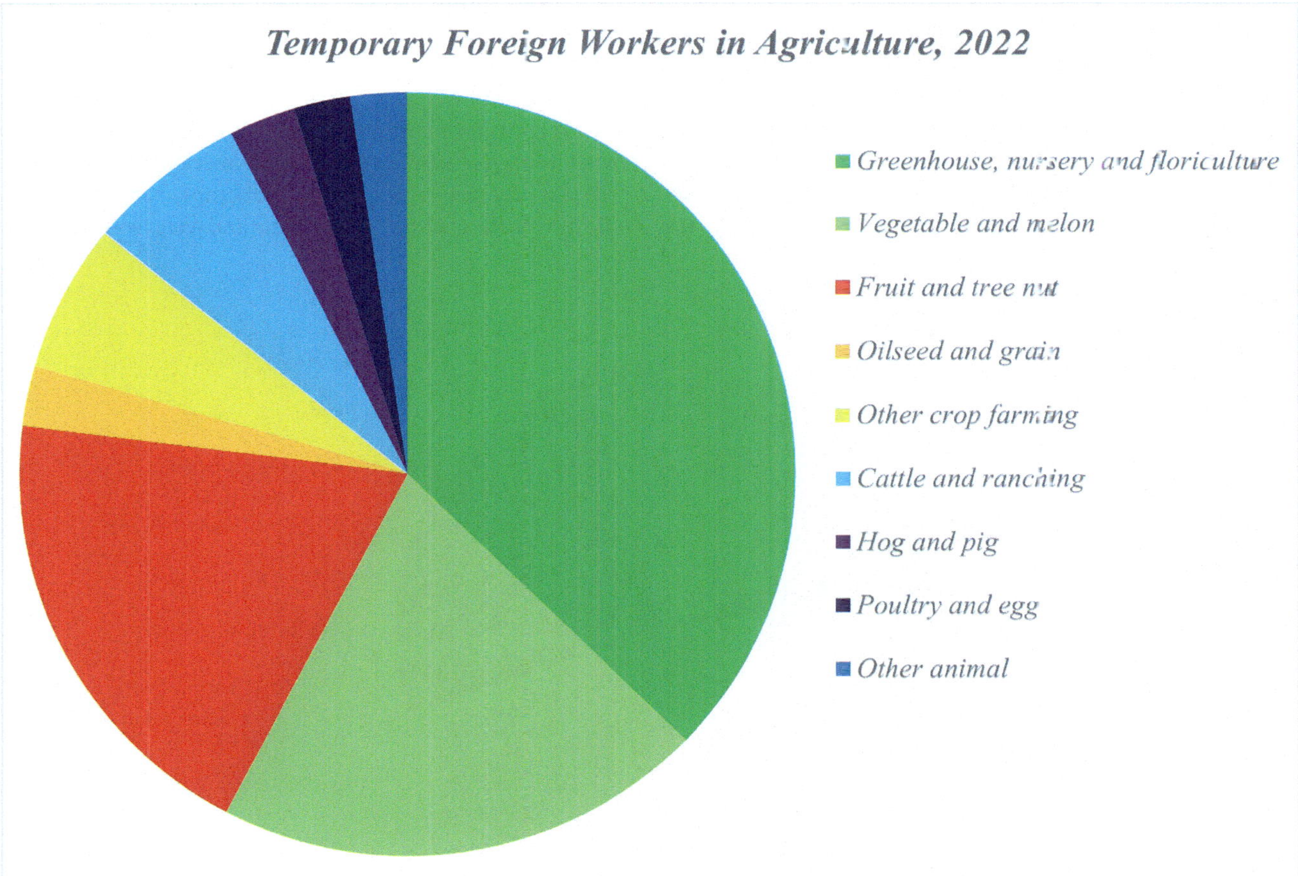

Fig. 16.7 Temporary foreign workers in agriculture, 2022. Workers involved in food and beverage processing and manufacturing are not included here (Statistics Canada 2023a)

16.4 Crop Types

(See Fig. 16.8).

16.4.1 Grains

16.4.1.1 Wheat

Wheat was the agricultural and political motivation for settlement of the Prairies between 1881 and 1914 (Kissinger and Rees 2009). Much effort resulted in development of numerous cultivars of hexaploid spring bread (common) wheat (*Triticum aestivum*), initially to overcome climate constraints and improve yield, and subsequently to cope with rust (*Puccinia sp.*) and wheat stem sawfly (*Cephus cinctus*) (McCallum and DePauw 2008). Cultivars of durum (*T. durum*) are increasingly grown. Approximately 60% of the Great Plains ecoregion of North America, including the Tallgrass and Midgrass Prairies of Canada, has been or is at high probability of being converted from grassland to commercial grain (Olimb and Robinson 2019). The use of spelt (*T. aestivum* var. *spelta*) was always limited on the Prairies, and recent increases in production are associated with smaller-scale and organic farms in several areas.

In recent years, however, wheat production has been surpassed by canola (Figs. 16.9, 16.10 and 16.11). Wheat ranks 2nd in MB and SK (behind canola) and 3rd in AB (behind beef and canola) (Agriculture and Agri-Foods Canada 2020a, 2022a; Statistics Canada 2022–23; Statistics Canada 2022b). This reflects a combination of international competition, climate change resulting in drier conditions in the southern part of the range (Warren and Lulham 2021), profitability, reliability of harvest, and individual producer decisions. Canada is no longer the "breadbasket of the world", with annual wheat production of 22–37 Mt superseded by China, India, Russia, and USA, and in some recent years by Australia, France, Pakistan, and Ukraine (FAO 2022).

Winter planting of bread wheat cultivars (6–10% of the total) is common in Ontario, Québec, and the Maritime Provinces (Statistics Canada 2023b). Soft winter wheat, used for cake and pastry flour, is more common. Hard winter wheat is used to make baguettes, flatbread, and noodles.

Production of durum wheat (5–6 Mt annually) has generally been increasing, showing lesser annual fluctuations than the combined wheat harvest (Statistics Canada 2022–23; Cereals Canada 2024a, b; Fig. 16.11). Durum is used to make pasta and semolina flour. Canadian consumption of pasta has been increasing (~8 kg y^{-1} per capita) at a more rapid rate than wheat bread consumption (~36 kg y^{-1} per capita). Durum exports also have increased, in contrast to the trend for spring (bread) wheat. Durum yields are typically 1.3–2.42 t ha^{-1}, compared to 2.8–4.0 t ha^{-1} for spring wheat (Cereals Canada 2024a, b; Statistics Canada 2022–23). Lower yields reflect more intense farming effort required, as well as losses from pests. Both spring and durum wheat require less fertilizer input per hectare than do canola or corn (maize).

Spring wheat has an average water footprint of 1800–2000 m^3t^{-1} (e.g. Multsch et al. 2016). International trade in wheat spreads its water footprint widely, with more than half of the virtual footprint attributed to North America and Australia.

Water footprints for particular crops vary regionally, reflecting differences in both climate and water sources. Anal-

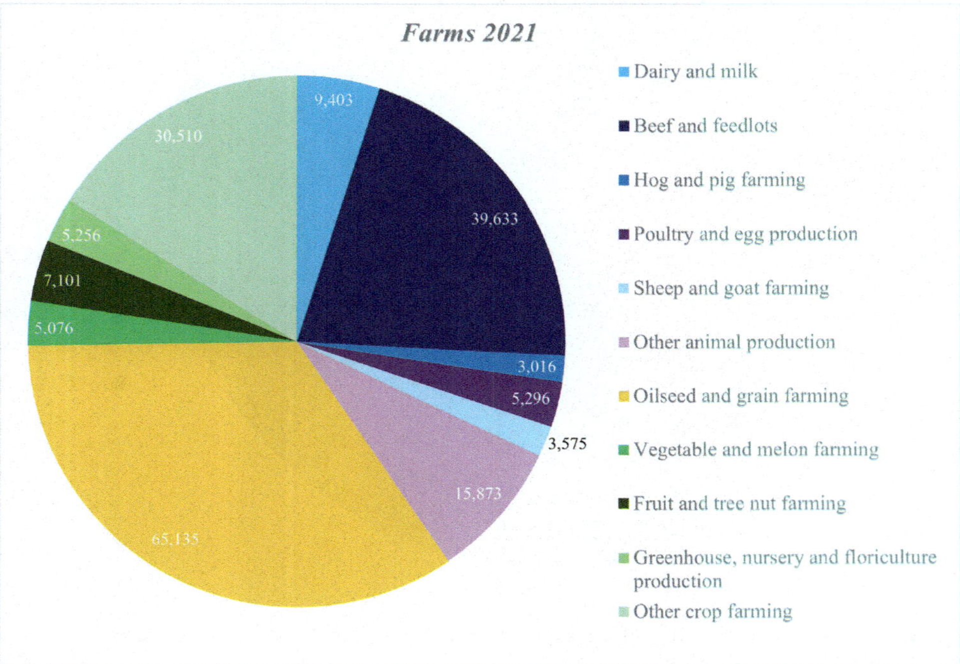

Fig. 16.8 Total numbers of farms in 2021, divided by product. Figures indicate number of farms (Census of Canadian Agriculture; Statistics Canada 2022–23)

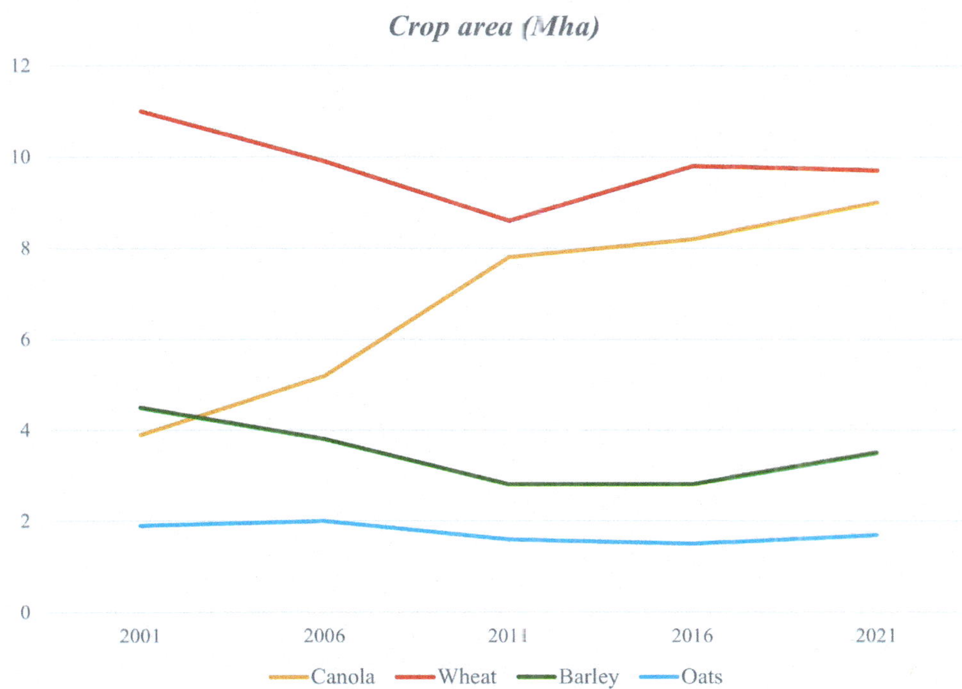

Fig. 16.9 Crop areas (millions of hectares), 2001 through 2021. The increase of canola hectares is matched by increased revenue from canola farming. Data from Census of Agriculture, taken at 5-year intervals (Census of Canadian Agriculture; Statistics Canada 2022–23)

Fig. 16.10 Wheat represents Saskatchewan's heritage, but economically it has been overtaken by canola. Author's photo

ysis of growing season climate indices and extremes over major growing regions for corn, wheat, and soybeans since 1951 indicates that crops have increasingly been exposed to extreme hot temperatures, above which yields decline (Zhu and Troy 2018). Rainfall trends are less consistent, indicating that precipitation by itself may not be an entirely accurate measure indicating potential water scarcity, increased water footprints, or agricultural drought.

The Canadian Grain Commission is the federal agency responsible for regulation of grain-handling, including interprovincial and international transport, quality monitoring and control, and food health (along with Health Canada). The Canadian Wheat Board, a federal agency which formerly controlled all sales and marketing of wheat and barley, was dissolved in stages between 2012 and 2015, largely as a result of farmer-induced political pressure. Since 2012, Canada

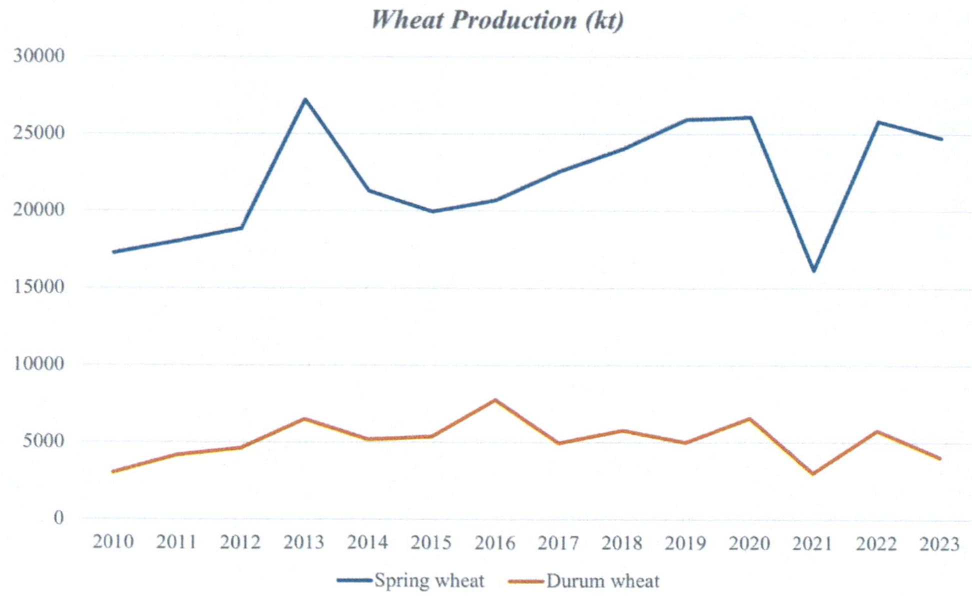

Fig. 16.11 Production of spring (bread) wheat and durum wheat, kt, 2010–2023. Drops in 2021 reflect adverse weather (Census of Canadian Agriculture; Statistics Canada 2022b)

Fig. 16.12 Grain elevators marked with wheat pool insignia, once symbolic of the prairies, have almost completely disappeared along with the Canadian Wheat Board. These elevators at Vulcan, Alberta, photographed by the author in 1995, have been demolished

has not had a supply management system for any grain (Fig. 16.12).

16.4.1.2 Oat

Oat*(Avena sativa)* production has shown sharp variations in recent years, reaching 5.2 Mt in 2022 before falling to 2.6 Mt in 2023 (Fig. 16.13), due to a combination of poor weather and abundant stockpiles (Statistics Canada 2022–23, 2023a). Modal yearly yields vary from 2.3 to 3.7 t ha^{-1}, with increased yields characteristic of wet winters followed by dry summers. Oat production is concentrated in the Midgrass Prairie of Alberta, Saskatchewan, and Manitoba. Oat production generally requires fewer agricultural inputs than does wheat (Menon et al. 2016). Most is used as animal feed, including specialty uses for racehorses. Annual consumption of oatmeal and rolled oats is approximately 60 kg y^{-1}, driven in part by increased demand for oat-based protein with its high nutritional value.

Harvested oat kernels (groats) are cleaned, dehulled, heated to enhance flavour and limit possible rancidity, and then dried for storage and to limit microbial activity. The oats are then steel-cut, milled, or ground to produce oatmeal or for use in breakfast cereal production. Rolled oats are steamed

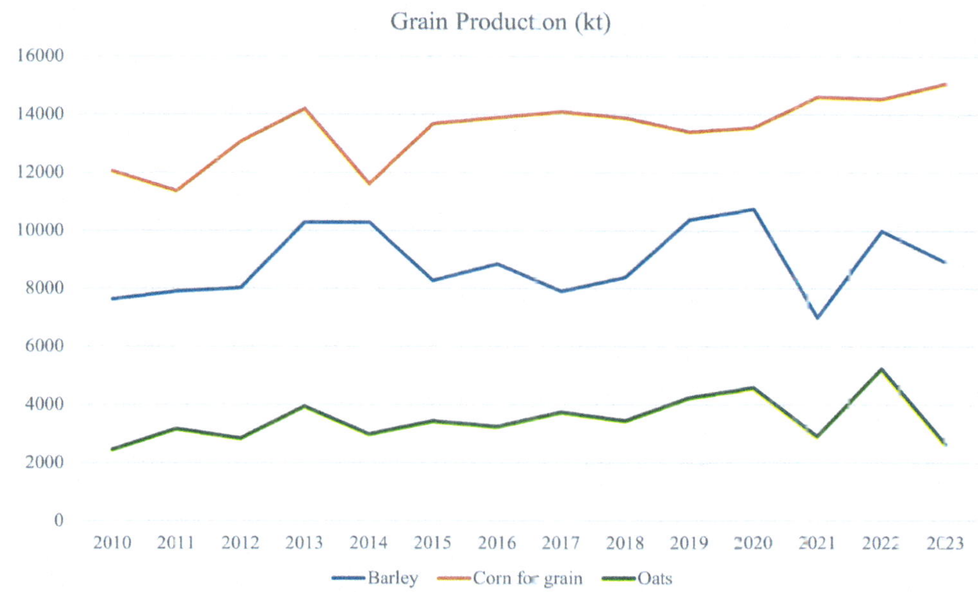

Fig. 16.13 Production of Barley, Corn (for grain), and Oats, 2010–2023. Drops in 2021 reflect adverse weather (Census of Canadian Agriculture; Statistics Canada 2022–23, 2022b, 2023a)

and rolled into flakes, with further rolling producing "quick" and "instant" oats.

Although oats and pure oatmeal are gluten-free, cross-contamination with wheat flour can occur. With at least 1.9% of Canadians practising dietary gluten avoidance (Mudryj et al. 2021), millers and retailers are offering certified gluten-free oat products. Canada is among the largest global producers and exporters of oats.

Because oatmeal is recognized as a healthy alternative to wheat flour (Menon et al. 2016), and with the desire to have gluten-free processing, organic production is of increasing interest, particularly in Ontario and Québec adjacent to larger urban markets. A life cycle assessment of oat production in Québec revealed that organic methods produced greater farm income per area. However, the necessary use of manure (including transport) had greater impacts on human and ecosystem health than conventional oat farming methods (Rodrigues Viana et al. 2022), due to eutrophication and enhanced N_2O emissions. Specific farming practices and regional context condition possible environmental benefits.

16.4.1.3 Barley

Barley (*Hordeum vulgare*) production also shows variations with weather conditions, existing stockpiles, and export markets, varying from 8 to 11 Mt annually (Fig. 16.13). Typical yields vary from 2.3 to 4.0 t ha^{-1}. The Midgrass Prairies of Alberta and Saskatchewan are the main producing areas, with lesser amounts in Manitoba and very little in Ontario, the original source of Canadian barley (Statistics Canada 2022b). Less water is required to produce barley in cooler northern Saskatchewan or Alberta (predominantly from precipitation) than in southern areas (from surface sources), due to reduced water consumption during growth (e.g. Ding et al. 2018).

Barley consumption is 180–200 kg y^{-1} per capita, including feed barley (for cattle and hogs), malting barley (for beer), and hull-less barley (for human consumption as grain). Recent trends show increased planting of feed barley, with decreases in malting barley, although the latter accounts for almost 50% of the hectares seeded (Canadian Grain Commission 2022). Canada exports about a third of annual production.

Beer consumption (occasionally to excess) is a Canadian stereotype which is increasingly losing validity. Beer's market share of alcohol consumption has declined from 50% in 2004 to about 35% today, with wine having an almost equal market share. Per capita consumption varies provincially, with BC the lowest (~60 l y^{-1}) and NL the highest (~90 l y^{-1}) (Beer Canada 2024a, b). Men consume 70% of production. About 8% of beer consumed was brewed in Canada, although this includes foreign beers produced in Canada under licence. Regulation of alcohol sales is a provincial responsibility. Restrictions exist on interprovincial transport of beer for sale or re-sale, although transport for personal use across provincial boundaries has been legal since 2017.

16.4.1.4 Rye

Rye (*Secale cereale*) is commonly planted in autumn, to vernalize for spring growth and seed production. Although rye is grown across Canada, most is produced on clayey Chernozemic soils in Manitoba and Saskatchewan (Statistics Canada 2022–23, 2023a). Yields increase in years with abundant snowfall and spring precipitation, followed by a dry summer, with good years seeing 3.1–3.4 t ha^{-1}. Canadian per capita annual consumption is approximately 10 loaves of rye bread. Although Canadian whiskey is commonly called "rye", most is made from corn.

Small quantities of rice and quinoa are grown locally. Wild rice (*Zizania sp.*, manoomin) is harvested from lakes in Manitoba and Ontario.

16.4.1.5 Corn

Corn (*Zea mays*, maize) is produced for grain, silage, and human consumption of ears and kernels. Grain production reached 15.1 Mt in 2023, the highest level recorded (Fig. 16.13). Yields varied from 380 to 460 bushels ha^{-1} (9.5–11.5 t ha^{-1}), from farms in Ontario and Québec (Mixedwood Plains), and Manitoba (former Tallgrass Prairie) (Statistics Canada 2022b, 2022–23). Grain corn is used for bread and tortillas, ethanol production, and animal feed. Silage production (10–12 Mt y^{-1}) is more widespread, with operations in all provinces, some in conjunction with dairy or beef farming. Sweet corn (170–220 t y^{-1}) is grown primarily in Ontario and Québec. Domestic availability is 5–6 kg y^{-1} per capita, approximately 60 ears.

The water demands of particular crops vary in seasonal sensitivity. Corn and soybeans are most sensitive to midsummer drought. Although corn has a lower water footprint (700–1100 m^3 t^{-1} yield) than wheat or canola (e.g. Multsch et al. 2016), the common use of water for irrigation often means that increased agricultural productivity cannot compensate for the long-term negative impact on water supply. This is a potential concern in southern Manitoba. Corn has a reputation as a water-demanding crop, but the problem is that it is grown in sandy soil and is commonly irrigated, rather than the volume of water used.

16.4.2 Oilseeds

16.4.2.1 Canola

Oilseeds grown in Canada are dominated by canola (Statistics Canada 2022b), produced in 1975 from cultivars of *Brassica napus* and *B. rapa* cross-bred to have lower concentrations of erucic acid (<2%). Canola was developed to replace rapeseed oil (from *B. rapa*), which had a bitter taste and high chlorophyll content giving it a green hue, in addition to being high in eutric acid (Bell 1982). The term "canola", an amalgam of "*Canadian oil low acid*", proved to be more palatable to the market than "rapeseed oil" was. Canola is conspicuous for its yellow flowers. Black seeds contain up to 45% oil. Processing involves extracting the seeds, heating and gentle milling (flaking), reheating, pressing, extraction of the oil, removal of lipids, gums, free fatty acids, and fine meal particles, filtration through clay to remove pigments, and steam distillation. Cold pressing, a less common alternative processing method, produces darker and stronger-tasting oil. Canola oil is the most popular for cooking and baking in Canada, due in part to an absence of trans-fats and lower saturated fat contents than corn, grapeseed, olive, palm, soybean, or sunflower oils. In addition to its food value, canola can be used for biodiesel production or blended with diesel.

Canola has become the dominant crop in the midgrass prairie and prairie parkland belt stretching from Winnipeg to Grande Prairie (AB) (Figs. 16.14 and 16.15). Smaller, less productive areas include the wettest parts of the Shortgrass Prairie, the dry plains west of Prince George (BC), and clay-rich plains in southern Ontario (Statistics Canada 2022b).

Fig. 16.14 Canola near Yorkton, SK. Author's photo

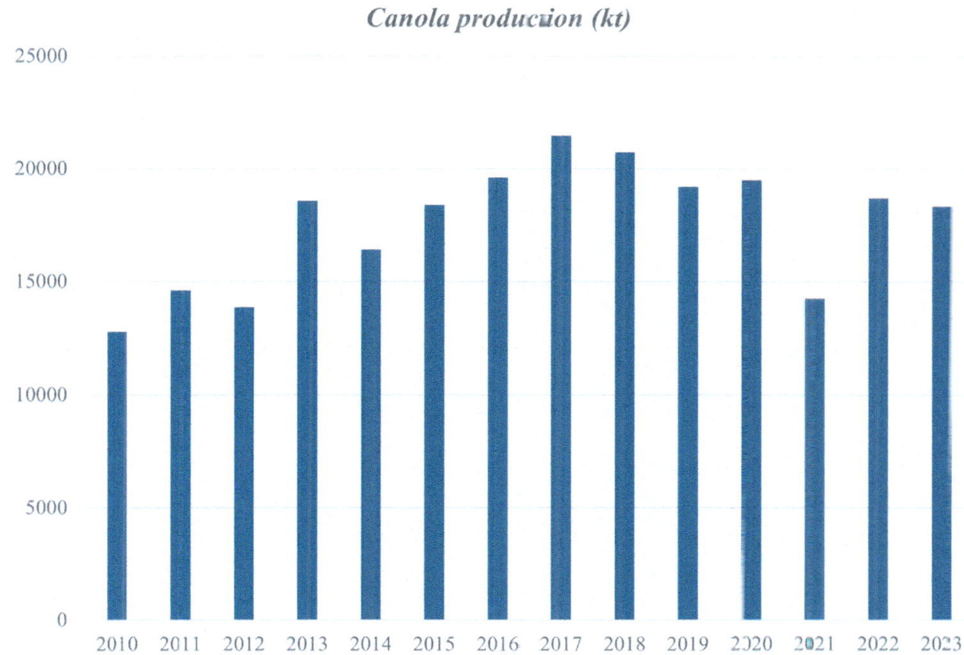

Fig. 16.15 Canola production (kt), 2010–2023. Drop in 2021 reflects adverse weather (Census of Canadian Agriculture; Statistics Canada 2022–23, 2022b)

Canola is the top agricultural commodity by dollar value in Saskatchewan and Manitoba and is second to beef in Alberta. Canada produces more than 20Mt of canola annually, with typical yields of 85–100 bushels ha^{-1} (2.0–2.3 t ha^{-1}). Yields are dependent on total harvested area and losses due to drought and insects (see White et al. 2019; Secchi et al. 2023).

Canola seeds genetically modified to resist the herbicide glyphosate, the most commonly used in Canada, were first introduced in 1995. More than 90% of canola grown today in Canada has some degree of genetic modification. Although initially controversial, genetically modified canola has been accepted by the majority of Canadian consumers. Differences between professed and actual behaviours are evident: although consumers say they mistrust GM foods, price and Canadian origin are the determining factors when purchasing (Sutherland et al. 2020; Macall et al. 2021; Yang and Arhonditsis 2022).

16.4.2.2 Other Oilseeds

Soybeans account for about 25% of total oilseed production, with 60% consumed domestically (Fig. 16.16). Yields are typically 100–110 bushels ha^{-1} (3 t ha^{-1}). Production is concentrated in Ontario, Manitoba, and Québec (Statistics Canada 2022b), primarily on loamy, stone-free alkaline luvisols. Although nitrogen fixation is critical in the growing process, root nodule N-fixing bacteria are sufficient, and the addition of fertilizer does not result in improved yields. By dollar value, soybeans are currently the third largest field crop in Canada (Statistics Canada 2022–23). About two-thirds of production is exported, primarily to Asian markets. Research has focused on breeding soybeans with desirable qualities for miso, sauce, and tofu.

Other oilseeds grown commercially include mustard (~160–170 kt y^{-1}, rapidly increasing since 2016, but substantially less than the 319 kt of 1994; Fig. 16.17), flax (240–680 kt y^{-1}), and sunflowers (80 Mt y^{-1}). Saskatchewan is the world's largest exporter of mustard seed, including yellow *(Sinapis alba)* and brown (*Brassica juncea*). Yields vary with weather conditions, ranging from 350 kg ha^{-1} to 1210 kg ha^{-1}, largely accounting for the variations in production. A global shortage of mustard has influenced prices. The large variation in flax (*Linum usitatissimum*) production for linseed oil reflects hot and dry weather during low yield years. Former flax producers have shifted to canola or pulses, which require less effort and produce higher yields. Most Canadian flax production is also concentrated in Saskatchewan.

Sunflowers are grown for snack food or use in baking (~60%), for oilseed (~40%), and as birdseed. Growing is concentrated in southern Manitoba, in the original Tallgrass Prairie. Sunflowers in Canada have water footprints ranging from 500 to 600 m^3t^{-1} yield (e.g. Ding et al. 2018).

16.4.3 Vegetables

Outdoor field vegetable production (not including potatoes) is concentrated in Ontario, Québec, and British Columbia. Field production occupies ~100 Mha and produces 2.0–2.5 Mt annually (Agriculture and Agri-Food Canada 2022b). Production is dominated by tomatoes, carrots, and onions

Fig. 16.16 Lentil and soybean production (kt), 2010–2023 (Census of Canadian Agriculture; Statistics Canada 2022–23)

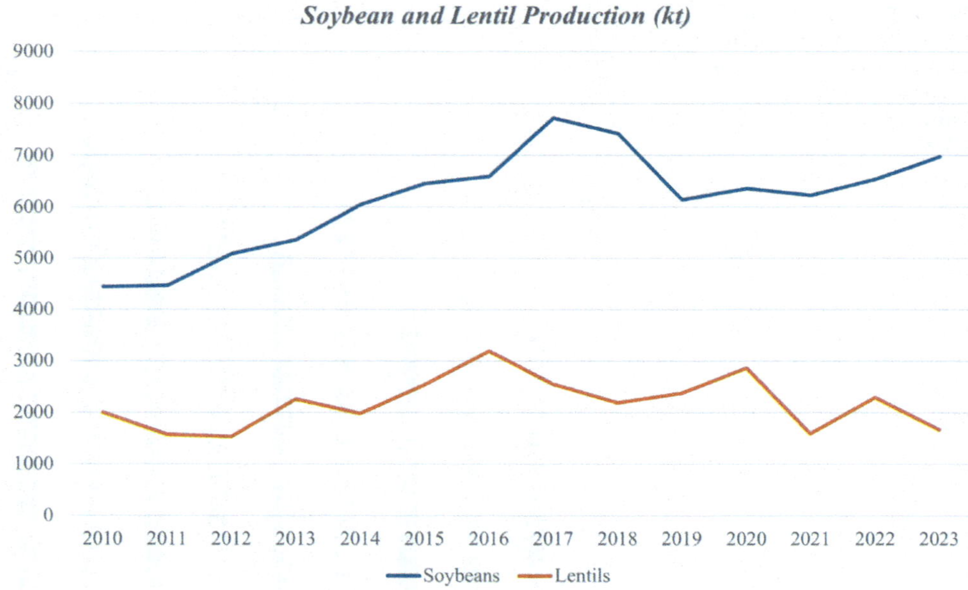

Fig. 16.17 Mustard and sunflower seed production (kt), 2010–2023 (Census of Canadian Agriculture; Statistics Canada 2022–23)

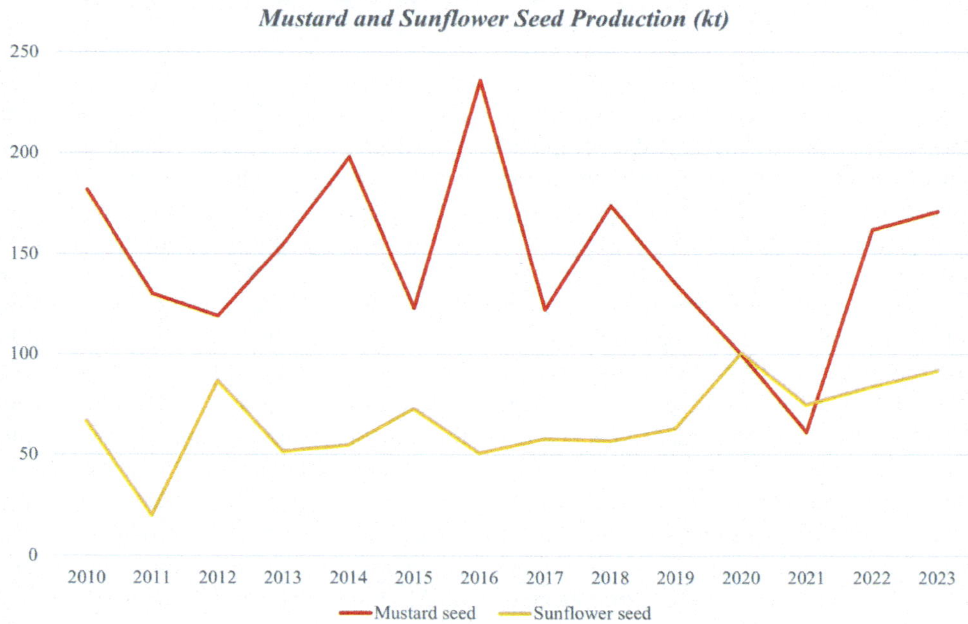

(Fig. 16.18). Other economically significant vegetables include cabbage, sweet corn, broccoli, peppers, lettuce, cauliflower, and asparagus. Climate makes vegetable and fruit importation a seasonal reality (e.g. Figure 16.19), with USA and Mexico being the primary sources (Dyer et al. 2011; Statistics Canada 2022–23).

Tomatoes only became a common part of Canadian diets in the 1870s (Munro and Small 1997). Today, they are second only to potatoes in global consumption. Canadian per capita consumption is ~20 kg y^{-1}, 60% from domestic sources. Sandy luvisols, such as the former glacial lake beaches of southwestern Ontario, are ideal growing sites. Tomatoes prefer daytime temperatures of 21–28 °C and night times ≥15 °C, characteristic of Dfa and Cfa climates. Most tomatoes are harvested by hand, and employment of temporary farm workers is common.

Carrots grow across Canada from mid-boreal regions south, with the exception of the dry Shortgrass Prairie (Munro and Small 1997). Optimum mean temperature during the growing season is 16–21 °C. They tolerate a wide variety of soil types, but prefer sandy or loamy soils. Including both fresh and packaged carrots, per capita annual consumption is ~10 kg.

Onion production of ~10 Mt y^{-1} is centred in southern Ontario (Agriculture and Agri-Food Canada 2022b). Organic-rich soils are preferred, and fertilization is usually necessary

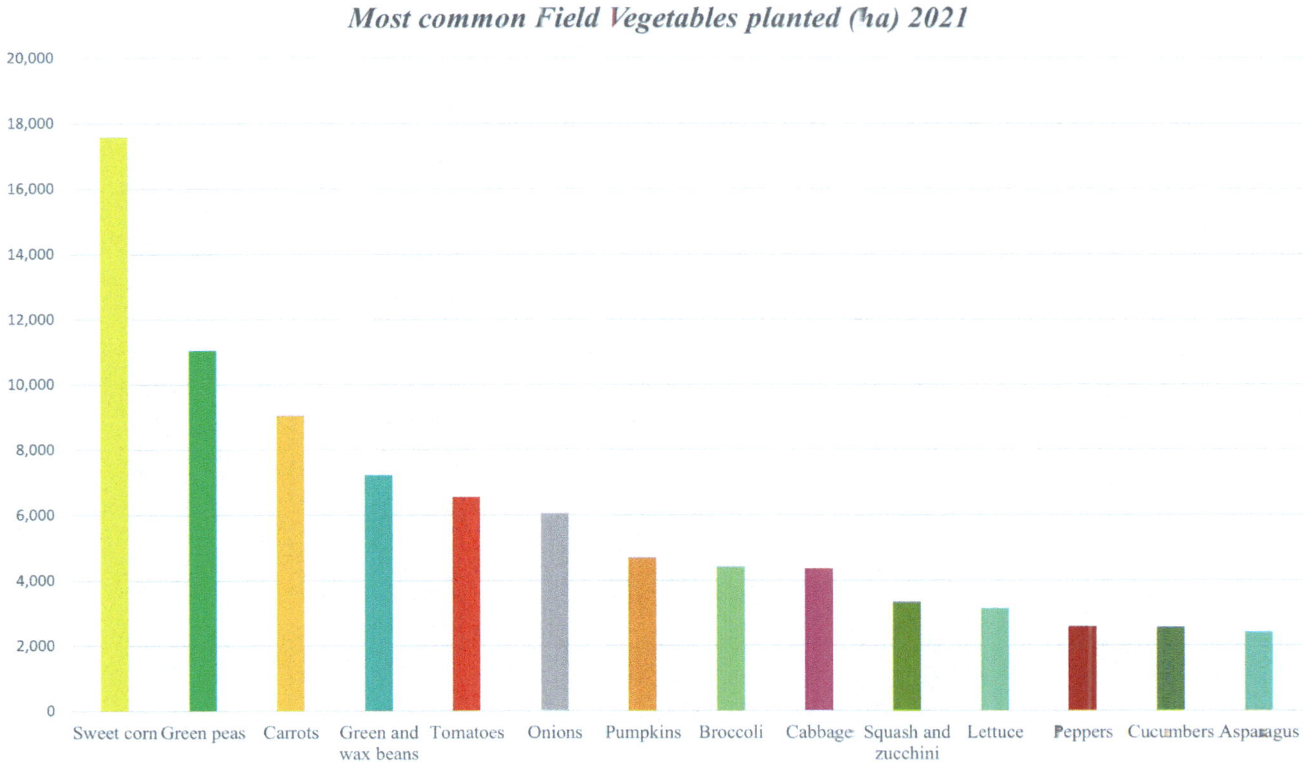

Fig. 16.18 Most common field vegetables planted by hectares, 2021. Field vegetables are planted outdoors, rather than in greenhouses (Agriculture and Agri-Food Canada 2022b)

for commercial production on other soil types. Commercial production usually involves transplanting, due to the short growing seasons. Onion annual consumption is currently ~ 11 kg y^{-1}, among the fastest rates of increase for vegetables in Canada (Helgi Library 2024a).

Pulses include beans, chickpeas, lentils (Fig. 16.16), and dry peas. Beans are grown in the St. Lawrence Lowlands, while chickpeas, lentils, and dry peas are grown on Dark Brown and Black Chernozems in the Midgrass and Shortgrass Prairies, primarily on large farms (Statistics Canada 2022–23). Dry peas generally occupy more northerly regions, extending north to Peace River (AB). All are sensitive to summer drought. Pulses are commonly grown as a part of a crop rotation with canola and wheat or oats. Canada is among the largest global producers of dry peas (3–4 Mt y^{-1}) and lentils (1.5–2.5 Mt y^{-1}). Fresh peas, including garden, snow, and sugar snap, are grown in southern Ontario and Québec.

16.4.4 Potatoes

Potatoes are the fifth largest primary crop, with production of 5–6.5 Mt y^{-1}. Production is increasingly split among provinces (Fig. 16.20), with Alberta, PEI, and Manitoba each holding slightly more than 20%, and New Brunswick and Québec 10–15% each (Statistics Canada 2022–23, 2022b). However, their socio-economic importance varies widely: 42% of the total farm cash receipts in Prince Edward Island (Figs. 16.21 and 16.22), and 18% in New Brunswick come from potatoes. O'Leary (PEI) and Florenceville-Bristol (NB) are the homes of the Canadian Potato Museum and a 4.3 m high potato statue, and the Potato World Museum and Hall of Recognition, respectively.

Total production is divided among the processing sector (65%), the fresh (table) sector with whole potatoes for market (20–25%), and the seed sector (10–15%) necessary for the production of subsequent generations. Approximately 400 varieties have been catalogued, and 150 are registered in Canada for seed commercial production. Prince Edward Island, Alberta, Quebec, Manitoba, and New Brunswick all produce seed potatoes commercially.

The approximately 20 potato varieties available to consumers are grouped by skin colour and best usage into four main categories (Barrett et al. 2023). White potatoes have white or light brown skin and white flesh and are used for boiling, frying, mashing, salads, and potato chips (crisps). From being the most popular 50 years ago, white potatoes now represent about 20% of the consumer market. Yellow potatoes (yellow-tan skin and flesh; 10% of market) are increasingly preferred for baking and mashing. Red potatoes (red

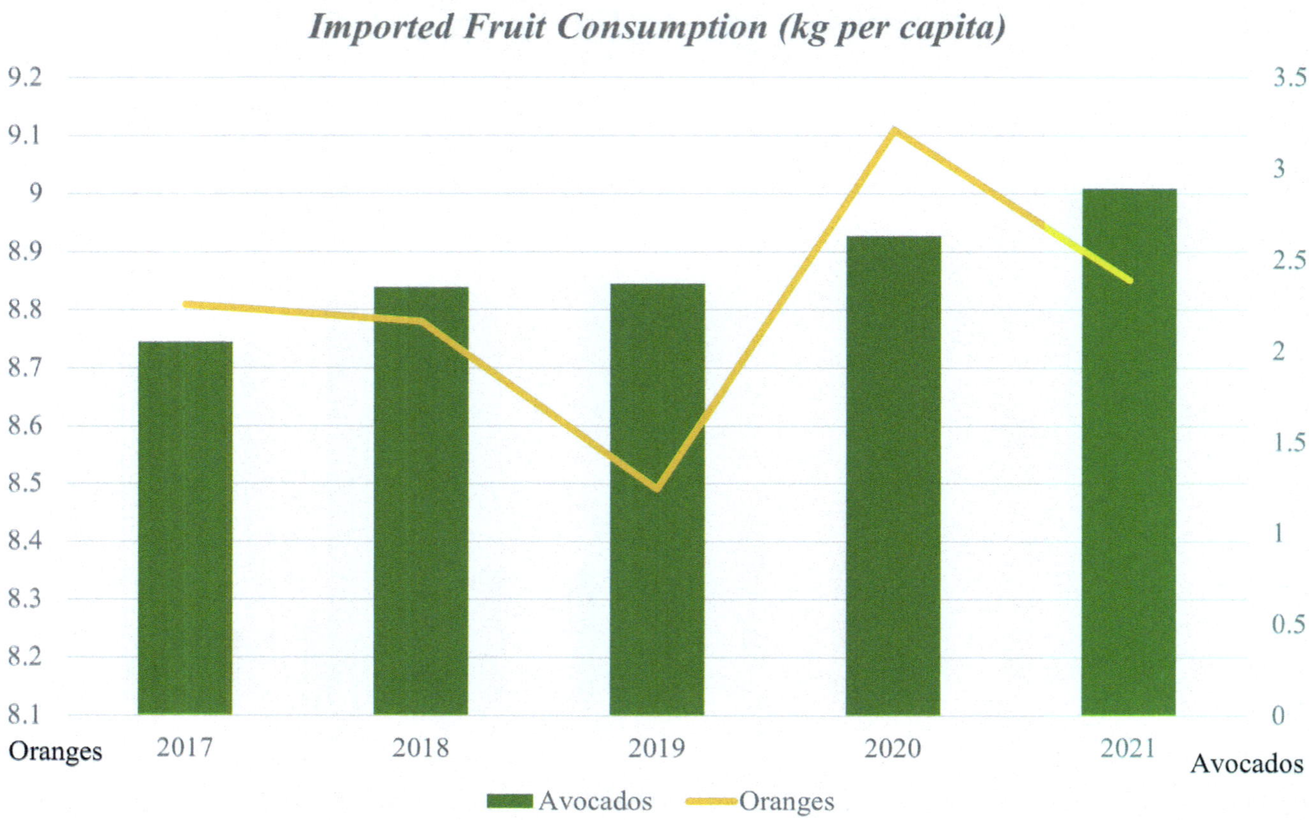

Fig. 16.19 Two examples of imported fruit consumption in Canada, 2017–2021. Avocado consumption reflects a steady rise from almost none in 2001 (Statistics Canada 2022–23)

Fig. 16.20 Potato production (kt) for the five provinces with the largest potato production (Census of Agriculture; Statistics Canada 2022–23, 2022b)

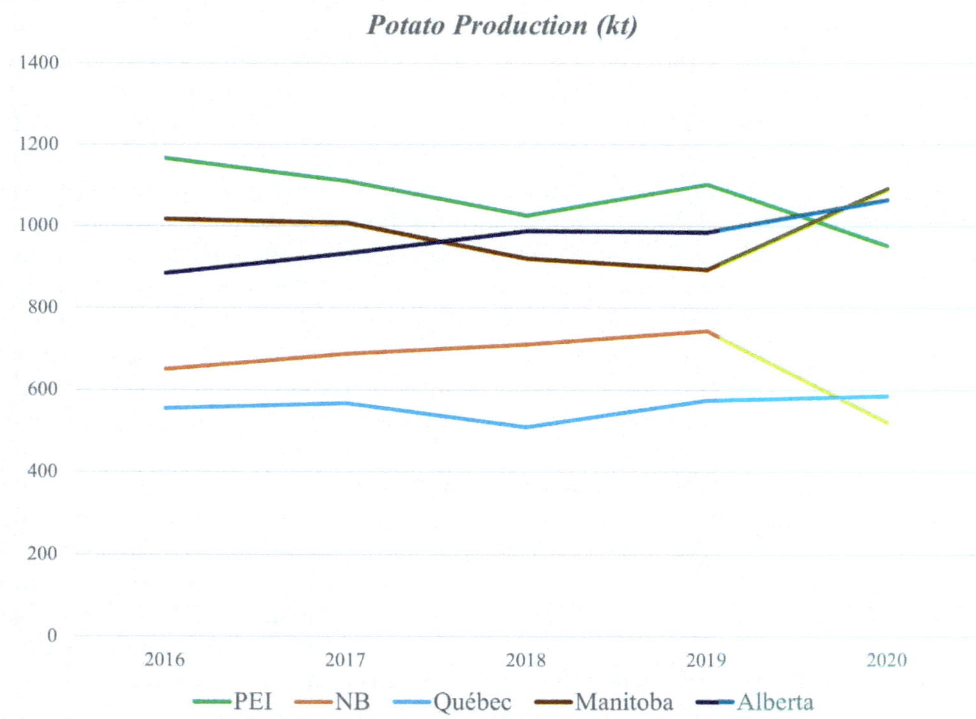

Fig. 16.21 Potatoes in flower, near Summerside PEI. Author's photo

Fig. 16.22 Potato fields in fallow (foreground) and planting (background), Miscouche PEI. Author's photo

skin, white flesh; 5% of market) are used for roasting and salads. Blue potatoes (blue-purple skin, white flesh) are only available locally, usually directly from farmers.

Russet potatoes (brown skin, white flesh) represent two-thirds of the consumer market. Russets have high starch and low moisture contents, making them superior for baking and commercial French fry manufacture. Their increasing dominance in the market reflects consumer choice of baked and French fries as preferred consumption styles. Consumer preference for consistency and industry preference for large-scale automated production favours russets, with consistent size and ease of processing. Potatoes are sorted, peeled by steaming, and blanched in cold water. Then, they are either machine-cut and cooked or cooked, mashed, and forced through sieves to produce consistent, thin, elongate fries, or placed in moulds to create different shapes.

Processing potatoes are used to produce French fries, chips, dehydrated flakes, frozen products, and starch and flour (Fig. 16.23). French fry processing plants are present in Alberta, Manitoba, New Brunswick, and Prince Edward Island. Ontario has the largest potato chip manufacturing capacity.

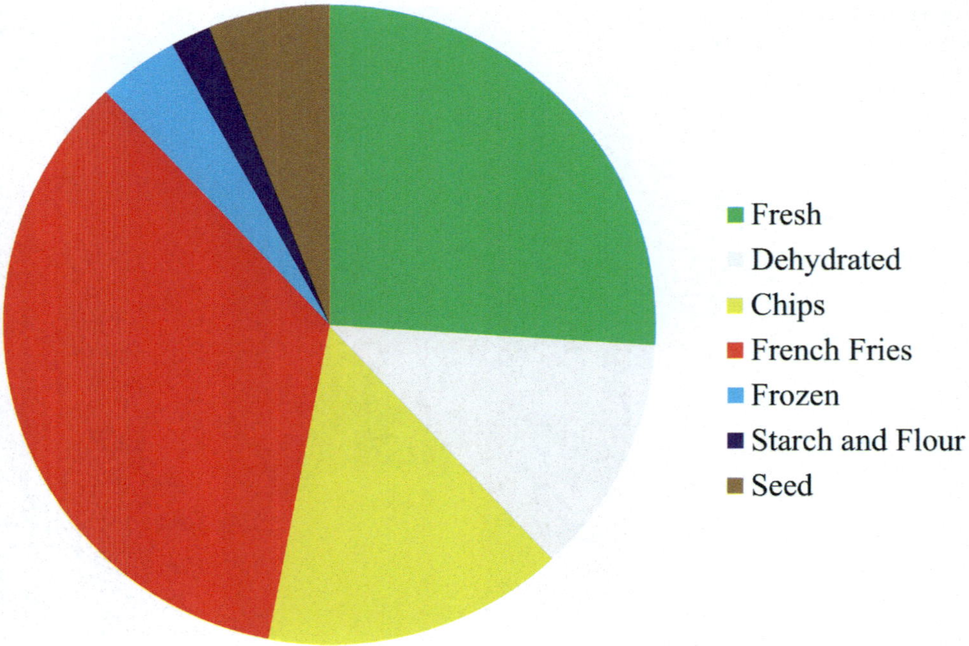

Fig. 16.23 Potato usage by product, Canada, 2008–2020. Seed potatoes are reserved for planting for next year's harvest. Data from various sources 2008–2020, including (Statistics Canada 2022–23, 2022b; Barrett et al. 2023; Williams and Marshall Strategy 2021; and others; compiled by the author)

Potato consumption is decreasing in most economically developed countries, but Canada is an exception, holding at ~60 kg y^{-1} per capita since 2011 (Helgi Library 2024b). Consumption includes an estimated 6.3 kg y^{-1} of French fries, approximately 1 medium order per week per person. Potato chip consumption per capita is about 3.7 kg y^{-1}, with approximately 85% of the population consuming some each year (Williams and Marshall Strategy 2021). Flavoured potato chips including ketchup, salt-and-vinegar, sour cream-and-onion, barbeque, and all-dressed (with all of the above flavours) are popular, and bacon-, lobster-, and maple-flavoured chips are locally available.

Poutine, invented at some time in the 1950s somewhere in Centre-du-Québec (there are many competing claims), has become popular throughout Canada, reaching stereotype status. Classic poutine consists of moderately thick, crispy outside, and soft inside fries, to which fresh (max. 1 day old) cheese curds and hot beef or chicken stick gravy are added immediately before serving.

16.4.5 Fruits

Soft fruits grown in Canada include cherries, peaches, pears, plums, nectarines, and apricots (Agriculture and Agri-Foods Canada 2023d). Frost-free durations in southern Ontario, the Okanagan Valley, Fraser Lowland, and southern Vancouver Island are sufficient for soft fruit maturation. Together, all soft fruits represent only 6% of the total hectares utilized for fruit production.

Fig. 16.24 Sunrise apples, among the most common types grown in Nova Scotia. Author's photo

Apples are the second most harvested fruit by hectares, and the largest by mass (350–400 kt y–1, 40% of total fruit production by mass; Agriculture and Agri-Foods Canada 2023d). Apple production in BC (e.g. Ambrosia, Gala, Granny Smith, Honeycrisp, Red Delicious) is severely affected by summer heat (notably in 2021). In contrast, Ontario apple production (e.g. Cortland, Empire, Gala, Macintosh, Spartan) has been aided in recent years by warmer summers and autumns. Cold springs and dry summers are unsuitable for Nova Scotia apples (e.g. Cortland, Gravenstein, Northern Spy, Spartan, Sunrise; Fig. 16.24).

In NS, a life cycle assessment (LCA) revealed that fertilizer (including manure) and diesel fuel use were major

Fig. 16.25 Wild lowbush blueberries, *Vaccinium angustifolium*. Author's photo

contributors to environmental impacts from both conventional and organic apple orchards, along with electricity required for refrigeration during storage (Keyes et al. 2015). Consuming locally produced apples and marine transport are more favourable than trucking.

Grapes (including both table (Labrusca) and wine grapes) show recent variations in production between 90 kt y^{-1} and 120 kt y^{-1}. Grapes are produced primarily in the Niagara region (ON) and the Okanagan valley (BC). Winter vine kill is a problem for viticulture and table grape operations in both regions.

Lowbush (wild) blueberries (Fig. 16.25), highbush (farmed) blueberries, and cranberries are all major fruit crops, produced in Maritime Canada and Québec. Recent cranberry production has varied from 155 to 210 kt y^{-1}. Highbush blueberry farms encompass 11–12 kha and 70–90 kt y^{-1} production. Lowbush blueberry production, more susceptible to variations in weather, insect activity, and harvesting effort, is 71–100 kt y^{-1}. Exports to USA account for more than 90% of highbush production, but less than half of lowbush production (Statistics Canada 2022–23).

Lowbush blueberries are pollinated by bumblebees and other native insects, but not by black flies (Hunter et al. 2000; Cutler et al. 2012; Moisan-Deserres et al. 2014). Native insect species are responsible for less than 20% of pollination, requiring the seasonal importation of "managed" pollinators, honeybees. Lowbush blueberries are concentrated along field margins. About 45% of the blueberry plants produce no pollen in a season. As blueberry pollen is relatively non-nutritious, honeybees relying on this "junk food pollen" may suffer malnourishment. Losses of 10% are typical for bee colonies seasonally imported from Ontario.

Other berries produced include strawberries (24–28 kt y^{-1}), raspberries (5–10 kt y^{-1}), and blackberries (Agriculture and Agri-Foods Canada 2023d). Lingonberries (partridgeberries), cloudberries (bakeapples), and bilberries are harvested from wild plants.

16.4.6 Tree Nuts

Twenty edible nut species are native to Canada. Black walnuts (*Juglans nigra*, southern Ontario), hazelnuts (*Corylus sp.*, filberts; southern Vancouver Island, Fraser Lowland, Okanagan (BC)), shagbark hickory nuts (*Carya ovata*, upper St. Lawrence Valley and Lake Ontario shoreline), and acorns from white oak (*Quercus alba*) are produced commercially. Acorns contain tannin, which causes a bitter taste that must be removed by grinding them into meal.

16.4.7 Cannabis

Since 2018, the production and sale of cannabis for recreational use have been legal in Canada. Cannabis is grown both indoors (1.9 Mm^2) and in open fields (685 ha). The economic impact on GDP between 2018 and 2021 is estimated at $43.8 billion (Deloitte 2022). Cannabis production was the largest agricultural industry by dollar value in New Brunswick in 2021 (Statistics Canada 2022–23). Regulation of cannabis is shared by federal, provincial, and territorial governments. Provinces and territories are responsible for regulating sales, distribution, further restrictions for personal (home) cultivation, restricting public use, and increasing the minimum legal age (18–21 in various jurisdictions).

16.4.8 Greenhouse Production

Vegetables are the major food type grown in commercial greenhouse operations (Agriculture and Agri-Foods Canada 2023c). More than 85% of production is destined for export to USA, and the operations are concentrated in proximity to the USA border in southern BC, Ontario (Dias et al. 2017), and Québec. Tomatoes and cucumbers each account for about one-third of sales, followed by bell peppers, strawberries, lettuce, eggplants, green beans, and herbs (Statistics Canada 2022–23; Agriculture and Agri-Foods Canada 2023c; Fig. 16.26).

Mushrooms are grown in composted agricultural by-products in closed, dark climate-controlled facilities, frequently with limited headroom. Harvesting mushrooms, which must be done by hand, is a labour-intensive, physically demanding process. Annual production is 130–140 kt, with about 55% in Ontario and 40% in BC (Agriculture and Agri-Foods Canada 2023c; Fig. 16.27). Proximity to the USA border is a dominant factor in mushroom operations. White,

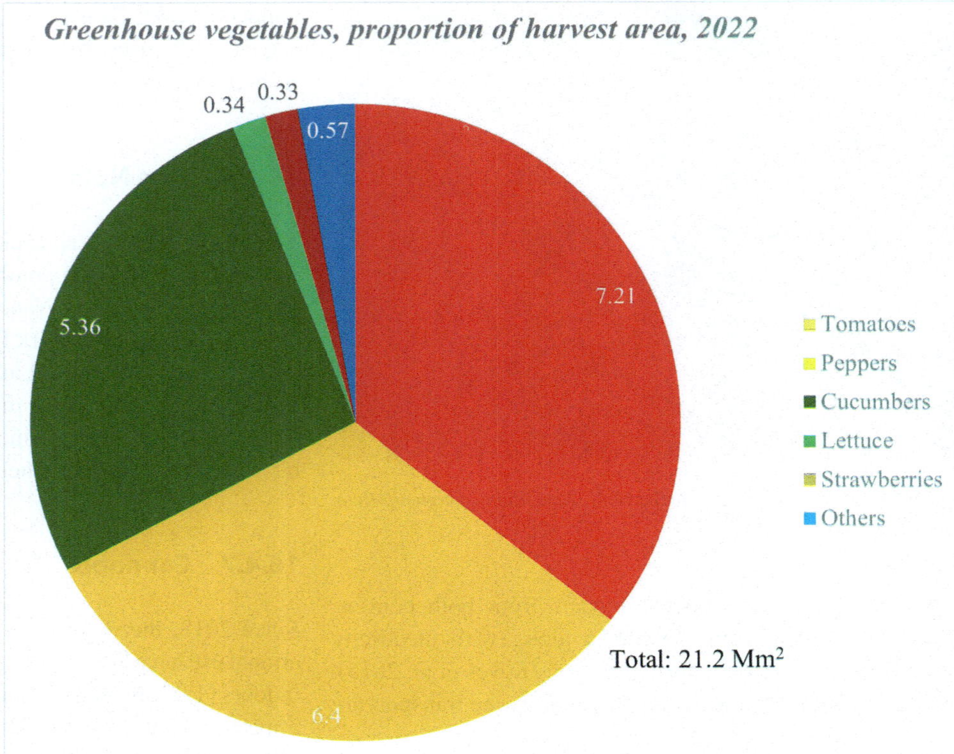

Fig. 16.26 Greenhouse vegetables, proportion of harvest area, 2022. Figures indicate millions of m^2. Compare with Fig. 16.18 (Agriculture and Agri-Foods Canada 2023c)

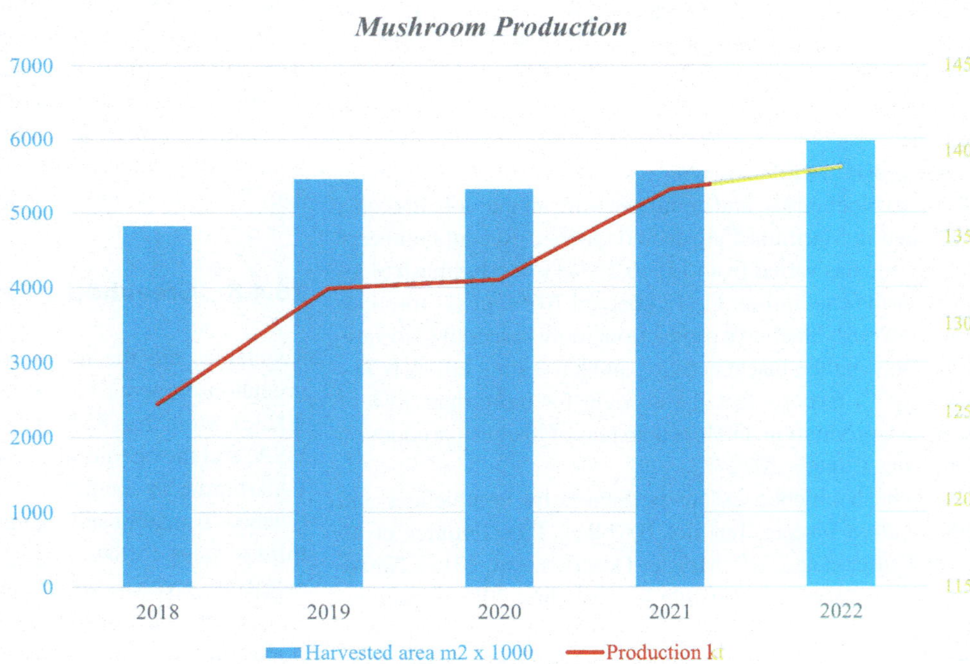

Fig. 16.27 Mushroom production, showing harvested area in 1000 s of m^2 (blue) and production in kt (red) (Agriculture and Agri-Foods Canada 2023c)

portobello, and shiitake mushrooms are the most commonly grown. Foraging is legal on Crown Land, although not in provincial or national parks.

16.4.9 Nurseries and Floriculture

Nursery production includes tree farming, ornamental plants and shrubs, Christmas tree growing, and turf (landscape sod) farming. Nurseries, including greenhouses devoted to floriculture, are concentrated in Ontario, British Columbia, Québec, and PEI. Floriculture represents about two-thirds of total sales of ~ $2.2 billion annually (Agriculture and Agri-Foods Canada 2020b).

Nurseries are the primary source of trees for urban areas. Trees planted on private lands are generally chosen by the residents. Cost, species availability, available space, growth rate, local climate (wind, snow cover, freezing rain), personal taste, neighbours' preferences all factor into these cultural decisions. Municipal and other initiatives to encourage planting of some species (typically native) and discourage others (typically non-native or invasive, such as Norway Maple (*Acer platanoides*) typically have limited or no effect on residents' decisions. As the choice reflects at least some elements of personal culture and values, residential trees are seldom planted specifically to reduce urban heat island effects (Almas and Conway 2016).

Augmentation of the urban forest is a priority in many cities (McGovern and Pasher 2016; Cheng et al 2021). The effects may not be uniform: affluent residents are more likely to plant private trees, and the resulting increases in real estate value, combined with gentrification, may accentuate disparities. Affluent neighbourhoods are also more likely to be aware of municipal or non-profit initiatives for urban greening. Lot size is also a factor: larger areas, whether public or private, tend to have more and healthier trees.

More than 1300 farms grow Christmas (or Holiday) trees over 2 kHa. Farms are spread across Canada, concentrated in Ontario, BC, Québec, and Nova Scotia (Agriculture and Agri-Foods Canada 2020b). More than 2 million are exported to USA annually. Preferred species vary by region (as well as personal taste), with balsam fir (*Abies balsamea*), Scots pine (*Pinus sylvestris*), and Fraser fir (*A. fraseri*) preferred in Atlantic Canada and Québec, western Canada, and southern Ontario and for export, respectively. As the latter two species are not native to Canada, large-scale plantations can create ecological issues.

16.4.10 Turf Farming

Turf farming is in an overall decline throughout the past 15 years, driven in part by summer watering restrictions and limitations in residential cosmetic pesticide use, combined with growing preference for hardscaped areas and natural gardens populated with native and pollinator-friendly species. Turf farming accounts for 5–6% of annual nursery-related sales (Agriculture and Agri-Food Canada 2020b). Kentucky bluegrass (*Poa pratensis*, not native to North America) and derivatives, commonly used for lawns, are C_3 plants, whereas crabgrass (*Digitaria sp.*), the bane of lawn owners, is C_4 species. As C_4 species do not increase photosynthetic activity as CO_2 levels rise and are warm-environment grasses driven by sunlight and not temperature, climate change affects C_3 and C_4 species differently.

16.5 Livestock

16.5.1 Beef (Fig. 16.8)

Beef production varies from 1.5 to 2.0 Mt annually. Cattle numbers vary substantially between and within years, but overall estimates of 10–12 million beef cattle and calves are reasonable (Agriculture and Agri-Foods 2023a). The average herd size is about 70 head, with 60% of operations having 47 or fewer, and only 1% having more than 500. Production is concentrated in Alberta ($5–6 billion economic value) and Saskatchewan ($1–2 billion), although all provinces have active beef operations (Statistics Canada 2022–23).

Common breeds include Angus (most registered head), Simmental, Charolais, Hereford, Limousin, Gelbvieh (in eastern Canada), and Shorthorn (in western Canada) (Agriculture and Agri-Foods 2023a). Breeding can occur naturally or by artificial insemination. Hormones have been in use since the 1960s to encourage muscle growth over fat production.

Summer grazing occurs on pastureland, open rangeland, or as part of a crop rotation system (often with grain or silage). In Alberta and Saskatchewan, much of the open rangeland of the Shortgrass Prairie, with solonetz soils, is not suitable for other forms of agriculture. During winter, cattle in the southern prairies may remain on open rangeland, while others are kept in woodlots (northern prairies, eastern Canada), or in open barns. Diets (rations) for indoor cows may be varied for particular animals and commonly include grains, fibre-rich roughage, and vitamin, mineral, oil, and fat supplements, in addition to hay. Earlier forage harvesting can improve quality.

Cattle are traded at auction markets. In western Canada, feeder calves (200–300 kg) destined for meat production are fed rations of hay with barley, corn, or wheat over 2–5 months at feedlots, until reaching market size of 600–700 kg. In eastern Canada and at smaller operations, many calves remain on the farm for this stage. Meat-packing plants are responsible for slaughtering, dressing, packaging, and distribution.

Practices show considerable regional variation across Canada (Sheppard et al. 2015). Prairie operations use earlier

Fig. 16.28 Beef cattle farming in Harricott NL differs substantially from the higher-intensity operations of western Canada. Author's photo

and shorter calving seasons, more winter grazing, more use of barley as feed, and autumn manuring than those in central and Atlantic Canada. However, these generalizations conceal a high degree of heterogeneity in both biophysical and socio-economic conditions for prairie beef producers (Pogue et al. 2018, 2020). Atlantic Canadian operations tend to be smaller, limiting local supply (Fig. 16.28). Newfoundland produces about 1 t of beef annually from 38 farms across the island and imports 14,000 t annually. Newfoundland is self-sufficient only in the supply-managed milk, chicken, and eggs, and not for beef, traded in free markets.

Globally, consumption of beef is projected to increase by 1.2% y^{-1} until 2050, driven by demand in eastern Asia (OECD-FAO 2020). Canadian beef exports account for about half of annual production, valued at $4.0–4.7 billion annually (Statistics Canada 2022–23). Exports to USA account for 75–80% of the total, along with Japan, Mexico, South Korea, and Vietnam. International trade in beef can be contentious, with health concerns complicated by lobbying from farmers in many countries, political events, linkage to other traded commodities, and public concerns. Exports to China and UK are among those which have been affected in recent years.

Beef consumption is declining in Canada, but currently is 16.9 kg y^{-1} per capita (Farm Credit Corporation 2023; Figs. 16.29 and 16.30). Ground beef is the most commonly consumed style: almost 50% of the population eats one beef burger patty at least weekly, and almost 90% consume at least one monthly (Technomic 2017). One in six buys fast-food daily (Bush 2024). A typical medium ground beef burger patty is 150 g prior to cooking. Calculations of calorie, carbohydrate, and fat content for burgers can be confusing, as most include the bun. Most of the almost 32,000 fast-food vendors and 400,000 workers in Canada serve beef burgers, which by law in most of Canada must be cooked as well-done. Fast-food and full-service restaurant sales were approximately equal in April 2023 at ~ $3.5 billion (Bush 2024).

16.5.2 Dairy

The dairy industry is the main economic agricultural activity in Québec, Ontario, British Columbia, Nova Scotia, and Newfoundland and Labrador and is also significant in PEI and NB. Dairy generates $6 billion economic activity in Québec and Ontario combined (Statistics Canada 2022a, b, 2022–23; Agriculture and Agri-Food Canada 2023b).

A dairy cow typically consumes about 30 kg of food and 100 l of water daily. Although all eat grass as part of a varied diet, "Grass-fed" cattle typically eat mixtures with at least 75% grazed or harvested herbaceous plants, including grass, legumes, brassicas, grain, and tree and shrub shoots. Feed for each cow varies depending on age, size, milk production, and stage of lactation. On many farms, each cow has an electronic dossier which is used to monitor diet and health. Cows typically relax for an hour or two after milking, but otherwise are either eating or ruminating. Farms may produce their own corn silage and hay, but most supplement on-farm sources with feed imported from other farms and provinces. Daily production is typically 30–40 l of milk per cow. Milk may be processed locally (typical for fluid milk) or may be shipped to larger Canadian processing facilities (typical for butter, ice cream, and yoghurt).

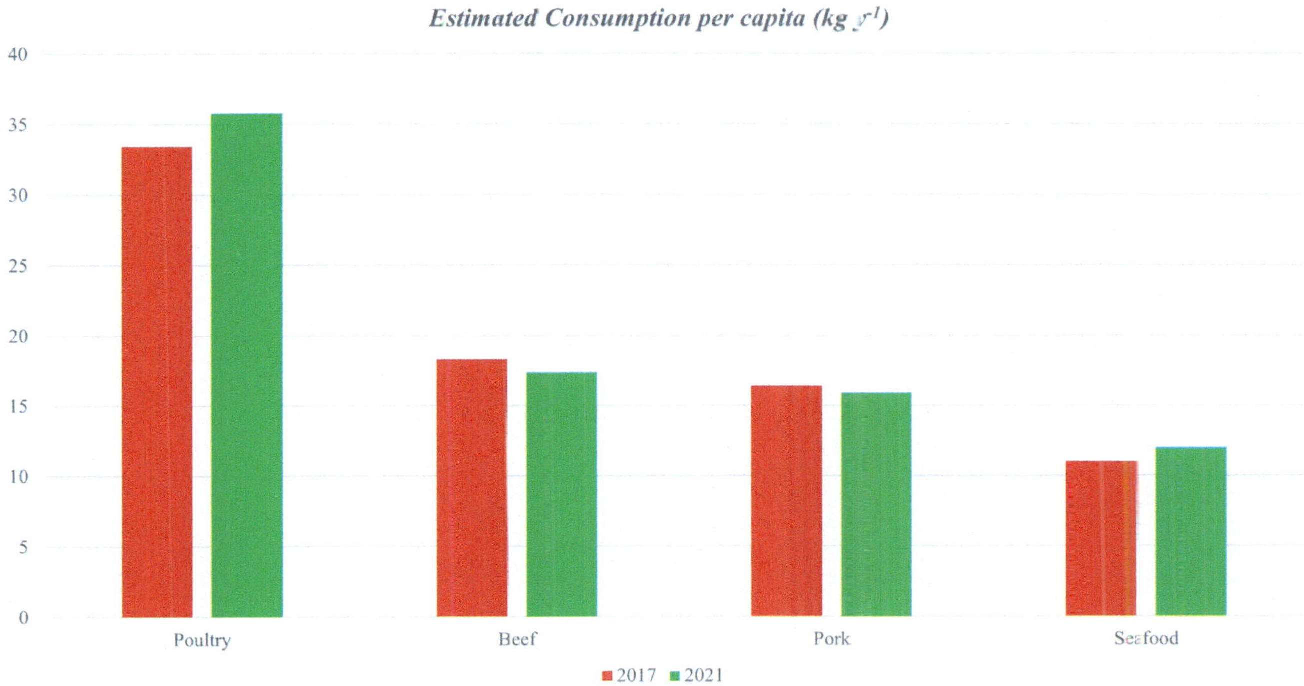

Fig. 16.29 Estimated consumption (kg y^{-1} per capita) of meats and seafood in Canada, 2017 and 2021. Beef and pork consumption have declined, while poultry and seafood consumption increased (Statistics Canada 2022–23; Farm Credit Corporation 2023)

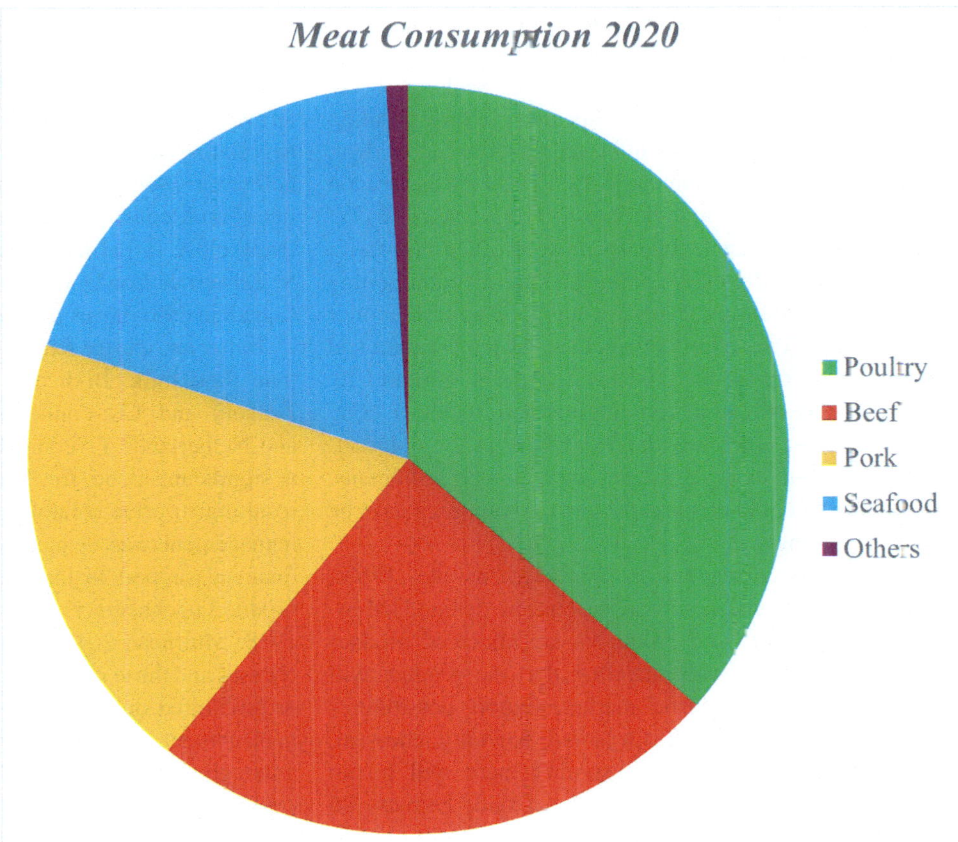

Fig. 16.30 Estimated proportions of meat consumption as a Canadian average, 2020 (compare with Fig. 16.29). "Others" includes sheep (the majority), goats, venison, and other wild game (Census of Agriculture; Statistics Canada 2022–23; Farm Credit Corporation 2023; and other sources compiled by the author)

Fig. 16.31 Total milk consumption (kg), 2012–2021. Although consumption of fluid milk has decreased, increased consumption of other milk products (e.g. cheese, yoghurt) and of industrial milk has increased (Census of Agriculture; Statistics Canada 2022–23)

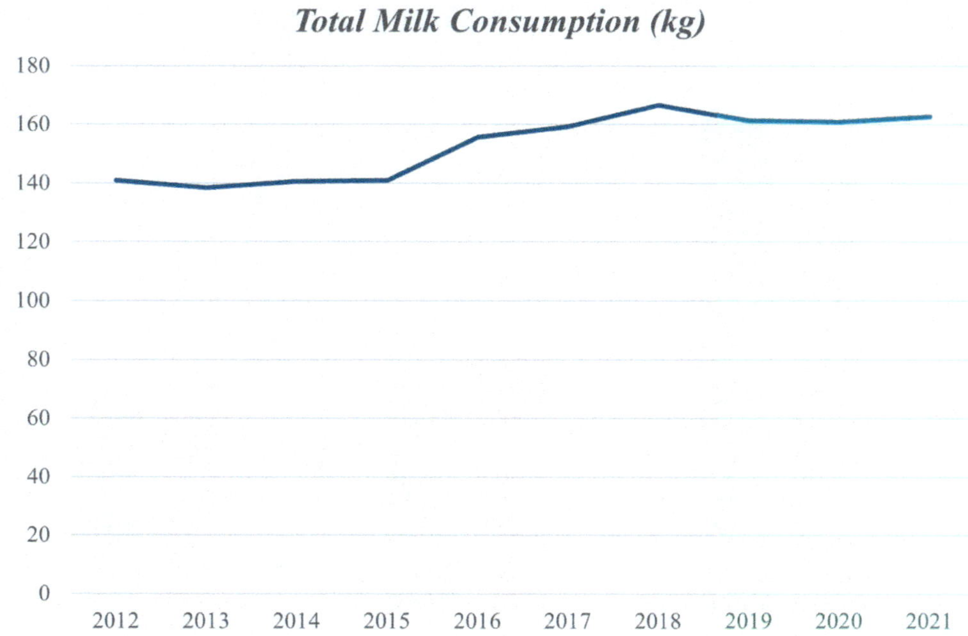

Dairy cows are most comfortable between 15 °C and 20 °C. Extreme heat events and longer warm periods have negative consequences for calf health, reduce weight gain, decrease reproduction, increase mortality, and cause declining milk production, and lower milk quality (Allen et al. 2015).

Free-ranging cattle attempt to adapt by seeking shade. Cattle subject to heat stress tend to spend more time standing than lying down. Differences in breed response are evident, with thicker-haired Holstein cattle, by far the most common in Canada, suffering more under heat (Ferrazza et al. 2017).

Most Canadian dairy cattle spend their lives indoors, a climatically dictated choice. Cooling fans and air conditioning are becoming ubiquitous. Barns are equipped with side roller curtains that are opened in hot weather. As temperature and humidity conditions inside the barn may differ substantially from those at an official meteorological station (Shock et al. 2016), farmers cannot rely on extrapolation from broadcast weather conditions. For dairy cattle as well as humans, local circumstances matter greatly, and the responses must be focused accordingly.

Dairy supply management, instituted in the 1960s, directly links domestic supply with domestic demand (Muirhead 2014). Farmers and producers are allocated production quotas, with the understanding that the product will have a guaranteed domestic market. Export possibilities are not considered (except for some specialty cheeses), because quotas are designated so that there will be no excess production beyond domestic demand. Farmers do not receive payments or subsidies, but the control on available supply means that costs are passed directly to retailers and consumers. The Canadian Dairy Commission is the federal Crown Corporation responsible for coordinating dairy policies, and administrates supply management of the Canadian dairy industry.

Supply management applies to all dairy products, including milk, butter, and cheese (Figs. 16.31 and 16.32). For liquid milk, which has limited transportability and durability, assuring local supply could mean supporting smaller dairy farms which otherwise might not be viable. It also limits the negotiating advantage dairies had over farmers, particularly before the 1960s. Practically, potential export competition in liquid milk could only come from USA dairies located relatively close to the large urban centres in the Great Lakes–St. Lawrence Lowlands and southern BC, areas where many Canadian dairy farms are located.

Butter and cheese are much more transportable and durable than fluid milk. Both can readily be exported from New Zealand and EU countries, including Germany, Ireland, and Netherlands. Oversupply from domestic producers was a significant issue for butter and non-specialty cheeses, resulting in price collapses. For these commodities, supply management was designed to eliminate gluts and provide financial support to farmers, as well as provide consistent products to consumers, a particular issue for butter (Chapman 1938; Muirhead 2014). Cheese imports, particularly "soft" cheeses and those made from raw milk, are severely restricted or prohibited on health grounds by Health Canada. Supply management does not apply to goat or sheep milk cheese, although much of the feta sold in Canada is produced from cows.

Although they all come from cows, milk, butter, and cheese could be argued to represent separate problems in domestic production and international trade. Supply management is applied to bulk fluid milk, mass-produced butter, and specialty

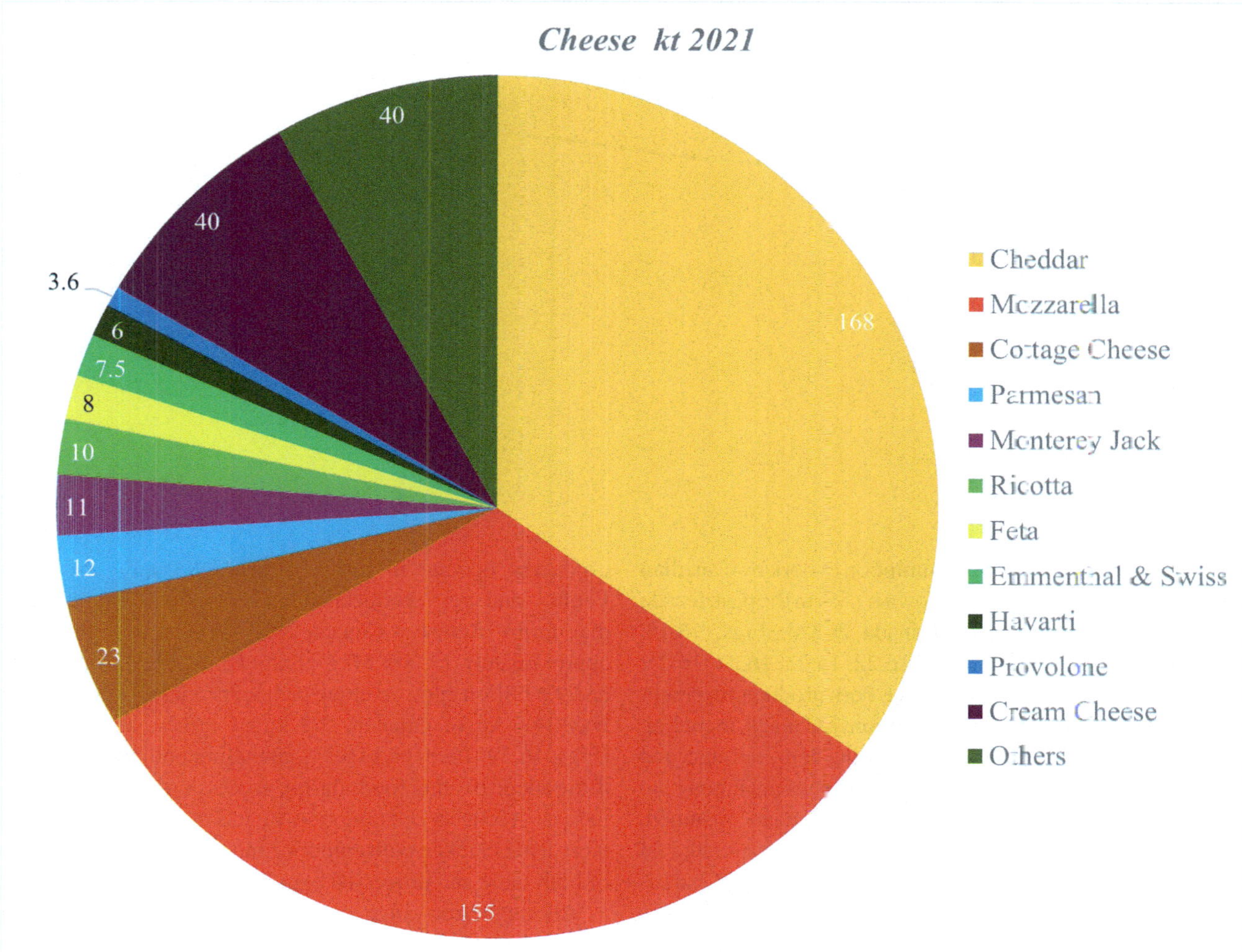

Fig. 16.32 Consumption of cheese produced in Canada, 2021. Figures indicate production in kt. Imported cheeses are not included (Census of Agriculture; Statistics Canada 2022–23)

cheeses with protected geographical status equally. This is a source of frustration to artisanal cheese producers in EU and UK, whose ability to export to Canada is bound with the broader international negotiations and the political pressure from farmers and dairies to retain supply management. Specialty cheeses are admitted only after high tariffs, and in limited quantities.

Dairy supply management continues to be a significant problem for Canada-USA trade negotiations, as well as others. Stresses include both economic and philosophical differences ("free market" neoliberal vs. government management principles). In contrast to the trend for consolidation evident elsewhere in Canadian agriculture, most dairy farms remain as family operations, typically with 50–100 cows. This structure gives both political strength and general public sympathy for dairy farmers' concerns.

16.5.3 Pork

Pork production is the fourth largest farming industry in Canada (Statistics Canada 2022–23). Production is concentrated in southeastern Manitoba, southern Québec, and southern Ontario with almost 14 million pigs on 7500 farms. Annual production of 2.3 Mt surpasses both beef and poultry by mass. Concentration and termination of provincial subsides has shuttered many smaller operations west of Winnipeg and east of Rivière-du-Loup, leaving environmental contamination problems to be resolved. Dependence on international trade, particularly exports to USA, China, Japan, and Mexico, has led to economic difficulties for several large vertically integrated operations and packing plants. About one-third of total production is consumed in Canada. Substantial imports come from USA, Germany, Denmark, and Spain.

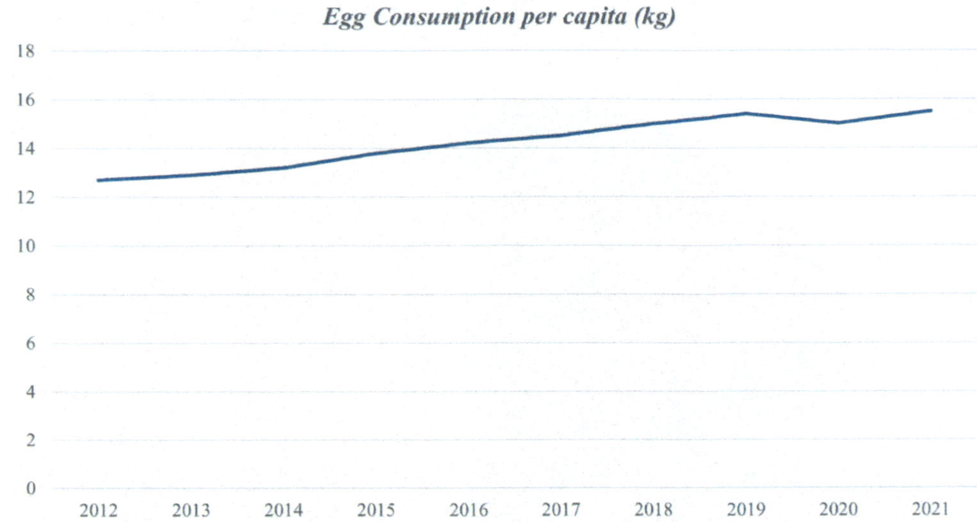

Fig. 16.33 Egg consumption per capita (kg), 2012–2021 (Census of Agriculture; Statistics Canada 2022–23)

Overall per capita annual consumption of pork by Canadian consumers was 15.9 kg in 2021, down by 5% in the past decade (Agriculture and Agri-Foods Canada 2022d; Figs. 16.16, 16.17, 16.18, 16.19, 16.20, 16.21, 16.22, 16.23, 16.24, 16.25, 16.26, 16.27, 16.28, 16.29 and 16.30). Pork products represent about 20–25% of total meat consumption in Canada, including poultry and fish. Cured products, including ham, sausage, and bacon account for about half. Ham sandwiches are among the most commonly reported form of processed meat consumed, and the largest contributor of energy, protein, and saturated fat among the red meat dishes cited (Shafiee et al. 2022). Charcuterie, using pork in galantine, pâté, rillette, and sausages, is a respected element of Québecois culinary practice. Bacon taken from a pig's back, termed "back bacon" or peameal bacon if rolled in finely ground corn meal in Canada, is marketed as "Canadian bacon" elsewhere.

Bacon consumption commonly appears on lists of stereotypical "Canadian" behaviour. A somewhat facetious 2009 survey commissioned by a major bacon processor suggested that 43% of Canadian men would choose eating bacon over sexual activity, 23% considered the smell of frying bacon to be their favourite fragrance, and 23% wondered if "my partner loves bacon more than me". Maple-flavoured and chocolate-flavoured bacon are available for sale. In reality, about three-quarters of Canadian households currently purchase bacon, with annual per capita consumption of 4.1 kg. Declining consumption reflects dietary constraints among newer Canadians, changing food preferences, health concerns, and rising costs.

16.5.4 Poultry and Eggs

Poultry production is a major agricultural industry in Québec southern Ontario, and the BC Lower Mainland. Local producers operate in the Atlantic and Prairie Provinces. Poultry and egg producers represent about 3% of Canadian farms (Statistics Canada 2022–23). Total production is approximately 13,000 Mkg of chicken, 150 Mkg of turkey, and 9.6 billion eggs. Average per capita consumption in 2021 was 34.4 kg of chicken, 3.3 kg of turkey, and 258 eggs (Figs. 16.29, 16.30 and 16.33), including both table eggs and those added to other foods during processing. Dietary changes indicate reduced consumption of turkey, stable or slightly increasing chicken consumption, and increased consumption of both table and processed eggs.

Although free-range eggs are commonly available and poultry less so, most production is factory-based (Pelletier 2018). Conventional cage housing systems account for more than 90% of egg production. General public concern for animal welfare has proven secondary to concerns about availability and price.

Poultry and egg production operates under a supply management framework similar to that for dairy, including the Egg Farmers of Canada, Turkey Farmers of Canada, Chicken Farmers of Canada, and the Canadian Hatching Egg Producers, overseen by the Farm Products Council of Canada. However, exporting and importing chicken and turkey is common. Ducks (about 1.7 million farmed) and geese (~70,000) are not subject to supply management.

16.5.5 Other Meats

Sheep and lamb production (about 1 million head) is concentrated in Ontario, Québec, and Alberta (Fig. 16.34). About 60% of the goat production (about 250,000 head) is located in Ontario (Statistics Canada 2022–23). Goat farmers in Ontario tend to be younger than the farming population as a whole. Both sheep and goats are raised for dairy, meat, and hides.

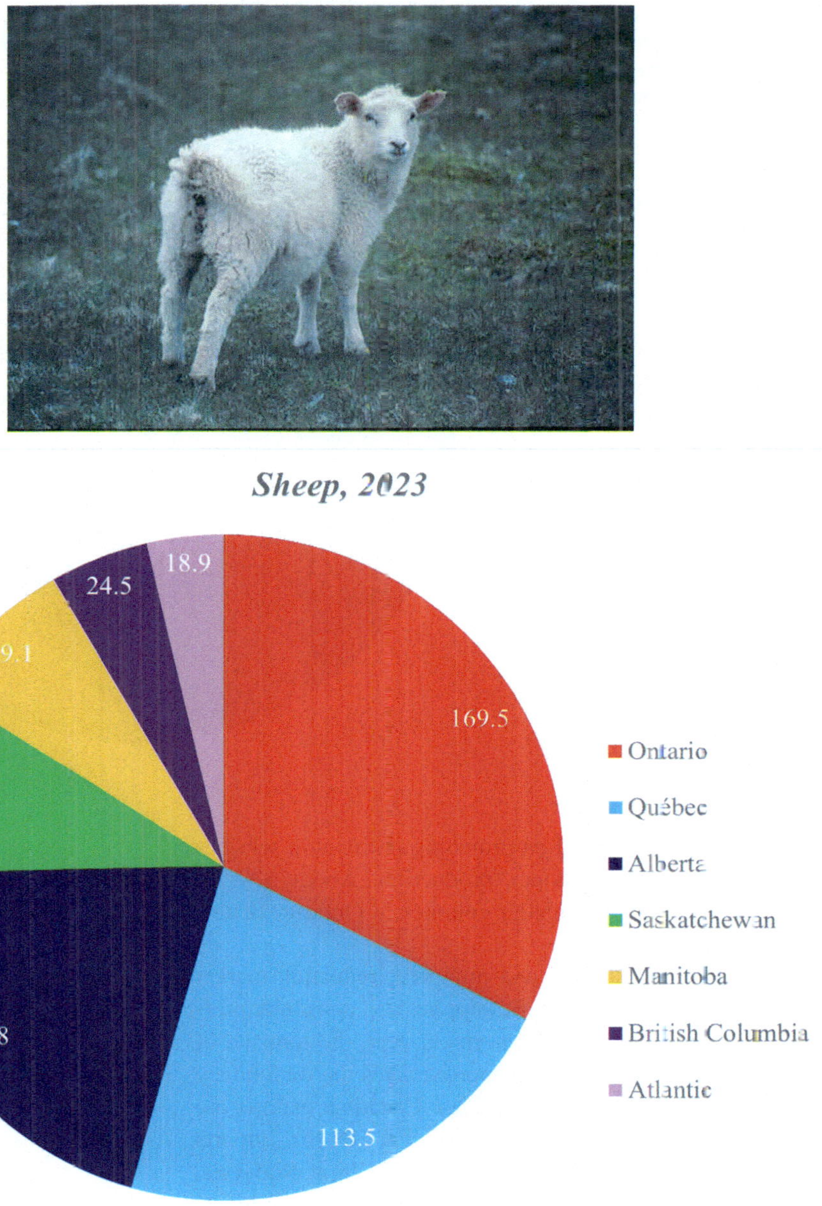

Fig. 16.34 Sheep numbers and proportions by province, 2023. Figures indicate sheep numbers in thousands (Statistics Canada 2022–23)

Other animals farmed for meat consumption include rabbits, wapiti, deer, and bison. Special licences are required to sell game meat in food stores.

16.6 Bees and Honey

Honeybees (*Apis mellifera*) are key components for agriculture and food, for honey production but especially for their role as pollinators (Agriculture and Agri-Food Canada, 2022c). Originating in Mediterranean Africa or western Asia, honeybees were introduced to North America ca. 1622 ce. Bees can survive as ferals in all but glacial, tundra, and taiga climates, but require human intervention in boreal climates. Typical maximum flight range is 3 km, so bees tend to utilize and pollinate plants within a short distance from the hive. The short flight range leads to distinctive honey characteristics from specific pollen. A worker bee will fly 800 km during its 6-week lifespan (Winston 1987; Nearman and van Engelsdorp 2022).

In addition to its value as human food, honey has potential for medical use due to its antioxidant and antimicrobial properties (Zammit Young and Blundell 2023). Beeswax is used for candles, furniture polish, shoe polish, crayons, lip balm, lipstick, hand cream, and automobile wax.

Canada produces 30–40 Mt of honey annually, with 80% from Alberta, Saskatchewan, and Manitoba (Statistics Canada 2022c, 2022–23; Fig. 16.35). Prairie colonies tend to be

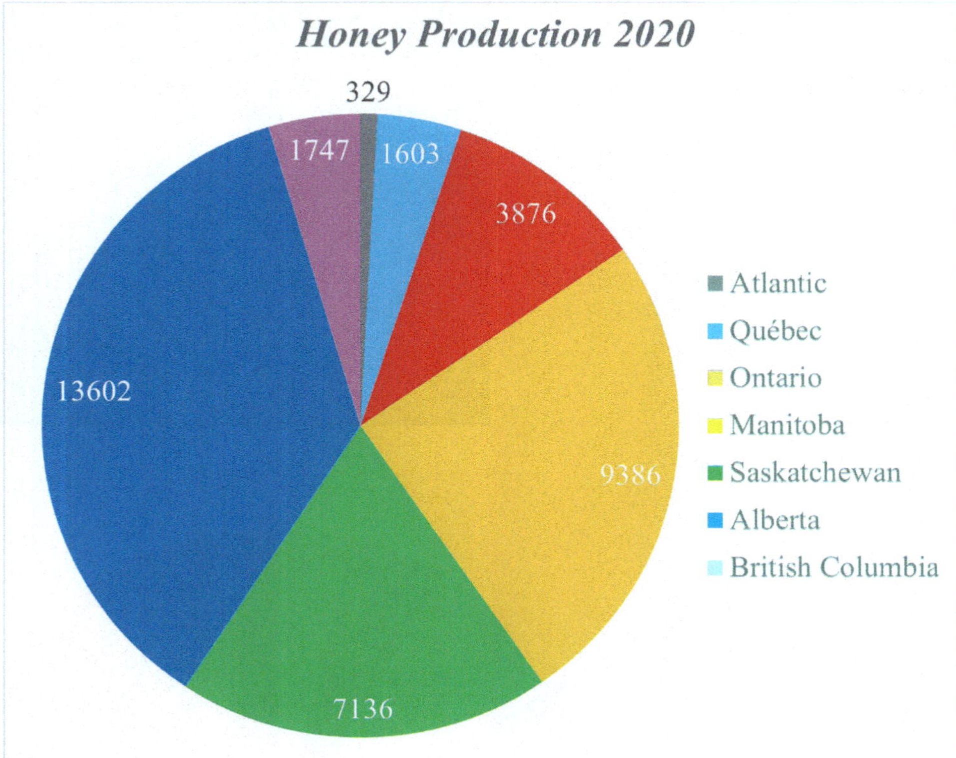

Fig. 16.35 Honey production and proportions by province, 2023. Figures indicate production in millions of kg (Statistics Canada 2022–23)

larger, with many focused on providing pollination services rather than strictly for honey production. About 60% of the beekeepers manage smaller colonies in Ontario, British Columbia, and Québec.

Honey is both exported and imported, reflecting market demand for inexpensive generic supplies and consumer preferences for specific types. In a typical year, the tonnage of imports and exports combined approximates the total honey production (Statistics Canada 2022c). Canadian exports are primarily directed to USA and Japan (Fig. 16.36). Imports come from a variety of sources, particularly China, Malaysia, Thailand, and USA. Trans-shipments from Asian producers are facilitated by the lower standards applied in USA, providing an incentive to import cheaper, less secure honey. Testing by the Canadian Food Inspection Agency between 2013 and 2018 revealed that ~7% of imported honey had been adulterated with corn syrup, detected by isotopic analysis of C_4 components (Canadian Food Inspection Agency 2022). The large Prairie colonies, producing honey greatly in excess of regional demand, are the major source of exports.

For many large beekeeping operations, the primary goal is providing pollination services, with honey production a by-product. The short flight range makes it profitable for crop growers to contract with beekeepers to bring colonies to their fields, so that bees will only pollinate their crop. Transported bees are used to pollinate blueberries (discussed above), apples, cherries, soybeans, and especially canola.

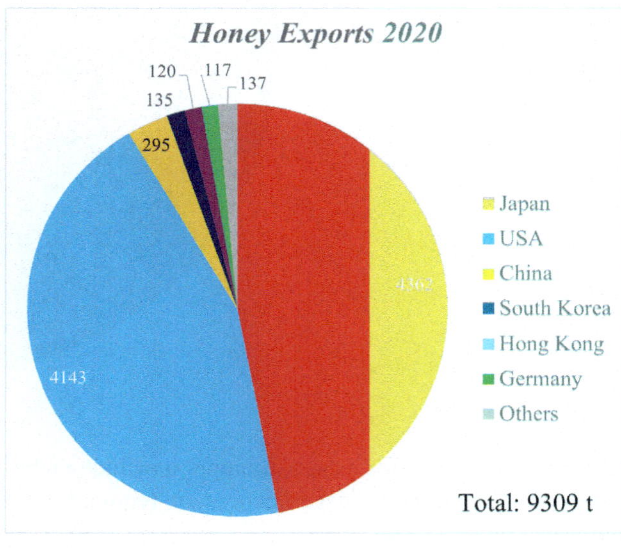

Fig. 16.36 Honey exports from Canada, 2020. Figures indicate exports in tonnes (Statistics Canada 2022c, 2022–23)

Hybrid canola can only be pollinated by honeybees and alfalfa leafcutter bees (Robinson et al. 2023). Bees are estimated to be responsible for ~50% of hybrid canola production, a growing industry. The total estimated dollar contribution of bee pollination to Canadian agriculture exceeds $7 billion annually, with hybrid canola accounting for more than half (Agriculture and Agri-Food Canada 2022a, b, c). In contrast to the cross-country circuits typical of many large USA operations, transport distances for bees involved in hybrid canola production are relatively short, with both bees and canola usually located in the same province. Restrictions on interprovincial bee movements vary, with NL imposing a total ban on importation.

In addition to weather, bee colonies are subject to infestation from varroa mites, foulbrood bacteria, chalkbrood parasitic fungi, and Moku virus (Canadian Association of Professional Apiculturists 2019). The presence of 3 varroa mites per 100 bees is fatal to an over-wintering colony. Colony collapse disorder, initiated by desertion of the colony by worker bees, has been attributed to insufficient genetic diversity, low quality or diversity of pollen sources, adverse weather conditions, stress resulting from transport, and insecticides (Ceparo et al. 2014). Dimethoate, Pyrethrin, and Neonicotinoid (neonic) insecticides are all toxic to bees. Health Canada, responsible for assessing the impacts of pesticides, has prohibited the use of neonicotinoids on orchard trees, berries, and fruiting vegetables, before or during blooming season.

16.7 Dietary Preferences and Agriculture

Dietary preferences and eating habits are changing. Approximately 10% of the population self-identifies as vegetarian or vegan (Charlebois 2018), with 2% as subject to celiac disease, and larger percentages restricting consumption of meat, gluten, sugar, carbohydrates, or other food types (Mudryj et al. 2021). These changes will inevitably impact food production destined for domestic consumption (e.g. Dyer et al. 2020).

Health Canada has issued the Canada Food Guide (various titles) periodically since 1942 (Health Canada 2019a, 2019b). Review of these reveals a progressive shift in recommended protein consumption, along with developing concerns about sugar, fat, and carbohydrates. The 1942 guide recommended eating kidney, liver, or heart once a week, along with 3–4 eggs weekly and a daily serving of tomatoes. The 1992 guide used a rainbow graphic, with "grain products" having the largest, most prominent components showing the greatest number of options, followed by "vegetables and fruits" and "milk products", with "meat and alternatives" having the least prominent position. Large ranges in recommended servings from each of the four groups were designed to account for differences in physiology, gender, and age. The 2007 version moved vegetables and fruits to the most prominent position.

The 2019 guide adopted a different approach, following widespread criticism of both the 2007 version and the initial proposed revision. The 2019 version does not include defined groups or suggested portion size but uses a photograph of a plate occupied by fruits and vegetables (half), whole-grain foods (a quarter), and a variety of proteins, deemphasizing red meat (a quarter), with water identified as the drink of choice (Health Canada 2019a). Public reaction was swift and highly varied.

In the interior of continental areas, consuming only plants may be insufficient to provide a person with the daily requirement of 110–130 μg per day of iodine. Grazing animals concentrate iodine in their flesh, so animal protein (including cheese and eggs) are other sources. In Canada, iodine is added to cattle feed, and iodine disinfectants are used by the dairy industry (Borucki Castro et al. 2011). In some affluent countries, the rate of iodine deficiency disorder has increased in recent years. Some research suggests that reduction in the consumption of dairy products and salt may be responsible, especially if people do not replace the iodine with other sources (Zimmermann and Andersson 2012).

Food waste is estimated at $31 billion annually, ~40% of annual production, of which 47% occurs in homes (National Zero Waste Council 2018, VCMI 2019; Figs. 16.37 and 16.38). Households in Canada waste ~ $30 worth of food each week ($1500 annually) on average, ~85% of which was edible (by mass) and ~ 80% was compostable. The average household food wastage is 0.5 kg d–1 (180 kg y^{-1}), with fruits and vegetables representing more than half. Avoidable home food waste is estimated to increase food costs by ~12%.

16.8 Greenhouse Gas Emissions

Agriculture and agri-food sector operations are responsible for 8–10% of Canada's GHG emissions of 70–75 Mt CO_2 equivalent (Fig. 16.39). The federal government has set a target to decrease emissions by achieving Canada's 30% below 2005 levels by 2030. Emissions result from fuel use in farming operations (~14 Mt), enteric fermentation (~24Mt) and crop production (~24Mt) (Environment and Climate Change Canada 2021).

Emissions from chemical fertilizer use (primarily N_2O and other nitrogen compounds) are 12–13 Mt. Increased N_2O emissions result from conversion of pasture to annual crop production, fertilizer intensive crops, and soil degradation releasing carbon and nitrogen (Karimi et al. 2020). N_2O emissions per hectare increased by almost 100% between 1981 and 2019 (Environment and Climate Change Canada 2021).

Hybrid canola production results in N_2O emissions associated with fertilizer manufacturing and application, crop residues, and leaching and volatilization. Additional CO_2 and CH_4 emissions result from fuel consumption. Changes in land use and management have decreased CO_2 emissions by increasing sequestration of C in less-disturbed soils (Shrestha et al. 2014). Although the intensity of emissions per hectare decreased, greatly increased canola production has resulted in higher total GHG emissions.

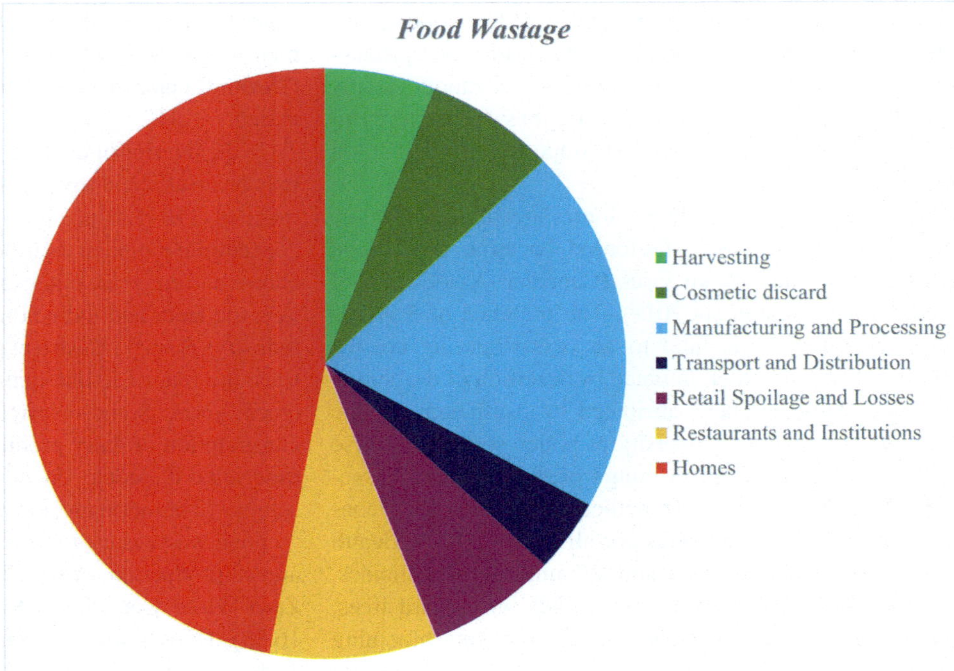

Fig. 16.37 Food wastage by sector proportion in Canada (National Zero Waste Council, 2018; VCMI 2019)

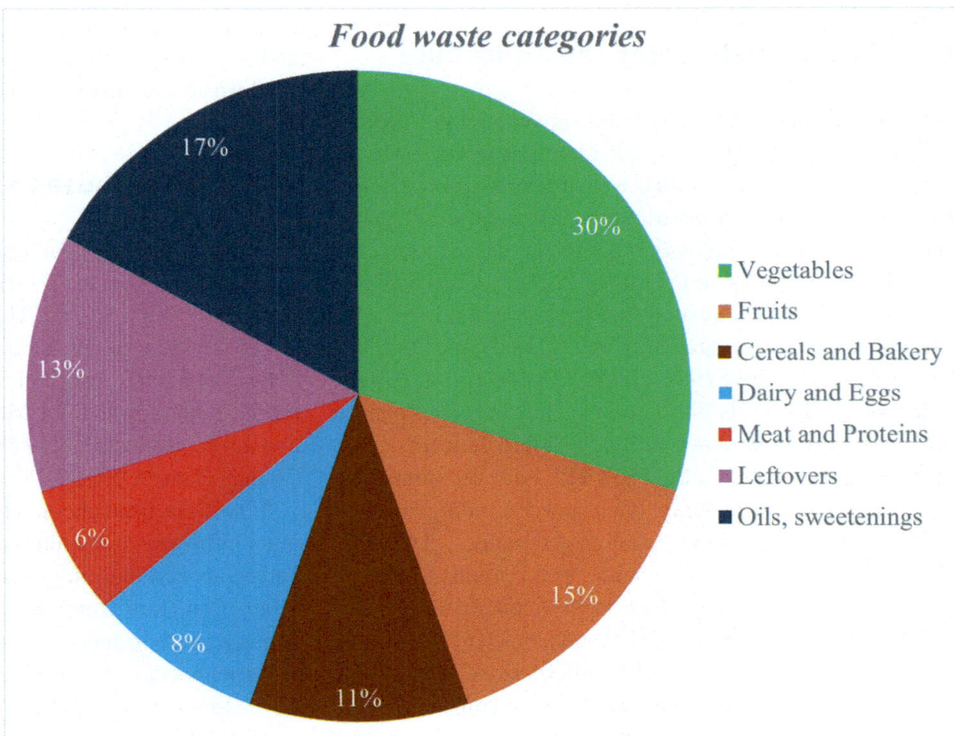

Fig. 16.38 Food wastage by category in Canada (National Zero Waste Council 2018; VCMI 2019)

Vegetable and fruit production results in emissions from planting, fertilization, irrigation, machinery operation, harvesting, cooling for storage, building maintenance, and product transport from farms (Dyer et al. 2011; Dyer and Desjardins 2018). Combined, vegetable and fruit production in Canada used 4.6 PJ between 2007 and 2016 (Environment and Climate Change Canada 2021). In a PEI study, nitrogen loading reached 44.4 kg ha^{-1} y^{-1} for soybean fields, commonly planted as an option or in rotation with potatoes. Potato rotated land was responsible for 84.5% of the annual nitrate load to the watershed (Liang et al. 2020).

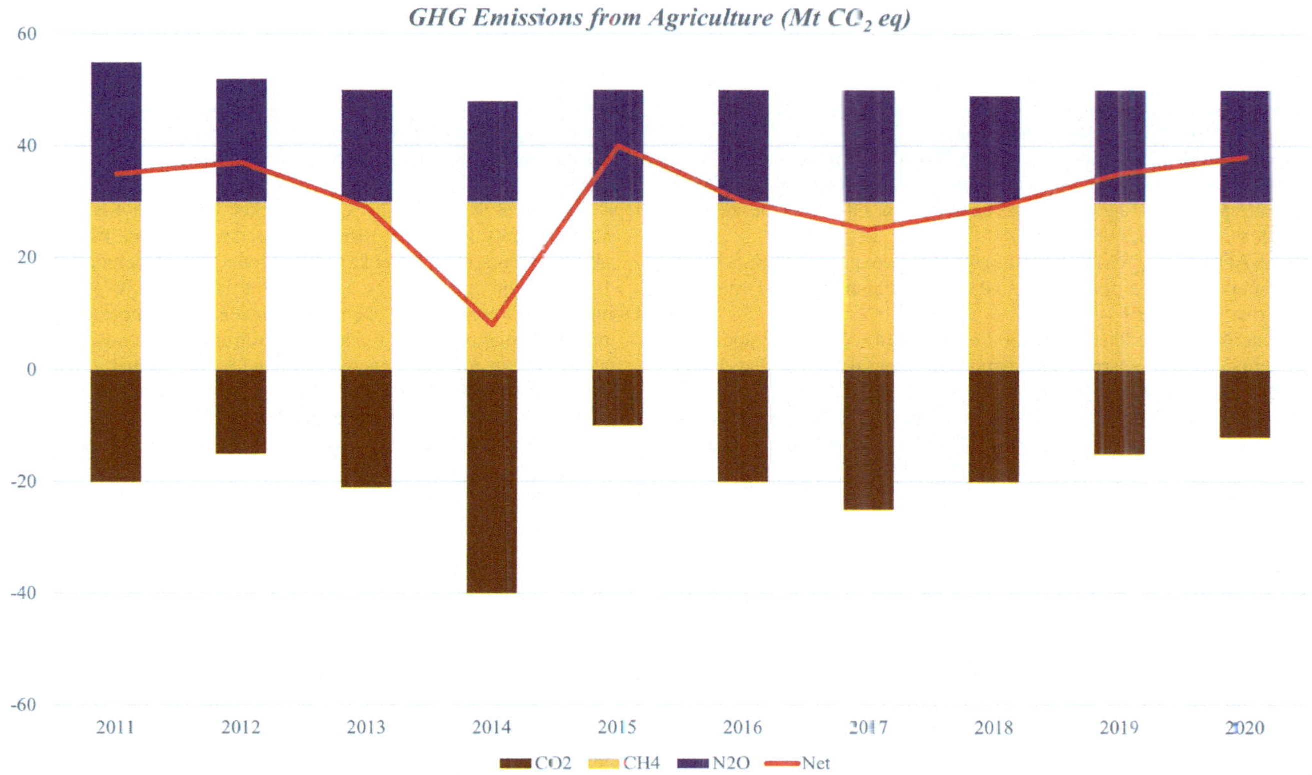

Fig. 16.39 GHG Emissions from agricultural operations, 2011–2020. Totals reflect emissions directly connected to agriculture (Environment and Climate Change Canada 2021)

GHG emissions are of particular concern in meat, dairy, and poultry operations (Vergé et al. 2016; Dimitrov and Wang 2019; Dyer et al. 2020; Oryschak and Beltranena 2020). Some progress has been made. Changes in egg production methods have resulted in a decrease in cradle-to-farm gate environmental impacts by about two-thirds of those estimated in 1962, including a 72% reduction in GHG emissions (Pelletier 2018).

GHG emissions from cattle are highly dependent on diet, in addition to herd size (Vergé et al. 2018). Use of higher-fibre diets has been shown to reduce CH_4 emissions, while crude proteins have the opposite effect (Zhao et al. 2022). Replacement of barley forage with corn reduces methanogenesis (Guyander et al. 2017).

References

Agriculture and Agri-Food Canada (2020a) Overview of the Canadian agriculture and agri-food sector. https://agriculture.canada.ca/en/sector/overview

Agriculture and Agri-Food Canada (2020b) Statistical overview of the Canadian ornamental industry 2019. Horticulture Section, Crops and Horticulture Division, Agriculture and Agri-Food Canada. https://agriculture.canada.ca/en/sector/horticulture/reports

Agriculture and Agri-Food Canada (2022a) Sector commodity breakdown. https://agriculture.canada.ca/sites/default/files/documents/2022-07

Agriculture and Agri-Food Canada (2022b) Statistical overview of the canadian field vegetable and mushroom industries 2020. Horticulture Section, Crops and Horticulture Division, Agriculture and Agri-Food Canada. https://agriculture.canada.ca/en/sector/horticulture/reports

Agriculture and Agri-Food Canada (2022c) Statistical overview of the Canadian Honey and Bee industry and the economic contribution of honey bee pollination 2021. Horticulture Section, Crops and Horticulture Division, Agriculture and Agri-Food Canada. https://agriculture.canada.ca/en/sector/horticulture/reports

Agriculture and Agri-Foods Canada (2022d) Canadian pork meat trends. https://agriculture.canada.ca/en/international-trade/market-intelligence/reports/customized-report-service-canadian-pork-meat-trends

Agriculture and Agrifood Canada (2023a) Beef Cattle Registrations in Canada. https://agriculture.canada.ca/sites/default/files/documents/2023-06/ap_gen_beefregcda_e.pdf

Agriculture and Agri-Food Canada (2023b) Overview of Canada's dairy industry. Dairy_industry_at_a_glance_2022

Agriculture and Agri-Food Canada (2023c) Statistical overview of the Canadian Greenhouse Vegetable and Mushroom Industry 2022. Horticulture Section, Crops and Horticulture Division, Agriculture and Agri-Food Canada. Report 13185E. https://agriculture.canada.ca/en/sector/horticulture/reports

Agriculture and Agri-Food Canada (2023d) Statistical overview of the Canadian fruit industry 2022. Horticulture Section, Crops and Horticulture Division, Agriculture and Agri-Food Canada. https://agriculture.canada.ca/en/sector/horticulture/reports

Agriculture and Agri-Foods Canada (2023e) What we heard report—fertilize emissions reduction. agriculture.canada.ca/en/department/transparency/public-opinion-research-consultations/share-ideas-fertilizer-emissions-reduction-target/what-we-heard-report-fertilizer-emissions-reduction

Allen JD, Hall LW, Collier RJ, Smith JF (2015) Effect of core body temperature, time of day, and climate conditions on behavioral patterns of lactating dairy cows experiencing mild to moderate heat stress. J Dairy Sci 98:118–127

Almas AD, Conway TM (2016) The role of native species in urban forest planning and practice: A case study of Carolinian Canada. Urban For Urban Green 17:54–62

Arnautovska U, McPhedran S, De Leo D (2014) A regional approach to understanding farmer suicide rates in Queensland. Soc Psychiatry Psychiatr Epidemiol 49:593–599

Barrett R, Robinson A, Thornton M, VanderZaag P (2023) Potato production in the United States and Canada. In: Caliskan ME, Bakhsh A, Jabran K (eds) Potato production worldwide, ch. 20, pp 365–379. Elsevier

Beaulieu MS (2014) Demographic changes in Canadian agriculture. Statistics Canada, 96-325-X-001

Beer Canada (2024) 2021 Industry trends. https://industry.beercanada.com/statistics

Bell JM (1982) From Rapeseed to Canola: A brief history of research for superior meal and edible oil. Poult Sci 61:613–622

Bialais C (2020) Organic agriculture in Canada. Library of Parliament, 2020-07-E

Borucki Castro SI, Lacasse P, Fouquet A, Beraldin F, Robichaud A, Berthiaume R (2011) Feed iodine concentrations on farms with contrasting levels of iodine in milk. J Dairy Sci 94:4684–4689

Burnett K, Skinner K, Hay T, Leblanc J, Chambers L (2017) Retail food environments, shopping experiences, first nations and the provincial Norths. Health Promot Chronic Dis Prev Can 37(10)

Bush O (2024) Fast food statistics in Canada. https://madeinca.ca/fast-food-canada-statistics/

Canadian Association of Professional Apiculturists (2019) Honey bee diseases and pests. University of Guelph

Canadian Food Inspection Agency (2022) Honey bee producer guide to the national bee farm-level biosecurity standard. Office Of animal biosecurity Canadian Food Inspection Agency, Ottawa, Ontario

Canadian Grain Commission (2022) Barley harvest annual report. https://www.grainscanada.gc.ca/en/grain-research/export-quality/cereals/maltingbarley/2022/barley-production.html

Cepero A, Ravoet J, Gómez-Moracho T, Bernal JL, Del Nozal MJ, Bartolomé C, Maside X, Meana A, González-Porto AV, de Graaf DC, Martín-Hernández R (2014) Holistic screening of collapsing honey bee colonies in Spain: a case study. BMC Research Notes 7:649. https://doi.org/10.1186/1756-0500-7-649

Cereals Canada (2024) Canada Western Amber Durum. Canadian Grain Commission. https://cerealscanada.ca/2022-wheat-crop/cwad

Chapman A (1938) The tariff and Canadian Butter. In: Innis HA (ed) Essays in political economy. University of Toronto Press, pp 59–74

Charlebois S (2018) The Conversation: Young Canadians lead the charge to a meatless Canada. Dalhousie University, Halifax NS. https://www.dal.ca/news/2018/03/16/the-conversation-young-canadians-lead-the-charge-to-a-meatless-c.html

Cheng Z, Nitoslawski S, van den Bosch CK, Sheppard S, Nesbitt L, Girling C (2021) Alignment of municipal climate change and urban forestry policies: A Canadian perspective. Environ Sci Policy 122:14–24

City of Yellowknife (2019) Yellowknife food and agriculture opportunities analysis. https://www.yellowknife.ca/en/exploring-yellowknife/resources/Foodies/Agriculture-Strategy

Cutler CG, Reeh KW, Sproule JM, Ramanaidu K (2012) Berry unexpected: Nocturnal pollination of lowbush blueberry. Can J Plant Sci 92:707711

Deloitte (2022) The economic and social impact of Canada's cannabis sector. https://www2.deloitte.com/content/dam/Deloitte/ca/Documents/consumer-business/ca-en-consumer-business-cannabis-annual-report-2021-AODA

Dias GM, Ayer NW, Khosla S, Van Acker R, Young SB, Whitney S, Hendricks P (2017) Life cycle perspectives on the sustainability of Ontario greenhouse tomato production: Benchmarking and improvement opportunities. J Clean Prod 140:831–839

Dimitrov DD, Wang J (2019) Geographic Inventory Framework for estimating spatial pattern of methane and nitrous oxide emissions from agriculture in Alberta, Canada. Environ Dev 32:100461. https://doi.org/10.1016/j.envdev.2019.100461

Ding D, Zhao Y, Guo H, Li X, Schoenau J, Si B 2018. Water Footprint for Pulse, Cereal, and Oilseed Crops in Saskatchewan, Canada. Water, 10:1609. https://doi.org/10.3390/w10111609

Dyer JA, Desjardins RL (2018) Energy use and fossil CO_2 emissions for the Canadian fruit and vegetable industries. Energy Sustain Dev 47:23–33

Dyer JA, Desjardins RL, Karimi-Zindashty Y, MConkey BG 2011. Comparing fossil CO_2 emissions from vegetable greenhouses in Canada with CO_2 emissions from importing vegetables from the southern USA. Energy Sustain Dev 15:451–459

Dyer JA, Worth DE, Vergé XPC, Desjardins RL (2020) Impact of recommended red meat consumption in Canada on the carbon footprint of Canadian livestock production. J Clean Prod 266:121785. https://doi.org/10.1016/j.jclepro.2020.121785

Environment and Climate Change Canada (2021) National Inventory Report 1990–2019: Greenhouse gas sources and sinks in Canada. Government of Canada, Ottawa

FAO (2022) Food and agriculture organization corporate statistical database. Food and Agriculture Organization, United Nations, Rome

Farm Credit Corporation (2023) 2023 economic trends: Meat products. www.fcc-fac.ca/en/knowledge/2023-economic-trends-meat-products

Ferrazza RA, Garcia HDM, Aristizabal VHV, Nogueria CS, Veríssimo CJ, Sartori JR, Sartori R, Ferreira JCP (2017) Thermoregulatory responses of Holstein cows exposed to experimentally induced heat stress. J Therm Biol 66:68–80

Galloway T, Li N (2023) Pass-through of subsidies to prices under limited competition: Evidence from Canada's Nutrition North program. J Public Econ 225:104971. https://doi.org/10.1016/j.jpubeco.2023.104971

Galloway T (2017) Canada's northern food subsidy nutrition north Canada: A comprehensive program evaluation. Int J Circumpolar Health 76

Government of Canada (2024) Nutrition north. https://www.nutritionnorthcanada.gc.ca/eng/1415385762263/1415385790537.

Guyander J, Little S, Kröbel R, Benchaar C, Beauchemin KA (2017) Comparison of greenhouse gas emissions from corn- and barley-based dairy production systems in Eastern Canada. Agric Syst 152:38–46

Health Canada (2019a) The new food guide. Health Canada, Office of Nutrition Policy and Promotion. Canada.ca/FoodGuide

Health Canada (2019b) History of Canada's food guides. Health Canada Publication 26–18–2165

Health Canada, 2021. Food and Nutrition Highlights 2020. https://www.canada.ca/content/dam/hc-sc/documents/services/publications/food-nutrition/highlights-faits-saillants-2020/highlights-2020.pdf

Helgi Library (2024a) Onion Consumption per capita in Canada. Helgi Library, Prague. https://www.helgilibrary.com/indicators/onion-consumption-per-capita/canada/

Helgi Library (2024b) Potato Consumption per capita in Canada. Helgi Library, Prague. https://www.helgilibrary.com/indicators/potato-consumption-per-capita/canada/

Hunter FF, Burgin SG, Woodhouse A (2000) Shattering the folklore: black flies do not pollinate sweet lowbush blueberry. Can J Zool 78. https://doi.org/10.1139/z00-133

Jones-Bitton A, Best C, MacTavish J, Fleming S, Hoy S (2020) Stress, anxiety, depression, and resilience in Canadian farmers. Soc Psychiatry Psychiatr Epidemiol 55:229–236

Karimi R, Pogue SJ, Kröbel R, Beauchemin KA, Schwinghamer T, Janzen HH (2020) An updated nitrogen budget for Canadian agroecosystems. Agr Ecosyst Environ 304:107046. https://doi.org/10.1016/j.agee.2020.107046

Kb KC, Green AG, Wassmansdorf D, Gandhi V, Nadeem K, Fraser EDG (2021) Opportunities and tradeoffs for expanding agriculture in Canada's North: an ecosystem service perspective. Facets 6:1728–1752. https://doi.org/10.1139/facets-2020-0097

Keyes S, Tyedmers P, Beazley K (2015) Evaluating the environmental impacts of conventional and organic apple production in Nova Scotia, Canada, through life cycle assessment. J Clean Prod 104:40–51

Kissinger M, Rees WE (2009) Footprints on the Prairies. Ecol Econ 68:2309–2315

Klöffel T, Young EH, Borchard N, Vallotton JD, Nurmi E, Shurpali NJ, Urbano Tenorio F, Liu X, Young GHF, Unc A (2022) The challenges fraught opportunity of agriculture expansion into boreal and Arctic regions. Agric Syst 203:103507. https://doi.org/10.1016/j.agsy.2022.103507

Koc M, MacRae R, Desjardins E, Roberts R (2008) Getting civil about food: The interactions between civil society and the state to advance sustainable food systems in Canada. J Hunger & Environ Nutr 3:122–144

Laforge J, Corkal V, Cosbey A (2021a) Farming the future: Agriculture and climate change on the Canadian Prairies. International Institute for Sustainable Development, Winnipeg. www.iisd.org

Laforge JML, Dale B, Levkoe CZ, Ahmed F (2021b) The future of agroecology in Canada: Embracing the politics of food sovereignty. J Rural Stud 81:194–202

Laforge J, Fenton AQ, Lavaleé-Picard V, McLachlan S (2018) New farmers and food policies in Canada. Can Food Stud 5:128–152

Landry V, Semsar-Kazerooni K, Tjong J, Alj A, Darnley A, Lipp R, Guberman GI (2021) The systemized exploitation of temporary migrant agricultural workers in Canada: Exacerbation of health vulnerabilities during the COVID-19 pandemic and recommendations for the future. J Migr Health 3:100035. https://doi.org/10.1016/j.jmh.2021.100035

Leblanc-Laurendeau O (2020) Food insecurity in Northern Canada: An overview. Library of Parliament Background Paper 2020-47-E

Liang K, Jiang Y, Qi J, Fuller K, Nyiraneza J, Meng F-R (2020) Characterizing the impacts of land use on nitrate load and water yield in an agricultural watershed in Atlantic Canada. Sci Total Environ 729:138793. https://doi.org/10.1016/j.scitotenv.2020.138793

Luby CD, Waldner C, Jelinski MD (2020) Update on demographics of the Canadian Dairy Industry for 2011 to 2016. Can Vet J 61:75–78

Lychuk TE, Moulin AP, Lemke RL, Izaurralde RC, Johnson EN, Olfert OO, Brandt SA (2021) Modelling the effects of climate change, agricultural inputs, cropping diversity, and environment on soil nitrogen and phosphorus: A case study in Saskatchewan, Canada. Agric Water Manag 252:106850. https://doi.org/10.1016/j.agwat.2021.106850

Macall DM, Williams C, Gleim S, Smyth SJ (2021) Canadian consumer options regarding food price decisions. J Hunger & Environ Nutr 3:100098. https://doi.org/10.1016/j.jafr.2020.100098

Magnan A (2015) The financialization of agri-food in Canada and Australia: Corporate farmland and farm ownership in the grains and oilseed sector. J Rural Stud 41:1–12

Malaj E, Freistadt L, Morissey CA (2020) Spatio-temporal patterns of crops and agrochemicals in Canada over 35 years. Front Environ Sci 8:556452. https://doi.org/10.3389/fenvs.2020.556452

McCallum BD, Depauw RM (2008) A review of wheat cultivars grown in the Canadian Prairies. Can J Plant Sci 88:649–677

McGovern M, Pasher J (2016) Canadian urban tree canopy cover and carbon sequestration status and change 1990–2012. Urban For Urban Green 17:222–232

Menon R, Gonzalez T, Ferruzzi M, Jackson E, Winderl D, Watson J (2016) Oats—from farm to fork. Adv Food Nutr Res 77:1–55. https://doi.org/10.1016/bs.afnr.2015.12.001

Mental Health Commission of Canada (2022) Fact sheet: Agriculture and suicide. Mental Health Commission of Canada. https://mentalhealthcommission.ca/resource/agriculture-and-suicide/

Moisan-Deserres J, Girard M, Chagnon M, Fournier V (2014) Pollen loads and specificity of native pollinators of lowbush Blueberry. J Econ Entomol 107 1156–1162

Mudryj A, Waugh A, Slater J, Duerksen DR, Bernstein CN, Riediger ND (2021) Dietary gluten avoidance in Canada: a cross-sectional study using survey data. Canadian Medical Association Journal Open 9. https://doi.org/10.9778/cmajo.20200082

Muirhead B (2014) Crying over split milk: The history of dairy supply management and its role in recent trade negotiations. Centre for International Governance Innovation, Special Paper 30. www.cigionline.org

Multsch S Pahlow M Ellensohn J Michalik T Frede H-G Breuer L (2016) A hotspot analysis of water footprints and groundwater decline in the High Plains aquifer region, USA. Reg Environ Chang 16:2419–2428

Munro DB Small E (1997) Vegetables of Canada. NRC Research Press

Mussell A (2021) Stratification in Canadian agriculture: Surveying the implications. Agri-Food Economic Systems, Policy Concepts Paper. http://www.agrifoodecon.ca/

Nasielski J, Srinivasan S, Jansen T, Akram-Lodhi AH (2024) Passion alone is not Sufficient What do we know about young farmers in Canada? In: Srinivasan S (ed) Becoming a young farmer. Rethinking Rural. Palgrave Macmillan, Cham. https://doi.org/10.1007/978-3-031-15233-7_2

National Zero Waste Council (2018) A food loss and waste strategy for Canada. https://nzwc.ca/Documents/NZWC-FoodLossWasteStrategy-EN.pdf

Natural Resources Canada (2022) Potash facts. https://natural-resources.canada.ca/our-natural-resources/minerals-mining/minerals-metals-facts/potash-facts/20521

Nearman A, van Engelsdorp D (2022) Water provisioning increases caged worker bee lifespan and caged worker bees are living half as long as observed 50 years ago. Sci Rep 12(1) https://doi.org/10.1038/s41598-022-21401-2

Newfoundland and Labrador (2023) NL Nutritious food basket. https://www.stats.gov.nl.ca/Statistics/Topics/personalfinance/PDF/NLNutritiousFoodBasket.pdf

O'Brien LV, Berry HL, Coleman C, Hanigan IC (2014) Drought as a mental health exposure. Environ Res 131:181–187

OECD-FAO (2020) Meat. OECD-FAO Agricultural Outlook 2022–2031. https://www.oecd-ilibrary.org/sites/ab129327-en/index.html?itemId=/content/component/ab129327-e

Olimb SK, Robinson B (2019) Grass to grain: Probabilistic modeling of agricultural conversion in the North American Great Plains. Ecol Ind 102:237–245

Oryschak MA, Beltranena E (2020) Reconsidering the contribution of Canadian poultry production to anthropogenic greenhouse gas emissions: returning to an integrated crop–poultry production system paradigm. Poult Sci 99:3777–3783

Pelletier N (2018) Changes in the life cycle environmental footprint of egg production in Canada from 1962 to 2012. J Clean Prod 176:1144–1153

Phillips PWB, Relf-Eckstein J-A, Jobe G, Wixted B (2019) Configuring the new digital landscape in western Canadian agriculture. NJAS—Wageningen J Life Sci 90–91:100295. https://doi.org/10.1016/j.njas.2019.04.001

Pogue SJ, Kröbel R, Janzen HH, Beauchemin KA, Legesse G, Maia de Souza D, Iravani M, Selin C, Byrne J, McAllister TA (2018) Beef production and ecosystem services in Canada's prairie provinces: A review. Agric Syst 166:152–172

Pogue SJ, Kröbel R, Janzen HH, Alemu AW, Beauchemin KA, Little S, Iravani M, Maia S, de Souza D, McAllister TA (2020) A socio-ecological systems approach for the assessment of ecosystem services from beef production in the Canadian prairie. Ecosyst Serv 45:101172. https://doi.org/10.1016/j.ecoser.2020.101172

Purc-Stephenson R, Doctor J, Keehn JE (2023) Understanding the factors contributing to farmer suicide: a meta-synthesis of qualitative research. Rural Remote Health 23: 8189. https://doi.org/10.22605/RRH8189

Rejeb A, Abdollahi A, Rejeb K, Treiblmaier H (2022) Drones in agriculture: a review and bibliometric analysis. Comput Electron Agric 198:107017. https://doi.org/10.1016/j.compag.2022.107017

Robinson SVJ, Cartar RV, Pernal SF, Waytes R, Hoover SE (2023) Bee visitation, pollination service, and crop yield in commodity and hybrid seed canola. Agr Ecosyst Environ 347:108396. https://doi.org/10.1016/j.agee.2023.108396

Rodrigues Viana L, Dessureault P-L, Marty C, Loubet P, Levasseur A, Boucher J-F, Paré MC (2022) Would transitioning from conventional to organic oat grains production reduce environmental impacts? A LCA case study in North-East Canada. J Clean Prod 349:131344. https://doi.org/10.1016/j.jclepro.2022.131344

Secchi MA, Fernandez JA, Stamm MJ, Durrett T, Vara Prasad PV, Messina CD, Ciampitti IA (2023) Effects of heat and drought on canola (*Brassica napus* L.) yield, oil, and protein: A meta-analysis. Field Crop Res 293:108848. https://doi.org/10.1016/j.fcr.2023.108848

Seguin R, Lefsrud MG, Delormier T, Adamowski J (2021) Assessing constraints to agricultural development in circumpolar Canada through an innovation systems lens. Agric Syst 194:103268. https://doi.org/10.1016/j.agsy.2021.103268

Shafiee M, Islam N, Randath DD, Vatanparast H (2022) Most frequently consumed red/processed meat dishes and plant-based foods and their contribution to the intake of energy, protein, and nutrients-to-limit among Canadians. Nutrients 14:1257. https://doi.org/10.3390/nu14061257

Sheppard SC, Bittman S, Donohoe G, Flaten D, Wittenberg KM, Small JA, Berthiaume R, McAllister TA, Beauchemin KA, McKinnon J, Amiro BD, MacDonald D, Mattos F, Ominski KH (2015) Beef cattle husbandry practices across Ecoregions of Canada in 2011. Can J Anim Sci 95:305–321

Shock DA, LeBlanc SJ, Leslie KE, Hand K, Godkin MA, Coe JB, Kelton DF (2016) Studying the relationship between on-farm environmental conditions and local meteorological station data during the summer. J Dairy Sci 99:2169–2179

Shrestha BM, Desjardins RL, McConkey BG, Worth DE, Dyer JA, Cerkowniak DD (2014) Change in carbon footprint of canola production in the Canadian Prairies from 1986 to 2006. Renew Energy 63:634–641

Soroka T (2022) Canada's Temporary Labor Migration Policy: The case of Mexican seasonal agricultural workers. Politeja 6(81):279–303

Statistics Canada (2022a) Volume of butter produced in Canada from 2005 to 2020. https://www150.statcan.gc.ca/t1/tbl1/en/tv.action?pid=3210011101

Statistics Canada (2022b) Thematic maps from the census of agriculture, 2021. Released 23–12–2022

Statistics Canada (2022c) Production and value of honey. https://www150.statcan.gc.ca/n1/daily-quotidien/221214/dq221214f-eng.htm

Statistics Canada 2022–2023 (2021) Census of agriculture. https://www.statcan.gc.ca/en/census-agriculture

Statistics Canada (2023a) Temporary foreign workers in the agriculture and agri-food sectors, by industry. Table: 32–10–0218–01

Statistics Canada (2023b) Winter Wheat: the grain that grows beneath the snow. https://www.statcan.gc.ca/o1/en/plus/3107-winter-wheat-grain-grows-beneath-snow

Sutherland C, Sim C, Gliem S, Smyth SJ (2020) Consumer insights on Canada's food safety and food risk assessment. J Agric Food Res 2:100038. https://doi.org/10.1016/j.jafr.2020.100038

Taylor NGA, Lunongo G, Jago E, Mah CL (2023) Observational study of population level disparities in food costs in 2021 in Canada: a digital National Nutritious Food Basket (dNNFB). Prev Med Rep. https://doi.org/10.1016/j.pmedr.2023.102162

Technomic (2017) Canadian burger consumer trend report. https://www.technomic.com/newsroom/burger-consumption-rebounds-canada

Thompson R, Hagen BNM, Lumley MN, Winder CB, Gohar B, Jones-Bitton A (2022) Mental health and substance use of farmers in Canada during covid-19. Int J Environ Res Public Health 19:13566. https://doi.org/10.3390/ijerph192013566

Valenzuela H (2023) Ecological management of the nitrogen cycle in organic farms. Nitrogen 4:58–84. https://doi.org/10.3390/nitrogen4010006

Vasseur L Catto NR (2008) Atlantic Canada. In: Lemmen, D, Warren, F, Lacroix J, Bush E (ed) From impacts to adaptation: Canada in a Changing Climate 2007. Government of Canada, Ottawa

VCMI (2019) The avoidable crisis of food waste: Technical Report. Value Chain Management International and Second Harvest. https://vcm-international.com/wp-content/uploads/2019/01/The-Avoidable-Crisis-of-Food-Waste-NR-January-17-2019.pdf

Vergé X, Maxime D, Desjardins RL, VanderZaag AC (2016) Allocation factors and issues in agricultural carbon footprint: a case study of the Canadian pork industry. J Clean Prod 113:587–595

Vergé X, VanderZaag AC, Desjardins RL, McConkey B (2018) Synergistic effects of complementary production systems help reduce livestock environmental burdens. J Clean Prod 200:858–865

Warren F, Lulham N (2021) Canada in a changing climate: national issues report. Government of Canada, Ottawa

Wen G, Ma B-L, Vanasse A, Caldwell CD, Smith DL (2022) Optimizing machine learning-based site-specific nitrogen application recommendations for canola production. Field Crop Res 288:108707. https://doi.org/10.1016/j.fcr.2022.108707

White J, Berg AA, Champagne C, Warland J, Zhang Y (2019) Canola yield sensitivity to climate indicators and passive microwave-derived soil moisture estimates in Saskatchewan, Canada. Agric for Meteorol 268:354–362

Williams and Marshall Strategy (2021) Canada: Potato Chips Market to Grow at CAGR of 1.90% until 2025. wm-strategy.com

Winston ML (1987) The biology of the honey bee. Harvard University Press, Cambridge, MA

Yang C, Arhonditsis GB (2022) What are the primary covariates of environmental attitudes and behaviours in Canada? A national-scale analysis of socioeconomic, political, and demographic factors. Eco Inform 69:101661. https://doi.org/10.1016/j.ecoinf.2022.101661

Zammit Young G-W, Blundell R (2023) A review on the phytochemical composition and health applications of honey. Heliyon 9:e12507. https://doi.org/10.1016/j.heliyon.2022.e12507

Zhang Y, Ostrovsky Y, Arsenault A (2021) Foreign workers in Canadian agriculture. Statistics Canada Economic and Social Reports 36–28–0001

Zhao K, Huang G, Luo B, Wu Y (2022) A factorial interval chance-constrained diet model for dairy farms under climate change: A case study for the Province of Saskatchewan, Canada. J Clean Prod 360:132059. https://doi.org/10.1016/j.jclepro.2022.132059

Zhu X, Troy TJ (2018) Agriculturally relevant climate extremes and their trends in the world's major growing regions. Earth's Future 6:656–672

Zimmermann MB, Andersson M (2012) Update on iodine status worldwide. Curr Opin Endocrinol Diabetes Obess 19:382–387

Energy

17

Abstract

Biomass represents the second largest source of renewable energy produced in Canada. Lignite (brown) and subbituminous coals, occurring in Cretaceous and Palaeogene deposits, are mined on the surface in northeast and southeast BC, Alberta, and Saskatchewan. Bituminous coals, occurring in Pennsylvanian (later Carboniferous) and Permian strata, are mined as surface mines in Alberta and BC, and until recently in underground collieries in Nova Scotia. Coal combustion generated 5.76% of electricity, 2.73 % of primary energy consumption, and 107 TWh used in Canada in 2022. Reduced use in electricity production represents responses to climate change concerns, along with the availability of alternatives and economic factors. Oil sands and bitumen represent more than 95 % of Canadian petroleum reserves. The estimated petroleum reserve for Canada increased by 1700% in 1999, due to technological advances, global price increase, and regulatory changes that facilitated use of the unconventional Athabasca Oil Sands. Petroleum reserves in Canada are estimated as 165 x 109 bbl, with more than 95% in the Athabasca Oil Sands, including bitumen and extra heavy oils. Marine petroleum production is currently confined to waters offshore of NL. Moratoria prevent offshore exploration and development in BC Pacific and Arctic waters. Licensing of offshore activity is a provincial responsibility, whereas environmental and safety regulations are federal. Pipelines are extensively used to move petroleum from Alberta westward to BC and the Pacific Ocean (Trans-Mountain), eastward to Ontario and Québec, and southward to USA. Natural gas forms a large component of the total energy mix in Alberta, Saskatchewan, Ontario, Nova Scotia, New Brunswick, and southern Yukon. Nationally, combustion of natural gas provided 11.16 % of electricity, 30.96 % of primary energy consumption, and 1216 TWh in 2022. Canada has 19 active nuclear reactors, with 1 in New Brunswick and the other 18 in southern Ontario. Hydroelectric impoundments represent the majority of the country's 1157 large dams. Hydro accounted for accounted for 61.48% of electricity, 26.44% of primary energy consumption, and 398 TWh of energy use in Canada in 2022. Wind generation represents the fastest growing renewable energy supply. Based on latitude and modal cloud cover, the maximum photovoltaic potentials for solar energy occur in the shortgrass prairies of southwestern Alberta and southern Saskatchewan. The experimental facility at Annapolis Royal (NS) is the sole operational tidal power plant.

Keywords

Biomass · Coal · Petroleum · Nuclear energy · Hydroelectricity · Renewable energy

All energy derived from organic material by humans utilizes primary productivity, the rate of new energy additions to the biosphere each year. Rapid storage and slow usage results in net accumulation, as occurs in fast-growing trees such as *Populus*. On the other hand, petroleum accumulates very slowly and is used rapidly. When primary productivity is less than consumption, the net result is the release of previously stored and collected energy. The choice of any fuel derived from organic matter involves balancing productivity and consumption, in addition to considerations of availability, economics, and emissions (Figs. 17.1 and 17.2).

17.1 Biomass Energy

17.1.1 Wood

Biomass represents the second largest source of renewable energy produced in Canada. Biofuels, including wood, have

N. Catto and G. R. Vincent, *Canada: Land and People*, World Regional Geography Book Series,
https://doi.org/10.1007/978-3-032-04391-7_17

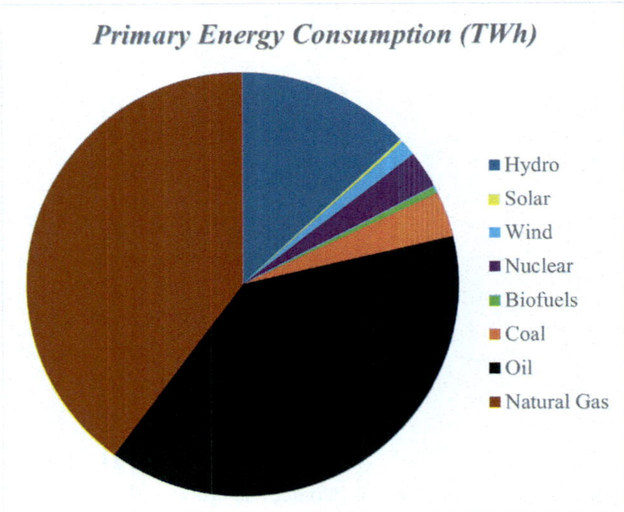

Fig. 17.1 Primary energy consumption (TWh) proportions by source (Canada 2022a, b). Primary energy consumption measures total domestic energy demand (Canada Energy Regulator 2023a)

been used by humans for at least 1.5 million years (Gowlett 2016). Wood burning remains common in Canada, with about 19% of Canadian households (3 million) burning wood at their primary dwelling in 2019 (Statistics Canada 2021b). About half of the 5% of Canadian households owning a cottage, cabin, or recreational farm burned wood at the secondary dwelling. Wood is typically sold in cords, 3.62 m^3, using 1.22 m (4') logs. Estimated consumption in 2019 was approximately 4 million cords (14.5 million m^3), 5 m^3 per household, producing about 23,000 MJ (Statistics Canada 2021b). By comparison, global mean fuel wood consumption in 2021 was approximately 0.3 m^3 per capita (Statista 2024). Birch and other hardwoods are the most popular, as burning conifers is less efficient (40% waste), has lower caloric value (10 MJ kg^{-1}), and produces more smoke, soot, and gases (e.g. Murray 2024; Francescato et al. 2008).

Burning wood at 500–600 °C produces black or grey smoke, with unburned fragments, firebrands, soot, ash, and gases. The particles and gases in this smoke produced by incomplete combustion may aggravate respiratory problems. Although smoke produced from burning vegetation is natural, it still can be very hazardous to human health (Stefanidou et al. 2008, Buss et al. 2021, Health Canada 2022a, b, 2023a, b). Wood smoke contains carbon monoxide, nitrous oxides, ozone, and several carcinogenic compounds including benzene and polycyclic aromatic hydrocarbons. Exposure to fine particulate matter concentrated in smoke may also impact cardiovascular health, leading to stress for people with heart problems (Henderson et al. 2011). In Canada, smoke allergies and sensitivities could affect one-third of households. Consequently, the Canadian Lung Association (2023) and other health organizations recommend not using residential wood stoves. Some municipalities, including Montreal (2023), have instituted restrictions or bans on residential wood stoves.

Wood pellets, formed by compressing waste sawdust, bark, fibre, and the residue from pulping, are also used as an energy

Fig. 17.2 Final energy consumption proportions by source (Canada 2022a, b). Final energy consumption refers to utilization by consumers (Canada Energy Regulator 2023a)

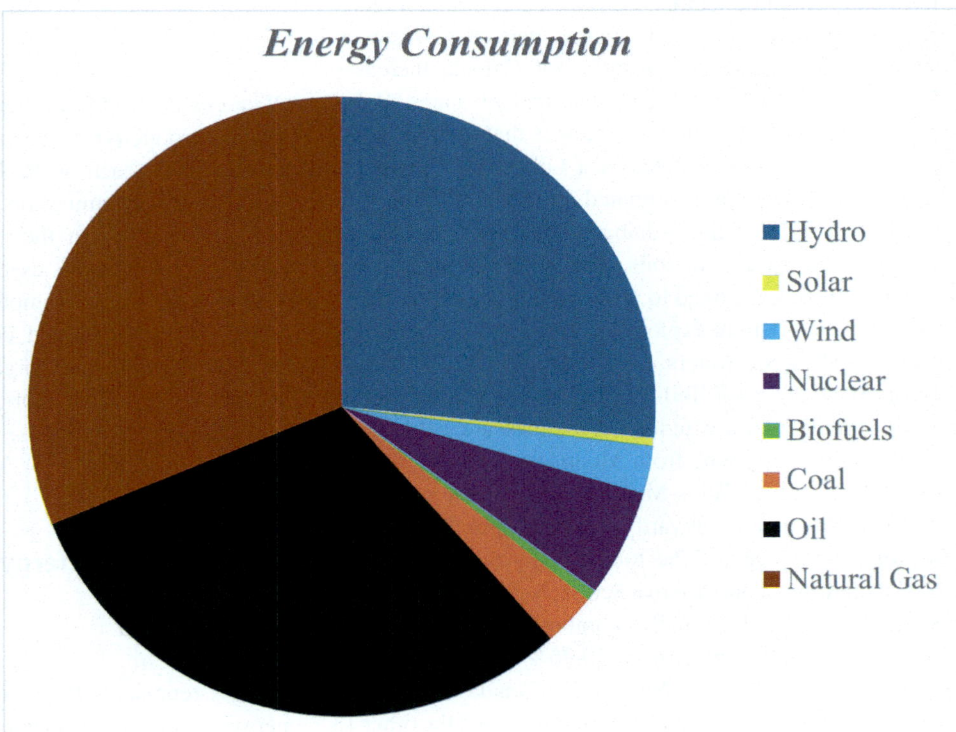

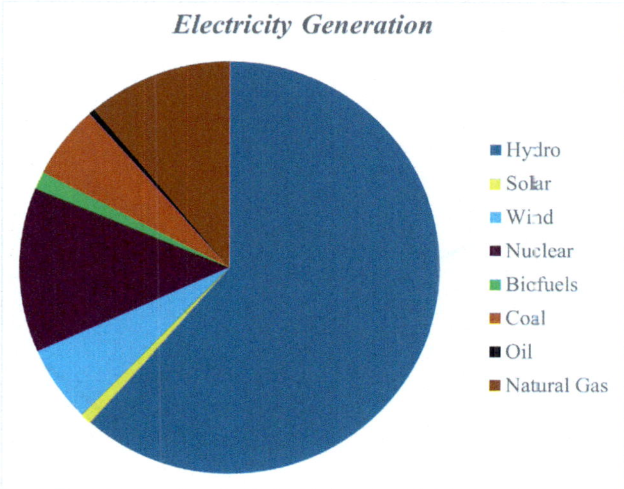

Fig. 17.3 Electricity generation proportions by source (Canada 2022a, b; Canada Energy Regulator 2023a)

source (Canadian Forest Service 2020, Wood Pellet Association of Canada 2023a, b). Typical sawmill operation leaves 20–25% of the original log as sawdust, with bark representing another 5–10%. Of the 3.5 million t produced annually, 85–90% is exported, particularly to the European Union and United Kingdom (USDA 2023). Pellet use replacing natural gas, petroleum, and coal has been identified as a key component to reduce GHG emissions, earning utilities carbon credits. However, the total amount of emission reductions calculated frequently does not include the impacts of Canadian production and transport. Wood pellet use within Canada is relatively low, and largely confined to on-site commercial operations. The 1% of Canadian households using wood pellets in 2019 consumed 196 million kg to produce 3.7 billion MJ (Statistics Canada 2021b). In total, biomass accounted for 0.6% of primary energy consumption and 1.39% of electricity generation in Canada during 2022 (Canada Energy Regulator 2023a; Figs. 17.1 and 17.3).

17.1.2 Peat

Peat is rarely used as a fuel source in Canada. Fibric or mesic peat accumulates at 0.1–4 mm y^{-1}. It produces about as much carbon dioxide per kg as burning bituminous coal, although considerably less sulphur dioxide (Environment and Climate Change Canada 2023a, b). Production from the 300 km^2 used (0.03% of Canada's total peat) is 1.3 million t annually, generating the equivalent of 2.1 Mt CO_2 emissions in 2021 (Environment and Climate Change Canada 2022a, b). Most of the harvest is used in commercial agriculture, both domestically and abroad.

17.2 Fossilized Biomass Energy

17.2.1 Coal

Under pressure over geologic time, peat is gradually converted to coal. In the process of conversion from fresh peat, through lignite, sub-bituminous, bituminous, to anthracite coal, the water content decreases from about 75% to less than 1% in an anthracite. Removal of water increases the carbon concentration by default, from ~60% in dry peat and lignite (not including water) and 70–76% in sub-bituminous to 76–89% in bituminous, and more than 90% in anthracite. Volatile contents also decrease, from as much as 70% in peat and 50–65% lignite to 15–40% in bituminous and typically < 10% in anthracite. The volatiles are the cause of coalmine explosions and miner asphyxiation. Sulphur contents tend to be lower in sub-bituminous coals than in either bituminous or lignite (Kentucky Geological Survey 2024).

Heavy metals, including toxic lead and cadmium, also tend to be higher in lignite and bituminous coal, reaching almost zero in anthracite (Schobert 2013). Extensive burning of coal for residential heating prior to 1950 resulted in heavy metal concentrations in air, water, and soil, in addition to coal-induced fogs and consequent respiratory illnesses (Fowler et al. 2020). Enhanced levels of lead and cadmium persist where coal dust settled, and where ashes and slag were used to infill ground, as in the old core of St. John's (NL), more than 70 years after widespread use ceased (Foley et al. 2011).

Heat contents in peats and lignites are 15–20 MJ kg^{-1}, similar to superior firewood, increasing to 18–23 MJ kg^{-1} in sub-bituminous, 25–30 MJ kg^{-1} in typical bituminous, and 32–33 MJ kg^{-1} in anthracite. Anthracite deposits represent less than 1% of global reserves, with the only large Canadian deposit located 160 km northeast of Stewart, BC (Coal Association of Canada 2024). Canada produces and exports lignite, sub-bituminous, and bituminous coal from 19 active mines in 2024 (Natural Resources Canada 2023 Figs. 17.4, 17.5, 17.6 and 17.7).

Lignite (brown) and sub-bituminous coals, occurring in geologically young Cretaceous (146–65.5 Ma) and Palaeogene (65.5–23 Ma) deposits, are mined on the surface in northeast and southeast BC, Alberta, and Saskatchewan (Coal Association of Canada 2024; Table 17.1). Strip mining involves progressive backfilling and reclamation as mining progresses, whereas reclamation only takes place after the conclusion of open-pit mining. Bituminous coals, occurring in Pennsylvanian (later Carboniferous, 323–299 Ma) and Permian (299–252 Ma) strata, are mined as surface mines in Alberta and BC, and until recently in underground collieries in Nova Scotia. Deposits in Nunavut and Yukon are not currently mined.

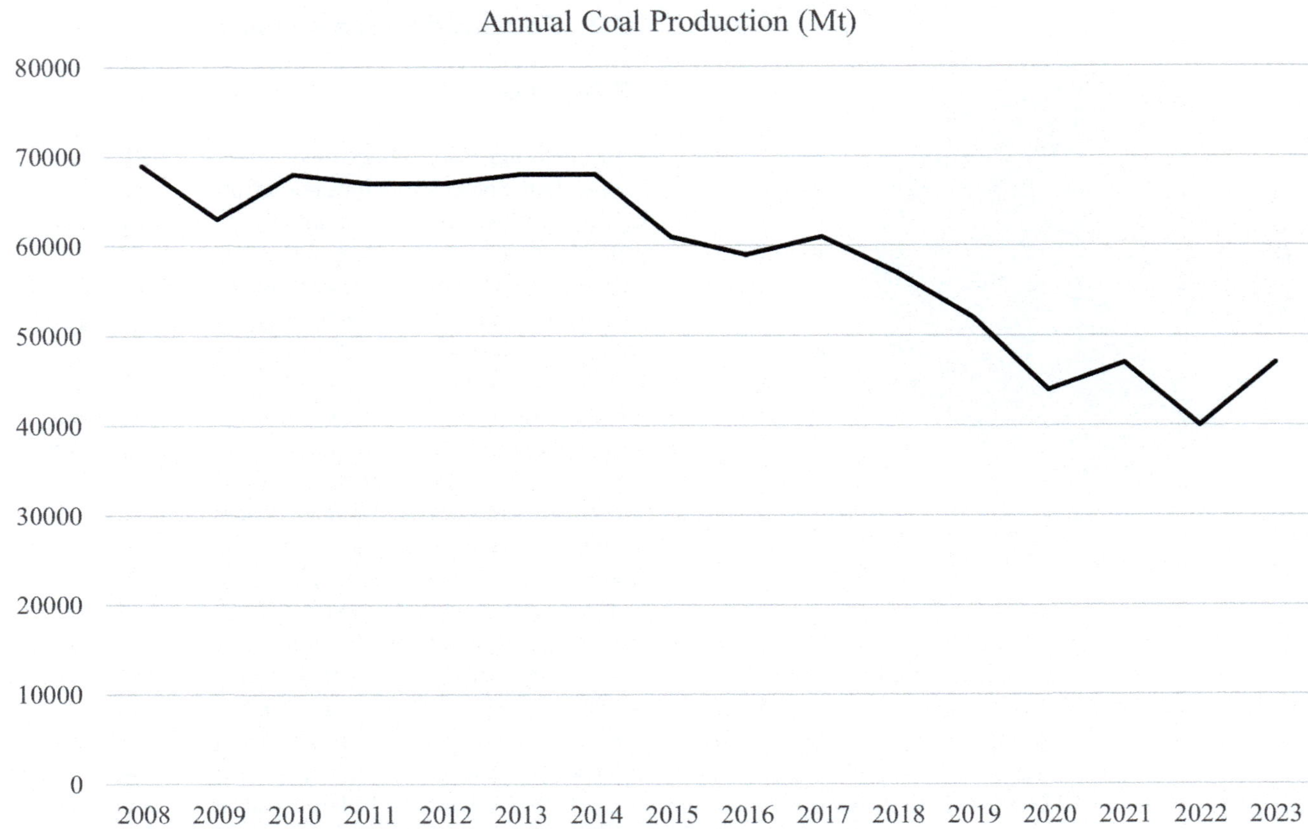

Fig. 17.4 Coal production (Mt), 2008–2023. (Canada Energy Regulator 2023a; Natural Resources Canada 2023, Statistics Canada Census of Canada)

In Alberta, thermal power plants were located next to large strip mines, minimizing the transport distance for sub-bituminous and bituminous coal. At Wabumun, west of Edmonton, coal was transported by conveyor belt under the Yellowhead Highway directly to the plant. The current plan of the Government of Alberta is to decommission all these coal plants by 2024, advanced from 2030, and to stop burning coal as a source of electricity (Alberta 2024b).

Lignite and sub-bituminous coals, increasingly exported rather than burned domestically, are used primarily to generate electricity (Coal Association of Canada 2024; Fig. 17.7). The low sulphur content of sub-bituminous coal makes it a preferred alternative to lignite for reducing acid rain and wastewater pollution, together with the higher caloric value. Bituminous coal is also used or exported to generate electricity, as well as in the production of coke for steel manufacture. A ready supply of metallurgic coal was critical in the development of the steel industry in Cape Breton (NS), prior to closure of the last undersea mine in 2001.

17.2.1.1 Consumption

Coal combustion generated 5.76% of electricity, 2.73% of primary energy consumption, and 107 TWh used in Canada in 2022 (Canada Energy Regulator 2023a; Figs. 17.1, 17.3 and 17.8). Reduced use in electricity production represents responses to climate change concerns, along with the availability of alternatives and economic factors. Ontario, which imported most of the coal used to generate 7,560 MW of electricity in 2003, had eliminated all coal-fuelled plants by 2015, with two converted to biofuel use. Coal is burned for electricity production in Alberta (7%; see Alberta 2024b), NB (13.9%), SK (41%), and NS (52%). The federal government, which does not have jurisdiction over resource management but is responsible for national climate change policy, has planned to phase out all use of coal for electricity production by 2030 (Canada 2023a, b). In Nova Scotia, political pressure to keep local mines operational, coupled with (perceived) lack of options, has resulted in discussion. Metallurgical coal, and coal for export, would continue to be mined.

17.2.2 Petroleum

Liquid hydrocarbon fuel options have been available to humans since petroleum was used in China ca. 347 ce, where the energy of petroleum was used to evaporate seawater

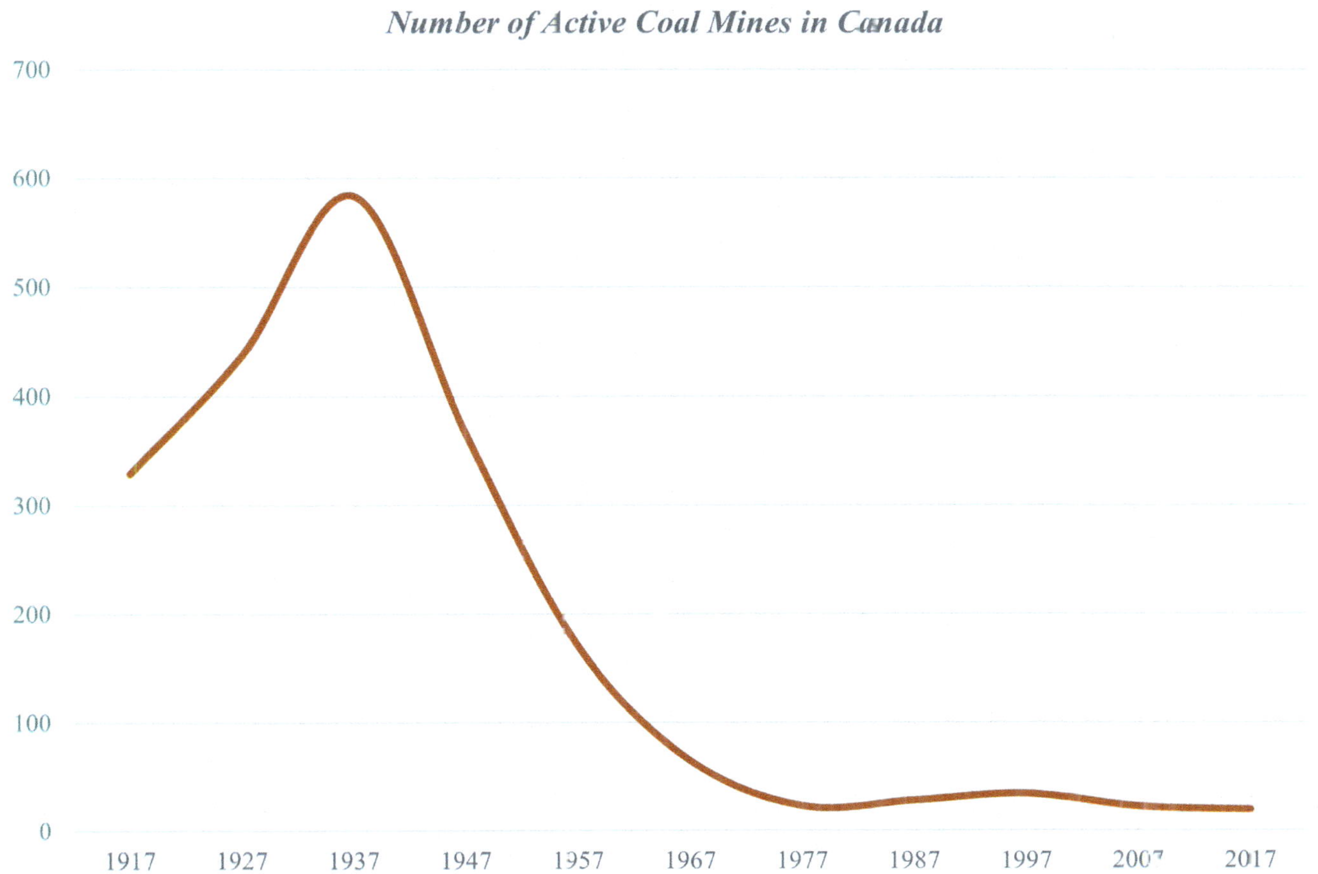

Fig. 17.5 Active coal mines in Canada, 1917–2017. Currently, there are 19 active mines (see Table 17.1). (Natural Resources Canada 2023, Statistics Canada Census of Canada)

to produce salt (Suihong 2016). The first commercial oil well was established in 1858 near Sarnia (Petroleum History Society 2023; Table 17.2). Gasoline, produced for the first time in 1860, surpassed declining whale oil production within 5 years.

Petroleum is a hydrocarbon with a variety of concentrations of carbon (82–85% by weight), along with hydrogen (10–14%), nitrogen, oxygen, and sulphur compounds, combined in alkanes (C_nH_{2n+2}, paraffins, 15–60%), naphthenes (C_nH_{2n}, cycloalkanes, 30–60%), aromatic hydrocarbons (including benzene and toluene, 3–30%), and viscous asphaltics ($C_nH_{1.2n}$, 0–10%) (Speight 1999; Schobert 2013). Dissolved bubbles of natural gas, and trace amounts of metals, including cadmium, lead, mercury, nickel, and vanadium, are also present. Petroleum is also categorized by sulphur content, with "sweet" oils having less and "sour" oils higher sulphur contents.

Bitumen consists dominantly of asphaltics remaining after the migration, diagenesis, or human refining of the other components, and varies from viscous liquids to waxy solids (Schobert 2013). The Athabasca Oil Sands and the LaBrea "tar pits" (Los Angeles USA) have natural bitumen deposits. Solidified natural asphalts are present in isolated areas, including the albertite at Albert Mines (NB). Kerosene was first distilled from an albertite-pitch mixture by NB chemist Abraham Gesner in 1846 (Black 1997; Table 17.2). Kerosene (termed "paraffin" in many countries) was initially used in lamps and is a component of many jet fuels.

Petroleum is produced through the decay of marine or lacustrine organisms, undergoing change or diagenesis over time, a process known as catagenesis (Schobert 2013). Eventually they produce kerogen, solidified organic matter. Higher concentrations of hydrogen in kerogen can result in petroleum generation, whereas lower concentrations are associated with natural gas. Burial to depths of 1.5–4 km and geothermal heat input (thermal maturation) produces pyrolysis, with the first droplets of petroleum forming roughly 10 million years after the first organisms decay. Petroleum is still forming today as marine organisms decay, fall to the ocean floor, and gradually undergo alteration, so in 10 million years' time, some petroleum will have formed. The greater the heat at depth, the more liquid and mobile the petroleum is, and the more likely it is then to flow through geological structures and to migrate into geological traps, or to reach the surface.

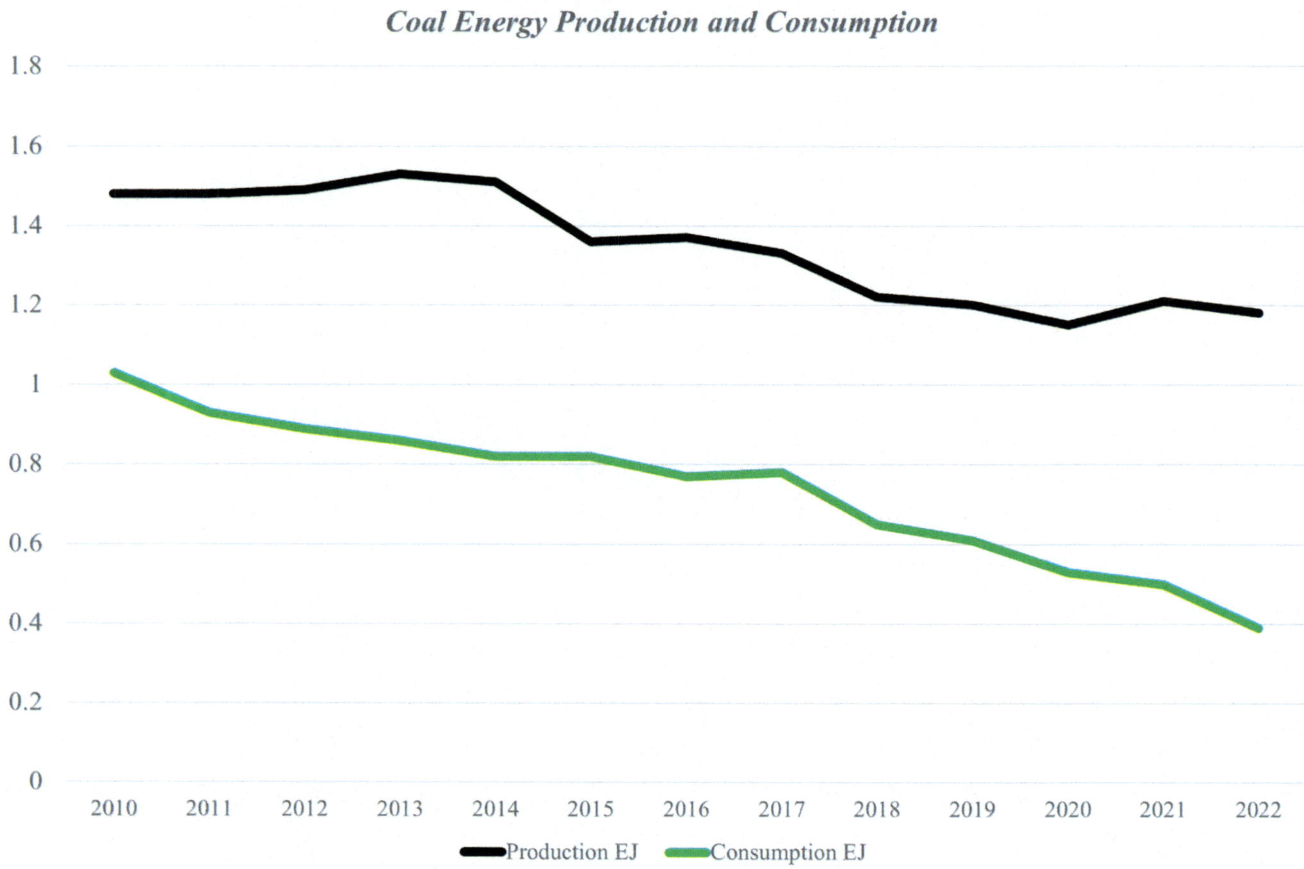

Fig. 17.6 Coal energy production and consumption (EJ, 10^{18} J), 2010–2022. Substantial amounts of coal are exported (Canada Energy Regulator 2023a; Natural Resources Canada 2023)

17.2.2.1 Processing

By convention, petroleum production in North America is measured in barrels (bbl). The barrel volume (159 l, or 42 US gallons) was standardized by producers in Pennsylvania in the 1860s, based on the English tierce used for wine. As the volume standard is based on specific temperature (15.6 °C) and standard atmospheric pressure conditions, a barrel of oil will have different volumes at different temperatures. Petroleum types with different densities ("light" and "heavy" oils) will expand and contract at different rates with temperature changes, complicating conversion from barrels to m^3, commonly used outside North America. Conversion factors are specific to the petroleum involved, typically 6.2898 bbl m^{-3} for "light" Alberta petroleum (Canada Energy Regulator 2016). Production is also measured in barrels per day, including the total volume of fluids recovered (petroleum, natural gas, and water or brine). Production of 1 barrel per day equates to ~ 50 t per year (Fig. 17.9).

Crude petroleum is refined ("cracked") by heating and fractional distillation (Canadian Fuels Association 2022, 2024). Typically, butane and propane are produced at low temperatures (20 °C), with gasoline (150 °C), kerosene (200 °C), diesel (300 °C), and heating fuel oil (370 °C) produced in sequence (Fig. 17.10). A single barrel can be refined to produce 70–85 l of gasoline, 10–15 l of kerosene (or jet fuel), 40–50 l of diesel, and 5–10 l of heating fuel oil. Some litres are lost during the refining process.

17.2.2.2 Unconventional Resources and Reserves

Resources and reserves are classified by the nature of the petroleum (Gordon 2012; SPE 2018; Radwan et al. 2023). Conventional petroleum, which flows readily when drilling allows it to reach the surface, represents ~33% of global and ~3.6% of Canadian reserves (bp 2023). Several types of "unconventional" oil comprise the remainder. "Heavy" oil and "extra heavy" oils, less readily flowing, represent a combined ~30% of global and ~5% of Canadian reserves. Deep source oils (extracted from $\geq$ 4600 m depth) represent ~2% of global reserves. Oil sands and bitumen represent ~33% of global and more than 95% of Canadian reserves (CAPP 2023; Alberta 2024a).

Technological advances in the extraction and processing of unconventional petroleum have resulted in a substan-

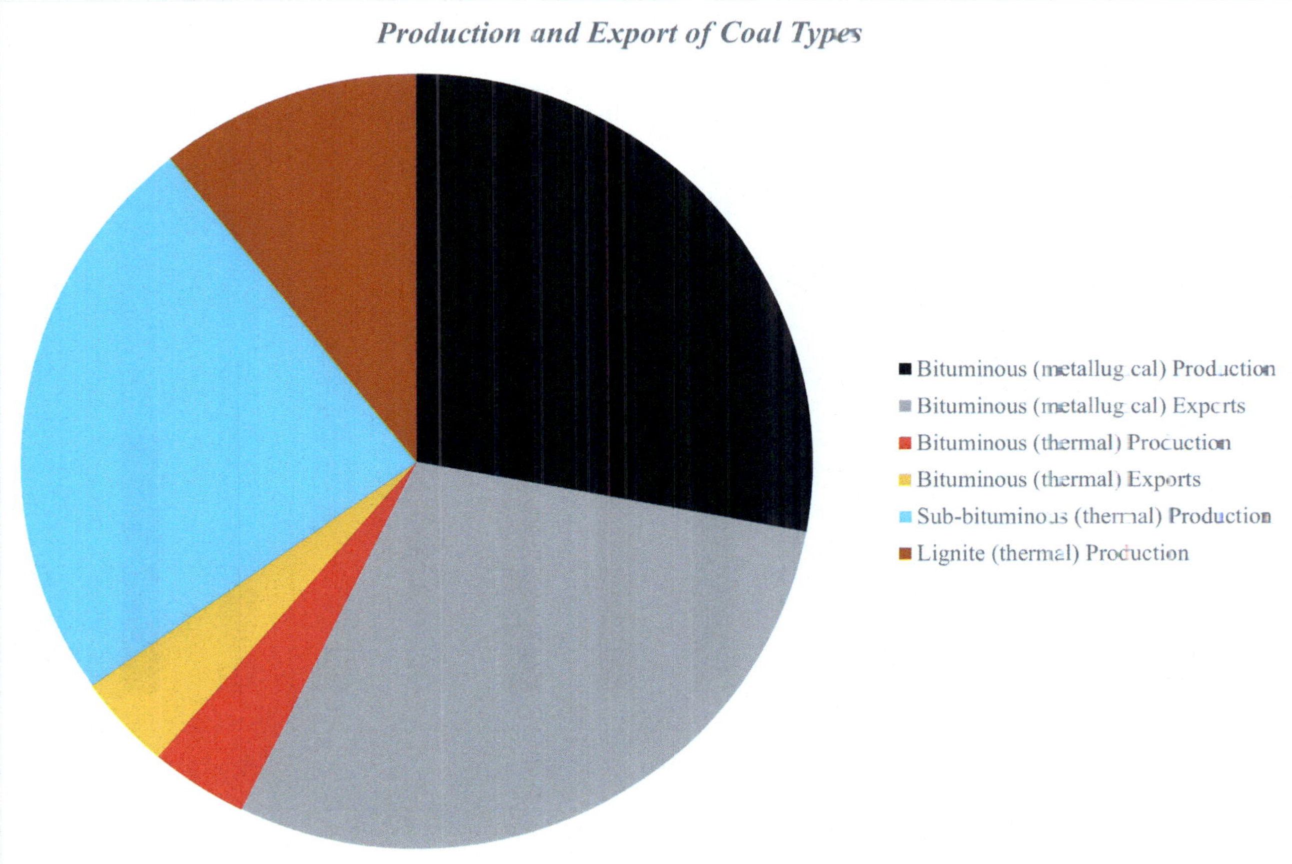

Fig. 17.7 Production and export of coal types, 2022 (Natural Resources Canada 2023, Statistics Canada Census of Canada)

Table 17.1 Coal mines in Canada 2022

Location	Annual production 2022	Coal type
Fording BC	8.5 Mt	Metallurgical bituminous
Elkview, Sparwood BC	7 Mt	Metallurgical bituminous
Elkford BC	5 Mt	Metallurgical bituminous
Line Creek, Sparwood BC	3.5 Mt	Metallurgical bituminous
Sparwood BC	2.7 Mt	Metallurgical bituminous
Tumbler Ridge BC	2.5 Mt	Metallurgical bituminous
Cardinal River AB	2 Mt	Metallurgical bituminous
Tumbler Ridge BC	1.5 Mt	Metallurgical bituminous
Willow Creek BC	1.2 Mt	Metallurgical bituminous
Donkin NS	Currently inoperative	Metallurgical bituminous
Coal Valley, Edson AB	2 Mt	Thermal bituminous
Genesee AB	4.7 Mt	Thermal subbituminous
Ryley AB	0.04 Mt	Thermal subbituminous
Estevan SK	6 Mt	Thermal lignite
Coronach SK	3.3 Mt	Thermal lignite

tial increase of proven reserves since 1995: not a larger oil resource, but more reserve technologically and financially available for processing (Radwan et al. 2023). Changes in reserves reflect technology and regulation more than exploration and geology. The estimated petroleum reserve for Canada increased by 1700% in 1999, due to technological advances, global price increase, and regulatory changes that facilitated use of the unconventional Athabasca Oil Sands

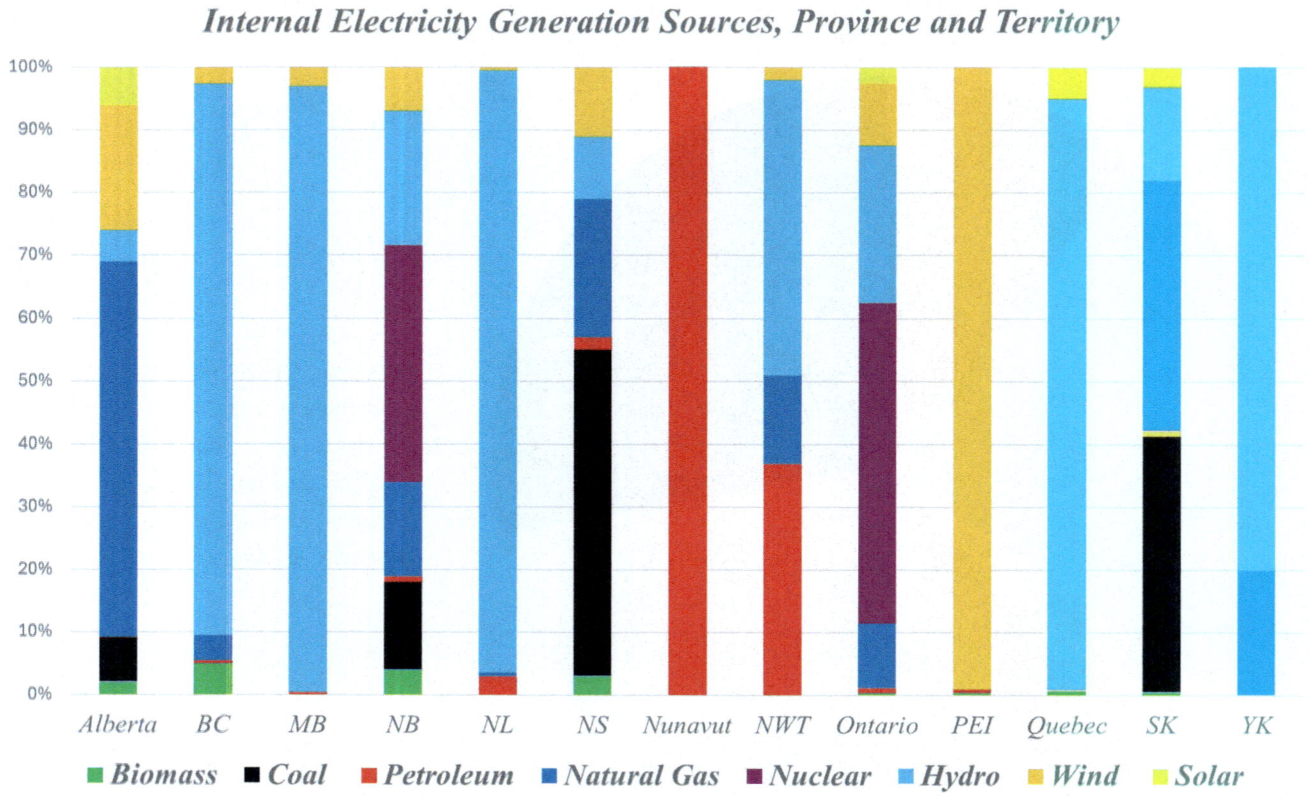

Fig. 17.8 Internal electricity generation sources, province and territory, 2022. Proportions do not include electricity transmitted across provincial lines. See text for discussion (Canada Energy Regulator 2023a)

(Fig. 17.11). Petroleum reserves in Canada are estimated as 165×10^9 bbl (2023), with more than 95% in the Athabasca Oil Sands, including bitumen and extra heavy oils (bp 2023; CAPP 2023; Alberta 2024a).

Oil seeps along the Athabasca River were identified by Indigenous residents shortly after their arrival ca. 9000 bce, and were first reported in writing ca. 1790 ce. In addition to the largest deposit north of Fort McMurray, other oil sands are located north of Cold Lake, along the Peace River, and in the Buffalo Head Hills (CAPP 2023). Petroleum originating from Devonian formations at depth has migrated upward to occupy void spaces in Cretaceous McMurray Formation sandstones and siltstones. The deposits are predominantly marine.

Petroleum production from oil sands is both water-intensive and energy-intensive, with the energy supplied by natural gas (Alberta 2024a). Production of 2.5–3 million bbl per day requires ~ 2 t of sand to produce 1 barrel of oil. Ex situ production (~40–45% of the total) involves excavation and heating the sand with warm water to liquify the bitumen, followed by discarding the processed sand and sending the remaining mud–oil slurry mixture to surface tailings ponds (CAPP 2023). Between 2 and 4.5 barrels of water are required to produce 1 barrel of synthetic petroleum. In situ production (~55–60% of the total), used for deposits ≥ 75 m depth, involves injection of steam or hot fluid to liquefy the bitumen at depth, along with hot air to stimulate combustion and get the petroleum to rise. In situ methods have higher efficiency and lower surface footprints than excavation, but are more costly.

Once bitumen is extracted, it is sent to upgraders (CAPP 2023; Alberta 2024a). Bitumen is too thick and viscous to flow through regular pipelines, and has too many contaminants to be refined as is conventional crude oil. Upgraders mix light hydrocarbons into bitumen, remove some impurities, and produce a refined product referred to as "synthetic crude". Extraction and upgrading are energy-intensive, with ~1 GJ required to produce 1 barrel of synthetic crude (delivering ~6.12 GJ).

Marine petroleum production is currently confined to waters offshore of NL (Table 17.2). Moratoria prevent offshore exploration and development in BC Pacific and Arctic waters. Licencing of offshore activity is a provincial responsibility, whereas environmental and safety regulations are federal (Environment and Climate Change Canada, Fisheries and Oceans Canada, Coast Guard). The joint Canada-NL Offshore Petroleum Board provides regulatory oversight. Annual production varies between 80 and 100 million bbl (Newfoundland and Labrador 2023),

Table 17.2 Events 1846–2023 in Canadian petroleum and natural gas industries

1846	A Gesner distills kerosene, NB
1851	CN Tripp forms company to produce asphalt and naphtha, Oil Springs ON
1858	JM Williams strikes oil in dug water well, Oil Springs ON
1859	HC Tweedle finds natural gas near Moncton NB
1860	JM Williams opens refinery in Hamilton ON
1864	More than 25 refineries operating in Petrolia and Sarnia ON
1866	Natural "sour" gas well at Port Colborne ON
1870	Oil springs resource exhausted
1883	Natural gas well at Langevin AB
1889	First producing gas well in Canada, Essex County ON
1897	Sarnia ON refinery begins production
1902	Oil well at Waterton Lakes AB
1904	Production from natural gas sand at Medicine Hat AB
1907	Ontario bans export of natural gas and electricity
1908	Production from natural gas well at Cochrane AB
1914	Turner valley AB oil field developed
1921	First oil drilling at Norman Wells NWT
1930	Alberta legislates to regulate gas wells
1934	Moose Jaw SK refinery begins production
1936	Burnaby AB refinery begins production
1938	Alberta Petroleum and Natural Gas Conservation Board formed (now ERCB)
1942	Canol pipeline and refinery at Norman Wells constructed
1947	Leduc AB discovery
1947	Lloydminster refinery begins production
1949	Construction of interprovincial pipeline begins
1953	Construction of trans-mountain pipeline spurred by Korean war
1954	Alberta gas trunk Line (now TransCanada) constructed
1955	Montreal refinery begins production
1956	Pembina AB field, 1500 wells
1960	Saint John NB refinery begins production
1966	Panarctic oil formed to exploit Arctic resources
1967	Commercial production from Athabasca oil sands begins
1970	*Arrow* spills 10,000 t of oil into Chedabucto Bay NS
1971	First Beaufort sea artificial island to support offshore exploration
1972	Federal government declares moratorium on BC offshore petroleum development
1973	Come-by-Chance NL refinery begins production
1975	Petro-Canada established
1976	Mackenzie valley pipeline inquiry: Justice Berger report recommends moratorium
1980	National energy program introduced
1980	Norman Wells-Zama (AB) pipeline approved
1982	Feb 15 Oil rig *Ocean Ranger* sinks offshore of NL, killing all 84 people aboard
1984	National energy program repealed with change of Federal government
1985	Panarctic oil production from Bent Horn, Cameron Island
1996	Bent horn decommissioned
1997	Hibernia platform installed offshore of NL
1999	Sable Island NS begins production
1999	Technological advances increase Athabasca oil sands reserves

(continued)

Table 17.2 (continued)

1846	A Gesner distills kerosene, NB
2002	Terra Nova offshore begins production
2005	White Rose offshore begins production
2013	Derailment and explosion at Lac Mégantic Québec kills 47 people
2015	Athabasca oil sands exceed 90% of total Canadian petroleum reserves
2016	Federal govt. declares moratorium on Arctic offshore petroleum development
2017	Mackenzie valley pipeline initiative abandoned on economic grounds
2017	Sable Island decommissioned
2017	Hebron offshore platform begins production
2020	Michigan gives notice of intent to decommission Enbridge Line #5
2023	Come-by-Chance refinery converted to biofuel production

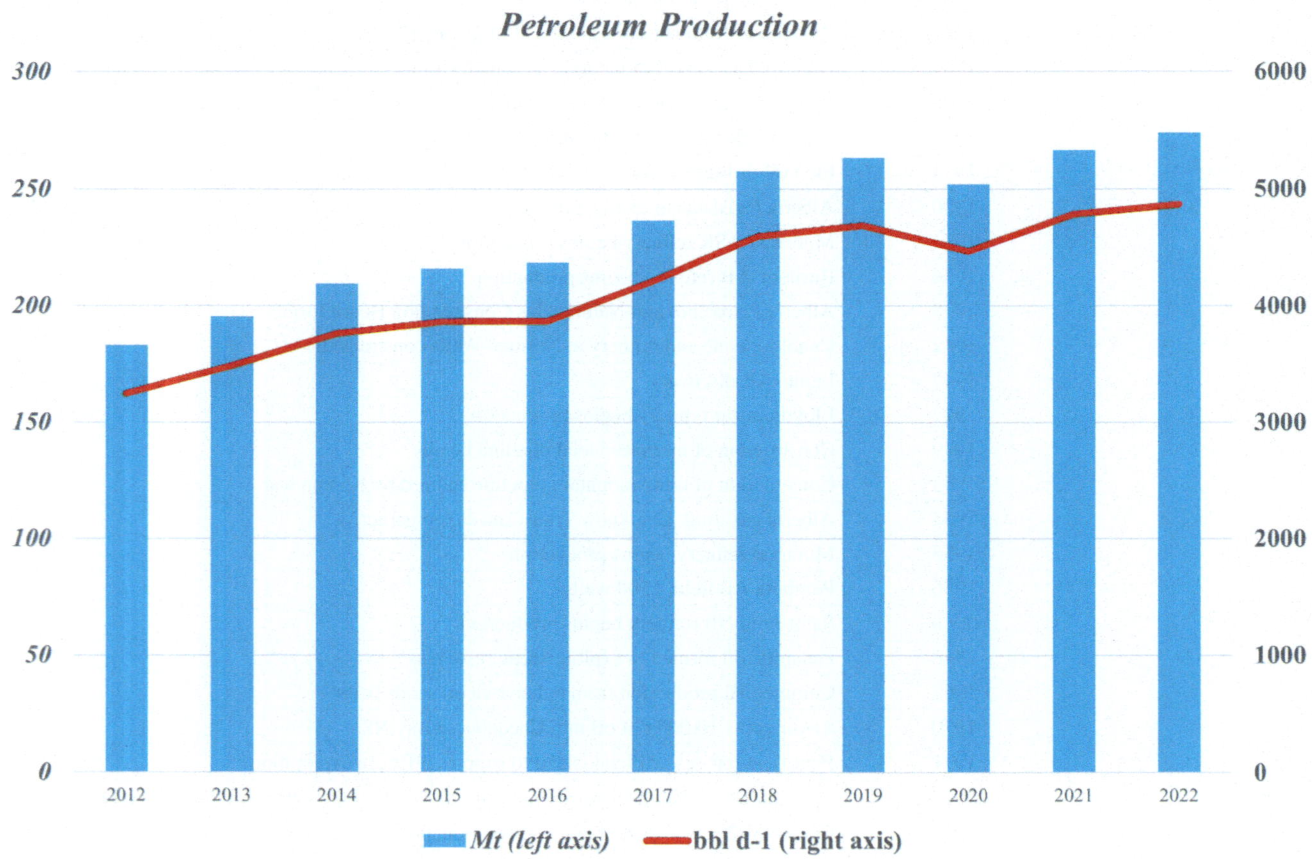

Fig. 17.9 Petroleum production in Mt and barrels d^{-1}, 2012–2022 (Canada Energy Regulator 2023a; Natural Resources Canada 2023)

Induced seismicity with small earthquakes has resulted from conventional hydrocarbon production, involving injection of wastewater into adjacent wells (USGS 2022). Over time, continued water injection will decrease the hazard, as systems become saturated. Unfortunately, the effectiveness of the technique in enhancing petroleum production also decreases. Small earthquakes can also be triggered by unconventional production techniques (including fracking).

17.2.2.3 Transport

Pipelines are extensively used to move petroleum from Alberta westward to BC and the Pacific Ocean (Trans-Mountain), eastward to Ontario and Québec, and southward to USA. Canadian oil represents 50–60% of total USA imports, and 97% of Canadian oil exports (Canada Energy Regulator 2023b; Fig. 17.12). While export of petroleum to USA is straightforward from a Canadian governmental perspective, movement across provincial boundaries is not,

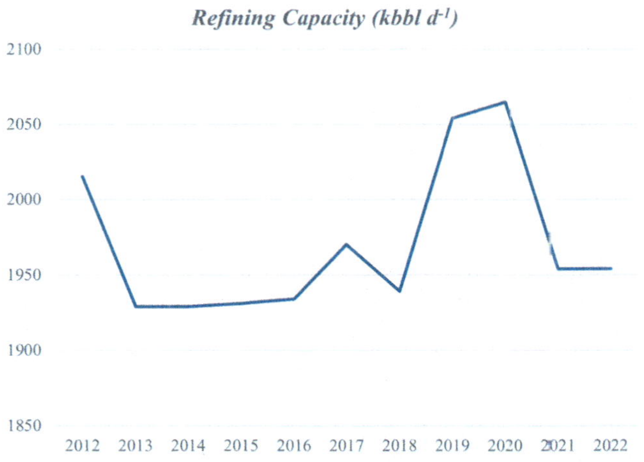

Fig. 17.10 Petroleum refining capacity in Canada, kbbl d^{-1} (Natural Resources Canada 2023; Canadian Fuels Association 2022, 2024)

involving federal-provincial areas of jurisdiction (see Plourde 2010). The proposed expansion of the Trans-Mountain line to carry 300,000 bbl d^{-1} provoked jurisdictional clashes between Alberta and BC, along with opposition by several environmental, citizen, and Indigenous groups, only partially resolved by the federal government purchase of the pipeline in 2018.

In eastern Canada, exploitation of petroleum from deposits in Devonian units east of Sarnia in 1858 led to construction of refineries and the concentration of petrochemical industries in the Lake Huron port. Subsequently, refineries to process imported petroleum were constructed in Québec and Saint John (NB) (Canadian Fuels Association 2022). When the Alberta petroleum industry began to develop in the late 1950s, a desire to protect it and simultaneously support existing markets led the federal government to regulate petroleum distribution in eastern Canada in 1961 (Plourde 2010). Alberta petroleum refined at Sarnia supplied western and central Ontario, whereas imported petroleum supplied the area east of the "Ottawa Valley Line" through eastern Ontario and the lower Ottawa River. A pipeline was constructed paralleling the St. Lawrence River and Lake Ontario west to Sarnia, designed so that petroleum could be moved in either direction. This decision solidified the pattern of petroleum transport in eastern Canada. Sarnia received petroleum from Alberta via pipelines through USA, constructed to bypass the difficult geology of the Canadian Shield.

A series of federal-provincial political controversies engulfed Canadian petroleum affairs throughout 1973–1985, involving federal regulation of exports and domestic prices (through taxation), disputes over exploration licencing, and the establishment of provincial and federal (Petro-Canada) petroleum producers (Halpern et al. 1988). The federal National Energy Program (1980), which embodied these policies and also gave preferential treatment to Petro-Canada including production rights in offshore developments, met

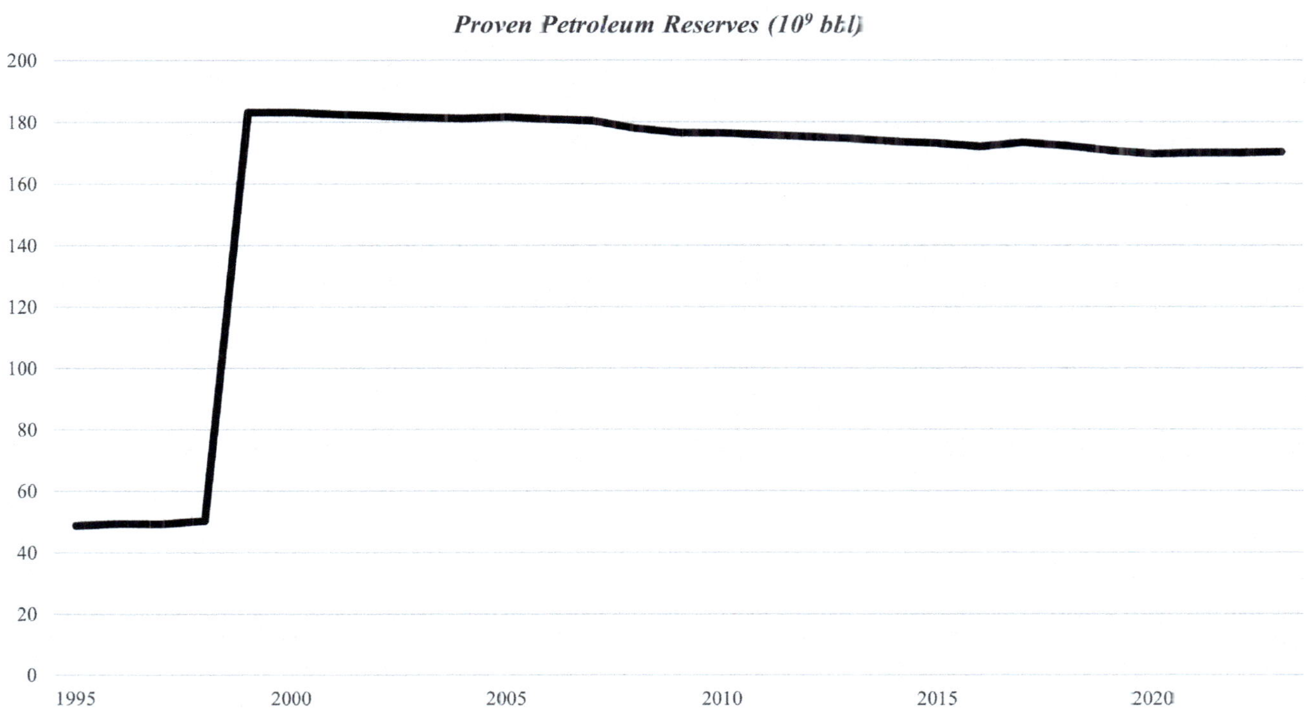

Fig. 17.11 Proven petroleum resources in Canada, 10^9 bbl, 1995–2023. The increase ca. 1999 reflects technological advances in Athabasca Oil Sands production, not new discoveries (bp 2023; CAPP 2023; Alberta 2024a)

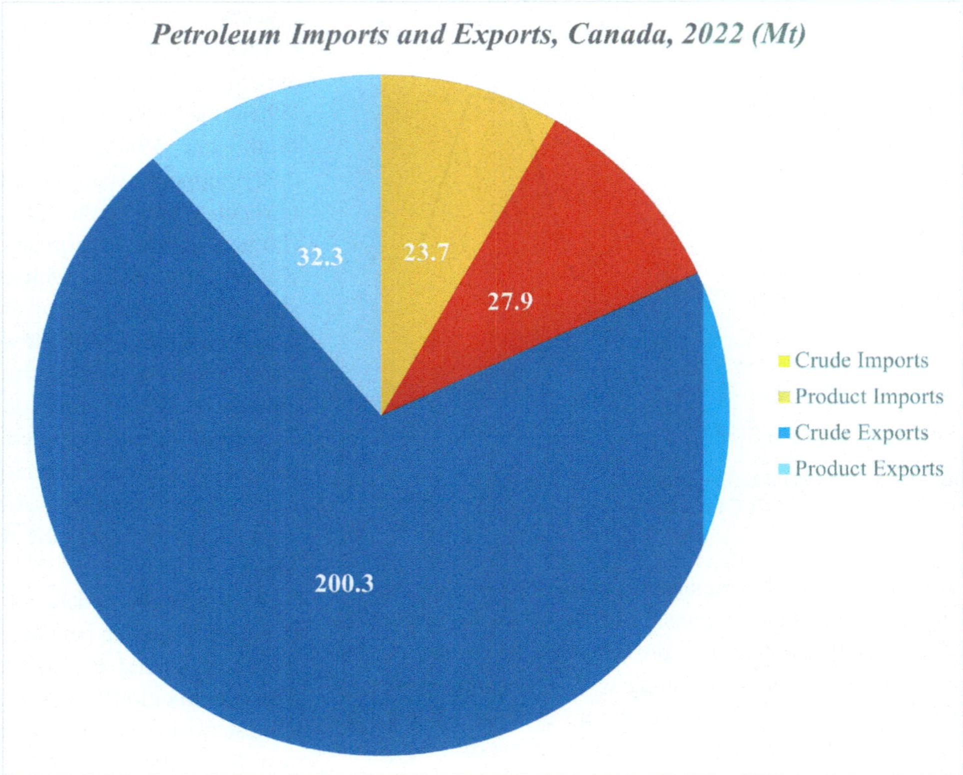

Fig. 17.12 Petroleum Imports and Exports, (Canada 2022a, b). Figures indicate values in Mt (Canada Energy Regulator 2023b)

with stiff resistance in Alberta and Saskatchewan (Table 17.2). Changes in market conditions and federal policy following the change in government following the 1984 election led to gradual federal disengagement, recognition of provincial jurisdiction, and the privatization of Petro-Canada (Plourde 2010). However, the Alberta-Federal rift remained strong in 2024.

Although restrictions on petroleum distribution disappeared, the geography of pipelines did not: Sarnia remained supplied largely from pipelines across Minnesota and Michigan, while Québec and Atlantic Canada primarily used oil imported from Venezuela and Saudi Arabia. No pipelines connect Atlantic Canada and Québec, and most NL offshore petroleum is exported unrefined. In 2020, Michigan gave notice of its desire to decommission Enbridge Pipeline #5, running under the Strait of Mackinac, cited environmental concerns. Decommissioning would significantly restrict petroleum supply to Sarnia, with no alternative pipeline and only rail transport as a theoretical option. As of 2024, this matter is currently under negotiation between governments.

Petroleum products are also transported by vessels, including international traffic and domestic movement of diesel and gasoline. Major spills have resulted from mishaps to the tankers *Arrow* (1970), *Odyssey* (1971), and *Kurdistan* (1979). *Arrow* spilled 10,000 t into the water and shoreline of Chedabucto Bay (NS) (Taylor 2019; Table 17.2).

Rail transport, predominantly exports to USA, has decreased to approximately 130,000 bbl d^{-1}. Accidents are a major concern with rail transport, with spills 4.5 times more likely to occur compared to pipeline flow (Green and Jackson 2015; Vragov et al. 2018; American Geosciences Institute 2019). On 6 July 2013, derailment of a train carrying 38,000 bbls for export at Portland (Maine) caused an explosion and fire that destroyed Lac-Mégantic, Québec, killing 47 people (Transportation Safety Board of Canada 2014).

17.2.2.4 Consumption

Global petroleum consumption currently outpaces exploration finds by about 10:1. Exploration in Canada is minimal at present, due to the available reserve of the Athabasca Oil Sands, the costs of NL offshore efforts, and moratoria in many regions. At present rates of domestic consumption, Canadian petroleum reserves would last 180–200 years, assuming no exporting or new discoveries. "Peak oil" is not a viable model for Canada.

Domestic consumption of oil was 1185 TWh (98.2 million t) in 2022 (Canada Energy Regulator 2023a; Figs. 17.1 and 17.13). Although this represents a drop since 2019, the decline is largely the result of COVID-related changes to travel. Since 2002, the modal rate of consumption has shown little change. Per capita consumption was 30,822 KWh, the third highest value globally, although declining by 25% since the 40,000 KWh peak in 2004 (Fig. 17.13). Including extraction, facility

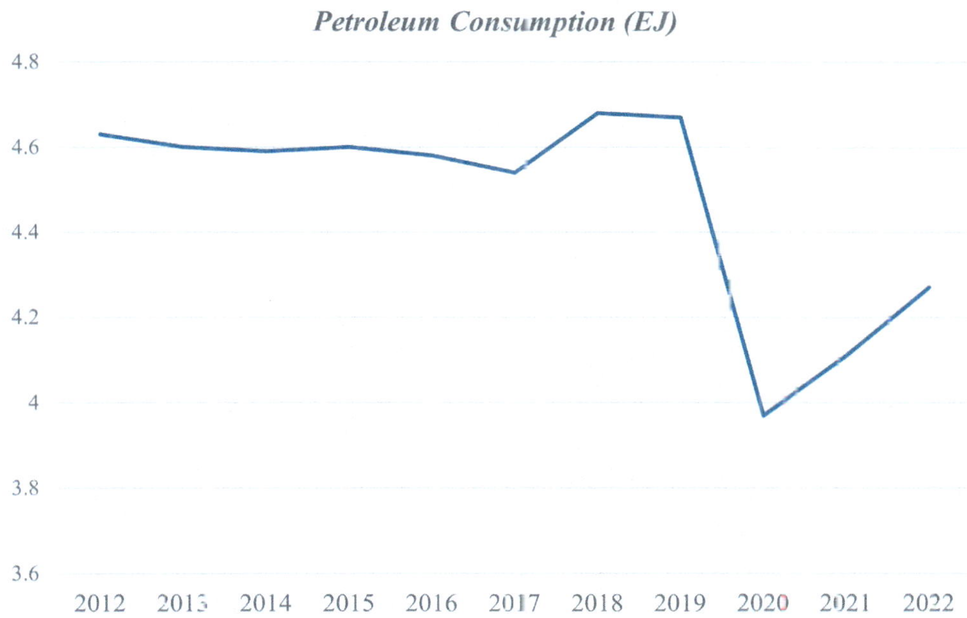

Fig. 17.13 Petroleum Consumption (EJ) in Canada, 2012–2022. Drop in 2020 reflects COVID-19 restrictions on travel and commercial activity (Canada Energy Regulator 2023a)

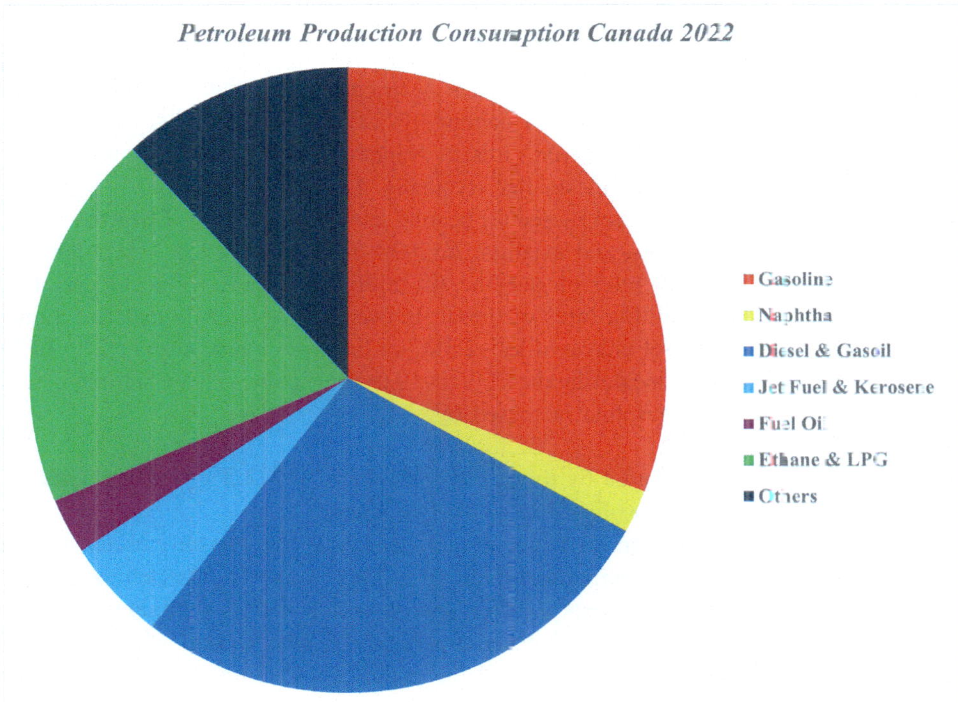

Fig. 17.14 Petroleum production consumption (Canada 2022a, b; Canada Energy Regulator 2023a; Statistics Canada 2023a)

construction, refining and distribution, light petroleum typically results in 210–240 g CO_2 equivalent emitted per KWh (Environment and Climate Change Canada 2023a, b).

Most of the petroleum is used for transportation (Fig. 17.14). Petroleum accounts for about 30% of primary energy consumption, representing 1185 TWy^{-1}, but only 0.47% of electricity generation, primarily as diesel fuel in northern and isolated communities. Use of oil for home heating has declined in most of Canada, with the exceptions of NL (19% of homeowners in 2021), NS (33%, and PEI (40%) (Statistics Canada 2023a). In NB, government incentives to convert have decreased the percentage of homeowner users of heating oil to 7%.

Gasoline represents 30–35% of petroleum usage in Canada. Distillates, including diesel, heavy oil for shipping, and heating oil also account for 30–35%. Jet fuel and asphalt represent 5–8% combined. As some gasoline and other fuel is used for industrial purposes, transportation encompasses about two-thirds of petroleum usage in Canada, industrial uses representing about 30%. With the exception of the period of COVID-19 travel and work-from-home restrictions, gasoline usage closely tracks overall petroleum consumption.

17.2.2.5 Northern and Arctic Communities

Approximately 250 communities depend exclusively on burning diesel fuel for electricity generation, including every community in Nunavut, Nunavik, and Nunatsiavut (Knowles 2016). In some areas, diesel fuel can be transported over winter roads, or by marine transport across open water. For coastal northern taiga and Arctic communities, marine transport of diesel fuel and other hydrocarbons is confined to the short open-water ice-free season (early July-late October in Nunavut). Costs are significantly influenced by operational concerns, including the requirement for ships to use lighter marine diesel oil rather than the former heavy fuel oil for environmental reasons.

Northern communities import and use five types of hydrocarbon fuel (Nunavut 2023). Premium 92-octane gasoline fuels automobiles, outboard motors ("kickers") for small boats, and snowmobiles. A-1 jet fuel (Avtur) is used for aircraft, and can also be burned to produce heat and electricity. Aviation gasoline (Avgas) is much less commonly used, in internal combustion propellor driven aircraft. Naphtha, used for camping fuel, is sold in smaller (typically 4-l) containers. Ultra-low sulphur diesel is used for heating, electricity generation, diesel trucks, and aviation. In 2020–21, Nunavut imported 24 million litres of gasoline, 40 million litres of jet fuel, and 137 million litres of diesel. Between 2016 and 2021, sales of diesel increased by ~2.3%, jet fuel by ~7% (influenced by a decline during the COVID-19 travel restrictions, while sales of gasoline increased by ~ 15.9% (Nunavut 2023).

Fuels are generally transported by truck, although short pipelines are used to move gasoline and jet fuel in some communities. Moving diesel or oil would require heated pipelines, which would induce permafrost melting. In Gjoa Haven, fuel pipelines have become exposed due to permafrost thawing and overland gullying flow. Exposed pipelines are crossed by roads and ATV tracks, increasing the potential hazards of collision or rupture.

Burning diesel fuel produces substantial GHG emissions, particularly nitrous oxides, in addition to particulates and heavy metals. Bioaccumulation (accumulation in individual organisms) and biomagnification (transfer of toxins to higher trophic levels) of mercury, cadmium, and lead are of concern for people consuming marine-based "country foods" diets, particularly as a source of zinc and vitamins-B6 and -D during winters. Unavailability or unsuitability of arctic char and seal leads to increased consumption of imported and expensive foods from the south, causing further dietary problems. At least half of Nunavut's population suffer from food insecurity.

Northern residents have fewer opportunities to reduce GHG emissions than do other Canadians. With its population of ~40,000, Nunavut currently produces ~600,000 t of CO_2 equivalent GHG, a per capita rate of 15.4 t approximately 13% below the Canadian average. In contrast, Nunavik's diesel generators may contribute as much as 43% of Québec's total CO_2 equivalent GHG, although they produce < 1% of the electricity produced by Hydro-Québec (Tarquti 2023). Québec's other electricity production is from renewable sources, and the per capita CO_2 equivalent GHG is 8.9 t, about half the Canadian average (Environment and Climate Change Canada 2022a, b). Climate and widely spaced communities require expenditures of fuel. In addition, the feeling that northern, particularly Indigenous, residents are paying for previous decisions made by southerners influences responses to some suggestions for fuel conservation from outsiders.

Prices charged for all fuels are subsidized by territorial, regional, and municipal governments, a necessity given high transportation costs, supply irregularities, low population numbers, and socio-economic challenges. There is little incentive for individual residents to adopt different patterns of energy use, even if opportunities were more available (e.g. Pinto and Gates 2022). Communities are increasingly exploring alternatives to diesel, particularly small-scale wind turbine installations.

17.2.2.6 Ethanol

Ethanol is a very common gasoline additive or substitute in Canada. Unlike biodiesel, ethanol can be manufactured from any organic material. Ethanol production in Canada stands at approximately 2×10^9 l y^{-1}, roughly 50% in excess of domestic consumption.

17.2.3 Natural Gas

Humans have used natural gas for about 3000 years. Hydrocarbon gases can be produced as a result of liquifying coal, or from refining petroleum. "Natural gas" is extracted from the ground without processing of pre-existing coal or oil (Natural Resources Canada 2020a). Natural gas varies in composition, including methane (CH_4); alkanes (C_nH_{2n+2}) including ethane (C_2H_6) and propane (C_3H_8); N_2, N_2O, and other nitrogen compounds; sulphur compounds ("sour gas"); and carbon monoxide (CO) (Schobert 2013; Faramawy et al. 2016). When extracted, natural gas is colourless and odourless: Natural gas has a variable intensity when burning, the origin. of the term "gaslighting". From the early nineteenth century, it was extensively used in factories, one of the major factors then that allowed the industrial revolution to take hold. In the 1870s, many people felt that natural gas would be the dominant fuel of the future in Canada. Although natural gas was largely replaced by electricity for lighting by 1905, it is

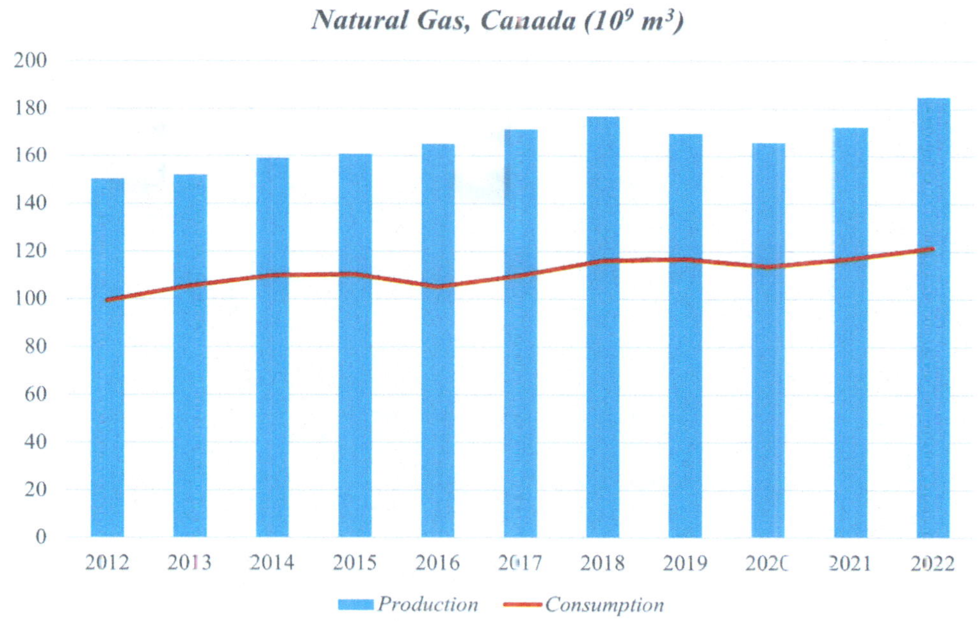

Fig. 17.15 Natural gas production and consumption, 10^9 m^3, 2012–2022 (Canada Energy Regulator 2023a), Statistics Canada Census of Canada

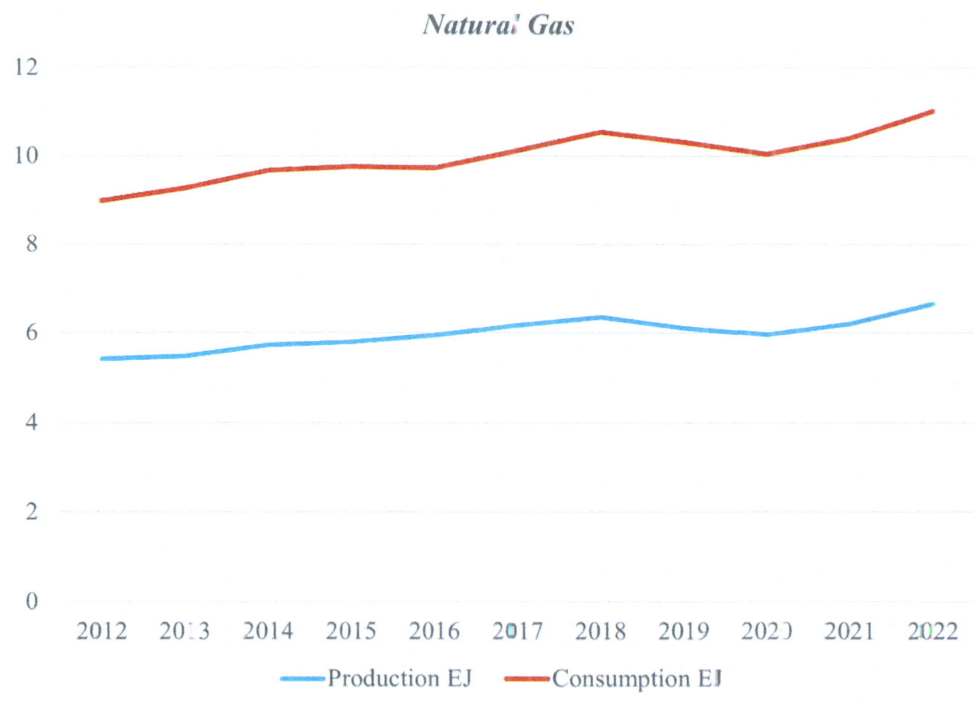

Fig. 17.16 Natural gas production and consumption, EJ, 2012–2022. Difference reflects exports to USA (Canada Energy Regulator 2023a, Statistics Canada Census of Canada)

still extensively used in much of Canada (Natural Resources Canada 2020a; Figs. 17.15 and 17.16). Nationally, combustion of natural gas provided 11.16% of electricity, 30.96% of primary energy consumption, and 1216 TWh in 2022 (Canada Energy Regulator 2023a; b; Figs. 17.1, 17.2 and 17.3).

The first recorded Canadian discovery of natural gas occurred near Moncton (NB) in 1859 (CAPP 2022; Table 17.2). The natural gas currently exploited in Canada is produced from the Western Canada Sedimentary Basin of northeastern BC, Alberta, and Saskatchewan (Fig. 17.17). Much production comes from shale and siltstone, and is referred to as "tight" reservoir gas (Natural Resources Canada 2020a).

Natural gas forms a large component of the total energy mix in Alberta, Saskatchewan, Ontario, Nova Scotia, New Brunswick, and southern Yukon (Canada Energy Regulator 2023a; Fig. 17.8). Usage has declined sharply in Québec, and was always low in Manitoba and NL. Industrial users are the largest consumers, followed by residential and commercial use for heating, and agriculture (Statistics Canada 2023b).

Fig. 17.17 Natural gas reserves (10^6 m^3) in Canada, 2002–2022. Increases since 2006 are largely technology-related (Natural Resources Canada 2020a, CAPP 2022, Canada Energy Regulator 2023a)

The large populations of Alberta and Ontario result in natural gas being used for ~50% of indoor space heating, ~65% of water heating, and by > 80% of businesses across Canada.

Nationally, natural gas is the third-largest source of electricity (trailing hydropower and nuclear energy). It accounts for 60% of electricity generation in Alberta (with a greater proportion used in Calgary and Edmonton), compared to ~40% in Saskatchewan, ~22% in Nova Scotia, ~10% in Ontario, ~4% in BC, and < 1% in Québec and NL. In NWT, Norman Wells uses natural gas generated from local sources.

Alberta and Ontario have different historical relations with natural gas. In Alberta, production developed as a result of petroleum exploitation. Natural gas originally was a by-product, and was frequently disposed of if inconvenient to capture. "Flaring" of gas from oil platforms still occurs, but although annoying from a conservation perspective, the 150 km^3 flared annually around the world compares with the 3400 km^3 consumed annually or the estimated ~200 trillion m^3 global reserve (bp 2023). Alberta produces > 98% of Canadian natural gas annually, much of which is exported the USA.

In contrast, Ontario initially used natural gas derived from the fields around Sarnia, long since exhausted (Table 17.2). Today, natural gas used in Ontario is imported from USA (Natural Resources Canada 2020a). Thus, although Canada is a net exporter of natural gas, by about 45–60 billion m^3 or 3:1, all the gas used in eastern Canada is imported. This phenomenon leads to west–east pricing and supply differences, occasionally fuelling acrimonious commentary and political tension. The different perceptions of natural gas as a problem or a solution in addressing climate change are another consequence, considered as an undesirable import by some and as a replacement for coal consumption by others.

In North America, natural gas is largely moved by pipeline (Fig. 17.18). Proposals for construction of liquified natural gas pipelines and terminals have met with substantial political and social opposition, particularly in Pacific Maritime BC and some Indigenous communities along the proposed routes. In 2024, there were four LNG liquefaction facilities, and two import terminals in Saint John NB and Hamilton ON (see Natural Resources Canada 2020a).

GHG emissions from natural gas differ considerably, due to the variations in gas composition, particularly the ratio of CH_4 to other gases. Emission figures of 125–180 g C equivalent per kWh are commonly quoted. Emission factors for consumed natural gas are approximately 1950 g CO_2 m^{-3} overall, and 0.037–1.9 g CH_4 m^{-3} and 0.033–0.060 g N_2O m^{-3}, varying by user sector (Environment and Climate Change Canada 2023a, b). Although natural gas usage by residents in different provinces varies substantially, annual mean values of 60 Mt CO_2 equivalent for heating with natural gas and an additional 25 Mt for electricity generation (which includes natural gas in some provinces) comprise a substantial proportion of the estimated 315 Mt y^{-1} mean for Canadian households in 2018–2020 (Statistics Canada 2022a, b,

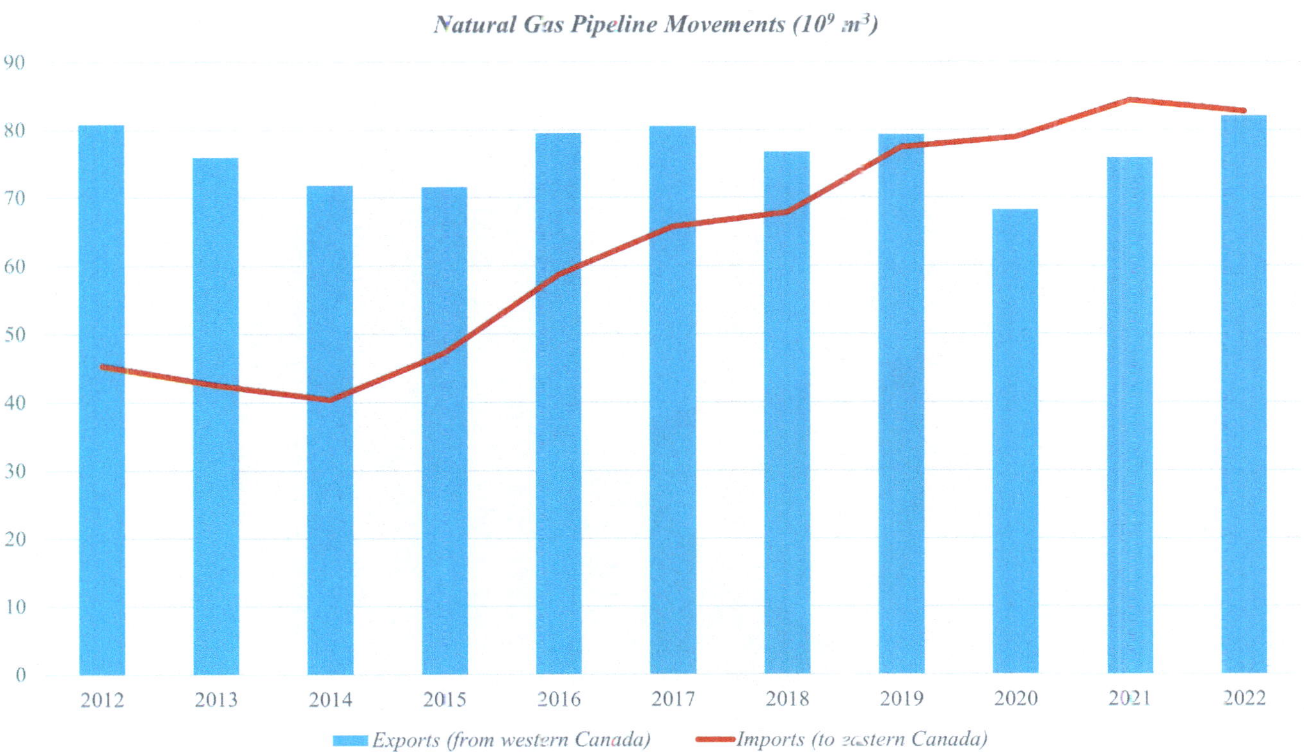

Fig. 17.18 Movements of natural gas by pipeline, 10^9 m^3, 2012–2022. See text for discussion (Natural Resources Canada 2020a; Canada Energy Regulator 2023a)

Fig. 17.19 Canadian household expenses (2021, Canadian $) (Statistics Canada 2022a, b, 2023b, and Census of Canada)

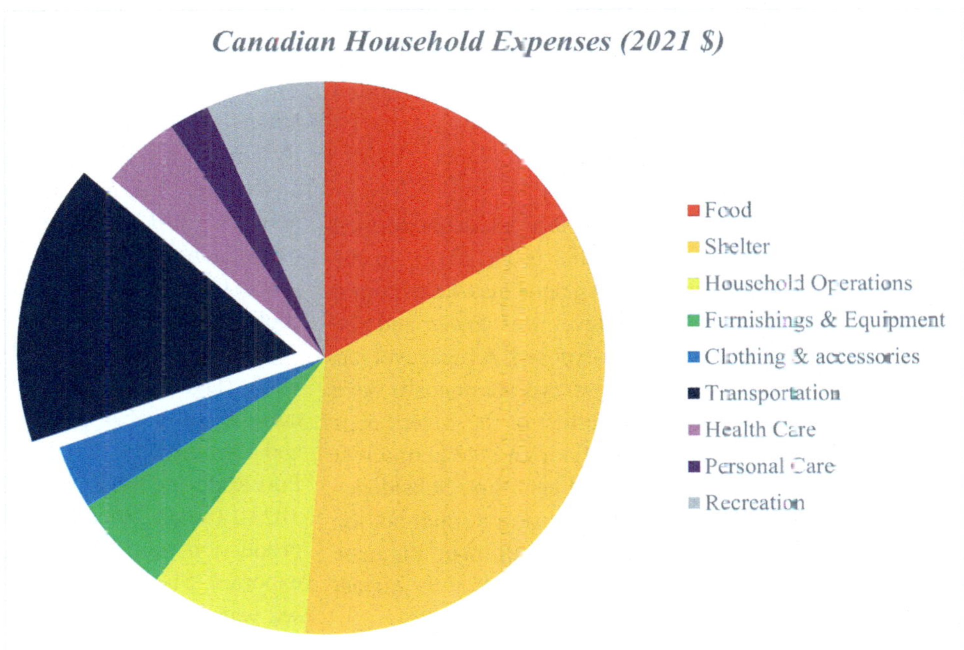

2023b; Figs. 17.19 and 17.20). At 8.61, equivalent CO_2 kg emissions per dollar spent are substantially higher for heating than for electricity generation (1.29), motor fuel usage (2.18), or other categories of household expenditure: fewer dollars are required to produce more heating-related emissions.

Global economic reserves of natural gas are currently estimated to last about 250 years (bp 2023). Canada's gas reserves in 2017 were estimated at 30.8 trillion m^3, 72% of which is from *unconventional* tight gas formations (Fig. 17.17).

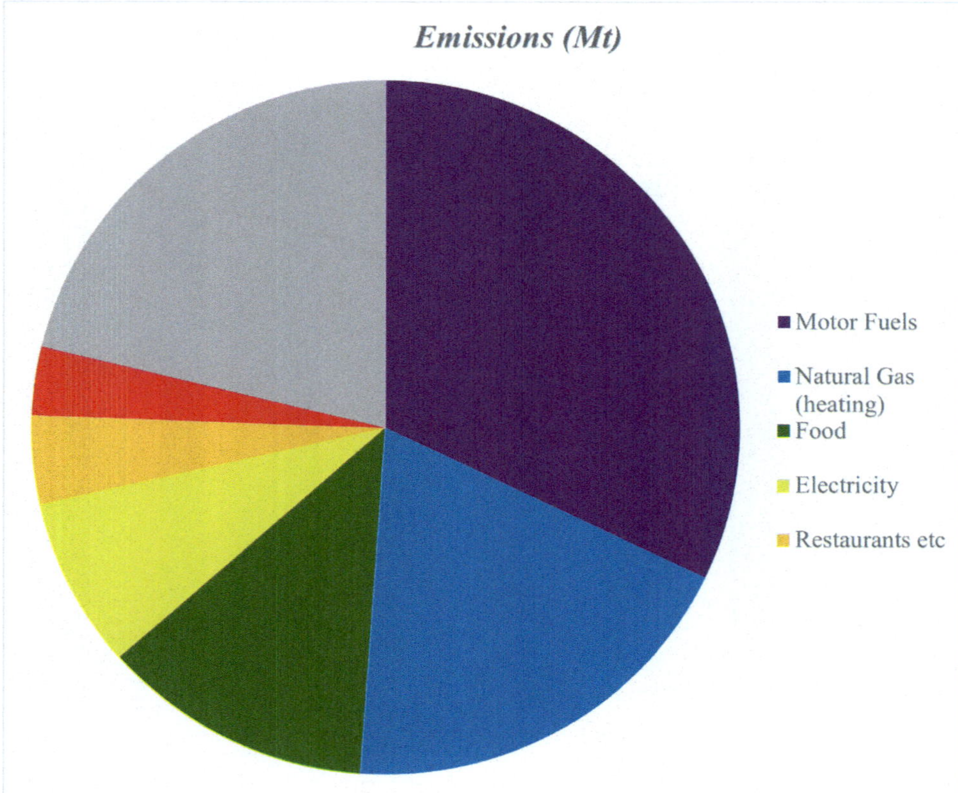

Fig. 17.20 GHG emissions (Mt) by Canadian households, 2018–2020 (Statistics Canada 2022a, b, 2023b)

Propane (first distilled in 1911) is also used to heat individual homes and businesses. Supplies are delivered by truck to exterior tanks. Approximately 2500 l of propane is required to heat a 150 m^2 space annually.

17.2.3.1 Gas Hydrates

Gas hydrates are deposits of solid to slushy methane, formed in areas of Subsea Permafrost (Riedel et al. 2017). In the southern Beaufort Sea during the last glacial maximum period (MIS 2, 25,000 years ago), sea level was lower and the southern parts of the sea floor were exposed. As sea level has risen, the flooded permafrost and gas hydrate deposits have become unstable, as the insulating effect of the water keeps temperatures above 0 °C on the sea floor and the permafrost melts through the input of geothermal heat flux. Subsidence of approximately 1 cm per year is occurring throughout the southern Beaufort Sea. Release of the 110 trillion m^3 methane trapped in the subsea permafrost could significantly impact GHG concentrations.

17.3 Nuclear Energy

Nuclear energy produced 12.89% of electricity, 5.52% of primary energy consumption, and 87 TWh in 2022 (Canada Energy Regulator 2023a; Figs. 17.1, 17.2, 17.3 and 17.21). Canada has 19 active reactors, with one in New Brunswick at Point Lepreau (constructed 1975, reactivated in 2012). The other 18 are located in southern Ontario, with 8 at the Bruce complex on Lake Huron (activated 1977–1987), 6 at Pickering on Lake Ontario (activated 1971–1986), and 4 at Darlington on Lake Ontario (activated 1992–1993, the last newly constructed power reactors in Canada). Each reactor produces between 540 and 878 MW. The last of the two reactors at Gentilly (Québec), along the St. Lawrence River, was shut down in 2012.

The Canadian-designed reactors use deuterium heavy water (D_2O) and natural "non-enriched" ^{238}U. Uranium is currently mined at McArthur River and Cigar Lake (SK), and formerly at Elliot Lake (ON), Uranium City (SK), and Port Radium (NWT) (Canadian Nuclear Safety Commission 2023). Canada accounts for about 20% of global uranium production (Fig. 17.22), and 85% of uranium mined is exported. The plan to produce D_2O at Glace Bay (NS) from sea water was a technical and commercial failure, but other plants using fresh water have been successful. Canada is the largest exporter of heavy water globally. Other smaller reactors are maintained for research and the production of medical radioisotopes. Proposals to develop Small Modular Reactors (100–300 MW) for local electricity and medical radioisotope production are currently under negotiation between various levels of government.

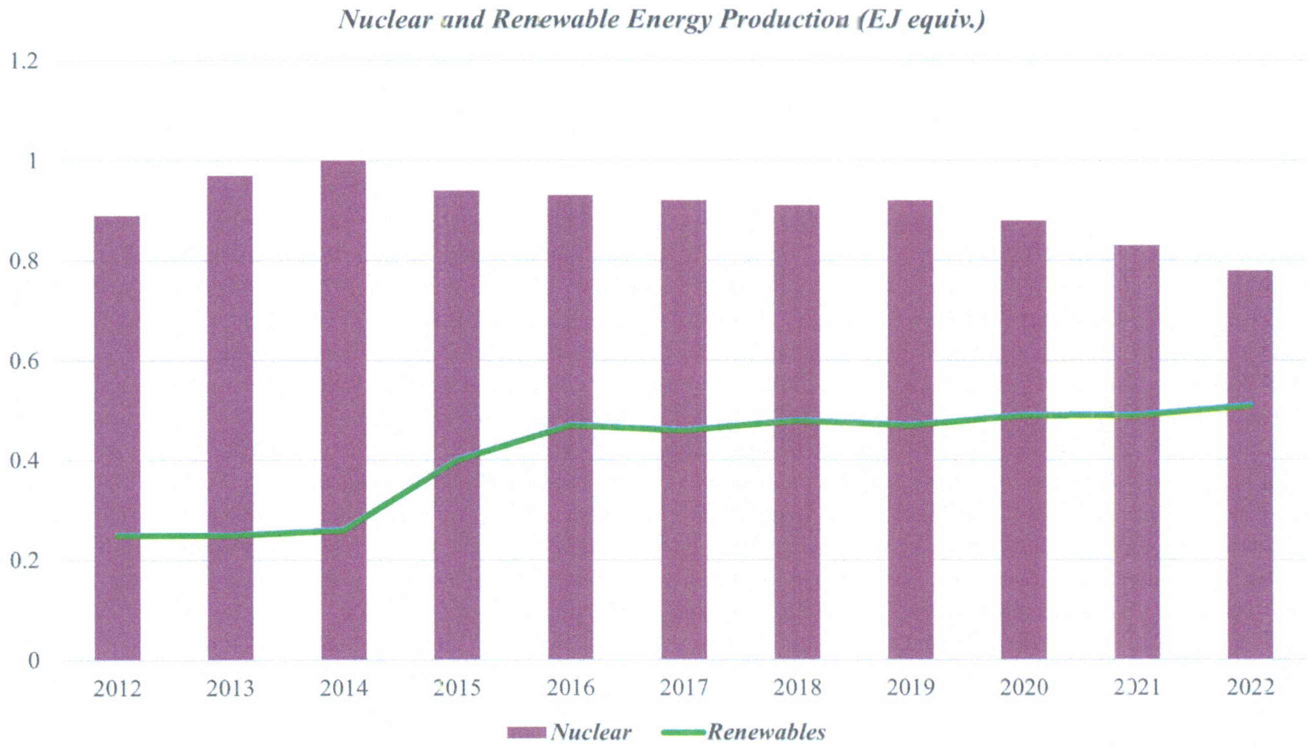

Fig. 17.21 Nuclear and renewable energy production (EJ equivalent), 2012–2022. Renewables are dominantly wind and solar energy; hydroelectricity is not included here (Canada Energy Regulator 2023a)

Fig. 17.22 Uranium production (kt), 2005–2023 (Canadian Nuclear Safety Commission 2023, and sources cited therein)

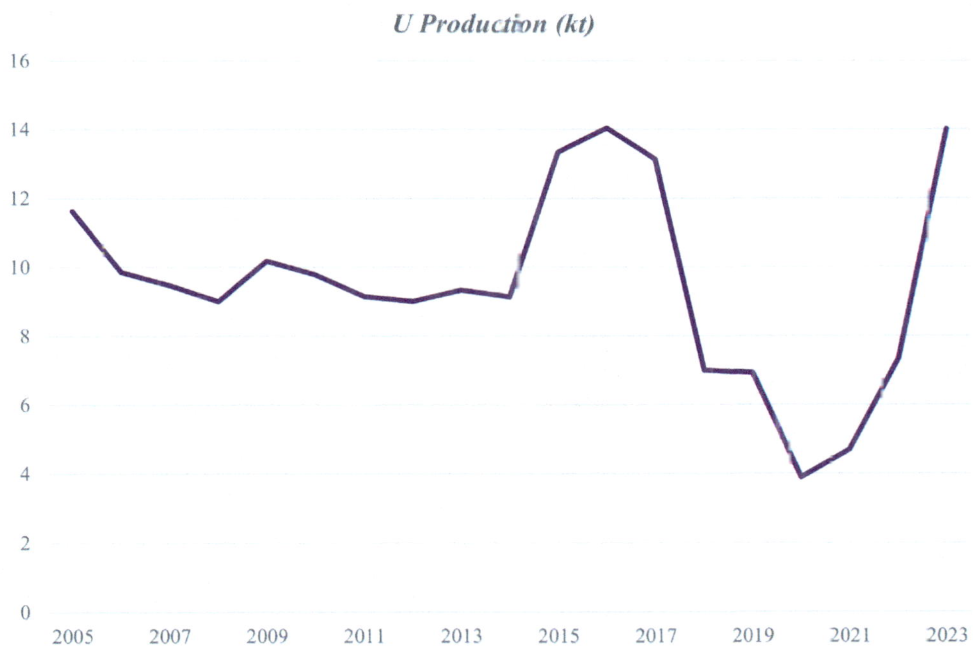

As elsewhere, uranium exploration, power plant development and operation, and nuclear waste disposal are highly controversial. Several provinces and territories have instituted moratoria or permanent bans on uranium exploration and mining. Accidents involving spillage of coolant D_2O have occurred, without loss of life or any detected environmental damage. Despite investigations of several locations in the Boreal Shield, no permanent nuclear waste repository has been authorized. The existing reactors have produced at least 52,000 t of high-level waste, mostly spent ^{238}U fuel bundles (Canadian Nuclear Safety Commission 2023). Waste is currently stored at the reactor sites, initially in water-filled fuel pools and subsequently in dry cask sites. Public opinion is generally more opposed (50%) than in favour (40%) of nuclear power development (e.g. Angus Reid 2023), percentages which did not change significantly after the Fukushima Daiichi plant failure in 2011.

17.4 Hydroelectricity

Hydroelectric power is generated by the kinetic energy of descending water passing through turbines. Energy production depends upon water volume and the difference in elevation, the head.

Run-of-river installations generate power either directly from or by partial diversion of the stream through turbines, without construction of a reservoir (Table 17.3). Large run-of-river installations utilize a diversion canal, leaving the main stream unimpaired by infrastructure. Advantages of run-of-river installations include the absence of potential CH_4 and heavy metal build-up in a reservoir, and no flooding upstream. Construction along a diversion canal maintains the aesthetics of the natural river.

Disadvantages include exposure to changes in precipitation patterns, particularly during dry summers or those resulting from climate change. Low stream levels add to potential impacts on migratory fish. Run-of-river systems typically generate less power than reservoir systems, unless the diversion canal brings large volumes of water and the difference in head is relatively great. For systems without diversionary canals, fish ladders may be required.

Most run-of-river plants are relatively small. The largest examples in Canada are the René-Lévesque (5400 GWh, R. Manicouagan, Québec); La Grande 1 (7900 GWh, R. La Grande, Québec); Limestone (900 GWh), Kettle (8500 GWh), and Long Spruce (7100 GWh), Nelson R., Manitoba; Beauharnois (11,700 GWh, St. Lawrence R.), and Sir Adam Beck (17,200 GWh, Niagara R.; Table 17.3). The site C dam on the Peace River (BC) will also be a run-of-river project. Beauharnois, fed by a 5 km-long canal delivering 8 million m^3s^{-1}, includes 36 turbines. Construction was completed in three phases from 1930 to 1961 (Hydro-Québec 2024). As with many contemporary installations, great care was paid to architectural design, with the façade exhibiting Art Deco style. The Sir Adam Beck station, originally opened in 1922 and most recently refurbished in 2013, draws from the Chippawa-Queenston Power Canal diversion of the *Welland River* (Ontario Power Generation 2021). Water is brought from the Welland River above Niagara Falls and diverted around the city to the power plant to the north. Diversion is regulated by the Canada-USA Niagara Treaty (1950), which requires flow over Niagara Falls of at least 2832 m^3s^{-1} during daylight hours in the tourist season, and 1,146 m^3s^{-1}

Table 17.3 Hydroelectricity sites

Site	Lat, long	Capacity (MW)	Type	River
Robert-Bourassa QC	53.8 °N 77.5 °W	5,616	Reservoir	La Grande
Churchill falls NL	53.5 °N 64.0 °W	5,428	Reservoir	Churchill
Mica BC	52.1 °N 118.5 °W	2,805	Reservoir	Columbia
La Grande-4 QC	53.9 °N 73.5 °W	2,779	Reservoir	La Grande
Gordon M. Shrum BC	56.0 °N 122.2 °W	2,730	Reservoir	Peace
Revelstoke BC	51.1 °N 118.2 °W	2,480	Reservoir	Columbia
La Grande-3 QC	53.7 °N 76.0 °W	2,417	Reservoir	La Grande
La Grande-2-A QC	53.8 °N 77.5 °W	2,106	Reservoir	La Grande
Beauharnois QC	45.3 °N 73.9 °W	1,853	Run-of-river	St. Lawrence
Manic-5 QC	50.6 °N 68.7 °W	1,528	Reservoir	Manicouagan
Sir Adam Beck II ON	43.1 °N 79.0 °W	1,499	Run-of-river w. Pool	Niagara
La Grande-1 QC	53.7 °N 78.5 °W	1,436	Run-of-river	La Grande
Limestone MB	56.5 °N 94.1 °W	1,350	Run-of-river	Nelson
René-Lévesque QC	49.7 °N 68.6 °W	1,326	Run-of-river	Manicouagan
Jean-Lesage QC	49.3 °N 68.3 °W	1,229	Run-of-river	Manicouagan
Kettle Rapids MB	56.4 °N 94.6 °W	1,220	Run-of-river	Nelson
Bersimis-1 QC	49.3 °N 69.6 °W	1,178	Reservoir	Betsiamites
Shipshaw QC	48.4 °N 71.2 °W	1,145	Reservoir	Saguenay
Manic-5-PA QC	50.6 °N 68.7 °W	1,064	Reservoir	Manicouagan
Robert H. Saunders ON	45.0 °N 74.8 °W	1,045	Reservoir	St. Lawrence
Outardes-3 QC	49.6 °N 68.7 °W	1,026	Run-of-river	Outardes

at all other times. Although many hydroelectric stations have topographic locations unsuitable for canal construction, other opportunities have been missed, notably at Churchill Falls NL.

17.4.1 Dams and Reservoirs

Throughout Canada, impoundments have been constructed for hydroelectric development, beginning in 1881 with Chaudière Falls in Ottawa-Gatineau (Catto 2020). Small reservoirs, such as Knox's along the Montague River PEI (rebuilt in 1917 for hydroelectricity), at Petty Harbour NL (1900), and the St. Margaret's Bay system (NS 1922), are found along many rivers in agricultural and coastal southern Canada. Québec has 1950 registered dams. All the provinces and territories, with the exception of Nunavut, have at least one dam constructed for power generation. There is a proposal to impound the Armshow River to supply power to Iqaluit.

Hydroelectric impoundments represent the majority of the country's 1157 large dams. Linear reservoirs are developed along the Exploits and Churchill (NL), Saint John (NB), Gatineau, St-Francois, and St-Maurice (QC), Ottawa (ON-QC), Madawaska and St. Lawrence (ON), Nelson (MB), Saskatchewan (SK), Bow, Brazeau, and North Saskatchewan (AB), Bridge, Columbia, Kootenay, Nechako, and Peace (BC) rivers. Dual purpose reservoirs are present along the St. Lawrence (hydroelectric combined with transportation, Long Sault, ON), Thames (flood control, Fanshawe Dam, ON), and South Saskatchewan (power, flood control, irrigation, recreation, Gardiner Dam, SK) rivers.

Twenty large reservoirs each cover at least 400 km^2, including raised impoundments of existing lakes. The Smallwood Reservoir, formed by impounding the Churchill River (NL) at Churchill Falls, Labrador, is the world's second largest in surface area (6527 km^2) and ranks 10th in surface area of all Canadian lakes. Southern Indian Lake, formed by the 85% impoundment of northern Manitoba's Churchill River and the associated diversion of 850 m^3s^{-1} into the Nelson River drainage, covers 2247 km^2. Lac Seul, a raised natural lake on the English River (ON) formed by damming at Ear Falls, covers 1657 km^2. Québec has 13 reservoirs exceeding 400 km^2, including Canipiscau (4539 km^2), Robert-Bourassa (2905 km^2), La Grande 3 (2451 km^2), Manicouagan (1788 km^2), and Gouin (1570 km^2). Willison Lake, formed by the impoundment of the Peace, Finlay, and Parsnip Rivers by the WAC Bennett Dam (BC), covers 1773 km^2.

Large nonlinear reservoirs can have significant impacts on the surrounding terrain and the lives of residents of all species. Many Canadian projects date from 1950–1975. The planners, engineers, builders, and operators were regarded by many in the general population as heroes. Frequently, the projects took on highly political and societal symbolic tones, particularly in Québec ('Les bâtisseurs d'eau'), but also elsewhere in Canada. Infrastructure was named for active or recently active politicians, or for the civil servants involved. Environmental assessments were cursory at best, and largely confined to geological and geotechnical concerns such as slope stability, reservoir loading, and construction challenges. When wildlife was considered, it was assumed to be capable of adaptation or relocation. Indigenous peoples received orders, not consultation: in some cases, they received no notice at all. The treatment of local non-indigenous residents was not much better. Much of the current reluctance or opposition to new hydroelectric proposals relates to the true horrors that have gradually been revealed in more recent years. One consequence is that the remaining 40% hydroelectric potential across Canada will almost certainly remain undeveloped in the foreseeable future, due to public resistance (e.g. International Energy Association 2010)

Environmental assessment practices have changed greatly since 1965, as has knowledge of environmental issues and impacts. Other large projects undertaken throughout the world in the 1960s, notably hydroelectric dam projects, received similarly limited environmental assessment prior to construction. Negative and unforeseen environmental effects resulted in many instances. History cannot be reversed, only appreciated, and future projects assessed on the basis of accumulated knowledge.

The Churchill Falls (Smallwood Reservoir) project (1965–1972) provides a convenient example of the shortcomings of environmental assessment and regard for basic human rights (see Past Due 2024). Construction of the dam was not preceded by a comprehensive Environmental Impact Assessment, as would be the case today. Archaeological surveys in the affected area were cursory, due to the shortage of time available and a lack of logistical and financial support. Many Innu residents received no notice of the impending flooding of the reservoir. Hunting lands and burial sites were flooded. Some Innu hunters lost all their personal belongings and have never received compensation. The Canadian Wildlife Service estimated that approximately 1,400 km^2 of waterfowl habitat were lost due to flooding, as numerous small, medium, and larger lakes were incorporated into the Smallwood Reservoir. In addition to inundated areas, areas subject to fluctuating water levels resulting from the requirements of dam management are also lost to waterfowl.

Caribou calving grounds were also inundated by the reservoir. Caribou generally shun deep valleys, such as the lower Churchill, preferring to travel across rolling uplands. Inundation of the terrain west of Churchill Falls has essentially closed this migration route, displacing the species to the west of the reservoir (Past Due 2024).

Trees killed by the flooding released mercury into the reservoir waters, resulting in contamination of fish and health problems for those who consume them. In 1977, 37% of the Sheshatshiu Innu population had elevated mercury levels.

Construction of reservoirs can result in slope failures as rising waters saturate sediment. On the Boreal and Taiga shield, the resistant bedrock limits slope failure. The shoreline of the flooded reservoir follows the bedrock-dominated topography, resulting in irregular configurations not aligned with respect to the prevailing wind, limiting erosion by wave action or seasonal ice shove. Sediment entrapment in the reservoir and subsequent sediment deprivation downstream are relatively low sediment. The most significant downstream impacts occur where streams are diverted into neighbouring systems to fill reservoirs, or where streams debouch from Shield areas into adjacent sedimentary lowlands.

In Cordilleran, Grassland Prairie, and Boreal and Taiga Plains locations, sediment erosion and slope failure are often serious concerns. Eroding sediment infilling the reservoir will reduce its volume and service life. Fine sediment in suspension can cause damage to turbines. Sediment impounded in the reservoir is not available to replenish bars and river banks downstream, leading to loss of habitat. These concerns were one factor in the change of the Peace River Site C project (BC) from an impoundment to a run-of-river design. Large slope failures in Cordilleran valleys could lead to catastrophic discharge of reservoir water, as occurred in the Vajont disaster (Italy, Semenza and Ghirotti 2000), requiring careful geotechnical assessment of the Mica Creek site (Columbia R, BC) prior to construction.

To avoid or reduce slope failure, one approach is to leave the vegetation and soil undisturbed during construction and reservoir filling, as was practised in the Peace and Columbia River projects. Unfortunately, vegetation submerged in a reservoir dies, sinks to the bottom in an anoxic environment, and decays to release CH_4 and bioaccumulated metals, including cadmium and mercury initially deposited in plants and soil from airborne pollution from distal sources. Initial flooding results in an increase in toxic methylmercury (CH_3Hg^+), formed by microbial methylation in CH_4-rich sediments. Methylmercury undergoes biomagnification, with increasing concentrations in aquatic species, particularly fish and waterbirds. Human health effects include memory loss, brain damage, sycosis, nervous system disorders, sensory impairment, cardiovascular diseases, autoimmune diseases, renal disfunction, chronic liver diseases, and irreversible damage to a foetus (Environment and Climate Change Canada 2016; Kim et al. 2016). Although concentrations gradually decline and reach background levels within 20 to 30 years, this is no consolation to those affected. Methylmercury concerns are a significant factor in opposition to large impoundments.

17.4.2 Consumption

Hydro accounted for accounted for 61.48% of electricity, 26.44% of primary energy consumption, and 398 TWh of energy use in Canada in 2022 (Canada Energy Regulator 2023a; Figs. 17.1, 17.2, 17.3, 17.23 and 17.24). Percentages of electricity generation vary from 96% (NL) and 94% (Québec) to none in PEI and Nunavut (Fig. 17.8). Initially, all provinces except Alberta formed government power corporations, although serval have now been privatized. Long-distance powerlines link Québec with Ontario and New Brunswick, Labrador with Newfoundland and Nova Scotia, and several provinces with USA (34 active lines). With the completion of the Newfoundland-Nova Scotia link under Cabot Strait, the system now connects all provinces to the North American power grid. Currently, high voltage transmission lines extend over 166,000 km, with more than 34,000 in Québec (Electricity Canada 2022a, b). Typically, 8–15% of electricity generated is lost in transit.

In addition to high winds and freezing rain (verglas, see Chap. 4), long-distance power lines can also be affected by variations in electromagnetic radiation from Sun ("space weather"). On 13 March 1989, a pulse of electromagnetic radiation generated strong electrical Direct Currents along the main power lines in Québec, causing six million people to lose power (Bolduc 2002). Hydro-Québec's explanation that failure was due to Sun, initially greeted with incredulity, was correct.

Standard household electricity is 120 V, 60-Hz, alternating current. Industry, particularly chemical production, hydrocarbon production and refining, pulp-and-paper manufacture, and primary metal production and refining, consumes ~40–45% of electricity produced. Residential users account for 30–35%, including home heating in provinces not using natural gas or petroleum for that purpose (Statistics Canada 2023a, b). Service sectors, including commerce, finance, government, and entertainment, account for 25–30% of usage. Use of electricity in transportation accounts for 0.3%.

17.5 Wind Energy

Wind turbines accounted for 5.97% of electricity, 2.49% of primary energy consumption, and 38 TWh of energy in 2022 (Canada Energy Regulator 2023a; Figs. 17.1, 17.2, 17.21 and 17.25). Wind generation represents the fastest growing renewable energy supply (Sadorsky 2021). In PEI, 99% of electricity generated on the island is produced by wind, although the province imports substantial amounts from New Brunswick in addition (Figs. 17.8 and 17.26).

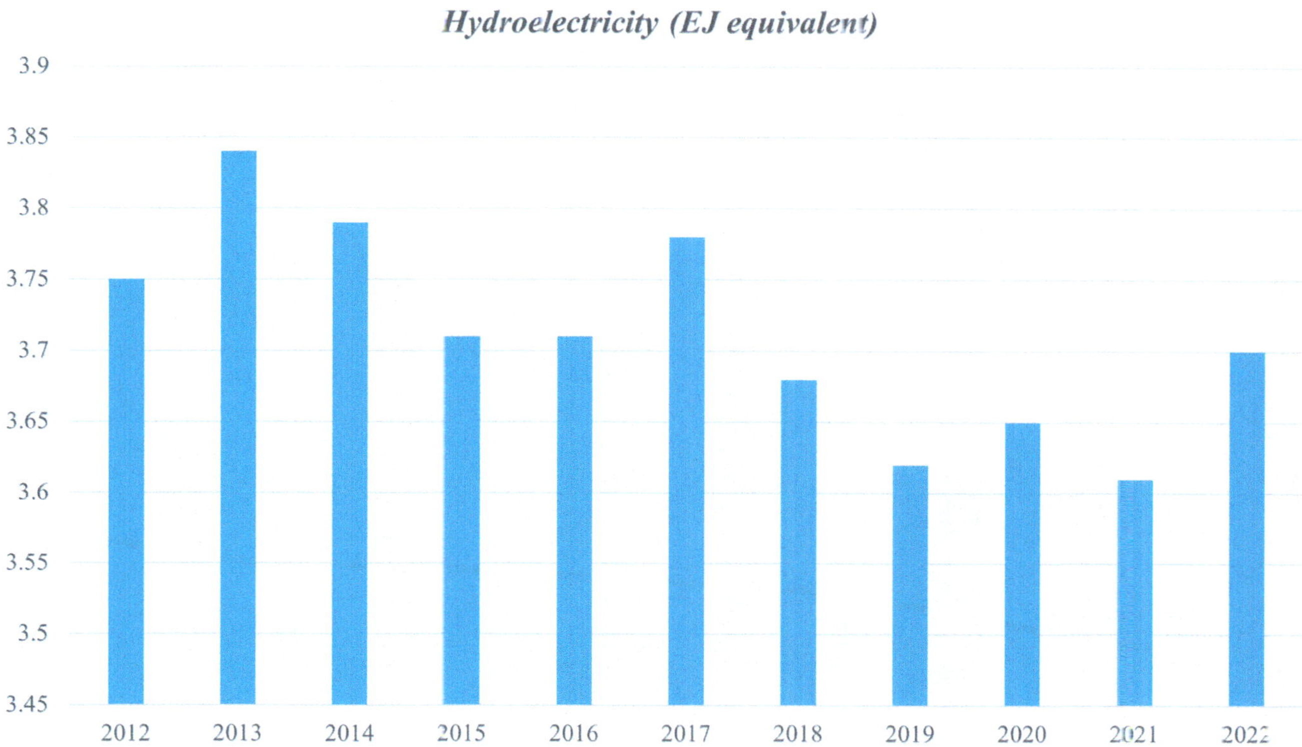

Fig. 17.23 Hydroelectricity production, EJ equivalent, 2012–2022 (Canada Energy Regulator 2023a)

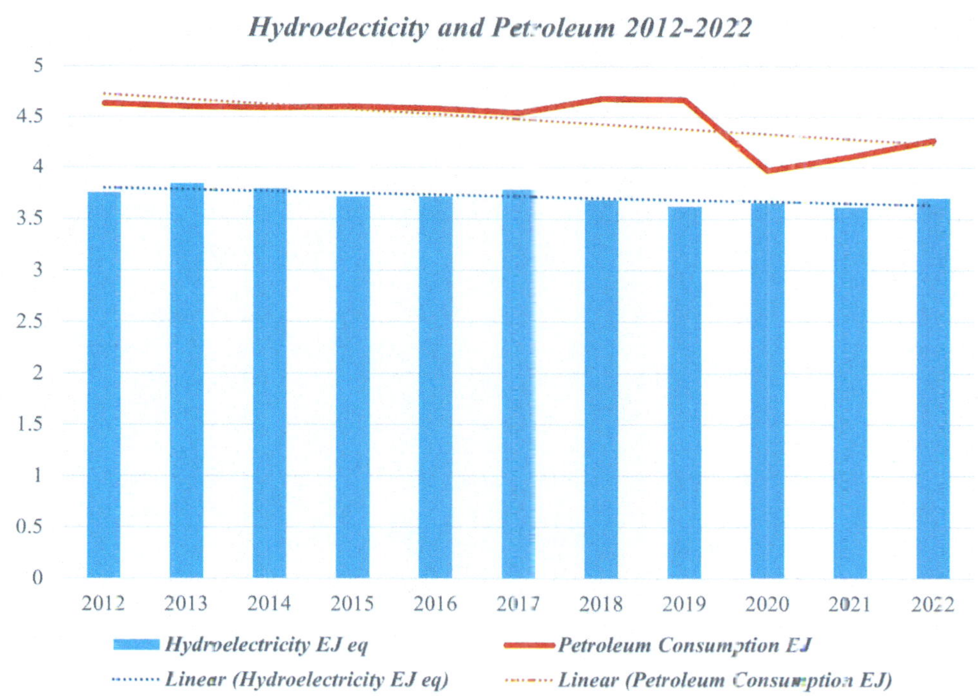

Fig. 17.24 Hydroelectricity and Petroleum Consumption (EJ eq.) compared, 2012–2022. Linear trends show a greater decline for petroleum consumption. Hydroelectricity consumption was largely unaffected by COVID-19 restrictions. Data from Canada Energy Regulator (2023a)

Wind power installations are generally practical where annual average wind speed exceeds 4.2 m s^{-1} (approximately 15 km h^{-1}; MacEachern and Yildiz 2018, Dehghani-Sanij et al. 2022). Modern turbines are designed to feather or cease rotating for wind speeds < 3.5 m s^{-1}, and when winds exceed 25 m s^{-1} (90 km h^{-1}) to prevent damage to the blades. Peak power generation typically requires 12–14 m s^{-1} (45–50 km h^{-1}). Suitably high wind speeds occur across NL, Nunavik, PEI, Cape Breton, along the Rocky Mountain foothills of Alberta, and over eastern Qikiqtaaluk, and in

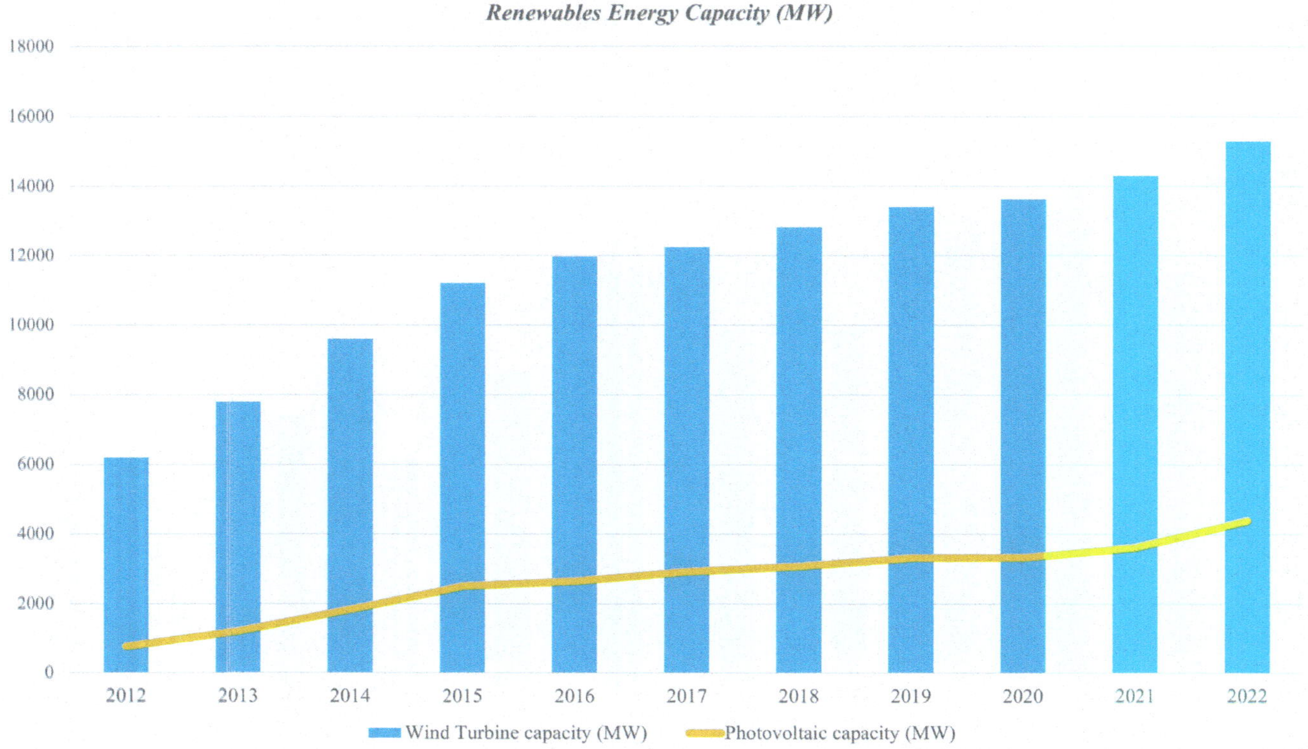

Fig. 17.25 Energy Capacity (MW) for wind turbines and photovoltaic arrays, 2012–2022 (Canada Energy Regulator 2023a)

Fig. 17.26 Wind turbines provide almost all the electricity generated on Prince Edward Island, although the province also imports substantial amounts from New Brunswick. See text for discussion. Author's photo

coastal areas and offshore of the Atlantic Provinces, British Columbia, and the Great Lakes (Canadian Wind Energy Association 2019). Much of Canada, therefore, has suitable physical conditions for wind turbine operations. In northern locations, LiDAR is used to measure wind velocity and turbulence to provide pre-installation data (NWT 2020). Future wind regimes in most areas have not been effectively modelled at present, although research is ongoing. The Atlantic regional study of Price et al. (2001) suggested that as much as a 10% reduction in summer winds could occur by ca. 2050, although winter wind strengths were anticipated to either remain unchanged or increase slightly. The impact of ice buildup on turbine blades appears limited (Wind Energy Institute of Canada 2024).

Wind speed increases proportional to the height above the ground, and turbines typically are mounted 50–80 m above ground level. In a coastal area, the approximate wind speed at Height z equals (speed at 1 m) × $(z)^{0.25}$, so a wind speed of 25 km h^{-1} at 1 m above the surface equates to 75 km h^{-1} at 75 m, although the directions may differ. To avoid turbulence, the sites should be unobstructed, with no structures or trees more than 50% of the height of the turbine, and there should be no significant obstructions within a radius of ten times the height. These restrictions pose significant constraints on onshore installations.

Most wind turbines have horizontally positioned rotors, although a few vertically oriented rotor systems exist (MacEachern and Yildiz 2018). As blade pitch can be adjusted, horizontal rotors are easier to control in areas with highly variable winds, especially for larger systems with blades exceeding 110 m length. Horizontal rotors also operate during lower wind speeds, and are generally more efficient power generators.

Offshore sites offer the advantages of unobstructed locations, but site selection is complicated by the needs and perceptions of nearshore fish harvesters and coastal residents, in addition to ecological concerns. Of Canada's more than 300 wind farms, only three are located offshore, in NB (2) and NS (1) (Wind Energy Institute of Canada 2024). The Gulf of St. Lawrence is a preferred location over open Atlantic coastal sites due to its consistent winds and lower mean significant and storm wave heights (Barrington-Leigh and Ouliaris 2017). The seasonal wind pattern suggests peak power generation would occur in winter months, the period of greatest demand in the boreal climate.

Onshore wind farms are concentrated in Ontario, Québec, and Alberta. On balance, operation of Alberta-installed turbines are more efficient than those elsewhere (capacity factors typically approaching 50%), reflecting consistent wind speeds. Installed capacity is sufficient to produce electricity to supply about 3.4 million homes annually. The largest installations with ≥300 MW capacity include Seigneurie de Beaupré, Rivière-du-Moulin, and Lac Alfred (Québec), Black Spring Ridge (Alberta), and Henvey Inlet (Ontario). Although PEI does not have large wind farms, 20% of the electricity it uses is produced from wind, with most of the reminder imported by submarine cable from hydrocarbon sources in NB. Wind power is also a possible solution for smaller communities: Rexton and Richibucto (NB) have each installed one wind turbine. Turbines could represent another avenue to allow isolated northern and Indigenous communities to stop burning diesel fuel (Knowles 2016).

Barrington-Leigh and Ouliaris (2017) suggested that if all suitable sites were developed, wind turbines could produce almost double the amount of electricity currently used in Canada. However, although public opinion in general favours increased turbine installation, significant local opposition to specific sites is frequent, based on concerns linked to noise, visual effects (especially if the turbines are illuminated), aesthetics, property values, and the effects on bats, birds, and other wildlife.

17.5.1 Wildlife Mortality

Bat mortality results from lung damage caused by changing air pressure from rotating blades, as well as direct collisions. Bats are attracted to turbines for foraging (Foo et al. 2017). Migratory, tree-dependent species suffer much greater mortality rates than other bats, particularly in summer (Goldenberg et al. 2021). In Québec, mortality decreased with increasing elevation, was greatest at sites to the southwest, and should no clear relation with installation size (MacGregor and Lemaître 2020). An estimate of total annual mortality of 47,000 bats from 4000 turbines in 2013 nationally (Zimmerman and Francis 2016) concurs with the estimate of 12 annual fatalities per installed turbine in Ontario.

Bird mortality varies by species and season, with raptors being particularly vulnerable (Katzner et al. 2016; Bose et al. 2018). Marine birds are most vulnerable at installations near breeding colonies, wintering sites, and areas where migrants concentrate (e.g. Thaxter et al. 2019). In some instances, birds appear to have adjusted their flight paths to benefit from enhanced wind flow above turbines (d'Entremont et al. 2017). A variety of studies have estimated bird mortality rates of 2–3 annually per turbine at Canadian sites, a rate approximately 5% of bird mortality at hydrocarbon producing sites (e.g. Sovacool 2013). Installations and roads also impact terrestrial species, including carnivores, rodents, and cervids (e.g. Sirén et al. 2017; Skarin et al. 2018). As with any human disturbance of terrain and environment, impacts on other life are inevitable.

17.6 Solar Energy

Solar energy accounted for 0.87% of electricity generation, 0.40% of primary energy consumption, and 6 TWh of energy in 2022 (Canada Energy Regulator 2023a; Figs. 17.1, 17.2, 17.3, 17.21 and 17.25). Based on latitude and modal cloud cover, the maximum photovoltaic potentials of 1300–1400 KWh/KWp occur in the shortgrass prairies of southwestern Alberta and southern Saskatchewan. Values of 1200–1300 KWh/KWp characterize N varying the northern and eastern Prairie Grasslands, and southwestern Ontario between Lakes Erie and Huron. Cloud cover and ambient moisture content are significant influences, with photovoltaic potentials along

49°N varying from maxima of 1370 KWh/KWp at Treelon (SK) to 1120 KWh/KWp at Roberval (Québec), 980 KWh/KWp at Vancouver, and 850 KWh/KWp at Gander (NL) (Natural Resources Canada 2020b). The Avalon Peninsula has the lowest photovoltaic potential of Canadian locations south of the High Arctic (Canada Energy Regulator 2018). Adjacent to the Arctic Circle, values vary from 900 KWh/KWp (northern Yukon) to 1050 KWh/KWp (Iqaluit).

Fixed solar panels must take advantage of the yearly average solar angle for a particular latitude. Moveable panels allow more efficient year-round collection. Panels at 49°N should be set at 75° at the December Solstice, shifted to 45° in June.

In 2024, there were approximately 43,000 personal residential and commercial installations, along with 80 arrays providing at least 10 MW. Most of the larger arrays are situated in Alberta and Ontario, with 3 in Saskatchewan and 1 in PEI (Martimanakis 2023). The Travers Solar Project at Vulcan (AB) is the largest, with 1.3 million panels producing 465 MW. Photovoltaic installations are seen as a viable alternative for small or isolated Indigenous and northern communities.

Larger solar panel arrays result in bird fatalities through collisions and solar-flux singeing of feathers. Birds appear to mistake the panel surfaces for water bodies. Walston et al. (2021) estimated annual total fatalities of 37,800 and 138,600 birds in USA, with more numerous and larger installations than in Canada. Although the mortality rate is low in comparison to other forms of human pressure, concern led to the halting of a proposed solar array in Frank Lake (AB) in 2023.

17.7 Tidal Energy

Human exploitation of tidal energy may date from the eleventh century grist mill at Carew Castle, Pembrokeshire, Cymru. In Canada, sites in Nunavik, Nunavut, British Columbia, and Atlantic Canada have drawn interest (Tarbotton and Larson 2006).

Potential tidal energy sites are present in Nunavik and Nunavut, including 9 of the 11 largest in Canada (Tarbotton and Larson 2006). Currently, diesel generators produce 100% of the energy generated in Nunavik and Nunavut. Ungava kangiqluk has macrotidal and hypertidal shorelines, with ranges reaching 16.8 m. Smoky Narrows (58.9 N 69.3 W) and Algemine Narrows (58.8 N 69.6 W) potentially could supply 2834 MW to the adjacent community of Tasiujaq (Leaf River) (Triton Consultants 2006). In Foxe Channel between Kingait (Cape Dorset), Qikiqtaaluk, and Ivujivik, Nunavik, three sites could produce 1972, 7584, and 10,426 MW respectively. North of Killiniq (Cape Chidley), the Gray Strait site has a potential of 6979 MW. Other smaller sites are present adjacent to Usqsuqtuuq (Gjoa Haven), Iqaluktuuttiaq (Cambridge Bay), Qingaut (Bathurst Inlet), and in embayments along Kangiqtualuk (Cumberland Sound) in coastal Qikiqtaaluk.

Although these sites have great theoretical potential, the challenges of development offshore in straits with prolonged ice cover are immense. Development would also have to avoid disruption of the seasonal ice cover, vital for both ecology and human access to transportation and marine hunting sites. Although Arctic communities would prefer to jettison high-polluting, inefficient, and expensive diesel fuel (Knowles 2016; Ouranos 2022; Nunavut 2023), with challenges of supply, tidal energy will not be involved in the near future.

Mesotidal conditions prevail along much of the British Columbia coastline, with tidal ranges locally exceeding 4 m restricted embayments including Skeena Estuary (Tarbotton and Larson 2006). Tidal currents with velocities in excess of 4 m s^{-1} occur in narrow channels in the Gulf Islands, notably Active Pass. In Johnstone Strait, tidal currents reach 7.7 m s^{-1}, with discharges of 3.2×10^5 m^3 s^{-1}, with flows at depth generally earlier and faster than surface movements (Sutherland et al. 2007). Several of these sites could generate power exceeding 300 M, with a theoretical maximum of 1335 MW in northwestern Johnstone Strait.

Although tidal power could replace the diesel fuel burned in isolated communities on Vancouver Island, Haida Gwaii, and the BC mainland coast, opposition to tidal development remains strong, due to concerns for the potential effects on orca, salmon migration, and nearshore fisheries. The BC government's goal to reduce diesel generation in isolated communities by 80% contrasts with the environmental concerns expressed by governments and citizens alike, particularly those pertaining to marine species.

The power of the tides of the Bay of Fundy have been noted by writers since the seventeenth century (Clark 1997; Catto 2020). The experimental 50 GWh/y facility at Annapolis Royal (NS), operational since 1984, is the sole operational tidal power plant. Here, an existing dam was retrofitted to allow increased tidal influx, resulting in modification of the harbour and Annapolis River.

Proposals for larger installations have been suggested for the Minas Channel (NS) (potential 1900 MW) and several sites in the Western Passage of Passamaquoddy Bay (NB-Maine) (Tarbotton and Larson 2006). Experience gained from tidal facilities in the Orkney Islands, South Korea, and elsewhere suggests that anchored in-stream turbines represent more feasible ecological and financial decisions than does barrage construction. As in British Columbia, opposition on both environmental and aesthetic grounds is significant. Nova Scotia currently generates power from hydrocarbons, along with imported hydroelectric energy from NL.

In the Minas Basin, sea ice cover can exceed 90%, with detached floes moving at 15 km h^{-1}. Marine salt and sediment inclusions causes floes to partially sink, leading to concerns they could be ingested into turbines. Large-scale construction may cause tidal current velocities to be reduced by up to 35% along the western head of Minas Basin, while increasing by a similar magnitude in the Southern Bight, east of Cape Split. In conjunction with rising relative sea level (~3.5 mm y^{-1}), in-stream Fundy tidal plants could result in changes in tidal amplitude throughout the USA Gulf of Maine (Pelling and Green 2013).

17.8 Geothermal Energy

Although geothermal energy is not at present used to generate electricity or by municipalities, geothermal heat pumps are used in residential and commercial buildings. Geothermal heat pumps use electricity to transfer heat to and from the subsurface, relying on the consistent temperature below the frost depth (Natural Resources Canada 2009). Energy is transferred through a flexible tube, either filled with a water–methanol mixture, or less commonly connected to another water well. The tubes can be aligned vertically, usually requiring 100 m depth, or as a horizontal matrix of similar extent at depths 3–5 m below ground surface. The space requirements limit the applicability of geothermal heat pumps for private buildings in built-up areas, and they cannot be used in continuous or deep permafrost.

Air source heat pumps act similarly, but depend on energy exchange with the open air rather than the subsurface. They will function effectively at temperatures above –30 °C. Air source heat pumps also require electricity to function. The ease of installation, minimal space requirements, and the option of installing small units (mini-splits) have made air source heat pumps an increasingly popular choice for residential and commercial buildings. In 2023, about 7% of Canadian homes used a heat pump, with air source models forming the majority of the 700,000 installed (Statistics Canada 2023a, 2023b). Government policies in the Maritime Provinces have encouraged installation: in NB, 32% of primary home heating systems used heat pumps in 2021.

17.9 Conclusion

Canada's total GHG emissions in 2021 were 670 megatonnes of carbon dioxide equivalent (Mt CO_2 eq), a 1.8% increase from 659 Mt CO_2 eq in 2020. From 2005 to 2021, Canada's GHG emissions decreased by 8.4% (62 Mt CO_2 eq) (Environment and Climate Change Canada 2022a, b, 2023a, b). Canada produces about 1.5% of global CO_2 eq emissions. Per capita GHG emissions were 14.25 t, with total emissions increasing by 117% since 1965. About 80% of GHG emissions in 2021 were related to the production and consumption of energy, mostly from the combustion of fossil fuels.

Canada's response to climate change is complicated by the constitutional division of federal-provincial powers, with the provinces responsible for energy production, municipalities, education (including postsecondary), roads, highways, and vehicle licencing, and responsibility for environmental matters split. Weather forecasting, meteorological investigations, and collation of climate data are the responsibility of the federal ministry Environment and Climate Change Canada (ECCC). ECCC has thus assumed responsibility for directing GHG mitigation, promotion conservation, and providing lead scientific representatives at international conferences.

Adaptation is handled at the federal level by Natural Resources Canada. From an initial focus in the early 1990s on geological concerns, including flooding, slope failures, and coastal erosion, efforts broadened to include attention to coping with extreme meteorological events. Provinces also have their own ministries responsible for climate change policy. Researchers and geographers within and working with the various government agencies are subject to political considerations, reassessments, and decisions.

Energy policy is a central component of any response to climate change. Varying spatial and socio-economic circumstances, political disagreements, public opinion, and personal history all play roles in any individual's position.

Successful responses require accurate assessments and appropriate, patient, and respectful communication with communities and individual people (e.g. Castro and Parewick 2008). Transforming a climate change assessment exercise into an effective plan for adaptation, mitigation, or (ideally) both requires intimate cultural understanding, something that geographers with training in both human and physical matters may be able to assist with. Substantial input from all research, administrative, socio-economic and cultural communities could lead to more appropriate and valuable adaptation and mitigation strategies. Community-driven communication and establishing effective working partnerships is essential.

References

Alberta (2024a) Oil sands 101. Government of Alberta. https://www.alberta.ca/oil-sands-101

Alberta (2024b) Phasing out emissions from coal. Government of Alberta, https://www.alberta.ca/climate-coal-electricity

American Geosciences Institute (2019) Transportation of oil, gas, and refined products. https://www.americangeosciences.org/geoscience-currents/transportation-oil-gas-and-refined-products

Angus Reid (2023) Public opinion poll: Angus Reid, Vancouver BC. https://angusreid.org/canada-energy-nuclear-power-oil-and-gas-wind-solar/

Barrington-Leigh C, Ouliaris M (2017) The renewable energy landscape in Canada: a spatial analysis. Renew Sustain Energy Rev 75:809–819

Black H (1997) Canadian scientists and inventors. Pembroke Publishers

Bolduc L (2002) GIC observations and studies in the Hydro-Québec power system. J Atmos Solar Terr Phys 64:1793–1802

Bose A, Dürr T, Klenke R, Henle K (2018) Collision sensitive niche profile of the worst affected bird-groups at wind turbine structures in the Federal State of Brandenburg, Germany. Sci Rep 8:3777. https://doi.org/10.1038/s41598-018-22178-z

bp (2023) bp Energy Outlook, 2023 edition. https://www.bp.com/content/dam/bp/business-sites/en/global/corporate/pdfs/energy-economics/energy-outlook/bp-energy-outlook-2023.pdf

Buss J, Mansuy N, Madrali S (2021) De-risking wood-based bioenergy development in remote and indigenous communities in Canada. Energies 14:2603. https://doi.org/10.3390/en14092603

Canada (2023) Powering past coal alliance: phasing out coal. Environment and Climate Change Canada. https://www.canada.ca/en/services/environment/weather/climatechange/canada-international-action/coal-phase-out.html

Canada Energy Regulator (2016) Energy conversion tables. https://apps.cer-rec.gc.ca/Conversion/conversion-tables.aspx

Canada Energy Regulator (2018) Market Snapshot: Residential solar is financially viable in some provinces and territories, but not in others. https://www.cer-rec.gc.ca/en/data-analysis/energy-markets/market-snapshots/2018

Canada Energy Regulator (2023a) Provincial and territorial energy profiles. https://www.cer-rec.gc.ca/en/data-analysis/energy-markets/provincial-territorial-energy-profiles/provincial-territorial-energy-profiles-canada.html

Canada Energy Regulator (2023b) Crude oil export summary. https://www.cer-rec.gc.ca/en/data-analysis/energy-commodities/crude-oil-petroleum-products/statistics/crude-oil-export-summary/index.html

Canadian Forest Service (2020) Forest bioenergy. Natural Resources Canada. https://natural-resources.canada.ca/our-natural-resources/forests/industry-and-trade/forest-bioeconomy-bioenergy-bioproducts/forest-bioenergy/13325

Canadian Fuels Association (2022) Canada's refining sector. 6003_FUEL_Refineries-Map-Canada-2022_Eng

Canadian Fuels Association (2024) Fuel production. https://www.canadianfuels.ca/our-industry/fuel-production/

Canadian Lung Association (2023) Residential Wood Burning. https://www.lung.ca/news/residential-wood-burning

Canadian Nuclear Safety Commission (2023) Uranium mines and mining. https://www.cnsc-ccsn.gc.ca/eng/waste/uranium-mines-and-millswaste/

Canadian Wind Energy Association (2019) Installed capacity. canwea-installed-capacity-map-canada-december-2019-e-web-v1

CAPP (2022) Canada's natural gas. Canadian Association of Petroleum Producers. https://www.capp.ca/wp-content/uploads/2022/04/Natural-Gas-Fact-Book-Web-404525.pdf

CAPP (2023) What are the oil sands? https://www.capp.ca/oil/what-are-the-oil-sands/

Catto NR (2020) Atlantic Canada's tidal coastlines: Geomorphology and multiple resources. In: Slaymaker O, Catto NR (eds) Landscapes and landforms of Eastern Canada. Springer International Publishers, Cham, Switzerland, pp 401–430

Catto NR, Slaymaker O, Kovanen DJ (2020) Human-modified Landscapes. In: Slaymaker O, Catto NR (eds) Landscapes and Landforms of Eastern Canada. Springer International Publishers, Cham, Switzerland, pp 475–507

Catto NR Parewick K (2008) Hazard and vulnerability assessment and adaptive planning: mutual and multilateral community–researcher communication, Arctic Canada. In: Liverman DGE, Pereira C, Marker B (eds) Communicating environmental geoscience. Geological Society, London, Special Publications, vol 305, pp 123–140

Clark RH (1997) Prospects for fundy tidal power. La Houille Blanche 83(3). https://doi.org/10.1051/lhb/1997019

Coal Association of Canada (2024) About the coal industry. https://www.coal.ca/coal-resources/about-the-coal-industry/

Dehghani-Sanij AR, Al-Haq A, Bastian J, Luehr G, Nathwani J, Dusseault MB, Leonenko Y (2022) Assessment of current developments and future prospects of wind energy in Canada. Sustainable Energy Technol Assess 50:101819. https://doi.org/10.1016/j.seta.2021.101819

d'Entremont MV, Hartley MI, Otter KA (2017) Comparing pre- versus postoperational movement of nocturnal migrants around a wind energy facility in northeast British Columbia, Canada. Avian Conserv Ecol 12:3. https://doi.org/10.5751/ACE-01046-120203

Electricity Canada (2022) Value of transmission. https://www.electricity.ca/knowledge-centre/the-grid/transmission/value-of-transmission/

Energy Institute (2023) Statistical review of world energy 2022. https://www.energyinst.org/statistical-review

Environment and Climate Change Canada (2016) Canadian mercury science assessment. En84–130/2–2016F-PDF

Environment and Climate Change Canada (2022) National Inventory Report 1990–2021: Greenhouse Gas Sources and Sinks in Canada. https://publications.gc.ca/collections/collection_2021/eccc/En81-4-1-2019-eng.pdf

Environment and Climate Change Canada (2023) Emission factors and reference values. https://www.canada.ca/en/environment-climate-change/services/climate-change

Faramawy S, Zaki T, Sakr AA-E (2016) Natural gas origin, composition, and pricing: a review. J Nat Gas Sci Eng 34:34–54

Foley R, Bell T, Liverman DGE (2011) Urban geochemical hazard mapping of St. John's, Newfoundland. Canada. Atlantic Geology 47:138–157

Foo CF Bennett VJ Hale AM Korstan JM Schildt AJ Williams DA (2017) Increasing evidence that bats actively forage at wind turbines. PeerJ 5:e3985. https://doi.org/10.7717/peerj.3985

Fowler D et al (2020) A chronology of global air quality. Philos Trans R Soc A 378. https://doi.org/10.1098/rsta.2019.0314

Francescato V, Antonini E, Bergomi LZ (2008) World Fuels Handbook. European Union, Biomass trade centres.

Goldenberg SZ, Cryan PM, Gorresen PM, Fingersh LJ (2021) Behavioral patterns of bats at a wind turbine confirm seasonality of fatality risk. Ecol Evol 11:4843–4853

Gordon D (2012) Understanding unconventional oil. Carnegie Papers

Gowlett JAJ (2016) The discovery of fire by humans: a long and convoluted process. Philos Trans R Soc B 371. https://doi.org/10.1098/rstb.2015.0164

Green KP, Jackson T (2015) Safety in the transportation of oil and gas: Pipelines or rail? Fraser Institute, Vancouver. fraserinstitute.org

Halpern PJ, Plourde A, Waverman L (1988) Petro-Canada: its role, control, and operations. Economic Council of Canada

Health Canada (2022) Health in a changing climate 2022. https://changingclimate.ca/health-in-a-changing-climate/chapter/5-0/

Health Canada (2023) Wood smoke. https://www.canada.ca/en/health-canada/services/air-quality/indoor-air-contaminants/avoid-wood-smoke.html

Henderson SB, Brauer M, MacNab YC, Kennedy SM (2011) Three measures of forest fire smoke exposure and their associations with respiratory and cardiovascular health outcomes in a population-based cohort. Environ Health Perspect 119:1266–1271

Hydro-Québec (2024) Beauharnois hydropower generating station. https://www.hydroquebec.com/facility-tours/tours-general-public/beauharnois-monteregie.html

International Energy Association (2010) Renewable energy essentials: Hydropower http://www.iea.org/publications/freepublications/publication/hydropower_essentials.pdf

Katzner TE et al (2016) Golden Eagle fatalities and the continental-scale consequences of local wind-energy generation. Conserv Biol 31:406–415

Kentucky Geological Survey (2024) Coal. University of Kentucky. https://www.uky.edu/KGS

Kim K-H, Kabir E, Jahan SA (2016) A review on the distribution of Hg in the environment and its human health impacts. J Hazard Mater 306:376–385

Knowles J (2016) Power shift: Electricity for Canada's remote communities. Conference Board of Canada, Ottawa

MacEachern C, Yildiz Y (2018) Wind Energy. In: Dincher I (ed) Comprehensive Energy Systems. Elsevier, Amsterdam, pp 667–701

MacGregor KA, Lemaître J (2020) The management utility of large-scale environmental drivers of bat mortality at wind energy facilities: The effects of facility size, elevation and geographic location. Glob Ecol Conserv 21:e00871. https://doi.org/10.1016/j.gecco.2019.e00871

Martimanakis S (2023) Ultimate list of the top five largest operational and under construction solar installations in Canada. https://pvbuzz.com/top-five-largest-solar-installations-canada/

Montréal (2023) Solid fuel-burning stoves and fireplaces. https://montreal.ca/en/topics/solid-fuel-burning-stoves-and-fireplaces

Murray D (2024) Firewood types and their heat output. https://forestry.com/wood/firewood-types-and-their-heat-output/

Natural Resources Canada (2009) Ground Source Heat Pumps (Earth Energy Systems). Natural Resources Canada, Office of Energy Efficiency. http://oee.nrcan.gc.ca/publications/infosource/pub/home/heating-heat-pump/gsheatpumps.cfm

Natural Resources Canada (2020a) Natural gas. https://natural-resources.canada.ca/energy/energy-sources-distribution/natural-gas/5639

Natural Resources Canada (2020b) Annual photovoltaic potential. https://ecosmartsun.com/solar-map-canada/

Natural Resources Canada (2023) Coal facts. https://natural-resources.canada.ca/our-natural-resources/minerals-mining/mining-data-statistics-and-analysis/minerals-metals-facts/coal-facts/20071

Newfoundland and Labrador (2023) The economy 2023. https://www.gov.nl.ca/fin/files/The-Economy-2023.pdf

Nunavut (2023) Community and Government Services, Petroleum Products Division 2018–21 Amalgamated Annual Report. Nunavut MaliGaliurvia. https://www.assembly.nu.ca/node/10935

NWT (2020) Wind monitoring in norman wells, NWT. NWT Research database. https://data.researchlicensing.ece.gov.nt.ca/Scientific/16677

Ontario Power Generation (2021) OPG's Sir Adam Beck I hydro station celebrates a century of clean power. https://www.opg.com/stories/opgs-sir-adam-beck-i-hydro-station-celebrates-a-century-of-clean-power/

Ouranos (2022) Executive summary: Impact of climate change on Nunavik's marine and coastal environment. https://www.ouranos.ca/sites/default/files/2022-07/proj-201419-enord-bleau-summary.pdf

Past Due (2024) Hydro-Québec: your bill is 50 years past due. https://50yearspastdue.ca/background-information/

Pelling HE, Green JAM (2013) Sea level rise, tidal power plants and resonance in the Gulf of Maine. J Geophys Res, Ocean 118:3462–3472

Petroleum History Society (2023) Six historical events in the first 100 years of Canada's petroleum industry. http://www.petroleumhistory.ca/history/wells.html

Pinto H, Gates ID (2022) Why is it so difficult to replace diesel in Nunavut? Renew Sustain Energy Rev 157:112030. https://doi.org/10.1016/j.rser.2021.112030

Plourde A (2010) Oil and gas in the Canadian Federation. University of Alberta, Department of Economics, Working Paper 2010–01. https://sites.ualberta.ca/~econwps/2010/wp2010-01.pdf

Price DT, McKenney DW, Caya D, Côté H (2001) Transient Climate change scenarios for high resolution assessment of impacts on canada's forest ecosystems. Report to the Climate Change Impacts and Adaptations Directorate, p 42

Radwan AE, Yin S, Hakimi M, Li Y (2023) Petroleum geology of conventional and unconventional resources: Introduction. Geol J 58. https://doi.org/10.1002/gj.4898

Riedel M Brent TA Taylor G Taylor AE Hong J-K Jin Y-K Dallimore SR (2017) Evidence for gas hydrate occurrences in the Canadian Arctic Beaufort Sea within permafrost-associated shelf and deep-water marine environments. Mar Pet Geol 81:66–78

Sadorsky P (2021) Wind energy for sustainable development: driving forces and future outlook. J Clean Prod 298:125779. https://doi.org/10.1016/j.jclepro.2020.125779

Schobert HH (2013) Chemistry of fossil fuels and biofuels. Cambridge University Press

Semenza E, Ghirotti M (2000) History of the 1963 Vaiont slide: the importance of geological factors. Bull Eng Geol Environ 59:87–97

Sirén APK, Pekins PJ Kilborn JR, Kanter JJ, Sutherland CS (2017) Potential influence of high-elevation wind farms on carnivore mobility. J Wildl Manag 81:1505–1512

Skarin A, Sandström P, Alam M (2018) Out of sight of wind turbines—Reindeer response to wind farms in operation. Ecol Evol 8:9906–9919

Sovacool BK (2013) The avian benefits of wind energy: A 2009 update. Renew Energy 49. https://doi.org/10.1016/j.renene.2012.01.074

SPE (2018) Petroleum resource management and system. Society of Petroleum Engineers, Tulsa

Speight JG (1999) The chemistry and technology of petroleum. Marcel Dekker, New York

Statista (2024) Production of wood fuel worldwide. https://www.statista.com/statistics/481654/wood-fuel-production-globally/

Statistics Canada (2021a) Households and the Environment Survey 2019. https://www150.statcan.gc.ca/n1/en/daily-quotidien/210601/dq210601c-eng.pdf?st=ssmu4NdO

Statistics Canada (2021b) Residential use of wood and wood pellets. https://www150.statcan.gc.ca/t1/tbl1/en/tv.action?pid=2510008301

Statistics Canada (2022) Canadian greenhouse gas emissions attributable to households, 2018. https://www150.statcan.gc.ca/n1/pub/11-627-m/11-627-m2022003-eng.htm

Statistics Canada (2023a) The heat is on: how Canadians heat their home during winter. https://www.statcan.gc.ca/o1/en/plus/2717-heat-how-canadians-heat-their-home-during-winter

Statistics Canada (2023b) Canadian social survey: Energy use. https://www150.statcan.gc.ca/n1/daily-quotidien/231030/dq231030b-eng.htm

Stefanidou M, Athanaselis S, Spiliopoulou C (2008) Health impacts of fire smoke inhalation. Inhalation Toxicol 20:761–766

Suihong S (2016) A perfect play of salt, bamboo, and gas: the first petroleum well (breakthrough) in the world. AAPG, Search and Discovery Article #70220

Sutherland G Foreman M Garrett C (2007) Tidal current energy assessment for Johnstone Strait, Vancouver Island. In: Proceedings, International Mechanical Engineers 221 Part A: Journal of Power and Energy, pp 147–157

Tarquti (2023). Electricity in Nunavik. https://tarquti.ca/learn/electricity-in-nunavik/

Tarbotton M Larson M (2006) Potential tidal energy resources. Canada Ocean Energy Atlas. Canadian Hydraulics Centre, Ottawa

Taylor GD (2019) Imperial standard: Imperial oil, Exxon, and the Canadian Oil Industry from 1880. University of Calgary Press

Thaxter CB, Ross-Smith VH, Bouten W, Clark NA, Conway GJ, Masden EA, Clewley GD, Barber LJ, Burton NHK (2019) Avian vulnerability to wind farm collision through the year: Insights from lesser black-backed gulls (*Larus fuscus*) tracked from multiple breeding colonies. J Appl Ecol 56:2410–2422

Transportation Safety Board of Canada (2014) Railway Investigation Report R13D0054: Runaway and main-track derailment Montreal, Maine & Atlantic Railway Freight train MMA-002 Mile 0.23, Sherbrooke Subdivision Lac-Mégantic, Quebec 06 July 2013. https://www.tsb.gc.ca/eng/rapports-reports/rail/2013/r13d0054/r13d0054.pdf

USDA (2023) Canada wood pellet annual. United States Department of Agriculture, CA 2023–002. https://apps.fas.usda.gov/

USGS (2022) Induced earthquakes. United States Geological Survey, http://earthquake.usgs.gov/research/induced

Vragov V, Ahmadi H, Grey K, Mclean J (2018) Quantitative comparison between pipeline and rail crude transportation cost to the society. https://www.ucalgary.ca/science/redevelop/files/redevelop/technical_paper_pipeline.pdf

Walston LJ, Szoldatits K, Lagory KE, Smith KP, Meyers SA (2021) A preliminary assessment of avian mortality at utility-scale solar energy facilities in the United States. Renew Energy 92:405–414

Wind Energy Institute of Canada (2024). https://www.weican.ca/, Tignish PEI

Wood Pellet Association of Canada (2023) Biomass is going mainstream: key learnings for Canada. https://www.canadianbiomassmagazine.ca/biomass-is-going-mainstream-new-report-outlines-key-learnings-for-canada/

Zimmerling JR, Francis CM (2016) Bat Mortality due to Wind Turbines in Canada. J Wildl Manag 80:1360–1369

Economic Development: From European Periphery to North American Periphery?

18

Abstract

Historically, economic development in Canada has been shaped by global and North American trends, local industries and natural resources, and political decisions. The nineteenth-century "National Policy" is a prime example, with ramifications that continue today. Canada has vast natural resources that have played a crucial role in its industrial development. The mining, forestry, and oil and gas industries have contributed significantly to the Canadian economy. Even if the size of its domestic market was limited, the development of transportation opened new territories to be explored and led to many mineral deposit discoveries with considerable international market potential or as development generator. Canada is highly likely to be influenced by what is happening in USA. Currents that hit the United States might eventually be felt in Canada. Industrialization, urbanization, and immigration that were observed south of the borders have typically been felt two or three decades later in Canada. Resource production involves hydrocarbons, forestry, grain agriculture, hydroelectricity, and aluminium. From a collection of smaller regional economies, each based on different resources, Canada's development has grown from a staple to a post-industrial economy. Canada's development has deeply been dependent on the physical and locational particularities of its regions. Each region of Canada has a different economic base, shaped by its natural resources, industries, and traditional activities. The Atlantic Provinces rely on fishing and natural resources. Central Canada is a major industrial and financial hub. The Prairies are known for their agriculture and energy. British Columbia benefits from a diversified economy, and the North represents a wealth of natural resources. Canada has been Britain's periphery, then the United States' periphery, before the Southern parts of Québec and Ontario, became a northern extension of the USA industrial core. The model can still be applied within Canada, with cores in Alberta and British Columbia among the western periphery.

Keywords

Core-periphery · Natural resources · Regional economies · Economic history and development

18.1 Introduction

Canada is still a young country. Within a little over four hundred years, this part of America went from a land inhabited by traditional societies who had practised subsistence economies to a country that has been colonized and developed into an industrial economy by waves of migrants In fact, development and its effects have spiralled over the centuries (Fig. 18.1). Landscapes that have evolved over centuries, or millennia in certain cases, have been completely transformed in less than four centuries. Many of these original societies were nomadic people that survived by hunting, fishing, foraging, and gathering. Some others became sedentary and developed some knowledge in agriculture that could support a large population. In that sense, a variety of lifestyles and mainly subsistence economy were influenced by their equally different environmental conditions throughout the continent. Different environmental conditions called for different ways to interact with the environment.

18.1.1 Indigenous People and Subsistence

Historical researchers (e.g. Kroeber 1923; also see Saku 1999, Chap. 11.1) identified five generalized Indigenous groups having lived in parts of Canada in pre-Columbian times. The fishers and hunters of the Arctic whose diet was made up of char, cod, seal walrus, caribou, and berries in summer, lived

N. Catto and G. R. Vincent, *Canada: Land and People*, World Regional Geography Book Series,
https://doi.org/10.1007/978-3-032-04391-7_18

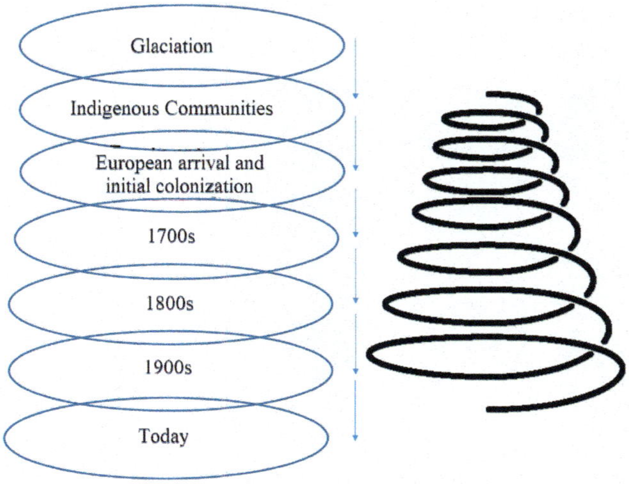

Fig. 18.1 Development spiral in Canada over time

in "snow houses" in winter and "skin tents" in summer. The fishers and hunters of the Pacific Northwest fished salmon and halibut, hunted forest deer, and gathered berries. They lived in large villages of planked houses covered with shingled roofs. The hunters of the boreal forests diet included caribou, elk, hare, freshwater fish, and berries. Their lodgings were double lean-to's of brush, teepees, and log cabins. Hunters from Laurentian-Appalachian woodlands feasted on deer, freshwater fish, waterfowl, wild rice, cultivated maize (corn), squash, and beans in the southeastern part of their region. Their shelters were made of birch bark teepees in the North part and lived in villages of bark longhouses over pole frames in the Southeast. Hunters of the Prairie and steppe chased buffalo, antelope, deer, rabbit, waterfowl, cooked pemmican, and harvested freshwater fish. They lived in skin tipis and villages of polygonal log houses roofed with sods (Kroeber 1923).

Subsequent research, including work by Indigenous scholars, has illustrated both the diversity and breadth of scope of pre-contact Indigenous economic activity throughout the current territory of Canada (e.g. Mann 2003; Doolittle 2004; Watson 2007; Brink 2008; Carlos and Lewis 2010; Jolicoeur 2020; Carlos 2023). Indigenous economies included specialized production, trade networks, societal (public) infrastructure including trail maintenance, standards and dispute resolution processes, mediums of exchange, and defined property rights (e.g. LeDressey et al. 2010, and references therein). Earlier researchers were unaware or underestimated the damaging effects of colonial activity on pre-existing Indigenous economies.

Institutional structures and financing limitations continue to be challenging for Indigenous people and communities (Schembri 2022), although economic outcomes continue to improve (Chernoff and Cheung 2023). Indigenous economic activity represented approximately 2.2% of Canada's GDP ($\$48.9 \times 10^9$) in 2020 (Ayotte and Bridger 2022). Compared with the Indigenous population of ~5% of Canada, this indicates that achieving economic reconciliation still requires effort (Chernoff and Cheung 2023).

18.1.2 European Explorations

European navigators have wandered on the Atlantic coast since at least the tenth century. The remains of a Norse settlement were identified on the tip of Newfoundland. Barraclough (2018) believes the settlement may have been used only sporadically. Basque, Portuguese, and French ships crossed the Atlantic to fish and whale on the Grand Banks. When Columbus accidentally discovered an unknown continent from the European perspective, numerous explorers navigated along the North American coast, some to try to find a passage to Asia, others to establish colonies. Most were economically motivated to tap new resources for European consumption or transformation. Thus, Imperial powers concurred in a "race around the world" to conquer various parts of "The New World". Eventually, Great Britain and France would colonize the Northern part of America, with British settling on the Atlantic coast and establishing trade posts in the Arctic North, and French on the Northern coast and later the St-Lawrence axis (see Chaps. 1 and 11). Numerous disputes emerged for territorial exclusivity. Consequently, boundaries had to be established or redefined on a regular basis.

18.2 Heartland and Hinterland

By then, European possessions were essentially becoming extensions of the European economic core, becoming the sites where the resources are found and exported to the European market. It is useful at this point to refer to the "Heartland and Hinterland model" (or core and periphery) to better understand how two regions interact with one another. The model is often used to analyse the relationships of power and dependence between different regions. It describes a spatial structure where a dominant central region exercises economic, political, and cultural control over dependent peripheral regions. The model is flexible. It can be applied at different geographic scales, from local to global, at different time intervals, and on a variety of spheres. Usually, it refers to economic exchanges, but it can also be applied to political, social, cultural interaction. It is therefore perfectly suited to document the evolution of rapports between Imperial powers and their colonies, but also how the role and the balance of power in distinct parts of a given area can change over time. As the following discussion will show, the application of the model can describe relations between regions that have grown more complex and sophisticated.

Simply, and very bluntly, the "Heartland" (or core) is where the power and control are, and the "Hinterland" (or periphery) is the area under the heartland's control or, as Friedmann (1973) put it, "a general theory of polarized growth". This power is expressed by the concentration of investment Capital, large labour force, adapted technology, and aggressive entrepreneurship or simply put, an industrial core is a developed region that will produce goods from resources coming from the periphery. The generated profits and wealth will then have multiplier effects (or spinoffs) that will induce growth. "Hinterland", on the other hand, in its basic role, is where resources are extracted to manufacture products in the Heartland. Over time, however, a resource producing region is not bound to necessarily stay that way. It can also create spinoffs and develop. Thus, it could be argued that core and periphery are interdependent.

18.2.1 Extension of European Systems

Figure 18.2 summarizes the mercantile model (Marshall 1991, for a mapped display of the model, McCann, 1998, for a graphical rendition of the model) which conveys adequately the evolution of the relation between Imperial powers (then core) with their North American possessions from the time of discovery (periphery). Colonies were then clearly the extension of Imperial urban and economic systems. Early on, European fishers would establish temporary seasonal first settlements along the American coasts, notably in Newfoundland, along the Gulf of St-Lawrence and on New-England shores. Fish would be dried, salted, and brought back to European markets at season's end. More explorations during the seventeenth century would eventually lead to permanent (or at least semi-permanent) settlements and the exploitation of other staples, specifically fur. Fur, especially beaver pelts, driven by French fashion industry, was in high demand after beaver having come close to extinction in Europe. Trading posts and depots multiplied, contributing to the stage of "expansion within colony", and pelts would be shipped to Europe from an expanding harbour town. Running along the St-Lawrence valley, throughout most of the Mississippi drainage basin, and deep in western territories, the French fur trade attained continental dimensions.

Britain wanted a part of this market. They started to trap beaver along the rivers. Hudson's Bay Company was then created in 1670 and was granted by the British Crown control of Rupert's Land that covered the entire watershed of Hudson's Bay (Brown 2017). Thus, an initial peripheral urban system pattern emerged. As a result of increased supply of furs, port activity intensified, and the fashion industry developed in Europe. New Capital circulated and fuelled the economy by creating multiplier effects.

Meanwhile, on the colonization front along the St-Lawrence River and on the Atlantic Coast, differences emerged between French and English settlements. British migration continued and added rapidly to the population, which also was becoming a considerable domestic market for its products and developing services (see Chap. 11). The colonies' economy still rested on exports of copious quantities of timber, fish, furs grains, rice, indigo, cotton, and tobacco and contributed enormously to the British economy, thus still being a profitable hinterland to the British Crown. Population growth, on the other hand, was comparatively slow in New France, generating mostly a subsistence economy. King Louis XIV was reluctant to "depopulate France to populate Canada" (Lacroix 2019). Still, Jean Talon succeeded to a certain extent in instilling an entrepreneurial climate. Tanneries, artisanal workshops, forestry, and a brewery now contributed to the colony's economy. Eventually, a shipyard would build vessels and the forges of Saint-Maurice would become the first industrial establishment in Canada. There were cotton mills here and there.

By then, colonization had been a long and slow process of development, especially by the French settlers. French migration and population growth had been slow compared to the British colonies. However, even with a smaller population count, France exercised an extensive control over large interior territories, contrasting with a limited but intensive grip on the coast by the British. The staples economy has established a network structure that would be, to a certain extent, the basis of a new and growing urban system.

The eighteenth-century St-Lawrence Valley was mainly agricultural. However, Montreal and Québec City slowly developed as commercial cities. Montreal became an important fur trade centre as a transfer station for pelts from the Upper St-Lawrence and Ottawa rivers, transferred from canoes or small boats to larger vessels for exports. Québec was more diversified as artisan workshops had developed, several shipbuilding sites were running, and it was the main trading connection between New France, France, and stations along the Gulf of Mexico and the West Indies (Harris 1987). Trois-Rivières, with a smaller fur hinterland, was much smaller. Nonetheless, these cities began to display an emerging urban form as commercial and trading activities would develop near the harbour.

Changes were coming with the end of the French Regime in 1763 as New France fell under British administration. British merchants started to flock to Québec City, as well as British loyalists following United States independence. From then on commerce was routed through Britain.

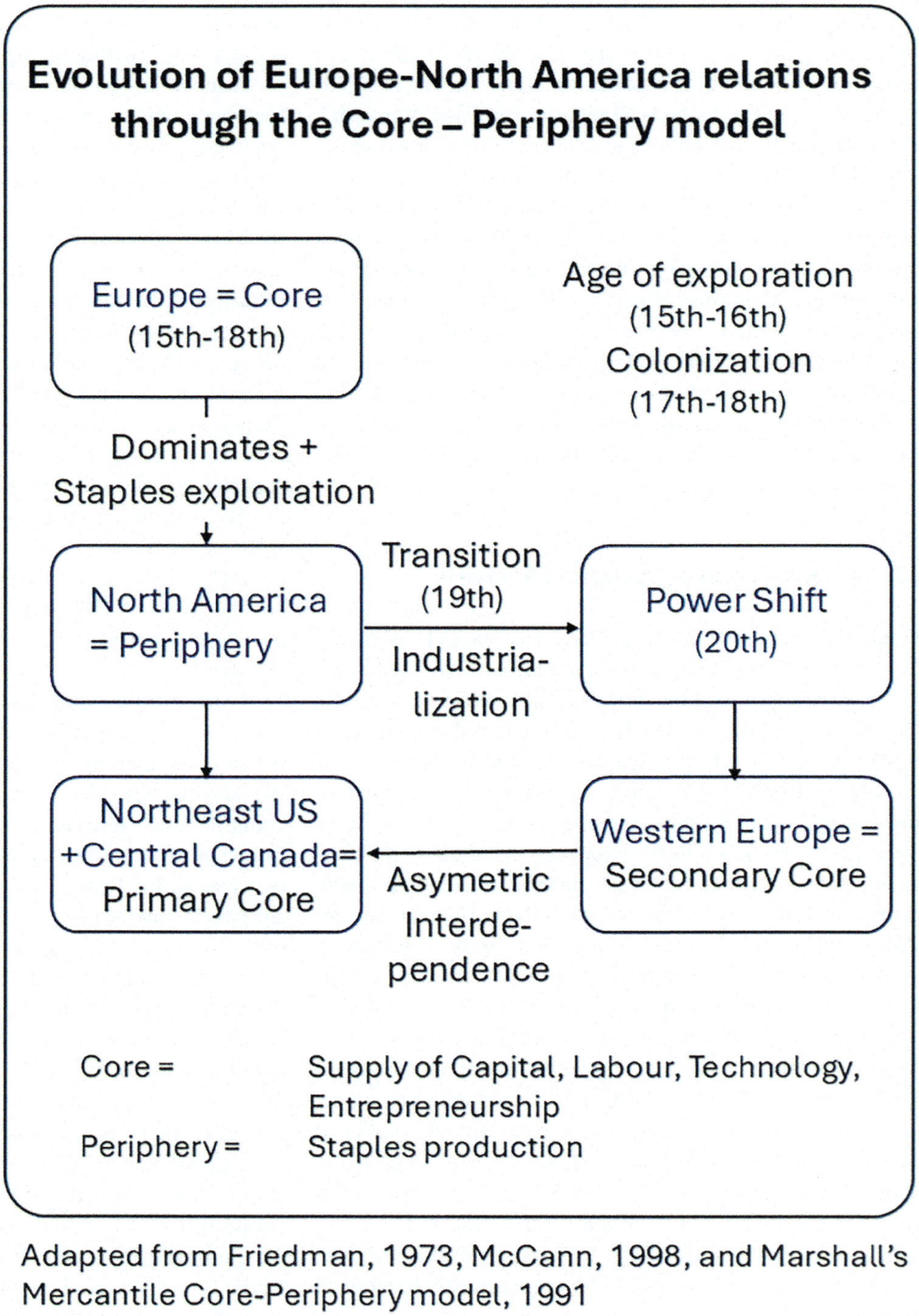

Fig. 18.2 Evolution of Europe-North America relations through the Core-Periphery model

18.2.2 Meanwhile, Something is Brewing in Great Britain…

The staples economy had been beneficial for both Imperial powers but probably more so for Great Britain. The variety of resources that could be exploited in the American colonies and the sheer quantity of its imports seem to have prompted the search for ways to transform massive quantities of resources much faster. This quest led to an avalanche of technological inventions and innovations during the eighteenth century that defines the term and fuelled the "Industrial Revolution". It transformed the way of producing goods. Mechanical first, and steam-powered machinery later in large factories, replaced manual artisan in small workshops. Improved agricultural instruments reduced the workforce needed in the fields. Repercussions were numerous and felt in many areas, including textiles and clothing, iron and steel, more efficient agriculture, food producing and packaging, and improved transportation.

The "magic ingredients" to the Industrial Revolution were coal and iron ore. Great Britain was blessed by its geography. It had large deposits of coal and iron ore mainly in the Northern half of its territory and in some other locations in Wales and Scotland. Manufacturing plants sprouted in cities adjacent to the deposits. Rural population flocked to the cities to get jobs. Cities' growth spiked. Their internal social structure was completely transformed and became geographically fragmented. Furthermore, the British population was turned upside down. From sparsely populated, its new distribution almost mimicked that of coalfields. From then on, it became the industrial core, as its American periphery doubled when New France was ceded. "See https://fulbrightworldhistory.weebly.com/start-of-the-industrial-revolution.html" or for illustrations.

18.2.3 Back to North America... Forests and Timber: River Basins and Deeper

Fur trade was now running further away and administered from London and shipped from Hudson Bay ports. Consequently, Canadian timber trade picked up during the Napoleonic Wars as Britain was barred from trading with European countries under Napoleon's continental blockade (Wynn 2001). Contrary to the continental span of fur trade timber trade had a fragmented configuration. It was carried along the riversides of forested areas along Saint John River in New Brunswick, the Montmorency, St-Maurice, and Ottawa rivers in Québec, and along Ontario lakes. Timbers would be floated to harbours for export. The unregulated cutting would progressively get deeper upstream, and the trees smaller (see Chap. 4). Eventually, sawmills or pulp and paper mills would develop near cutting areas to process the resource and cut transportation costs.

Technologically, North America was nearly one hundred years behind Europe at the beginning of the nineteenth century. Pre-industrial economy prevailed. The urban landscape exhibited an essentially mercantile city portrait, but some early manufacturing of entrepôt-type industry was noticeable. This type of manufacturing involves the primary transformation of resources being performed before shipping overseas (or somewhere else) for manufacturing of finished goods. Processing of raw material transported from the hinterland including milling, meatpacking, distilling, brewing, tanning, and refining, contributed to gradually transform the waterfronts and induce a geographical separation of land uses, especially in USA cities. They were still compact walkable cities. In that sense, the peripheral role of hinterland became organized. A complementary system of "industrialization at home" cities developed from the field to a market, to partial processing, then to export site. This stage occurred a little later in a still insular Canada.

The United States enjoyed a period of commercial prosperity by being neutral during the Napoleonic Wars. The USA could export some of their staples, including much in demand cotton in Britain, reexport commodities from the West Indies, and European manufactured goods throughout the world (Groves 2001). Economic mercantile effervescence led to significant infrastructure works that would connect an emerging industrial core to its extended periphery and drive settlements westward. Connections through rivers, canals, roads, and bridges allowed better communication so resources and goods could move. For the first six decades of the nineteenth century the United States experienced a population growth rate of no less than 32%, increasing from 5.3 million to 31.4 million through natural increase and immigration. An urbanization rate from six to twenty per cent accompanied this trend. Although the Northeast proportion of the total USA population declined from 50 to 34%, the region was the innovative pivot (United States Census 2020). Through immigration and domestic skills and expertise, innovative technology would develop rapidly. Steamboats, rail, industrial economies, and specialty clusters combined and connected the developing industrial Northeast, south of the Canada-USA border. The 1854 Reciprocity Treaty with Canada (East and West, not including NB and NS) extended the United States periphery by guaranteeing access to staples until 1866. Therefore, the North American market became a lucrative market. Moreover, it marked a shift from a transatlantic East–West core-periphery configuration to a continental North–South formation. For that matter, the ex-British colony achieved the stage of "economic independence". In Canada, however, urban life was still limited to Québec City and Montreal. The economy was one of subsistence along the St-Lawrence shores and a pre-industrial one in the port cities. To that effect, Canada became the periphery of the American industrial core.

18.2.4 1800s: An Important Century for Canada as Well

The nineteenth century proved to be an important one to Canada. The difference between Canada in 1800 and in 1891 is striking. Changes were profound physically, economically, and socially. Disparate parts of a wild land puzzle were brought to constitute a mostly domesticated landscape stretched and pieced together from sea to sea by a steel ribbon. Modern Canada was born in 1867. There was still much to explore and to discover further inland apart from spaces emptied of beavers by the fur trade. Recent history proved that Canada still had a considerable wealth of untapped resources on its territory!

New France, and then British North America's economy, had relied essentially on fish and fur trade until the early 1800s. The few sporadically dispersed settlements survived on subsistence agriculture. The population totalled 382,000 in 1800, the most important part (225,000, 60%) being in Lower Canada, along the St-Lawrence River. Atlantic Canada was home to approximately 100,000 and Upper Canada, 50,000 people (Statistics Canada 2010). Those numbers had swollen by 155% following the American Independence with the influx of British Loyalists (see Chap. 11).

In 1891 by contrast, Canada was in an industrializing phase. Its population had risen to 4.7 million. The southern parts were connected by an East–West railway line along which a string of towns and rural areas were attached. Ontario and Québec provinces played a dominant role. They had become the Canadian core, where 75% of the population resided, mainly in the St-Lawrence axis, and where a concentration of industrial and manufacturing activities was located. Agriculture was practised in rural areas of the Maritimes, the St-Lawrence axis, southern Manitoba, central Alberta, and central British Columbia. Mining, forestry, and fishing resources were generally peripheral but connected to the industrial regions.

18.2.5 Early Canada: A Partial Core or Periphery?

The French and British colonies had previously been the European periphery. However, Britain abolished the economic policy of protectionism which aimed to protect the economy of its colonies in 1846. In the meantime, railways slowly made their appearance in Canada and facilitated circulation of goods within Canada and the Maritimes colonies (see Chap. 14). Canada also turned south toward the United States to sell its raw materials and became part of the USA industrial hinterland by providing resources under the Reciprocity Treaty from 1854 to 1866. Lower Canada was still the most populated part of the Union in the first half of the century. With the previous Loyalist influx, settlement expanded away from the traditional settlement pattern toward the Eastern Townships but remained predominantly rural at 80%. If Lower Canada dominated the first half of the nineteenth century, Upper Canada ruled during the second half. Natural increase combined with Irish and Scottish immigration to increase population and surpass that of Lower Canada. Commercial farming, wheat for a short while, was very efficient and allowed rapid industrialization and trade with Upstate New York. A handful of cities emerged but, together, accounted for a mere 10% of the population.

The relative proximity to Europe that Québec City had over Montreal gradually slipped. Because of the obligatory transhipment necessitated by the Lachine Rapids, Montreal gained prominence as a major service centre for wholesaling and entrepôt-type manufacturing, and St-James Street was the prime location of financial activities. Québec City did retain an edge in the lumber trade and shipbuilding. However, the rise of Montreal as an international export gateway was short circuited by the numerous canal openings in the United States during the first half of the nineteenth century. From then on, Montreal was in direct competition with New York to have access to the Great Lakes.

18.2.6 Southern Attraction (or Temptations) Vortex?

The opening of the Erie Canal in 1825, the Oswego in 1826, and the Welland in 1829 indeed allowed a somewhat safer route between the Midwest to New York through Buffalo. The British colonies, later provinces, had retained the right to manage their resources. They had developed commercial ties with their adjacent American neighbours. Along with the first railway lines, their orientation would be mainly North–South. Keen to protect their own resources in a much larger domestic market, USA businesses started to look across the Northern border. This would eventually divert business strategic position to Southern Ontario, particularly the Toronto region. Hence, numerous industrial clusters developed in the USA North-East manufacturing belt and would eventually spread westward to "overflow" in Canada, to a certain extent, in Southern Québec and ultimately in Southern Ontario.

Nonetheless, Montreal was still a national transportation and an industrial platform. Indeed, the Lachine Canal, heralded as the industrialization cradle of Canada, would become a very industrial hub by the second half of the nineteenth century. Its location by the rapids was the initial factor for the hydraulic energy potential it provided. Over the years, hundreds of industrial manufacturers ranging from sugar refineries, paint manufacturing, tanneries, foundries, wood and textile mills, shoemaking, and bridge making would locate along the canal corridor. Added to that concentration, the Grand Trunk Railway (later Canadian National) had already made its way to Montreal and established its maintenance shop nearby in 1856. Other such shops were also built in Montreal (Angus shops), Moncton, Toronto, Stratford, Winnipeg, and Calgary, among many others.

18.3 The Confederation: Toward a Canadian Integration

The Dominion of Canada was born on an agreement, among other things, to link its constituents by rail. The idea was to build an economy based on the national market, even though the large territory was inhabited by a population of approximately three million people. Under the Constitution each province is responsible to manage its own natural resources (see Chaps. 1 and 15). There also exist some complicated provincial dispositions to protect their consumers that act as trade barriers between the provinces (Union des consommateurs). New Brunswick and Nova Scotia joined in 1867. Canada acquired Rupert Land (most of the Prairies and the Canadian North) from the Hudson's Bay Company in 1870. Manitoba (essentially the Red River settlement) entered the federation in 1870, British Columbia in 1871, and Prince Edward Island followed in 1873. The Intercolonial Railway was completed in 1876 between Halifax and Québec City (at Charny) where it connected to the Grand Trunk, allowing rail transportation to Sarnia. The Canadian Pacific Railway reached Winnipeg in 1882. The transcontinental line was completed in 1885 when "the last spike" was driven to reach the segment that started from Port Moody on the West Coast (see Chap. 14). The western terminal would be moved to Vancouver the next year. PEI was connected by a ferry service to satisfy its BNA constitutional condition to be connected to the rest of Canada.

Early on, the newly formed Canada faced a headwind. The world was plunged in a long-lasting economic depression (1873–1883). International trade was in decline. In times of hardship, states often tend to isolate. Consequently, exports of Canadian raw materials decreased. The Confederation "railway project" and the ongoing industrialization were decelerating. Out migration to the USA was on the rise. John A. Macdonald, Canada's first prime minister, ousted in 1873, returned to power in 1878 with the proposal of the protectionist National Policy.

18.3.1 Trying to Tie This Huge Land Together

Trying to unify a fragmented country into an internal market appeared to be a challenging task. Because of its large geography, Macdonald was convinced that Canada had to rely on transportation infrastructure to tie the various parts of the territory together. Equally, this would accelerate the ongoing gradual transition from a staples economy to an industrial one. The Conservative party under Macdonald proposed an electoral programme that would improve economy, discourage emigration, stimulate industry, and combat North–South "temptations" by encouraging East–West trade by building a transcontinental railway system (Charbonneau et al. 2018).

The "National Policy" intended to break the "natural" connections with American regions across the borders and favour East–West connections within Canada by the introduction of protectionist tariffs and promoting the building of a railway line across the land. According to Macdonald, the set of measures included in the programme would contribute to create "a union of interest, a union in trade, and a union in feeling" (Wynn 2001).

Amid a long-lasting economic depression, the National Policy started to be applied in 1879 to combat emigration to the US (essentially to create spinoffs). The policy would contain three main complementary and interlocking components: the instauration of protective tariffs; the construction of railway segments that would cross the continent from Halifax to Vancouver; and western colonization, especially in the Prairies.

The tariffs were used to help and protect Canadian industry by imposing custom tariffs on foreign products, particularly those coming from the USA. USA industry was experiencing the benefits of industrialization. Productivity and diversity of production were on the rise. Having long been kept in the role of raw materials exporter, Canada's industry was developing more slowly. The tariffs charged by the Canadian government will help finance the construction of a railway line that will connect the existing network in the east to the western part of the Canadian territory. Under this policy, for example, a steel manufacturer in southern Ontario would ideally have its coal shipped 2,000 km from Cape Breton Island instead of using nearer, and better-quality, Pennsylvanian coal. Equally, iron ore would have to be mined within Canada, probably around Peterborough (Marmora), and shipped by train to the producing plant, instead of coming by boat from Michigan or Minnesota.

"The Canadian Pacific Railway Company ("CPR") was initially formed from an effort to physically unite Canada from coast to coast following the nation's confederation on July 1, 1867" (CPR 2006). The construction of a railway line to cross Northern Ontario, the Prairies, and the mountain ranges to reach the Pacific Coast, Macdonald believed, would create a transcontinental railway to connect the distinct parts of Canada and make Canada enter the "Steel Era". It would have the double outcome of creating an internal market and establishing an integrated economy in which products and materials would circulate mostly in an East–West fashion as opposed to a North–South way. Rail, it was believed, or hoped, would play an integrator role in forging a Canadian character more than simply connecting parts of the land. One portion of the whole picture would be the part that the Prairies would play.

Part of the National Policy plan, paired with the existing Dominion Lands Act of 1872, was that the construction of the transcontinental railway would "open" the western part of Canada and facilitate the settlement of immigrants in the Prairies. The government used an aggressive promotional campaign nationally, continentally and in Europe with a host of colourful and inviting prospectuses designed to attract immigrants who would cultivate the land and develop agricultural settlements (see Chap. 11). The main objective was to encourage cereal and grain farming that could be moved to eastern and western rail terminals and exported internationally, thus contributing to Canada's economic growth.

18.3.2 Were the Outcomes as Expected?

The context of the implementation of the National Policy was difficult. There had been a long recession. The railway line to Vancouver was completed but the cost was high. Immigration suffered as well. The effects of the policy were not as strong as intended. Liberal Prime Minister Wilfrid Laurier reinstated Macdonald's wish in 1900 to "consolidate Confederation and bring our people… gradually to become a nation (Wynn 2001). The results were coming and would be visible in the long term.

Immigration was low, ranging between 25,000 and 40,000 from 1870 to 1880, and could barely offset the number of people leaving for the United States. However, numbers would pick up at the beginning of the twentieth century. Indeed, a little over 2.8 million people immigrated to Canada between 1900 and 1914 to boost Canada's population to 7.8 million, 1913 being the peak year with 400,900 (see Chap. 11). Many newcomers settled in "immigrant blocs" forming an "ethnic group tapestry" in the Prairies, sometimes occupying a whole township, sometimes more than one.

18.3.3 Multiplying Effects?

Generally, the National Policy re-enforced the role of Central Canada as the economic core and extended its periphery. Located at "break-of-bulk" points, Montreal, and especially Toronto at the crossroad of several railway lines, benefited from the induced development. Both cities experienced an increased concentration of industrial activities. The Lachine Canal in Montreal was already heavily industrialized, and Toronto saw its waterfront attract numerous new industries ranging from foundries, abattoirs, mills, furniture makers, breweries, rail workshops, railway yards, and port transhipment facilities.

The construction of the Grand Trunk had already prompted the establishment of foundries producing rolling stock, locomotives, and agricultural implements principally intended for a regional market. However, the construction of the CPR and the opening of the West stimulated the iron and the steel manufacturing industry that would supply almost everything related to railway building, including spikes, wheels, engines, shafts, and hosts of different parts, carriage and wagon makers, bridge parts, and also increased the demand for agricultural implements. Several establishments located in Toronto, including a young company named Massey that would become later Massey-Ferguson. Gad (1994) identified twelve such establishment (under "Engine builders, etc.) in the Toronto area in 1882. There were also "Machinery manufacturers" (27) likely supplying the variety of manufacturing plants in the region. Moreover, the steel industry developed in Hamilton and Sault Ste. Marie during the same period.

Unforeseen developments also occurred. The Canadian Shield and the west mountain ranges offered new challenges to railway builders. The demand created in explosives was so high that a "sophisticated supply chain" had to be developed to enable work to progress accordingly. The building of three dynamite factories was even necessary to satisfy demand (Wall 2017).

The construction of the railway line led to the discovery of minerals while blasting rock on the Canadian Shield in Ontario, including nickel, copper, and iron near Sudbury. Consequently, prospecting expeditions were conducted that resulted in further discoveries and the exploitation of gold, cobalt, platinum, silver, and many more.

The Canadian Pacific construction also opened new frontiers to access forest products and establish sawmills in Northern Ontario. Logging would also supply wood for railroad ties and bridges. The railway opened the Prairies to agriculture in the 1880s. Settlement was timid in the beginning but intensified at the turn of the twentieth century with record numbers of immigrants arriving in Canada. The land has been canvassed by multiple secondary railway lines and created a system that would collect wheat for it travel by rail to Thunder Bay, then transshipped on vessels to journey through the St-Lawrence network for international destinations. Equally, a hierarchy of cities, towns, and villages, each with their grain elevators, emerges to provide farmers with an increasing variety of services according to their size. The Prairies urban system is a good example of the central place theory pertinence.

But, surprisingly, the cost of rails was high since they, as financial reports indicate, were imported! Canadian foundries, including the Toronto Rolling Mill, produced railway tracks when they were made of iron, but with the development of steel production these firms could not, or did not, convert to the Bessemer process in their manufacturing. Steel rail production seemed to appear only from the early 1900s in Canada. Canada had to rely on British steel rail manufacturers who supplied the CPR: Barrow Hematite Steel Company (Barrow-in-Furness, Lancashire (now Cumbria)) heralded as the largest rail manufacturer in the world, Ebbw Vale Company (Ebbw Vale, South Wales), Mersey Steel and Iron Co (Liverpool), West Cumberland Iron and Steel Company (Workington), and Patent Nut and Bolt Co (near Birmingham). This fact earned Canada the following scathing remark, "There is no other country in the world with TEN THOUSAND MILES OF RAILWAY (author's emphasis) that does not make its own rails" (Bartlett 1885). Bartlett went further in arguing that such an extensive use of steel would justify the investments in steel production facilities. In that regard, although Canada was industrializing rapidly, it had not reached its full maturity yet. According to Rostow's stages of growth framework, Canada was showing "preconditions to take-off". The United States had reached the take-off stage in the 1840s. Benefitting from its proximity to the United States, Canada's economy reached that stage in the 1890s (Firestone 1960, 1968). In both cases, the two countries, being young, maintained a dependent role toward UK as a resource provider for a long time, longer for Canada, although their economic growth was remarkably fast. The lack of a large domestic market has been a deterrent for Canada. Because provinces can manage their own natural resources and a huge market lies so close, the East–West market was not as appealing.

18.3.4 The National Policy: Perverse Effects or Silver Lining?

Despite these efforts to keep the economic exchange within, the National Policy and its successors did not discourage foreign investments. Indeed, foreign companies, most of them from the United States, established subsidiaries in Canada ("branch plants") to avoid tariffs. This way, they could manufacture their goods in Canada and avoid their products being hit by high protective tariffs. Of course, establishing production plants in Canada necessitated large investments but had most probably been weighted before doing so. It could be argued that this way it did contribute to grow the domestic economy by employing a local workforce. One could wonder what would have happened if not?

Canada has welcomed many such ventures. There were 1,350 United States affiliates companies in Canada in 1934. Five percent (67) of them were established before 1900 (Marshall et al. 1936). For instance, carmaker Ford subsidiary Ford Motor Company of Canada, Limited, built an assembly plant in Walkersville ON (now part of Windsor) in 1899. This way it could access Canadian and British Empire markets. General Motors in 1918 and Chrysler in 1925 followed suit. Still, Canadian consumers and auto workers were disadvantaged by the North American market conditions. This eventually led to the Auto Pact (Automotive Products Trade Agreement) signed in 1965.

The North American Auto Pact aimed at creating a continental market for automobiles. Under this agreement, American and Canadian production of the "Big Four" (General Motors, Ford, Chrysler (now Stellantis), and defunct American Motors) and already present Studebaker (closed in 1966) and Volvo in Halifax (closed in 1998) would be integrated by: eliminating customs fees on parts and automobiles manufactured in Canada and the United States; guaranteeing production in Canada; allowing assembly plants to specialize in producing a limited range of models, thus realizing economies of scales and increasing market size to the continental scale (for instance GM similar Chevrolet Camaro and Pontiac Firebird were the last models produced at the Ste-Thérèse QC GM plant before closing). It essentially created an auto free-trading bloc before being terminated after the WTO ruled it illegal. However, the majority of its aspects were included in the recent NAFTA. Eventually, foreign car import rose, and manufacturing plants would open in Canada, concentrating mainly in Southern Ontario (see Chap. 12).

Similarly, to comply with a Province of Québec law that forced foreign corporation to invest in the province, with an exclusive contract binding them to the New York Times, a group of American businessmen from Utica NY, also active in Baie St. Paul Lumber Co., built a paper mill in Donnacona QC (near Québec City) to supply the daily publisher. They acquired extensive logging concessions to feed their production. Later Donnacona Papers had contracts with Hearst Corporation (Vallières 2012), but attempted to diversify their offering by adding construction "insulboard" made from wood waste to their production. Such dispositions enabled a North–South type on interactions.

A recent reverse trend is the buying of established Canadian companies by foreign interests (Figs. 18.3 and 18.4). For instance, Stelco and Dofasco, steel works that developed in early 1900s, are now under foreign control of US Steel in Pittsburgh and Arcelar Mittal in Belgium respectively (Hamilton Spectator 2012). United Kingdom's Rio Tinto acquired Alcan aluminium operations in 2007 (Financial Times 2007). Anglo Swiss Glencore-Xstrata bought Falconbridge copper, nickel, aluminium, lead, and zinc in 2006 (Glencore Canada 2024). Labatt, Canada's largest brewery, has been owned by Anheuser-Busch InBev since 2016.

18.4 Canada: Still a Rich Resource Economy

Canada has vast natural resources that have played a crucial role in its industrial development. The mining, forestry, and oil and gas industries have contributed significantly to the Canadian economy. Even if the size of its domestic market was limited, the development of transportation opened new territories to be explored and led to many mineral deposit

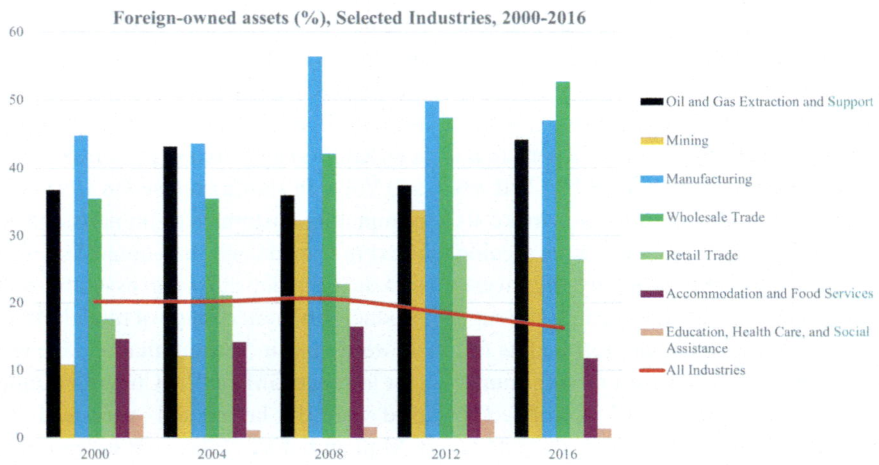

Fig. 18.3 Foreign-owned assets (%) in selected industries, 2000–2016. Concentrations are notable in oil and gas extraction, Manufacturing (particularly in automotive industries, as discussed in text), and wholesale trade. Increased involvement of multinational (particularly USA-based) companies in Retail Trade has occurred since 2000. Variation in Mining reflects acquisitions and mergers, including activity by large Canadian-owned firms. Educational, Health Care, and Social Assistance, along with Communication and Cultural Industries, have remained dominantly Canadian-owned (Data from Statistics Canada; latest available in December 2024)

Fig. 18.4 Percentage ownership of Economic Assets in Canada 2022, by country where ownership or head office is located (Data from Statistics Canada 2024)

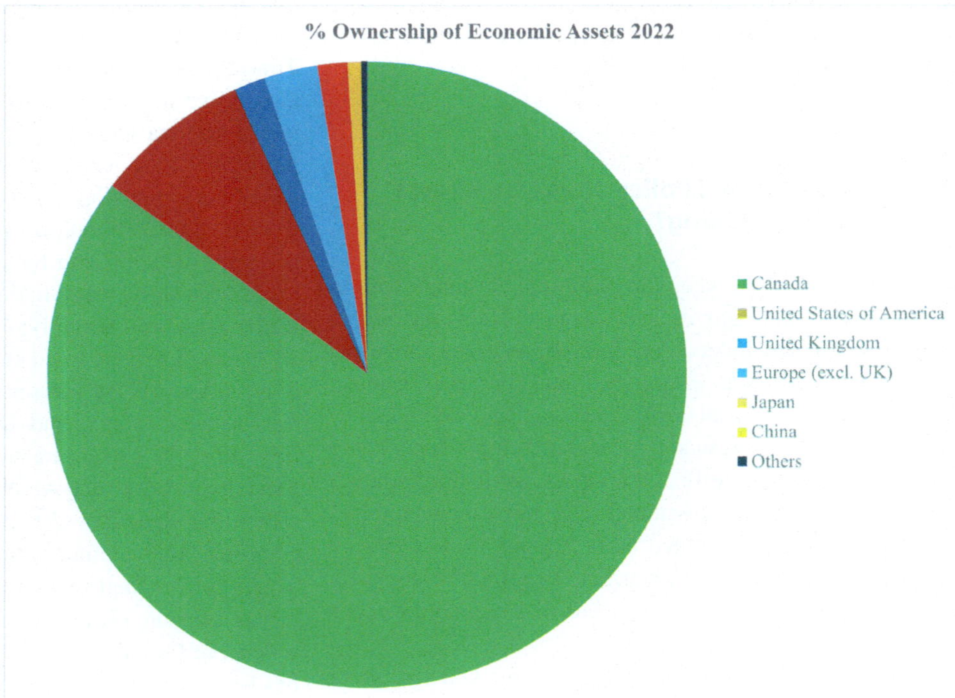

discoveries with considerable international market potential or as development generator. The richness of huge, forested areas also showed enormous potential. In numerous occasions, they led to resource or company towns. These towns have an ephemeral life span and are as randomly distributed as the resources are located. Most of them are mono-industrial, that is, extracting is their economic base.

Once the resource is depleted or no more profitable, activities cease, workers leave, services will eventually close as well. Some others will try to reorient or diversify their economic base. Elliot Lake ON is a typical example. Uranium mining operations started in 1955 with a contract to supply United States Atomic Energy Commission that was terminated in 1962 (Canadian Nuclear Safety Commission 2015). Contracts were signed with Britain, Japan, and Ontario Hydro in the 1970s. Global demand vanished in the early 1990s and mines closed. The town council opted to convert the community into a retirement village.

For a long time, Canada's raw materials wealth has contributed to industrialization in Europe and United States with minimal use domestically. Iron and coal were favoured within Canada in the first phase of industrial development. They were used early in Nova Scotia and New Brunswick before Confederation and evolved rapidly in the Montreal region and ultimately concentrated in Southern Ontario. However, iron became surpassed by steel which Canadian industry took a while to convert to. However, the discoveries of minerals (zinc, gold, nickel, silver, copper, cobalt, potash, among others) while performing earthworks for railway construction and the consequential prospecting works led to the development of international trade and investment networks, particularly with the United States. Their use has been valuable in various domains as in manufacturing as well as the benefits of coal, gold, lithium, titanium, and uranium in medicine for instance.

18.5 Twentieth Century: Modernization

It is already clear that by being situated directly north of the United States, Canada is highly likely to be influenced by what is happening south of the border. For instance, currents that hit the United States might eventually be felt in Canada. Indeed, waves like industrialization, urbanization, and immigration that were observed south of the borders have typically been felt two or three decades later in Canada. Likewise, a magnitude of one to ten can be applied to certain numbers: Canada's population is (very) roughly one tenth of the USA population, as its GNP is, for instance.

As it did in Britain a century before, the mushrooming of factories created a vacuum of people leaving rural areas to find work in cities, as well as attracting immigrants. Industrialization was already being felt in second part of the nineteenth century. Continued dependence on raw materials led to creation of resource or company towns throughout Canada. Manufacturing cores developed in regions with favourable proximity to large markets and accessible resource material supply. So did the USA industrial north-east around mid-century by transforming resources harvested in the United States but also gradually importing Canadian raw materials until commercial laws encouraged American industrial investments in Canada. Canadian-fuelled industrialization developed as well. From then on, adjacent industrial regions in Canada developed in the Montreal region and especially in the Ontario Great Lakes region.

Manufacturing, the process of adding value to raw materials, may consist of primary processing of raw materials, such as refining, such as oil, and derivative products such as tar, asphalt, or the assembly and production of finished goods, such as appliances and cars. A market needs to be large enough to realize economies of scale. A small domestic market limits economies of scale and may need reliance on export for industries to survive and to grow. The Maritime provinces have gone through this during the eighteenth century before companies were swallowed by larger firms that moved operations to Central Canada. Thus, Canada had a loosely scattered population of five million in the early twentieth century that made large-scale manufacturing uneconomical. Hence, Canada needed an open and/or a larger domestic market to grow, especially with such a large neighbour directly south.

The opening of the Prairies to settlement and farming, railway expansion into new regions, immigration waves and population increase that made population double within three decades, USA investments and technological advance, the introduction of scientific management, and the abundance of cheaper hydroelectricity than fossil-powered machinery to fuel new manufacturing plants all contributed to substantial manufacturing increase and Canadian economic modernization. The domestic market size, coupled with that of the United States, and gradual decrease in tariffs then made Canada transition from an essentially resource exporter to mass manufacturing.

18.5.1 Highs and Lows

The first half of the twentieth century was marked by major events which had opposite effects on the Canadian economy, two world wars and the Great Depression. Being unscathed by World War I hostilities, demand for production of military equipment surged. The early stage of industrialization limited Canada's participation in the war effort to produce equipment for Canadian and British soldiers including clothing and boots, riffles and ammunitions, shovels, etc. (Balakrishnan et al. 2007). Exports to Europe continued for goods while countries were rebuilding. Meanwhile, Canada's economy kept on growing. Then the Great Depression hit.

In time of crisis, economies tend to contract. Raw material and manufactured goods exports dropped. The market was then limited to a much smaller Canadian East–West market, and, consequently, dismal revenues. Regions with low economic diversity relying on timber and wheat, for instance, were hit hard. Larger manufacturers, such as steel mills, could trade with British Empire countries to try to recoup lost business but were running at far less than their capacity (Balakrishnan et al. 2007). A need to diversify was felt, especially in the Prairies, and efforts to do so were made to develop an array of manufacturing facilities (Balakrishnan et al. 2007).

A more developed industrial Canada contributed heavier and more technologically sophisticated military equipment during World War II (Balakrishnan et al. 2007). Economic recovery from the Great Depression was well underway.

Expansion of services and the constant creation of new jobs, notably office jobs, improved productivity, characterized the Western World after the Second War and Canada was no exception.

18.5.2 Since World War II

The momentum started during the World War II was prolonged to become a mutually reinforced economic and demographic boom. The baby boom, between 1946 and 1964, led to meteoric population growth from twelve to twenty million in Canada. A period of unprecedented effervescence followed. Urban population grew from 53% in 1941 to 66% in 1971. Most of this growth resulted in a new way of life, detached houses in suburbia. Since the 1980s it has stabilized at around 80%. The sudden and sustained increase in births for two decades resulted in increased demand for a large variety of consumer goods, for the construction industry and building materials, for the manufacturing and sales of furniture, appliances, cars (thanks to the Auto Pact), oil, and other forms of energy.

The public sector swelled. With increased needs in the education system, new schools were built, and new teachers were hired to welcome, and educate, this new generation. Pressure on the healthcare system and public services in general was felt as well. With such demands on goods and services, the private sector has grown and changed accordingly. Manufacturing firms continued to invest in production methods. Increased automation resulted in gradually lower blue-collar workers and more managerial and specialized jobs. Large infrastructure projects were conducted in the 1950s and 1960s as an extension of the postwar economic boom. Highways were built to carry traffic, the St-Lawrence seaway was built to carry cargo freight, hydroelectric projects were built to carry electricity to expanding cities and industries.

The Baby Boom generation had a considerable impact on society at every stage of their life. As they represented a massive proportion of the total population, as consumers they contributed to create significant new demands at every stage of their life. Equally, as they reached adolescence, the prospect of university education was imminent. Consequently, new universities were created. The 1960s is viewed as a Golden Age for education and postsecondary education in Canada as "universities ballooned in size and gradually became more research-intensive" (Usher 2024). In 1965–66, twenty-two were either new or institutions had their status modified (Charbonneau 2016) as well as many programmes continued to evolve or were created. A host of technical and general education colleges (CÉGEPs in Québec) were added to the system. As a result, baby boomers reached a higher level of education than previous generations. Hence, they entered the workforce in the 1960s and 1970s with new and diverse sets of skills that contributed to the rise of new sectors of activities.

18.5.3 From Fields to Offices

Economic activities have been divided into three main categories from primary to tertiary sectors which describe quite well the gradual detachment from the environment in its natural state, the primary being close to the earth, the following sectors generally moving away from it. At the same time, a trend of less primary jobs and more tertiary occupations was well under way (Fig. 18.5). The Clark-Fisher Economic model shows the trends that developed countries have followed from the pre-industrial to the postindustrial eras (Fig. 18.6). What is striking at first sight is the opposite trends of the primary (raw material extraction) and tertiary (services) sectors. The secondary sector (processing and manufacturing) has experienced an increase of the proportion of workers followed by a decrease. The trends shown in the model are consistent with what Canada's industrial sectors experienced between 1935 and 2009 (Bank of Canada 2012).

Primary activities have benefited throughout the last centuries from innovations that reduce the number of people required to conduct tasks, but which also reduce the muscular and physical effort. Ever more efficient tools and machinery have replaced the ploughs, picks, and shovels and increased worker productivity, productivity being the economic output produced by an individual. This can be measured in value produced by an activity or in quantity of units produced by an individual. For instance, by felling and processing trees, a "harvester" is at the "cutting edge of forestry technology". Similarly, precision farming allows farmers to better manage crop production by using agriculture dedicated information technology. Contemporary mining equipment does not compare with picks and shovels used a century ago. The results have been comparable in each domain: the work of one machine operator has replaced that of tens or hundreds of workers. However, Statistics Canada differentiates agriculture and "other" primary activities (Fig. 18.5). Agriculture jobs account for approximately two percent (1.9% in 2009 vs 1.3% in 2023) of total jobs, illustrative of modern and highly mechanized extensive and high-performing agriculture (Statistics Canada 2015, 2023). There is some discrepancy on sector definition data between the Bank of Canada and Statistics Canada, but the latter estimates the "industry" sector to account for 18.5% (17.6% in 2023) of total jobs in 2009. Still, the trend is compatible with the deindustrialization trend shown in Clark's model.

The secondary sector is the sector by which raw materials are transformed and manufactured into good. Firm executives want to improve the efficiency of their means of production. This includes the manner of accomplishing or dividing the work and the machinery used. Henry Ford innovated by applying scientific principles of production management put forward by Taylor using the assembly line, along which line are aligned workers who have a specific task to accomplish

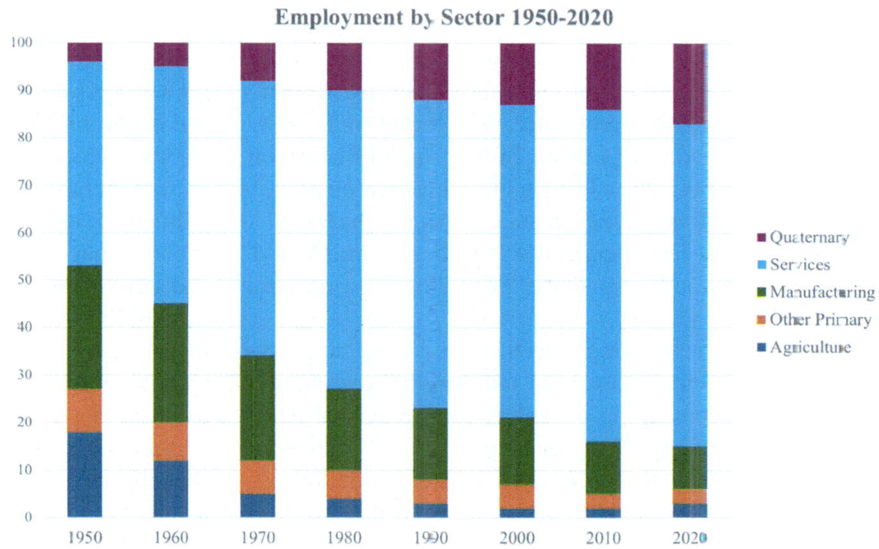

Fig. 18.5 Employment (%) by Economic Sector, 1950–2020. Agriculture is separated from other primary industries, following Statistics Canada practice. Sharp declines in agricultural and manufacturing employment since 1950 are evident (Data from Statistics Canada)

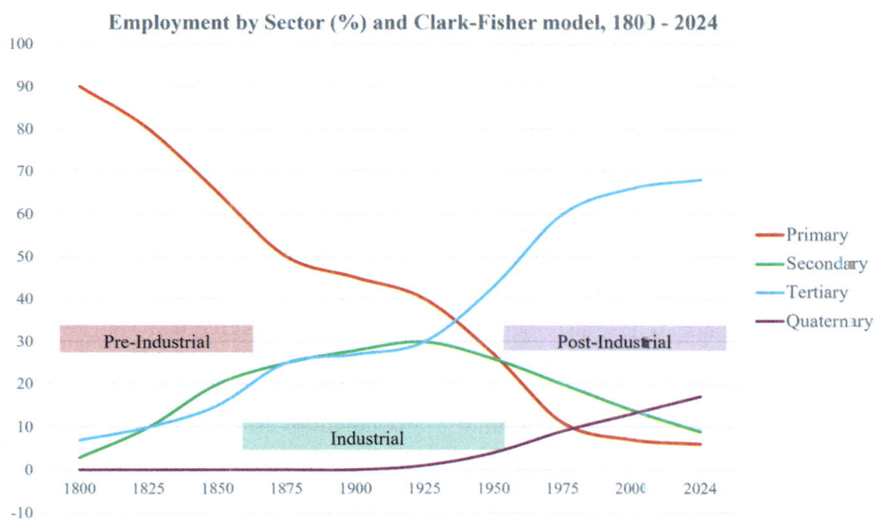

Fig. 18.6 Employment by Sector (%) and Clark-Fisher model of economic development, 1800–2024. Relatively high levels of Tertiary employment prior to 1925 (compared to secondary) reflect impact of governmental, administrative, and military employment for Canada's relatively small population. Industrialization accelerated over the period 1870–1930. After a recovery from decline during World War II, de-industrialization began in the 1950s. Tertiary employment increased rapidly between 1950 and 1980. Since 1990, employment in the Quaternary sector has seen the greatest rate of growth (Data from Statistics Canada, and from Marshall (1936) for years prior to 1934)

throughout their shift. The production equipment has evolved over the years, requiring ever less labour by entrusting part of the task to robotic devices which accomplish precise tasks without tiring to attain peak production around the 1960s–early 1970s. Since then, and in conjunction with the development of large firms, faster and more efficient transportation modes allowed these firms to conquer markets, resources, and labour internationally. This became particularly apparent from the 1970s when several long-established factories in older industrial regions reconsidered their options and opted to relocate, or close, their activities in developing countries. This scenario was noticeable in the American industrial core and its Canadian extent. Montreal and Toronto waterfronts, for instance, were vacated from industrial use.

The Beauce region, just south of Québec City, had developed in the 1980s a highly competitive and flexible highly automated garment industry that specialized in jeans clothing and was subcontracting international production for all jeans brands. Most of these manufacturing plants were closed by 2010 after trade liberalization in the 1990s and the lifting of tariff barriers in 2005 as brands turned to Asian or Latino-American subcontractors. Some of the manufacturers that remain in the regions have opted to attack the upscale denim market. At least one could promptly change its production to manufacture medical uniforms during the COVID-19 pandemic. In another instance, five textile mills in the town of Huntingdon (population 2,600), south of Montreal, closed in December 2004 leaving eight hundred employees without work. Efforts have been made to reverse the trend but the region is still struggling and is "Québec's Third World" according to the coordinator of the "Pacte de rue" organization (Lacoursière 2024).

The tertiary sector brings together everything that is not contained in the first two sectors and therefore constitutes a very heterogeneous group, including jobs in several fields, incomes from a few dollars per hour to several million per year. The tertiary sector is referred to as the service sector.

The direct link with the first two sectors is that the products resulting from production are transported to be marketed. This includes transportation, distribution, and direct retailing. However, this marketing also involves research in product design, market studies, advertising, production management, personnel management, etc.

More systematically, the tertiary sector can typically be divided into three types of activities, all of which are intended to offer: (1) personal or consumer services (retail, education, health and social services, entertainment, specialized consultants in various domains); (2) business services (finance, law, consultants and various professionals; (3) public services which are offered by the different levels of government (civil protection, public health, defence, roads, also includes health, education). However, more recently, the tertiary sector has been divided into new subcategories: the quaternary (or knowledge-based), sector linked to the development of innovative technologies and information, activities engaged in research, consulting, leasing, or concession of high value-added grey matter services, and the quinquennial sector, being that of decision-makers and positions exercising leadership as business leaders or senior public administration officials.

With around eighty per cent of its workforce in the tertiary sector, Canada is comparable to other developed countries (Fig. 18.3). This tertiarization has evolved since Canada started to industrialize with the timely addition necessitated by the economic transformations brought about by the development of the secondary sector, for example supervision (foremen) on the production floor, accountants and bookkeepers, managers of the staff, production managers, etc. At the same time, new concerns are creating new, or at least multiplying, jobs like architects, publicists, notaries and lawyers, town planners, designers, analytical specialists, and consultants of all kinds in a wide range of fields, different tasks in office work (reception, telephony, filing, secretarial work, etc.), R&D sectors (research and development) in most firms and domains. The fields of entertainment and tourism are also developing to produce their share of new jobs. The public sector, especially since the Second World War, has seen a growth in government jobs at all levels, as well as in education and health sectors. What is remarkable for Canada (and for the United States as well) is that it developed from essentially a raw material economy to a diversified modern economy in a little over one century. It is still strongly tied to a resource economy but considerable diversification and R&D (research and development) efforts have been, and still are, invested in these domains to either improve the extraction process, to minimize the adverse effects on the environment (like carbon capture), to improve the quality of agricultural yields, to manage the efficient exploitation of the forests, among other things to improve competitiveness, which illustrates that there are jobs that are directly and indirectly related to Canadian natural resources. They generate 14.9% of the national GDP but they also represent 45% of manufacturing production and close to one job out of ten (Cross and Mintz 2024).

18.5.4 Next Step: Continental Integration?

Of course, sharing the continent and being economic partner with the world's largest economy has been an extraordinary advantage. Canada and United States are mutual predominant trade partners. Trade between the two countries has intensified to become "partners" in CUSFTA (Canada–United States Free Trade Agreement) in 1989, in NAFTA (North American Free Trade Agreement) in 1994 when Mexico was added, and in CUSMA (Canada, United States, and Mexico Agreement) when renegotiated in 2018. The current agreement is a contentious issue at time of publishing. The trade agreement has to be renegotiated in 2026. However, the American administration has proved to be reluctant to pursue such an accord since its election in 2024. The natural, irresistible North–South attraction that the National Policy had tried to repress by favouring East–West linkages has been strengthened by such agreements. Canada is also member of fifteen trade agreements involving fifty-one countries (Government of Canada 2023).

18.5.5 A Low Dollar? Good for Export or Double-Edged Sword for Oil?

The value of the Canadian dollar is intrinsically linked to crude oil price, and it fluctuates accordingly. When the oil price is high, so is the Canadian dollar value. However, exports of manufactured goods potentially suffer because they become more expensive. Conversely, Canadian manufactured goods become more attractive under a devaluated dollar because of lower oil prices (Carbone and McKenzie 2016) and have the effect of boosting export revenues.

18.6 Sequential Eras of Resource Production

Agricultural, forestry, and energy resources and production are discussed in detail in Chaps. 15, 16, and 17, respectively. This section provides a concise overview in the context of economic history.

18.6.1 Black Gold: Coal and "First" Oil Era

The term Black gold refers to both coal and oil because of their colour. There are two uses for coal: thermal coal to fire electricity power plants; and metallurgical coal used as fuel in iron and steel manufacturing.

Canadian coal mining was first performed for its use in iron works on Cape Breton Island in Nova Scotia. It led to the regional economic development for a while but could hardly progress beyond that scale. Nova Scotia had developed ties with USA Atlantic Coast port cities before Confederation. Even though the intent of the National Policy was that coal from Nova Scotia would fuel the development of Central Canada's industry, transportation costs proved too high and American competition and quality, superior.

Coal deposits were later discovered during excavation work in Alberta along the CPR route in the western mountain ranges, and on Vancouver Island, too far from where the coal was needed. In 2022, Canadian coal mines extracted forty-seven million tonnes of coal and exported thirty-six million tonnes. British Columbia is the dominant producer with half of Canadian mined coal, while Alberta and Saskatchewan extract 28% and 13%, respectively. Nova Scotia, the early great producer, lags far behind with 0.1% (Natural Resources Canada 2020). Of the 11 million tonnes used domestically, 59% was metallurgical coal and 41% for thermal use. Alberta, with 51%, and Saskatchewan, 30%, generate the largest proportion of their electricity through coal fuelled thermal power plants. Nova Scotia and New Brunswick use more diversified sources of electricity. Consequently, coal account for respectively 15% and 4% of their generated electricity. Canada is phasing out thermal coal use by 2030.

Oil exploitation is an important industry in Canada, contributing significantly to the country's economy. Canada has the third-largest oil reserves in the world, after Venezuela and Saudi Arabia, mainly concentrated in the Alberta oil sands (Natural Resources Canada 2019).

The history of petroleum development in Canada is a compelling tale spanning more than 160 years, marked by unforeseen discoveries, major technological innovations, periods of dazzling economic growth, and growing environmental challenges.

Exploratory drilling led to the discovery of oil fields at Oil Springs, Ontario, and Turner Valley, Alberta, laying the foundations for a fledgling industry. Oil Springs exploitation was marginal and did not last long. The presence of oil was discovered in Waterton and Turner Valley southwest of Calgary around 1900. Pumping of conventional oil lasted from 1914 to 1985. When Turner Valley production started to slow down, a much larger oil field was found in Leduc in 1947, just south of Edmonton which propelled Alberta, and Canada, as a major oil producer. Canada soon became the tenth-largest oil producer in the world. Leduc ceased production in 1984. However, Athabasca oil sands exploitation, which required a completely different process, had started further north in 1967.

18.6.2 Green Gold: Forestry

Canadian forestry comprises three subsectors: solid wood transformation into construction products, pulp and paper, and logging. Softwood is, and has been for a long time, an important contributor to the forest product economy. However, lingering issues with the United States persist. For decades, the United States accused Canada of subsidizing softwood lumber producers, thus violating Canada's international commitments and being unfair to the USA softwood industry. In response, the USA has applied punitive duties to imported Canadian softwood lumber.

Is the boreal forest burning? Forest fire frequency has and is bound to increase with climate change. At times, fires can endanger other activities, notably oil production in the Fort McMurray area as in 2016, forcing the evacuation of nearly 90,000 people and production for approximately two weeks.

Paper production was a large part of the forest industry for the first three quarters of the twentieth century. However, it has faced several challenges since the 1980s. Already in the late 1970s, international competition was hurting Canadian paper producers. For instance, Brazil could compete at cheaper cost and by using eucalyptus trees which attain maturity in approximately ten years, much faster than Canadian coniferous trees, in the paper production process. Gradual media movement to electronic formats since the early twenty-first century, the introduction of digital books, digital filing of records, and even "teaching without paper" has hit the paper industry. R&D and scientific management have attempted to recoup by promoting substitutes and the use of production residues for biofuels.

18.6.3 Yellow Gold: Wheat Era

Wheat farming began as subsistence agriculture along the St-Lawrence River by early French settlers. During the first half of the nineteenth century, Southern Ontario, then Upper Canada, wheat farming took on commercial dimensions exporting to Great Britain through the St-Lawrence, and through New York after canal openings. However, soil exhaustion reduced yields, which coincided with the Canadian Pacific Railway line construction and the opening of the Prairies to immigration and farming. Slow at the beginning, Prairie wheat farming and increased mechanization allowed wheat production to increase by 714% between 1880 and 1920. In the late 1920s, aided by the global context, record yields propelled Canada's wheat exports to 50% of world's exports (Statistics Canada 2015). In 2021, Canada was the world's fourth-largest wheat exporter.

18.6.4 White Gold: Hydro and Aluminium Era

Enormous quantities of electricity are required in the aluminium manufacturing process. Bauxite (or alumina) importation has varied over the years. In the early 1900s, France was one of the very few bauxite exporters. Later, bauxite used in Canada has come from Jamaica, Surinam, Guyana, and Guinea, but has recently come from the United States, Brazil, and China.

In the early 1900s, aluminium manufacturers became interested by Québec's hydroelectric potential. They first tended to locate near watercourses to build dams because of tremendous electricity demands occurring during the electrolysis process. First in Shawinigan, along the St-Maurice River, then in Alma and Arvida, along the Saguenay River, the company that would become Alcan built hydroelectric power plants to feed its aluminium smelters. The electricity excess was then sold to Hydro Québec. Out of the ten aluminium smelters in Canada, the only outside Québec is in Kitimat BC, a planned town, smelter, dam, and hydro powerhouse that Alcan also built to start producing in 1954. The newer smelters in Bécancour and Deschambault are fuelled by a direct power line from La Grande powerhouse from James Bay and a combination of natural gas and solar power, respectively. In 2022, Canada was the world's fourth-largest aluminium producer.

Another notable example is the textile industry that tended to locate close to rivers to tap the hydraulic power potential. Some located to be close to hydroelectric power as did the Montmorency Cotton Manufacturing Company in 1889, later Dominion Textile, at the foot of the Montmorency Falls near Québec City, where there was a Quebec and Levis Electric Light Company hydroelectric plant. Similarly, the textile industry developed in the Eastern Townships and the Montréal region for the same reasons.

18.6.5 Blue Gold: "The Water Builders' Era"

Blessed by a generous natural hydrographic system, the Province of Québec is the champion in producing hydropower thanks to the mammoth giant hydro projects a generation of "water builders" engaged in during the 1960s and 1970s (Hydro Québec 2001, 2024). The Manicouagan project on the Côte Nord region (North Shore) is a complex built between 1959 and 1978 that contains seven powerhouses which, together, produce 5,595 megawatts of renewable energy. Even more impressive is La Grande complex in the James Bay area with eight powerhouses, built between 1971 and 1996, with a capacity of 15,240 megawatts. These projects were built with the intention of selling electricity to the United States. By far, Québec has the biggest hydropower capacity with a total capacity of 37,000 mw, which is twice as much as BC, which in turn is twice as much as Ontario and Newfoundland and Labrador.

Challenges are appearing on electricity infrastructures. The transition to a carbon reduced economy is already increasing demand on electricity production. Hydro Québec has indicated the need for adding power generating facilities, be it wind farms or more dams. The same should be felt by other provinces, also considering that Canada has planned to phase out coal use by 2030 and new internal combustion engine cars sales by 2035. The provinces of Alberta and Saskatchewan are showing reluctance to transitioning.

Furthermore, climate change might have perverse effects on hydroelectricity infrastructure. For instance, Hydro-Québec's earnings in winter 2024 decreased 30% from the same quarter of the previous year due to lower exports, lower precipitation, and mild temperatures (Hydro Québec 2024).

18.6.6 Steel Blue Era (Newfoundland and Labrador) Era?

Offshore oil fields were discovered 315 kms from the coast of Newfoundland in 1979. Hibernia platform extraction started in 1997. The Terra Nova and White Rose floating production vessels, respectively in 2002 and 2005 and Hebron platforms (2017) were later added to produce together approximately 5% of Canada's oil production, behind Alberta (84%) and Saskatchewan (10%). Platform extraction is the typical practice of the FIFO (Fly in fly out) model as teams alternate three-week work shifts, in a combined working and living environment, and three weeks at home time.

18.6.7 "Gooey" (or Sticky) Gold: Bituminous Sands Era

Oil sands exploitation started in 1967. In 2024, the main oil sands production lie in the Albertan Cold Lake, Peace River, and Athabasca areas. Oil sands have replaced conventional oil extraction from Turner Valley and Leduc.

The oil extraction method is different from conventional pumping. Because it involves the need to isolate bitumen from soil, combined with higher wages, the process is more complex and more expensive. Oil prices are volatile: consequently, when global oil is high, oil sands exploitation is more profitable.

Because of its intensive water and energy use, oil sands exploitation has acquired the reputation of being referred to as "dirty oil" by environmentally conscious groups. However, through research, "friendlier" ways to improve its efficiency and acceptability, the industry has developed methods such as "steam-assisted gravity drainage, solvent-assisted extraction, and direct contact steam generation (Natural Resources Canada 2022) techniques (Report Linker 2024) that prove to be somewhat more efficient as well as more sustainable. In doing so, these factors contributed to reducing the operating cost so the break even point of production cost has gone down from 77$ to 45$/barrel between 2015 and 2022 (Canadian Energy Centre 2023). As a result, oil sands production strongly influences the exchange rate of the Canadian dollar. In this regard, Alberta could be seen as the Canadian economic thermometer.

Nonetheless, a contentious issue remains, "How to get it out of Canada?" Pipeline stories (or nightmares) have become a divisive issue among the different regions of Canada. Export has limited pipeline access to international markets other than the United States because of political and environmental concerns. The proposed Keystone XL to Nebraska, Energy East to Saint John NB, and Northern Gateway to Kitimat BC projects have been stopped. However, a complex network of pipelines still runs essentially North–South to the United States, an exception being the Trans Mountain pipeline from Alberta to Burnaby BC.

18.7 Canada Today: Diversifying and Postindustrializing

Since Jacques Cartier set foot, and Champlain founded Québec, Canada has rapidly come a long way, especially since Confederation in 1867. Favoured by its geography, Canada has capitalized on its natural resources and its locational advantage directly north of its powerful American neighbour which became its principal trade partner.

From a collection of smaller regional economies, each based on different resources, Canada's development has grown from a staple to a postindustrial economy. However, Canada remains close to its resource base. Part of its development stems from an enduring will to enhance the performance of its primary sector by continuing to add value to its resource base. Lumber, grain, electricity, and oil exports have supported the development of increasingly efficient expertise. Primary processing, tools and machinery, productivity, R&D, product diversification, precision, technological advancement, management, and engineering have all, at the same time, contributed and benefited from resource trading. Simultaneously, a knowledge-based, or value-added service sector has developed to become an important part of Canada's economy by contributing 12.1% of the total national GDP in 2022 (EDC 2023). Hence, Canada has transitioned from the industrial age to a postindustrial society, marked by a decline in the manufacturing sector and a rise in the service and information sectors. Nested in a global economic restructuring by which goods manufacturing is performed in developing regions, this change has had a profound impact on the country's economy, society, and urban landscape. The urban fabric has been modified by generating gentrification pockets throughout Canadian cities.

Throughout the years, Canada has developed internationally renowned expertise including, among others: engineering firm SNC-Lavalin (now AtkinsRéalis) who has, through its different iterations, an impressive worldwide list of achievements of infrastructure projects. Bombardier now focuses on business aircraft design and building after having developed and sold its recreational vehicles and rail transit branches. Agriculture machinery manufacturer Massey-Ferguson (now in USA); and Research in Motion which developed the Blackberry smartphone and has specialized in cybersecurity since Apple and Samsung took the cell phone market over; and the defunct Nortel that developed network equipment and semiconductors. Their presence in their respective ecosystems has, by attracting complementary businesses, generated multiplying effects and industry ancillary development that turned into innovation clusters.

Relying on fundamental and applied scientific research, the information and communications sector (ICTs) has grown rapidly and is playing an increasingly key role in the Canadian economy which has led to an increase in the demand for skilled, creative, and highly educated workers that foster economic and urban development. Higher education institutions have and are playing a significant role in not only producing a specialized workforce, but also by being involved, along with private firms, industry, research centres, suppliers, and governmental organizations in the symbiotic formation of "knowledge clusters". Porter (1998) defines clusters as "critical masses in one place of linked industries and institutions" that position them to be globally competitive. It is a somewhat updated version of Marshall's (1927) concept of "agglomeration effect" expressing the magnetic effect the establishment of companies can have on part or service suppliers. California's Silicon Valley and Hollywood are long-standing examples of specialized clusters in high technology and entertainment, respectively. The automobile industry in the Detroit region, the wine industry in Napa Valley (California), and the bicycle industry in Taiwan are other examples. The difference between Marshall's and Porter's approaches being that agglomeration will have the effect of attracting complementary manufacturing activities, whereas clusters will include agencies such as R&D labs, government organizations, higher education institutions, and venture capital.

This passage to a postindustrial phase has favoured a collection of large metropolitan centres, home of corporate head offices, universities, and colleges, which contribute to the development and maintaining of a "knowledge culture". This culture could be simply summarized as a "know-how" culture and the main ingredients being knowledge and innovation.

According to 2021 data, Canada is 41st on the Harvard Growth Lab's Economic Complexity Index (ECI) with 0.58 and has been constantly slipping from 22nd since 2000 (Harvard University 2021). For comparison, the USA sat 14th at 1.40. Japan has been constantly first with an above 2.26 index during the same period. "Know-how!" is the expression that best describes the index. In that sense, it would still put Canada in the international core but in a rather peripheral position. Its continued reliance on resource export such as minerals, grains and cars would partly explain the comparative lower index for a developed country.

Indeed, the Canadian GDP per capita, which was comparable to that of the United States until 2013, has not kept the pace of its southern neighbour. Economists argue that a decrease in productivity due to weak investments in machinery and technology has been an issue in Canada since the mid-2010s (BMO 2023; TD 2023).

18.7.1 Why Has Canadian Productivity Declined During the 2010s?

During the 2010s, Canadian productivity experienced a notable decline due to a combination of economic, structural, and investment-related factors especially when compared to the United States to which it had consistently been close to (Worldbank.org for a graphic comparison). One of the primary reasons was insufficient business investment, particularly in key areas such as machinery, equipment, and intellectual property (Rogers 2024). Compared to other advanced economies, Canadian firms have historically underinvested in research and development (R&D), limiting innovation and efficiency gains. This reluctance to adopt new technologies has contributed to sluggish productivity growth.

Another significant factor was global economic shocks, including the aftermath of the 2008–2009 financial crisis and the 2015 oil price collapse. The latter had a profound impact on Canada's resource-dependent economy, leading to reduced capital investment in the energy sector. Given that a substantial portion of Canada's economic growth is tied to commodities, lower oil prices weakened overall economic performance and slowed productivity improvements (Rogers 2024).

Moreover, declining terms of trade played a role in limiting productivity growth. While Canada benefitted from high commodity prices in the early 2000s, the stabilization and subsequent decline of these prices in the 2010s meant that real income growth slowed, revealing underlying weaknesses in productivity (Rao 2011). Additionally, global trends in productivity stagnation affected Canada, as many advanced economies experienced similar slowdowns due to structural shifts in labour markets, ageing populations, and reduced innovation momentum (Erber et al. 2017).

Overall, Canada's productivity decline during the 2010s was the result of underinvestment in key industries, economic disruptions from external shocks, and a broader slowdown in global productivity trends. Addressing these challenges requires a stronger focus on business investment, technological adoption, and R&D spending to drive sustainable long-term growth.

Recent geopolitical developments, such as the imposition of United States tariffs on Canadian goods in early 2025, pose significant risks to Canada's economic stability. These tariffs threaten to disrupt economic recovery, potentially leading to increased consumer prices, higher unemployment, and a possible recession. The manufacturing sector has already experienced contractions due to tariff uncertainties, highlighting the vulnerability of Canada's productivity to external shocks.

Despite having a highly educated workforce, Canada faces challenges in fully capitalizing on this asset. Ensuring that investments in education translate into productivity gains is crucial for future economic growth (RBC 2024).

18.8 The "Cluster Era"

Canada's development has deeply been dependent on the physical and locational particularities of its regions. Each region of Canada has a different economic base, shaped by its natural resources, industries, and traditional activities. The Atlantic Provinces rely on fishing and natural resources; Central Canada is a major industrial and financial hub; the Prairies are known for their agriculture and energy; British Columbia benefits from a diversified economy; and the North represents a wealth of natural resources.

Typical associated activities closely related to local geographical properties started to naturally emerge along Canada's "Main Street" (the axis between Québec City and Windsor). Large, diversified metropolitan regions such as Montreal and Toronto already had significant industrial concentrations at the turn of the century. We will insist here on some very illustrative examples of "natural" industrial agglomerations that present the characteristics to be qualified as "clusters". The Cluster Atlas of Canada (Spencer 2014) offers a more detailed list of clusters in Canada. We take a closer look at a few industries that evolved to become clusters in the provinces of Québec, Ontario and Alberta.

For one, the Montreal and St-Lawrence Lowlands have experienced over the years the establishment of firms that would eventually form clusters. Nonetheless, as it was the case along the Lachine Canal in Montreal, the type of manufacturing performed was rather eclectic. However, concentrations of manufacturing activities started to appear. For instance, Canadian Vickers, subsidiary of aeroplane and vessel manufacturer British Vickers Limited, Canadian General Electric, division of General Electric, and USA's Alco's Montreal Locomotive Works were located a bloc apart from one another along St-Lawrence port installation and railway lines. Like most heavy industries in Montreal, they closed, and Vickers and MLW installations demolished. CGE buildings were converted into French retailer Decathlon's distribution centre in 2019.

Pratt and Whitney opened in Longueuil in 1928 as a repair shop close to then Montreal's St-Hubert airport. The opening of airports in Dorval in 1941 and Mirabel in 1975 fuelled the development of other related firms which contributed to the agglomeration effect of the industry in the Montreal's metropolitan region (Ben Hassen et al.) Research centres and higher education institutions eventually got involved to upgrade Marshall's definition of "agglomeration" to that of Porter's "cluster". The current mandate of this cluster is work for a sustainable aerospace industry by reducing continuously its environmental footprint by, among other things, reducing aircraft weight (Carrier 2023) thus relying on the development of alloys such as aluminium.

Being a Canadian leader in aluminium production with nine smelters and one refinery located along the St-Lawrence River and its tributaries, the Province of Québec government launched the so-called AluQuébec cluster in 2013 with the mandate to support innovation in manufacturing solutions and the development of products suitable for a variety of applications. As such aluminium production is a complement to the automobile and aerospace industry.

Similarly, a clothing industry developed downtown along St-Laurent Boulevard in Montreal and concentrated in the St-Laurent-Chabanel district to be called "Le quartier de la guenille" (pejorative term that textually means "rag district" but used to mean the garment district). The district suffered the same fate as the Beauce region cited earlier in the late 1990s–early 2000s as the number of firms and employees decreased significantly to the point that a change in orientation was the only option. The response was the creation of "La cite de la mode", an urban revitalization project designed to attract investments and upscale and designer fashion businesses. However, the title of the project was revised to be more inclusive and marketed as "Pôle du design" and "Marché central" to attract complementary businesses in design, urban manufacturing, and technology. Urban agriculture projects jumped on the opportunity to revive the district. Lufa Farms, for instance, started to use the building rooftops to establish greenhouse farming in 2011. In 2024, they had "colonized" five rooftops totalling over 50,000 m^2 producing a variety of vegetables, including cucumbers, tomatoes, peppers, and lettuce.

Agriculture is still present in the St-Lawrence axis as a "peripheral" activity that is often sacrificed to suburban and industrial uses. At places, including Prince Edward County and the Niagara Valley, traditional agriculture has been converted into a more added-value venture which is grape growing and wine making which pairs well with tourism.

Further west along the Canadian "Main Street" is the Toronto and Golden Horseshoe region where heavy industry such as steel, auto, agriculture implements, and machinery has converged for many decades. The twenty-first-century development is a combination of old and new. From 1904 when Ford built an assembly plant in Walkerville, the automobile industry has been an extension of the American automobile industry core. Then, international competition copied the model by building assembly plants. The industry is a well-integrated system, comprising car assembly plants, parts manufacturers, primary parts suppliers, diversified parts suppliers, machinery and production equipment, shop supply materials, automotive technology, raw materials manufacturers (i.e. steel), and other materials including rubber, plastics, and glass, totalling approximately 133,300 assembly and parts workers across 515 establishments in 2018 (Focal Initiative 2023), and an estimated 462,000 spin-off and supported jobs in 2022 (Unifor 2023).

The Toronto-Waterloo Innovation Corridor is a 100-km stretch that has also been nicknamed Silicon North because of the strong presence of an estimated 15,000 high tech firms employing over 200,000 tech workers, which equates to a startup density only surpassed by the Silicon Valley in California. Instrumental to this concentration of knowledge-based firms is the presence of numerous higher education institutions. Indeed, there are no less than twenty colleges and universities within a 100-km radius from Toronto, some of them closely tied to the needs of a rapidly changing economy. One particular area within this corridor which has attracted attention is the region delimited by Kitchener, Cambridge, and Guelph, also referred to as Canada's knowledge intensive Technology Triangle. A long and diverse industrial presence coupled with institutions such as the University of Waterloo (2024) who act as innovation incubator and offer co-op programmes and startup support to facilitate knowledge transfer to private firms, or wherever it is needed, like in the auto industry for instance, for which the University hosts a "Centre for Automotive Research". Its webpage says it all: "Partnering automotive industry and cutting-edge research"!

Incidentally, capitalizing on the auto industry long presence in the area and the more recent high tech development node cohabiting, the provincial government launched its Ontario Vehicle Innovation Network (2024) to contribute to the advancement of the auto industry into the smart mobility era. Part of the industrial agglomeration thus became a cluster in the sense that Porter defined it.

A perfect example of Startup from University of Waterloo and the next case in point is the development of the SolarPass patented technology to reduce the environment impact of water treatment after use in the oil sands extraction process. The method reduces the need for enormous quantities of energy and use of hazardous chemicals. Other endeavours in the Alberta oil extraction industry include research destined to produce energy while finding ways to reduce greenhouse gas emissions by improving extraction technology, by capturing carbon emissions, and by mitigating environmental risk of the industry among other research avenues. University of Alberta (2024) and University of Calgary (Research Money 2024) are engaged in such ways "to energize the world—without compromising the earth". On the Pacific Coast, the Vancouver region has attracted the American entertainment industry.

Of course, the examples covered here are not exhaustive. They illustrate just a sample of several clusters that can exist in a region. Moreover, they are illustrative of larger economic regions. Smaller regions can also either be included in larger cluster networks or form their own smaller development ecosystems by, for instance, promoting startups or on-site training. Graduates from diverse programmes can integrate already present expertise nodes or start new ventures.

18.8.1 Canada's Clusters Support Global Innovation Clusters Programmes

In 1991, Canada relied on MIT's Michael Porter to evaluate its place in the global economy. His report stated that Canada was still too depended on its resource-based industries and the United States economy and efforts should be put in R&D, equipment and productivity improvement (Balkrishnan et al. 2007). The rise of the dollar around 2010 seems to have stifled the urge to follow Porter's recommendations, and the same situation occurred in the mid-2010s. Subsequently in 2017, (a response to increasing "innovation gap" and GDP per capita gap?) the government put in place the Innovative Superclusters Initiative (ISI) designed to increase the competitiveness of Canadian firms and strengthen clusters out of industrial agglomerations already in place in the Canadian landscape and meshing them with related information, knowledge, and technological agencies (Doloreux and Frigon 2022; Innovation, Science, and Economic Development 2022).

Although included in the cluster programme, the Province of Québec had already embarked in a similar approach in the 1980s by investing heavily in aerospace (as discussed previously), pharmaceuticals, IT, metal and mineral refinery, and hydroelectric production and transmission (Institute of competitiveness and prosperity).

There are five identified clusters that are generally associated, but not exclusively, to five geographically defined regions: Oceans in Atlantic Canada, targeting at "growing the ocean economy by harnessing emerging technologies that digitize and optimize marine operations and increase marine safety"; Scale AI in Québec, especially in the Montreal region where AI works have been progressing rapidly, "to build intelligent supply chains and make Canada a global export leader by capturing new market opportunities"; Advanced manufacturing, automation and robotics, to establish "innovation solutions in manufacturing and their commercialization" worldwide, in Ontario; Protein Prairies, aiming at "increasing value of key Canadian crops"; and Digital health and geographical analytics, focusing on "applications in natural resources, precision health and manufacturing sectors", in British Columbia (Canadian Innovation Space 2024). Projects for each cluster are mapped (Innovation, Science, and Economic Development 2022).

At the time of writing (June 2024), no explicit and systematic assessment was available, but the programme was reconducted in 2022 for a five-year period. The automotive battery sector is projected to be booming soon with the planned construction of battery manufacturing plants in Québec and Ontario, and the necessary electricity infrastructure upgrades. Nonetheless, the expected results are that Canada will increase its competitiveness on global markets, and that the "innovation gap" with competitors will decrease. Widespread economic impact, even in rural areas at the lower steps of the ladder, is also expected (or wished for?).

18.9 Rare Earth and Silicon Gold? R&D&C (for Creativity) and Knowledge Era

Canada has come a long way, capitalizing from natural to (increasingly specialized) human resources throughout its development! Canada has been Britain's periphery, then the United States' periphery, before the Southern parts of Québec and Ontario became a northern extension of the USA industrial core. The model can still be applied within Canada, but cores (or semi-cores?) can be found in Alberta and British

Columbia among the western periphery. Colleges and universities were instrumental in creating a "Silicon North", especially on Canada's "Main Street" with an important concentration in Southern Ontario. Diversification even gave international notoriety in entertainment and creativity by being called Hollywood North, by exporting Cirque du Soleil shows, and Moment Factory immersive experiences all over the world, and much more.

References

Ayotte C, Bridger J (2022) Indigenous peoples economic account: Methodology and preliminary results. Income and Expenditure Accounts Technical Series. Statistics Canada Catalogue No. 13–604-M2022001

Balakrishnan J, Eliasson JB, Sweet TRC (2007) Factors affecting the evolution of manufacturing in Canada: An historical perspective. J Oper Manag 25:260–283

Bank of Canada (2012) 75 ans d'Histoire en chiffres: 1935–2010. https://publications.gc.ca/collections/collection_2012/banque-bank-canada/FB4-11-2010-fra.pdf

Barraclough ER (2018) Gudrid Thorbjarnardóttir (c980–?): 'Far-travelled' Icelandic Pioneer to America. BBC World Histories 11 (August/September, 2018), pp 52–53

Bartlett JH (1885) Manufacture consumption and production of iron, steel, and coal, in the dominion of Canada. Dawson, Montreal

BMO (2023) A tale of two countries. https://bmofundcentral.com/articles/a-tale-of-two-countries-how-the-u-s-and-canada-ended-up-on-different-economic-paths/

Brink JW (2008) Imagining head smashed. In: Aboriginal Buffalo Hunting on the Northern Plains. Edmonton: Athabasca University Press

Brown JHS (2017) An Ethnohistorian in rupert's land: Unfinished conversations. Athabasca University Press, Edmonton

Canadian Energy Centre (2023) Canadian upstream oil sector supply costs continue to decline. https://www.canadianenergycentre.ca/canadian-upstream-oil-sector-supply-costs-continue-to-decline/

Canadian Innovation Space (2024) Canada's Global innovation clusters. https://canadianinnovationspace.ca/partners/canadas-global-innovation-clusters/

Canadian Nuclear Safety Commission (2015) History of uranium mining in the Elliot Lake region of Ontario and associated effects on water quality and fish intended for human consumption. https://archives.bape.gouv.qc.ca/sections/mandats/uranium-enjeux/documents/NAT32_VA.pdf

McKenzie CJC (2016) Going Dutch? The impact of falling oil prices on the Canadian economy. Can Public Policy 42:168–180

Carlos AM (2023) The country they built: Dynamic and complex indigenous economies in North America before 1492. J Econ His 83:319–358

Carlos AM, Lewis FD (2010) Commerce by a Frozen Sea: Native Americans and the European Fur Trade. University of Pennsylvania Press, Philadelphia

Carrier C-E (2023) Quand la collaboration propulse le Savoir. https://www.ledevoir.com/economie/797098/criaq-quand-la-collaboration-propulse-le-savoir

Charbonneau K, de Munnik D, Murphy L (2018) Canada's Experience with trade policy. Bank of Canada Staff Discussion Paper 2018–1. https://www.bankofcanada.ca/wp-content/uploads/2018/01/sdp2018-1.pdf

Charbonneau L (2016) A look back at universities at the time of Canada's Centennial through the pages of UA. https://universityaffairs.ca/features/feature-article/look-back-universities-time-canadas-centennial-pages-university-affairs/

Chernoff A Cheung C (2023) An overview of the indigenous economy in Canada. Bank of Canada, Staff Discussion Paper 2023–25. https://www.bankofcanada.ca/wp-content/uploads/2023/10/sdp2023-25.pdf

CPR (2006) Fact book. https://s21.q4cdn.com/736796105/files/doc_downloads/resources/cp-2006-fact-book.pdf

Cross P Mintz JM (2024) Resources are still our golden goose. Financial Post. https://financialpost.com/opinion/resources-still-canada-golden-goose-need-protect-them

Doloreux D, Frigon A (2022) The Innovation Superclusters initiative in Canada: a new policy strategy? Science and Public Policy 49:148–158

Doolittle WE (2004) Permanent vs. shifting cultivation in the Eastern woodlands of North America Prior to European Contact. Agric Hum Values 21:181–189

EDC (2023) Service exports boost Canadian Economy and Global Brand. https://www.edc.ca/en/article/service-exports-boost-canadian-economy.html

Erber G, Fritsche U, Harms PC (2017) The global productivity slowdown: Diagnosis, causes and remedies. Intereconomics 52 45–50

Financial Times (2007) Rio Tinto in $44bn takeover of Alcan. https://www.ft.com/content/3664a78c-3005-11dc-a68f-0000779fd2ac

Firestone OJ (1960) Development of Canada's economy 1850–1900. In: Trends in the American economy in the nineteenth century. Princeton University Press, pp 217–252

Firestone OJ (1968) Education and economic development: the Canadian Case. https://www.roiw.org/1968/341.pdf

Focal Initiative (2023) Automotive industry labour market analysis. https://www.futureautolabourforce.ca/wp-content/uploads/2020/06/Regional-Profiles_Provincial-Ontario.pdf

Gad G (1994) Location patterns of manufacturing: Toronto in the Early 1880s. Urban History Review 22:113–138

Friedman J (1973) Urbanization, Planning and National Development, Beverly Hills: Sage

Glencore Canada (2024) Who we are. https://www.glencore.ca/en/who-we-are/our-history

Government of Canada (2023) International Trade Agreements. https://www.international.gc.ca/trade-commerce/trade-agreements-accords-commerciaux/agr-acc/index.aspx

Groves PA (2001) The Northeast and regional Integration, 1800–1860. In: Mitchell RD Groves PA (eds.) North America: The historical geography of a changing continent. Rowman and Littlefield, Totowa NJ, pp 198–219

Hamilton Spectator (2012) Hamilton's steel industry from birth to boom and beyond. https://www.thespec.com/news/hamilton-region/hamiltons-steel-industry-from-birth-to-boom-and-beyond/article_4dec b079-39c2-50d5-b456-916ca473ec21.html

Harris RC (1987) Historical Atlas of Canada: Volume I: From the beginning to 1800. University of Toronto

Harvard University (2021) Country and product complexity ratings. https://atlas.cid.harvard.edu/rankings

Hydro Québec (2008) Synthèse des connaissances Environnementales. https://www.hydroquebec.com/data/developpement-durable/pdf/pop_06_08.pdf

Hydro Québec (2024) Un avenir à bâtir. https://www.hydroquebec.com/data/documents-donnees/pdf/rapport-annuel-2023-hydro-quebec.pdf

Innovation, Science, and Economic Development (2022) Canada's cluster projects: Maps. https://www.ic.gc.ca/app/scr/cmb-dgcm/spi-pi/mp.html?tpcId=341&lang=eng

Jolicoeur PC (2020) Detecting early widespread Metal use in the Eastern North American Arctic around AD 500–1300. Am Antiq 86:111–132

Kroeber A (1923) Anthropology. Harcourt Brace, New York

Lacoursière A (2024) Pacte de Rue: Intervention. https://www.pactderue.org/en

Lacroix J-M (2019) La Nouvelle-France (1608–1713), in Histoire du Canada—Des origines à nos jours. Tallandier, Paris, p 528

Le Dressay A, Lavallee N, Reeves J (2010) First Nations trade, specialization, and market institutions: A historical survey of first nation market culture. Aboriginal Policy Research Consortium International (APRCi), p 72. https://ir.lib.uwo.ca/aprci/72

Mann CC (2003) 1491: New revelations of the Americas before Columbus. Alfred A. Knoff, New York

Marshall A (1927) Industry and trade. Macmillan, London

Marshall B (1991) The Real World: Understanding the Modern World Through New Geography Boston: Houghton Mifflin

Marshall H, Southard FA, Taylor KW (1936) Canadian-American Industry: A study in international investment. Yale University Press, New Haven

McCann, L (1998) Heartland and Hinterland: A Regional Geography of Canada Scarborough ON: Prentice-Hall

Natural Resources Canada (2019) Oil resources. https://natural-resources.canada.ca/our-natural-resources/energy-sources-distribution/fossil-fuels/crude-oil/oil-resources/1808

Natural Resources Canada (2020) Coal facts. https://natural-resources.canada.ca/our-natural-resources/minerals-mining/mining-data-statistics-and-analysis/minerals-metals-facts/coal-facts/20071

Natural Resources Canada (2022) Oil sands scientific research. https://natural-resources.canada.ca/our-natural-resources/energy-sources-distribution/fossil-fuels/crude-oil/oil-sands-science-research/18095

Ontario Vehicle Innovation Network (2024) Ontario's ecosystem. https://ovin-navigator.ca/automotive-mobility-sector/ontarios-ecosystem/

Porter ME (1998) Clusters and competition: new agendas for companies, government, and institutions. In: On competition. Harvard Business Review Books, Cambridge, MA

Rao S (2011) Cracking Canada's productivity conundrum. Institute for Research in Public Policy, https://irpp.org/research-studies/cracking-canadas-productivity-conundrum/

RBC (2024) Canada's growth challenge: Why the economy is stuck in neutral. https://thoughtleadership.rbc.com/canadas-growth-challenge-why-the-economy-is-stuck-in-neutral/

Report Linker (2024) Canada oil sands market overview 2024–2028. https://www.reportlinker.com/market-report/Oil/526548/Oil-Sands

Research Money (2024) Technology developed at University of Calgary is poised to transform oil sands industry. https://researchmoneyinc.com/article/technology-developed-at-university-of-calgary-is-poised-to-transform-oil-sands-industry

Rogers C (2024) The productivity problem. https://www.bankofcanada.ca/2024/03/productivity-problem/

Saku JC (1999) Aboriginal census data in Canada: a research note. Can J Nativ Stud 19:365–379

Schembri L (2022) Economic Reconciliation: Supporting a return to indigenous prosperity. Remarks at the National Aboriginal Capital Corporations Association, Gatineau, Quebec. https://www.bankofcanada.ca/2022/05/economic-reconciliation-supporting-a-return-to-indigenous-prosperity/

Spencer GM (2014) Cluster analysis of Canada. Munk School of Global Affairs, University of Toronto. https://clustercollaboration.eu/sites/default/files/international_cooperation/cluster-atlas.pdf

Statistics Canada (2010) Progress of population in Canada. https://www65.statcan.gc.ca/acyb02/1867/acyb02_1867001803-eng.htm

Statistics Canada (2015) The evolution of natural resource production. https://www150.statcan.gc.ca/n1/pub/11f0027m/2012079/part-partie1-eng.htm#h2_4

Statistics Canada (2023) Labour force characteristics by industry, annual. https://www.150.statcan.gc.ca/t1/tbl1/en

TD (2023) Mind the gap. https://economics.td.com/ca-falling-behind-standard-of-living-curve

Unifor (2023) Quick facts on Canada's auto industry. https://autotalks.uniforautohub.ca/quick_facts

Union des Consommateurs (2015) Levée des Barrières au commerce intérieur et protection du consommateur: l'exemple de l'Union européenne. https://www.uniondesconsommateurs.ca/wp-content/uploads/2015/11/R01-UC-barriere-commerce-interieur-rapport-F-v2-FR.pdf

United States Census (2020) Historical centre of population 1790–2020. https://www.census.gov/data/tables/time-series/dec/center-pop-text.html

University of Alberta (2024) Research: Energy and environment. https://www.ualberta.ca/research/our-research/energy-and-environment.html

University of Waterloo (2024) Centre of Automotive Research. https://uwaterloo.ca/centre-automotive-research/

Usher A (2024) Alex Usher on Canadian Higher Education. https://higheredstrategy.com/alex-usher-on-canadian-higher-education/

Vallières M (2012) Portneuf, Québec. In: Les régions du Québec—histoire en bref. Les Presses de l'Université Laval

Wall D (2017) East and west, connected by rail. https://canada.constructconnect.com/joc/news/projects/2017/03/east-and-west-connected-by-rail-1022562w

Watson PJ (2007) Early plant cultivation in the Eastern Woodlands of North America. White P, Denham T. Routledge, The Emergence of Agriculture. London, pp 134–153

https://data.worldbank.org/indicator/NY.GDP.PCAP.CD?locations=CA-US

Wynn G (2001) Forging a Canadian Nation. In: Mitchell RD Groves PA (eds) North America: The historical geography of a changing continent. Rowman and Littlefield, Totowa NJ, pp 357–78

https://onlinelibrary.wiley.com/doi/10.1111/j.1541-0064.2011.00384.x

Winter in Canadian Cities 19

Abstract

Throughout the twentieth century, Canada was portrayed by and to international audiences as a land of cold and snow, largely in rural settings, and the stereotype persists today. This chapter focuses on urban areas, emphasizing winter life in Canadian cities. As well, the chapter looks at outdoor winter recreation, as pursued mainly by urban residents. Canadian cities are generally cold in winter, with mean January temperatures below 0°C in most and temperatures below -40°C recorded in prairie cities. Most frostbite injuries in urban settings can be attributed to insufficient protective clothing and lack of preparation for weather conditions, rather than extreme cold. This has led to a focus on wind chill warnings as a preventative measure. Cold remains a significant hazard for socio-economically disadvantaged people. Chronic hypothermia can result from exposure to cold over long periods. Wind chills of -10 (or less) may be sufficient to induce hypothermia, especially among populations with other comorbidity factors, particularly homelessness. Seasonal affective disorder (SAD) affects 15-20% of Canada's residents, with 2-5% seriously affected in their lifetime. SAD suffers represent about 10% of those diagnosed with depression in Canada. All areas of Canada are exposed to at least some snowfall (and all its consequences), although the amount of snow cover and its duration vary significantly across the country. Property damage and economic losses due to disruption of business remain significant, both from blizzards and less intense snowfall events. Snow ploughing and ice removal are of significant expense for Canadian cities. Ironically, the most popular Canadian spectator winter sports (hockey, figure skating, curling) are conducted indoors under climate-controlled conditions. Winter-related injuries and deaths are related to occupational, socio-economic, and recreational events. In Canada, most casualties from winter conditions involve people engaged in recreational activities. Despite the stereotypes, winter is generally unpopular in Canada. Directly connecting climate change to winter conditions may miss the mark by assuming stereotypical attitudes.

Keywords

Stereotypes and attitudes · Urban areas · Cold · Snow · Seasonal affective disorder · Outdoor recreation

Stereotypically, Canadians are considered to be cold-adapted and winter-loving. Winter may be regarded as "inescapable and defines us in some ways as a people and country" (National National Post 2023a, b). A casual perusal of quotes about Winter in Canada (e.g. LineUp Media 2023) reveals an emphasis on cold, snow, and embracing and rising above the same. Internationally, the phenomenon dates to Champlain ("Every Canadian winter was a mortal challenge to its habitants"), Voltaire ("Quelques arpents de neige"), and others. Throughout the twentieth century, Canada was portrayed by and to international audiences as a land of cold and snow, largely in rural settings (e.g. Berton 1975), and the stereotype persists today (e.g. Toronto Sun 2022).

Elsewhere in this volume, we have discussed winter climate conditions in the Ecoregions (Chaps. 2–9), freezing rain and verglas events (Chap. 4), impacts on agriculture (Chap. 5), avalanches (Chap. 6), icebergs (Chap. 8), and winter roads (Chap. 14). This chapter focuses on urban areas, emphasizing winter life in Canadian cities. As well, the chapter looks at outdoor winter recreation, as pursued mainly by urban residents.

19.1 Cold

Canadian cities are generally cold in winter, with mean January temperatures below 0 °C in most and temperatures below −40 °C recorded in prairie cities (Table 19.1). Although

N. Catto and G. R. Vincent, *Canada: Land and People*, World Regional Geography Book Series,
https://doi.org/10.1007/978-3-032-04391-7_19

Table 19.1 January Mean and Minimum temperature recorded, Canadian and selected international cities. Canadian data from Climate Normals, Environment and Climate Change Canada; International Data from https://en.climate-data.org/

	January mean °C	Minimum °C
Harbin	−18.5	−53.0
Winnipeg	−16.4	−45.0
Saskatoon	−15.5	−50.0
Regina	−14.7	−50.0
Québec	−12.8	−36.1
Edmonton	−12.1	−48.3
Ottawa	−10.3	−36.1
Montreal	−9.7	−37.8
Fredericton	−9.4	−37.2
Minneapolis	−9.0	−37.0
Moncton	−8.9	−32.2
Moscow	−8.0	−42.2
Charlottetown	−7.7	−30.5
Calgary	−7.1	−45.0
Halifax	−5.9	−28.5
Toronto	−5.5	−31.3
St. John's	−4.5	−23.8
Stockholm	−2.0	−28.2
Vancouver	4.1	−17.8
Victoria	4.6	−15.6

cold winters are not unique to Canada and are disliked by many residents (e.g. Leger 2023b, National Post 2023a, b) they are considered to have significant influences on many aspects.

19.1.1 Frostbite

Frostbite results from freezing of skin and underlying tissues and can occur at any temperature below 0 °C (Handford et al. 2017; Egro et al. 2024). Exposure to cold does not result in either heightened sensitivity or resistance to frostbite (Burström et al 2017). Previous frostbite injury reduces ability to sense cold, and multiple exposures over time appear to increase the likelihood of severe injury (Handford et al. 2017). Lack of previous experience and limited perception of cold (Sugg et al. 2019) can lead to enhanced risk, primarily resulting from failure to take precautions.

Frostbite severity ranges from frostnip affecting only the epidermis; through superficial frostbite affecting the upper dermis and causing rigidity, brittleness, and blistering; to deep frostbite, penetrating through the entire skin to muscle, tendon, and bone. Deep frostbite is marked by waxy skin appearance, stiffness in joints, severe blistering, and gangrene. In severe cases, loss of digits and limbs can occur, or amputation may be required.

Traditionally, treatment for frostbite involved applying cold water or snow, partially to reduce the likelihood of gangrene. Today, rapid warming, thrombolytic therapy (blood-clotting inhibiting drugs), and parenteral iloprost (used to treat pulmonary arterial hypertension) are the preferred treatments for severe frostbite (Shenaq and Gottlieb 2017; Sheridan et al. 2022; Regli et al. 2023). In some cases, eliminating the potential for cardiac arrest must take priority over the treatment of the frostbite injuries.

Several studies have focused on frostbite in military personnel, in victims engaged in recreational or winter sport activities (Fudge 2016), and in residents in northern and Arctic Canada (Do et al. 2013). In more southerly and urban settings in Canada, the incidence of frostbite appears to be substantially underestimated, as many cases of frostnip in particular go unreported or are subject only to first aid treatment. However, frostbite occurrence in the general population and homeless people in particular has been studied in urban settings, including Montreal (Koutsavlis and Kosatsky 2003), Toronto (Gower and Bassil 2014), and Vancouver (Henderson et al. 2012).

Alcohol or illicit drug consumption inhibits shivering, which can result in frostbite at temperatures only slightly below 0 °C (Endorf and Nygaard 2022). Factors which inhibit blood flow to the extremities, including use of β-blocker medication and diabetes, also influence frostbite (Egro et al. 2024). Other factors include a history of smoking, psychiatric disturbance, and working with NO_2, freon, or solid CO_2 (Sever et al. 2010). Age is also a factor, although statistically men aged 35–60 are most likely to be affected, as a result of age in combination with other factors.

Most frostbite injuries in urban settings can be attributed to insufficient protective clothing and lack of preparation for weather conditions, rather than extreme cold. This has led to a focus on wind chill warnings as a preventative measure.

19.1.2 Wind Chill

Wind chill is a perceived effect of wind on exposed skin (Osczevski 1995, Environment Canada 2017, Langford and Fox 2021). Wind affects perceived temperature by removing sweat and disrupting the thin boundary layer of air directly above the skin. Relative humidity also affects human perception of cold: many state that the "dry cold" of Prairie Canada is more bearable than cold in a more humid environment. As with measures of perceived heat (humidex), wind chill is not an actual temperature. Different people, as well as different animals, will sense cold differently (Mader et al. 2010; Provençal et al. 2016; de Freitas and Grigorieva 2017).

Laboratory tests quantifying heat loss from skin under non-extreme conditions can be compared with human perceptions (Ahmad et al. 2016).

The current version of the Wind Chill Temperature (WCT) chart for Canada was adopted in 2005 (Maarouf and Bitzos 2000, Osczevski and Bluestein 2005, Environment Canada 2017). The chart compares wind speed with reported air temperature to calculate a WCT. WCT values of −27 or lower are considered to be extremely likely to induce frostbite. Hypothermia-inducing values are not specified, as hypothermia involves reduction of core body temperature, not exposed skin temperature. Although attempts to use different regional scales for Prairie and Atlantic Canada proved unsuccessful, different values are used by Environment Canada to trigger wind chill warnings in different regions.

WCT values can be used to estimate freezing times for exposed cheeks or fingers. The differences in rate reflect capillary density in human skin, as well as human circulation. Elevated cheek and facial temperatures, while enhancing protection against frostbite and other cold-related injuries, also increase heat loss to the surrounding cold environment. Paradoxically, elevated facial temperatures increase the chance of frostbite.

The wind and temperature values used to estimate WCT are recorded at meteorological stations. Instruments typically are mounted 10 m above the surface, where temperatures may be either higher or lower than those at ground level. Wind speed in unobstructed situations increases exponentially with height, so that wind speeds measured at 10 m elevation can be used to calculate the speed experienced by a cheek or finger for a human standing on the ground, depending on the height of the person. The current WCT formula emphasizes wind speed over temperature, and does not consider humidity. Under identical wind and temperature, the “dry cold” may be perceived as more comfortable, although the WCT and frostbite susceptibility may be identical. Greater relative humidity values result in perceived colder conditions.

The estimated WCT does not account for individual variations in perception or reaction to cold, or for progressive numbness as the duration of exposure increases (Shitzer 2007; Shitzer and Tikuisis 2010). They are also based on exposed cheeks and fingers, rather than on scientifically more precise whole-body models (see Ben Shabat et al. 2013). However, if the primary objective is to communicate hazard to the general public in a readily understood fashion, to encourage protective action, a simpler designation is the most effective approach. Environment Canada (2017) surveys suggest that more than 80% of Canadians use wind chill data when planning outdoor activities.

The term “toque” (“tuque”, “touque”) for a knitted cap has been used since the mid-1860s in Canadian English and dates to the initial coureurs de bois in Québec. Although used in other cold climates under a variety of names, toques stereotypically are considered iconically Canadian. Toques are frequently sold or given away in campaigns to promote mental health, combat homelessness, and other causes. Although heat loss through an unclothed head is roughly in proportion to the 10% of body mass, it represents (see Vreeman and Carroll 2008), wearing clothing on any body part is advisable in cold weather.

19.1.3 Hypothermia

Since 1953, days with at least one hourly wind chill value below -30 have decreased significantly at 76% of Canadian weather stations (Mekis et al. 2015, Vincent et al. 2015). However, cold remains a significant hazard for socio-economically disadvantaged people (Koutsavlis and Kosatsky 2003; Gasparrini et al. 2015; Stares and Kosatsky 2015). Hypothermia kills approximately 80 to 95 people in Canada each year, many in urban areas (Gower and Bassil 2014, Environment Canada 2017). Days do not have to reach extreme cold to cause hypothermia (Day 2006). Chronic hypothermia can result from exposure to cold over long periods (e.g. by living in an unheated place). Wind chills of −10 (or less) may be sufficient to induce hypothermia, especially among populations with other comorbidity factors, particularly homelessness (Gower and Bassil 2014).

Hypothermia has long been recognized as a serious medical hazard (Guly 2011; Ireland et al. 2011; Paal et al. 2016; Shenaq and Gottlieb 2017; Peiris et al. 2018). Hypothermia occurs when a person loses heat faster than it can be produced within their body.

Four degrees of hypothermia are recognized. In mild hypothermia, body temperature drops below 36 °C. Sufferers are lethargic but conscious and shivering, with pale and cool extremities and numbness in limbs or digits. Accelerated heart rate and rapid, shallow breathing may occur.

Under moderate hypothermia, respiratory rate, blood pressure, and pulse are all depressed. The person may exhibit slurred speech, slower reflexes, or lose consciousness and will stop shivering. The core body temperature will be below 34 °C.

Severe hypothermia occurs when the body temperature drops below 30 °C. Victims may be unresponsive, suffer irregular heartbeat, shallow breathing, muscle rigidity, and have extremely cold skin. This is a life-threatening state. Approximately 30% of patients with severe hypothermia who receive treatment nevertheless succumb, although survival is possible in rare cases for profound hypothermia, with body temperatures below 20 °C. The preferred technique for treatment in cases of cardiac arrest is extracorporeal life support, involving slow warming and oxygenation of patients.

Hypothermia is a hazard for those engaged in outdoor recreation, and in cases where vehicle breakdown leaves motorists stranded. Motorists are advised to stay with their vehicle, put out flares or marker pyramids, move arms and legs frequently, and not to eat snow, which reduces the internal body temperature. If the motor is operational, run it for 10 min in each hour, while keeping windows slightly open to prevent carbon monoxide poisoning.

In British Columbia, hypothermia has been the primary or contributing cause in an annual average of 25 deaths since 1998 (Henderson et al. 2012; Stares and Kosatsky 2015). Although the highest death rates affected those over 75, the typical victim is under 50. For almost two-thirds of victims, consumption of alcohol and/or other psychoactive substances were contributing factors. For those under age 75, male deaths outnumber female deaths by 4:1. Statistically, victims were more likely to have lower socio-economic status, and live in rural areas.

Homeless people are more subject to cold, creating enhanced risk of hypothermia in most areas other than relatively warm Vancouver and Victoria. Toronto's homeless population has a mean age of ~42 years, with about two-thirds male, and hypothermia is a significant issue (Gower and Bassil 2014). In Calgary and other cities, shelter usage by the homeless is not directly correlative with cold but peaks on weekdays and days with precipitation (Jadidzadeh and Kneebone 2015). Calgary's cold, dry winters may correlate with lesser use of shelter, under conditions when the danger of hypothermia is elevated.

19.2 Seasonal Darkness

Seasonal darkness results from both high latitude and cloud cover, coupled with additional particulate matter and SO_2 aerosols in urban environments. Annually, insolation received at the surface in Toronto is less than 50% of the potential value (e.g. Rouse 1993). Shorter wavelengths are more subject to reflection and adsorption, with ultraviolet input reduced by $\geq$75% in winter. Thermal infrared wavelengths are relatively unaffected, resulting in lesser temperature reductions along with lower amounts of ultraviolet radiation input. Although excessive exposure to UV radiation is harmful, too little exposure can also produce adverse health consequences, including reduction of Vitamin-D production (Neale et al. 2023). Recent evidence also suggests that limited amounts of UV radiation can result in improved autoimmune function, with reduction of related diseases and infections (Neale et al. 2023), possibly including COVID-19 (To et al. 2021).

Cloud cover further reduces the amount of insolation received at the surface. During a typical winter, approximately one-third of daylight hours in Edmonton (53.5°N) have bright sunlight, decreasing to 20% or less in Halifax (44.6°N) and Toronto (43.6°N). Overcast conditions, with Sun not visible, characterize 50% or more of winter daylight hours in cities throughout Atlantic Canada, southern Québec, southern Ontario, and Vancouver.

Seasonal affective disorder (SAD) affects 15–20% of Canada's residents, with 2–5% seriously affected in the lifetime (CMHA 2013; Lukmani et al. 2019). SAD suffers represent about 10% of those diagnosed with depression in Canada (Levitt et al. 2000; Lukmani et al. 2019). Younger and middle-aged adults (20–50 y) and women are more likely to experience SAD (Magnusson and Boivin 2003). SAD appears to be related to both the biophysical effect of decreased exposure to insolation (including UV) and the psychological effect produced by overcast weather. The additional stress and enforced social isolation resulting from COVID-19 also contributed to SAD incidence in Canada throughout 2019–2020 (e.g. BC Doctors 2020).

On occasion, UV radiation in bright sunlight and high albedo combine to produce snow blindness (photokeratitis; American Academy of Ophthalmology 2023). Children are particularly susceptible. Fortunately, the effects are usually temporary and can be alleviated by rest in a darkened location. The Inuit invention of sun goggles (Canadian Innovation Space 2017, Parks Canada 2021) is an adaptive response to the problem in Arctic environments. In urban areas, snow blindness is limited by restricted UV radiation and the lower albedo of urban snow with dirt, other contaminants, and recrystallized ice.

19.3 Snow

All areas of Canada are exposed to at least some snowfall (and all its consequences), although the amount of snow cover and its duration vary significantly across the country (Table 19.2). Sensitivity, the potential impact of a hazard, depends upon human, community, or societal conditions. Familiarity with the hazard, or measures taken to reduce the impact, can result in a reduction of sensitivity.

St. John's, NL, receives 335 cm of snow annually, with as much as 90 cm in one day (Environment Canada 2020; Fig. 19.1, Table 19.2). Based on experience, residents have developed coping strategies, involving shovels, snowploughs, snow tyres, and snowmobile suits. Weather forecasting has improved to allow "snow days" to be announced one to two nights in advance, letting people prepare and make alternate arrangements for work, school, and social commitments. Although exposure to snowfall is high, sensitivity is low.

"Snow days", with closure of schools and government offices, have become an established phenomenon. Although definitive statistical evidence is lacking, the predicted amount of snow required to initiate advance closure has decreased over the past twenty years, along with increases in the amount of advance notice given, from essentially none to up to 3 days. (This phenomenon has been noted in informal records kept

Table 19.2 Snowfall data, selected Canadian Cities. Area data from Census of Canada; snowfall data from Climate Normals, Environment and Climate Change Canada; Budget data from websites for each city

City	Area (km^2)	Annual mean (cm)	Days with >0.2 cm snow	Snow clearing budget (10^6 y^{-1})
St. John's	446.1	335	79	$ 20–25
Québec	454.1	303	70	$ 45–55
Moncton	140.7	282	50	$ 5–8
Montreal	365.1	210	59	$ 170–200
Ottawa	2778.0	175	52	$ 75–95
Halifax Reg. municipality	5475.6	154	25	$ 25–30
Calgary	825.3	128	54	$ 50–55
Toronto	630.2	122	41	$ 120–140
Winnipeg	464.1	113	53	$ 80–87
Regina	178.8	100	56	$ 8–15
Vancouver	115.0	45	9	$ 4–5

Fig. 19.1 Typical urban snowscape

by Catto over a career teaching at Memorial University in St. John's.) This reflects improved weather forecasting, particularly with respect to snowfall amount and estimated time of arrival. Other factors may include an increased emphasis on safety (especially for schoolchildren), increased ability and inclination to work from home, use of weather-related terminology which emphasizes hazard (e.g. "weather bomb"), and a general societal increase in risk-aversion. The latter factor perhaps explains why the Trip Advisor (2019) survey respondents aged 55 and older were more likely to categorize snow days as unnecessary or overreactive ("pretty lame"), based on their childhood reminiscences. "Weather bomb", initially used by the Bergen School of Meteorology during World War II to describe rapidly intensifying marine storms and subsequently applied to extratropical cyclones by Sanders and Gyakum (1980), came into general use in Canadian weather forecasts ca. 2015–2018.

19.3.1 Extratropical Cyclones

Extratropical (mid-latitude) cyclones are the dominant weather pattern in boreal areas. Although they form throughout the year, they represent most of the cold-weather storm events in Canada at latitudes of 30°–60°N.

An extratropical cyclone has a low-pressure centre, with converging and ascending air associated with interaction and energy exchange between warm- and cold-weather fronts (Fig. 19.2). Air spirals inward counterclockwise, and there is no central calm "eye". The cyclone draws energy from

Fig. 19.2 Extratropical cyclone, 30 October 2010. http://www.crh.noaa.gov/mpx/?n=10oct2627

both water and land and does not depend on a warm ocean energy source. Extratropical cyclone diameters may exceed 1000 km, and the area simultaneously affected is therefore much larger than that beneath a tropical cyclone (hurricane). Radar images show a comma-like pattern, with the tail formed by strong southwest-to-northeast winds. Although most develop in stages over seven to ten days, bomb cyclones can form and intensify much more rapidly.

Before development begins, adjacent cold and warm air masses are present along a (quasi-) stationary front. Precipitation develops as the rising warm air mass as it cools with altitude. Cyclogenesis occurs if the cold air mass begins to move, wedging under the warm air and generating further uplift. Distinctive cold and warm fronts begin to develop, moving in counterclockwise patterns.

During the second, open-wave stage, the cold front moves more rapidly than the warm air mass, forming an inverted-V pattern on the weather map. The warm air is now elevated on two fronts: the cold front, along the leading edge of the cold air, and a warm front, as the warm air slides upwards over cool air to the northeast.

During the third, occlusion stage, continued movement of the cold front elevates the warm air, eventually isolating if from the surface. A thermally inverted pattern results. Condensation within the warm air mass results in development of substantial precipitation, typically as snow during the winter months. Rapid condensation results in the intensified development and enhanced precipitation characterizing a "bomb". An occluded front, symbolized on a weather map by alternating cold and warm front symbols, is the diagnostic configuration of a mid-latitude cyclone.

As the extratropical cyclone continues to move, the warm air mass becomes completely isolated from uplift. The storm loses energy during the final, dissipation stage, as the warm air gradually cools and amalgamates with the surrounding cooler air in the upper troposphere.

Extratropical cyclonic storm tracks vary seasonally, typically shifting towards the pole in summer and moving south during the winter (Phillips 1990). Across North America, cyclones undergo cyclogenesis across the prairies in the summer and autumn ('Alberta Clipper'), with occlusion and precipitation in eastern Canada. During the winter, cyclogenesis shifts to the Gulf of Mexico and the southern USA, with occlusion and storm activity over New England and Atlantic Canada. The combination of atmospheric circulation, Jet Stream movement, and the Appalachian Mountains focuses storm activity over eastern Québec and Atlantic Canada throughout the year. Québec and St. John's both receive more than 300 cm of snow annually, with more than 70 days of measurable snowfall. Climate change is anticipated to reduce the number of extratropical cyclones, although the intensity of individual events may increase (Basu and Sauchyn 2024).

In contrast to tropical cyclones, winter storms do not receive official names from Environment and Climate Change Canada or the USA National Weather Service. The broad and highly differential impact of extratropical cyclones and the rapid development of "bombs" (in contrast to the more predictable North Atlantic hurricanes) are cited as reasons for not naming these storms (Palmer 2013). Experience with difficulty in the use of human names for tropical cyclones is also involved. The use of human names is intended to make the storms more memorable, both prior to and after arrival, allowing them to be readily identified. However, anthropomorphizing a storm can cause people to assign human attributes, with unfortunate results. Research and some experience indicate that people may not consider storms with female, "friendlier", "unusual", or "non-aggressive" names as threats comparable to storms with more "powerful" or "forceful"-sounding names and may therefore underestimate the need to prepare (e.g. Catto and Batterson 2011, Skilton 2013, Jung et al. 2014; also see Rainear et al. 2017)

Unofficial names may be assigned by commercial broadcasters (e.g. Weather Network's use of names from classical mythology) or locally (Poloncarz 2010, Rainear et al. 2017) or are designated in retrospect by those impacted. "White Juan", a major extratropical cyclone affecting Nova Scotia in January 2004, was locally designated after the fact in association with the officially named Hurricane Juan, which had arrived in September 2003. "Snowmageddon" has been used for at least 10 storms in Canada and northeast USA since 1996.

Major extratropical cyclones such as "White Juan" (February 2004), January 2015 ("Juno"), January 2016 ("Jonas"), January 2020 ("Snowmageddon", Avalon Peninsula; Larochelle-Côté, 2023), and February 2024 ("Snowmageddon", Nova Scotia) bring snowfall totals of 90–160 cm to Atlantic Canada, with drifts reaching 15 m high. The eight-day closure of St. John's resulting from the mid-latitude cyclone of mid-January 2020 tested the residents' adaptive capacity. All Canadian cities have experienced significant snowfall events in recent years, including Toronto (55 cm, 17 January 2022) and Victoria (65 cm, 28–29 December 1996).

Environment Canada defines a blizzard as a weather event lasting a minimum of 4 h (6 h north of the taiga—tundra tree line) with sustained winds at least 40 km/h, visibility less than 1 km, and wind chill of -25 or lower. A whiteout is a blizzard accompanied by strong downdrafts that obscure the horizon. Blizzards have caused loss of human lives in recent times, including 76 people killed in March 1941 in Saskatchewan and Alberta, 21 in Southern Ontario in December 1944 and 12 in Montreal in March 1971.

Weather forecasting, improved governmental and public services, and general preparedness have made winter storm events somewhat less deadly in the past forty years, although individual fatalities unfortunately continue. Canadian urban residents, including those in Toronto and Victoria, generally have good preparedness or adaptation strategies. The March 1993 "Storm of the Century" brought 100–120 km h^{-1} winds and between 5 and 60 cm of snow from Alabama to Massachusetts, causing more than 300 deaths and $18 billion (USA, 2024 equivalent) damage. Unfortunately, effective precautions and emergency response were lacking in some areas. Many television and radio stations did not air weather forecasts, or downplayed predicted snowfall amounts. In comparison, the February 2004 "White Juan" storm brought comparable winds and higher snowfall totals (up to 95 cm) to Maritime Canada, but no deaths resulted. The exposure and magnitude of the events were similar, but effective preparation and response greatly reduced sensitivity and vulnerability to "White Juan". The January 2020 event resulted in one death due to exposure on the Avalon Peninsula, in addition to two deaths attributed to snow shovelling (CBC 2020). The February 2024 storm in Nova Scotia resulted in one death from a propane explosion (CBC 2024).

Property damage and economic losses due to disruption of business remain significant, both from blizzards and less intense snowfall events. The December 1996 (southern British Columbia) and January 1999 (Toronto) occluded front events cost $450 million and $300 million (2024 equivalents) respectively. The January 2015 and January 2016 storms cost about $250 million each across New England and Maritime Canada. The total economic impact of the January 2020 and January 2024 events exceeded $100 million in direct and indirect costs.

Statistics Canada coordinated a study of the social aspects of the January 2020 storm in northeast Avalon (Larochelle-Côté 2023), revealing the significance of demographic characteristics. People aged 65 and older represented 15% of the overall population (17% in St. John's), and about 30% of the more than 22,200 people living alone. The demography of the physically oldest urban core directly north and northeast of the downtown business district contrasts with suburban Paradise (10.3% aged ≥ 65, mean age 37.3 y, 7.7% living alone, 11.3% with 2021 income ≥ $100,000; Statistics Canada 2021–23). In the older neighbourhoods of St. John's, dating from the 1940s or earlier, more than 9% of dwellings are in need of major repairs, compared with 4% in Paradise and the northeast Avalon as a whole. These neighbourhoods are also characterized by lower household incomes and employment, and higher levels of residential instability, marked by frequent changes of dwelling, temporary renters, boarding houses, and (since 2022) visible homelessness.

Although Paradise received 91 cm of snow, more than downtown St. John's (~75 cm), a full State of Emergency lasted 3 days in the suburban community and 8 days in St. John's. Snow removal was difficult among the steep streets and closely spaced houses flanking the harbour (Fig. 19.3), necessitating deployment of Canadian Armed Forces personnel to assist in relief. Military personnel also assisted in Nova Scotia (2004, 2024), Toronto (1999), and in other Canadian cities.

19.3.2 Lake Effect Snow

Some of the highest snowfall totals are associated with marine and lake shorelines. Lake effect snow particularly impacts the Bruce Peninsula on the eastern shore of Lake Huron, increasing snowfall by 20–50% (Liu and Moore 2004; Lake

Fig. 19.3 Clearing snow in downtown St. John's, January 2020

Huron Weather 2024). Similar events can occur along the Atlantic coastline.

Lake effect snow is produced when cold air travels over relatively warm lake water (Wiley and Mercer 2020). Advection results in evaporation, causing the formation of low-lying clouds and fog. If wind speed is low, the moisture-laden advection fog persists over the water, obstructing vision but not producing substantial precipitation. Open water is required: ice cover reduces moisture exchange, precluding formation of advection fog or lake effect precipitation.

If the wind velocity is higher and directed towards land, uplift and advection result in incorporation of more moisture, which cools and condenses at higher altitudes. Heavy snowfall begins near the shoreline, where the moisture-laden air is cut off from the warm source. The snow belts are concentrated near the shoreline and do not extend far inland. Lake effect snow is often developed under relatively warm conditions in late autumn or early winter, especially following mild autumns which warm lake water and slow formation of lake ice cover.

19.3.3 Roof Loading

Failure of building roofs due to snow loading is attributed to design, construction, and maintenance problems (Zmigrodzki 2011, Geis et al. 2012, Bass and O'Rourke 2017). Estimation of the maximum snow load for a roof is based on 60–80% of the maximum ground load (snow on the ground, considering both thickness and snow density; Taylor 1980). For lower buildings, the calculation must also include the possibility of snowdrift cover. Rain-on-snow events must also be considered, as rain is denser than snow or ice pellets. Recommended design snow loads vary from 0.9 kPa (Edmonton) to 3.8 kPa (Québec). Most collapse incidents involve commercial buildings and warehouses, particularly of metal construction. However, collapses involving residences and schools have occurred (CTV 2023).

19.3.4 Snow Clearance

Snow ploughing and ice removal is a significant expense for Canadian cities (Table 19.2). Snow removal is a municipal responsibility in urban areas and commonly represents ~5% of a city's annual budget. Economic costs are largely confined to clean-up efforts: lost productivity and business activity is generally made up by activity in subsequent days, although longer disruptions can have significant effects. For hospitality and entertainment-related businesses, a snowfall on the wrong day (e.g. New Year's Eve, Valentine's Day) can result in losses that are not subsequently recovered. As only about 20% of Canadians were working from home in November 2023, decreased from the previous year (Statistics Canada 2024), clearing streets and sidewalks on normal business days remains a priority for municipalities and citizens alike. University operations have become less affected in recent years, with increased use of online teaching.

Most cities use diesel snowploughs (Campbell and Langevin 1995a, 1995b). A typical plough, with a 3.5–4 m wide, 2 t front blade, can move snow accumulations up to 1 m height. Attached supplemental wing ploughs, designed to clear arterial roads with high traffic volumes, are lighter (~1 t). Large capacity snowblowers and front-end loaders are used in Eastern Canadian cities with greater amounts or greater densities of snow, where successive snowfalls can result in banks in excess of 3 m height. Road graders are used in Western cities, where snow densities are lower. Smaller tracked ploughs are used to clear sidewalks. Snow which does not subsequently melt or is displaced by property owners is trucked to waste disposal areas, isolated from bodies of water to prevent contamination by road salt, heavy metals, and petrochemicals.

Diesel consumption for a typical large snowplough is 2.0–2.5 km l^{-1}, with higher rates for snowblowers. Few alternatives exist: trials with electric and autonomous ploughs generally have not been successful (see Lindquist 2024), due to capacity issues and the inherent difficulties in autonomous plough–nonautonomous vehicle interactions. GHG emissions specifically related to snowclearing equipment generally have not been tracked, although diesel heavy duty vehicles contribute substantially (c.f. Lukman et al. 2018). Environment and Climate Change Canada's (2021) discussion paper on vehicle emissions does not mention snowclearing equipment, perhaps because alternatives are not available.

Emphasis has focused on more efficient use of equipment, both to reduce emissions and enhance residents' satisfaction with operations. As up to 25% of a snowplough's operation time may be spent idling (Minnesota Department of Transportation 2023), operational measures have the potential to reduce fuel consumption. GIS technology is increasingly used to manage operations (Liu et al. 2014; Rasul et al. 2022), in addition to allowing residents the ability to track the location of equipment in many municipalities.

A truism of winter urban life is that almost nobody is always happy with municipal snowclearing operations (e.g. National Post 2023a, b, Canadian Press 2024), including this author. Municipalities give priority to clearing collector and major streets, used by public transit and by snowploughs leaving depots, followed by more general street clearance and sidewalk clearance. In some (a minority of) municipalities, residents are responsible for clearing sidewalks in front of their property, and are liable to fines if they do not. Businesses are responsible for providing access to their operations, as are public services such as Canada Post. Residents are responsible for clearing their own driveways and walkways, including hard-packed windrows discourteously deposited by unnecessary use of wing ploughs. "Snow rage" incidents involving disputes with plough drivers or between neighbours (e.g. Global News 2024) are reported every winter.

Recent discussion has focused on addressing the balance between clearing streets and clearing sidewalks (e.g. Gray and Cook 2021). Initially framed as a discriminatory practice against women (e.g. CBC 2018; Roach 2023), discussion has evolved to recognize the negative effects more generally on non- or occasional drivers, including physically challenged, elderly (of diverse genders), recent immigrants, university and college students, homeless, and economically disadvantaged people. In many cases, the contrast lies in the differences in automobile use and prevalence between suburban residents (of all genders) and inner-city residents (of all genders). The socio-economic aspects of clearing snow after a blizzard (e.g. Larochelle-Côté 2023) apply to more general snow removal efforts. Cities where substantial percentages of non- or occasional drivers are also users of public transit face the necessity of clearing general-use streets for buses on a priority basis.

Because different types of equipment are used to clear sidewalks and streets, assigning priority to either would not necessarily affect clearance of either. However, the indiscriminate use of wing ploughs on medium- and low-traffic volume streets results in the infilling of sidewalks with displaced street snow, frequently creating banks too high and too compacted for the tracked sidewalk ploughs to negotiate. A comprehensive policy would avoid such duplication and excess effort, in addition to reducing GHG emissions and creating better conditions for all residents. Unfortunately, although some cities have attempted to make operational changes (e.g. Gray and Cook 2021), much remains unaccomplished.

19.3.5 Road Salt

Ice clearance generally is accomplished by using road salt. NaCl is most effective at temperatures >−9.5 °C. At −3 °C, 1.8 kg of ice melt per kg of NaCl applied results after 15 min. For 1 mm-thick ice, that clears about 2 m^2, so clearing 1 km of standard two-way street (width 7 m) under a bare-pavement policy would require about 3500 kg of NaCl. The estimated annual consumption of NaCl for road clearance exceeds 5 million tonnes (Environment and Climate Change Canada 2018). Existing technologies for more efficient application could reduce usage by ~20%. $CaCl_2$ is effective to −29.1 °C and is more commonly used in Prairie Canada, but it is also more expensive than NaCl.

NaCl road salt has been recognized as a pollutant (Schuler et al. 2017; Moulding et al. 2022), as well as causing freezing point depression in freshwater, delaying formation of seasonal ice cover. Suggested alternatives (Sustainable Technologies Evaluation Program 2020) include use of $CaCl_2$, $MgCl^2$ (minimum effective temperature −15 °C), calcium acetate (minimum effective temperature −7 °C), beet juice, waste liquid from breweries, and starch residue (minimum effective temperatures −17 °C to −23 °C, commonly added to NaCl salt before application), and sand, which is non-toxic but adds sediment to freshwater bodies and does not melt ice. Some of the alternatives also may have adverse environmental effects (Gillis et al. 2021).

19.3.6 Shovelling as a Hazard

Shovelling snow is hard work (CCOHS 2024). A typical shovelful contains 3–5 kg of snow (~1 million snowflakes); 1 m^3 of partially compacted driveway snow represents between 200 and 1000 kg (Fig. 19.4). Ten minutes of shovelling and ten minutes of running produce equivalent effects on heart rates (Franklin et al. 1995). Use of different types of shovels or snow pushers does not appear to affect the amount of work performed, energy exerted, or heart rate (Smolander et al. 1995; Huang and Paquet 2002; McGorry et al. 2003).

Snowfall events result in increased emergency room visitation associated with myocardial infarction (heart attack) and angina, but decreases in asthma-related visits (Blindauer et al. 1999). The likelihood of having an acute coronary event related to snow shovelling appears to be greater for males, particularly those with a family history of premature death from coronary causes, are over 60, who have a history of stable angina, or who are taking coronary medications (Spitalnic et al. 1996, Nichols et al. 2012, Sauter et al. 2015, Toukola et al. 2015, Auger et al. 2017, Baranchuk 2017, Simon 2019). Many victims did not have a previous diagnosis of heart disease prior to the shovelling-related heart attack. People who are unaccustomed to exercise are particularly susceptible, especially if they approach the snowfall as a challenge to be immediately mastered (see Sauter et al. 2015, Auger et al. 2017, Simon 2019, Strauss-Mazzullo 2020) or are elderly (e.g. CBC 2020; CTV 2024). Although most victims are male, women also suffer heart-related illnesses due to shovelling. An estimated 1200 cardiac arrests and 100 deaths are attributed to snow shovelling in North America each year.

Snow shovelling can also cause shoulder and back injuries (Tanaka et al. 2015). The Ontario Chiropractic Association (2003) estimated that approximately 70% of back injuries in Ontario are related to shovelling or using personal snowblowers. Estimated numbers of shovelling injuries in the USA annually vary from 10,000 to 70,000. Other common causes of injuries are slips and falls and being struck with a shovel (~1700 annually in USA; Watson et al. 2011). The use of ergonomic-designed bent handle shovels results in significant reductions to back strain and the possibility of injury, although not affecting the total amount of work required. Attempts to clear snow from residential roofs result in additional injuries and fatalities, primarily from falls (Aulanko et al. 2012; Bylund et al 2016).

19.4 Falls

Falling is a significant winter hazard (Levy et al. 1998, Bischoff-Ferrari et al. 2007, Morency et al 2012, Hsu et al 2016), especially for the ever-increasing proportion of elderly urban residents. Replacement of door delivery with more remote or central communal services (e.g. Sibalija et al 2015) has increased the potential for injury. Under some circumstances, municipalities can be liable for injuries resulting from falls on inadequately cleared sidewalks (Supreme Court of Canada 2021). Several incidents of people being killed or severely injured by falling ice from roofs or rock walls have been recorded in the

Fig. 19.4 Shovelling (again) in St. John's. See text for discussion

past decade (e.g. Canadian Occupational Safety 2022; CBC 2023).

19.5 Winter Driving

Winter driving poses enhanced hazards (Kelsall and Redelmeier 2016). Although more intense precipitation results in increased chance of crash and injury (Andrey et al. 2003; Brubacher et al. 2016), the type of winter precipitation does not markedly influence accident rates (Andrey 2010; Black and Mote 2015). Long-term studies of Canadian motorists indicate that the accident probability during snowfall events has not changed (Andrey 2010). Over the winter, drivers do not become acclimatized to weather or less likely to have accidents. The posted speed limit is a significant factor, with greater rates in rural areas (Andrey et al. 2013).

Weather forecasts and warnings are often intended for drivers. Less-experienced, female, and elderly drivers are more likely to consult weather forecasts prior to setting out (Kilpeläinen and Summala 2007; Drobot et al 2014). Long trips are more likely to be postponed than are shorter ones, and commercial traffic tends to be less affected than personal trips (Andrey et al. 2013; Roh et al. 2016). Truckers may take advantage of improved snow clearing on routes normally used largely by commuters (Roh et al. 2016). Several studies have revealed that although drivers tended to consult weather forecasts before leaving, their driving behaviour did not change, influenced primarily by their perceptions at the moment (Unrau and Andrey 2006; Kilpeläinen and Summala 2007; Andrey 2010). Drivers generally considered the driving conditions to be better than those forecast. As with other studies of drivers illusory superiority (wishful thinking), the tendency to rate one's competence at performing a task at an unreasonably high level, appears to be involved.

19.5.1 Potholes

Potholes are a sign of forthcoming spring in many cities, although any thaw is likely to see their formation or reappearance (Fig. 19.5). Water infiltrates into pavement cracks formed by bottom-up fatigue cracking, top-down cracking, frost heave, or combinations of these. An ice lens develops by subsequent freezing along the substrate (subbase)-asphalt boundary, forcing the asphalt upward. Upon warming, the thawed ice lens leaves a void, subject to vertical compressive stress from traffic pressure, lateral tensile strain at depth in the subbase, and vertical compressive stress at the base of the subbase. Disconnected asphalt then falls into the hole.

Potholes are ubiquitous. Toronto filled more than 172,000 annually between 2020 and 2023, with 201,054 in the last year (Toronto 2024). The City of Ottawa filled more than a million potholes from 2018 to March 2023. Montreal filled an average of 176,000 over 2017–2019. An April 2016 survey by the Canadian Automobile Association (Automotive News Canada 2017) suggested an annual $1 billion in pothole-related damage to vehicles. Cost per incident ranges from $360 to $2100.

Filling or repairing potholes costs $25–50 for materials (asphalt) and $100–$300 labour. As most potholes are filled in winter, the expedient cold asphalt method ("throw-and-go")

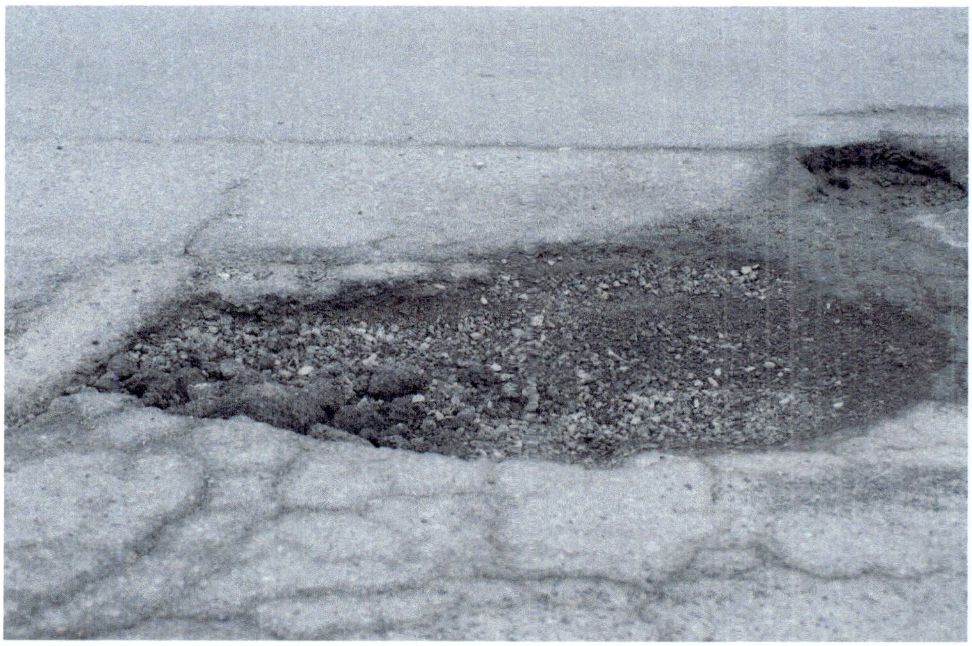

Fig. 19.5 A pothole, one of 15 over a 100 m extent of street, April 2018. See text for discussion

is generally used (Biswas et al 2016; Hafezzadeh et al. 2021). Cold asphalt repairs generally last less than one year.

19.6 Outdoor Winter Recreation

Ironically, the most popular Canadian spectator sports (hockey, figure skating, curling) are conducted indoors under climate-controlled conditions. This section considers outdoor recreation, including skating, skiing, snowboarding, snowmobiling, and ice fishing, focusing on economic aspects and vulnerability to climate change.

Winter-related injuries and deaths are related to occupational, socio-economic, and recreational events. In Canada, most casualties from winter conditions involve people engaged in recreational activities. Detailed discussions of the hazards involved in winter sport activities are provided by Tator (2008).

Outdoor recreational skating under climate change has received particular attention (Brammer et al 2015; SIRC 2023). Backyard rinks, maintained by individual homeowners for family and neighbours, are monitored through several social media groups and websites (Mekis et al. 2020). Declining numbers and shortening seasons in Eastern Canadian cities are regarded as tangible symbols of the impact of winter warming.

The Rideau Canal Skateway, maintained by the National Capital Commission through the centre of Ottawa, is also considered iconic and symbolic (Agrawal and Jahanandish 2019). Although subject to yearly variability (Fig. 19.6), the overall pattern shows a decline in the number of skating days (when ice thickness is suitable) from 2001. In 2023, the Skateway was unable to open at all. Between 1970 and 1995, the Skateway was open an average of 63 days per season. Between 1998 and 2022, the average season was 49 days.

Skiing and snowboarding generate ~$10 billion in Canada annually. The market has largely recovered from the effect of COVID-19 restrictions, leaving snow quality and quantity as the main problem. Recent trends suggest that declines in the amount of snowfall (as much as 70%) in low elevation sites) are not matched by declines in visitation (~10%) or revenue (~stable) at ski resorts. Resorts increasingly are diversifying from strict emphasis on downhill skiing to include snowboarding, cross-country, and other non-skiing attractions, a process which has been ongoing for several decades (see Hudson 2000).

Skiing and snowboarding are dependent on snow cover. Although artificial snow can be constructed, it is dependent on cold temperatures and abundant free water (Scott et al 2017), limiting a hill's ability to cope with long-term adverse conditions.

Regional and local conditions will significantly influence a hill's future, in addition to overall patterns of climate change (Scott et al 2017; Hagenstad et al 2018). Low elevation resorts, dependent on larger individual snowfall events, are more susceptible to adverse effects than are higher elevation resorts, where snowfall is more consistent (Figs. 19.7 and 19.8).

Recreational snowmobiling has grown in popularity in recent decades (Reich 1999, Canadian Council of Snowmobile Organizations 2019, Rutty et al 2023) and is now a $15

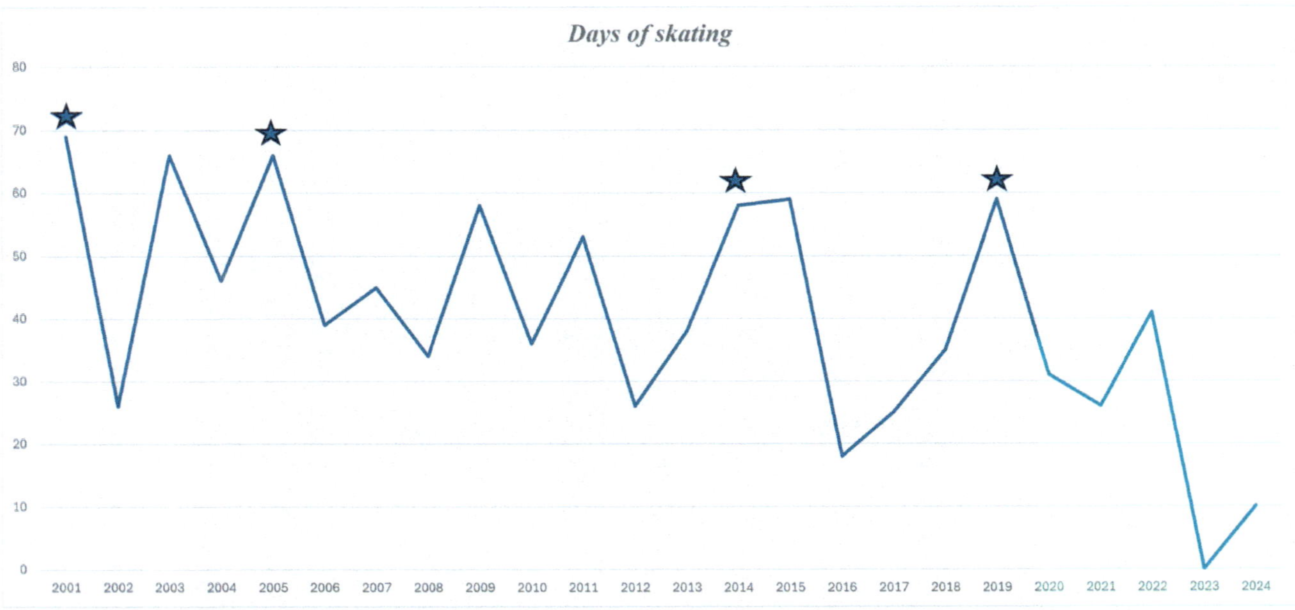

Fig. 19.6 Days of skating on the Rideau Canal, Winter 2001–2024. Stars indicate years when the Canal was open for skating during the previous December. Data from Agrawal and Jahanandish (2019; National Capital Commission)

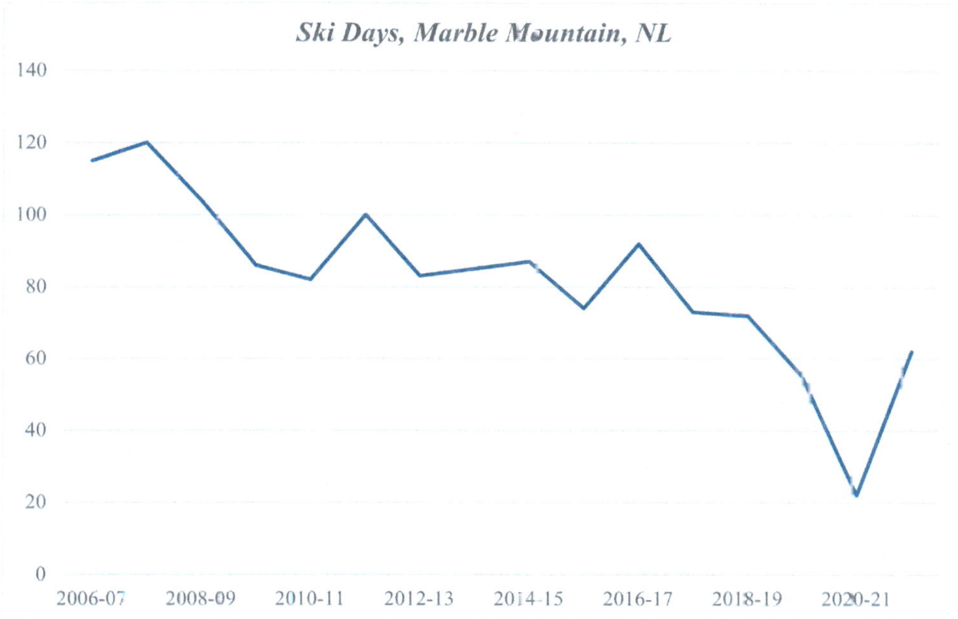

Fig. 19.7 Ski Days, Marble Mountain, NL. Marble Mountain has a base elevation of 1660 m asl, top elevation 2730 m asl. Data from Marble Mountain Development Corporation, Annual Reports, Government of Newfoundland and Labrador

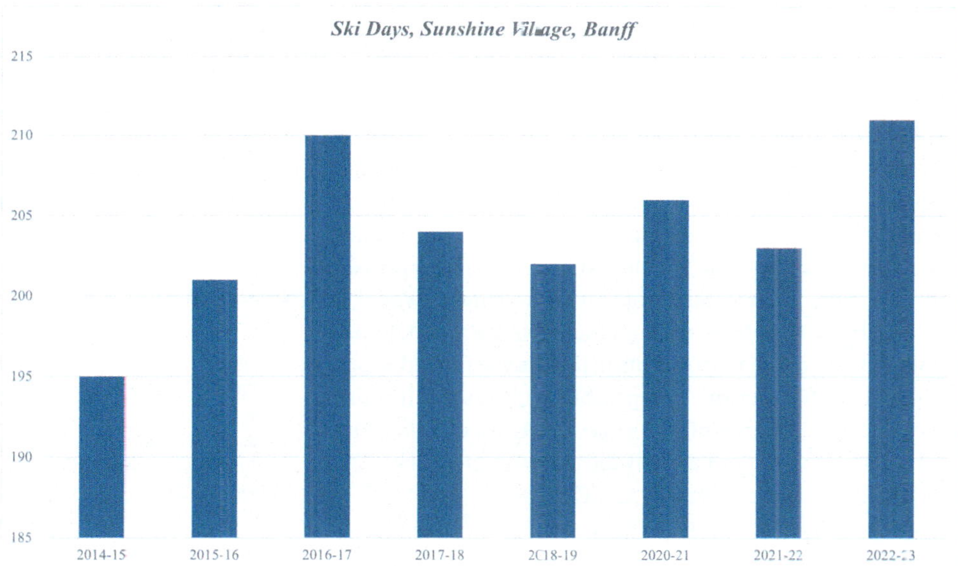

Fig. 19.8 Ski Days, Sunshine Village, Banff. Sunshine Village has a base elevation of 10 m asl, top elevation 546 m asl. 2019–2020 season is excluded due to early closure under COVID-19 restrictions. Data from https://www.snowpak.com/alberta/banff/sunshine-village/opening-dates

billion industry throughout North America. There are at least 618,000 recreational snowmobilers in Canada. In Ontario during 2022–2023, snowmobiling generated $1.5 billion million expenditures, $1.8 billion in total GDP impacts, 9307 equivalent full-year jobs, and $538 million in tax revenue (Harry Cummings and Associates 2023).

Predictions prior to 2010 suggested that snowmobiling would be eliminated in southern and central Ontario as a result of climate change, as well as encountering significant restrictions in Atlantic Canada and Québec (e.g. Vasseur and Catto 2008). More recent work undertaken in direct consultation with snowmobile associations has emphasized temporal and spatial variability of snowfall conditions across Ontario (Rutty et al. 2023), as well as in other areas. As most recreational snowmobilers come from urban communities, they are able to shift from region to region to take advantage of snow conditions, in addition to considering snowmobiling as a day-to-day pursuit. As a result, overall declines in snow pack thickness and duration in specific areas, which adversely affect local residents (especially those using snowmobiles for hunting and fishing, see Chaps. 8 and 9) as well as professional snowmobilers engaged in long-distance racing, have much lesser impact on recreational users. These snowmobilers are highly resilient and able to take advantage of marginal conditions, moving their recreation to areas within driving distance which have snow. Effects on local service businesses are profound, but overall expenditures (vehicles, gasoline) remain relatively unaffected.

Snowmobile use generates a small fraction (~0.1%) of total GHG emissions from transportation and motorized vehicle use. CO emissions are a problem with older machines with two-stroke engines. The most significant environmental impact comes from noise pollution.

Ice fishing is a recreational activity generally practised by local residents and hence tends to be underestimated as an economic activity. Lake Winnipeg is one area which attracts ice fishers from outside the local region, allowing estimation of impacts. In 2016–17, there was approximately $221 million in direct spending by anglers on Lake Winnipeg, generating $102 million GDP, $44.5 million in wages in local businesses, and $52 million in tax revenue (MB Wildlife Federation 2018).

Ice fishing requires a minimum thickness of 25 cm (35 cm if vehicles are used). The absence of suitable conditions throughout the winter of 2023–2024 in southern Ontario and Québec has resulted in economic losses in smaller communities (e.g. Orillia Matters 2024; Canada's National Observer 2024).

19.7 Commentary

In response to the Leger March 2023 public opinion survey (National Post 2023a, b) commissioned by the Association of Canadian Studies (ACS), the ACS president expressed surprise of the limited attention given to winter, which is "inescapable and defines us in some ways as a people and country". Only 5% of respondents in this survey ranked winter as their favourite season, comparable with results in other surveys. Less than half of the respondents agreed that they liked winter, and more than half did not like snow. The Trip Central (2019) survey ranked "cold", "snow", and "ice" as the most disliked elements. Winter driving was considered as the most unpleasant aspect, followed by scraping ice off vehicle windshields, slush, overcast, and short days. Elderly respondents were more prone to dislike winter than was the group as a whole.

Another survey (Leger 2023b) during the generally anomalously warm December 2023 detailed respondents' attitudes to climate change. Most linked climate change to the warm weather, and although 66% felt some concern as a result, more than 80% appreciated not having to shovel or brave cold. Residents of Québec (76%) and those aged 18 to 34 (75%) were more likely to attribute the warm weather to climate change. Approximately two-thirds of respondents stated that they had not been directly impacted by climate change.

One common Canadian response to winter is to run away to a warmer place (Trip Advisor 2019). Online surveys and informal polls (e.g. Leger 2023a) suggest that more than 40% of respondents plan winter holidays in southern locations. Although not representative of the population as a whole, this does indicate the prevalence of winter travel. Canadians represented more than half of international visitors to Cuba in 2023. A similar pattern is evident in tourists to Canada, with winter consistently the least preferred season among potential visitors (e.g. Theckedath 2021).

These data highlight both the weaknesses in the cold-adapted, winter-loving stereotype of Canadians, and the difficulties sometimes encountered in discussions of climate change. Fairly common responses to concerns about winter weather and outdoor recreation include, "I don't like winter, so less of it is good"; "I don't skate /ski / snowboard"; "I always go to Florida / Cuba / Arizona / etc.", and "I drive, I like driving, and I don't like anything that interferes with driving". Emphasis on the impact on winter sports (e.g. Malik et al. 2020; SIRC 2023) is unconvincing to those who are uninterested in them, or whose winter sports take place indoors (e.g. hockey, figure skating, curling). Fewer, albeit somewhat larger, events (cf. Basu and Sauchyn 2024) may be preferred to dealing with multiple repetitive snowfalls. The tendency to blame human shortcomings for loss of life during natural winter events (e.g. Catto 2012) suggests that the issues would not exist had the humans involved performed better. In short, directly connecting climate change to winter conditions may miss the mark by assuming stereotypical attitudes.

References

Agrawal N, Jahanandish A (2019) Is climate change impacting Rideau Canal Skateway, the world's longest skating rink? Nat Hazards 98:91–101

Ahmad T, Rashid T, Khawaja H, Moatamedi M (2016) Study of wind chill factor using infrared imaging. Int J Multiphys 10:325–341

American Academy of Ophthalmology 2023. https://www.aao.org/eye-health/diseases/photokeratitis-snow-blindness

Andrey J (2010) Long-term trends in weather-related crash risks. J Transp Geogr 18:247–258

Andrey J, Mills B, Leahy M, Suggett J (2003) Weather as a chronic hazard for road transportation in Canadian cities. Nat Hazards 28:319–343

Andrey J, Hambly D, Mills B, Afrin S (2013) Insights into driver adaptation to inclement weather in Canada. J Transp Geogr 28:192–203

Auger N, Potter BJ, Smargiassi A, Bilodeau-Bertrand M, Paris C, Kosatsky T (2017) Association between quantity and duration of snowfall and risk of myocardial infarction. Can Med Assoc J 189:E235–E242. https://doi.org/10.1503/cmaj.161064

Aulanko M, Handolin L, Söderlund T, Pajarinen J (2012) Accidental falls related to shovelling snow from rooftops: analysis of injuries in an extraordinary epidemic in southern Finland. Scand J Surgery 101:271–274

Automotive News Canada (2017) The $1 billion hidden cost of potholes keeps dealers busy. https://canada.autonews.com/article/20170426/CANADA/170429879/the-1-billion-hidden-cost-of-potholes-keeps-dealers-busy

Baranchuk A (2017) The cardiovascular risk of snowfall and snow shovelling in Canada. Can Med Assoc J 189:E545. https://doi.org/10.1503/cmaj.732903

Bass RAD, O'Rourke M (2017) Observations of snow load effects on four school buildings in New England. Struct Congress 2017. American Society of Civil Engineers, pp 302–313

Basu S, Sauchyn D (2024) Winter storm activity across Canada at the end of the century: a CMIP5 multi-model projection. Atmos Ocean 62:183–191. https://doi.org/10.1080/07055900.2023.2277710

Ben Shabat Y, Shitzer A, Fiala D (2013) Modified wind chill temperatures determined by a whole body thermoregulation model and human-based facial convective coefficients. Int J Biometeorol. https://doi.org/10.1007/s00484-013-0698-z

Berton P (1975) Hollywood's Canada: the Americanization of our national image. McLelland and Stewart, Toronto

Bischoff-Ferrari HA, Orav JE, Barrett JA, Baron JA (2007) Effect of seasonality and weather on fracture risk in individuals 65 years and older. Osteoporos Int 18:1225–33

Biswas S, Hashemian L, Hasanuzzaaman M, Bayar A (2016) A study on pothole repair in Canada through questionnaire survey and laboratory evaluation of patching materials. Can J Civil Eng 43:443–450. https://doi.org/10.1139/cjce-2015-0553

Black AW, Mote TE (2015) Effects of winter precipitation on automobile collisions, injuries, and fatalities in the United States. J Transp Geogr 48:165–175

Blindauer KM, Rubin C, Morse DL, McGeehin M (1999) The 1996 New York blizzard: impact on noninjury emergency visits. Am J Emerg Med 17:23–27

Brammer JR, Samson J, Humphries MM (2015) Declining availability of outdoor skating in Canada. Nat Clim Chang 5(1):2–4. https://doi.org/10.1038/nclimate2465

Brubacher JR, Chan H, Erdelyi S, Schuurman N, Amram O (2016) The Association between regional environmental factors and road trauma rates: a geospatial analysis of 10 Years of road traffic crashes in British Columbia, Canada. Plos ONE 11(4):e0153742. https://doi.org/10.1371/journal.pone.0153742

Burström L, Björ B, Nilsson T, Pettersson H, Rödin I, Wahlström J (2017) Thermal perception thresholds among workers in a cold climate. Int Arch Occup Environ Health. https://doi.org/10.1007/s00420-017-1227-x

Bylund P-O, Johansson J, Albertsson P (2016) Injuries sustained during snow removal from roofs resulting in hospital care. Int J Inj Contr Saf Promot 23(1):105–109. https://doi.org/10.1080/17457300.2014.992349

Campbell JF, Langevin E (1995a) Operations management for urban snow removal and disposal. Transp Res Part A: Policy Pract 29:359–370

Campbell JF, Langevin E (1995b) The snow disposal assignment problem. J Oper Res Soc 48:919–929

Canada's National Observer (2024) Quebec forced to cancel ice fishing villages for first time due to warm weather. https://www.nationalobserver.com/2024/02/13/news/quebec-cancel-ice-fishing-villages-first-time-warm-weather

Canadian Innovation Space (2017) Snow goggles. https://canadianinnovationspace.ca/snow-goggles/

Canadian Occupational Safety (2022) Government inspected site weeks before fatal ice fall incident. https://www.thesafetymag.com/ca/news/general/government-inspected-site-weeks-before-fatal-ice-fall-incident/424563

Canadian Press (2024) Heaps of snow left at bottom of driveways by Toronto plows could be your problem next winter. https://www.cp24.com/news/heaps-of-snow-left-at-bottom-of-driveways-by-toronto-plows-could-be-your-problem-next-winter-1.6733646.

Canadian Council of Snowmobile Organizations (2019) Economic impact of snowmobiling in Canada

Catto NR, Batterson MJ (2011) Igor and other hurricane and extratropical transitions in Newfoundland: geomorphologic and landscape impacts. GeoHydro, Quebec City, Quebec

Catto NR (2012) Natural hazard and vulnerability assessment in Atlantic Canada: review, progress, and challenges. In: Haque CE, Etkin D (eds.) Disaster risk and vulnerability. McGill-Queen's University Press, pp 235–274

CBC (2018) How Bill Morneau may use Sweden's gender-balanced snow clearing to adjust Canadian budgets. https://www.cbc.ca/news/politics/gender-analysis-budget-snow-sweden-1 4494640

CBC (2020) 2 dead in Harbour Grace following historic Newfoundland storm, Mayor says. https://www.cbc.ca/news/canada/newfoundland-labrador/shovelling-deaths-1.5432959

CBC (2023) NWT power company was aware of falling ice issue years before worker death. https://www.cbc.ca/news/canada/north/ntpc-fine-follow-1.6889154

CBC (2024) Senior dies in Nova Scotia after explosion caused by snow-damaged propane line. https://www.cbc.ca/news/canada/nova-scotia/historic-storm-woman-dies-snow-on-propane-line-1.7112945

CCOHS (2024) Ergonomics—shovelling snow. Canadian Council on Occupational Health and Safety. https://www.ccohs.ca/oshanswers/ergonomics/snow_shovelling.html

CMHA (2013) Seasonal affective disorder Canadian Mental Health Association, British Columbia division. https://bc.cmha.ca/documents/seasonal-affective-disorder-2/

CTV (2023) Roof of school collapses on Manitoba First Nation. https://winnipeg.ctvnews.ca/heavy-snow-causes-school-roof-to-collapse-on-manitoba-first-nation-1.6350466

CTV (2024) Deaths in Renfrew County highlight risks of shovelling snow. https://ottawa.ctvnews.ca/deaths-in-renfrew-county-highlight-risks-of-shovelling-snow-1.6726991

Day MP (2006) Hypothermia: a hazard for all seasons. Nursing 2006, December, 45–47. www.nursing2006.com

de Freitas CR, Grigorieva EA (2017) A comparison and appraisal of a comprehensive range of human thermal climate indices. Int J Biometeorol 61:487–512

Do MT, Fréchette M, McFaull S, Denning B, Ruta M, Thompson W (2013) Injuries in the North – analysis of 20 years of surveillance data collected by the Canadian Hospitals Injury Reporting and Prevention Program. Int J Circump Health 72:21090. https://doi.org/10.3402/ijch.v72i0.21090

Doctors of BC (2020) Coping with Seasonal Affective Disorder amid COVID-19. https://www.doctorsofbc.ca/together-health/coping-seasonal-affective-disorder-amid-covid-19

Drobot S, Anderson ARS, Burghardt C, Pisano P (2014) US Public Preferences for weather and road condition information. Bulletin, American Medical Society June 2014, pp 849–859

Endorf FW, Nygaard RM (2022) Socioeconomic and comorbid factors associated with frostbite injury in the United States. J Burn Care Res 43:646–651

Environment and Climate Change Canada (2018) Road salts: frequently asked questions. https://www.canada.ca/en/environment-climate-change/services/pollutants/road-salts/frequently-asked-questions.html

Environment and Climate Change Canada (2021) Discussion paper on heavy-duty vehicles and engines in Canada: transitioning to a zero-emission future. https://www.canada.ca/en/environment-climate-change/services/canadian-environmental-protection-act-registry/heavy-duty-vehicle-engines-zero-emission-future-discussion-paper.html

Environment Canada (2017) Canada's Wind Chill Index. https://www.canada.ca/en/environment-climate-change/services/weather-health/wind-chill-cold-weather/wind-chill-index.html

Environment Canada (2020) Canadian Climate Normals 1981-2010. https://climate.weather.gc.ca/climate_normals/index_e.html

Egro FM, Roy E, Friedstat J (2024) Update on cold-induced injuries. Clin Plastic Surg 51:303–311

Franklin BA, Hogan P, Bonzheim K, Bakalyar D, Terrien E, Gordon S, Timmis GC (1995) Cardiac demands of heavy snow shovelling. J Am Med Assoc 273:880–882

Fudge J (2016) Exercise in the cold: preventing and managing hypothermia and frostbite injury. Sports Health 8(2). https://doi.org/10.1177/1941738116630542

Gasparrini A et al (2015) Mortality risk attributable to high and low ambient temperature: a multicountry observational study. Lancet 386:369–375

Geis J, Strobel K, Liel A (2012) Snow-induced building failures. J Perform Constr Facil 26:377–388

Gillis PL, Salerno J, Bennett CJ, Kudla Y, Smith M (2021) The relative toxicity of road salt alternatives to freshwater mussels: examining the potential risk of eco-friendly de-icing products to sensitive aquatic species. Am Chem Soc EST Water 1:1628–1636. https://doi.org/10.1021/acsestwater.1c00096

Global (2024) Snow shovelling dispute leads to assault. https://globalnews.ca/news/10226442/snow-shovelling-dispute-leads-to-assault-arrest-peterborough-county-opp/

Gower S, Bassil K (201) Toronto public health. Health Impacts of Cold Weather. Technical Report. June 2014. www.toronto.ca/health/reports

Gray B, Cook T (2021) Cost of providing harmonized snow clearing for the entire City of Toronto for the 2021/2022 Winter Season including an equity lens. Budget Briefing Note, City of Toronto

Guly H (2011) History of accidental hypothermia. Resuscitation 82:122–125

Hafezzadeh R, Autelitano F, Giuliani F (2021) Asphalt-based cold patches for repairing road potholes—an overview. Constr Build Mater 306:124870. https://doi.org/10.1016/j.conbuildmat.2021.124870

Hagenstad M, Burakowski E, Hill RL (2018) The economics of snow for recreation in a changing climate. Proceedings international snow science workshop, Innsbruck, Austria, pp 531–536

Handford C, Thomas O, Imray CHE (2017) Frostbite. Emergency medicine clinic north. America 35:281–299

Harry Cummings and Associates (2023) The economic impact of snowmobile trails in Ontario. Ontario Federation of Snowmobile Clubs

Henderson S, Wan V, Kosatsky T (2012) A preliminary examination of the association between cold weather mortality in British Columbia between 1986 and 2010: final report. BC Centre for Disease Control, Vancouver

Hsu J, Shaw R, Novak A, Li Y, Ormerod M, Newton R, Dutta T, Fernie G (2016) Slip resistance of winter footwear on snow and ice measured using maximum achievable incline. Ergonomics 59(5):717–728

Huang C-t, Paquet V (2002) Kinetic evaluation of two snow-shovel designs. Int J Ind Ergon 29:319–330

Hudson S (2000) Snow business: a study of the international ski industry. Cassell, London

Ireland S, Endacott R, Cameron P, Fitzgerald M, Paul E (2011) The incidence and significance of accidental hypothermia in major trauma—a prospective observational study. Resuscitation 82:300–306

Jadidzadeh A, Kneebone R (2015) Shelter from the storm: weather-induced patterns in the use of Emergency Shelters. University of Calgary, School of Public Policy Research Papers 8(6). www.policyschool.ca

Jung K, Shavitt S, Viswanathan M, Hilbe JM (2014) Female hurricanes are deadlier than male hurricanes. Proc Natl Acad Sci 111:8782–8787

Kelsall D, Redelmeier DA (2016) Winter road safety is no accident. Can Med Assoc J Editor 188:241

Kilpeläinen M, Summala H (2007) Effects of weather and weather forecasts on driver behaviour. Transport Res Part F 288–299

Koutsavlis A, Kosatsky T (2003) Environmental-temperature injury in a Canadian metropolis. J Environ Health 66(5):40–45

Lake Huron Weather (2024) Lake effect Snow. https://www.lakehuronweather.org/wxSnowLakeEffect.php

Langford HV, Fox LR (2021) The wind-chill index: lessons from history. Wilderness Environ Med 32:392–399

Larochelle-Côté S (2023) Impact of the January 2020 snowstorm in northeast Avalon Peninsula, Newfoundland and Labrador, 2020. Statistics Canada. https://www150.statcan.gc.ca/n1/pub/16-002-x/2020001/article/00001-eng.htm

Leger (2023a) Canadians' Winter Travel Plans. https://leger360.com/surveys/canadians-winter-travel-plans/

Leger (2023b) Climate change in Canada. https://leger360.com/surveys/climate-change-in-canada/

Levitt AJ, Boyle MH, Joffe RT, Baumal Z (2000) Estimated prevalence of the seasonal subtype of major depression in a Canadian community sample. Can J Psychiat 45(7):650–654

Levy AR, Bensimon DR, Mayo NE, Leighton HG (1998) Inclement weather and the risk of hip fracture. Epidemiology 9:172–177

Lindquist E (2024) Do robot snowplows have a future in Canada? https://www.nationalobserver.com/2024/01/12/analysis/do-robot-snowplows-have-future-canada

LineUp Media (2023) Canada winter quotes. https://lineupmedia.fm/canada-winter-quotes

Liu AQ, Moore GWK (2004) Lake-effect snowstorms over southern Ontario, Canada, and their associated synoptic-scale environment. Mon Weather Rev 132:2595–2609

Liu G, Ge Y, Oiu TZ, Soleymani HR (2014) Optimization of snow plowing cost and time in an urban environment: a case study for the City of Edmonton. Can J Civ Eng 41:667–675

Lukman RK, Cerinšek M, Virtič P, Horvat B (2018) Improving efficient resource usage and reducing carbon dioxide emissions by optimizing fleet management for winter services. J Clean Prod 177:1–18

Lukmani A, Williams JVA, Bulloch AGM, Bhattarai A, Patten SC (2019) Seasonal variation in symptoms of depression: a Canadian population based study. J Affect Disorders 1:255:142–149

Maarouf A, Bitzos M (2000) Windchill indices: a review of science, current applications and future directions for Canada. Meteorological Service of Canada, Environment Canada, Technical Report En56–152/2000, p 30

Mader TL, Johnson LJ, Gaughan JB (2010) A comprehensive index for assessing environment stress in animals. J Anim Sci 88:2153–2165

Magnusson A, Boivin D (2003) Seasonal affective disorder: an overview. Chronobiol Int 20:189–207

Malik K, McLeman R, Robertson C, Lawrence H (2020) Reconstruction of past backyard skating seasons in the Original Six NHL cities from citizen science data. Can Geogr 64:564–575

Manitoba Wildlife Federation (2018) Economics of the recreational fishery on Lake Winnipeg. https://mwf.mb.ca/archives/674

Orillia Matters (2024) Ripple effect: local tourism industry devastated by lack of snow. https://www.orilliamatters.com/local-news/ripple-effect-local-tourism-industry-devastated-by-lack-of-snow-8047403

McGorry RW, Dempsey PG, Leamon TB (2003) The effect of technique and shaft configuration in snow shoveling on physiologic, kinematic, kinetic, and productivity variables. Appl Ergon 34:225–231

Mekis E, Vincent LA, Shephard MW, Zhang X (2015) Observed trends in severe weather conditions based on humidex, wind chill, and heavy rainfall events in Canada for 1953–2012. Atmos Ocean 53:383–397

Minnesota Department of Transportation (2023) Using advanced data analytics to reduce snowplow fuel consumption. https://mdl.mndot.gov/index.php/_flysystem/fedora/2023-08/202303ts.pdf

Morency P, Voyer C, Burrows S, Goudreau S (2012) Outdoor falls in an urban context: winter weather impacts and geographical variations. Can J Public Health 103:218–222

Moulding BJG, King GKK, Shenton M, Bray JP, Nichols SJ, Kefford BJ (2022) Assessing the relative toxicity of different road salts and effect of temperature on salinity toxicity: LCx values versus no-effect concentration (NEC) values. Arch Environ Contam Toxicol 82:281–293

National Post (2023a) Canadians hate winters and we have the numbers to prove it. https://nationalpost.com/news/canada/canadians-hate-winters-and-we-have-numbers-to-prove-it

National Post (2023b) Snow plow operators and residents clash in GTA cities this winter. https://nationalpost.com/news/canada/snow-plow-operators-and-residents-clash-in-some-gta-cities-this-winter

Neale RE, Lucas RM, Byrne SN, Hollestein L, Rhodes LE, Yazar S, Young AR, Berwick M, Ireland RA, Olsen CM (2023) The effects of exposure to solar radiation on human health. Photochem Photobiol Sci 1:1–37

Nichols RB, McIntyre WF, Chan S, Scogstad-Stubbs D, Hopman WM, Baranchuk A (2012) Snow-shoveling and the risk of acute coronary syndromes. Clin Res Cardiol 101:11–15

Ontario Chiropractic Association (2003) Snow shovelling number one cause of back pain. https://www.greatlifechiro.com/canadian-poll-says-snow-shoveling-number-one-cause-of-winter-back-pain/

Osczevski R (1995) The basis of wind chill. Arctic 48:372–382

Osczevski R, Bluestein M (2005) The new wind chill equivalent temperature chart. Bull Am Meteorol Soc 86:1453–1458

Paal P, Gordon L, Strapazzon G, Maeder MB, Putzer G, Walpoth B, Wanscher M, Brown D, Holzer M, Broessner G, Brugger H (2016) Accidental hypothermia—an update. Scand J Trauma, Resuscit Emergency Med 24:111. https://doi.org/10.1186/s13049-016-0303-7

Palmer R (2013) What's in a storm name? Weather Channel policy draws critics, but catches on. IBT, 8 February, http://www.ibtimes.com/whats-storm-name-weather-channel-policy-drawscritics-catches-1073016

Parks Canada (2021) Iggaak (ilgaak) Inuit snow goggles. https://parks.canada.ca/culture/cseh-twih/202124

Peiris AN, Jaroudi S, Gavin M (2018) Hypothermia. J Am Med Assoc 319(12):1290

Phillips D (1990) The Climates of Canada. Environment Canada. 176 p

Poloncarz MC (2010) Review of the Erie County storm fund. Erie Country Report, pp 15. https://www.erie.gov/comptroller/pdfs/October_2006_Storm_Report_FINAL.pdf

Provençal S, Bergeron O, Leduc R, Barrette N (2016) Thermal comfort in Quebec City, Canada: sensitivity analysis of the UTCI and other popular thermal comfort indices in a mid-latitude continental city. Int J Biometeorology 60:591–603

Rainear AM, Lachlan KA, Lin CA (2017) What's in a #Name? An experimental study examining perceived credibility and impact of winter storm names. Weather Climate Soc 9:815–822

Rasul A, Seo J, Xu S, Kwon TJ, MacLean J, Brown C (2022) Optimization of snowplow routes for real-world conditions. Sustainability 14:13130. https://doi.org/10.3390/su142013130

Regli IB, Oberhammer R, Zafren K, Brugger H, Strapazzon G (2023) Frostbite treatment: a systematic review with meta-analyses. Scand J Trauma, Resusc, Emerg Med 31(96). https://doi.org/10.1186/s13049-023-01160-3

Reich LS (1999) Ski-dogs, pol-cats, and the mechanization of winter: the development of recreational snowmobiling in North America. Technol Cult 40:484–516

Roach N (2023) Snow clearance isn't gender neutral. https://www.womeninurbanism.ca/words/snow-clearance-gender

Roh H-J, Sahu PK, Sharma S, Datia S, Mehran B (2016) Statistical investigations of snowfall and temperature interaction with passenger car and truck traffic on primary highways in Canada. J Cold Regions Eng 30:04015006. https://doi.org/10.1061/(ASCE)CR.1943-5495.0000099

Rutty M, Cardwell F, Gunn G (2023) Snowmobiling and climate change: exploring shifts in snowmobile activity using a temporal analogue approach in Ontario (Canada). Tour Hosp 4:604–617

Sanders F, Gyakum JR (1980) Synoptic-dynamic climatology of the "Bomb." Mon Weather Rev 108:1589–1606

Sauter T, Haider DG, Eicklin ME, Exadaktylos AK (2015) The snow, the men, the shovel, the risk? ER admissions after snow shovelling: 13 winters in Bern. Swiss Med Weekly 2015(145):w14104

Schuler MS, Hintz WD, Jones YDK, Lind LA, Mattes BM, Stoler AB, Sudol KA, Relyea RA (2017) How common road salts and organic additives alter freshwater food webs: in search of safer alternatives. J Appl Ecol 54:1353–1361

Scott D, Steiger R, Rutty M, Pons M, Johnson P (2017) The differential futures of ski tourism in Ontario (Canada) under climate change: the limits of snowmaking adaptation. Curr Issue Tour 22. https://doi.org/10.1080/13683500.2017.1401984

Sever C, Kulahci Y, Acar A, Karabacak E (2010) Unusual hand frostbite caused by refrigerant liquids and gases. Turk J Trauma Emerg Surg 16:433–438

Shenaq DS, Gottlieb LJ (2017) Cold Injuries. Hand Clin 33:257–267

Sheridan RL, Goverman JM, Walker TG (2022) Diagnosis and treatment of frostbite. N Engl J Med 386(23):2213–2220

Shitzer A (2007) Paradox: increased blood perfusion to the face enhances protection against frostbite while it lowers wind chill equivalent temperatures. Int J Biometeorol 51:383–393

Shitzer A, Tikuisis P (2010) Advances, shortcomings, and recommendations for wind chill estimation. Int J Biometeorol 56:495–503

Sibalija J, Ciminsky ML, Fuchigami K, He M-H LS, Trevithick J, Bickerton G, Zecevic A (2015) Potential implications of mail delivery conversion on older adults' risk of falls in the Winter.

Simon H (2019) Snow shoveling—caution, men at risk. Harvard Medical School Commentaries on Health. https://search.credoreference.com/articles/16929

SIRC (2023) The impact of shorter winters on Canadian sport. Sport Information Resource Centre, Ottawa. Canada. https://sirc.ca/wp-content/uploads/2023/07/Winter-Sport-Lit-Review.pdf

Skilton EC (2013) Camille was no lady but Katrina was a bitch: Gender, hurricanes and popular culture. PhD dissertation, Tulane University, New Orleans

Smolander J, Louhevaara V, Ahonen E, Polark J, Klen T (1995) Energy expenditure and clearing snow: a comparison of shovel and snow pusher. Ergonomics 38:749–753

Spitalnic SJ, Jagminas L, Cox J (1996) An association between snowfall and ED Presentation of Cardiac Arrest. Am J Emer Med 14:572–573

Stares J, Kosatsky T (2015) Hypothermia as a cause of death in British Columbia,1998–2012: a descriptive assessment. Can Med Assoc J Open 3:E352–E358 https://doi.org/10.9778/cmajo.20150013

Statistics Canada (2024) Working from home in Canada. https://www150.statcan.gc.ca/n1/daily-quotidien/240118/dq240118c-eng.htm

Strauss-Mazzullo H (2020) Shovelling snow in Finnish Lapland: social and aesthetic perspectives on an everyday activity. Polar Rec 56(e32):1–14. https://doi.org/10.1017/S0032247420000339

Sugg MM, Stevens S, Runkle JD (2019) Estimating personal ambient temperature in moderately cold environments for occupationally exposed populations. Environ Res 173:497–507

Supreme Court of Canada (2021) Nelson (City) v. Marchi, 2021 SCC 41 (CanLII). https://www.canlii.org/en/ca/scc/doc/2021/2021scc41/2021scc41

Sustainable Technologies Evaluation Program (2020) Alternatives to Salt. www.sustainabletechnologies.ca

Tanaka M, Kobayashi T, Seko T, Shida M, Kumamoto T (2015) The influence of throwing height on muscle fatigue in repetitive snow-shoveling movements. Rigakuryoho Kagaku 30(4):609–614

Tator CH (2008) Catastrophic injuries in sports and recreation: causes and prevention—a Canadian study. University of Toronto Press

Taylor DA (1980) Roof snow loads in Canada. Can J Civ Eng 7:1–18

Theckedath D (2021) Canada's tourism economy. Library of Parliament, 2021-33-E

To T, Zhang K, Maguire B, Terebessy E, Fong I, Parikh S, Zhu J, Su Y (2021) UV, ozone, and COVID-19 transmission in Ontario, Canada, using generalized linear models. Environ Res 194:110645. https://doi.org/10.1016/j.envres.2020.110645

Toronto Sun (2022) Canada still seen as a land of snow, ice, and maple syrup. https://torontosun.com/news/national/canada-still-seen-as-a-land-of-snow-ice-and-maple-syrup

Toronto (2024) Potholes. https://www.toronto.ca/services-payments/streets-parking-transportation/road-maintenance/potholes/

Toukola T, Hookana E, Junttila J, Kaikkonen K, Tikkanen J, Perkiömäki J, Marja-Leena Kortelainen M-L, Huikuri HV (2015) Sudden cardiac death during physical exercise: characteristics of victims and autopsy findings. Ann Med 47:263–268

Trip Central (2019) What Canadians hate about winter. https://media.tripcentral.ca/staticCMS/content/pdfs/WhatCanadiansHateAboutWinter-SunVacations.pdf

Unrau D, Andrey J (2006) Driver response to rainfall on urban expressways. Transp Res Rec 1980:24–30

Vasseur L, Catto N (2008) Atlantic Canada. In: Lemmen DS, Warren FJ, Lacroix J, Bush E (eds) From Impacts to Adaptation: Canada in a Changing Climate 2007. Government of Canada, Ottawa, pp 119–170

Vasseur L, Thornbush M, Plante S (2015) Gender-based experiences and perceptions after the 2010 Winter Storms in Atlantic Canada. Environ Res Public Health 12:12518–12529

Vincent LA, Zhang X, Brown RD, Feng Y, Mekis E, Milewska EJ, Wan H, Wang XL (2015) Observed trends in Canada's climate and influence of low-frequency variability modes. J Clim 28:4545–4560

Vreeman R, Carroll A (2008) Seasonal medical myths that lack convincing evidence. British Medical Journal 337: a2769 https://doi.org/10.1136/bmj.a2769

Watson DS, Shields BJ, Smith GA (2011) Snow shovel-related injuries and medical emergencies treated in US EDs, 1990 to 2006. Am J Emerg Med 29:11–17

Wiley J, Mercer A (2020) An updated synoptic climatology of Lake Erie and Lake Ontario Heavy Lake-Effect Snow Events. Atmosphere 11:872. https://doi.org/10.3390/atmos11080872

Zmigrodzki S (2011) Investigations following structural failure of a roof in Canada. Proc Instit Civil Eng 164:125–138

Patterns, Lines On and Around the Land

20

Abstract

Boundaries within Canada, and between Canada and the USA, evolved as European settlement and political negotiations progressed. These boundaries were not precisely defined as areas of European settlement were relatively remote from one another. Wording in treaties was frequently vague and lacked precision, especially when based on maps that were not as accurate as modern maps. Indigenous communities were not considered or consulted during boundary discussions. The British North America Act (1867) officialized the creation of the Dominion of Canada and its governing structure. Powers were split between two levels of government. The federal government would take care of matters that are of national and international concern, while the provincial governments would have jurisdiction over regional specificities such as property, contracts and torts, local work, undertakings, and businesses. People that have settled the land throughout Canada since ca. 1620 have left persistent traces of their presence by the different land managements they used, which have left a varied mosaic of patterns on the land. They left clear, sometimes imbricated or superimposed, imprints on the ways pieces of land were divided that are still easily noticeable. The French seigneurial system left imprints of long narrow lots. The British left traces of townships. After 1867, Canada surveyed the remaining land using the township model used by the American Homestead Act to execute the Dominion Land Survey. As it is flanked by three oceans, Canada has rights and responsibilities over maritime territory. As a coastal country, Canada has to demarcate its maritime zones as defined by the United Nations Convention on the Law of the Sea. Canada's Arctic sovereignty is a complex and strategic issue of growing national and international importance. Canadian sovereignty over the Arctic is particularly at stake today when climate change, national security, and natural resources have become paramount issues.

Keywords

Boundaries · Settlement · Surveying · Maritime zones · Arctic sovereignty

20.1 The Contour of Canada

While British merchants were still pursuing ventures along the Mediterranean coasts, Jacques Cartier sailed across the Atlantic and discovered Canada (from a European perspective) in 1534. Despite failed attempts to establish French colonies along the North American coast, nothing much had changed when Samuel de Champlain laid the foundations of Acadia in 1604 and founded Québec City four years later. Then, the bases of a new country were laid out. Gradually, immigrants crossed the ocean to arrive in New France. In the meantime, as Britain's economy started to industrialize, its merchants followed suit and turned to North America in search of new markets for its manufactured goods as well as new raw material pools. British colonies were soon established on the Atlantic coast of what is now the United States. Great Britain became interested in acquiring the French possessions as well. Hence, an extended period of battles between France and Great Britain and, later, after France finally capitulated between Great Britain and the United States followed.

What was New France then, approximately as far west as the Mississippi River, became Province of Québec under British rules after the 1763 Treaty of Paris. Great Britain relinquished the southernmost part of the province after the American Revolutionary War under the 1783 Treaty of Paris. Eight years later, in 1791, the province of Québec became divided into Upper Canada and Lower Canada, respectively

N. Catto and G. R. Vincent, *Canada: Land and People*, World Regional Geography Book Series,
https://doi.org/10.1007/978-3-032-04391-7_20

the southern parts of today's Ontario and Québec. They were reunited in 1841 as the United Province of Canada, but differentiated as Canada West and East, until the creation of the Dominion of Canada in 1867 when they were separated again to become the provinces of Ontario and Québec along with New Brunswick and Nova Scotia.

It is necessary at this point to go back to events that predate the creation of Canada. At the time of signing the 1783 Treaty of Paris, the wording of the boundary's definitions was rather vague and not precisely stated, if at all. Several issues arose and had to be settled.

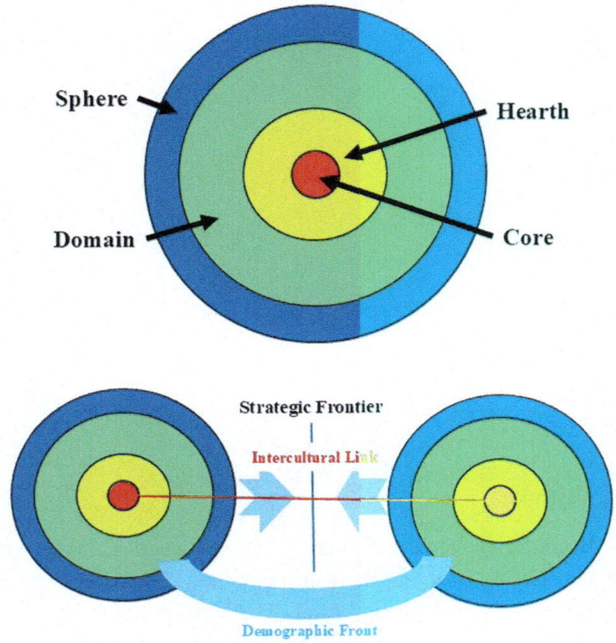

Fig. 20.1 Core-domain model

20.2 Types of International Boundaries

We can look at three theoretical main types of boundary definitions that are used to delimit political regions: consequent, geometric, and natural boundaries (Poole 1964). Examples of those three categories of boundaries have been used to draw lines around and within Canada throughout its history.

Consequent boundaries are, as the term indicates, consequent to (or resulting from) early or previous settlements. However, as simplistic as it may seem, difficulties and questioning may arise in using them to cut a region in different jurisdictions such as which and how early settlements should be taken into account? Should permanent or temporary settlements be considered as previous settlements? Settlements that were isolated from each other may have grown to meet and even mix, thus making the establishment of a line difficult to set.

Geometric boundaries are superimposed "artificial" lines that may appear to have been drawn arbitrarily to divide a region into different political entities. Such lines as parallels, meridians, and lines joining features such as lakes, rivers, or other lines often set remotely on maps are examples of such boundaries.

Natural, or physiographic boundaries refer to water bodies (river, lake, bays, sea) or mountains (watershed divide) used as dividing lines.

20.3 Early Settlements and Blurry Lines

Boundaries were not precisely defined as areas of European settlement were relatively remote from one another. Moreover, wording in the treaties was frequently vague and lacked precision, especially when based on maps that were not as accurate as modern maps. Similar to Meinig's core-domain-sphere model, where the core is the main nucleus of a given group whose power over space diminishes with distance but still exerts some influence, albeit subject to competing authority, the sphere was the blurry line that divided opposing powers (Fig. 20.1). Indigenous communities were not considered or consulted during boundary discussions.

20.3.1 Geography of New France and New England

During an era of maritime transportation that led to the discovery of the "New World", North America was a continental block turned inward with a complex hydrographic network, critical for both fur trade and navigation, and with rather difficult access. European explorers navigated along the coasts in search for an "entry door" to the interior, which Cartier and then Champlain found first, allowing France to colonize the Saint Lawrence Valley (Fig. 20.2).

Colonization of Acadia did not enable a deep penetration of the continent. Moreover, part of Acadia, as a peninsula on the Atlantic coast, and situated directly north of the British colonies, was very exposed to confrontations. Therefore, the Saint Lawrence River was a crucial axis that allowed penetration and control over a large part of the continent, which was the envy of Britain.

British colonies developed on the Atlantic coast and were stuck between the sea on the east and the Appalachian range on the west in an era when navigation was the main transportation mode. Furthermore, the Saint Lawrence Valley had to be accessed by bypassing Acadia and going through the Gulf and then navigating upstream to Québec. This is primarily why the British relentlessly attacked Québec. They wanted access to the interior, what was difficult from the east coast. Alterna-

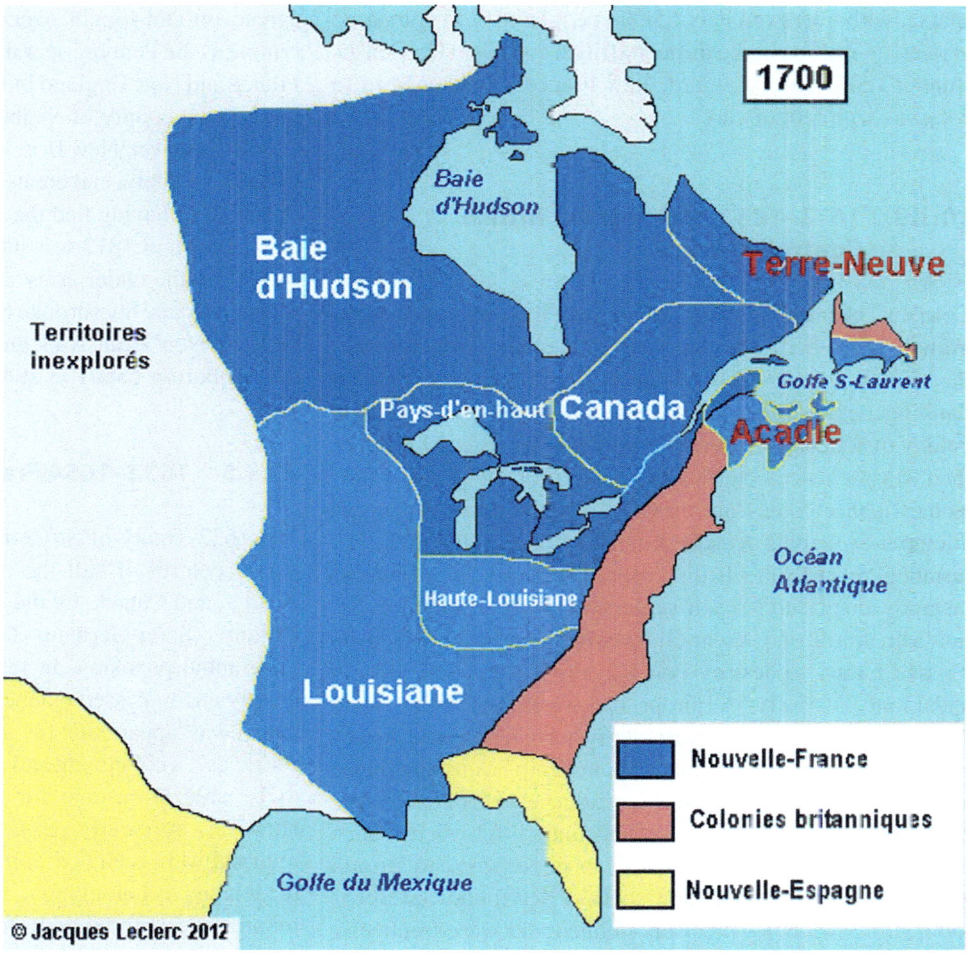

Fig. 20.2 Nouvelle-France ca. 1700. (Wikimedia commons, public domain)

tively, the continent could be accessed through Hudson Bay, what the British eventually did.

20.3.2 1604–1613 Foundation of Acadia and New France

After a failed attempt at Ile Ste-Croix in 1604, French settlements were established across the Bay of Fundy, first in Port Royal in 1605, and subsequently along the Bay of Fundy from Pentagouet (Penobscot in present Maine) to Port Royal, and Cap-Sable on southernmost tip of Nova Scotia. For a population of 800 in 1680, such a large territory was difficult to defend. However, Mi'kmaq people lived in the Maritimes, with Wolastoqiyik (Maliseet) groups along the Saint John River and Northwest Maine. Abenakis (Wapánahki) lived in Maine and along the St Lawrence south shore and were strategically located as they acted as a buffer between British and French colonies. Since French colonists developed alliances with all of the Indigenous peoples, they could resist better to eventual attacks from British colonies wanting to extend control.

Meanwhile, from 1608 French settlements slowly developed as agricultural and self-sufficient seigneuries spread along the St Lawrence River, accounting for 90% of New France's population in the mid-eighteenth century. However, for the first few decades New France revolved around fur trade, with Québec City as the main trading post. Networks were organized along water routes: Sainte-Croix, Saint John, Pentagouët (today's Penobscot), and Kennebec Rivers in Acadia. The Saint Lawrence River and its tributaries was by far the most fertile territory. Fur trade allowed New France to expand further west and south by establishing trade posts along water routes. At its peak it covered a large part of North America along the St. Lawrence axis, the Great Lakes and the Mississippi River totalling 8,000,000 km^2 (Havard 2003). Divided in "regions", in 1700 New France comprised Acadie, Terre-Neuve (Newfoundland), Canada/Laurentian Valley, Upper Country (Great Lakes), Hudson Bay, Upper (Illinois Country), and Lower Louisiana. These vast territorial claims were based on the presence of beaver. However, apart from settlements regularly spread along the Saint-Lawrence River, the trade posts commonly located at river mouths represented little population. The total population, as a result, was sparse, with approximately 65,000 people in 1755 compared to over 1.4 million in the thirteen British colonies (US Census Bureau 1909). In that regard, New France was likened to “a colossus with feet of clay”.

20.3.3 1613–1632—19 Years of British Control in Acadia

Colonies became nervous when neighbouring population growth intensified. The Saint Lawrence Valley was far from the British colonies. Furthermore, the mountainous terrain flanking the south side was more difficult to cross. The French colony of Acadia was an easier prey as it lay directly north of then Massachusetts North (now Maine) of the original thirteen British colonies and has been the target of numerous attempts by Britain to take control of part of the area. For instance, in 1613 the British captain Samuel Argall almost instantly displaced French settlers when they settled in St-Sauveur, north of present Bar Harbor ME, after which he crossed Fundy to destroy Port Royal (Thorp 1996). Britain and France’s rivalry in Europe had repercussions in North America. Acadia had a particularly turbulent early history as it was regularly tossed like a “poker chip” as its allegiance alternated numerous times through a succession of exchanges between French and Britain crowns until 1763 when France relinquished all its North American possessions to Britain except St-Pierre and Miquelon, south of Newfoundland. Territorial limits were not precisely defined and were frequently adapted or fought to as settlement progressed.

20.3.4 1621 Colony of Nova Scotia (New Scotland)

King James I of England granted Acadia to Sir William Alexander in 1621 (Nova Scotia Royal Charter 1621) “to establish a Scottish colony in North America” to be named Nova Scotia (Parks Canada 2022). The new colony included present-day Nova Scotia, New Brunswick, Gaspésie Peninsula, and part of Maine.

> “The lands and islands within the promontory of Cape Sable, westward to the roadstead of St. Mary, crossing its entrance or mouth of that roadstead to the St. Croix River, following to its remotest source, from that indefinite place, direct north to the St. Lawrence; eastward along the south shore of that river to Cape Gaspé, then south-southeast to the right of the Bacalaos Isles, onward to the mouth of the Gulf at the northernmost point of Cape Breton and from there southward to and including Sable Island, and to the starting point of that Cape” (Nova Scotia Royal Charter 1621).

Because of the wording and inaccurate mapping techniques, borders were not precisely defined and often not agreed on and fought over. The western limits could be as remote as the Penobscot or the Kennebec rivers, between New France and New England in 1687 (Griffith 2005). The western limit of the county of Sunbury, established in 1765, was the Penobscot River. New Brunswick would eventually be carved from Nova Scotia and created in 1784 to administer the influx of Loyalists having fled the American Revolution. The battle of Hampden in 1812 was intended to regain the northeastern part of Maine under army officer and colonial administrator Sir John Coape Sherbrooke (Burroughs 2024). The exact limit between New Brunswick and Maine would be finalized with the Ashburton Treaty in 1842.

20.3.5 1632–1654 France Regains Control

The 1632 Treaty of St-Germain-en-Laye allowed France to regain control of “all the places occupied in New France, Acadia, and Canada by the subjects of His Majesty of Great Britain” (Saint-Germain Treaty, Article 3) with no more geographic precision. In reality, the British occupation had mainly encompassed Québec and Port Royal. The aim of the treaty was apparently for the French company “Les Cent-Associés”, a colonizing and trading company formed in 1626, to be able to resume fur trading as quickly as possible. However, no clearly defined limits to the territories to be returned were included in the treaty and therefore implicitly recognized as belonging to the French. There was little need for precise territorial limits to revive the trade, which gives the impression that trade, posts and routes prevailed over territorial possessions (Paireau 2012). The fur trade was the main (or “only”) economic base, contrasting with more diversified and populous New England.

20.3.6 1654–1710 Neglected Acadie, in Unsettling Times

For as long as French people have been present in North America, their presence had been subjected to confrontations with British powers. Acadia has often paid the price. Isolated from the Saint Lawrence Valley colony, France did not appear keen to populate or defend it. For instance, the “King’s daughters (Filles du Roi)” were sent in New France to instil natural increase in the colony (see Chap. 11). None of them were sent in Acadia. Acadia’s population grew from 800 to 1,700 between 1680 and 1710, whereas the Laurentian region population went from 9,700 to 16,000 during the same period. Although its growth rate was superior, the total number remains meagre compared to the Saint Lawrence Valley and, particularly, to New England, its immediate neighbour New England whose numbers went from 155,000 in 1680 to 357,000 in 1710, let alone more economically diversified than relying solely on fur trading as the French did. Furthermore, the military presence was insignificant. There were never more than a few hundred soldiers in all Acadia in 1710 to resist the constant and an outnumbering aggressive British threat. Consequently, Acadia was an easier prey for the British Crown. Thus, French and British territorial claims were continually readjusted.

As a result, after Argall's attack in 1613, Acadia continued to alternate control from one crown to another: returned to France by the Treaty of St-Germain-en-Laye, attacked again by the English of Boston under Sedgwick in 1654; given back to France by the Treaty of Breda in 1667 with no specific mention of its geography; attacked once more by Phipps at Port Royal in 1690. The Treaty of Ryswick of 1697 returned Plaisance (southern half of Newfoundland), Hudson Bay (no real agreement reached but British kept Fort Albany and French Fort Bourbon (Fort York)) and Acadia to France. England kept its colony in Newfoundland (northern half). Nicholson's English attack resulted in the definitive conquest of Port Royal, renamed Annapolis Royal, in 1710 (Parks Canada 2022).

Under the Treaty of Utrecht in 1713 that followed, the colony of Acadia ceased to exist as a governmental entity, even though most Acadian people stayed. France lost "historical" Acadia (today's Nova Scotia Peninsula) but kept Ile-Royale (Cape Breton Island), and Ile-Saint-Jean (Prince Edward Island). Sparsely inhabited then, and mostly by Acadians, France also held present-day New Brunswick, which Britain contested as well. Because of unprecise wording regarding territorial limits, claims on space differed. France stretched Acadia's limits as far west as the Kennebec River, while Britain claimed first the Penobscot, then the Ste-Croix River as its western limit. As Baron Saint-Castin reigned over the Penobscot River's mouth until 1700 (incidentally Castine ME was named after him), Acadia's governor Villebon claimed in 1698 that Kennebec River was the boundary (Williamson 1832), or even further west along the St. George River. Such a definition of Acadia and New France would guarantee year-long access to the Atlantic. For Britain, the Acadia granted to William Alexander to found New Scotland (or Nova Scotia) was part of the "deal". Indigenous communities were not considered or consulted during these discussions.

20.3.7 1670 Rupert's Land

In the meantime, further inland, fur traders and explorers Radisson and Des Groseilliers travelled north and west beyond the Great Lakes and the French fur trading area in 1658–1859 and discovered that beaver pelts were plentiful, and this vast area had the potential to be very lucrative. Unable to convince the French Crown to extend its fur trading westward, the two entrepreneurs sought support from the British Crown. Seduced by the idea, Prince Rupert started to gather a company of investors and organized financing to prepare an exploratory expedition that proved to be successful. Rupert then convinced King Charles II to establish a fur trading company. Consequently, "The Company of Adventurers" was granted a commercial monopoly. Still unclaimed, the Hudson's Bay watershed, a huge territory of 3.9 million km^2 (a third of present-day Canada), was given to the Hudson's Bay Company in 1670. Moreover, the prince was named the first governor of the newly established Hudson's Bay Company, and this new land would bear his name. Again, limits were set with the cartographic skills and techniques of the time. That vast piece of land would later become valuable in the history of Canada (see Chap. 1).

20.4 British Conquest and the Royal Proclamation of 1763 (First British Constitution of the Province of Québec)

Ranging from 1755 to 1763, the Seven Years' War proved to be difficult for France. The capture of Louisbourg in 1758, the defeat of General Montcalm at the Plains of Abraham in 1759, and the capitulation of Montreal in 1760 almost marked the end of France in North America. Almost, because France regained Saint-Pierre-et-Miquelon, despite losing it temporarily between 1793 and 1802. After the British conquest of New France, the Treaty of Paris of 1763 officialized the ruling of French possessions by Britain. The large colony that was New France had become too much to oversee and a financial drain for the French Crown.

The Royal Proclamation delineated new boundaries in North America. Part of New France, populated mainly by settlers of French origin, became the Province of Quebec. Its territory was restricted to the St. Lawrence Valley. The former New France was reduced to the St. Lawrence Valley, whose eastern limit was set to Saint-Jean River near Havre-Saint-Pierre and became a new British colony. Western boundaries were drawn west of the thirteen British colonies despite their request to extend them across the Appalachian Range. Territories west of the thirteen colonies boundaries to the Mississippi River and north to Rupert's Land and the Province of Québec would be left to Indigenous peoples (designated "Indian Territories") considered as hunting grounds where no settlement was permitted without approval from the "Indian Department" and controlled by the British Military (Crown-Indigenous Relations 2011). The British colony of Nova Scotia expanded to include New Brunswick, Prince Edward Island, and Cape Breton. Newfoundland included the continental area east of Saint-Jean River, Anticosti, and Iles-de-la-Madeleine (Magdalen Islands).

After the fall of the French regime, French settlers became British subjects under British civil and criminal law. Nevertheless, the English population was comparably very small, approximately 1,000 or 1.3% of the total population (Dickinson 2007) of around 70,000 (Couture 2007), consisting mostly of military personnel and a handful of merchants residing mostly in Québec City and Montreal.

20.4.1 Québec Act 1774: Defining Québec Distinctiveness

The 1774 Québec Act was the United Kingdom's response to the demands of the Canadian population and British merchants. The Crown revoked the assimilative policies introduced by the Royal Proclamation of 1763. As a way to gain loyalty from the still majority French-Canadian population, their property rights and traditional customs were restored. Hence, they could freely speak their language and practice their religion. French civil law was reinstated in Québec (although not in the former Acadia), and seigneurial tenure of land maintained. Although several members of the anglophone minority had hoped for a true British colony, French ancestry remained and would eventually have geographical ramifications in the making of Canada.

Most critically, the northwest part of the "Indian Territories" from the junction of the Ohio and Mississippi rivers was ceded to the Province of Québec. The considerably enlarged Québec territory included the Great Lakes region that had a rich fur trade potential, the fishery-rich Labrador, Anticosti, and Magdalen Islands, while substantially reducing the territory reserved for Indigenous peoples. Québec's northern frontier was extended to the James Bay watershed line south of Mistassini Lake, and included Labrador.

Several irritating decisions from the British parliament since 1763 led to increased dissatisfaction among the Thirteen Colonies along the Atlantic seaboard. For instance, taxes imposed on stamps, sugar, and tea to pay for war debts angered colonists. Furthermore, although their population was increasing rapidly, their demands to extend their territories across the Appalachian Mountains were left unanswered. Furthermore, instead of extending the limits of the colonies beyond the mountains, the British Crown assigned the territory to Indigenous nations. Similarly, the policy changes between the British Conquest of New France and the Québec Act were viewed as too authoritarian towards the British minority as well as too lenient towards the French population. The British Parliament reacted by passing the oppressive Coercive Acts of 1774, referred to by the colonies as the Intolerable Acts. This succession of events resulted in the Declaration of Independence in 1776 and the subsequent war, ending in victory for the United States of America as recognized by the Treaty of Paris signed in 1783.

20.4.2 1783: Treaty of Paris Consequences

Meanwhile, everything was relatively calm in the Province of Québec... until after the USA Revolution. United Empire Loyalist settlers fled the United States to find sanctuary and security in the northern colony of Québec where they hoped they could live like British citizens. However, the Loyalists were no longer British but were also not US citizens. They felt they did not "fit in" in either the United States, the reason they fled, and in the French world of the Province of Québec.

Under the treaty, Britain surrendered regions south of the Great Lakes and the "Indian territories" to the USA, reducing the total area of the Province of Québec and pushing Indigenous populations further north and west. Because of the loss of these valuable fur trapping grounds, Montreal's merchants founded the Northwest Company that would compete with the Hudson's Bay Company and exploit the western part of Rupert's Land and beyond in regions as far as the Pacific Coast.

Following the 1783 Treaty of Paris, 100,000 loyalists escaped the United States. Approximately half of them did retreat to the British Isles. The other half migrated to North American British colonies (see Chap. 11). Nova Scotia welcomed 35,000 loyalists and, for doing so, London carved New Brunswick and Cape Breton colonies out of Nova Scotia in 1784 where most of them were directed to. Incidentally, the border between Nova Scotia and New Brunswick is the Missaguash River, almost exactly halfway between Fort Lawrence and Fort Beausejour (Fort Cumberland after 1755). Roughly, 10,000 migrated to the "Province of Québec" on the north shore of Baie des Chaleurs in Gaspésie, along the Richelieu River leading to the Eastern Townships, along the St. Lawrence River west of Montreal and the established seigneuries, and on the shores of Lakes Ontario and Erie, setting them geographically apart from settlers from French descent (see Chap. 11). To accommodate their demands, the British parliament worked on a new constitution.

20.4.3 Constitutional Act 1791: A Balancing Act

By the late seventeenth century, the Province of Québec's population along the St. Lawrence Valley had grown to approximately 160,000. The population west of Montreal was comparatively much lower, around 15,000 (Coats 1923). However, most importantly, of these 160,000 people, 90% were French-speaking and Catholic while 90% of the 15,000, most of them Loyalists, were English speaking and Protestant living in a British territory under French customs. The British Parliament became sympathetic to their demands for a new constitution. Consequently, a new constitution was created in 1791, trying to satisfy the two distinct realities.

Fundamentally, the Constitutional Act of 1791 set apart the French Catholic population from the English Protestant majority by dividing the Province of Québec into the (French) Lower Canada and the (English and loyalist) Upper Canada proclaimed by some as "the child of the American Revolution" (Wilson n.d.) with each having its parliament. The dividing line was established at the westernmost seigneurie, further west of Ottawa River and St-Lawrence River junction, the southern border along 45°N and between St-Lawrence and Atlantic watersheds. Under this act, Lower Canada would keep what was obtained in the Québec Act of 1774. British laws would be reestablished in Upper Canada. Seigneurial system would remain along with township cadastral system in Lower Canada while the latter would prevail in Upper Canada.

20.4.4 The Act of Union 1840—Takes Effect in 1841

Population increased considerably between 1791 and 1840. In Lower Canada, natural increase was the main reason for growth from 160,000 to 600,000. This rate of growth was unsustainable for the already settled area along the St. Lawrence River and created overpopulation conditions. Therefore, new regions in Gatineau, Las Saint-Jean and Eastern Townships were opened for settlement. Some people migrated to cities or emigrated to the United States in search of a better life. Upper Canada population grew to 456,000 in 1840 and to 1,396,000 in 1860–1861 (Census of Canada, Statistics Canada) with immigration from the British Isles as the main factor. Most of the immigrants went to Upper Canada.

However, unsatisfied with the way London treated Lower Canada, rebellion movements organized in 1837–1838 which, backed by the Durham Report of 1839, led the British Parliament to pass the Act of Union to reunify Upper and Lower Canada (and to become Canada West and Canada East, respectively) into the Province of Canada (or Unified Canada). A goal was to reduce Lower Canada's political power.

20.5 The British North America Act of 1867—The Birth of Modern Canada

In the meantime, industry timidly developed in Canada. After the elimination of protectionist tariffs with the United Kingdom, Canada signed in 1854 a reciprocity treaty with the United States that lasted until 1866. Canada then needs a new market for its products.

Politicians opted to bet on a domestic market with North American British colonies. A need for a union of British colonies was considered mandatory, triggering the proposal of the Canadian Confederation by members of "The Great Coalition", the Canadian politicians George Brown, George-Étienne Cartier and John A. MacDonald in 1864. The British North America Act was presented and adopted by the British Parliament in 1867. Canada then became a British Dominion which translates into more autonomy, for instance with the ability to write (for the first time) its own constitution for approval by London, while remaining a member of the British Empire (see Chap. 1 for further discussion). The next step would be to incorporate colonies into the union, populate the land, and build a domestic market large enough to commercialize Canada's resources and products.

The Province of Canada was divided again into Ontario and Québec this time. However, the northeastern coast (with no clearer definition) of Québec remained part of the British colony of Newfoundland. The colonies of Nova Scotia and New Brunswick agreed to be part of the new confederation, while PEI and Newfoundland declined. Located between Ontario and Québec, Ottawa is established as the federal capital.

The governing structure of Canada would be a federation which divides powers between the central (federal) government and its constituents, the provinces which are proportionally represented in the government. The central government takes care of matters that cater to Canada as a whole and what is not covered by provinces while provinces manage affairs that are more specific to their region and population such as education, healthcare, and resources (including energy). However, some jurisdictions are shared by both levels of government, including agriculture, economic development, and immigration. Québec retained the "distinctiveness" it wished and obtained under the Québec Act of 1774. The division of responsibilities between the federal and provincial governments was contentious in 1866–67 and remains a source of controversy and political activity. Please see Chap. 1 for a detailed discussion, Chaps. 15, 16, and 17 for discussion of impacts on resource sectors, and this Chapter for additional commentary.

20.6 Other Parts of British North America in the 1800s

Before exploring the territorial evolution of Canada from 1867, a look at how several borders were settled during the nineteenth century involving the United Kingdom, the United States, and Russia is needed. As a reminder, Québec (Upper and Lower Canada), Nova Scotia, New Brunswick, and Newfoundland were still settlements under British colonial governments in 1800.

The early nineteenth century was marked by the Anglo-French (or Napoleonic) Wars and the Continental blockade imposed by Napoleon on the United Kingdom, who retaliated by preventing the United States from trade with France with blockades of British ships along the coast. Infuriated by this, the United States, still in a fighting spirit and strong from having obtained independence struck back by targeting British North American positions in Canada between 1812 and 1814 (British–Indigenous–USA War). However, the USA armies underestimated the opposition, and these campaigns all failed (see Chap. 1 for more details). A string of military forts built over the years by France and Britain to resist potential territorial aggressions (Henry, Ingall, George, Erie, York, Niagara, Chambly, and Carillon/Ticonderoga (the last two built by France and lost to Britain), among others, are silent remnants of the strategic significance of the border between the United States and Canada.

On the Atlantic Coast, the "Labrador Strip (Coast of Labrador)" (loosely and imprecisely defined) was detached from Lower Canada and assigned to Newfoundland Colony in 1809.

20.6.1 New Brunswick-USA Border

While the western part of the continent was being divided, the 1842 Webster-Ashburton Treaty (Yale Law School 2008), named after United States Secretary of State Daniel Webster and British diplomat Alexander Baring, 1st Baron Ashburton, aimed at correcting imprecisions issued from previous boundary definitions in the Treaty of Paris (Article 2d; US National Archives 2022) and to: 1. settle the dispute and define the boundary between the state of Maine and the colony of New Brunswick in the Aroostook region; 2. settle the border between the Lake Superior and Lake of the Woods; 3. confirm the 49th parallel as the boundary from the Lake of the Woods to the Rocky Mountains. Additionally, the Treaty would confirm the shared use of the Great Lakes. Their negotiation sessions would be spread between 1839 and 1842, when it was agreed by both parties.

Canada and New Brunswick claimed control of the Aroostook Valley region. Both Québec and New Brunswick offered concessions to settlers. In 1830, Great Britain intervened and established the frontier at the Madawaska River, leaving the western part of the coveted territory to Quebec.

The meaning of Article 2 of the 1783 Treaty of Paris was interpreted differently by the two countries. As detailed as the 2.0 m by 1.4 m Mitchell map (1755) that was used might have been, it contained significant inaccuracies. The Treaty of Paris defined the Maine-New Brunswick boundary "… from the Northwest Angle of Nova Scotia, viz. that Angle which is formed by a Line drawn due North from the Source of St. Croix River to the Highlands; along the said Highlands which divide those Rivers that empty themselves into the river St. Lawrence, from those which fall into the Atlantic Ocean, to the northwesternmost Head of Connecticut River;…" and further in the same article: " East, by a Line to be drawn along the Middle of the river Saint Croix, from its Mouth in the Bay of Fundy to its Source, and from its Source directly North to the aforesaid Highlands, which divide the Rivers that fall into the Atlantic Ocean from those which fall into the river Saint Lawrence;…" (Article 2d; US National Archives 2022), surprisingly with no map attached to the treaty (Osher Map Library 2022).

Consequently, until 1839, both British and USA authorities laid claim to a vast territory comprising the valleys of the Saint John, Madawaska, Aroostook, and Sainte-Croix rivers. Both claims corresponded roughly to the delimitation of the Saint John River watershed. However, the British claim considered the southernmost limit of the watershed (towards the Aroostook River close to the border), while the USA claim interpreted the northern limit of the watershed to be the border, aggressively extended further northeast in 1792. Important to the British was the necessity to keep a military road between Québec and Halifax, which the USA claim made difficult.

Both sides acted upon their comprehension of the treaty. American lumbermen received concessions in the region. Maine even started to run a census in what appeared as a frontier advancing front. British subjects had settled in almost all of eastern Maine during the 1812 war, which London intended to annex. Both parties accused the other of trespassing and tension escalated to the point that militia on both sides were mobilized in the Aroostook War (Lang 2001). King William of the Netherlands was named as arbitrator to the disagreement and proposed a boundary as a compromise which followed the Saint John and Saint-Francis rivers. After a rejection by both parties of the solution proposed, Washington and London agreed to refer the dispute to a joint boundary commission. Agreement was concluded in 1839 and the Webster-Ashburton Treaty was ratified in 1842, awarding Canada 2331 km^2 more than the King of the Netherlands had suggested and viewed as a "pure diplomatic compromise" (Poole 1964).

20.6.2 Northwestern Ontario—USA

The border between the United States and Canada between Lake Superior and Lake of the Woods was also based on Mitchell's 1755 map (and its inaccuracies). Geographic details were limited, especially in the sparsely populated regions in the mid-eighteenth century. Wording of the 1783 Treaty of Paris read "through Lake Superior Northward of the Isles Royale & Phelipeaux to the Long Lake," then along "the Water Communication between it & the Lake of the Woods … to the most Northwestern Point thereof, and from thence on a due West Course to the River Mississippi" (US National Archives 2022).

Interpretations of what the treaty intended were difficult at times, but common ground was reached. The intent of the treaty was to establish the boundary in the middle of Lake Superior up to Isle Royale, then "to the Long Lake", corresponding to, it was agreed, Pigeon River, through the river and a complex series of lakes (in a region of many lakes) on which were defined mid-water lines, along Rainy River to Lake of the Woods. However, later surveys proved the shape of the lake was different, and the boundary is skewed north in favour of the United States.

Nonetheless, what is more peculiar about the boundary between Lake Superior and Lake of the Woods is the location of the northwesternmost point and the oddity it created. Lake of the Woods shape is also inaccurately depicted on Mitchell's (1755) map. The "Northwesternmost point" cited in the 1783 Treaty of Paris proved to be the trickiest one to find.

The Mississippi River does not extend sufficiently far north as to be "due west" of Lake of the Woods". Rather, it takes its source at Lake Itasca, almost due south of the Lake of the Woods, and not present on the Mitchell map. David Thompson, former Hudson's Bay Company explorer and later British surveyor in chief, identified Lake Bemidji as the northernmost source of the Mississippi in 1797, also south of Lake of the Woods. An 1824 map by Thompson identified four possibilities for the point. A survey by Tiarks in 1825 first identified a different shape of Lake of the Woods, and of two possibilities, he identified "the northwesternmost" point on Angle Inlet (Lass 1980). Reaffirmed by the Webster-Ashburton Treaty, this was later documented to be at 49.39°N 95.153°W, creating a geographical oddity (see below). Once the northwesternmost point was established and agreed upon, south to 49°N and west to the Rocky Mountains was straightforward enough to be reaffirmed by the treaty.

20.6.3 Western North America

Further inland, fur trade in the west by the Hudson's Bay Company and later, the Northwest Company, had kept the United States from expanding northward, in the spirit of the Manifest Destiny (as termed in the 1840s). The Convention of 1818 Treaty addressed issues arising from the 1783 Treaty of Paris and firmly established the boundary between British North America and the USA at 49°N, for the sake of simplicity, running from Lake of the Woods to the then Stony (now Rocky) Mountain Divide (U.S. Department of State 2009), roughly separating the watersheds of Hudson Bay and Gulf of Mexico. Both parties ceded territory, with Britain losing more including part of the Red River settlement extending south into present-day northwestern Minnesota and northeastern North Dakota. The parties also agreed to share the administration of the land further west, which the USA called the Oregon Country and the British the Columbia District, for the next ten years (Nicholson 2006). The agreement seemed at the time an amicable one. However, this turned into a severe dispute later, the Oregon Border dispute of 1846.

The Columbia District or Oregon Country was disputed territory over time. Previously, Russian fur traders, hunters, and trappers had made their way along the Pacific coast from Alaska to as far south as Fort Ross (phonetically derived from Rus), 150 kms north of San Francisco. Spain also explored the Pacific Northwest, leaving names such as Cordova, Valdés AK, Haro and Juan De Fuca Straits, Cortés, San Juan, and Lopez Islands, Spanish Banks (see *Richthofen* for extended list). There had been waves of American, British, and French-Canadian settlers who were used as pawns to eventually revendicate parts of the area. Britain's claims were based on exploration and fur trade and the presence of loyal Canadian farmers at Fort Vancouver (present-day Vancouver WA) on the Columbia River. The USA based its claim on the presence of American settlers, albeit later arrivals than British and French-Canadian traders through the Hudson's Bay Company long-time monopoly over the area.

Claims from both sides ranged from 42°N (present-day Oregon southern border) by the British, based on the 1803 Nootka Treaty between Spain and Britain (Stein 2008), to 54°40' °N (Jacobs 2011) by USA, the southern boundary of Alaska under the third article of the USA-Russian Treaty of 1825 (Explore North n.d.). Russia had initially claimed land north of 55°N, adjusted to 54°40' N to include the totality of Prince of Wales Island.

The Manifest Destiny ideology was by then firmly rooted in the USA belief that transcended into a territorial quest across the continent stemming from the immense Louisiana Purchase (1803). James K. Polk, campaigning for the United States presidency at the 1844 election, ran an insistent expansionist platform mainly targeting Texas, California, and Oregon. Consequently, the "54–40 or fight" slogan emerged. Polk won the election but became occupied with the Mexico-USA War started in 1846 over the Texas annexation. Fighting two wars would have been difficult. Somehow, his position softened, and Polk agreed on a compromise to extend the existing 49°N border to the Strait of Georgia in 1846. Vancouver Island would remain a British Territory. However, Point Roberts remains to this day, as does the Northwest Angle (in Minnesota), one of the few borderland oddities between Canada and the United States created by the 49°N boundary.

While most European explorers sailed west, Russians travelled inland on a challenging eastward venture, then crossed the Bering Strait as early as mid-eighteenth century to establish the colony of Russian America. The Russian-American Company was created in 1799 (Roth 2023) in search of a fertile ground for fur bearing animals as Siberia had been overhunted. Settlements were established in Alaska and along the Pacific Coast to hunt seals and sea otters.

Britain and Russia signed the St. Petersburg Treaty in 1825 establishing the boundary between Alaska, then a Russian possession, and the British claim at 141°W. Alaska proved to be a difficult and costly venture needing complicated organization, combined with the severe depletion of the fur resources. Consequently, Russia sold it to the United States in 1868 for $7.2 million USD ($153 million in 2024 equivalent). From then, the future British Columbia shared a western and a southern boundary with the United States. But, in the spirit of the Manifest Destiny, there was an American reaction asking for more and wanted to annex the Columbia District and therefore have a continuous access to the northwest Pacific Coast. It was thought among expansionist advocates that annexation of British Columbia would be the natural next step (Shi 1978).

The ambition intensified when gold was found in the Cariboo region in 1858. A flood of people followed, with many Americans among them. London reacted immediately by creating the Crown colony of British Columbia and designating Fort Langley, on the Fraser River, as its capital. Nonetheless, north–south economic and social interactions across the borders were more "natural," even more so after the gold frenzy faded. Vancouver Island remained a separate colony until 1866 when it was annexed to the colony of British Columbia (McGillivray 2005). In the meantime, discussions led to the creation of the Canadian Confederation in 1867 with Ontario, Québec, New Brunswick, and Nova Scotia as the four initial partners.

Instrumental in the purchase of Alaska, US Secretary of State William H. Seward was particularly active in promoting the annexation of British Columbia (Shi 1978). Amid discontent and feeling of being neglected, growing support (truly or not) for annexation was reported, notably stirred by American agitators. After all, the colony of British Columbia was very isolated from the other continental British colonies, and even more from London. Washington wanted the incorporation of British Columbia in retribution for England supplying warships to the South and causing damages during the Civil War, known as the *Alabama* (ship) Claim. The claim was lodged in 1869, and Britain paid $15.5 million ($315 million 2024 eq.) compensation in 1872 (Office of the Historian n.d.).

Both the United States and Canada alongside London courted British Columbia to be part of their federation. British Columbia agreed to join Confederation under the condition of being connected to eastern Canada by railroad. The United States, similarly, tried to seek entrepreneurs to build a Northern Pacific Railway line and capital to buy Rupert's Land from the Hudson's Bay Company. Although the Hudson's Bay Company governors estimated their possession at $40 million value, they transferred it for $1.5 million to the British Government who then ceded it to Canada in 1869 (McIntosh and Smith 2006). The Rupert's Land acquisition, including the North-Western Territory that the company was administering, by Canada helped seal the process and the legislative council of the British Columbia colony finally voted to adhere to the Canadian Confederation in 1870, effective in 1871. The eastern boundary was set to run from the intersection of 49°N along the Great Divide to 120°W. The northern border was 60°N, which still defines the divide between provinces and territories, as well as north and south in contemporary Canada.

However, the Alaska panhandle boundary dispute was finally settled, with award of the territory to USA in 1903. Washington had declined Ottawa's proposal to determine the exact location of the border between Alaska and British Columbia. The discovery of gold in 1896 in Yukon rushed the need for a ruling. Canada wanted access to the Pacific which was refused by an international tribunal in which a British delegate "representing" Canada, Lord Alverstone, Lord Chief Justice of England, infuriated Canada by supporting the USA claim (Farr and Block 2006).

20.6.4 Canada-USA Boundary from Northeast New Brunswick to Lake Ontario

The easternmost part of the Canada-USA Boundary was included in the New Brunswick-Maine settlement (1842). The remaining segment of the border with New Hampshire, Vermont, and New York, covered in Article 2 of the 1783 Treaty of Paris, has also been finetuned by the Webster-Ashburton Treaty. Indian Stream NH and Fort Blunder NY, an American fortification wrongly built north of the border, hence the name, were issues that were rapidly settled.

This part of the border is an example of "surprising technicity" (Sanguin 1974) that has been set using the three types of international boundaries, subsequent, geometric, and natural boundaries. The purpose here is to show the complex application of these diverse types.

The Québec–Maine boundary westward from New Brunswick follows the median line of (natural) LeBeau Lake, then 44 km along St-Francis River (or Kelly Rapids) with a "complicated" median lane because of abandoned meanders to Estcourt, south of Pohénégamook. A 103-km-long geometric straight line from Estcourt to Lac Frontière, shifting southward for 32 km in a (geometric) straight line until it meets Saint John River; then 50 km along the (geometric) median lane of the Saint John River Southwest Branch to Little Saint John Lake. From Little Saint John Lake to the New Hampshire border, it follows exactly the (natural) watershed between the Atlantic Coast and St. Lawrence slopes for 190 km and extends for a further 68 km along the Quebec-New Hampshire border along Halls Stream, where it meets 45°N at Beecher Falls VT. The boundary follows 45°N until it meets the St. Lawrence River at Akwesasne. From there, to Lake Superior, the boundaries are rivers (St. Lawrence, Niagara, Détroit, Saint Clair) and lakes (Ontario, Erie, St. Clair, Huron, and Superior) through which lines are cut in the middle of the lakes or channels.

Before Confederation and benefitting from Imperial tariffs, the Atlantic colonies were still British colonies having commercial ties with Great Britain, West Indies, and New England under the reciprocity treaty from 1854 to 1866. They had developed a wind, wood (especially in NB), and water economy, supplying timber, wooden ship building, and sea export and were not originally interested in joining Confederation originally. Atlantic Canada was enjoying prosperity, then known as Golden Age of Sail. Nevertheless, New Brunswick and Nova Scotia entered the Confederation in 1867, and Prince Edward Island in 1873. Still maintaining strong ties with the United Kingdom and relying on fisheries, Newfoundland kept its back turned on Canada for even longer (see Chap. 1).

20.7 Internal Boundaries 1867–2001

The progressive changes to internal boundaries in Canada are summarized in Chap. 1 and Table 1.11 (also see Atlas of Canada 2017). This section highlights some particular points.

The British North America Act (1867) officialized the creation of the Dominion of Canada and its governing structure. Powers were split between two levels of government. The federal government would take care of matters that are of national and international concern, while the provincial governments would have jurisdiction over regional specificities such as property, contracts and torts, local work, undertakings, and businesses (see Chap. 1). Some areas would be shared by both levels and, dear to Québec, it would retain its Civil Law. Ottawa was established as the Federal capital.

Although having been the site of the first round of negotiation in 1864 for the creation of Canada in 1867, Prince Edward Island, the "Birthplace of Confederation" only joined the federation in 1873. As was the case with British Columbia, the United States courted the Island to join the union (Canadian Archives 2020). The colony's legislature had hoped that the construction of a "trans-island" railway would bolster its economy, but soon realized the cost was beyond its capacity to pay (see Chap. 14). However, Canada had a new round of discussions with the Island and offered terms that would entice its incorporation. Canada would assume the British pressures and Prince Edward Island's railway debt and facilitate transfer of land from absentee owners.

When British Columbia became a British colony, the contour lines of the country were set. British Columbia would join the four initial provinces in Confederation under the condition of being connected by rail. British Columbia had continued growth, based especially on its mining, forest, agriculture, and fisheries. After entry to Confederation in 1871, these catered mostly to its domestic market until the promised railway connection was completed in 1885. From then, resources could be exported eastward and growth accelerated.

Canada acquired an immense swath of land between Ontario and BC, a land bank. Rupert's Land and the North-Western Territory were to be transferred to Canada in 1869, but the arrangement only became effective in 1870, slowed by the Métis protest under Louis Riel to resist Canada's takeover of their land in the Red River Valley. The area would become the Northwest Territories, an immense land reserve from Alaska to Northern Québec, later subdivided into new and enlarged provinces entering the Canadian federation. The partition of the Northwest Territories started almost as soon as Rupert's Land was transferred to Canada, with the creation of a very small Manitoba ("postage stamp") in 1870. Canada viewed the southern part of the Northwest Territories as having enormous potential for agricultural land for expansion and settlement that would result from the completion of the Canadian Pacific Railway. The Atlantic coast northeast of the 1867 extent of Québec was still part of the British colony of Newfoundland, albeit very vaguely defined territorially. At that time, only the fishery resources of the Atlantic Coast were recognized as important.

Although a series of eight Indigenous-Crown Treaties were negotiated covering the land area of the Northwest Territories (Crown-Indigenous Relations 2011), the subsequent subdivisions into Districts and Provinces did not follow the treaty boundaries or respect the pre-existing Indigenous territories. Traditional Indigenous territories focused on natural features, particularly rivers and areas preferentially occupied by *Bison*. None of the internal or the Canada-USA geometric boundaries established between 1818 and 1912 bore any relation to the Indigenous populations and traditional territories.

Sequentially, in 1874, Ontario expanded northward to 51°N, with its western border extending north from "Hunter's Island" in Quetico Provincial Park. The District of Keewatin was separated from the Northwest Territories in 1876, to facilitate subsequent expansion of Ontario and Manitoba. The Arctic Archipelago was transferred to Canada by Great Britain in 1880, and assigned to the NWT. In 1881, Manitoba, now accessible by train and with 66,000 settlers (Association of Manitoba Land Surveyors 2023), was expanded to the North, West, and East, reaching five times its original size. Winnipeg rapidly became a boom town, a gateway to the west. Townships filled, towns and villages sprouted. The eastern border was disputed by Ontario, but the boundary dispute would be resolved in 1889 when Ontario was extended to Lake of the Woods (95°50' W) and north to the Albany River (Government of Canada 1889).

In 1882, four "provisional districts" were established in the Prairie regions of NWT: Saskatchewan (central part of modern Saskatchewan, expanded in 1886 eastward into the district of Keewatin to follow the Nelson River, Cross Lake, Rosa Lake, and Lake Winnipeg); Assiniboia (southern part of modern Saskatchewan); Alberta (southern and central parts of modern Alberta); and Athabasca (northern parts of modern Alberta and Saskatchewan, with the Athabasca River as its northeastern limit, along with northwestern Manitoba). The CPR traversed the Districts of Assiniboia and Alberta. The northern part of Alberta and Saskatchewan Districts lay along the projected railway routes later utilized by the Canadian Northern and Grand Trunk Pacific Railways (see Chap. 14). Riel led a second resistance (rebellion) further west, in Batoche SK (Saskatchewan District), until he was defeated in 1885. Athabasca Districts, extending from 55°N to 60°N, lay beyond the region of nineteenth– early twentieth-century railway construction.

In 1895, the District of Keewatin expanded eastward to James Bay. The remaining Northwest Territories were divided to create four new districts: Yukon (westernmost part); Mackenzie (mainland between Yukon and Keewatin; Franklin (Arctic Archipelago); and Ungava (modern northern Québec, Nunavik, with its northern limit to be extended to Eastmain River in 1898). The District of Athabasca was enlarged to the western limit of Keewatin at 100°W. In 1897, the mainland Boothia and Melville peninsulas were transferred to the District of Franklin, while Southampton Island was re-assigned to the District of Keewatin.

Yukon became a separate territory in 1898, with a series of border adjustments with NWT until 1901. Its eastern border consists, for the most part, of the drainage divide between the Yukon and Mackenzie River drainage basins in the Mackenzie Mountains. Most of Yukon Territory is defined by the Yukon River watershed east of the Alaskan border.

In 1905, the provinces of Alberta (Alberta, western Athabasca, and westernmost Saskatchewan Districts) and Saskatchewan (Assiniboia and most of the eastern Athabasca and Saskatchewan Districts) were established, separated by 110°W and extending from 49°N to 60°N. The Manitoba-Saskatchewan border was approximately 102°W, with jogs resulting from the survey correction lines (see below). In 1912, Manitoba absorbed the southwestern part of Keewatin District, to 60°N; Ontario absorbed the southeastern part of Keewatin District, extending to Hudson Bay; and Québec absorbed Ungava District. The remanent Keewatin District north of 60°N and east of 102°W (from 1920), Mackenzie District, and the Arctic Archipelago in Franklin District, remained with the Northwest Territories until 1999.

20.7.1 Nunavut

Nunavut is the Inuktitut word for "our land". Its population is over 80% Inuit. The creation of Nunavut in 1999 was the result of the Canada-Inuit land claim initiative and better governance options started in the 1960s and 70 s with the ultimate goal of obtaining a separate territory. The Nunavut Land Claims Agreement was signed in 1992 and submitted to a referendum, accepted by 84.7% of the voting population. Including the former Northwest Territories districts of Keewatin, Franklin, and the northeastern part of the Mackenzie District, Nunavut became Canada's third territory in 1999. The mainland Nunavut-NWT boundary, although geometric, lies between ethnographic (Inuit-Inuktitut and Dene) and natural (tundra-taiga) communities.

The only territory still not accessible by ground transportation, only boats or planes allow access to Nunavut. Furthermore, communication remains difficult for such a large territory of over 2 million km^2 and a population of 36,858 (Statistics Canada 2021) spread in 25 communities in a challenging environment. Consequently, its decentralized governing structure is conceived accordingly by having Nunavut divided in three administrative regions: Kivalliq (mainland west of Hudson Bay, much of the former Keewatin, along with Shugliaq (Southampton Island); regional centre at Kangiqtiniq (Rankin Inlet); Kitikmeot (southwestern part, including terrain flanking Coronation Gulf, eastern Kitlineq (Victoria Island), and Kingngailap Nunanga (Boothia Peninsula), regional centre at Iqaluktuuttiaq (Cambridge Bay); and Qikiqtaaluk (northern and eastern Arctic Archipelago, *Sanirajak* (Melville Peninsula), and islands in southern Hudson and James Bays, regional centre at Iqaluit). Iqaluit is the overall capital, along with the 2 other main regional and 8 other centres, assumes administrative functions (Wyman 2011).

20.7.2 Newfoundland and Labrador

In 1912, the District of Ungava became exclusively part of Québec, doubling its size. However, the coastal islands, difficult to define properly, were kept under Ottawa's control for geostrategic reasons. The geographic definition of Labrador, part of the Colony of Newfoundland, was still vaguely termed as "coast" without any precise inland boundary. As both Canada and Newfoundland had diverging views on the question, the Labrador territory dispute was referred to the United Kingdom Privy Council to settle the continuing dispute between Québec and Newfoundland (Heritage Newfoundland and Labrador 2013). The "strip along the coast," never precisely defined before, was defined as all the terrain draining into the Atlantic Ocean, essentially making the boundary the height of land between the open Atlantic and Hudson Bay watersheds. The 1927 judgement, final and without right to appeal, was in favour of Newfoundland. The whole Labrador Sea watershed (and more) was assigned to Newfoundland, as well as the area between the watershed and 52°N. The decision granted Newfoundland an area of almost 300,000 km^2, rich in mineral resources and hydropower potential (see Chap. 8).

L'actualité (2022) argues that Canada defended its case very poorly. Québec never officially recognized the ruling, citing the boundary as "non définitif" on many maps. After 1927, Newfoundland made two attempts to sell Labrador to Canada (see Chap. 1). Canada's objections became moot after 1949.

The coast of Newfoundland was the first region of Canada to be visited by Europeans (beginning with Leif Eirikson ca. 1015). However, it was the last province to enter the Canadian Confederation in 1949. Newfoundland had spurned previous Canadian offers in 1864–68 and 1892, but the financial and political collapse of 1934 resulted in loss of independent political status and the UK colonial Commission of Government (see Chap. 1). In the aftermath of World War II, the UK indicated that Newfoundland could not remain as a British colony: the options were independence, amalgamation with the USA, or joining Canada. Following two close referendums in 1948, Newfoundland entered the Canadian federation. Notably, the Avalon Peninsula that contains St. John's voted against Confederation in both 1948 referendums, supporting Confederation by only 25% and 34% of the votes in the first and second referenda, respectively (Hiller 1997).

20.8 Patterns on the Land: Colonial Cultural Footprints

People that have settled the land throughout Canada since ca. 1620 have left persistent traces of their presence by the different land managements they used, which have left a varied mosaic of patterns on the land. They left clear, sometimes imbricated or superimposed, imprints on the ways pieces of land were divided that are still easily noticeable. The French seigneurial system left imprints of long narrow lots. The British left traces of townships. After 1867, Canada surveyed the remaining land using the township model used by the American Homestead Act to execute the Dominion Land Survey.

20.8.1 French Seigneurial System

The seigneurial system has had a profound structural and still visible influence on the geographic pattern of land occupied by French settlers throughout the continent (Fig. 20.3). Everywhere where there were French settlements this long narrow lot type of land tenure was used and still perceived (e.g. Red River in Manitoba, Acadia, Detroit, New Orleans). Traces of the seigneurial patterns are still noticeable even in urban settings through the street pattern orientation.

The seigneurial system was a social structure that was spatially transposed to organize, establish settlement, and develop the colonial economy. It was intended to ensure, above all, the settlement and agricultural colonization in New France. Seigneuries were large pieces of land, divided in "cen-

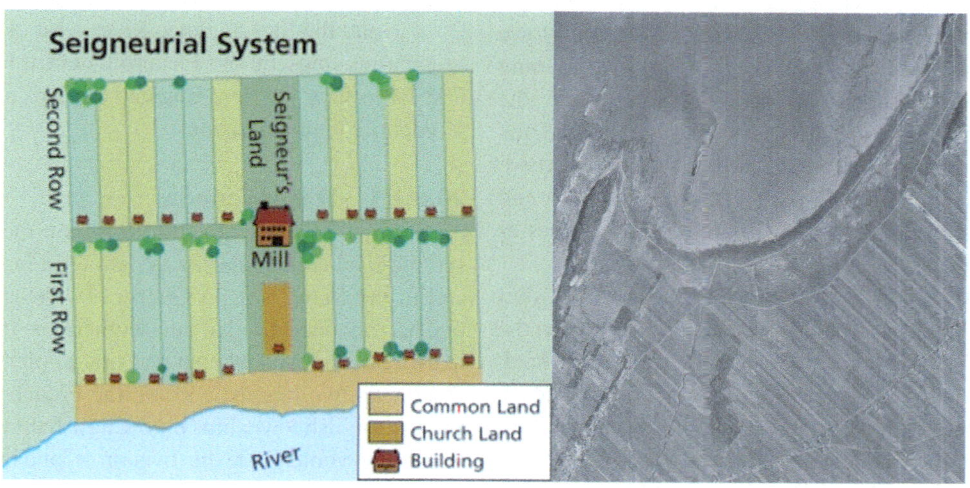

Fig. 20.3 Seigneurial survey system. Left: Diagram showing alignment of lots at 90° to main river transportation route. Additional ranges were gradually developed inland over time. Lots were frequently divided through inheritance to produce greatly elongated properties, each with river access. The seigneurial system was used throughout the St. Lawrence Valley of Québec, along the Detroit River (Windsor, Ontario), and in parts of Acadia. In modified form, it was used in Métis communities along the Red River in Manitoba. Right: the seigneurial system is still evident in land use patterns in many areas of Québec, as in this example from the Kamouraska area. (National Air Photo Library)

sives" (lots), offered by the King through the governor to entrepreneurs, who would be called seigneurs, and had the task to administer, develop and have the land worked (cultivated) by their "censitaires". The King of France remained the proprietor of all New France land. Seigneurs were nobles (merchants, militaries, professionals, religious communities) who granted concessions of land in long narrow strips to censitaires. Censitaires were peasants, French immigrants, or sons from already established families in New France, who rented their piece of land, which they had to clear, to build their house and utilities on, and to maintain by cultivating and farming. Harvested grains would be crushed at the seigneurial mill. Some censitaires would engage in fur trade away from home during winter.

Seigneuries, usually rectangular pieces of land, were divided by the colonial authorities into long and narrow strips (censives) at 90° to waterways. The Saint Lawrence River was the main natural artery along which seigneuries were aligned, as well as its tributaries (e.g. Batiscan, Jacques-Cartier, Chaudière, St-Maurice, L'Assomption, Richelieu). Seigneuries were made up of a number of long lots with a seigneurial domain consisting of space reserved for a flour mill and/or a sawmill, a church and its presbytery, a cemetery, the seigneur's manor, a quay, its only source of external contact, and eventually a school. When the first row of lots were filled, a second row, called "rang" would be opened, and a second, third one, or more. A "montée" would permit to access to those "rangs", and a road (a "route de rang") would allow the circulation between rangs. For instance, there were approximately 7,000 "censives" (long lots) in 177 existing seigneuries in 1725 showing a remarkable geographical regularity (Laberge 2010, p.50).

The development of these long strips showed a uniform pattern: meadows graced the riverfront; then a road running parallel to the watercourse in front of the lots to link the land together and permitted access to neighbours and the seigneurial domain; houses would be built along the road; utilities at the back along with ploughed land; pastures; and wooded land that gradually receded with the use of trees as building material and as firewood.

The first peasants on a given piece of land would start from zero: clear the land, build a house and utilities from the wood and rocks on their land, cultivate and maintain the land, live from their land, and typically would own horses, cows, pigs, chicken, and grow vegetables. The concession would be bequeathed to the eldest son. Any other sons would have to acquire a concession from the seigneur of his or another seigneurie. With rare exceptions, daughters did not inherit land.

The seigneurial system was essentially a widespread low-density subsistence economy operating in independent "seigneurial silos". It failed to be attractive as its progress was slow and was an obstacle to economic development. It more or less maintained the peasants in a state of constant dependence on their seigneurs and prevented them from developing a market for their products and accumulating wealth. Moreover, to ensure equal development of the terroir, it was necessary to encourage a fair distribution of tenant farmers on the land to be developed, and to avoid the creation of dense settlements until the first concessions had been properly developed (Boileau 1999). Eventually, natural concentrations would develop around churches and slowly evolve into villages. Exceptionally, Jean-Talon had Jesuits' plans implemented to develop three radial villages in the Québec region. One developed as such is today known as Charlesbourg or Trait-Carré. Bourg-Talon developed in the late twentieth century, keeping the same radial lines, but bears the name Bourg-Royal.

Later, new agricultural waves of colonization on the Canadian Shield facilitated by newly completed railways (Gatineau, Laurentians, Mauricie, Matapedia, Saguenay-Lac St-Jean, then peripheral regions including Abitibi-Temiskaming, Upper Mauricie, Lower North Shore) followed the same pattern, sometimes in environments too stubborn for profitable agriculture. As a result, efforts then turned to forestry and mining or migration to cities or to the United States.

20.8.2 Loyalist Townships

After the British conquest of New France, authorities were instructed to establish a system of townships of sufficient size, freehold of land tenure, and extent in such places as may be needed, instead of maintaining the seigneurial system. A first sequence of townships was surveyed on the shores of Lake Ontario and the St. Lawrence River west of the westernmost seigneuries (present-day Eastern Ontario) to fit the needs of first large flow of New England settlers unhappy with the seigneurial tenure divisions.

20.8.2.1 Lower Canada

As opposed to the seigneurial long lot, the township was a symmetrical rectangular lot system used in Québec, Ontario, and in the Maritimes. In Québec, it was applied to regions that had not been divided into seigneuries. In fact, it did more than divide the land when it was first implemented. It divided Québec in two linguistic blocs: the French seigneurial bloc, and the English townships bloc. English thinking of order and symmetry would guide the division of land into an orderly and rigid fashion.

Following the 1791 Constitutional Act, the English administration opened the large territory between the seigneuries of the St. Lawrence Valley and the United States border for colonization to accommodate British immigrants and Loyalists who fled the American Revolution. This vast territory was divided into townships, of approximately 10 miles (16.1 km) by 10 miles, except for those located on the banks of a navigable waterway, which were 9 miles (14.5 km) wide and 12 miles (19.3 km) deep containing 11 ranges of 28 lots for a total of 308 lots each, with English toponymy (Québec 2004). Discharged officers and soldiers who fought in North America, along with settlers from the British Isles, were initially granted land and settled in the townships bordering the seigneuries. The Irish were the largest group of British immigrants, followed by the Scots and English (see Chap. 11). French-Canadian migration to the townships, largely a consequence of the overpopulation of the St. Lawrence lowlands and the declining efficiency of the seigneuries agricultural system, would start in mid-nineteenth century. Part of the region that constitutes the first block of townships on the Québec territory bears the name of Eastern Townships (Cantons-de-l'Est; Fig. 20.4). Townships are also present in the neighbouring regions of Centre-du-Québec, Chaudière-Appalaches, and Montérégie.

The township model was also applied in Abitibi (Fig. 20.5) and Lac Saint-Jean, but with the distinction of having twice as many lots within them for a comparable size (Québec 2004). It is also present in Lower Laurentians, Ottawa Valley, Gaspésie, and Lower North Shore (Côte-Nord) hinterland.

20.8.2.2 Upper Canada

The concept of township was officially applied to Upper Canada immediately following the 1791 Constitutional Act. However, Cataraqui Township, present-day Kingston, had been quickly surveyed in 1783 and opened in 1784 to provide Loyalist settlers a safe haven on land bought from the Mississauga (Ross Geiger 2001). This first Ontario township was 6 sq miles (15.5 km^2), which the new settlers were accustomed to, containing 25 lots wide and seven concessions deep for a total of 175 lots. Its baseline, and that of the neighbouring townships, was parallel to the waterfront and its lots were at right angles. Consequently, shifts of orientation according to waterfront contours would be inevitable at places, resulting in somewhat irregular shapes in subsequent rows of townships (Figs. 20.6, 20.7). Furthermore, a reorientation of baselines proved necessary at places, especially in the southern tip of Ontario.

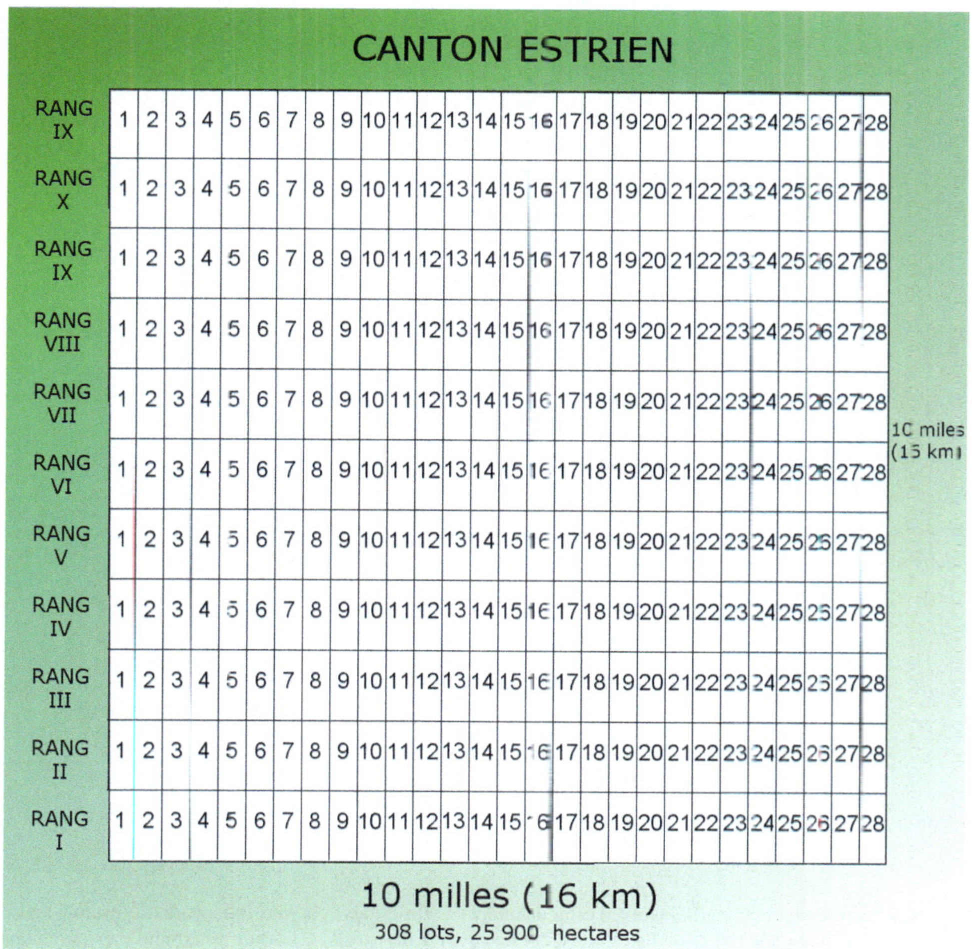

Fig. 20.4 Typical Canton estrien (Eastern township) survey, southern Québec. (Wikimedia commons, public domain)

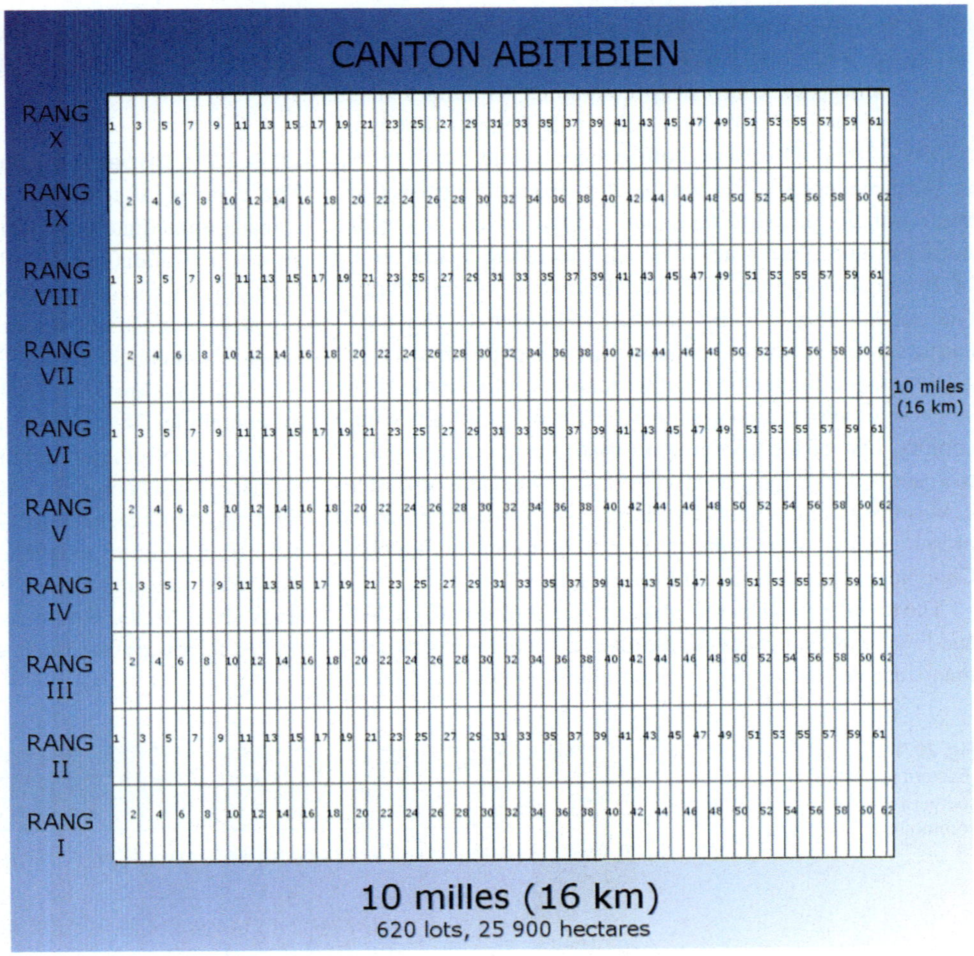

Fig. 20.5 Typical Canton abitibien (Abitibi township) survey, Québec. (Wikimedia commons, public domain)

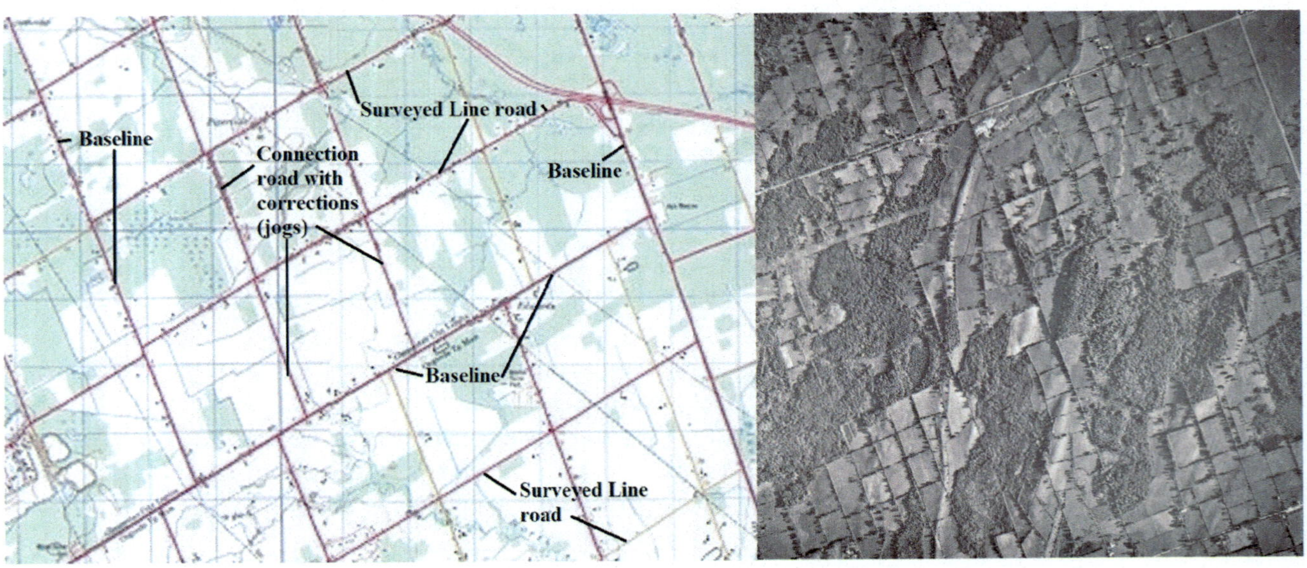

Fig. 20.6 Surveying in Ontario involved establishment of "Baselines", usually at 90° to major rivers or lakeshores, followed by Line roads surveyed at right angles to the baseline. Connection roads between the Line roads were marked by corrections (jogs) at each intersection. In contrast to the Dominion Land Survey in western Canada, Baselines were independently established along county, township, or other borders, leading to irregular patterns. Left: an example from the eastern part of Ottawa (formerly Carleton County) and adjacent Russell County. Right: baselines and connection roads, Douro area, east of Peterborough. (National Air Photo Library)

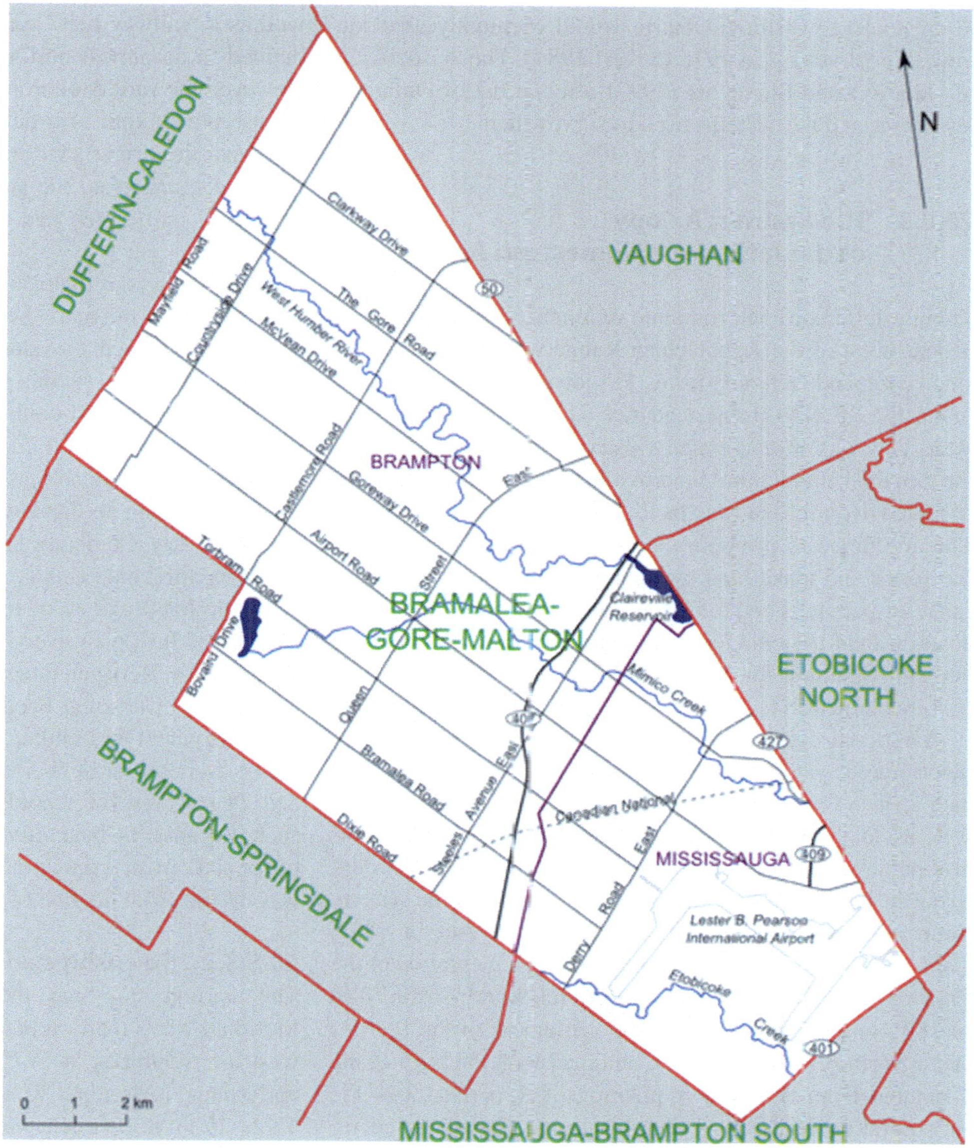

Fig. 20.7 "Gore" (also spelled "Gower") triangular patterns result from differing orientation of baselines in response to shoreline alignment. West of Toronto, the shoreline of Lake Ontario is aligned NE-SW, resulting in the Gore pattern of the Bramalea area, west of Etobicoke (Toronto), as shown in the Federal electoral district of Bramalea-Gore-Malton. (Wikimedia Commons, public domain)

Township and lot sizes have been revised from time to time from 6 to 10 sq miles (15.5 to 25.9 km^2). Seven survey systems (Weaver 1962) were applied to Ontario's land. Technical details for each type of survey method can be found online (Ontario Laws 1990).

The "Single-front" township constitutes the first and oldest generation of Ontario township on the eastern north shore of Lake Ontario. Most of the 103 townships of this type were 6 sq. miles, surveyed between 1783 and 1818. Because of its generous allowance for roads, only 13 "Front-and-rear" townships were surveyed (8 regular and 5 "special") from 1787 to 1818, located in the Niagara Peninsula. The "Double-front" 96 townships, implemented mostly between 1815 and 1829, account for approximately 18% of Southern Ontario and are typically located in the Peterborough region. There were 26 "2,400-acre sectional" townships surveyed from 1829 to 1851. They are relatively widespread but there seems to be a slight concentration in the western tip of Southern Ontario. The "1,000-acre sectional" is representative of a period of growth between 1835 and 1906 as 238 townships of this type were surveyed from the Georgian Bay to the Ottawa River region. The "640-acre sectional", the American type 6-mile square township, was adopted in 1859 and implemented along the north shore of Lake Huron, north of Sault Ste. Marie and Near Fort William (present-day Thunder Bay), albeit in three different patterns: Pattern 1, with road allowance, was used to survey 22 townships; Pattern 2, without road allowance in 42; and Pattern 3 adopted in 1874, with single fronts, was extensively used to create 431 townships in the northern part of Ontario. Finally, the "1,800-acre" was adopted in 1906 consisting of a 9-mile square township, to be applied in Clay Belt region of Northern Ontario, unsuitable for agriculture, to create 53 townships. Ontario would eventually abandon township surveying in 1935 (Sebert 1984). The Association of Ontario Land Surveyors (1997) offers a list of Ontario's townships and the systems used to survey them.

20.8.3 The Prairies: A Copy of the Jefferson's Homestead Act

Thomas Jefferson had a romantic vision that the large expanse of land west of the Appalachian Range would be settled by small independent family farms. His ideas were later reflected in the 1862 (USA) Homestead Act. The intent of the Homestead Act was to encourage western migration and settlement in the United States by providing settlers with 160 acres (64.7 ha) of public land and facilitate an agricultural economy. The act aimed to promote the development of the western territories and make land ownership accessible to a broader range of people. The Homestead Act opened up millions of acres of land to courageous farmers who would settle, live, and cultivate their lot. The Homestead Act would be signed by Lincoln in 1862.

A rational system of allocation of space was created to determine a systematic layout to measure, divide, and allocate space. The result is still clearly visible in the USA as well as in the Canadian Prairies with a regular quadrilateral pattern uniformly applied from east to west, and south to north. The "township" became the basic unit of the system. Soon after the adoption of the Homestead Act in 1862, Canada followed suit by beginning the implementation of the Prairie Dominion Land Survey (McKercher and Wolfe 1986) in 1871 to encourage and expand settlement in the Prairies (then Northwest Territories). Combined with the birth of the Canadian Dominion, the acquisition of Rupert's Land, the incorporation of British Columbia in the Canadian federation, and the project of building a railway line across the continent, the Dominion Land Survey was a precipitated, nonetheless colossal, engineering endeavour to affirm Canadian sovereignty over these lands. The technical aspects, executed in challenging conditions at times, over 478,000 km^2 with rudimentary equipment, were rather similar to the USA Homestead Survey.

20.8.3.1 Dividing the Land: The Grid System

The application of the survey on the land created a rigid grid system notwithstanding the topography. For instance, lines will run through rivers, lakes, elevations, or any other features. Parts of British Columbia, because of their exceptionally mountainous landscape, were exempted from surveys. The land that had been submitted to previous forms of land division was left untouched, as in the case of the long lots along the Red River, federal parks, mountainous regions, and Indigenous reserves. As a result, the Prairies provinces, as well as a "railway belt" corridor in British Columbia, were divided in an almost uniform patchwork quilt grid pattern. However, the variety of crops that have been grown, from the intense blue colour of a field of flax, the deep green leaves of crops, the bright shades of canola, to the golden wheat fields have enhanced the geometrical aspect of the agricultural land into a rich and mixed palette of colours in the landscape.

Specific parts of the resulting grid were allotted to particular uses from the start. Sections 11 and 29 of each township were reserved for educational purposes. Although the building of school would not occupy a full section, the remainder of the land could be sold to finance construction and school operations. Equally, exchanges of land could be completed for practical reasons.

When selling its land to the Dominion of Canada, the Hudson's Bay Company had a clause retaining ownership of 5% of the land. This condition was met by granting the company lot 8 and ¾ of lot 26 of each township, and the entirety of lot 26 in every fifth township (McKercher and Wolfe 1986). Being an integral part of western Canada development, the Canadian Pacific was given land as a form of partial payment for building the railway line that would allow access to homesteaders. Odd-numbered sections, except 11 and 29 reserved for schools, of townships were granted as such. Because the land allocated to the CPR had to be "fairly fit for settlement", much of the land allocated to the railway was distant from the line of track.

20.8.3.2 Townships and Sections

The starting step was the establishment of the "Prime Meridian West" that was determined to be 10 miles (16.1 km) west of Winnipeg at 97^0 27′ 28.4″W, marked by the aptly named Meridian Road (McKercher and Wolfe 1986; Table 20.1). Subsequent meridians occurred at 4° W increments, from the second meridian at 102°W to the seventh meridian at 122°W. The meridians run from the "baseline", set at the 49th parallel to the 60th parallel. There are baselines at every four townships. From that point, land was divided into "ranges", counting upward from 1 westward. Ranges are 6 miles (10 km) apart. Township lines run northward every 6 miles, thus creating a square of 6 × 6 miles, identified as a "township". A township was divided into 36 "sections" of 1 × 1 mile (1.6 × 1.6 km), each identified by number from 1 to 36, starting at the southeast (or bottom right) corner moving west (or left) then north in a "Snakes and ladders way" (i.e. from 6 it moves directly up one square and right; Figs. 20.8, 20.9). This way of counting was adopted to minimize the pedestrian distance the surveyors would have to cover, being more time efficient. Incidentally, although the execution is the same as in the United States, the location of township 1 in the US township system was in the top right (Northeast) corner, then "snaking" west and south.

Table 20.1 Principal Meridians of western Canada

Principal meridian	Longitude (W)	Notes
First	97°27'28" (97.46°)	West of Winnipeg, intended to be west of any previously surveyed Métis community land
Second	102°	West of MB-SK border 49–56°N; along border 56–60°N.
Third	106°	Passes east of Saskatoon SK
Fourth	110°	Designated AB-SK border, 1905
Fifth	114°	Passes through Calgary AB
Sixth	118°	Surveyed in AB and along CPR line in BC
Seventh	122°	Surveyed in Fraser Lowland, BC
Second (east)	94.5°	Theoretical in northeastern Manitoba; not surveyed

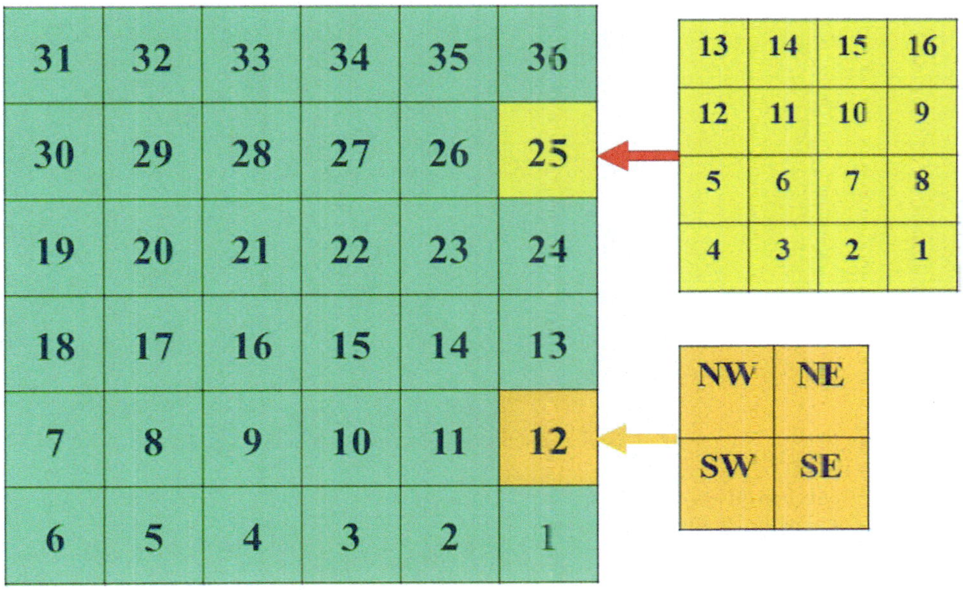

Fig. 20.8 Dominion land survey system used in Prairie Canada. Each township block (green) is divided into 36 sections. Sections in turn can be divided into Legal Sub-Divisions (LSD) numbered 1 through 16 (yellow), or into quarter sections (orange). Quarter section divisions were commonly allotted to nineteenth-century homesteaders

Sections can be subdivided into smaller portions of land, from half (1 × 0.5 miles), quarter, to sixteenth Sects. (0.25 × 0.25 miles), and all possibilities between. Quarter Sects. (0.5 × 0.5 miles, 64.7 ha), identified as their geographic location (NE, SE, SW, or NW), are the most common form of subdivision. Under the Dominion Lands Act (1872), homesteaders could apply for a quarter section.

20.8.3.3 Correction Lines

Longitude lines converge northward. Consequently, for each township to remain the same size, adjustments need to be made. For every four townships from the 49th to the 60th parallel, a baseline was added (Fig. 20.10). Then, a correction line was inserted in the middle of two baselines. At each correction line a westward 225 ft (68.6 m) "jog" is inserted from a meridian, thus shifting the west side of the township to the north 225 ft further west. The jog increments by 225 ft (225, 450, 675, etc.) when meeting every township west line, until the system resets when reaching the next meridian. Due to the equipment used, with 66 ft (20.1 m) surveying chains, the distance of 225 ft was approximate.

The 225 ft incremental jogs result in gradually smaller townships north of baselines along the next meridian. This situation is noticeable mostly in Alberta. Thus, instead of dividing the townships in smaller sections, the same size of sections was retained (1 mile per 1 mile), but their number was reduced. The typical township of 36 sections might be reduced to 30, 24, 18, 12, or 6. However, the numbering system remained. Then, as the "exceeding" western range is removed, so are section numbers 6, 7, 18, 19, 30, and 31 for instance. If two are removed, the column of Sections 5 to 32 is eliminated, and so on. The same applies to the townships east of the prime meridian but in the opposing direction. Sections in a fractional

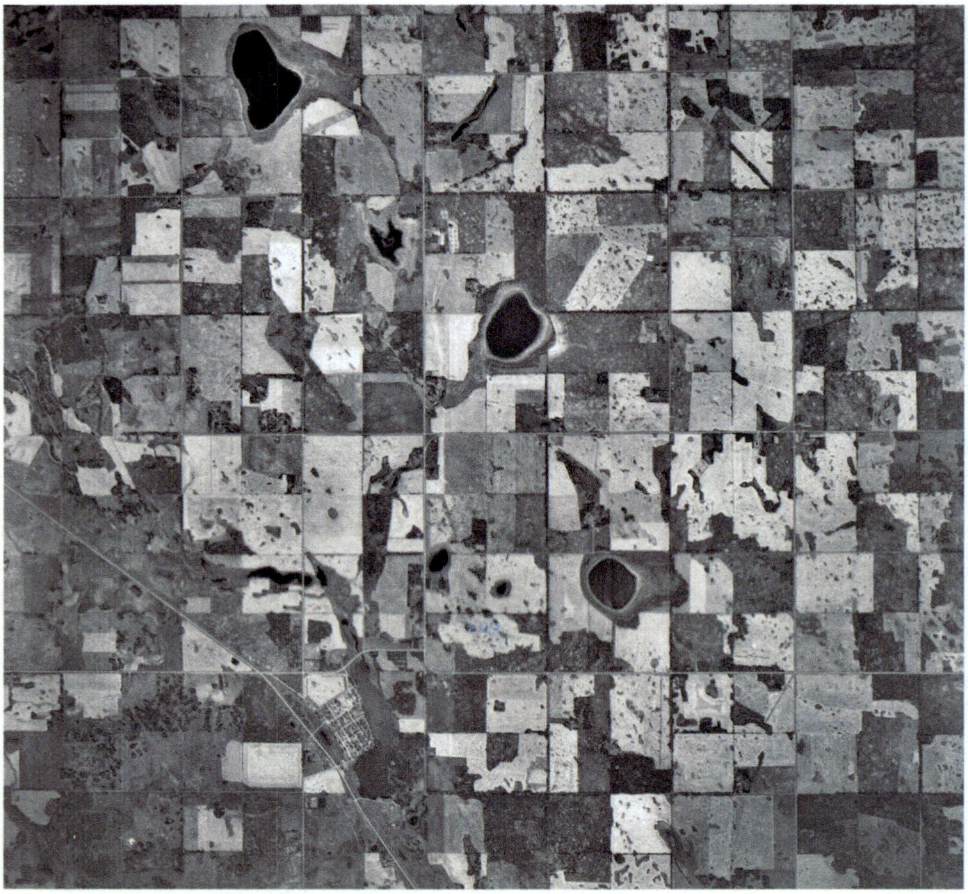

Fig. 20.9 Aerial photograph of Saltcoats SK. Roads follow township and range lines. Fields, indicated by different tones from agricultural activities, show the pattern of quarter sections. The grid pattern does not follow topography. (National Air Photo Library)

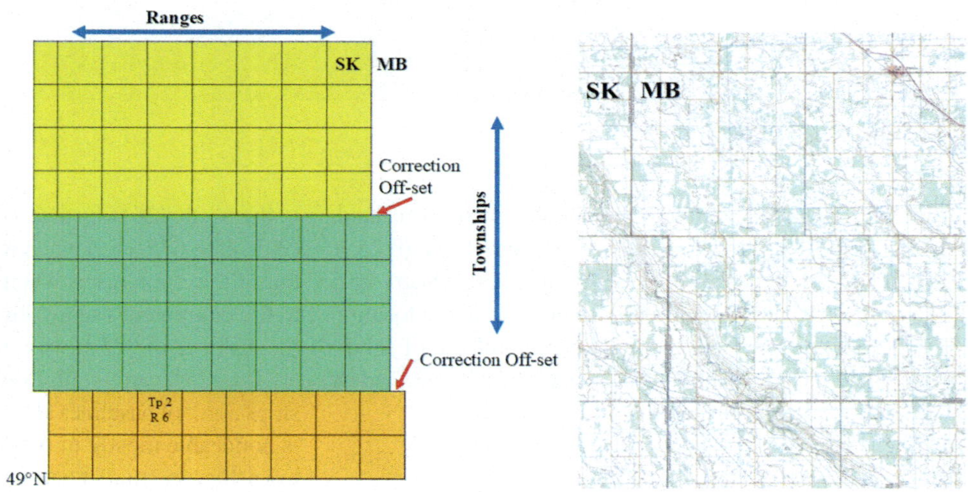

Fig. 20.10 Correction offsets for ranges, typical of the DLS in the southern Prairies. Offsets occurred at 2 townships north of 49°N, and at every 4 townships to the north. Inset map shows the offset Manitoba-Saskatchewan border near Elkhorn (NTS map 62 F/14)

township are numbered the same as though the township was a full one (McKercher and Wolfe 1986, Barnett 2001).

20.8.3.4 A Relative Uniformity

The Dominion Land Survey has resulted in a uniform division of land in the Prairies and parts of British Columbia with only slight differentiations between systems. The first and second systems were applied in southern Manitoba, a small part of southeast Saskatchewan, and the Prince Albert Region between 1871 and 1879. They included an extensive road allowance adjoining each section. To minimize "left-over land" against the next meridian and attempt to have all the sections approximately the same size, lines were run "true north–south" in the second system. The result was sections were a few feet wider at the southern edge and slightly "wedge shaped" (McKercher and Wolfe 1986).

The third survey was by far the largest. It covered the part of Manitoba north of the first survey and the main settled areas of Saskatchewan and Alberta after 1881. It featured a shrunken road allowance, from 1 ½ to 1 chain (99 × 66 ft, 30.2 × 20.1 m), and east–west roads only at every second township. Consequently, the townships are not perfectly square, being 486 chains on their base (1.596 km) and 483 chains (1.586 km) sidelines.

The fourth and fifth surveys were used only in only two parts of British Columbia: the Canadian Pacific "railway belt", and the westward extension of Alberta's Peace River region. Because of the difficult nature of terrain, a "theoretical" survey was performed virtually by applying the Dominion rules through mathematical projection along most of the railway belt (McKercher and Wolfe 1986). However, the Lower Fraser Valley had been surveyed by the Province of British Columbia previously. Although the numbering system differs, it became considered as part of the Dominion Survey.

20.8.3.5 A Recent Trend: Circling the Squares

Agricultural practices on the land have contributed to somewhat change the rigid quadrilateral aspect of parts of the Prairies. Circular irrigation is a system in which equipment rotates around a pivot and crops and sprays water through a series of sprinklers located along a long arm (tower). It was invented and developed in 1940 by Colorado farmer Frank Zybach to provide a more efficient use of water as it allows an adjustable water distribution from the centre to the end of the tower (NASA). He later licenced his patent to Valley Manufacturing in Nebraska (Valley Manufacturing demo). Valley engineers spent the next decades refining Zybach's innovation to a state-of-the-art implement that is less labour intensive, customizable to cultivated crops, fully adaptable to the environmental conditions and remotely controlled that contributes to perform "precision farming". The self-propelled device circles around the central pivot that feeds the individually adjustable sprinklers flow rate along the tower" that distributes the water to evenly cover the crops from centre to extremity. On the field it can distribute water and/or fertilizer from the centre on a quarter section and up to a full section of a township in a circular pattern, thus creating a mosaic of different patterns.

Although Agriculture Canada (1975) estimated the cost to be comparable to other systems, since it did not cover a whole square by missing corners, it was believed to be more costly to use by surface unit. However, newer systems have an "end gun" sprinkler fitted so it can also irrigate the neglected corners, although not as efficiently. In Canada, it was first installed in Burdett, Alberta, in 1962 by Valley Manufacturing. The wide use of circular irrigation, clearly visible on satellite imagery, appears to be well adapted to a dry region like southern Alberta and an illustration of "contagious diffusion" that has "pivoted" from the village, incidentally, neatly fitted on a quarter section, to a large surrounding area. The region of Outlook, Saskatchewan, is another, although less extensive, example.

20.8.4 British Columbia: An Amalgam from One System to Another

British Columbia, from before it was even a colony, has had a form of land survey. No less than nine systems of surveys have been applied to different parts of British Columbia throughout its history (Taylor 1975).

> British Columbia has a long and colorful history of trying different land surveying systems and methods. There are wonderful descriptions of the way BC landscapes and politics challenged 9 systems that were tried. There were also additional methods specific to areas in the Kootenays and near railways that didn't fit into any system (Peace River Forage Association 2015)

Even though the United States and the British government had agreed that the Oregon country, to which the British Columbia colony was part of, was open to both countries, London had granted the Hudson's Bay Company a hunting and trading licence. The company started laying boundaries to protect its hunting grounds (Taylor 1975). Their first system was implemented in 1851 and consisted of Metes and Bounds in the Fort Victoria District. Relying on topographical features such as trees, rock, rivers as natural boundaries, the end result was an irregular of mostly rectangular pattern of lots of sizes that varied from 12 to 840 acres (4.8–340 ha) (Taylor). Nonetheless, the resulting pattern showed some regularity.

The second system was run from 1858 to 1860 on Vancouver Island, which became necessary with the influx of miners and prospectors from California to be outfitted on their way to the Fraser Valley to search for gold, somewhat to the distaste of Hudson's Bay Company agents who feared the disrupting of the fur trade. Sections of 20 chains by 50 chains (400 m- 1 km) were drawn in the Saanich Peninsula and the island's east coast north to Duncan and Nanaimo, showing a much more regular grid pattern along the lines of a township system.

Directly linked to the Gold Rush, a third survey system was initiated in the Fraser Valley in 1858. It was decided to adopt the same system as in the United States, and Chief Commissioner of Lands and Work Colonel R.C. Moody wrote to the Surveyor General in Olympia WA to obtain copies of the "Books of Instructions" (Taylor 1975). Consequently, a 36 section township in New Westminster was changed to a 3 miles square township (23.3 km^2), in this case named "block", and the system adopted by the United States was applied to Sea and Lulu islands (present-day Vancouver Airport and Richmond, respectively) using the American numbering system, that is, opposite to the Canadian system that was used later in the Canadian Prairies. As mentioned above, this survey would eventually be kept "as is" under the Dominion Land Survey.

The fourth system, implemented in the district between the North Arm of the Fraser and Burrard Inlet comprising present-day Vancouver and Burnaby from 1859 to 1863, produced variable sized and shaped lots. A large part granted of the central Vancouver area was granted to the Canadian Pacific Railway, which applied its own survey to eventually develop the upscale Shaughnessy neighbourhood.

The fifth system, started in 1873, established 27 townships in the New Westminster District, conformed to the Dominion Land Survey and fell within the Railway Belt, but used a different "disastrous numbering system" (Taylor 1975).

The sixth British Columbia Land Survey System, also known as the Provincial Townships System, adopted in 1879, consisting of a township survey system inspired by the American model, without road allowances and starting from the southeast corner. The survey ran north and west, numbering sections from the southeast, similar to the Dominion system. In areas where there was insufficient space to create at least four complete townships, a different system was used, involving the creation of rectangular district lots. This alternative system was employed when the standard township system couldn't be applied due to limited space or irregular geography. However, this system did not account for the earth's curvature, resulting in true quadrilateral townships (Taylor 1975).

The seventh British Columbia Land Survey System, also known as the Island Townships System, was specifically designed for surveying land on Vancouver Island and the Gulf Islands. Similarly to other systems, it divided land into sections, but with adaptations to suit the island geography. Under this system, land was surveyed into sections approximately one mile square. However, unlike other systems used in British Columbia, no townships were officially created due to the relatively small size and irregular shape of the islands. The surveyors adapted the land division method to suit the geography of the islands better, focusing on subdividing land into sections rather than larger townships, which was more typical in mainland areas. The land survey was implemented as if a large grid was applied over the Georgia Straight and 1 × 1-mile squares (1.6–1.6 km^2) that were on or touched an island were numbered as sections (Taylor 1975).

The standard District Lot System (the 8th system) had its beginnings in 1861, but its statutes were changed in 1874 to allow more flexibility by offering different size options, to which more options were authorized in 1952. However, the Land Act was amended in 1953 to further simplify the system by only requiring surveyed lots to be rectangular in shape with no boundary surpassing 80 chains (1.61 km) in length (Taylor 1975). This process is still widely used and is more compliant and easier to use (or "fit") in unsurveyed land.

The ninth system was applied along a railway corridor and the Peace River block. This railway belt is a strip of land 40 miles wide on both sides of the Canadian Pacific Railway mainline, extending from the British Columbia coast. After British Columbia joined Confederation, this land was transferred to Canada's federal government. The Dominion Land Survey (DLS) was used to survey this area, specifically employing the 4th System of Survey. The 40-mile (64.3 km) railway belt and Peace River Block were significant areas of land surveyed by the Dominion Land Survey using systems that did not initially account for road allowances. Over time, these lands were returned to provincial control in 1930, and more flexible surveying methods, such as the District Lot system, have been adopted in areas with irregular terrain. For small lots along highways and bodies of water, the survey lines follow early practices, with side lines parallel to roads or natural features, continuing a tradition of aligning surveys with infrastructure.

20.8.5 Atlantic Provinces

In the Atlantic Provinces, land subdivision was most extensive in agricultural regions and areas where settlement resulted from government initiatives and sales. Prince Edward Island was divided into 67 rectangular and square lots, all but two of which had ocean frontage (Fig. 20.11). Areas of New Brunswick settled by Loyalists after 1784 were surveyed into rectangular and less-regular polygonal lots, generally with access to the Saint John River or Fundy Coast (Southwest). In the Acadian areas, no comprehensive land subdivision system was imposed (Fig. 20.12). The same applies to Nova Scotia, with systematic subdivision confined to the Annapolis River valley and areas of Loyalist settlement.

On the island of Newfoundland outside of the larger communities, surveying until ca. 1950 was largely the responsibility of individual property owners. Records are frequently absent, and disputes over title between long-time residents and the Government of Newfoundland and Labrador still are common (Law Society of Newfoundland and Labrador 2021).

Fig. 20.11 Field pattern, Winter river watershed, PEI. Surveying did not follow local topographic features, and was similar in style to that used in southern Ontario. (Norm Catto photo)

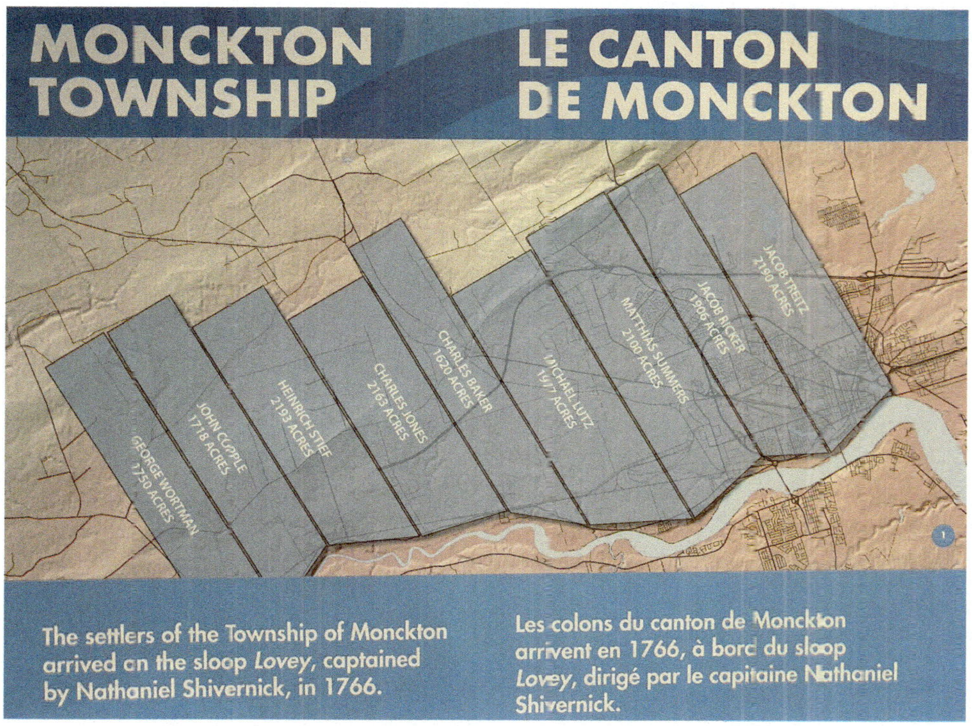

Fig. 20.12 Survey of Monckton (now Moncton) NB followed a modified seigneurial system. (Courtesy Moncton Resurgo Museum)

20.9 Maritime Boundaries

The territorial waters of a coastal country represent a strip of sea adjacent to its coastline over which the sovereignty of that state extends. They can be considered to be the extension of its territory onto the sea. Canada has a large land territory. As it is flanked by three oceans, Canada has rights and responsibilities over maritime territory (Fig. 20.13). As a coastal country, Canada has to demarcate its maritime zones as defined by the United Nations Convention on the Law of the Sea (UNCLOS).

Maritime zones of a country are measured starting from a "baseline" that runs along the coastline. Article 5 of the United Nations Convention on the Law of the Sea defines the baseline as the low-water line along the coast as marked on large-scale charts officially recognized by the coastal state (UNCLOS 1994). The definition of a baseline can be complicated by the nature of its coastline, the presence of islands, and deep indentations. The idea is to connect the coastline points in straight lines at low water to roughly form a smooth continuing and homogenous contour of the land area. Therefore, each body of water, such as rivers, lakes, and bays, that falls within the baseline is defined as "Internal waters". "Archipelagic waters" are included to draw the baseline as is the case of the Canadian Arctic Islands.

Fig. 20.13 Canada's Oceans Estate includes all areas with the Exclusive Economic Zone. Note the exclusion extending from St. Pierre towards the southeast, off the coast of Newfoundland. Offshore of the Grand Banks and in the northern Beaufort Sea, the continental shelf extends beyond the EEZ boundary. (Fisheries and Oceans Canada)

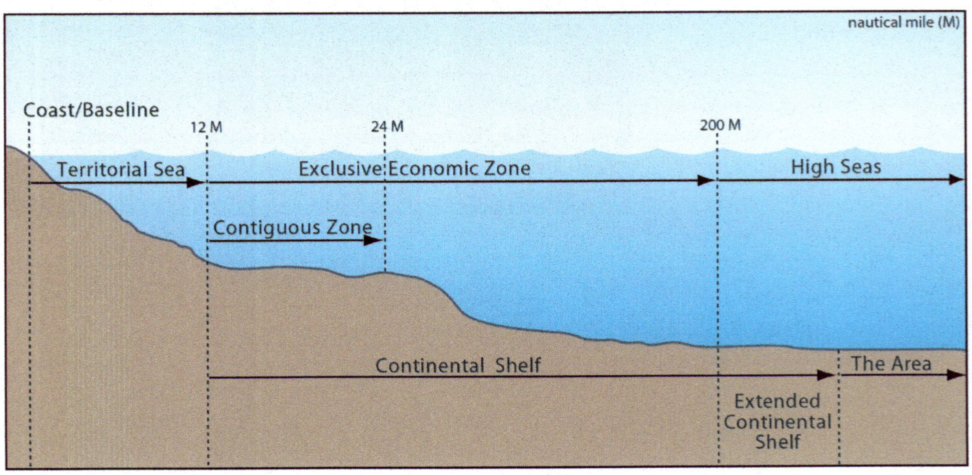

Fig. 20.14 Schematic representation of marine boundaries. (National Oceanographic and Atmospheric Administration)

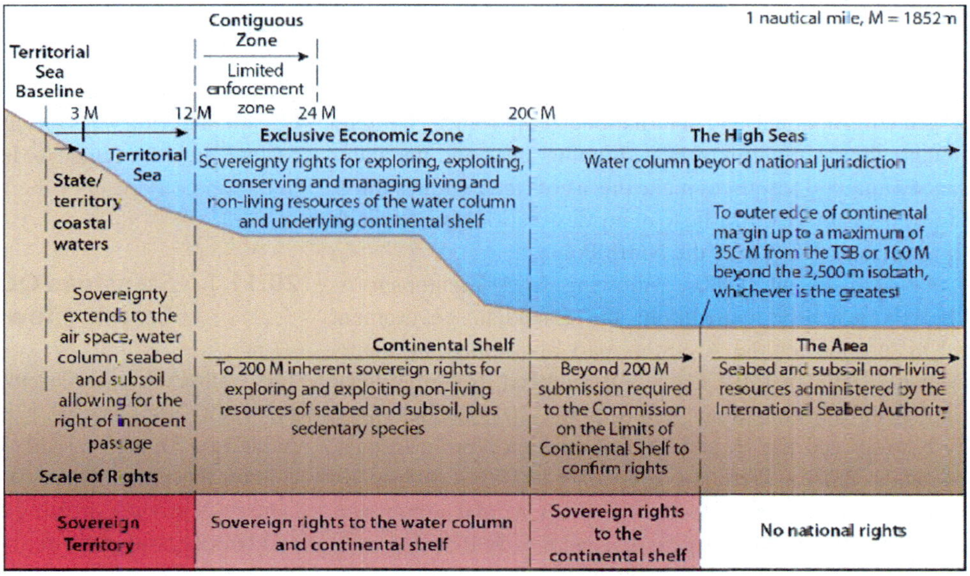

Fig. 20.15 Scale of sovereign rights of coastal states in marine zones. (National Oceanographic and Atmospheric Administration)

"Territorial sea" refers to coastal waters that extend "to a limit not extending 12 nautical miles" (22.2 km) at all points of the baseline (Figs. 20.14 and 20.15). It is more or less an extension of the sovereign land territory. If the territorial sea limit meets a neighbouring country's limit, as is the case with the United States on the East and West Coasts (see Chaps. 2 and 7), and Denmark's Greenland in the north, the limit is set at the median point. Hans Island happens to be almost exactly midpoint between Greenland and Canada in the Kennedy Channel. The island was claimed by both countries until 2022 when they both agreed to divide the island in two equal parts (see Chap. 9). A "contiguous zone" is an added 12 nautical miles to the territorial sea or 24 nautical miles from the baseline.

The exclusive economic zone (EEZ) extends to a maximum of 200 nautical miles (374.4 km). Resources contained in the EEZ are under sovereign rights of the coastal country to explore, exploit, conserve, and manage natural resources: fisheries in the Grand Banks off Newfoundland are among the richest fishing grounds, known for cod, haddock, and other species (Chap. 2), and the waters off British Columbia are abundant in salmon, halibut, and shellfish (Chap. 7).

Oil and Gas in the Atlantic Ocean offshore fields, including Hibernia, Terra Nova, and White Rose, located 315 km east of St. John's, relatively close to the zone's limit, are significant sources of crude oil (see Chap. 2). Arctic Ocean potential reserves in the Beaufort Sea have been explored (see Chap. 9). The seabed may as well contain valuable minerals like nickel, copper, and rare earth elements. The Atlantic and Pacific Ocean coasts have potential for tidal and offshore wind energy projects (see Chap. 17). All of these zones considered, Canada has an ocean estate of approximately 7.1 million km^2.

20.9.1 Saint-Pierre-and-Miquelon Islands

Saint-Pierre-and-Miquelon Islands are all that is left of all the possessions France had in North America. Their history is a relatively epic one for such a small piece of territory and illustrates its strategic location on many regards.

Colonized by Breton, Norman, and Basque fishermen from the beginning of the sixteenth century, Jean Talon had annexed the islands to New France in 1670. They were deserted for some time and were ceded to Britain following the Treaty of Utrecht in 1713. Under the Treaty of Paris in 1763, France lost all its possessions in North America but were granted fishing rights along the coast of Newfoundland and was handed back Saint-Pierre-and-Miquelon. They were successively invaded for France participation in the American Revolution, and in 1793 during the French Revolutionary Wars, during which colonists were deported to France while British settlers were moved in. Following another sequence of back-and-forth attacks and exchanges between France and Great Britain, France regained the then deserted island in 1815. Fishing allowed the islands to be resettled during the nineteenth century. It became French Overseas Territory after World War II and is a now a French overseas collectivity, a little piece of Europe left in North America.

France claimed an EEZ of over 12,000 km^2 awarded in 1992, 24 nautical miles around the islands. An arbitration tribunal awarded France a 200 nautical mile long narrow corridor southward within Canada's EEZ. This zone allows France to exploit marine resources, including fishing and potential underwater minerals (Sovereign Limits n.d.). This zone might prove to be strategic for fishing as it has right to an extended, although narrow, southward corridor.

20.10 Canadian Claims in the Arctic

Canada's Arctic sovereignty is a complex and strategic issue of growing national and international importance. Climate change, the opening up of new shipping lanes and the abundance of natural resources make this a coveted region, subject to new pressures. Air traffic to many destinations is also much shorter when flying over the North Pole.

As early as the 1880s, following USA Lieutenant A. Greely's scientific expedition, the Canadian government began to consider the Arctic as part of its internal waters (Vandal 2019). In 1909, Canada formulated its formal legal claims under international law (Vandal 2019). Challenged by Norway and the United States views on Nordic Islands, Canada's claim to Arctic was revived in 1925, basing its privilege on its history of exploration, settlement, and administration of Nordic territories by French and mostly British power. Several RCMP posts were established throughout the Arctic islands. The presence of Indigenous peoples in the Arctic can also be viewed as a fundamental element of Canadian sovereignty. Unfortunately, the forced relocation of Indigenous people from northern Québec to the vastly different environment of southern Umingmak Nuna (Ellesmere Island), in a bid to establish sovereignty by community presence, was a shameful consequence (see Chap. 9). Today, Indigenous rights and traditional knowledge are recognized and considered in the management of this region. Canada has ratified various international treaties asserting its sovereignty over the Arctic, such as the United Nations Convention on the Law of the Sea.

Canadian sovereignty over the Arctic is particularly at stake today when climatic change, national security, and natural resources have become paramount issues. The impacts of climate change on the Arctic are rapid and profound (see Chap. 9). Melting sea ice is opening up new shipping lanes that Europeans have searched for centuries that incidentally led to the discovery of the Americas. This foreseeable possibility has sparked increased interest in exploiting natural resources and increasing maritime traffic, posing challenges in terms of maritime safety and environmental protection. The U.S. icebreaker *Polar Sea's* crossing of the Northwest Passage in 1985 raised questions about Canadian sovereignty (Pharand 1986). The Canadian Arctic is rich in natural resources such as hydrocarbons, minerals and seafood, potential exploitation of which raises complex issues relating to environmental protection, Indigenous rights, and the distribution of benefits. Economic development in the Arctic must be sustainable, considering environmental and social issues. However, the Arctic is a strategic region for the defence of Canadian territory and the protection of its national interests.

20.11 Oddities

This set of sometimes complex lines around and within the Canadian territory has produced some oddities that the United States Center for Land Use Interpretation (2014) has compiled in its Linear Portrait of the USA/Canada Border.

20.11.1 Stanstead QC/Beebe VT, Border Crossing Town

The Canada–US border in this area was established after the Treaty of Paris in 1783 and refined by the Webster-Ashburton Treaty of 1842. The borderline follows 45°N and cuts through the towns of Stanstead (in Québec) and Derby Line (in Vermont), creating a situation where some properties and buildings straddle both countries or streets run along the border.

The Haskell Free Library and Opera House was intentionally built between 1901 and 1904 on the border by a wealthy Canadian-American couple, Martha Stewart Haskell and her husband Horace Haskell. Their goal was to create a shared cultural space for people on both sides of the border. They deliberately positioned the building so that it would serve both communities: the entrance is in the United States (Vermont), but most of the library's books and the opera stage are located in Canada (Québec). The building was designed to symbolize the friendship and cooperation between the two nations. It was meant to be accessible to both Americans and Canadians, allowing them to share cultural and educational resources. Visitors can enter the library from the USA or Canada but technically cross an international boundary when inside the building. There is a black line running through the floor of the library, marking the exact location of the border. Special arrangements exist with border patrols to allow people from both sides to access the building without needing to go through full customs procedures, although the area is closely monitored.

Canusa Street, a contraction of Canada and USA, is a road that straddles the Canada–United States border and runs along the boundary between Stanstead, Québec, and Derby Line, Vermont. The reason for this unique situation lies in the historical development of the area and the placement of the international boundary. The towns of Stanstead and Derby Line were developed as closely knit communities with shared infrastructure, despite being in two different countries. As the towns grew, urban planning did not strictly adhere to the boundary. Instead, streets including Canusa Street were built following the natural layout of the area, effectively straddling the border because the boundary line happened to divide these towns.

Historically, the border between Stanstead and Derby Line was a soft border with minimal physical barriers, allowing residents to move freely between the two countries. While security has increased over time, especially after 9/11, Canusa Street remains a unique example of a place where the border is not marked by fences or walls but runs down the middle of a residential street.

20.11.2 American Enclaves in Canada

The establishment of parallels to run through a large part of the continent created several American enclaves in Canada and some surprising oddities. Point Roberts, WA is located at the southern tip of the Tsawwassen Peninsula. This part of Washington State is cut off from the rest of the U.S. and can only be accessed by land through Canada.

Lake of the Woods was drawn as oval on the 1755 Mitchell map, and the boundary would cut in the middle, resulting in several pieces of isolated USA territory. The "Northwest Angle" is isolated from Minnesota and is only reachable either by water or through Manitoba by land. When signed, the Webster–Ashburton Treaty of 1842 utilized the exact location of the northwesternmost point of the lake, which proved difficult to establish and needed experts to evaluate. When found, the placement of the northwestern corner and shape of the lake created a section of the United States that is isolated and is more easily accessed from Canada. Elm Point, Minnesota suffers the same fate.

Piney Pinecreek "Border Airport" is probably a particularly rare and surprising oddity as the terminal and facilities are in the United States and the runway is almost half and half in Canada and the United States. However, it is not a major commercial airport but is valuable for the local aviation community and cross-border travel.

Alburgh Tongue and Province Point, Vermont are separated from the rest of the U.S. by Lake Champlain by the crossing of the 45th parallel and are more easily accessed from Canada although "the Tongue" is also accessible through bridges. Province Point, a very small, wooded enclave, is completely isolated from the United States.

20.11.3 On the Wrong Side of History?

Île Sainte-Croix (now Dochet Island) is in Maine due to historical treaties, particularly the Treaty of Paris (1783) and the Webster-Ashburton Treaty (1842), which established the St. Croix River as part of the boundary between the United States and Canada. Although the island holds deep historical significance for both nations, more so for Acadia, it is officially part of the U.S. state of Maine. The Saint Croix Island International Historic Site is administered by the U.S. National Park Service, while Parks Canada also maintains an interpretive site on the Canadian side of the river, celebrating the shared history of Île Sainte-Croix.

20.11.4 Sitting on an Interprovincial Border

The Dominion Land Survey was accomplished when Manitoba was still a "postage stamp" size province. Manitoba was slightly enlarged in 1882 sharing its western border with the North-West Territories, then again in 1912, this time sharing its border with Saskatchewan.

Since its first expansion in 1882 the border between Manitoba and first the Northwest Territories followed the Dominion Land Survey grid from the Canadian-US border at 101° 30'W on a road allowance between ranges 29 and 30 west of prime meridian. From there, a correction line occurs at the second township and every fourth township north of 49°N to 56°N, shifts the border 1.5 mile (2.5 km) west every 24 miles (38.6 km). North of 56°N, the border follows along 102°W to 60°N. The western border of Saskatchewan is a set meridian along 110°W (a straight line), passing through Lloydminster.

Flin Flon, a fairly isolated zinc and copper mining city of approximately 5,000 people founded in 1927, is located directly north of a correction line, with most of its area being in Manitoba. The remainder of Greater Flin Flon is in Saskatchewan, south of the border, not west. Several municipal services will cover both sides of the borders while provincial services will be provided by their respective governments. Consequently, while Saskatchewan does not observe Daylight Saving, Greater Flin Flon does, meaning that Flin Flon, SK is in a different time zone than the rest of the province during the summertime.

References

Agriculture Canada (1975) Irrigation on the Prairies. https://publications.gc.ca/collections/collection_2014/aac-aafc/agrhist/A53-1488-1975-eng.pdf

Association of Manitoba Land Surveyors (2023) The first boundary extension. https://amls.ca/wp-content/uploads/2023/02/amls_The-first-boundary-extension-New.pdf

Association of Ontario Land Surveyors (1997) List of subdivided townships in the Province of Ontario with dates and systems of survey. https://www.aols.org/site_files/content/pages/practice-manual/historical-information/aols_pm_list_of_subdivided_townships.pdf

Atlas of Canada (2017) The Atlas of Canada—Territorial Evolution from 1867 to 2017. https://atlas.gc.ca/ette/en/index.html

Barnett D (2001) Early Surveys and Settlements in Central Alberta. University of Alberta

Boileau G (1999) Réflexion sur les villages du Québec. Histoire Québec 4(2):17–21

Burroughs P (2024) Sherbrooke, Sir John Coape. Dictionary of Canadian Biography. https://www.biographi.ca/en/bio.php?id_nbr=3132

Canadian Archives (2020) Prince Edward Island 1873. https://www.bac-lac.gc.ca/eng/discover/politics-government/canadian-confederation/Pages/prince-edward-island-1873.aspx

Coats RH (1923) The growth of population in Canada. Ann Am Acad Polit Soc Sci 107(1):1–6. https://doi.org/10.1177/000271622310700102

Center for Land Use Interpretation (2014) United divide: a linear portrait of the USA/Canada border. https://clui.org/projects/united-divide

Couture C (2007) Québec. The Canadian Encyclopedia. https://www.thecanadianencyclopedia.ca/en/article/quebec

Crown-Indigenous Relations (2011) History of treaty-making in Canada. Crown-indigenous relations, Government of Canada. https://www.rcaanc-cirnac.gc.ca/eng/1314977704533/1544620451420

Dickinson JA (2007) The english-speaking minority of quebec: a historical perspective. Int J Sociol Lang 185:11–24

Explore North (n.d.) Convention between the United States of America and His Majesty the Emperor of All the Russias, relative to navigating, fishing, Etc., in the Pacific Ocean. https://www.explorenorth.com/library/history/bl-rusus1825.htm

Farr DML, Block N (2006) Alaska Boundary Dispute. https://www.thecanadianencyclopedia.ca/en/article/alaska-boundary-dispute

Government of Canada (1889) Ontario Boundaries Act, 1889. https://www.justice.gc.ca/fra/pr-rp/sjc-csj/constitution/loireg-lawreg/p1t101.html

Griffiths NES (2005) From migrant to acadian: A North American border people, 1604–1755. McGill-Queen's University Press, Montreal and Kingston

Havard G, Vidal C (2003) Histoire de l'Amérique française. Flammarion, Paris

Heritage Newfoundland and Labrador (2013) Labrador Boundary Dispute. https://www.heritage.nf.ca/articles/politics/pdf/labrador-boundary-dispute-documents.pdf

Hiller J (1997) The 1948 referendums. Heritage Newfoundland and Labrador. https://www.heritage.nf.ca/articles/politics/referendums-1948.php

Jacobs F (2011) A Not-so-Straight Story. New York Times. https://archive.nytimes.com/opinionator.blogs.nytimes.com/2011/11/28/a-not-so-straight-story/

L'Actualité (2022) Le Québec s'est-il vraiment fait « voler » le Labrador ? https://lactualite.com/societe/le-quebec-sest-il-vraiment-fait-voler-le-labrador/

Laberge A (2010) Portraits de campagnes: la formation du monde rural laurentien au xviiie siècle. Université Laval.

Lang N (2001) Une population séparée à jamais: Le confit de la frontière internationale au Madawaska (1830–1842). Les Cahiers du GERHICO 2"133–144. https://demelerlespinceaux.umce.ca/

Lass WE (1980) Minnesota's boundary with Canada: its evolution since 1783. Minnesota Historical Society, St. Paul

Law Society of Newfoundland and Labrador (2021) Refortming the Law on Adverse Possession. https://lsnl.ca/wp-content/uploads/2021/05/Reforming-the-Law-on-Adverse-Possession_CBA-NL-Final-Report-May-2021.pdf

McGillivary B (2005) Geography of British Columbia. UBC Press.

McIntosh A Smith SA (2006) Rupert's Land. The Canadian Encyclopedia. https://www.thecanadianencyclopedia.ca/en/article/ruperts-land

McKercher RB, Wolfe B (1986) Understanding Western Canada's dominion land surveying system. University of Saskatchewan, Saskatoon. https://static1.squarespace.com/static/6148b130f68d7f1bb745938c/t/6185ab073b28f4218e2ed168/1636150031489/Understanding%2B_DL_Survey_System.pdf

Mitchell J (1755) A map of the British colonies in North America with the roads, distances, limits, and extent of the settlements, humbly inscribed to the right honourable the earl of Halifax. https://oshermaps.org/map/789.0001

Nicholson NL (2006) Convention of 1818. Canadian Encyclopedia. https://www.thecanadianencyclopedia.ca/en/article/convention-of-1818

Nova Scotia Royal Charter (1621). https://gutenberg.ca/ebooks/frasera-novascotiaroyalcharter/frasera-novascotiaroyalcharter-00-h-dir/frasera-novascotiaroyalcharter-00-h.html

Office of the Historian (n.d.) The Alabama Claims, 1862–1872. United States Department of State. https://history.state.gov/milestones/1861-1865/alabama

Osher Map Library (2022) The Mitchell map: introduction and overview. https://oshermaps.org/special-map-exhibits/mitchell-map/introduction

Ontario Laws (1990) Survey Methods. https://www.ontario.ca/laws/regulation/901029

Paireau V (2012) Le traité de Saint-Germain en 1632: ébauche d'une diplomatie "colonial"? Actes des Congrès Nationaux des Sociétés Historiques Et Scientifiques Année 2012(133–6):14–26

Parks Canada (2022) Fort Anne Historic Site. https://parks.canada.ca/lhn-nhs/ns/fortanne/culture/charles-fort/histoire-history

Peace River Forage Association (2015) Understanding legal land locations. https://peaceforage.bc.ca/wp-content/uploads/2021/06/FF_91_Legal_Land_Description_SB2.pdf

Pharand D (1986) La souveraineté du Canada dans l'Arctique. Revue Québécoise De Droit International 3:289–306. https://doi.org/10.7202/1101420ar

Poole AFN (1964) The Boundaries of Canada. Canadian Bar Review 42: 100–139. CanLIIDocs 7. https://canlii.ca/t/t64l

Québec (2004) Répertoire des cantons du Québec 2004. Ressources naturelles, faune, et parcs, Gouvernment du Québec. https://mrnf.gouv.qc.ca/documents/foncier/cantons.pdf

Ross Geiger DA (2001) Ontario's first township. The Ontario Land Surveyor, Summer 18–22. https://www.krcmar.ca/resource-articles/2001_Summer_Ontario's%20First%20Township_1.pdf

Roth M (2023) Early encounters on a Western frontier: the search for Svyatoy Nikolai (1807–1808). National Oceanic and Atmospheric Association. https://oceanexplorer.noaa.gov/explorations/23svyatoy-nikolai/features/russian-america/russian-america.html

Sanguin A-L (1974) La frontière Québec-Maine: quelques aspects limologiques et socio-économiques. Cahiers De Géographie Du Québec 18(43):159–185. https://doi.org/10.7202/021180ar

Sebert LM (1984) The Survey of the Sidelines of Concession Lots. Ontario Land Surveyor Spring 18–19. https://www.krcmar.ca/resource-articles/1984_Spring_The%20Survey%20of%20the%20Sidelines_1.pdf

Shi DE (1978) Seward's Attempt to Annex British Columbia, 1865–1869. Pacific Historical Review 47:217–238

Statistics Canada (2021) Census of Canada, Population. https://www12.statcan.gc.ca/census-recensement/2021/dp-pd/prof/i

Stein M (2008) How the States got their shapes. https://www.libraryaware.com/Records/ViewFirstChapterLegacy?recordID=7249991

Sovereign Limits (n.d.) Canada-France (St. Pierre & Miquelon). https://sovereignlimits.com/boundaries/canada-france-maritime

Taylor WA (1975) A History of Survey Systems. Ministry of Sustainable Resource Management, Government of British Columbia. https://ltsa.ca/wp-content/uploads/2020/10/Crown-Lands-A-History-of-Survey-Systems-1.pdf

Thorp DB (1996) Equals of the king: the balance of power in early acadia. Acadiensis 25(2):3–17

UNCLOS (1994) United Nations Convention on the Law of The Sea. https://www.un.org/depts/los/convention_agreements/texts/unclos/unclos_e.pdf

US Census Bureau (1909) Population in the Colonial and Continental periods. https://www.census.gov/library/publications/1909/decennial/century-populaton-growth.html

US Department of State (2009) Rush-Bagot Pact, 1817, and Convention of 1818. https://2001-2009.state.gov/r/pa/ho/time/jd/91716.htm

US National Archives (2022) Treaty of Paris 1783. https://www.archives.gov/milestone-documents/treaty-of-paris

Vandal G (2019) Une menace à la souveraineté canadienne en Arctique. La Tribune. https://www.latribune.ca/2019/06/27/une-menace-a-la-souverainete-canadienne-en-arctique-ec82a02668bbf6180ce7780fc0dfe650/

Valley Manufacturing (n.d.) What is a center pivot? https://www.youtube.com/watch?v=2bILpvH3EuQ

Weaver WF (1962) Crown Surveys in Ontario. Department of Lands and Forests, Government of Ontario. https://www.aols.org/site_files/content/pages/public-awareness/crown-surveys-in-ontario.pdf

Williamson WD (1832) The History of the State of Maine. Hallowell: Glazier, Masters, and Company

Wilson WR (n.d.) Historical Narratives of Early Canada. http://www.uppercanadahistory.ca/contact.html

Wyman O (2011) A functional review of decentralization. Government of Nunavut. https://assembly.nu.ca/library/GNedocs/2011/000533-e.pdf

Yale Law School (2008) The Webster-Ashburton Treaty. https://avalon.law.yale.edu/19th_century/br-1842.asp

21 Commentary: Three Geographical Issues, 2024 and Beyond

Abstract

This chapter offers thoughts on some geographical issues currently impacting Canada: demography and immigration; international relations; and environment and climate change.

Keywords

Demography · Immigration · International relations · Enviornment

Prediction of the future is highly uncertain. Extrapolation based on previous trends can easily go awry if environmental, cultural, or political factors are sharply altered. This chapter offers thoughts on some geographical issues currently impacting Canada, in the sure knowledge that these trends identified in December 2024 will be subject to future correction.

21.1 Demography and Immigration

Over the lifetime of a person or country, change is inevitable. Demographically, Canada in 2024 differs substantially from the country that existed 30 or 60 years previously (Table 21.1). Between 1961 and 2021, the population doubled. The percentage of immigrants increased by 277%, and the Indigenous population increased from 219,000 to more than 1.8 million. The birth rate, rate of natural increase, and total fertility rate all fell markedly. Changes in median age and age distribution indicate a transition from a youth-dominated to an older society. Growth in population in recent decades is largely due to immigration. These demographic changes are paralleled by societal changes, as is evident by comparing Statistics Canada's (1961) Official Handbook of Present Conditions and Recent Progress with the information presented in the 2021 Census of Canada (Statistics Canada 2021–2023).

In 2021, the Baby Boomer Generation represented the largest segment of Canada's population (Table 21.2). Lower birth rates for subsequent generations result in lower population totals. Projected increases for 2031 assume that most population growth will continue to be from immigration, accounting for the increased numbers of Generations X, Millennials, and Z, based on the likely ages of the prospective immigrants (Fig. 21.1). The average age at retirement, 65 years, lies within the latter half of the Baby Boomer Generation. The median ages for Public Service Executives (50), Members of Parliament (53), physicians (49), lawyers (45), and university professors (51) all fall within Generation X.

Future immigration to Canada is dependent on conditions in other countries, the attractiveness of Canada as an initial and final destination, and the quotas set by the federal government in response to policy decisions, public opinion, and political will. Although the demographic significance of high immigration numbers is commonly understood, public opinion and political policies are always subject to change. Reaction to perceived increases in irregular migrants, concern about the availability of social services (particularly medicine), and shortage of affordable housing represent public and political obstacles to high immigration quotas. Although none of these factors are related to officially sanctioned immigration (see Chap. 11), the higher rates of immigration since 2018 were generating public and political reservations in mid-late 2024.

21.2 International Relations

Canada has had a tradition of involvement with the United Nations and other international bodies since the early 1950s. Approximately 35% of official development assistance is channelled through multilateral organizations (Canada 2023). In general, attitudes towards the United Nations and World

N. Catto and G. R. Vincent, *Canada: Land and People*, World Regional Geography Book Series,
https://doi.org/10.1007/978-3-032-04391-7_21

Table 21.1 Comparative demographic data, 1961–2021. Data from Statistics Canada (2021–2023)

Parameter	1961	1991	2021
Population	18,238,247	27,296,859	36,991,981
Population km^{-2}	7.5	5.0	3.7
CMA population %	41	61	74
Indigenous population %	1.2	3.7	4.9
Crude birth rate	25.8/1000	13.8/1000	10.2 / 1000
Crude death rate	7.8/1000	7.1/1000	8.1/1000
Population growth % p.a	2.7	1.2	0.8
Rate of natural increase %	1.9	0.7	0.1
Total fertility rate	3.8	1.7	1.4
Infant mortality rate	28.4/1000	6.7/1000	4.4/1000
Mean childbearing age	27.8 y	27.8 y	31.4 y
Mean life expectancy	70.9 y	77.4 y	82.2 y
Female: male ratio	0.98	1.02	1.02
Median age	26.3 y	33.5 y	41.0 y
Population 0–14 y %	33.9	20.9	15.6
Population $\geq$ 65 y %	7.6	11.6	18.8
Immigrant population %	8.3	15.8	23.0

Table 21.2 Generations in Canada. Data from statistics Canada 2021–2023

Generation	Total 2021	Projection 2031	Projection 2046
Greatest (born before 1928)	135,560	25,000	–
Interwar (1928–1945)	2,716,910	1,220,000	–
Baby boomers (1946–1965)	9,212,640	8,145,200	4,102,700
X (1966–1980)	7,069,355	7,198,200	6,776,800
Millennials (1981–1996)	7,926,575	8,750,000	9,127,900
Z (1997–2012)	6,736,525	7,945,700	9,220,200
Alpha (2013 and later)	3,194,415		

Bank are positive, with mixed feelings concerning the International Monetary Fund (IPSOS 2021).

Since 1956, more than 125,000 have served in various capacities in UN-sanctioned peacekeeping operations, most recently in Rwanda (1993), East Timor (1999), Haiti (2004), and Mali (2013). In April 2024, however, Canadian involvement in peacekeeping operations has declined to less than 60 personnel, deployed in the Democratic Republic of Congo, Mali, South Sudan, Haiti, Cyprus, and the Middle East. Support for deployment of military personnel in peacekeeping operations is strongest among those over age 55 (Boomers, with memories of previous, perceived successful operations) and weakest among those under 35 (Boucher 2020a, 2020b).

From a geographical perspective, recent immigration; differences between large CMA, smaller CMA and CA, and rural areas; political preferences among areal groups of citizens; and reaction to specific international issues and conflicts

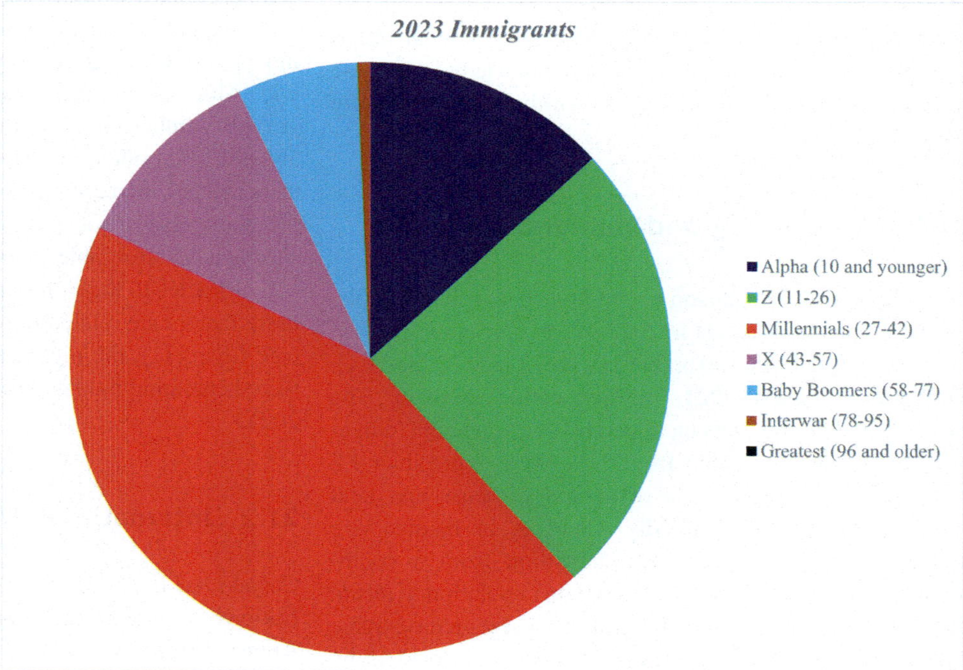

Fig. 21.1 Immigrants by generation, 2023. Data from Statistics Canada (2023)

have substantial influences on attitudes towards international organizations and involvement (e.g. MLI 2022). In general, international affairs are of far lesser concern compared to cost-of-living, housing affordability, healthcare, and economic issues. Under Canada's federal system of government, many of these responsibilities fall within provincial jurisdiction.

Geographic proximity, trade and economic connections, cultural and media influences, marketing and commercial activity, and family connections all lead to Canadians' views of international relations being strongly conditioned by attitudes towards USA. Perception of the USA political administration at any particular time significantly influences attitudes. In general, Canadian residents are likely to describe USA as prosperous and a nice place to visit. However, Canadians in general and women in particular are also likely to say that USA is racially divided, without a caring society or a good system of government, and that it is not a safe country overall (e.g. Angus Reid 2023). The Angus Reid (2023) survey also revealed that majorities of those Canadians surveyed considered Canada to be a caring society, prosperous, not racially divided, a positive player in world affairs, and a country they're proud to live in. Political differences strongly influenced the perception of the system of government. Perceptions of Canada as a caring society, not racially divided, also vary demographically and with experience (see Chap. 11).

21.3 Environment and Climate Change

Climate change is a substantial concern among Canadians, particularly the linkage with wildland fire, extreme heat events, and flooding. Impacts on winter are also of concern to many, although not all (see Chap. 19).

There is general agreement that some measures should be taken. However, there is less agreement as to what those measures should be, and how rapidly that should be implemented, particularly those that impact cost-of-living and personal transportation. In Alberta and Saskatchewan, with substantial hydrocarbon production and economic revenue, federally imposed regulations on GHG emissions are regarded as intrusions on provincial political jurisdiction, as well as poor economic policy and failure to acknowledge residents' concerns. Energy policy has been a subject of dispute between provincial and federal governments for decades (see Chap. 17), leading to general mistrust on many other issues.

The original Department of the Environment (now Environment and Climate Change Canada) was established in 1971. Canada was an original signatory to the UN Framework Convention on Climate Change (1992) and the Kyoto Protocol (2002), both commitments of the Federal Government. The National Round Table on the Environment and Economy, intended to promote sustainable development, was established in 1994. A series of national action programmes and initiatives on climate change began in 1995, along with provincially, regionally, and locally focused initiatives. Responsibility for GHG mitigation rested with Environment Canada, whereas adaptation was assigned to Natural Resources Canada. Between 1988 and April 2024, Canada developed at least 11 national climate plans, coupled with 9 GHG reduction targets.

The Kyoto commitment to reduce GHG emissions from 1990 levels by 6% by 2012 was not met, with emissions increasing by more than 30% over that period. Canada also signed the Copenhagen Accord (2009), committing to a reduction of 17% from 2005 levels by 2020, a commitment which also was not met.

Climate change policy and participation in international accords became a significant political issue in the interim. After a change in federal administration, Canada formally withdrew from the Kyoto Protocol in 2012, citing the non-participation of China and USA. After another change in federal administration, Canada signed the Paris Agreement in 2015. The current GHG emissions plan (2018) aims for a 40% cut from 2005 levels by 2030 (ECCC 2020, 2023). Significant differences in preferred policy exist among the federal political parties, and between the federal and most provincial governments. By law, federal elections occur no more than 4 years apart. The most recent election, in April 2025, produced a minority government.

References

Angus Reid (2023) Views from (both sides of) the 49th: Canadians take more pride in their country than Americans do in their own. https://angusreid.org/us-canada-border-crossing-customs/#gsc.tab=0

Boucher JC (2020a) Public opinion and Canadian defence policy. In Lagassé P, Vucetic S, Juneau T (eds) Canadian defence policy. Palgrave MacMillan, New York

Boucher JC (2020b) Peacekeeping? It's an age thing. Canadian Institute of global affairs. https://www.cgai.ca/peacekeeping_its_an_age_thing

Canada (2023) Partnership organizations. https://www.international.gc.ca/world-monde/international_relations-relations_internationales/partnerships_organizations-partenariats_organisations.aspx?lang=eng

Dominion Bureau of Statistics (now Statistics Canada) (1961) Official handbook of present conditions and recent progress. https://publications.gc.ca/collections/collection_2017/statcan/11-203/CS11-203-1961-eng.pdf

ECCC (2020) National inventory report: greenhouse gas sources and sinks in Canada. Environment and climate change Canada. https://publications.gc.ca/site/eng/9.506002/publication.html

ECCC (2023) Canada's greenhouse gas and air pollutant emissions projections. Environment and climate change Canada. https://publications.gc.ca/collections/collection_2023/eccc/En1-78-2023-eng.pdf

IPSOS (2021) Canada's Reputation Shines internationally and at home. https://www.ipsos.com/en-ca/news-polls/canada-reputation-shines-internationally-and-at-home

MLI (MacDonald-Laurier Institute) (2022) Canadians support a stronger, more principled foreign policy. https://macdonaldlaurier.ca/canadians-support-a-stronger-more-principled-foreign-policy-mlis-annual-foreign-policy-and-international-affairs-survey/

Statistics Canada (2021–2023) Census of Canada 2021. https://www12.statcan.gc.ca/census-recensement/2021/as-sa/98-200-X/2021003/98-200-X2021003-eng.cfm

Statistics Canada (2023) Immigrant status and period of immigration by gender and age, July 1, 2022 to June 30, 2023. https://www150.statcan.gc.ca/t1/tbl1/en/tv.action?pid=9810034701

The manufacturer's authorised representative in the EU is Springer Nature Customer Service Centre GmbH, Europaplatz 3, 69115 Heidelberg, Germany. If you have any concerns regarding our products, please contact ProductSafety@springernature.com

Printed and bound by CPI Group (UK) Ltd, Croydon, CR0 4YY
16/09/2026
02217809-0001